D1379063

INORGANIC CHEMISTRY

INORGANIC CHEMISTRY

A GUIDE TO ADVANCED STUDY

by

R. B. HESLOP

Senior Lecturer in Chemistry
The University of Manchester Institute of
Science and Technology

and

K. JONES

Lecturer in Chemistry
The University of Manchester Institute of
Science and Technology

A Completely Revised Successor to Heslop/Robinson, Inorganic Chemistry

ELSEVIER SCIENTIFIC PUBLISHING COMPANY
Amsterdam — Oxford — New York 1976

ELSEVIER SCIENTIFIC PUBLISHING COMPANY
335 Jan van Galenstraat
P.O. Box 211, Amsterdam, The Netherlands

AMERICAN ELSEVIER PUBLISHING COMPANY, INC.
52 Vanderbilt Avenue
New York, New York 10017

ISBN: 0-444-41426-6

Printed in The Netherlands

Preface

In April 1960, one of us published, in collaboration with the late Professor P.L. Robinson, a text-book of inorganic chemistry with the same title as this present volume. A second edition appeared in 1963 and a third edition in 1967. These facts and much encouraging comment suggest that the choice of material and its manner of presentation have met with wide approval.

In writing this book we have tried to maintain the original aim of providing a text-book of moderate size which does not presume to be a substitute for oral teaching or seek to deal with specialised topics which may be features of courses in some colleges and universities. We have therefore concentrated our attention on presenting the chemistry of the elements by treatment of the groups in the Periodic Table, and have omitted, after due consideration, the separate treatment of topics such as biochemical or environmental aspects of inorganic chemistry.

We have, however, given greater attention to the physical methods used for the determination of structure and the understanding of chemical bonding. We have introduced ideas of molecular and orbital symmetry, and made use of them throughout the text. Another important change is the exclusive use of SI units with the appropriate alterations in physical formulae when the international system makes them necessary.

We have also added, on the suggestion of many helpful critics, a number of problems based on the material in the book, together with answers, and we hope this innovation will prove useful to the students for whom the book is primarily intended.

We should like to thank Mrs. H.M. Shaw, Mrs. J.I. Jones, and particularly Mrs. M.M. Latta for typing the manuscript and helping to prepare it for publication.

Contents

1 Modern Inorganic Chemistry

The ambit of inorganic chemistry is well established; it comprises the natural occurrence and artificial preparation of the elements, their properties and reactions, and those of their compounds, together with a rational correlation and theoretical interpretation of the phenomena.

Although the oldest discipline in chemistry, inorganic chemistry continues to excite increasing interest as a subject of pure and applied research and is finding manifold applications in industry. It is the foundation of vast endeavour in fine and heavy chemicals, in ceramics, in the production of glass and building materials, and in extraction metallurgy. There is hardly an element which is not being investigated from some aspect at the present time. It is concerned in the preparation of catalysts and has made, and is making, contributions which facilitate the production and exploitation of atomic energy. Inorganic chemists have continued to call attention to much in this vast field which awaits exploration, and this book will have failed in its purpose if something of the enthusiasm and confidence which inspires them is not reflected in its pages. A few of the exciting themes dealt with are: current interpretation of chemical bonding and structure with the modern physico-chemical evidence, non-stoichiometry, electron-deficient compounds, complex compounds, organometallic compounds, materials with exceptional properties (such as great hardness, high melting points, corrosion resistance), noble gas compounds. Most of the compounds of carbon with non-metallic elements lie outside inorganic chemistry, but organometallic compounds (Chapter 18) form one of its most important current topics.

The advances of the last three decades have taken inorganic chemistry right out of the stage of being a mass of arbitrarily chosen facts brought into some relationship by the Periodic Table. Although modern inorganic chemistry remains in a measure descriptive and pictorial (as do the other branches of the subject) the pictures have grown in precision and the descriptions have become more quantitative. Hence an historical approach is now inappropriate because the subject has developed to a point which makes entry from the level of present theoretical and physico-chemical knowledge desirable; such an entry has the advantage of being both easier and more lively. For this reason earlier work on the structure of the atom, culminating in 1913 with Bohr's atomic model, and other classical material has not been included.

1.1. Approach to the Subject

Quantum theory has provided a picture of the atom that allows an immediate approach to valency and molecular structure while furnishing a far more secure basis for the periodic arrangement of the elements than selected physical and chemical properties. This is not to overlook the achievements of Thomson, Rutherford and Bohr in atomic structure, or of Newlands, Lothar Meyer, Mendeleev, Bohr and others in evolving the Periodic Law, but rather to seize upon developments which their prescient work made possible. Readers will find it easier to assimilate and remember the facts embodied in the Periodic Classification when these are seen to emerge from the systematic development of electronic structure with increasing atomic number.

The description of atoms and molecules which is used — that of one or more positive nuclei surrounded by a cloud of electrons which, for many purposes, is equivalent to a smeared-out negative charge — presents in pictorial form the results of thirty years of quantum mechanics. All available evidence suggests that this general picture is unlikely to suffer substantial modification. Theoretical chemistry accepts the Schrödinger equation and is largely concerned with finding the most direct mathematical path to a unified explanation of the physico-chemical properties of the molecules. This description has been adopted and used in the interpretation of valency theory, including metal—ligand bonding, and in discussing structure generally. A brief account of the nucleus is included because not only does it determine the extra-nuclear structure of the atom, but is also the seat of radioactivity and the source of nuclear energy. Some knowledge of it is necessary in order to follow present ideas on the synthesis of elements taking place in the stars.

The kinetic and thermodynamic aspects of reactions are stressed wherever the data are available, and examples are worked out which should enable the reader to understand and acquire facility in these essential calculations.

So much of inorganic matter is crystalline that once a means of investigating the structure of solids became available it was eagerly applied. The result has made it necessary to include a short account of the solid state. With this goes a little about the growth of crystals and the way atoms in their lattices suffer dislocation.

1.2. Suggestions on the Use of this Book

This book is a brief epitome of modern inorganic chemistry and the reader may usefully begin anywhere, since liberal cross-references provide a constant link with definitions and the underlying theory. These appear mainly in the earlier chapters which need not be mastered before the rest of the book can be understood, and their assimilation by repeated reference will be repaid by a fuller appreciation of what follows. It is idle to suppose that a real appreciation of any branch of chemistry can be had without two things—adequate theory and sufficient facts. The best way to grasp theory is to apply it constantly, and the easiest way to remember facts is to seek their relation to theory.

Appended to every chapter is a short list of references which will be found useful in extending the reader's knowledge of particular aspects of the subjects treated. Because a student's time is limited and his requirement is a wide, balanced view of the subject, books, monographs and review articles, rather than original papers, have mainly been cited.

1.3. Symbols and Abbreviations

Chemical symbols and formulae have been used extensively in place of names to save space and often to secure clarity. Symbols which appear frequently are:

Atomic number	Z	Magnetic flux density	$\mathbf{B}$
Mass number	A	Entropy	S
Isotopic mass	m	Frequency	ν
Relative atomic mass (atomic weight)	A_r	Wavelength	λ
		Concentration of X	[X]
Relative molar mass (molecular weight)	M_r	Activity of X	a_X
		Specific reaction rate	k
Enthalpy	H	Equilibrium constant	K
Gibbs function	G	Magnetic moment	μ

Some of the abbreviations which often appear are:

$I(1)$	ionisation energy (first)	u.v.	ultraviolet
r_M	single-bond covalent radius of M atom	i.r.	infrared
r_{M^+}	ionic radius of M^+	n.m.r.	nuclear magnetic resonance
emf	electromotive force	e.s.r.	electron spin resonance
E^0, M^+/M	standard redox potential for M^+/M couple	n.q.r.	nuclear quadrupole resonance
AO	atomic orbital	ln x	logarithm of x to base e
MO	molecular orbital	h.c.p.	hexagonal close-packed
b.p.	boiling point	c.c.p.	cubic close-packed
m.p.	melting point	b.c.c.	body-centred cubic

Other abbreviations in the text are explained where they occur.

1.4. Units

The Système Internationale (SI), which was adopted in 1960, is based on the metre (m) as the unit of length, the kilogram (kg) as the unit of mass, the second (s) as the unit of time, the ampere (A) as the unit of electric current, the kelvin (K) as the unit of temperature and the mole (mol) as the unit of amount of substance. There is a further basic unit, that of luminous intensity, the candela, which is not required in this book, but the first six units, and other units derived from them, will be used when appropriate.

The principal changes introduced with the SI are:

(1) The metre and kilogram replace the centimetre and the gram as the coherent units, although the latter remain as submultiples.

(2) The unit of energy is the joule ($J = kg\ m^2\ s^{-2}$); thus the variously defined calories and non-metric units of energy are superseded.

(3) 'Electrostatic' and 'electromagnetic' units are replaced by SI electrical units.

For definitions of the basic units and for further information on SI the student should consult one of the first two references given at the end of Section 1.5. The word *mole* however is worthy of a little further treatment here. A mole is defined as that amount of substance which contains the same number of molecules, ions, atoms, electrons, as the case may be, as there are atoms of carbon in exactly 0.012 kg of carbon-12. The word is not a synonym for 'gram-molecule' which can be applied only to substances which exist as discrete molecules at a specified temperature. A gram-molecule of oxygen refers to 32.00 g of oxygen in the form of O_2 molecules. The expression 'a mole of oxygen' is ambiguous, because 'mole' can be applied to an amount of O_2 molecules or an amount of O atoms, and the former obviously has twice the mass of the latter. Similarly a mole of Al_2Br_6 has twice the mass of a mole of $AlBr_3$. It is always safest, therefore, to use a formula to specify exactly what species are being considered. The Avogadro constant, $N_A = 6.0220 \times 10^{23}$ mol^{-1}, expresses the number of specified particles in one mole, for example the number of O_2 molecules per mole of O_2. Some other important constants are given in Table 1.1. The unified atomic mass constant, m_u, is defined as one-twelfth of the mass of an atom of carbon-12, and is used in the calculation of atomic weights (2.2).

The use of an internationally recommended system of electric and magnetic equations founded on four dimensionally independent quantities, length, mass, time and electric current, requires the introduction of two

TABLE 1.1

VALUES OF SOME PHYSICAL CONSTANTS (5 SIGNIFICANT FIGURES)

Physical constant	Symbol	Value
Speed of light in a vacuum	c_0	2.9979×10^8 m s^{-1}
Permeability of a vacuum	μ_0	$4\pi \times 10^{-7}$ kg m s^{-2} A^{-2} exactly
Permittivity of a vacuum	ϵ_0	8.8542×10^{-12} kg^{-1} m^{-3} s^4 A^2
Unified atomic mass constant	m_u	1.6605×10^{-27} kg
Planck's constant	h	6.6262×10^{-34} J s
Gas constant	R	8.3143 J K^{-1} mol^{-1}
Faraday's constant	F	9.6487×10^4 C mol^{-1}

physical constants, the permittivity of a vacuum, ϵ_0, and the permeability of a vacuum, μ_0. In defining these quantities in the SI the opportunity was also taken to *rationalise* the equations, that is to ensure that the factors 2π or 4π occur only where the equation relates to a field of cylindrical or spherical symmetry. Thus, in order to be able to express Ampere's law in a rationalised form in which the quantities can all be expressed in SI units, it becomes necessary to give μ_0 the value $4\pi \times 10^{-7}$ kg m s^{-2} A^{-2} exactly. The law is then written

$$F = \frac{\mu_0 I_1 I_2 l}{2\pi d}$$

where F is the force between two parallel conductors of negligible cross-section and length l, carrying currents of I_1 and I_2, respectively, and separated by a distance d in a vacuum.

Similarly Coulomb's law in its four-quantity rationalised form is expressed as

$$F = \frac{Q_1 Q_2}{4\pi\epsilon_0 r^2}$$

in which F is the force exerted by two point charges Q_1 and Q_2 separated by a distance r in a vacuum. The value of ϵ_0 is defined by the equation

$$\epsilon_0 = \mu_0^{-1} c_0^{-2} ,$$

where c_0 is the velocity of light in a vacuum, and is given by $(8.854\,185 \pm 0.000\,006) \times 10^{-12}$ kg^{-1} m^{-3} s^4 A^2.

TABLE 1.2

DERIVED SI UNITS WITH SPECIAL NAMES

Physical quantity	Name of SI unit	Symbol	Definition
Frequency	hertz	Hz	s^{-1}
Force	newton	N	kg m s^{-2}
Energy	joule	J	N m
Power	watt	W	J s^{-1}
Pressure	pascal	Pa	N m^{-2} = J m^{-3}
Electric charge	coulomb	C	A s
Electric potential difference	volt	V	J A^{-1} s^{-1}
Electric resistance	ohm	Ω	V A^{-1}
Electric conductance	siemens	S	Ω^{-1}
Electric capacitance	farad	F	A V^{-1} s
Magnetic flux	weber	Wb	V s
Inductance	henry	H	V A^{-1} s
Magnetic flux density	tesla	T	V s m^{-2}

TABLE 1.3
APPROVED PREFIXES USED IN SI

Prefix	Symbol	Fraction	Prefix	Symbol	Multiple
deci	d	10^{-1}	deka	da	10
centi	c	10^{-2}	hecto	h	10^2
milli	m	10^{-3}	kilo	k	10^3
micro	μ	10^{-6}	mega	M	10^6
nano	n	10^{-9}	giga	G	10^9
pico	p	10^{-12}	tera	T	10^{12}
femto	f	10^{-15}			
atto	a	10^{-18}			

The changes in these basic equations of electrostatics and electromagnetism are significant when one needs to relate magnetic and electric properties to other derived SI units. Worked examples at appropriate places in the text will illustrate the use of these equations (e.g. 3.2.4 and 4.2.3).

There are several units which are used in this book which have special names and symbols in the international system. A list of these, with their definitions, is given in Table 1.2.

Table 1.3 gives the prefixes which have been agreed internationally for fractions and multiples of SI units. Thus 1 pm is 10^{-12} m and 1 kJ is 10^3 J. It will be noted that, apart from the range 10^{-2} to 10^2, all the indices are divisible by 3. Thus the measure of a very large quantity or a very small one can always be expressed as a number between 1 and 1000 if a prefix is used. A typical ionic radius of 1.45×10^{-10} m can be conveniently expressed as 145 pm, for example.

1.5. Tabulation of Numerical Values

To avoid needless repetition of units, it is convenient to place only figures in the body of a table. When this is done, the headings must also be dimensionless, and this is achieved by showing the symbol for the physical quantity (in italics) divided by the unit in which it is expressed (in ordinary upright print). Thus the figure 525, in a column headed $I(1)$/kJ mol^{-1} means that the value of the first ionisation energy there tabulated is 525 kJ mol^{-1}. A similar convention is used for labelling the axes of graphs.

SI units have been used almost exclusively in the presentation of tables and in graphs. Thus, for example, all m.p. and b.p. are expressed in kelvins and all pressures in pascals. In a very few cases other units such as the electronvolt and the Bohr magneton have been used, but they are defined in the text in terms of SI units. Furthermore the coherent SI units are used in all calculations so that the advantages of the method known as quantity calculus (see references 2 and 3) are obtained.

References

1. The Royal Society, Quantities, units and symbols, London, 1971.
2. M.L. McGlashan, Physico-chemical quantities and units, Royal Institute of Chemistry Monograph for Teachers No. 15, 2nd edition, London, 1971.
3. R.B. Heslop and Gillian M. Wild, SI units in chemistry, Applied Science Publishers Ltd., London, 1971.
4. Policy for NBS usage of SI units, J. Chem. Ed., 48, 469 (1971).
5. M.L. McGlashan, Internationally recommended names and symbols for physico-chemical quantities and units, Ann. Rev. Phys. Chem., 24, 51 (1973).

1.6. Symmetry

The shapes and spatial arrangements of orbitals, atoms, and molecules provide a simple fundamental basis on which a great deal of Inorganic Chemistry can be rationalised. Such a study requires an elementary knowledge of symmetry so that the shape of any object such as a molecule can be measured quantitatively and described unambiguously.

Symmetry elements are the units of symmetry, and their precise definitions are best understood in conjunction with their associated symmetry operations. Thus a symmetry operation can be defined as doing something (actually or notionally) to an object (whose symmetry one wishes to detail) which leaves the object in an indistinguishable (not necessarily identical) situation. Hence a symmetry operation is an action which demonstrates the existence of a symmetry element. There are five types which can be described as:

1. *Rotation* (the symmetry operation) about an *axis* of symmetry, C (the symmetry element). The order n of the rotation C_n is defined as 360° divided by the angle of rotation.
2. *Reflection* (operation) through a *plane* of symmetry, σ (the symmetry element).
3. *Reflection* (operation) through a *centre of inversion*, i (the symmetry element). This is essentially a three-dimensional analogue of the two-dimensional plane of symmetry. Any point which can be described by the co-ordinates $+x$ $+y$ $+z$ will have a corresponding point at $-x$ $-y$ $-z$.
4. *Rotation followed by reflection* in a plane perpendicular to the axis (operation) defines the element of symmetry known as an *improper* axis of symmetry, S.
5. Finally, the operation of *rotation* through 360° (or alternatively doing nothing) is the symmetry element of *identity*, E. Such an operation must result in an identical situation to the starting position.

Consider the octahedral SF_6 molecule. Rotation through 90° about the vertical axis through F—S—F would leave the molecule with a configuration in space indistinguishable from its original one. This proper axis of symmetry is called a four-fold axis and labelled C_4 (Fig. 1.1). Rotation about the same axis through 180° is a C_2 axis while other C_2 axes bisect F—S—F bond angles

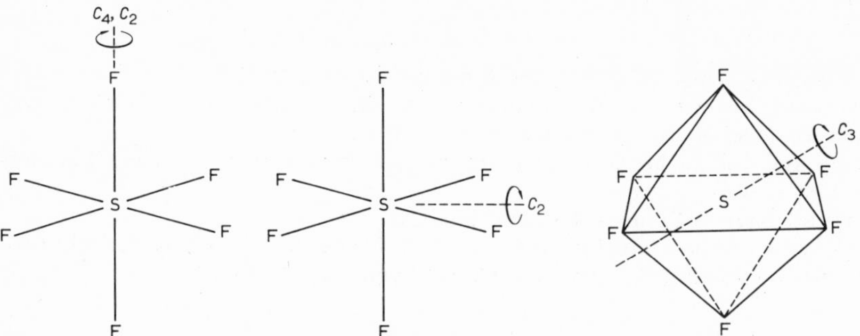

Fig. 1.1. Proper axes of symmetry in the SF_6 molecule.

and C_3 axes pass through the centre of each face of the octahedron formed by the F atoms. The order of rotation n is determined by dividing 360° by the angle of rotation.

The principal axis is that with the highest order of rotation (the C_4 axis in SF_6) which is usually drawn vertically in a representation of the molecule and designated the z axis in Cartesian coordinates.

Planes of symmetry σ are usually further defined as vertical, horizontal, and dihedral with respect to the principal axis. In SF_6, there are three planes (3 σ_h) horizontal to the principal axis and six dihedral planes (6 σ_d) which bisect the different two-fold axes (Fig. 1.2).

SF_6 has a centre of symmetry at S, because any F atom moved from its origin through S to a position diametrically opposite and equidistant from S finishes in the position of another F atom.

The improper rotation axis S_6 arises when a rotation through 60° is carried out along the axis through the centre of any face of an octahedron (if SF_6 is drawn in a cube, the F atoms lie at the centre points of each face, the S_6 axes are the four cube diagonals), followed by reflection in a plane perpendicular to the axis (the plane contains the centre points of four diametrically opposed edges) (Fig. 1.3).

The total number of symmetry operations and their associated elements

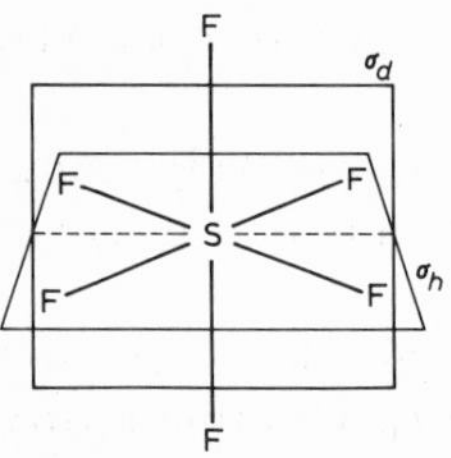

Fig. 1.2. SF_6 molecule, illustrating σ_d and σ_h planes of symmetry.

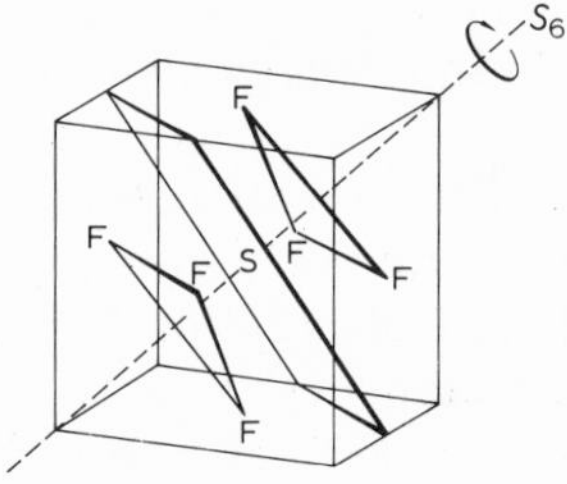

Fig. 1.3. SF_6 molecule, illustrating an S_6 axis.

of symmetry can be used to classify objects into point groups. The Schoenflies system of point group symbols provides a convenient shorthand notation by which objects of similar symmetry can be recognised. Apart from the special point groups I_h, O_h, T_d, which refer to icosahedral, octahedral, and tetrahedral symmetry, most inorganic molecules are described by the C and D point groups. Classification into these groups can be achieved simply by determining which axes of rotation and planes of symmetry are present. The presence of an n-fold axis indicates the point group C_n; if in addition there is a horizontal plane of symmetry, the group is C_{nh}, but if there are only vertical planes, the group is C_{nv}. The presence of 2-fold axes perpendicular to the principal axis C_n is the basis of the D_n group. Again, if in addition there is a horizontal plane, the group is D_{nh}, but if there are dihedral planes (planes which bisect the 2-fold axes), the group is D_{nd}.

The absence of axes of symmetry indicates lower orders of C group symmetry such as C_s, which contains only one plane of symmetry, C_i, which has a centre of inversion as its only element of symmetry, and C_1 which has no elements of symmetry other than identity. S groups are less common and contain both proper and improper axes of rotation but no planes.

TABLE 1.4

Examples of some common point groups

Point group	Elements of symmetry	Molecules
C_1	E	CHFClBr
C_i	E, i	CHClBr—CHClBr (staggered)
C_s	E, σ	SO_2BrF, HOCl
C_2	E, C_2	H_2O_2
C_{2v}	$E, C_2, 2\,\sigma_v$	$SnCl_2$, SiH_2Cl_2
C_{3v}	$E, C_3, 3\,\sigma_v$	PCl_3, H_3GeCl
C_{4v}	$E, C_4, 4\,\sigma_v$	IF_5, SF_5Cl
$C_{\infty v}$	$E, C_\infty, \infty\,\sigma_v$	NO, HCN
C_{2h}	E, C_2, σ_h, i	*trans*-N_2F_2
C_{3h}	E, C_3, σ_h, i	$B(OH)_3$
D_{2d}	$E, C_2, 2\,C_2$ (perp), $2\,\sigma_d, S_4$	$H_2C{=}C{=}CH_2$
D_{3d}	$E, C_3, 3\,C_2, 3\,\sigma_d, i, S_6$	C_2Cl_6 (staggered)
D_{4d}	$E, C_4, 4\,C_2, 4\,\sigma_d, S_8$	S_8
D_{2h}	$E, C_2, 2\,C_2$ (perp), $2\,\sigma_v, \sigma_h, i$	B_2Cl_4, B_2H_6
D_{3h}	$E, C_3, 3\,C_2, 3\,\sigma_v, \sigma_h, S_3$	BCl_3, PCl_5
D_{4h}	$E, C_4, 4\,C_2, 4\,\sigma_v, \sigma_h, i, S_4$	XeF_4, $PtCl_4^{2-}$
$D_{\infty h}$	$E, C_\infty, \infty\,C_2, \infty\,\sigma_v, i$	CO_2, Cl_2

Experience at recognising symmetry elements leads to a speedy way of determining point groups (Table 1.4). However the complete symmetry description of a point group is found in its character table.

Point group	Symmetry operations	
Irreducible representations or symmetry species	Characters in rows COLUMNS	Rotational, translational assignments and Cartesian transformations

For example:

D_{3d}	E	$2C_3$	$3C_2$	i	$2S_6$	$3\sigma_d$		
A_{1g}	1	1	1	1	1	1		x^2+y^2, z^2
A_{2g}	1	1	−1	1	1	−1	R_z	
E_g	2	−1	0	2	−1	0	(R_x, R_y)	$(x^2-y^2, xy)\ (xz, yz)$
A_{1u}	1	1	1	−1	−1	−1		
A_{2u}	1	1	−1	−1	−1	1	T_z	
E_u	2	−1	0	−2	1	0	(T_x, T_y)	

Character tables have been constructed for all point groups and those of some chemically important symmetry groups are included for reference (Appendix I). However, chemists will want to use character tables rather than dwell on the mathematical subtleties of matrix algebra and group theory to arrive at their fundamental derivation, although the rules and conditions which such tables uniquely satisfy can be readily demonstrated. As the name implies, the table is made up of characters (which are usually integers), the significance of which will be made clear in later applications (5.3.4), arranged in vertical columns under each type of symmetry operation which the particular point group possesses. A horizontal row of characters constitutes an irreducible representation (which describes the symmetry of a particular part of an object described by the point group), collectively known by the symbol Γ but designated individually by a symbol *A*, *B*, *E*, or *T* and often referred to as a symmetry species. *A* and *B* both refer to one-dimensional representations, *E* and *T* refer to two- and three-dimensional representations (doubly and triply degenerate species), respectively, as indicated by the character in the identity column. The subscripts *g* and *u* (German gerade and ungerade, in this context meaning symmetric and antisymmetric) are included in terms for the irreducible representations of shapes which have a centre of symmetry, and indicate the sign of the character in the inversion centre column. Other subdivisions are designated by subscript numerals, primes and double primes.

To the right of the character table, it has become the convention to indicate the irreducible representations to which the rotational R_x R_y R_z and translational T_x T_y T_z modes correspond, as well as showing how the binary functions of x y z such as xy, x^2-y^2 transform. Such information is useful when discussing the symmetry of orbitals in bonding as well as in interpreting infrared and Raman spectra. The application of symmetry and group theory to each of these topics will be explained in the appropriate section.

Reference

G. Davidson, Introductory Group Theory for Chemists, Elsevier, Amsterdam, 1971.

1.7. Suggestions for Further Reading

Appended to every chapter in the book is a short list of references which will be found useful in extending the reader's knowledge of particular aspects of the subjects treated in that chapter. More generally there are two recent publications containing reviews of inorganic chemistry written by experts in their respective fields: *Comprehensive Inorganic Chemistry*, Executive Editor, A.F. Trotman-Dickenson, 5 vols., Pergamon Press, Oxford, 1973, and *Inorganic Chemistry: Series One and Two* (*M.T.P. International Review of Science*), Consultant Editor, H.J. Emeléus, 10 vols., Butterworths, London, 1973 and 1975.

Useful review articles will be found in *Advances in Inorganic Chemistry and Radiochemistry*, Eds. H.J. Emeléus and A.G. Sharpe, published annually since 1959 by Academic Press, New York, and in *Progress in Inorganic Chemistry*, also published annually since 1959 by Interscience, New York. The *Chemical Reviews* have been published by the American Chemical Society since 1901; the *Quarterly Reviews of the Chemical Society* were published by the Chemical Society of London from 1947 to 1971 and have been succeeded by the *Chemical Society Reviews* which also appear quarterly. The reader's attention is also drawn to *Accounts of Chemical Research*, published since 1968 by the American Chemical Society, and to the international edition in English of *Angewandte Chemie*, published since 1962 by Verlag Chemie at Weinheim. This journal contains less specialised articles of particular value to undergraduate students, as do the *Journal of Chemical Education*, published monthly by the American Chemical Society and *Education in Chemistry*, published six times a year by the Chemical Society of London.

More specialised articles of greater interest to research workers will be found in *Advances in Organometallic Chemistry*, Eds. F.G.A. Stone and R. West, published annually from 1963 by Academic Press, New York, and *Coordination Chemistry Reviews*, Ed. A.B.P. Lever, published since 1966 by Elsevier, Amsterdam.

Practical aspects of inorganic chemistry are well covered in *Inorganic Syntheses*, McGraw-Hill, New York, of which fourteen volumes have appeared since 1939, and in *Techniques of Inorganic Chemistry*, Eds. H.B. Jonassen and A. Weissberger, Interscience, New York, of which six volumes have been published. Smaller books on practical inorganic chemistry include W.G. Palmer, *Experimental Inorganic Chemistry*, Cambridge, 1954, G. Pass and H. Sutcliffe, *Practical Inorganic Chemistry*, 2nd edition, Chapman and Hall, London, 1974, and C.F. Bell, *Synthesis and Physical Studies of Inorganic Compounds*, Pergamon, Oxford, 1972.

For the student seeking further to test his understanding by attempting problems, there are B.J. Aylett and B.C. Smith, *Problems in Inorganic Chemistry*, English Universities Press, London, 1965, which contains general problems, mainly non-numerical, and R.B. Heslop, *Numerical Aspects of Inorganic Chemistry*, Elsevier, London, 1970, which contains mainly numerical problems, and makes use of SI units and the methods of quantity calculus.

There are many reference books containing thermodynamic and spectral data. Two small books which have limited but useful tables of such information are *Tables of Physical and Chemical Constants*, 14th edition, Longmans, London, 1973, and W.E. Dasent, *Inorganic Energetics*, Penguin Books, Harmondsworth, 1970. Finally, the definitive authority on the naming of inorganic compounds is I.U.P.A.C., *Nomenclature of Inorganic Chemistry*, 2nd edition, Butterworths, London, 1970.

The Atomic Nucleus: Genesis of the Elements

2

The chemical properties of an element depend upon the nature of its atoms; many of the advances in chemistry during the present century are directly attributable to knowledge of atomic structure provided by the study of atomic physics.

An atom consists of a positive *nucleus*, with a radius of a few femtometres, surrounded by an electron cloud with a radius of a few hundred picometres. The nucleus consists of positively charged *protons* and uncharged *neutrons* (Table 2.1). In a free atom the number of protons in the nucleus equals the number of *electrons* in the electron cloud. This number, the *atomic number*, Z, fixes the place of the element in the Periodic Table and determines its chemical character. An atom with eight protons in its nucleus is oxygen, one with seventeen protons is chlorine. The *mass number*, A, of an atom is the sum of protons and neutrons in the nucleus; they are known collectively as *nucleons.*

An atom has three noteworthy features: (i) the enormous density of the nucleus, about 5×10^{12} times that of uranium metal, (ii) the small volume of the nucleus, about 10^{-15} that of the whole atom, (iii) the very high proportion of the total mass which is due to the nucleus.

2.1. Isotopes

A particular *nuclide* (i.e. a specific kind of atom) is designated by the name of the element followed by the mass number. Thus carbon-12 is the

TABLE 2.1

PROPERTIES OF SOME FUNDAMENTAL PARTICLES

Particle	Symbol	m_a/m_u	m_a/kg [a]	*charge*/C
Proton	^1_1p	1.007277	1.67262×10^{-27}	$+1.602 \times 10^{-19}$
Neutron	^1_0n	1.008665	1.67492×10^{-27}	0
Electron	e	0.000548	9.1096×10^{-31}	-1.602×10^{-19}

[a] For the definition of m_a, the isotopic mass, see Section 2.2.

name given to those atoms of carbon for which $A = 12$ ($Z = 6$, and N the *neutron number* = 6). Carbon-13, however, has $Z = 6$ and $N = 7$. Nuclides of the same element, like carbon-12 and carbon-13, are *isotopes* of that element; isotopes have the same atomic number but different mass numbers. The symbols are $^{12}_{6}C$ and $^{13}_{6}C$, the mass number and atomic number being kept to the left to leave space on the right for charge numbers and atomicities. Thus $^{14}_{7}N_3^-$ refers to an azide ion containing three atoms of nitrogen-14 and carrying a charge equal to that of one electron.

Isotopes were first recognised in the natural radioactive series which comprise heavy elements, but positive-ray analysis soon showed that light elements also had isotopes, and the development of the mass spectrograph enabled all the elements to be examined for isotopes. The operation of the mass spectrograph is based on the deflection of collimated beams of positively charged particles, in electric and magnetic fields of known strength. The ions are formed as an anode ray of positively charged particles, by evaporating from a hot filament, or, as in Aston's earliest apparatus, by passing a discharge through a vapour. Adjustment of the strength of the fields enables particles of the same charge to mass ratio to be focussed. With such a beam from an element possessing isotopes, the several images are brought to different foci, and from their positions the masses of the individual isotopes may be determined.

Molecules have rotational and vibrational energies (5.3.2) which are quantised, taking only certain discrete values which depend in magnitude on the masses of the atoms involved, and are therefore different for molecules containing different isotopes of the same element, for instance $^{1}H^{35}Cl$ and $^{1}H^{37}Cl$. Changes in rotational energy are characterised by the absorption of radiation in the far infrared, and in vibrational energy by absorption in the near infrared; they give rise to lines in these spectral regions. Particular lines which occur singly when one nuclide is present appear in groups when there are several nuclides. From the small differences in frequency both the presence and relative masses of the nuclides may be inferred. By this sensitive method Giauque and Johnson (1929) showed that oxygen contained molecules of $^{16}O^{17}O$ and $^{16}O^{18}O$.

2.2. Natural Isotopic Ratios — Atomic Weights

The mass spectrometer is a mass spectrograph in which the positive ions are produced at a steady rate, and the photographic plate, by means of which the positions of the ion images were formerly observed, is replaced by a slit behind which is a collector connected to devices for amplifying and measuring the ion current. The slit scans the spectrum and the ion current shows a series of peaks. Every peak indicates an isotope and its height represents the relative number of ions. The design of these instruments now allows an accuracy of 0.001%, making the method better than that of chemical analysis for the determination of atomic weights.

High-resolution mass spectrometers enable the *isotopic mass*, m_a, of any nuclide to be compared with the mass of the carbon-12 atom. Thus if carbon-12 is assigned a mass of 12.000000 m_u, nitrogen-14 has a mass of 14.003074 m_u and oxygen-16 a mass of 15.994914 m_u. The *atomic weight* of an element, more correctly called the *relative atomic mass*, A_r, is the weighted mean of the masses of the naturally-occurring isotopes divided by m_u. For example, natural carbon is 98.89% ^{12}C and 1.11% ^{13}C for which $m_a = 13.00335\ m_u$. Thus A_r for carbon is $(0.9889 \times 12.000000\ m_u + 0.0111 \times 13.00335\ m_u)/m_u = 12.011115$. It is important to realise that A_r is a dimensionless quantity. However, the *molar mass*, *M*, of carbon, i.e. the mass of 6.022×10^{23} carbon atoms containing the natural distribution of ^{12}C and ^{13}C, is 12.011115 g mol^{-1}.

2.3. Separation of Isotopes

2.3.1. Gaseous diffusion

Gases of different density, ρ, diffuse at different rates:

$$\text{rate} \propto \rho^{-1/2}$$

Uranium hexafluoride made from natural uranium contains one part of $^{235}UF_6$ in 140 parts of $^{238}UF_6$, for which the corresponding ratios of densities is 1.0086 and of diffusion rates 1.0043 respectively. Thus separation can be achieved by a process of successive diffusions through a series of porous partitions, after which the lighter fraction becomes the feed for a later stage while the heavier fraction is returned to an earlier point in the sequence.

2.3.2. Thermal diffusion

When two gases of different densities are in a vertical tube which is cool and has an electrically heated wire down the axis, the lighter gas diffuses preferentially towards the hot wire where it rises in a convection current, while the heavier streams downwards on the surface of the tube. The principle is applicable to isotopes and, with a column 30 m long and a temperature difference of 600 K, almost complete separation of ^{35}Cl from ^{37}Cl was effected as early as 1938.

2.3.3. Electromagnetic method

A perfect separation is possible with a mass spectrograph in which the positive ions of the different isotopes pass into collectors, but the quantities are small. About 1943, however, a large-scale apparatus, the calutron, was constructed in which an ideal rigorous separation was sacrificed in order to increase the yield, since the ions in beams of heavy current repel one another and cause spreading. The method has been further refined and is extensively used.

2.3.4. Molecular distillation

In 1921 the isotopes of mercury were partially separated by distillation in a high vacuum with a distance between the evaporating and condensing surfaces equal to about the mean free path of the atoms. The rate at which the isotopes evaporate is inversely proportional to the square roots of their masses so that the condensate is a little richer in the lighter isotopes.

2.3.5. Chemical methods

Isotopes are not absolutely identical in chemical properties, but the difference is observable only in the light elements where the ratio of the mass of the isotopes is large, for instance ^{1}H and ^{2}H. Thode and Urey (1939) concentrated nitrogen-15 in ammonium nitrate, to the extent of 70%, by allowing a solution of the salt to flow down a column against a counter current of ammonia. The equilibrium constant for

$$^{15}NH_3(g) + {}^{14}NH_4^+(aq) \rightleftharpoons {}^{14}NH_3(g) + {}^{15}NH_4^+(aq)$$

has a value of about 1.033.

2.3.6. Gas chromatography

This technique enables gases to be separated by the selective adsorption of one or more from a mixture on a suitable packing in a column. The adsorbed gas, should that be required, may be subsequently recovered by elution. Though only partially successful when applied to the isotopes of neon because the difference in their adsorption coefficient on charcoal is small, the method has enabled deuterium to be separated from a 1 : 1 deuterium—hydrogen mixture. The column is packed with palladium-black on asbestos which preferentially adsorbs the hydrogen and allows the deuterium to pass on.

2.3.7. Ionic migration

The migration of isotopic ions under the influence of an electric potential difference has been studied in aqueous solution, in ionic crystals, in fused salts and in molten metals. In practice the best separations are achieved in aqueous solution and in fused salts. Klemm (1954) was able to obtain LiCl enriched in lithium-7 from the anode compartment of an apparatus in which current was passed through the fused salt, the cathodic deposition of Li being prevented by the continuous introduction of chlorine.

2.4. Nuclear Binding Energy

Except for hydrogen-1, the mass of an atom is less than the mass of the protons, neutrons and electrons which compose it. For oxygen-16, m_a =

15.99491 m_u, but the sum of the masses of 8 protons, 8 neutrons and 8 electrons is 16.13193 m_u:

Total mass of 8 protons and 8 electrons (i.e. 8 $^1_1$H atoms) = 8 × 1.007825 m_u = 8.06260 m_u.
Total mass of 8 neutrons = 8 × 1.008665 m_u = 8.06932 m_u.
Total mass of protons, neutrons and electrons composing the $^{16}_{8}$O atom = 16.13193 m_u.

There is therefore, a loss of 0.13702 m_u when the nucleons are bound in the nucleus. The Einstein equation:

$$\Delta E = \Delta m c_0^2$$

where ΔE is the energy release, Δm is the loss of mass, and c_0 is the velocity of light in a vacuum, relates energy and mass. Strictly the mass of a particle depends on its velocity, but it is sufficient here to equate the masses of the various particles to their rest masses, that is their masses at zero velocity.

Using Einstein's equation, the mass lost in the formation of one atom of $^{16}_{8}$O from its fundamental particles is equivalent to

$$\begin{aligned}\Delta E &= 0.13702 \times 1.660 \times 10^{-27}\ \text{kg} \times (2.998 \times 10^{8}\ \text{m s}^{-1})^2 \\ &= 2.05 \times 10^{-11}\ \text{kg m}^2\ \text{s}^{-2} \\ &= 20.5\ \text{pJ}\end{aligned}$$

This quantity is the *binding energy* of the $^{16}_{8}$O nucleus, the energy released when it is formed from its protons and neutrons, or, alternatively, the energy which would be required to split the nucleus itself into separate nucleons.

A quantity of particular interest for a nucleus is the binding energy per nucleon. As there are 16 nucleons in oxygen-16, the *B.E.* per nucleon is (20.5/16) pJ = 1.28 pJ. The *B.E.* per nucleon is highest for atoms of atomic number near 25. Consequently energy can be released if very heavy atoms break into smaller ones, as in uranium fission (2.9.4.3), and also if very light atoms are converted by a fusion process into heavier ones, the source of energy in stars (2.11).

The nuclear binding forces are strong attractive forces, independent of charge and of very short range. Just as an individual extra-nuclear energy level holds two electrons (Pauli principle (3.3)) so an individual nuclear energy level appears to hold two protons and two neutrons. This probably accounts for the stability of the α-particle, which is the $^4_2$He nucleus, and those nuclides with even numbers of protons and neutrons. Oxygen-16 and carbon-12 are outstanding examples, and the majority of the stable (non-radioactive) nuclides are further instances. There are 164 stable nuclides containing even numbers of both protons and neutrons, 104 for which either Z is even and N is odd or vice versa, and only four, $^2_1$H, $^6_3$Li, $^{10}_{5}$B and $^{14}_{7}$N, for which both Z and N are odd numbers.

2.5. Nuclear Models

Nuclear models have been devised as an aid in the discussion of nuclear properties; they have been very useful but must not be taken as an accurate picture of real nuclei, about the structure of which much remains to be known. Two important models are the *Shell model* in which the emphasis is on the individual particles, and the *Liquid Drop model* in which there is the idea of a surface tension that must be overcome for fission to take place.

In the shell model it is assumed that the energy of a system of nucleons is capable of taking only certain discrete values, i.e. it is quantised; the possible values of the energy cannot be reliably calculated by quantum mechanics, however, because the nature of the nuclear forces is not fully known. It is further assumed that the spins of protons and of neutrons are paired, each to each, and moreover that there is a coupling between the orbital angular momentum and the spin leading to a splitting in levels. These levels account for the experimental fact of the magic numbers of either protons or neutrons (2, 8, 20, 28, 50, 82, 126), so called because they represent especially stable nuclei. There is a broad resemblance between the shell model of the nucleus and the electron shells of the extra-nuclear structure, but the nucleons are extremely closely packed. Acceptance of the shell model implies an assumption that each nucleon moves in an average central field provided by all the others, rather as the extra-nuclear electrons move in the electrostatic field of the nucleus.

2.6. Nuclear Spin

Measurement of the energy emitted in radioactive changes (2.9) and of the magnetic properties of nuclei indicate that some nuclei possess spin (2.7) which depends on the energy state. Since in nuclei there is pairing, only nuclei which have an unpaired proton or neutron show spin.

The nuclear shell model has allowed the prediction of energy levels and spins for nuclei in which N or Z is near one of the magic numbers. For example, ${}^{19}_{9}F$ has an even number of neutrons which produce no net spin, but the odd proton as expected gives nuclear spin for the nucleus in its ground state in agreement with experiment.

The shell model is less successful in predicting energy levels for nuclei as N and Z depart from the magic numbers. Bohr and Mottelson, in their *Collective model* of the nucleus (a kind of combination of Shell and Liquid Drop models), interpret the deviations as due to the existence of low-energy vibrations and rotations in non-spherical nuclei.

2.7. Nuclear Magnetic Resonance Spectroscopy

2.7.1. Physical principles

Those atomic nuclei which have an odd number of either protons or neutrons, or odd numbers of both, possess intrinsic magnetic moments. The nuclear spin quantum number, I, analogous to the electron spin quantum number s (3.2.4) differs from it in being able to take various integral and half-integral values.

The laws governing the vector addition of nuclear spins are not known, hence while ^1H is unambiguous with $I = \frac{1}{2}$, the value of I for ^2H could be 1 or 0 depending on whether the spins are aligned or opposed. In fact ^2H has $I = 1$, but for ^4He the observed spin is zero. Observed spins have been rationalised, and empirical rules have been formulated which can be used to evaluate I:
(1) nuclei with p and n even (charge and mass even) have zero spin,
(2) nuclei with p and n odd (charge odd and mass even) have integral spin,
(3) nuclei with odd mass have half-integral spins.
Examples are:

$I = 0$ $\quad {}^4_2He, {}^{12}_6C, {}^{16}_8O, {}^{32}_{16}S$

$I = \frac{1}{2}$ $\quad {}^1_1H, {}^{13}_6C, {}^{15}_7N, {}^{19}_9F, {}^{29}_{14}Si, {}^{31}_{15}P$

$I = 1$ $\quad {}^2_1H, {}^{14}_7N$

$I = \frac{3}{2}$ $\quad {}^9_4Be, {}^{11}_5B, {}^{33}_{16}S, {}^{35}_{17}Cl, {}^{79}_{35}Br, {}^{81}_{35}Br$

$I = 2$ $\quad {}^{36}_{17}Cl$

$I = \frac{5}{2}$ $\quad {}^{17}_8O$

$I = 3$ $\quad {}^{10}_5B$

$I = \frac{7}{2}$ $\quad {}^{51}_{23}V$

Magnetic nuclei placed in a magnetic flux of flux density $\boldsymbol{B}$, acquire an energy of magnetic interaction

$$E = \mu_z \boldsymbol{B}$$

where μ_z is the component of the nuclear moment in the direction of the magnetic flux (Fig. 2.1). Quantum mechanics (3.1) restricts the orientations to those for which

$$\mu_z = \frac{m\mu}{\sqrt{I(I+1)}} \text{ and } \theta \text{ is therefore } \cos^{-1} \frac{m}{\sqrt{I(I+1)}}$$

where $m = I, I-1, I-2 \ldots -I$. It also restricts the energy changes, which can arise when the nucleus interacts with electromagnetic radiation, to those for which $\Delta m = \pm 1$.

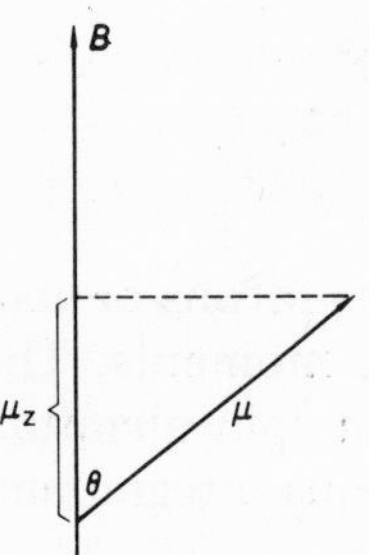

Fig. 2.1. Magnetic moment, μ, represented as a vector.

Thus for the absorption of radiation with a frequency of ν

$$E = h\nu = \frac{\mu}{\sqrt{I(I+1)}} B \therefore \nu = \left\{ \frac{\mu}{h\sqrt{I(I+1)}} \right\} B$$

The fraction in braces is characteristic of a particular nucleus.

For a magnetic flux density of 1 tesla, absorption frequencies are in the range 1—50 MHz, so that measurements are in the radiofrequency range. The energy which is absorbed causes the nuclear magnets, which precess round the direction of the applied magnetic flux, to tilt, so that θ increases in value. The precession may be indicated as shown in Fig. 2.2. Thus identical nuclei, irradiated at a particular frequency, would be expected to absorb energy at the same strength of applied flux.

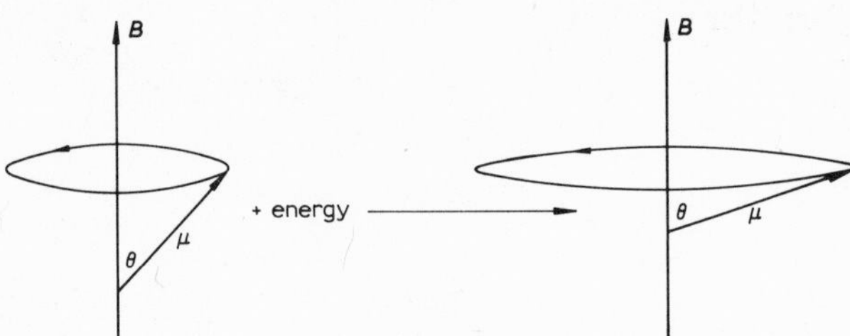

Fig. 2.2.

Chemical interest in n.m.r. spectroscopy arises from the fact that the flux density $\boldsymbol{B}$ round the nucleus depends on the flux due to its surrounding electrons and to adjacent nuclei as well as on the external flux produced by the laboratory magnet.

$$\boldsymbol{B} = \boldsymbol{B}_a + \boldsymbol{B}_b + \boldsymbol{B}_c + \boldsymbol{B}_d$$

$\boldsymbol{B}_a$ = The flux density due to the laboratory magnet;

$\boldsymbol{B}_b$ = the flux density due to currents induced by $\boldsymbol{B}_a$ in electrons around the nuclei;

$\boldsymbol{B}_c$ = the flux density due to neighbouring magnetic nuclei, large in solids, but absent in liquids and gases where random molecular motion occurs;

B_d = the flux density due to neighbouring nuclei in the same molecule but unlike B_c in being indirect, depending on currents induced in electrons by the adjacent nuclei, and also in not being averaged to zero by random molecular motion in liquids and gases.

2.7.2. Measurement of internuclear distances in solids

In solids, B_b and B_d may be neglected in relation to B_c. The strength of the applied flux B_a required to produce resonance absorption in a solid is therefore related to B and the spectrum has the form shown in Fig. 2.3.

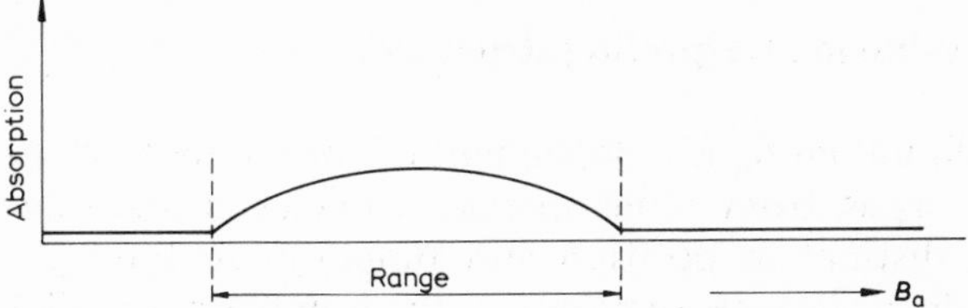

Fig. 2.3. Nuclear magnetic resonance spectrum for a solid.

The shape of the absorption curve reflects the range and distribution of local magnetic fluxes which depend on the distances of nuclear separation. From it, nuclear distances may be determined with a higher degree of precision than by other methods: for instance in NH_4Cl, r_{N-H} = (103.8 ± 0.4) pm by n.m.r. spectroscopy and (103 ± 2) pm by neutron diffraction.

2.7.3. Application to molecular structure determination

Nuclear magnetic resonance spectroscopy is used predominantly in the elucidation of molecular structure in liquids and gases. B_b and B_d (unlike B_c) take discrete values, and the plot of absorption against B_a shows fairly sharp bands. Usually B_b has a larger value than B_d, and spectra obtained at low resolution can be interpreted in terms of B_b variations. The ratio B_b/B_a is characteristic of a particular environment in a molecule. The value of $B_b/B_a \times 10^6$, referred to standard environment structure is termed the 'chem-

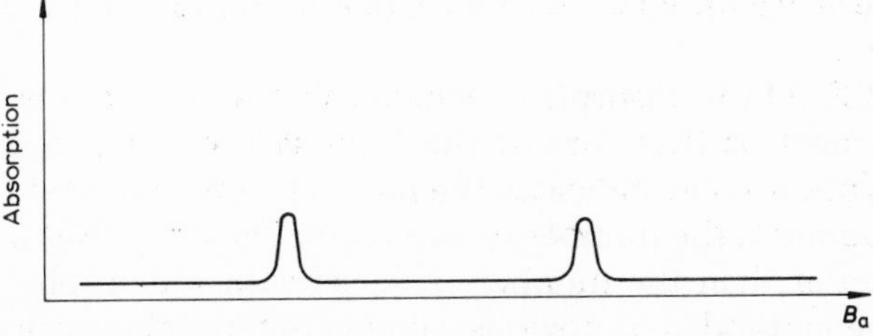

Fig. 2.4. Low resolution ^{31}P resonance spectrum of $P_4O_{13}^{6-}$ ion.

ical shift' characteristic of a particular environment. In the ^{31}P resonance spectrum of the tetrapolyphosphate ion, for example, there are two absorption bands of equal intensity (Fig. 2.4). One arises from the two central P atoms, which both have the same type of environment; the other from the two terminal P atoms, which both lie in a rather different environment:

$$\left[\begin{array}{c} \text{O}\quad\ \text{O}\quad\ \text{O}\quad\ \text{O} \\ |\qquad\ |\qquad\ |\qquad\ | \\ \text{O—P—O—P—O—P—O—P—O} \\ |\qquad\ |\qquad\ |\qquad\ | \\ \text{O}\quad\ \text{O}\quad\ \text{O}\quad\ \text{O} \end{array}\right]^{6-}$$

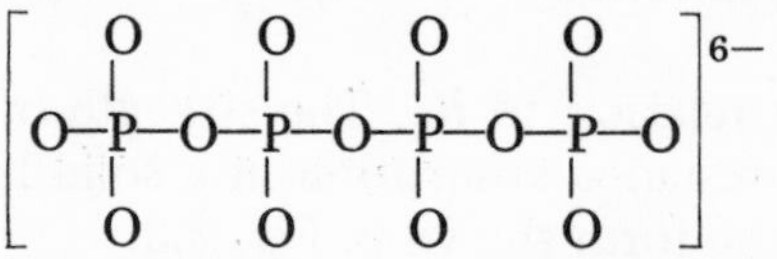

2.7.4. Spin coupling and its value in interpreting structure

The flux density $\boldsymbol{B}_d$, which unlike $\boldsymbol{B}_b$ is independent of the strength of the externally applied flux $\boldsymbol{B}_a$, arises from other magnetic nuclei in the same molecule, particularly ones distinct in position and function. In these the adjacent nuclear magnets induce circulatory currents; the number of possible local fluxes depends on the orientations of the neighbouring nuclear magnets. In the T-shaped ClF_3 molecule, two F atoms (F_a) have one kind of environment, and the other (F_b) has a different environment

F—Cl—F
(a) | (a)
F
(b)

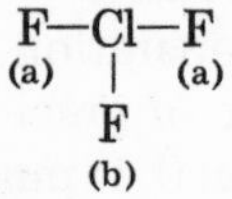

Each ^{19}F ($I = \frac{1}{2}$) may have two effective orientations (i.e. averaged over all precession angles)

$\rightarrow$
or
$\leftarrow$

F_b experiences three different $\boldsymbol{B}_d$ fluxes due to the two nuclei F_a (I and II)

$\rightarrow$ I $\rightarrow$ II	$\rightarrow$ I $\leftarrow$ II $\leftarrow$ I $\rightarrow$ II	$\leftarrow$ I $\leftarrow$ II
$\frac{1}{4}$ *of the molecules*	$\frac{1}{2}$ *of the molecules*	$\frac{1}{4}$ *of the molecules*

but the two F_a atoms experience only two $\boldsymbol{B}_d$ fluxes due to F_b, according as to whether its spin is $\rightarrow$ or $\leftarrow$.

At a low temperature (220 K) the absorption, which can only be recorded by a high resolution instrument, is therefore of the form shown in Fig. 2.5.

The area under an absorption band indicates the number of paramagnetic nuclei in a particular environment; the number of peaks (for the $I = \frac{1}{2}$ case) in a multiplet band is one greater than the number of nuclei with which spin—spin interaction occurs. The method is of obvious value in determining structure.

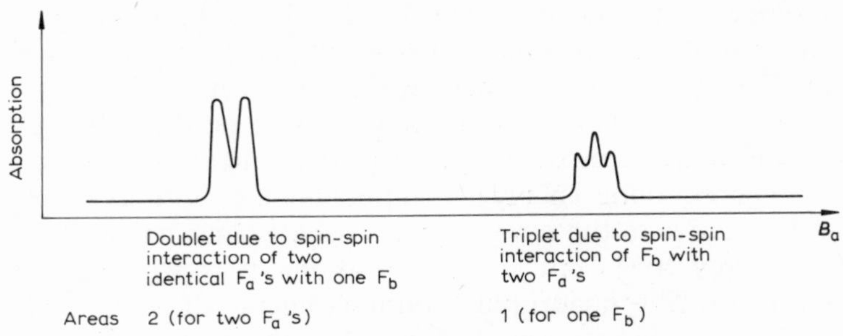

Fig. 2.5. High-resolution ^{19}F resonance spectrum of ClF_3 at 220 K.

2.7.5. Applications to study of fast exchange reactions

At about 330 K, the n.m.r. spectrum of ClF_3 has a single ^{19}F band because of rapid exchange of fluorine atoms between the two environments. As the temperature is decreased the band broadens and eventually splits into the doublet and triplet shown above. From this behaviour, information on the kinetics of the fluorine exchange may be deduced. The techniques can often be applied to other processes such as this which involve no net chemical change.

2.8. Nuclear Quadrupole Coupling

2.8.1. Principles

Where the spin quantum number I of a nucleus exceeds $\frac{1}{2}$, the distribution of its positive charge is non-spherical. The electric quadrupole moment, Q, measures the extent to which the distribution of positive charge deviates from spherical symmetry (Fig. 2.6).

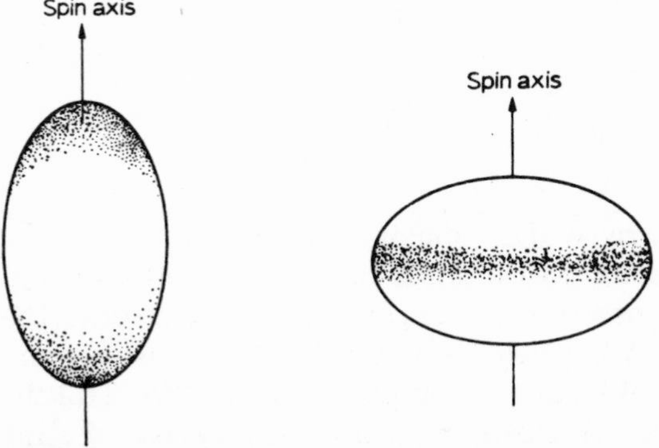

Fig. 2.6. Left: Prolate spheroid of +ve charge. Greatest charge density at poles. Right: Oblate spheroid of +ve charge. Greatest charge density at equator.

In atoms and molecules the nuclei are enveloped in an electronic cloud. When the electrons round a quadrupolar nucleus produce a non-spherical charge distribution, there is interaction between the nuclear field and the field due to the electrons; as a result the nucleus has a potential energy which depends on the orientation of its quadrupole moment to the electronic field (quadrupole coupling energy). The possible orientations of the spin axis of a nucleus relative to the rotation axis of the molecule as a whole represent quantised energy levels and produce rotational lines with nuclear quadrupole fine structure. The quadrupole coupling energy of a single nucleus is proportional to eQq; where e is the charge on the proton; Q is the nuclear quadrupole moment; q is the field gradient, being equal to $\partial^2 V/\partial z^2$ where V is the potential at the nucleus arising from all the charges outside it and z is a co-ordinate referred to a fixed axis. The nuclear quadrupole coupling constant eQq/h, where h is Planck's constant (3.1), can be calculated from the hyperfine structure of rotational lines in the microwave range — for gases — or from the nuclear quadrupole resonance spectrum of solids in the radiofrequency range. In the latter case only the modulus of eQq/h is obtained.

2.8.2. Applications

Since s electrons and filled inner shells have spherical symmetry, and since d and f electrons do not penetrate close to the nucleus, the field gradient depends largely on the p electrons in the valence shell. Quadrupole coupling constants therefore provide information about the character of chemical bonds; a quadrupolar nucleus can be imagined as a kind of built-in probe giving information about the electronic field around it.

Some values for eQq/h of ^{35}Cl in chlorine compounds are:

	$\frac{eQq}{h}$/MHz
BrCl	−103.6
ICl	−82.5
SiF_3Cl	−43.0
TlCl	−15.8
KCl	+0.04

(In the case of solids, the sign is arrived at by analogy.)

The numerical value of eQq/h for KCl is small because the chloride ion, with an s^2p^6 structure, has a symmetrical arrangement of p electrons. The large value for BrCl indicates that the field due to the p electrons is asymmetric; some of the p electrons take part in covalent bonding. For TlCl the field due to the p electrons of the Cl is evidently only slightly asymmetric; hence the compound is essentially ionic in character.

The quadrupole coupling constants of ^{14}N in some of its compounds are:

	$\frac{eQq}{h}$/MHz
HCN	−4.58
CH_3CN	−4.40
NH_3	−4.08
CH_3NC	+0.5

The low numerical value for the ^{14}N coupling constant in CH_3NC indicates a fairly uniform p-electron distribution around the N atom — it is therefore essentially quadricovalent — compared with the asymmetric arrangement around the tervalent nitrogens in the other compounds.

Clearly the measurement of nuclear quadrupole coupling constants is valuable in elucidating the nature of valency bonds. Furthermore, being particularly sensitive to the electronic distribution near the nucleus the coupling constants provide information about the adequacy of wave functions in that region. By contrast, bond energies and dipole moments give information mainly about the overlap region of the valence bond.

2.9. Radioactivity

The manifestations of radioactivity are the emission from the nucleus of α-particles ($^4_2$He nuclei moving with high kinetic energies), β^--particles (electrons of nuclear origin) and β^+-particles (positrons, 2.9.3). Particle emission is often accompanied by energy in the form of electromagnetic waves of very high frequency known as γ-radiation. The energies involved in a radioactive change are so great that the change is unaffected in rate by alterations of temperature of several thousand Kelvins. Individual emissions are governed by statistical laws, and are generally accompanied by a characteristic radiation.

The probability of a nucleus disintegrating in unit time is the decay constant, λ. For N nuclei

$$\frac{dN}{dt} = -\lambda N \text{ and } N = N_0 e^{-\lambda t}$$

where N_0 is the original number of nuclei and N the number left after t (in convenient units).

A useful constant is the half-life, $t_{1/2}$, the time taken to halve the number of atoms of the nuclide.

$$\text{Substituting } \frac{N}{N_0} = \tfrac{1}{2} \text{ above: } \lambda = \frac{\ln 2}{t_{1/2}} = \frac{0.693}{t_{1/2}}$$

2.9.1. α-Emission

The α-particle is a helium nucleus. This corresponds to a helium atom which has lost two electrons and become He^{2+}. Its emission depends on the probability of the α-particle penetrating the potential energy barrier arising from the nuclear interactions. Thus the greater the energy of the particle in relation to the energy barrier the higher the decay constant. The energies of α-particles, though not identical for a particular species, usually lie within a narrow range and are high, generally greater than 0.7 pJ. Because of their size and charge, α-particles do not penetrate far into matter. Their energies can be roughly assessed by observing their penetration, or can be measured from their response to magnetic and electric fields.

2.9.2. β^--Emission

A β^--particle is an electron of nuclear origin. Its emission from a nucleus is always accompanied by the emission of an uncharged particle of negligible mass called an antineutrino. Thus ${}^{14}_{6}C$ changes to ${}^{14}_{7}N$ by a β^--emission:

$${}^{14}_{6}C \rightarrow {}^{14}_{7}N + {}^{0}_{-1}\beta + \text{antineutrino}$$

The total release of energy can be calculated from the isotopic masses of carbon-14 and nitrogen-14.

$$\begin{aligned} m_a({}^{14}_{6}C) &= 14.003242\ m_u \\ m_a({}^{14}_{7}N) &= 14.003074\ m_u \\ \hline \Delta m_a &= 0.000168\ m_u \\ &= 24.8\ \text{fJ} \end{aligned}$$

The antineutrino has spin which enables the angular momentum to be conserved in the process of disintegration. The energy of 24.8 fJ which is released in the above example is shared between the β^--particle and the antineutrino, thus the actual kinetic energy of the β^--particle from a particular nuclide can vary, and the energy calculated above represents the *maximum β-energy*, that is the largest energy which the β^--particle produced by that particular process can have. Energies of particles derived from a nuclear change are often expressed in mega-electronvolts, MeV. The electronvolt is not an SI unit, in fact it is not strictly a unit at all, being the product of a true unit, the volt, and a physical quantity, the charge on the electron:

$$\begin{aligned} 1\ \text{electronvolt} &= 1\ \text{V} \times 1.6021 \times 10^{-19}\ \text{C} \\ &= 1.6021 \times 10^{-19}\ \text{J} \\ \therefore 1\ \text{MeV} &= 1.6021 \times 10^{-13}\ \text{J} \end{aligned}$$

Thus the maximum β^--energy above could also be expressed as

$$\frac{24.8 \times 10^{-15}}{1.6021 \times 10^{-13}}\ \text{MeV} = 0.155\ \text{MeV}$$

The emission of a particle, either α or β, is sometimes followed by the emission of a γ-ray, similar to an X-ray but of shorter wavelength:

$^{24}_{11}Na \xrightarrow{\beta^-}$ unstable nuclide $\xrightarrow{\gamma} {}^{24}_{12}Mg$

The probable explanation is that an immediate rearrangement of protons and neutrons, causing a release of energy, occurs after the nucleus has lost a particle.

Alternative processes often occur, such as those illustrated in Fig. 2.7 in which chlorine-38 produces three β-particles, in the proportions and of the energies shown. Conversion to a stable argon atom is completed by low energy β-emission being followed by γ-radiation; but this is not necessary after high energy β-emission.

Frequently disintegration processes are far more complicated.

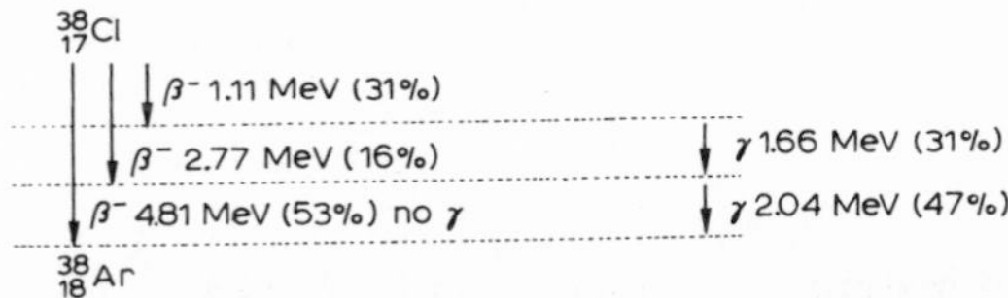

Fig. 2.7. Disintegration scheme for chlorine-38.

2.9.3. Positrons: β^+-emission

Dirac (1928) showed theoretically that a particle of the mass of an electron may have an energy less than $-mc_0^2$ or greater than $+mc_0^2$ but not of an intermediate value, and proposed the following hypothesis. The negative states are normally all filled and give an all-pervading, uniform density of negative charge whose presence cannot be detected experimentally: for otherwise ordinary electrons would spontaneously disappear by falling into the lower energy states and taking on the properties of particles of negative mass (negative kinetic energy). Such particles would, for instance, move *against* an applied force instead of with it. A perfect vacuum is thus a *sea* of electrons in a negative energy state (not to be confused with a negative charge), and only *changes* in this background are observable. If, however, a particle in the sea is given enough energy ($> 2mc_0^2$), it can go into a positive energy state and is observed as an ordinary electron. When an electron goes into this state it leaves a 'hole' and this is also observed, but as something with positive energy (for to fill the 'hole', and make it disappear, negative energy must be added) and with positive charge (being a missing negative charge in the uniform sea). Excitations of this kind are actually observed with energy absorption of at least $2mc_0^2$ or 1.02 MeV, and the process is described as *pair production.* The 'hole' behaves like a positive 'electron' and is called a *positron.* The reverse process also occurs, an electron and a positron annihilating each other with the liberation of the same amount of energy as that required in pair production. Both processes require the presence of a nucleus — as 'catalyst' — in order that momentum conservation conditions are satisfied.

Anderson (1932) discovered the positron in a study of cosmic rays, using the Wilson cloud chamber. Blackett (1933) confirmed his findings.

Curie and Joliot obtained positron—electron pairs from heavy metals bombarded with high energy (5 MeV) γ-rays derived from beryllium mixed with polonium. The average life of the positron is about 10^{-9} s. On colliding with an electron both are annihilated and γ-radiation — the annihilation radiation — is emitted. The formation and annihilation of a positron—electron pair may be represented as in Fig. 2.8.

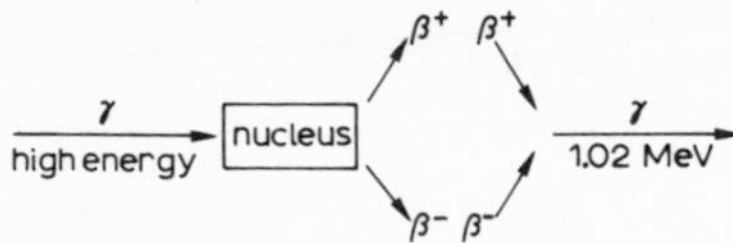

Fig. 2.8.

The energy of a β^+-particle at the instant of emission can be found by considering copper-64. This nuclide, in one of its transitions, emits a positron with a maximum energy of 106 fJ, giving nickel-64.

For ^{64}Cu	$m_a = 63.92976\ m_u$
^{64}Ni	$m_a = 63.92796\ m_u$
Difference	$\Delta m_a = \ \ 0.00180\ m_u$
	$= 269$ fJ
Energy required to create a β^+	$= 163$ fJ
Maximum kinetic energy available to the positron	$= 106$ fJ

A nuclide which is neutron-rich is normally a β^--emitter, one which is neutron-deficient may be a β^+-emitter. Thus sodium-24, with one neutron more than the stable sodium-23, emits a β^- particle and changes to magnesium-24, but sodium-22, with one neutron fewer than the stable isotope, loses a β^+-particle and becomes neon-22. Since 1.02 MeV, i.e. 163 fJ, is necessary to create a positron—electron pair however, a neutron-deficient nuclide can only be a β^+-emitter if at least that amount of energy is available from the decay process. The usual alternative, where there is insufficient energy for pair production, is the capture by the nucleus of an extra-nuclear electron. The process is called K-capture because the electron which is captured will normally come from the K shell which is nearest to the nucleus.

2.9.4. Nuclear reactions

2.9.4.1. α-Induced reactions

Rutherford (1919) noticed that high-energy α-particles passing through nitrogen produced protons of long range:

$$^{14}_{7}N + ^{4}_{2}He \rightarrow ^{17}_{8}O + ^{1}_{1}H$$

This is a *nuclear reaction.* As the particle which initiates it is an α-particle and the particle which is liberated is a proton, it is designated an (α, p) reaction.

In 1932 Chadwick produced neutrons by bombarding beryllium with α-particles:

$$^{9}_{4}Be + ^{4}_{2}He \rightarrow ^{12}_{6}C + ^{1}_{0}n$$

This is an (α, n) reaction; notice that the sum of the mass numbers and the sum of the atomic numbers is unchanged in any nuclear reaction.

(α, n) and (α, p) reactions can be used in the production of radionuclides of light elements, e.g.:

$$^{24}_{12}Mg\ (\alpha, n)\ ^{27}_{14}Si\ (\beta^{+},\ t_{1/2} = 42\ s)$$

2.9.4.2. Proton-induced reactions

Cockroft and Walton (1932) disrupted lithium-7 with protons energised in a linear accelerator:

$$^{7}_{3}Li + ^{1}_{1}H \rightarrow 2\ ^{4}_{2}He$$

This (p, α) reaction liberates 17.3 MeV of energy, there being a mass reduction of 0.0186 m_u in the process.

2.9.4.3. Neutron-induced reactions

The neutron, because it has no charge, can penetrate nuclei which repel α-particles and protons, and it is the most useful bombarding particle for the production of radionuclides:

$$^{23}_{11}Na\ (n, \gamma)\ ^{24}_{11}Na\ (\beta^{-},\ t_{1/2} = 15.1\ h)$$

$$^{32}_{16}S\ (n, p)\ ^{32}_{15}P\ (\beta^{-},\ t_{1/2} = 14.3\ d)$$

$$^{27}_{13}Al\ (n, \alpha)\ ^{24}_{11}Na$$

The (n, γ), or neutron-capture reaction, in which a gamma ray but no particle is emitted, yields a product isotopic with the target. It is of particular importance in the production of heavy nuclides, including the transuranium elements (31.1).

The (n, f) reaction, or neutron-induced fission, is exemplified by

$$^{235}_{92}U + ^{1}_{0}n \rightarrow ^{92}_{36}Kr + ^{142}_{56}Ba + 2\ ^{1}_{0}n$$

The target atom, uranium 235, breaks into one particle of mass about 95 m_u, another of mass about 140 m_u and two or three neutrons. Nuclear reactors are designed so that these neutrons set up a steady chain reaction. The primary fission products have high A/Z ratios and lose β^{-}-particles, e.g.:

$$^{92}_{36}Kr \xrightarrow[3s]{\beta^-} ^{92}_{37}Rb \xrightarrow[5s]{\beta^-} ^{92}_{38}Sr \xrightarrow[2.7h]{\beta^-} ^{92}_{39}Y \xrightarrow[3.6h]{\beta^-} ^{92}_{40}Zr\ \text{(stable)}$$

A fission product of particular interest is technetium-99 ($t_{1/2} = 2.2 \times 10^5$ y), an element which does not occur in nature, as any present when the Earth's crust was formed has presumably decayed away.

2.9.4.4. Reactions induced by high-energy γ-rays

A well-known example is

$$^{9}_{4}\text{Be} + \gamma \rightarrow {}^{8}_{4}\text{Be} + {}^{1}_{0}\text{n}$$

It enables a mixture of beryllium powder with an emitter of high-energy γ-rays such as antimony-124 to be used as a laboratory neutron source.

2.10. Mössbauer Spectroscopy

There are some stable nuclei, that of ^{57}Fe is an example, which have an excited state with a very short lifetime which is only a few keV in energy above the ground state. For iron-57 the first excited state of the nucleus ($I = \frac{3}{2}$) is 14.4 keV (2.31×10^{-15} J) above the ground state ($I = \frac{1}{2}$) and its lifetime is about 10^{-7} s. Because the lifetime is so short, absorption of energy by the nucleus is limited to the range ΔE predicted by the Uncertainty Principle (3.1.1):

$$\Delta E \Delta t \sim h$$

$$\therefore \Delta E \sim \frac{h}{\Delta t} = \frac{6.626 \times 10^{-34}\ \text{J s}}{1 \times 10^{-7}\ \text{s}} = 6.6 \times 10^{-27}\ \text{J}$$

Thus there is a very narrow energy range, representing only about 10^{-12} of the difference between the $I = \frac{1}{2}$ state and the $I = \frac{3}{2}$ state, for which resonant absorption is possible.

γ-radiation of the correct energy for resonant absorption by iron-57 nuclei (2.19% of natural iron) is obtainable from cobalt-57, a nuclide which decays, with a half-life of 270 days, by capturing an extra-nuclear electron. The decay scheme is shown in Fig. 2.9.

Although it was realised that the 14.4 keV γ-radiation from ^{57}Co should suffer resonance absorption by ^{57}Fe, the phenomenon was not observed until

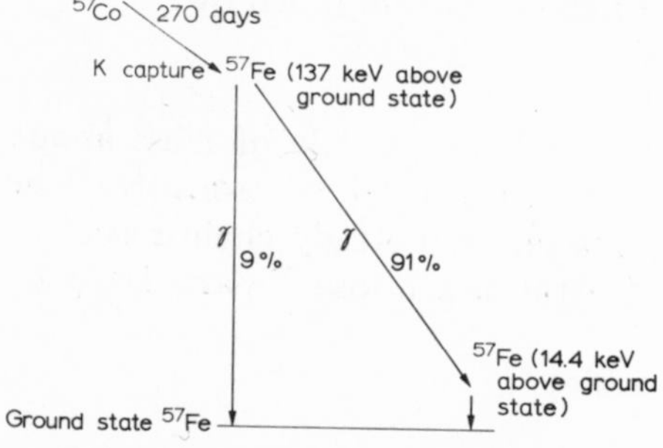

Fig. 2.9. Decay scheme for cobalt-57.

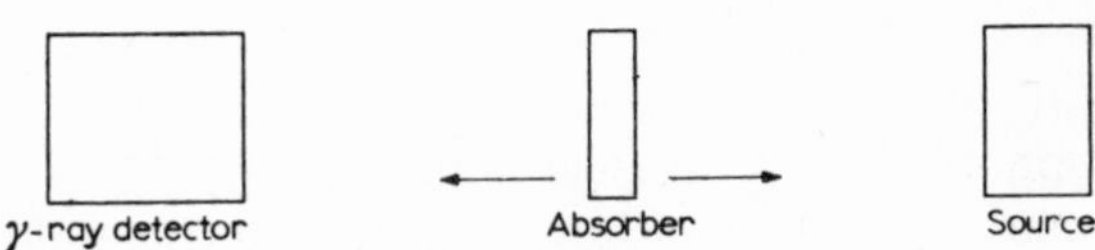

Fig. 2.10. Arrangement of source, absorber and detector for Mössbauer spectrometry.

1958, when Mössbauer succeeded in producing satisfactory sources by incorporating ^{57}Co in stainless steel to minimise the reduction of the γ-energy caused by loss of energy to daughter atoms on recoil.

The apparatus for Mössbauer spectroscopy consists of a source such as ^{57}Co in steel, an absorber containing the daughter nuclide, ^{57}Fe in this particular case, and a γ-ray detector (Fig. 2.10). If γ-radiation from the source strikes an absorber in which the ^{57}Fe atoms have the same electronic environment as in the daughter atoms produced in the source itself, there is resonance absorption by which nuclei in the absorber are raised to the $I = \frac{3}{2}$ state. But if the ^{57}Fe atoms in the absorber are in a different state of chemical combination from those produced by ^{57}Co decay in the source, the absorber must be moved at a steady speed relative to the source to fulfill the conditions for resonance absorption, the Doppler shift in frequency being just sufficient to compensate for the change induced in the target nucleus by its electronic environment. The speed at which the absorber must move to cause maximum energy absorption is called the *isomer shift*, δ, and it can be positive or negative (Fig. 2.11). The size of δ depends particularly on the density of s electrons (3.2.1) at the nucleus. In iron-57 an increase in electron density causes a negative isomer shift; since d electrons (3.2.2) tend to shield the nucleus slightly from the s electrons the value of δ falls as the number of d electrons in the iron atom falls. Mean values of δ for some oxidation states of iron are:

	Fe^{I}	Fe^{II}	Fe^{III}	Fe^{IV}	Fe^{V}
$\delta/mm\ s^{-1}$	+2.3	+1.5	+0.7	+0.2	−0.6

Thus a δ value can be used to determine the oxidation state of an element in the solid compound used as a Mössbauer absorber. For example, Fe^{+} in a solid solution of $FeCl_2$ in NaCl, Fe^{3+} in nonstoichiometric FeO and Fe^{4+} in Sr_2FeO_4 have been confirmed, and the mixed oxide $FeVO_3$ has been shown to contain Fe^{3+} but not Fe^{2+}.

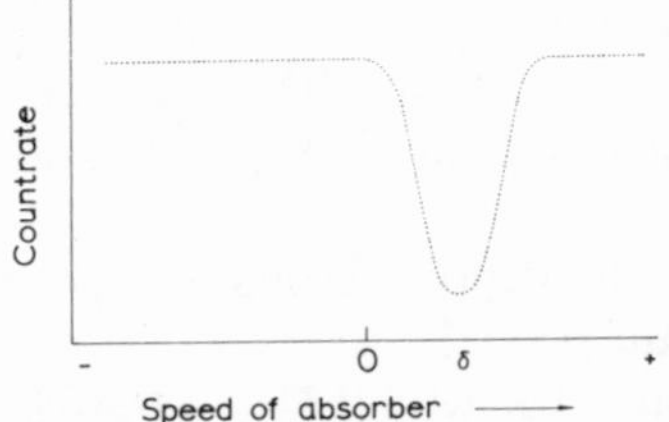

Fig. 2.11. Mössbauer spectrum. No quadrupole interaction.

2.10.1. Quadrupole interactions

A nucleus is not necessarily spherical. Any nuclear state with $I > \frac{1}{2}$ has a quadrupole moment Q which can align itself either with an electric field gradient or across it. For example an $I = \frac{3}{2}$ state becomes, in a field of gradient q, a doublet of separation ΔE_Q (Fig. 2.12a(ii)). As ΔE_Q is proportional to Qq it is possible to calculate the field gradient around the nucleus from the distance between the two absorption maxima (Fig. 2.12b).

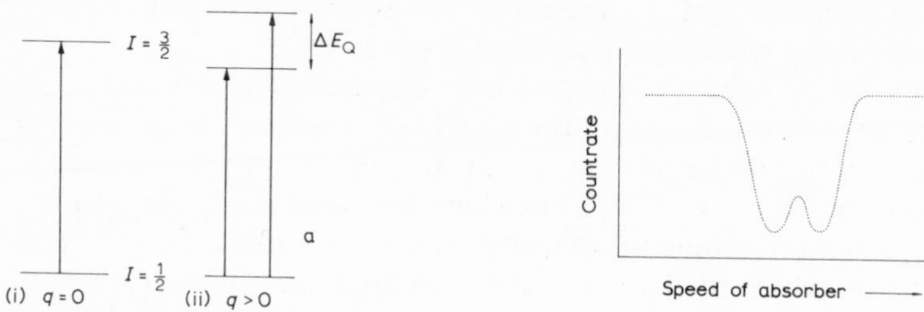

Fig. 2.12a. Mössbauer absorption in target Fe-57 atom with (i) spherically symmetrical electronic environment, (ii) unsymmetrical environment.

Fig. 2.12b. Mössbauer spectrum for case (ii) in Fig. 2.12a.

In the $Fe(CN)_6^{4-}$ ion there is a low-spin d^6 arrangement (6.3.1) surrounded octahedrally by CN^- ligands. There is thus identical electron density in the x, y and z directions, $q = 0$, and there is no quadrupole splitting. But if the symmetry of the electron density is destroyed by introducing a heterogroup into the co-ordination sphere, as in $Fe(CN)_5NO^{2-}$, a large ΔE_Q is observed. $Fe(CN)_6^{3-}$ is quite different from $Fe(CN)_6^{4-}$ because the low-spin d^5 ion has an unpaired electron, and quadrupole splitting is observed for it. The situation of Fe^{II} and Fe^{III} in high-spin complexes is the reverse; in this case the field gradient, and therefore ΔE_Q, is small in the Fe^{III} case ($t_{2g}{}^3e_g{}^2$) and large when there is a less symmetrical arrangement of d electrons in Fe^{II} ($t_{2g}{}^4e_g{}^2$) (see 6.3).

Mössbauer spectroscopy can be used to study the influence of different ligands on the s-electron density around a metal atom. In the series $Fe(CN)_5L^{-n}$ the isomer shifts have been found to decrease in the series L = $NH_3 > PPh_3 > SO_3^{2-} > CN^- > CO$ because back-donation of d electrons to the ligand (6.2) reduces the shielding of the s electron from the Fe nucleus and hence reduces the isomer shift.

Mössbauer studies have been extended to other elements with low-lying excited states of short lifetime. Target nuclei include iridium-191, which absorbs the 129 keV γ-rays from osmium-191 (β, 16d), and also ^{61}Ni, ^{119}Sn and ^{197}Au.

The method has some advantages over nuclear quadrupole resonance spectroscopy for the study of field gradients around nuclei:

(i) It can be applied to nuclei with ground states for which $I = \frac{1}{2}$ whereas for n.q.r. the ground state must have $I > \frac{1}{2}$.

(ii) The technique is simpler.
(iii) It is possible to obtain the sign, as well as the magnitude, of q, the field gradient, in solid absorbers.

2.11. Stellar Energy

The recognition of the enormous energy set free in nuclear reaction has suggested the source from which the stars draw their energy. At the exceedingly high temperatures prevailing (10^8 K) the nuclei are stripped of electrons and attain velocities comparable with those of particles from the cyclotron, thus making possible thermonuclear reactions. Weizsacker and Bethe independently (1938) proposed as the energy source of stars the carbon cycle in which four protons are converted into one alpha particle with the release of 4.8 pJ (30 MeV) (Fig. 2.13).

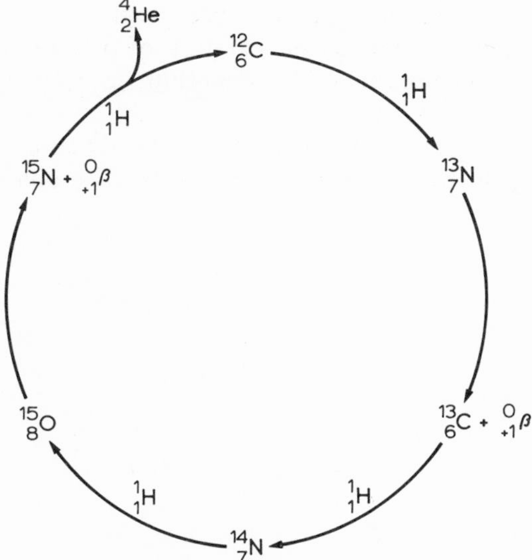

Fig. 2.13. The stellar carbon cycle.

The energy of cooler stars like the sun (10^7 K) appears to emerge from a proton—proton cycle:

$$^{1}_{1}H + {}^{1}_{1}H \rightarrow {}^{2}_{1}H + \beta^{+} + 0.07 \text{ pJ}$$

$$^{1}_{1}H + {}^{2}_{1}H \rightarrow {}^{3}_{2}He + \gamma + 0.88 \text{ pJ}$$

$$^{3}_{2}He + {}^{3}_{2}He \rightarrow {}^{4}_{2}He + 2\,{}^{1}_{1}H + 2.05 \text{ pJ}$$

Four protons are converted into an α-particle with the release of 4.1 pJ, including the annihilation energy of a positron—negatron pair. In the hottest stars more complicated cycles than either the carbon cycle or the proton cycle must occur (vide infra).

2.12. Genesis and Abundance of the Elements

Many estimates have been made of the relative abundance of the elements in the Universe, notably by Goldschmidt (1931), Brown (1949) and Urey (1952). The logarithm of the estimated abundance, taking $\log A_{Si} = 6$ as standard, is plotted against atomic number in the diagram (Fig. 2.14). The points of interest are:

(i) the higher abundance values for the even elements,
(ii) the rapid fall in abundance with atomic number up to element 45 after which the variations are smaller,
(iii) the surprisingly low abundance of Li, Be and B,
(iv) the high values for elements with Z around 26, 54 and 78,
(v) the abnormal abundance of iron.

Various theories have been advanced to explain the relative abundance of the elements. In recent years knowledge has accumulated of the types of nuclear transformations occurring in stars. A continuous process of synthesis and consumption of elements (Burbidge, Burbidge, Fowler and Hoyle, 1957) accounts for the observed differences in composition of stars of different ages and also such abnormalities as the presence of technetium in S-type stars.

Except in catastrophic phases a star has a self-governing energy balance, the temperature attained depending on the nuclear fuel available. The con-

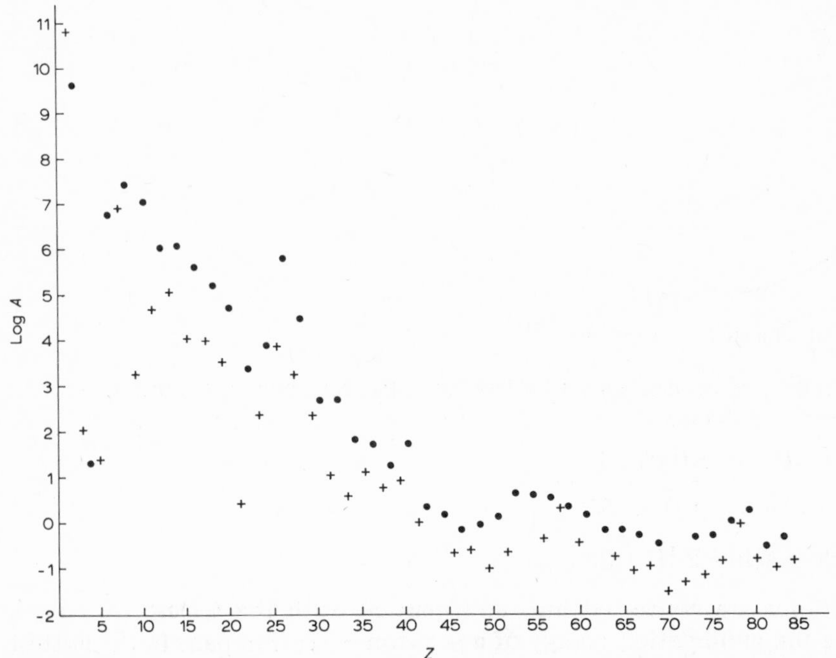

Fig. 2.14. Cosmic abundance of elements compared with silicon (10^6) (Suess and Urey). (●) Even atomic numbers. (+) Odd atomic numbers.

sumption of hydrogen is responsible for the production of energy on a large scale, and the nuclear reactions employed in the changes do not require such a high temperature as other stellar processes. Cyclic processes, like the proton cycle and the carbon cycle, synthesise He and those isotopes of C, N, O, F, Ne and Na not produced by He consumption or α-processes.

The conversion of helium to ^{12}C, ^{16}O and ^{20}Ne occurs in stars whose temperatures are about 10^8 K.

$$^{4}He + {}^{4}He \rightarrow {}^{8}Be,$$

$$^{8}Be + {}^{4}He \rightarrow {}^{12}C^{*} \rightarrow {}^{12}C + \gamma \text{ etc.}$$

At temperatures of about 2×10^8 K the so-called α-process also becomes possible. Under these conditions the γ-rays produced in the helium consumption are sufficiently energetic to bring about the change

$$^{20}Ne + \gamma \rightarrow {}^{16}O + \alpha$$

and for the α-particle released to have the energy necessary for such reactions as $^{20}Ne(\alpha,\gamma)^{24}Mg$ and $^{24}Mg(\alpha,\gamma)^{28}Si$, in which the α-addition goes on certainly as far as ^{40}Ca, and possibly to ^{48}Ti. The high abundance of nuclei with mass numbers divisible by four compared with that of their neighbours is due to the remarkable efficiency of the He consumption and α-processes. The C,N cycle belongs to this stage; it results in the catalytic conversion of hydrogen into helium until the hydrogen is exhausted.

Elements of the first transitional period are synthesised at about 3×10^9 K in the so-called e-process. The nuclear reactions operating are of many types (γ,α) (γ,p) (γ,n) (α,γ) (p,γ) (n,γ) (p,n) but they always lead to preponderance of elements belonging to the iron group. The strongly-marked peak in the abundance curve at ^{56}Fe is due to the e-process, which occurs when stellar evolution is at an advanced stage.

Two nuclear reactions involving neutron capture are responsible for the production of the majority of nuclides. The s-process, in which (n,γ) reactions occur on a time-scale of 10^2 to 10^5 years per neutron capture, gives rise to most of the nuclides in the mass-range 23—46 which are not formed in the α-process, together with a large proportion of those in the range 63 to 209. The s-process is responsible for peaks in the abundance curve at mass numbers 90, 138 and 208. The r-process also involves (n,γ) reactions but with a time-scale of 0.01 to 10 s per neutron capture. It is responsible for very many nuclides in the mass range 70 to 209, and for Th and U. Peaks in the abundance curve are produced at mass numbers 80, 138 and 194. The advanced stage of stellar evolution necessary to build up elements of such high mass numbers is uncommon in the Universe and accordingly the heavier elements are relatively rare.

The p-process is one of proton capture (p,γ) or of gamma-ray absorption with neutron emission (γ,n), and is responsible for many proton-rich nuclides of low abundance, generally derived from iron-group elements. Finally, an x-process is responsible for the synthesis of D, Li, Be and B which are unstable at the temperatures reached inside stars, where they are converted to helium by a large number of processes including $^2D(p,\gamma)^3He$, $^6Li(p,\alpha)^3He$ and $^{11}B(p,\alpha)^8Be \rightarrow 2\ ^4He$.

The rarity of D, Li, Be and B compared with H, He, C, N, and O is due partly to the inefficiency of their production and partly to their consumption in processes such as those just set out.

The composition of the Earth differs from that of the Universe by having a much lower proportion of hydrogen and helium; these elements comprise about 90% and 9% respectively of the currently accepted grand total of matter. The small planets, Venus, Mercury and Mars, appear to have compositions similar to that of the Earth. Presumably their masses are too small for them to attract and retain the lighter particles in the form of an atmosphere as they move through space. Jupiter, however, has a core of iron and siliceous material surrounded by ice, solid methane and ammonia and finally by hydrogen and helium.

Apart from the small values for hydrogen and helium, the Earth's composition is not remarkable, the percentages by weight of the commonest elements, taking hydrosphere and lithosphere together, are:

O	49.20	Mg	1.93	C	0.08
Si	25.67	H	0.87	S	0.06
Al	7.50	Ti	0.58	Ba	0.04
Fe	4.71	Cl	0.19	N	0.03
Ca	3.39	P	0.11	F	0.03
Na	2.63	Mn	0.08	Sr	0.02
K	2.40			All others	0.47

The composition of the atmosphere at sea level in moles per million moles is:

N_2	7.8×10^5	Ne	1.8×10^1	N_2O	0.5
O_2	2.1×10^5	He	5.2	H_2	0.5
Ar	9.3×10^3	CH_4	1.5	O_3	0.4
CO_2	3×10^2	Kr	1	Xe	0.08

2.13. Man-made Heavy Elements

Elements of atomic number greater than 92 have been made either by neutron-bombardment of uranium in a reactor or by nuclear reactions between cyclotron-accelerated ions such as C^{6+} and heavy atoms like uranium-238 (31.5). In general the half-lives of the nuclides so formed tend to fall as atomic numbers increase. For example, the longest-lived isotope of element 104 has a half-life of 70 s. Seaborg has predicted, however, that some isotopes of elements with atomic numbers in the range 110—114 might be much more stable, not only to spontaneous fission but also to α-emission; half-lives as high as 10^8 years have been suggested. The search for such elements in nature has not been successful, however, and recent calculations of fission barriers and neutron separation energies for such nuclei indicate that the production of superheavy atoms either in the astrophysical r-process or in man-made nuclear transformations must be considered unlikely.

Further Reading

B.H. Bransden, The elementary particles, Q. Rev., 22 (1967) 474.

A. Weiss, Nuclear magnetic resonance in solid-state chemistry, Angew. Chem. Internat. Ed., 11 (1972) 607.

N.N. Greenwood, Applications of Mössbauer spectroscopy to problems in solid-state chemistry, Angew. Chem. Internat. Ed., 10 (1971) 716.

K.E. Zimen, Nuclear energy reserves and long-term energy requirements, Angew. Chem. Internat. Ed., 10 (1971) 1.

H.D. Fryer and K. Wagener, Electrochemical processes for the enrichment of isotopes, Angew. Chem. Internat. Ed., 6 (1967) 757.

R.L. Mössbauer, Gamma resonance spectroscopy and chemical bonding, Angew. Chem. Internat. Ed., 10 (1971) 462.

G.T. Seaborg, Prospects for further considerable extension of the Periodic Table, J. Chem. Ed., 46 (1969) 626.

W.M. Howard and J.R. Nix, Production of superheavy nuclei by multiple capture of neutrons, Nature, 247 (1974) 27.

J.A.S. Smith, Nuclear quadrupole resonance spectroscopy, J. Chem. Ed., 48 (1971) 39.

G.M. Bancroft, Mössbauer spectroscopy — An introduction for inorganic chemists and geochemists, McGraw-Hill, New York, 1973.

N.N. Greenwood and T.C. Gibb, Mössbauer spectroscopy, Chapman and Hall, 1971.

T.C. Gibb, Principles of Mössbauer spectroscopy, Chapman and Hall, 1975.

H. Götte and G. Kloss, Nuclear medicine and radiochemistry, Angew. Chem. Internat. Ed., 12 (1973) 712.

H.A.C. McKay, Principles of radiochemistry, Butterworth, London, 1971.

Electronic Structures of Atoms. The Periodic Table

3

3.1. Wave Mechanics

The subject called wave mechanics arose in 1926 as a result of two main sets of experimental data which were inconsistent with established concepts of classical physics.

(i) Changes in the internal energy of an atom accompanying emission or absorption of light are not continuous but occur in discrete amounts. Moreover spectroscopic evidence shows that an atom exists only in certain discrete energy states characteristic of the element. The change from an energy state E to one E' is accompanied by emission (when $E > E'$) or absorption (when $E < E'$) of light, whose frequency ν is related to the energy change thus:

$$|E - E'| = h\nu$$

where h (Planck's constant) = 6.6262×10^{-34} J s.

(ii) A beam of electrons shows interference effects exactly as does a beam of light. A diffraction pattern is obtained when light is reflected or diffracted from a grating. The regularly spaced atoms in a crystal provide a grating which diffracts not only X-rays but also electron beams. Hence wave properties must be associated with the electron.

The occurrence of discrete energy states of a 'bound' electron in an atom and the co-existence of wave and particle properties in a free electron, can be accounted for on the basis of wave mechanics. Observation of the diffraction patterns produced when electrons of known energy encounter crystals of known atomic spacing shows that the wave length, λ, associated with an electron of velocity, v, is given by

$$\lambda = h/mv$$

Wave and particle properties can then be reconciled by the following argument.

Normally, a wave motion is a disturbance which varies at any point with a *frequency*, ν, and is propagated with some *phase velocity*, u. The *wavelength*,

λ, is then given by the relationship $\lambda\nu = u$, and represents the distance, along the direction of propagation, between points where the variation is exactly similar or *in phase*. A train of waves of this kind extends over space as long as the source goes on emitting energy; it describes a steady state. But if a wave is to be associated with a localised particle, it is necessary to consider a travelling *pulse* or *wave packet*. Such a pulse can be described only by combining a group of wave trains whose phase velocities (and hence also wave lengths) depend on frequency, varying slightly about their mean values: it travels with a *group velocity*, v_g, given by

$$v_g = \frac{d\nu}{d\left(\frac{1}{\lambda}\right)}$$

Now for the electron, $\lambda = h/mv$ (experimentally) and therefore

$$v_g = \frac{h}{m}\frac{d\nu}{dv}$$

but if the wave packet, which carries the energy, is always to stay with the particle, v_g must be the same as the actual particle velocity v and the frequency ν must then satisfy the equation

$$\frac{d\nu}{dv} = \frac{m}{h}v$$

The solution is simply $\nu = E/h$, where $E\ (= \frac{1}{2}mv^2)$ is the energy of the particle, say an electron.

The equivalence $E = h\nu$ expresses Planck's law relating the energy of a pulse of radiation, a *light quantum*, to its frequency. Thus the law acquires a more general significance and enables us to associate a wave packet with a *material particle*, its frequency being similarly related to the energy of the particle. Such a wave has to be associated with a particle simply in order to describe its behaviour; it is not necessarily real in the same sense as an electromagnetic wave. Its interpretation must now be examined.

Let ψ be the amplitude, varying from point to point in space, of the wave associated with a moving particle. Relative amplitudes at different points can then be calculated just as in optical diffraction theory; and in all diffraction experiments the degree of darkening of a photographic plate at different points depends on the relative values of ψ^2, the *intensity*, at these points. Now, on a particle picture, this effect must depend on the *number of particles* landing at the different points, and since one particle can arrive only at one point, it is necessary to make a *statistical* interpretation. The *probability* of a particle being found in a given small region is proportional to the square of the amplitude of the associated *wave function* in that region. Since this probability is also proportional to the size of the region the actual probability must be $\psi^2 d\tau$, where $d\tau$ is the volume element and ψ^2 is a *probability density*.

In the diffraction of a free particle, the probability density changes with time as the wave packet progresses, but the interpretation turns out to be quite general. In other situations, where the particle is bound within a certain region of space, *stationary patterns* occur (like the standing waves on a vibrating string, where the actual profile is fixed, instead of travelling along the string as it does in a whip), and again ψ^2 determines the probability of finding the particle at a given point. Stationary patterns are of supreme importance in molecular theory since they describe an electron bound in an atom or molecule. They are invoked throughout this book and their calculation is taken up in a later section.

3.1.1. The Uncertainty Principle

The need to associate a wave packet with a moving particle has another important consequence. In the one-dimensional case of plane waves representing a particle moving in the x-direction, the extension of the wave packet, Δx, is determined by the range of wavelengths admitted in the packet and this, in turn, by the momentum range, Δp_x. It can then be shown that

$$\Delta x \Delta p_x \sim h$$

This is one consequence of the famous Heisenberg Uncertainty Principle, which relates the simultaneous uncertainties in certain pairs of measurable quantities with the universal Planck's constant, h. Similar relationships exist for the y and z co-ordinates and momentum components. If the wave packet associated with the electron is very compact (Δx small) it must be very 'impure', containing a mixture of wave trains corresponding to a considerable range of momenta (Δp_x large); it is then known fairly accurately where the electron is at a given time, but only roughly what momentum to associate with it. When, however, the momentum of the electron is measured more and more precisely, the width of the wave packet must increase, and there is a corresponding loss of precision in possible knowledge of its position. However, since h is very small, these uncertainties, except in atomic physics, are often negligible. The position of a particle weighing 1 g can be determined optically without sensibly disturbing its momentum, but even in principle this is not true for an electron.

3.1.2. The wave equation

Any disturbance Ψ (e.g. the displacement of an elastic string or the height of a water wave) which is propagated along the x-direction with velocity u, satisfies a simple partial differential equation, the *wave equation*:

$$\frac{\partial^2 \Psi}{\partial x^2} = \frac{1}{u^2} \cdot \frac{\partial^2 \Psi}{\partial t^2}$$

Here Ψ depends on both position and time, $\Psi = \Psi(x, t)$. More generally for a three-dimensional disturbance,

$$\nabla^2\Psi = \frac{\partial^2\Psi}{\partial x^2} + \frac{\partial^2\Psi}{\partial y^2} + \frac{\partial^2\Psi}{\partial z^2} = \frac{1}{u^2} \cdot \frac{\partial^2\Psi}{\partial t^2}$$

where $\Psi = \Psi(x, y, z, t)$.

Proceeding immediately to the stationary patterns, a particular type of one-dimensional solution is, for example,

$$\Psi(x, t) = \psi(x) \sin 2\pi\nu t$$

where $\psi(x)$ is the amplitude, which depends only on position, and the disturbance at a point varies between $+\psi(x)$ and $-\psi(x)$, oscillating ν times per unit time owing to the time-dependent factor which has extreme values ± 1. In this case substitution in the wave equation shows that $\psi(x)$ must satisfy a time-independent amplitude equation:

$$\frac{d^2\psi}{dx^2} = \frac{4\pi^2\nu^2}{u^2}\psi$$

It is customary to call ψ the 'wave function', though strictly it is only the amplitude of the wave function Ψ, the time-dependent quantity. Provided the medium is uniform, so that u is constant, the last equation has solutions of the form

$$\sin \frac{2\pi\nu}{u}x \text{ and } \cos \frac{2\pi\nu}{u}x$$

and these stationary patterns (standing waves) are repeated when x increases by $\lambda = u/\nu$, the wavelength.

Returning to the case of an electron with kinetic energy E, the associated *wave function* is known to have a wavelength

$$\lambda = \frac{h}{mv} = \frac{h}{\sqrt{2mE}}$$

and the amplitude equation then becomes

$$\frac{d^2\psi}{dx^2} + \frac{8\pi^2 mE}{h^2}\psi = 0$$

This equation applies only to a particle with constant kinetic energy. Nevertheless, it shows with unexpected clarity the origin of *quantisation* – namely, the fact that a particle bound within a certain region of space can exist only in certain discrete energy states. For if the particle is now confined to a certain region, by means of barriers a distance l apart, ψ must vanish at and outside the end points and these 'boundary conditions' require that the region contains an integral number of half waves, i.e. $n(\frac{1}{2}\lambda) = l$. Substituting for λ, the only possible values of E are

$$E = \frac{h^2}{8ml^2} n^2 \qquad \text{where } n = 1, 2, 3, \ldots$$

Thus a particle moving back and forth within definite limits can exist in states described by a stationary probability pattern only for certain discrete values of the energy.

The Schrödinger equation represents a generalisation to the case in which there is an external potential field and the kinetic energy is no longer constant. The kinetic energy, which might be expected to take the place of E in the above equation, is now (classically) $E—V$, where E is the total energy and $V = V(x)$ is the potential energy, which in general varies from place to place. The final wave equation for determining the stationary wave patterns associated with a particle in an arbitrary potential field is then

$$\nabla^2\psi + \frac{8\pi^2 m}{h^2}(E - V)\psi = 0$$

Although the historical derivation sketched above depends largely on suggestive analogies, more profound derivations have since been given, and the equation itself is now regarded as entirely satisfactory so long as extremely small relativistic effects are neglected. On physical grounds (and ψ^2 has a physical meaning) it is assumed that the only solutions which mean anything are well-behaved. That is they are everywhere finite, smoothly varying, and vanishing at infinity. This requirement alone leads automatically to the quantisation so foreign to classical physics.

Sometimes the wave equation is rewritten in the form

$$\left[\frac{1}{2m}\left(\frac{-h^2}{4\pi^2}\right)\nabla^2 + V(x, y, z)\right]\psi = E\psi$$

The quantity in square brackets is an 'operator' — the Hamiltonian operator, allowing the wave equation to be written in the simpler form:

$$H\psi = E\psi$$

All *eigenvalue equations* have this form; and there are solutions ψ only for certain *eigenvalues* of the factor E on the right hand side. This form immediately suggests the final generalisation from 1 to n electrons. Since the classical energy expression is

$$E = \tfrac{1}{2}mv^2 + V(x, y, z) = \frac{1}{2m}(p_x^2 + p_y^2 + p_z^2) + V(x, y, z)$$

where $p_x = mv_x$ etc. are components of momentum, the 1-electron equation may be obtained by replacing p_x^2 etc. by the differential operators

$$-\frac{h^2}{4\pi^2} \cdot \frac{\partial^2}{\partial x^2} \text{ etc.}$$

letting the resultant operator (in square brackets above) operate on a wave function ψ, and equating the result to $E\psi$. The same recipe, applied to a system of N particles with masses m_1, m_2, m_3 ... m_N, gives

$$\left[\frac{1}{2m_1}\left(\frac{-h^2}{4\pi^2}\right)\nabla_{(1)}^2 + \frac{1}{2m_2}\left(\frac{-h^2}{4\pi^2}\right)\nabla_{(2)}^2 + \ldots V\right]\psi = E\psi$$

where V and ψ are now functions of the co-ordinates of all the particles and there is a ∇^2 for every particle. $\psi^2 d\tau_1$, $d\tau_2$ ... $d\tau_n$ is then the probability of simultaneously finding particle 1 in volume element $d\tau_1$, particle 2 in volume element $d\tau_2$, etc.

3.2. Application of Wave Mechanics to the Hydrogen Atom

In the hydrogen atom, an electron of charge $-e$ moves about a proton of charge $+e$. If the distance between them is r, its potential energy is expressed by

$$V = -\frac{e^2}{4\pi\epsilon_0 r}$$

and

$$\nabla^2\psi + \frac{8\pi^2 m}{h^2}\left(E + \frac{e^2}{4\pi\epsilon_0 r}\right)\psi = 0$$

One type of solution follows on taking $\psi = f(r)$ and assuming that, as the potential is spherically symmetrical, there must be a solution dependent only on r; in which case

$$\frac{\partial\psi}{\partial x} = \frac{d\psi}{dr}\frac{\partial r}{\partial x} \text{ etc.}$$

Then since $r = \sqrt{x^2 + y^2 + z^2}$

$$\left(\frac{\partial r}{\partial x}\right)_{yz} = \frac{1}{2}\frac{2x}{\sqrt{x^2 + y^2 + z^2}} = \frac{x}{r}$$

and

$$\frac{\partial\psi}{\partial x} = \frac{x}{r}\frac{d\psi}{dr}$$

Differentiating again:

$$\frac{\partial^2\psi}{\partial x^2} = \frac{1}{r}\frac{d\psi}{dr} - \frac{x^2}{r^3}\frac{d\psi}{dr} + \frac{x^2}{r^2}\frac{d^2\psi}{dr^2}$$

$\partial^2\psi/\partial y^2$ and $\partial^2\psi/\partial z^2$ may be found similarly. Adding, and using $x^2 + y^2 + z^2 = r^2$,

$$\nabla^2\psi = \frac{3}{r}\frac{d\psi}{dr} - \frac{1}{r}\frac{d\psi}{dr} + \frac{d^2\psi}{dr^2} = \frac{2}{r}\frac{d\psi}{dr} + \frac{d^2\psi}{dr^2}$$

As $\nabla^2\psi$ reduces to this form if ψ is a spherically symmetrical function, it follows that spherically symmetrical solutions must satisfy the equation

$$\frac{d^2\psi}{dr^2} + \frac{2}{r}\frac{d\psi}{dr} + \frac{8\pi^2 m}{h}\left(E + \frac{e^2}{4\pi\epsilon_0 r}\right) = 0$$

A standard trial solution is $\psi = e^{-ar}$, for the exponential factor remains on differentiation, and can be extracted as a common factor. Thus

$$\frac{d\psi}{dr} = -ae^{-ar} \text{ and } \frac{d^2\psi}{dr^2} = +a^2e^{-ar}$$

and substitution gives

$$e^{-ar}\left[a^2 - \frac{2}{r}a + \frac{8\pi^2 m}{h^2}\left(E + \frac{e^2}{4\pi\epsilon_0 r}\right)\right] = 0$$

If this is to be true for all values of r, the constant term and the term in $1/r$ must both vanish: thus

$$a^2 + \frac{8\pi^2 mE}{h^2} = 0, \text{ and } -2a + \frac{2\pi me^2}{\epsilon_0 h^2} = 0$$

$$\therefore a = \frac{\pi me^2}{\epsilon_0 h^2}$$

and the E value for this state is

$$E_1 = -\frac{a^2h^2}{8\pi^2 m} = -\frac{me^4}{8\epsilon_0^2 h^2}$$

Two important points arise. First, E_1 is negative and hence the electron is bound within the atom, and, secondly E_1 is expressed in terms of universal constants and is itself a useful constant. The energy, $-E_1$, is the *ionisation energy* necessary to remove the electron entirely from the hydrogen atom, according to the equation

$$H \rightarrow H^+ + e^-$$

Its actual value derived from spectroscopic measurements agrees with that calculated from the equation above

$$-E_1 = \frac{me^4}{8\epsilon_0^2 h^2}$$

$$= \frac{9.109 \times 10^{-31}\text{ kg} \times (1.602 \times 10^{-19}\text{ C})^4}{8 \times (8.854 \times 10^{-12}\text{ kg}^{-1}\text{ m}^{-3}\text{ s}^4\text{ A}^2)^2 \times (6.626 \times 10^{-34}\text{ J s})^2}$$

$$= 2.18 \times 10^{-18}\text{ kg m}^2\text{ s}^{-2}$$

$$= 2.18\text{ aJ}$$

3.2.1. Radial probability

We have seen that the value of $\psi^2 \mathrm{d}r$ is proportional to the probability of finding one electron in an element of volume d*r*. The probability of one electron being somewhere is obviously one, and it is necessary to multiply ψ by a constant so that

$$\int_{-\infty}^{+\infty} \psi^2 \mathrm{d}r = 1$$

The wave amplitude, ψ, is then said to be *normalised.* For the case under discussion, where ψ depends only on *r*, the probability of an electron being at a distance between *r* and *r* + d*r* from the nucleus (i.e. in a shell of radius *r* and thickness d*r*, and therefore of volume $4\pi r^2 \mathrm{d}r$) must be $4\pi^2\psi^2\mathrm{d}r$. When the radial probability density, $4\pi r^2\psi^2$, is plotted against *r* (Fig. 3.1), there is a

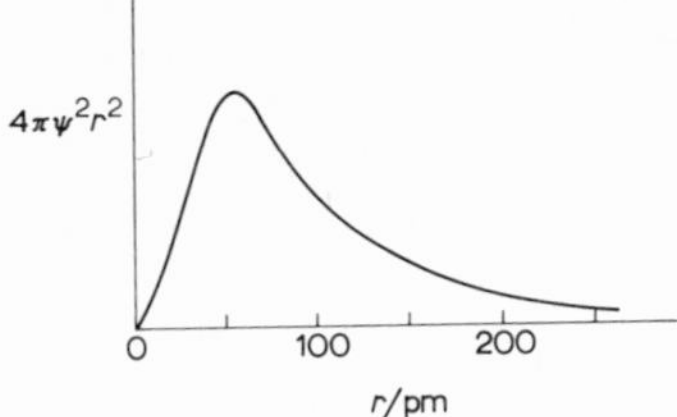

Fig. 3.1. Variation of $4\pi\psi^2 r^2$ with *r* for 1s orbital in H atom.

maximum at $r = 1/a$. This is the radius of the thin spherical shell which has a greater probability of holding the electron than any equally thin spherical shell of larger or smaller radius. The radius, $1/a$, can be calculated in terms of ϵ_0, *m*, *e* and *h*.

$$\begin{aligned}
\frac{1}{a} &= \frac{\epsilon_0 h^2}{\pi m e^2} \\
&= \frac{(8.854 \times 10^{-12}\ \mathrm{kg^{-1}\ m^{-3}\ s^4\ A^2}) \times (6.626 \times 10^{-34}\ \mathrm{J\ s})^2}{(3.142 \times 9.109 \times 10^{-34}\ \mathrm{kg}) \times (1.602 \times 10^{-19}\ \mathrm{C})^2} \\
&= 5.29 \times 10^{-11}\ \mathrm{m} \\
&= 52.9\ \mathrm{pm}
\end{aligned}$$

This is the Bohr radius of the semiclassical approach (1913) and is often adopted as the atomic unit of length. The electron has a maximum probability of being at this distance from the nucleus but a good chance of being anywhere within a very considerable volume.

The probability density, ψ^2, is commonly represented pictorially by density of shading (Fig. 3.2):

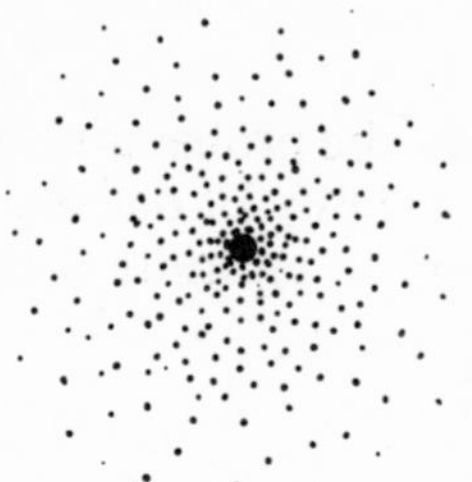

Fig. 3.2. Electron probability density represented by density of shading.

or, more simply, by sketching a boundary contour (on which all points have this same value of ψ^2) enclosing, say 95% of the density (Fig. 3.3):

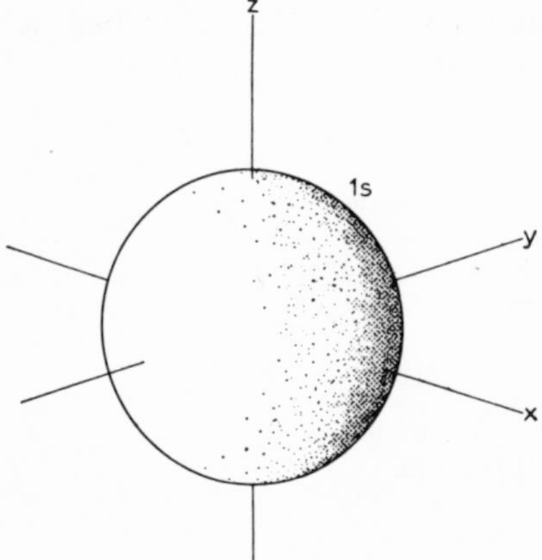

Fig. 3.3. Atomic orbital represented by spherical bounding surface.

The ψ-function, indicated by this boundary surface, is called an *atomic orbital*, being the wave-mechanical counterpart of the classical orbit. An electron, like that in the unexcited hydrogen atom, whose orbital is spherically symmetrical is called an s electron.

In addition to the state of lowest energy E_1, there are excited states with greater energies E_2, E_3 ... E_n for which it can be shown that

$$E_n = \frac{-me^4}{8\epsilon_0^2 h^2} \times \frac{1}{n^2} = \frac{E_1}{n^2}$$

n being an integral quantum number. When n is very large, E_n approaches zero. For the free electron $E > 0$.

The energy set free when an electron falls from an energy level n to the energy level 2 is

$$E_n - E_2 = \frac{-me^4}{8\epsilon_0^2 h^2}\left(\frac{1}{n^2} - \frac{1}{2^2}\right) = \frac{+me^4}{8\epsilon_0^2 h^2}\left(\frac{1}{2^2} - \frac{1}{n^2}\right)$$

In this form the equation relates the wave number of lines in the visible hydrogen spectrum (Balmer, 1885) to the change in energy state of the electron. The *wave number*, $\frac{1}{\lambda}$, is simply the number of waves per unit length

$$\frac{1}{\lambda} = \frac{\nu}{c_0}$$

where c_0 is the velocity of light in a vacuum. Since

$$E_n - E_2 = \Delta E = h\nu = hc_0/\lambda$$

$$hc_0/\lambda = \frac{me^4}{8\epsilon_0^2 h^2}\left(\frac{1}{2^2} - \frac{1}{n^2}\right)$$

and

$$\frac{1}{\lambda} = \frac{me^4}{8\epsilon_0^2 h^3 c_0}\left(\frac{1}{2^2} - \frac{1}{n^2}\right) = R\left(\frac{1}{2^2} - \frac{1}{n^2}\right)$$

The constant R agrees, within the limits of experimental error, with the empirical one found by Rydberg (1889)

$$R = \frac{(9.109 \times 10^{-31}\ \text{kg})(1.602 \times 10^{-19}\ \text{C})^4}{8(8.854 \times 10^{-12}\ \text{kg}^{-1}\ \text{m}^{-3}\ \text{s}^4\ \text{A}^4)^2(6.626 \times 10^{-34}\ \text{J s})^3(2.998 \times 10^8\ \text{m s}^{-1})}$$

$$= 1.097 \times 10^7\ \text{m}^{-1}$$

The radial density distribution of the charge of an electron of energy E_2 is shown by plotting $4\pi\psi^2 r^2$ against r, as in Fig. 3.4. Pictorial representations

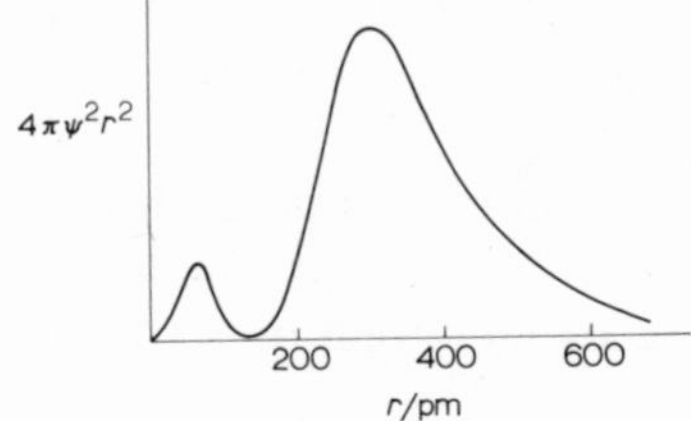

Fig. 3.4. Variation of $4\pi\psi^2 r^2$ with r for 2s orbital in H atom.

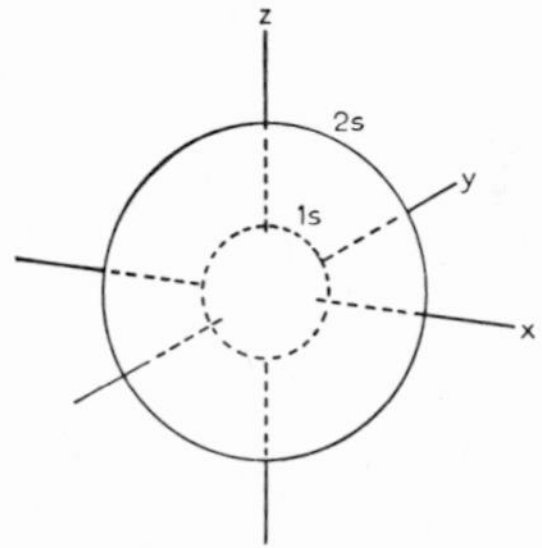

Fig. 3.5. Boundary surface for 2s electron, compared with that for 1s.

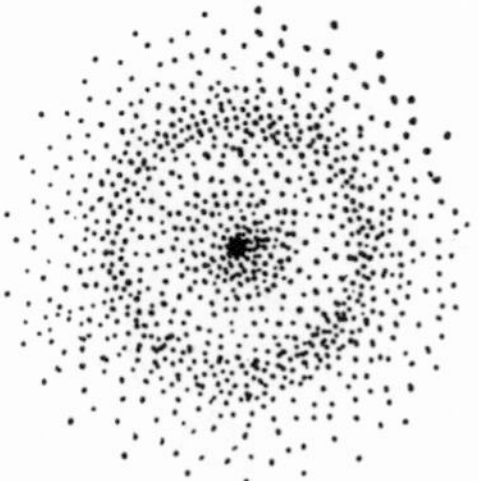

Fig. 3.6. Representation of probability density for electron of energy E_2.

of both ψ and ψ^2 are given in Fig. 3.5 and Fig. 3.6, the former obscuring the details but indicating schematically the general size and shape of the orbital. In the latter the density of the dots is intended only to show where ψ is large.

3.2.2. p, d and f functions

Though the potential V in the Schrödinger equation depends only on r, there are solutions which depend on x, y and z separately, e.g. $\psi = xf(r)$. Wave functions of this type are p functions and there are three solutions for the same energy value. As a result of the spherical symmetry of the potential V, the orbitals which are not themselves spherically symmetrical occur in groups, the members of a given group having the same value of E but quite independent ψ-patterns. Solutions in the same group are described as *degenerate*, their number being the *degree of degeneracy*, and different orbitals of the same group are often said to be *equivalent*, that is energetically alike.

For the special case where $V(r) = -e^2/4\pi\epsilon_0 r$, substitution in the Schrödinger equation shows the p-type solution of lowest energy to give

$$E = \frac{-me^4}{8\epsilon_0^2 h^2} \cdot \frac{1}{2^2} = E_2$$

E_2 is the energy of the second-lowest s state. For the single-electron hydrogen atom this energy level is fourfold degenerate, corresponding to one s and three p orbitals.

All other elements have atoms with more than one electron, and accordingly, an electron in them moves, roughly speaking, in the field created by the nucleus and the other electrons together. This field is not of a simple inverse distance form; and in such atoms s electrons differ in energy from p electrons and from d electrons when these are present.

The number of higher energy solutions is infinite, their 'atomic' character disappearing for $E \to 0$, when the electron becomes free. Some of the higher atomic orbitals are represented by boundary surfaces including most of the

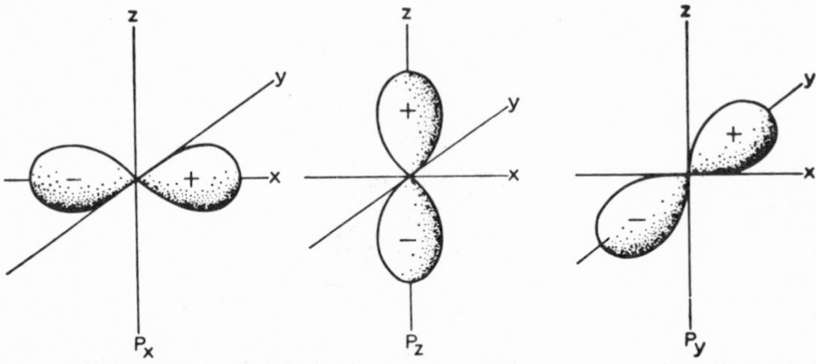

Fig. 3.7. Boundary surfaces for p electrons.

probability density, in Fig. 3.7 for p electrons and in Fig. 3.8 for d electrons. The figures do not strictly show surfaces of constant ψ (ψ vanishes at the nucleus) but rather their general disposition in relation to one another, and for clarity the different parts of the separate lobes are joined up into a single

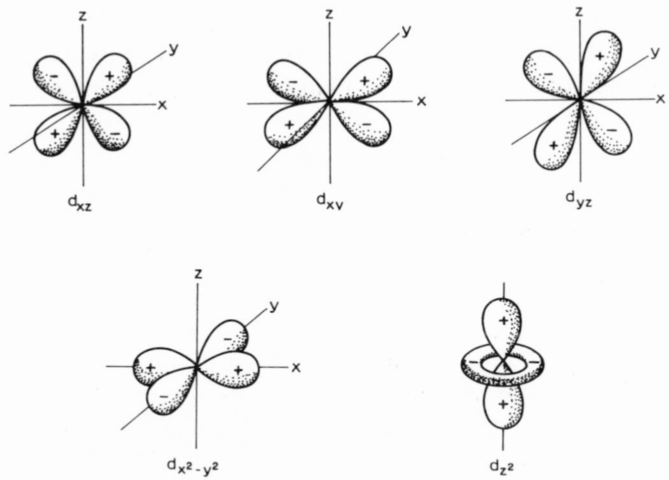

Fig. 3.8. Boundary surfaces for d electrons.

pattern. The pattern indicates simply the sign of ψ and the regions where it is large, the different regions being separated by nodal surfaces in which $\psi = 0$. The sevenfold degenerate f states are less easy to represent in a two-dimensional drawing.

The terminology s, p, d, f, ... is a legacy from the Bohr theory which attempted to classify the 'sharp', 'principle', 'diffuse' and 'fundamental' series of classical spectroscopy.

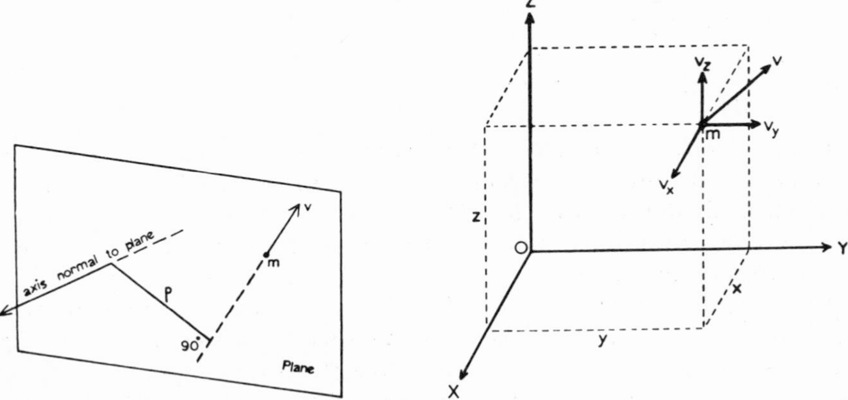

Fig. 3.9. Illustrating angular momentum. Fig. 3.10. Components of angular momentum.

3.2.3. Orbital angular momentum

The *angular momentum* of a particle about a point O is a quantity $M = pmv$ where p is the perpendicular distance from O to the line of motion and mv is the ordinary momentum; the quantity is associated with an axis through O, namely the normal to the plane in which the motion is taking place (Fig. 3.9). The angular momentum is thus a vector quantity and can be described in terms of components parallel to the three axes. Its units are J s (the dimensions of action and also those of h, Planck's constant). The magnitude of the vector is

$$M = \sqrt{M_x^2 + M_y^2 + M_z^2}$$

and the direction (i.e. the axis) makes angles

$$\cos^{-1}\frac{M_x}{M}, \cos^{-1}\frac{M_y}{M}, \cos^{-1}\frac{M_z}{M}$$

with the three co-ordinate axes. Here for instance $M_x = y(mv_z) - z(mv_y)$ and is seen from Fig. 3.10 to be the total momentum of the y and z momentum components about the x axis, the contributions being counted positive for anticlockwise and negative for clockwise motion. M_y and M_z follow by cyclic rearrangement, $xyz \rightarrow yzx \rightarrow zxy$.

Now M and its three components are all observable dynamical quantities just as is the energy E. It turns out that, in a definite energy state, M and *one* of its components can simultaneously have definite, quantised values. Thus if a particular direction in space is singled out, by means of an applied magnetic field for example, the component in this direction can take only certain quantised values, while the other two components must remain uncertain. It is customary to call the particular axis the z axis. The definite allowed values of M_z are then integral multiples of $h/2\pi$

$$M_z = m_l(h/2\pi) \text{ where } m_l = 0, \pm1, \pm2, \dots \pm l$$

those of M itself being

$$M = \sqrt{l(l+1)}\,\frac{h}{2\pi} \text{ where } l = 0, 1, 2, \dots$$

The s, p, d, f electrons have angular momenta corresponding to $l = 0, 1, 2, 3$, respectively, and the different degenerate states can be classified according to the simultaneously definite values of M_z.

Thus there are three p orbitals, corresponding to $m_l = -1, 0, +1$; five d orbitals, corresponding to $m_l = -2, -1, 0, +1, +2$; and in general $(2l + 1)$ orbitals, corresponding to $m_l = -l, \dots 0 \dots +l$, for an angular momentum M with quantum number l. This classification suggests a *vector model* (Fig. 3.11) in which different states correspond to different settings of a vector of length l, representing the angular momentum. This model goes back to the early days of quantum theory, when it was found empirically necessary to

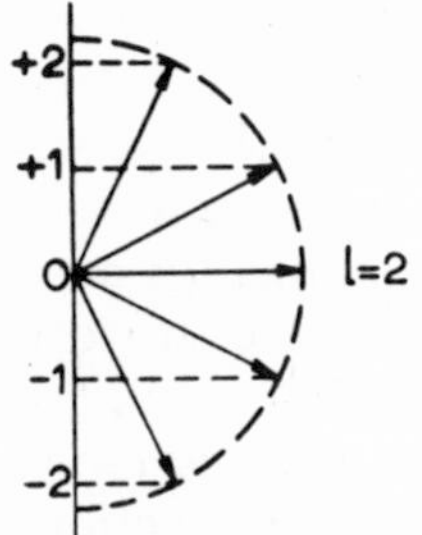

Fig. 3.11. Representation of angular momenta.

assume the rather curious angular momentum value $\sqrt{l(l+1)}$ (in units $h/2\pi$) instead of the l employed in the semi-classical vector diagram. The wave mechanical treatment is in fact capable of bringing theoretical and experimental results into perfect agreement.

The quantisation of angular momentum components in a given direction is experimentally realised in the Zeeman effect. Here the unique direction is fixed by an applied magnetic field; the states with different electronic angular momentum components in this direction correspond to slightly different energy values since the circulating charge behaves like a small magnet, whose component along the field is proportional to m_l. The splitting of the initially degenerate level, due to the different coupling energies, is spectroscopically observable. Quantitative discussion of this magnetic effect is deferred (3.4.4.3). The Zeeman effect does, however, provide a very powerful tool in the interpretation of atomic spectra (see 3.2.4).

At this stage, it must be stressed that any mixture of two or more degenerate solutions is also a solution of the wave equation, and is an equally satisfactory description of an electron with a given energy and angular momentum; but, for example, an equal mixture of solutions with common E and l but with $m_l = +1$ and -1, respectively, would describe a state in which M_z was uncertain and, if observed, would be equally likely to yield either result. This arbitrariness disappears only if the spherical symmetry of the problem is destroyed by fixing an axis (e.g. by an applied field); for when this is done the only stationary states are those compatible with this axis of symmetry, and such states correspond to definite values of the third characteristic, M_z. To make this clear, consider the three p orbitals, describing electrons with the same energy, and with $l = 1$, but with $m_l = -1, 0, +1$. These orbitals, which are called p_{-1}, p_0 and p_{+1}, describe the stationary states when a feeble magnetic field (just sufficient to reduce the symmetry) is applied along the z axis; but they may be *mixtures* of the p orbitals p_x, p_y, and p_z, which were introduced before any reference was made to symmetry or angular momentum. In fact p_z which is symmetrical about the z axis, does coincide with p_0: but p_{-1} and p_{+1} are mixtures of p_x and p_y (and vice versa) which can be interpreted, somewhat naively, as permitting the 'jumping' of an electron from one orbital to the other, so as to give respectively clockwise or anticlockwise circulation. On the other hand, p_x and p_y correspond to standing waves (this time 'rotational') built up from p_{-1} and p_{+1} which are oppositely directed travelling waves. The same situation arises with the d electrons. Five 'travelling wave' solutions can be used which all give electron distributions completely symmetrical about a particular axis, or five standing wave solutions which are less symmetrical but more easily visualized. This ambiguity need cause no alarm; the freedom of choice simply means it is possible to work with those orbitals most suitable for a given application. In describing the spectroscopic states of free atoms (spherical symmetry), and of linear molecules (axial symmetry), it is often advantageous to employ the functions which correspond to definite M_z values; but this is not so in systems of lower symmetry in which there is no such thing as quantised angular momentum. And in discussing molecular structure where the atomic orbitals are employed only as 'building bricks' for constructing molecular wave functions, it is usually quite immaterial which type of orbital is used. The mixtures such as p_x, p_y, p_z have, however, an important practical advantage. Unlike the travelling wave solutions, they can be directly represented by stationary patterns with a fixed orientation in space, and permit a pictorial approach to the construction of molecular wave functions.

3.2.4. Electron spin

The interpretation of the Zeeman effect provided by the theory of orbital angular momentum is incomplete. Anomalous Zeeman splittings are observed (indeed, the so-called 'normal' splitting is found to be the exception) and require a further development of the theory. The difficulty first arises in the one-electron (hydrogen-like) system already considered, for even an s electron, with *no* angular momentum, is found to have two accessible energy states in the presence of a magnetic field. This led Goudsmit and Uhlenbeck (1925) to postulate an intrinsic angular momentum due to *spin* of the electron itself. Treating this on the same footing as orbital angular momentum the observed *doublet splitting* of an s state requires a spin quantum number, $s = \frac{1}{2}$, such that there would be two possible z components (separated by an integral multiple of $h/2\pi$) namely $+\frac{1}{2}$ and $-\frac{1}{2}$ (cf. Fig. 3.11).

The general interpretation of the Zeeman effect for a one-electron system then calls for an extension of the vector picture. Consider an electron in a p state, denoting z components of the orbital and spin angular momenta, in units of $h/2\pi$, by m_l and m_s respectively. The possibilities are $m_l = -1, 0, +1$ and $m_s = -\frac{1}{2}, +\frac{1}{2}$; thus the possible values of the total angular momentum z components are $(-1 \pm \frac{1}{2})$, $(0 \pm \frac{1}{2})$, $(+1 \pm \frac{1}{2})$. These correspond to different z components of total angular momentum with quantum number j, where j can take the values $1 + \frac{1}{2} = \frac{3}{2}$ and $1 - \frac{1}{2} = \frac{1}{2}$. The two vector diagrams for the allowed states are shown in Fig. 3.12, along with the composition of the orbital and spin vectors, the spin being parallel or anti-parallel to the orbital vector. The method of composition is applicable for arbitrary orbital angular momentum with quantum number l, j taking values $l + \frac{1}{2}$ and $l - \frac{1}{2}$.

The alternative couplings of orbital and spin angular momentum give rise

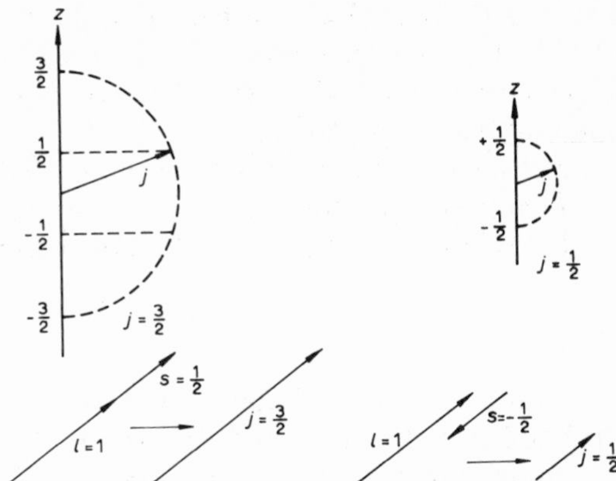

Fig. 3.12. Vector addition of orbital and spin angular momenta.

to another experimentally observable phenomenon. The magnetic field associated with l performs in effect a Zeeman experiment on the spinning electron and, even in the absence of an external field, the two couplings correspond to slightly different energies which are revealed in the fine structure of spectral lines. Thus p, d, f ... states are always resolved into doublets by the electron spin, while the s state is single, being unresolved because the orbital magnetic moment is then zero. Further resolutions should occur when an external field is applied and these are in complete accord with the observed Zeeman splittings. A rigorous justification of this semi-classical vector model comes from quantum mechanics. The pictorial interpretation must not be taken too literally (in getting quantitative results, for instance, angular momenta with quantum numbers, l, j etc. have magnitudes $\sqrt{l(l+1)}$, $\sqrt{j(j+1)}$ etc.); nevertheless, the pictorial model is extremely useful in the general classification and understanding of spectra.

3.3. Application of Wave Mechanics to Many-electron Systems

The wave equation for several electrons in the presence of a central nucleus may be taken over from the final equation of 3.1.2 where now $m_1 = m_2 = \ldots = m$, the electron mass. Abbreviating the co-ordinates of the electrons from (x_1, y_1, z_1), (x_2, y_2, z_2), ... to 1, 2 ... the wave function is then represented by $\Psi(1, 2, \ldots N)$, and $V(1, 2, \ldots N)$ contains the potential energy of each electron in the field of the nucleus together with interaction terms representing the mutual repulsion of the electrons. This partial differential equation cannot be solved in closed form, that is to give formulae for wave functions and energy levels, even for the helium atom. But highly developed approximation methods yield numerical values of the energy in essentially perfect agreement with experiment (1 part in 50,000), which leaves little doubt of the general validity of the wave equation itself.

When approximate solutions are extended to heavier atoms and to molecules, the accuracy is rather lower (e.g. 1 to 2%). The most fruitful approximation method is that of Hartree (1928), later justified and refined by Slater (1930) and Fock (1930). This is suggested by first neglecting the electron repulsion and observing that the probability function $P(1, 2, \ldots N)$ must then be approximated by the product $P_1(1)\ P_2(2)\ \ldots\ P_N(N)$, for then the probability of any configuration of the electrons is a product of the probabilities of N independent events. Since $P = \Psi_1^2$, this means the many-electron wave function must also be a product,

$$\Psi(1, 2, \ldots N) = \psi_1(1)\psi_2(2) \ldots \psi_N(N)$$

in which electron 1 is described by orbital ψ_1 and probability pattern $P_1 = \psi_1^2$, electron 2 by orbital ψ_2 and probability pattern $P_2 = \psi_2^2$, and so on. Pictorially, each electron occupies its own orbital, as if it alone were being considered. The probability of finding electron 1 at a given point (x, y, z) is proportional to $P_1(x, y, z)$, that of finding electron 2 at the same point is proportional to $P_2(x, y, z)$ and by addition the probability P of finding an electron (i.e. any of the N electrons) at the point (x, y, z) is

$$P(x, y, z) = P_1(x, y, z) + P_2(x, y, z) \ldots + P_N(x, y, z)$$

Thus the probability density in the system, to this approximation, is obtained simply by pictorially superimposing contributions from individual electrons, each described by its own orbital. The orbitals $\psi_1, \psi_2 \ldots \psi_N$ are, however, not necessarily the orbitals of the one-electron problem in which all electrons but one are literally ignored, for it is possible to choose all the orbitals in such a way that the resultant wave function closely approaches the exact solution. These best orbitals then represent the motion of an electron in an *effective* field. This is the Hartree field produced by the nucleus together with the remaining electrons, severally regarded as a smeared-out negative charge with a density equal to that of its probability function.

Certain restrictions are imposed on this picture by the fact that electrons are indistinguishable and by the existence of electron spin. Indistinguishability implies that $P(1, 2 \ldots N)$ shall be unchanged in value if the positions of any two particles are interchanged, and this in turn implies that ψ can at most change sign. For electrons, the wave function must be an antisymmetric sum of products,

$$C[\psi_1(1)\psi_2(2)\psi_3(3) \ldots \psi_N(N) - \psi_1(2)\psi_2(1)\psi_3(3) \ldots \psi_N(N) + \ldots]$$

whose sign reverses in any interchange (the constant C being chosen to normalise the probability function so that the chance of finding all particles somewhere is unity). Fortunately, this does not destroy the simplicity of the picture; the electron probability density $P(x, y, z)$ is given by exactly the same sum-rule as for a single product. But there is now a fundamental restriction on the choice of orbitals. It might have been expected that every electron added to a nucleus would, in the lowest energy or ground state, be described by the lowest energy solution of the one particle wave equation. However the requirement of antisymmetry rules out this possibility because an antisymmetric Ψ vanishes if any two orbitals (e.g. ψ_1, ψ_2 above) are identified. The orbitals occupied by the electrons must therefore all be different and in the ground state the tendency will be for N electrons to occupy the first $\frac{1}{2}N$ orbitals, in ascending energy order, of the one-electron problem.

The existence of spin calls for a further slight modification. An orbital such as ψ_1 describes the spatial distribution of an electron in a given energy state. If, in addition to given energy, a particle has one of two given spin components, this must also be indicated in its orbital description by adding a factor α, which is unity if the particle has spin $s_z = \frac{1}{2}$ and vanishes otherwise, or β, which is likewise unity or 0 according as $s_z = -\frac{1}{2}$ or not. The set of orbitals is therefore in effect duplicated, becoming $\psi_1\alpha$, $\psi_1\beta$, $\psi_2\alpha$, $\psi_2\beta$ ... At the same time $\psi_1\alpha$ and $\psi_1\beta$, having the same spatial part ψ_1, lead to identical contributions to the electron density.

The effect of indistinguishability, which denies the occupation of two orbitals which are alike in every respect (including now the spin), may consequently be stated in modified form. Any orbital ψ can be empty or occupied by either 1 or 2 electrons, appearing once with each spin factor in the second instance, but not by more than 2. The conclusion is that the electron density is of the form

$$P(x, y, z) = n_1P_1(x, y, z) + n_2P_2(x, y, z) + \dots n_NP_N(x, y, z)$$

where each of the 'occupation numbers' n_1, n_2, ... may be 1 or 2, and an orbital is said to be 'filled' when it is 2. This is the wave-mechanical form of the Pauli exclusion principle.

3.4. Electronic Configurations of Atoms

In the ground states of atoms, the lowest energy orbitals are doubly occupied but some of those of higher energy, which may be degenerate, may be only singly occupied. Later it will be seen that the occupied orbitals of highest energy are the *valence orbitals*, responsible for the chemical behaviour of the atom. Whereas the lower energy orbitals are tightly localised

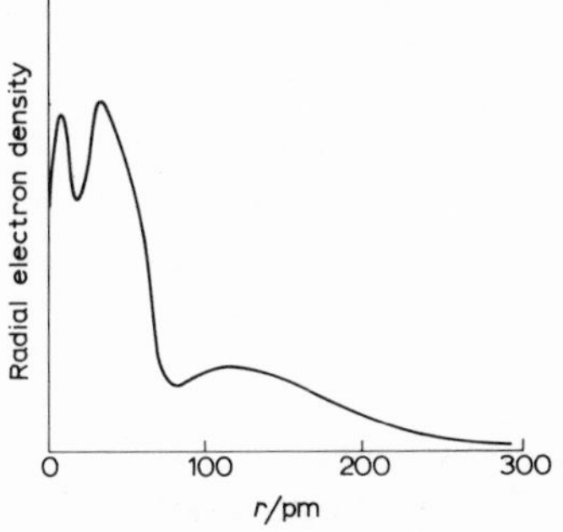

Fig. 3.13. Shell structure. Charge distribution in argon (after Hartree).

about the nucleus, the higher ones are much more diffuse, and when a whole set of orbitals is occupied the resultant electron density falls into 'shells' (Fig. 3.13), giving the wave-mechanical analogue of the shells of Bohr theory. The shells are determined by the *principal* quantum number, n, and orbitals in the individual shells are distinguished by the *subsidiary* quantum numbers l and m_l. (When the orbitals are not strictly hydrogen-like, n can still be defined as the number of nodes in an orbital minus one). This classification is general, depending only on the spherical symmetry of the system, but it must be stressed that orbitals in the same shell but with different l values do not have quite the same 'spread', and for $n \geqslant 3$ the division into shells is not very clear-cut.

The nature of departures from the hydrogen-like form must now be discussed. It is the essence of the Hartree method that each orbital describes one electron in the field of the nucleus *and* all the other electrons (each regarded as 'smeared out' according to its orbital). Thus when there are two electrons in a 1s orbital and one in a 2s orbital, outside a nucleus of charge $+3e$, the 2s electron will 'feel', roughly speaking, an effective nuclear charge of little more than $+1e$. The simplest way of improving hydrogen-like orbitals, in order to take into account electron repulsion, is to use 'screening rules' (Slater, 1931) to estimate an effective nuclear charge for each orbital. In better approximation, the orbitals are obtained by numerical solution of the equations given by Hartree and Fock. But since each depends, through the field arising from the charge distribution, upon the forms of *all* the occupied orbitals, tedious repetition is necessary to get a *self-consistent field.* Fortunately, the main features of atomic orbitals can be understood in terms of the screening due to various groups of electrons and the non-coulombic form of the resultant field. In particular, this is responsible for a separation of the energy levels of orbitals in the same shell but with different l values, the order becoming s < p < d < f ... .

The effective nuclear charge, Z^*, acting on an electron is given by $Z^* = Z - \sigma$, where Z is the atomic number and σ is a screening constant.

To determine σ, divide the electrons into the respective orbital groups:

1s
2s and 2p
3s and 3p
3d
4s and 4p
4d
4f
5s and 5p

Then σ is the sum of the contributions:

(a) Zero from any orbital group outside the one considered.

(b) 0.35 in general, but 0.30 in the case of the 1s, from every other electron in the orbital group considered.

(c) 0.85 from every electron in the quantum level immediately below (nearer to the nucleus than) the electron considered, and 1.00 from every electron in levels still nearer the nucleus, provided that the electron considered is in an s or a p orbital. If the electron considered is in a d or f orbital, every electron in lower orbital groups contributes 1 towards the value of σ.

Thus for an electron in the 3s or 3p level in silicon which has the configuration $1s^2$, $2s^2$, $2p^6$, $3s^2\ 3p^2$ ($Z = 14$)

$$\begin{aligned} \sigma &= 3 \times 0.35 \text{ (for the 3 other electrons in 3s and 3p)} \\ &\quad +8 \times 0.85 \text{ (for the 8 electrons in 2s and 2p)} \\ &\quad +2 \times 1 \text{ (for the 2 electrons still nearer the nucleus)} \\ &= 9.85 \\ \therefore Z^* &= Z - \sigma = 4.15 \end{aligned}$$

The method is useful for orbitals with principal quantum numbers up to 4, but for orbitals of higher energy it is progressively less accurate.

It is now possible to predict the electron configurations of simple atoms by adding Z electrons to a central nucleus, filling the available orbitals in ascending energy order in accordance with the Pauli principle. This is because the total electronic energy is a sum of the characteristic energies of the occupied orbitals (counted once or twice according to occupation) along with a smaller electron interaction term. Consequently the normal or *ground state* occurs when the electrons occupy the orbitals of lowest energy, with not more than 2 in each. Thus, the atoms from H to B have electron configurations which may be indicated by H ($1s^1$), He ($1s^2$), Li ($1s^2\ 2s^1$), Be ($1s^2\ 2s^2$), B ($1s^2\ 2s^2\ 2p^1$). At this point a difficulty arises. In the next atom, carbon, there are 2 electrons outside the $1s^2\ 2s^2$ group, but there are three 2p orbitals all of equal energy into which the electrons may go. This ambiguity is admitted by writing the configuration C ($1s^2\ 2s^2\ 2p^2$) without specifying which of the equivalent 2p orbitals are occupied. Corresponding to such a configuration there are several states whose energies differ somewhat. We are interested primarily in the lowest of these, which is the ground state, and in its spectroscopic character. Before proceeding further in 'building up' the atoms of the Periodic Table, it is therefore necessary to classify the states associated with a given electron configuration and to decide which is lowest.

3.4.1. Spectroscopic states

After specifying the configuration in terms of the occupation of orbitals of given (n, l), without reference to m_l which denotes different equivalent orbitals, the possible states may be classified according to angular momentum. We shall therefore summarise and extend the vector scheme (3.2.3), using the carbon $1s^2\ 2s^2\ 2p^2$ configuration in illustration.

(1) s, p, d, ... orbitals describe electrons with angular momentum quantum numbers 0, 1, 2 .... For any value of l there are $2l + 1$ equivalent orbitals, corresponding to z-components of $m_l = l, l-1, \ldots -l$.

(2) Each electron has a spin with quantum numbers $s = \frac{1}{2}$ and possible z-components $m_s = +\frac{1}{2}, -\frac{1}{2}$.

(3) The orbital and spin vectors may be separately combined to give *total* angular momenta with orbital and spin quantum numbers (for 2 electrons) $L = (l_1 + l_2), (l_1 + l_2 - 1) \ldots (l_1 - l_2)$ and $S = s_1 + s_2$ $(= 1)$, $s_1 + s_2 - 1$ $(= 0)$; and z-components $M_L = L, L-1, \ldots -L$ and $M_s = S, S-1, \ldots -S$. The possibilities are indicated, for carbon, in Fig. 3.14. The

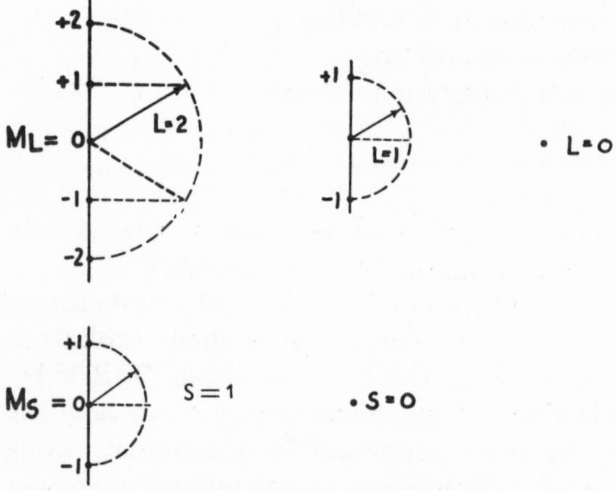

Fig. 3.14. Resultant orbital and spin angular momentum for configuration $2p^2$.

greatest possible L is simply the maximum value of M_L, which is $m_{l_1} + m_{l_2}$; but the latter has a maximum value $l_1 + l_2$ and the greatest possible L is thus $L = l_1 + l_2 = 2$. The other L values correspond to different couplings in which $M_{L(\max)}$ and hence L are reduced by unit steps. The allowed total angular momenta therefore correspond to $L = 2, 1, 0$, each with its $2L + 1$ possible z-components. Atomic states are classified according to L value in essentially the same way as orbitals in terms of l value. Those in which L has the value 0, 1, 2, 3, ... are called S, P, D, F, ... states. For instance, S, P and D states are associated with the $1s^2\, 2s^2\, 2p^2$ configuration of carbon.

(4) The total orbital angular momentum and the total spin, described by L and S, may be coupled exactly as in the one-electron case (3.2.4). Their resultant total angular momentum is described by a quantum number J, with possible values $L + S, L + S - 1, \ldots L - S$; and for each J there are $2J + 1$ possible z-components $M_J = J, J-1, \ldots -J$. There are $2S + 1$ possible couplings of L and S if $S < L$ (or $2L + 1$ if $L < S$) and, since these introduce slightly different coupling energies, a state of given L is always a *multiplet*, with multiplicity $2S + 1$ (or $2L + 1$ if $L < S$). Since for carbon, $S = 1$ or 0, we expect S, P and D states of multiplicity 1 and 3 to appear. It is customary to indicate these singlet and triplet states by ^1S, ^1P, ^1D and ^3S, ^3P, ^3D. Not all these states occur. Thus ^3D would require two electrons in the same orbital with the same spin and this, by

the Pauli principle, is inadmissible. In fact only ^1S, ^3P and ^1D are found. The individual states of a multiplet may be labelled (when necessary) by adding their J values as subscripts. The states comprising the triplet P level of carbon are $^3P_2$, $^3P_1$, $^3P_0$. States of given J are not resolved further (according to M_J) unless a magnetic field is applied, when each shows a Zeeman splitting into $2J + 1$ components.

By the use of this vector model it is possible to recognise and classify observed atomic states and to associate them with definite electron configurations. It will be noticed that mention has not been made of the doubly occupied orbitals of closed shells; these together give a zero contribution to the angular momenta, and consequently spectral data provide direct information about the outer, singly-occupied orbitals. Later this will be seen to be of the utmost importance (4.1.1). Briefly, the valence of an atom, that is the number of *covalent* bonds in which it can participate, is the number of singly occupied outer orbitals, and this is shown by the fact that a configuration with n such orbitals gives rise to states with a maximum multiplicity of $p = 2S + 1 = n + 1$. Hence the valency of an atom whose electron configuration gives states of maximum multiplicity p is $p - 1$. Even at this stage, important conclusions can be drawn. Thus, the lowest energy configuration of carbon is observed to give states of maximum multiplicity $p = 3$. Carbon is not however bivalent; and the implication is that the normal quadrivalence must in some way involve an excitation, from a ground state with *two* singly-occupied p orbitals (the theoretically expected $1s^2\ 2s^2\ 2p^2$) to an excited state with *four* (namely, $1s^2\ 2s\ 2p^3$). The energy order of the various states cannot be predicted from the vector model, but rules are available for this purpose.

Finally, it must be remembered that the vector model gives only an approximate description of the electronic structure. According to (4) the present picture is one of L–S (or Russell–Saunders) coupling. For heavy atoms this picture is often inadequate, the L–S coupling being broken (3.4.4.3).

3.4.2. Hund's rules. The Landé *g*-factor

The order of the energy levels arising from any configuration can be predicted, when L–S coupling is operative, by rules proposed by Hund (1925):

(1) The state of maximum multiplicity lies lowest. The electrons then occupy equivalent orbitals singly, as far as possible, with parallel spins.

(2) For a given multiplicity, the state of maximum L lies lowest.

(3) For a given S and L, the state of maximum or minimum J lies lowest, according as the set of equivalent orbitals is more or less than half filled. When the set is exactly half-filled J is equal to S.

The first two rules are of greatest chemical importance. From them it follows immediately that the lowest energy states of the carbon atom lie in the order ^3P, ^1D, ^1S; carbon has a ^3P ground state.

Ground states for the elements following carbon are tabulated in Table 3.1.

At a later stage (3.4.4.3) we shall also need to know precisely what mag-

TABLE 3.1

GROUND STATES FOR ELEMENTS 7 TO 10

	Electronic configuration	Unpaired electrons	S	$2S+1$	L	J	Spectroscopic state
N	$1s^2 2s^2 2p^3$	3	$\frac{3}{2}$	4	0	$\frac{3}{2}$	$^4S_{3/2}$
O	$1s^2 2s^2 2p^4$	2	1	3	1	2	$^3P_2$
F	$1s^2 2s^2 2p^5$	1	$\frac{1}{2}$	2	1	$\frac{3}{2}$	$^2P_{3/2}$
Ne	$1s^2 2s^2 2p^6$	0	0	1	0	0	$^1S_0$

netic moment is associated with an atom in a given angular momentum state, this determining the Zeeman splitting of the levels. The magnetic moment is proportional to the angular momentum, but experiment indicates that the proportionality factor is twice as great for spin as for orbital angular momentum. When both contributions occur, the formula for the effective magnetic moment, μ, is consequently less simple: it is

$$\mu = g\sqrt{J(J+1)}\ \mu_B$$

where

$$g = 1 + \frac{J(J+1) + S(S+1) - L(L+1)}{2J(J+1)}$$

and

$$\mu_B = \frac{eh}{4\pi m} = 9.274 \times 10^{-24}\ \mathrm{A\ m^2}$$

is the *Bohr magneton.* This *g-factor* was first proposed on experimental grounds by Landé and later derived theoretically.

3.4.3. The Aufbau Principle and the Periodic Table

It is now possible to work through the Periodic Table, predicting the electronic configurations of the ground states, by adding to each successive nu-

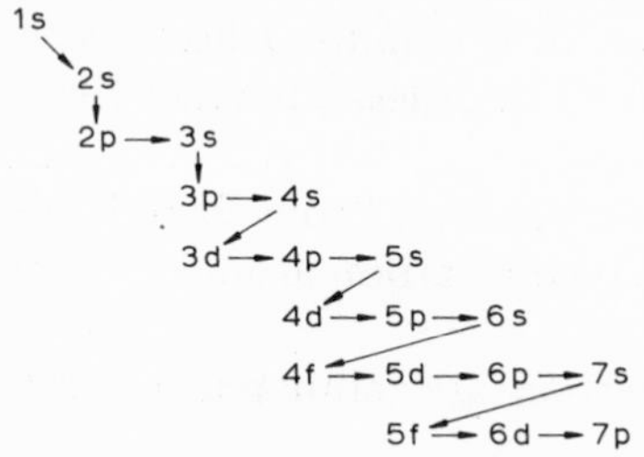

1s
2s
2p → 3s
3p → 4s
3d → 4p → 5s
4d → 5p → 6s
4f → 5d → 6p → 7s
5f → 6d → 7p

Fig. 3.15. Order of filling atomic orbitals.

cleus the number of electrons needed to produce a neutral atom, filling the available orbitals in ascending order of energy and observing the Pauli Principle and Hund's rules. The principles already discussed, upon which this process is based, are known collectively as the Aufbau Principle. For this it is

3.4.3.1. *Electron configurations of the elements*

Ground states of the free atoms

Atomic No.	Element	K	L		M			N			
		1s	2s	2p	3s	3p	3d	4s	4p	4d	4f
1	H	1									
2	He	2									
3	Li	2	1								
4	Be	2	2								
5	B	2	2	1							
6	C	2	2	2							
7	N	2	2	3							
8	O	2	2	4							
9	F	2	2	5							
10	Ne	2	2	6							
11	Na	2	2	6	1						
12	Mg	2	2	6	2						
13	Al	2	2	6	2	1					
14	Si	2	2	6	2	2					
15	P	2	2	6	2	3					
16	S	2	2	6	2	4					
17	Cl	2	2	6	2	5					
18	Ar	2	2	6	2	6					
19	K	2	2	6	2	6		1			
20	Ca	2	2	6	2	6		2			
21	Sc	2	2	6	2	6	1	2			
22	Ti	2	2	6	2	6	2	2			
23	V	2	2	6	2	6	3	2			
24	Cr	2	2	6	2	6	5	1			
25	Mn	2	2	6	2	6	5	2			
26	Fe	2	2	6	2	6	6	2			
27	Co	2	2	6	2	6	7	2			
28	Ni	2	2	6	2	6	8	2			
29	Cu	2	2	6	2	6	10	1			
30	Zn	2	2	6	2	6	10	2			
31	Ga	2	2	6	2	6	10	2	1		
32	Ge	2	2	6	2	6	10	2	2		
33	As	2	2	6	2	6	10	2	3		
34	Se	2	2	6	2	6	10	2	4		
35	Br	2	2	6	2	6	10	2	5		
36	Kr	2	2	6	2	6	10	2	6		
		2	8		18						

Electron configurations of the elements (continued)

Ground states of the free atoms

Atomic No.	Element	K	L	M	N				O				P
					4s	4p	4d	4f	5s	5p	5d	5f	6s
37	Rb	2	8	18	2	6			1				
38	Sr	2	8	18	2	6			2				
39	Y	2	8	18	2	6	1		2				
40	Zr	2	8	18	2	6	2		2				
41	Nb	2	8	18	2	6	4		1				
42	Mo	2	8	18	2	6	5		1				
43	Tc	2	8	18	2	6	6		1				
44	Ru	2	8	18	2	6	7		1				
45	Rh	2	8	18	2	6	8		1				
46	Pd	2	8	18	2	6	10						
47	Ag	2	8	18	2	6	10		1				
48	Cd	2	8	18	2	6	10		2				
49	In	2	8	18	2	6	10		2	1			
50	Sn	2	8	18	2	6	10		2	2			
51	Sb	2	8	18	2	6	10		2	3			
52	Te	2	8	18	2	6	10		2	4			
53	I	2	8	18	2	6	10		2	5			
54	Xe	2	8	18	2	6	10		2	6			
55	Cs	2	8	18	2	6	10		2	6			1
56	Ba	2	8	18	2	6	10		2	6			2
57	La	2	8	18	2	6	10		2	6	1		2
58	Ce	2	8	18	2	6	10	1	2	6	1		2
59	Pr	2	8	18	2	6	10	3	2	6			2
60	Nd	2	8	18	2	6	10	4	2	6			2
61	Pm	2	8	18	2	6	10	5	2	6			2
62	Sm	2	8	18	2	6	10	6	2	6			2
63	Eu	2	8	18	2	6	10	7	2	6			2
64	Gd	2	8	18	2	6	10	7	2	6	1		2
65	Tb	2	8	18	2	6	10	9	2	6			2
66	Dy	2	8	18	2	6	10	10	2	6			2
67	Ho	2	8	18	2	6	10	11	2	6			2
68	Er	2	8	18	2	6	10	12	2	6			2
69	Tm	2	8	18	2	6	10	13	2	6			2
70	Yb	2	8	18	2	6	10	14	2	6			2
71	Lu	2	8	18	2	6	10	14	2	6	1		2
					32								

necessary to know the energy order of the orbitals, shown in Fig. 3.15. The diagram is slightly over-simplified; for example one 5d electron is found in the ground state of lanthanum and only then does the 4f level begin to fill, similarly actinium has one 6d electron and thorium has two, so that the 5f level does not begin to fill until protactinium is reached.

Electron configurations of the elements (continued)

Ground states of the free atoms

Atomic No.	Element	K	L	M	N	O				P			Q
						5s	5p	5d	5f	6s	6p	6d	7s
72	Hf	2	8	18	32	2	6	2		2			
73	Ta	2	8	18	32	2	6	3		2			
74	W	2	8	18	32	2	6	4		2			
75	Re	2	8	18	32	2	6	5		2			
76	Os	2	8	18	32	2	6	6		2			
77	Ir	2	8	18	32	2	6	9					
78	Pt	2	8	18	32	2	6	9		1			
79	Au	2	8	18	32	2	6	10		1			
80	Hg	2	8	18	32	2	6	10		2			
81	Tl	2	8	18	32	2	6	10		2	1		
82	Pb	2	8	18	32	2	6	10		2	2		
83	Bi	2	8	18	32	2	6	10		2	3		
84	Po	2	8	18	32	2	6	10		2	4		
85	At	2	8	18	32	2	6	10		2	5		
86	Rn	2	8	18	32	2	6	10		2	6		
87	Fr	2	8	18	32	2	6	10		2	6		1
88	Ra	2	8	18	32	2	6	10		2	6		2
89	Ac	2	8	18	32	2	6	10		2	6	1	2
90	Th	2	8	18	32	2	6	10		2	6	2	2
91	Pa	2	8	18	32	2	6	10	2	2	6	1	2
92	U	2	8	18	32	2	6	10	3	2	6	1	2
93	Np	2	8	18	32	2	6	10	4	2	6	1	2
94	Pu	2	8	18	32	2	6	10	6	2	6		2
95	Am	2	8	18	32	2	6	10	7	2	6		2
96	Cm	2	8	18	32	2	6	10	7	2	6	1	2
97	Bk	2	8	18	32	2	6	10	9	2	6		2
98	Cf	2	8	18	32	2	6	10	10	2	6		2
99	Es	2	8	18	32	2	6	10	11	2	6		2
100	Fm	2	8	18	32	2	6	10	12	2	6		2
101	Md	2	8	18	32	2	6	10	13	2	6		2
102	No	2	8	18	32	2	6	10	14	2	6		2
103	Lr	2	8	18	32	2	6	10	14	2	6	1	2
104	Rf	2	8	18	32	2	6	10	14	2	6	2	2

Four types of electronic structures can be distinguished.

(i) The Group 0 gases: In the atoms of these elements all the sets of s, p, d, and f orbitals which occur are filled to capacity. Helium has the structure $1s^2$ and all the others have the outer electronic arrangement $ns^2\ np^6$. These atoms all have spherically symmetrical electron distributions and ^1S ground states.

(ii) The main-group elements forming Groups IA, IIA and Groups IIB to VIIB: These elements have atoms in which singly occupied orbitals are confined to the outermost shell. All the underlying levels are filled to capacity. This type of electronic structure is found in all the non-metals.

(iii) The transition elements forming Groups IIIA through to IB: In these, of which there are three series, the hitherto empty d orbitals of an inner shell begin to fill. The elements would be expected to have an outer electronic structure ranging from $(n{-}1)d^1\ ns^2$ to $(n{-}1)d^9\ ns^2$, but they do not show a rigid adherence to the ns^2 arrangement. A Cr atom in its ground state has $3d^5\ 4s^1$ and a Cu atom has $3d^{10}\ 4s^1$. This is because a shell which is half-filled or completely filled is particularly strongly exchange-stabilised.

The quantum mechanical *exchange energy* of an atom is directly related to the number of ways electrons of parallel spin and equivalent energy can exchange with one another. In a d shell containing five electrons of parallel spin ten such exchanges are possible. But in a d shell containing only four electrons the ways in which they can exchange are reduced to six. In the section which follows other examples will be noted of the tendency for d and f shells to gain electrons from outer orbitals as they fill up. For example, Pd has the outer electron configuration $4d^{10}$ rather than $4d^9\ 5s^1$ or $4d^8\ 5s^2$.

(iv) The inner transition elements: These comprise two series, the lanthanides and actinides. They not only have unfilled d orbitals but also underlying partly filled f orbitals. The similarity of the configurations of the outermost energy levels causes the inner transition elements to have very similar properties (30.1).

3.4.4. The Periodic classification

The above considerations give rise to the Periodic Law, a general regularity in the succession of properties of the elements, established empirically by Mendeleev. They determine the modern form of the Periodic Table (Table 3.2).

This particular tabulation is one of many that have been proposed and has the advantage over some of affording quick reference to the periods (horizontal with Arabic numerals) and the groups (vertical with Roman numerals) and of showing the relation of the transition elements to the rest. Major features to note are:

(i) There are only two elements, hydrogen and helium, in the first period. Two electrons are sufficient to fill the first quantum level. Because hydrogen can form both a cation and an anion it is placed at the head of Group I and also, in parenthesis, at the head of Group VII. Helium falls naturally in Group 0 above neon.

(ii) Group 0, the noble gases, lie between the halogens which give uninegative ions and the alkali metals which give unipositive ions.

TABLE 3.2.

THE PERIODIC TABLE

The first three Periods

s orbitals filling			p orbitals filling					
Group	I	II	III	IV	V	VI	VII	0
Period								
1	H						(H)	He
2	Li	Be	B	C	N	O	F	Ne
3	Na	Mg	Al	Si	P	S	Cl	Ar

All the Periods

			d orbitals filling										p orbitals filling					
Group	IA	IIA	IIIA	IVA	VA	VIA	VIIA		VIII		IB	IIB	IIIB	IVB	VB	VIB	VIIB	0
Period																		
1	H																(H)	He
2	Li	Be											B	C	N	O	F	Ne
3	Na	Mg											Al	Si	P	S	Cl	Ar
4	K	Ca	Sc	Ti	V	Cr	Mn	Fe	Co	Ni	Cu	Zn	Ga	Ge	As	Se	Br	Kr
5	Rb	Sr	Y	Zr	Nb	Mo	Tc	Ru	Rh	Pd	Ag	Cd	In	Sn	Sb	Te	I	Xe
6	Cs	Ba	La*	Hf	Ta	W	Re	Os	Ir	Pt	Au	Hg	Tl	Pb	Bi	Po	At	Rn
7	Fr	Ra	Ac†	Rf														
				f orbitals filling														
6	* Lanthanides			Ce	Pr	Nd	Pm	Sm	Eu	Gd	Tb	Dy	Ho	Er	Tm	Yb	Lu	
7	† Actinides			Th	Pa	U	Np	Pu	Am	Cm	Bk	Cf	Es	Fm	Md	No	Lr	

(iii) Groups IA, IIA and Groups IIB to VIIB are the non-transition elements and they flank the transition elements which occupy the centre of the Table. They include all the non-metals, except the noble gases.

(iv) The transition metals include the lanthanides and actinides and are all in the centre of the Periodic Table from Periods 4 to 7. Of the inner transition elements only lanthanum and actinium actually appear in the Table, to show where their respective series begin.

The Periodic Table is a useful basis from which to begin a comparison of the properties of individual elements because those elements with similar electronic configurations (3.4.3.1) are found in the same groups. For example, the configurations of the atoms C, Si, Ge, Sn, Pb, excluding the filled inner shells, are $2p^2$, $3p^2$, $4p^2$, $5p^2$ and $6p^2$ respectively; all the elements have the same spectroscopic ground state as carbon, namely ^3P, and the similarity of their outermost structure is responsible for a family relationship in chemical properties.

3.4.4.1. Ionisation energies

The energy required to withdraw an electron from an atom against the attraction of the nuclear charge is the *first ionisation energy* of the atom.

The value of the ionisation energy depends on:

(i) the quantum level of the electron of highest energy in the ground state of the atom,

(ii) the effective nuclear charge acting on that electron.

The first ionisation energies of elements 1 to 36 (H to Kr) are shown diagramatically in Fig. 3.16.

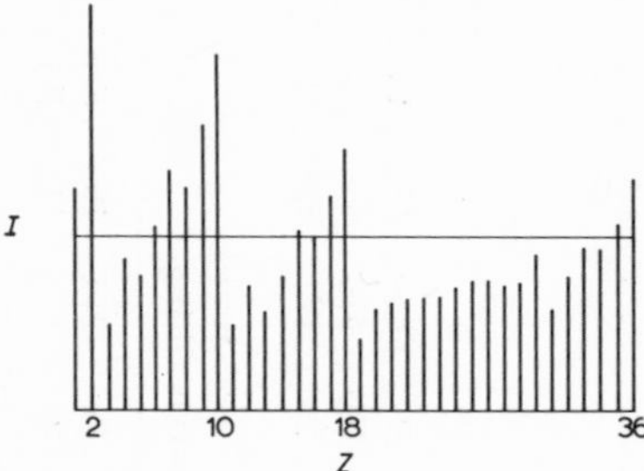

Fig. 3.16. First ionisation energies of elements 1 to 36 (horizontal line represents 1 MJ mol^{-1}).

Periodicity in the values of the first ionisation energies is evident; the changes arise from the following causes:

(i) Much energy is required to remove an electron from a completely filled outer quantum shell, for instance, from the atom of a Group 0 gas.

(ii) The first ionisation energy of an alkali metal is low because the outer s electron is heavily screened from the Z positive charges on the nucleus by $Z-1$ electrons in relatively compact orbitals. The screening constant is high and the effective nuclear charge is low.

(iii) The effective nuclear charge on the outer shell increases through a period and with it the first ionisation energy also increases, but the upward trend is not smooth. When the removal of an electron leaves an ion with a spherically symmetrical shell, for instance in the following examples:

$B(1s^2 2s^2 2p^1) \rightarrow B^+(1s^2 2s^2) + e$
$O(1s^2 2s^2 2p^4) \rightarrow O^+(1s^2 2s^2 2p^3) + e$

the energy required is usually somewhat less than the ionisation energy of the preceding element. The 'staggered' rise in ionisation energies through a short period is supporting evidence for the existence of s- and p-type orbitals.

(iv) For a particular group, an electron is more easily withdrawn from an atom in a later period than from an atom in an early period:

$Ne \rightarrow Ne^+ + e \qquad I = 2.08$ MJ mol^{-1}
$Ar \rightarrow Ar^+ + e \qquad I = 1.51$ MJ mol^{-1}

In this example the ionisation energy is lower in the second case largely because the outermost electrons of Ar belong to a higher quantum level, and therefore have a higher energy, than those of Ne.

The energy required to remove a second electron from a unipositive ion is the second ionisation energy. For Group IIA metals it is about twice the first ionisation energy.

$Be \rightarrow Be^+$	$I = 0.90$ MJ mol^{-1}	$Be^+ \rightarrow Be^{2+}$	$I = 1.76$ MJ mol^{-1}
$Sr \rightarrow Sr^+$	$I = 0.55$ MJ mol^{-1}	$Sr^+ \rightarrow Sr^{2+}$	$I = 1.06$ MJ mol^{-1}

This difference is because the effective nuclear charge has been increased by the removal of the first electron. For Group IA elements the second ionisation energies are very large indeed. For instance:

$Li \rightarrow Li^+$	$I = 0.52$ MJ mol^{-1}	$Li^+ \rightarrow Li^{2+}$	$I = 7.25$ MJ mol^{-1}
$Rb \rightarrow Rb^+$	$I = 0.40$ MJ mol^{-1}	$Rb^+ \rightarrow Rb^{2+}$	$I = 2.64$ MJ mol^{-1}

This is because of (a) the lower energy-state and (b) the far weaker shielding of the electrons in the complete s^2p^6 shell.

For transition and inner transition elements there is little variation (0.72 ± 0.06 MJ mol^{-1} for a transition series and 0.57 ± 0.03 MJ mol^{-1} for the lanthanides). The metals following the lanthanides are exceptions to the general rule of decreasing ionisation energy down a group. In them the screening effect of the rather diffuse d shell is outweighed by the increasing nuclear charge.

Ag	0.76 MJ mol^{-1}	Cd	0.87 MJ mol^{-1}
Au	0.89 MJ mol^{-1}	Hg	1.01 MJ mol^{-1}

Elements of B subgroups have higher ionisation energies than those of the corresponding A subgroups.

K	0.42 MJ mol^{-1}	Cu	0.74 MJ mol^{-1}
Ca	0.59 MJ mol^{-1}	Zn	0.91 MJ mol^{-1}

The increased binding of the electron due to the penetration of its orbital into orbitals nearer the nucleus is greater with 18 than with 8 electrons immediately beneath it, since it comes into the region of a positive charge greater by 10 units.

There is a simple mathematical relation between the ionisation energies of the members of a series of isoelectronic species of consecutive atomic number. The energy required to remove the electron completely from a system consisting only of a nucleus of charge Ze and a single electron is

$$E = \frac{Z^2me^4}{8\epsilon_0^2h^2}$$

Thus

$$\mathrm{d}E/\mathrm{d}Z = \frac{2Zme^4}{8\epsilon_0^2h^2}$$

and

$$\mathrm{d}^2E/\mathrm{d}Z^2 = \frac{2me^4}{8\epsilon_0^2h^2} \equiv 2.62\ \mathrm{MJ\ mol^{-1}}$$

The same value of $\mathrm{d}^2E/\mathrm{d}Z^2$ is obtained for the series of atoms and ions with the $1s^2$ configuration (Table 3.3).

TABLE 3.3

Classification	Process	I/MJ mol^{-1}	$\frac{\Delta I}{\Delta Z}$/MJ mol^{-1}	$\Delta\left(\frac{\Delta I}{\Delta Z}\right)$/MJ mol^{-1}
$I(1)$ (He)	$He \rightarrow He^+ + e$	2.37		
			} 4.93	
$I(2)$ (Li)	$Li^+ \rightarrow Li^{2+} + e$	7.30		} 2.62
			} 7.55	
$I(3)$ (Be)	$Be^{2+} \rightarrow Be^{3+} + e$	14.85		} 2.62
			} 10.17	
$I(4)$ (B)	$B^{3+} \rightarrow B^{4+} + e$	25.02		} 2.64
			} 12.81	
$I(5)$ (C)	$C^{4+} \rightarrow C^{5+} + e$	37.83		

For a series in which the electron which is ionised is in the second quantum level the value of $\Delta(\frac{\Delta I}{\Delta Z})$ is $2.62/2^2$ MJ mol^{-1}, and for a series in which an electron of principal quantum number 3 is removed the value is $2.62/3^2$ MJ mol^{-1}. This law,

$$\frac{\mathrm{d}^2E}{\mathrm{d}Z^2} = \frac{2.62}{n^2}\ \mathrm{MJ\ mol^{-1}}$$

is known as the *law of second differences* and it is a useful principle for predicting ionisation energies, certainly for values of n up to 3. For $n > 3$ the law cannot be tested so successfully because the data are incomplete.

The law also enables estimates to be made of electron affinities. For example, extrapolation of the figures in Table 3.3 gives the value

$$I = 2.37 - (4.93 - 2.62)\ \text{MJ mol}^{-1}$$
$$= 0.06\ \text{MJ mol}^{-1} = 60\ \text{kJ mol}^{-1}$$

for the process

$H^- \rightarrow H + e$

and this is also the electron affinity of the H atom, the energy *released* in the reverse process.

3.4.4.2. Electron affinities

The energy *released* when an extra electron is taken up by an atom is its electron affinity, *A*. This can be estimated indirectly by applying the Born—Haber cycle to ionic compounds (4.2.3) and by other methods.

In the halogens, except for fluorine, the smaller the atom the greater the electron affinity and the ease of anion formation:

F 0.36 Cl 0.39 Br 0.36 I 0.33 MJ mol^{-1}

In Group VI the values for two electrons are

O —0.70 S —0.33 Se —0.41 MJ mol^{-1}

Once an electron has been received by an atom the uninegative ion which is formed repels further electrons; hence the negative affinities displayed by O, S and Se in forming binegative ions.

3.4.4.3. Magnetic properties

The *permeability*, μ, of a substance is defined by the equation

$$\mu = B/H$$

where B is the magnetic flux density in the material when it is placed in a field of strength H. The permeability of a vacuum is $\mu_0 = 4\pi \times 10^{-7}$ kg m s^{-2} A^{-2} exactly.

The relative permeability, μ_r, of a substance is the ratio of the permeability, μ, for that substance to μ_0:

$$\mu_r = \mu/\mu_0$$

It is a dimensionless quantity, as is the magnetic susceptibility for the substance, χ_m, defined by the equation:

$$\chi_m = \mu_r - 1$$

Diamagnetism and paramagnetism. Most substances are *diamagnetic*, with χ_m negative. In a diamagnetic substance the flux density in a particular magnetic field is lower than the flux density in a vacuum in the same field and the substance will tend to move towards the weakest part of an inhomogeneous magnetic field. Diamagnetism arises from the fact that the movement of electrons within a substance, when placed in a magnetic field, has the effect of opposing the applied field.

The numerical value of χ_m for a diamagnetic substance is small and it is not affected by temperature.

Paramagnetic substances are those for which χ_m is positive. The flux density in such a substance is greater than that in a vacuum in the same field, and a paramagnetic material tends to move to the strongest part of an inhomogeneous field. When paramagnetism occurs it completely overshadows the feeble diamagnetism also present; it arises from the presence of unpaired electrons whose spin and angular momenta interact with the field. The measurement of paramagnetic susceptibility is therefore of value in gaining information about the numbers of unpaired electrons in a molecule or ion.

The Gouy balance. The susceptibility of a solid or liquid is most commonly determined using a Gouy balance. A sample of substance contained in a long silica tube of uniform cross-section is suspended from a balance beam so that the lower end is opposite the pole pieces of an electromagnet. If a magnetic field is applied by energising the magnets, the paramagnetic material is drawn downwards and additional weights have to be added to the opposite pan of the balance to restore the sample to its original position. For a sample with a cross-section of 1 cm^2 in an external flux density of 1 tesla (the SI unit of flux density defined as kg s^{-2} A^{-1}) the weight required is a few hundred mg, thus an ordinary analytical balance can be adapted for the purpose.

The value of χ_m is obtained from

$$\chi_m = \frac{2F\mu_0}{AB_0^2}$$

where F is the force acting on the sample with cross-section A in an external flux of density B_0. The value of B_0 is usually obtained by calibration with a substance for which χ_m is known.

The magnetic moment of a particle. In general, paramagnetism arises from a permanent magnetic moment, not an induced one, which is associated with a definite state of angular momentum. The *magnetic moment*, μ, of an atom or ion, arising from total angular momentum with quantum number J is given by

$$\mu = g\sqrt{J(J+1)}\,\mu_B$$

where g is the Landé-factor and μ_B the Bohr magneton:

$$\mu_B = \frac{eh}{4\pi m} = 9.274 \times 10^{-24} \text{ A m}^2$$

For atoms in S states, $L = 0$ and $J = S$, therefore $g = 2$ (3.4.4.4) but in all other states $g < 2$. Consequently observed values of g indicate whether the paramagnetic moment arises from a combination of orbital and spin angular momenta or from spin momentum only. In the latter case, the observation of the maximum value of J arising from a given electron configuration indicates the number, n, of unpaired spins, since $J_{max} = S_{max} = \frac{1}{2}n$.

The relation between magnetic moment and magnetic susceptibility. Since the magnetic moment of an assembly of atoms depends on statistical averaging over all possible z-components, and regular orientation is hindered by thermal agitation, the relation between μ and χ_m must involve temperature. For ordinary temperatures

$$\chi_m \times \frac{M}{\rho} = \frac{\text{a constant}}{T} \qquad \text{(Curie's law)}$$

where M is the molar mass of the substance, ρ is its density and T is the thermodynamic temperature.

More accurately $\chi_m \times (M/\rho)$ is proportional not to $1/T$ but to $1/T - \theta$, a result due to Weiss, where θ is the *Curie temperature*. As θ is small for most paramagnetic substances it is customary to use Curie's law in its original form and calculate an *effective magnetic moment*, μ_{eff} for a specified temperature:

$$\mu_{eff} = \sqrt{\chi_m \times \frac{M}{\rho} \times T \times \frac{3k}{N_A \mu_0}}$$

where N_A is the Avogrado constant and k the Boltzmann constant. When other small effects such as the small opposing diamagnetism are taken into account, magnetic moments determined from measurements of susceptibility are found to be in good agreement with those inferred from spectroscopy.

Magnetic moments of transition-metal ions. For a free gaseous ion in which the separation between states with different J values is large compared with kT, only the lowest J state is occupied, there is strong L—S coupling (3.4.1), and the value of μ should be given by

$$\mu = g\sqrt{J(J+1)}\ \mu_B$$

There are no data for gaseous ions to test this equation but in lanthanide compounds the unpaired 4f electrons are so little affected by their chemical environment, being shielded from surrounding ions by the 5s and 5p electrons, that the experimental magnetic moments agree well with those calculated from the equation.

For transition-metal ions of the Period 4 the arguments above do not apply because (a) the spacing between J states is about equal to kT at normal temperatures and (b) the unpaired d electrons are influenced by their surroundings. Agreement with the formula above is very poor and the values of

μ are close to

$$\mu = 2\sqrt{S(S+1)}\ \mu_B = \sqrt{n(n+2)}\ \mu_B$$

which is equivalent to putting $L = 0$ in the more general equation above. In the second statement of the equation, n is the number of unpaired electrons. The fact that these ions comply with this 'spin-only' formula implies that the orbital contributions to the magnetic moment are nullified, or nearly so. This 'quenching' of the orbital contribution can be explained as follows.

An electron can have orbital angular momentum about a particular axis only when its orbital may be transformed into an equivalent, degenerate orbital by rotation about that axis. Thus a d_{xy} orbital is transformed into $d_{x^2-y^2}$ by a rotation of 45° about the z axis in a free ion, and a d_{xz} into d_{yz} by a rotation of 90°. In a free ion the z axis represents any direction, since all directions are equivalent.

In, say, an octahedral field produced by ligands around the ion (6.3) the d_{z^2} and $d_{x^2-y^2}$ (e_g) orbitals are raised in energy relative to the d_{xy}, d_{xz} and d_{yz} (t_{2g}). Thus the d_{xy} and $d_{x^2-y^2}$ are no longer equivalent, though the d_{xz} and d_{yz} remain so. Furthermore, the d_{z^2} has no orbital momentum around the z axis, thus no orbital contribution to the magnetic moment can arise from e_g electrons, and if the t_{2g} orbitals are either half-filled or completely filled their total contribution is zero. Thus in the following configurations in an octahedral field the orbital contributions are completely quenched and the spin-only formula applies:

$t_{2g}^3 e_g^x$ where $x = 0, 1$ or 2
$t_{2g}^6 e_g^y$ where $y = 0, 1, 2$ or 3

For other ground-state configurations some orbital contribution is expected

TABLE 3.4

PARAMAGNETIC MOMENTS OF SOME TRANSITION-METAL IONS

Ion	Number of unpaired electrons	μ_s/μ_B	Observed values/μ_B
Ti^{3+}	1	1.73	1.65—1.79
V^{4+}	1	1.73	1.68—1.78
V^{3+}	2	2.83	2.75—2.85
Cr^{3+}	3	3.87	3.70—3.90
Mn^{3+}	4	4.90	4.90—5.00
Mn^{2+}	5	5.92	5.65—6.10
Fe^{3+}	5	5.92	5.70—6.0
Fe^{2+}	4	4.90	5.10—5.70
Co^{3+}	4	4.90	4.3
Co^{2+}	3	3.87	4.30—5.20
Ni^{2+}	2	2.83	2.80—3.50
Cu^{2+}	1	1.73	1.70—2.20

in a truly octahedral environment, but when tetragonal distortion is present (6.3.4) the degeneracy of the t_{2g} orbitals is split and the possibility of an orbital contribution is further reduced.

An argument similar to that for the octahedral field can be applied to the tetrahedral case, and in these compounds of Period 4 transition metals the paramagnetic moments are close to those predicted by the spin-only formula (Table 3.4). It will be noticed that the tendency is for experimental values to be a little smaller than the calculated μ_s when the d shell is less than half-filled, and sometimes appreciably larger than μ_s when the d shell is more than half-filled.

Transition-metal compounds of Periods 5 and 6, with the exception of those of the platinum group, generally have low paramagnetic moments; this is almost certainly due to the metal—ligand bonding being much more covalent in character than in the Period 4 compounds.

3.4.4.4. Electron spin resonance

General principles. Any atom or molecule with an incompletely paired system of electrons has a magnetic moment. In a magnetic field the moment acquires an interaction energy which depends upon its orientation relative to the field. In a flux of density one tesla the interval between the quantised energy levels is about 10^{-23} J, and transitions may be induced by irradiation with microwaves of wavelength 0.5 to 5 cm. The consequent absorption spectrum arises from electron spin resonance.

In a non-linear molecule containing one unpaired electron, the electronic orbital motion makes no contribution to the magnetic moment, and

$$\mu = g\sqrt{S(S+1)}\ \mu_B$$

μ_B is the Bohr magneton; S is the spin quantum number which is $\frac{1}{2}$ for a system containing one unpaired electron; and g is the Landé splitting factor. Here we must use a more exact value derived from relativistic quantum mechanics, in which g is 2.00229 in this idealised case.

In a magnetic flux of density B, the energy of interaction is

$$E = \mu_z B$$

μ_z being the component of the magnetic moment vector in the field direction. The values of μ_z are restricted to $gM\mu_B$, with M equal to S, $S-1$, $S-2$, ... $-S$. A transition between states is permitted if $\Delta M = \pm 1$. As absorptive transitions to higher energy states are more numerous than radiative transitions downwards, measurements are made for the $\Delta M = +1$ case

$$\therefore\ \Delta E = gB\mu_B = h\nu$$

$$\therefore\ \nu = \frac{gB\mu_B}{h}$$

To determine their electron spin resonance, the molecules are irradiated with monochromatic microwaves of frequency about 10^{10} Hz, while the magnetic field applied on them is varied. According to the derivation above, all paramagnetic species should absorb at exactly the same flux density when irradiated at equal frequencies; for example 0.357 tesla for 10^{10} Hz radiation:

$$B = \frac{h\nu}{g\mu_B} = \frac{6.626 \times 10^{-34}\ \mathrm{J\,s} \times 10^{10}\ \mathrm{s^{-1}}}{2.0023 \times 9.274 \times 10^{-24}\ \mathrm{A\,m^2}}$$

$$= 0.357\ \mathrm{kg\,A^{-1}\,s^{-2}} = 0.357\ \mathrm{T}$$

But if this were the case, e.s.r. spectroscopy would be without chemical interest.

However there are deviations from constant ν/B because g can vary from the value 2.00229. For paramagnetic molecules, deviations are small but usually sufficient to separate the absorptions of different species, but for paramagnetic ions g can vary considerably. Not only is the position of the absorption variable, but so also is its nature. Absorption in a single band is rare; there is usually a symmetrical multiplet.

Applications

(1) The spectrum of a chloroform solution of $KSO_3 \cdot NO \cdot SO_3K$ (containing one unpaired electron) consists of three bands of equal intensity, equally spaced by 1.3 mT. The middle one corresponds to $g = 2.0054$, which is rather higher than the spin-only value because of incomplete quenching of an orbital contribution to the magnetic moment. There are three bands because one of the nuclei present, ^{14}N, is paramagnetic (2.7) with nuclear spin quantum number $I = 1$ (M_N is therefore +1, 0 or −1). As the moment of the electron can have two values represented by $M_e = +\frac{1}{2}$ or $-\frac{1}{2}$, six possible energy states arise from the possible combinations of electronic and nuclear moments (Fig. 3.17).

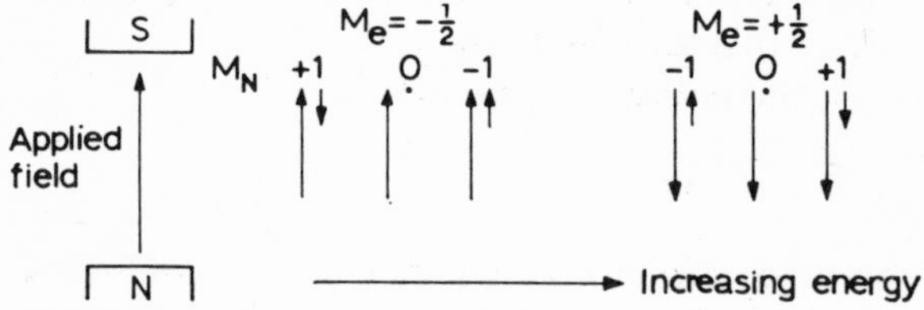

Fig. 3.17. Possible combinations of electronic (M_e) and nuclear (M_N) moments for nitrogen-14 atoms.

An antiparallel arrangement of moments represents a lower energy than a parallel one, the extent of the difference increasing with the probability of finding the unpaired electron at the nitrogen. The energy level diagram has the form shown in Fig. 3.18.

The selection rules are $\Delta M_e = +1$ and $\Delta M_N = 0$. The three possible energy

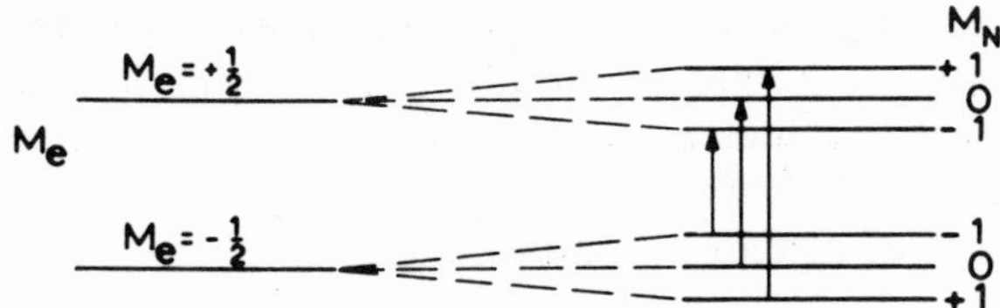

Fig. 3.18. Energy levels and allowed transitions for nitrogen-14 atom in a negative field.

transitions are as shown. From the splitting, the unpaired electron density at the nitrogen can be calculated.

(2) The absorption due to transition metal ions also shows multiplet structure due to nuclear magnetism. For example, the spectrum of a benzene solution of bis(cyclopentadienyl)vanadium has eight equally intense bands associated with ^{51}V, which has $I = \frac{7}{2}$, and $2I + 1 = 8$ components of magnetic moment.

(3) More important than the nuclear splitting, however, is the observed deviation from $g = 2.00229$, arising from the orbital contribution to the electronic magnetic moment which provides information on crystal-field and spin—orbital coupling effects. In free atoms or monatomic ions

$$\mu = g\sqrt{J(J+1)}\,\mu_B$$

and

$$g = 1 + \frac{J(J+1) + S(S+1) - L(L+1)}{2J(J+1)} \tag{3.4.2}$$

For Ti^{3+} (d^1) the values are:

$S = \frac{1}{2}$, $L = 2$ and $J = \frac{3}{2}$

$$\therefore g = 0.8$$

But when the Ti^{3+} ion is surrounded by other ions in a crystal the d orbital energy levels are split (6.3); the effect of this is to destroy the significance of L and the orbital contribution is quenched by the crystal field or, of course, by molecule formation. If $L = 0$, then $S = J$ and g is therefore 2, the spin-only value; actually 2.00229 after relativistic correction. A further modification to the energy states and to the wave functions, which has the effect of restoring some significance to the orbital contributions to magnetic moments, is made by spin—orbit coupling.

Spin—orbit coupling. An electron tends to align its spin anti-parallel to its orbital angular momentum. Thus, when it has orbital angular momentum, this is maintained by being weakly coupled to the spin; furthermore, the spin itself tends to generate orbital angular momentum. As a result, there is competition between the quenching effect of the ligands and the sustaining effect of the spin—orbit coupling. The spin—orbit coupling constant, λ, expressed in energy units, depends on the atomic number, on the number and type of outer electrons, and on the values of L and S for the ion in the particular state.

For an octahedral arrangement of ligands around a central d^1 ion:

$$g = 2.00229 - \frac{8\lambda}{\Delta_0}$$

where Δ_0 is the ligand-field splitting (6.3) and λ is the *spin—orbit coupling constant.*

The moment of an ion in which spin—orbit coupling is effective is given by:

$$\mu = \mu_s \left(1 - \alpha \frac{\lambda}{\Delta_0}\right)$$

where μ_s is the spin-only moment and α is a constant which depends upon the spectroscopic state of the free ion. For high-spin ions of the first transition series:

$\alpha = 0$ if $L = 0$ (d^5)
$\alpha = 2$ if $L = 2$ (d^1, d^4, d^6, d^9)
$\alpha = 4$ if $L = 3$ (d^2, d^3, d^7, d^8)

The value of λ is positive for ions with fewer than five d electrons, and negative for those with six or more.

TABLE 3.5

VALUES OF THE SPIN—ORBIT COUPLING CONSTANT λ FOR SOME TRANSITION-METAL IONS

Ion	Occupancy of d orbitals	λ/kJ mol^{-1}
Ti^{2+}	d^2	+0.73
V^{2+}	d^3	+0.67
Cr^{2+}	d^4	+0.68
Mn^{2+}	d^5	zero
Fe^{2+}	d^6	−1.22
Co^{2+}	d^7	−2.16
Ni^{2+}	d^8	−3.76

The first group in Table 3.5 have observed moments slightly lower than the spin-only value; the last have moments significantly larger, the effect being greatest when the crystal field is weakest. Table 3.5 above should be compared with Table 3.4.

Further Reading

H. Eyring, J. Walter and G.E. Kimball, Quantum chemistry, Wiley, New York, 1944.
G. Herzberg, Atomic spectra and atomic structure, 2nd edition, Dover, New York, 1944.
H.A. Kramers, Quantum mechanics, Interscience, New York, 1957.
J.C. Slater, Quantum theory of atomic structure, McGraw-Hill, New York, 1960.
R.S. Berry, Atomic orbitals, J. Chem. Ed., 43 (1966) 283.
I.B. Cohen and T. Bustard, Atomic orbitals, J. Chem. Ed., 43 (1966) 187.
R.P. Feynman, R.B. Leighton and M. Sands, The Feynman lectures on physics, Quantum mechanics, Addison—Wesley, Reading, Mass., 1965.
A. Hermann, From Planck to Bohr, The first fifteen years in the development of the quantum theory, Angew. Chem., Internat. Ed., 9 (1970) 34.

4 The Electronic Theory of Chemical Bonding

In principle, the existence of stable compounds could be predicted by solving the wave equation for systems of nuclei (masses m_1, m_2, ...) and electrons (mass m_e). Solutions would occur for certain most probable nuclear configurations, and would correspond to lower energy values than when the systems were separated into neutral atoms at rest. The large mass ratio (usually $> 10{,}000 : 1$) of nuclei to electrons makes it permissible to discuss separately the electron distribution and the relatively sluggish motion of the nuclei. To an exceedingly good approximation, the nuclei may be regarded as moving in a 'cloud' of electrons whose distribution is determined by a wave function depending on the instantaneous positions of the nuclei. This general picture is adopted universally in the electronic interpretation of valency, the stability of different configurations being discussed as though the nuclei were at rest.

Another guiding principle of great value was derived theoretically by Feynman and Hellmann (1939). It states that the forces which hold together the positive nuclei in a molecule or crystal are just those which they would experience if they were embedded in a static distribution of negative charge of density P, $P(x, y, z)$ being the probability density for finding an electron at the point x, y, z. This principle at once provides immediate physical insight into the nature of bonding and its classification. Briefly, all bonds can be interpreted electrostatically in terms of a charge cloud, provided this is determined wave-mechanically rather than classically. Usually, the form of the charge cloud is inferred from experimental, intuitive and theoretical considerations.

When internuclear distances are large compared with molecular distances, every nucleus has its normal complement of electrons fairly tightly localised about it. At molecular distances, the inner shells remain in this condition, but the outermost electrons are affected in varying degrees, ranging from a slight distortion of their orbitals to an electron transfer from one atom to another.

Such situations correspond in the broad view to covalent and ionic descriptions of bonding. Thus, although there is no sharp boundary between the two approaches, it is convenient to consider covalent molecules first.

4.1. The Covalent Bond

4.1.1. Molecular orbital theory and the LCAO-MO method

The simplest of all molecules is the ion H_2^+ since it possesses two nuclei and but one electron. If orbitals can be found to describe the states of the one electron in a two-centre field of this kind, it should be possible to develop an aufbau theory of diatomic molecules, exactly as in the one-centre instance. This can then be extended to multi-centre situations. The one-electron orbitals, which extend over all nuclei, are called *molecular orbitals*; or in the case of a crystal, *crystal* or *Bloch* orbitals. In such an approach, the available molecular orbitals are filled in ascending energy order and the resultant electron density is, in this approximation, again just a sum of the orbital contributions. And again, as in the Hartree method, the picture will be a good one when each orbital is chosen so as to take account of the various nuclear charges, the screening effect of electrons in inner orbitals, and the average disposition of electrons in the other molecular orbitals.

In 1927 Burrau determined accurately the molecular orbital of lowest energy for the system H_2^+ and found, by considering a range of internuclear distances, that at 106 pm there was an energy minimum which was 268 kJ mol^{-1} lower than that of a system comprising a normal H atom and a distant proton. This bond length and the theoretical dissociation energy agree excellently with spectroscopic values. The electron density calculated by Burrau is indicated in Fig. 4.1; it clearly substantiates the Feynman—Hellmann principle by showing a considerable piling up of charge in the bond region. Burrau's results form the true starting point of an aufbau theory of molecular structure. They were, however, almost immediately overshadowed by those of Heitler and London, who considered the normal hydrogen molecule and later developed a general theory of *two*-electron bonds so closely in accord with the accepted ideas of G.N. Lewis (1916) that it won universal popularity. It was not for some time that the relationship between the two approaches was appreciated.

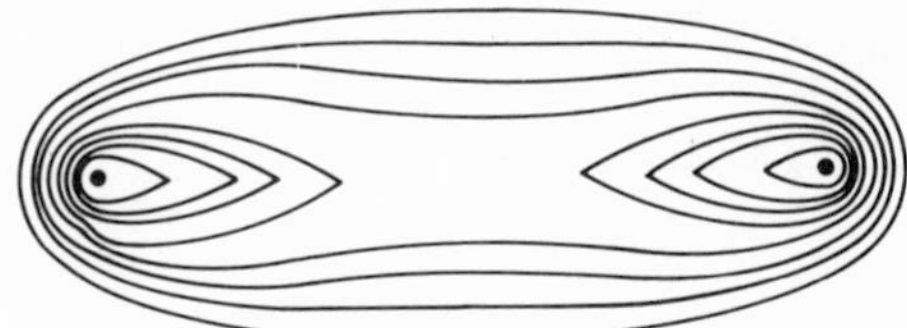

Fig. 4.1. Contours of electron density in H_2^+ (after Burrau).

Since accurate calculations are not usually feasible, it is necessary to find means of securing fair approximations to molecular orbitals and of estimating their relative energies. The simplest way is to build them up out of atomic orbitals, taking:

$$\psi = c_1\varphi_1 + c_2\varphi_2 + c_3\varphi_3 + \ldots$$

where φ_1, φ_2 etc. are suitable atomic orbitals put together with numerical coefficients c_1, c_2, c_3. This is called the *linear combination of atomic orbitals* (LCAO) approximation.

For H_2^+, such an approximate MO would be $\psi = c_a\varphi_a + c_b\varphi_b$ where φ_a and φ_b are 1s orbitals of the nuclei a and b. Since the ψ-function of an AO falls off exponentially, ψ at a point near nucleus a will be essentially $c_a\varphi_a$, which is a solution of the wave equation for an electron associated with nucleus a. ψ behaves in the same way when the electron is near b, giving $c_b\varphi_b$. Thus it remains to determine the values of c_a and c_b. There are standard mathematical methods for this, which give sets of coefficients and energies, not only for the lowest state, but for as many states as there are atomic orbitals. Here in H_2^+ where the two centres are identical, the probability of finding the electron, $P = \psi^2$, will be symmetrical, which means that $c_a = \pm c_b$. Hence the two solutions are

$$\psi_1 = N_1(\varphi_a + \varphi_b) \text{ and } \psi_2 = N_2(\varphi_a - \varphi_b)$$

where the constants N_1 and N_2 are chosen so that the functions are correctly normalised. The values of φ_a and φ_b along the axis through the nuclei a and b obtained by the LCAO method are shown graphically in Fig. 4.2 together

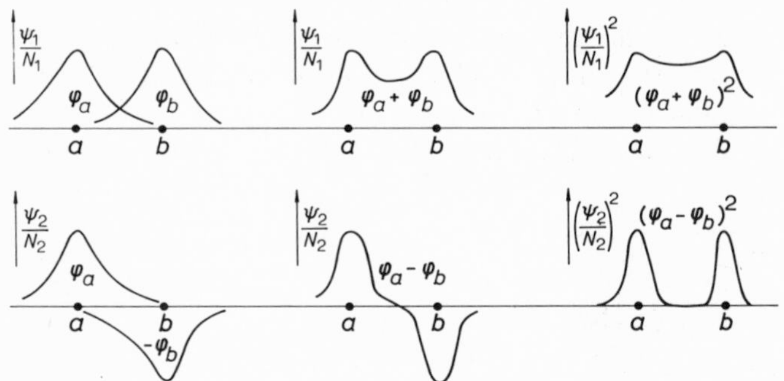

Fig. 4.2. Linear combination of atomic orbitals.

with the corresponding electron density distributions. Clearly there is considerable electron density, proportional to ψ_1^2, between nuclei a and b, but zero density mid-way between a and b for ψ_2^2.

The shapes of these molecular orbitals are indicated in Fig. 4.3, and it is clear that ψ_1 must approximate to Burrau's ground state solution and ψ_2 represents an excited state. In fact ψ_1, in error by only a few percent, still suffices to predict a stable molecule; but the excited state, giving repulsion at all internuclear distances, indicates spontaneous dissociation. The two molecular orbitals are described as *bonding* and *anti-bonding* partners; they are designated by σ1s and σ^*1s, being built out of 1s atomic orbitals, where σ indicates their symmetry about the molecular axis and the asterisk distinguishes the anti-bonding from the bonding MO.

As with atomic orbitals, the set of solutions of the *two*-centre wave equation is infinite; σ1s and σ*1s are the two of lowest energy. Some of the solutions would lead to certain molecular orbitals with more than one node across the bond, others with nodal planes through both nuclei, and so on. The main features of all these molecular orbitals can be quite well reproduced in LCAO approximation, as is indicated schematically in Fig. 4.3, where the names of the orbitals are also given.

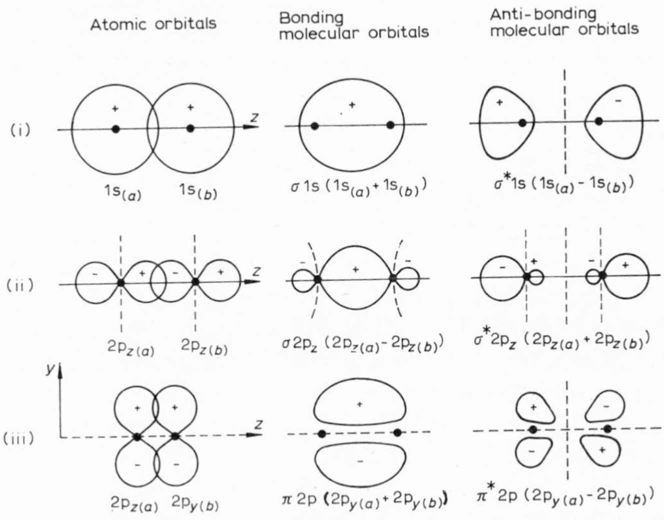

Fig. 4.3. Molecular orbitals in LCAO approximation (nodes indicated by broken lines).

The symmetry of an MO is specified by the number of nodal planes containing the molecular axis. The types with 0, 1, 2, ... such planes are designated by σ, π, δ, ... by analogy with the AO notation s, p, d, ... Again, these numbers correspond to quantised values of the angular momentum — momentum around the molecular axis — $\lambda(h/2\pi)$, where $\lambda = 0, \pm 1, \pm 2, \ldots$. Orbitals other than those of σ type thus occur in degenerate pairs, corresponding to circulation in one sense or the other; and molecular orbitals such as $\pi 2p_x$ and $\pi 2p_y$ are mixtures of equal parts of those representing states of definite angular momentum ($\lambda = \pm 1$), and are the counterparts of the atomic orbitals $2p_x$ and $2p_y$ which were adopted in place of $2p_{+1}$ and $2p_{-1}$.

In the hydrogen molecule H_2, the two lowest energy molecular orbitals of such a homonuclear diatomic molecule are σ1s and σ*1s. The relative energies of the molecular orbitals and the atomic orbitals from which they arise are shown in Fig. 4.4. The valence orbitals of the combining atoms, H_a and H_b, which have the same coulombic energy, are placed directly opposite one another. The energies of the molecular orbitals appear in the middle of the diagram. The σ1s MO is more stable than the combining 1s valence orbitals, the σ*1s is correspondingly less stable.

According to the aufbau approach, the electronic configuration of the normal molecule H_2 would be written $H_2[(\sigma 1s)^2]$. The charge cloud would be

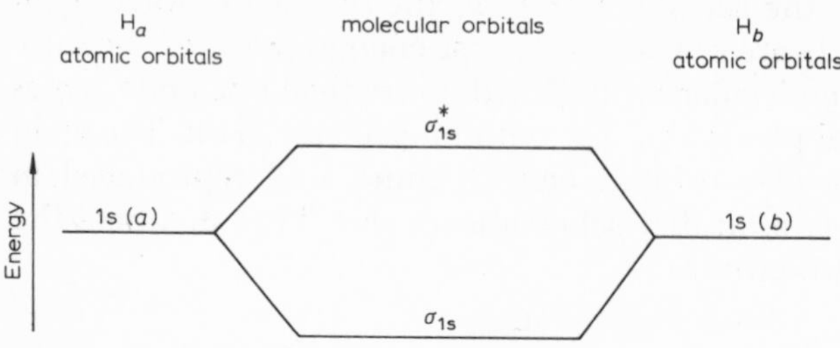

Fig. 4.4. Relative energies of molecular orbitals and the 1s atomic orbitals from which they are formed.

very similar in shape to that in H_2^+, only twice as dense and, by exerting a correspondingly greater attraction upon the nuclei, would lead to a shorter and stronger bond. But, on the other hand, further addition of electrons would weaken the bond. Thus the hypothetical molecule He_2 would have the structure $He_2[(\sigma 1s)^2(\sigma^* 1s)^2]$ and the bonding effect of the $\sigma 1s$ electrons would be offset by the anti-bonding effect of the others. If the charge clouds (cf. Fig. 4.3) are superimposed, it turns out that there is no accumulation of charge in the bond region, the final density being roughly that of two unmodified helium atoms. More precisely, there is a slight repulsion which grows rapidly when the closed shells are pushed together. The saturation property of chemical binding is thus immediately interpreted: an atom with one or more *singly* occupied orbitals has a corresponding number of valencies, but one with only *doubly* occupied orbitals is inert.

Most chemical bonds, whether σ, π, or δ bonds, are two-electron bonds, and in the aufbau picture both electrons occupy a bonding MO whose anti-bonding partner is empty. All two-electron bonds can be explained in this way, but their characters and strengths vary widely according to the form of the charge density. The bonding MO, whatever its type, gives a density contribution

$$\psi^2 = c_a{}^2\varphi_a{}^2 + c_b{}^2\varphi_b{}^2 + 2c_a c_b \varphi_a \varphi_b$$

where the first two terms are AO charge clouds with 'weight factors' c_a^2 and c_b^2. The third term is best described as an overlap density, since it can be large only in the 'overlap region' between the atoms where both φ_a and φ_b are large. Now when c_a and c_b are roughly equal (they are exactly so for all homonuclear systems) the charge will spread more or less evenly over the two atoms and the binding will arise mainly from the attraction exerted by the overlap density between them. This density depends jointly on the *amount* of charge described by the overlap function $\varphi_a\varphi_b$ and upon its weight factor $2c_ac_b$; the amount is clearly greater the greater the overlap of the atomic orbitals, but the weight-factor (fixed so that ψ is normalised) is not very sensitive to overlap. Consequently, in a 'homopolar' situation ($c_a \simeq c_b$), the strength of a bond involving two given atomic orbitals is determined largely by their overlap, the larger the overlap the stronger the bond. This conclusion, embodied in the *principle of maximum overlap*, is corroborated by a great weight of chemical evidence. When, however, the bonded atoms have very different electron affinities, c_a and c_b may be of different orders of magnitude, describing the much greater chance of finding an electron on one nucleus rather than on the other. The change of bond type can be seen from the limiting case $c_a \to 1$, $c_b \to 0$: for then the MO degenerates into an AO φ_a and this acquires both electrons of the two-electron bond; the overlap density of the covalent bond disappears and the charge cloud becomes that of two oppositely charged ions. There is clearly a perfectly smooth transition from covalent to ionic binding, the amount of ionic character being reflected in the disparity between the MO coefficients and in a 'lopsided' bond orbital.

To summarise: two-electron σ bonds occur when the singly occupied atomic orbitals of different atoms overlap in pairs; these bonds are strong when the overlap is large. The combination of atomic orbitals which are doubly occupied, or which overlap only weakly, need not in general be considered. The bonding molecular orbitals lean towards the more electronegative atom, approaching the form of an AO of that atom in extreme cases. These principles, as will be seen later, apply with little modification to polyatomic molecules and crystals, the only difference being that strong overlapping is not confined to *one pair* of atomic orbitals.

4.1.1.1. Simple diatomic molecules

The electron configuration of simple diatomic molecules can be readily discussed in terms of the molecular orbitals described above. It is only necessary, in the aufbau approach, to know the energy order of various orbitals, and for homonuclear molecules where the difference in energy between the 2s and 2p orbitals is large (taking the bond as the z axis) this is:

$$\sigma 1s < \sigma^* 1s < \sigma 2s < \sigma^* 2s < \sigma 2p_z < \begin{pmatrix} \pi 2p_x \\ \pi 2p_y \end{pmatrix} < \begin{pmatrix} \pi^* 2p_x \\ \pi^* 2p_y \end{pmatrix} < \sigma^* 2p_z \ldots$$

Where the energy difference between 2s and 2p orbitals in the atoms is small, the 2s and $2p_z$ orbitals must be considered together in an LCAO scheme. The effect of this *sp hybridisation* is to stabilise the $\pi 2p_x$ and $\pi 2p_y$ orbitals relative to the $\sigma 2p_z$. In the B_2, C_2 and N_2 molecules, for example, the π2p orbitals lie lower in energy than the σ2p.

With the notation and the principles explained above, electronic structures of some homonuclear diatomic molecules may be described along the following lines:

Lithium. $2Li(1s^2 2s^1) \to Li_2\{KK(\sigma 2s)^2\}$

Here the two K-shell (1s) electrons on each nucleus are indicated in the molecular configuration simply by the two K's, the inner shells being more or less undisturbed, and the two valence electrons give rise to a single bond. The bond is a weak one (104 kJ mol^{-1} against 436 kJ mol^{-1} for H—H) because the 2s orbital of Li is larger and more diffuse than the 1s orbital of H and consequently allows less effective overlap. Furthermore, interaction between the filled K shells makes the bond long and weak.

Beryllium
The electron configuration of a Be_2 molecule would have to be $KK(\sigma 2s)^2(\sigma^*2s)^2$ according to the principle outlined above. As the σ^*2s orbital is the antibonding partner of the $\sigma 2s$ orbital the net interaction would be zero, and Be_2 has not been observed.

Nitrogen. $2N(1s^2\ 2s^2\ 2p^3) \rightarrow N_2\ \{KK(\sigma 2s)^2(\sigma^*2s)^2(\pi 2p)^4(\sigma 2p)^2\}$
Here the occupancy of both $\sigma 2s$ and σ^*2s by pairs of electrons means that bonding is due entirely to overlap of 2p atomic orbitals which give rise to a $\sigma 2p_z$ MO and two degenerate, that is equivalent, $\pi 2p$ molecular orbitals. These π orbitals describe electron densities which vanish on the N—N axis but have maxima some distance to either side of it (Fig. 4.5). There are

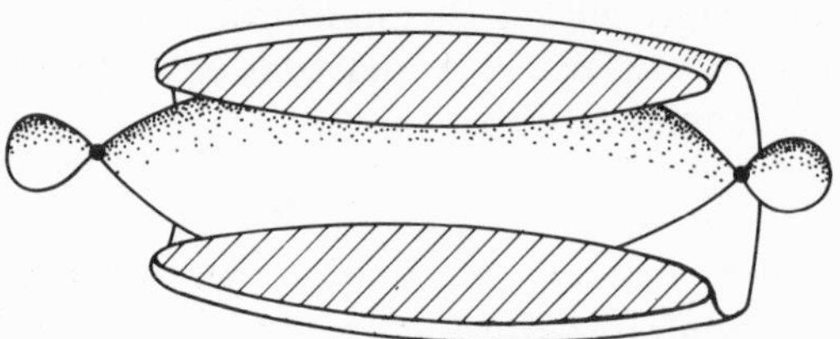

Fig. 4.5. Diagram showing schematically the general disposition of charge in N_2 (cylindrical density due to two π bonds cut away to reveal σ bond).

three net bonding interactions and the *bond order* is three, the maximum value for a diatomic molecule. The electronic arrangement accounts for the very high thermal stability of N_2, 940 kJ mol^{-1}, the very short intermolecular distance, 110 pm, and the diamagnetism. The $\pi 2p$ orbitals in the N_2 molecule lie lower in energy than the $\sigma 2p$, as is indicated in the paragraph heading.

Oxygen. $2O(1s^2\ 2s^2\ 2p^4) \rightarrow O_2\ \{KK(\sigma 2s)^2(\sigma^*2s)^2(\sigma 2p)^2(\pi 2p)^4(\pi^*2p)^2\}$
The O_2 molecule contains two electrons more than does N_2. They are accommodated, in accordance with Hund's rules, separately in the two π^*2p orbitals. Thus there is a σ bond and two 'half' π bonds, giving a bond order of two. Since the two odd electrons have parallel spins the molecule is paramagnetic. The bond length is 121 pm and its strength is 493 kJ mol^{-1}. When an electron is removed to give O_2^+ the bond length decreases to 112 pm, consistent with the removal of an antibonding electron and an increase of the bond order to $2\frac{1}{2}$, but addition of an electron to give O_2^- increases the bond length to 126 pm and reduces the bond order to $1\frac{1}{2}$, since the extra electron must enter an antibonding orbital.

Fluorine. $2F(1s^2\ 2s^2\ 2p^5) \rightarrow F_2\{KK(\sigma 2s)^2(\sigma^*2s)^2(\sigma 2p)^2(\pi 2p)^4(\pi^*2p)^4\}$
In this molecule both the bonding π orbitals and their antibonding partners have their full complement of electrons. Thus the only effective bond is that due to overlap of $2p_z$ atomic orbitals. This electron structure is consistent with the weakness and length of the F—F bond (155 kJ mol^{-1}; 142 pm) and the diamagnetism of fluorine.

Neon
The molecule Ne_2 has not been observed. All the antibonding molecular orbitals would be filled as well as the bonding molecular orbitals and the bond order would be zero.

Energy level diagrams
The energy levels of the molecular orbitals in the oxygen molecule and of the atomic orbitals from which they are formed, together with the occupancy of the orbitals, are shown schematically in Fig. 4.6. The energy levels are

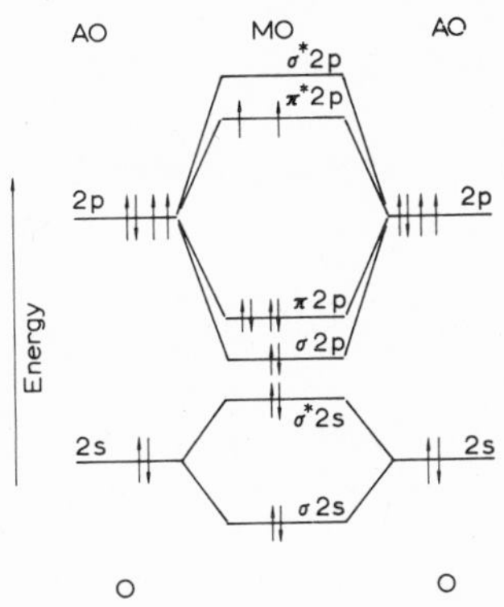

Fig. 4.6. Molecular orbital energy diagram for O_2.

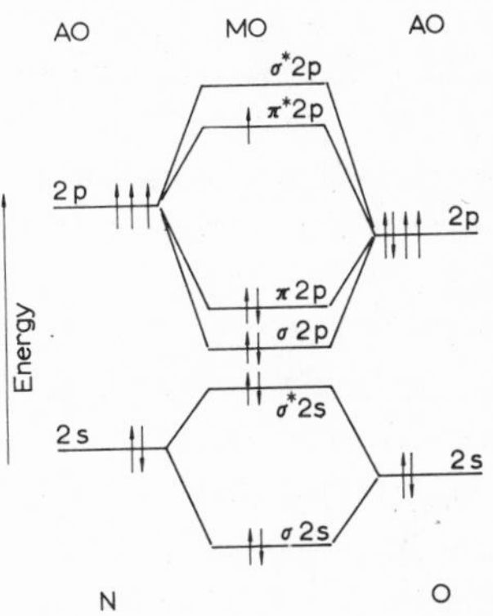

Fig. 4.7. Molecular orbital energy diagram for NO.

not drawn to scale; the σ2p and π2p orbitals have very similar energies.

For the molecule NO (Fig. 4.7) the energy levels of the various atomic orbitals are shown to be lower for the oxygen atom than for the nitrogen atom, in accordance with the relative electronegativities. Because there is one electron fewer than in O_2 only one of the π^*2p orbitals contains an unpaired electron. The bond order is $2\frac{1}{2}$ and the gas is paramagnetic.

For the molecule LiH, Li has 2s, $2p_x$, $2p_y$, and $2p_z$ atomic orbitals in its valence level able to combine with the 1s AO of hydrogen. Thus a σ MO can be formed by overlap of H_{1s} with Li_{2s} and also by H_{1s} with Li_{2p_z}. As the energy of Li_{2s} is lower than that of Li_{2p_z} it is likely that the H_{1s}—Li_{2s} will be the more important bonding contribution although both give rise to corresponding antibonding orbitals to make up the same number of molecular orbitals as there are atomic orbitals.

Interaction of H_{1s} with Li_{2p_x} and Li_{2p_y} does not lead to any net overlap, as the positive overlap of one lobe is cancelled by the negative overlap of the other (Fig. 4.8). Hence Li_{2p_x} and Li_{2p_y} remain as non-bonding orbitals.

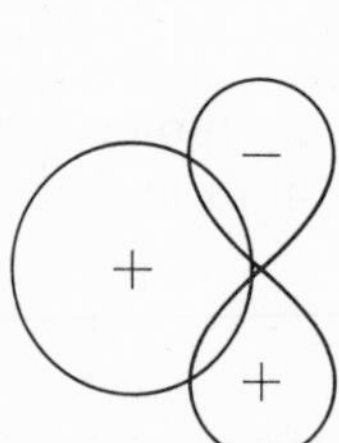

Fig. 4.8. Cancelling of positive and negative overlap.

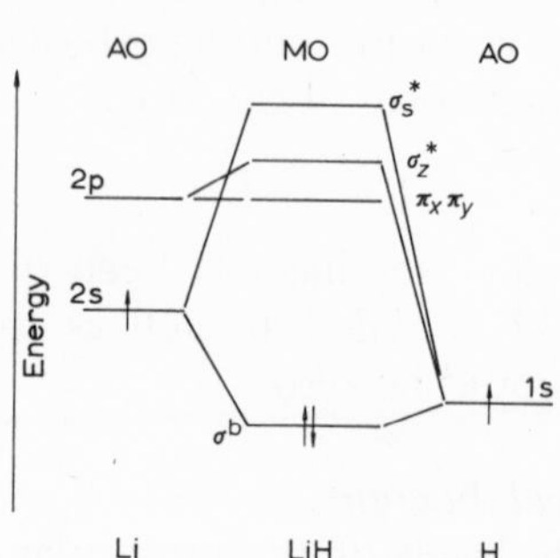

Fig. 4.9. Orbital energy diagram for LiH molecule.

Thus the two electrons from each atom will enter the lowest energy orbital, σ^b, giving a bond order of 1. As the σ^b MO is closer in energy to the H_{1s} AO it is reasonable to assume that electrons spend most of their time closer to the H nucleus than the Li and the description $Li^{\delta+}$ $H^{\delta-}$ might be appropriate.

In HF, omitting the fluorine K shell, H_{1s} can combine with the F_{2s}, $2p_x$, $2p_y$, $2p_z$ valence orbitals. As before, interaction of H_{1s} with F_{2p_x} and F_{2p_y} does not lead to any net overlap, and although both F_{2s} and F_{2p_z} have the correct symmetry to interact with H_{1s}, their energies are very different. Thus F_{2s} is too strongly bound to the F atom to combine with H_{1s}, and to a first approximation remains as a non-bonding orbital leaving F_{2p_z}—H_{1s} as the principal bonding interaction. Hence the eight valence electrons of H and F enter three non-bonding orbitals, and the σ-bonding MO, being closer in energy to the F_{2p_z} AO, implies that the electrons spend more time closer to F than H, making the description $H^{\delta+}$ $F^{\delta-}$ appropriate.

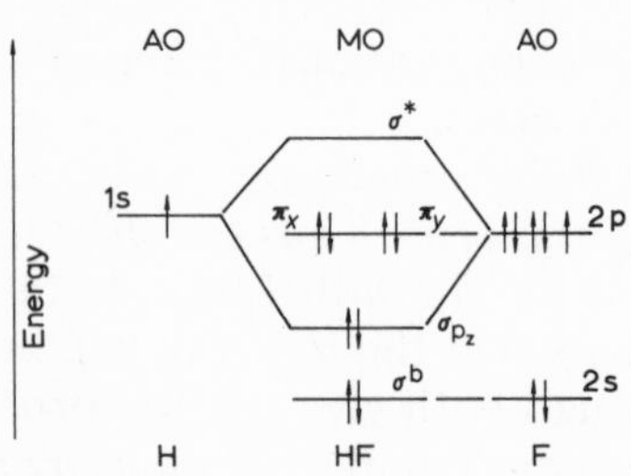

Fig. 4.10. Orbital energy diagram for the HF molecule.

4.1.1.2. Hybridisation

Although simple pairing of the atomic orbitals on different atoms suggests MO forms which account surprisingly well for the general properties of many diatomic molecules, it must be remembered that the best molecular orbitals are solutions of a wave equation and that simple LCAO forms are rather rough approximations only. Thus, by building an MO out of a number of atomic orbitals instead of just a pair, a better approximation can be obtained. It will appear later that this refinement termed hybridisation is often quite indispensable, even in qualitative descriptions. The HF molecule can be used to illustrate the principles and provide an alternative description of its bonding.

The Mulliken notation is a useful one for indicating the order of energy (ascending z, y, x, w, v ...) and symmetry type (σ, π ...) of molecular orbitals without specifying the atomic orbitals which combine to form them.

Order of energy	Mulliken symbol	Atomic orbitals contributing
Lowest	$z\sigma$	$1s_H$, 2s, $2p_z$
Second lowest	$y\sigma$	2s, $2p_z$
Third lowest	$x\pi$ and $w\pi$	$2p_x$ and $2p_y$

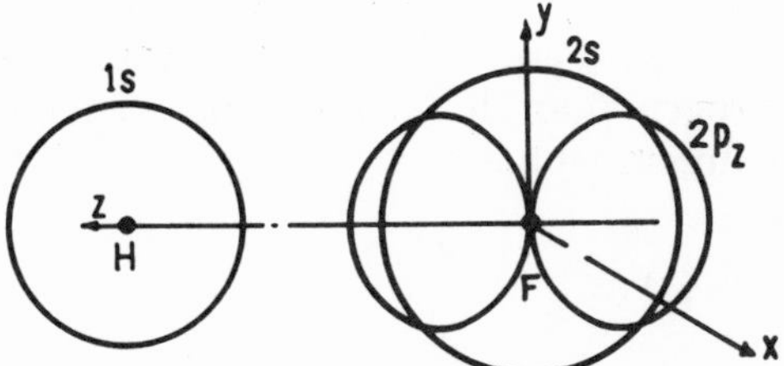

Fig. 4.11. Hydrogen fluoride with the 2s + $2p_z$ atomic orbitals of fluorine shown. The $2p_x$ and $2p_y$ atomic orbitals, normal to the z axis, are omitted for clarity.

The $z\sigma$ MO is of bonding type, concentrating charge between the nuclei, though it may lean more towards the fluorine atom. The $y\sigma$ MO contains an inappreciable amount of $1s_H$ and is hardly a molecular orbital at all; it leans to the 'rear' of the fluorine atom, away from the hydrogen. The other two orbitals are pure fluorine atomic orbitals each of which has its distinctive symmetry. The structure is

$$H(1s) + F(1s^2\ 2s^2\ 2p^5) \rightarrow HF(K\ z\sigma^2\ y\sigma^2\ x\pi^2\ w\pi^2)$$

There is thus a normal σ bond, although it cannot be well represented by less than 3 atomic orbitals, which is densest at the fluorine end, and 3 doubly occupied non-bonding orbitals. The latter give a striking picture of the three lone pairs of the octet on the fluorine atom; one projects to the rear while

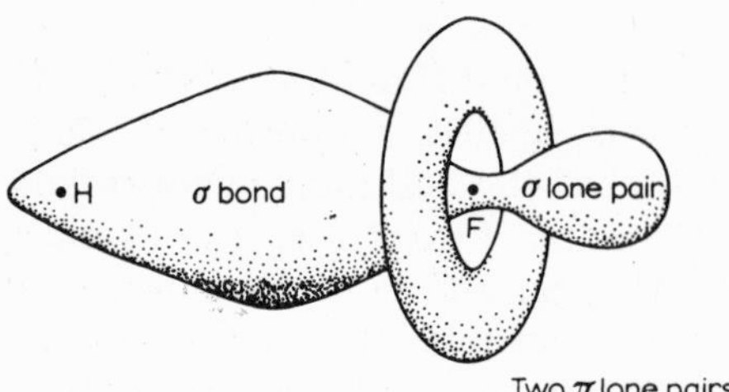

Fig. 4.12. Principal regions of high charge density in HF (schematic).

the other two combine to give an axially symmetrical charge distribution, in form something like a doughnut with the fluorine nucleus in the hole (Fig. 4.12).

By examining the molecular orbitals more carefully it is now possible to retrieve the simple picture in which molecular orbitals are approximated by overlapping suitable pairs of orbitals derived from the different atoms. For clearly the $z\sigma$ MO could be regarded as being formed by combining the 1s orbital of the hydrogen with a modified fluorine orbital which is itself a *mixture* of the 2s and $2p_z$ on the same (fluorine) atom. Such modified orbitals, which are still essentially atomic though not pure atomic orbitals, are called *hybrid* atomic orbitals. They are an immense aid to description because they make it possible to retain the pair picture in which bonds are associated with strongly overlapping pairs of atomic orbitals (now including hybrid as well as pure atomic orbitals), one on each of two atoms. The general forms of the hybrid orbitals which can be achieved by mixing are easily inferred. Thus the 2s and $2p_z$ atomic orbitals in the hydrogen fluoride example yield an infinite range of hybrid pairs, the most symmetrical being

$$h_1 = \frac{1}{\sqrt{2}}(\varphi_{2s} + \varphi_{2p_z}) \text{ and } h_2 = \frac{1}{\sqrt{2}}(\varphi_{2s} - \varphi_{2p_z})$$

These are illustrated in Fig. 4.13 and are called *digonal* sp hybrids; they are exactly similar or 'equivalent' except in orientation. Clearly, a plausible de-

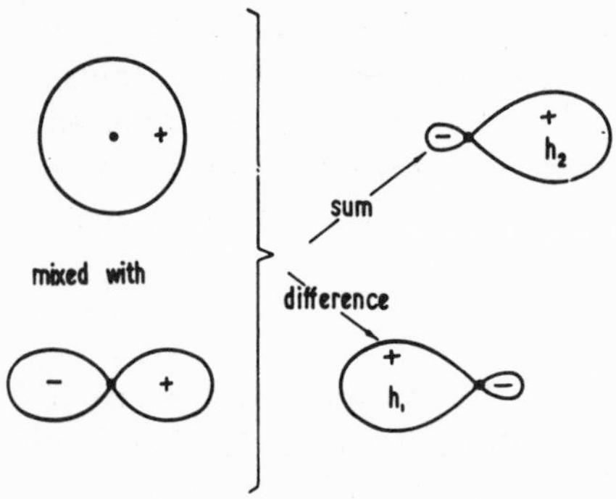

Fig. 4.13. Representing combination of s and p atomic orbitals to give sp hybrids.

scription of the $z\sigma$ MO in HF could be obtained by overlapping h_1 with the hydrogen 1s AO and using h_2 as the lone pair orbital, $y\sigma$. This would not be the best description because there is no reason, here, for choosing a pair of exactly similar hybrids; but it is still qualitatively useful. More generally, departures from symmetry are permissible, but the most acceptable pairs are related so that they overlap as little as possible. If the 2s content of h_1 is increased and its $2p_z$ content decreased, then the 2s content of h_2 must decrease and its $2p_z$ content increase.

It is probable that sp hybridisation of this kind is significant in a large number of diatomic molecules, but its importance depends on a number of delicately balanced factors. It is favoured by the fact that overlap in a bond region (that is where the electrons are attracted by *two* nuclei) may be increased. It occurs most easily when the p orbital is not much higher in energy than the s but is opposed by, for instance, an increasing p character in a lone pair orbital. Hybridisation also accounts most readily for the fact that electrons repel one another and tend to stay apart, particularly (owing to the exclusion principle) when each has the same spin. It will be seen later that the forms and disposition of hybrid orbitals can be discussed in terms of repulsions between the different pairs of bonding and lone pair electrons. This more detailed analysis of the electronic energy helps to give some understanding of the factors determining molecular geometry.

4.1.1.3. Valence states. Promotion

In discussing bonds involving hybrid atomic orbitals it is necessary to introduce 'atomic valence states'. Accepting the above description (4.1.1.2) we can discuss the energy of the hydrogen fluoride molecule. The bond is formed by overlapping h_1 and $1s_H$, each contributing one electron, where the atoms, supposing they could be taken apart without changing the orbital forms, would be in the respective states

$$H(1s^1) \text{ and } F(1s^2\ 2p_x^2\ 2p_y^2\ h_2^2\ h_1^1).$$

The hydrogen would be in a 'true atomic' state and the fluorine in a 'valence' state. The latter state is hypothetical, because h_1 and h_2 would, in fact, go smoothly into 2s and $2p_z$ orbitals as the nuclei were being separated; it does, however, allow the energy changes to be visualised. As Fig. 4.14 shows, the net binding energy is the 'gross binding energy' minus the 'valence state excitation energy'. The latter, which is characteristic of the valence state considered and constant from one molecule to another, accordingly forms an exceedingly useful datum. The high excitation energy involved is often completely offset by the more satisfactory overlap of the hybrid atomic orbitals, so that the net binding energy becomes considerably larger than could be accounted for without invoking hybridisation.

There are many instances in which a mixing of the atomic orbitals occupied in the ground state of an atom does nothing to improve its capacity for bond formation. Beryllium, for example, has the ground state structure $Be(1s^2\ 2s^2)$ and, being without singly occupied orbitals, should be zerovalent.

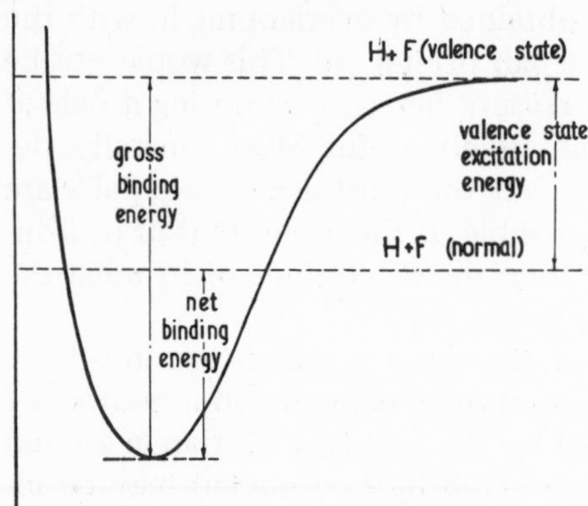

Fig. 4.14. Energy relationships in HF molecule (energy vs. distance between nuclei).

Carbon, with the structure $C(1s^2\ 2s^2\ 2p^2)$, should be bivalent; mixing the singly occupied 2p orbitals is found merely to change their orientation. The bonds formed by these atoms can best be understood in terms of valence states in which 'electron promotion' has occurred. The promotions envisaged are

$$Be(1s^2\ 2s^2) \rightarrow Be(1s^2\ 2s^1\ 2p_x^1)$$

$$C(1s^2\ 2s^2\ 2p^2) \rightarrow C(1s^2\ 2s^1\ 2p_x^1\ 2p_y^1\ 2p_z^1)$$

where the x, y, and z directions are chosen in terms of the molecular environment of the atom. These valence states would serve to describe bivalent beryllium and quadrivalent carbon but, since two types of AO appear, would not account for the formation of the sets of *identical* bonds found in the linear molecule $BeCl_2$ and the tetrahedral molecule CH_4. The difficulty disappears when the possibility of hybridisation is introduced; for mixing a 2s and 2p orbital has already been seen to yield, among other possibilities, precisely equivalent hybrid orbitals (Fig. 4.13) pointing in opposite directions. Promotion and hybridisation are employed freely to set up atomic valence states which give the best account of bond formation in specific molecular situations. With the help of these ideas it is possible usefully to discuss the bonding in really complicated polyatomic molecules.

4.1.1.4. Polyatomic molecules

In Fig. 4.3, the atomic orbitals of atoms were paired, according to symmetry and degree of overlap, to yield molecular orbitals of various type, all describing a two-electron bond. Essentially similar considerations may be employed in dealing with polyatomic molecules; but then the bonds, instead of being superimposed (e.g. σ and π bonds), may lie in different regions of space, uniting different pairs of atoms. If there are 'obvious' pairs of strongly overlapping atomic orbitals, or if such pairs can be formed by invoking not unreasonable valence states, it is possible to obtain a fairly satisfactory picture of the molecular electronic structure in terms of the 'localised' molecular orbitals which result. As with the diatomic molecule (4.1.1.1), this description succeeds in explaining the accumulation of electron density in a number of localised bond regions. The application of these ideas is best illustrated by reference to specific molecules.

Beryllium chloride, $BeCl_2$. This is a linear molecule in which both Be—Cl bonds are equivalent. The appropriate valence state of the Be atom is thus $Be(1s^2\, h_1{}^1\, h_2{}^1)$, where h_1 and h_2 are the digonal hybrids of Fig. 4.13. The relevant orbitals on the chlorine atoms are the singly occupied $3p_z$ atomic orbitals (z referring to the molecular axis), 3s, $3p_x$ and $3p_y$ containing lone pairs. It is likely also that some digonal hybridisation would occur at the chlorine atoms, the incorporation of some 3s character strengthening the overlap in the bonds and pushing one of the lone pairs to the rear of each chlorine. The resulting situation, shown in Fig. 4.15, should be compared with that in HF (Fig. 4.12).

Heavier atoms with the same ns^2 ground state configurations, for example zinc, cadmium and mercury, form linear dichlorides and dibromides whose electronic structures must be closely similar.

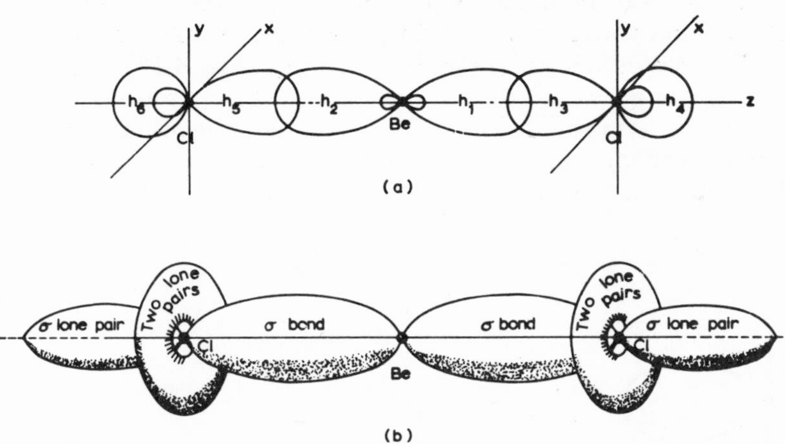

Fig. 4.15. Beryllium chloride: (a) Hybrid orbitals of the σ bonds and σ lone pairs ($2p_x$ and $2p_y$ orbitals omitted), (b) regions of maximum charge density (schematic).

Acetylene, C_2H_2. This molecule is also linear. Again digonal valence states are appropriate, each carbon having the prepared configuration $C(1s^2\, h_1{}^1\, h_2{}^1\, 2p_x{}^1\, 2p_y{}^1)$. A σ C—C bond and two σ C—H bonds, all collinear, arise from obvious pairings (Fig. 4.16); the singly occupied carbon 2p atomic orbitals overlap laterally to give two π bonds, exactly as in the diatomic molecule N_2. Consequently, there is a carbon—carbon triple bond.

Water, H_2O. This is a non-linear molecule, the H—O—H angle being about 105°. Hybridisation occurs somewhat less easily in oxygen than in carbon.

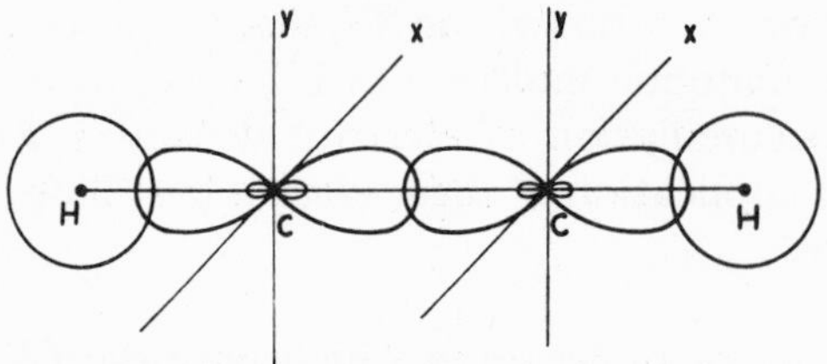

Fig. 4.16. Hybrid orbitals of the σ bonds in C_2H_2 ($2p_x$ and $2p_y$ orbitals omitted).

By neglecting it altogether, it is still possible to get a rough explanation of the molecular shape. A strong overlap occurs between the singly occupied oxygen 2p orbitals and the hydrogen 1s atomic orbitals and suggests localised molecular orbitals at about 90°. But the overlap can be improved and the lone pairs better separated, thus lowering their repulsion energy, by admitting a fair amount of 2s—2p mixing.

Boron trichloride, BCl_3. Here the molecule is planar, with the boron atom at the centre of an equilateral triangle of chlorine atoms (Fig. 4.17). The valence state must be described in terms of three similar hybrid atomic orbitals pointing towards the corners of the triangle. Such orbitals can be formed by mixing the 2s AO and two 2p atomic orbitals, say $2p_y$ and $2p_z$; they lie in the *yz* plane of the latter and are precisely equivalent (Fig. 4.18). If the so-called *trigonal* hybrids are denoted by h_1, h_2 and h_3, the appropriate boron valence state must be $B(1s^2\ h_1^1\ h_2^1\ h_3^1)$. The hybrid atomic orbitals overlap chlorine 3p atomic orbitals, directed towards the boron atom, to form localised molecular orbitals similar to those in beryllium chloride.

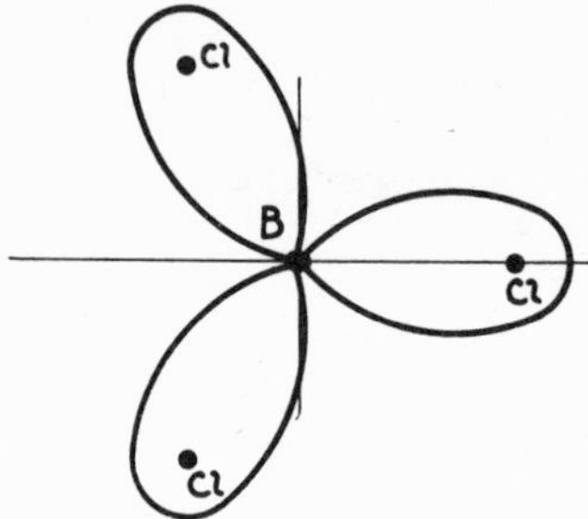

Fig. 4.17. Localised molecular orbitals in BCl_3.

Methyl radical, CH_3. The trigonal hybridisation just mentioned is indicated whenever an atom forms three identical bonds directed at 120° to each other. In carbon the valence state entails promotion and also hybridisation: it is $C(1s^2\ 2p_x^1\ h_1^1\ h_2^1\ h_3^1)$, where the third 2p AO, mainly $2p_x$, is not involved in the mixing and lies perpendicular to the plane of the hybrids. Three C—H bonds in the planar methyl radical are accounted for by an overlap of h_1, h_2

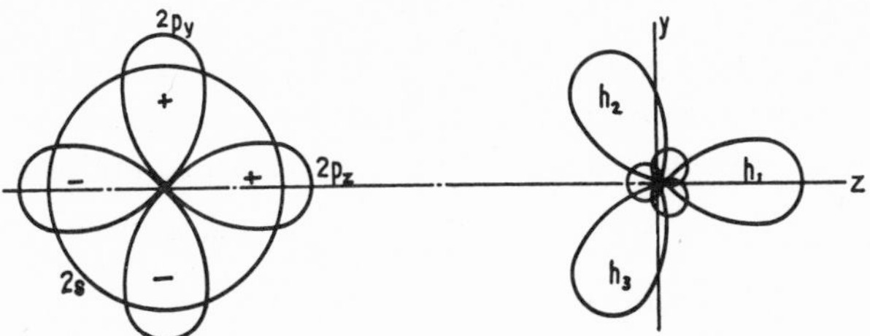

Fig. 4.18. Mixing of 2s, $2p_y$ and $2p_z$ atomic orbitals to form three equivalent hybrid atomic orbitals, h_1, h_2, h_3. (Cf. Fig. 4.13, except that here the possibilities are superimposed instead of being shown separately.)

and h_3 with the three hydrogen 1s atomic orbitals. The free radical character of the system arises from the singly occupied $2p_x$ orbital.

Ethylene, C_2H_4. This molecule is also flat, and the bonds from each carbon atom make very nearly 120° with each other. Assuming the same trigonal valence state as in the methyl radical, four C—H bonds and a central C—C bond are readily accounted for; and, when the two CH_2 groups are rotated about the C—C bond until their singly occupied 2p atomic orbitals are parallel, they overlap laterally to give a normal π bond (Fig. 4.19). The planar configuration is thus stabilised by π bonding and any twisting between the two CH_2 groups would lead to a reduction in binding energy.

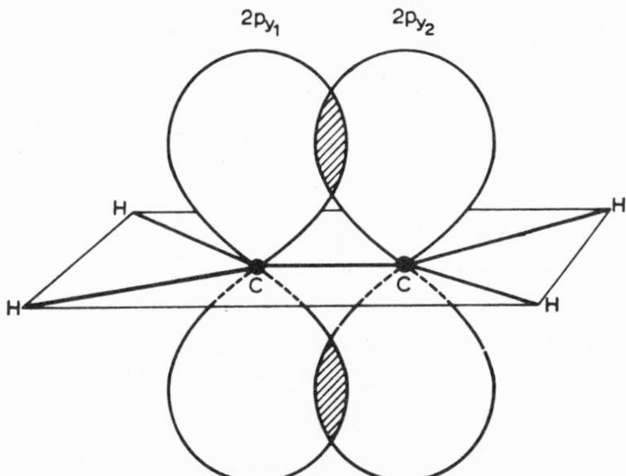

Fig. 4.19. Formation of a π bond in ethylene (σ bonds omitted). If the two 2p orbitals are parallel there is substantial lateral overlap giving a π bond: rotation of the CH_2 groups lessens the binding energy.

Methane, CH_4. The hydrogen atoms are at the corners of a regular tetrahedron, with the carbon at the centre. A carbon valence state with four identical, singly occupied hybrid atomic orbitals can be set up by allowing the mixing of 2s and *all* the 2p atomic orbitals. When they are denoted by h_1, h_2, h_3, and h_4, the carbon valence state is $C(1s^2\, h_1{}^1\, h_2{}^1\, h_3{}^1\, h_4{}^1)$. The pairing of atomic orbitals, describing four identical strong bonds, is then obvious (Fig. 4.20(a) and (b)). This *tetrahedral* hybridisation occurs whenever carbon forms four single bonds; thus it is appropriate to ethane and indeed to all the paraffins and their derivatives. The possibility of free rotation about the bonds leads to long, flexible chains.

4.1.1.5. The wide occurrence of hybridisation

Consideration of the bonds formed by atoms with s and p electrons in their valence shells has brought to light three principal types of valence state. Although their realisation may require promotion and hybridisation, it often leads to strongly directed bonds with a very high net binding energy.

The principal types of hybrid which occur are:

(i) Digonal or sp hybrids, formed from an s and *one* p orbital and pointing in opposite directions along the axis of the p orbital;

(ii) trigonal or sp^2 hybrids, formed from an s and *two* p orbitals and inclined at 120° in the plane of the 2p orbitals;

(iii) tetrahedral or sp^3 hybrids, formed from an s and *three* p orbitals and inclined at about 109.5° (the tetrahedral angle).

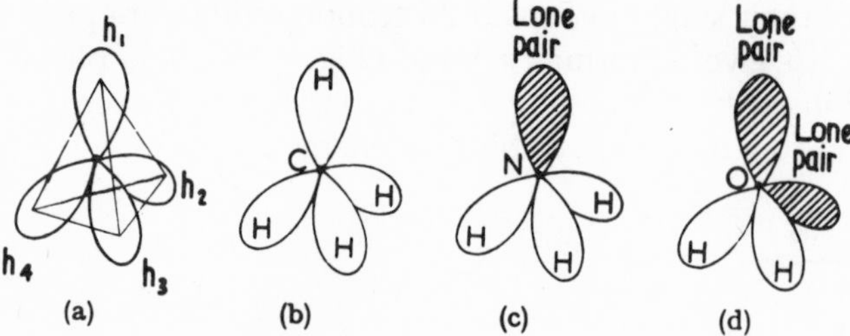

Fig. 4.20. Tetrahedral hybridisation in a series of iso-electronic molecules.

It is important to remember that, as already recognized, these types are exactly appropriate only in symmetrical situations (e.g. $BeCl_2$, BCl_3, CH_4) and that departures from them are frequent. It is now generally believed that hybridisation is much more widely appropriate than was at first thought. It is interesting to consider, for example, what would happen if one of the protons in CH_4 were transferred to the carbon nucleus. The resulting system, NH_3, is iso-electronic with the original, differing from it only in an increase of 1 in the central charge and a loss of symmetry. But the four C—H molecular orbitals have become three N—H molecular orbitals together with a lone pair orbital, h_1 say. The tetrahedral valence state is therefore a plausible description of the nitrogen atom in ammonia, although at first sight hybridisation does not appear to be necessary; for the pyramidal form of the molecule (Fig. 4.20(c)) could be accounted for simply by overlapping three hydrogen orbitals with the three singly occupied 2p orbitals of the nitrogen ground state. The H—N—H angle is, however, 106.7° which is neither the tetrahedral nor the right angle. The hybrid h_1 has increased its s content, thereby lowering the energy of the lone pair electrons (which are no longer attracted by another nucleus), and at the same time decreasing the s content of the other hybrids. This allows their approach towards the right-angled set of three pure p orbitals — an effect which is helped by repulsion from the concentrated lone-pair charge. The equilibrium configuration results from a quite fine energy balance, involving the mutual repulsions of all four electron pairs.

A tetrahedral disposition of four hybrids appears to give a much better description of the actual electronic structure than could be achieved without hybridisation, largely because it puts individual electron pairs in different regions of space. The same is true for H_2O where the system can be imagined as having been formed from the iso-electronic CH_4 by the shrinking of two protons into the carbon nucleus. Then a tetrahedral oxygen valence state becomes more reasonable, the two lone pairs being well separated (as in Fig. 4.20(d)) instead of being superimposed in one 2s and one 2p orbital.

The electron-donor property of nitrogen and the hydrogen-bonding property of oxygen are both connected with the existence of localised and strongly directed lone pairs of this kind. Later, it will be seen how a great deal of stereochemistry may similarly be interpreted in terms of a theoretically reasonable distribution of electron pairs.

4.1.1.6. Covalency maxima

Earliest ideas on covalence were based largely on the view that the atom under consideration attained a Group 0 gas structure by sharing electrons. Faced with such compounds as PCl_5 and SF_6, in which sulphur and phosphorus share 10 and 12 electrons respectively, Sidgwick added the suggestion that every element had a certain maximum covalency which depended on atomic number (see Table 4.1).

TABLE 4.1

MAXIMUM COVALENCY AS IT DEPENDS ON ATOMIC NUMBER

Atomic number	Period	Maximum number of shared pairs
1	1	1 (2 electrons)
3—9	2	4 (8 electrons)
11—35	3 and 4	6 (12 electrons)
37—92	5, 6 and 7	8 (16 electrons)

The maximum covalency is by no means always reached, being shown by many elements only in their fluorides, and clearly represents the number of orbitals in the outermost, partially filled, shell which might be invoked in forming valence states. It is 1 for hydrogen, 4 for elements of the second period, 6 for those of the third and fourth periods and 8 for the rest of the elements. The last group of elements, having s, p and d orbitals $(1 + 3 + 5 = 9)$ available, might be expected to exhibit a maximum covalency of 9. But the outermost electronic shells become less defined with increasing principal quantum number and the amount of promotion which can take place is restricted. A covalency of 4 is common in the third period, though there is a limited d orbital participation in spite of the d orbitals being unoccupied in the ground states.

4.1.1.7. Hybridisation with d orbitals

With the admission of d orbitals the number of principal types of hybridisation is increased from 3 (digonal, trigonal, tetrahedral) to over 40; fortunately only a few of these lead to systems of strong bonds.

Bipyramidal, sp^3d. A typical bipyramidal molecule is phosphorus pentachloride, PCl_5, which is monomeric in the vapour and has the form shown in Fig. 4.21. The hybridisation in the central plane could be sp^2, giving the three bonds at 120° with one another. And the third p orbital could be mixed with the appropriate d orbital, since both are symmetrical about the central axis, to give one hybrid directed up and another down as in Fig. 4.22.

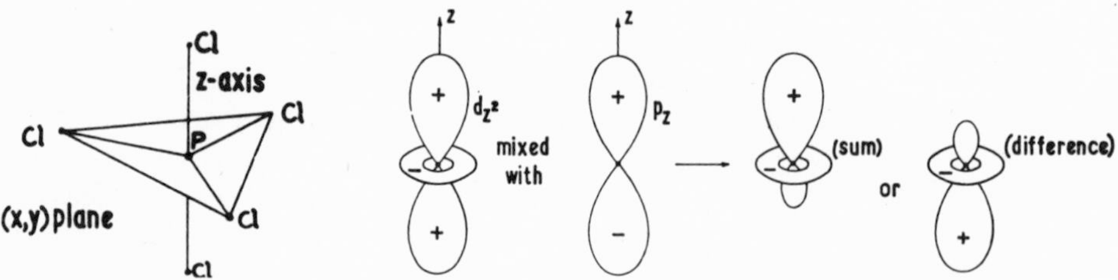

Fig. 4.21. Bipyramidal structure of PCl_5.

Fig. 4.22. One type of pd hybridisation.

In these circumstances two chlorine atoms would be expected to be bound somewhat differently from the other three; the bond above and below the plane are in fact slightly longer. The corresponding element of the first short period, nitrogen, forms no such compound because there are no 2d atomic orbitals and the 3d atomic orbitals are too high in energy to allow of appreciable mixing. Thus NF_3 is the highest fluoride.

Octahedral, sp^3d^2. A molecule of this kind is SF_6, which has four bonds lying at right angles in the plane of the sulphur and the other two pointing up and down from the plane (Fig. 4.23). The two vertical bonds are formed essentially as in a bipyramidal molecule, but the in-plane bonds are best described in terms of hybrids incorporating a second d orbital. The mode of formation of one of the in-plane hybrids is indicated in Fig. 4.24. Here the four hybrids so formed would not be quite equivalent to the other two, but, if slight mixing is permitted, a strictly equivalent set of six can be found; these point to the corners of a regular octahedron.

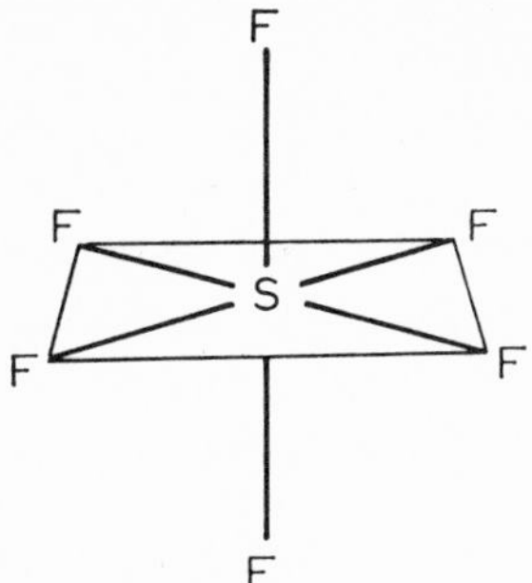

Fig. 4.23. Octahedral structure of SF_6.

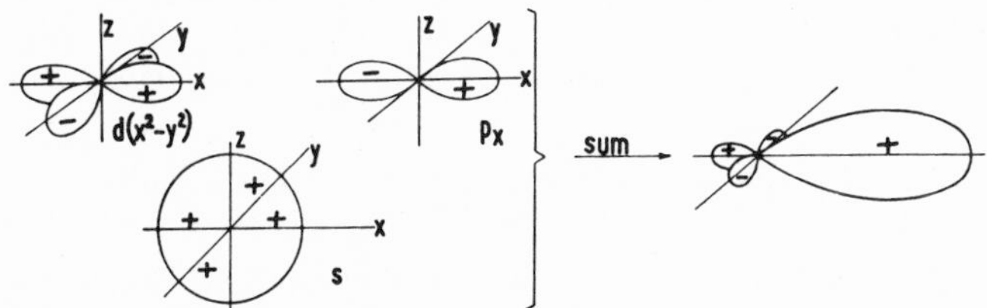

Fig. 4.24. One type of spd hybridisation.

Square planar, dsp^2. Certain ions have four planar orbitals directed towards the corners of a square without atoms above or below the plane. Examples of 'square' ions are $Ni(CN)_4^{2-}$ and $PtCl_4^{2-}$. The hybridisation shown in Fig. 4.24 is again appropriate. In planar ions of this kind, however, it is possible for the primary system of hybrid bonds to be supplemented by a secondary system involving the remaining orbitals of the valence shell. In the plane square instance, the secondary system can be formed from the out-of-plane orbitals (one p and three d) as in Fig. 4.25. Such orbitals have a node in the molecular plane and therefore have π character. They can overlap with p orbitals on the atoms to which they point, to give a π bond superimposed on the primary σ bond. Double bonding of this kind is almost certainly important in planar complexes when sufficient electrons are available.

It should be noted that the orbitals which are mixed belong, in the first two examples, to the same quantum shell (3s, 3p, 3d). Often, however, the

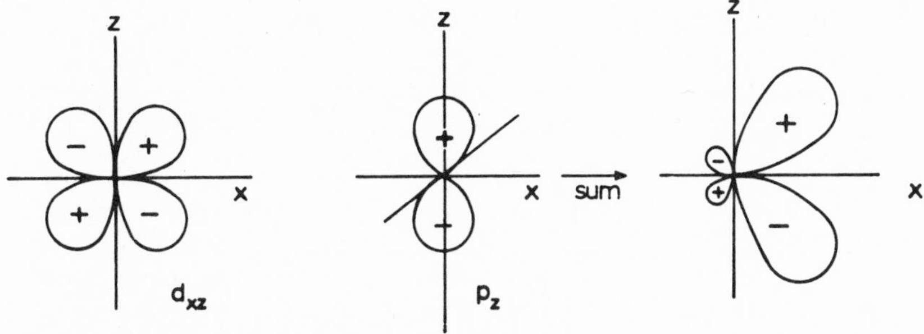

Fig. 4.25. Secondary hybrid suitable for π-bond formation.

d orbital belongs to a lower-quantum shell (e.g. 4s, 4p, 3d) and this is indicated by writing the d orbital first. In the last example (Ni, Pt) we therefore refer to dsp^2 hybridisation.

Compounds in which the maximum covalency is exhibited are naturally very stable. Thus, for instance, CCl_4 and SF_6 are unreactive towards water, whereas $SiCl_4$ and PCl_5 are easily hydrolysed because the valency of each central atom is capable of being increased.

4.1.1.8. Hybridisation and symmetry

That such hybrid orbitals are involved in the bonding can be demonstrated perhaps more convincingly by the use of symmetry. Clearly the shape of any molecule will determine which orbitals have the right symmetry to account for the bonding in that particular shape. For example take the ethylene molecule. The point group is D_{2h} which can be determined by enumerating what symmetry elements planar ethylene possesses. These will of course coincide with the symmetry operations listed in the D_{2h} character table (Table 4.2). As there are three independent C_2 axes of symmetry it is necessary to define the axes along the Cartesian co-ordinates and label the atoms (Fig. 4.26). Thus in order to determine the symmetry of the orbitals available in ethylene, carry out the symmetry operations of D_{2h} on all the orbitals of carbon atoms A and B and hydrogen atoms C, D, E and F. For each operation determine the appropriate character according to the notation: +1 for symmetrical behaviour, −1 for antisymmetrical, and 0 if the operation causes the orbital to change into something different (but still indistinguishable from the original).

Consider the 2s orbital of carbon atom A. Carrying out the identity operation E leaves the 2s orbital in an indistinguishable situation, to give the character +1. The $C_2(z)$ operation changes $C_A(2s)$ into $C_B(2s)$, which, being different, gives the character 0. The $C_2(y)$ operation is similar, but $C_2(x)$ would give +1. The complete row would be

D_{2h}	E	$C_2(z)$	$C_2(y)$	$C_2(x)$	i	$\sigma(xy)$	$\sigma(xz)$	$\sigma(yz)$
+ C_A (2s)	1	0	0	1	0	1	1	0

If we examine the rows of characters in the Character table of D_{2h} (Table 4.2) we find that none correspond with $C_A(2s)$. It becomes apparent that the

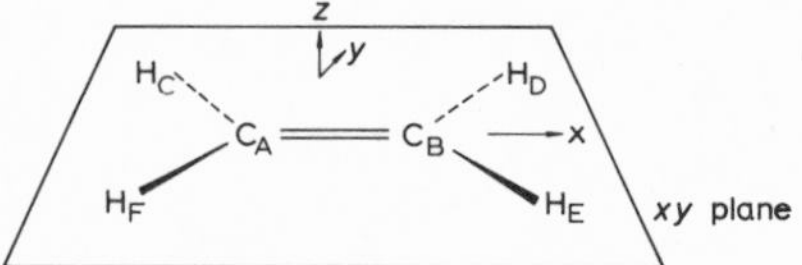

Fig. 4.26. Ethylene molecule, with atoms labelled to illustrate the symmetry operations.

TABLE 4.2

CHARACTER TABLE OF D_{2h}

	D_{2h}	E	$C_2(z)$	$C_2(y)$	$C_2(x)$	i	$\sigma(xy)$	$\sigma(xz)$	$\sigma(yz)$	
	A_g	1	1	1	1	1	1	1	1	
	B_{1g}	1	1	−1	−1	1	1	−1	−1	
	B_{2g}	1	−1	1	−1	1	−1	1	−1	
	B_{3g}	1	−1	−1	1	1	−1	−1	1	
	A_u	1	1	1	1	−1	−1	−1	−1	
	B_{1u}	1	1	−1	−1	−1	−1	1	1	
	B_{2u}	1	−1	1	−1	−1	1	−1	1	
	B_{3u}	1	−1	−1	1	−1	1	1	−1	
$C_A(2s)$ +	$C_B(2s)$	2	0	0	2	0	2	2	0	$A_g + B_{3u}$
$C_A(2p_z)$ +	$C_B(2p_z)$	2	0	0	2	0	2	2	0	$B_{3g} + B_{1u}$
$C_A(2p_y)$ +	$C_B(2p_y)$	2	0	0	−2	0	2	−2	0	$B_{1g} + B_{2u}$
$C_A(2p_x)$ +	$C_B(2p_x)$	2	0	0	2	0	2	2	0	$A_g + B_{3u}$
4 H(1s)		4	0	0	0	0	4	0	0	$A_g + B_{1g} + B_{2u} + B_{3u}$

2s orbital of C_A cannot be considered in isolation. It is found however that the resulting character representation obtained by considering the 2s orbitals of carbon atoms A and B together will be reducible i.e. made up of a unique combination of rows of characters from D_{2h}. The rows are labelled by symbols (A_g, B_{1g} etc.) and describe the symmetry properties of the orbitals under consideration. Continuing, it can be shown that the remaining orbitals of carbon must be considered in pairs and the 1s orbitals of all four hydrogen atoms must be considered together.

Hence the 1s orbitals of hydrogen will combine with carbon orbitals of the same symmetry as denoted by the symmetry symbol to form proper molecular orbitals. Thus the carbon 2s, $2p_x$ and $2p_y$ transform in the same way as the hydrogen 1s i.e. $A_g + B_{1g} + B_{2u} + B_{3u}$, so that the formation of sp^2 hybrid orbitals at each carbon atom is indicated.

To obtain both a picture and a quantitative description of the molecular orbitals involved, carry out the operations of D_{2h} on an individual $H_C(1s)$ orbital and note its transformations:

D_{2h}	E	$C_2(z)$	$C_2(y)$	$C_2(x)$	i	$\sigma(xy)$	$\sigma(xz)$	$\sigma(yz)$
$H_C(1s)$	C	E	D	F	E	C	F	D

Multiply this result by the characters appropriate to the symmetry species of interest and sum:

$$A_g = C + E + D + F + E + C + F + D$$

$$\therefore \ \psi(A_g) = \tfrac{1}{2}(\sigma_C + \sigma_D + \sigma_E + \sigma_F)$$

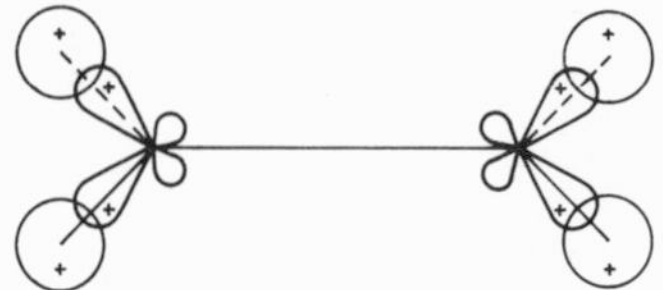

$$B_{1g} = C + E - D - F + E + C - F - D$$

$$\therefore \ \psi(B_{1g}) = \tfrac{1}{2}(\sigma_C - \sigma_D + \sigma_E - \sigma_F)$$

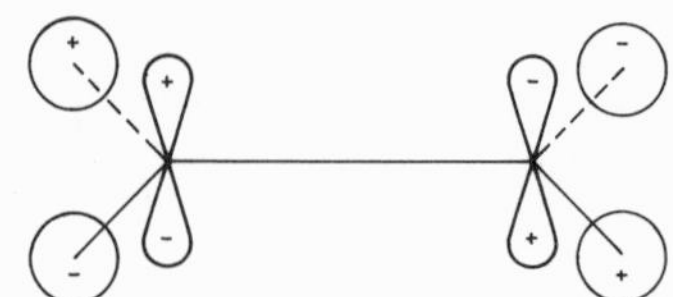

$$B_{2u} = C - E + D - F - E + C - F + D$$

$$\therefore \ \psi(B_{2u}) = \tfrac{1}{2}(\sigma_C + \sigma_D - \sigma_E - \sigma_F)$$

$$B_{3u} = C - E - D + F - E + C + F - D$$

$$\therefore \ \psi(B_{3u}) = \tfrac{1}{2}(\sigma_C - \sigma_D - \sigma_E + \sigma_F)$$

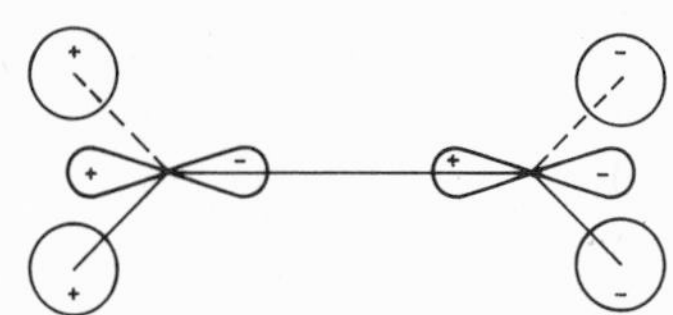

After considering carbon to hydrogen bonding, the remaining orbitals represent the orbitals bonding carbon to carbon i.e. A_g + B_{3g} + B_{1u} + B_{3u}. The A_g and B_{2u} are involved in σ bonding as sp^2 hybrid orbitals while the B_{1u} and B_{2g} are the π-contribution of p_z orbitals.

4.1.2. Covalent radii

The nature of the electron cloud about a nucleus makes it difficult to define the size of an atom. However, the distance between a nucleus and its nearest neighbour is a precisely measurable quantity, and from it the atomic radius may be deduced. Internuclear distances in solids and bond lengths in molecules can be determined by a variety of methods, such as X-ray or electron diffraction and band spectroscopy. Internuclear distances in solids depend to some extent on the way the atoms are packed. In diamond, however, the C—C distance (154.2 pm) is very nearly the same as it is in a saturated hydrocarbon (152 to 155 pm). The covalent radius of carbon for tetrahedral bonding is thus 77 pm. In silicon the Si—Si distance is 226 pm, indicating a covalent radius of 113 pm. Thus the C—Si distance in a compound should be (77 + 113) pm = 190 pm, which is close to the measured value in tetramethylsilane, $Si(CH_3)_4$, 187 pm.

For hydrogen, the covalent radius varies considerably when the element is in different states of combination, from, for instance, 28 pm in the halogen halides to 37 pm in the hydrogen molecule, H_2.

From a comparison of a large number of covalent bond distances we can assign to atoms single-bond covalent radii which, when added together, successfully predict internuclear distances for bonds of this type. The radii depend to some extent on the number and spatial distribution of the bonds, but the variations thus caused are usually less than 5 pm.

4.1.2.1. Double-bond and triple-bond radii

The existence of π-bonding as well as a sigma bond between two nuclei shortens the internuclear distance. In alkenes the distance between the 'doubly-bound' carbon atoms is (134 ± 2) pm; hence, carbon has been assigned a double-bond radius of 67 pm. The carbon triple-bond radius deduced from the spectra of alkynes is 60 pm. For nitrogen the triple-bond radius is 55 pm. In HCN, the C—N distance is 115 pm which is the sum of the carbon and nitrogen triple-bond radii. It should be emphasised, however, that many compounds have bonds which cannot be described, even to a first approximation, simply as single, double or triple. The description implied by the formulation N ≡ C—C ≡ N for the bonding in cyanogen is a poor one

TABLE 4.3

SINGLE-BOND COVALENT RADII/pm

Group IA	IIA	IIIA	IVA	IB	IIB	IIIB	IVB	VB	VIB	VIIB
Li 134	Be 90					B 82	C 77	N 75	O 73	F 72
Na 154	Mg 130					Al 118	Si 113	P 106	S 102	Cl 99
K 196	Ca 174	Sc 144	Ti 136	Cu 138	Zn 131	Ga 126	Ge 122	As 119	Se 116	Br 114
Rb 211	Sr 192	Y 162	Zr 148	Ag 153	Cd 148	In 144	Sn 141	Sb 138	Te 135	I 133
Cs 225	Ba 198	La 169	Hf —	Au 150	Hg 149	Tl 148	Pb 147	Bi 146	Po —	At —

because the C—C distance is only 137 pm. Clearly the electron density in the region between these two atoms must nearly be as high as that in the double bond of an alkene.

4.1.3. Bond energies

Strictly, what has been said in the discussion of valency refers to an idealised situation in which all the atomic nuclei are at rest in equilibrium positions. Such a situation is not realised even at the absolute zero of temperature, because there still remains the zero-point energy of vibration. The electronic binding energy, that is the energy required to separate a molecule into its component atoms in their ground states and at rest, is thus not directly measurable. What can be measured is the 'dissociation energy', which is the energy required for separation of the atoms under specified conditions (e.g. referred to standard temperature and pressure). The two quantities differ by the energies of vibration, rotation and translation; the latter two are usually small, but the vibrational energy may amount to a substantial fraction of the binding energy.

The *bond dissociation energy* of a bond R—X in a molecule RX, written $D(\mathrm{R{-}X})$, is given by

$$D(\mathrm{R{-}X}) = \Delta H_f{}^\circ(\mathrm{R}) + \Delta H_f{}^\circ(\mathrm{X}) - \Delta H_f{}^\circ(\mathrm{RX})$$

where the ΔH_f° values refer to enthalpies of formation per mole of the gaseous species under defined standard conditions. Thus for the O—H bond in H_2O:

$$D(\mathrm{HO{-}H}) = \Delta H_f{}^\circ(\mathrm{OH}) + \Delta H_f{}^\circ(\mathrm{H}) - \Delta H_f{}^\circ(\mathrm{H_2O})$$

Bond dissociation energies can be determined by calculations from thermochemical data. The bonds in the H_2O molecule will serve as examples.

The equilibrium

$$2\,\mathrm{H_2O(g)} + \mathrm{O_2(g)} \rightleftharpoons 4\,\mathrm{OH(g)}$$

has been studied at high temperatures, the equilibrium concentration of OH free radicals being determined in this case by absorption spectrometry. Knowledge of the equilibrium constants for the reaction at two temperatures allows the calculation of the mean enthalpy of the reaction using the Van't Hoff equation and hence of the standard enthalpy, ΔH_r°, using the Kirchhoff equation. The bond dissociation energies of H_2 and O_2 being determined similarly and the standard enthalpy of formation of gaseous H_2O being also known, the bond dissociation energy $D(\mathrm{O{-}H})$ can be calculated as follows:

			Enthalpy change
(a)	$\mathrm{H_2(g)}$	$\rightarrow 2\,\mathrm{H(g)}$	$D(\mathrm{H{-}H})$
(b)	$\mathrm{O_2(g)}$	$\rightarrow 2\,\mathrm{O(g)}$	$D(\mathrm{O{=}O})$
(c)	$\mathrm{H_2O(g)}$	$\rightarrow \mathrm{H_2(g)} + \frac{1}{2}\,\mathrm{O_2(g)}$	$-\Delta H_f(\mathrm{H_2O})$
(d)	$2\,\mathrm{OH(g)}$	$\rightarrow \mathrm{H_2O(g)} + \frac{1}{2}\,\mathrm{O_2(g)}$	$-\frac{1}{2}\Delta H_r$

Adding: $2\,\mathrm{OH(g)} \rightarrow 2\,\mathrm{H(g)} + 2\,\mathrm{O(g)}$ $\qquad 2\,D(\mathrm{O{-}H})$

Thus:

$$D(\text{O—H}) = \tfrac{1}{2}\, D(\text{H—H}) + \tfrac{1}{2}\, D(\text{O=O}) - \tfrac{1}{2}\Delta H_f(\text{H}_2\text{O}) - \tfrac{1}{4}\Delta H_r$$

The value of D(O—H) at 298 K is 429 kJ mol^{-1}. Furthermore, addition of $2D$(H—H), D(O=O) and $-2\Delta H_f(H_2O)$, gives ΔH for the reaction:

$$2\ H_2O(g) \rightarrow 4\ H(g) + 2\ O(g)$$

(i.e. 2 × Eqn (c) + 2 × Eqn (a) + Eqn (b) above).

The enthalpy change for

$$H_2O(g) \rightarrow 2\ H(g) + O(g)$$

is thus found to be 927 kJ mol^{-1} at 298 K. This is the energy needed to break both bonds in the H_2O molecule. Thus D(H—OH) is (927—429) kJ mol^{-1} = 498 kJ mol^{-1}.

It will be noticed that D(O—H) and D(H—OH) have similar, but not equal, values. The *mean bond energy* or the *bond energy term*, E(O—H), for the O—H bond is $\tfrac{1}{2}$ (927 kJ mol^{-1}) = 463.5 kJ mol^{-1}.

It is found experimentally that the total dissociation energy of a polyatomic molecule can be fairly accurately represented as the sum of the mean bond energies of the individual bonds and, moreover, that the values of E for bonds between particular elements do not vary much from one molecule to another. This is a remarkable empirical fact, for the electronic energy of a molecule must depend considerably on its size and shape, and the existence of characteristic bond properties must, in a general way, be a consequence of the strong localisation of the orbitals describing individual bonds. Tables of mean bond energies have been drawn up and have been of considerable value, more particularly in organic chemistry, where molecules often contain large numbers of bonds but relatively few different types (very common ones are C—H, C—C, C—O, C—N). However, the additivity rule breaks down when systems whose atoms adopt different valence states are compared; for the gross or 'intrinsic' bond energy will be offset by different excitation energies in

TABLE 4.4

Mean bond energies for some X—X single bonds/kJ mol^{-1}

B—B	331	N—N	159	O—O	138	F—F	155
C—C	348	P—P	172	S—S	264	Cl—Cl	243
Si—Si	200	As—As	134	Se—Se	184	Br—Br	193
Ge—Ge	159	Sb—Sb	126	Te—Te	138	I—I	151
Sn—Sn	142	Bi—Bi	105				

Some values for $E(H\text{—}X)$/kJ mol^{-1}

H—H	436	H—N	390	H—Se	276
H—C	416	H—P	318	H—F	565
H—Si	293	H—As	247	H—Cl	431
H—Ge	289	H—O	464	H—Br	368
H—Sn	251	H—S	339	H—I	297

different situations. In these circumstances the atomic valence state excitation energies should be subtracted from the sum of the intrinsic bond energies in calculating the total dissociation energy of the molecule; unfortunately valence state data are at present limited.

4.1.4. Electronegativity scales

A homopolar molecule like Cl_2 has zero dipole moment (5.3.7) because the electrons are shared equally between the two nuclei. But in the HCl molecule the chlorine atom attracts the electrons more strongly than does the hydrogen atom. Consequently the molecule is polar; its chlorine end is negative and its hydrogen end is positive.

The *electronegativity* of an atom was defined by Pauling as the power of that atom within a molecule to attract electrons to itself. Various methods have been used to construct *electronegativity scales* for the comparison of the electronegativities of the elements. Scales based on dipole moment values, stretching force constants (5.3.2), ionisation energies and electron affinities have been proposed. Some of the methods are of limited application; for example the Mulliken scale which takes the absolute negativity of an atom X as $\frac{1}{2}(I_X + A_X)$ has the defect that electron affinities are known for very few atoms.

4.1.4.1. The Pauling scale

Two scales of electronegativities will be considered more fully. That due to Pauling is based on the empirical relation between the electronegativities of two atoms and the dissociation energy of the bond formed between them. The bond dissociation energies of H_2, Cl_2 and HCl are:

$$D(\text{H—H}) = 436 \text{ kJ mol}^{-1}$$
$$D(\text{Cl—Cl}) = 243 \text{ kJ mol}^{-1}$$
$$D(\text{H—Cl}) = 431 \text{ kJ mol}^{-1}$$

A quantity Δ_{HCl}, the *ionic stabilisation energy* of the HCl molecule, is defined as

$$\Delta_{HCl} = D(\text{H—Cl}) - \sqrt{D(\text{H—H}) \times D(\text{Cl—Cl})}$$

Its value is 106 kJ mol^{-1}.

Then

$$\chi_{Cl} - \chi_H = \sqrt{\Delta_{HCl}/96.5 \text{ kJ mol}^{-1}}$$

where χ_{Cl} and χ_H are the electronegativities of Cl and H atoms respectively. The constant 96.5 kJ mol^{-1} was taken because it is equivalent to one electronvolt per molecule. Thus

$$\chi_{Cl} - \chi_H = \sqrt{106/96.5} = 1.05$$

Pauling used $\chi_H = 2.1$ as a standard so that all electronegativities would be positive. Thus $\chi_{Cl} = 2.1 + 1.05 = 3.15$. Pauling himself used $\chi_{Cl} = 3.16$; bond energy values have been redetermined since the scale was first constructed.

4.1.4.2. The Allred—Rochow scale

In this method the electron-attracting power of an atom is considered to be proportional to Z^*/r^2 where Z^*e is the effective nuclear charge (3.4) exerted on the outermost electrons of the atom and r is its single-bond covalent radius (4.1.2).

In order to obtain a set of electronegativity values comparable with Pauling's, Allred and Rochow plotted Pauling values of χ against Z^*/r^2 for a large number of elements. They found Pauling values to lie near the straight line

$$\chi = \frac{3.59 \times 10^3 \, Z^* \, \text{pm}^2}{r^2} + 0.744$$

and they used this equation to obtain their electronegativity scale (Table 4.5). The great advantage of their method is that the required data, i.e. electron configurations and single-bond covalent radii, are known for most of the elements.

The actual electron-attracting power of an atom depends on its environment in the molecule; accordingly the electronegativity ascribed to an ele-

TABLE 4.5

ELECTRONEGATIVITIES ON (a) THE ALLRED—ROCHOW SCALE, (b) THE PAULING SCALE

	a	b		a	b		a	b
F	4.10	3.98	Bi	1.67	2.02	Mo	1.30	2.16
O	3.50	3.44	Zn	1.66	1.65	Hf	1.23	—
N	3.07	3.04	Fe	1.64	1.83	Mg	1.23	1.31
Cl	2.83	3.16	Mn	1.60	1.55	Zr	1.22	1.33
Br	2.74	2.96	Cr	1.56	1.66	U	1.22	1.38
C	2.50	2.55	Pb	1.55	2.33	Pu	1.22	1.28
Se	2.48	2.55	Ir	1.55	2.20	Sc	1.20	1.36
S	2.44	2.58	Os	1.52	—	Lu	1.14	1.27
I	2.21	2.66	In	1.49	1.78	Th	1.11	—
As	2.20	2.18	Be	1.47	1.59	La	1.08	1.10
H	2.20	2.10	Re	1.46	—	Nd	1.07	1.14
P	2.06	2.19	Cd	1.46	1.69	Pr	1.07	1.13
Ge	2.02	2.01	Rh	1.45	2.28	Ce	1.06	1.12
Te	2.01	—	V	1.45	1.63	Ca	1.04	1.00
Sb	1.81	2.05	Pt	1.44	2.28	Na	1.01	0.93
Ga	1.82	1.81	Tl	1.44	2.04	Sr	0.99	0.95
Ni	1.75	1.91	Ag	1.42	1.93	Li	0.97	0.98
Cu	1.75	1.90	Ru	1.42	—	Ba	0.97	0.89
Si	1.74	1.90	W	1.40	2.36	K	0.91	0.82
Sn	1.72	1.96	Ta	1.33	—	Rb	0.89	0.82
Co	1.70	1.88	Ti	1.32	1.54	Cs	0.86	0.79

ment in a scale of values can only be an average one for its most common valency state and oxidation state. Clearly the higher the charge number of an element in a compound the more strongly its atom attracts electrons.

4.1.4.3. Effect of electronegativity difference on bond length

Where two atoms connected by a bond differ considerably in electronegativity the bond length is shorter than the sum of the covalent radii. Thus in ClF the bond length is 163 pm, that is 8 pm shorter than the sum of the radii of Cl and F.

The empirical equation: Internuclear distance

$$r_{A-B} = r_A + r_B - 9|\chi_A - \chi_B| \text{pm}$$

due to Schomaker and Stevenson, is an attempt to relate bond length to bond polarity. It should be stressed, however, that the shortening of a bond can rarely be proved to be entirely due to its polar nature; there is usually the possibility of some double-bond character which would contribute to the shortening.

4.1.5. Non-localised orbitals

It has been possible to represent all the bonds thus far considered by using localised molecular orbitals built up from pairs of atomic orbitals (or hybrid

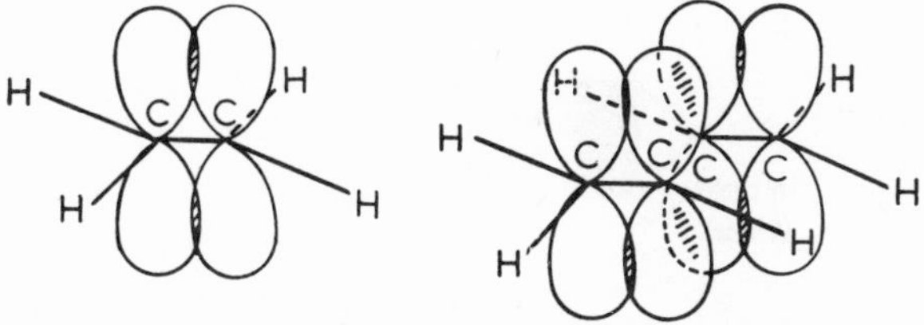

Fig. 4.27. Singly occupied 2p orbitals in ethylene and butadiene. The σ-bonded framework is omitted. (For clarity the greater extent of the actual overlap has not been shown.)

atomic orbitals). But Fig. 4.27 shows that unique pairs cannot always be found. In ethylene there is no ambiguity, the highest occupied orbital providing a normal π bond; but in butadiene, where there is a very similar 'σ-bonded framework', the electrons of the 2p atomic orbitals normal to the molecular plane are less easily accommodated. In instances of this kind, which occur throughout organic chemistry, every such AO overlaps *two* or *three* neighbours (end atoms excepted), and if *one* is admitted in a π-type MO then *all* must be admitted. Hence the so-called 'π electrons' must be accommodated in molecular orbitals which extend over the whole molecule, the restricted 'pairing' approximation being no longer valid.

The coefficients of the individual atomic orbitals, in the LCAO approximation to any π-type MO, can, as always, be determined by standard methods. Fig. 4.28 indicates their values in the butadiene molecular orbitals and

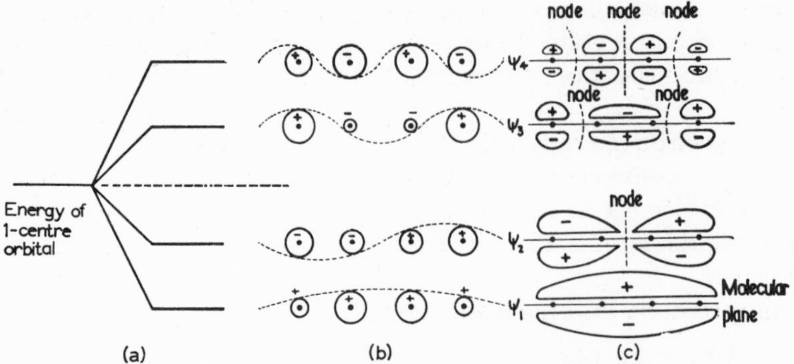

Fig. 4.28. Energy levels and π-orbitals for a chain of 4 atoms. (a) MO energies, relative to that of electron confined to one centre. (b) Coefficients in LCAO approximation — indicated by size of circles. (c) Regions where ψ is greatest, separated by nodes.

illustrates some general properties of *non-localised molecular orbitals.* Clearly the orbitals may be classified as bonding (ψ_1 and ψ_2) or anti-bonding (ψ_3 and ψ_4) on energetic grounds, and their character is reflected in their charge density contributions — the highest energy MO putting a node across every bond, and the lowest leading to a pile-up of charge in every bond. The wave-like pattern followed by the coefficients is also typical; the non-localised orbitals for any long chain would be similar and the allowed energies would behave as in Fig. 4.29.

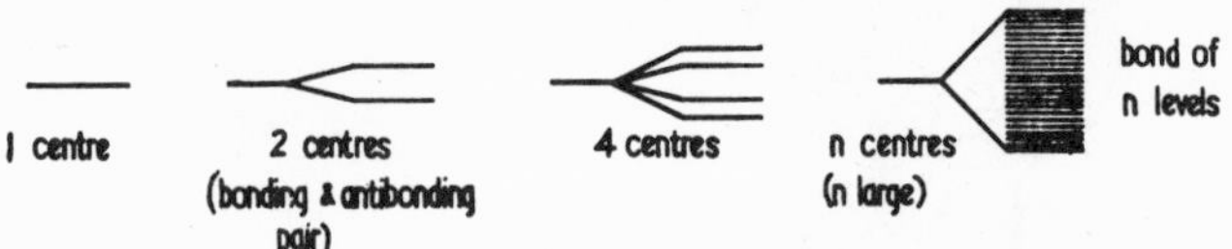

Fig. 4.29. Energy levels for a chain of atoms, showing origin of energy bands.

Generally, for every energy state of an electron moving about one centre there are n related states in a crystal of n such centres. The energies of these states lie within a *band* whose width depends on the strength of the interaction between neighbours.

In foregoing sections the one-centre orbitals described have fallen into *pairs*, interaction between different pairs being so small that each pair could be considered by itself, but in conjugated molecules and in metals there is no alternative but to admit complete non-localisation. This non-localisation is essential to an understanding of the characteristic properties of such systems.

4.1.6. Electronic structure of conjugated molecules

Butadiene shows features common to all 'conjugated' molecules, that is those in which the double bonds of classical chemistry cannot be uniquely allocated. According to the aufbau approach, its electronic structure, apart from that of the σ-bonded framework, would be $(\psi_1)^2(\psi_2)^2$. The extra stabilisation due to these 'mobile' π electrons is in fact greater than that which could be provided by two ethylenic π bonds. The difference in the stabilisation which accompanies the increased delocalisation of the same number of electrons, is the 'resonance' energy. If the system is twisted about its centre link it does indeed, from the point of view of the π electrons, break into two ethylenic halves, because interaction between the central 2p atomic orbitals diminishes to zero as they are rotated (cf. Fig. 4.19). It is accordingly the resonance energy which keeps such systems flat.

More insight into the π-electronic structure can be gained from the coefficients indicated in Fig. 4.28. As stated earlier, the squares of these coefficients, and therefore the *areas* of the circles in Fig. 4.28(b), represent the amounts of charge associated with corresponding atoms in each MO. The electrons in ψ_1 and ψ_2 are associated mainly with internal and end atoms, respectively; but, on adding up the contributions, each atom gets just the same share, one electron, and the charge is thus uniformly distributed. This is true for a large class of hydrocarbons though non-uniformities arise when hetero-atoms are introduced and when ionisation occurs; thus, for example, the removal of an electron from ψ_2 would give a net positive charge mainly on the *end* atoms. On the other hand, the electron density *between* the atoms is determined by products of adjacent AO coefficients. More precisely, $p_{1-2} = 2\sum c_1c_2$ (for the occupied molecular orbitals) measures the π-bond density in link 1—2; it has the value 1 for ethylene (with 1 π bond), but is fractional in other molecules. For butadiene, the π bond orders are about 0.45 for the central, and 0.89 for the outer bonds. These substantial differences in π-bonding in different regions of the molecule are reflected in the extent to which the underlying σ bonds are strengthened and shortened. It has in fact proved possible to interpret both bond lengths and bond properties in terms of bond orders.

4.1.7. Electronic structure of metals

Metals are generally distinguished from non-metals by (a) their excellent thermal and electrical conductivity and (b) their great mechanical strength and ductility. These properties are a direct result of the non-localised nature of the bonding in metals; the electrons are mobile, like the π electrons in butadiene and, furthermore, in a true metal there are no underlying directed bonds.

Lorentz first suggested that a metal consists of an array of cations in a sea of free electrons, and Sommerfeld put the idea on a wave-mechanical basis, describing the electrons by standing waves. Fig. 4.28, which refers to the prototype of a one-dimensional crystal, shows that the standing wave description is not inappropriate even when the presence of the cations is explicitly recognised. But a more accurate picture shows more. The energies of the allowed 'standing waves', the Bloch orbitals, fall into characteristic *bands*, one associated with each atomic level, whose widths and positions can be correlated with a range of non-structural (e.g. electric and magnetic) properties.

The true metals comprise the elements of Groups IA, IB, IIA, and the transition elements, including lanthanides and actinides. With the exception of manganese and uranium they all have one of the three simple structures:

(i) Body-centred cubic (Fig. 4.30), e.g. Na, K, Mo, Fe
(ii) Face-centred cubic (Fig. 4.31, 4.33), e.g. Cu, Ag, Au, Fe
(iii) Close-packed hexagonal (Fig. 4.32, 4.33), e.g. Be, Mg, Zr

In (i) the number of nearest neighbour atoms, viz. the crystal co-ordination number, is 8. Obviously there are insufficient electrons to account for the binding in terms of normal electron-pair covalencies. For instance, the one valence electron of sodium cannot, in any conceivable way, provide covalent bonds with its near neighbours. The necessary sharing, which results in 'partial' bonds, is strictly comparable with that which occurs in butadiene (4 electrons giving 3 partial π bonds) or benzene (6 electrons giving 6 partial π bonds); but the spreading out must be much more complete in a metal and the crystal bond orders must be relatively small.

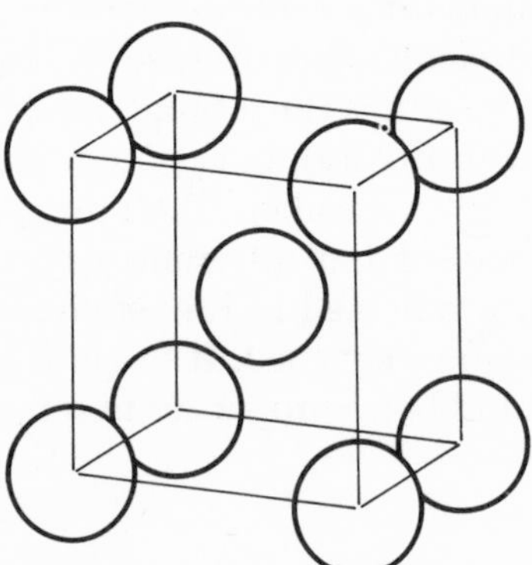

Fig. 4.30. Body-centred cubic packing.

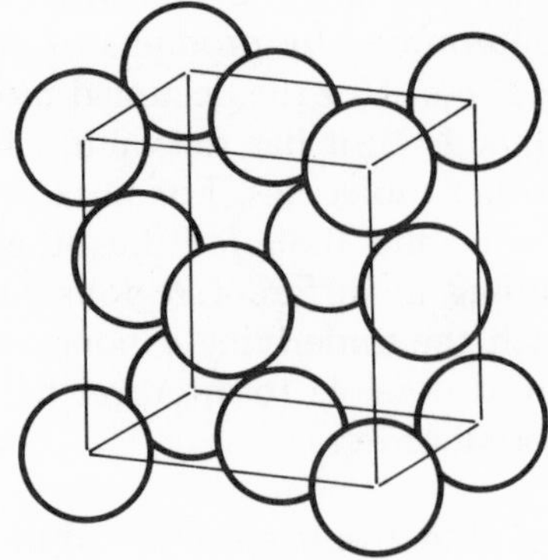

Fig. 4.31. Cubic close-packing (face-centred cubic).

Fig. 4.32. Hexagonal close-packing.

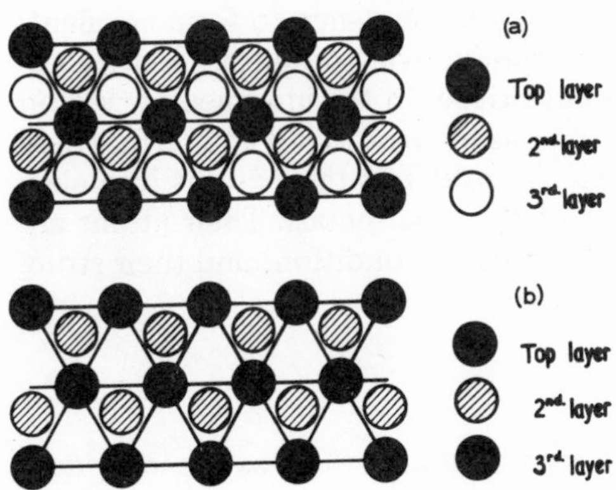

Fig. 4.33. (a) Face-centred cubic packing (cubic close-packing) (shown in depth). (b) Hexagonal close-packing (shown in depth).

4.1.7.1. Passage from metals to non-metals across the Periodic Table

It is largely because the formation of a close-packed metallic structure demands a fairly easy removal of electrons that the metals lie at the left hand side of the Periodic Table. Such elements pool their valence electrons readily, having low ionisation potentials, and the large de-localisation energy then easily leads to a net binding. Proceeding to the right along every period, metallic properties become feebler and the tendency towards covalently-bonded structures increases. Carbon, for instance, does not allow its valence electrons to escape but readily shares them covalently with 4 neighbours. The open tetrahedral structure of diamond, with its strongly directed bonds (which make it brittle), is thus energetically preferred to a close-packed metallic structure. In diamond, carbon completes its octet in the usual sense — a sense which entirely breaks down in the true metals — and in this context is said to satisfy the (8—N) *rule.* Every atom is bound covalently to 8—N others, N being the number of its group in the Periodic Table.

Silicon, germanium and grey tin have diamond-like structures, and although the latter two are metallic to a considerable degree, their bonding is largely covalent, the localisation merely being less complete than in, say, diamond. Other structures are also possible. Graphite consists of well-separated layer planes in which every carbon has three neighbours (as in a giant polycyclic hydrocarbon), but the absence of a fourth covalent bond is compensated for by non-localised π bonding which makes graphite into a 'two-dimensional metal' (the conductivity being primarily along the single planes). Other elements conforming to the (8—N) rule are arsenic, antimony and bismuth (N = 5) with three nearest neighbours, and selenium and tellurium (N = 6) with only two. In the latter, the atoms are connected in spiral chains which are held together by much weaker forces. In iodine (N = 7) the essential units are simply diatomic molecules.

Thus a gradual breakdown of metallic properties is observed in passing along the various periods, together with a growing tendency to form covalently bonded units. Thermal and electrical conductivity diminish, density decreases, and the materials become hard but brittle. In the intermediate region of the Periodic Table are included the metals of Groups IIB and IIIB, where there is still a tendency to obey the (8—*N*) rule, but their atoms lose electrons with a readiness approaching that of the true metals. Their atoms are sometimes said to be in an 'incompetely ionised' condition, and their structures are distorted forms of the simple lattices.

4.1.8. Long-range electrostatic bonds

The forces which hold together well-separated units such as the individual molecules in ice or naphthalene are weak compared with covalent bonds. The strongest of them are often termed 'electrostatic' bonds, signifying that the forces can be attributed to an interaction of the unmodified, static charge distributions of the separate systems. Examples are the *ion—dipole* bonds

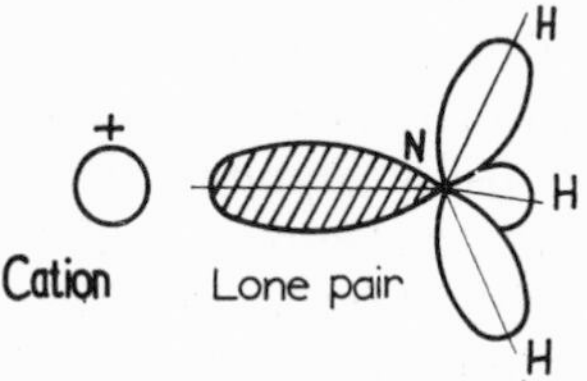

Fig. 4.34. Ion—dipole interaction.

which occur in hydrates and ammines. These are usually weaker than ordinary covalent bonds and break relatively easily on heating. The binding force arises (Fig. 4.34) from the attraction between the positive ion and the lone-pair electron density. Appreciable mixing of the lone pair orbitals with those on the other atom is not required, overlap being small. The attraction arises mainly from a very asymmetrical charge distribution *within* the separate systems, but some mixing (covalency) also contributes to it.

4.1.9. Hydrogen bonds

Similar forces account for the *hydrogen bond* between hydrogen, already bonded to one atom, and the lone pair on a strongly electronegative element such as fluorine, oxygen or nitrogen. Of these, the H ···· F bond is the strongest, with an energy about one tenth that of a σ bond and a length about twice as great. Such bonds are responsible for the open tetrahedral structure of ice (11.8.1), in which every hydrogen (with a net positive charge) attaches itself through the lone pair to a neighbouring oxygen atom. It is the persistence of this structure, in the liquid state, which makes the boiling point of water higher than that of other non-metallic hydrides of Group VI. The boiling points of ammonia and hydrogen fluoride are likewise high in relation to other hydrides of Groups V and VII for the same reason. Similar O ···· H bonds occur in many polymers; thus, in non-polar solvents, they bind carboxylic acids into dimers (Fig. 4.35). In *o*-hydroxy aromatic aldehydes and *o*-nitrophenols (Fig. 4.36) *intra*molecular O ···· H bonds prevent the characteristic O—H vibrations which are normally revealed in infrared spectra. Hydrogen bonding also reduces the solubility of these compounds in water. The *m*- and *p*-isomers which give the normal O—H frequencies are much more soluble.

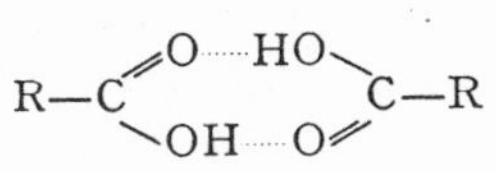

Fig. 4.35. Dimeric bonding in carboxylic acids.

Fig. 4.36. Prevention of O—H vibration by O ···· H bonds.

4.1.10. Polarisation and dispersion forces

Electrostatic interactions between the highly symmetrical atoms of a Group 0 gas are exceedingly small, but weak long-range forces do operate in both gases and liquids. To account for these *Van der Waals* forces, it is necessary to introduce a more refined treatment which recognises the disturbance of one charge density by another. Each electronic system can induce a weak dipole in the other, giving a *polarisation* force which depends on the polarisability of the system and falls off as the sixth power of the distance. At the same time a correlation of the electronic motions of the two systems provides further attractive force. London first suggested that the latter was like the force between rapidly varying electric dipoles. He called it a *dispersion force*. It is now known that dispersion forces operate over considerable distances and that they are strongest when (a) one system is in an excited state and (b) the systems are identical (the case of 'resonance' interaction). Van der Waals forces between atoms or small molecules are well accounted for in this way; those between large molecules are, as yet, but partly understood. Forces of this type always occur as a 'higher order' effect, and also play their part in determining the cohesive properties of molecular crystals and many other non-metallic solids. These forces are usually completely masked by those associated with true chemical valence and will not be considered further.

4.2. The Ionic Bond

4.2.1. Energy changes in the formation of ionic bonds

Suppose that in their power of attracting electrons (their electronegativities (4.1.4)) two atoms *a* and *b* are so different that the singly occupied atomic orbitals φ_a and φ_b give a bonding MO which is almost purely φ_a. The electron configuration in the molecule then contains φ_a^2 instead of $\varphi_a^1\varphi_b^1$. A typical ionic bond arising in this way is indicated in Fig. 4.37. The binding force in these circumstances comes predominantly from the attraction of oppositely charged ions, whose outer orbitals are all doubly occupied. The equilibrium position occurs when these filled orbitals begin to interpenetrate, the situation then resembling that in a hypothetical molecule such as He_2. The repulsive force which arises is of short range, corresponding to a potential energy which can be represented fairly well by a term $a/4\pi\epsilon_0 r^n$, where r is the internuclear distance, a is a constant, ϵ_0 is the permittivity of a vacuum and n, the Born exponent, is a number which depends on the electron configuration of the ion:

Type of electron configuration	Born exponent, n
He	5
Ne	7
Ar and Cu^+	9
Kr and Ag^+	10
Xe and Au^+	12

If cation and anion are of different types, for example Na^+ (Ne type) and Cl^- (Ar type) the mean value, 8 in this case, is taken. Because when one atom loses z_1 electrons and the other gains z_2 electrons the Coulomb potential energy is $-z_1z_2e^2/4\pi\epsilon_0 r$, the net *interaction energy* of the two ions at distance r is

$$E_{\text{int}} = \frac{\alpha}{4\pi\epsilon_0 r^n} - \frac{z_1 z_2 e^2}{4\pi\epsilon_0 r}$$

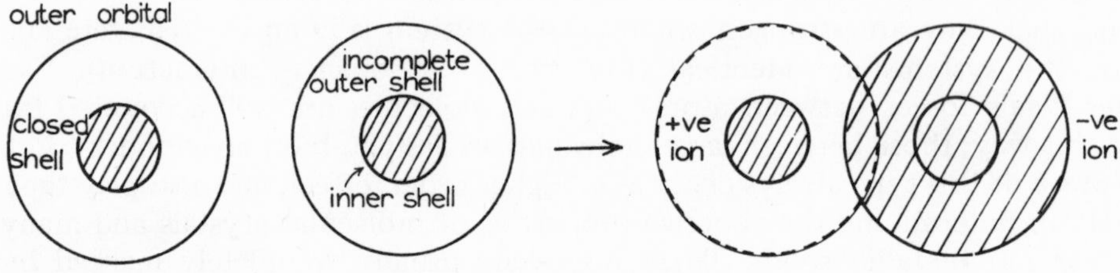

Fig. 4.37. Ionic binding.

4.2.2. Factors determining the formation of ions

A simple anion is formed when a non-metal accepts one or two electrons:

$X \rightarrow X^-$; (X = H, F, Cl, Br, I),
$X \rightarrow X^{2-}$; (X = O, S, Se).

Anions formed from single atoms all have the Group 0 gas structures $1s^2$ or $ns^2\ np^6$. On the other hand, cations are formed when a metal loses one, two, three or even four electrons. Various electron configurations occur (see Table 4.6).

TABLE 4.6

CONFIGURATIONS OF CATIONS

Kinds of ion			Electron configuration (term in brackets indicates a Group 0 gas inner shell)			
(i)	No electrons	H^+				
(ii)	Two electrons	Li^+, Be^{2+},	$1s^2$			He type
(iii)	'8-electron' structures	Na^+, Mg^{2+}, Al^{3+}	$1s^2$	$2s^2$	$2p^6$	Ne type
		K^+, Ca^{2+}, Sc^{3+}	(Ne)	$3s^2$	$3p^6$	Ar type
(iv)	'18-electron' structures	Cu^+, Zn^{2+}, Ga^{3+}	(Ar)	$3d^{10}$		
		Ag^+, Cd^{2+}, In^{3+}	(Kr)	$4d^{10}$		
		Au^+, Hg^{2+}, Tl^{3+}	(Xe)	$5d^{10}$		
(v)	'Inert pair' structures	In^+, Sn^{2+}, Sb^{3+}	(Kr)	$4d^{10}$	$5s^2$	
		Tl^+, Pb^{2+}, Bi^{3+}	(Xe)	$5d^{10}$	$6s^2$	
(vi)	Transition ions	V^{3+}	(Ar)	$3d^2$		
		Fe^{3+}	(Ar)	$3d^5$		
		Fe^{2+}	(Ar)	$3d^6$		
		Ni^{2+}	(Ar)	$3d^8$		

Both anions and cations are stable when they have a closed shell (or subshell, e.g., nd^{10}) of electrons, since the electrons outside such a shell are heavily screened from the nuclear charge and therefore loosely bound. Moreover, as the effective nuclear charge upon the outer electrons is increased by the removal of electrons and decreased by their addition, successive ionisations become increasingly difficult.

4.2.2.1. Types of cation

Of the various types of cation, those with the Group 0 gas structure are by far the most stable. They do not show appreciable tendency to share their electrons with other atoms or ions. Cations with the '18-electron' configuration are moderately stable, although the Cu^+ ion can lose a second electron to form Cu^{2+}, a transition type of coloured ion with $3s^2\ 3p^6\ 3d^9$ structure. But the $3d^{10}$ group is stabilised by an increase of nuclear charge, and the further ionisation of zinc from Zn^{2+} to Zn^{3+} does not occur. The unexpected, though not great, stability of the 'inert pair' structures is due to the outer s electrons penetrating the rather diffuse 18-electron shell immediately beneath them. The effect increases with higher shells; thus the ions of the third long period show greater stability than those of the second; Pb^{2+} is, for instance, much more stable than Sn^{2+}. However, large singly charged ions with an inert pair, such as Tl^{+}, show some of the properties of the alkali metals, though the bonds they form are by no means purely ionic. The least stable ions are those of the transition metals; in these variable charge number is common. Cobalt ($3s^2\ 3p^6\ 3d^7\ 4s^2$) forms Co^{2+} and, less easily, Co^{3+} ions, but nickel ($3s^2\ 3p^6\ 3d^8\ 4s^2$) only Ni^{2+}, the tendency to lose 3d electrons falling with increasing nuclear charge. The more highly charged ions, particularly of the transition elements, are frequently invoked in describing molecular complexes; but then the purely ionic picture is not very satisfactory as the bonds often possess considerable covalent character.

4.2.3. Lattice energies of ionic crystals

The energy of a system of spherical ions is at a minimum when each is surrounded by as many others of opposite charge as possible (7.2.6); this number of nearest neighbours is the *crystallographic co-ordination number* of the ion. Thus in sodium chloride (Fig. 7.11) the ions are in a regular lattice and every ion has the co-ordination number six.

When N_A ions (one mole) of each kind are brought together to form a crystal the interaction energy is clearly not just that of N_A pairs of ions. The formula in 4.2.1 is replaced by

$$E_{int} = \frac{BN_A}{4\pi\epsilon_0 r^n} - \frac{N_A A_a z_1 z_2 e^2}{4\pi\epsilon_0 r}$$

where the first term represents the total energy of repulsion, and A_a, the Madelung constant, in the second term, depends only on the geometry of the crystal.

The value of E_{int} is a minimum when $r = r_0$, the equilibrium distance between neighbouring cation and anion in the crystal.

Putting

$$\left(\frac{\partial E_{int}}{\partial r}\right)_{r=r_0} = \frac{-nBN_A}{4\pi\epsilon_0 r_0{}^{n+1}} + \frac{N_A A_a z_1 z_2 e^2}{4\pi\epsilon_0 r_0{}^2} = 0$$

$$\therefore B = \frac{A_a z_1 z_2 e^2 r_0^{n-1}}{n}$$

The fall in energy on assembling the ions in the crystal, the *lattice energy*, $U = -E_{int}$ is then given by the Born—Landé equation:

$$U = \frac{N_A A_a z_1 z_2 e^2}{4\pi\epsilon_0 r_0}\left(1 - \frac{1}{n}\right)$$

Values of the Madelung constant for some common types of lattice (7.2.1) are given in Table 4.7.

TABLE 4.7

SOME MADELUNG CONSTANTS

Structure type	A_a
CsCl	1.76267
NaCl	1.74756
Wurtzite	1.64123
Blende	1.63805
Fluorite	5.03879
Rutile	4.816

The inter-ionic distance r_0 and the type of lattice are found from determinations of crystal structure (7.1.1).

The lattice energy may be deduced from experimental thermochemical data by considering a suitable cycle of changes (Born and Haber, 1919). The cycle for formation of sodium chloride, with the energy changes occurring at each step shown alongside the arrows, is:

$$\begin{array}{ccc} \text{Na(solid)} + \tfrac{1}{2}\,\text{Cl}_2\,\text{(gas)} & \xrightarrow{\Delta H_s + \frac{1}{2}\Delta H_d} & \text{Na(g)} + \text{Cl(g)} \\ \downarrow \Delta H_f & & \downarrow I_{Na} \quad \downarrow -A_{Cl} \\ \text{NaCl(s)} & \xleftarrow{-U} & \text{Na}^+\text{(g)} + \text{Cl}^-\text{(g)} \end{array}$$

It may also be conveniently set out thus:

			Energy requirement per mole
	$Na(s)$	$\rightarrow Na(g)$	ΔH_s
	$Na(g)$	$\rightarrow Na^+(g) + e$	I_{Na}
	$\frac{1}{2}\,Cl_2(g)$	$\rightarrow Cl(g)$	$\frac{1}{2}\,\Delta H_d$
	$Cl(g) + e$	$\rightarrow Cl^-(g)$	$-A_{Cl}$
	$Cl^-(g) + Na^+(g)$	$\rightarrow NaCl(s)$	$-U$
Adding:	$Na(s) + \frac{1}{2}\,Cl_2(g)$	$\rightarrow NaCl(s)$	ΔH_f

ΔH_s is the enthalpy of sublimation of sodium to free atoms, ΔH_d the enthalpy of dissociation of chlorine molecules to atoms, ΔH_f is the enthalpy of formation of sodium chloride from the elements; these, with the ionisation energy of sodium, I_{Na}, and the electron affinity of chlorine, A_{Cl}, refer to one mole of material. Thus:

$$\Delta H_f = \Delta H_s + \tfrac{1}{2}\Delta H_d + I_{Na} - A_{Cl} - U$$

Therefore

$$U = \Delta H_s + \tfrac{1}{2}\Delta H_d + I_{Na} - A_{Cl} - \Delta H_f$$

4.2.4. Ionic radii

The meaning and determination of ionic radius can be illustrated by reference to the alkali-metal halides. Their crystal structure and internuclear distance are found by X-ray analysis (7.1.1). Most of them have the same type of crystal lattice as sodium chloride (Fig. 7.11) but their inter-ionic distances differ. Cubic arrangement of ions in NaCl indicates 6-co-ordination of both anions and cations. The internuclear distance, r_0, for a number of similar lattices is given:

Internuclear distance		Internuclear distance	
KCl	314 pm	NaCl	276 pm
KF	269 pm	NaF	231 pm

These distances represent the sum of the radii of neighbouring, spherical ions. The method most generally used to deduce ionic radii from internuclear distance is that due to Pauling (1927), and is applicable only to isoelectronic pairs. To obtain the ionic radius of the singly charged, isoelectronic ions K^+ and Cl^- (both with Ar configuration) in potassium chloride the distance between their nuclei is divided in the inverse ratio of the effective nuclear charge exerted on the outer electron shell of each ion. The screening constant σ (3.4) is given by:

$$\begin{aligned} \sigma = {} & 8 \times 0.35 && \text{(3s and 3p electrons)} \\ & + 8 \times 0.85 && \text{(2s and 2p)} \\ & + 2 \times 1.0 && \text{(1s)} \\ = {} & 11.6 \end{aligned}$$

Since $Z^* = Z - \sigma$ (3.4), thus for K^+ ($Z = 19$) $Z^* = 7.4$, and for Cl^- ($Z = 17$) $Z^* = 5.4$. The radius of the Cl^- ion, r_{Cl^-}, is given by:

$$\frac{7.4}{5.4 + 7.4} \times 314 \text{ pm} = 181 \text{ pm}$$

and that of K^+ is

$$\frac{5.4}{5.4 + 7.4} \times 314 \text{ pm} = 133 \text{ pm}$$

Similar treatment of the data for NaF gives

$$r_{F^-} = \frac{6.5}{6.5 + 4.5} \times 231 \text{ pm} = 136 \text{ pm}$$

$$r_{Na^+} = \frac{4.5}{6.5 + 4.5} \times 231 \text{ pm} = 95 \text{ pm}$$

The sum of r_{Na^+} and r_{Cl^-} should therefore be (95 + 181) pm = 276 pm, in agreement with the X-ray determination.

4.2.4.1. Pauling's method for ions of charge greater than one

For an ion of charge z, Pauling's method is to calculate first the hypothetical 'univalent' radius, r_1, that the ion would possess provided it retained its electron configuration but suffered coulombic attraction as though it carried unit charge. Then r_z for the actual ion is computed from the equation

$$r_z = \frac{r_1}{z^{2/(n-1)}}$$

where n is the Born exponent (4.2.1).

TABLE 4.8

IONIC RADII

Radii in crystals with 6-co-ordination/pm

		H^-	Li^+	Be^{2+}	B^{3+}		
		150	60	31	20		
N^{3-}	O^{2-}	F^-	Na^+	Mg^{2+}	Al^{3+}	Si^{4+}	P^{5+}
171	140	136	95	65	50	41	34
P^{3-}	S^{2-}	Cl^-	K^+	Ca^{2+}	Sc^{3+}	Ti^{4+}	V^{5+}
212	184	181	133	99	81	68	59
			Cu^+	Zn^{2+}	Ga^{3+}	Ge^{4+}	As^{5+}
			96	74	62	53	47
As^{3-}	Se^{2-}	Br^-	Rb^+	Sr^{2+}	Y^{3+}	Zr^{4+}	Nb^{5+}
222	198	195	148	113	93	80	70
			Ag^+	Cd^{2+}	In^{3+}	Sn^{4+}	Sb^{5+}
			126	97	81	71	62
Sb^{3-}	Te^{2-}	I^-	Cs^+	Ba^{2+}	La^{3+}	Ce^{4+}	
245	221	216	169	135	115	101	
			Au^+	Hg^{2+}	Tl^{3+}	Pb^{4+}	Bi^{5+}
			137	110	95	84	74

As stated, the values in Table 4.8 are for 6 co-ordinated ions. To calculate the ionic radius for an ion in a crystal when its co-ordination number is a, the following formula is used

$$\frac{r_a}{r_6} = \left(\frac{a}{6}\right)^{1/(n-1)}$$

Thus for 8-co-ordinated Cl^- ($n = 9$)

$$r_8 = r_6 \times (\tfrac{8}{6})^{1/8}$$

$$= 181 \text{ pm} \times 1.038 = 188 \text{ pm}$$

and for 4-co-ordinated Cl^-

$$r_4 = r_6 \times (\tfrac{4}{6})^{1/8} = 172 \text{ pm}$$

4.2.5. Trends in ionic and covalent radii

An ion may be looked upon as a sphere with a radius which depends on (a) the quantum level of its outermost electrons and (b) the effective nuclear charge acting on those electrons. Ionic radii can be derived as indicated in 4.2.4, and they show clear trends through the Periodic Table. Within a group there is an increase in radius with atomic number which is due to the increasing energies of the successive outermost electrons, the effective nuclear charge acting on these electrons remaining about the same:

Li^+	Na^+	K^+	Rb^+	Cs^+
60	95	133	148	169 pm

F^-	Cl^-	Br^-	I^-
136	181	195	216 pm

Within a period the sizes of the positive ions decrease as the effective nuclear charges increase with atomic number, the quantum level of the outermost electrons remaining the same:

Cs^+	Ba^{2+}	La^{3+}
169	135	115 pm

For corresponding reasons a 2-positive ion is larger than a 3- or 4-positive one:

Fe^{2+} 76 Fe^{3+} 64 pm
Pb^{2+} 120 Pb^{4+} 84 pm

Across a transition series, for a series of ions of equal charge, the increase in effective nuclear charge with atomic number as electrons enter an inner d shell is rather small, and the radii of the ions remain about constant:

Mn^{2+}	Fe^{2+}	Co^{2+}	Ni^{2+}
80	76	78	78 pm

The minor variations are due to ligand field effects (27.10).

In the lanthanides an increase of 14 in atomic number occurs as the 4f shell is filled up; these f electrons exert only weak screening and the ionic radii fall from 115 (La^{3+}) to 93 (Lu^{3+}) pm. This, the lanthanide contraction, affects the size of the ions of the elements which follow in the Periodic Table.

However, the sizes of atoms cannot be compared as easily as those of ions. The *Van der Waals radius* is that of the spherical, non-bonded atom in a crystal of the solid element; it can be measured only for the noble gases, although estimates of its value can be made for some other atoms. Pauling has assumed that in directions other than that of a covalent bond the electrons of an atom are in almost the same environment as those of the corresponding anion: thus, for instance, since the anion F^- has a radius of 136 pm, the Van der Waals radius of the fluorine atom must be about the same. The single-bond covalent radius of fluorine is, of course, much smaller. Thus a fluorine molecule can be represented as a pair of spherical atoms fused together (Fig. 4.38):

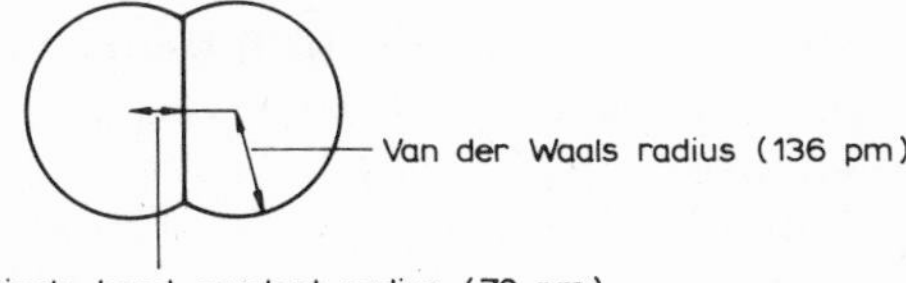

Fig. 4.38. Diagrammatic representation of F_2 molecule.

In view of what has been said (4.1.1) about the nature of covalent bonding, this picture obviously gives an over-simplified representation of the electron densities in the molecule. Nevertheless, the Van der Waals radii can account satisfactorily for effects such as steric hindrance between parts of a molecule not directly bonded to each other.

As Van der Waals radii can be measured only for the noble gases, and estimated only for those elements which form simple anions, some other criterion of size is necessary for the atoms of metals. The metallic radius obtained by halving the internuclear distance in the solid metal is not the same as a Van der Waals radius because neighbouring atoms cannot be considered as non-bonded. The concept of metallic radius is of limited value; in many metals (e.g. zinc, 39.2) the internuclear distances differ in the different crystal planes. Nevertheless two generalizations can be made.

(a) When a metal forms compounds in which the bonds can be described as covalent single bonds, the covalent radius assigned to the metal is less than its metallic radius. Thus for aluminium the single-bond covalent radius in the alkyls is inferred to be 118 pm, whereas the metallic radius in the face-centred cubic lattice of the metal is 143 pm. But the ionic radius of Al^{3+}, 50 pm, is very much less than either the metallic radius or the covalent radius, as would be expected from the larger effective nuclear charge acting on the outer electrons.

(b) Trends in metallic radii, like those in single-bond covalent radii and also in Van der Waals radii as far as these are known, are similar to the trends in ionic radii. Sizes increase down a group in the Periodic Table:

	Li	Na	K	Rb	Cs
Metallic radius	155	190	235	248	267 pm
Covalent radius	134	154	196	211	225 pm

But sizes decrease along a period:

	Na	Mg	Al
Metallic radius	190	160	143 pm
Covalent radius	154	130	118 pm

As for ions, the trends can be explained in terms of the energies of the outermost electrons and the effective nuclear charges acting upon them.

4.2.6. Interpretation of lattice energies

The experimental (Born—Haber) and theoretical (Born—Landé) estimates of lattice energy agree well for typically ionic solids such as the Group IA halides, and show that the theoretical picture of complete electron transfer is satisfactory. Values for salts of '18-electron' type cations, however, often show considerable discrepancies. Representative differences between U_{expt} and U_{calc} are: RbI, 17 kJ mol^{-1}; CdI_2, 360 kJ mol^{-1}; PbO_2, 890 kJ mol^{-1}. These indicate a gradual departure from the purely ionic condition found in NaCl; the small cations such as Pb^{4+} tend to retain some hold on their electrons and the binding acquires considerable covalent character.

Deviations of this kind used to be ascribed to the 'polarisation' of the large anions by the small cations, the charge density of the former being deformed by the strong electric field of the latter. Although such effects undoubtedly occur, the deviations can most readily be accounted for along the lines already indicated. The electron transfer from a small cation which has a correspondingly great electron-attracting power is never complete, and the bonds that such a cation forms may exhibit considerable covalent character. Nevertheless, the term 'polarisation' is convenient, and continues to be used to describe this effect. Moreover, the effect, always accompanied by an abnormally large negative enthalpy of formation and high lattice energy, is not confined to '18-electron' ions. Thus although there is no discrepancy between U_{expt} and U_{calc} for BaO, the difference for MgO is 83 kJ mol^{-1}, in accord with the smaller size and higher electron-attracting power of the Mg^{2+} ion.

4.2.7. Kapustinskii's equation

Using a semi-empirical approach, Kapustinskii developed an equation for calculating lattice energies of compounds which contain non-spherical ions and for which the computation of Madelung constants is not possible. The equation can be written in the form:

$$U/\text{kJ mol}^{-1} = 1.202 \times 10^{-7}\ \text{m}\ \frac{\nu z_c z_a}{r_c + r_a}\left(1 - \frac{3.45 \times 10^{-11}\ \text{m}}{r_c + r_a}\right)$$

where ν is the number of ions in the formula, 2 for NaCl, 3 for $CaCl_2$ and so on, and r_c and r_a are the formal 6-co-ordinate radii (Table 4.8) of cation and anion respectively. This equation has proved remarkably accurate in cases where it can be checked against precise calculations of lattice energy. The equation can also be used, where the lattice energy of a compound can be calculated from thermochemical data, to calculate the apparent radii, the so-called *thermochemical radii*, of complex ions.

For $KClO_4$, U is 591 kJ mol^{-1}. Thus if $r_c + r_a = r_0$

$$\frac{1.202 \times 10^{-7}\ \text{m} \times 2 \times 1 \times 1}{r_0}\left(1 - \frac{3.45 \times 10^{-11}\ \text{m}}{r_0}\right) = 591$$

Solving the quadratic equation:

$r_0 = 3.69 \times 10^{-10}$ m = 369 pm

As r_c for K^+ = 133 pm, ∴ r_a for ClO_4^- = 236 pm, which is the thermochemical radius of the ClO_4^- ion. Values for some other complex ions are given in Table 4.9.

TABLE 4.9

THERMOCHEMICAL RADII OF SOME IONS/pm

NO_3^-	189	SO_4^{2-}	230
BrO_3^-	191	CrO_4^{2-}	240
IO_4^-	249	BeF_4^{2-}	245
BF_4^-	228	BO_3^{3-}	191
CO_3^{2-}	185	PO_4^{3-}	238

4.2.8. Water as a solvent for ionic compounds

Water consists of polar molecules with a strong tendency to form hydrogen bonds (4.1.9). The liquid has a high relative permittivity; ϵ_r = 78. Its molecules orient themselves around cations with the oxygen atoms pointing inwards, the complex so formed being stabilised by a *hydration energy*, more correctly an enthalpy of hydration, arising from electrostatic attraction between the cation and the excess of negative charge on the oxygen. Around anions the H_2O will be oriented with hydrogen towards the centre.

The coulombic attraction between hydrated cations and anions is low because ϵ_r for water is so large.

Water is, in consequence, a good solvent for ionic compounds but a poor one for non-electrolytes whose molecules are without strongly polarised charges.

4.2.8.1. Hydration energies of ions

For a salt MX the sum of the enthalpies of hydration of the gaseous M^+ and X^- ions can be obtained by determining (a) the enthalpy of hydration, ΔH_h(MX), (b) the lattice energy, U for MX, and then applying a thermochemical cycle:

MX(s) + aq	→ M^+(aq) + X^-(aq)	ΔH_h(MX)
M^+(g) + X^-(g)	→ MX(s)	$-U$
M^+(g) + X^-(g) + aq	→ M^+(aq) + X^-(aq)	$\Delta H_h(M^+) + \Delta H_h(X^-)$

Clearly the value of $\Delta H_h(X^-)$ can then be calculated if $\Delta H_h(M^+)$ is known. Various methods have been used to apportion the contributions of the separate ions. One procedure which gives results very similar to those obtained from the most sophisticated methods is to determine $\Delta H_h(M^+) + \Delta H_h(X^-)$ for two large ions of small charge which do not attract water molecules strongly and then make the assumption that $\Delta H_h(M^+) = \Delta H_h(X^-)$. The enthalpies of hydration tabulated in Table 4.10 are calculated from

$$\Delta H_h(Cs^+) = \Delta H_h(I^-) = -280 \text{ kJ mol}^{-1}$$

TABLE 4.10

ENTHALPIES OF HYDRATION OF GASEOUS IONS (expressed as $-\Delta H_h$/kJ mol^{-1})

H^+	1110	Be^{2+}	2520	Al^{3+}	4700	F^-	485
Li^+	530	Mg^{2+}	1960	Ga^{3+}	4740	Cl^-	350
Na^+	420	Ca^{2+}	1615	Tl^{3+}	4240	Br^-	320
K^+	340	Sr^{2+}	1475	Sc^{3+}	4000	I^-	280
Rb^+	315	Ba^{2+}	1340	La^{3+}	3340	OH^-	510
Cs^+	280	Zn^{2+}	2080	Fe^{3+}	4460	CN^-	345
Ag^+	490	Cd^{2+}	1840			NO_3^-	310
Tl^+	345	Hg^{2+}	1850			ClO_4^-	225

Trends in the figures indicate that small ions of high charge, such as Al^{3+}(g), release the greatest energy on hydration. Another feature is that a d^{10} ion such as Cd^{2+} forms slightly stronger bonds with water molecules than a d^0 ion of similar size such as Ca^{2+}.

4.2.8.2. The solubility of ionic compounds in water

From the foregoing cycle it is clear that ΔH_h(MX), the enthalpy of hydration of the solid, is given by

$$\Delta H_h(MX) = U + \Delta H_h(M^+) + \Delta H_h(X^-)$$

The value of ΔH_h(MX) is close to zero for many salts because small, highly charged ions which confer a high lattice energy are also those that release most energy on hydration. The solubility of a salt MX in water depends not only on ΔH_h(MX) but also on the entropy change of the process of hydration. Thus a salt can be appreciably soluble, because of the generally favourable entropy factor, even when it dissolves endothermically. However, when the cations are sufficiently polarising to confer a good deal of covalent character on the binding in the crystal, the large lattice energy becomes the dominant factor, ΔH_h(MX) is strongly positive, and the compound is sparingly soluble.

Further Reading

H.B. Gray, Electrons and chemical bonding, Benjamin, New York, 1964.

C.A. Coulson, Valence, 2nd edition, Oxford University Press, 1961.

A.L. Allred and E.G. Rochow, A scale of electronegativity based on electrostatic force, J. Inorg. Nucl. Chem., 5 (1958) 264.

L.F. Phillips, Basic quantum chemistry, Wiley, New York, 1965.
J.N. Murrell, S.F.A. Kettle and J.M. Tedder, Valence theory, Wiley, London, 1965.
C.J. Ballhausen and H.B. Gray, Molecular orbital theory, Benjamin, New York, 1964.
R.H. Hochstrasser, Molecular aspects of symmetry, Benjamin, New York, 1966.
D.S. Urch, Orbitals and symmetry, Penguin Books, London, 1970.
L. Pauling, The nature of the chemical bond, 3rd edition, Cornell University Press, Ithaca, N.Y., 1960.
T.C. Waddington, Lattice energies and their significance in inorganic chemistry, in Adv. Inorg. Chem. and Radiochem., Vol. 1, Academic Press, New York, 1959.
K.H. Stern and E.S. Amis, Ionic size, Chem. Rev., 59 (1959) 1.
T.H. Cottrell, The strengths of chemical bonds, 2nd edition, Butterworths, London, 1958.
F.A. Cotton, Chemical applications of group theory. 2nd edition, Interscience, New York, 1971.
H.H. Jaffé and M. Orchin, Symmetry in chemistry, Wiley, New York, 1965.
M.L. Huggins, 50 years of hydrogen bond theory, Angew. Chem. Internat. Ed., 10 (1971) 147.
M. Klessinger, Polarity of covalent bonds, Angew. Chem. Internat. Ed., 9 (1970) 500.
R.J. Gillespie, Electron-pair repulsions and molecular shape, Angew. Chem. Internat. Ed., 6 (1967) 819.
L.D. Pettit, Multiple bonding and back-co-ordination in inorganic compounds. Quart. Rev., 25 (1971) 1.
M. Orchin and H.H. Jaffé, Symmetry operations and their importance for chemical problems, J. Chem. Ed., 47 (1970) 247.
R.J. Gillespie, The electron-pair repulsion model for molecular geometry, J. Chem. Ed., 47 (1970) 18.
R.M. Gavin, Jr., Simplified molecular orbital approach to inorganic stereochemistry, J. Chem. Ed., 46 (1969) 413.
C.A. Coulson, The shape and structure of molecules, Clarendon Press, Oxford, 1973.
J.E. Ferguson, Stereochemistry and bonding in inorganic chemistry, Prentice-Hall, New Jersey, 1974.
B.M. Gimarc, Applications of qualitative MO theory, Accounts Chem. Res., 7 (1974) 384.

Bonding and Structure in Compounds of Non-transition Elements

5

5.1. General Principles

The forces holding atoms together in molecules and crystals have been interpreted in terms of located electron density. There are further theoretical principles which make it possible to account for the shapes of molecules.

Localised molecular orbitals

Although the forces on the nuclei depend on the charge density P (that is the probability density for finding an electron at the given point), their equilibrium configuration is determined by the total energy of the system, and to understand the factors on which it depends, the *interaction* between different electrons must be examined. The interaction energy depends on the probability of two electrons being a given distance apart and accordingly on how their motions are *correlated.* Fortunately, this correlation is mainly associated with the exclusion principle, which prevents electrons of like spin from occupying the same orbital or, more generally, the same region of space. If the various doubly occupied orbitals are chosen so as to be localised essentially in different parts of space, this correlation requirement is automatically met — for then electrons with the same spin are obviously always well separated. Thus, by working in terms of localised molecular orbitals, it is possible to get a picture not only of where an electron is *on the average* (the charge density) but also of where different electrons are *at the same instant.*

The water molecule provides a good example. In the true molecular orbital approach, the electrons responsible for bonding would occupy orbitals extending over the whole molecule, but in the earlier discussion of valency the electrons were supposed to occupy localised molecular orbitals. The two descriptions are, however, exactly equivalent, provided the two types of MO are related by

$$\psi_1 = \frac{1}{\sqrt{2}}(\varphi_a + \varphi_b) \text{ and } \psi_2 = \frac{1}{\sqrt{2}}(\varphi_a - \varphi_b)$$

It is easily seen, for instance, that they give the same charge density contribution; for, when the orbitals are each doubly occupied, the density is

$$2\psi_1^2 + 2\psi_2^2 = (\varphi_a^2 + 2\varphi_a\varphi_b + \varphi_b^2) + (\varphi_a^2 - 2\varphi_a\varphi_b + \varphi_b^2) = 2\varphi_a^2 + 2\varphi_b^2$$

But, by allocating the electrons to the localised molecular orbitals, we can describe, without mathematical analysis, the fact that two electrons with the same spin tend at any instant to occupy different regions of space. Different electron pairs have, in fact, an interaction energy which can be closely estimated by regarding each as a smeared-out charge.

5.2. The Orientation of Bonds around Non-transitional Atoms

The energy of interaction of different electron pairs is a minimum when the different pairs (either lone pairs or bond pairs) are as far apart as possible. N.V. Sidgwick and H.M. Powell (1940) drew attention to the use of this simple principle for predicting the shapes of small molecules. As was mentioned in 1.6, the symmetry properties of such shapes are described exactly by their point groups.

The pairs of valence electrons around a non-transitional atom are arranged, according to their number, to be as far apart as possible. Two pairs are arranged diametrically opposite one another, thus the $BeCl_2$ molecule is linear because the Be atom uses its only two valence electrons in forming two σ bonds with Cl atoms.

Boron, with three valence electrons, forms a planar molecule BCl_3 with Cl—B—Cl angles of 120°; the CCl_4, PF_5 and SF_6 molecules are respectively tetrahedral, trigonal bipyramidal and octahedral. These are the shapes which enable the electron pairs to interact as little as possible.

The arrangement of electron pairs around an atom is basically the same whether they are σ-bonding pairs or lone pairs. The Sn atom in $SnCl_2$ has three pairs of electrons around it but one of these is a lone pair. The individual molecules observed in the vapour are V-shaped belonging to point group C_{2v} (Fig. 5.1), the Cl—Sn—Cl angle being somewhat less than 120° because the lone pair repels the bonding pairs a little more strongly than they repel one another. In a similar way nitrogen, with five valence electrons, uses only three of them to form σ-bonds with H atoms in the NH_3 molecule and therefore has one lone pair remaining. The molecule is consequently a trigonal pyramid (C_{3v}) (Fig. 5.2.b) because the three bond pairs and the lone pair all repel one another; the H—N—H angles (106.7°) are rather less than the

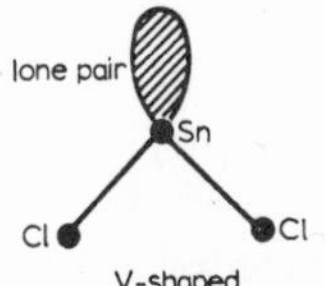

Fig. 5.1. Single molecule of tin(II) chloride.

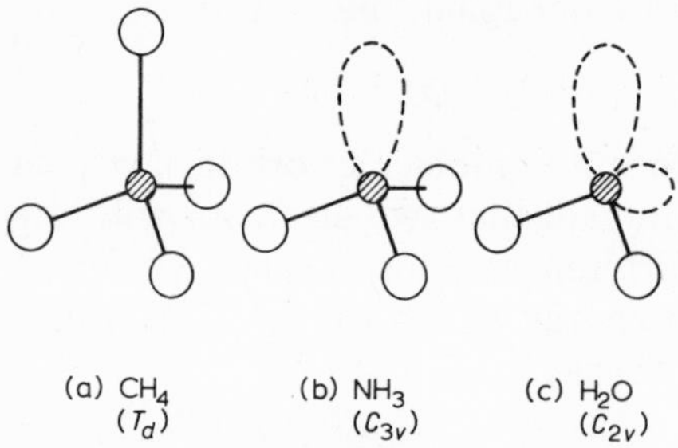

Fig. 5.2. Shapes of methane, ammonia and water molecules.

H—C—H angles in the tetrahedral molecule CH_4 (T_d) (109.5°) because the lone pair again repels the bond pairs rather more strongly than they repel one another. From similar considerations it can be predicted that the H_2O molecule will be V-shaped (C_{2v}) (Fig. 5.2.c) with a still smaller bond angle, 104.5° in fact.

The shapes of molecules containing a central (non-transitional) atom surrounded by five electron pairs are shown in Fig. 5.3. In fact, all the covalent compounds of non-transitional elements with five electron pairs which have been investigated have the trigonal bipyramidal structure or one of these derived forms. Other shapes are possible but are ruled out if lone-pair—bond-pair repulsions are assumed to be stronger than those between bond pairs. Thus, in $TeCl_4$, lone-pair—bond-pair repulsion is minimised when the lone pair lies at about 120° to two of the bonds and about 90° to the other two. In the trigonal pyramid (Fig. 5.4) the lone pair would be close to three (instead of two) bond pairs and electron repulsion energy would be higher; this alternative does not occur.

For similar reasons ClF_3 is T-shaped (C_{2v}), not triangular (D_{3h}). The F—Cl—F angle is actually only 87.5° because of the strength of the repulsion between the lone pair and the bond pairs.

The arrangement of six pairs of electrons about the atom of a non-transi-

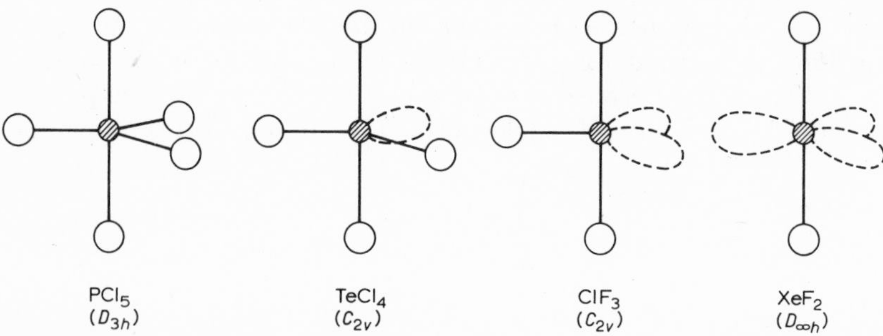

Fig. 5.3. Shapes arising from the disposition of five electron pairs round a central atom. (In $TeCl_4$ and ClF_3 the two bonds shown vertical are actually inclined away slightly from the lone pairs.)

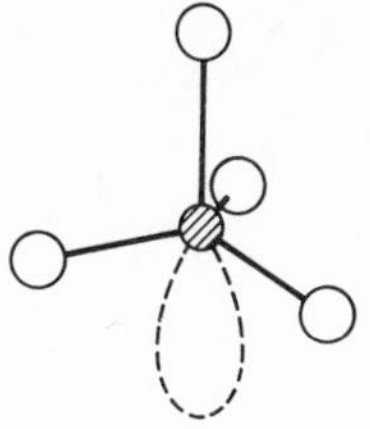

Fig. 5.4. Trigonal pyramid (C_{3v}).

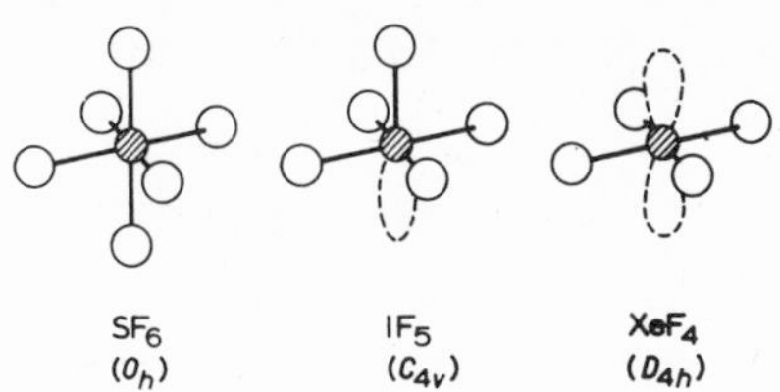

Fig. 5.5. Shapes arising from the disposition of six electron pairs round a central atom.

tional element is basically octahedral, but there again lone pairs tend to adopt positions in which their repulsive energy is a minimum (Fig. 5.5).

5.2.1. Shapes of ions

The shapes of ions can be predicted similarly. Taking the ICl_4^- ion as an example, the number of electron pairs around the iodine atom can be computed as follows:

Valence electrons of iodine atom	7 electrons
Charge of −1 on ion	1 electron
	8 electrons
1 electron for each I—Cl bond	4 electrons
Remainder attached as lone pairs	4 electrons

Thus there are four σ-bonds and two lone pairs and the ICl_4^- ion is square planar (D_{4h}).

5.2.2. Factors influencing bond angles

Bond angles depend first of all on the number of pairs of electrons around the central atom and on whether they are lone pairs or bond pairs. Interaction between lone pairs is stronger than that between a lone pair and a bond pair, which is in turn stronger than that between bond pairs.

Another factor is the electronegativity of the central atom. Thus the bond angle in H_2O is 104.5° but in H_2S only 92.5°, corresponding to a fall in electronegativity from 3.5 in oxygen to 2.5 in sulphur. The bond pairs lie nearer to the O in H_2O than to the S in H_2S, so that in H_2O the repulsion between the bond pairs is greater and the effect of the lone pairs on the angle between the bonds less than in H_2S. In terms of the molecular-orbital theory there is a greater p-orbital contribution to the bonding orbitals.

TABLE 5.1

SHAPES OF MOLECULES WITH NON-TRANSITIONAL ELEMENT AS CENTRAL ATOM (No π-BONDING)

Number of electron pairs in valence shell	Bonding pairs	Lone pairs	Shape of molecule	Point group	Example
2	2	0	Linear	$D_{\infty h}$	$BeCl_2$ (vap.)
3	3	0	Triangular	D_{3h}	BCl_3
	2	1	V-shaped	C_{2v}	$SnCl_2$(vap.)
4	4	0	Tetrahedral	T_d	CH_4
	3	1	Trigonal pyramidal	C_{3v}	NH_3
	2	2	V-shaped	C_{2v}	H_2O
5	5	0	Trigonal bipyramidal	D_{3h}	PCl_5(vap.)
	4	1	Distorted tetrahedron	C_{2v}	$TeCl_4$
	3	2	T-shaped	C_{2v}	ClF_3
	2	3	Linear	$D_{\infty h}$	XeF_2
6	6	0	Octahedral	O_h	SF_6
	5	1	Square pyramidal	C_{4v}	IF_5
	4	2	Square	D_{4h}	XeF_4

5.2.3. Molecules containing π-bonds

The prediction of molecular shape using the simple principle of electron-pair repulsion can be extended to molecules which contain π-bonds as well as σ-bonds. In molecules containing only non-transitional atoms the existence of π-bonding has hardly any influence on the shape. When calculating the number of lone pairs left on the central atom after the bonds have been formed, it is sufficient to transfer to each of the outer atoms sufficient electrons to provide it with a noble gas configuration (i.e. 8 electrons in its valence shell or two electrons in the case of hydrogen).

Consider the case of the SO_2Cl_2 molecule. The two oxygen atoms have 6 valence electrons each and can therefore each accept two more, the chlorine atoms have seven and can accept only one each. Thus the electrons required from the S atom are:

Required by two oxygen atoms	4 electrons
Required by two chlorine atoms	2 electrons
	6 electrons

Since the S atom has only 6 electrons in its valence shell it is left without a lone pair when the bonds are formed. The two oxygens and the two chlorines should consequently be arranged tetrahedrally around it. In fact the

Cl—S—O angle is 106.5° and the Cl—S—Cl angle is 111°, both within 3° of the angle at the centre of a regular tetrahedron but as the atoms are different the point group is C_{2v}. Clearly this extremely simple approach, in which the effect of π-bonding is ignored, is sufficient to provide an almost perfect prediction of molecular shape. The same principle can be extended even to cases where there is extensive π-bonding. Consider, for example, the N_3^- ion, and imagine the middle N atom to carry the negative charge, giving it 6 electrons in all. Since three are needed by each of the other N atoms to give them noble gas configurations, the first N atom is left without a lone pair and the N_3^- ion should be linear ($D_{\infty h}$), as it is in fact.

5.2.4. Isoelectronic molecules and ions

The ions PO_4^{3-}, SO_4^{2-} and ClO_4^- are *isoelectronic*, i.e. they contain equal numbers of electrons. It follows from the principles outlined above that they must have the same structure, tetrahedral (T_d) in this case. Similarly the isoelectronic CO_3^{2-} and NO_3^- ions are triangular (D_{3h}).

The principle can be extended to atoms and molecules with equal numbers of *valence* electrons. Thus ICl_4^- and XeF_4 are both square planar (D_{4h}), SO_3^{2-} and XeO_3 are both pyramidal (C_{3v}).

5.2.5. Shapes of larger molecules and ions

Although the concept of electron-pair repulsion can be applied in principle to larger molecules and ions in which, unlike the cases quoted above, there is no single central atom, interactions between electron pairs on adjacent atoms naturally make the problem of prediction more complicated. The molecules of H_2O_2 (24.1) and N_2H_4 (11.7.2) illustrate this point.

5.3. Experimental Methods for Determining the Shapes of Molecules

Inorganic chemicals can roughly be divided into molecular and crystalline materials. The word molecule is usually understood in the context of discrete and finite structures as opposed to the infinite arrays which occur in ionic crystals. Clearly there are many materials which are borderline.

As the non-transition elements give rise to most molecular compounds, experimental techniques which provide information about the structures of molecular species will be discussed in this section. Techniques which have been more widely applied to ionic materials will be discussed in connection with transition-metal chemistry in Chapter 6.

5.3.1. Electron diffraction

The technique is based on the fact that any electron in a beam of electrons exhibits wave properties, the wavelength being dependent on the excitation voltage. Thus electrons of about 5×10^4 volts ($\lambda \sim 5$ pm) can be diffracted by gas or vapour at low pressure. The atoms in the molecules scatter electrons in much the same way that they diffract X-rays (7.1.1) and although the molecules will have all possible orientations the motion of any one molecule is negligible during the time of its interaction with the electron beam. The diffraction pattern is recorded photographically as a series of diffuse concentric rings around the undeflected beam. The relative intensities of these rings are measured by means of a microdensitometer and the intensity due to coherent molecular scattering is obtained. From this information, interatomic distances (including all the non-bonding distances) can be computed, and accurate bond lengths and bond angles can be calculated.

The technique is ideally suited to small or highly symmetrical molecules, but as the number of atoms in the molecule increases assumptions have to be made to enable a structure to be worked out. Light atoms such as hydrogen cannot be located accurately in molecules containing much heavier ones, and thermal vibrations lead to uncertainty of position. Apart from the accumulation of accurate structural parameters for small molecules such as O_2, Br_2 and NH_3, electron diffraction has played an important role in determining the bonding in various compounds. Diborane was shown to contain two bridging hydrogen atoms (Fig. 11.1), and $(H_3Si)_3N$ was shown to be planar at N, giving strong evidence in support of $p\pi$—$d\pi$ bonding.

5.3.2. Infrared absorption spectra

Infrared radiation is of lower energy and longer wave length than ordinary light. Absorption in the range 2 μm to 200 μm, which is from 50 cm^{-1} to 10^4 cm^{-1} on the wave-number scale, is due to molecules absorbing quanta of energy and thereby increasing their rotational and vibrational energy.

The fundamental absorption band of gaseous HBr in the infrared (Fig. 5.6)

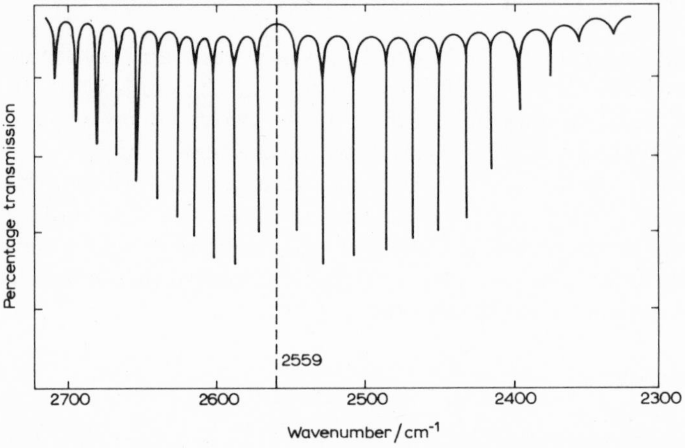

Fig. 5.6. Fundamental absorption band for HBr gas, showing rotational fine structure.

shows the many simultaneous changes brought about in the vibrational and rotational energy of that molecule. The fine structure of the spectrum is associated with changes in rotational energy.

Generally, for a molecule of moment of inertia I, the energy of rotation

$$E = \frac{h^2}{8\pi^2 I} \times J(J+1)$$

where J is a rotational quantum number 0, 1, 2, 3... Thus the energy increase between two lines in the fine structure represents

$$\Delta E = \frac{2h^2}{8\pi^2 I} = h^2/4\pi^2 I$$

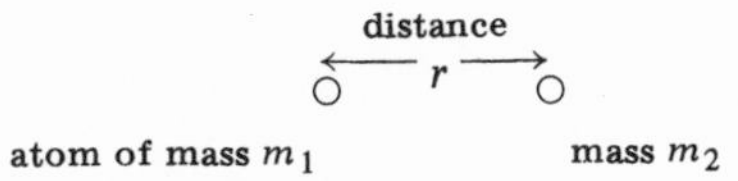

Fig. 5.7. Representation of heteronuclear diatomic molecule.

For a diatomic molecule comprising atoms of mass m_1 and m_2 separated by distance r (Fig. 5.7), the moment of inertia is the product of the reduced mass μ and the distance squared:

$$I = \mu r^2 \text{ where } \mu = \frac{m_1 m_2}{m_1 + m_2}$$

Thus if m_1, m_2 and I are known, the internuclear distance r can be found. For molecules with more atoms, such as NH_3, both bond angles and internuclear distances have to be considered and the calculation is more involved.

The middle of the absorption band for HBr, corresponding to 2559 cm^{-1}, represents the energy, ΔE, absorbed in lifting the vibrational energy of the H—Br bond from its lowest quantum level ($v = 0$) to the next ($v = 1$). This is the *fundamental frequency* of the bond. A quantity called the *force constant* of the bond, k, which is a measure of the resistance of the bond to stretching, is given by

$$k = \left\{\frac{2\pi(\Delta E)}{h}\right\}^2 \mu$$

A comparison of these force constants (Table 5.2) indicates that the HF bond resists stretching more than does the bond in HCl and the other hydrogen halides. Furthermore the π-bonded molecules NO and CO show greater rigidity than the single-bonded molecules.

The value of a stretching frequency depends on two factors:
(a) the masses of the atoms,
(b) the strength of the bond.
The frequency of the infrared absorption increases as the masses are reduced

TABLE 5.2

FUNDAMENTAL VIBRATIONAL ABSORPTIONS AND FORCE CONSTANTS FOR SOME DIATOMIC MOLECULES

Molecule	Fundamental absorption band/cm^{-1}	Force constant/N m^{-1}
HF	2907	970
HCl	2886	480
HBr	2559	410
HI	2230	320
NO	1876	1530
CO	2143	1840

and the bond strengths are increased:

>C—H stretch ~2900 cm^{-1}
>C—O— stretch ~1100 cm^{-1}
>C=O stretch ~1750 cm^{-1}

In molecules with three or more atoms, bond-bending vibrations also occur:

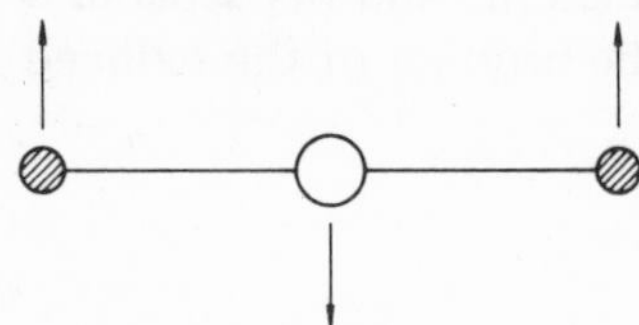

These are of lower frequency than the stretching vibrations.

For a molecule to be infrared-active the vibrational change must cause a change in dipole moment (5.3.7). This can be illustrated by reference to the CO_2 and H_2O molecules, both of which have symmetric and asymmetric stretching frequencies.

For the CO_2 molecule the former is without net change in dipole moment and is infrared inactive; there is a single peak near 2400 cm^{-1}.

(a) Symmetric stretching	(b) Asymmetric stretching
$\overleftarrow{\mathrm{O}}$—C—$\overrightarrow{\mathrm{O}}$	$\overleftarrow{\mathrm{O}}$—$\overrightarrow{\mathrm{C}}$—$\overleftarrow{\mathrm{O}}$
No net change in dipole moment	A net change in dipole moment
Infrared inactive	Infrared active

But for the angular H_2O molecule there is a net change in dipole moment, whether both hydrogen atoms move away from the oxygen atom, or one hydrogen atom moves towards it as the other moves away from it.

(a) Symmetric stretching (b) Asymmetric stretching

Accordingly, H_2O has a double peak in the stretching frequency absorption in the region of 3700 cm^{-1}.

Various groups present in molecules give rise to characteristic frequencies for the stretching and bending of bonds. These frequencies are altered only slightly by the presence of other groups and can therefore be used for diagnostic purposes. The numerous 'fingerprinting' applications of infrared spectroscopy lie outside the scope of this text but the way in which the shapes of simple molecules can be determined from their vibrational spectra will be outlined (5.3.4).

5.3.3. Raman spectroscopy

When monochromatic light is passed through a material, some of it is scattered. Most of this scattered light has the same frequency as the incident light and is known as Rayleigh scattering. A small fraction of the scattered light has a different frequency, because quanta are absorbed by molecules and, having lost some of their energy by inducing vibrational or rotational change in these molecules, are emitted as quanta of lower energy than the incident radiation. This is known as Raman scattering, and such spectra supply similar information to infrared absorption spectra. The major difference arises from their origin, in that for infrared activity the motion of the nuclei must give rise to a change in dipole moment whereas Raman activity results from a change in polarisability, i.e. the separation of positive and negative charge within the molecule. Furthermore it is easier to obtain data pertaining to the far infrared region (low frequencies) by means of Raman spectroscopy, as emission will occur close to the incident radiation which can itself be selected in a region convenient for measurements. In practice, the mercury arc lamps which were formerly used to provide a very intense source of monochromatic light (near u.v.) to examine liquids or concentrated colourless solutions have been largely superseded by laser beams. Intense, monochromatic sources of radiation are thus conveniently available to investigate solids and even coloured materials, and this development has been largely responsible for the recent upsurge in the use of this spectroscopic technique. In addition to the recording of the frequency of a Raman band another parameter can be measured; this is the degree of depolarisation, ρ, defined as the ratio of the intensity of the scattered light polarised perpendicularly to a plane, to the intensity of the scattered light polarised parallel to the plane.

$$\rho = \frac{I_{\perp}}{I_{\parallel}}$$

Totally symmetrical vibrations give rise to polarised Raman bands with a depolarisation ratio of less than 0.857. This was formerly measured by recording spectra with polaroid film placed between the light source and sample, first in one position, then a position perpendicular to the first one.

However, laser light is plane polarised anyway so all that is now required is an Arhens prism to rotate the plane of polarisation through 90°.

5.3.4. The deduction of structure by vibrational spectroscopy

By way of example consider how the shape of the SF_4 molecule could be determined by vibrational spectroscopy. The procedure, which is general and applicable to all such situations, is to determine the symmetry of the normal modes of vibration for particular shapes by means of the selection rules, then to compare this prediction with the fundamental vibrations actually recorded. Possible structures which will be considered are:

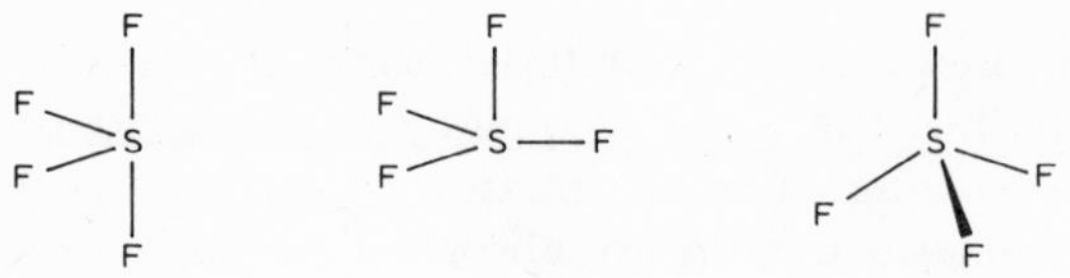

distorted tetrahedron C_{2v} | trigonal pyramid C_{3v} | tetrahedral T_d

To determine how the atoms of SF_4 behave in C_{2v} symmetry, for which the orientation in Cartesian co-ordinates and character table are as follows,

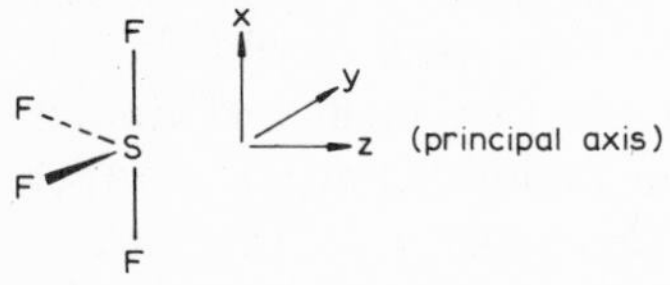

x, y, z (principal axis)

C_{2v}	E	C_2	$\sigma(xz)$	$\sigma(yz)$	
A_1	1	1	1	1	T_z α_{x^2} α_{y^2} α_{z^2}
A_2	1	1	−1	−1	R_z α_{xy}
B_1	1	−1	1	−1	T_x R_y α_{xz}
B_2	1	−1	−1	1	T_y R_x α_{yz}

carry out each symmetry operation of this point group on all the atoms, and in each case record how many atoms do not change position as a result of that operation. For example, performing the C_2 operation (rotation through 180°) will cause all F atoms to change position but S to remain the same, whereas reflection in a plane of symmetry will cause two F atoms to change while S and the other two F atoms remain unchanged.

C_{2v}	E	C_2	$\sigma(xz)$	$\sigma(yz)$	
Unshifted atoms	5	1	3	3	

The total number of degrees of freedom of a polyatomic molecule is equal to three times the number of atoms, so it is necessary to multiply this result by a factor to give the total representation. The factor for proper operations is given by the equation $C_n^k = 2\cos(2\pi k/n) + 1$, and for improper operations by the equation $S_n^k = 2\cos(2\pi k/n) - 1$ which reduce as follows: $E = 3$, $C_2 = -1$, $C_3 = 0$, $C_4 = 1$, $C_6 = 2$, $\sigma = 1$, $i = -3$, $S = -2$, $S_4 = -1$, $S_6 = 0$. Thus the total representation is

C_{2v}	E	C_2	$\sigma(xz)$	$\sigma(yz)$
	5	1	3	3
X	3	−1	1	1
Γ_{3N}	15	−1	3	3

and represents some unique combination of the irreducible representations A_1, A_2, B_1, B_2 which make up the C_{2v} point group.

In order to determine how many of each irreducible representation comprise the total representation Γ_{3N}, we can use the equation:

$$n_\Gamma = \frac{1}{g} \sum g_R \gamma_\Gamma^R \gamma_{DP}^R$$

where n_Γ is the number of each individual irreducible representation, g is the group order, g_R is the coefficient of the symmetry operation, γ_Γ^R is the character of the species in the character table; γ_{DP}^R is the direct product in total representation. Using this equation, the values are as follows:

$$n_{A_1} = \tfrac{1}{4}(1 \cdot 1 \cdot 15 + 1 \cdot 1 \cdot -1 + 1 \cdot 1 \cdot 3 + 1 \cdot 1 \cdot 3) = 5$$
$$n_{A_2} = \tfrac{1}{4}(1 \cdot 1 \cdot 15 + 1 \cdot 1 \cdot -1 + 1 \cdot -1 \cdot 3 + 1 \cdot -1 \cdot 3) = 2$$
$$n_{B_1} = \tfrac{1}{4}(1 \cdot 1 \cdot 15 + 1 \cdot -1 \cdot -1 + 1 \cdot 1 \cdot 3 + 1 \cdot -1 \cdot 3) = 4$$
$$n_{B_2} = \tfrac{1}{4}(1 \cdot 1 \cdot 15 + 1 \cdot -1 \cdot -1 + 1 \cdot -1 \cdot 3 + 1 \cdot 1 \cdot 3) = 4$$

Thus the total representation is made up of $5A_1 + 2A_2 + 4B_1 + 4B_2$, which can be checked to see if it gives the direct product.

C_{2v}	E	C_2	$\sigma(xz)$	$\sigma(yz)$
$5A_1$	5	5	5	5
$2A_2$	2	2	−2	−2
$4B_1$	4	−4	4	−4
$4B_2$	4	−4	−4	4
Γ_{3N}	15	−1	3	3

Any discrepancy in addition or any non-integer value indicates an error in the working out.

Selection rules predict that non-linear molecules give rise to $3n-6$ normal modes of vibration ($3n-5$ for linear molecules) where n is the number of atoms. The six (or 5 for linear) modes to be deducted are due to translational and rotational modes. The symmetry of these modes are indicated to the right of the character table and for C_{2v} are $A_1 + B_1 + B_2$ for the three translational, and $A_2 + B_1 + B_2$ for the three rotational modes. Thus the $3n-6$ normal modes of vibration for SF_4 in the C_{2v} point group are

$$\begin{array}{rl} & 5A_1 + 2A_2 + 4B_1 + 4B_2 \\ - & A_1 + A_2 + 2B_1 + 2B_2 \\ \hline & 4A_1 + A_2 + 2B_1 + 2B_2 \qquad (3n-6=9) \end{array}$$

Using the terms to the right of the character table, the translational terms also indicate which species involve a change in dipole moment, and hence infrared activity, while the term α refers to species which involve a change in polarisability necessary for Raman activity. (The symbol α is often omitted in such tables.) Thus $4A_1 + 2B_1 + 2B_2$ will give 8 infrared active modes, while all modes $4A_1 + A_2 + 2B_1 + 2B_2$, totalling 9, are Raman active, of which the $4A_1$, being totally symmetric, will be polarised.

This procedure can be repeated for the other geometries. For C_{3v} the orientation and character table are:

z

F

F—S<F F

C_{3v}	E	$2C_3$	$3\sigma_v$	
A_1	1	1	1	T_z $\alpha_{x^2} + \alpha_{y^2}$ α_{z^2}
A_2	1	1	−1	R_z
E	2	−1	0	T_z T_y R_x R_y α_{xz} α_{yz} $\alpha_{x^2-y^2}$ α_{xy}

The number of unshifted atoms when an operation of each type is carried out is:

C_{3v}	E	C_3	σ_v
Unshifted atoms	5	2	3
×	3	0	1
Γ_{3N}	15	0	3

This total representation is made up of

$$n_{A_1} = \tfrac{1}{6}(1\cdot1\cdot15 + 2\cdot1\cdot0 + 3\cdot1\cdot3) = 4$$
$$n_{A_2} = \tfrac{1}{6}(1\cdot1\cdot15 + 2\cdot1\cdot0 + 3\cdot{-1}\cdot3) = 1$$
$$n_E = \tfrac{1}{6}(1\cdot2\cdot15 + 2\cdot{-1}\cdot0 + 3\cdot0\cdot3) = 5$$

$4A_1 + A_2 + 5E = 15$ ($3n$ modes) allowing for the double degeneracy of the E species. Deduct the three translational ($A_1 + E$) and three rotational ($A_2 + E$)

modes which leaves $3A_1 + 3E$ i.e. 6 infrared modes and 6 Raman modes of which the $3A_1$ being totally symmetric will be polarised.

Finally for the tetrahedral model it is often easier to visualise some operation of the T_d point group by orienting the molecule in a cube.

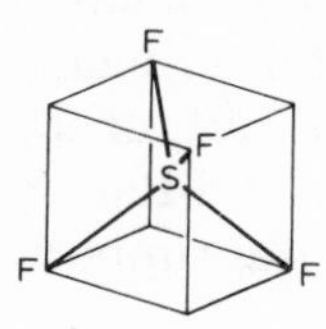

T_d	E	$6S_4$	$3C_2$	$8C_3$	$6\sigma_d$	
A_1	1	1	1	1	1	$\alpha_{x^2} + \alpha_{y^2} + \alpha_{z^2}$
A_2	1	−1	1	1	−1	
E	2	0	2	−1	0	$\alpha_{x^2} + \alpha_{y^2} - \alpha_{2z^2}$ $\alpha_{x^2} - \alpha_{y^2}$
T_1	3	1	−1	0	−1	$R_{x,y,z}$
T_2	3	−1	−1	0	1	$T_{x,y,z}$ α_{xy} α_{yz} α_{xz}

Once again determine the number of unshifted atoms when an operation of each type from the T_d point group is carried out:

T_d	E	S_4	C_2	C_3	σ_d
Unshifted atoms	5	1	1	2	3
X	3	−1	−1	0	1
Γ_{3N}	15	−1	−1	0	3

This total representation is made up of:

$$n_{A_1} = \tfrac{1}{24}(1 \cdot 1 \cdot 15 + 6 \cdot 1 \cdot -1 + 3 \cdot 1 \cdot -1 + 8 \cdot 1 \cdot 0 + 1 \cdot 6 \cdot 3) = 1$$
$$n_{A_2} = \tfrac{1}{24}(1 \cdot 1 \cdot 15 + 6 \cdot -1 \cdot -1 + 3 \cdot 1 \cdot -1 + 8 \cdot 1 \cdot 0 + 6 \cdot -1 \cdot 3) = 0$$
$$n_E = \tfrac{1}{24}(1 \cdot 2 \cdot 15 + 6 \cdot 0 \cdot -1 + 3 \cdot 2 \cdot -1 + 8 \cdot -1 \cdot 0 + 6 \cdot 0 \cdot 3) = 1$$
$$n_{T_1} = \tfrac{1}{24}(1 \cdot 3 \cdot 15 + 6 \cdot 1 \cdot -1 + 3 \cdot -1 \cdot -1 + 8 \cdot 0 \cdot 0 + 6 \cdot -1 \cdot 3) = 1$$
$$n_{T_2} = \tfrac{1}{24}(1 \cdot 3 \cdot 15 + 6 \cdot -1 \cdot -1 + 3 \cdot -1 \cdot -1 + 8 \cdot 0 \cdot 0 + 6 \cdot 3 \cdot 1) = 3$$

$A_1 + E + T_1 + 3T_2 = 15$ ($3n$ modes) allowing for the double and triple degeneracy of the E and T species respectively. Deducting the three translational (T_2) and three rotational (T_1) modes leaves $A_1 + E + 2T_2$ of which the $2T_2$ are infrared active and all are Raman active with A_1 polarised.

Thus summarising and comparing with experimental data:

SF_4	Calculated			Found
	C_{2v}	C_{3v}	T_d	
I.r. active	8	6	2	5 (or 7)
Raman active	9	6	4	5 (or 8)
Polarised	4	3	1	1

These results are typical of such calculations and highlight the major problem, that of band assignment. What can safely be inferred from the above data is that SF_4 is not tetrahedral, as more fundamentals have been found than T_d can account for. The reasons why insufficient bands have been assigned could be due to several reasons such as accidental degeneracy (two bands occurring at the same frequency), poor resolution, low intensity or simply the inability to identify fundamentals from overtones or combination bands. In such situations where further evidence is necessary for an unambiguous assignment, assistance from a spectroscopist skilled in the art, or the utilisation of some other technique is required. For SF_4, the i.r. and Raman spectra were interpreted on the basis of C_{2v} symmetry and this was supported by its ^{19}F n.m.r. spectrum (5.3.6).

5.3.5. Microwave spectra

Very accurate measurements of the moments of inertia, and hence the bond angles and bond lengths of gaseous molecules, can be obtained by observing absorption in the microwave (radar wave) region of the spectrum, around 1 cm in wavelength. For this purpose, the gas is put in a long tube through which microwaves pass to a receiver which records the absorption at particular frequencies. These frequencies correspond to transitions between possible rotational states of the molecule. The equation

$$h\nu = \Delta E = \frac{h^2}{4\pi^2 I}$$

enables the moments of inertia for all the possible rotational modes of the molecule to be determined, and hence, in suitable cases, the bond lengths and bond angles.

5.3.6. Nuclear magnetic resonance

First-order spectra will occur if the chemical shifts of the different nuclei are relatively large compared with the spin—spin coupling constants between those nuclei. This consideration is not so important in inorganic chemistry, as chemical shifts for elements other than hydrogen are also much larger. The situation can also be influenced by increasing the field strength, as chemical shifts are field dependent whereas spin—spin coupling constants are field independent.

First-order spectra usually show the following features:

a. A separate line or group of lines is observed for each distinct magnetic nucleus.
b. Multiplets are well separated, with intensity proportional to the number of each type of magnetic nucleus.
c. Spacing within a multiplet gives the spin—spin coupling constant.
d. The number of lines and the relative intensities within each multiplet are related to the number and type of equivalent nuclei.

e. Resonances are not split by spin interactions between chemically equivalent nuclei.

Apart from the recording of nuclear magnetic parameters to enlarge the characterisation of a compound, much of the above data can indicate structural information. Theoretical spectra can be predicted for various shapes, and compared with the recorded spectrum. Consider again possible structures for SF_4:

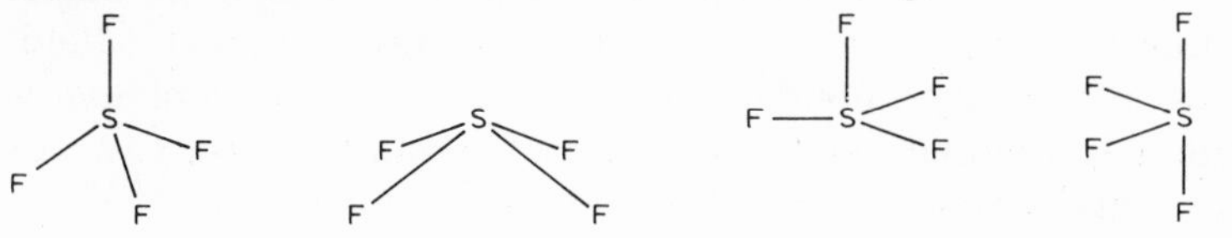

tetrahedral	square based pyramid	trigonal pyramid	distorted tetrahedron
T_d	C_{4v}	C_{3v}	C_{2v}

The ^{19}F spectrum for both T_d and C_{4v} would be a single line. C_{2v} having two sets of two equivalent fluorines, would give rise to two sets of 1:2:1 triplets. C_{3v}, having one fluorine in a different environment from the other three, would give rise to a 1:3:3:1 quartet and a 1:1 doublet. At room temperature the recorded spectrum shows a single band suggestive of T_d or C_{4v} but at 175 K two pairs of triplets are observed, indicative of C_{2v} symmetry. The single band at room temperature arises as a result of rapid exchange of fluorines owing to thermal excitation.

Several elements possess more than one isotope with spins of $I = \frac{1}{2}$. Each isotope should be regarded separately, including any with spin zero which are present in the compound. The relative intensities due to each isotope will be proportional to their natural abundance, but occasionally materials enriched with a particular isotope might be encountered.

Mention should also be made of isotopes with a spin greater than $\frac{1}{2}$. Spin—spin interactions give rise to $2I + 1$ peaks. Thus in $^{14}NH_3$ (^{14}N, $I = 1$), the ^{14}N n.m.r. spectrum shows a quartet of relative intensity 1:3:3:1 because the ^{14}N signal is split by three equivalent hydrogen atoms while the ^{1}H n.m.r. spectrum shows two peaks of equal intensity arising from the splitting by ^{14}N of the equivalent hydrogen resonance.

5.3.7. Electric dipole moments

5.3.7.1. Electric permittivity

The force between two electric charges Q_1 and Q_2 separated by a length l in a medium of permittivity ϵ is

$$F = \frac{Q_1 Q_2}{4\pi\epsilon l^2} \qquad \text{(cf. 1.4)}$$

The ratio $\epsilon/\epsilon_0 = \epsilon_r$, the *relative permittivity* of the medium, is a number greater than one. The permittivity of a vacuum, $\epsilon_0 = 8.854 \times 10^{-12}$ F m^{-1}.

The capacitance of a condenser is proportional to ϵ_r for the material between the plates.

5.3.7.2. Polarisation of molecules in an electric field

Electrically, a molecule can be considered as a collection of negative charges located on electrons and positive charges located on atomic nuclei. In a homopolar diatomic molecule like Br_2 the centres of positive and negative charge are both situated at the mid-point of the line joining the nuclei unless the molecule is placed in an electric field. However, in such a field, between the plates of a charged condenser, the centre of positive charge is drawn towards the negative plate and the centre of negative charge towards the positive plate; polarisation is induced in the molecule. The *induced polarisation*, P_D, is independent of temperature.

In a molecule like HBr, however, the centre of negative charge lies nearer the bromine nucleus than does the centre of positive charge even in the absence of a field. The molecule has a *permanent dipole moment.* Such molecules, in an electric field between condenser plates, exhibit not only induced polarisation but also *orientation polarisation*, P_μ, because they tend to align with their bromine atoms pointing towards the positive plate. The orientation polarisation decreases with temperature because the thermal agitation destroys the alignment. For a gas of molecular mass M and density ρ in which the molecules are polar, P_M, the *molar polarisation*, is the sum of P_D and P_μ,

$$P_M = P_D + P_\mu$$

Substituting:

$$\left(\frac{\epsilon_r - 1}{\epsilon_r + 2}\right)\frac{M}{\rho} = \frac{N_A\alpha}{3\epsilon_0} + \frac{N_A p_e^2}{9\epsilon_0 kT}$$

α is the *polarisability* of the molecule and p_e is its *dipole moment.*

If ϵ_r and ρ are measured for a particular gas at a series of temperatures and the values of $\left(\frac{\epsilon_r - 1}{\epsilon_r + 2}\right)\frac{M}{\rho}$ are plotted against the reciprocal of the thermodynamic temperature, a straight line is obtained for which the gradient is $N_A p_e^2/9\epsilon_0 k$ and the intercept on the axis $K/T = 0$ is $N_A\alpha/3\epsilon_0$. For HBr the slope of the line is 3.85×10^{-3} m^3 K mol^{-1}. Thus p_e^2 for HBr is

$$\frac{3.85 \times 10^{-3}\,\text{m}^3\,\text{K mol}^{-1} \times 9 \times 8.854 \times 10^{-12}\,\text{F m}^{-1} \times 1.38 \times 10^{-23}\,\text{J K}^{-1}}{6.02 \times 10^{23}\,\text{mol}^{-1}}$$

$$= 7.04 \times 10^{-60}\,\text{C}^2\,\text{m}^2$$

and $p_e = 2.65 \times 10^{-30}$ C m

A non-SI unit of dipole moment still in common use is the debye (D), 3.33×10^{-30} C m. Thus the dipole moment of HBr is 0.79 D.

Some examples from Table 5.3 illustrate the kind of information which

TABLE 5.3

DIPOLE MOMENTS OF SOME MOLECULES

Molecule	p_e/D	$10^{30} p_e$/C m
HF	1.91	6.35
HCl	1.04	3.45
HBr	0.79	2.65
HI	0.42	1.39
H_2O	1.85	6.17
H_2S	0.94	3.12
NH_3	1.47	4.88
PH_3	0.55	1.83
BCl_3	0	0
PF_5	0	0
SO_2	1.63	5.42
CO_2	0	0

dipole moment measurements provide:

(i) In HCl the internuclear distance is 127.5 pm. As the charge on a proton is 1.60×10^{-19} C the dipole moment of a completely ionised arrangement

$$\overset{H^+}{\bullet} \qquad\qquad \overset{Cl^-}{\bullet}$$
$$\longleftarrow 127.5\ \text{pm} \longrightarrow$$

would be $1.60 \times 10^{-19}\ \text{C} \times 127.5 \times 10^{-12}\ \text{m} = 20.2 \times 10^{-30}$ C m. Since the actual dipole moment is only about one-sixth of that, the H—Cl bond can be said to have about one-sixth ionic character. Similarly the figure for HI shows the H—I bond to have only 5% ionic character.

(ii) The zero dipole moment of CO_2 must arise from the polarities of the two bonds cancelling one another exactly. This implies a linear molecule O=C=O. By the same token, H_2O and SO_2 must be angular as they both have considerable dipole moments. BCl_3 has zero dipole moment because the molecule is planar with Cl—B—Cl angle of 120°; whereas NH_3, with its appreciable moment, has a pyramidal molecule.

For larger molecules, permittivity measurements are made on dilute solutions in non-polar solvents. One important application is to distinguish between isomers. For example, consider the compound $PtCl_2 \cdot 2Et_3As$ which has two forms: one is white, *m.p.* 415 K, sparingly soluble in benzene; the other is yellow, *m.p.* 394 K, very soluble in benzene. The white isomer has a dipole moment of about 10 D and could be either tetrahedral or *cis*-planar. The yellow one is practically without dipole moment and must be *trans*-planar, thus:

```
          Cl
          |
Et3As—Pt—AsEt3
          |
          Cl
```

Further Reading

P.J. Wheatley, The determination of molecular structure, 2nd edition, Oxford, 1968.

K. Nakamoto, Infrared spectra of inorganic and coordination compounds, 2nd edition, Wiley, New York, 1970.

S.D. Ross, Inorganic infrared and Raman spectra, McGraw-Hill, London, 1972.

P.J. Wheatley, The chemical consequences of nuclear spin, North-Holland, Amsterdam, 1970.

G.E. Bacon, Neutron diffraction, 2nd edition, Oxford University Press, London, 1962.

A. Abragam, The principles of nuclear magnetism, Oxford University Press, London, 1961.

C.H. Townes and A.L. Schawlow, Microwave spectroscopy, McGraw-Hill, New York, 1955.

J.W. Smith, Electric dipole moments, Butterworth, London, 1955.

R.S. Drago, Physical methods in inorganic chemistry, Reinhold, New York, 1965.

G. Barrow, Introduction to molecular spectroscopy, McGraw-Hill, New York, 1962.

A. Anderson (Ed.), The Raman effect, Marcel Dekker, New York, 1972.

G.J. Moody and J.D.R. Thomas, Dipole moments in inorganic chemistry, Arnold, London, 1971.

B. Schrader, Chemical applications of Raman spectroscopy, Angew. Chem. Internat. Ed., 12 (1973) 884.

R.S. Tobias, Raman spectroscopy in inorganic chemistry, J. Chem. Ed., 44 (1967) 1 and 70.

J.R.L. Swain, Dipole moments in inorganic chemistry, Ed. Chem., 8 (1971) 105.

H.E. Hallam, Infrared and Raman spectra of inorganic compounds, RIC Reviews, 1 (1968) 39.

I.R. Beattie, Helium-neon laser Raman spectroscopy, Chem. Brit., 3 (1967) 347.

S.K. Freeman, Applications of laser Raman spectroscopy, Wiley, New York, 1973.

Bonding in Transition-metal Complexes

6

6.1. Electron Configurations

The non-transitional elements are characterised by electron configurations of the type

[(filled shells)ns^y, np^z]

in which the total number of valence electrons, $y + z$, varies from 1 to 8 and hybridisation involves ns and np orbitals and, less commonly, nd orbitals. The metals with $y = 1$ or 2, $z = 0$, have small electronegativities and readily form ionic bonds. As z increases electrons are lost less easily, ionic character diminishes, and the atom holds near itself up to 4 electron pairs; this number may be increased to 6 by d hybridisation and acceptance of electrons.

In the transition elements the electron configuration is

[(filled shells)$(n-1)d^x$, ns^y]

in which $y = 1$ or 2 and x goes from 1 to 10. They resemble the Group I and Group II metals in having low electronegativity; but, owing to the incomplete d shell, the number of electrons available for bonding is potentially much larger. The $(n-1)$d, ns, np and even the nd orbitals may lie so close together that promotion and hybridisation occur freely. Often, however, all the d orbitals are not free to participate in hybridisation; they may describe a stable sub-shell, not affected by bonding. The valence properties of a transition metal are thus controlled to some extent by the stability of incomplete d shells. This in turn is determined largely by the strength and symmetry of the electric field imposed by the attached atoms, ions or groups, known collectively as ligands.

The nature of the ligands also affects the magnetic properties of transition-metal complexes. For example the magnetic moment of $(NH_4)_2[Mn(H_2O)_6](SO_4)_2$ is 5.88 μ_B but that of $K_4[Mn(CN)_6] \cdot 3H_2O$ is only 2.18 μ_B. Clearly the dipositive manganese has five unpaired electrons in the former complex and only one unpaired electron in the latter. The colours of complexes are also affected by the ligands which surround the metal atom. Thus the aquated nickel(II) ion is green, whereas the ion $Ni(en)_3^{2+}$ is lilac. A satisfactory theory of metal—ligand bonding must explain these three phenomena of variable molecular symmetry, variable colour and variable paramagnetism encountered in transition-metal chemistry.

A molecular orbital treatment of metal—ligand bonding has been developed (6.6) which is of particular value in giving a picture of what is happening in highly covalent complexes such as carbonyls; but although it is the most fundamental approach, it is not always easy to apply. For many complexes, however, a sufficiently accurate theoretical explanation of the observed phenomena may be obtained by allocating electrons to atomic orbitals. This is the basis of the electrostatic *crystal field theory* and its modification the *ligand field theory*.

6.2. Ligand Field Theory

In transition-metal ions all the orbitals except the highest d orbitals are either completely full or quite empty. A metal cation, an ionic crystal or a mononuclear complex is symmetrically surrounded, as nearest neighbours, either by negative ions or dipoles both of which direct electrons towards it. The five d orbitals described in 3.2.2, although distinct, all have the same energy in a free atom or ion — they are degenerate. But the presence around the cation of negative charges due to ligands reduces this degeneracy, and it is clear that the symmetry properties of the complex can be used to determine how the d orbitals behave.

6.2.1. Octahedral complexes

The most common geometrical arrangement in complexes is octahedral. By observing the effect of performing the symmetry operations of point group O_h on each orbital and recording +1 for a symmetrical result, −1 for an antisymmetrical result, and 0 for something different, it is possible to determine the characters appropriate to each symmetry operation. Particular note must be made of the signs of the lobes when considering d orbitals. When the operations of point group O_h are performed on the d_{xy} orbital alone, several operations leave d_{xy} in some other orientation, as shown in Fig. 6.1 below:

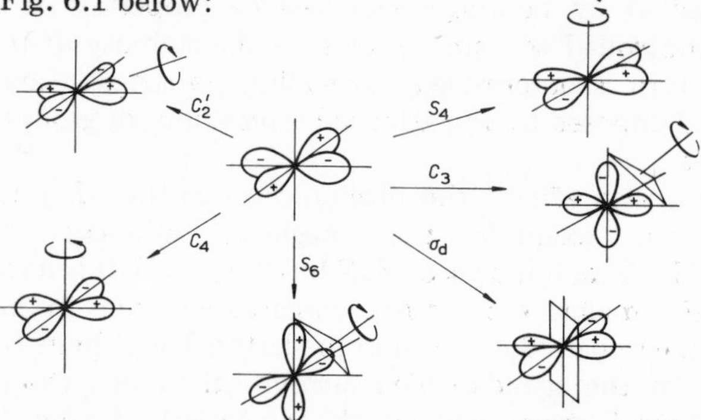

Fig. 6.1. Operations of point group O_h performed on d_{xy} orbital.

This results in the representation:

O_h	E	C_3	C_2	C_4	C_4^2	i	S_4	S_6	σ_h	σ_d
d_{xy}	1	0	−1	−1	1	1	−1	0	1	−1

which does not coincide with an irreducible representation of the point group O_h.

It is clear that with some operations such as C_3, d_{xy} becomes d_{xz} or d_{yz}. Thus it is necessary to consider these three orbitals together and to sum the three combinations, for which we obtain the T_{2g} irreducible representation:

O_h	E	C_3	C_2	C_4	C_4^2	i	S_4	S_6	σ_h	σ_d
d_{xy} d_{xz} d_{yz}	3	0	1	−1	−1	3	−1	0	−1	1

A further set of characters can be obtained for $d_{x^2-y^2}$ and d_{z^2}. In order to recognise the relation between these two orbitals it is necessary to regard d_{z^2} as $d_{2z^2-x^2-y^2}$, that is, as a linear combination of $d_{z^2-x^2}$ and $d_{z^2-y^2}$. For example the C_4 rotation, involving a 90° rotation about the z axis, leaves the d_{z^2} orbital unchanged (+1) but changes the signs of $d_{x^2-y^2}$ (−1), hence the character is zero. Likewise the horizontal plane σ_h leaves both orbitals appearing the same, giving a character of 2. In the C_3 operation, $d_{x^2-y^2}$ becomes $d_{z^2-x^2}$, in which half of the orbital retains its original direction (the x portion) but with opposite sign ($-\frac{1}{2}$). Likewise $d_{z^2-x^2}$ and $d_{z^2-y^2}$ become $d_{y^2-z^2}$ and $d_{y^2-x^2}$, each retaining half of its original directional properties also of opposite sign, which will amount to $-\frac{1}{2}$ for the d_{z^2} contribution and giving by addition a character of −1 (Fig. 6.2).

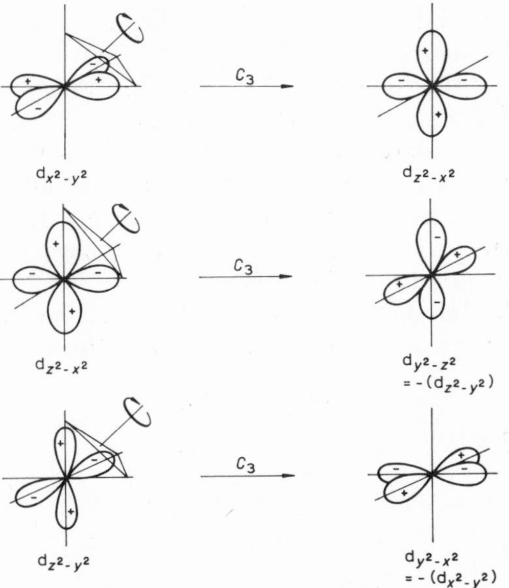

Fig. 6.2. C_3 operations on $d_{x^2-y^2}$, $d_{z^2-x^2}$ and $d_{z^2-y^2}$ orbitals.

Thus the total representation for d_{z^2} and $d_{x^2-y^2}$ is:

O_h	E	C_3	C_2	C_4	C_4^2	i	S_4	S_6	σ_h	σ_d
$d_{x^2-y^2}$, d_{z^2}	2	−1	0	0	2	2	0	−1	2	0

giving the irreducible representation E_g.

These results are also indicated at the right hand side of the character table opposite the symmetry species, and as similar calculations have been detailed for all point groups it is not necessary to work through this derivation on each occasion.

Thus the five degenerate d orbitals of the free ion are split into two groups, the high energy e_g and lower energy t_{2g} (Fig. 6.3). (Lower case letters

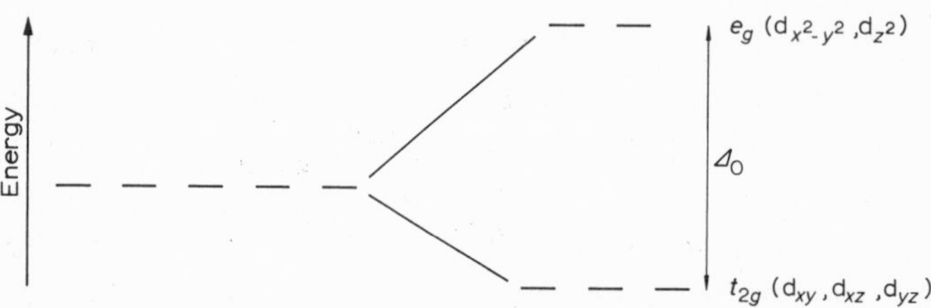

Fig. 6.3. Splitting of d orbitals in an octahedral ligand field.

e and t are used when referring to energy states, whereas upper case E and T are reserved for symmetry species). The energy difference Δ_0, the *ligand field splitting* in the octahedral field, is in the range 100—400 kJ mol^{-1} depending on the metal ion and the ligands (1 kJ mol^{-1} is equivalent to 83.7 cm^{-1}, in which units Δ_0 is often measured (6.2.1.1)).

In its earliest form the crystal field theory was applied by Bethe in the following way to explain the splitting of the energies of the d orbitals. Imagine the ligands on the x-, y- and z-axes to act as points of negative charge. These negative charges will interact more strongly with orbitals lying along the axes than with the orbitals which lie at 45° to them. Thus, in the octahedral field, the energies of the e_g orbitals will be higher than those of the t_{2g}. This attractively simple picture is unsatisfactory because even the smallest ligands like F^- ions do not remotely resemble point negative charges.

A realistic physical picture of the interaction between ligand and metal must be based on the recognition that the atomic orbitals necessarily become

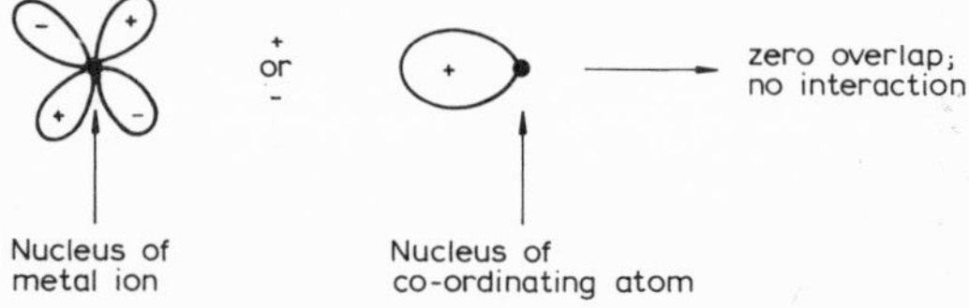

Fig. 6.4. Zero overlap of a t_{2g} orbital with a σ orbital on one of the co-ordinate axes.

mixed to form molecular orbitals when the nuclei approach to within a few hundred picometres of one another. It is immediately apparent that there can be no net overlap between a t_{2g} orbital on the metal and a σ orbital of a ligand lying on one of the co-ordinate axes (Fig. 6.4).

However the σ orbital of the ligand can combine with the e_g orbital to give a pair of σ molecular orbitals, one bonding and the other antibonding (4.1.1) (Fig. 6.5).

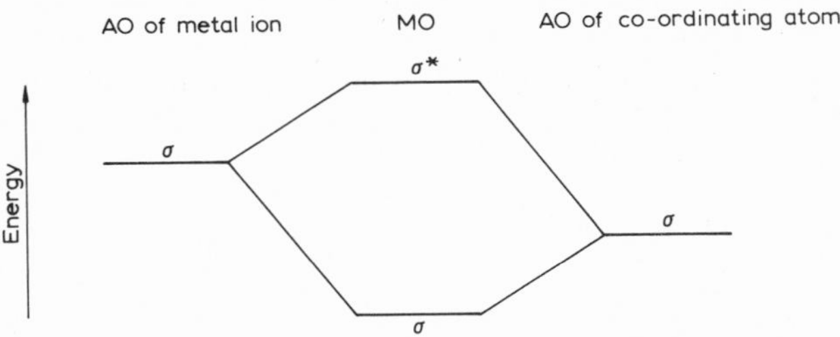

Fig. 6.5. Combination of e_g orbital of metal with sigma orbital of ligand.

Since there is no net interaction between the t_{2g} orbital lying nearest the ligand and the σ orbital of the co-ordinating atom, the complete energy level diagram which results from the combination of atomic orbitals is shown in Fig. 6.6.

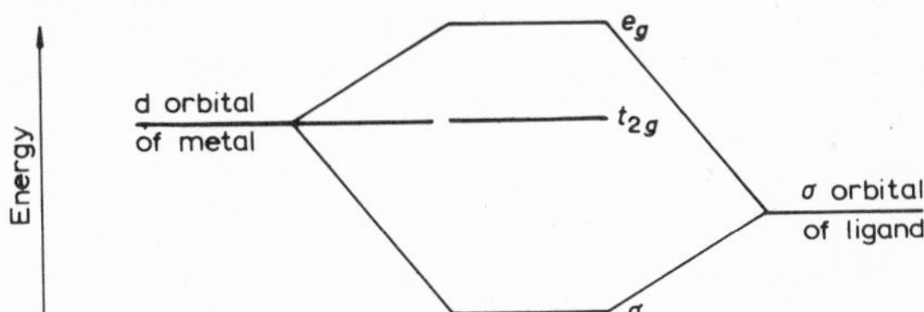

Fig. 6.6. Energy diagrams for combination of e_g and t_{2g} orbitals with σ orbital of ligand.

Although it is important to recognise that the point-charge model of Bethe's crystal field theory is quite without physical reality, it is possible to use the idea of coulombic interactions to give results which are qualitatively correct. The term *ligand field theory* will be used here in this sense, though it is used by some authors to describe a *quantitative* adjustment of the crystal field theory using parameters derived from considerations of the spin—orbital coupling and inter-electronic repulsion.

6.2.1.1. The measurement of Δ_0

The principles involved in the measurement of the ligand-field splitting are most simply explained when there is one d electron. This occurs in the $Ti(H_2O)_6{}^{3+}$ ion the purple colour of which is due to an absorption of light energy to raise the electron, normally located in a t_{2g} orbital, to an e_g orbital (Fig. 6.7).

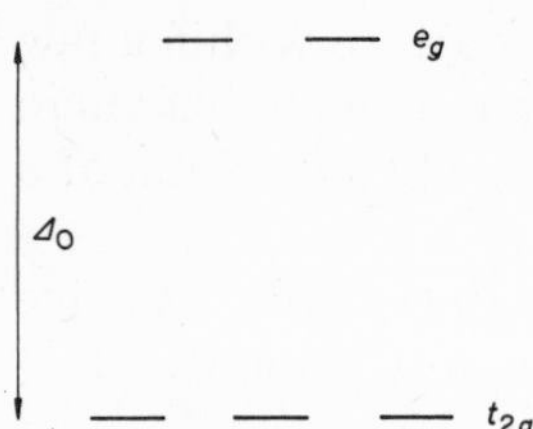

Fig. 6.7. Promotion of electron from t_{2g} orbital to e_g orbital.

The absorption band for the $Ti(H_2O)_6^{3+}$ ion in aqueous solution has the form shown in Fig. 6.8.

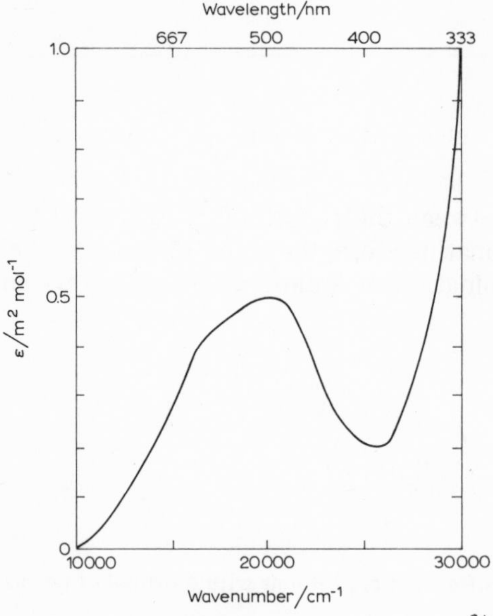

Fig. 6.8. Absorption spectrum of $Ti(H_2O)_6^{3+}$ ion.

The band with a maximum absorption at 490 nm (20,400 cm^{-1}) represents the energy absorbed in promoting the d electron from a t_{2g} orbital to an e_g orbital. [The strong absorption in the near ultraviolet is probably due to a charge-transfer process (6.6.3)]. The absorption at 490 nm is not a thin line but a band, because electronic excitation is accompanied by vibrational excitations spread over several thousand wave numbers. Octahedral complexes of Ti^{III} formed by ligands other than the water molecule show maximum absorption at different wavelengths. The size of the energy difference Δ_0 is very simply related, in this one-electron case, to the wave-number of the absorption maximum. Transition from t_{2g} to e_g in a ligand field of strength Δ_0 gives rise to an absorption maximum at the frequency $\nu = \Delta_0/h$.

6.2.1.2. The spectrochemical series

The strength of the field depends on the intensity of the electron density placed by the ligands into the regions covered by the different d orbitals. It will be large when the lone pairs offered by the ligands occupy well directed orbitals, but smaller for those ligands, such as halide ions, without strongly directed lone-pair electrons. For a given cation, the extent of the splitting increases with ligands along the sequence:

$I^- < Br^- < Cl^- < F^- < OH^- < H_2O < NH_3 < CN^-$

This is a representation of the *spectrochemical series* containing some of the commonest ligands. The position of an absorption band of a compound shifts towards the blue (shorter wave length) as one ligand is replaced by a successor which exerts a stronger ligand field.

With more than one d electron, the factors governing the positions of absorption maxima are more complicated (6.4.1) but, nevertheless, the ligand-field strengths can usually be determined from the spectra. The spectrochemical series is almost independent of the choice of central ion, although the precise value of Δ_0 does, of course, depend on this choice.

6.2.1.3. Other factors affecting ligand-field strengths

Three empirical rules on d-orbital splitting (indicating ligand-field strength) emerge:

(i) For a given ligand and a given oxidation state of the cation, it has a small range for metals in the same period. For $M(H_2O)_6{}^{2+}$ in Period 4, Δ_0 varies from 7,800 cm^{-1} (90 kJ mol^{-1}) for Mn^{II} to 11,000 cm^{-1} (131 kJ mol^{-1}) for Cr^{II}.

(ii) For a given ligand, it is larger for a higher oxidation state than for a lower oxidation state of the same metal.

Δ_0 for $Fe(H_2O)_6{}^{2+}$ is 10,400 cm^{-1} (124 kJ mol^{-1})
Δ_0 for $Fe(H_2O)_6{}^{3+}$ is 13,700 cm^{-1} (164 kJ mol^{-1})

(iii) For a given ligand and a given oxidation state, it is about 30% larger for metals in Period 5 than for metals in Period 4. And for metals in Period 6, Δ_0 is about 30% larger again.

6.3. High-spin and Low-spin Complexes

6.3.1. Octahedral complexes

In an octahedral complex of Ti^{III}, the single d electron occupies a t_{2g} (low energy) orbital in the unexcited ion. In a similar Cr^{III} complex, the three d electrons, their spins orientated parallel, occupy the three t_{2g} orbitals, and the system is stable for three reasons:

(i) the electrons occupy the d orbitals of lowest energy;

(ii) the electrons are in different d orbitals and thus the coulombic repulsion between them is small;

(iii) the electrons have parallel spins and thus their quantum mechanical exchange energy is low.

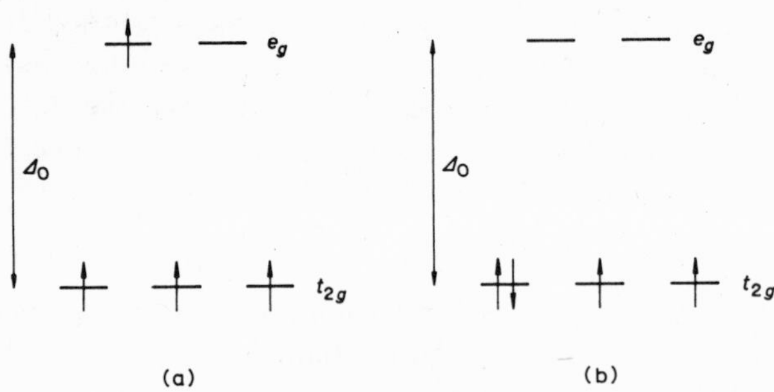

Fig. 6.9. Two possible distributions of four d electrons in an octahedral field.

In a d^4 ion such as Mn^{3+} there are two possible orientations (Fig. 6.9).

In (a) the coulombic repulsion is a minimum, and the existence of six pairs of electrons with parallel spin further stabilises the ion. But there is one electron in a high-energy e_g orbital.

In (b) the coulombic repulsion is greater because two electrons occupy the same orbital, and there are only three pairs with parallel spins. But the high-energy e_g orbital is unoccupied.

Whether in a Mn^{III} complex the electrons are arranged as in (a) or as in (b) depends on the value of Δ_0. In the strong ligand field exerted by six CN^- ions, the energy drop occasioned when the electron falls from an e_g to a t_{2g} orbital more than compensates for the increase in coulombic energy and exchange energy. Hence, the d orbitals in $Mn(CN)_6{}^{3-}$ are occupied as in (b), the magnetic moment is that due to two unpaired spins only, and the complex is a *low-spin complex*. The ligand acetylacetone, however, exerts a weaker field, Δ_0 is small, and the complex is a *high-spin complex* with a magnetic moment due to four unpaired spins as shown in (a). The energy drop if the e_g electron fell into a t_{2g} orbital would not be large enough to compensate for the increase in coulombic and exchange energy.

In a d^5 ion such as Fe^{3+} a strong octahedral ligand field will cause both e_g electrons to fall to t_{2g} orbitals (Fig. 6.10). The orbital energy released is $2\,\Delta_0$.

The number of parallel pairs falls from 10 to 4, a difference of 6; accordingly there is a considerable loss of exchange stabilisation and the field has to be a strong one to induce spin-pairing. Cyanide ions exert a field

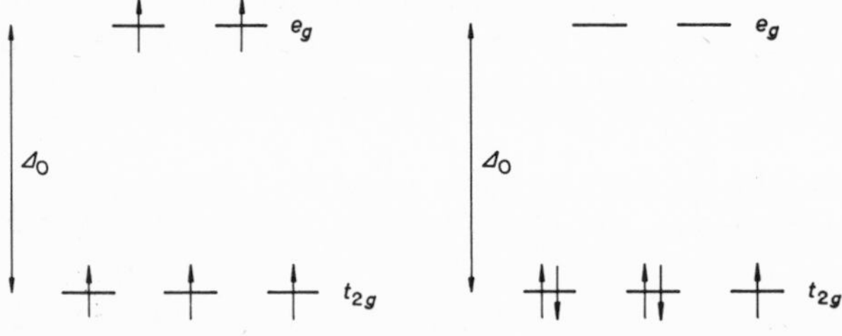

Fig. 6.10. Spin pairing induced by strong ligands placed octahedrally around a d^5 ion.

strong enough to cause spin-pairing and the $Fe(CN)_6^{3-}$ ion has a magnetic moment due to 1 unpaired electron. It is a *low-spin complex*. But the field exerted by water molecules is too weak to cause spin-pairing and the hydrated ion $Fe(H_2O)_6^{3+}$ has a magnetic moment indicative of five unpaired spins. (Note that the intermediate case of 3 unpaired spins never occurs here.)

Ions with d^6 structure can also form high- and low-spin complexes (Fig. 6.11). The release of orbital energy is again $2\Delta_0$; but, as the number of

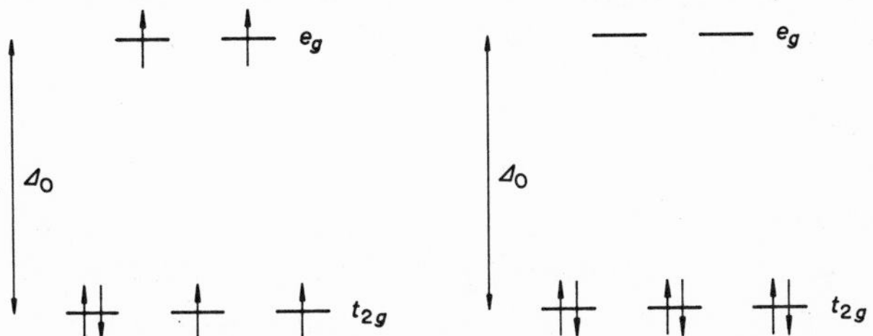

Fig. 6.11. Two possible arrangements of six d electrons in an octahedral ligand field.

parallel pairs is reduced by only four (10—6), the value of Δ_0 required to force spin-pairing does not need to be so large, and even the NH_3 molecule exerts a strong enough field to do this in the $Co(NH_3)_6^{3+}$ ion, which is diamagnetic.

Ions with d^7 configurations also form both high-spin and low-spin octahedral complexes (Fig. 6.12). Octahedral d^8 ions of necessity have two elec-

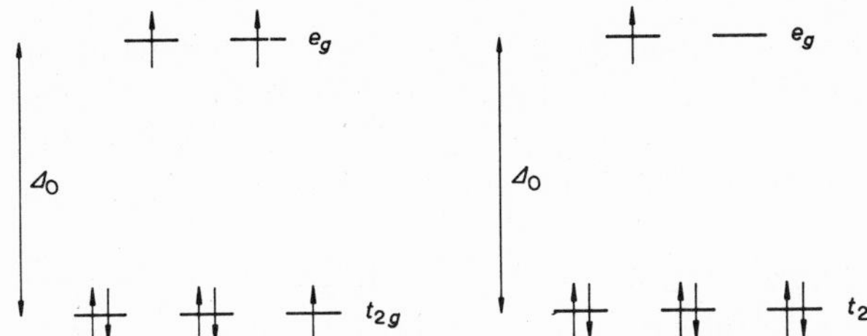

Fig. 6.12. Two possible arrangements of seven d electrons in an octahedral ligand field.

trons in the e_g orbitals and d^9 ions have three; because of this the strength of the ligand field is without effect on the extent of spin-pairing.

6.3.2. Summary of the magnetic effect of the ligand field

The effect of the ligand field on the magnetic properties of a transition metal ion can now be summarised:

(i) For a given metal ion there is a critical value of Δ_0 at which the state of lowest energy ceases to be the high-spin form and becomes the low-spin form.

(ii) Since for a given ligand Δ_0 is greatest for Period 6 metals and for high oxidation states, low-spin complexes are most likely to be formed by highly charged ions of heavy transition metals.

(iii) Because there is a comparatively small decrease in exchange stabilisation in the $t_{2g}^4\, e_g^2 \rightarrow t_{2g}^6$ transition, spin-pairing in d^6 ions is caused by comparatively weak fields. Ions containing four or five d electrons undergo spin-pairing less readily than d^6 ions, and d^7 ions still less readily.

Thus, according to the simple theory, spin-pairing is most likely to be caused by ligands to the left of the spectrochemical series in the octahedral complexes of ions such as Ir^{3+} (d^6) and, particularly, Pt^{4+} (d^6).

6.3.3. Tetrahedral complexes

When ligands are arranged tetrahedrally around an ion the five degenerate d orbitals are split into two groups, t_2 comprising d_{xy}, d_{xz}, d_{yz} which have higher energy by virtue of their directions lying closer to the ligands than do those of e, d_{z^2} and $d_{x^2-y^2}$ (Fig. 6.13).

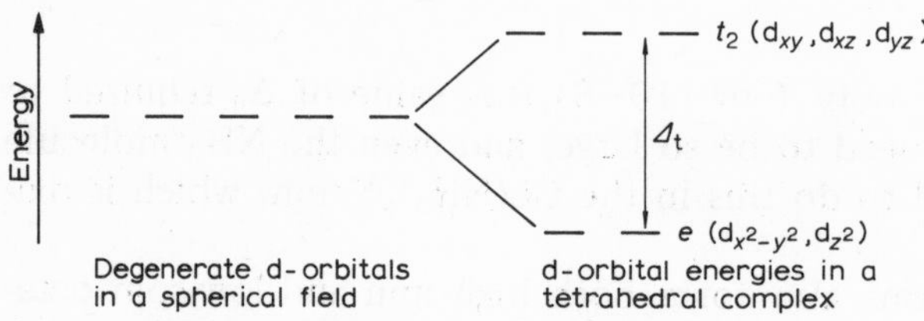

Fig. 6.13. Splitting of d-orbital energies in a tetrahedral field.

That this is so can again be confirmed by carrying out the operations of point group T_d on each d orbital and recording how each is affected. The result agrees with the irreducible representations E and T_2 (not E_g and T_{2g}, because there is no centre of symmetry in point group T_d) as indicated in the character table for T_d.

According to simple electrostatic theory, Δ_t, the ligand field splitting energy of the d electrons for a tetrahedral field should be 4/9 Δ_0 (based on the square of the ligand interactions $4^2/6^2 = 16/36 = 4/9$) where Δ_0 is the splitting produced by an octahedral field exerted by the same ligands at the same metal ion.

6.3.4. Tetragonal complexes

Symmetrical non-linear molecules cannot remain in equilibrium in an orbitally degenerate state but must be distorted in some way which will destroy the degeneracy. This effect, which is found whenever the electron configuration of such a central atom contains unequally occupied, degenerate orbitals, was predicted by Jahn and Teller in 1937.

Consider the Cu^{2+} ion (d^9) and suppose that the $d_{x^2-y^2}$ orbital contains only one electron while the d_{z^2} contains a pair. Then we have the simple physical picture that the two ligands on the z axis in an octahedral Cu^{2+} complex are more strongly shielded from the nuclear charge of the copper than the four ligands on the x- and y-axes. Consequently the Cu—ligand distances on the x- and y-axes tend to become shorter than the Cu—ligand distance on the z-axis. The octahedron has become distorted because the d_{z^2} and the $d_{x^2-y^2}$ orbitals of the metal ion are not equally occupied. This is a Jahn—Teller effect. Distortion is not to be expected for octahedrally co-ordinated ions which contain symmetrical arrangements of d electrons, for example $t_{2g}^3e_g^0$, $t_{2g}^3e_g^2$, $t_{2g}^6e_g^0$ and $t_{2g}^6e_g^2$, but for $t_{2g}^3e_g^1$, $t_{2g}^6e_g^1$ and $t_{2g}^6e_g^3$ arrangements, where the e_g orbitals facing the ligands are not symmetrically occupied, large distortions are likely to occur.

As a consequence, the octahedral disposition of ligands described by point group O_h changes spontaneously to one of tetragonal form, D_{4h} symmetry. Examination of the character table for D_{4h} will reveal that the degeneracy of octahedral t_{2g} and e_g orbitals are split into b_{2g}, e_g and b_{1g}, a_{1g} respectively (Fig. 6.14). As an example, CuF_2 has a distorted rutile structure in which

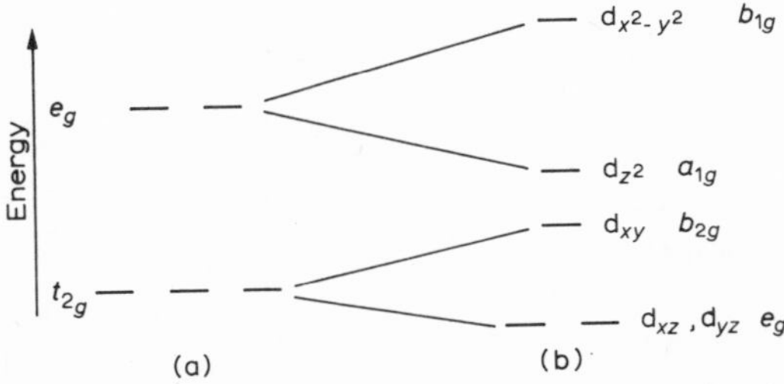

Fig. 6.14. Splitting of d-orbital energies in (a) octahedral, (b) tetragonal fields.

there are two F^- ions at 227 pm and four at 193 pm from the Cu^{2+}. The effect on the orbital energies is to split the degeneracy of the t_{2g} and e_g orbitals (Fig. 6.14). In a very strong ligand field the energy of the d_{z^2} orbital may fall below that of the d_{xy} and the order of energies is the same as for a square complex (6.3.5). In fact the ligands on the z axis are sometimes withdrawn altogether to give square complexes; an example of this is the $[Cr(H_2O)_4]^{2+}$ ion ($t_{2g}{}^3e_g{}^1$).

Jahn—Teller distortion of octahedral complexes usually produces tetragonal arrangements with one long (z) axis and two short (x and y) axes, but there are instances where one axis is short and the two others are longer. At present it is not possible to predict which of these arrangements should be the more stable.

Jahn—Teller effects are also manifest in absorption spectra, as can be appreciated from Fig. 6.14, because a splitting into t_{2g} orbitals of low energy and e_g of high energy is replaced by a much more complicated system of energy levels. (It will be seen that splitting of the e_g is accompanied by some splitting of the t_{2g} levels when distortion occurs. This is to be expected because both the d orbitals in the xy plane will be raised somewhat by the closer approach of the ligands in that plane.) Furthermore absorption spectra can indicate the existence of distortions in *excited states*, illustrated by the

Ti^{3+}aq discussed in 6.2.1.1. Suppose the single electron in a t_{2g} orbital of Ti^{3+} is promoted to the e_g level (Fig. 6.7) the excited state so produced has an unsymmetrical occupancy of the d_{z^2} and $d_{x^2-y^2}$ orbitals and their energies are split. It is this splitting which accounts for the hump or "shoulder" near 600 nm on the absorption spectrum of $Ti(H_2O)_6^{3+}$ (Fig. 6.8). The effect is even more marked in the spectrum of the CoF_6^{3-} ion in which the symmetrical ground state $t_{2g}^4e_g^2$ gives rise to the unsymmetrical excited state $t_{2g}^3e_g^3$. Here there is a clear division of the two high-energy orbitals, giving rise to a twin peak (Fig. 6.15).

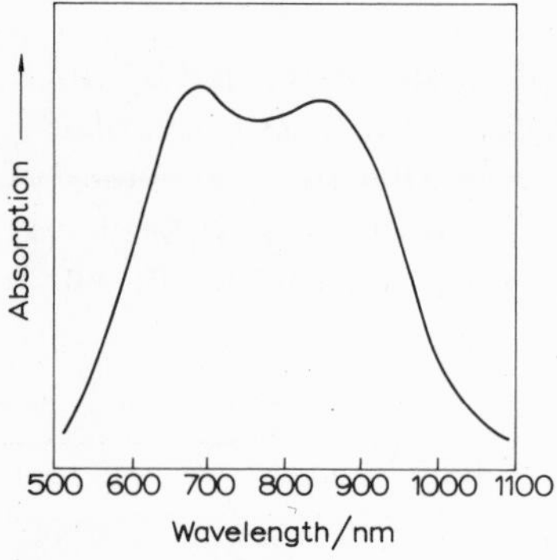

Fig. 6.15. Absorption spectrum of CoF_6^{3-} ion.

6.3.5. Square complexes

A square complex ML_4 can be thought of as an octahedral arrangement ML_6 in which tetragonal distortion has occurred to such an extent (say along the z axis) that two ligands have been totally removed. Thus in the field exerted by ligands lying on x and y axes of a central ion, the energy of the $d_{x^2-y^2}$ orbital is raised to a high level (6.3.3). The energy release in the transition of an electron from this orbital to a d_{xy} orbital is about the same as in an octahedral field of the same strength. Ions with d^8 configuration (Ni^{2+}, Pd^{2+}, Pt^{2+}) tend to form square complexes because the unfavourably situated $d_{x^2-y^2}$ orbital can be left empty while considerable energy is released as the electron falls to the d_{xy} level (Fig. 6.16). Such a splitting is consistent with the symmetry of the D_{4h} point group.

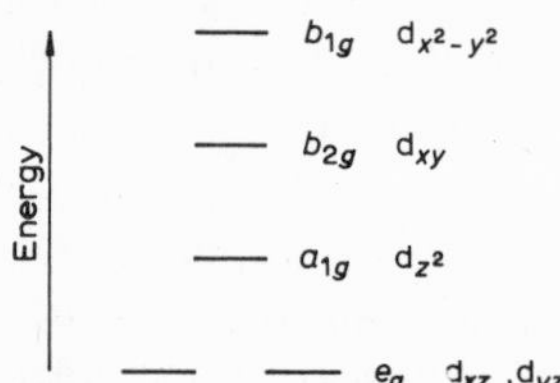

Fig. 6.16. Order of energies in a square complex.

6.3.6. Distortion of tetrahedral complexes

Equally filled degenerate orbitals are also present in the perfectly tetrahedral complexes $[FeO_4]^{2-}$(e^2), $[FeCl_4]^-$ ($e^2t_2^3$), $[CoCl_4]^{2-}$ ($e^4t_2^3$), $[ZnCl_4]^{2-}$ and $[Zn(acac)_2]$ ($e^4t_2^6$). Slightly distorted tetrahedra occur less frequently, probably, for example, in the vanadium tetrahalides (e^1). Usually a complete transition to the planar square form takes place. A spin-free arrangement $e^2t_2^2$ in a tetrahedral field would incur strong distorting forces; these are completely relieved in the square system, both degenerate pairs of orbitals being uniformly filled.

6.4. Absorption Spectra of Transition-metal Complexes

A solution of a transition-metal compound owes its colour to the absorption of light at certain wave-lengths in the visible range. The absorption for a particular wavelength is expressed as a molar extinction coefficient (ϵ):

$$\epsilon = \frac{1}{cd} \log_{10} \frac{I_0}{I}$$

where c is the concentration of the solution, I_0 is the intensity of the incident light and I the intensity of the light transmitted after passing through a length d of the solution.

The radiant energy absorbed by these compounds is used to promote electrons from low-lying d orbitals to higher ones. The principles involved in the simplest case of an octahedral d^1 complex have already been described (6.2.1.1). Because the wave-lengths of absorption bands are of importance for the determination of ligand field strengths, we shall discuss the general principles applying to complexes with more than one d electron.

As already explained under atomic spectra (3.2.1), not all electron transitions are allowed by quantum mechanics. Conditions under which certain forbidden electron transitions may occur are indicated below:

(i) Transitions in which the multiplicity, $(2S + 1)$, changes are forbidden. Since $S = n/2$, where n is the number of unpaired electron spins, another way of expressing this is to say that transitions which cause a change in the number of unpaired spins are forbidden. Nevertheless, weak absorption ($\epsilon_{Max} <$ 0.1 m^2 mol^{-1}) can occur, even with high-spin d^5 (for example Mn^{2+}) complexes, where any transition between d orbitals must involve a change in n, because of the weak coupling of spin and orbital angular momenta.

(ii) According to Laporte's selection rule, for any electron transition $\Delta l = \pm 1$. Thus transitions from one d orbital to another ($\Delta l = 0$) are 'Laporte forbidden'. Although this rule applies to free gaseous ions, there is always some low-intensity absorption ($\epsilon_{Max} < 5$ m^2 mol^{-1}) by ions in complexes because the d orbitals are to some extent mixed with p or f orbitals.

(iii) Actually, nearly all transition metal compounds owe their colour to absorption due to 'forbidden' transitions. At shorter wave-lengths in the ultraviolet, there are usually intense absorption bands (ϵ_{Max} up to 10^3 m^2 mol^{-1}) arising from 'allowed' transitions. But in a few instances these 'allowed' transitions cause strong absorption in the visible spectrum; an example is the thiocyanatoiron(III) complex formed in the thiocyanate test for Fe^{3+} ions.

6.4.1. Energy level diagrams for ions with more than one d electron

6.4.1.1. d^2 ions

In an ion containing two d electrons the energy levels are affected not only by the field due to the ligands but also by the interaction between the two electrons themselves under the influence of the field. We shall consider the effect of ligands on the Russell—Saunders states (3.4.1) of the free ion.

The free V^{3+} ion has as its lowest energy state a ^3F state (Hund's rules, 3.4.2). Ligands octahedrally round the metal ion split the energy of the ^3F state into three components, denoted by the symbols $^3A_2$, $^3T_2$ and $^3T_1$.

The excited states lying next above the ^3F state in a free d^2 ion are ^1D and ^3P. We shall consider here only the triplet state, because we are concerned with transitions between the lowest energy state, a triplet state in this case however strong the field, and states of the same multiplicity; other transitions are strongly forbidden and give rise to very weak absorption. The energy-level diagram for a d^2 ion in an octahedral field has the form shown in Fig. 6.17.

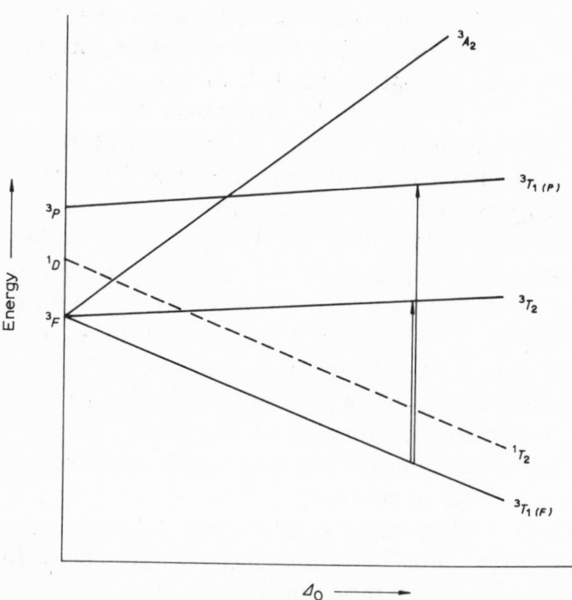

Fig. 6.17. Energy diagram for d^2 ion in an octahedral field. The two transitions which are illustrated give rise to two absorption bands.

Only the lowest singlet state ^1D is drawn, merely to show that it always lies above the lowest triplet state. The ^3P state gives rise to only one state, $^3T_1(P)$ in the field.

The spectrum of $[V(ox)_3]^{3-}$ has two absorption bands at 17000 and 24000 cm^{-1}. As the energy diagram shows, these energies correspond to transitions from a $^3T_1$ state to $^3T_2$ and $^3T_1(P)$ respectively, at a point where the ligand-field splitting, Δ_0 is 17800 cm^{-1}. The transition $^3T_1 \rightarrow {}^3A_2$ is unlikely because it represents a movement of both electrons simultaneously.

6.4.1.2. d^9 ions

The visible absorption spectrum of a d^9 system in an octahedral field can be interpreted by means of an energy level diagram which is effectively the d^1 diagram inverted. The d^9 ion can be imagined as a d^{10} ion with a hole where an electron is missing. Thus, as far as differences between the two energy levels are concerned, the treatment is the same as for a system containing one positron capable of being excited from an e_g orbital to a t_{2g} orbital by energy equivalent to Δ_0.

In fact the absorption spectrum of a Cu^{II} complex consists of two bands because the e_g and t_{2g} levels are split in Jahn—Teller distortion (6.3.4), much more so than in the $Ti^{3+}(d^1)$ case.

6.4.1.3. d^8 ions

The absorption spectra of Ni^{II} compounds have excited considerable attention. The spectrum of the $Ni(en)_3{}^{2+}$ ion has three almost symmetrical absorption peaks (Fig. 6.18). The energy diagram for Ni^{II} in an octahedral field, shown in Fig. 6.19, with singlet states omitted, resembles that of the d^2 system V^{III}.

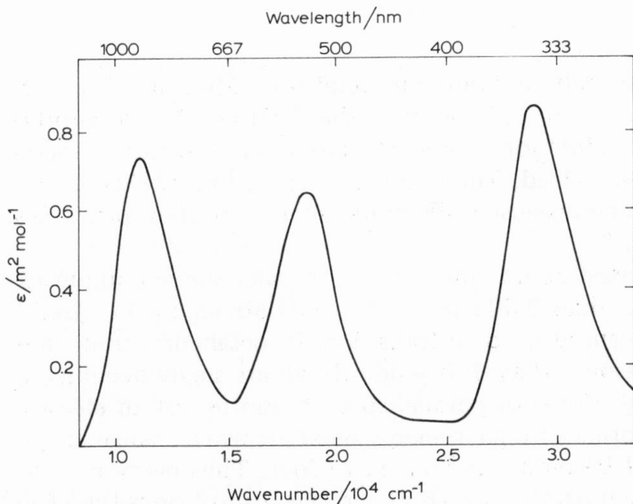

Fig. 6.18. Absorption spectrum of $Ni(en)_3{}^{2+}$

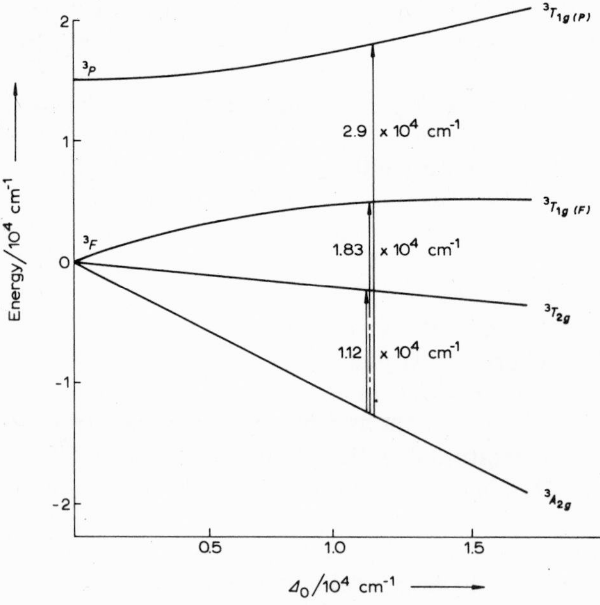

Fig. 6.19. Energy as a function of Δ_0 for $Ni^{II}(d^8)$ system.

The ^3F state splits in the field into three components but their energies are inverted in relation to those in a d^2 system. The ^3P, as in the d^2 ion, gives only a $^3T_1$ state in the ligand field, and as before (6.4.1.1) the ^1D state always lies above the lowest triplet state.

Thus the three absorption bands are due to transitions from the $^3A_2$ state to the $^3T_2$, the $^3T_1(F)$ and the $^3T_1(P)$ states respectively. For energy transitions of 11200, 18300 and 29000 cm^{-1} the value of Δ_0 is 11200 cm^{-1}.

6.4.1.4. d^5 ions

Hydrated manganese(II) salts contain the octahedral $Mn(H_2O)_6^{2+}$ ion. In common with other high-spin d^5 complexes — the $Fe(H_2O)_6^{3+}$ ion is another example — the hydrated Mn^{2+} ion is almost colourless; absorption in the visible range is very weak indeed. The absorption spectrum of this ion has been thoroughly studied, and because it illustrates some important principles, it is reproduced in Fig. 6.20.

The first point of interest about this spectrum is that the maximum extinction coefficient is less than 0.004 $m^2\ mol^{-1}$, only about one-hundredth of the usual absorption caused by d—d transitions in octahedral transition-metal complexes. The reason is that all five 3d orbitals are singly occupied in the ^6S ground state by electrons of parallel spin. Promotion of an electron from one d orbital to another of higher energy must involve a change in spin and the multiplicity must be reduced from six to four. Thus every possible d—d transition is multiplicity-forbidden (6.4) and the absorption is therefore very weak.

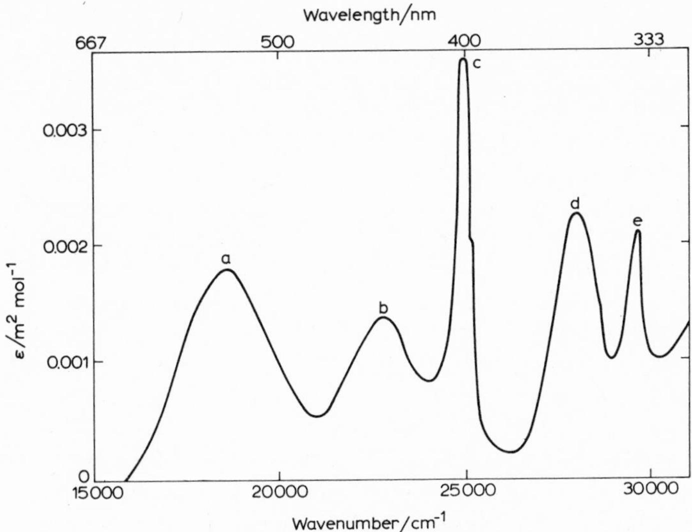

Fig. 6.20. Visible and near-ultraviolet spectrum of $Mn(H_2O)_6{}^{2+}$ ion in aqueous solution.

The second point of interest is that there are many bands in the spectrum (cf. $Ti(H_2O)_6{}^{3+}$ and $Ni(en)_3{}^{2+}$). The reason is apparent from the energy diagram, Fig. 6.21, which shows the energies of the four quadruplet states

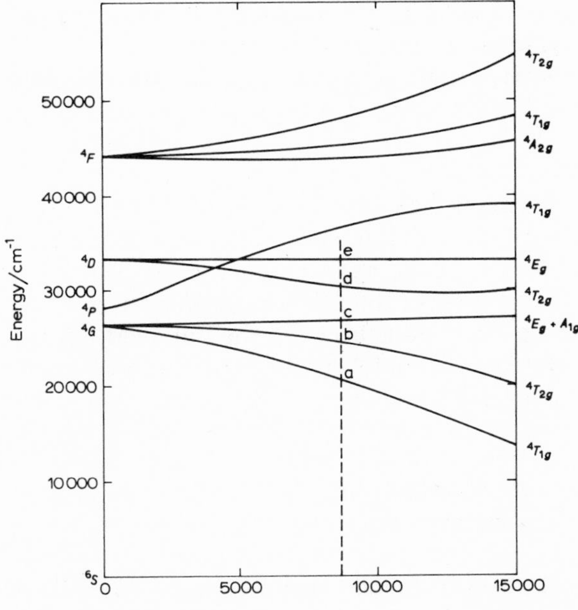

Fig. 6.21. Energies of the quadruplet states of $Mn(H_2O)_6{}^{2+}$, relative to the ^6S state of the free gaseous Mn^{2+} ion.

relative to the ^6S state in the free, gaseous Mn^{2+} ion and in ligand fields of increasing strength. There are numerous energy states lying sufficiently close together to cause a number of absorptions in the visible and the near ultraviolet. The energy transitions from ^6S which are responsible for the absorption peaks are lettered for easy comparison with Fig. 6.21, the value of Δ_0 which fits the observed bands is ~ 8600 cm^{-1}.

The third point of interest is that bands c and e in Fig. 6.20 are much sharper than the others. Comparison with Fig. 6.21 indicates that these bands are due to d—d transitions in which the energy difference is affected very little by ligand field strength — the lines (*e.g.* ^4D—^4E_g) are about parallel to the ^6S base line. In these transitions the electron 'jump' brought about by the absorption of energy is from one t_{2g} orbital to another. In an octahedral field the difference in energy between two t_{2g} orbitals is affected much less by the field strength than is the difference between a t_{2g} and an e_g orbital. For transition of the t_{2g}—t_{2g} type the energy absorbed is therefore affected very little by the lengths of the vibrating metal—ligand bonds at the moment the electron jumps. Unlike the t_{2g} orbitals which lie between the octahedrally disposed ligands, the e_g orbitals lie opposite them and their energies are sensitive to changes in metal—ligand bond length (*i.e.* to changes in Δ_0) which occur as the bonds vibrate. In t_{2g}—e_g transitions the size of the energy absorption therefore depends a good deal on the lengths of the metal—ligand bonds at the moment the electron jumps; there is a wide spread of energy transition and a broad absorption band (a, b and d, Figs. 6.20 and 6.21).

The fourth point of interest in the absorption spectrum is the 'shoulder' on the peak at 25000 cm^{-1}. The ^4G state of the free Mn^{2+} ion gives rise to four symmetry states in the complexed ion. Of these the ^4E_g and $^4A_{1g}$ are almost, but not exactly degenerate; the line in Fig. 6.21 should actually divide very slightly.

6.4.1.5. Other d^n ions

The spectra of other d^n systems are not treated in detail; but, in general, it may be said that an octahedral d^n system has an energy level diagram resembling that of an octahedral d^{10-n} system, except that the order of levels coming from each Russell—Saunders state of the free ion is reversed.

The energy level diagram for a d^n system in a tetrahedral field has the same form as that for a d^{10-n} system in an octahedral field. The values of Δ_t can be determined in a similar way to those of Δ_0.

6.5. Evidence for Covalent Bonding in Transition-metal Complexes

Although the explanation just given of metal—ligand bonding in terms of electrostatic interaction is sufficient to account for the general form of the spectra arising from d—d transitions, the frequencies of the absorption peaks do not always coincide with those calculated from the theory. Furthermore, a model based purely on coulombic attractions cannot account for the order of ligands in the spectrochemical series. This is because there is always a degree of covalent bonding between metal and ligand. The presence of covalence can be shown by several methods.

(i) How the Landé splitting factor, g, varies under the influence of crystal-field and spin—orbital coupling effects has been described (3.4.4.4). A d electron which spends only some of its time on the transition-metal atom in a complex will contribute only in part to the orbital component of magnetic moment. From an experimental determination of the g-value, the mean distribution of the electrons which cause the magnetism can be calculated. The result shows that the d-electrons in transition metals are always delocalised to some extent. Even in octahedral fluorocomplexes, where a purely electrostatic interaction of M^{n+} and F^- ions might be expected to give an accurate representation of the bonding, the evidence is that the d electrons of the metal spend up to 5% of their time on the fluorines.

(ii) How the unpaired electron density round an atom can be found from the nuclear splitting in the e.s.r. spectrum has been described for an odd-electron molecule (3.4.4.4). The same principles can be applied to the finding of electron densities round atoms in paramagnetic ions. For instance, the e.s.r. spectrum of the interesting $[(NH_3)_5CoO_2Co(NH_3)_5]^{5+}$ complex shows eleven lines in the hyperfine splitting, indicating that the unpaired electron is in an MO which gives it an equal probability of being on each cobalt. Moreover, all metal nitrosyls have three-line e.s.r. spectra due to nitrogen splitting ($I = 1$ for nitrogen), indicating considerable covalent character in the M—N bond.

(iii) An unpaired electron in an orbital about a paramagnetic nucleus has a very large effect on the magnetic field at that nucleus. Measurement of the ^{19}F chemical shift (2.7.4) in the n.m.r. spectra of fluorocomplexes of Mn^{II} and Ni^{II} shows that the t_{2g} and e_g electrons of the metal spend more than a negligible part of their time around the fluorines.

(iv) Shifts in the n.q.r. spectra (2.8.1) of halide ions in various complexes can be related to the so-called percentage ionic character of the bonds. Analysis of the quadrupole resonance spectra of square and octahedral complexes of Pd and Pt shows that the metal—halogen bonds are only partly ionic (Table 6.1)

TABLE 6.1

PARTIAL IONIC CHARACTER OF METAL—HALIDE BONDS FROM n.q.r. SPECTRA

Complex	Shape	Ionic character of M—X (%)
$PdBr_4^{2-}$	square	60
$PtBr_4^{2-}$	square	57
$PdBr_6^{2-}$	octahedral	37
$PtBr_6^{2-}$	octahedral	38

(Note: the greater the oxidation number of the metal the more covalent the metal—halogen bond.)

(v) The energy-level diagrams above are based on the assumption that the energy separations between Russell—Saunders states (3.4.1) in a complexed M^{n+} ion are the same as in a free, gaseous M^{n+} ion. In fact, the actual positions of lines in the d—d absorption spectra always indicate that there is less interaction between the d electrons themselves in a complexed ion than in a free gaseous ion. Presumably the ligands provide outlets by which d electrons can to some extent escape from the metal; indeed the ligands effectively increase the size of the d-electron cloud. An arrangement of ions and molecules in order of their 'cloud-expanding' or nephelauxetic effect:

$$F^- < H_2O < NH_3 < SCN^- < Cl^- < CN^- < Br^- < I^-$$

is almost independent of the central metal ion involved, as also is the arrangement in the spectrochemical series (6.2.1.2). But the order of ligands for the nephelauxetic effect, unlike that for the spectrochemical series, appears to be closely related to the increasingly covalent character of the metal—ligand bond.

(vi) In some solids which contain paramagnetic ions, the moments of the separate ions tend to align themselves parallel. However, the effect is small above the Curie temperature (3.4.4.3) of the solid because the thermal vibrations are sufficiently great to upset the alignment, but below this point the susceptibility rises rapidly as temperature falls. Such solids are termed *ferromagnetic*. In another class of solids, called *antiferromagnetic*, which also exhibit normal paramagnetic behaviour at high temperatures, the moments of the separate ions in the crystal lattice tend to align themselves antiparallel; for these materials, at temperatures below a characteristic temperature called the Néel point, the susceptibility falls rapidly because thermal agitation is insufficient to prevent antiparallel alignment of the ionic moments.

Some transition-metal oxides such as MnO and NiO are antiferromagnetic for another reason. In them the metal ions retain their moments but these line up in the NaCl type crystal lattice with one half in one direction and one half in the other, as is shown in the diagram on the next page. The moments of the metal ions almost certainly do not interact directly with one another but through the oxide ions.

The antiferromagnetism can be explained if we suppose some sharing of electrons between O^{2-} and adjacent Mn^{2+} ions (Fig. 6.22). Single electron occupancy of a metal d orbital makes the electron with which it pairs align with antiparallel spin and all the other unpaired d electrons align similarly to maintain the maximum exchange stabilisation (6.3.1). Thus the spins, and the magnetic moments due to them, are aligned in opposition on two metal ions separated by an oxygen. The immediate point of this is that the explanation of antiferromagnetism implies a certain amount of covalent character in the bonding of these metal oxides.

Obviously when the delocalisation of metal d-electrons is slight, the results from crystal field theory need only slight modification to give calculated values of Δ which agree with experimental results. However, a more satis-

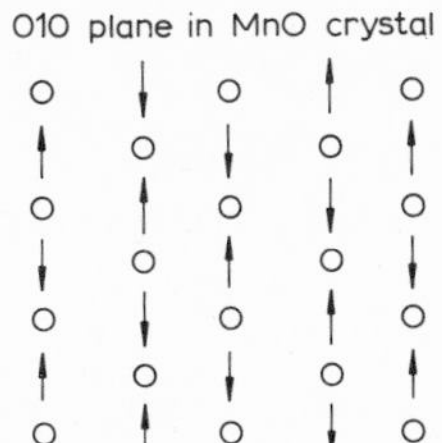

Alignment of magnetic moments of Mn^{2+} ions in MnO crystal
↑ Mn^{2+} showing direction of its magnetic moment
O = O^{2-} ion

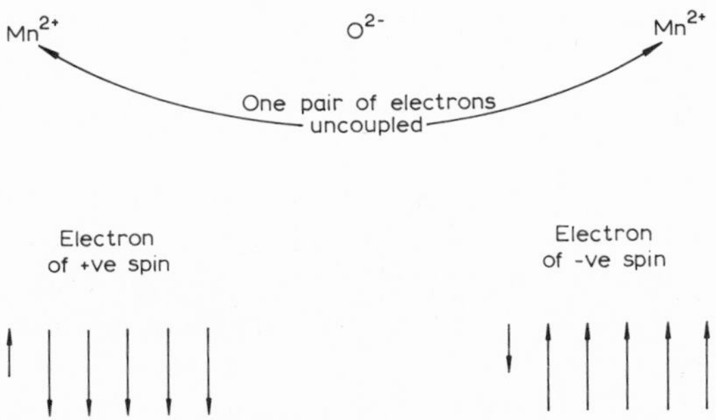

Fig. 6.22. Representation of the effect of an uncoupled electron pair from an oxide ion upon neighbouring Mn^{2+} ions.

factory energy level diagram can be constructed if we accept that (a) the spin—orbit coupling constant (3.4.4.4) and (b) the energy differences between the Russell—Saunders states are not the same for a complexed ion as for a free gaseous ion. Generally λ is 15—30% less in a complex than in a free ion. The energy difference between two states of the same symmetry arising from two Russell—Saunders states of the free ion, for example the ${}^3T_{1g}(F)$ and ${}^3T_{1g}(P)$ states in the d^8Ni^{2+} system (Fig. 6.18), is usually about 30% less than for the transition (${}^3F \rightarrow {}^3P$ in this case) in the free ion.

When the smaller values for λ and for the separations of the energies of the Russell—Saunders states are used, the simple electrostatic theory can give good agreement between the calculated and experimental values of Δ. This is the method of the ligand field theory; in it a basically electrostatic theory is modified to allow for electron delocalisation, in other words, for some covalent character in the metal—ligand bonding. When, however, the metal—ligand bonding in a complex is largely covalent, the best treatment is provided by the molecular orbital theory.

6.6. Molecular Orbital Theory of Metal—Ligand Bonding

6.6.1. Octahedral complexes

In general the LCAO-MO method can be applied to complexes in the same way as for simple molecules (4.1.1). There are three steps.

The first step. In this we seek to pick out the valency atomic orbitals of the atoms in the molecule under consideration. For a transition metal of Period 4, the valence atomic orbitals are 3d, 4s and 4p. The ligands may have σ and π valence orbitals; for example in fluorine the valence orbital is composed of 2s and $2p_z$, the π valence orbitals are $2p_x$ and $2p_y$.

The second step. The next stage is to construct proper linear combinations of valence atomic orbitals for the molecule. For this a knowledge of the symmetry of the various orbitals is required in order that proper combinations can be made.

The third step. This final step is to place the molecular orbitals in order of energy.

It has already been shown (6.2.1) how the 3d orbitals are split by the influence of an octahedral environment into two groups, d_{z^2} and $d_{x^2-y^2}$ of the e_g symmetry species and d_{xy} d_{xz} and d_{yz} of the t_{2g} symmetry species. Using the same procedure, the 4s orbital transposes as a_{1g} and the three 4p orbitals form the triply degenerate t_{1u} species.

The symmetry description of the six ligands can be determined by performing the operations of point group O_h and recording how many ligands remain unchanged in position as a result of each operation. For example, the C_4 operation involving a 90° rotation about any axis results in the two ligands on that axis remaining in the same position while the other four ligands perpendicular to the axis change positions. Likewise reflection in a horizontal plane leaves the four ligands in the plane the same while the other two ligands change positions. In all, this leads to the following reducible representation:

O_h	E	C_3	C_2	C_4	C_4^2	i	S_4	S_6	σ_h	σ_d
6 ligands	6	0	0	2	2	0	0	0	4	2

By means of the equation

$$n_\Gamma = \frac{1}{g}\sum g_R \gamma_\Gamma^R \gamma_{DP}^R$$

this representation can be reduced to $A_{1g} + E_g + T_{1u}$.

O_h	E	C_3	C_2	C_4	C_4^2	i	S_4	S_6	σ_h	σ_d
A_{1g}	1	1	1	1	1	1	1	1	1	1
E_g	2	−1	0	0	2	2	0	−1	2	0
T_{1u}	3	0	−1	1	−1	−3	−1	0	1	1
	6	0	0	2	2	0	0	0	4	2

Thus it is possible for linear combinations of ligand orbitals to bond with the t_{1u}, e_g and a_{1g} metal orbitals. The relationship between d^2sp^3 hybridisation and the symmetry of the orbitals involved is apparent. For example, the linear combination which has the same symmetry as the $4p_x$ metal orbital has a positive value in the $+x$ direction and a negative value in the $-x$ direction (Fig. 6.23). This is the combination $\sigma_1 - \sigma_3$. Similarly the combination

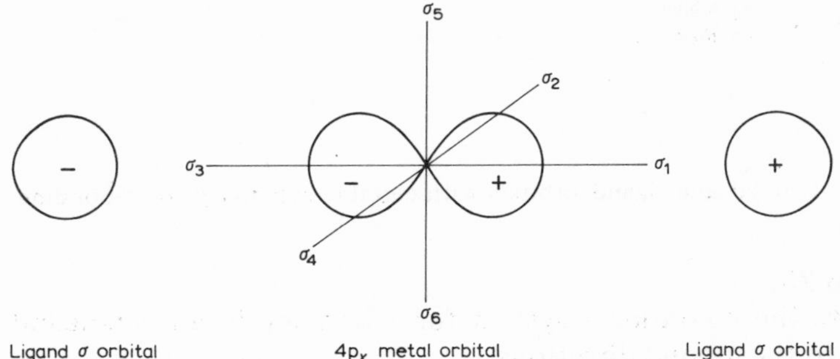

Fig. 6.23. Linear combination of $4p_x$ orbital on metal with $\sigma_1 - \sigma_3$ ligand orbitals.

$\sigma_2 - \sigma_4$ has the correct symmetry for overlap with $4p_y$ and the combination $\sigma_5 - \sigma_6$ for overlap with $4p_z$. The 4s metal orbital, positive in all directions, combines with $\sigma_1 + \sigma_2 + \sigma_3 + \sigma_4 + \sigma_5 + \sigma_6$. The $3d_{x^2-y^2}$ orbital, the lobes of which have alternately positive and negative ψ values, matches the $\sigma_1 - \sigma_2 + \sigma_3 - \sigma_4$ in symmetry (Fig. 6.24). The ligand σ orbital combination for $3d_{z^2}$

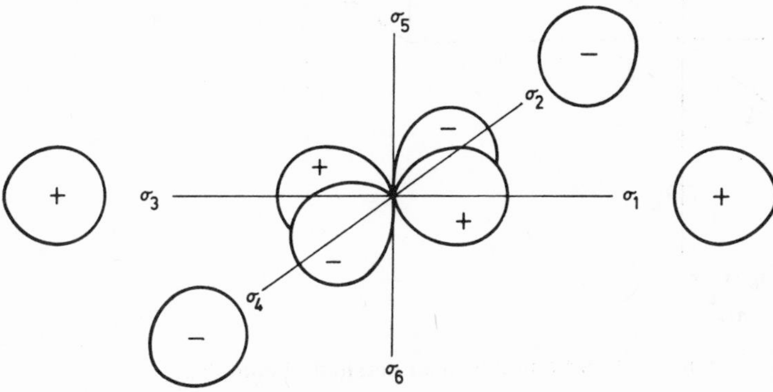

Fig. 6.24. Combination of $d_{x^2-y^2}$ with $\sigma_1 - \sigma_2 + \sigma_3 - \sigma_4$.

presents more difficulty; it is in fact $2\sigma_5 + 2\sigma_6 - \sigma_1 - \sigma_2 - \sigma_3 - \sigma_4$.

For π-bonding the appropriate metal orbitals are

$3d_{xy}$, $3d_{xz}$, $3d_{yz}$, (t_{2g}),
$4p_x$, $4p_y$, $4p_z$, (t_{1u}).

The 4p orbitals are involved in both σ- and π-bonding. The proper $t_{1u}(\pi)$ and $t_{2g}(\pi)$ ligands can be written down as for the σ linear combination of atomic

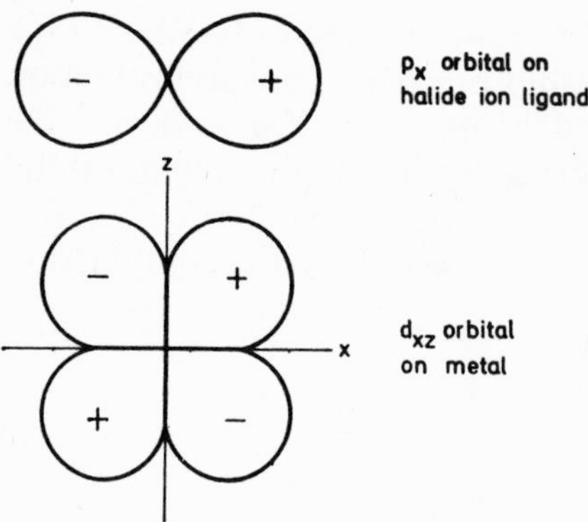

Fig. 6.25. Example of metal and ligand orbitals with suitable symmetry for π-bonding.

orbitals (see Fig. 6.25).

Fig. 6.26 shows the co-ordinate system for π-bonding in an octahedral complex, relative to the σ-bond directions.

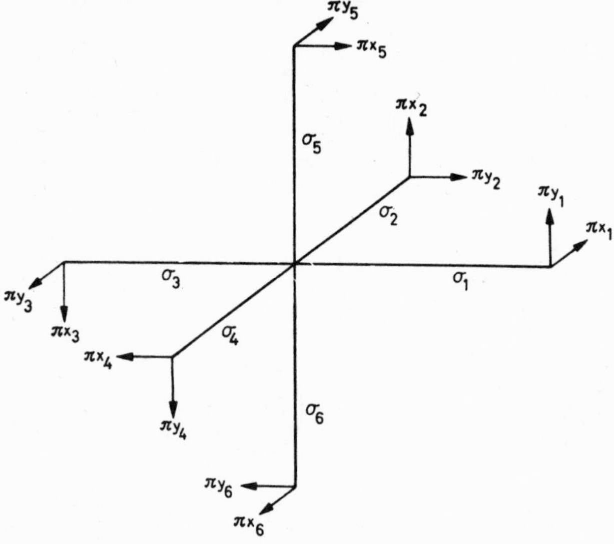

Fig. 6.26. Co-ordinate system for σ and π bonds in an octahedral complex.

In addition to the six ligand π-orbital combinations given in Table 6.2, there are six more made up of the same π orbitals as the t_{1u} and t_{2g} but with the signs of the second and fourth terms changed. These are the t_{2u} and t_{1g} respectively. These are non-bonding because there are no metal orbitals of the same symmetry, so that overlap is not possible.

Finally, to place the molecular orbitals in order of energy, reasonable approximations for the coulombic and exchange energies may be obtained by following the three rules:

TABLE 6.2

PROPER METAL- AND LIGAND-ORBITAL COMBINATIONS FOR OCTAHEDRAL COMPLEXES

Symbol	Metal orbital	Combination of ligand orbitals
(i) For sigma bonds		
a_{1g}	s	$\frac{1}{\sqrt{6}}(\sigma_1 + \sigma_2 + \sigma_3 + \sigma_4 + \sigma_5 + \sigma_6)$
e_g	d_{z^2}	$\frac{1}{2\sqrt{3}}(2\sigma_5 + 2\sigma_6 - \sigma_1 - \sigma_2 - \sigma_3 - \sigma_4)$
	$d_{x^2-y^2}$	$\frac{1}{2}(\sigma_1 - \sigma_2 + \sigma_3 - \sigma_4)$
t_{1u}	p_x	$\frac{1}{\sqrt{2}}(\sigma_1 - \sigma_3)$
	p_y	$\frac{1}{\sqrt{2}}(\sigma_2 - \sigma_4)$
	p_z	$\frac{1}{\sqrt{2}}(\sigma_5 - \sigma_6)$
(ii) For π bonds		
t_{1u}	p_x	$\frac{1}{2}(\pi_{y_2} + \pi_{x_5} - \pi_{x_4} - \pi_{y_6})$
	p_y	$\frac{1}{2}(\pi_{x_1} + \pi_{y_5} - \pi_{y_3} - \pi_{x_6})$
	p_z	$\frac{1}{2}(\pi_{y_1} + \pi_{x_2} - \pi_{x_3} - \pi_{y_4})$
t_{2g}	d_{xz}	$\frac{1}{2}(\pi_{y_1} + \pi_{x_5} + \pi_{x_3} + \pi_{y_6})$
	d_{yz}	$\frac{1}{2}(\pi_{x_2} + \pi_{y_5} + \pi_{y_4} + \pi_{x_6})$
	d_{xy}	$\frac{1}{2}(\pi_{x_1} + \pi_{y_2} + \pi_{y_3} + \pi_{x_4})$

(i) The order of coulombic energies is taken to be

$$\sigma_{\text{ligand}} < \pi_{\text{ligand}} < 3d < 4s < 4p$$

The values are lower for ligand orbitals because the ligand is more electronegative than the metal.

(ii) The extent of mixing of metal atomic orbitals and ligand orbitals in the molecular orbitals is proportional to the overlap of the atomic orbitals with ligand orbitals and inversely proportional to their coulombic energy differences.

(iii) Bonding σ molecular orbitals are more stable than bonding π molecular orbitals and antibonding σ^* are less stable than antibonding π^*.

The relative energies of molecular orbitals for an octahedral complex are illustrated in Fig. 6.27.

Points to note about Fig. 6.27 are:

(i) A bonding MO has a lower energy and an antibonding MO a higher energy than either of the orbital types contributing to it; a non-bonding MO has the same energy as the orbital type from which it is formed.

(ii) The energies of bonding orbitals generally lie near those of the ligand orbitals, and the energies of antibonding orbitals near those of the metal orbitals.

(iii) There is accommodation for 12 electrons in σ-bonding orbitals, for

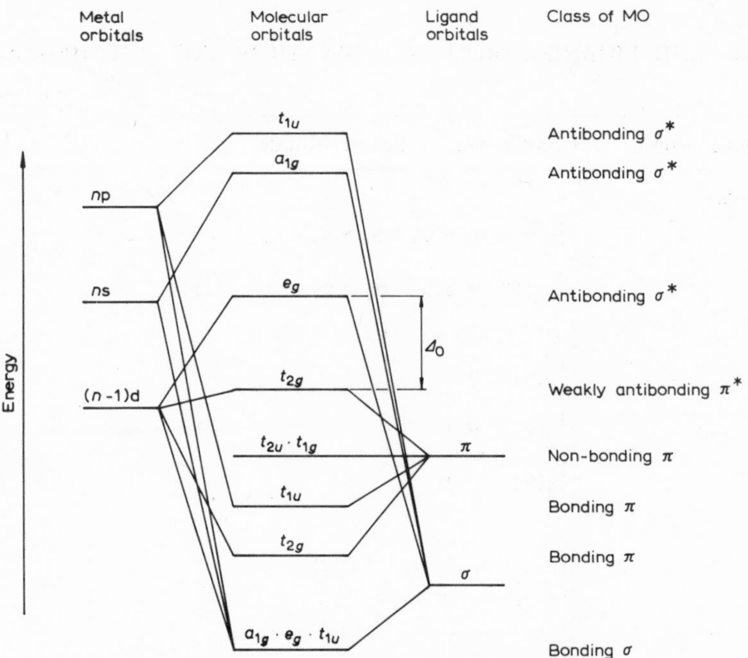

Fig. 6.27. Energies of orbitals involved in the formation of an octahedral complex.

12 electrons in π-bonding orbitals and for 12 electrons in non-bonding π-orbitals.

The advantage of the molecular orbital approach to metal—ligand bonding is that we can picture the filling of molecular orbitals with electrons in the order of their energy, irrespective of the origin of the electrons. By dispensing with an ionic model, which is no longer necessary, the various energy levels can be represented much more realistically.

Consider for instance an octahedral complex formed by Cl^- ions with a metal ion having three d electrons. Disregarding those electron pairs of the chlorine ions which are directed away from the metal, we have $(6 \times 6) + 3$ valence electrons available, 39 in all. Of these 12 are needed to fill the sigma bonding orbitals, and 24 for the bonding and non-bonding π orbitals t_{2g}, t_{1u}, t_{2u} and t_{1g}. The last three of the 39 electrons are accommodated in the weakly antibonding t_{2g} orbitals. The form of the energy level diagram shows that these t_{2g} orbitals, although formed largely from metal d orbitals, also contain a contribution from ligand π orbitals (Rule (ii) above). In the present instance the ligands have sufficient electrons to fill their π orbitals; and the t_{2g} level remains high above the ligand π level. But when there are empty π orbitals on the original ligands the t_{2g} electrons tend to escape towards them. The result is to lower the energy of the t_{2g} orbitals.

The cyanide ion, although it is without empty π orbitals of low energy, does have empty π* orbitals which lie only slightly higher (Fig. 6.28). The π* orbitals are of suitable symmetry to combine with the $t_{2g}(\pi^*)$ of the metal;

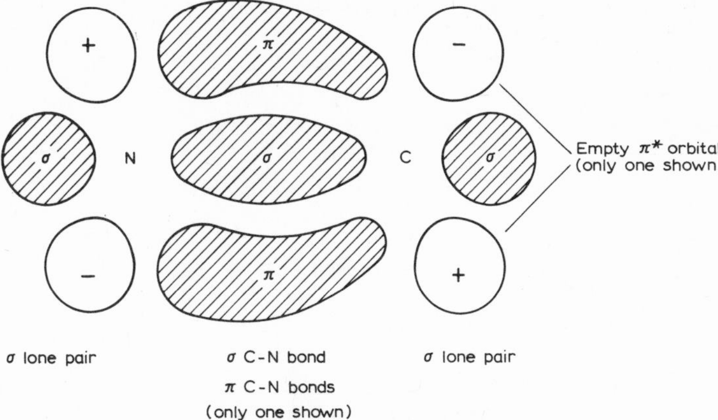

Fig. 6.28. Molecular orbitals in CN^- ion (KK $z\sigma^2 y\sigma^2 x\sigma w\pi^4$).

electrons in the t_{2g} which are released towards them are said to undergo π-back-donation to the ligand. The ligand itself is known as a π-acceptor ligand. The resultant fall in the energy of the orbital increases the gap between its energy and that of the $e_g(\sigma^*)$ orbital (*i.e.* the value of Δ_0 increases). Thus the MO theory is able to explain why the CN^- ion lies so far towards the high-energy end of the spectrochemical series.

Clearly the effect of π-back-donation is to increase the covalent character of the bond between the metal and the co-ordinating atom, whereby the bond is also strengthened. In the present example the creation of π bonding between metal and carbon increases the strength of the M—C bond; at the same time however the C—N bond is weakened a little because of the partial filling of an orbital which is antibonding with respect to C—N. Manifestly, the strength of the C—N bond in cyanocomplexes of different metals must be different and this can be demonstrated by infrared spectroscopy (5.3.2).

6.6.2. Tetrahedral complexes

In the building-up of the molecular orbitals in a tetrahedral complex, the only σ orbital that is entirely of the metal is the *n*s (a_1). The $(n-1)$ d_{xy}, d_{xz} and d_{yz} (t_2) and the np_x, p_y and p_z (t_1) orbitals may be used in both σ and π bonding. The $(n-1)$ $d_{x^2-y^2}$ and d_{z^2} are pure π orbitals (Fig. 6.29). There is maximum π-bonding in a tetrahedral complex when all π-bonding *e* and t_2 orbitals (five altogether) are filled and all the π* are empty. For example, the permanganate ion, MnO_4^- has six electrons in the π-bonding t_2 and four in the slightly higher *e* orbitals, making five π-bonds.

6.6.3. Charge-transfer spectra

A solution of iodine in carbon tetrachloride has the same violet colour as iodine vapour; this is due in both instances to an absorption of light charac-

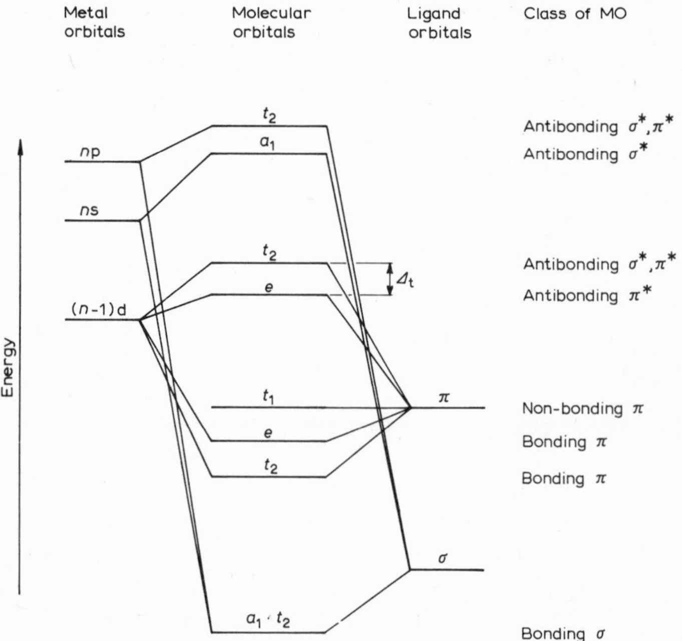

Fig. 6.29. Energy levels of orbitals involved in the formation of a tetrahedral complex.

teristic of I_2 molecules. But in a basic solvent, such as alcohol or pyridine, iodine gives a brown solution, although appreciable reaction has not occurred between the I_2 and the solvent molecules. The difference is because in such solvents every energy quantum absorbed transfers an electron from a molecule of the base to a high-energy, empty orbital of the I_2 molecule. This can be considered as an intermolecular electron transfer. The charge-transfer absorption gives rise to charge-transfer spectra.

There are also intramolecular transitions, particularly in complex ions. Electronic transitions from an orbital localised mainly on one atom to an orbital localised mainly on another atom usually occur at much higher energies than d—d transitions. The resulting *charge-transfer spectrum* consists of strong absorption, usually in the ultraviolet region. Such charge-transfer processes are of two types: ligand to metal, abbreviated L → M, in which an electron is transferred from an orbital localised mainly on the ligand to one localised mainly on the metal, and metal to ligand in which the electron is transferred in the reverse direction, abbreviated M → L.

The L → M charge transfer is shown by complexes of O^{2-}, Cl^-, Br^- and of I^- particularly. For d^0 complexes, where a d—d transition is not possible, Δ can be estimated only from charge-transfer absorption frequencies. An example is MnO_4^-, the purple colour of which is due to a low-energy $O^{2-} \rightarrow Mn^{7+}$ charge transfer in the tetrahedral complex. The bands at 1.85, 3.22 and 4.44×10^4 cm^{-1} in this absorption spectrum have been assigned to transitions $t_1(\pi) \rightarrow e(\pi^*)$, $t_2(\pi) \rightarrow e(\pi^*)$ and $t_1(\pi) \rightarrow t_2(\sigma^*, \pi^*)$ respectively (Fig. 6.29). Thus Δ is found to be 2.59×10^4 cm^{-1} by difference $e(\pi^*) \rightarrow t_2(\sigma^*, \pi^*)$.

The M $\rightarrow$ L charge transfer occurs in complexes containing ligands which have stable but empty orbitals. Complexes of π-acceptor ligands such as NO, CO and CN^- often have such absorption bands.

Charge-transfer transitions, unlike d—d transitions, are often neither multiplicity-forbidden nor Laporte-forbidden; consequently they give strong absorption bands with extinction coefficients in the range 50—200 m^2 mol^{-1}. This fact enables charge-transfer absorption to be distinguished from d—d transitions for which ϵ_{Max} is usually below 10 m^2 mol^{-1}.

A charge-transfer process is obviously closely related to an oxidation—reduction process. In fact, with ligands of low electronegativity a charge-transfer process can become a redox reaction. Thus the dark colour of a solution containing Cu^{2+} and Cl^- ions is attributed to strong charge-transfer (L $\rightarrow$ M) absorption in a Cu^{II} chloro-complex. Bromide ions cause an even stronger colour than Cl^-. But I^- ions lose electrons to orbitals centred entirely on the copper; as a result the Cu^{2+} cations are reduced to Cu^+ and the iodide anions are oxidised to iodine:

$$2Cu^{2+} + 4I^- \rightarrow 2CuI + I_2$$

6.6.4. Factors influencing structure in transition-metal complexes

The shape of a complex ion or molecule is governed largely by two types of interaction:

(i) The ligands themselves exert a mutual coulombic repulsion, greatest for large, monatomic anions such as Br^- and I^-, smallest for small, uncharged ligands like NH_3 and H_2O.

(ii) The electrons in the π^* and σ^* orbitals, by their number and distribution, stabilise some arrangements of ligands but not others; stabilisation happens particularly when the ligands themselves create strong fields.

Most transition-metal complexes have one of the three structures; octahedral, square or tetrahedral. The octahedral structure, which utilises all six of the strong σ orbitals $(n-1)d_{x^2-y^2}$, $(n-1)d_{z^2}$, ns, np_x, np_y and np_z, is the preferred arrangement unless interactions (i) and (ii) above make this an unstable structure.

Weak-field ligands with large mutual repulsions (*e.g.* I^-, Br^- and very large molecules) tend to form tetrahedral complexes because the coulombic repulsions are reduced when there are few ligands and they are not crowded together.

In complexes with strong-field ligands the number of d electrons contributed by the metal ion is of great importance. A d^6 ion with strong-field ligands forms low-spin, undistorted octahedral complexes in which six electrons are accommodated in the weakly antibonding $\pi^* t_{2g}$ orbitals. But in a d^7 complex one electron must enter the strongly antibonding $\sigma^* e_g$ orbital. The usual result is that one ligand is lost and an ML_5 complex, with one unpaired electron in a potentially σ-bonding orbital like the d_{z^2}, is formed; this ML_5 monomeric complex then dimerises to M_2L_{10}. Thus d^7 ions with strong-field ligands such as CO and CN^- often form dimeric complexes with metal—metal bonds; examples are the compound $Mn_2(CO)_{10}$ and the anion $[Co_2(CN)_{10}]^{6-}$.

The d^8 complexes have already been discussed under ligand-field theory (6.4.1). For them the molecular-orbital picture is effectively the same, a square arrangement of strong-field ligands enables eight electrons to be accommodated in relatively stable orbitals. In instances where the metal is in a low oxidation state, however, a trigonal bipyramidal structure is possible (*e.g.* $Fe(CO)_5$), because back-donation both strengthens the π-bonding, which is favoured in the trigonal-bipyramidal arrangement, and relieves the concentration of negative charge at the metal.

For d^9 metal ions, both octahedral and square arrangements of ligands have electrons in strongly antibonding σ^* orbitals. Tetragonal arrangements are often more stable, but dimeric complexes with metal—metal bonds are also possible. Complexes of d^{10} metal ions are tetrahedral. The arrangement enables all the electrons to occupy low-energy orbitals (6.6.2).

For complexes of weak-field ligands with small mutual repulsions, there is usually no reason for the structure to be other than octahedral unless the metal ion is either a d^9 or a d^{10} one. The d^9 complexes are usually tetragonal, the d^{10} almost always tetrahedral.

In the complexes of Period 5 and Period 6 transition metals, the higher values of Δ make the occupancy of antibonding orbitals a more important factor in determining structure. At the same time the greater distances between the ligands, owing to longer metal—ligand bonds, reduce their mutual repulsion. Thus $PdBr_4^{2-}$ and $PtBr_4^{2-}$ are planar, which is the preferred symmetry for d^8 strong-field, weak-repulsion ligands; whereas $NiBr_4^{2-}$ is tetrahedral, which is the preferred symmetry for d^8 weak-field, strong-repulsion ligands.

Further Reading

T.M. Dunn, The visible and ultraviolet spectra of complex compounds, in J. Lewis and R.G. Wilkins (Eds.), Modern coordination chemistry, Interscience, New York, 1960.

C.K. Jørgensen, Absorption spectra and chemical bonding in complexes, Pergamon, London, 1961.

J.S. Griffith, The theory of transition metal ions, Cambridge University Press, London, 1961.

C.K. Jørgensen, The nephelauxetic series, in F.A. Cotton, Progress in inorganic chemistry, Vol. 4, Interscience, New York, 1962, p. 73.

C.J. Ballhausen, Intensities of spectral bands in transition-metal complexes, in F.A. Cotton, Progress in inorganic chemistry, Vol. 2, Interscience, New York, 1960, p. 251.

R.S. Drago, Physical methods in inorganic chemistry, Reinhold, New York, 1965.

J.S. Griffith and L.E. Orgel, Ligand field theory, Quarterly Reviews, 11 (1957) 381.

L.R. Orgel, Spectra of transition-metal complexes, J. Chem. Phys., 23 (1955) 1004.

C. Nordling, ESCA, Angew. Chem. Internat. Ed., 11 (1972) 83.

D.R. Eaton and K. Zaw, Magnetic resonance methods in the study of transition metal complexes. Co-ord. Chem. Rev., 7 (1971) 197.
F.A. Cotton, Ligand field theory, J. Chem. Ed., 41 (1964) 466.
D.W. James and M.J. Nolan, Vibrational spectra of transition-metal complexes and the nature of the metal—ligand bond. Prog. Inorg. Chem., 9 (1968) 195.
B.A. Goodman and J.B. Raynor, Electron spin resonance of transition-metal complexes, Adv. Inorg. and Radiochem., 13 (1970) 136.

7

The Solid State

7.1. Experimental Evidence on Structure

The term *structure* when applied to solids refers to the arrangement of the atoms, ions or molecules which compose them. This arrangement in a solid may be the ordered one of the crystal or the random one of amorphous material. In fact crystals are probably never perfectly ordered nor amorphous material completely random, but this does not detract from the practical usefulness of the distinction. Evidence on which our current knowledge of the structure of solids is based has been obtained from a study of their diffraction of X-rays, electrons or neutrons. These methods are very briefly outlined below.

7.1.1. X-Ray diffraction

Friedrich (1912), at the suggestion of Von Laue, demonstrated that X-rays produce diffraction patterns when passed through crystals. In the same year W.L. Bragg investigated the reflection of monochromatic beams of X-rays from the surfaces of crystals such as NaCl and ZnS. He found that, for a particular salt, there were certain angles between the incident beam and the surface which gave rise to strong reflection. Fig. 7.1 illustrates his explanation: the lines AB, A′B′ and A″B″ represent parallel rays reflected from successive planes of atoms or ions in a crystal. It should be noted, however, that the interaction occurs between X-rays and the extra-nuclear electrons of the atoms or ions. The path differences between rays AB and A′B′, reflected from planes separated by a distance d, is $ED = 2d \sin \theta$, since $B'B = B'D$, and $BD = 2d$. Reinforcement of reflection can occur only if the waves along BC are in phase, that is if the difference in path length represents an integral number of wavelengths.

Thus the condition for reinforcement of reflection is

$$n\lambda = 2d \sin \theta$$

in which λ is the wavelength of the X-rays and n is a small whole number. The intensity of the beam falls sharply to zero when the glancing angle is altered slightly from that required for strong reflection, because the reflected waves are then out of phase.

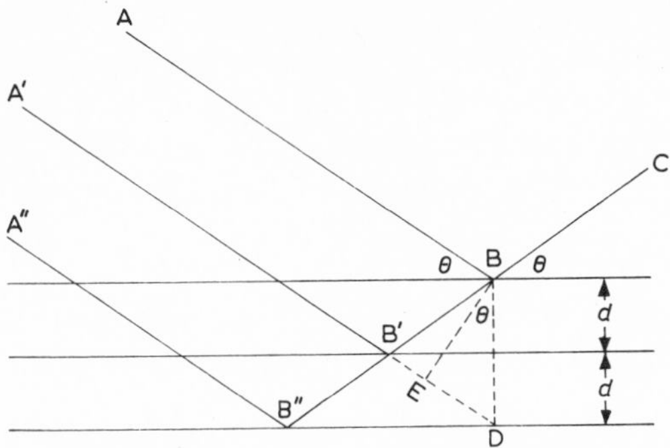

Fig. 7.1. Illustrating the derivation of the Bragg equation.

Once it became possible to measure distances between planes in a crystal, the positions of different atoms relative to one another could be determined. In a three-dimensional array of atoms there will be many possible planes. In Fig. 7.2 the small circles represent atoms in a crystal, spaced out and confined to a two-dimensional representation for clarity. Obviously the layers of atoms joined by solid lines are separated by greater distances than those joined by dotted lines: $d = d'\sqrt{2}$.

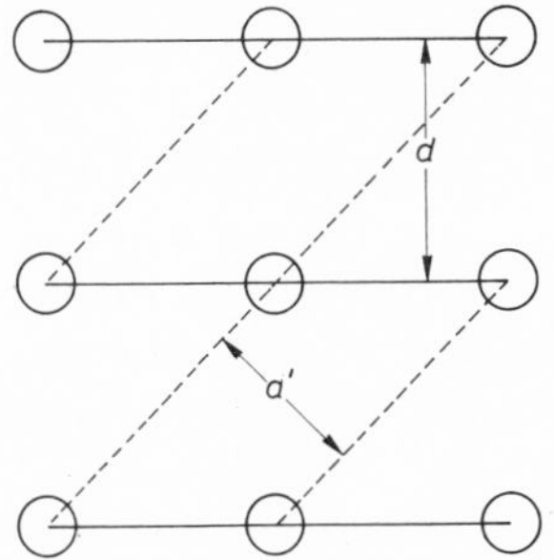

Fig. 7.2. 100 and 110 planes in a cubic crystal (circles locate positions of nuclei).

The ways that atoms can be arranged in a true metal have already been discussed (4.1.7). Figs. 7.8, 7.9 and 7.10 show the planes labelled 100, 110 and 111 which occur in a cubic system. In the b.c.c. structure (Fig. 4.30) the ratios of the distances between 100, 110 and 111 planes are

$$d_{100} : d_{110} : d_{111} = 1 : \sqrt{2} : \frac{1}{\sqrt{3}}$$

whereas in the c.c.p. structure (Fig. 4.31) the corresponding ratios are

$$d_{100} : d_{110} : d_{111} = 1 : \frac{1}{\sqrt{2}} : \frac{2}{\sqrt{3}}$$

Clearly it is possible to distinguish between a b.c.c. structure and a c.c.p. structure by measuring the distances between layers of atoms for various orientations of a crystal relative to the incident beam.

In ionic crystals like NaCl (Fig. 7.11) some planes will contain Na^+ ions only, some Cl^- only and some both Na^+ and Cl^-. As the intensities of the reflections depend not only on the spacing of atoms in a plane but also on the number of electrons per atom, we should expect a layer of Cl^- ions, with 18 orbital electrons each, to reflect X-rays more strongly than a parallel layer of Na^+ ions with only 10 electrons each. In general, by measuring both the distances between various planes and the intensities of the reflections from them, it is possible to determine relative positions of atoms and ions within crystalline solids, though analysis of the results often presents a complex mathematical problem.

7.1.1.1. Powder method

The atoms in crystals form a diffraction grating for X-ray and other short-wave radiation. Because it is a three-dimensional grating, diffraction occurs only at certain orientations of the crystal with respect to the direction of an incident beam. Debye and Scherrer (1917) used the diffraction pattern recorded for a monochromatic beam of X-rays falling on a powder of small crystals, or crystal fragments, to deduce Bragg angles and hence structure. The *powder method*, in which finely divided material, compressed or held in the form of a rod, is rotated in a beam of near-monochromatic X-rays and the diffractions from it are recorded on a photographic film mounted in a circular camera of known dimensions, has been developed to provide diffraction angles and intensities with great precision.

The method is much used in chemical investigation because of its applicability to powders and to material of poor crystallinity. It readily shows the presence of crystalline components in mixtures and hence has been an instrument for both qualitative and quantitative analysis. For these purposes the diffraction pattern of the sample is compared with the patterns of known materials. Provided the symmetry of the unit cell (7.2.1) is high, powder photographs can be indexed without great difficulty, but for unit cells of low symmetry a powder photograph will often contain a number of overlapping lines. In such cases it is more useful to examine single crystals rather than powders.

7.1.1.2. Single crystal method

Schiebold (1919) showed that a single crystal fixed in a narrow beam of monochromatic X-rays gives diffractions which may be recorded as dark spots on a photographic plate. Provided the geometry of the system is known the Bragg angles may be found and the structure deduced. The *single crystal method* has also been much developed and, where a good crystal is available, it now provides the most powerful means of determining structure, particularly elaborate structure.

7.1.2. Neutron diffraction

The diffraction of X-rays and electrons is due to interaction with the orbital electrons of the atoms they encounter. The diffraction of neutrons springs from different causes:

(i) Nuclear scattering is brought about by interaction with protons or neutrons in the nucleus and depends upon (a) nuclear *size* (which increases only slowly with atomic weight) and (b) nuclear *structure* (a dependence which appears to vary arbitrarily from one element to the next). Thus the hydrogen atom scatters neutrons as well as the potassium and better than the cobalt. This is in marked contrast to X-rays where the scattering increases smoothly with the number of extra-nuclear electrons in the atom.

(ii) Magnetic scattering arises from interaction between the magnetic moment of the neutron and that of the atom or ion in question. Thus the Fe^{3+} ion, with its unpaired electrons, gives additional scattering superimposed upon the nuclear scattering. Magnetic scattering is used to investigate the magnetic properties of alloys.

Neutrons emerging from an atomic reactor have a range of energies. They are collimated and monochromated to give a narrow pencil of neutrons with an energy corresponding to a wave length of about 100 pm. After scattering by the specimen, usually a single crystal, the diffraction pattern is obtained by means of a counter, filled with $^{10}BF_3$, connected to a pen-recorder.

The diffraction of neutrons provides a way of locating hydrogen atoms in compounds and is used to complement X-ray study of crystals especially by locating and characterising water molecules in hydrates.

7.2. Structure and Properties

The properties of a solid depend not only on the number and kind of atoms composing it but also on their arrangement. The empirical formula reveals nothing of this, for a compound AB_2 can exist in the solid:

(i) as separate molecules (B—A—B in solid carbon dioxide);

(ii) as an infinite layer (Fig. 7.3) (cadmium iodide, CdI_2, contains puckered layers of this form);

(iii) in various three-dimensional structures (Fig. 7.4) (fluorite, CaF_2, has cubic crystals with one of these arrangements).

In (i) every carbon atom is connected to two oxygen atoms, in (ii) every cadmium atom has six iodine atoms as nearest neighbours, and in (iii) every calcium atom has eight fluorine atoms as nearest neighbours. The chemical

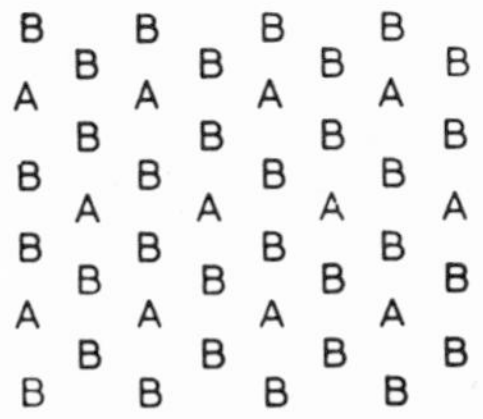

```
B     B     B     B
   B     B     B     B
A     A     A     A
   B     B     B     B
B     B     B     B
   A     A     A     A
B     B     B     B
   B     B     B     B
A     A     A     A
   B     B     B     B
B     B     B     B
```

Fig. 7.3. Infinite layer of formula AB_2.

formula of a solid should therefore be considered in relation to its crystal structure. Even two solids as similar in formulation as PCl_5 and PBr_5 differ structurally; the former consists of equal numbers of PCl_4^+ ions and PCl_6^- ions, the latter of equal numbers of PBr_4^+ and Br^- ions.

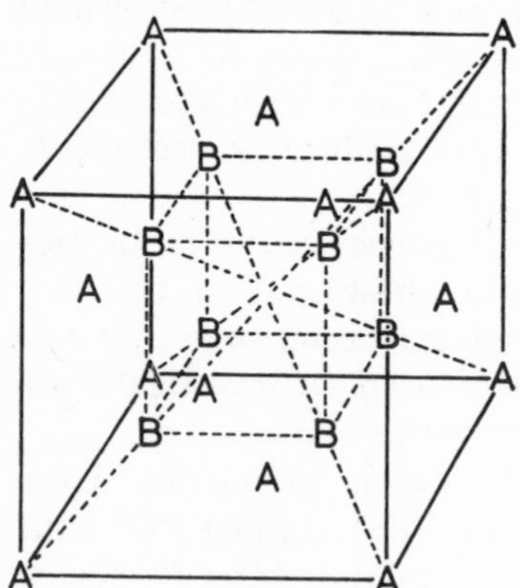

Fig. 7.4. Cubic structure of formula AB_2.

7.2.1. The unit cell

A crystalline solid consists of atoms (or ions) packed regularly in a three-dimensional arrangement. For any point in the pattern it is possible to find other points possessing exactly the same environment. Such points then define a regular *lattice*: one lattice point can be reached from another by taking a suitable number of primitive translations (steps) along each of three suitable directions. Clearly the whole pattern can be built up by translating a block of essentially different points lying in a certain *unit cell*. The nature of the solid is determined by the size, shape and content of its unit cell.

The size and shape of a unit cell is defined by the lengths (a, b, c) of three intersecting edges and the angles (α, β, γ) between them (Fig. 7.5).

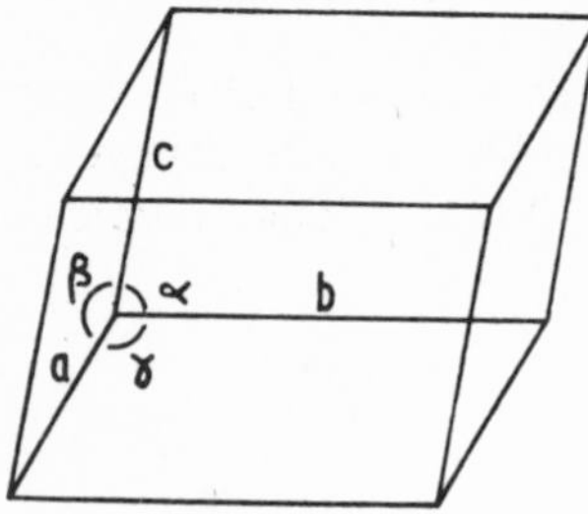

Fig. 7.5. Unit cell.

There are seven types of unit cell (Table 7.1) and therefore seven *simple* or *primitive* lattices with one unit of pattern at each cell corner.

TABLE 7.1

THE SEVEN TYPES OF UNIT CELL

System	Cell edges	Cell angles	Example
Cubic	$a = b = c$	$\alpha = \beta = \gamma = 90°$	Rock salt
Tetragonal	$a = b \neq c$	$\alpha = \beta = \gamma = 90°$	White tin
Orthorhombic	$a \neq b \neq c$	$\alpha = \beta = \gamma = 90°$	Mercuric chloride
Monoclinic	$a \neq b \neq c$	$\alpha = \gamma = 90°, \beta \neq 90°$	Potassium chlorate
Triclinic	$a \neq b \neq c$	$\alpha \neq \beta \neq \gamma \neq 90°$	Potassium dichromate
Hexagonal	$a = b \neq c$	$\alpha = \beta = 90°, \gamma = 120°$	Silica
Rhombohedral	$a = b = c$	$\alpha = \beta = \gamma \neq 90°$	Calcite

7.2.2. Bravais lattices

If the *contents* of a unit cell containing a number of units of pattern (atoms, molecules) have symmetry, the number of distinct types of space

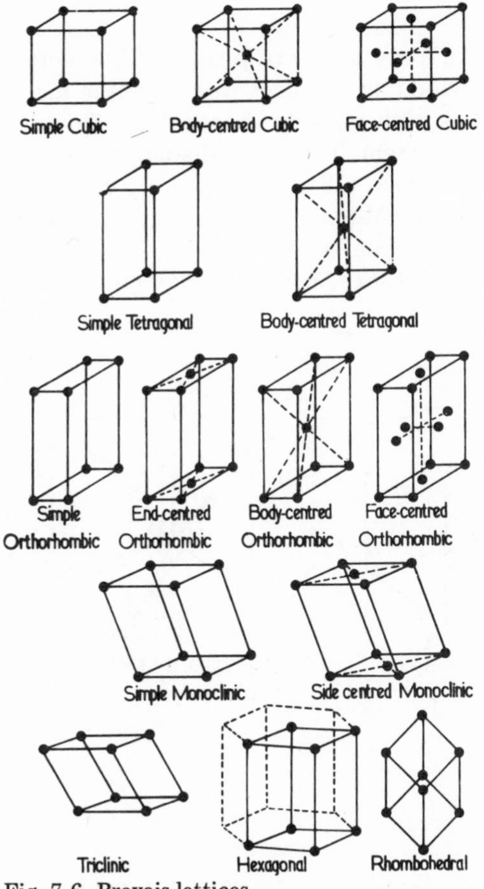

Fig. 7.6. Bravais lattices.

lattice, as Bravais noted in 1848, becomes *fourteen* (Fig. 7.6). And when other symmetry operations are recognised (e.g. rotation of the lattice) there are found to be 230 distinct varieties of crystal symmetry.

A compound lattice that contains body-, face-, or end-centred points can always be regarded as a *simple* lattice with a smaller unit cell. Thus, the body-centred tetragonal lattice is a special case of the triclinic lattice (Fig. 7.7). The centred lattices describe most readily the higher symmetry.

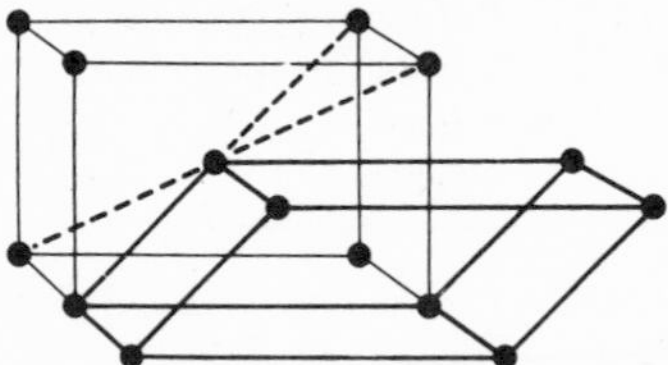

Fig. 7.7. Relation of body-centred tetragonal to triclinic.

7.2.3. Miller indices

For the convenient naming of faces of a crystal, or of planes within a crystal, a system was introduced in 1839 which uses indices which are the reciprocals of the intercepts the plane makes with suitably chosen axes. The idea is most easily understood for the cubic system. The plane shaded in Fig. 7.8 is a 100 plane, the intercepts made on the OA, OB and OC axes by the plane being 1, ∞, and ∞, respectively, so that the Miller indices are $\frac{1}{1}$, $\frac{1}{\infty}$, $\frac{1}{\infty}$, which are represented by the figures 100. The plane shaded in Fig. 7.9 is a 110 plane, the intercepts on the OA, OB and OC axes being 1, 1 and ∞, and that shaded in Fig. 7.10 is a 111 plane.

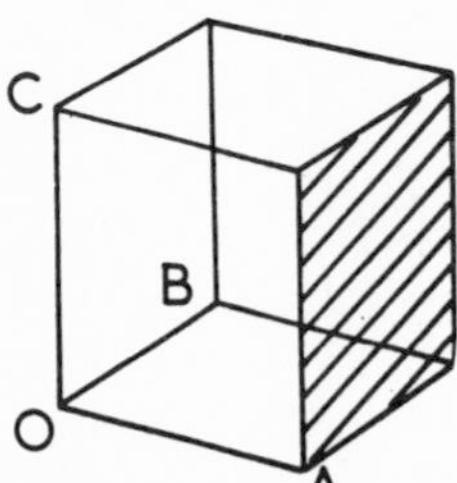

Fig. 7.8. 100 plane in cubic system.

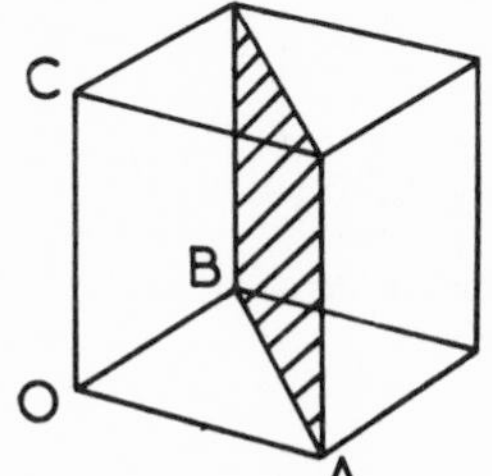

Fig. 7.9. 110 plane in cubic system.

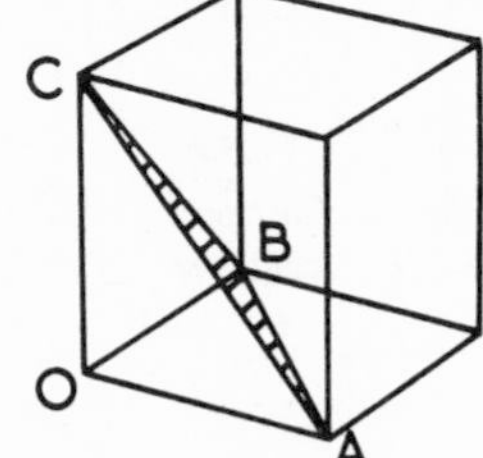

Fig. 7.10. 111 plane in cubic system.

7.2.4. Cohesive forces in crystals

Although the sub-division of valency bonds into types is not rigid, four

TABLE 7.2

HOMODESMIC SOLIDS

Type of binding		Examples
(i)	Ionic	Rock salt, fluorite
(ii)	Homopolar	Diamond, zinc blende
(iii)	Metallic	Copper, manganese
(iv)	Van der Waals	Argon

main types are distinguishable in solids. When all the bonds are of the same type the solid is *homodesmic* (Table 7.2). Should the bonds be all of equal strength the solid is *isodesmic*, otherwise it is *anisodesmic*.

Heterodesmic solids contain more than one type of bond; two examples are given:

(i) Solid carbon dioxide has homopolar bonds in the individual molecules; but the molecules are held to one another by Van der Waals' forces.

(ii) Antimony consists of puckered sheets of covalently bound atoms with metallic binding holding the sheets together.

7.2.5. Ionic binding

Monatomic ions may be considered as incompressible spheres which are usually only slightly polarised by the ions of opposite charge around them. There are two conditions for an arrangement of ions in space to have minimum potential energy:

(i) The larger ions (usually the anions) round a smaller ion of opposite sign must all 'touch' it.

(ii) The co-ordination number of an ion, that is the number of nearest neighbours of opposite sign, must be as large as possible, subject to condition (i) being observed.

The co-ordination numbers 5, 7, 9, 10 and 11-are excluded by geometry if the ionic arrangement is to form a regular spatial pattern; in addition, the co-ordination number 12 is excluded by the requirement that positive and negative charges balance one another. Only 6 Cl^- ions (181 pm) can be accommodated round a Na^+ ion (95 pm) so that they all 'touch' the cation; very many more Na^+ ions could be accommodated round the Cl^- ion. But electrical neutrality must be achieved, and as the system is stable only when the cation touches all the anions surrounding it, the co-ordination in sodium chloride is 6 : 6. Any single Na^+ ion has as its nearest neighbours 6 Cl^- ions arranged octahedrally; similarly any Cl^- has a corresponding arrangement of Na^+ ions around it (Fig. 7.11).

The Cs^+ ion (169 pm) can accommodate about it 8 Cl^- ions and the co-ordination in caesium chloride is 8 : 8 (Fig. 7.12).

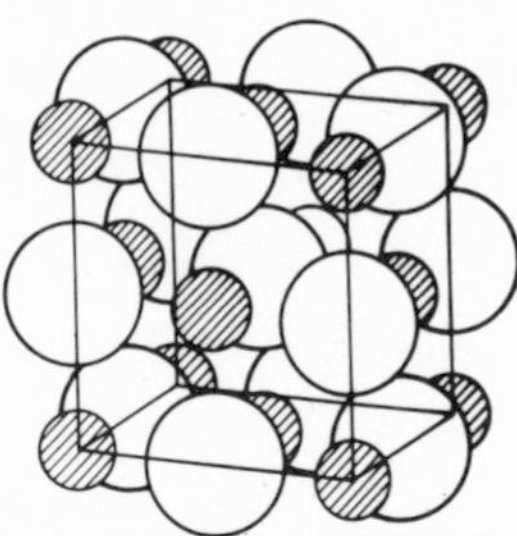

Fig. 7.11. Unit cell of NaCl. In this and subsequent diagrams the shaded circles represent the metallic (or less electronegative) constituent.

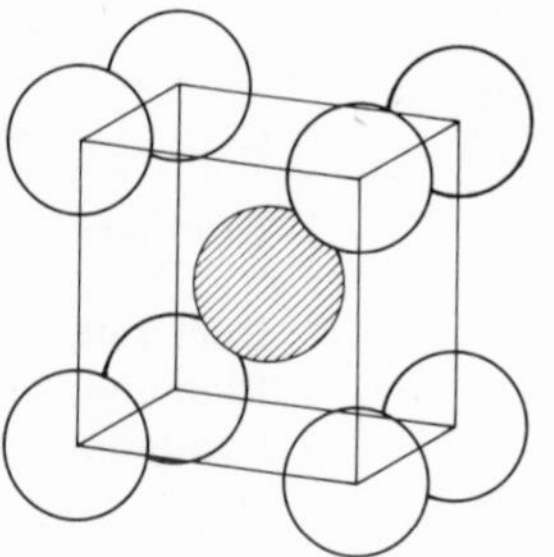

Fig. 7.12. Unit cell of CsCl.

7.2.6. Radius ratios

The number of anions surrounding the cation in the crystal, the *crystallographic co-ordination number* of the cation, depends on the ratio r_c/r_a, of the cationic and anionic radii. Fig. 7.13 represents the smallest cation which is capable of holding apart four anions in the same plane. Its radius $r_c = \sqrt{2}(r_a - r_c)$ and $r_c/r_a = \sqrt{2} - 1 = 0.414$. This is the smallest *radius ratio,* r_c/r_a for which 6-co-ordination of the cation is possible — in addition to the four anions in the plane one anion can be accommodated above, one below the cation. For lower values of r_c/r_a, 6-co-ordination of the cation is not possible because the anions touch one another but not the cation. The limiting values of r_c/r_a for various co-ordination numbers are given in Table 7.3.

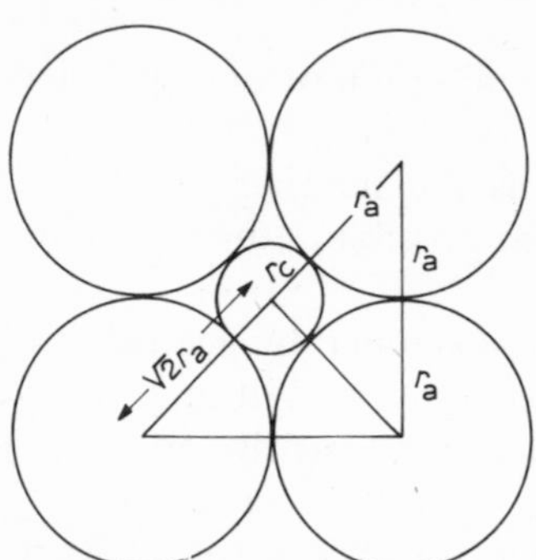

Fig. 7.13. Illustrating the radius ratio rule.

In crystals of a compound AB_2 the co-ordination number of A is twice that of B and the structure is determined by the co-ordination number of the smaller ion. The *radius ratio rule* is clear-cut only with simple ions. Dimorphism is much commoner where complex ions are involved. Even the relatively simple CO_3^{2-} ion can co-ordinate in different ways without causing any change in the empirical formula of the compound; thus $CaCO_3$ exists as

TABLE 7.3

CORRESPONDENCE BETWEEN r_c/r_a VALUES, CO-ORDINATION NUMBER OF CATION, AND THE ARRANGEMENT OF IONS

r_c/r_a	Co-ordination number of cation	Arrangement
1 → 0.73	8	At cube corners
0.73 → 0.41	6	Octahedral
0.41 → 0.22	4	Tetrahedral
Square, triangular and linear arrangements are possible in anisodesmic crystals:		
0.73 → 0.41	4	Square
0.22 → 0.16	3	Triangular
0.16	2	Linear

both the minerals calcite and aragonite. In general, the rule is most likely to break down where large, easily polarised anions are present.

7.2.7. Strength of ionic binding

As the energy required to separate a pair of ions is approximately proportional to

$$\frac{z_1 z_2}{r_c + r_a}$$

where z_1 and z_2 are the respective charges, the hardness (Table 7.4) and melting point (Table 7.5) of crystals with similarly charged ions should vary inversely as the inter-ionic distance. Where the inter-ionic distance is about the same, the relative hardness and melting point become dependent on the ionic charges (Table 7.6).

TABLE 7.4

$(r_c + r_a)$-VALUES AND HARDNESS OF SOME OXIDES

	MgO	CaO	SrO	BaO
$(r_c + r_a)$/pm	205	240	253	275
Hardness (Mohs' scale)	6.5	4.5	3.5	3.2

TABLE 7.5

$(r_c + r_a)$-VALUES AND MELTING POINTS OF THE SODIUM HALIDES

	NaF	NaCl	NaBr	NaI
$(r_c + r_a)$/pm	229	276	290	314
M.p./K	1261	1074	1013	933

TABLE 7.6

INFLUENCE OF IONIC CHARGE ON STRENGTH OF IONIC BINDING

	NaF	CaO
$(r_c + r_a)$/pm	229	240
$z_1 \times z_2$	1	4
Hardness (Mohs' scale)	3.2	4.5
M.p./K	1261	2843

Crystals with strong ionic bonds always have low coefficients of thermal expansion. Though the solids are non-conducting because the electrons are all firmly bound in atomic orbitals, conductance by migration of ions occurs in the fused state. Moreover, ionic bonds are often broken by media of high relative permittivity such as water ($\epsilon_r = 78$). Binary simple salts, exemplified by NaCl, are generally more easily dissociated in aqueous solution than such compounds as MgO where the interionic attraction is large.

Certain principles are evident in the building up of ionic crystals. These, known as Pauling's rules, are useful even when the ionic picture is not strictly valid and considerable covalent character must be conceded:

(i) A co-ordinated polyhedron of anions is formed about each cation, the cation—anion distance being determined by the radius sum, and the co-ordination number by the radius ratio.

(ii) In a stable co-ordination structure, the total electrostatic strength of the bonds which reach an anion from all the neighbouring cations is equal to the charge on the anion. The term electrostatic strength is defined as follows. If a cation of charge $+ze$ is surrounded by n anions, the electrostatic strength of the bonds from the cation to a neighbouring anion is z/n. The rule therefore implies that charges in ionic crystals are neutralised locally.

(iii) The existence of edges, and particularly faces, common to two anion polyhedra decreases the stability of the structure.

(iv) Cations with a high charge number and a small co-ordination number tend not to share polyhedron elements with one another.

(v) The environments of all chemically similar anions in a structure tend to be similar.

These rules are particularly valuable in their application to more complicated structures.

7.2.8. The Jahn—Teller effect in crystals

So far it has been assumed that the field due to a monatomic ion is spherically symmetrical, but for cations which have certain d-electron configurations this is untrue. For instance a regular octahedral arrangement of anions is not possible around the Cu^{2+} ($t_{2g}^6\ e_g^3$) ion which has a degenerate ground state, with either the d_{z^2} or the $d_{x^2-y^2}$ orbital containing only one electron. If it is the $d_{x^2-y^2}$ then the charge on the Cu nucleus is more shielded along the z-axis than in the x- and y-axial directions. As a result the equilibrium positions of anions in the xy-plane will be closer to the metal than those on the z-axis. Consequently, the arrangement of anions round the cation is distorted from the octahedral, even when the radius ratio is favourable for 6 : 6 co-ordination.

Thus CuO has a structure which is unlike the monoxides of other transition metals of Period 4 such as MnO and NiO, which have cubic NaCl structures. The lack of distortion in NiO suggests that the Ni^{2+} ion exists in the symmetrical high-spin form of the $t_{2g}^6\ e_g^2$ state. The magnetic properties of NiO are in accordance with this electronic configuration (6.5). But on the other hand PdO has a tetragonal unit cell; the crystal field exerted by the oxygen ions is evidently sufficient to cause spin-pairing of the electrons in the e_g level of the Period 5 ion. There is actually a square arrangement of oxygen atoms around the Pd (Fig. 37.9) as expected for a low-spin d^8 cation.

7.2.9. Homopolar crystals

Unlike ionic structures, where crystallographic co-ordination numbers are determined by the radius ratio rule and the bonds are not directed, homopolar structures have directed bonds. An element in the Nth group of the Periodic Table can form 8—N bonds per atom and co-ordination numbers are small, usually four or less.

Few crystals are perfectly homopolar. An example is diamond in which the individual carbon atoms are held tetrahedrally to 4 others by bonds similar to those between the carbon atoms in aliphatic compounds; accordingly the crystal is a truly homopolar, homodesmic solid. Silicon and germanium are similar. Zinc blende, ZnS, has like character, but the 4 bonds about the individual atoms are formed from 2 electrons derived from the zinc and 6 from the sulphur atoms. The valency electrons tend to be drawn towards the sulphur atoms, conferring some polarity on the bonds. This illustrates the fact that there is no sharp demarcation between homopolar and ionic crystals. Zinc blende would normally be considered covalently bonded — a low co-ordination number usually implies directed bonds — but in this, and indeed in many other instances, it is convenient to regard the structure formally as being composed of Zn^{2+} and S^{2-} ions, the latter distorted, or polarised, in the field of the former. Such an approach is possible because in solids the exigencies of packing and electrical neutrality are determining factors. But divergence from ionic character means that bond lengths are shortened and co-ordination numbers reduced below those required by the radius ratio rule.

7.2.10. Metals

The true metals comprise those of Groups IA and IIA and the transition groups: they are designated by T. Atoms of a T metal are closely packed and without directional bonds. The metals are relatively soft and malleable since the atoms glide easily over one another.

B sub-group metals are rather more covalent in character, those of Groups IIB and IIIB being nearer true metals than Se, Te, As, Sb and Bi. Zinc and cadmium, for example, have distorted, close-packed hexagonal arrangements of atoms in which the axial ratios are about 1.87 instead of the ideal 1.63. Aluminium and indium have approximately c.c.p. lattices, and thallium has a close-packed hexagonal one. In Group IVB, white tin possesses a structural character between that of lead and silicon; its co-ordination number is 6. Grey tin has the diamond structure and a co-ordination number of 4. Lead is c.c.p. and behaves as a true metal.

7.2.11. Substitutional alloys

These are obtained by starting with metal X and gradually replacing its atoms with those of metal Y. Two true metals yield a TT alloy. But TB and BB alloys are also possible.

A typical TT combination is formed by silver and gold, both of which have a face-centred cubic lattice and atoms of very nearly the same size (Au, 144.2 pm; Ag, 143.2 pm). They give a continuous series of solid solutions with random distribution of gold and silver atoms. An alloy of this type is possible when the radius ratio of the atoms does not exceed 1.14 and their charge numbers are alike.

Copper has a metallic radius of 128.8 pm. At a high temperature, gold forms a continuous solid solution with it. Rapid quenching retains this condition, but slow annealing allows the atoms to segregate. At a composition Au : Cu = 1 : 1, alternate sheets of gold and copper appear and the symmetry is converted from cubic to tetragonal. With Au : Cu = 1 : 3, the ordered form is cubic, Au being at the cube corners and Cu at the face centres (Fig. 7.14). The disordered form of this alloy of gold and copper is brittle; the ordered form is more malleable, more ductile, and a better conductor. At the *order-disorder transition temperature* the specific heat increases markedly, thermal energy being necessary to disorder the atoms and increase the entropy.

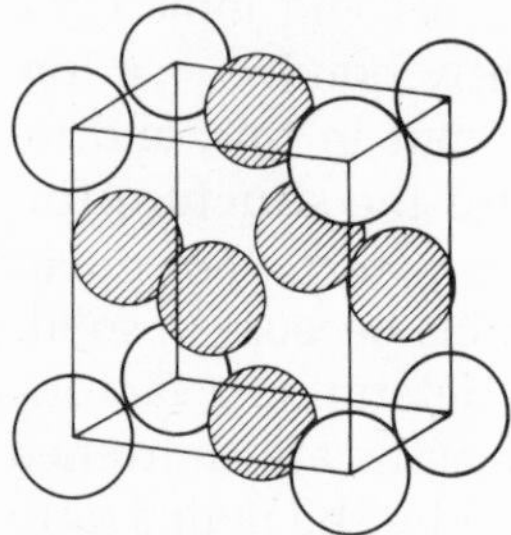

Fig. 7.14. Unit cell of the ordered form of Cu_3Au.

7.2.11.1. Electron compounds

An example of a TB alloy is found in the system silver—cadmium. Silver has a cubic close-packed structure, cadmium an elongated close-packed hexagonal one like zinc (39.2). Five phases occur. The random α-phase contains less than 42% Cd. At about this composition a β-phase begins to be formed, and at 50% Cd the structure is ordered body-centred cubic. At 57% Cd a complicated γ-phase appears, and from 70—82% Cd a close-packed hexagonal ϵ-phase. Finally, at high Cd concentrations a η-phase is formed which is a solid solution of silver in cadmium in close-packed hexagonal arrangement. Many other TB systems show a similar sequence of intermediate phases. In the zinc, aluminium and tin alloys of copper, β-, γ- and ϵ-phases appear at certain compositions (Table 7.7). Hume—Rothery (1926) pointed

TABLE 7.7

COMPOSITION AND PHASE OF SOME ALLOYS

β	γ	ϵ
CuZn	Cu_5Zn_8	$CuZn_3$
Cu_3Al	Cu_9Al_4	Cu_5Al_3
Cu_5Sn	$Cu_{31}Sn_8$	Cu_3Sn

out that the appearance of a particular phase depended on the ratio of valency electrons to atoms (Table 7.8). For a β-phase it is 3/2. For a γ-phase the ratio is 21/13 and for an ϵ-phase 7/4. The fractions are most easily remembered as β 21/14, γ 21/13 and ϵ 21/12. The phases themselves are sometimes referred to as *electron compounds.*

TABLE 7.8

RATIO OF VALENCY ELECTRONS TO ATOMS FOR SOME ALLOYS

Cu (1) + Zn (2)	$\frac{3}{2}$
3 Cu (1) + Al (3)	$\frac{6}{4}$
5 Cu (1) + Sn (4)	$\frac{9}{6}$

An alloy of a T metal with a B element of Groups VB and VIB often has intermediate phases quite different from those of the constituent elements. They frequently have structures like NiAs (Fig. 7.18); examples are the arsenides, antimonides, bismuthides, sulphides, selenides and tellurides of many T metals. The binding in these is partly covalent and partly ionic.

7.2.12. Interstitial alloys

The borides, carbides and nitrides of the transition metals have metallic properties. Only atoms with small covalent radii are capable of occupying the interstices in relatively close-packed arrangements:

	B	C	N
r/pm	82	77	75

These interstitial alloys are of great technical importance; for instance all steels are interstitial alloys of carbon and iron.

When the radius ratio is less than 0.59 the alloy is 'normal' and the metal—interstitial atom arrangement is c.c.p., h.c.p. or b.c.c. The 'complex' interstitial alloys have a radius ratio greater than 0.59 and are less stable. Carbon and nitrogen always occupy octahedral holes in interstitial alloys. In c.c.p. and h.c.p. lattices there are as many octahedral holes as there are metal atoms and twice as many tetrahedral holes.

Austenite, an iron—carbon phase in steel, has a c.c.p. arrangement of iron atoms and a maximum carbon content corresponding to the filling, quite at random, of one twelfth of the octahedral interstices. Rapid quenching of austenitic steel, which is stable at high temperature, causes the iron atom arrangement to change to body-centred, but the interstitial carbon atoms hold the metal atoms apart and allow them to assume a tetragonal, rather than a cubic, configuration. This supersaturated solid solution is *martensite*. Tempering precipitates the carbon as an ϵ-iron carbide with a h.c.p. structure at low temperatures, but at higher temperatures *cementite*, Fe_3C, is produced. During these last transformations the iron positions alter only slightly, the changes being due to the diffusion and re-arrangement of the carbon atoms.

The nature of the valency forces in interstitial alloys has been variously explained. It is clear that such alloys are restricted to metals with incompletely filled d orbitals. Electrons may be donated by the interstitial atoms, leaving these as positive ions and resulting in binding of a metallic nature.

7.2.13. Clathrates

When quinol is crystallised from aqueous solution in the presence of argon at 4 MPa, the solid has the properties of quinol but contains argon which is set free when the quinol is melted or dissolved. The gas molecule is trapped inside a cage of hydrogen-bonded quinol molecules. Clathrates are formed

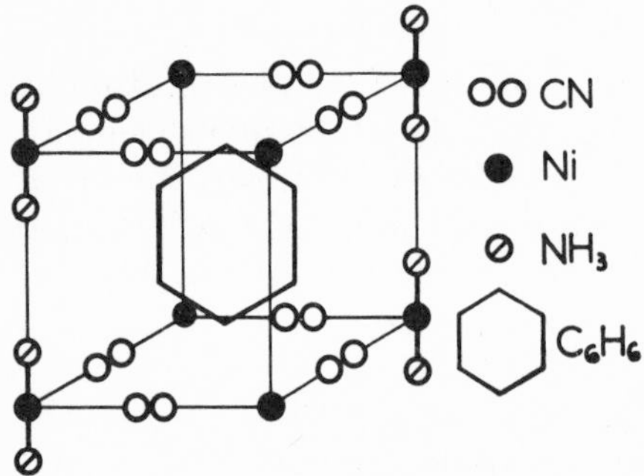

Fig. 7.15. Unit cell of benzene ammino-nickel cyanide clathrate.

by krypton, xenon and such gases as HCl, HBr, SO_2 and CO_2 with quinol, the general formula being $[C_6H_4(OH)_2]_3X$, where X represents a molecule of the gas.

An ammoniacal solution of nickel cyanide produces a pale mauve clathrate when shaken vigorously with benzene. The formula is $Ni(CN)_2NH_3 \cdot C_6H_6$ and the structure as shown in Fig. 7.15.

7.2.14. AB structures

There are three principal types among isodesmic crystals of this empirical formula:

(i) 8 : 8 co-ordination as in caesium chloride (Fig. 7.12); e.g. CsBr, CsI, RbF, TlCl, TlBr, TlI, NH_4Cl, NH_4Br, AgLi, AlFe.

(ii) 6 : 6 co-ordination as in rock salt (Fig. 7.11); e.g. Li, Na and K halides, NH_4I, RbCl, RbBr, RbI, the oxides MgO, CaO, SrO, BaO, CaO, MnO and NiO, and the compounds LaSb and CeBi.

(iii) 4 : 4 co-ordination as in zinc blende (Z) (Fig. 7.16) and wurtzite (W) (Fig. 7.17); e.g. (Z) AgI, AlSb, BeS, BeSe, SiC, CdSe, CuBr, HgS; (W) AgI, BeO, CdS, NH_4F, ZnO, MgTe.

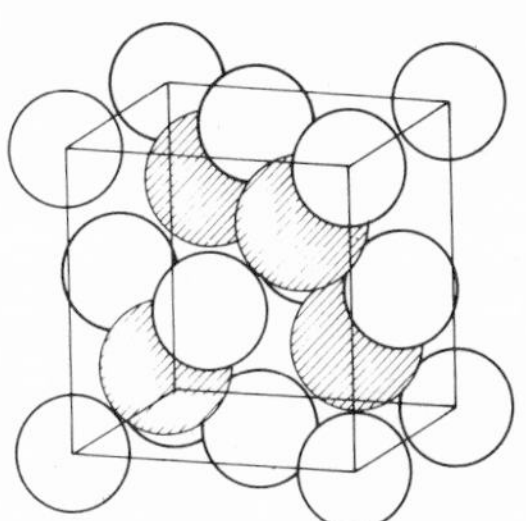

Fig. 7.16. Zinc blende structure.

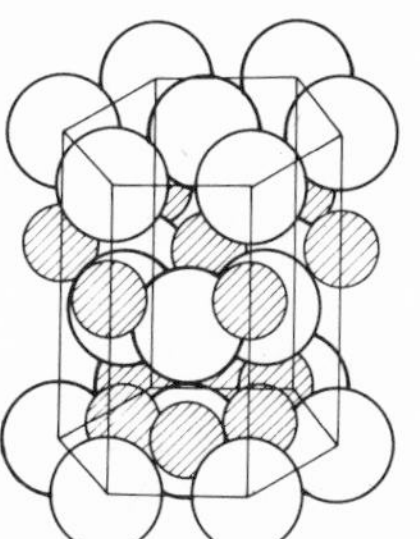

Fig. 7.17. Wurtzite structure (3 unit cells).

Zinc forms a bipositive ion with an 18-electron structure and a moderately small radius (74 pm). The large S^{2-} ion (184 pm) is considerably polarised by it and the two forms of ZnS are predominantly homopolar, since both atoms can adopt sp^3 hybridisation and pool their valence electrons. Blende is cubic, wurtzite hexagonal, there being an average of 4 electrons per atom for bond formation as in diamond. Homopolar compounds of this type are formed by many pairs of elements whose valency electrons total 8. They are called *adamantine compounds* and have either a blende or wurtzite structure (see Table 7.9).

The *Grimm—Sommerfeld rule* for adamantine compounds states that if the sum of the atomic numbers is constant and the number of valency electrons is constant the inter-atomic distances are constant.

The non-adamantine lattice of nickel arsenide, NiAs, is another common

TABLE 7.9

EXAMPLES OF ADAMANTINE COMPOUNDS

Formulae	Valency electrons	Atomic numbers	Inter-atomic distances
CuBr (Z)	1 + 7 = 8	29 + 35 = 64	246 pm
ZnSe (Z)	2 + 6 = 8	30 + 34 = 64	245 pm
GaAs (Z)	3 + 5 = 8	31 + 33 = 64	244 pm
GeGe (Z)	4 + 4 = 8	32 + 32 = 64	246 pm

AB structure (Fig. 7.18); it is also displayed by FeS, FeSe, CoS, CoSe, CoTe, NiS, NiSe, NiTe, NiSb, MnSb, CoSb. Each metal atom is surrounded by a slightly distorted octahedron of six Group VB or VIB atoms, which themselves form nearly regular tetrahedra without metal atoms, the structure being anisodesmic.

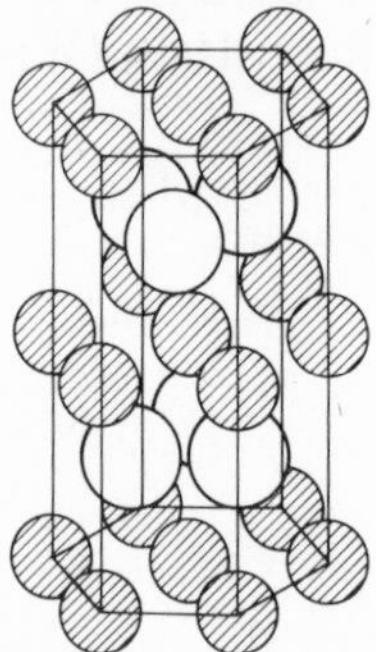

Fig. 7.18. Nickel arsenide structure (3 unit cells).

Symmetry in a crystal may be affected by some metal ions tending to form low-level excited states with asymmetrical electron distributions. Examples among AB compounds are HgO and PbO.

Thus the structure of HgO is quite different from that of ZnO and CaO. There are two short, collinear Hg—O bonds and two weaker bonds perpendicular to them. The lower symmetry can be explained in terms of electronic-promotion energies. For Zn^{2+} and Cd^{2+} the promotion $(n{-}1)d^{10} \rightarrow (n{-}1)d^9ns^1$ requires about 960 kJ mol^{-1}, but for Hg^{2+} only 500 kJ mol^{-1} is needed. Linear complexes employing sd hybrid orbitals in bond formation are very common among Hg^{II} compounds, whereas Zn^{II} and Cd^{II} usually occur in highly symmetrical arrangements with metal ion co-ordination numbers 4 or 6.

Another type of metal ion which normally has a distorted co-ordination sphere in a crystal is the 'inert pair' type with ns^2 ground state. In the Pb^{2+} ion, for example, there is a low-lying excited state, $ns^1\ np^1$; so that in PbO, the ion is best described as 3-co-ordinated, being based on a tetrahedral arrangement with one missing corner, and having a rather indefinite number of next nearest neighbours.

7.2.15. AB_2 structures

Five types of AB_2 structure are exemplified below.

(i) The fluorite structure shows 8 : 4 co-ordination, every Ca^{2+} ion being surrounded by eight F^- ions arranged at the corners of a cube and every F^- by four Ca^{2+} arranged tetrahedrally. It is effectively the CsCl structure with the alternate diagonal pairs missing (Fig. 7.19). Other compounds with the fluorite lattice are SrF_2, BaF_2, $SrCl_2$, CdF_2, PbF_2 and ThO_2.

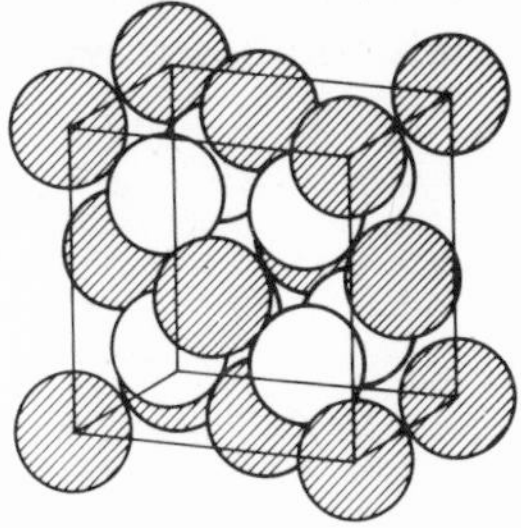

Fig. 7.19. Unit cell of CaF_2, fluorite.

The related antifluorite structure, exemplified by the monoxides and sulphides of Na, K and Rb (but not Cs), has every anion surrounded by eight cations in an arrangement which is the reverse of the CaF_2 lattice.

(ii) In rutile, a form of TiO_2, the co-ordination is 6 : 3, every Ti being surrounded by six O atoms, arranged roughly octahedrally, and each O by three Ti atoms. The unit cell is tetragonal (Fig. 7.20). Other substances with this unit cell structure are ZnF_2, MnF_2, CoF_2, SnO_2, TeO_2, MnO_2.

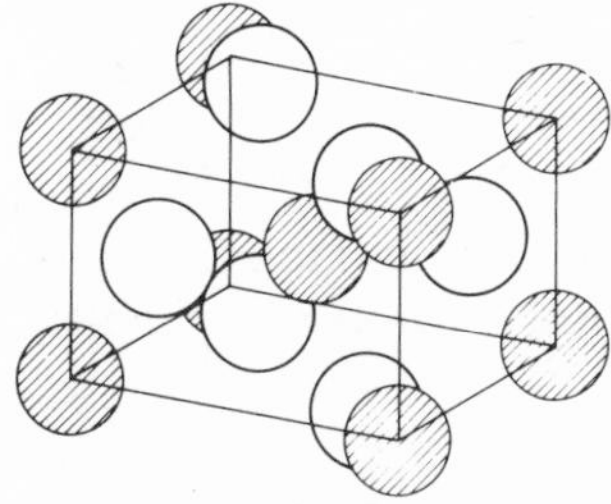

Fig. 7.20. Unit cell of TiO_2, rutile.

Copper(II) fluoride has a distorted rutile structure in which there are two F^- ions at 227 pm and four at 193 pm from the Cu^{2+} ion. This is typical of the Jahn—Teller distortion expected for a d^9 ion. But when the metal ion has 5, 6, 7, 8 or 10 d electrons, its difluoride has an essentially undistorted structure.

(iii) Cristobalite, a form of SiO_2, shows 4 : 2 co-ordination. BeF_2 also has this structure.

(iv) Cadmium iodide, CdI_2, has a layer lattice; its relation to the NiAs structure can be seen by comparing Fig. 7.21 with Fig. 7.18. The force of attraction between layers is weak and the crystal can be cleaved easily into sheets. Examples of this structure are CaI_2, $MgBr_2$, $FeBr_2$, FeI_2; typically, bipositive metals combined with easily polarisable uninegative ions. Many hydroxides $M(OH)_2$ have this structure but in these the layers are bound by rather stronger forces — the so-called hydroxyl bonds between OH^- ions of adjacent layers.

(v) Cadmium chloride, $CdCl_2$, is structurally related to CdI_2 (Fig. 7.21). Successive layers of halide ions in the latter are arranged like alternate layers in hexagonal close packing, ABAB (4.1.7), whereas in the former they are arranged as in cubic close packing: ABCABC (Fig. 7.22). Examples of this structure are $MgCl_2$, $MnCl_2$, $NiCl_2$; typically, bipositive metals combined with uninegative ions of moderate polarisability.

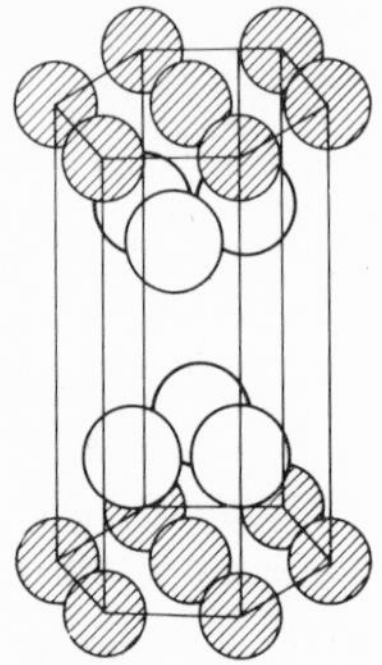

Fig. 7.21. Structure of CdI_2 (3 unit cells).

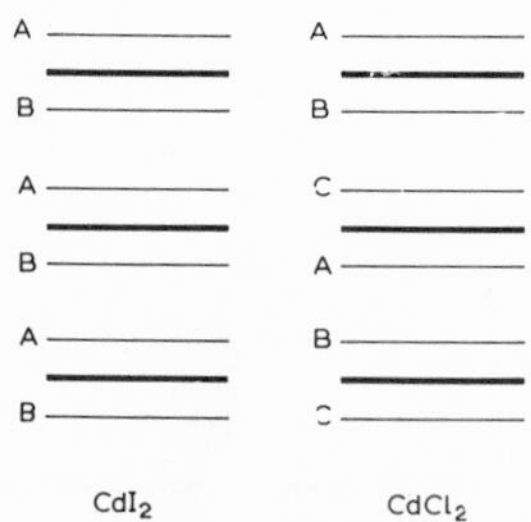

Fig. 7.22. The packing of successive layers of halide ions in CdI_2 and $CdCl_2$. Thick lines represent layers of Cd^{2+} ions.

7.2.16. AB_3 structures

The very simple ReO_3 structure (Fig. 7.23) is rather uncommon. None of the trioxides of Group VI metals has quite such a simple structure, evidently

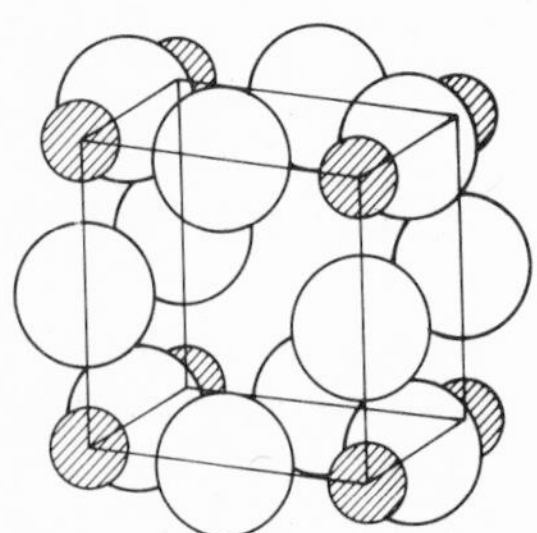

Fig. 7.23. Unit cell of ReO_3.

because the M^{6+} cations are too small for perfect octahedral co-ordination by oxygen ions. Of these trioxides, WO_3 has an almost regular ReO_3 structure but MoO_3 is distorted; it has one oxygen further from the Mo than the other five thus making what is effectively a square pyramidal arrangement of oxygens round the molybdenum. Many trifluorides, for instance AlF_3, ScF_3 and FeF_3, are similar to ReO_3, but the co-ordination in them is rarely perfectly octahedral; the terpositive cations are often just too small to fill the octahedral space completely.

Manganese(III) fluoride is an interesting example of Jahn—Teller distortion: in it there are three different Mn—F distances (Fig. 7.24). The magnetic

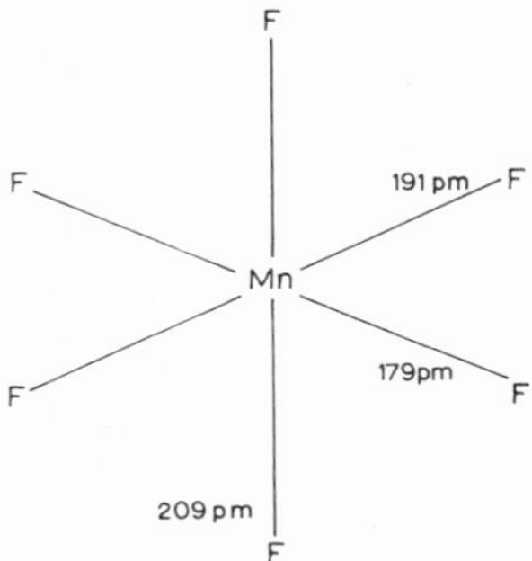

Fig. 7.24. Mn—F distances in MnF_3 crystal.

moment of 4.9 Bohr magnetons implies that MnF_3 has four unpaired d electrons in the Mn^{3+}; three of these are in t_{2g} orbitals and the fourth is possibly in the d_{z^2}. This arrangement accounts for the long bonds on the z-axis (vertical) but not for the unequal Mn—F distances along x- and y-axes. However, neighbouring MnF_6 arrangements are oriented in the crystal so that a fluoride ion, common to both, is at either 191 pm from each Mn^{3+} ion or at 209 pm from one and 179 pm from the next Mn^{3+} ion, thus allowing close-packing of distorted octahedra. It is the packing of the anions in the crystal, in addition to Jahn—Teller distortion, that determines the symmetry.

Other trifluoride structures which are of interest are the BiF_3 structure (Fig. 7.25), which is related to CaF_2, and the tysonite (LaF_3) structure which contains hexagonal nets like those in boron nitride (Fig. 15.6) with two further fluorines above and below each La, making each metal atom five-co-ordinate.

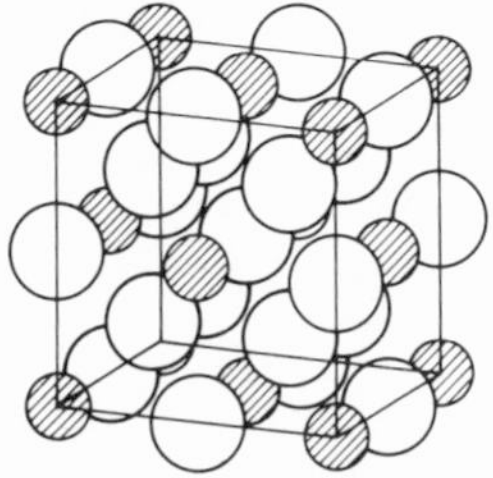

Fig. 7.25. Unit cell of BiF_3.

Trichlorides, tribromides and tri-iodides usually have layer lattices. Their structures are treated in greater detail in Chapter 26.

7.2.17. Complex ions in crystals

Solids containing complex ions often have quite simple structures: for example K_2PtCl_6 consists of K^+ ions and octahedral $PtCl_6^{2-}$ ions in which the chlorine atoms are bound covalently to the platinum. The structure resembles that of fluorite (Fig. 7.19) with the complex ions occupying places in the cubic lattice corresponding to those of the Ca^{2+} ions, and the K^+ ions places corresponding to the F^- ions (Fig. 7.26).

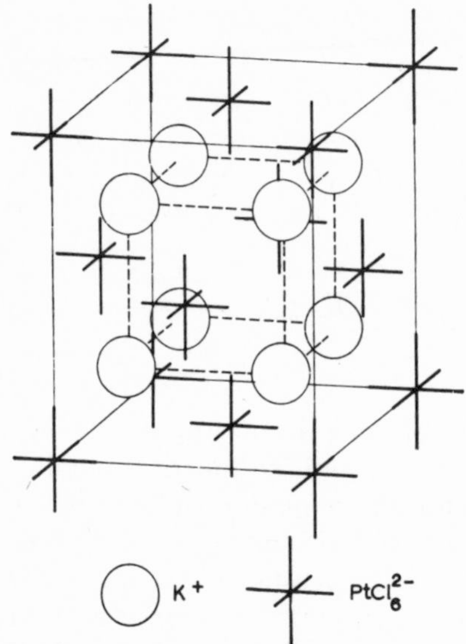

Fig. 7.26. Potassium chloroplatinate lattice.

7.2.18. Perovskites

Not all ternary compounds contain complex ions. Many with the empirical formula ABC_3 have the cubic perovskite ($CaTiO_3$) structure (Fig. 7.27).

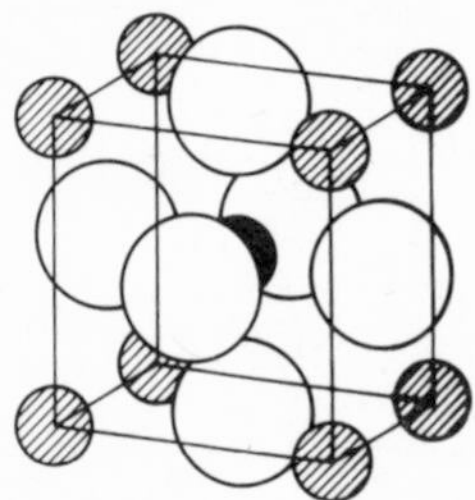

Fig. 7.27. Unit cell of perovskite, $CaTiO_3$ (black atom is Ti).

Examples are $CaZrO_3$, $LaAlO_3$, and $KMgF_3$. The charges on the ions are not important provided the packing is close and the electrical neutrality is maintained. The larger cation occupies the Ca^{2+} position in all compounds of this structure:

$(1 + 5)^+\ 6^-$	$(2 + 4)^+\ 6^-$	$(3 + 3)^+\ 6^-$	$(1 + 2)^+\ 3^-$
Na W O_3	Ca Ti O_3	La Al O_3	K Mg F_3

Particularly interesting perovskite structures are the tungsten bronzes. The ideal composition is $NaWO_3$, the charge-type being $(1 + 5)^+\ 6^-$, as shown above. However, the compound can be crystallised with varying deficiencies of Na^+, provided that, for every missing Na^+ ion, a 5-positive tungsten ion becomes 6-positive. These sodium-deficient materials are known as *incomplete lattice* defect structures. When the deficiency is complete the limiting formula of WO_3 is reached.

7.2.19. Spinels

The spinels are minerals with the empirical formula AB_2O_4. There are 2 : 3 spinels containing A^{2+} and B^{3+} ions, and 4 : 2 spinels containing A^{4+} and B^{2+} ions (Table 7.10). All spinel structures have a c.c.p. array of anions (Fig. 7.28).

TABLE 7.10

SPINEL STRUCTURES

	2 : 3 spinels	4 : 2 spinels
Normal	$MgAl_2O_4$ $MgMn_2O_4$ $Co^{II}(Co^{III})_2O_4$	
Inverse	$Fe^{III}(Fe^{II}Fe^{III})O_4$ $Fe(NiFe)O_4$	$Zn(TiZn)O_4$ $Co(SnCo)O_4$

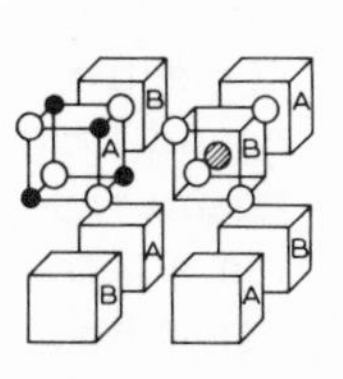

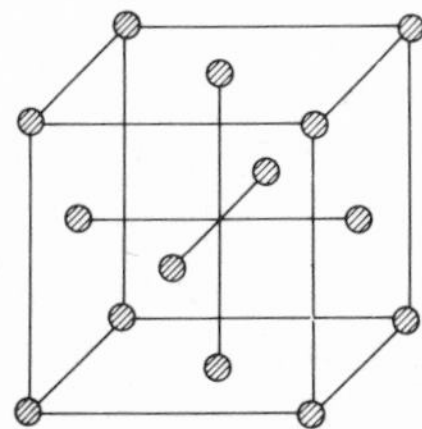

Fig. 7.28. The structure of spinel, $MgAl_2O_4$. The structure contains four MgO_4 tetrahedra (B) and four Al_4O_4 cubes (A) arranged as shown in the eight octants of a face-centred cube of Mg atoms (right). The unit cell contains 32 oxygen atoms arranged as in cubic close-packing with 8 Mg in tetrahedral spaces and 16 Al in octahedral spaces.

In a *normal* spinel the A^{2+} cations occupy one-eighth of the tetrahedral interstices (positions where a cation can be surrounded by four anions) and B^{3+} cations occupy half the octahedral spaces. In an *inverse* spinel such as $NiFe_2O_4$ all the A^{2+} and half of the B^{3+} cations are in octahedral holes, the other half of the B^{3+} cations are in tetrahedral holes. These inverse spinels are therefore best represented as $B(AB)O_4$; for instance the above example is $Fe^{III}(Ni^{II}Fe^{III})O_4$. The 4 : 2 spinels all have inverse structures.

The spinels are of interest structurally because of the several factors which influence the arrangement of the ions.

(i) As the 3+ ion in a 2 : 3 spinel is usually smaller than the 2+ ion, it would be expected to confer greater stability on the lattice by occupying the smaller, tetrahedral, sites. This is realised when $r_{B^{3+}} \ll r_{A^{2+}}$, the tendency being towards the inverse form.

(ii) The Madelung constant and therefore the lattice energy for the structure is greatest when the cation of greater charge has the higher co-ordination number. This factor favours the occupancy of octahedral sites by 3+ ions (i.e. the normal spinel lattice) in a 2 : 3 spinel. It is notable that the factors (i) and (ii) both favour the $B^{II}(A^{IV}B^{II})O_4$ structure for 4 : 2 spinels, and all of them have this form.

(iii) Certain ions tend to occupy octahedral holes because the arrangement confers the maximum ligand field stabilisation. For example the Ni^{2+} ion (d^8) always displaces other ions from octahedral sites, and spinels containing it are always inverse. The Mn^{3+} ion (d^3) displaces all other ions except Ni^{2+} from octahedral holes, consequently all its spinels except $Mn(NiMn)O_4$ are normal.

This tendency of d^3 and d^8 ions to occupy octahedral holes has been explained in terms of ligand field theory in the following way. Let us consider a d^3 ion. In an octahedral crystal field its five degenerate d orbitals are split into three t_{2g} and two e_g orbitals.

In the lowest energy state of the ion the three electrons will all occupy t_{2g} orbitals. If they were randomly distributed between e_g and t_{2g} however, the mean occupancy of the t_{2g} would be only $\frac{3}{5}$ of 3, i.e. 1.8 electrons. As the t_{2g} orbital lies lower by Δ_0 than the e_g the reduction in potential energy caused by all three occupying the lower energy level is $(3.0-1.8)\Delta_0 = 1.2\Delta_0$. This is the *ligand field stabilisation energy*.

For a d^3 ion in a tetrahedral field a random distribution of electrons would give a mean occupancy of the lower energy level of $\frac{3}{5}$ of 2, i.e. 1.2 electrons whereas the level is capable of holding two electrons. Thus the ligand field stabilisation in this case is $0.8\Delta_t$. As Δ_0 is about twice Δ_t, the ligand field stabilisation of a d^3 ion is far greater in an octahedral field than in a tetrahedral field. Examples of ligand field stabilisation energies for other high-spin d^n ions are shown in Table 7.11 which indicates that, on consideration of ligand field stabilisation, d^3 and d^8 ions will have a particularly strong tendency to displace others from octahedral sites in the spinel structure.

(iv) A tendency to covalency can also affect the arrangement but this is not usually important because the O^{2-} ion is not easily polarisable.

TABLE 7.11

LIGAND FIELD STABILISATION ENERGIES FOR TRANSITION METAL IONS

	Octahedral field	Tetrahedral field
d^1 and d^6	$0.4\ \Delta_0$	$0.6\ \Delta_t$
d^2 and d^7	$0.8\ \Delta_0$	$1.2\ \Delta_t$
d^3 and d^8	$1.2\ \Delta_0$	$0.8\ \Delta_t$
d^4 and d^9*	$0.6\ \Delta_0$	$0.4\ \Delta_t$
d^5 and d^{10}	0	0

* These ions have substantial additional stabilisation because of Jahn—Teller distortion.

When the various factors are close to counterbalancing one another there can be a completely random arrangement of metal ions among the eight tetrahedral and sixteen octahedral sites. This is the case with $FeMn_2O_4$.

7.2.20. Silicates

In silica and the silicates oxygen atoms are arranged tetrahedrally round silicon atoms. These tetrahedra may be

(i) separate,
(ii) linked in chains or rings of 2, 3, 4 or 6 units,
(iii) linked in long single or double chains,
(iv) linked in sheets,
(v) joined in 3-dimensional frameworks.

The tetrahedra always share corners, never edges or faces (Pauling's rules 7.2.7).

Silicates may contain other oxygen ions besides those forming the tetrahedra. These O^{2-} anions can be replaced by OH^- and F^- provided electrical neutrality is maintained by the replacement of some of the cations by other cations of lower charge. Similarly cations are replaceable, without changing the structure of the silicate, by others of about equivalent size and the same charge, or by different numbers of other cations providing the same net charge; for instance Ca^{2+} may be replaced by $2Na^+$. The cation Al^{3+} is particularly important, since the ratio $r_{Al^{3+}}/r_{O^{2-}} \sim 0.43$ is close to the transition ratio from 4 co-ordination to 6 co-ordination arrangement. Aluminium thus fits into either a tetrahedral group, $AlO_4{}^{5-}$, or an octahedral group, $AlO_6{}^{9-}$. The $AlO_4{}^{5-}$ group has roughly the same size as the $SiO_4{}^{4-}$ tetrahedron and can replace it provided electrical neutrality is maintained by an adjustment of positive charge elsewhere in the structure.

In the orthosilicates, separate $SiO_4{}^{4-}$ tetrahedra are linked only by cations. An example is olivine, Mg_2SiO_4, essentially a packing of SiO_4 tetrahedra and MgO_6 octahedra. The oxygen—silicon ratio higher than 4 found in some orthosilicates arises from oxygen atoms which are co-ordinated only to the metal atoms; for example in cyanite, Al_2SiO_5 (Fig. 7.29).

Two SiO_4 tetrahedra may be linked by sharing one corner (Fig. 7.30), giv-

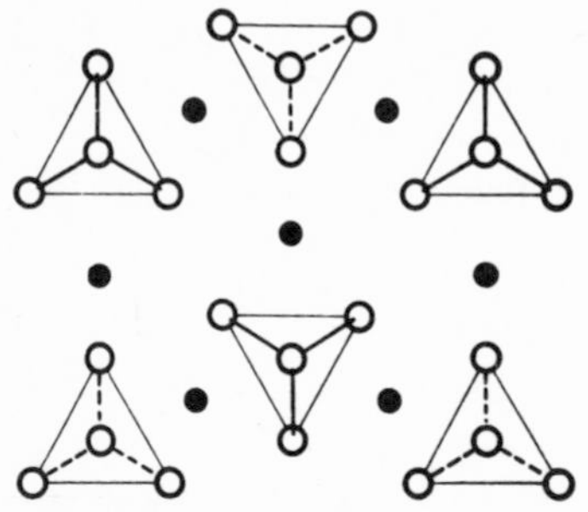

Fig. 7.29. Arrangement of Al_2O^{4+} ions and SiO_4^{4-} tetrahedra in Al_2SiO_5.

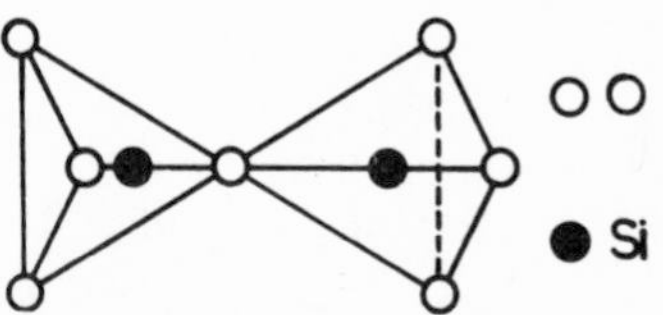

Fig. 7.30. Two SiO_4 tetrahedra with one shared corner.

ing a separate Si_2O_7 unit. Thorveitite, $Sc_2Si_2O_7$, is an example.

Closed rings of SiO_4 tetrahedra give an O : Si ratio of 3; found, for example, in beryl, $Be_3Al_2(Si_6O_{18})$. Each Be is co-ordinated by 4 oxygens from 4 different six-membered rings, and each Al by 6 oxygens from 6 different rings. Relatively wide, empty channels pass through the structure (Fig. 7.31).

Pyroxenes are silicates in which the SiO_4 tetrahedra share two corners to form long chains. The O : Si ratio again is 3, and diopside, $CaMg(SiO_3)_2$, is an example (Fig. 7.32).

Amphiboles consist of double chains, exemplified by tremolite, $(F, OH)_2Ca_2Mg_5Si_8O_{22}$. The additional OH^- and F^- anions cannot be linked to Si atoms; they are co-ordinated around the cations. Up to a quarter of the SiO_4 tetrahedra can be replaced by AlO_4^{5-}, electrical neutrality being maintained by the replacement of Mg^{2+} by Al^{3+}, or the addition of Na^+ or Ca^{2+} cations (Fig. 7.33).

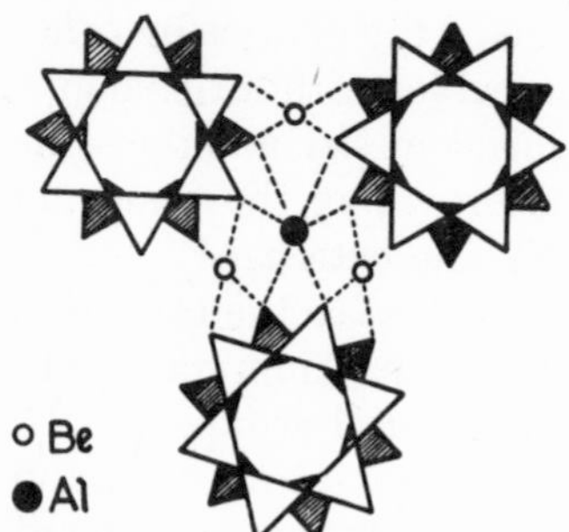

Fig. 7.31. Structure of beryl, showing channels running through superimposed rings of SiO_4 tetrahedra.

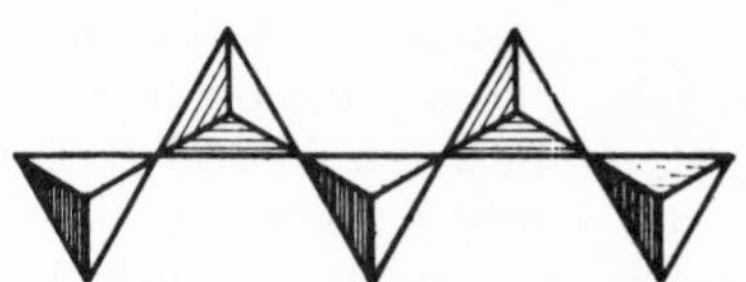

Fig. 7.32. Arrangement of SiO_4 tetrahedra in a pyroxene.

Micas are made up of sheets of SiO_4 tetrahedra (Fig. 7.34), usually with AlO_4 tetrahedra replacing some of the SiO_4 units. In muscovite the sheet-

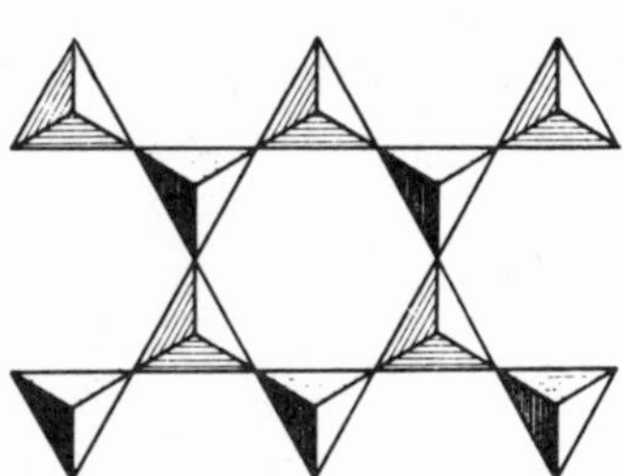

Fig. 7.33. Double-chain arrangement of SiO_4 tetrahedra in an amphibole.

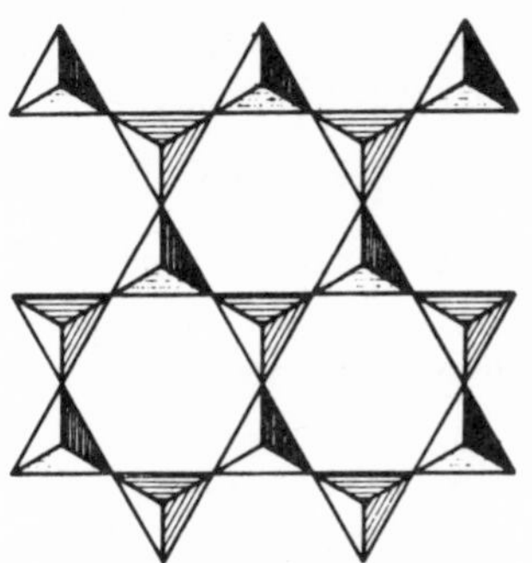

Fig. 7.34. Part of the sheet of SiO_4 tetrahedra in a mica.

composition is $(Si_3AlO_{10})^{5-}$, one quarter of the Si atoms being replaced by Al atoms. Two such sheets, with their tetrahedral vertices inwards, are linked by Al^{3+} ions whose octahedral co-ordination is completed by OH^- ions. The double sheets which have in consequence the composition $[Al_2AlSi_3O_{10}(OH)_2]^-$ are stacked one upon another with sufficient K^+ ions between them to maintain electrical neutrality. The cleavage of the mica is due to weakness along these layers of K^+ ions. A wide variety of replacement is possible; thus the 'binding' aluminium may be replaced by magnesium and the 'sandwich' potassium by sodium. The structure of talc is somewhat similar, except that the sheets are electrically neutral. In clays, the sheets are held together by hydroxyl bonds.

Silica, in its various crystalline forms, is a three-dimensional framework of SiO_4 tetrahedral units which share all their corners. Cristobalite for instance has SiO_4 tetrahedra arranged as are the atoms in zinc blende; tridymite as are those in wurtzite.

The felspars have very open structures, shown two-dimensionally in Fig. 7.35 for simplicity. The cations are accommodated in holes running through

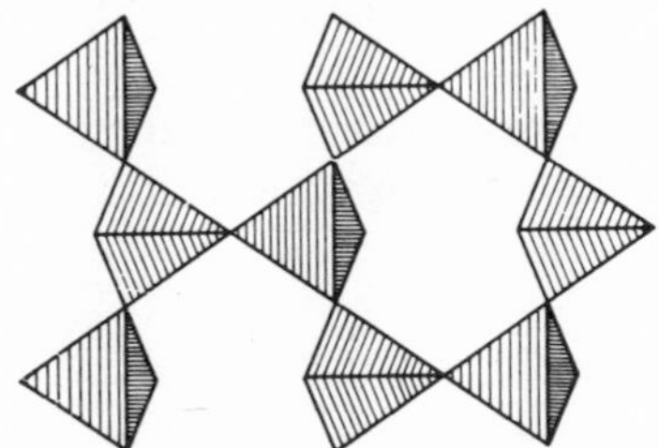

Fig. 7.35. Linking of SiO_4 tetrahedra in a felspar, showing very open structure.

the structure. Orthoclase, $KAlSi_3O_8$, has one quarter of the SiO_4 tetrahedra replaced by AlO_4 tetrahedra and, to preserve neutrality, requires one K^+ ion for every Al^{3+} ion present.

Zeolites have an even more open framework. They contain water molecules which can be removed without the framework collapsing. Because cations are easily exchanged, the materials have been used extensively as 'base exchangers' for water-softening:

$$Na_3CaAl_5Si_5O_{20} \cdot 6\ H_2O + Ca^{2+} \rightleftharpoons NaCa_2Al_5Si_5O_{20} \cdot 6\ H_2O + 2\ Na^+$$

The solid reverts to the sodium form when treated with brine.

7.2.21. Water of crystallisation

Salt hydrates hold water molecules as
(i) co-ordinated water,
(ii) anion water,
(iii) lattice water, or
(iv) zeolite water.

The term water of constitution is a misnomer often applied to the compounds wrongly formulated as $Fe_2O_3 \cdot H_2O$ and $MgO \cdot H_2O$. These are actually the true hydroxides $FeO(OH)$ and $Mg(OH)_2$, and most of this class of compound have a layer lattice consisting of sheets of OH^- and O^{2-} ions with cations between them.

Co-ordinated water. Many cations form complex ions with four or six molecules of water co-ordinated to the metal:

$$[Co(H_2O)_6]Cl_2, \quad [Mg(H_2O)_6]Cl_2, \quad [Be(H_2O)_4]SO_4$$

The metal ion has usually charge number +2 or +3 and, being small, has a high complexing power. Small ions such as Be^{2+} (31 pm) and Mg^{2+} (65 pm) give hydrated ions much larger than themselves, thus $Be(H_2O)_4{}^{2+}$ has about the same dimensions as $SO_4{}^{2-}$, consequently $BeSO_4 \cdot 4\ H_2O$ has its hydrated cations and its anions arranged as in the CsCl structure. The magnesium compound $[Mg(H_2O)_6]Cl_2$ has a very slightly distorted fluorite lattice.

Anion water. This is not common, but certainly occurs in $CuSO_4 \cdot 5\ H_2O$ and probably in $ZnSO_4 \cdot 7\ H_2O$. When copper sulphate pentahydrate is heated the water of the complex ion is lost in two stages giving first $CuSO_4 \cdot 3\ H_2O$ and then $CuSO_4 \cdot H_2O$; the anion water however remains tenaciously held up to 520 K.

In $CuSO_4 \cdot 5\ H_2O$ the copper atoms have the environment shown in Fig. 7.36. The fifth water molecule is hydrogen-bonded to oxygen atoms of

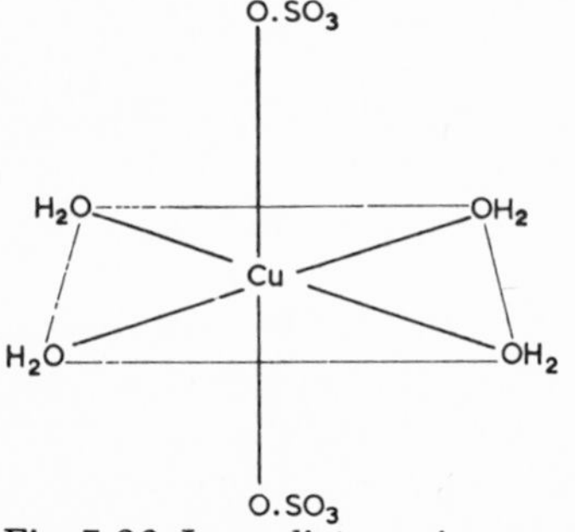

Fig. 7.36. Immediate environment of Cu^{2+} ion in crystal of copper sulphate pentahydrate.

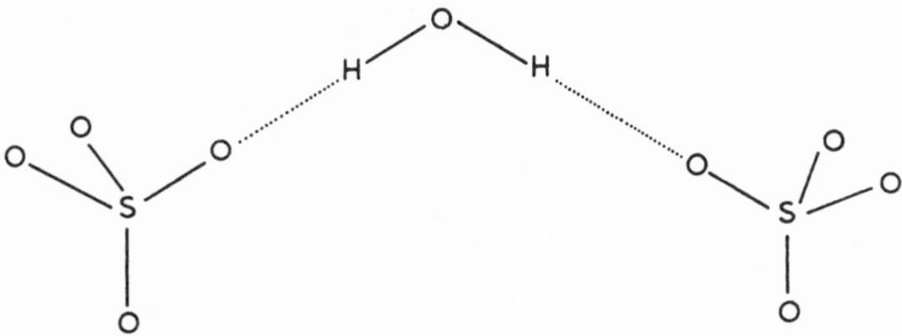

Fig. 7.37. Hydrogen-bonding of SO_4^{2-} tetrahedra by water molecule in $CuSO_4 \cdot 5\ H_2O$.

neighbouring tetrahedral SO_4^{2-} ions (Fig. 7.37).

The water molecule appears to behave in some crystals very much as it does in ice (11.8.1), attaching itself to other molecules by approximately tetrahedrally disposed hydrogen bonds provided atoms of sufficiently high electronegativity are present (Fig. 7.38). It is attached to F or O atoms in various ways; Fig. 7.39(b) shows the one employed in crystalline $CuSO_4 \cdot 5\ H_2O$. In some instances the water molecule is attached directly to both anion and cation (Fig. 7.40).

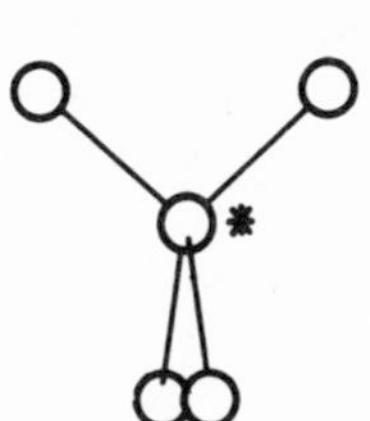

Fig. 7.38. Environment of a water molecule (*) in ice.

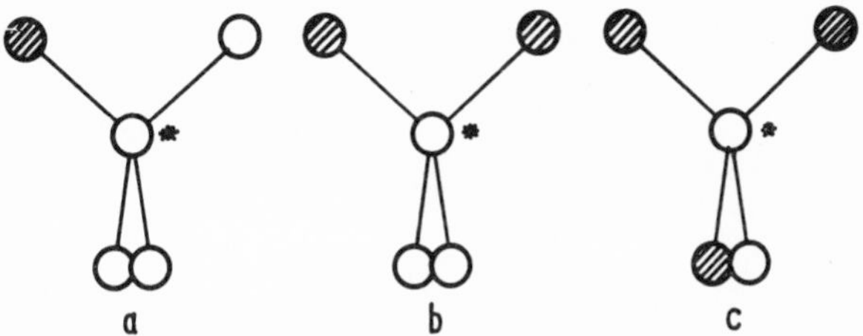

Fig. 7.39. Environment of water molecule (*) attached to oxygen atoms of anions (shaded circles) and also to other water molecules.

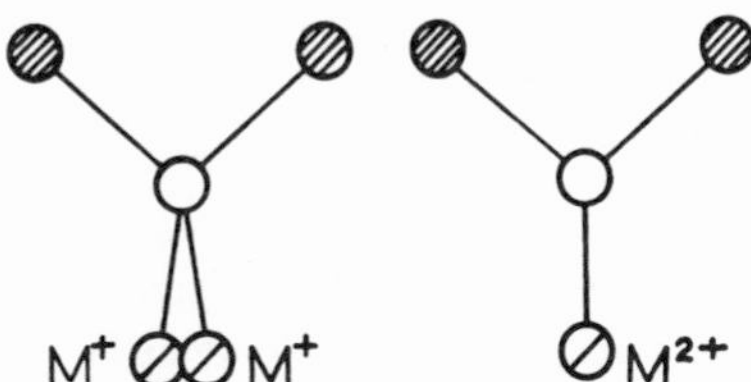

Fig. 7.40. Water molecules attached to both anions (shaded circles) and cations (circles with bar).

Lattice water. In some crystals, water molecules occupy definite lattice positions but are not associated directly with either anion or cation. In the alums six of the water molecules are co-ordinated round the 3+ cation and the other six are arranged at a much greater distance about the unipositive cation.

Zeolite water. This is the water found between the layers of a crystal lattice and in the interstices. It cannot be removed stepwise as can the water from $CuSO_4 \cdot 5\,H_2O$. Moreover, the lattice is only slightly affected by its removal.

Finally, it should be emphasised that structures of hydrates can never be predicted from chemical compositions. The hexahydrate $CaCO_3 \cdot 6\,H_2O$, for example, does not contain $Ca(H_2O)_6^{2+}$ ions. Every Ca^{2+} is surrounded by eight oxygen atoms, six belonging to H_2O molecules and two belonging to CO_3^{2-} ions.

7.2.22. Inherent thermodynamic defects in crystals

Crystal lattices always contain imperfections. Those which are inherent in the thermodynamics of the solid state occur to some extent in all crystals; at any temperature above zero on the Kelvin scale there is some displacement of atoms (or ions) from lattice sites. In stoichiometric ionic crystals two types of defect arise from thermal causes. A *Schottky defect* is due to the movement of an anion and a cation to the surface of the crystal (Fig. 7.41A). Since the displacement involves a pair of ions, the stoichiometry and electric neutrality are maintained. A *Frenkel defect* is due to the displacement of a cation or an anion to an interstitial position (Fig. 7.41B) thereby leaving a vacant lattice site. But the stoichiometry and electric neutrality are maintained as before.

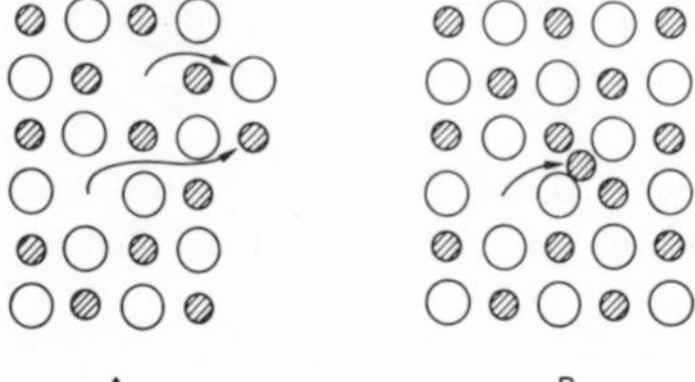

Fig. 7.41. Two-dimensional illustration of the formation of (A) a Schottky defect, (B) a Frenkel defect.

In any particular ionic compound there is one type of defect which is dominant. Schottky defects are associated with compounds in which co-ordination numbers are high and space for interstitial ions is consequently small. In compounds like ZnS where co-ordination numbers are low, Frenkel defects are more likely.

By applying statistical thermodynamics to the systems and ignoring the changes in vibrational entropy which accompany the production of lattice defects, equations can be derived for the number of lattice defects in an ionic crystal AB at a temperature T. For Schottky defects

$$n_s/N = \exp\,(-W_s/2RT)$$

where n_s/N is the ratio of defects to the ideal number of ion pairs and W_s is the energy required to produce one mole of defects. For Frenkel defects

$$n_f/\sqrt{NN^*} = \exp(-W_f/2RT)$$

where n_f is the number of vacant sites, N the ideal number of lattice sites, N^* the total number of interstitial positions and W_f the energy required to produce one mole of Frenkel defects.

For NaCl, W_s is about 190 kJ mol^{-1}, roughly one-quarter of the lattice energy. For the crystal at 670 K, n_s/N is about 5×10^{-8}, and at 1073 K, one degree below the melting point, n_s/N is about 3×10^{-5}, which can be interpreted to mean that about one position in thirty is vacant along any line of ions. Values of W_s and W_f vary a great deal from one compound to another: the lower the values the greater the proportion of defects at a particular temperature.

7.2.22.1. Specific defect structures

As well as defects which are inherent in the thermodynamics of the solid state and therefore occur to some extent in all crystals there are sometimes *specific defects* characteristic of particular stoichiometric compounds.

The simplest type of specific defect is that due to place exchange (Fig. 7.42).

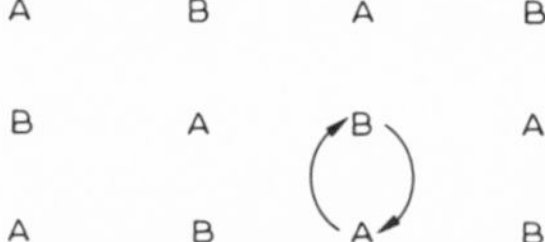

Fig. 7.42. Disordering of a lattice by place exchange.

Its occurrence is limited to alloys because in ionic compounds the exchange would give rise to large electrostatic repulsions. An example of the phenomenon is Cu_3Au (7.2.11) which has an ordered form (Fig. 7.14) below about 700 K and a disordered form above that temperature. Another example is the alloy of composition FeAl, which has a CsCl lattice at low temperatures but completely random occupation of the sites at high temperatures.

The random distribution of cations can introduce specific defects into many types of regular crystalline array. For example MgO has the NaCl structure, but if the Mg^{2+} ions (radius 65 pm) are replaced by equal numbers of Li^+ (radius 60 pm) and Fe^{3+} (radius 64 pm) a disordered NaCl structure of composition $LiFeO_2$ is obtained when the solid is quenched from 1000 K. Specific defects of this type are very common in the spinels.

Random distribution of anions can also occur. An example is β-YOF which has a defect CaF_2 structure in which the cation sites are occupied by Y^{3+} and the anion sites contain randomly distributed O^{2-} and F^-.

In the examples of specific defects mentioned above, all the lattice sites are occupied, but it is also possible to make stoichiometric compounds which have incomplete lattices. For example one mole of MnO_2 fused with six moles of MgO gives rise to a compound $MnMg_6O_8$ which has a defect NaCl structure in which two Mg^{2+} ions are replaced by a Mn^{4+} ion and a cation vacancy. The Mn^{4+} ions and the vacancies are arranged in an ordered manner in a cubic unit cell of edge 838 pm; the cubic cell containing eight MgO in pure magnesium oxide has an edge of 842 pm. This phenomenon is called *superlattice ordering* of defects.

7.2.23. Non-stoichiometric compounds

The concept that a chemical compound must have a constant composition was one of the foundations of nineteenth-century chemistry, but increasingly in recent years it has become apparent that any solid phase can have a range of composition, usually small, but in some cases quite large.

There are three ways in which stoichiometric imbalance can be incorporated in a solid. *Substitutional incorporation* occurs in alloys; atoms of one component substitute atoms of the other component:

A	B	A	B		A	B	A	B
B	A	B	A	→	B	A	A	A
A	B	A	B		A	B	A	B
Formula A_6B_6 (i.e. AB)					Formula A_7B_5			

An example is β-brass, nominally CuZn, which can vary from $CuZn_{0.65}$ to $CuZn_{1.16}$ at room temperature without a change of phase.

Compounds which contain supernumerary atoms or ions in interstitial sites, effectively unbalanced Frenkel defects, display *interstitial incorporation*:

A^+	B^-	A^+	B^-		A^+	B^-		
B^-	A^+	B^-	A^+	$\xrightarrow[\text{atom of B}]{\text{loses an}}$	B^-	A^+	B^-	A^+
						A		
A^+	B^-	A^+	B^-		A^+	B^-	A^+	B^-

A familiar example is zinc oxide, which turns yellow on heating because some oxygen is lost, and zinc atoms which are left in the lattice move into interstitial positions. The deviation from stoichiometry is small in this case; at 1100 K the excess of Zn over O is only 7×10^{-3}%. Of course, an increase in the pressure of oxygen reduces the deviation. *Subtractive incorporation* occurs in compounds in which one component has a complete lattice whereas the other, the deficit component, has vacant lattice sites:

		B^-	A^+		B^-	$\square_+$	B^-	A^+
A^+	B^-	A^+	B^-	$\xrightarrow[\text{atom of B}]{\text{gains an}}$	A^{2+}	B^-	A^+	B^-
B^-	A^+	B^-	A^+		B^-	A^+	B^-	A^+

An example is iron(II) oxide which has the almost constant composition $Fe_{0.95}O$ over the temperature range 900—1300 K. It should be noted that the formula is written so as to indicate which sub-lattice is incomplete. The existence of vacant cation sites ($\square_+$), in effect unbalanced Schottky defects, means that electric neutrality has to be achieved by some of the Fe^{2+} ions losing electrons to become Fe^{3+}:

$$2Fe^{2+} + O \rightarrow 2Fe^{3+} + O^{2-} + \square_+$$

Deviations from ideal stoichiometry in ionic crystals imply changes in the charge numbers of some of the ions, usually the cations. Deviations usually occur in the direction of another stable oxidation state of the metal concerned; thus copper(I) oxide tends to be metal deficient because Cu^{2+} ions can be incorporated to balance vacant cation sites. On the other hand, iron(III) oxide tends towards metal excess, i.e. towards Fe_3O_4 which contains Fe^{2+} ions as well as Fe^{3+}. There are three conditions for a compound to be stable over a range of composition.

(i) the energy needed to produce defects must be low (7.2.22);

(ii) the energy differences between oxidation states must be small;

(iii) the ions of higher and lower oxidation state must not differ much in size, otherwise the lattice suffers great distortion when an ion gains or loses an electron.

The problem of predicting the range of composition over which a particular phase is stable is a matter for specialist books, but some semiquantitative generalisations can be made which arise from considerations of lattice energies.

(i) Deviations on the side of metal excess are usually very small because the presence of interstitial atoms, or cations of reduced charge, reduces the lattice energy. However, the metal excess can be incorporated more easily if the stoichiometric compound has an incomplete lattice (e.g. γ-Fe_2O_3).

(ii) Deviations on the side of metal deficiency tend to be larger because the incorporation of cations of higher charge and smaller size increases lattice energy by increasing coulombic interactions.

A further generalisation is that the range of composition for a phase increases with the polarisability of the anions. Iodides tend to show much greater deviation from ideal stoichiometry than fluorides, for example, and tellurides and selenides show greater variation of composition than do oxides.

Non-stoichiometric compounds are of particular importance in catalysis and in semiconductors. These and other aspects will be treated at appropriate places later in the text.

7.2.23.1. Semiconductivity

Suppose N atoms or ions in the gas phase condense to form a crystal, the discrete energy levels of the atoms interact and spread into bands of allowed energy separated by energy gaps; each bond can accommodate $2nN$ electrons, where n is the degeneracy of the atomic energy level which contributes to the band. The band structure of electronic energy levels in a solid is a consequence of the exclusion principle (3.3), and can be deduced from the solution of the Schrödinger equation for the system, but instead of the spherical field potential appropriate to a simple atom a varying potential function has to be used which has the periodicity of the crystal lattice. The whole system of bands can be looked upon as a series of very large delocalised orbitals (4.1.7) extending in three dimensions.

Fig. 7.43a illustrates energy bands in a pure element or stoichiometric compound which is an insulator. The highest occupied band of electrons, the valence band, contains its total complement of $2nN$ electrons, and the band above it, the conduction band, contains none. There is no mechanism for the movement of electrons, i.e. electrical conduction, through the solid, unless electrons in the valence band can be promoted to the conduction band. However the energy gap ΔE between the two bands is too great for thermal excitation to effect the promotion.

Fig. 7.43b illustrates the energy levels in an element like silicon or germanium.

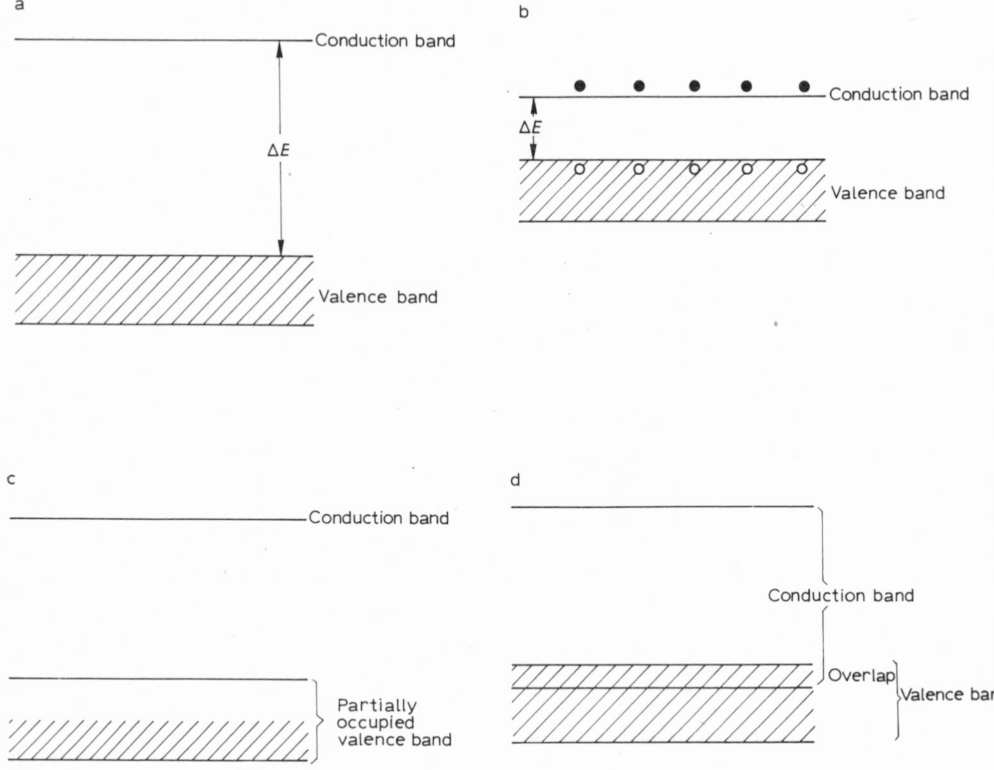

Fig. 7.43. (a) Showing the energy gap, ΔE, between a filled valence band (shaded) and an empty conduction band in an insulator. (b) Energy gap ΔE between valence band and conduction band in a semi-conductor. As the temperature is raised, some electrons (shown black) are promoted to the conduction band, leaving holes in the valence band. (c) Showing partial occupation of the highest occupied energy level in a metal such as sodium. (d) Showing overlapping energy bands leading to immediately accessible energy levels for conduction electrons in a metal like magnesium.

TABLE 7.12

ENERGY-BAND GAPS IN GROUP IV ELEMENTS

	ΔE/kJ mol^{-1}
Diamond	480
Silicon	117
Germanium	76
α-Tin	7.5

Compared with an insulator ΔE is small (Table 7.12), the promotion of electrons from the valence band to the conduction band can occur when the temperature is raised, and the substance becomes a semiconductor with a conductance which increases exponentially with temperature:

$$\kappa = \kappa_0 \exp(-\Delta E/RT)$$

The electrons in the conduction band and the positive holes in the valence band all contribute to the conductivity; their numbers increase with temperature as more electrons are promoted to the high energy level. A pure substance like silicon, which exhibits semiconductivity simply because ΔE is small, is called an *intrinsic* semiconductor. Its conductance is large compared with that of an insulator but small compared with that of a typical metal:

Conductivities at 293 K	
Sodium chloride	10^{-12} S cm^{-1}
Silicon	10^{-2} S cm^{-1}
Gold	5×10^{5} S cm^{-1}

Figs. 7.43c and d illustrate the energy bands in metallic conductors. In a metal like sodium with one 3s electron per atom the highest valence band contains only N electrons, i.e. it is only half-filled (Fig. 7.43c).

In a metal like magnesium the valence band is full but it is overlapped in energy by the conduction band (Fig. 7.43d). In both cases there are energy levels so close to the highest occupied levels that electrons can move easily into them under the influence of an applied field, and the metals are therefore good conductors.

Impurity-doped crystals and nonstoichiometric solids also exhibit semiconductivity. Nonstoichiometric oxides as semiconductors are treated in 23.6.1, where the general principles of *extrinsic* or *impurity* semiconductors are explained.

The conductivity of elements like silicon and germanium can be enhanced by doping with very small amounts of elements like phosphorus, arsenic or aluminium. A small number of phosphorus atoms, for example, can be accommodated in the silicon lattice without disrupting it, but as P has 5 valence electrons and Si only 4, every P atom donates an additional electron which cannot be accommodated in the valence band. Such an electron is attached to the P^+ centre much as the electron of the H atom is attached to the proton, but its orbit is very large because the interaction between it and the P^+ is reduced by the large permittivity of the silicon. For the electron attached to the proton, the Bohr radius r_0 is given by

$$r_0 = \frac{h^2 \epsilon_0}{\pi m e^2} = 53 \text{ pm}$$

but in the present case

$$r = \frac{h^2 \epsilon}{\pi m e^2}$$

and as ϵ for silicon = 11.8 ϵ_0

$r = 11.8 \times 53 \text{ pm} = 624 \text{ pm}$

Thus the orbit of this electron extends over a sphere of radius 624 pm containing about 50 Si atoms. It is effectively free from the P^+ centre and consequently acts as a conducting electron when an electric potential difference is applied across the crystal. Silicon doped with phosphorus is an *n-type* semi-conductor since conduction is due to an excess of *negative* charge which cannot be accommodated in the valence band.

A small number of aluminium atoms in silicon produce a *p-type* semiconductor, the valence band has rather too few electrons to fill it because the Al atom has only 3 valence electrons.

The so-called positive holes in the valence band make it possible for electrons to move through the band under the influence of an applied potential.

7.2.24. Crystal growth

An increased precision in the location of atoms in the crystal lattice has shown that some irregularity in their position is very common. This in turn has contributed an understanding of the way in which a crystal grows and of the changes brought about in a crystal lattice when it is bombarded by particles from without or when some of its atoms suffer radioactive disintegration within.

Crystal growth from vapour, as in sublimation for example, makes the simplest approach to the subject, and the principles involved apply to the more complicated situation in solutions. Consider the flat 001 surface of a cubic crystal in contact with its vapour and partly covered by an incomplete layer (Fig. 7.44).

When the vapour pressure is raised by an amount Δp above the vapour—solid equilibrium vapour pressure, the layer grows with a speed proportional to Δp until the surface is covered. But in order to start a new layer a two-dimensional nucleus must be formed (Fig. 7.45) for which the rate of nucleation has been found to be proportional to $e^{-A/\Delta p}$, A being a constant.

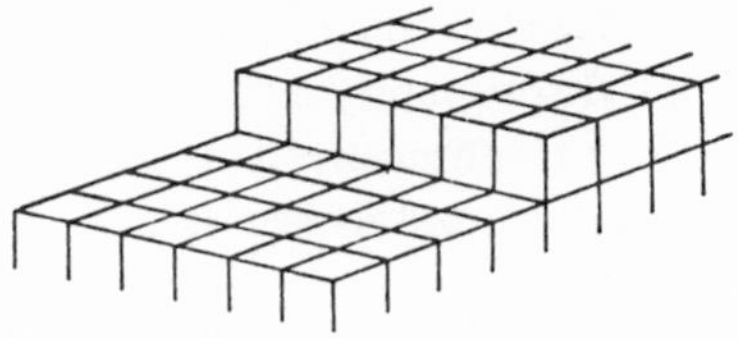

Fig. 7.44. 001 crystal surface partly covered with another layer.

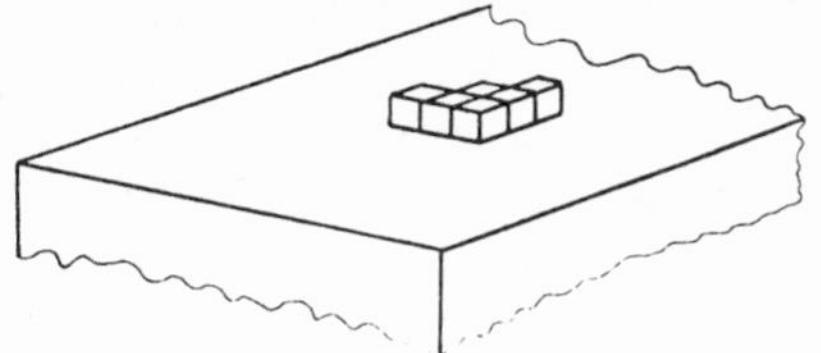

Fig. 7.45. Nucleus for the start of a new layer.

Clearly the stepwise character of a 110, 101, 011 or 111 surface makes nucleation easy to occur, but these surfaces disappear in the course of their growth leaving 100, 010 and 001 surfaces only (Fig. 7.46).

For surfaces of low indices such as the 100, the rate of nucleation predicted by the $e^{-A/\Delta p}$ formula is minutely small compared with the experimental rate. This great discrepancy between theory and practice is best explained by assuming the growing crystal surface always to have random imperfections and, indeed, never to have been covered at any time by an unbroken, perfect layer.

Unit displacement of a *dislocation* need not be normal to the dislocation lines and can be parallel to it, giving rise to *screw dislocations* as shown in Fig. 7.47. When a screw dislocation terminates in the exposed face of a crystal, there is a permanently exposed 'cliff' of atoms; the addition of a further layer to the surface simply perpetuates the conditions and the need for two-dimensional nucleation disappears. The crystal grows up a series of spiral stairways.

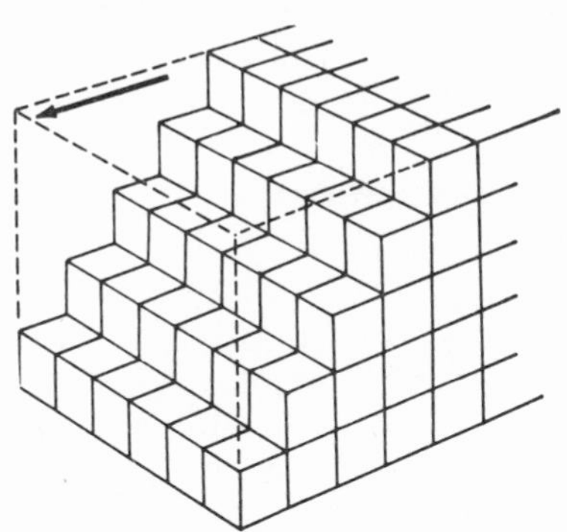

Fig. 7.46. Growth of 001 surface from 101 plane.

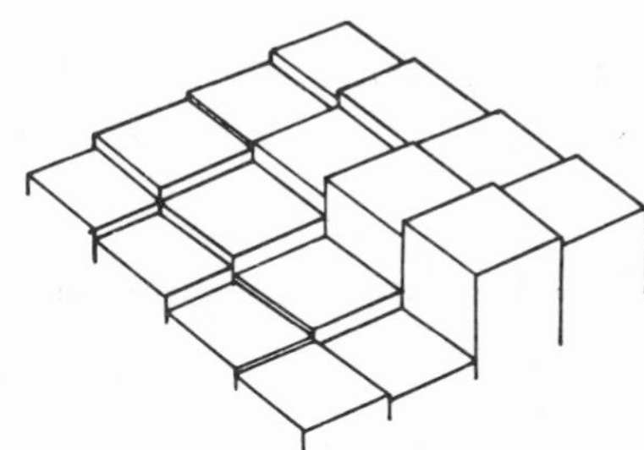

Fig. 7.47. Screw dislocation in crystal.

At low supersaturations, crystals do not grow unless they contain dislocations, but these must not be too close to one another or growth is inhibited because the surface then approaches a perfect form. Perfection is, of course, impossible according to this theory, and is very unlikely to be realised in practice. This explanation of crystal growth is supported by study of the markings present on crystal faces and by the great discrepancy between the strength deduced theoretically for perfect crystals and the measured strengths of actual crystals; imperfections often make the actual crystals weaker by a factor of 100.

The growth of a crystal in solution probably proceeds similarly to the growth in vapour. In all instances a very high supersaturation (1.5 times) is normally needed to initiate growth.

7.2.25. Reactions between solids

The rate of a chemical reaction between two grains of solid in contact with one another depends upon two factors which are of little importance in reactions between molecules:

(i) the mean length of the diffusion path, say half the radius of the average-sized grain, several μm at least,

(ii) the slow rate of diffusion through the solids, probably less than one-millionth of the rate of diffusion which would be due to normal motion in a liquid at a comparable temperature.

A useful working rule about the rate of reactions between solid grains (Tammann's rule) is that significant reaction will not occur until the thermodynamic temperature is two-thirds of the *m.p.* of the lower-melting solid. An increase in temperature affects diffusion rates by increasing the concentration of imperfections such as vacant lattice sites and dislocations. The obvious way of increasing imperfections by doping is rarely used in preparative work however because the product of reaction is contaminated thereby.

The length of the diffusion path in a solid-state reaction can be reduced in favourable cases to atomic dimensions by making a mixed precipitate, a suitable complex or a solid solution suitable for thermal decomposition, e.g.:

(i) Nb^{V} and Ti^{IV} in solution $\xrightarrow{OH^-}$ $TiNb_2O_7 \cdot n\ H_2O$ (precipitate)

$$\downarrow \text{heat}$$

$$TiNb_2O_7$$

(ii) $2\ NH_4^+ + Mg^{2+} + 2\ CrO_4^{2-}$ in solution

$\rightarrow (NH_4)_2Mg(CrO_4)_2 \cdot 6\ H_2O$ (crystals)

$$\downarrow \text{heat}$$

$$MgCr_2O_4$$

(iii) $MgC_2O_4 \cdot aq + 2\ FeC_2O_4 \cdot aq$

$\rightarrow MgFe_2(C_2O_4)_3 \cdot 6\ H_2O$ (solid solution)

$$\downarrow \text{heat}$$

$$MgFe_2O_4$$

A similar principle is used in reactions between solids and gases. Thus a mixture of $BaCl_2$ and $NiCl_2$ reacts with F_2 at 600 K to give a mixture of

BaF_2 and NiF_2, but the compound $Ba[Ni(CN)_4]$ reacts with F_2 at that temperature to give $BaNiF_5$.

7.2.25.1. Thermal decomposition of solids

There are two principal reaction paths for a decomposition formulated

$$AB(s) = A(s) + B(g)$$

(i) Release of B leaves the phase AB supersaturated with respect to A. The nuclei of A then grow by diffusion within the body of the solid.

(ii) Release of B at the surface of a grain is followed by the growth of nuclei of A in the surface.

Surface diffusion is often susceptible to the influence of the gas in contact with the solid. Thus the reaction

$$VI_3(s) \rightarrow VI_2(s) + \tfrac{1}{2}I_2(g)$$

proceeds at a lower temperature and produces more ordered crystals of VI_2 under a pressure of 10^5 Pa of argon than in a vacuum. The iodine molecules in equilibrium with the surface evidently improve the mobility of the V atoms in the surface layer. Similarly the recrystallisation of nickel at 1300 K is encouraged by a trace of I_2 and that of Fe_2O_3 by a little HCl gas.

7.2.25.2. Chemical transport in the gas phase

If one of the two interacting solids is transportable in the gas phase there are two important advantages:

(i) Mechanical contact between reactants is improved, the substance transported as a gas becoming available over the *whole* surface of the grain of the other solid and not merely at a single point of contact.

(ii) Exact stoichiometric proportions are unnecessary because isothermal transport ceases when one of the starting materials is used up.

Some examples of reactions with gaseous transport are:

(i) the formation of Ca_2SnO_4 in the presence of carbon monoxide:

$$SnO_2(s) + CO \rightarrow SnO(g) + CO_2$$

$$SnO(g) + 2\ CaO(s) + CO_2 \xrightarrow{1200\ K} Ca_2SnO_4(s) + CO$$

(ii) the formation of $NiCr_2O_4$ from Cr_2O_3 and NiO in the presence of O_2:

$$Cr_2O_3(s) + \tfrac{3}{2}O_2 \rightarrow 2\ CrO_3(g)$$

$$2\ CrO_3(g) + NiO \rightarrow NiCr_2O_4(s) + \tfrac{3}{2}O_2$$

Advantages also accrue if the product of reaction is chemically transportable.

(i) No troublesome protective layer is formed.

(ii) The product can be separated along a temperature gradient, often as a single crystal suitable for X-ray analysis.

(iii) The use of a gas phase of suitable composition makes it possible to prepare a substance with a specific composition even in a case where complicated phase relations normally exist.

The synthesis of Al_2S_3 in the presence of I_2 is a reaction in which vapour-phase transport along a temperature gradient facilitates the production of pure crystals. Aluminium reacts very slowly with sulphur vapour at 1100 K because the layer of Al_2S_3 which is formed acts as a diffusion barrier. But when aluminium is heated to 1100 K in one end of a sealed quartz tube containing sulphur, large crystals of Al_2S_3 form at the other end of the tube, maintained at 1000 K. The probable course of reaction is:

$$2\ Al(l) + 3\ I_2(g) \rightarrow Al_2I_6(g)$$

$$Al_2I_6(g) + \tfrac{3}{2}\ S_2(g) \rightarrow Al_2S_3(s) + 3\ I_2$$

7.2.25.3. Reactions leading to products structurally similar to the solid reactant

The intercalation compounds of graphite (17.10) are examples of compounds in which the main structural features of the original solid reactant, graphite in this case, are preserved in the product. Similarly the compounds MoS_2 and WS_2 which, like graphite, have layer lattices, react with solutions of Na, K, Rb and Cs in liquid ammonia to give intercalation compounds like $Cs_{0.5}MoS_2$ in which Cs atoms occupy planes between the MoS_2 layers. The compounds NbO_2F and TaO_2F, which both have the ReO_3 structure, absorb HF, possibly as H^+ and F^- ions accommodated in "perovskite holes". In all of these reactions there is advantage in working at low temperature to avoid decomposition of the products; in such reactions transport effects are probably of small importance.

7.2.25.4. The influence of lattice defects upon the nature of the product

A few cases are known where the manner of preparation of a solid influences the way it reacts subsequently. Niobium(V) oxide has a number of crystalline forms. One of them, called T-Nb_2O_5, can be made in three ways:

(i) by oxidising NbO_2 in air at 850 K,
(ii) by treating NbO_2 with Cl_2 at 550 K and then heating to 750 K in air,
(iii) by heating niobic acid in air at 850 K.

The samples are indistinguishable in X-ray analysis but they behave differently when they are heated. The first changes sharply to B-Nb_2O_5 at 1090 K, the second changes to the B-form very slowly, reaction being incomplete after a week at 1130 K, and the third is converted to the M-form without intervening formation of B-Nb_2O_5. Evidently the lattice imperfections which are characteristic of the method of preparation themselves control the course of further reaction.

Further Reading

H. Schäfer, Preparative solid state chemistry, The present position, Angew. Chem. Internat. Ed., 10 (1971) 43.

R. Hoppe, The co-ordination number — an "inorganic" chameleon, Angew. Chem. Internat. Ed., 9 (1970) 25.

H.G. Hertz, Structure of the solvation shell of dissolved particles, Angew. Chem. Internat. Ed., 9 (1970) 124.

J. Nölting, Disorder in solids (ionic crystals and metals), Angew. Chem. Internat. Ed., 9 (1970) 489

G. Will, Crystal structure analysis by neutron diffraction, Angew. Chem. Internat. Ed., 8 (1969) 356 and 950.

R. Hoppe, Madelung constants, Angew. Chem. Internat. Ed., 5 (1966) 95.

H. Heyer, The kinetics of crystal growth, Angew. Chem. Internat. Ed., 5 (1966) 67.

H.J. Goldschmidt, Interstitial alloys, Butterworth, London, 1968.

N.N. Greenwood, Ionic crystals, lattice defects and nonstoichiometry, Butterworth, London, 1968.

A.F. Wells, Structural inorganic chemistry, 4th edition, Oxford University Press, London, 1975.

R. Hoppe, Madelung constants as a new guide to the structural chemistry of solids. Adv. Fluorine Chem., 6 (1970) 387.

S.F. Lincoln, Solvent co-ordination numbers of metal ions in solution, Co-ord. Chem. Rev., 6 (1971) 309.

P.F. Weller, An introduction to principles of the solid state, J. Chem. Ed., 47 (1970) 501 and 48 (1971) 831.

E.F. Gurnee, Fundamental principles of semiconductors, J. Chem. Ed., 46 (1969) 81.

K.S. Førland, Lattice defects and nonstoichiometric solids, Ed. Chem., 8 (1971) 139.

S.J. Bass, Chemistry of semiconductors, Chem. Brit., 5 (1969) 100.

D.J.M. Bevan and P. Hagenmuller, Non-stoichiometric compounds: Tungsten bronzes, vanadium bronzes and related compounds, Pergamon, Oxford, 1975.

D.W. Adams, Inorganic solids, Wiley, London, 1974.

8

Oxidation-Reduction: Redox Reactions

8.1. Definitions

A great many chemical reactions can be broadly classified in terms of oxidation and reduction. Such reactions involve basically a transfer of electrons. Thus when sodium burns in chlorine the overall reaction can be regarded as the result of two processes:

$$\left.\begin{array}{l} 2\,Na - 2\,e \rightarrow 2\,Na^+ \\ Cl_2 + 2\,e \rightarrow 2\,Cl^- \end{array}\right\} 2\,Na + Cl_2 \rightarrow 2\,Na^+\,Cl^- \text{ (an ionic solid)}$$

The sodium atom loses an electron easily and is said to be *oxidised* by the chlorine to a sodium ion; and the chlorine atom, which accepts an electron readily is an *oxidising agent.* By accepting an electron the chlorine atom is *reduced* to a chloride ion and the sodium behaves towards the chlorine as a *reducing agent.* Normally the processes occur simultaneously, but in electrolysis they take place at different electrodes. One electrode puts electrons into, the other takes them out of the electrolyte.

Generally speaking the metallic elements are reducing agents and the non-metallic elements, with the exception of the noble gases, oxidising agents. But the terms are relative. Two oxidising agents may compete for electrons as happens when fluorine is bubbled through molten sodium chloride in which are the ions Na^+ and Cl^-,

$$\left.\begin{array}{l} 2\,Cl^- - 2\,e \rightarrow Cl_2 \\ F_2 + 2\,e \rightarrow 2\,F^- \end{array}\right\} F_2 + 2\,Cl^- \rightarrow 2\,F^- + Cl_2$$

Here fluorine, the stronger oxidising agent, takes the electrons and oxidises the chloride ion, which is just the reverse of the conversion of molecular chlorine to the ion by sodium described above. In the second instance the chloride ion acts as the reducing agent and is oxidised to the normal molecule. The Cl atom and the Cl^- ion are said to form an *oxidation—reduction couple.* The relative strengths of oxidising and reducing agents clearly depend on the free energy changes in the processes. These can be studied under controlled conditions in electric cells.

8.2. Electrodes

Although every electrode process is one of either oxidation or reduction, it is convenient to recognise three main types of electrode:

(i) The cation electrode very readily liberates cations into the solution and is usually a metal, M. The reaction is $M \rightarrow M^{z+} + z\,e$; the electrons flow away through the external circuit and the cations migrate into solution, the metal being oxidised by the solution. But the direction of the reaction can be reversed by driving electrons in at the electrode (electrolysis), under these conditions the cations approach the electrode, the *cathode*, and there receive electrons and are reduced to metal.

One electrode of this kind is the *hydrogen electrode*, in which pure hydrogen passes over platinised platinum. Such a surface is thermodynamically equivalent to one of gaseous hydrogen, cations being produced according to $\frac{1}{2}H_2 \rightarrow H^+ + e$.

(ii) The anion electrode very readily liberates anions and is commonly a metal coated with one of its insoluble salts. Silver coated with silver chloride, for example, accepts electrons from the external circuit and produces chloride anions which migrate into the solution,

$$AgCl + e \rightarrow Ag + Cl^-$$

Silver ions, Ag^+, in the solid chloride are reduced to the metal. When the reaction is made to proceed in reverse, electrons are extracted from the electrode to give silver ions, $Ag - e \rightarrow Ag^+$. These silver ions combine with chloride ions from the solution to form silver chloride on the surface of the electrode.

(iii) The oxidation—reduction electrode is one in which the electrode itself is inert (e.g. pure platinum) and the solution contains ions of variable charge, say iron(II) and iron(III) ions. Supply of electrons to this system causes reduction of the higher charged ion to the lower, $Fe^{3+} + e \rightarrow Fe^{2+}$, and removal of electrons causes oxidation of the lower charged ion to the higher, $Fe^{2+} - e \rightarrow Fe^{3+}$.

The free energy changes which accompany the flow of electrons always depend on the chemical potentials of the ions taking part in the reactions and hence on the *ionic activities.* The nature of the electrolyte is irrelevant except in so far as it determines these activities. Each electrode in its ambient solution constitutes a *half-cell.* By putting two half-cells together, preventing mixing if the two solutions differ, a complete cell is formed, with a natural direction of operation determined by the free energy changes. It is customary to represent such an arrangement (here the hydrogen electrode on the left and the Fe^{3+}/Fe^{2+} electrode on the right) thus:

$$Pt, H_2|H^+|Fe^{2+}, Fe^{3+}|Pt$$

In such a cell the electric potential difference is equal in sign and magnitude to the potential of the electrode on the right minus that on the left. In this particular case, unless $[Fe^{2+}]$ is very large indeed compared with $[Fe^{3+}]$, the cell has a positive potential, thus a wire connecting the two electrodes would convey electrons from the electrode on the left to that on the right. The two electrode reactions are

$\frac{1}{2} H_2 \rightarrow H^+ + e$
$e + Fe^{3+} \rightarrow Fe^{2+}$

and the cell reaction is

$\frac{1}{2} H_2 + Fe^{3+} \rightleftharpoons H^+ + Fe^{2+}$

If the cell described above has as its left-hand half-cell a *standard hydrogen electrode*, one in which pure hydrogen gas at 298 K and 101 325 Pa is in contact at a platinised platinum surface with hydrogen ions at unit activity, the *emf* of the cell is called the *electrode potential* or *redox potential* of the Fe^{3+}/Fe^{2+} couple. The *activity* a_i of an ion i (e.g. H^+) in aqueous solution is given by the equation

$$a_i = \frac{\gamma_i m_i}{\text{mol kg}^{-1}}$$

in which m_i = amount of i/mass of water, and γ_i is the activity coefficient of i, a dimensionless quantity which depends on the strength of the solution and the charge carried by the ion. For methods of determining γ text-books of physical chemistry should be consulted.

In the special case where the Fe^{2+} and Fe^{3+} ions are in their standard states, i.e., $a_{Fe^{2+}} = a_{Fe^{3+}} = 1$, the *emf* of the afore-mentioned cell is the *standard redox potential* of the Fe^{3+}/Fe^{2+} *couple*, written E^0, Fe^{3+}/Fe^{2+} according to the I.U.P.A.C. "Stockholm convention" (1953). Its value is +0.77 V.

For the cell

Pt, $H_2|H^+|Zn^{2+}|Zn$

in which the elements H_2 and Zn are in their standard states (i.e. pure substances at 298 K and 101,352 Pa) and the ions H^+ and Zn^{2+} are at unit activity, the *emf* is —0.76 V, thus E^0, Zn^{2+}/Zn = —0.76 V.

It follows that for the cell

$Zn|Zn^{2+}|Fe^{2+}, Fe^{3+}|Pt$

the *standard emf* is +0.77 V — (—0.76) V = +1.53 V and the cell reaction is

$2\ Fe^{3+} + Zn \rightleftharpoons 2\ Fe^{2+} + Zn^{2+}$

The standard equilibrium constant K_a for the electron-transfer reaction is related to the standard *emf* E by the equation

$$\ln K_a = \frac{zF\Delta E}{RT}$$

where zF is the charge transferred in one mole of reaction. A table of stan-

TABLE 8.1

STANDARD REDOX POTENTIALS (pH 0)

Electrode Reaction		Couple	E^0/V
F_2 + 2 e	= 2 F^-	F_2/F^-	+2.87
O_3 + 2 H^+ + 2 e	= H_2O + O_2	O_3/O_2	+2.07
$S_2O_8^{2-}$ + 2 e	= 2 SO_4^{2-}	$S_2O_8^{2-}/SO_4^{2-}$	+2.01
Ag^{2+} + e	= Ag^+	Ag^{2+}/Ag^+	+1.98
Co^{3+} + e	= Co^{2+}	Co^{3+}/Co^{2+}	+1.82
H_2O_2 + 2 H^+ + 2 e	= 2 H_2O	H_2O_2/H_2O	+1.77
MnO_4^- + 4 H^+ + 3 e	= MnO_2 + 2 H_2O	MnO_4^-/MnO_2	+1.69
2 HClO + 2 H^+ + 2 e	= Cl_2 + 2 H_2O	$HClO/Cl_2$	+1.63
Ce^{4+} + e	= Ce^{3+}	Ce^{4+}/Ce^{3+}	+1.61
2 HBrO + 2 H^+ + 2 e	= Br_2 + 2 H_2O	$HBrO/Br_2$	+1.59
MnO_4^- + 8 H^+ + 5 e	= Mn^{2+} + 4 H_2O	MnO_4^-/Mn^{2+}	+1.51
PbO_2 + 4 H^+ + 2 e	= Pb^{2+} + 2 H_2O	PbO_2/Pb^{2+}	+1.45
2 HIO + 2 H^+ + 2 e	= I_2 + 2 H_2O	HIO/I_2	+1.45
Cl_2 + 2 e	= 2 Cl^-	Cl_2/Cl^-	+1.36
$Cr_2O_7^{2-}$ + 14 H^+ + 6 e	= 2 Cr^{3+} + 7 H_2O	$Cr_2O_7^{2-}/Cr^{3+}$	+1.33
2 HNO_2 + 4 H^+ + 4 e	= N_2O + 3 H_2O	HNO_2/N_2O	+1.29
ClO_2 + H^+ + e	= $HClO_2$	$ClO_2/HClO_2$	+1.28
$N_2H_5^+$ + 3 H^+ + 2 e	= 2 NH_4^+	$N_2H_5^+/NH_4^+$	+1.275
Tl^{3+} + 2 e	= Tl^+	Tl^{3+}/Tl^+	+1.25
MnO_2 + 4 H^+ + 2 e	= Mn^{2+} + 2 H_2O	MnO_2/Mn^{2+}	+1.23
O_2 + 4 H^+ + 4 e	= 2 H_2O	O_2/H_2O	+1.21
2 IO_3^- + 12 H^+ + 10 e	= I_2 + 6 H_2O	IO_3^-/I_2	+1.195
ClO_4^- + 2 H^+ + 2 e	= ClO_3^- + H_2O	ClO_4^-/ClO_3^-	+1.19
Cu^{2+} + 2 CN^- + e	= $Cu(CN)_2^-$	Cu^{2+}:$CN^-/Cu(CN)_2^-$	+1.12
N_2O_4 + 2 H^+ + 2 e	= 2 HNO_2	N_2O_4/HNO_2	+1.07
Br_2 + 2 e	= 2 Br^-	Br_2/Br^-	+1.065
HNO_2 + H^+ + e	= NO + H_2O	HNO_2/NO	+1.00
Pd^{2+} + 2 e	= Pd	Pd^{2+}/Pd	+0.99
NO_3^- + 3 H^+ + 2 e	= HNO_2 + H_2O	NO_3^-/HNO_2	+0.94
2 Hg^{2+} + 2 e	= Hg_2^{2+}	Hg^{2+}/Hg_2^{2+}	+0.92
Cu^{2+} + I^- + e	= CuI	Cu^{2+}:I/CuI	+0.86
Rh^{3+} + 3 e	= Rh	Rh^{3+}/Rh	+0.8
Ag^+ + e	= Ag	Ag^+/Ag	+0.799
Hg_2^{2+} + 2 e	= 2 Hg	Hg_2^{2+}/Hg	+0.79
Fe^{3+} + e	= Fe^{2+}	Fe^{3+}/Fe^{2+}	+0.77
O_2 + 2 H^+ + 2 e	= H_2O_2	O_2/H_2O_2	+0.68
MnO_4^- + e	= MnO_4^{2-}	MnO_4^-/MnO_4^{2-}	+0.57
I_2 + 2 e	= 2 I^-	I_2/I^-	+0.54
Cu^+ + e	= Cu	Cu^+/Cu	+0.52
VO^{2+} + 2 H^+ + e	= V^{3+} + H_2O	VO^{2+}/V^{3+}	+0.36
$Fe(CN)_6^{3-}$ + e	= $Fe(CN)_6^{4-}$	$Fe(CN)_6^{3-}/Fe(CN)_6^{4-}$	+0.36
Cu^{2+} + 2 e	= Cu	Cu^{2+}/Cu	+0.34
SO_4^{2-} + 4 H^+ + 2 e	= H_2SO_3 + H_2O	SO_4^{2-}/H_2SO_3	+0.17
Cu^{2+} + e	= Cu^+	Cu^{2+}/Cu^+	+0.15
Sn^{4+} + 2 e	= Sn^{2+}	Sn^{4+}/Sn^{2+}	+0.15
S + 2 H^+ + 2 e	= H_2S	S/H_2S	+0.14
TiO^{2+} + 2 H^+ + e	= Ti^{3+} + H_2O	TiO^{2+}/Ti^{3+}	+0.1

TABLE 8.1 (continued)

Electrode Reaction		Couple	E^0/V
$P + 3H^+ + 3e$	$= PH_3$	P/PH_3	+0.06
$UO_2^{2+} + e$	$= UO_2^+$	UO_2^{2+}/UO_2^+	+0.05
$2H^+ + 2e$	$= H_2$	H^+/H_2	0.00
$Pb^{2+} + 2e$	$= Pb$	Pb^{2+}/Pb	−0.126
$Sn^{2+} + 2e$	$= Sn$	Sn^{2+}/Sn	−0.136
$Mo^{3+} + 3e$	$= Mo$	Mo^{3+}/Mo	−0.2
$N_2 + 5H^+ + 4e$	$= N_2H_5^+$	$N_2/N_2H_5^+$	−0.23
$Ni^{2+} + 2e$	$= Ni$	Ni^{2+}/Ni	−0.25
$V^{3+} + e$	$= V^{2+}$	V^{3+}/V^{2+}	−0.255
$H_3PO_4 + 2H^+ + 2e$	$= H_3PO_3 + H_2O$	H_3PO_4/H_3PO_3	−0.276
$Co^{2+} + 2e$	$= Co$	Co^{2+}/Co	−0.28
$Tl^+ + e$	$= Tl$	Tl^+/Tl	−0.34
$In^{3+} + 3e$	$= In$	In^{3+}/In	−0.34
$Cd^{2+} + 2e$	$= Cd$	Cd^{2+}/Cd	−0.40
$Cr^{3+} + e$	$= Cr^{2+}$	Cr^{3+}/Cr^{2+}	−0.41
$Fe^{2+} + 2e$	$= Fe$	Fe^{2+}/Fe	−0.44
$Sb + 3H^+ + 3e$	$= SbH_3$	Sb/SbH_3	−0.51
$Ga^{3+} + 3e$	$= Ga$	Ga^{3+}/Ga	−0.53
$As + 3H^+ + 3e$	$= AsH_3$	As/AsH_3	−0.60
$U^{4+} + e$	$= U^{3+}$	U^{4+}/U^{3+}	−0.61
$Te + 2H^+ + 2e$	$= H_2Te$	Te/H_2Te	−0.72
$Cr^{3+} + 3e$	$= Cr$	Cr^{3+}/Cr	−0.74
$Zn^{2+} + 2e$	$= Zn$	Zn^{2+}/Zn	−0.763
$TiO^{2+} + 2H^+ + 4e$	$= Ti + H_2O$	TiO^{2+}/Ti	−0.89
$Nb^{3+} + 3e$	$= Nb$	Nb^{3+}/Nb	−1.1
$Mn^{2+} + 2e$	$= Mn$	Mn^{2+}/Mn	−1.18
$Ti^{2+} + 2e$	$= Ti$	Ti^{2+}/Ti	−1.63
$Al^{3+} + 3e$	$= Al$	Al^{3+}/Al	−1.66
$U^{3+} + 3e$	$= U$	U^{3+}/U	−1.80
$Be^{2+} + 2e$	$= Be$	Be^{2+}/Be	−1.85
$Np^{3+} + 3e$	$= Np$	Np^{3+}/Np	−1.86
$Sc^{3+} + 3e$	$= Sc$	Sc^{3+}/Sc	−2.08
$H_2 + 2e$	$= 2H^-$	H_2/H^-	−2.25
$Lu^{3+} + 3e$	$= Lu$	Lu^{3+}/Lu	−2.25
$Am^{3+} + 3e$	$= Am$	Am^{3+}/Am	−2.32
$Mg^{2+} + 2e$	$= Mg$	Mg^{2+}/Mg	−2.37
$Gd^{3+} + 3e$	$= Gd$	Gd^{3+}/Gd	−2.40
$La^{3+} + 3e$	$= La$	La^{3+}/La	−2.52
$Na^+ + e$	$= Na$	Na^+/Na	−2.71
$Ca^{2+} + 2e$	$= Ca$	Ca^{2+}/Ca	−2.87
$Sr^{2+} + 2e$	$= Sr$	Sr^{2+}/Sr	−2.89
$Ba^{2+} + 2e$	$= Ba$	Ba^{2+}/Ba	−2.90
$Ra^{2+} + 2e$	$= Ra$	Ra^{2+}/Ra	−2.92
$Rb^+ + e$	$= Rb$	Rb^+/Rb	−2.925
$K^+ + e$	$= K$	K^+/K	−2.925
$Li^+ + e$	$= Li$	Li^+/Li	−3.04

dard redox potentials is obviously of value in inorganic chemistry to predict the extent of electron-transfer reactions.

Suppose, for instance, we wish to know what happens when a solution containing Sn^{4+} ions and Sn^{2+} ions is mixed with one containing Fe^{3+} ions and Fe^{2+} ions.

For $Sn^{4+} + 2\,e \rightarrow Sn^{2+}$, $E^0 = +0.15$ V
For $Fe^{3+} + e \rightarrow Fe^{2+}$, $E^0 = +0.77$ V

Thus for the reaction

$2\,Fe^{3+} + Sn^{2+} \rightleftharpoons 2\,Fe^{2+} + Sn^{4+}$

at 298 K

$$\ln K_a = \frac{zF(E_1^0 - E_2^0)}{RT}$$

$$\therefore \ln K_a = \frac{2 \times 96.5\ \text{kC mol}^{-1} \times (0.77 - 0.15)\ \text{V}}{8.314\ \text{J K}^{-1}\ \text{mol}^{-1} \times 298\ \text{K}}$$

$$= 48.3$$

and $K_a = e^{48.3} = 10^{21}$

The Fe^{3+} ion is clearly a much stronger oxidising agent than the Sn^{4+} ion, and the Sn^{2+} ion is a much stronger reducing agent than the Fe^{2+} ion. Quite generally a couple of high positive potential will oxidise one of low potential, and the greater the difference in the two potentials the larger the equilibrium constant for the electron-transfer reaction.

8.3. Factors Governing the Size of Redox Potentials

The redox potential of a metal ion/metal couple depends on the free-energy change of the process whereby one mole of the metal at 298 K is converted into ions at unit activity in aqueous solution. To calculate this energy the process is separated into three stages. In the worked examples, heats of reaction are used in place of free energies because the entropy changes are not all known.

M(solid) $\rightarrow$ M(gas)	ΔH = enthalpy of sublimation
M(gas) $\rightarrow$ M^{n+}(gas) + n e	ΔH = sum of first n ionisation energies
M^{n+}(gas) + aq $\rightarrow$ M^{n+}aq (unit activity)	ΔH = (enthalpy of hydration of the ion)
Adding: M(solid) + aq $\rightarrow$ M^{n+}aq + n e	ΔH = enthalpy change for the half-reaction

Applied to zinc and copper we have:

	ΔH^0/kJ mol^{-1}
$Zn(s) \rightarrow Zn(g)$	131
$Zn(g) \rightarrow Zn^{2+}(g) + 2\,e$	2650
$Zn^{2+}(g) + aq \rightarrow Zn^{2+}aq$	—2931
$Zn(s) + aq \rightarrow Zn^{2+}aq + 2\,e$	—150

	ΔH^0/kJ mol^{-1}
$Cu(s) \rightarrow Cu(g)$	341
$Cu(g) \rightarrow Cu^{2+}(g) + 2\,e$	2705
$Cu^{2+}(g) + aq \rightarrow Cu^{2+}aq$	—2987
$Cu(s) + aq \rightarrow Cu^{2+}aq + 2\,e$	+59

Thus the enthalpy of reaction for the process

$Zn(s) + Cu^{2+}aq \rightarrow Zn^{2+}aq + Cu(s)$

is given by

$$\begin{aligned}\Delta H &= (-150 - 59) \text{ kJ mol}^{-1} \\ &= -209 \text{ kJ mol}^{-1}\end{aligned}$$

As the entropy of the products is probably not greatly different from that of the reactants,

$\Delta G \sim \Delta H$

Thus for the two-electron transfer:

$$E^0, \text{Cu}^{2+}/\text{Cu} - E^0, \text{Zn}^{2+}/\text{Zn} = \frac{-\Delta G}{2F} = \frac{+209 \text{ kJ mol}^{-1}}{2 \times 96.5 \text{ kC mol}^{-1}} = 1.09 \text{ V}$$

in near agreement with the differences between the two experimental values: +0.34 — (—0.76) = +1.10 V.

There are two major causes of 'noble' character (that is, high E^0, M^{n+}/M values) in a metal:

(a) High sublimation energy, usually associated with high *b.p.* (cf. Zn, *b.p.* 1180 K and Cu, *b.p.* 2583 K).

(b) High ionisation energy, associated particularly with transition metals which form cations of high charge.

The effect of (b) is usually partly reduced in small, highly-charged ions by their larger heats of hydration.

For non-metals such as the halogens, the value of E^0, X_2/X^- depends to a great extent on the hydration energy of the X^- ion. For instance with fluorine and chlorine we have:

	ΔH^0/kJ mol^{-1}
$\frac{1}{2}F_2(g) \rightarrow F(g)$	+78
$F(g) + e \rightarrow F^-(g)$	−360
$F^-(g) + aq \rightarrow F^-aq$	−485
$\frac{1}{2}F_2 + aq + e \rightarrow F^-aq$	−767

	ΔH^0/kJ mol^{-1}
$\frac{1}{2}Cl_2(g) \rightarrow Cl(g)$	+122
$Cl(g) + e \rightarrow Cl^-(g)$	−390
$Cl^-(g) + aq \rightarrow Cl^-aq$	−350
$\frac{1}{2}Cl_2(g) + aq + e \rightarrow Cl^-aq$	−618

Thus the calculated enthalpy of the reaction

$$\frac{1}{2}F_2 + Cl^-aq \rightarrow \frac{1}{2}Cl_2 + F^-aq$$

is given by $\Delta H = (-767 + 618)$ kJ mol^{-1}.

Using the approximation above, that $\Delta G = \Delta H$

$$E^0, F_2/F^- - E^0, Cl_2/Cl^- = \frac{-\Delta H}{F}$$

$$= \frac{+149 \text{ kJ mol}^{-1}}{96.5 \text{ kC mol}^{-1}} = 1.55 \text{ V}$$

This is again in reasonable agreement with the difference given in Table 8.1, E^0, F_2/F^- − E^0, Cl_2/Cl^- = (+2.87 − 1.36) V = 1.41 V. (The error introduced by ignoring the entropy change of reaction will be larger in this case than in the previous example ($\Delta G = \Delta H - T\Delta S$).

The relative importance of the factors which determine the redox potentials of couples which include oxoanions (e.g. MnO_4^-/Mn^{2+}) are difficult to assess. The oxidations which they bring about involve a number of steps for which the energies cannot be evaluated accurately.

8.4. Oxidation States

A convenient, formal concept used to characterise the state of oxidation of an element in a compound, and also in nomenclature (28.5), is that of *oxidation number*. Atoms in a free element have zero oxidation number. In a molecule or ion, the oxidation number is equal to the charge the atom might be expected to have if the molecule were made entirely of ions, the anions among these being assumed to have noble gas structures. It follows that the sum of the oxidation numbers of the atoms forming a species (molecule or ion) is equal to the charge on that species.

Thus in SO_2, oxygen, the more electronegative element, is imagined to exist as O^{2-} (neon structure) ions, and the sulphur as an S^{4+} ion. The sulphur is in an oxidation state +4, designated S^{IV}. In SO_3, the sulphur has the oxidation state S^{VI}, as it also has in the SO_4^{2-} ion:

$$S^{6+} + 4\,O^{2-} \equiv SO_4^{2-}$$

Oxygen in most classes of compounds is in the −2 state but there are two notable exceptions to this general rule:

(i) Peroxides in which oxygen is conventionally in the −1 state, as is illustrated by $2\,H^+ + O_2^{2-} \equiv H_2O_2$

(ii) Oxygen fluoride in which fluorine is the more electronegative element and oxygen is in a +2 state: $O^{2+} + 2\,F^- \equiv F_2O$.

But these exceptions should not be allowed to confuse the general rule that the oxygen atom assumes the −2 state in molecules and ions.

Oxidation or reduction of a compound always involves a change in the oxidation state of at least one atom in the compound:

S^{-II}	S^0	S^{IV}	S^{VI}
H_2S	S	SO_2	SO_3

$\longrightarrow$ oxidation

$\longleftarrow$ reduction

In these sulphur compounds, the oxidation state of the sulphur is raised by oxidation, lowered by reduction.

The number of electrons which must be transferred to effect a certain redox process is apparent when the change in oxidation number is known. A familiar example is oxidation by the permanganate ion MnO_4^- ($\equiv Mn^{7+} + 4\,O^{2-}$) which contains Mn^{VII}. The reduction of MnO_4^- to Mn^{2+} (Mn^{II}) clearly requires the addition of 5 electrons:

$$MnO_4^- + 8\,H^+ + 5\,e \rightarrow Mn^{2+} + 4\,H_2O$$

The oxidation of Sn^{2+} to Sn^{4+} requires the removal of 2 electrons.

$$Sn^{2+} \rightarrow Sn^{4+} + 2\,e$$

When this oxidation is effected by MnO_4^- in acid solution the balanced equation is obtained by imagining a ten-electron transfer:

$$2\,MnO_4^- + 16\,H^+ + 5\,Sn^{2+} \rightarrow 5\,Sn^{4+} + 2\,Mn^{2+} + 8\,H_2O$$

It must be remembered that this transfer is done not in one step but in many steps.

8.5. Elements with Several Oxidation States

Most of the non-metals and the transition metals (27.5) demonstrate in their chemistry their ability to exist in several oxidation states. It is useful to be able to summarise the relationships between the states by a visual aid like a redox potential diagram. One method due to W.T. Latimer, but given here with the I.U.P.A.C. sign convention, is illustrated in Fig. 8.1 for the common oxidation states of manganese.

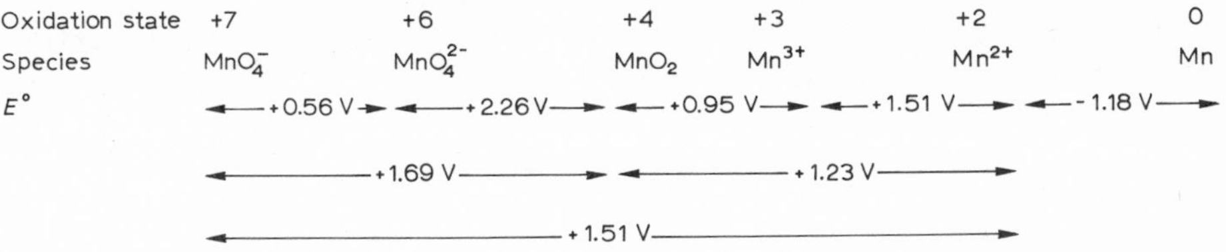

Fig. 8.1. Latimer diagram for manganese at pH 0.

Latimer diagrams summarise the numerical relationships between various oxidation states by indicating the standard redox potential measured for the half cell comprising the two species joined by the line. For example +0.56 V refers to the standard redox potential for the 1-electron transfer from MnO_4^- to MnO_4^{2-} (*a*), E^0 for the 2-electron transfer MnO_4^{2-} to MnO_2 is +2.26 V (*b*) resulting in E^0, MnO_4^-/MnO_2 = +1.69 for a 3-electron transfer (*c*). Notice that the redox potential for the three-electron transfer (*c*) is equal to $(2\,b + a)/3$. This result is to be expected because the Gibbs free-energy change in the half-reaction $MnO_4^- \rightarrow MnO_2$ at pH 0 is the same regardless of the path of reaction:

$$\Delta G = -3\,\text{e} \times 1.69\text{ V} = -(\text{e} \times 0.56\text{ V}) - (2\,\text{e} \times 2.26\text{ V})$$

As all E^0 values are compared with E^0, H^+/H_2 = 0, the half-reaction here really refers to the reduction reaction with hydrogen at standard pressure in a solution for which $a_{H^+} = 1.00$, but in what follows we shall use the expression "half-reaction" for convenience.

The figures also provide information about the possible disproportionation reaction in which manganate would be converted to a mixture of permanganate and manganese dioxide:

$$3\,\text{Mn}^{\text{VI}} \rightarrow 2\,\text{Mn}^{\text{VII}} + \text{Mn}^{\text{IV}}$$

Addition of the half-reactions,

	$2\,\text{Mn}^{\text{VI}} \rightarrow 2\,\text{Mn}^{\text{VII}} + 2\,\text{e}$	$\Delta G^0 = +2 \times 0.56$ eV
and	$\text{Mn}^{\text{VI}} + 2\,\text{e} \rightarrow \text{Mn}^{\text{IV}}$	$\Delta G^0 = -1 \times 2.26$ eV
gives	$3\,\text{Mn}^{\text{VI}} \rightarrow 2\,\text{Mn}^{\text{VII}} + \text{Mn}^{\text{IV}}$	$\Delta G^0 = -1.14$ eV

Since $\Delta G^0 = 1\,RT \ln K$, and 1 eV = 9.65×10^4 J mol^{-1}, therefore

$$\log_{10} K = \frac{1.14 \times 9.65 \times 10^4\text{ J mol}^{-1}}{8.314\text{ J K}^{-1}\text{ mol}^{-1} \times 298\text{ K} \times 2.303} = 19.28$$

and

$$K = \frac{a_{\text{Mn}^{\text{IV}}} \times (a_{\text{Mn}^{\text{VII}}})^2}{(a_{\text{Mn}^{\text{VI}}})^3} = 10^{19.28} = 4.5 \times 10^{19}$$

Thus the high value of the equilibrium constant $K = 4.5 \times 10^{19}$ at 298 K indicates that manganate(VI) ions disproportionate almost completely to MnO_4^- ions and MnO_2 in aqueous solution at pH 0.

A graphical method of illustrating redox relations between the oxidation states of an element has been suggested by E.A.V. Ebsworth. Instead of redox potentials the free energies of the various half reactions

$$M^0 \rightarrow M^{z+} + z\,e$$

strictly the reactions

$$M^0 + z\,H^+ \rightarrow M^{z+} + \tfrac{1}{2}z\,H_2$$

are plotted against the oxidation state (z). The ordinates represent the Gibbs free energies of the various oxidation states, relative to the metal at pH 0. Fig. 8.2 for manganese may be compared with the Latimer diagram above. The Gibbs free-energy change for the *oxidation* of Mn to Mn^{2+}

$$Mn^0 + 2\,H^+ \rightarrow Mn^{2+} + H_2$$

i.e., the half-reaction

$$Mn^0 \rightarrow Mn^{2+} + 2\,e$$

is $-2\,e \times +1.18\,V = -2.36$ eV, sometimes written as -2.36 volt equivalents. Thus the ΔG value for Mn^{2+} lies 2.36 eV below that for Mn^0, taken as zero. The slope of the line drawn between this point and 0.0 is +2.36/2, that is $-E^0$, Mn^{2+}/Mn.

Other points are plotted similarly from the data given on the Latimer diagram for manganese. Thus the point for Mn^{III} is 1.51 eV above that for Mn^{II},

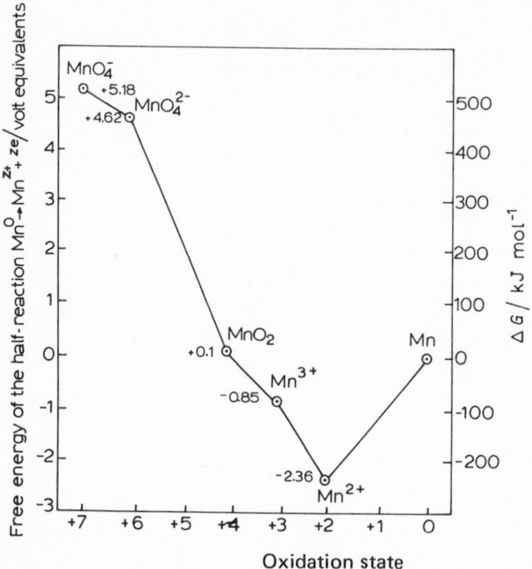

Fig. 8.2. Oxidation state diagram for manganese at pH 0.

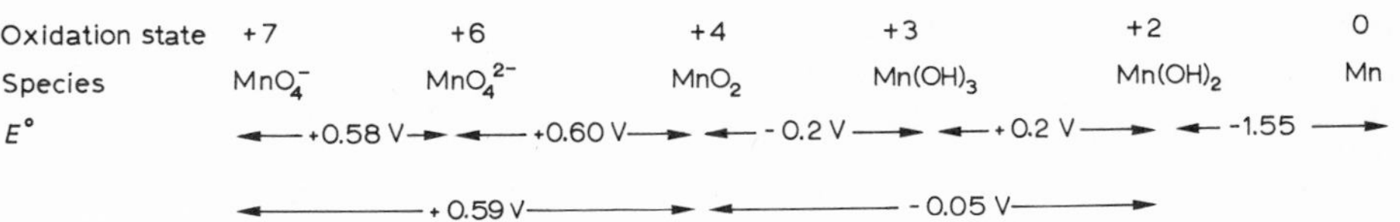

Fig. 8.3. Latimer diagram for manganese at pH 14.

and so on. The points are plotted from right to left to give an order of oxidation states which follows the I.U.P.A.C. convention:

M(oxidised) + electron(s) → M(reduced)

This graphical method provides a clear picture of the redox-solution chemistry of an element. For the case illustrated, the minimum at Mn^{2+} indicates that this state is thermodynamically stable with respect to Mn^0 and to all the oxidation states of manganese from Mn^{III} to Mn^{VII}. The fact that Mn^{VI} lies higher than a straight line drawn from Mn^{VII} to Mn^{IV} indicates that Mn^{VI} will disproportionate largely into the Mn^{VII} and Mn^{IV} states. But for the equilibrium

$$3\ Mn^{IV} \rightleftharpoons Mn^{VI} + 2\ Mn^{III}$$

K will be very small indeed since the Mn^{IV} point lies well below the straight line from Mn^{VI} to Mn^{III}. However, for

$$2\ Mn^{III} \rightleftharpoons Mn^{IV} + Mn^{II}$$

K will be large, as Mn^{III} lies at a 'convex' point.

The diagrams given above refer to the reactions of manganese compounds at pH 0 (normal acidity). For alkaline solutions (pH 14) the reactions are different, and the Latimer diagram and Ebsworth graph are also different.

For the free energies of half-reaction the graph has the form shown in Fig. 8.4. Under alkaline conditions, $Mn(OH)_3$ disproportionates almost completely into $Mn(OH)_2$ and MnO_2:

$$K = \frac{a_{Mn(OH)_2} \times a_{MnO_2}}{a_{Mn(OH)_3}} \sim 10^5$$

Both $Mn(OH)_2$ and MnO_2 are thermodynamically stable species (there are

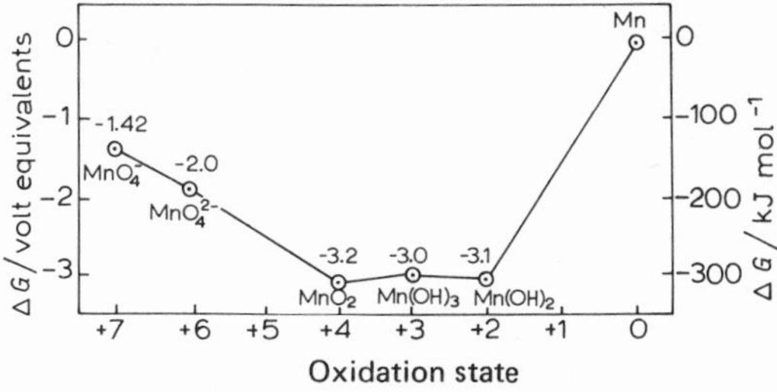

Fig. 8.4. Oxidation state diagram for manganese at pH 14.

minima at both the +2 and +4 states). The manganate(VI) ion has less tendency to disproportionation in alkaline than in acid solution; the line through the +7, +6 and +4 states is almost straight, meaning that

$$K \text{ for } \frac{(a_{\mathrm{Mn^{VII}}})^2 \times a_{\mathrm{Mn^{IV}}}}{(a_{\mathrm{Mn^{VI}}})^3} \sim 1$$

Thus these three oxidation states of manganese can exist together in appreciable concentrations in aqueous alkalis; whenever three or more species are joined by a straight line on an oxidation state diagram they are capable of existing in such an equilibrium with one another.

Because the MnO_4^{2-} ion does not disproportionate in strong alkalis, an alkaline solution of K_2MnO_4 can be made by oxidising MnO_2 in strong aqueous KOH, but the addition of even a very weak acid causes the manganate(VI) to disproportionate to permanganate and MnO_2:

$$3\,MnO_4^{2-} + 4\,H^+ \rightarrow 2\,MnO_4^- + MnO_2 + 2\,H_2O$$

Oxidation-state diagrams will be used throughout the book to illustrate the relative thermodynamic stabilities of oxidation states for those elements which have several. Unless there is special interest in the case for the alkaline solution, only the diagram for pH 0 will be given.

8.6. Redox Potentials of Complex Couples

In the foregoing section we have seen that the redox potential of the Mn^{III}/Mn^{II} couple depends on the pH of the solution. For $Mn(H_2O)_6^{3+}/Mn(H_2O)_6^{2+}$ at pH 0, $E^0 = +1.51$ V, but for $Mn(OH)_3/Mn(OH)_2$ at pH 14, $E^0 = +0.10$ V. The oxidising power of Mn^{III} is clearly dependent on its chemical form. In a perfectly general way the chemical form of a particular oxidation state of a metal can be modified by the addition of complexing agents. For example, manganese(III) becomes a very weak oxidising agent in the presence of a large excess of CN^- ions, the value of E^0, $Mn(CN)_6^{3-}/Mn(CN)_6^{4-}$ being -0.22 V compared with +1.51 V for $Mn^{3+}aq/Mn^{2+}aq$. An energy cycle explains this change in E^0:

(a) $Mn(H_2O)_6^{3+} + e \rightarrow Mn(H_2O)_6^{2+}$ $\qquad \Delta G^0 = 1 \times 1.51\ \mathrm{V} \times 96.5\ \mathrm{kC\ mol^{-1}}$

(b) $Mn(H_2O)_6^{2+} + 6\,CN^- \rightarrow Mn(CN)_6^{4-} + 6\,H_2O$ $\qquad \Delta G^0 = -RT \ln K_{II}$

(c) $Mn(CN)_6^{3-} + 6\,H_2O \rightarrow Mn(H_2O)_6^{3+} + 6\,CN^-$ $\qquad \Delta G^0 = +RT \ln K_{III}$

(d) $Mn(CN)_6^{3-} + e \rightarrow Mn(CN)_6^{4-}$

For (d) $\Delta G^0 = -(1.51 \times 96.5)\ \mathrm{kJ\ mol^{-1}} + RT(\ln K_{III} - \ln K_{II})$, where K_{II} = the cumulative stability constant for $Mn(CN)_6^{4-}$, i.e.,

$$\frac{a_{Mn(CN)_6{}^{3-}}}{a_{Mn(H_2O)_6{}^{3+}} \times (a_{CN^-})^6}$$

and K_{III} = the cumulative stability constant for $Mn(CN)_6{}^{3-}$. (see 28.6.2)

The constant K_{III} is much larger than K_{II}; this is because the absence of one antibonding electron from $Mn(CN)_6{}^{3-}$ very greatly strengthens the bonding in the complex. Thus ΔG^0 is far more positive for reaction (d) than for reaction (a), and E^0 is consequently greatly reduced.

The effect of strong-field ligands on the redox potential of an M^{III}/M^{II} transition-metal couple is a fairly general one. Thus reference to Table 8.1 indicates that the $Fe(H_2O)_6{}^{3+}$ ion is a considerably stronger oxidising agent than the $Fe(CN)_6{}^{3-}$ ion. However, there are a few cases where the addition of a strong-field ligand enhances the oxidising power of iron(III). This is the case when orthophenanthroline is added to Fe^{3+}. In this system the cumulative stability constant of the $[Fe(phen)_3]^{2+}$ ion is greater than that of the $[Fe(phen)_3]^{3+}$ because the former, unlike the latter, is a low-spin complex. Consequently, as reference to the foregoing equation will show, E^0 for the Fe^{III}/Fe^{II} system is increased in the presence of orthophenanthroline, E^0, $[Fe(phen)_3]^{3+}/[Fe(phen)_3]^{2+}$ being +1.14 V.

Further Reading

W.L. Latimer, The oxidation states of the elements and their potentials in aqueous solutions, 2nd edition, Prentice-Hall, Englewood Cliffs, N.Y., 1952.

A.G. Sharpe, The principles of oxidation and reduction, Monographs for Teachers Series, No. 2, The Royal Institute of Chemistry, London, 1959.

E.A.V. Ebsworth, A graphical method of representing the free energies of oxidation—reduction systems, Education in Chemistry, 1 (1964) 123.

G. Kortüm, Treatise on electrochemistry, 2nd edition, Elsevier, Amsterdam, 1965.

R.M. Lawrence and W.H. Bowman, Electrochemical cells for space power, J. Chem. Ed., 48 (1971) 359.

A.K. Vijh, Electrochemical principles involved in a fuel cell, J. Chem. Ed., 47 (1970) 681.

Acids and Bases

9

9.1. Introduction

Early studies of acids and bases were restricted to aqueous solutions and were made with an inadequate understanding of the peculiarities of water. The compounds were defined as substances which dissociated in water, the acids to give hydrogen ions and the bases hydroxide ions (Arrhenius, Ostwald, 1887). This definition does not express present views about aqueous solutions and cannot be applied to solutions in other solvents.

Thermodynamics shows that free H^+ ions cannot exist in appreciable concentration in water itself, and spectroscopy that the hydration of the proton is a strongly exothermic reaction. Accordingly the dissociation of an acid in water leads to the hydrated proton H^+ aq., often represented by H_3O^+ and referred to as the hydroxonium ion:

$$H^+ + H_2O \rightarrow H_3O^+ \qquad \Delta H \sim -1210 \text{ kJ mol}^{-1}$$

The hydrogen ion in aqueous solutions is almost certainly associated with more than one water molecule (for the structure of water see 11.8.1). However H_3O^+ itself exists in some ionic lattices, for instance in $H_3O^+ClO_4^-$ which is isomorphous with $NH_4^+ClO_4^-$.

On the other hand, an acid dissociates in anhydrous ethyl alcohol to give the cation $C_2H_5OH_2^+$ and in liquid ammonia to give the cation NH_4^+.

Turning to bases, we find something similar. For instance, sodium hydroxide is a strong base in water, giving the hydroxide ion OH^-. But it is less ionised in anhydrous alcohol than is sodium ethoxide, C_2H_5ONa, which gives the anion $C_2H_5O^-$. This anion also appears in solutions of amines in ethyl alcohol. In liquid ammonia the anion present is NH_2^-. The liquids exemplify solvents in which the anion is not the hydroxide ion.

Clearly acids and bases cannot be defined simply as substances respectively producing by dissociation the ions H^+ and OH^-.

9.2. Lowry—Brönsted Theory of Acids and Bases

Lowry and Brönsted (1923) independently defined an acid as a compound or ion with a tendency to lose a proton, and a base as a compound or ion with a tendency to gain a proton. The proton will be associated with a solvent molecule but may for convenience be represented formally by H^+. An acid A by losing a proton becomes a base B. The base B will tend to regain the proton and revert to the acid A. There will be an equilibrium:

$$\underset{\text{acid}}{A} \rightleftharpoons \underset{\text{base}}{B} + \underset{\text{proton}}{H^+}$$

Thus for acetic acid:

$$CH_3COOH \rightleftharpoons CH_3COO^- + H^+$$

Clearly the acetate ion is by definition a base. Moreover, the acetic acid and the acetate ion are together known as a *conjugate acid—base pair.* The term is applied generally to any two species related to each other as A is to B in the general equation.

Acids

The definition allows for the existence of three types of acid:

(i) Molecular acids, such as hydrochloric, sulphuric and acetic acids, which lose protons to give, as conjugate bases, the chloride, bisulphate and acetate ions respectively:

$$HCl \rightleftharpoons Cl^- + H^+$$
$$H_2SO_4 \rightleftharpoons HSO_4^- + H^+$$
$$CH_3COOH \rightleftharpoons CH_3COO^- + H^+$$

(ii) Anion acids, such as the bisulphate HSO_4^- and bi-oxalate $C_2O_4H^-$ ions, which lose protons to give, as conjugate bases, the sulphate SO_4^{2-} and oxalate $C_2O_4^{2-}$ ions respectively:

$$HSO_4^- \rightleftharpoons SO_4^{2-} + H^+$$
$$HC_2O_4^- \rightleftharpoons C_2O_4^{2-} + H^+$$

(iii) Cation acids, such as the hydroxonium H_3O^+, ammonium NH_4^+, and the anilinium $C_6H_5NH_3^+$ ions, which lose protons to give water, ammonia and aniline, respectively, as conjugate bases:

$$H_3O^+ \rightleftharpoons H_2O + H^+$$
$$NH_4^+ \rightleftharpoons NH_3 + H^+$$
$$C_6H_5NH_3^+ \rightleftharpoons C_6H_5NH_2 + H^+$$

Bases

Two types of base are possible:

(i) Molecular bases, such as ammonia and ethylamine, which combine with protons to give cations:

$$CH_3NH_2 + H^+ \rightleftharpoons CH_3NH_3^+$$

(ii) Anion bases, such as the hydroxide and acetate ions, which combine

with protons to give uncharged molecules:

$$OH^- + H^+ \rightleftharpoons H_2O$$
$$CH_3COO^- + H^+ \rightleftharpoons CH_3COOH$$

The equilibrium $A \rightleftharpoons B + H^+$ is purely formal because of the non-existence of the proton in water and other solvents. When an acid loses its proton it does so to a base. The acid is thereby converted to its conjugate base and the base to its conjugate acid, as exemplified in the general and particular expressions:

$$A_1 + B_2 \rightleftharpoons B_1 + A_2$$
$$\underset{\text{acid}}{CH_3COOH} + \underset{\text{base}}{H_2O} \rightleftharpoons \underset{\text{base}}{CH_3COO^-} + \underset{\text{acid}}{H_3O^+}$$

The latter represents the transfer of a proton from an acetic acid molecule to a water molecule and the consequent formation of two ions. Written in the opposite sense,

$$CH_3COO^- + H_3O^+ \rightleftharpoons CH_3COOH + H_2O$$

the equilibrium is seen to depend upon the transfer of a proton from a hydroxonium ion to an acetate ion. This process is responsible for the 'buffer action' (9.4) of an acetate when a strong acid is added to the solution.

9.3. Acid—Base Equilibria in Water

Reactions in which a proton is transferred are known as *protolytic reactions.* The conjugate pair, H_3O^+/H_2O is used as a standard for comparing the strengths of other proton acids and their conjugate bases.

For the reaction

$$A + H_2O \rightleftharpoons B + H_3O^+$$

the *dissociation constant* of the acid A is given by

$$K_a = \frac{a_B \times a_{H_3O^+}}{a_A}$$

Thus for H_2SO_4 in the reaction $H_2SO_4 + H_2O \rightleftharpoons H_3O^+ + HSO_4^-$:

$$K_a = \frac{a_{HSO_4^-} \times a_{H_3O^+}}{a_{H_2SO_4}}$$

The relative activity of an ion is obtained from the molality m_i by the equation:

$$a_i = \frac{\gamma_i m_i}{\text{mol kg}^{-1}}$$

and γ_i by the Debye—Hückel equation. Thus K_a is a dimensionless quantity. The acid exponent pK_a, defined as $-\log_{10}K_a$, is used as a convenient measure of acid strength (Table 9.1).

TABLE 9.1

THE STRENGTHS OF SOME ACIDS

Acid			pK_a
Hydrochloric acid	$HCl + H_2O$	$\rightleftharpoons Cl^- + H_3O^+$	−6.1
Perchloric acid	$HClO_4 + H_2O$	$\rightleftharpoons ClO_4^- + H_3O^+$	−4.8
Nitric acid	$HNO_3 + H_2O$	$\rightleftharpoons NO_3^- + H_3O^+$	−1.37
Sulphuric acid	$H_2SO_4 + H_2O$	$\rightleftharpoons HSO_4^- + H_3O^+$	−3.1
Orthophosphoric acid	$H_3PO_4 + H_2O$	$\rightleftharpoons H_2PO_4^- + H_3O^+$	+2.15
Dihydrogenphosphate ion	$H_2PO_4^- + H_2O$	$\rightleftharpoons HPO_4^{2-} + H_3O^+$	+7.21
Monohydrogenphosphate ion	$HPO_4^{2-} + H_2O$	$\rightleftharpoons PO_4^{3-} + H_3O^+$	+12.3
Ammonium ion	$NH_4^+ + H_2O$	$\rightleftharpoons NH_3 + H_3O^+$	+9.25
Aquated iron(III) ion	$Fe(H_2O)_6^{3+} + H_2O$	$\rightleftharpoons Fe(H_2O)_5OH^{2+} + H_3O^+$	+2.34

The strength of a base can be expressed similarly. Thus for the reaction

$$NH_3 + H_2O \rightleftharpoons NH_4^+ + OH^-$$

$$K_b = \frac{a_{NH_4^+} \times a_{OH^-}}{a_{NH_3}}$$

However it has become the custom to express the dissociation constants of bases by tabulating the pK_a values of their conjugate acids. For the NH_4^+

$$K_a = \frac{a_{NH_3} \times a_{H_3O^+}}{a_{NH_4^+}}$$

and the product of K_a for NH_4^+ and K_b for NH_3 is thus $a_{H_3O^+} \times a_{OH^-}$, which is the *autoprotolysis constant* for water, K_w, so called because it refers to the transfer of a proton from one water molecule to another:

$$H_2O + H_2O \rightleftharpoons H_3O^+ + OH^-$$

The value of K_w at 298 K is 1.0×10^{-14}. Thus for NH_3, pK_b is $14 - 9.25 =$ 4.75 (Table 9.1).

An acid for which pK_a is small, e.g. HCl, is a strong acid, i.e. it ionises strongly in water; in consequence its conjugate base, Cl^-, is a very weak base indeed:

$$pK_b(Cl^-) = 14 - (-6.1) = 20.1$$

9.4. Buffer Action

The hydroxonium ions in a solution of a strong acid such as HCl are largely converted to undissociated acetic acid molecules when added to a solution of acetate ions:

$$H_3O^+ + CH_3COO^- \rightleftharpoons CH_3COOH + H_2O$$

$$\text{Since } K_a(\text{acetic acid}) = \frac{a_{H_3O^+} \times a_{CH_3COO^-}}{a_{CH_3COOH}}$$

$$\text{therefore } a_{H_3O^+} = K_a \times \frac{a_{CH_3COOH}}{a_{CH_3COO^-}}$$

The pH of the solution is given by $pH = -\log_{10} a_{H_3O^+}$

$$\text{Hence } pH = pK_a - \log_{10} \frac{a(\text{acid})}{a(\text{conjugate base})}$$

The ratio a(acid)/a(base) is the *buffer ratio* for the particular solution. Thus the pH of a solution containing a weak acid such as acetic acid and its conjugate base is not affected by dilution because neither pK_a nor the buffer ratio is changed. Furthermore, small additions of H_3O^+ or OH^- alter the buffer ratio, and therefore the pH, only slightly. Clearly the buffering effect is most efficient when a(acid) is approximately the same as a(base) because that is the condition for the conversion of a small amount of acid to its conjugate base to have least effect on the ratio.

9.5. Methods of Studying Acid—Base Equilibria in Water

For an equilibrium such as

$$CH_3COOH + H_2O \rightleftharpoons CH_3COO^- + H_3O^+$$

in which ions appear on one side of the equation only, the equilibrium constant is conveniently obtained from the electrical conductance. For equilibria such as

$$NH_4^+ + H_2O \rightleftharpoons NH_3 + H_3O^+$$

where conductance changes are small, the concentration of H_3O^+ ions may be obtained from potential measurements. The concentration of hydroxonium and hydroxide can also be inferred from their respective catalytic effect in specific reactions. Sometimes the equilibrium may be investigated colorimetrically. The concentration of an uncharged species such as ammonia may, on occasion, be determined from partition coefficients with another phase, say chloroform.

9.6. Acids and Bases in Non-aqueous Protonic Solvents

A protonic solvent is one which contains hydrogen in the molecule and which is thus capable of acting, under certain conditions, as a proton donor. The behaviour of a solute in a solution depends on whether the solvent (i) tends to gain or lose a proton, (ii) admits recombination of ions.

Water, being both a proton donor and a proton acceptor, is said to be *amphiprotic.* It has a considerable dipole moment and a very high dielectric constant so that solvation of the ions, together with the dielectric effect (4.2.8.2), discourages their recombination. These characteristics are responsible for the simple form taken by the ionic theory before the study of non-aqueous solvents revealed the more complex relationship between concentration and degree of ionisation.

9.7. Amphiprotic Solvents Resembling Water

Alcohols resemble water as solvents, but their *relative permittivities,* ϵ_r (dielectric constants) are smaller. For methanol, at 298 K, $\epsilon_r = 32.6$ and for ethanol $\epsilon_r = 24.3$, whereas for water $\epsilon_r = 78.5$.

For equilibria like

$$NH_4^+ + EtOH \rightleftharpoons NH_3 + EtOH_2^+$$

where ions are featured on both sides of the equation, the permittivity of the solvent has comparatively little effect on the equilibrium constant. The value of $a_{NH_3} \times a_{EtOH_2^+}/a_{NH_4^+}$ is about six times greater than that of $a_{NH_3} \times a_{H_3O^+}/a_{NH_4^+}$ because EtOH is a rather stronger base than H_2O.

But for equilibria such as

$$RNH_2 + EtOH \rightleftharpoons RNH_3^+ + EtO^-$$

in which both ions are on one side, the lower permittivities of the alcohols cause the equilibrium constants to be much lower than in the corresponding reactions in aqueous solution.

The electrical free energy of a pair of ions of charges $+e$ and $-e$ separated by a distance r in a medium of permittivity ϵ is given by

$$G = \frac{e^2}{4\pi\epsilon r}$$

Since $\epsilon = \epsilon_r \times \epsilon_0$, the difference between the electrical free energy in water and that in ethanol is given by the Born equation:

$$\Delta G = \frac{N_A e^2}{4\pi\epsilon_0 r}\left(\frac{1}{\epsilon_r(\text{water})} - \frac{1}{\epsilon_r(\text{ethanol})}\right)$$

For charges separated by 200 pm:

$$\Delta G = \frac{6.022 \times 10^{23}\ \text{mol}^{-1} \times (1.602 \times 10^{-19}\ \text{A s})^2}{4\pi \times 8.854 \times 10^{-12}\ \text{kg}^{-1}\ \text{m}^{-3}\ \text{s}^4\ \text{A}^2 \times 2 \times 10^{-10}\ \text{m}} \left(\frac{1}{78.5} - \frac{1}{24.3}\right)$$

$$= -1.97 \times 10^4\ \text{kg m}^2\ \text{s}^{-2}\ \text{mol}^{-1}$$
$$= -19.7\ \text{kJ mol}^{-1}$$

The ratio of the equilibrium constants in water and alcohol for an acid which dissociates to give two singly-charged ions is obtained from

$$\ln \frac{K_a \text{ in water}}{K_a \text{ in ethanol}} = \frac{-\Delta G}{RT}$$

$$= \frac{1.97 \times 10^4\ \text{J mol}^{-1}}{8.314\ \text{J K}^{-1}\ \text{mol}^{-1} \times 298\ \text{K}}$$

$$= 7.95$$

Whence K_a in water/K_a in ethanol ~3000. This is of the same order as the observed figures for such cases, suggesting that the difference in behaviour is due largely to differences in inter-ionic attraction.

9.8. Basic or Protophilic Solvents

Ammonia (*b.p.* 240 K, $\epsilon_r = 22.4$) is, like water, an unsymmetrical molecule with a lone pair and a fairly high dipole moment. The autoprotolysis reaction is slight; $a(NH_4^+) \times a(NH_2^-) \sim 10^{-32}$ at the boiling point. The protonic acid, NH_4^+, formed by ammonium salts in solution in ammonia, liberates hydrogen with such metals as calcium:

$$2\ NH_4^+ + Ca \rightarrow 2\ NH_3 + Ca^{2+} + H_2$$

and decomposes salts of weaker acids:

$$4\ NH_4^+ + Mg_2Si \rightarrow 2\ Mg^{2+} + 4\ NH_3 + SiH_4$$

Metallic amides, imides and nitrides behave as bases in the solvent. Thus potassium amide turns phenolphthalein red and can be titrated conductometrically with an ammonium salt:

$$KNH_2 + NH_4Cl \rightleftharpoons KCl + 2\ NH_3$$

Heavy metal amides, imides or nitrides are precipitated by potassium amide:

$$AgNO_3 + KNH_2 \rightarrow AgNH_2 + KNO_3$$
$$PbI_2 + 2\ KNH_2 \rightarrow PbNH + 2\ KI + NH_3$$
$$BiI_3 + 3\ KNH_2 \rightarrow BiN + 3\ KI + 2\ NH_3$$

These correspond with the precipitation, sometimes of the hydroxide:

$$CaCl_2 + 2\ KOH \rightarrow Ca(OH)_2 + 2\ KCl$$

and sometimes of the oxide:

$$MgCl_2 + 2\ KOH \rightarrow MgO + 2\ KCl + H_2O$$

in aqueous potassium hydroxide solution. Furthermore, zinc amide reacts with potassium amide in liquid ammonia to give potassium ammonozincate:

$$Zn(NH_2)_2 + 2\ KNH_2 \rightleftharpoons K_2[Zn(NH_2)_4]$$

a reaction formally similar to that between zinc hydroxide and aqueous potassium hydroxide:

$$Zn(OH)_2 + 2\ KOH \rightleftharpoons K_2[Zn(OH)_4]$$

Ammonia, a strongly basic solvent and therefore a strong proton acceptor, encourages the dissociation of acids:

$$NH_3 + CH_3COOH \rightleftharpoons NH_4^+ + CH_3COO^-$$

Acids of $pK_a \sim 5$ in water dissociate almost completely in liquid ammonia. But the acids which dissociate extensively in water do not stand out so markedly in this respect in ammonia because the ionisation of acids weak in water, such as the carboxylic acids, is so much greater in liquid ammonia. This basic solvent is said to have a *levelling* effect on the strength of the acids. However, as the acidic properties (proton donating powers) of ammonia are very weak, strong bases dissociate in it only slightly and weak bases scarcely at all.

The solubilities of salts in liquid ammonia are markedly different from their solubilities in water. Ammonium salts are generally very soluble, as are also metal nitrates and iodides, including silver iodide, but fluorides and chlorides have low solubilities. Ammonia is not, in general, a good solvent for highly ionic compounds, possibly largely because of its low dielectric constant. But it is a better solvent than water for non-polar compounds, and for those with highly polarisable anions such as iodides. Non-metals like iodine, sulphur, selenium and phosphorus show a moderate solubility.

The alkali metals dissolve in ammonia, without the evolution of hydrogen, giving blue, strongly conducting solutions. In dilute solutions the cation appears to be Na^+, probably solvated as $Na(NH_3)_m^+$, and the anion NH_3^- or $(NH_3)_n^-$, which also contributes the colour. Metals such as platinum and iron catalyse the decomposition of the solution with the formation of sodium amide and the liberation of hydrogen:

$$2\ Na + 2\ NH_3 \rightarrow 2\ NaNH_2 + H_2$$

A solution of sodium in liquid ammonia furnishes a useful reducing agent.

9.9. Acidic or Protogenic Solvents

(i) Acetic acid has a strong levelling effect on the strength of bases, such as aliphatic amines and alkylanilines, which do not ionise greatly in water:

$RNH_2 + CH_3COOH \rightleftharpoons RNH_3^+ + CH_3COO^-$

Inorganic acids which are strong acids in water ionise less readily in acetic acid, and differences between their strengths become more obvious in that medium; for instance the dissociation constants of nitric, hydrochloric, sulphuric, hydrobromic and perchloric acids are spread in the ratios 1 : 9 : 30 : 160 :400. Because the dissociation of perchloric acid in acetic acid:

$$\underset{\text{acid}}{HClO_4} + \underset{\text{base}}{CH_3COOH} \rightleftharpoons \underset{\text{acid}}{CH_3COOH_2^+} + \underset{\text{base}}{ClO_4^-}$$

is so considerable it may be used to titrate, potentiometrically, such compounds as amides and oximes which ionise fairly strongly as bases in the solvent.

(ii) As a solvent, formic acid might be expected to behave as does acetic acid; but, having a dielectric constant (ϵ_r 62) almost 10 times greater than that of acetic acid (probably through hydrogen bonding in the liquid), it shows a particularly high autoprotolysis constant:

$$a(HCOOH_2^+) \times a(HCOO^-) = 10^{-6} \text{ at } 298 \text{ K}$$

(iii) Sulphuric acid also displays a high degree of autoprotolysis

$$a(H_3SO_4^+) \times a(HSO_4^-) = 2.4 \times 10^{-4} \text{ at } 298 \text{ K}$$

As however, the other dissociation of sulphuric acid:

$$2\,H_2SO_4 \rightleftharpoons H_3O^+ + HS_2O_7^-$$

occurs to only a slightly less extent, the interpretation of conductance and e.m.f. measurements is difficult. The dielectric constant is not known, but it is probably high, and interionic forces should be correspondingly small. The low volatility and the high viscosity are also consistent with a high dielectric constant.

Sulphuric acid is so strongly protogenic that most compounds of oxygen and nitrogen accept protons from it to some extent. Not only amines, but ethers and ketones, give twice the normal freezing point depression, suggesting that such reactions as

$$R_2O + H_2SO_4 \rightarrow R_2OH^+ + HSO_4^-$$

go virtually to completion. Many carboxylic acids dissociate as bases in sulphuric acid:

$$RCOOH + H_2SO_4 \rightleftharpoons RCOOH_2^+ + HSO_4^-$$

And even perchloric acid ionises like a weak acid in sulphuric acid:

$$a(H_3SO_4^+) \times a(ClO_4^-)/a(H_2SO_4) \times a(HClO_4) \sim 10^{-4} \text{ at } 298 \text{ K}$$

Certain substances, however, undergo reactions more complicated than proton transfer when dissolved in sulphuric acid. Nitric acid, for instance, gives the nitronium ion which is the active agent in aromatic nitration:

$HNO_3 + 2\ H_2SO_4 \rightleftharpoons NO_2^+ + H_3O^+ + 2\ HSO_4^-$

(iv) Hydrofluoric acid has a surprisingly low conductance in the anhydrous liquid condition; but it has a very high dielectric constant (ϵ_r 84), a high dipole moment (1.9 D) and a strong tendency to associate. For simplicity, it is treated here as a monobasic acid, rather than the dibasic acid, H_2F_2, which it really is. It is similar to sulphuric acid in its effects on acids and bases, the only acids showing measurable dissociation in it being perchloric and per-iodic acids. It dissolves water, ethers, ketones, aliphatic acids and even nitric acid, all of them functioning as bases:

$H_2O + HF \rightleftharpoons H_3O^+ + F^-$
$Et_2O + HF \rightleftharpoons Et_2OH^+ + F^-$
$HNO_3 + HF \rightleftharpoons H_2NO_3^+ + F^-$

Some non-metal fluorides dissolve to give acid solutions:

$AsF_5 + 2\ HF \rightleftharpoons H_2F^+ + AsF_6^-$

Alkali-metal chlorides, bromides, iodides and cyanides dissolve in hydrogen fluoride to give the free acids:

$NaCl + HF \rightarrow Na^+ + F^- + HCl(gas)$

Many solutes react with anhydrous hydrogen fluoride, an example being sulphuric acid which produces fluorosulphuric acid:

$H_2SO_4 + 2\ HF \rightarrow HSO_3F + H_3O^+ + F^-$

Mixtures of hydrogen fluoride and boron trifluoride are very powerful proton donors: the substances react as shown.

$BF_3 + 2\ HF \rightleftharpoons H_2F^+ + BF_4^-$

In this liquid mixture even aromatic hydrocarbons behave as bases; for instance hexamethylbenzene is largely converted into $C_6HMe_6^+BF_4^-$ and forms a strongly conducting solution.

9.10. Superacid Media

In research to discover solvents which will enable even compounds like alkanes to behave as bases, Gillespie and his co-workers have developed so-called superacid media. Pure HSO_3F is a stronger protonating agent than any pure acid other than $H_2S_2O_7$, but the addition of SbF_5 or mixtures of SbF_5 and SO_3 greatly enhances the acidity. With SbF_5 alone the HSO_3F gives rise to the equilibria

$$SbF_5 + 2HSO_3F \rightleftharpoons [SbF_5(OSO_2F)]^- + H_2SO_3F^+$$

and

$$2\,SbF_5 + 2\,HSO_3F \rightleftharpoons [F_5Sb{-}O{-}S(F)(=O){-}O{-}SbF_5]^- + H_2SO_3F^+$$

The still stronger acid obtained when SO_3 is added as well as SbF_5 is probably due to the reaction:

$$3\,SO_3 + SbF_5 + 2\,HSO_3F \rightleftharpoons H_2SO_3F^+ + [SbF_2(SO_3F)_4]^-$$

because the conductivity of the solution increases with addition of SO_3 up to the point where the SO_3/SbF_5 ratio becomes 3. This solution is exceptionally effective in generating carbonium ions, e.g.:

$$R_3CH + H_2SO_3F^+ \rightleftharpoons R_3CH_2^+ + HSO_3F \rightleftharpoons H_2 + R_3C^+ + HSO_3F$$

a reaction in which donation of a proton by the superacid ion to the alkane acting as a base is followed by evolution of H_2 and the production of a carbonium ion.

9.11. Aprotic Solvents

Hydrocarbons and their halogen derivatives have no tendency to gain or lose protons; they are inert and exhibit no levelling effect. The dielectric constants are very low (ϵ_r 2 to 6) and the ions associate, rendering conductance measurements of no value for determining the extent of protolysis. Protolytic equilibria are also complicated by association of the uncharged molecules themselves; carboxylic acids, for example, exist as dimers in benzene. The same factors reduce greatly the solubilities of acids and bases in these solvents.

9.12. Comparison of Acid Strengths in Various Solvents

Because the dissociation of an acid depends in a complex way on the chemical properties, molecular dipole and dielectric constant of the solvent in which it is dissolved, attempts to define absolute acid strength independently of the solvent have been unavailing. Nevertheless, relative strengths are independent, within a power of 10, of the nature of the solvent provided the acids belong to the same charge type, whether that be molecular, anionic or cationic. The independence of the nature of the solvent shown by acids of the same chemical character is even more marked.

9.13. Acid Strength and Molecular Structure

(i) Simple hydrides of Groups V to VII show the approximate pK_a values given in Table 9.2.

Where there are several acids of a similar type the pK_a values vary regularly down the series. The changes in value are an expression of the overall effect of the energies of (a) dissociation of the hydrides to atoms, (b) ionisation of the hydrogen atom, (c) hydration of the hydrogen ion, (d) electron affinity of the other atom or radical, and (e) the solvation of the anion.

TABLE 9.2

pK_a VALUES OF HYDRIDES OF GROUPS V, VI, VII

Group V		Group VI		Group VII	
NH_3	23	H_2O	14	HF	3
		H_2S	7	HCl	−6
		H_2Se	4	HBr	−8
		H_2Te	3	HI	−9

These stages are set out below and give by addition the heat of dissociation of a binary acid into ions in aqueous solution.

	Stage		Heat required
	(a) HX(gas)	$\rightarrow$ H(g) + X(g)	D
	(b) H(g)	$\rightarrow H^+(g) + e$	I_H
	(c) $H^+(g)$ + aq	$\rightarrow H^+_{aq}$	ΔH_{aqH^+}
	(d) X(g) + e	$\rightarrow X^-(g)$	$-A_X$
	(e) $X^-(g)$ + aq	$\rightarrow X^-_{aq}$	ΔH_{aqX^-}
Adding:	HX(g) + aq	$\rightarrow H^+_{aq} + X^-_{aq}$	$\Delta H_{\text{(acid dissociation)}}$

Thus for acid dissociation

$$\Delta H = D - A_X + \Delta H_{aqX^-} + I_H + \Delta H_{aqH^+}$$

As the last two terms (I_H and ΔH_{aqH^+}) are constants for all binary acids, the change of heat of acid dissociation from one acid to another must depend on the first three terms. The dissociation constants depend on ΔG, not ΔH, of course, but for most of the cases which have been thoroughly studied the entropy differences are not large enough to affect the trends in the change of acid strength through a series of binary acids.

In the series NH_3, H_2O and HF, increasing electron affinity (A) is the likeliest cause of increasing acid strength; but in the group HF to HI the determining factor is probably the decreasing dissociation energy (D).

(ii) Oxoacids of the general formula H_nXO_m fall into four fairly well-defined classes depending on the ratio of oxygen to hydrogen atoms in the molecule:

(a) Where $m = n$, as in HOCl, HOBr, H_3BO_3 and H_6TeO_6, the pK value for the first ionisation is between 7 and 11.

(b) Where $m = n + 1$, as in $HClO_2$, HNO_2, H_2SO_3 and H_3PO_4, the pK value is approximately 2.

(c) Where $m = n + 2$, as in HNO_3, $HClO_3$ and H_2SO_4, the pK value is between -1 and -3.

(d) Where $m = n + 3$, as in $HClO_4$, the pK value is approximately -8.

Clearly, telluric acid, about which there was doubt, fits into this scheme as H_6TeO_6, not as $H_2TeO_4 \cdot 2\,H_2O$. However, phosphorous and hypophosphorous acids, with pK 2 and 1 respectively, behave as $HPO(OH)_2$ and $H_2PO(OH)$, in accordance with their known basicities. Increase in acid strength appears to be associated with the stabilisation of the anion. The greater the charge number of the central atom, the greater its electronegativity, and the lower the energy of the molecular orbitals which can accommodate the electron to which the anion owes its charge.

(iii) Aliphatic acids have a pK_a of about 5. The substitution of hydrogen in alkyl groups by elements with high electronegativity (F, Cl) increases acid strength, particularly when more than one hydrogen atom is replaced, and when substitution is made at the carbon atom nearest the carboxylic group. In the dichloracetate ion, for example, stability is conferred through the chlorine increasing the electronegativity of the alkyl C and thereby lowering the energy of the whole ion by attracting away some of the negative charge conferred on the other C by the oxygen atoms.

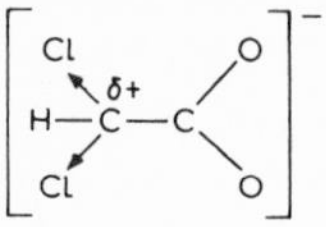

9.14. Generalised Acid—Base Theory

The disadvantage of the Lowry—Brönsted theory is that it applies only to acids containing hydrogen.

A more general concept of acids and bases was developed by G.N. Lewis, who defined a base as an atom, molecule or ion with at least one pair of electrons not shared in a covalent bond, and an acid as a species in which some atom, the acceptor atom, has a vacant orbital which can accommodate a pair of electrons. The typical acid—base reaction, in the Lewis scheme, is

$$A + :B \rightleftharpoons A:B$$

The species A:B may be called a *co-ordination compound*, an *adduct*, or an *acid—base complex*.

The following reactions, with the acid formulated first, illustrate the principle:

$$BF_3 + :NMe_3 \rightleftharpoons F_3BNMe_3$$
$$Ag^+ + 2(:NH_3) \rightleftharpoons Ag(NH_3)_2^+$$
$$FeCl_3 + :Cl^- \rightleftharpoons FeCl_4^-$$
$$CH_3CO^+ + :OCH_3^- \rightleftharpoons CH_3CO_2CH_3$$
$$H^+ + :Br^- \rightleftharpoons HBr$$

All cations are Lewis acids and all anions are bases. Since H^+ is classified as an acid, there is no difference between a base in this scheme and a base in the Lowry—Brönsted scheme, but a proton acid like HBr is considered an acid—base adduct by Lewis.

Clearly, it is possible to classify a large proportion of chemical reactions as interactions between generalised acids and bases. But the attempts which have been made to develop rules about the stabilities of adducts have proved only moderately successful. One empirical rule which has been suggested is the *principle of hard and soft acids and bases*, the HSAB principle.

9.15. The HSAB Principle

Pearson (1963) defined a soft base as one in which the valence electrons are easily polarised, and a hard base as one in which the valence electrons are tightly held. Typical soft bases are large anions like I^- and H^- and molecules like $(C_6H_5)_3As$ in which the donor atom is large; typical hard bases are small anions like F^-, symmetrical oxo-anions like ClO_4^-, and molecules like NH_3 in which the donor atom is small.

A hard acid is one in which the electron-accepting atom is small, has a high positive charge and does not have valence electrons with easily distorted orbitals; a soft acid is one in which the acceptor atom is large, carries a low positive charge or none at all, and has electrons in orbitals which are easily distorted. Examples of hard acids are ions like Al^{3+} and compounds like BF_3; typical soft acids are metal atoms, ions like Hg_2^{2+} and compounds like $InCl_3$.

The HSAB principle can be stated in the form: Hard acids tend to form adducts by co-ordination to hard bases rather than to soft bases, and soft acids tend to co-ordinate to soft bases rather than to hard bases.

Table 9.3 shows a number of Lewis acids which are classified as hard, intermediate or soft, and treats some bases similarly. The tables are not meant to imply a quantitative scale of hardness or softness; in fact the disadvantage of the HSAB principle, and of the whole concept of generalised acids and bases, is that it has at present no sound thermodynamic basis. However, Pearson has suggested ways in which progress might be made towards predicting the thermodynamic stabilities of adducts.

To define strength in a simple way, consider the equilibria:

$$A' + A{:}B \rightleftharpoons A + A'{:}B$$
$$B' + A{:}B \rightleftharpoons B + A{:}B'$$

If the two reactions proceed far to the right it means that A′ is a stronger acid than A, and B′ is a stronger base than B. In principle it should be possible from a study of such reactions to make a list of Lewis acids in order of strength and a list of bases similarly. Then we should expect the equilibrium constant, K, for the reaction

$$A + {:}B \rightleftharpoons A{:}B$$

TABLE 9.3

THE HSAB PRINCIPLE

Hard acids	
	H^+, Li^+, Na^+, K^+,
	Be^{2+}, Mg^{2+}, Ca^{2+}, Sr^{2+}, Mn^{2+},
	Al^{3+}, Cr^{3+}, Co^{3+}, As^{3+},
	BF_3, $Al(CH_3)_3$, Al_2Cl_6
Intermediate acids	
	Fe^{2+}, Co^{2+}, Ni^{2+}, Cu^{2+}, Zn^{2+}, Pb^{2+}, Sn^{2+},
	$B(CH_3)_3$, SO_2, NO^+, $C_6H_5^+$, R_3C^+
Soft acids	
	Pd^{2+}, Cd^{2+}, Pt^{2+}, Hg^{2+},
	Cu^+, Ag^+, Tl^+, Hg_2^{2+},
	$GaCl_3$, GaI_3, B_2H_6,
	RO^+, RS^+, RSe^+ (R = Alkyl),
	metal atoms
Hard bases	
	H_2O, OH^-, F^-, Cl^-, CO_3^{2-}, ClO_4^-, NO_3^-,
	ROH, RO^-, R_2O, NH_3, RNH_2
Intermediate bases	
	$PhNH_2$, C_5H_5N, N_3^-, Br^-, NO_2^-
Soft bases	
	R_2S, RSH, RS^-, I^-, SCN^-, R_3P, R_3As,
	CN^-, RNC, CO, C_2H_4, H^-, R^-

to be related to strength factors, S, for the acid and base respectively by

$$\ln K = S_A S_B$$

Unfortunately it is quite impossible to construct a table of strengths of bases which is independent of the nature of the reference acid. Ahrland, Chatt and Davies (1958) pointed out that among transition-metal ions, class *a* acceptors (later named hard acids by Pearson) form their most thermodynamically stable adducts with donor atoms in Period 2 of the Periodic Table whereas class *b* acceptors (metal ions now classified as soft acids) co-ordinate best with donor atoms in the later periods (Table 9.4). Hard acids co-ordinate most strongly to the smallest, least polarisable donor atoms while soft acids co-ordinate most strongly to larger, more polarisable donor atoms. In Group VII the order of donor strength is exactly reversed for soft acids as compared with hard ones. In Group V, however, the very polarisable bases (e.g. alkylarsines and stibines) do not form particularly strong adducts because they act as weak bases towards all acids. Nevertheless they do form stronger complexes with soft acids than with hard acids.

The implication of the argument is that the relation above should be modified. One possibility would be

TABLE 9.4
COMPARISON OF DONOR STRENGTH OF ATOMS TOWARDS ACIDS

	(a) Hard acids:	(b) Soft acids:
Group V donors	N > P > As > Sb > Bi	N < P > As > Sb > Bi
Group VI donors	O > S > Se > Te	O < S ~ Se ~ Te
Group VII donors	F > Cl > Br > I	F < Cl < Br < I

$$\ln K = S_A S_B + \sigma_A \sigma_B$$

where σ_A and σ_B are softness factors for acid and base respectively. The σ values could be made negative for hard acids and bases and positive for soft acids and bases. Thus the strength of a complex would depend not only on the absolute strengths of the acid and base producing it but also on their softness factors. Consequently an adduct formed from a strong acid A and a strong base B may be quite thermodynamically unstable if A is very hard and B is very soft because the product $\sigma_A \sigma_B$ will be negative and ln K will be significantly less than the product $S_A S_B$.

Although the HSAB principle does describe much of chemical reactivity in a qualitative way, and has useful predictive power, it has not yet enabled tables of K values for Lewis acids and bases to be constructed.

9.16. Non-protonic Solvents

Attempts to relate the behaviour of non-protonic ionising solvents to generalised acid—base theory have been only moderately successful. In many cases there is strong evidence for self-ionisation to produce a positive ion which is the acid and a negative ion which is the base characteristic of the system.

Thus the reactions of liquid N_2O_4 (*b.p.* 294 K) can be rationalised on the basis that it appears to ionise, perhaps only to a small extent:

$$N_2O_4 \rightleftharpoons NO^+ + NO_3^-$$

Zinc dissolves in the liquid, liberating NO, and forming the ions NO^+ and $[Zn(NO_3)_4]^{2-}$. The solution can be titrated with a solution of a nitrate in the same solvent.

$$2\,NO^+ + 2\,NO_3^- \rightleftharpoons 2\,N_2O_4$$

The pale-yellow BrF_3 (*b.p.* 405 K) appears to ionise:

$$2\,BrF_3 \rightleftharpoons BrF_2^+ + BrF_4^-$$

AgF dissolves to give a strongly conducting solution containing Ag^+ and BrF_4^-, while SbF_5 produces the ions BrF_2^+ and SbF_6^-. The reaction between the two solutions:

$Ag^+ + BrF_4^- + BrF_2^+ + SbF_6^- \rightleftharpoons AgSbF_6 + 2\ BrF_3$

has been followed conductometrically. Substances which give rise to BrF_2^+ ions in this solvent behave as acids and those producing BrF_4^- act as bases.

Liquid NOCl (*b.p.* 267 K) has a rather high relative permittivity of 18.2 at the boiling point. It dissolves metal halides, e.g.:

$NOCl + FeCl_3 \rightleftharpoons NO^+ + FeCl_4^-$

This particular solution has been titrated conductometrically with Me_4NCl in the solvent:

$NO^+ + FeCl_4^- + Me_4N^+ + Cl^- \rightleftharpoons Me_4NFeCl_4 + NOCl$

In the three examples given above the solvent appears to produce by self-ionisation a positive ion which acts as an acid and a negative ion which acts as a base. However liquid SO_2 (*b.p.* 263 K) with a relative permittivity of 14 does not appear to ionise to give SO^{2+} and SO_3^{2-}. This is hardly surprising, as the force of attraction between doubly charged ions in a solvent of moderate permittivity would obviously be high.

Although sulphites and thionyl halides dissolve in liquid SO_2, and increase conductance, isotopic labelling shows that the sulphur in a thionyl compound does not exchange with sulphur in the solvent molecules.

The idea that it is the ionisation of a solvent which is primarily responsible for the way it behaves when Lewis acids and bases are dissolved in it has recently been strongly criticised. R.S. Drago has drawn attention to the importance of the solvating action of solvents. It is unrealistic to consider that the generalised acid—base interaction in a solvent will be unaffected by the nature of that solvent, because acid, base and adduct will all be solvated and the enthalpies of solvation may well be disparate. Consider the application of an enthalpy cycle to the reaction between solvated acid and base:

A(solv) + B(solv) $\rightleftharpoons$ AB(solv)

The value of ΔH for this reaction will depend upon four enthalpy changes:

(i)	A(solv) → A(g) + solvent	$-\Delta H_a$
(ii)	B(solv) → B(g) + solvent	$-\Delta H_b$
(iii)	A(g) + B(g) → AB(g)	ΔH_c
(iv)	AB(g) + solvent → AB(solv)	ΔH_{ab}

∴ A(solv) + B(solv) = AB(solv) $\quad \Delta H_c + \Delta H_{ab} - \Delta H_a - \Delta H_b$

ΔH_a, ΔH_b and ΔH_{ab} are respectively the enthalpies of solvation of acid, base and complex, and ΔH_c is the enthalpy of the reaction between A and B in the gas phase. Clearly it is only ΔH_c which can be predicted by the HSAB method and its value will not be similar to the overall enthalpy of the reaction in solution unless $\Delta H_{ab} - \Delta H_a - \Delta H_b$ is negligible.

Further Reading

R.J. Gillespie, Fluorosulphuric acid and related superacid media, Acc. Chem. Res., 1 (1968) 202.

R.P. Bell, Acids and bases, Their quantitative behaviour, 2nd edition, Chapman and Hall, London, 1969.

D.P.N. Satchell and R.S. Satchell, Quantitative aspects of the Lewis acidity of covalent metal halides and their organo derivatives, Chem. Rev., 69 (1969) 251.

R.G. Pearson, Hard and soft acids and bases, Chem. Brit., 3 (1967) 103.

R.S. Drago and B.B. Wayland, A double-scale equation for correlating enthalpies of Lewis acid—base interactions, J. Amer. Chem. Soc., 87 (1965) 3571.

S. Ahrland, J. Chatt and N. Davies, The relative affinities of ligand atoms for acceptor molecules and ions, Quart. Rev., 12 (1958) 265.

A. Albert and E.P. Serjeant, Ionization constants of acids and bases, Methuen, London, 1962.

D.D. Perrin, Dissociation constants of inorganic acids and bases in aqueous solution, (I.U.P.A.C.) Butterworth, London, 1969.

J.F. Coetzee and C.D. Ritchie, Solute—solvent interactions, Marcel Dekker, New York, 1969.

L.F. Audrieth and J. Kleinberg, Non-aqueous solvents, Wiley, New York, 1953.

R.G. Pearson, Hard and soft acids and bases, J. Chem. Ed., 45 (1968) 581 and 643.

Hydrogen

10

10.1. Occurrence and Physical Properties

Terrestrially, except in the upper atmosphere, hydrogen occurs almost entirely in compounds; of these water and the hydrocarbons are the most abundant. Its importance is, however, greatly beyond what might be suggested by its terrestrial abundance (0.81%) because it combines with nearly every other element and is, moreover, an essential constituent of all living matter. It is a colourless, odourless gas which is less dense than any other gas and is almost insoluble in water.

10.1.1. Isotopes

Hydrogen has three isotopes. The atom of deuterium, ${}^{2}_{1}H$, is about twice as heavy as that of ordinary hydrogen; tritium, ${}^{3}_{1}H$, three times as heavy. These uniquely large mass ratios are responsible for a difference in chemical properties between the isotopes of hydrogen far greater than that shown by the isotopes of any other element. However, so little deuterium and tritium are present in natural hydrogen (1.6×10^{-4} and 10^{-18} by weight respectively) that its properties are substantially those of ${}^{1}_{1}H$ itself.

10.1.2. The hydrogen ions H^+ and H^-

With electron configuration $1s^1$, hydrogen nearly always forms covalent bonds, but positive and negative singly charged ions have been recognised. Loss of the electron leaves the proton H^+, the ionisation energy of the hydrogen atom being 1.31 MJ mol^{-1}. The proton exerts so strong a positive field that it is unable to exist alone in the presence of polarisable species. Thus the

TABLE 10.1

PHYSICAL PROPERTIES OF HYDROGEN

B.p./K	*M.p.*/K	*Critical temp.*/K
20.4	14.0	33.2

'hydrogen ion' in water becomes H_3O^+, in ammonia NH_4^+, the proton being bonded to the molecule by a lone pair of electrons (9.1).

The addition of an electron to the hydrogen atom produces the hydride ion, H^-, the electron affinity being 68 kJ mol^{-1}. This quantity and the dissociation energy of the hydrogen molecule, $H_2 \rightarrow 2\ H$, namely $\Delta H = 436$ kJ mol^{-1}, are of importance in a consideration of the heat of formation of metal hydrides. The figures can be compared with those for fluorine:

$\frac{1}{2}\ H_2 \rightarrow H$	$\Delta H = +218$ kJ mol^{-1}	$\frac{1}{2}\ F_2 \rightarrow F$	$\Delta H = +80$ kJ mol^{-1}
$H + e \rightarrow H^-$	$\Delta H = -68$ kJ mol^{-1}	$F + e \rightarrow F^-$	$\Delta H = -360$ kJ mol^{-1}

As the H^- ion is comparable in size with the F^- ion, the lattice energies of ionic hydrides and fluorides of the same structural type must be similar. Applying the Born—Haber cycle we can see that the formation of an ionic fluoride will release far more energy than will the formation of an ionic hydride. Nevertheless, H^- ions exist in alkali-metal hydrides. The hydride ion is large because of the mutual repulsion of the electrons, which offsets the nuclear attraction, but variable (from 126 pm in LiH to 154 pm in CsH) because the rather diffuse electron cloud is easily polarised.

The H^- ion is an extremely strong base; the reaction

$$NaH + H_2O \rightarrow NaOH + H_2$$

when represented as

$$H^- + H_2O \rightarrow OH^- + H_2$$

indicates that H^- is a stronger base than OH^- in aqueous solution. Similarly,

$$NaH + NH_3 \rightarrow NaNH_2 + H_2$$

or

$$H^- + NH_3 \rightarrow NH_2^- + H_2$$

shows H^- to be a stronger base than NH_2^- in liquid ammonia.

10.2. *Ortho* and *Para* Hydrogen

10.2.1. Reason for the two forms

Two hydrogen atoms combine to form the very stable hydrogen molecule, $2H(1s^1) \rightarrow H_2(\sigma 1s^2)$. Heisenberg showed, however, that when nuclear spins are taken into account there are, in effect, two observable 'isomers'. These result from parallel coupling of the nuclear spins, with three possible quantum states; and antiparallel coupling, with only one. At room temperature, the different quantum states are about equally probable, but spontaneous transitions between them have a very low probability and can be ignored. Consequently, ordinary hydrogen behaves as though it were a mixture of 3 vol. *ortho* hydrogen (spins parallel) and 1 vol. *para* hydrogen (spins antiparallel). But, owing to symmetry, the allowed states of rotation of the molecule as a whole differ in the two instances, the lowest state being somewhat lower for the *p*- than for the *o*-form. Thus at very low temperatures, where the molecules tend to go into their lowest quantum states, the proportion of *p*-hydrogen tends towards 100%, when there would be true thermodynamic equilibrium. However, transition from one nuclear spin state to another is so slow (the collision 'half-life' at room temperature is several years) that the 3:1 proportion persists in metastable equilibrium during cooling. These conclusions are confirmed by measurements of conductance and specific heat, which are distinctly greater for the *p*- than the *o*-form (in contrast with the *b.p.* and *m.p.* which are slightly less). True equilibrium at any temperature is achieved in the presence of a catalyst: (i) activated charcoal at low temperature, or a transition metal at room temperature; (ii) atomic hydrogen, (iii) a paramagnetic substance such as O_2 or NO. Thus, at 20 K in the presence of active charcoal, 99.7% pure *p*-hydrogen results. On the other hand, concentration of *o*-hydrogen beyond 75% is impossible and its tabulated properties are those inferred from the mixture.

10.2.2. Conversion of forms and its application

The mechanism of conversion, which is exothermal in the direction $o \rightarrow p$, involves dissociation of the molecule and recombination of the atoms, during which the nuclear spins re-couple, parallel or antiparallel, in equilibrium proportions. This occurs, for example, on collision (the high temperature mechanism); and probably in chemisorption, when the atoms are separated by going into different lattice sites and subsequently recombine with nuclear spins oppositely coupled.

Knowledge of the two forms of molecular hydrogen has found industrial application. As stated, the change $o\text{-}H_2 \rightarrow p\text{-}H_2$ is exothermal, but so slow that it takes about a month for normal 25% *p*-liquid to be converted to 90% *p*-liquid. The energy released by the change is sufficient to evaporate 64% of the original liquid. Hence to keep liquid hydrogen without loss constant refrigeration is necessary. However a rapid and effective catalyst, a hydrous iron(III) oxide, has been developed which enables normal hydrogen to be converted to 99% *p*-hydrogen during liquefaction. The change greatly facilitates the storage of liquid by making it unnecessary to refrigerate to prevent loss by boiling.

10.2.3. *Ortho* and *para* deuterium

Deuterium also exists in *ortho* and *para* forms, but the *ortho* form is the more stable at low temperatures. The equilibrium mixture at elevated temperatures contains 33.3% of *para* deuterium.

The *ortho* forms of both hydrogen and deuterium have a small magnetic moment due to the nuclear spins of the two being of the same sense; the *para* forms have none. Nonetheless, hydrogen is essentially diamagnetic because it is without unpaired electrons, and the magnetic moment of a nucleus is very much less than that of an electron.

Strictly, hydrogen and deuterium are not unique, and nuclear spins should be recognised in discussing other molecules. But only with a low moment of inertia is the separation of rotational energy states large enough to give one form an appreciable preference, and then only at low temperatures. *Ortho* and *para* forms of F_2, Cl_2 and N_2 have been distinguished at very low temperatures, but generally the 'high temperature' mixture is the equilibrium form for all readily accessible temperatures, accordingly the nuclear spins may be disregarded.

10.3. Reactions of Hydrogen

Hydrogen is relatively unreactive at ordinary temperatures. Apart from fluorine, which spontaneously explodes with hydrogen in the dark, the strongly electronegative elements react with hydrogen only on heating or on irradiation. Combination with chlorine, which is started by light or heat, is a simple chain reaction:

Cl_2	$\xrightarrow{h\nu} 2\,Cl^*$	Initiation
$Cl^* + H_2$	$\rightarrow HCl + H^*$	Chain propagating steps
$H^* + Cl_2$	$\rightarrow HCl + Cl^*$	

(* indicates activated atom)

With bromine and with oxygen the chain reactions are more complex.

Combination with nitrogen and the hydrogenation of gaseous hydrocarbons occur on surface-catalysts. Only a few reactions of hydrogen are known to involve H_2 molecules and for these the activation energies (E_A) amount to about 25—30% of the sum of the bond energies, the usual value for reactions of this type.

The comparatively unreactive character of hydrogen in the absence of catalysts arises largely from the strength of the H—H bond, which also accounts for the low thermal stabilities of many hydrides, for instance:

PH_3 $\quad \Delta G_f = +18\ \text{kJ mol}^{-1}$
H_2Se $\quad \Delta G_f = +70\ \text{kJ mol}^{-1}$

Of the metals, most of those in Groups IA, IIA, IIIA, IVA and VA and also the lanthanides and actinides combine exothermally with hydrogen under suitable conditions.

Hydrogen reduces many metal oxides to metal at moderate temperatures:

$$CuO + H_2 \rightarrow Cu + H_2O$$
$$Fe_3O_4 + 4\,H_2 \rightarrow 3\,Fe + 4\,H_2O$$

But it is not as effective a reducing agent as carbon at high temperatures.

Hydrogen reacts with carbon monoxide to give a variety of products, depending on the conditions; the most important of these reactions is that used for the production of methanol:

$$CO + 2\,H_2 \rightarrow CH_3OH \qquad \Delta H = -125\ \mathrm{kJ\ mol^{-1}}$$

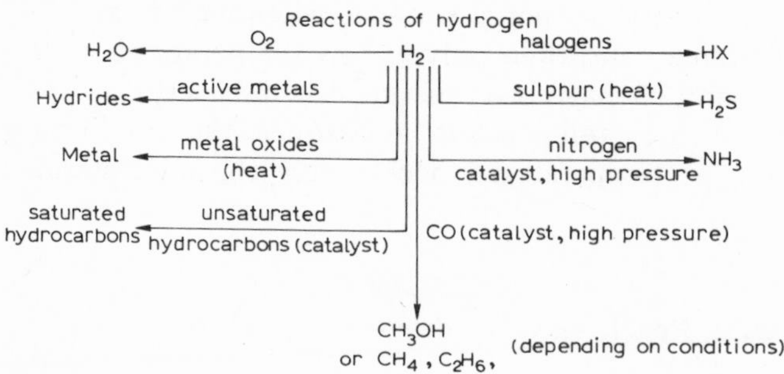

10.4. Atomic Hydrogen

The hydrogen molecules are dissociated into atoms by heat, by radiation of a specific wavelength, and by an electric discharge at low pressure. The atoms do not combine on collision because the energy set free, 436 kJ mol^{-1}, causes immediate redissociation when converted into vibrational energy. For recombination to take place a third body must be present to absorb the excess of energy. The lifetime of a hydrogen atom at room temperature, even at low pressures, is short on account of the catalytic combination of atoms on the walls of the vessel.

Atomic hydrogen is a stronger reducing agent than H_2: unlike the latter it combines directly with Ge, Sn, As, Sb, Te, and reduces some oxoanions in solid compounds:

$$As + 3\,H \rightarrow AsH_3$$
$$BaSO_4 + 8\,H \rightarrow BaS + 4\,H_2O$$

Atomic hydrogen torch

Langmuir showed that hydrogen passed through an arc between tungsten electrodes is partly dissociated into atoms even at ordinary pressure. If this stream of gas is directed on to a metal surface, the heat of combination is sufficient to produce a high temperature. This is the principle of the hydrogen torch, used in welding metals, which in action surrounds the weld with an atmosphere of the molecular hydrogen and thus prevents surface oxidation. It is especially applicable to such metals as aluminium which are readily oxidised but which do not easily absorb hydrogen.

10.5. Deuterium

The isotope ^2_1H was observed spectroscopically in 1931 and the preparation of nearly pure D_2O was achieved two years later by prolonged electrolysis of an 0.5 *M* NaOH solution between nickel electrodes. Three factors favour the liberation of H_2 at the cathode rather than D_2:

(i) H_2O is more readily reduced than D_2O:
$H_2O + e \rightarrow OH^- + H$;
(ii) the activation energy for $H + H \rightarrow H_2$ is less than for $D + D \rightarrow D_2$;
(iii) HD exchanges with the water: $HD + H_2O \rightarrow HDO + H_2$.

Differences in overvoltage (the potential arising from the non-reversible nature of the electrode processes) are probably not of major importance since the material of the cathode does not affect the separation factor.

The concentration of deuterium in ordinary water was attained by successively reducing the original volume to about one seventh by electrolysis. As the electrolysis proceeds the proportion of D_2 in the evolved gas rises. When it reaches 0.02%, the gas is burnt in oxygen and the H_2O/D_2O mixture added to the electrolyte of an earlier stage. Such an electrolytic separation has produced most of the considerable quantities of D_2O already in use for 'moderating' fast neutrons in heavy-water atomic reactors.

Heavy water is also produced by the Spevack method which uses the exchange reaction:

$$D_2O(\text{in ordinary water}) + H_2S \rightleftharpoons D_2S + H_2O \text{ (or DHS + DHO)}$$

A temperature of 373 K favours the forward reaction. The equilibrium mixture of sulphides is passed into water at 298 K, when the reverse reaction is favoured. Continued cycles lead to a liquid containing ~2% D_2O which is concentrated to 90% by fractional distillation and to 99.8% by electrolysis.

The physical properties of D_2O differ slightly from those of H_2O (Table 10.2). Its dielectric constant is about 2% below that of H_2O, making the liquid a slightly poorer ionising solvent; the autoprotolysis constant, $a(D_3O^+) \times a(OD^-)$, is 3×10^{-15} at 298 K, and the solubilities of electrolytes are corre-

TABLE 10.2

PHYSICAL PROPERTIES OF H_2O AND D_2O

	B.p./K	*M.p.*/K	ρ/g cm^{-3} at 293 K
H_2O	373	273	0.998
D_2O	374.4	276.8	1.106

TABLE 10.3

PHYSICAL PROPERTIES OF H_2 AND D_2

	B.p./K	*M.p.*/K	*Latent heat of fusion*/J mol^{-1}	*Heat of dissociation*/ kJ mol^{-1}
H_2	20.4	14.0	117	436
D_2	23.5	18.65	217	438

spondingly less than in water. For instance, the solubility of NaCl at 298 K in H_2O is 0.359 w/w, in D_2O, 0.309 w/w.

Deuterium gas is made by the electrolysis of D_2O containing some P_2O_5. The *b.p.*, *m.p.*, latent heats of fusion and evaporation, and heat of dissociation of D_2 are all higher than those of H_2 (Table 10.3).

Deuterium compounds are exactly analogous to those of hydrogen, and are often obtained directly from D_2O:

$$Mg_3N_2 + 6\,D_2O \rightarrow 3\,Mg(OD)_2 + 2\,ND_3$$
$$CaC_2 + 2\,D_2O \rightarrow Ca(OD)_2 + C_2D_2$$
$$SO_3 + D_2O \rightarrow D_2SO_4$$

Deuterium is somewhat less reactive than hydrogen. Activation energies for reactions such as

$$D_2 + Br_2 \rightarrow 2\,DBr \qquad \text{and} \qquad 3\,D_2 + N_3 \rightarrow 2\,ND_3$$

are greater than for those with H_2. In keeping with this, the adsorption of deuterium on an active surface is slower than that of hydrogen.

10.6. Exchange Reactions

When H_2 and D_2 are mixed at a sufficiently high temperature, exchange occurs:

$$H_2 + D_2 \rightarrow 2\,HD$$

Many such exchange reactions are known; the one above is almost certainly an atomic reaction with a chain mechanism:

$$D + H_2 \rightarrow HD + H; \qquad H + D_2 \rightarrow HD + D$$

Exchange on a catalytic surface is also common. Deuterium replaces hydrogen in water, ammonia, methane and other simple hydrides on the surface of metals such as platinum and nickel which are good catalysts for hydrogenation. The exchange of adsorbed deuterium with gaseous hydrogen on finely divided nickel is almost certainly an adsorption—desorption process:

$$D_2 + 2\,Ⓢ \rightleftharpoons 2\,D(\text{adsorbed}), \qquad H_2 + 2\,Ⓢ \rightleftharpoons 2\,H(\text{adsorbed})$$
$$D(\text{adsorbed}) + H(\text{adsorbed}) \rightleftharpoons HD(\text{gas}) + 2\,Ⓢ$$

(Ⓢ represents an empty site)

Heavy water exchanges deuterium with compounds containing labile hydrogens:

$$NH_4{}^+ + D_2O \rightleftharpoons NH_3D^+ + HDO$$
$$CH_3NH_2 + D_2O \rightleftharpoons CH_3NHD + HDO$$
$$CH_3OH + D_2O \rightleftharpoons CH_3OD + HDO$$

The suggested mechanism of exchange between D_2O and the hexa-amminecobalt(III) ion is

$$Co(NH_3)_6^{3+} + D_2O \rightleftharpoons Co(NH_3)_5NH_2^{2+} + HD_2O^+$$
$$Co(NH_3)_5NH_2^{2+} + D_2O \rightleftharpoons Co(NH_3)_5NH_2D^{3+} + OD^-$$
$$H_3O^+ + OD^- \rightleftharpoons HDO + H_2O$$

But the hydrogen atoms of alkyl groups do not exchange readily unless enol—keto tautomerism occurs:

$$(CH_3)_2C{=}O \rightleftharpoons CH_3\overset{\overset{\displaystyle OH}{|}}{C}{=}CH_2$$

$$CH_3\overset{\overset{\displaystyle OH}{|}}{C}{=}CH_2 + D_2O \rightleftharpoons [CH_3\overset{\overset{\displaystyle O}{|}}{C}{=}CH_2]^- + HD_2{}^+O$$
$$\rightleftharpoons CH_3\underset{\underset{\displaystyle O}{\|}}{C}{-}CH_2D + HDO$$

The rate of this reaction is measured by isolating some of the reaction product in the pure state, burning it to produce H_2O and HDO, and determining the deuterium present by density or with the mass spectrometer. Here deuterium is, in fact, being used as a non-radioactive tracer.

10.7. Tritium

The heaviest isotope of hydrogen, tritium, was first made (1934) by bombarding deuterium compounds with deuterons:

$${}^2_1D + {}^2_1D \rightarrow {}^3_1H + {}^1_1H$$

Certain other nuclear reactions lead to tritium formation:

$${}^9_4Be + {}^2_1H \rightarrow 2\,{}^4_2He + {}^3_1H$$

$${}^6_3Li + {}^1_0n \rightarrow {}^4_2He + {}^3_1H$$

The second of these, the bombardment of lithium-6 with neutrons in an atomic reactor, has been extensively used in its production; the tritium is adsorbed by uranium metal and released by heating. The nuclide is a low-energy β-emitter (0.018 MeV) with a half-life of 12.5 years:

$${}^3_1H \xrightarrow[\text{12.5 years}]{\beta^-} {}^3_2He$$

Tritium has proved a useful tracer, since the amount of it in a compound can be deduced from its β-activity. For the measurement of such low-energy radiation either gas-counting or liquid-scintillation counting is used.

10.8. Industrial Production and Uses of Hydrogen

Hydrogen is manufactured principally by the steam—hydrocarbon process, based on the mixture of hydrocarbons discarded in the 'cracking' of crude petroleum. This material is freed from sulphur by passing it over bauxite at 670 K to convert sulphur compounds to H_2S which is removed by a caustic scrubbing. The vapour along with steam then goes over a nickel catalyst at 1100 K and reactions occur exemplified by:

$$C_3H_6 + 3\,H_2O \rightarrow 3\,CO + 6\,H_2$$

To the hot gases leaving the furnace, steam is added in order to reduce the temperature to 670 K, when a second reaction occurs:

$$3\,CO + 3\,H_2O \rightarrow 3\,H_2 + 3\,CO_2$$

The hydrogen is freed from CO_2 by scrubbing the gases with aqueous ethanolamine

$$2\,HO{-}CH_2{-}CH_2{-}NH_2 + H_2O + CO_2 \underset{390\,K}{\overset{340\,K}{\rightleftharpoons}} (HO{-}CH_2{-}CH_2NH_3)_2CO_3$$

From this the CO_2 is subsequently recovered by heating the solution to 390 K. In industrial practice, further $CO \rightarrow CO_2$ stages are necessary to complete the removal of CO.

Vast quantities of hydrogen are produced in refining processes associated with the petroleum industry and designed to increase the octane number of the product. Typical reactions are the dehydrogenation of naphthalenes and paraffins to aromatic hydrocarbons:

$$C_6H_{12} \rightarrow C_6H_6 + 3\,H_2; \qquad C_6H_{14} \rightarrow C_6H_6 + 4\,H_2$$

Industrial hydrogen is used mainly in catalytic hydrogenation processes:
(i) $N_2 \rightarrow NH_3$ (Haber process);
(ii) $CO \rightarrow CH_3OH$ (Methanol process);
(iii) Unsaturated vegetable oils → saturated fats (particularly for margarine).
(iv) The 'oxo' process: the hydroformylation of alkenes to aldehydes and to alcohols. In this a mixture of H_2 and CO is added to the alkene in the presence of cobalt catalyst:

$$R{-}CH{=}CH_2 + CO + H_2 \xrightarrow[20\,MPa]{390\,K} R{-}CH_2{-}CH_2{-}CHO$$

If the temperature is increased to 460 K the main product is the alcohol:

$$R{-}CH_2{-}CH_2{-}CHO + H_2 \rightarrow R{-}CH_2{-}CH_2{-}CH_2OH$$

The active catalyst in the reaction is probably $Co_2(CO)_8$. Besides other synthetic purposes, hydrogen finds use in the removal of sulphur, nitrogen and oxygen from petroleum feed stocks by the catalytically promoted reactions:

$$C_4H_4S + 4\ H_2 \rightarrow C_4H_{10} + H_2S$$

$$C_5H_5N + 5\ H_2 \rightarrow C_5H_{12} + NH_3$$

$$C_6H_5OH + H_2 \rightarrow C_6H_6 + H_2O$$

Further Reading

W.E. Jones, S.D. MacKnight and L. Teng, The kinetics of atomic hydrogen reactions in the gas phase, Chem. Rev., 73 (1973) 407.

E.A. Evans, Tritium and its compounds, Butterworth, London, 1966.

R.P. Bell, The proton in chemistry, 2nd edition, Chapman and Hall, London, 1973.

J.B. Levy and B.K.W. Copeland, The kinetics of the hydrogen—fluorine reaction, J. Phys. Chem., 72 (1968) 3168.

W.K. Hall, F.E. Lutinski and J.A. Hall, Rates of *para*-hydrogen conversion and the H_2—D_2 exchange, Trans. Faraday Soc., 60 (1964) 1823.

A. Farkas, *Ortho*hydrogen, *para*hydrogen and heavy hydrogen, Cambridge University Press, 1935.

The Hydrides

11

11.1. Types of Hydrides

There are broadly four types of binary hydrides, i.e. compounds in which only one other element is combined with hydrogen:

(a) Saline hydrides such as LiH, CsH, BaH_2 and EuH_2 are salt-like solids containing hydride ions and metal ions in ionic lattices.

(b) Metallic hydrides are metallic in appearance and are usually nonstoichiometric, with formulae which depend somewhat on the method of preparation. They are more brittle than the metals from which they are made, they are conductors or semiconductors of electricity, and they are bound partly by ionic and partly by metallic bonds.

(c) Hydrides containing multicentre bonds are exemplified by the polymers Be_nH_{2n}, Al_nH_{3n} and the boranes (11.5). The bonding is predominantly covalent, but the molecules contain insufficient valence electrons to permit all the atoms to be joined by electron-pair bonds.

(d) Covalent molecular hydrides such as Si_2H_6, SnH_4, SbH_3 and H_2Te are gases or volatile liquids. Elements in Groups IVB, VB, VIB and VIIB form compounds of this kind with hydrogen. Although these are not called hydrides when the element is more electronegative than hydrogen, for example HCl is by convention hydrogen chloride not chlorine hydride, all such covalent binary compounds of hydrogen will be grouped together in this chapter.

The proceeding classification is not rigid. It is possible to argue for the inclusion of MgH_2 in (a), (b) or (c). Furthermore many of the lanthanides have hydrides which are largely metallic at composition LnH_2 but become predominantly ionic when the composition is close to LnH_3.

11.2. The Saline Hydrides

The ionic hydrides can be made by heating the metals in hydrogen. Those of the alkali metals have the NaCl lattice, SrH_2, BaH_2 and EuH_2 are less regular. All have stoichiometric compositions. Those stable in the fused state like LiH, or those like BaH_2 which can be dissolved in a suitable melt, yield hydrogen at the anode on electrolysis in accordance with Faraday's laws; the ionisation is therefore

$$MH \rightleftharpoons M^+ + H^-$$

The hydride ion in these compounds has a radius of about 145 pm, rather greater than that of the F^- ion in NaF. The alkali metal hydrides are much denser than the parent metals. For CsH, ρ = 3.42 g cm^{-3} whereas for Cs itself, ρ = 1.90 g cm^{-3}. All the ionic hydrides are oxidised by air. RbH and CsH ignite spontaneously at room temperature, and all react vigorously with water. The reaction, which is essentially

$$H^- + H_2O \rightarrow OH^- + H_2$$

demonstrates incidentally that the H^- ion is a very strong base.

The hydrides of lithium and sodium are of commercial importance because they are used for making the valuable reducing agents lithium aluminium hydride (11.4) and sodium borohydride (11.5.3). Sodium hydride is not made easily in bulk simply by passing hydrogen over molten sodium because the NaH which is formed tends to form a crust which inhibits further reaction. One method is to pass hydrogen at high pressure and 550 K into liquid sodium dispersed in kerosene containing a surface active agent like magnesium stearate.

The ionic hydrides are all exothermic compounds:

$$Li + \tfrac{1}{2} H_2 \rightarrow LiH, \qquad \Delta H^0 = -90 \text{ kJ mol}^{-1}$$
$$Cs + \tfrac{1}{2} H_2 \rightarrow CsH, \qquad \Delta H^0 = -55 \text{ kJ mol}^{-1}$$
$$Sr + H_2 \rightarrow SrH_2, \qquad \Delta H^0 = -177 \text{ kJ mol}^{-1}$$

Calcium hydride is almost certainly saline in type. It has a stoichiometric formula and is an exothermic compound:

$$Ca + H_2 \rightarrow CaH_2, \qquad \Delta H^0 = -183 \text{ kJ mol}^{-1}$$

but it is unlike the other dihydrides in two respects, it reacts rather quietly with cold water and it does not combine with molten LiH to give a compound $LiMH_3$ of antiperovskite structure as do SrH_2, BaH_2 and EuH_2. Ytterbium hydride, YbH_2, resembles CaH_2 in this respect and in its structure, and it too must be considered as on the borderline of the ionic class.

11.3. The Metallic Hydrides

The transition metals Ti, Zr, Hf, V, Nb, Ta, Cr and Pd all combine with hydrogen to give compounds which are to some extent non-stoichiometric. For example the palladium hydride made by absorption of hydrogen at room temperature has a maximum hydrogen content represented by the formula $PdH_{0.6}$. It was formerly thought that compositions TiH_2, ZrH_2 and HfH_2 could not be reached, but recent work with very pure samples of the metals and hydrogen has enabled the dihydrides to be made. They have tetragonal lattices at room temperature but cubic CaF_2 structures at rather higher temperatures. ZrH_2 and HfH_2 have enthalpies of formation similar to that of SrH_2, but TiH_2 is less thermally stable. At compositions below MH_2 the solids contain several phases. Their appearance does not vary greatly with composition; they are brittle, black, metallic solids.

The usual products of the reaction of the Group VA metals with hydrogen are monohydrides somewhat deficient in hydrogen. In this group the α-phase, the so-called solution phase, exists for a wide range of compositions, certainly up to $TaH_{0.15}$ for tantalum, unlike Group IVA where the disposition of atoms in the metal is altered by a much smaller intake of hydrogen. The thermodynamic data which are available suggest that VH, NbH and TaH are only weakly exothermic. The only other hydrides of d-block elements which have been unambiguously characterised are CrH, which has an anti-NiAs structure, and $PdH_{0.6}$, which has an NaCl structure with a high proportion of vacant sites.

Because these compounds are metallic in appearance, it was once thought that the hydrogen was merely occluded as hydrogen atoms into interstitial positions in the metal lattice, but this explanation is clearly unsatisfactory. Except in the Group VA elements quite small percentages of hydrogen are sufficient to cause changes in the arrangement of the metal atoms in the lattice. In the case of Pd the absorption of hydrogen increases the lattice parameter of the c.c.p. crystals from 383.3 pm to 389.4 pm without a change of phase, but any further absorption so strains the lattice as to cause a sudden increase to 401.8 pm with the appearance of a β-phase.

There are two theories of bonding in the transition-metal hydrides. One is that hydrogen atoms lose electrons to the conduction band (7.2.23.1) of the metal and become H^+ ions. Thus the metallic hydride is considered to be an alloy. The other theory is that hydrogen atoms gain electrons from the conduction band, which is thereby depleted but not denuded of electrons and can still confer metallic character. The second of these theories is supported by the greater weight of evidence:

(i) The expansion of the metallic lattice is consistent with the inclusion of ions about 130 pm in radius, not much less than the value of r_{H^-} in the Group IIA hydrides.

(ii) Estimates of lattice energies on this basis agree with the known thermal stabilities. For example it is possible to predict that the hydrides of Group VIIA should be unstable, and none has yet been made.

(iii) Paramagnetic susceptibility falls as the hydrogen content increases, as expected if electrons of parallel spin in the conduction bands are used up in H^- formation. Conductance also falls; if the hydrides were alloys, both paramagnetism and conductance should increase with hydrogen content.

The lanthanide hydrides are of particular interest because their properties are just those we should predict using the theory that conduction bands are depleted by hydride formation. Both Eu and Yb form dihydrides; EuH_2 is undoubtedly ionic, and no higher hydride of the metal is formed; YbH_2 is predominantly ionic, but an unstable $YbH_{2.5}$ is known. These are the two lanthanides whose M^{2+} ions contain respectively 7 and 14 electrons more than the Xe atom, ions we should therefore expect to be particularly strongly exchange-stabilised. All the other lanthanides form both dihydrides and trihydrides. The cubic CaF_2 phase persists for the lighter lanthanides up to MH_3 but for the heavier ones it is replaced by a hexagonal phase. In both cases the extra hydrogens are accommodated in octahedral sites. The MH_3 phases require higher temperatures and pressures for their production. But the most significant difference between the dihydride and the trihydride is that the latter loses its paramagnetism and its conductance as the ideal composition MH_3 is approached; it ceases to be a metallic hydride and becomes effectively an ionic hydride — the conduction band has been denuded.

The actinides up to americium are all known to form hydrides. Uranium forms only UH_3, which has two crystalline forms. ThH_2 is similar to HfH_2 but PuH_2 has the CaF_2 structure; in the composition range PuH_2 to $PuH_{2.75}$ the additional hydrogen occupies octahedral sites, but above $PuH_{2.75}$ a hexagonal PuH_3 phase appears.

The lanthanide and actinide hydrides clearly occupy the border between ionic and metallic hydrides, the ionic character increasing with hydrogen content.

In many instances one element forms both molecular and polymeric hydrides, which makes it convenient to treat both types under the heading of covalent hydrides. We shall deal with them by Periodic Groups.

11.4. The Hydrides of Beryllium, Magnesium, Aluminium and Zinc

Beryllium hydride can be made by pyrolysis of its etherate, which is obtained by the action of $LiAlH_4$ on $(CH_3)_2Be$ in ether. Another method is to pyrolyse ditertiarybutylberyllium:

$$[(CH_3)_3C]_2Be \rightarrow BeH_2 + 2\ CH_3\underset{\displaystyle CH_3}{\underset{|}{C}}{=}CH_2$$

The white, insoluble solid reacts slowly with water but releases hydrogen rapidly in dilute acids. It decomposes on heating to 470 K. The structure is in doubt, but infrared data are consistent with a hydrogen-bridge, polymeric character:

$$>Be\langle{}^{H}_{H}\rangle Be\langle{}^{H}_{H}\rangle Be\langle{}^{H}_{H}\rangle Be<$$

There are three satisfactory preparative routes to the light-grey solid MgH_2:

$$EtMgI \text{ or } Et_2Mg \xrightarrow{450\ K} MgH_2 + C_2H_4 + \text{other products}$$

$$LiAlH_4 + Me_2Mg \xrightarrow{\text{ether}} MgH_2$$

$$Mg + H_2 \xrightarrow[\text{pressure, } MgI_2 \text{ as catalyst}]{\text{high temperature and}} MgH_2$$

The solid made by the third method is less reactive with air and water than that made by the first two. Neutron-diffraction studies indicate a rutile structure with r_{H^-} = 131 pm, but infrared studies suggest bridge-bonding. A three-dimensional lattice incorporating hydrogen bridges is not impossible.

Aluminium hydride is best made by the action of LiH on an excess of Al_2Cl_6 in ether:

$$6\ LiH + Al_2Cl_6 \rightarrow 6\ LiCl + \frac{2}{n}\ Al_nH_{3n}$$

The solution is filtered quickly to remove LiCl, and the white, polymeric hydride then precipitates slowly, probably as the result of the slow decomposition of an etherate. If the same reactants are mixed but with LiH in excess, the product is the ether-soluble lithium tetrahydridoaluminate, $LiAlH_4$, more usually called lithium aluminium hydride, which is a valuable reducing agent. It converts aldehydes and ketones to alcohols, nitriles to primary amines and many inorganic halides to hydrides:

$SnCl_4 \rightarrow SnH_4$; $SiCl_4 \rightarrow SiH_4$

Aluminium hydride is similar in its reducing properties but it is not soluble in diethyl ether; however it is reported to dissolve as a monomer in tetrahydrofuran. The polymer is thought to have a hydrogen-bridged structure.

Although adducts such as trimethylaminegallane, Me_3NGaH_3, have been prepared, attempts to make digallane by treating them with BF_3 have proved unsuccessful. However, trialkylgallanes, R_3Ga, can be made by treating gallium trihalides with aluminium alkyls, and some salts of the GaH_4^- ion have been prepared.

The preparation of a polymer In_nH_{3n} has been claimed, but the existence of this compound is in doubt. Mass spectrometry of hydrides formed by atomic reaction in a fast flow system indicates that InH_3 should be even less thermally stable than GaH_3.

Zinc dihydride has been made as a white precipitate by the action of $LiAlH_4$ in dry ether on dimethylzinc at 230 K. It decomposes slowly at 290 K and rapidly at 350 K. A white compound, CdH_2, is reported to have been made at 195 K by a similar method, but it decomposes into its elements at about 260 K.

11.5. Hydrides of Boron (Boranes)

The structures and reactions of the hydrides of boron are of great interest. They have been the object of much research, and several new hydrides have been made relatively recently. Six of them were discovered by Stock (1914–20) whose starting material was the gaseous product of the reaction between magnesium boride and dilute hydrochloric acid. They were B_2H_6, B_4H_{10}, B_5H_9, B_5H_{11}, B_6H_{10} and $B_{10}H_{14}$. Stock's remarkable success was due to a vacuum technique which he developed for handling compounds sensitive to oxygen and moisture. Monomeric BH_3 was not found; there would be insufficient electrons in such a molecule to stabilise the bonding in the way it is stabilised in the monomeric boron halides (15.3).

Most of the boranes which have been fully characterised belong to either the B_nH_{n+4} or the B_nH_{n+6} series:

(i) B_nH_{n+4} series (the stable boranes):

B_2H_6 (gas); B_5H_9, B_6H_{10}, B_8H_{12} (liquids); $B_{10}H_{14}$, $B_{18}H_{22}$ and iso-$B_{18}H_{22}$ (solids)

(ii) B_nH_{n+6} series (the unstable boranes):

B_4H_{10}, B_5H_{11}, B_6H_{12}, B_9H_{15}(liquids) $B_{10}H_{16}$, $B_{20}H_{26}$(solids)

In addition to the above, the structure of the solid, $B_{20}H_{16}$, has been determined and a thermally unstable B_8H_{18} (*b.p.* 211 K) has been identified.

The usual terminology is di-, tetra-, penta- (etc.) borane to signify the number of boron atoms, followed by a numeral to show the number of hydrogen atoms: thus $B_{10}H_{14}$ is written decaborane-14.

Generally speaking, the stabilities of the hydrides increase with their molar mass. However, members of the B_nH_{n+4} series are more stable than those of the B_nH_{n+6} series which have about the same mass. Thus pentaborane-9 decomposes slowly even at 420 K, and is hydrolysed by water only on prolonged heating, whereas pentaborane-11 decomposes rapidly at 300 K and is hydrolysed immediately by cold water.

The boranes, with the exception of B_6H_{10}, are no longer made by Stock's method. The starting material for them is diborane, B_2H_6, for which the best method of large scale preparation is to add the boron trifluoride—ether complex slowly to a suspension of lithium hydride in ether and gently reflux the mixture:

$$6\ LiH + 8\ Et_2O \cdot BF_3 \rightarrow 6\ LiBF_4 + B_2H_6 + 8\ Et_2O$$

Ethane, usually present as an impurity, is removed by passing the products into dimethyl ether at 193 K, to form a solid complex, $BH_3 \cdot Me_2O$, from which the ethane may be pumped. The complex is decomposed by warming and the diborane purified by fractional distillation.

The higher boranes listed above, with the exception of the two $B_{18}H_{22}$ isomers, are made by heating diborane alone or with hydrogen. Thus when passed through a tube at 388 K, diborane is largely converted to pentaborane-11. This, warmed with hydrogen at 373 K, gives tetraborane with some diborane:

$$2\ B_5H_{11} + 2\ H_2 \rightarrow 2\ B_4H_{10} + B_2H_6$$

When, however, diborane and hydrogen are passed through a tube at 470 K, pentaborane-9 is obtained. Kinetic studies suggest the changes involve a radical mechanism with perhaps borine, BH_3, as an intermediate, though direct evidence of this entity is wanting. At higher temperatures the boranes give non-volatile products of variable composition, $(BH_x)_n$. At 970 K, the breakdown of these materials to boron and hydrogen is complete.

When kept dry at room temperature, diborane suffers about a 10% decomposition per year, but with water it hydrolyses rapidly to boric acid and hy-

TABLE 11.1

REACTIONS OF DIBORANE

	Reagent / conditions	Product
B_2H_6	$\xrightarrow{H_2 + \text{heat}}$	higher boranes
	$\xrightarrow{H_2O}$	$H_3BO_3 + H_2$
	$\xrightarrow{NaH}$	$Na^+BH_4^-$
	$\xrightarrow{BMe_3}$	$B_2H_2Me_4$
	$\xrightarrow[390\ K]{NH_3}$	$(NH_3)_2BH_2^+BH_4^-$
	$\xrightarrow{NMe_3}$	$Me_3N^+BH_3^-$
	$\xrightarrow{BX_3}$	B_2H_5X (X = Cl and Br)
	$\xrightarrow{\text{alkenes}}$	alkyl boranes
	$\xrightarrow[370\ K,\ 2\ MPa]{CO}$	H_3BCO

drogen. Like the two pentaboranes it burns spontaneously in air.

In general, the reactions of boranes fall into two principal classes:

(i) Those in which a BH_3 group is removed from the borane, often by a nucleophile to give first an adduct, which then dissociates or decomposes

$$B_2H_6 + 2\,ROH \rightarrow 2\ \begin{matrix} R \\ H \end{matrix}\!\!>\!O \cdot BH_3$$

$$\downarrow \text{dissociates}$$

$$ROBH_2 + H_2$$

$$\downarrow \text{disproportionates}$$

$$(RO)_2BH + (RO)_3B + B_2H_6$$

Related reactions occur with molecules whose donor power is insufficient to allow stable adducts to be formed:

$$B_2H_6 + AsH_3 \rightarrow H_2 + H_2AsBH_2 \text{ polymer}$$

With unsaturated ligands, irreversible decomposition of the adduct often occurs by the transfer of hydrogen from a boron atom to the ligand:

$$2\,CH_3{-}CH{=}O + B_2H_6 \rightarrow 2\,CH_3{-}CH_2{-}O{-}BH_2$$

$$\downarrow \text{disproportionates}$$

$$(EtO)_3B + B_2H_6$$

(ii) Those in which hydrogen-bridge bonds are cleaved unsymmetrically. This probably occurs in the ammonia—diborane reaction:

$$H_2B(\mu\text{-}H)_2BH_2 + 2\,NH_3 \longrightarrow \left[H_2B(NH_3)_2\right]^+ + BH_4^-$$

11.5.1. Structures

Considerable interest has centred on the structures of the boranes. They are all electron-deficient compounds; that is they have too few valency electrons to permit every one of the adjacent atoms to be held together by electron-pair bonds. Diborane has a double hydrogen-bridge at right angles to the plane of the other four hydrogen atoms (Fig. 11.1). The presence of two

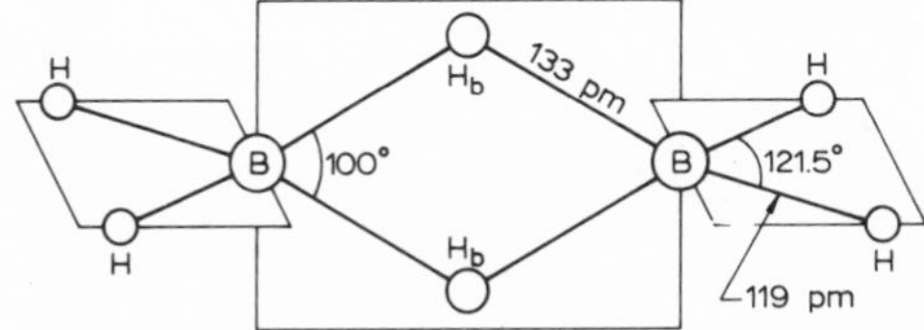

Fig. 11.1. Hydrogen-bridged structure of diborane.

kinds of boron—hydrogen bond is indicated both by the Raman spectrum and by the chemical evidence that four only of the hydrogen atoms in the molecule are replaceable by methyl groups. Electron diffraction leads to the parameters: B—H 119 pm, B—H_b 133 pm, B—B 177 pm, ∠HBH 121.5° and ∠H_bBH_b 100°. Raman and infrared spectra of the tetramethyl compound suggest an absence of terminal hydrogen atoms, and electron diffraction shows the four carbon atoms and two boron atoms to be coplanar. Although a double hydrogen bridge is certain, the precise nature of the bonds involved is still uncertain. Clearly they are abnormal and a suggestion is that they are formed by an overlap of the sp^3 tetrahedral hybrids and hydrogen 1s orbitals. Each electron pair is then less localised than usual, extending over three centres (Fig. 11.2). This accounts for the symmetry and absence of free rotation between the boron atoms.

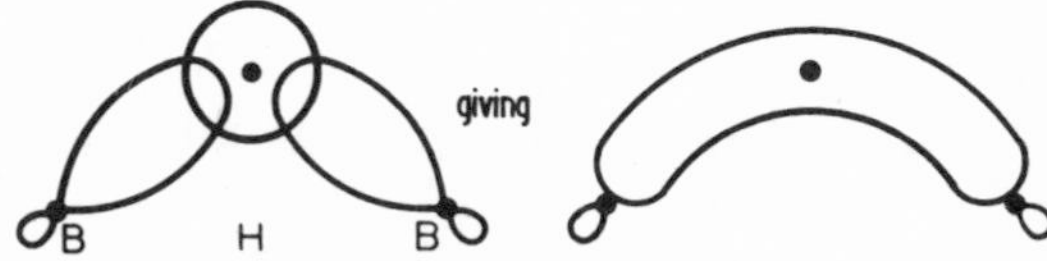

Fig. 11.2. Contributions to molecular orbitals in diborane.

The boron skeletons of the higher boranes can be considered as fragments of octahedra (B_5H_9 and $B_{10}H_{16}$) or of icosahedra. The $B_{12}H_{12}{}^{2-}$ ion present in $K_2B_{12}H_{12}$ is almost perfectly icosahedral with I_h symmetry (Fig. 11.3). The structural relation of $B_{10}H_{14}$ and B_6H_{10} to the $B_{12}H_{12}^{2-}$ ion are shown in the diagram (Fig. 11.4). At each corner of the figures is a boron atom which is attached to a hydrogen atom 120—130 pm away; the hydrogen-bridge bonds are shown as curves. The actual structures are somewhat less regular than in the simplified illustrations. In $B_{10}H_{14}$, for example, the B—B distances vary from 171 pm for the 1—3 distance to 201 pm for the 5—10 and 7—8 distances. The straight lines do not represent bonds; the nature and disposition of the bonds are discussed below.

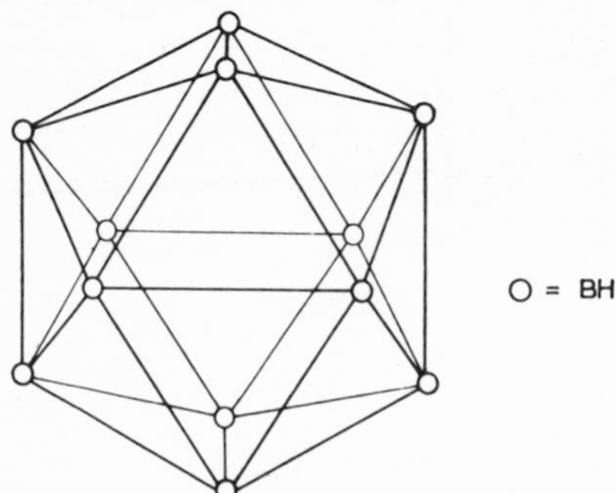

Fig. 11.3. The $B_{12}H_{12}^{2-}$ ion.

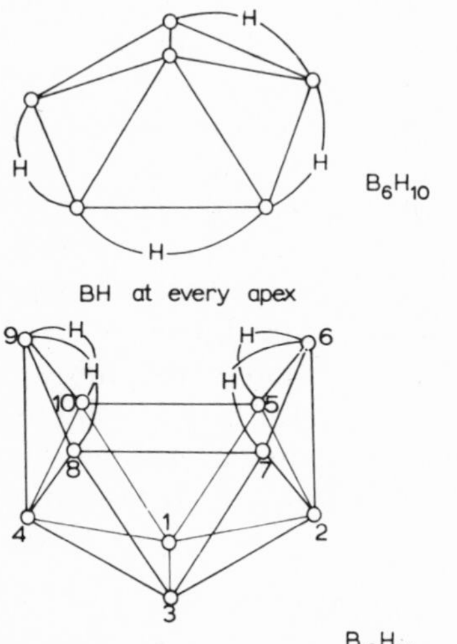

Fig. 11.4. Structures of B_6H_{10} and $B_{10}H_{14}$ in which the circles indicate BH and ⌒H⌒ a hydrogen bridge bond (cf. $B_{12}H_{12}^{2-}$ in Fig. 11.3).

The structure of $B_{20}H_{16}$, the only borane yet characterised which has fewer hydrogen atoms than boron atoms in the molecule, consists of two large isocahedral fragments fused together (Fig. 11.5).

There are two isomeric boranes of formula $B_{18}H_{22}$. The relation of their structures to that of $B_{10}H_{14}$ is shown in the form of planar projections (Fig.

11.7) of the molecules, which can be considered as $B_{10}H_{14}$ molecules (Fig. 11.6) joined at two boron atoms originally hydrogen-bridged.

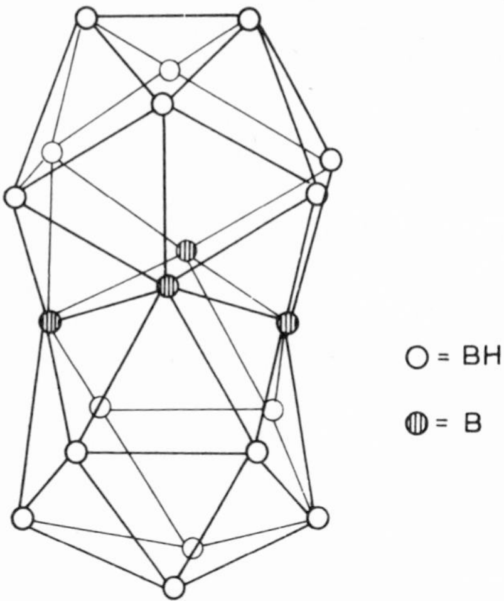

Fig. 11.5. Structure of $B_{20}H_{16}$.

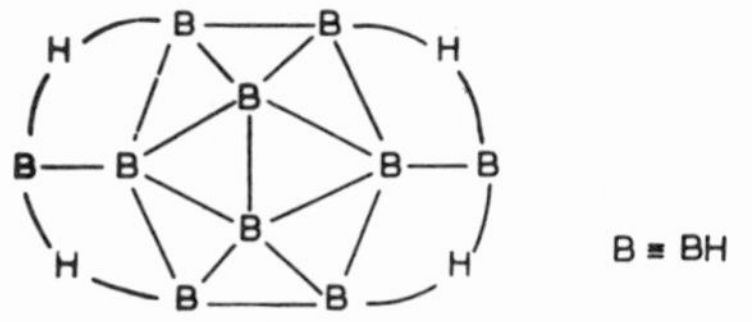

Fig. 11.6. Planar projection of $B_{10}H_{14}$ molecule (B indicates BH).

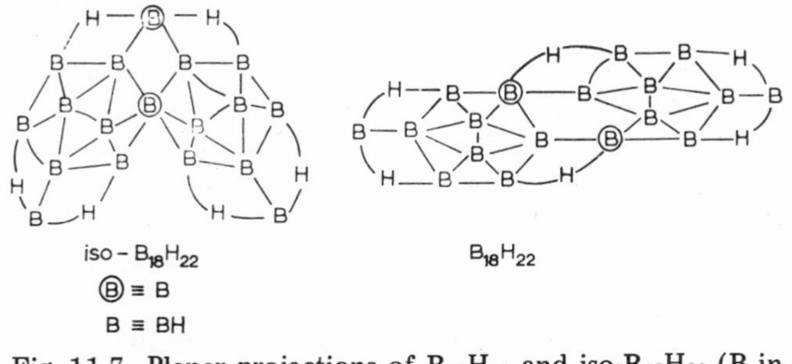

Fig. 11.7. Planar projections of $B_{18}H_{22}$ and iso-$B_{18}H_{22}$ (B in circle indicates B, and B indicates BH).

11.5.2. Some other borane structures

The structure of B_4H_{10} is shown in Fig. 11.8.

$B_{10}H_{16}$ consists of two square pyramids like those in Fig. 11.9 joined by a B—B bond at their apices.

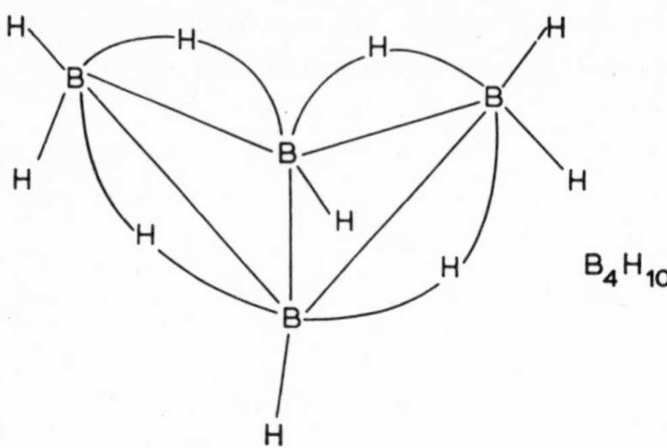

Fig. 11.8. The structure of B_4H_{10}. The bond angles are such that the boron skeleton can be considered to be either an octahedral or an icosahedral fragment.

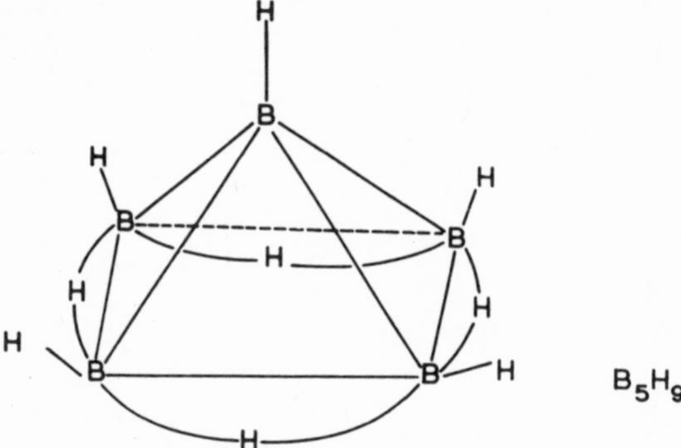

Fig. 11.9. The structure of B_5H_9.

Many of the atoms in borane molecules are obviously joined by bonds quite different in type from the sigma- and pi-bonds used in describing carbon compounds. The 'end' B—H bonds can be considered as normal, two-centre bonds in which two electrons are used to make a bonding MO. But to describe hydrogen-bridge bonding and also the boron—boron bonds, the concept of three-centre bonds is useful. In these bonds three atoms are bound together by two electrons. The three-centre bonds are considered to be of three types:

(a) The hydrogen-bridge bonds described above in which a hydrogen s-orbital overlaps hybrid orbitals of two boron atoms, each of which contributes the equivalent of one-half electron:

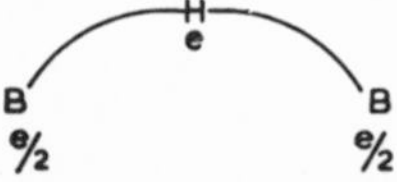

(b) Closed three-centre bonds formed by the overlap of boron hybrid orbitals containing both s and p contributions:

(c) Open three-centre bonds formed by the overlap of hybrid orbitals on two of the boron atoms with a p orbital on the third atom:

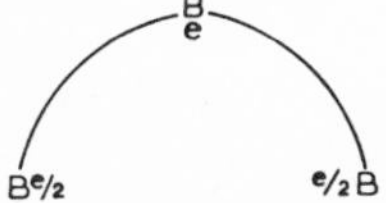

The structure of $B_{10}H_{14}$, described in bonds of these types, is shown in Fig. 11.10.

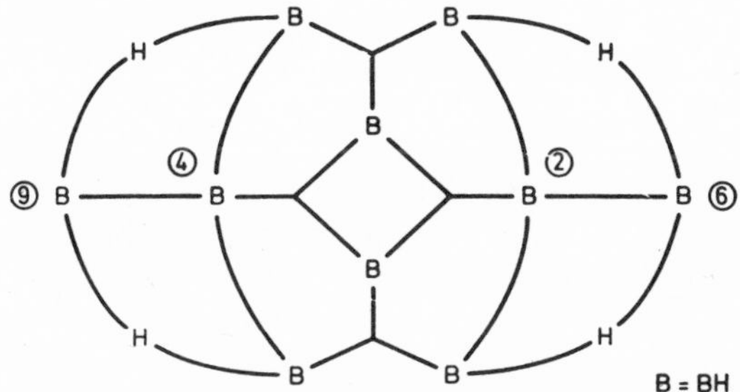

Fig. 11.10. Topological drawing of $B_{10}H_{14}$ (B indicates BH).

This topological diagram is something of an oversimplification. A more realistic approach would be to imagine the boron atoms in such large icosahedral fragments as being bound by 'multicentre' bonds. Nevertheless the three-centre bond approach does predict correctly that atoms 6 and 9 should constitute the most positive positions in $B_{10}H_{14}$ — these are the points at which nucleophiles attack the molecule — and that 2 and 4 should be the most negative, again in agreement with experiment.

11.5.3. Borohydrides

Sodium borohydride results from the reaction

$$4\,NaH + B(OMe)_3 \rightarrow NaBH_4 + 3\,NaOMe$$

which proceeds rapidly at 520 K, and is extracted from the products with isopropylamine. Lithium borohydride is best made by passing diborane into an ethereal solution of lithium hydride. These borohydrides are salt-like; the sodium compound has a face-centred cubic lattice of discrete Na^+ and tetrahedral BH_4^- ions, and the lithium compound is only a little less regular. They are involatile and unaffected by dry air. Lithium borohydride is particularly useful for making other borohydrides:

$$2\ LiBH_4 + BeBr_2 \rightarrow Be(BH_4)_2 + 2\ LiBr$$

Beryllium, aluminium and some transition metals such as thorium differ from the alkali metals in forming volatile borohydrides which constitute the most volatile compounds of these elements. On pyrolysis they decompose giving hydrogen and non-volatile residues. Electron diffraction and infrared studies suggest they possess bridge structures.

In addition to the BH_4^- ion there are $B_3H_8^-$, made by the reaction

$$NaBH_4 + B_2H_6 \rightarrow NaB_3H_8 + H_2$$

at 370 K in diglyme, and an important series $B_nH_n^{2-}$ (n = 6 to 12) of *closo*-borane anions (*closo* = cage-like) with structures based on polyhedra of B atoms.

Salts of $B_{12}H_{12}^{2-}$ can be prepared in good yield by reactions such as

$$2\ NaBH_4 + 5\ B_2H_6 \rightarrow Na_2B_{12}H_{12} + 13\ H_2$$
$$2\ R_3N + 6\ B_2H_6 \rightarrow (R_3NH)_2B_{12}H_{12} + 11\ H_2$$

and those of $B_{10}H_{10}^{2-}$ by the reaction

$$2\ R_3N + B_{10}H_{14} \rightarrow (R_3NH)_2B_{10}H_{10} + H_2$$

The structures of the $B_{12}H_{12}^{2-}$ and the $B_{10}H_{10}^{2-}$ ions are shown (Figs. 11.3 and 11.11). Their compounds are much more thermodynamically stable than

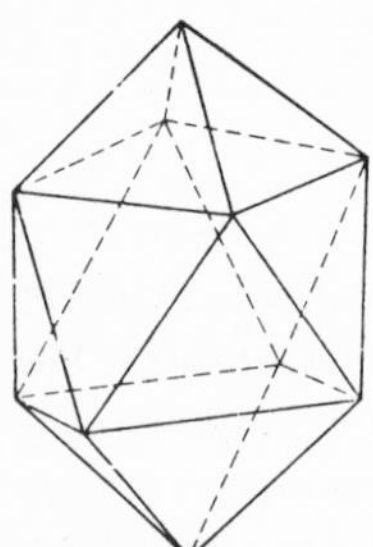

Fig. 11.11. Structure of the $B_{10}H_{10}^{2-}$ ion (BH at every apex).

those of the other $B_nH_n^{2-}$ ions. The reactions of these two ions have been extensively studied; attack by electrophiles such as RCO^+, $C_6H_5N_2^+$ and Br^+ in strongly acidic media gives rise to a great variety of substitution products.

The $B_{10}H_{10}^{2-}$ ion can be oxidised by Fe^{3+} in aqueous solution to give an ion $B_{20}H_{18}^{2-}$ which can then be reduced by Na in liquid NH_3 to a $B_{20}H_{18}^{4-}$ ion consisting of two bicapped antiprisms like $B_{10}H_{10}^{2-}$ joined by a boron—boron bond.

11.5.4. Compounds derived from boranes

11.5.4.1. Carboranes

Closely related in structure to the *closo*borane anions are the polyhedral carboranes of general formula $B_{n-2}C_2H_n$. Those most extensively studied are

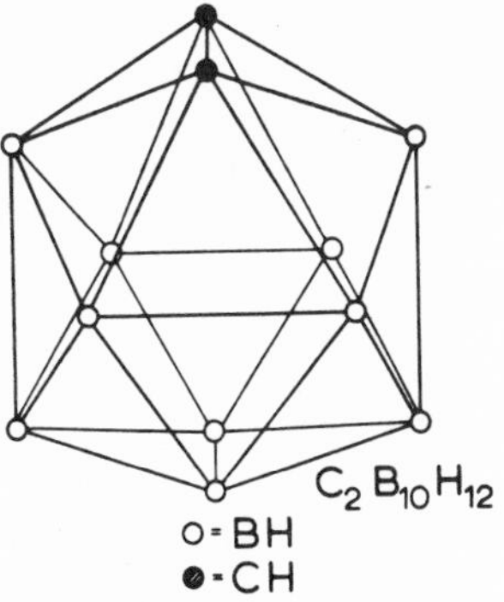

Fig. 11.12. The structure of 1,2-dicarba*closo*decaborane.

the dicarba*closo*decaboranes, $B_{10}C_2H_{12}$. The 1,2-isomer (Fig. 11.12) has an icosahedral structure in which two CH groups replace adjacent isoelectronic BH^- groups of the $B_{12}H_{12}^{2-}$ ion (Fig. 11.3). It is made by the reactions:

$$B_{10}H_{14} + 2\,R_2S \rightarrow B_{10}H_{12}(R_2S)_2 + H_2$$
$$B_{10}H_{12}(R_2S)_2 + CH{\equiv}CH \rightarrow B_{10}C_2H_{12} + 2\,R_2S + H_2$$

Carbon-substituted derivatives can be made similarly from alkynes other than C_2H_2:

$$R'C{\equiv}CR'' + B_{10}H_{12}(R_2S)_2 \rightarrow B_{10}H_{10}C_2R'R'' + 2\,R_2S + H_2$$

When the 1,2-derivatives are heated to 700 K they isomerise smoothly to 1,7-derivatives in which the two carbon atoms occupy corners separated by one BH corner. A 1,12-isomer, in which the two carbon atoms are at diametrically opposite corners, can be obtained only by more severe heat treatment and with considerable loss due to decomposition.

The hydrogen atoms of the CH groups are easily replaced by lithium atoms when 1,2- and 1,7-$B_{10}C_2H_{12}$ are treated with C_2H_5Li. These lithium compounds can be used to reach other derivatives, as represented below:

$$B_{10}C_2H_{10}Li_2 \xrightarrow{CO_2} B_{10}C_2H_{10}(CO_2H)_2$$
$$B_{10}C_2H_{10}Li_2 \xrightarrow{NOCl} B_{10}C_2H_{10}(NO)_2 \qquad B_{10}C_2H_{10}Li_2 \xrightarrow{I_2} B_{10}C_2H_{10}I_2 \qquad B_{10}C_2H_{10}Li_2 \xrightarrow{H\cdot CHO} B_{10}C_2H_{10}(CH_2OH)_2$$

The other *closo*carboranes are of less importance. Two *nido*carborane anions (*nido* = nest-like) with the formula $B_9C_2H_{12}^-$ can be obtained by the action of ethoxide ions in ethanol upon the 1,2- and 1,7-dicarba*closo*decaboranes:

$$B_{10}C_2H_{12} + EtO^- + 2\,EtOH \rightarrow B_9C_2H_{12}^- + B(OEt)_3 + H_2$$

These isomeric ions (Fig. 11.13) can be considered to be derived from the parent carboranes by the removal of BH^{2+} units, leaving $B_9C_2H_{11}^{2-}$ ions which capture protons to yield $B_9C_2H_{12}^-$. Thus an open face is created by the loss of one of the corner atoms of the icosahedron; it is not known how the twelfth hydrogen is bound, but it is probably highly labile.

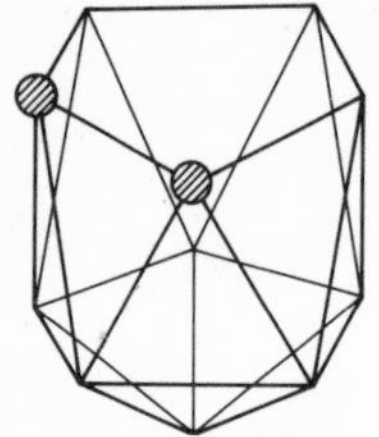
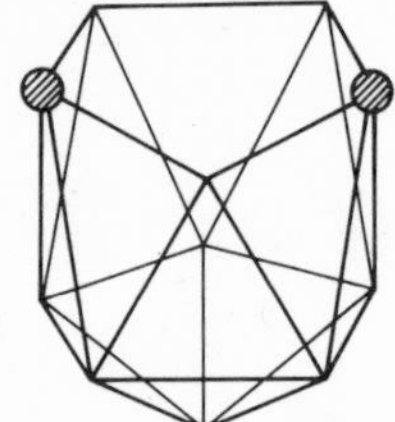

Fig. 11.13. Structures of the isomeric $B_9C_2H_{12}^-$ ions (shaded circles = CH).

11.5.4.2. Dicarbollide complexes

Hawthorne (1964) recognised the strong resemblance of the pentagonal open face of a $B_9C_2H_{12}^-$ ion to the pentagonal $C_5H_5^-$ ion. He therefore attempted to make transition-metal complexes analogous to dicyclopentadienyls (18.2.5). He generated $B_9C_2H_{11}^{2-}$ ions by treating $B_9C_2H_{12}^-$ ions with a very strong base.

$$B_9C_2H_{12}^- + NaH \rightarrow Na^+ + B_9C_2H_{11}^{2-} + H_2$$

These ions were found to react easily with some transition-metal ions, particularly d^6 ions like Fe^{2+} and Co^{3+}, to give structures with D_{5d} symmetry (Fig. 11.14). The ions such as $(B_9C_2H_{11})_2Fe^{2-}$ and $(B_9C_2H_{11})_2Co^-$ have become known as dicarbollide complexes; the trivial name dicarbollide ion for $B_9C_2H_{11}^{2-}$ is derived from the Spanish *olla* (= a pot), because of the shape. Complexes containing one dicarbollide ion and a π-bonding ligand such as $C_5H_5^-$, e.g. $(\pi\text{-}C_5H_5)Fe(B_9C_2H_{11})^-$, have also been made.

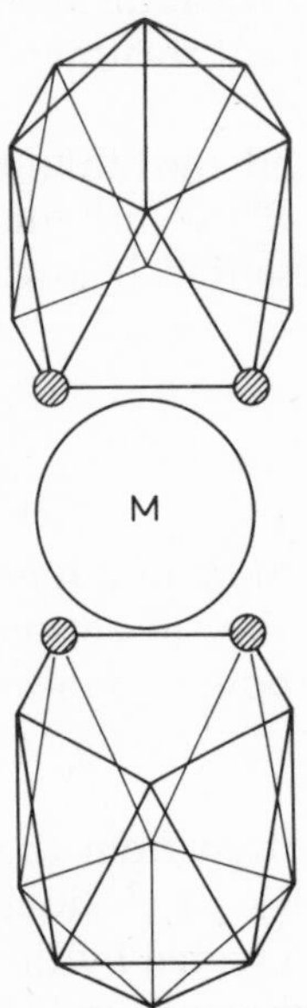

Fig. 11.14. Sandwich structure of $(B_9C_2H_{11})_2Fe^{2-}$ and $(B_9C_2H_{11})_2Co^-$ ions.

Another *nido*carborane anion $B_7C_2H_9^{2-}$, called the dicarbazapide ion because of its shoe-like shape, from the Spanish *zapato* (= a shoe), also forms complexes with transition-metal ions. An example is $(B_7C_2H_9)_2Co^-$ in which the cobalt atom lies between the openings of the 'shoes'.

11.5.4.3. Boranes containing other hetero-atoms

A few compounds have been made in which boron atoms in the framework are replaced by sulphur atoms. The thiaborane ion $B_{10}SH_{10}^{2-}$ reacts with transition-metal ions to form complexes similar to the bis(dicarbollyl)-metal ions, e.g.:

$$2\ B_{10}SH_{10}^{2-} + Fe^{2+} \rightarrow (B_{10}SH_{10})_2Fe^{2-}$$

Three isomeric carbaphosphaboranes of formula $B_{10}CPH_{11}$ have been made. The 1,2-isomer is made by the reaction

$$Na_3B_{10}CH_{11} + PCl_3 \rightarrow 1,2\text{-}B_{10}CPH_{11} + 3\ NaCl$$

Like $B_{10}C_2H_{12}$, the corresponding 1,7- and 1,12-isomers can be made by heat treatment.

11.5.4.4. Nitrogen derivatives

The reaction between ammonia and diborane depends upon the conditions. Excess of ammonia at low temperatures produces the salt-like diammoniate, $[(NH_3)_2BH_2]^+BH_4^-$, because with NH_3 there is an unsymmetrical cleavage ($BH_2^+ + BH_4^-$) in contrast to the symmetrical one ($BH_3 + BH_3$) resulting from reactions with amines. The monomer $BH_3 \cdot NH_3$ has indeed been made by the action of an ammonium halide on a borohydride:

$$NH_4X + MBH_4 \rightarrow BH_3 \cdot NH_3 + H_2 + MX$$

At higher temperatures excess of ammonia gives boron nitride, BN. When, however, the proportions are two molecules of ammonia to one of diborane, the volatile cyclic borazine is produced in yields up to 45%. In this compound the boron and nitrogen atoms are all in trigonal (sp^2) valence states,

```
  /BH—NH\
HN       BH
  \BH—NH/
```

Fig. 11.15. Borazine.

the 2p lone pairs on the three nitrogens providing 6 π electrons as in benzene. The boron—nitrogen and carbon—carbon analogues are isoelectronic and compounds of this kind show certain similarities, but because the boron—nitrogen bond is between atoms of different electronegativity and has an unsymmetrical charge cloud it is decidedly weaker. This is true even in borazine, although the bonds in it are non-localised. Consequently borazine is considerably more reactive than benzene, forming addition compounds with hydrogen halides, methanol, water and methyl iodide.

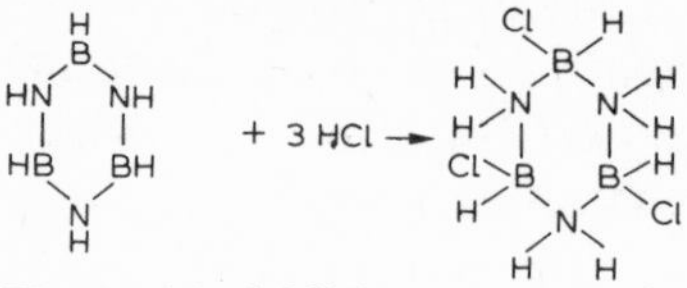

Fig. 11.16. Addition compound of borazine with HCl.

The hydrochloride loses hydrogen at 320 K giving the symmetrical trichloroborazine:

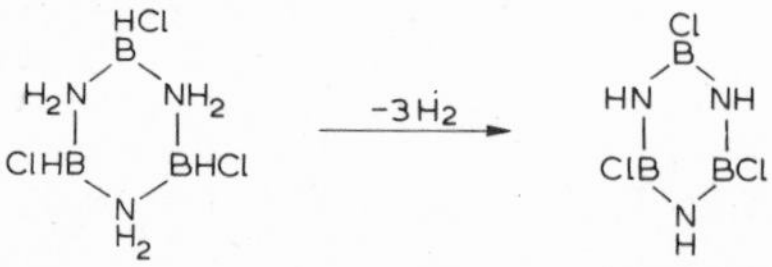

Fig. 11.17. Dehydrogenation of the hydrochloride of borazine.

11.5.4.5. Phosphorus and arsenic derivatives

Development from aminoboron chemistry led to phosphino- and arsinoboron compounds. Diborane and phosphine react at 163 K to give $B_2H_6 \cdot 2\,PH_3$, which is much less stable than the diammoniate. Trimethylamine displaces phosphine from it quantitatively at 233 K, suggesting the ionic structure $H_3P^+ \cdot BH_3^-$:

$$B_2H_6 \cdot 2\,PH_3 + 2\,Me_3N \rightarrow 2\,PH_3 + 2\,Me_3N \cdot BH_3$$

At ordinary temperatures diborane reacts slowly with phosphine to form hydrogen and a non-volatile white polymer of the approximate composition $(H_2P \cdot BH_2)_x$. It behaves similarly with arsine. Methyl substituted phosphine and arsine yield more stable compounds than the hydrides, some being recrystallisable from organic solvents even when they are exposed to air.

11.5.4.6. Oxygen and sulphur derivatives

A number of oxygen and sulphur compounds are sufficiently powerful electron-pair donors to form borine adducts by reaction with diborane. Of these the unstable solid dimethyl ether borine, $Me_2O^+ \cdot BH_3^-$, used in the purification of diborane, is well known.

11.6. Hydrides of Carbon, Silicon, Germanium and Tin

No element forms such a variety of hydrides as carbon. In addition to the numerous known alkanes, C_nH_{2n+2}, structural isomerism predicts many more not yet isolated. For example, over thirty six million isomeric forms of $C_{25}H_{52}$ can be formulated. To these an even larger number of possible unsaturated hydrocarbons may be added.

The hydrides of silicon and germanium are less thermally stable than those of carbon but compounds analogous with the alkanes have been made containing up to ten Si or Ge atoms per molecule. For members of these series containing four or more Si or Ge atoms, isomeric forms with branched chains are possible and their existence has been demonstrated by gas chromatography for the higher silanes and germanes. Many of the isomers have been characterised by the use of n.m.r. spectra.

A mixture containing all the silanes is obtained when magnesium silicide is dropped into dilute hydrochloric acid in an enclosed, low-pressure system. Fairly pure monosilane itself is made in good yield by adding magnesium silicide to ammonium bromide in liquid ammonia, in which the ammonium salt behaves as a strong acid. The pure gas results from the action of lithium aluminium hydride on silicon tetrachloride:

$$LiAlH_4 + SiCl_4 \rightarrow LiCl + AlCl_3 + SiH_4$$

Monosilane can be converted into a mixture of higher silanes by circulating the gas through a silent electric discharge. The homologous series of silanes resembles that of alkanes in showing a gradation of physical properties (Table 11.2).

TABLE 11.2

M.P. AND *B.P.* OF THE LOWER SILANES

	SiH_4	Si_2H_6	Si_3H_8	n-Si_4H_{10}
m.p./K	88	141	156	189
b.p./K	161	259	328	380

The thermal stabilities of the silanes are much lower than those of the corresponding alkanes. The higher silanes decompose on moderate heating to give lower silanes and solid unsaturated hydrides:

$$Si_5H_{12} \rightarrow Si_2H_6 + SiH_4 + \tfrac{2}{x}(SiH)_x$$

Above 770 K, decomposition to the elements is complete. The silanes are spontaneously inflammable and explosive in air; they are not hydrolysed at a pH of less than 7, but in water containing a trace of alkali hydrolysis is rapid and complete. They have strong reducing properties; permanganates are reduced to manganese dioxide, and mercury(II) to mercury(I) ions and mercury. Unlike the alkanes, the silanes enter into substitution reactions with halogen acids (other than hydrofluoric) on warming in the presence of the corresponding aluminium halide, e.g.:

$$SiH_4 + HCl \xrightarrow{Al_2Cl_6} SiH_3Cl + H_2$$

Mean bond energies for M—H bonds in Group IV are

E(C—H) = 416 kJ mol^{-1}	E(Si—H) = 293 kJ mol^{-1}
E(Ge—H) = 289 kJ mol^{-1}	E(Sn—H) = 251 kJ mol^{-1}

11.6.1. Silyl radical and silyl compounds

As the electronegativity of silicon (1.74) is less than that of carbon (2.50), the silyl radical SiH_3 should be a less powerful electron acceptor than the methyl group. This would be so were it not for the vacant d orbitals of the silicon atom which enable π-bonding to take place. Theory, supported by experimental evidence, indicates that there would be overlap between a vacant silicon d_π orbital and a p_π orbital on an atom of a Group V, VI or VII element already attached to the silicon atom by a σ bond. The π bond is generally stronger the more electronegative the acceptor atom, but silicon attracts electrons more strongly than its accepted electronegativity would suggest. This kind of bonding cannot occur with carbon because its d orbitals (3d) are too high in energy to contribute appreciably. The fact also accounts nicely for the strength of the C—H bond and the weakness of the Si—H bond.

This point is illustrated by the silyl halides, whose properties are greatly affected by the presence of a silicon—silicon bond. The preparation of the chloride, bromide and iodide of monosilane is described above; the fluoride is made by the action of the chloride on antimony fluoride:

$$3\ SiH_3Cl + SbF_3 \rightarrow 3\ SiH_3F + SbCl_3$$

The iodide is liquid at room temperature, the rest are gaseous. Surprisingly only the bromide is spontaneously inflammable in air. The reaction of the fluoride with water is not recorded; the others are hydrolysed immediately to disilyl ether, a colourless gas:

$$2\ SiH_3X + H_2O \rightarrow (SiH_3)_2O + 2\ HX$$

On the other hand, hydrolysis with aqueous alkalis is complete, producing hydrogen and silicates. Silyl iodide gives a Wurtz-type reaction with sodium, which affords a useful path to disilane:

$$2\ SiH_3I + 2\ Na \rightarrow Si_2H_6 + 2\ NaI$$

Silyl chloride and ammonia give amines, the most stable being the liquid trisilylamine; this is spontaneously inflammable in air and vigorously decomposed by water into silica, ammonia and hydrogen. Silyl iodide may be converted into several other silyl compounds by means of a silver salt.

$$SiH_3I \rightarrow (SiH_3)_2Se \rightarrow SiH_3Br \rightarrow SiH_3Cl \rightarrow SiH_3NC \rightarrow SiH_3NCS \rightarrow$$

$$\rightarrow SiH_3NCO \rightarrow (SiH_3)_2O \rightarrow SiH_3F$$

The sequence implies that a compound may be converted into one coming later in the series by means of the appropriate silver salt, although all the changes have not been tested.

In general, silyl compounds present a contrast to their methyl analogues in that they more readily enter into reactions in which the identity of the radical is maintained. This is exemplified in the instantaneous conversion of the silyl halides by water into disilyl ether, and in their rapid reaction with silver salts. The principal causes of this reactivity are

(i) the ease with which the co-ordination number of the silicon atom can be raised from four to six;

(ii) the larger size, and consequently greater vulnerability to attack, of the silicon atom;

(iii) the appreciable polarity of the $Si^+—H^-$ bond (polarity is almost absent in the C—H bond) which renders it more reactive towards nucleophilic reagents.

These three factors also favour the formation of complexes.

11.6.2. Germanes and stannanes

Germanium hydrides are made in ways similar to those used for making boranes and silanes. A mixture of mono-, di- and trigermanes results from the action of dilute hydrochloric acid on magnesium germanide. Monogermane itself is conveniently made by reducing germanium tetrachloride with ethereal lithium aluminium hydride. Germanes from Ge_2H_6 to $Ge_{10}H_{22}$ have been made by circulating monogermane at 50 kPa through an ozoniser electric discharge tube at 195 K.

The germanes decompose at lower temperatures than the silanes, but are less inflammable and much less easily hydrolysed; monogermane is not attacked even by 30% caustic soda. Halogenation may be effected as with the silanes. An amorphous, yellow polymer $(GeH_2)_x$ is obtained when calcium germanide, Ca_2Ge, is treated with acid. Between 390 and 490 K it decomposes, the three volatile germanes being among the products.

TABLE 11.3

MELTING POINTS AND BOILING POINTS OF THE LOWEST GERMANES

		M.p./K	*B.p.*/K
Monogermane	GeH_4	108	183
Digermane	Ge_2H_6	164	302
Trigermane	Ge_3H_8	167	383

Tin forms the gaseous hydride SnH_4. It is best made by reducing tin(IV) chloride with ethereal lithium aluminium hydride. Stannane decomposes, at room temperature, into tin and hydrogen, but it is not hydrolysed by 15% caustic soda. This hydride is formed by the action of atomic hydrogen on metallic tin.

There is considerable doubt about the existence of a lead hydride; certainly no trace of it appears when metallic lead is treated with atomic hydrogen.

11.7. Group VB Hydrides

11.7.1. Ammonia

Ammonia is made in the laboratory by heating an ammonium salt with a base:

$$2\ NH_4^+ + CaO \rightarrow Ca^{2+} + 2\ NH_3 + H_2O$$

or by treating a nitride with water:

$$Mg_3N_2 + 6\ H_2O \rightarrow 3\ Mg(OH)_2 + 2\ NH_3$$

It is manufactured industrially by passing nitrogen and hydrogen over an iron catalyst at ~750 K and 20—100 MPa:

$$N_2 + 3\ H_2 \rightarrow 2\ NH_3 \qquad \Delta H = -92\ \text{kJ mol}^{-1}$$

Because the reaction is exothermic, the higher the temperature the more unfavourable the equilibrium; below 750 K however the rate is inconveniently low. A conversion of 10—14% in the Haber process operating at 30 MPa, and 40% in the Claude process operating at 90 MPa is usual. The ammonia is removed from the mixed gases and the residual mixture of nitrogen and hydrogen is recycled through the converter.

The ammonia molecule has been shown by infrared and microwave studies to be pyramidal with the H—N—H angle ~107°. This is because the valency electrons round the nitrogen atom can be roughly described as using sp^3 hybrid orbitals, one containing two electrons (the lone pair) and the other three being singly occupied. Overlap of these with the 1s orbitals of the three hydrogen atoms should give three molecular orbitals (at the tetrahedral angle 109.5°) but the hybridisation is not strictly sp^3. The lone-pair orbital is more electron-repellant than the others, thus forcing the bonds together and reducing the angle between them to 107°.

The ammonia crystal has an approximately face-centred cubic lattice, the lone pair of electrons being used in hydrogen bonding to three other molecules. The free inversion displayed by the gaseous molecule thus does not occur in the solid as infrared spectroscopic evidence shows.

Ammonia, like water and hydrogen fluoride, is associated in the liquid

TABLE 11.4

MELTING POINTS AND BOILING POINTS OF GROUP VB HYDRIDES

		M.p./K	*B.p.*/K
Ammonia	NH_3	195	240
Phosphine	PH_3	141	184
Arsine	AsH_3	157	211
Stibine	SbH_3	185	256
Hydrazine	N_2H_4	275	386
Diphosphine	P_2H_4	174	325

state, as is evident from its melting point, boiling point, latent heat and surface tension considered in relation to those of phosphine and arsine. The association is due to hydrogen bonding which occurs in liquid ammonia but not in liquid phosphine and arsine.

Liquid ammonia has a significant dipole moment (1.49 D) and is an ionising solvent. Through its lone pair, the molecule is a strong proton acceptor, and the liquid facilitates the extensive dissociation of weak acids; thus acetic acid is almost as completely dissociated in liquid ammonia as a mineral acid is in water:

$$NH_3 + CH_3 \cdot COOH \rightleftharpoons NH_4^+ + CH_3 \cdot COO^-$$

Liquid ammonia dissolves many of the active metals (Na, Ca) to give blue solutions in which there are solvated metal ions and solvated electrons. The solutions are fairly stable but have strong reducing properties.

Gaseous ammonia burns in oxygen with a low-temperature flame to produce nitrogen and water:

$$4\,NH_3 + 3\,O_2 \rightarrow 2\,N_2 + 6\,H_2O$$

At about 1050 K, on a platinum catalyst, ammonia and air readily react to give nitric oxide and water

$$4\,NH_3 + 5\,O_2 \rightarrow 4\,NO + 6\,H_2O$$

When ammonia and air are passed through a hot platinum gauze, the reaction provides the first step in the manufacture of nitric acid. The NO produced is oxidised to NO_2, as it cools, by the excess of air which is present. Finally successive reactions with water and oxygen from the air give a solution of nitric acid:

$$3\,NO_2 + H_2O \rightarrow 2\,HNO_3 + NO$$
$$2\,NO + O_2 \rightarrow 2\,NO_2$$

Ammonia is very soluble in water; there the molecule is partly hydrated and partly converted to ammonium ion:

$$NH_3 + H_2O \rightleftharpoons NH_4^+ + OH^- \qquad K_b = 1.8 \times 10^{-5}$$

The formation of an ammonium hydroxide is very improbable in these aqueous solutions.

The gas reacts with acids to give ammonium salts containing the tetrahedral NH_4^+ ion:

$$NH_3 + HCl \rightarrow NH_4^+Cl^-$$

Ammonium halides dissociate on heating; salts of some of the oxo-acids decompose:

$$NH_4Cl \rightleftharpoons NH_3 + HCl$$
$$NH_4NO_3 \rightarrow N_2O + 2\,H_2O$$

Ammonium salts are usually water-soluble. They resemble those of potassium and rubidium in this respect, the NH_4^+ ion being about the same size as the Rb^+ ion. Except where hydrogen bonding affects the structure (26.3.1) ammonium salts are isostructural with potassium salts.

11.7.2. Hydrazine

Hydrazine is still manufactured (Raschig, 1907) by oxidising aqueous ammonia, present in a large excess, with sodium hypochlorite. Two reactions occur:

$$NH_3 + NaOCl \rightleftharpoons NH_2Cl + NaOH$$
$$2\,NH_3 + NH_2Cl \rightleftharpoons NH_2 \cdot NH_2 + NH_4Cl$$

The reactants are mixed at a low temperature and rapidly heated to promote reaction of the chloramine with ammonia. Glue or gelatine is used to promote the slow reaction and to inhibit the secondary reaction:

$$N_2H_4 + 2\,NH_2Cl \rightarrow N_2 + 2\,NH_4Cl$$

This the glue does by chelating such metal ions as Cu^{2+} which catalyse the secondary reaction. Commercially the hydrazine is recovered as the hydrate.

To prepare anhydrous hydrazine, sulphuric acid and alcohol are added to the hot solution of the hydrate. From this crystals of hydrazine sulphate separate on cooling:

$$NH_2 \cdot NH_2 + H_2SO_4 \rightleftharpoons (NH_2 \cdot NH_3^+)HSO_4^-$$

Distillation with concentrated NaOH solution gives anhydrous hydrazine, *b.p.* 386.5 K, which is thermally stable but is very reactive and burns in air.

It forms a monohydrate, $N_2H_4 \cdot H_2O$, and is a weak base. The existence of the hydrazinium cation $N_2H_6^{2+}$ has been confirmed in the compounds $N_2H_6(SbCl_6)_2$ and $N_2H_6(BF_4)_2$. Both the base and its salts are strong reducing agents, converting iodates to iodides, iron(III) salts to iron(II) salts, and gold(III) salts to colloidal gold.

Hydrazine bears the same relation to ammonia as hydrogen peroxide to water. That the molecule is similar to hydrogen peroxide is shown by the Raman spectrum and high dipole moment (1.83 D) of the monomeric vapour. Like the hydroxyl groups in hydrogen peroxide, the NH_2-groups in hydrazine are without free rotation.

In the solid (*m.p.* 275 K) the molecules, apparently hydrogen-bonded to one another, are arranged in zig-zag chains.

Anhydrous hydrazine burns spontaneously in dry oxygen and reacts readily with halogens:

$$2\,I_2 + N_2H_4 \rightarrow 4\,HI + N_2$$

It sets free ammonia from ammonium chloride and decomposes when heated:

$$3\,N_2H_4 \rightarrow N_2 + 4\,NH_3$$

The anhydrous liquid is a good solvent for sulphur, selenium, phosphorus, and arsenic.

Aqueous solutions of hydrazine, like those of hydrogen peroxide, show both oxidising and reducing properties. In acids the redox potential is high and suggests that hydrazine should be a strong oxidising agent:

$$N_2H_5^+ + 3\,H^+ + 2\,e \rightleftharpoons 2\,NH_4^+ \qquad E^0 = +1.27\ V$$

The reaction is slow, however, with all but the strongest reducing agents such as Ti^{3+}. Hydrazine is easily oxidised in either acids or alkalis; the reactions are complicated, nitrogen being the commonest product:

$$N_2 + 5\,H^+ + 4\,e \rightarrow N_2H_5^+ \qquad E^0 = -0.23\ V$$
$$N_2 + 4\,H_2O + 4\,e \rightarrow N_2H_4 + 4\,OH^- \qquad E^0 = -1.15\ V$$

The four-electron change necessary for the quantitative conversion to nitrogen occurs only within certain limits of pH, concentration and temperature. Chlorine, bromine, iodine and iodates bring about this reaction quantitatively at ~pH 7. Dissolved molecular oxygen oxidises aqueous hydrazine to nitrogen in a series of stages, so that hydrazine is an effective deoxidant for boiler-water. Several metal ions, particularly copper, catalyse the reaction.

11.7.3. Phosphines

Phosphine, like ammonia, is pyramidal with the H—P—H angle 93° (cf. NH_3, 107°). It is much less soluble and a much weaker base than ammonia, but a much stronger reducing agent. Phosphonium salts are decidedly less stable than those of ammonium. They dissociate as do ammonium salts on heating; PH_4Cl is stable at 200 K and completely decomposed at ~220 K, the corresponding temperatures for PH_4Br and PH_4I being ~220 K, 270 K and 270 K, 335 K, respectively. The same order is found for the ammonium salts, but with temperatures in the range 500—650 K. The difference in behaviour is ascribable to the lower electronegativity of phosphorus (2.06 against 3.07 for nitrogen) as also is the ready disruption of the phosphonium ion by water:

$$PH_4^+I^- + H_2O \rightarrow PH_3 + H_3O^+ + I^-$$

Alkyl and aryl substituted phosphines are similar to the amines in structure but highly inflammable. The quarternary phosphonium bases are, like the corresponding nitrogen compounds, very strongly ionised.

The unstable, colourless, liquid diphosphine, P_2H_4, (*m.p.* 174 K, *b.p.* 324.7 K) is a minor by-product of the hydrolysis of phosphides (Ca_3P_2) which give mainly phosphine; it is separated from the latter by freezing it out. Unlike hydrazine it is without basic properties. But like hydrazine it is readily oxidised and is a strong reducing agent; it has the same structure. There the similarity ends because its lone pairs are quite ineffective, so that it is insoluble in water and without trace of basic character. It is photo- and heat-sensitive giving phosphine and phosphorus:

$$6\,P_2H_4 \rightarrow 8\,PH_3 + P_4$$

The liberated phosphorus adsorbs some of the phosphine to form a polymer, $(P_2H)_x$. The yellow solid is odourless and insoluble in cold dilute hydrochloric acid, but is decomposed by water giving hydrogen and a phosphorous acid.

11.7.4. Other Group V hydrides

Like phosphine, the trihydrides of arsenic and antimony are not formed by direct combination with molecular hydrogen. They are usually prepared by reducing arsenic or antimony compounds with atomic hydrogen produced at a zinc surface dissolving in dilute hydrochloric acid:

$$Zn + 2\,H^+ + 2\,Cl^- \rightarrow Zn^{2+} + 2\,H + 2\,Cl^-$$

Both are strong reducing agents, without basic properties and easily decomposed by heat; the decomposition of arsine begins at 500 K.

Phosphorus, arsenic and antimony are all readily attacked by atomic hydrogen to yield their respective trihydrides. There is evidence that a hydride of bismuth is not formed in this way.

11.8. Hydrides of Group VIB

The bond angles in water vapour, hydrogen sulphide and hydrogen selenide are respectively 104.5°, 92.3° and 90°. Roughly speaking the 'prepared' oxygen atom is well described as using tetrahedral hybrid sp^3 orbitals, two doubly occupied and two singly, and the water molecule results from the overlap of each of the latter with an s orbital of hydrogen. Less hybridisation occurs with sulphur and the other elements, which have a much lower electronegativity than oxygen and the final form is determined by various factors. Oxygen probably adopts the symmetrical (tetrahedral) configuration because in this way it is better able to 'draw in' the hydrogen electrons and approach the condition of having two electrons in each hybrid.

The Group VI hydrides exhibit an appreciable increase in acid strength with increasing molecular weight, the pK_a values of aqueous solutions being approximately as given in Table 11.5. This is mainly due to a fall in dissociation energies; solvation energies and electron affinities increase in the reverse direction. There is no connection between acid strength and electronegativity, for the dipole moments decrease from water to hydrogen telluride.

TABLE 11.5

FIRST pK_a VALUES FOR GROUP VI HYDRIDES

H_2O	H_2S	H_2Se	H_2Te
14.0	7.0	3.8	2.6

11.8.1. Water

Upon the singular properties of water all biochemistry, much of geochemistry and a great deal of general chemistry depends. Ice at 90 K has a rigid arrangement of atoms in which oxygen is tetrahedrally co-ordinated by sp^3 orbitals with four hydrogens, two closely (100 pm) and two more remotely (176 pm). This leads to the very open wurtzite structure with the oxygen atoms 276 pm apart, and the hydrogen atoms at points one third along this distance. It accounts for ice being less dense than water.

The hydrogen bonding in ice suffers a progressive break-down as the temperature rises and there is increasing freedom of movement of the H_2O units, but the structure remains sufficiently open for the density of melting ice at 273 K to be less than that of water at 273 K, when the liquid still retains

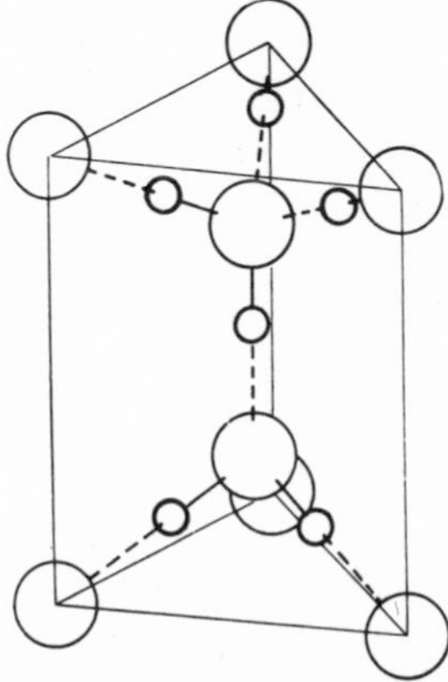

Fig. 11.18. Wurtzite-like structure of ice.

some of the open character of the solid. As the liquid warms, the density is affected by (a) the breaking of hydrogen bonds leading to a closer packed structure, and (b) the thermal expansion. At 277 K these just balance one another, and water is at its maximum density. Above 277 K thermal expansion takes charge and the density continues to fall to the boiling point, but at 323 K about half the hydrogen bonding still remains and even boiling water retains some.

Water is the most remarkable of all solvents. It has a very high dielectric constant (78 at 298 K), has a considerable dipole moment (1.85 D), and is amphoteric. As an ionising medium it acts, upon occasion, as either a donor or an acceptor of protons: thus it facilitates the dissociation of both bases and acids:

$$H_2O + \underset{\text{crystal}}{Na_2O} \rightarrow 2\,Na^+ + 2\,OH^-$$

$$H_2O + \underset{\text{gas}}{HCl} \rightarrow H_3O^+ + Cl^-$$

Because water molecules are dipolar they cluster round both positive and negative ions; these hydrate sheaths greatly reduce the ionic attraction between the oppositely charged ions and hence their tendency to recombine.

The chemical versatility of water is further increased by its ability either to oxidise or reduce. For the equilibrium in pure water at 298 K,

$$H^+(10^{-7}\,M) + e \rightleftharpoons \tfrac{1}{2}\,H_2 \qquad E = -0.414\text{ V}$$

Water gives hydrogen with reducing agents possessing redox potentials more negative than this, but the process is very slow when the hydrogen overvoltage is high. Thus zinc (E^0, $Zn^{2+}/Zn = -0.763$ V) is passive in pure water, though it sets free hydrogen from dilute acids where E, H^+/H_2 is much nearer zero.

The reducing action of the water is summarised in the equation

$$\tfrac{1}{2}\,O_2 + 2\,H^+(10^{-7}\,M) + 2\,e \rightleftharpoons H_2O \qquad E = +0.815\text{ V}$$

In consequence of this, oxygen is liberated only by strong oxidising agents. The redox potential decreases as the OH^- ion concentration increases, and alkaline solutions are more easily oxidised than pure water.

11.8.2. Hydrogen sulphides

By contrast, hydrogen sulphide is without hydrogen bonding; the solid is close-packed, and the liquid density has a normal temperature dependence. The gas burns in air; there is considerable decomposition within the flame, which consequently deposits sulphur on a cold surface, and complete combustion at its outer edge. The catalysed oxidation by air on an iron(III) oxide surface can be made to release the whole of the sulphur:

$$H_2S + \tfrac{1}{2}\,O_2 \rightarrow H_2O + S$$

Hydrogen sulphide is moderately soluble in water in which it is a weak dibasic acid:

$$H_2O + H_2S \rightleftharpoons H_3O^+ + HS^- \qquad pK_a = 7.0$$
$$H_2O + HS^- \rightleftharpoons H_3O^+ + S^{2-} \qquad pK_a = 13.9$$

Like carbon dioxide, sulphur dioxide and ammonia it is completely expelled from boiling water. It acts as a reducing agent with, generally, the liberation of sulphur, but in strong oxidising agents, such as concentrated nitric acid, the sulphur may be converted into SO_2 or SO_3, or their derivatives:

$$S + 2\,H^+ + 2\,e \rightleftharpoons H_2S \qquad E^0 = +0.141\text{ V (in acid)}$$
$$S + 2\,e \rightleftharpoons S^{2-} \qquad E^0 = -0.508\text{ V (in alkali)}$$

Sulphur also forms yellow, liquid hydrides H_2S_x where x = 2–6. H_2S_2 and H_2S_3 can be made by dissolving sulphur in aqueous Na_2S, pouring the solution of polysulphides into concentrated HCl at 263 K, and fractionating the yellow oil which separates. Mixtures of the higher sulphides, also called sulphanes, can be made by the action of aqueous H_2S on the lower chlorides of sulphur (22.5):

$$S_xCl_2 + 2\ H_2S \rightarrow 2\ HCl + H_2S_{x+2}$$

All the liquids readily decompose into H_2S and sulphur. They dissolve in solvents like benzene and chloroform. Their structures involve chains of sulphur atoms.

$$H{-}S{-}S{-}S{-}S{-}S{-}H$$

11.8.3. Hydrogen selenide and hydrogen telluride

Hydrogen selenide, made by decomposing aluminium selenide with water, is much less thermally stable than its sulphur analogue:

$$Al_2Se_3 + 3\ H_2O \rightarrow 3\ H_2Se + Al_2O_3\ \text{(hydrated)}$$

It is slowly oxidised to selenium by moist oxygen. The corresponding compound of tellurium decomposes rapidly at room temperature and is even more easily oxidised. Both have strong reducing properties; thus H_2Se gives free selenium in addition to the selenide when passed into a solution of a heavy metal cation. There are no hydrogen polyselenides or polytellurides, nor would these be expected in view of the weakness of the Se—Se and Te—Te bonds and of the bonds between these elements and hydrogen. The last point is exemplified by a marked increase in ΔH_f^0 of the simple hydrides in passing down Group VI (Table 11.6). This increase is also observed in the covalent hydrides of Groups IV, V and VII.

TABLE 11.6

ENTHALPIES OF FORMATION OF HYDRIDES OF GROUP VI

	H_2O	H_2S	H_2Se	H_2Te
ΔH_f/kJ mol^{-1}	−286	−20	+86	+154

The hydrides of selenium and tellurium, like those of tin and antimony, are strongly endothermic, probably owing to the high dissociation energy of the hydrogen molecule and the small electronegativity of the hydrogen atom.

11.9. Hydrides of Group VIIB

As was found in the two previous groups, the boiling point of the first member of Group VII is again notably high because of association.

Hydrogen chloride shows a considerable fall from the value for hydrogen fluoride; thence the boiling points increase with molecular weight (Fig. 11.19).

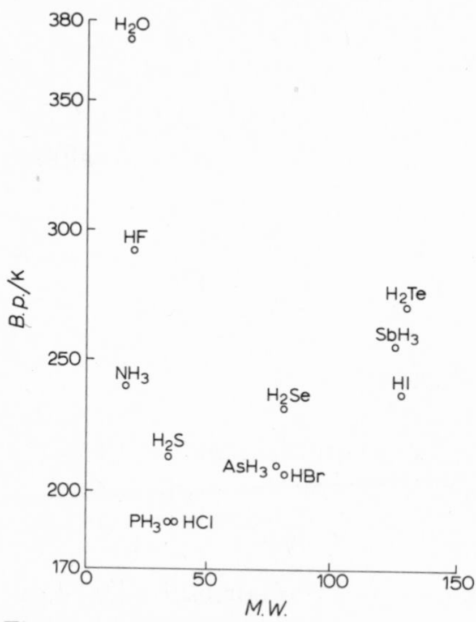

Fig. 11.19. Relation between *b.p.* and molecular weight for hydrides of Groups V, VI and VII.

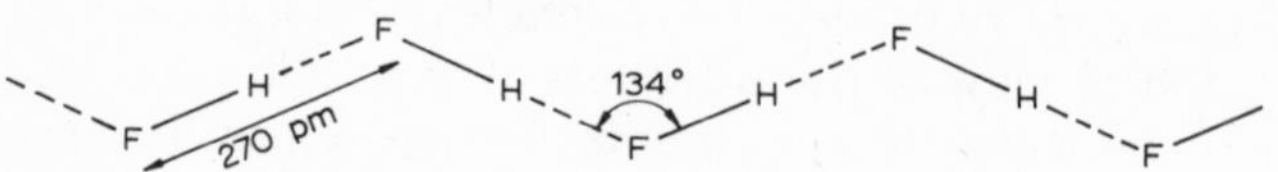

Fig. 11.20. Chain of hydrogen fluoride in the solid.

Hydrogen fluoride is very strongly hydrogen-bonded; its entropy of vaporisation at the boiling point is high (104 J K^{-1} mol^{-1}), but not as high as that of water (109 J K^{-1} mol^{-1}). Solid hydrogen fluoride contains zig-zag chains (Fig. 11.20). The hydrogen bonding with fluorine, though the strongest known, is considerably less than the 290 kJ mol^{-1} typical of a σ covalent bond.

TABLE 11.7

PROPERTIES OF THE HYDRIDES OF GROUP VIIB

	HF	HCl	HBr	HI
B.p./K	292	188	206	238
ΔH_f/kJ mol^{-1}	−271	−92	−36	+26
D(H—X)/kJ mol^{-1}	565	431	368	297
pK_a	3	−6	−8	−9

Hydrofluoric acid is made by treating CaF_2 with H_2SO_4. It is the principal starting material for the preparation of other fluorine compounds. Although the liquid attacks glass it can be kept in copper or Monel metal vessels or in certain fluorocarbon polymers like PTFE. It is a useful solvent for both organic and inorganic compounds. For the autoprotolysis

$$2\ HF \rightleftharpoons H_2F^+ + F^- \qquad K = 10^{-10}$$

Unlike HCl, HBr and HI it ionises rather weakly in aqueous solution (Table 11.7).

Gaseous HCl is made conveniently by the action of H_2SO_4 on a chloride:

$$H_2SO_4 + Cl^- \rightarrow HSO_4^- + HCl$$

but HBr and HI are better prepared by direct combination on a platinum catalyst.

The decrease in ionic character from hydrogen fluoride to hydrogen iodide is marked and due to increasing symmetry in the distribution of the bonding electrons. At first sight, the bond is a simple one, compounded of the 1s and the p_x atomic orbitals of hydrogen and halogen respectively.

$$\psi = a\psi(\mathrm{H{:}1s}) + b\psi(\mathrm{Cl{:}3p_x})$$

In a molecular orbital of this kind, b/a always exceeds unity, but the ratio is much larger for fluorine than for iodine, indicating that the charge cloud 'leans' more heavily towards the fluorine than towards the iodine atom in the respective hydrides.

This simple picture is inadequate because some sp hybridisation occurs on the halogen atom, one hybrid being engaged in the bond and the other pointing out behind the halogen and containing a lone pair; the lone pair plays a large part in determining the dipole moment of the halogen halides.

Disparity between the MO coefficients, a and b, indicates ionic character in every instance; the dipole moment of hydrogen chloride is 1.03 D. Its infrared absorption spectrum gives a bond length of 126 pm, so that, were the molecule completely ionic (i.e. H^+Cl^-), the value of μ would be 6.05 D. In fact, the bonding is mainly covalent in the free molecule. If the degree of polarity is measured by 1.03/6.05 = 0.169, it could be said that the bond in HCl is about 83% covalent. On this basis hydrogen iodide would be 95% covalent, the corresponding polarity being 0.05.

Dissociation of the halogen halides in water is due to the breaking of the hydrogen—halogen bond through the approach of a lone pair on the water molecule to a hydrogen atom in the hydrogen halide causing a retraction of the molecular orbital towards the halogen and thereby allowing the water molecule to capture the liberated proton.

$$H_2O + HCl \rightarrow H_2O \cdots H^+Cl^- \rightarrow H_3O^+Cl^-$$

The necessary energy for the process is provided by the enthalpy of formation of the aquated H^+ and halide ions. The increase in acidity from hydrogen fluoride to hydrogen iodide arises almost entirely from the decreasing strength of the hydrogen—halogen bond, since the other enthalpy and entropy terms are largely self-cancelling.

The monohydrate formed by HCl at low temperatures has been shown by

X-ray diffraction at 238 K to be $H_3O^+Cl^-$, every hydrogen atom being bonded by hydrogen bonds to the nearest chlorine atom. The hydrogen bonds OH···Cl are 295 pm in length.

Further Reading

K.M. Mackay, Hydrogen compounds of the metallic elements, Spon, London, 1966.

E. Wiberg and E. Amberger, Hydrides, Elsevier, Amsterdam, 1971.

G.G. Libowitz, The solid-state chemistry of binary metal hydrides, Benjamin, New York, 1965.

W.N. Lipscomb, Boron Hydrides, Benjamin, New York, 1963.

E.L. Muetterties and W.H. Knoth, Polyhedral boranes, Marcel Dekker, New York, 1968.

R.N. Grimes, Carboranes, Academic Press, New York, 1970.

M.F. Hawthorne, Carborane chemistry, Accounts Chem. Res., 1 (1968) 281, 6 (1973) 118.

K. Wade, Electron deficient compounds, Nelson, London, 1971.

F.G.A. Stone, Hydrogen compounds of the Group IV elements, Prentice-Hall, Englewood Cliffs, New Jersey, 1962.

F. Franks (Ed.), Water. A comprehensive treatise, 4 vols., Plenum Press, New York, 1973.

R.A. Horne (Ed.), Water and aqueous solutions, Wiley-Interscience, New York, 1972.

W.M. Mueller, J.P. Blackledge and G.C. Libowitz, Metal hydrides, Academic Press, New York, 1968.

D.J.G. Ives and T.H. Lemon, Structure and properties of water, RIC Reviews, 1 (1968) 62.

J.N. Murrell, The hydrogen bond, Chem. Brit., 5 (1969) 107.

B.L. Shaw, Inorganic hydrides, Pergamon, Oxford, 1967.

The Noble Gases — Group 0

12

12.1. The Elements

The gases helium, neon, argon, krypton, xenon and radon constitute Group 0 of the Periodic Table. The first five were isolated by Ramsay and his associates (1894—98); radon was discovered in the disintegration products of radium (1900). With the exception of helium ($1s^2$) their atoms have ns^2np^6 ground states; consequently they have no tendency to form diatomic molecules (4.1.1.1). The gases are monatomic, and the ratio C_P/C_V is in every case close to 5/3, the theoretical value for an ideal monatomic gas. Ionisation energies are high (Fig. 3.16), and compound formation is limited to a few fluorine and oxygen derivatives of the three heaviest members of the group. Before 1962 the elements were called the inert gases, but since the discovery of fluorocompounds of xenon in that year it has become more appropriate to describe them as noble gases.

TABLE 12.1

ATOMIC PROPERTIES OF THE NOBLE GASES

	He	Ne	Ar	Kr	Xe	Rn
Z	2	10	18	36	54	86
$I(1)$/MJ mol^{-1}	2.36	2.07	1.51	1.34	1.17	1.03
Van der Waals radius/pm	120	160	190	200	220	

The gases are difficult to liquefy because interactions between the molecules are so slight, being due entirely to dispersion forces (4.1.10). The elements have therefore much lower *b.p.* and polarisabilities than compounds of similar molar mass in which less symmetrical arrangements of electrons give rise to polarisation forces. Thus neon (M = 0.020 kg mol^{-1}) has a *b.p.* 65 K below that of methane (M = 0.016 kg mol^{-1}). Xenon (M = 0.131 kg mol^{-1}) has a polarisability (5.3.7.2) of only 4.5×10^{-40} F m^2 whereas Br_2 (M = 0.160 kg mol^{-1}) has a polarisability almost ten times greater: α_{Br_2} = 4.20×10^{-39} F m^2. Helium has a negative Joule—Thomson coefficient above

40 K — spontaneous expansion causes the gas to warm up — and preliminary cooling is necessary before the gas can be liquefied by controlled expansion. A notable characteristic of all the elements is the narrow range of temperature over which they are liquid (Table 12.2). Helium has to be compressed before it will solidify.

TABLE 12.2

PHYSICAL PROPERTIES OF THE NOBLE GASES

	He	Ne	Ar	Kr	Xe	Rn
M.p./K	0.9*	24	84	116	161	202
B.p./K	4.2	27	87	120	166	211
ΔH_{vap}/kJ mol^{-1}	0.08	1.8	6.7	9.6	13.6	18.0

* At 2.6 MPa.

Helium II

When helium-4 is cooled to 2.18 K at 10^2 kPa pressure (the λ-point) a remarkable liquid, helium II, is obtained. It has

(i) very high heat conductance — 600 times that of copper at room temperature;
(ii) very low viscosity — about one thousandth that of hydrogen gas;
(iii) ability to flow up the surface of the containing vessel.

The liquid is produced only from $^4_2$He, not from $^3_2$He.

12.2. Helium

The α-particles set free in radioactive disintegration take up electrons to form atoms of helium-4. The gas is therefore to be found associated with minerals containing α-emitters; pitchblende, which contains uranium, and monazite, which contains thorium, are examples. Helium-3, though a stable nuclide, comprises only 1.4×10^{-3}% of terrestrial helium; it is a product of the radioactivity of tritium, itself a result of the action of cosmic rays on deuterium (10.5):

$$^3_1\text{H} \xrightarrow[12.5\ \text{years}]{\beta} {}^3_2\text{He (stable)}$$

Helium-5 and helium-6 are radioactive nuclides of short half-life.

Helium is much less common on earth than on larger planets or stars (2.12). Its high molecular velocity, and the large mean free path available, enable it to escape readily from the upper atmosphere, where it is relatively abundant, because of the weakness of the gravitational field at that height. The principal source of the element is natural gas (up to 8% He) found in the south of the U.S.A., the other gases, mainly hydrocarbons, being separated from it by liquefaction. Attempts have been made to isolate it from the atmosphere by making use of its ability to diffuse through heated, thin-walled silica capillaries.

12.3. Separation of Noble Gases from Air

The non-radioactive noble gases are all present in the atmosphere. Their volume percentages at sea level are: argon 0.93%, neon 1.8×10^{-3}%, helium 5.2×10^{-4}%, krypton 1.1×10^{-4}%, xenon 9×10^{-6}%.

Liquid air is, to a first approximation, a ternary mixture of nitrogen (*b.p.* 77 K), argon (*b.p.* 87 K) and oxygen (*b.p.* 90 K). In the fractional distillation of liquid air there is a point on the column where the mid-boiling fraction reaches a maximum concentration, and from which a side-cut may be taken to give a liquid containing mainly oxygen and argon. This cut is fractionated separately into crude argon and oxygen; the oxygen is returned to the column. The crude argon, which has up to 20% oxygen, is mixed with hydrogen and sparked to convert the oxygen into water; the unused hydrogen is later oxidised by hot CuO.

Neon is not condensed during the main distillation of liquid air; it accumulates on the nitrogen side and is withdrawn. The nitrogen in the extracted gas is removed, first by low-temperature liquefaction and, finally, by charcoal adsorption.

Krypton and xenon remain dissolved in the liquid oxygen; they can be separated by selective low-temperature adsorption.

12.4. Uses

Helium, formerly used for airships, is now employed, like argon, to provide an inert gaseous shield during the welding of Mg, Al, Ti and stainless steel. It has a future in gas-cooled atomic reactors as a material for transferring heat, since it is inert and does not become active under irradiation. An oxygen—helium mixture is used in the treatment of asthma as it diffuses more rapidly than air through constricted lung passages. A similar mixture is supplied to deep-sea divers because helium, being less soluble than nitrogen, does not cause caisson sickness or 'bends' by bubbling out of the blood when the pressure is released. Helium is a suitable gas for low-temperature gas thermometry, because of its low boiling point and also its near-ideal behaviour.

Argon is chiefly employed in welding and other operations which require both a non-oxidising atmosphere and the absence of nitrogen. Since about 1920, it has been used in gas-filled electric bulbs to reduce the rate of evaporation from the tungsten filament, and also more recently in thermionic tubes (thyratons) and fluorescent lamps. Krypton replaces argon in high-efficiency filament lamps, such as miners' cap lamps. Xenon is employed in some electronic flash tubes for high-speed photography. The first four members of the group, particularly neon, but krypton increasingly, are used in the low-pressure discharge tubes for coloured signs.

The helium—neon laser

The first excited state (^3S) of the He atom lies 1.9 MJ mol^{-1} above the ground state. If a helium atom of this energy collides with a neon atom, an energy transfer can occur which raises the Ne to a metastable state with the $1s^22s^22p^54s^1$ configuration and provides a small increment of translational energy. A helium—neon laser consists of a discharge tube of a few mm bore containing He and Ne in the molar ratio 10 : 1 at a pressure of about 200 Pa, closed at the ends with mirrors, one of which allows a fraction of the light to pass through it. Excitation of the He atoms with a radiofrequency discharge exceeding 27 MHz, followed by transfer of energy from the He atom to Ne atoms, produces a situation where nearly all the neon atoms in the tube are in the metastable state. Suppose that one of the metastable Ne atoms loses energy in the form of light emission — the transition is to a $1s^22s^22p^53p^1$ configuration — the light photon will tend to induce emission from another metastable Ne rather than be absorbed, simply because there are so few Ne atoms in lower energy states to absorb it. Furthermore the light emission which is induced is *coherent* with that which induces it, i.e. the waves have the same wavelength, the same direction of propagation and the same phase angle. Such waves reflected from a mirror at one end of the tube will multiply the induced emissions on their return path and eventually, after many reflections, produce an intense beam of coherent radiation, some of which passes through the partial mirror at one end. A continuous radiofrequency discharge can maintain the rate of excitation of He atoms at a level which ensures a continuous laser beam. (Laser = light amplification by stimulated emission of radiation.) The advantage of coherent radiation is that a narrow beam diverges hardly at all, thus intense radiation can be directed with great accuracy. Laser beams are being used increasingly in chemistry; their application has re-vitalised Raman spectroscopy (5.3.3).

12.5. Radon

The isotopes of radon (element 86) result from the α-particle activity of the radium isotopes which belong respectively to the U-238, Th-232 and U-235 natural radioactive series. These isotopes, formerly called radon, thoron and actinon, are all α-emitters of short half-life:

$$^{222}_{86}\mathrm{Rn} \xrightarrow[3.8\ \mathrm{d}]{\alpha} {}^{218}_{84}\mathrm{Po}\ (\alpha,\ 3\ \mathrm{min})$$

$$^{220}_{86}\mathrm{Rn} \xrightarrow[55\ \mathrm{s}]{\alpha} {}^{216}_{84}\mathrm{Po}\ (\alpha,\ 0.15\ \mathrm{s})$$

$$^{219}_{86}\mathrm{Rn} \xrightarrow[3.9\ \mathrm{s}]{\alpha} {}^{215}_{84}\mathrm{Po}\ (\alpha,\ 1.8 \times 10^{-3}\ \mathrm{s})$$

Element 86 is a noble gas and has physical and chemical properties appropriate to its position as the last member of Group 0.

12.6. Compounds of the Noble Gases

Pre-1962

Recognition of the stable electronic structure of the noble gases (3.4.3) although contributing enormously to the interpretation of valency, probably restricted attempts to prepare chemical compounds of the elements. Pauling in 1933 predicted their formation, but up to Bartlett's work in 1962 there were only the compounds observed in discharge tubes, some unstable hydrates and the clathrates.

Hydrates and deuterates have been made by compressing the gases with water and D_2O. Those formed by the heavier elements are the most stable and contain six H_2O or D_2O molecules to one inert gas atom; as for example in $Xe \cdot 6\,H_2O$ where the xenon atom is evidently polarised by the strong dipole of the water molecule. In keeping with this, the water-solubility of the gases increases down Group 0.

TABLE 12.3

ABSORPTION COEFFICIENTS OF NOBLE GASES IN WATER AT 293 K

He	Ne	Ar	Kr	Xe	Rn
0.0097	0.014	0.05	0.11	0.24	0.51

Quinol forms clathrates (7.2.13) with argon, krypton and xenon when crystallised from benzene or water under a considerable pressure of the noble gas. The noble gas atoms are caged inside groups of quinol molecules which are joined together by hydrogen bonds. The argon clathrate contains about 9% by weight, corresponding to about one argon atom to three quinol molecules. The molecular ratios for the krypton and xenon clathrates are nearly the same as for clathrates holding molecules such as CO, CO_2 and SO_2, which are about the same size as Kr and Xe.

Post-1962

The history of the discovery that the noble gases actually do form stable, conventional compounds is an interesting and exciting one. Bartlett had found that platinum hexafluoride was a strong enough oxidising agent to convert molecular oxygen to the oxygenyl compound $O_2^+PtF_6^-$. From this he reasoned that xenon should be similarly converted by PtF_6 to a cation, since the first ionisation energy of xenon is about the same as that of the oxygen molecule (i.e. $O_2 \rightarrow O_2^+ + e$). When he mixed the deep-red vapour of PtF_6 with an excess of xenon at room temperature, they reacted immediately to give a yellow solid which he identified as $Xe^+PtF_6^-$. In this way a wide new field of investigation was opened up; some of the results from this will now be described.

It is now clear that when Xe reacts with PtF_6 the compound $XePtF_6$ is not the only product. The adduct can contain more platinum, and then has the composition $Xe(PtF_6)_x$ where x is between 1 and 2. Similar adducts, $Xe(RuF_6)_x$ and $Xe(RhF_6)_x$, are also formed when xenon reacts with the appropriate hexafluoride at room temperature.

12.6.1. Fluorides

Xenon tetrafluoride

The first fluoride of xenon, XeF_4, was prepared in 1962 by Claasen, Selig and Malm who heated 1 vol. of fluorine with 5 vol. of xenon in a metal vessel which they afterwards cooled rapidly. It has since been synthesised by simply passing the two gases through a heated nickel tube and condensing out the product. The tetrafluoride is a colourless, crystalline solid which readily sublimes, the vapour being also colourless. It can be stored in Pyrex, is soluble without reaction in hydrogen fluoride, and in iodine pentafluoride.

Some of the reactions of XeF_4 are summarised below in the form of equations.

$$XeF_4 + 2\,H_2 = Xe + 4\,HF$$
$$XeF_4 + 4\,HCl = Xe + 4\,HF + 2\,Cl_2$$
$$XeF_4 + 4\,I^-aq = Xe + 4\,F^-aq + 2\,I_2$$
$$6\,XeF_4 + 12\,H_2O = 24\,HF + 4\,Xe + 2\,XeO_3 + 3\,O_2$$
$$XeF_4 + 2\,CH_2{=}CH_2 = CH_2F \cdot CH_2F + CH_3 \cdot CHF_2 + Xe$$
$$XeF_4 + 2\,CH_3 \cdot CH{=}CH_2 = 2\,CH_3 \cdot CH_2 \cdot CHF_2 + Xe$$
$$XeF_4 + 4\,NO = 4\,NOF + Xe$$
$$XeF_4 + 4\,NO_2 = 4\,NO_2F + Xe$$
$$3\,XeF_4 + 4\,BCl_3 = 4\,BF_3 + 6\,Cl_2 + 3\,Xe$$
$$XeF_4 + 2\,SbF_5 = XeSb_2F_{12} + F_2$$
$$XeF_4 + 2\,SF_4 = Xe + 2\,SF_6$$

The equations do not imply exact stoichiometry. With aqueous alkalis the reaction of XeF_4 is more complex than with water; the disproportionation reaction gives some Xe^{VIII} in the form of XeO_6^{4-}.

Xenon difluoride

This is produced by heating the streaming elements in a short nickel tube (short to reduce the formation of XeF_4), by subjecting them to an electrical discharge, by heating a mixture of xenon, oxygen and fluorine and in many other ways. Like the tetrafluoride the difluoride is a colourless, crystalline solid; it is the least volatile fluoride of xenon and melts about 400 K. The liquid and vapour are also colourless. It dissolves in hydrogen fluoride without reaction, is reduced by hydrogen to Xe and HF, and reacts with SbF_5 to give $Xe(SbF_6)_2$.

Like XeF_4 the difluoride reacts with olefins to give difluoroalkanes. Its reaction with water is expressed in the equation

$$2\,H_2O + 2\,XeF_2 = 2\,Xe + 4\,HF + O_2$$

Fluorine atoms in both XeF_2 and XeF_4 can be replaced by other very electronegative groups e.g.

$$XeF_2 + HOSO_2F = FXeOSO_2F + HF$$
$$XeF_2 + HOClO_3 = FXeOClO_3 + HF$$

Xenon hexafluoride

This is prepared by heating xenon and fluorine under pressure. As would be expected, it is the most volatile of the three fluorides: the white crystals melt at 322.5 K. The hexafluoride also dissolves in hydrogen fluoride, but differs from XeF_2 and XeF_4 by giving a conducting solute:

$$HF + XeF_6 \rightarrow XeF_5^+ + HF_2^-$$

It is reduced by hydrogen to Xe and HF and, under a controlled hydrolysis, yields the oxofluoride $XeOF_4$ or XeO_3 according to conditions.

Liquid XeF_6 dissolves CsF to give the salt $CsXeF_7$.

$$CsF + XeF_6 \rightarrow CsXeF_7$$

This can be separated as a yellow solid which is stable up to 330 K. Above that temperature it loses XeF_6 to give colourless Cs_2XeF_8, stable up to 670 K:

$$2\,CsXeF_7 \rightarrow Cs_2XeF_8 + XeF_6$$

Although RbF is only slightly soluble in XeF_6, it reacts with it to give colourless $RbXeF_7$, stable below 293 K, and Rb_2XeF_8, also colourless, stable up to 670 K. KF and NaF form only the octafluoroxenates K_2XeF_8 and Na_2XeF_8, and these are less thermally stable than those of Rb and Cs.

Xenon hexafluoride attacks silica to produce $XeOF_4$ and SiF_4:

$$2\,XeF_6 + SiO_2 \rightarrow 2\,XeOF_4 + SiF_4$$

Oxofluorides

The compound $XeOF_4$ can be made as a colourless, mobile liquid by condensing into a nickel can at 77 K equimolar amounts of XeF_6 and H_2O. The oxofluoride is produced when the mixture is allowed to come to room temperature:

$$H_2O + XeF_6 \rightarrow XeOF_4 + 2\,HF$$

The HF and any unchanged XeF_6 can be removed by adding NaF which converts HF to $NaHF_2$ and XeF_6 to Na_2XeF_8.

Colourless crystals of XeO_2F_2 can be made by the reaction between XeO_3 and $XeOF_4$:

$XeO_3 + XeOF_4 \rightarrow 2\ XeO_2F_2$

Oxofluorides of formula XeO_2F_4 and XeO_3F_2 have also been characterised, the latter is made by treating Na_4XeO_6 (12.6.3) with XeF_6.

12.6.1.1. Fluoride structures

Xenon difluoride has a linear molecule in both gaseous and solid phases (cf. Fig. 5.3). In the solid these are packed parallel to one another (Fig. 12.1).

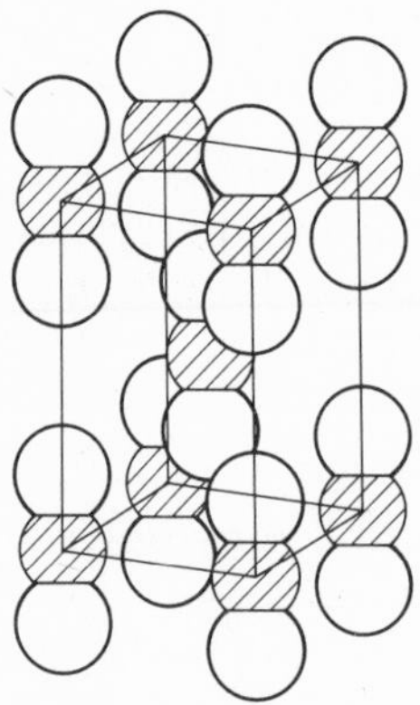

Fig. 12.1. Arrangement of molecules in XeF_2 crystal.

The bonding may involve only one σ5p AO of the Xe and the σ5p atomic orbitals of the fluorines, leaving three lone pairs symmetrically placed in a plane through the Xe atom and at right angles to the F—Xe—F line.

If we consider three atomic orbitals, the $5p_z$ on the xenon atom and the $2p_z$ on each of the fluorine atoms, we see that these can be combined to give three molecular orbitals, of which one is bonding, one is non-bonding and one is anti-bonding. The symmetry and overlap of the atomic orbitals is represented in Fig. 12.2.

F_a	Xe	F_b	
(−)(+)	(−)(+)	(−)(+)	anti-bonding
(−)(+)		(+)(−)	non-bonding
(−)(+)	(+)(−)	(−)(+)	bonding

Fig. 12.2. Atomic orbitals available for MO formation in XeF_2.

The non-bonding MO receives no contribution from the xenon $5p_z$ AO because this orbital has the wrong symmetry for combination, as indicated by the algebraic signs of the ψ values.

The energies and occupancies of the molecular orbitals are represented in Fig. 12.3.

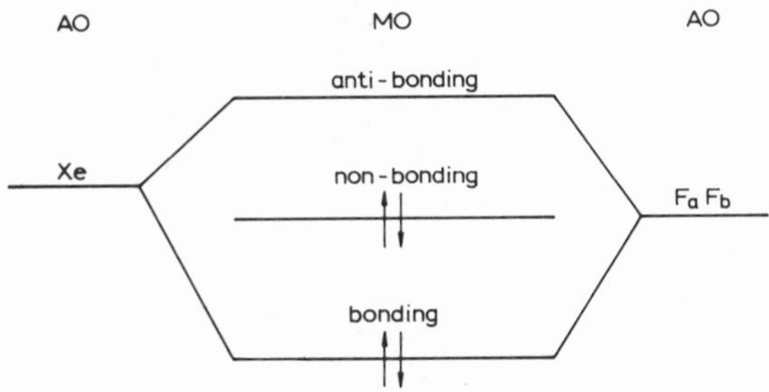

Fig. 12.3. Energies of molecular orbitals in XeF_2.

XeF_2, in common with the other xenon compounds, is an electron-excess compound, in contrast with the electron-deficient compounds such as the boranes (11.5).

The molecular orbital theory predicts that stability is enhanced by high electronegativity of the outer atoms and lower ionisation energy of the central atom, in agreement with the known stabilities of noble gas compounds. The theory also predicts that a compound like XeF_2 will have considerable ionic character, with a charge of about $+e$ on the xenon and about $-\frac{1}{2}\ e$ on each fluorine. If we accept that the bond order is 0.5 we can calculate that the Xe—F distance for a bond order of 1.0 should be 183 pm, which is about the distance between Xe and the terminal F in $XeSb_2F_{12}$:

F —184 pm— Xe —235 pm— F (147°) — Sb_2F_{10}

in which the other Xe—F distance is so long that the compound can be considered as very close to the ionic model $XeF^+(FSb_2F_{10})^-$.

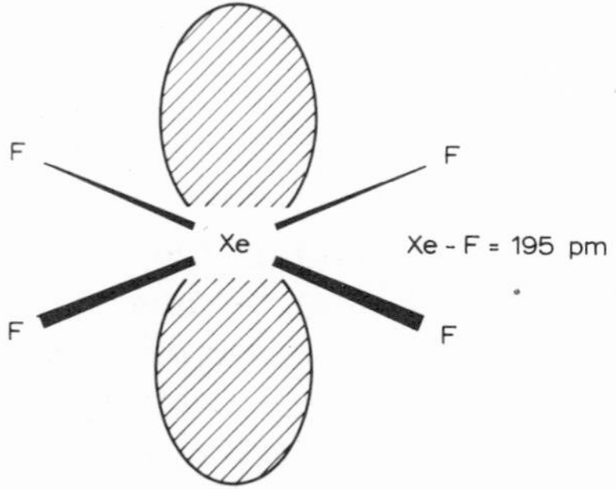

Fig. 12.4. Bonds and lone pairs in XeF_4 molecule.

Xenon tetrafluoride has a planar D_{4h} structure which is considered to be due to the use of two 5p atomic orbitals of the xenon, and to have two lone pairs above and below the xenon atom (Fig. 12.4).

The various methods which have been used to study the structure of XeF_6 have produced conflicting conclusions. Thus electron diffraction data show the molecular symmetry to be a considerable distortion of O_h, but the very small dipole moment suggests that the molecule is very symmetrical. Simple theory predicts a distorted molecule, because if the molecular orbitals are considered to be formed from the three σ5p atomic orbitals of the xenon and the σ5p atomic orbitals of the fluorine atoms, there remains a lone pair of electrons on the xenon. Four different crystalline forms of XeF_6 have been distinguished by X-ray analysis. The cubic form which is stable nearest to the *m.p.* contain F^- ions and XeF_5^+ ions which are isostructural and isoelectronic with the $XeOF_4$ molecule (Fig. 12.5). A possible reason for the conflicting evidence on the structure of XeF_6 in the vapour phase is that the molecule is fluxional (18.2.5). Infrared analysis shows that the force constant of the Xe—F bond is small, only 330 N m^{-1}, suggesting a bond order of little more than 0.5, consonant with considerable ionic character in the molecule.

$XeOF_4$ has been shown by infrared and Raman spectroscopy to have a square pyramidal C_{4v} structure (Fig. 12.5). The Xe and the F atoms are almost coplanar. Analysis of chemical shift data obtained from the Mössbauer spectrum indicates that the oxygen atom withdraws approximately twice as much electron density from the Xe as does any of the fluorine atoms.

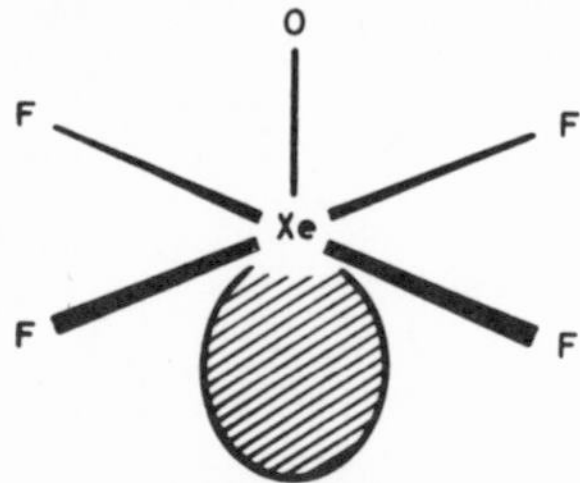

Fig. 12.5. Structure of $XeOF_4$ molecule; it has one σ lone pair.

12.6.2. Fluorocomplexes

The ability of the xenon fluorides to donate F^- ions might be expected to decrease from XeF_2 to XeF_6 because of the increasing effective nuclear charge of the Xe. However, although XeF_4 is a weak fluoride ion donor, XeF_6 is a rather stronger one than is XeF_2; this effect may be due to the crowding of occupied orbitals around the xenon atom in XeF_6. Ionic formulations have been suggested for some of the compounds of XeF_6 with Lewis acids, e.g. $(XeF_5)^+AsF_6^-$, mainly on the evidence of infrared spectra.

XeF_2 adducts with Lewis acids MF_5 (M = P, As, Sb, Pt, Ir, Os, Ru, Rh) have been studied conductometrically in BrF_3, and X-ray studies have also been made. There are three principal types of adduct, 1 : 1 adducts of general formula $XeMF_7$, 2 : 1 adducts Xe_2MF_9 and 1 : 2 adducts XeM_2F_{12}. They have a high degree of ionic character; ionic formulations $(XeF)^+MF_6^-$, $(Xe_2F_3)^+MF_6^-$ and $(XeF)^+M_2F_{11}^-$ have been suggested for some of them.

The $Xe_2F_3^+$ ion in $(Xe_2F_3)^+SbF_6^-$ has the structure:

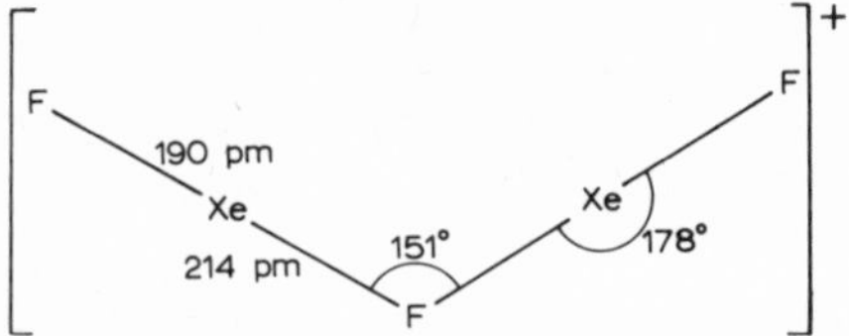

12.6.3. Oxides

Xenon trioxide

Xenon tetrafluoride does not react with dry silica, but it undergoes a succession of reactions with water of which the approximate results may be represented thus:

$$6\ XeF_4 + 12\ H_2O \rightarrow 4\ Xe + 2\ XeO_3 + 3\ O_2 + 24\ HF$$

The trioxide, a non-volatile, highly explosive solid, can be separated from this mixture. It can be made more directly by the slow hydrolysis of XeF_6:

$$XeF_6 + 3\ H_2O \rightleftharpoons XeO_3 + 6\ HF$$

Xenon trioxide, as X-ray analysis shows, has a trigonal pyramidal structure

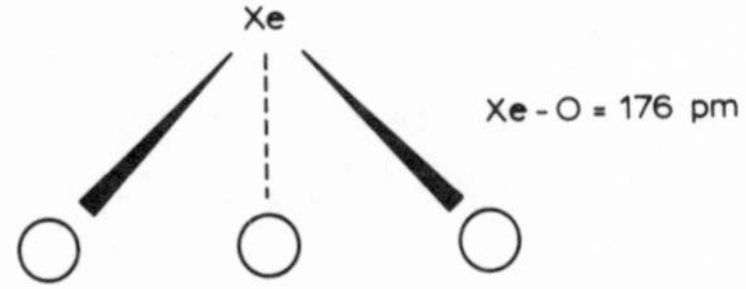

Fig. 12.6. Pyramidal structure of XeO_3 molecule.

shown in Fig. 12.6, similar to the isoelectronic IO_3^- ion, with a lone pair of electrons directly above the Xe atom.

An aqueous solution of XeO_3 is a very strong oxidising agent. It will oxidise Cl^- to Cl_2, Br^- to BrO_3^- and Mn^{2+} to MnO_4^- in strongly acidic conditions, and it converts alcohols and carboxylic acids to CO_2 and water. Alkalis cause disproportionation to Xe^{VIII} and Xe^0:

$$2\ XeO_3 + 4\ OH^- \rightarrow XeO_6^{4-} + Xe + O_2 + 2\ H_2O$$

The addition of $Ba(OH)_2$ to aqueous XeO_3 causes precipitation of barium perxenate, $Ba_2XeO_6 \cdot 1.5\ H_2O$. Perxenates such as $Na_4XeO_6 \cdot 6\ H_2O$ and $K_4XeO_6 \cdot 9\ H_2O$ can also be made; a satisfactory method is to pass ozone into solutions of XeO_3 and then neutralise with alkali. They are remarkably thermally stable and can be heated to 450 K without decomposition.

Xenon tetroxide

When Ba_2XeO_6 is added slowly to sulphuric acid at 267 K the thermally unstable gas XeO_4 is produced. Infrared measurements show the molecule to be tetrahedral.

12.7. Compounds of Radon and Krypton

Radon reacts spontaneously at room temperature with F_2 and also with all the halogen fluorides except IF_5. The product is RnF_2.

Krypton is much more difficult to oxidise. The thermally unstable KrF_2 has been made by a number of high-energy, low-temperature methods, for example by passing a silent discharge through a krypton/fluorine mixture at 90 K and 2 kPa and then cooling suddenly to 20 K. It is a colourless solid which can be stored in Kel-F containers at 230 K. Its molecule has $D_{\infty h}$ symmetry like XeF_2. It hydrolyses much more rapidly than XeF_2 in neutral and acidic solutions.

12.8. Compounds of Xenon with Halogens other than Fluorine

Very small amounts of $XeCl_2$ have been made by the action of electric discharges on Xe/Cl_2 mixtures followed by trapping of the product in inert matrices at 20 K. Evidence for the formation of $XeCl_2$, $XeBr_2$ and $XeCl_4$ has been obtained by an interesting application of Mössbauer spectroscopy. A typical experiment utilised a source containing $^{129}ICl_2^-$ and an absorber consisting of a xenon clathrate. The chemical shift which was observed indicated the existence of Xe—Cl bonds in the β-decay product of the $^{129}ICl_2^-$.

12.9. The Thermochemistry of Noble-Gas Compounds

Xenon difluoride is an exothermic compound; $\Delta H_f = -117$ kJ mol^{-1}. Although the first ionisation energy is so large, other energy factors, particularly the release of electrostatic energy when a xenon cation combines with a fluoride ion, are sufficient to compensate for it. The enthalpy change of the reaction

$$XeF^+ + F^- \rightarrow (XeF)^+F^-$$

calculated from the known dimensions of the ions, is -695 kJ mol^{-1}. Thus it is possible to calculate, from an energy cycle, the so-called resonance energy of the XeF_2 molecule, i.e. $-\Delta H$ for the conversion of the ion-pair $(XeF)^+F^-$ to the covalent XeF_2.

The first five quantities on p. 301 have been experimentally determined. If the resonance energy of KrF_2 is assumed to be the same as that of XeF_2, and the electrostatic energy of the $(KrF)^+F^-$ pair is calculated, an energy cycle which utilises experimental values for the ionisation energy of Kr and for the $Kr^+ + F \rightarrow KrF^+$ reaction enables ΔH_f to be calculated for KrF_2; the value is +25 kJ

			ΔH/kJ mol^{-1}
	Xe	$= Xe^+ + e$	+1170
	XeF_2	$= Xe + F_2$	+117
	F_2	$= 2\ F$	+155
	F + e	$= F^-$	−334
	$F + Xe^+$	$= XeF^+$	−196
	$XeF^+ + F^-$	$= (XeF)^+F^-$	−695
Adding:	XeF_2	$= (XeF)^+F^-$	+217

mol^{-1}. Although the energy of electrostatic interaction between KrF^+ and F^- is large, it is not sufficient to compensate for the very large ionisation energy of krypton. Similar calculations show that ArF_2 and $XeCl_2$ must be endothermic compounds which are unlikely to be prepared on anything but a small scale and under high-energy, low-temperature conditions.

Similar energy cycles have been used to explain why XeO_3 and XeO_4 can be prepared but XeO and XeO_2 have not yet been made. In the case of XeO_4, the energy needed to produce an Xe^{4+} ion, more than 10 MJ mol^{-1}, is very largely compensated for in the formation of an ion cluster $Xe^{4+}(O^-)_4$. The release of energy in the process $Xe^+ + O^- \rightarrow (Xe)^+O^-$ is however very much smaller than $I(1)$ for Xe.

12.10. Uses for Noble-Gas Compounds

It has been suggested that the danger from radon gas in uranium mines might be reduced by circulating the air over a halogen fluoride which would convert the radon to RnF_2. Xenon compounds are likely to find increasing application as oxidising agents; they are 'clean' in their action because the reduction product, xenon, is so easy to remove.

Further Reading

J.G. Malm, H. Selig, J. Jortner and S.A. Rice, The chemistry of xenon, Chem. Rev., 65 (1965) 199.

J.H. Holloway, Noble-gas chemistry, Methuen, London, 1968.

H. Selig, Fluoride chemistry of the noble gases, in Halogen chemistry, Vol. 1, Ed. V. Gutmann, Academic Press, London and New York, 1967.

J.H. Wolfenden, The noble gases and the periodic table, Telling it like it was, J. Chem. Ed., 46 (1969) 569.

G.A. Cook (Ed.), Argon, helium and the rare gases (2 vols.), Interscience, New York, 1961.

N. Bartlett, Noble-gas compounds, Endeavour, 31 (1972) 107.

R. Davies, Lasers in chemistry, Ed. Chem., 9 (1972) 92.

N.K. Jha, Recent advances in the chemistry of noble gas elements, RIC Reviews, 4 (1971) 147.

I.R. Beattie, Helium—neon laser Raman spectroscopy, Chem. Brit., 3 (1967) 347.
N. Bartlett, The chemistry of the noble gases, Elsevier, Amsterdam, 1971.
A.H. Cockett, K.C. Smith and N. Bartlett, The chemistry of monatomic gases, Pergamon, Oxford, 1975.

The Alkali Metals — Group IA

13

13.1. The Elements

The atom of an alkali metal contains one s electron outside a noble gas core (Table 13.1); the outermost electron is so heavily shielded from the attractive force of the nucleus that the ionisation energy is very low. The first ionisation energy of caesium is lower than that of any other element for which a measurement has been made. The second ionisation energies are high, about ten times greater than the first, thus the alkali metals form unipositive ions in chemical reactions and exhibit the oxidation state +1 in all their compounds.

Another consequence of the low effective nuclear charge acting on the outermost electrons is that the atoms are large. Furthermore the unipositive ions are also large and, with the exception of the smallest of them, Li^+, have very weak polarising power indeed. Thus the chemistry of the alkali metals is almost entirely the chemistry of their +1 cations.

The low densities of the metals are also clearly attributable to the large internuclear distances in the solids. These have body-centred cubic structures in which the outer electrons available for metallic bonding are sufficient to fill only half the vacancies in the conduction band. In consequence the bonding is weak, the metals are soft, their *m.p.* low, and they are exceptionally good conductors of electricity. Except for Li, the *b.p.* are also low compared

TABLE 13.1

ATOMIC PROPERTIES OF THE ALKALI METALS

	Li	Na	K	Rb	Cs	Fr
Z	3	11	19	37	55	87
Electron configuration	[He]$2s^1$	[Ne]$3s^1$	[Ar]$4s^1$	[Kr]$5s^1$	[Xe]$6s^1$	[Rn]$7s^1$
$I(1)$/kJ mol^{-1}	526	501	426	409	380	
r_{M^+}/pm	60	95	133	148	169	
Metallic radius/ pm	155	190	235	248	267	

TABLE 13.2
PHYSICAL PROPERTIES AND ELECTRODE POTENTIALS OF THE ALKALI METALS

	Li	Na	K	Rb	Cs
M.p./K	453.7	370.5	336.5	312.2	301.5
B.p./K	1620	1156	1030	961	978
ρ/g cm^{-3}	0.53	0.97	0.86	1.53	1.90
E^0, M^+/M/V	−3.02	−2.71	−2.92	−2.99	−3.02

with those of most other metals. The vapours contain small amounts of M_2 molecules; their dissociation energies are small, the largest, *D*(Li—Li) being only 114 kJ mol^{-1}. In these molecules the bonding is believed to be due largely to overlap of the outer s-orbitals of the atoms, as in H_2; the difference is that the atomic nuclei are heavily shielded, unlike those of hydrogen, and interaction between such a nucleus and a valence electron on an adjacent atom is therefore much less than in hydrogen.

13.1.1. Occurrence and properties

Sodium (2.63% of lithosphere) and potassium (2.40% of lithosphere) are high in abundance among the elements in the earth's crust; their amounts in sea water are respectively 1.14% and 0.04%. Sodium and its compounds are chiefly derived from sodium chloride; the principal source of potassium is carnallite, $KMgCl_3 \cdot 6\,H_2O$. The other elements of the family are much less common. Lithium (0.0065% of lithosphere), mainly in aluminosilicates such as petalite, $(Li,Na)AlSi_4O_{10}$, and spodumene, $LiAl(SiO_3)_2$, has an abundance less than rubidium (0.031% of lithosphere) and about ten times greater than caesium. Francium occurs naturally only in minute amounts and, of its 21 known isotopes — all radioactive — the longest lived is ^{223}Fr (half-life 22 min). Its progenitor is actinium-227, and the nuclide is itself a β-emitter:

$$^{227}_{89}\mathrm{Ac} \longrightarrow {}^{223}_{87}\mathrm{Fr} + {}^{4}_{2}\mathrm{He}$$

$$^{223}_{87}\mathrm{Fr} \xrightarrow{22\ \mathrm{min}} {}^{223}_{88}\mathrm{Ra} + \beta^-$$

It is co-precipitated with Rb and Cs perchlorates or chloroplatinates.

Both potassium and rubidium have natural active isotopes. Potassium-40 is a feeble β-emitter with a half-life of 1.3×10^9 years; it makes up about 0.012% of the natural element. Of the two natural isotopes of rubidium, one, ^{87}Rb, which accounts for nearly 28%, is a weak β-emitter, half-life 6×10^{10} years.

Separation of the isotopes of lithium

Lithium occurs as $^6_3$Li (7.3%) and $^7_3$Li (92.7%) with one isotope nearly 17% greater in mass than the other. The isotopes show a slight difference in chemical behaviour, and when lithium amalgam falls through a methanol solution of lithium chloride an equilibrium is established:

$$^{7}_{3}\text{Li (in amalgam)} + {}^{6}_{3}\text{Li}^{+}\text{ (in methanol)} \rightleftharpoons {}^{6}_{3}\text{Li (in amalgam)} + {}^{7}_{3}\text{Li}^{+}\text{ (in methanol)}$$

Lithium-6 has a slightly less negative electrode potential; and as a result the proportion of $^{6}_{3}Li$ in the amalgam increases.

Lithium-6 ions are preferentially adsorbed by a zeolite in its sodium form and some separation of $^{6}Li^{+}$ from $^{7}Li^{+}$ can be attained by means of very tall cation-exchange columns. The separation of lithium-6 is important because of its use in the preparation of tritium in nuclear reactors (10.7).

13.1.2. The metals

Lithium and sodium are both made by electrolysis, either of a fused LiCl/KCl mixture in the former case or of fused NaCl in the latter. However, K, Rb and Cs are made by reduction of their chlorides at high temperature and low pressure in reactions such as:

$$KCl + Na \rightarrow NaCl + K$$
$$2RbCl + Ca \rightarrow CaCl_2 + 2Rb$$

Thus a sodium—potassium alloy can be obtained by passing sodium vapour through molten KCl. The potassium can then be separated by fractional distillation of the alloy. Caesium can be made most conveniently by reducing the compound $CsAlO_2$ which is the product obtained when caesium alum is calcined:

$$2\,CsAlO_2 + Mg \rightarrow MgAl_2O_4 + 2\,Cs$$

The pure metals are silvery-white.

Reactivity with oxygen and water increases down the group. Lithium does not react readily with oxygen below 400 K, but Na and K combine slowly and Cs ignites even at room temperature. Lithium reacts quietly with cold water, sodium reacts rapidly but the hydrogen which is evolved is rarely ignited, potassium reacts more violently, and the heaviest congeners explosively. With H_2, Li reacts at 700 K to give LiH; the hydrides of the other members can be made at lower temperatures but they are less thermally stable. Lithium, however, is the most reactive of the metals towards nitrogen and carbon: it combines with N_2, even at room temperature, to give Li_3N, and with carbon above 1100 K to give Li_2C_2.

Behaviour with liquid ammonia

The metals dissolve in liquid ammonia to form metastable, coloured solutions; and the solubilities can be as high as 5 *M*. With Li, Na and K it is possible, above a certain concentration, to have two immiscible solutions in equilibrium, a heavier, blue, dilute metal phase and a lighter, bronze, concentrated metal phase. This is because dissolving the metal causes a considerable increase in liquid volume. The blue solutions have approximately the same absorption spectra, irrespective of the particular metal present.

The constitution of the solution changes with concentration:

(i) In extremely dilute solutions the metal atoms form M^+ ions surrounded by NH_3 molecules with their N atoms directed towards the metal. The electrons which are set free by the ionisation form centres in the liquid around which NH_3 molecules are arranged with hydrogens directed towards the electron. The presence of these electrons has been shown by paramagnetic resonance (3.4.4.4).

(ii) In moderately concentrated solutions ($M/20$ to M) ammoniated metal ions are bound by paired electrons into small clusters of two, three or four ions.

(iii) In more concentrated solutions ($>M$) ammoniated metal ions are bound together by unpaired electrons in much the same way as are the metal ions in a molten metal.

These systems are remarkably stable and the solutions liberate hydrogen only very slowly:

$$2\ M + 2\ NH_3 \rightarrow 2\ MNH_2 + H_2$$

But traces of transition metal ions catalyse the reaction. The solutions have strong reducing properties and are used in preparative chemistry.

Uses of the metals

Lithium is used for the production of its alkyls, which are important reagents in the synthesis of many organic chemicals. Up to 0.1% Li is added to Al, Mg and Zn alloys to make them harder. An alloy of Na and Pb is used in the manufacture of tetraethyllead which is used as an anti-knock in petrol and is one of the most commercially important organometallic compounds. Sodium is among the cheapest non-ferrous metals; it is increasingly used as a reducing agent in metal extraction, e.g.:

$$TiCl_4 + 4\ Na \rightarrow 4\ NaCl + Ti$$

Liquid sodium is circulated through the cores of some modern nuclear reactors such as the fast breeder reactor to transfer heat from them. Caesium alloys are used in photoelectric cells.

13.2. Halides

The alkali-metal halides, MX, form colourless, cubic crystals. Only CsCl, CsBr and CsI have the caesium chloride lattice (Fig. 7.12); the others all have the NaCl structure at ordinary temperatures. If the radius-ratio rule were obeyed we should expect KF, RbF, RbCl, and CsF to have 8:8 co-ordination and LiCl, LiBr and LiI to have 4:4 co-ordination. The reason for this marked deviation from the rule in this family of salts is not known. These halides are generally markedly ionic, though, as expected, LiI is least so, for I^- is the largest and most easily polarised halide ion, and Li^+, the smallest alkali-metal cation, possesses the strongest polarising power.

Gross non-stoichiometry does not occur in these compounds, but when alkali-metal halides are heated in the vapour of the metal they become highly coloured as a result of the incorporation of small numbers of metal atoms. These atoms appear to be converted to M^+ ions which occupy normal cation sites in the crystal lattice; the electrons so released occupy vacant anion sites and are known as F centres. Similar colorations are induced in the crystals by irradiation with X-rays or neutrons; in all cases the coloured crystals give perfectly colourless aqueous solutions.

Lithium halides, and to a lesser extent sodium halides, form stable solid complexes with some nitrogen ligands. For example, LiI reacts with NH_3 to give $Li(NH_3)_4I$, and NaCl can be precipitated from aqueous solution with *p,p'*-methylenedianiline (MDA) as $(MDA)_3NaCl$. The chloride, bromide and iodide of lithium are much more soluble in alcohol and ether than those of the other alkali metals; solvation of the small Li^+ ions is evidently stronger than that of the larger unipositive ions. Lithium fluoride is much less soluble in water than any of the other alkali-metal halides, in spite of the large hydration energy of Li^+; the high lattice energy of the salt is evidently responsible.

Polyhalides

Many polyhalides of alkali metals are known, the most important being the tri-iodides and the very stable mixed pentahalides of the type $MICl_4$. Lithium and sodium do not form tri-iodides. The potassium compound exists as a monohydrate stable below 298 K, but dehydration is accompanied by loss of iodine. The rubidium and caesium compounds are anhydrous and crystalline. The stronger positive fields round the lithium and sodium ions curtail the spread of negative charge from the iodide ion to an iodine molecule.

$$I^- + I_2 \rightarrow (I\text{—}I\text{—}I)^-$$

The tribromides $RbBr_3$ and $CsBr_3$ are also known, and mixed trihalides such as KIBrCl and CsIBrF have been made. In all of these compounds the triatomic anions are linear, in accordance with the simple electron-pair repulsion theory. These anions all carry a single negative charge, thus the metals in these compounds exhibit their normal oxidation state of +1.

13.3. Oxides

The oxides of the alkali metals fall into three classes, the normal oxides, M_2O, containing the O^{2-} ion, the peroxides, M_2O_2, containing the O_2^{2-} ion, and the superoxides, MO_2, containing the paramagnetic O_2^- ion. Direct combination of the metals with oxygen yields Li_2O from lithium, mainly Na_2O_2 from sodium and mainly the orange superoxides from K, Rb and Cs. The normal oxides of sodium and potassium can be made by heating the nitrates with the metal or its azide, e.g.:

$$2\,NaNO_3 + 10\,Na \rightarrow 6\,Na_2O + N_2$$
$$NaNO_3 + 5\,NaN_3 \rightarrow 3\,Na_2O + 8\,N_2$$

or by thermal decomposition of the higher oxides. Rb_2O and Cs_2O are best made by burning the metals in a limited supply of O_2 and then distilling off any excess of metal. All the normal oxides except Cs_2O have the antifluorite lattice (Fig. 7.19). The orange crystals of Cs_2O have the anti-$CdCl_2$ lattice (Fig. 26.1).

The peroxide of lithium is best made by precipitation with H_2O_2 from alcoholic LiOH:

$$2\,LiOH + 2\,H_2O_2 + H_2O \rightarrow Li_2O_2 \cdot H_2O_2 \cdot 3\,H_2O$$

The precipitate can be dried over P_2O_5 to give the white, anhydrous Li_2O_2, but the compound is decomposed by warming. Evidently the strong positive field around the Li^+ ion tends to polarise the O_2^{2-} ion, with the effect of splitting it rather easily into O^{2-} and an oxygen atom. The weaker fields around the larger cations of the heavier congeners permit O_2^{2-} and even O_2^- ions to exist in the same crystal with them. Pale-yellow Na_2O_2 is the main product of the direct combination of sodium and oxygen. The coloured compounds K_2O_2, Rb_2O_2 and Cs_2O_2 can be made by thermal decomposition of the superoxides; Cs_2O_2 for example is obtained from CsO_2 at 560—600 K.

A yellow solid containing up to 45% of LiO_2 has been made by passing ozone into a suspension of Li_2O_2 in Freon-12 at 200 K, but the compound reverts to Li_2O_2 at 240 K. The compounds KO_2, RbO_2 and CsO_2 have the tetragonal lattice of CaC_2 (Fig. 14.6). The compounds are paramagnetic because the O_2^- ion contains an unpaired electron.

Li_2O dissolves quietly in water to form aqueous LiOH. Lithium hydroxide monohydrate crystallises from the solution. Structurally every lithium ion is surrounded tetrahedrally by two hydroxide ions and two water molecules. Every tetrahedral group has an edge and two corners to produce double chains held laterally by hydrogen bonds (shown by broken lines) between the hydroxide ions and the water molecules. Every water molecule has four

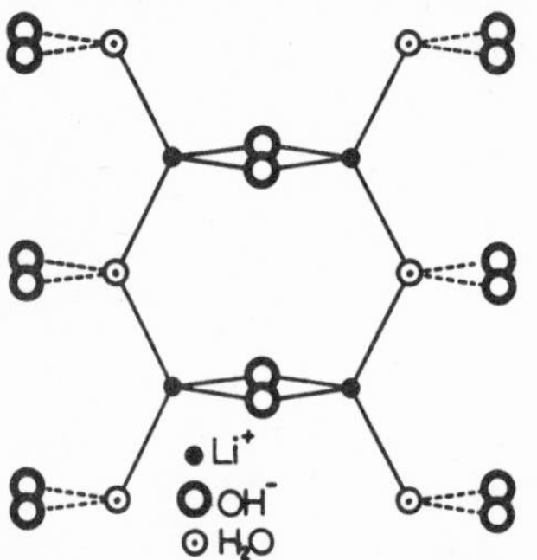

Fig. 13.1. Structure of $LiOH \cdot H_2O$.

near neighbours, two lithium ions of the same chain and two hydroxide ions, one from each of two other chains (Fig. 13.1).

The other oxides dissolve energetically in water; the peroxides and superoxides liberate oxygen in the process. Phase rule studies show that both NaOH and KOH form several hydrates.

13.4. Sulphides

All the alkali metals form hydrosulphides MHS, monosulphides M_2S, and a series of polysulphides M_2S_x, where x has values from 2 to 6.

The hydrosulphides are prepared by saturating an aqueous or aqueous alcoholic solution of the appropriate hydroxide with hydrogen sulphide, and may be crystallised as hydrates therefrom, e.g. $NaHS \cdot 3\,H_2O$ and $KHS \cdot \frac{1}{2}\,H_2O$. With the exception of LiSH which is thermally unstable, these may be dehydrated or even fused without extensive decomposition.

The monosulphide and polysulphides are formed by burning the metals in sulphur vapour, by the action of sulphur on the metals dissolved in liquid ammonia, and by the action of the molten metals on sulphur dissolved in toluene. Hydrates or alcoholates and, in some cases, the anhydrous compounds may be prepared by dissolving sulphur in hot solutions of the hydrosulphides or monosulphides. Potassium, rubidium and caesium give all the sulphides where $x = 1, 2, 3, 4, 5$, or 6; sodium only up to the pentasulphide, and lithium only those for which $x = 1$, 2 and 4. All the metals form two polysulphides of relatively outstanding stability; one is invariably the disulphide, and the other is the tetrasulphide in the case of lithium and sodium, and the pentasulphide in the case of potassium, rubidium or caesium. The amount of water of crystallisation and the solubility decrease with increase in atomic number of the metal, the gradation being most marked between sodium and potassium.

13.5. Oxoacid Salts

Lithium carbonate decomposes to the oxide when heated in a stream of hydrogen at 1070 K:

$$Li_2CO_3 \rightarrow Li_2O + CO_2$$

The strong polarising action of the small cation on the large complex anion, CO_3^{2-}, presumably assists the reaction. Lithium oxide has the antifluorite structure (Fig. 7.19), and the reaction proceeds because (i) the highly polarisable carbonate ion readily loses an oxygen ion and (ii) the gain in lattice energy resulting from the substitution of a smaller oxide ion for the larger carbonate ion enables the centres of charge to approach more closely.

Lithium carbonate is so sparingly soluble as to be precipitated from solutions of lithium salts by sodium carbonate. It dissolves in water containing carbon dioxide but there is no direct evidence that a solid bicarbonate can be formed. The bicarbonates of the other alkali metals may all be obtained in solid form. Those of rubidium and caesium are more soluble and thermally stable than the sodium and potassium salts.

The decahydrate $Na_2CO_3 \cdot 10\,H_2O$ has been shown to contain $Na_2(H_2O)_{10}^{2+}$ ions in which there is approximately octahedral co-ordination of the sodium ions by water molecules two of which are shared by the two metal ions.

Lithium nitrate decomposes on heating in a similar way to calcium nitrate to give the oxide:

$$4\,LiNO_3 \rightarrow 2\,Li_2O + 2\,N_2O_4 + O_2$$

The other alkali-metal nitrates yield the nitrites and oxygen. Only the lithium and sodium nitrates are deliquescent and hydrated, the rest are anhydrous.

Lithium orthophosphate, Li_3PO_4, is precipitated when sodium phosphate is added to an alkaline solution of a lithium salt. Phosphates of the other metals are soluble.

The sulphates of potassium, rubidium and caesium are isomorphous and anhydrous. The solubility of sodium sulphate depends upon the composition of the solid phase. It increases rapidly with temperature up to 305.54 K, and within this range crystallisation gives the decahydrate. Above that temperature the solubility falls slightly with temperature and the anhydrous salt separates as fine crystals. Lithium sulphate crystallises as the monohydrate, $Li_2SO_4 \cdot H_2O$. It is doubtful whether Li_2SO_4 forms an alum, although such compounds are formed by the elements from sodium to caesium (e.g. $KAl(SO_4)_2 \cdot 12\,H_2O$).

13.6. Nitrogen Compounds

All the alkali metals react when heated in ammonia to form white, crystalline amides, MNH_2, insoluble in organic solvents, and decomposed immediately by cold water:

$$NH_2^- + H_2O \rightarrow OH^- + NH_3$$

Only lithium forms an imide; it is made by heating the amide, $LiNH_2$, to near its *m.p.* under reduced pressure:

$$2\,LiNH_2 \rightarrow Li_2NH + NH_3$$

Dark-red lithium nitride, Li_3N, is made by heating the metal in nitrogen, but for nitrides of the other alkali metals indirect methods must be used. Na_3N can be made by dissolving sodium and sodium azide in liquid ammonia, allowing the NH_3 to evaporate, and then warming:

$$NaN_3 + 8\,Na \rightarrow 3\,Na_3N$$

The red solid decomposes at 423 K to the metal and N_2, and reacts with hydrogen at 390 K to give sodium hydride and ammonia:

$Na_3N + 3\ H_2 \rightarrow 3\ NaH + NH_3$

It is evident that lithium forms more stable nitrogen compounds than do the other Group I metals.

13.7. Organometallic Compounds

An organometallic compound is one in which there is direct bonding between carbon and a metal, or, in more general usage, between carbon and an element of lower electronegativity. The organometallic compounds of the alkali metals belong to three classes:

(i) Colourless alkyls and aryls, of largely covalent character.

(ii) Coloured ionic compounds such as $(C_6H_5)_3C^-Na^+$ and $C_6H_5CH_2^-Na^+$.

(iii) Colourless ionic compounds of hydrocarbons, like acetylene and cyclopentadiene, which have acidic character (18.2.1).

The alkyls and aryls can be made by adding the metal to the alkyl or aryl halides or to an organomercury in benzene or ether:

$$2M + RX \rightarrow RM + MX$$
$$2M + R_2Hg \rightarrow 2RM + Hg$$

The lithium compounds, except for CH_3Li, C_2H_5Li and C_6H_5Li which are solids, are liquid at room temperature. In solution in hydrocarbon C_2H_5Li is hexameric and t-BuLi is tetrameric, with Li and C atoms occupying the vertices of interpenetrating tetrahedra. In general it appears that unbranched alkyllithiums are hexamers in cyclohexane and benzene, but branching at the α- or β-carbon atom favours tetramer formation. Mass spectrometry shows the vapour of ethyllithium to contain mainly the hexamer and tetramer; the bonding is probably of the multicentre type.

An X-ray analysis of the structure of CH_3Li has shown that the solid contains $(CH_3Li)_4$ units as shown in Fig. 13.2. The lithium atoms are arranged in a regular tetrahedron, the carbon atoms in an interpenetrating tetrahedron which has its vertices opposite to the face-centres of the first. The C—Li distance is 228 pm. Every carbon atom can be considered to form a four-centre bond with three lithium atoms.

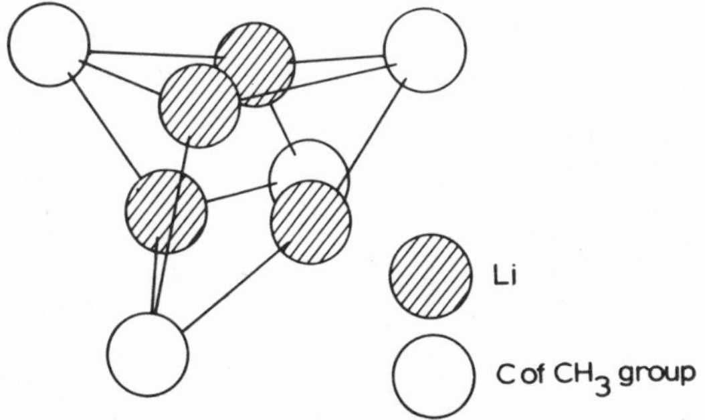

Fig. 13.2. Tetrameric methyllithium.

The alkyls and aryls of the other alkali metals are colourless, amorphous solids, insoluble in organic solvents. Like the lithium compounds, they are thermally unstable, strongly hydrolysed by water, and most of them are spontaneously inflammable in air. The compounds initiate polymerisation in styrene and similar substances.

Structure determinations on the alkyls indicate that ionic character in these compounds is greatest when the metal atom is large and the alkyl group is small. Thus CH_3K, made from Me_2Hg and a sodium—potassium alloy in pentane, has a structure of the NiAs type, whereas C_2H_5Na, made from sodium and Et_2Hg, can crystallise in a rhombohedral form which has the layer structure shown in Fig. 13.3.

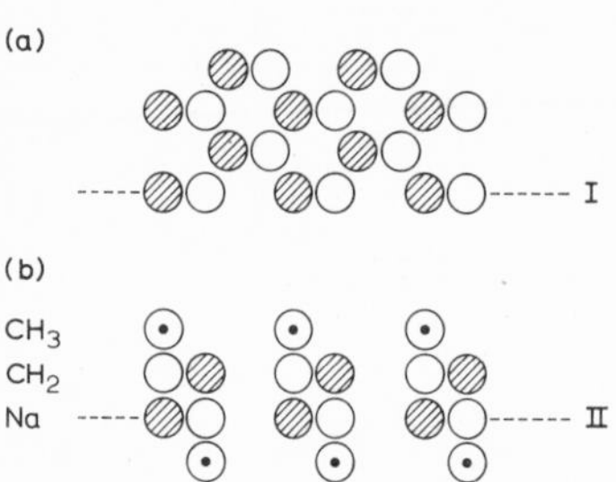

Fig. 13.3. Layer lattice of rhombohedral NaEt. (a) Plan of layer through II in (b). (b) Elevation of layer through I in (a).

The coloured ionic compounds such as $Ph_3C^-Na^+$ can be made by treating the halide with the metal:

$$Ph_3CCl + 2Na \rightarrow Ph_3C^-Na^+ + NaCl$$

They usually give conducting solutions when dissolved in ether. The anion is almost certainly stabilised by a delocalisation of the negative charge over the aromatic system.

The metals react with acidic hydrocarbons in solvents such as liquid ammonia or dimethylformamide:

$$2RC{\equiv}CH + 2Na \rightarrow 2RC{\equiv}C^-Na^+ + H_2$$
$$2C_5H_6 + 2Na \rightarrow 2C_5H_5^-Na^+ + H_2$$

The products are colourless solids.

13.8. Complex Compounds

Though the sizes and electronic configurations of the alkali-metal cations are not conducive to co-ordination with ligands, chelating groups do impose some acceptor properties on them. Thus when sodium hydroxide is added to salicylaldehyde the salt produced takes up a further molecule of salicylaldehyde to form a compound which is covalent.

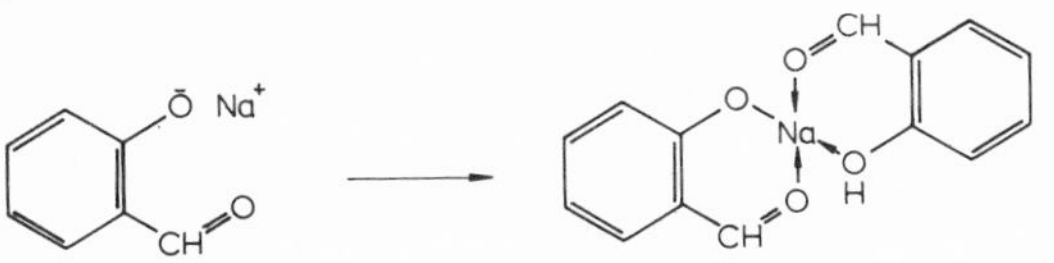

A similar result is sometimes achieved by hydration. The anhydrous sodium derivative of acetylacetone is salt-like but the dihydrate shows covalency, being soluble in non-polar solvents like toluene.

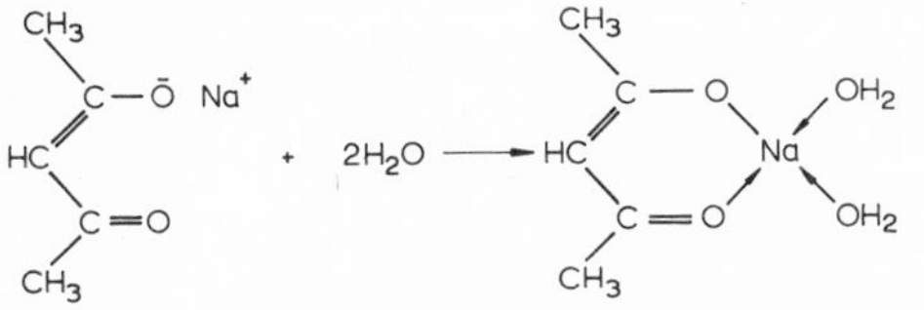

Complexing agents have recently been developed which are capable of making alkali-metal salts dissolve in non-polar solvents like chloroform; they act by surrounding the metal ion and effectively hiding it from the solvent.

The cyclic polyether known as dibenzo-18-crown-6 (Fig. 13.4) reacts with RbCNS to give a 1:1 complex which has been shown to have the structure shown in Fig. 13.5, in which the Rb atom is co-ordinated to six coplanar oxygen atoms encircled by the organic molecule. Not only salts of alkali metals but also most of the alkaline earth metals and some transition metals can give 1:1 complexes with macrocyclic polyethers.

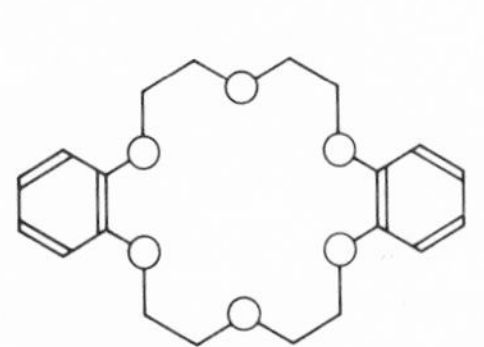

Fig. 13.4. Dibenzo-18-crown-6.

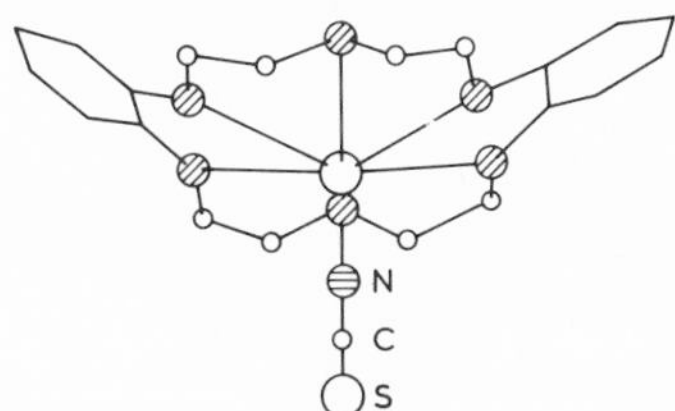

Fig. 13.5. 1:1 complex of RbCNS with dibenzo-18-crown-6.

Macrocyclic systems containing atoms other than oxygen are of even greater interest because they permit the production of ligands which will completely surround the cations rather than merely encircle them. The complexes which are formed are called cryptates (Greek *kryptus* = hidden) because the metal ion is effectively hidden from the solvent by the ligand which encloses it. Thus the compound $N(CH_2CH_2OCH_2CH_2OCH_2CH_2)_3N$, shown diagrammatically in Fig. 13.6, will enable salts of alkali metals and alkaline-earth metals to be extracted rapidly into chloroform as cryptates.

Rubidium thiocyanate reacts with the compound which is formulated above to produce a crystalline solid $[RbC_{18}H_{36}N_2O_6]SCN \cdot H_2O$. The rubidium ion in the cryptate has been shown to be surrounded by six O atoms at 289—291 pm, and two N atoms at 300 pm. These eight atoms form the

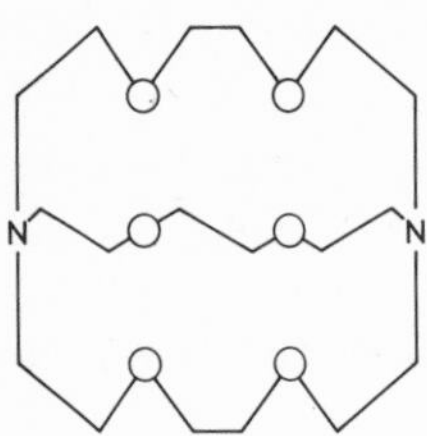

Fig. 13.6. Cryptate-forming ligand $N(CH_2CH_2OCH_2CH_2OCH_2CH_2)_3N$.

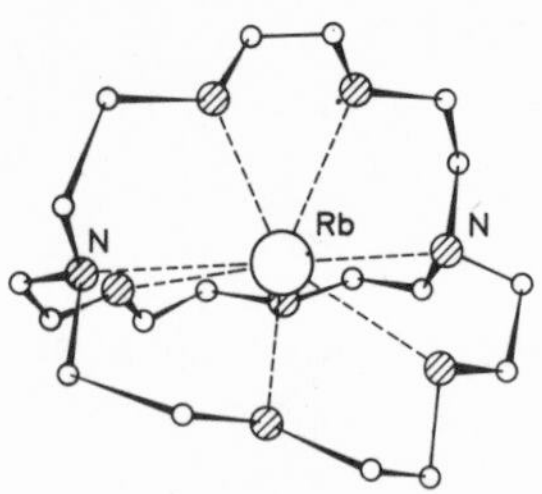

Fig. 13.7. Environment of the Rb^+ ion in the cryptate $[RbC_{18}H_{36}N_2O_6]SCN \cdot H_2O$.

corners of a slightly distorted bi-capped trigonal prism which encloses the rubidium (Fig. 13.7).

13.9. Similarities of Lithium and Magnesium

A feature of the second period of the Periodic Table is the similarity the chemistry of an individual element shows to that of the element a group higher in the third period: *diagonal similarities* as they are often termed:

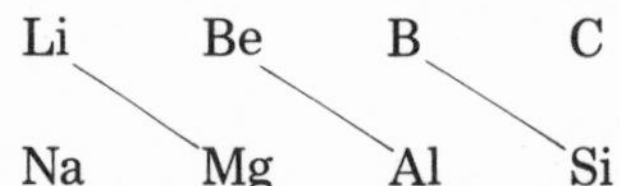

In general the polarising powers of cations increase across a period because of the increasing charge, but diminish down a group because of the increasing ionic radius. When both movements are made simultaneously, one effect, in part at least, compensates for the other and there may be no marked change in properties. In illustration of this the similarity between lithium and magnesium may be cited; this is exemplified by:

(i) the formation of normal oxides, not peroxides, when the metal is burnt;
(ii) the instability of the carbonates and nitrates;
(iii) the formation of the carbides and nitrides by direct combination of the elements;
(iv) the insolubility of the carbonates and phosphates;
(v) the strong hydration of the ions;
(vi) the solubility of salts such as chlorides in organic solvents;
(vii) the high solubility of most of the alkyls in organic solvents.

Further Reading

Chemical Society Special Publication No. 22, The alkali metals, London, 1967.
F.M. Perel'man, Rubidium and caesium, Pergamon, Oxford, 1965.

M.C.R. Symons, Nature of metal solutions, Quart. Rev., 13 (1959) 99.
M.C.R. Symons and W.T. Doyle, Colour centres in alkali halide crystals, Quart. Rev., 14 (1960) 62.
D.S. Laidler, Lithium and its compounds, Royal Institute of Chemistry, Monograph No. 6, London, 1957.
R.J.P. Williams, The biochemistry of sodium, potassium, magnesium and calcium, Quart. Rev., 24 (1970) 331.
M.R. Truter, Crystal chemistry of d^0 cations, Chem. Brit., 7 (1971) 203.
K.E. Chave, Chemical reactions and the composition of sea water, J. Chem. Ed., 48 (1971) 149.
W.A. Hart, O.F. Beumel, Jr. and T.P. Whaley, The chemistry of lithium, sodium, potassium, rubidium, caesium and francium, Pergamon, Oxford, 1975.
P.B. Chock and E.O. Titus, Alkali-metal ion transport and biochemical activity, Prog. Inorg. Chem., 18 (1973) 287.

Beryllium, Magnesium and the Alkaline Earth Metals — Group IIA

14

14.1. The Elements

In both their physical and chemical properties beryllium and magnesium stand apart from the later members of the group, which are known collectively as the alkaline earth metals. Unlike the latter, which corrode rapidly in moist air, Be and Mg can be used as major constituents of light alloys. Furthermore these two metals form many covalent compounds, whereas the compounds of the alkaline earths are essentially those of their bipositive ions.

Atomic properties of the elements are shown in Table 14.1. The effective nuclear charge on a 2s electron in Be is much greater than in Li, consequently the atom is small and the first ionisation energy is large. The extreme smallness of the Be^{2+} ion makes it so strongly polarising that Be^{II} compounds are almost entirely covalent. Furthermore, provided electron-rich ligands are available, the number of covalent bonds can be increased to four by donation and the use of sp^3 hybrid orbitals. In this respect Be is very different from the other members of the group. For the elements Ca to Ra, however, the first two ionisation energies are small and, though the M^{2+} ions are smaller than the corresponding M^+ ions in Group IA, they have low polarising power, and the compounds are predominantly ionic. The Mg^{2+} ion is inter-

TABLE 14.1

ATOMIC PROPERTIES OF THE ELEMENTS

	Be	Mg	Ca	Sr	Ba	Ra
Z	4	12	20	38	56	88
Electron configuration	$[He]2s^2$	$[Ne]3s^2$	$[Ar]4s^2$	$[Kr]5s^2$	$[Xe]6s^2$	$[Rn]7s^2$
$I(1)$/kJ mol^{-1}	898	736	589	548	503	508
$I(2)$/kJ mol^{-1}	1762	1449	1144	1060	960	975
Ionic radius of M^{2+}/pm	31	65	99	113	135	148
Metallic radius/pm	112	160	197	216	222	

TABLE 14.2

PHYSICAL PROPERTIES AND ELECTRODE POTENTIALS OF THE ELEMENTS

	Be	Mg	Ca	Sr	Ba	Ra
Density/g cm^{-3}	1.85	1.74	1.54	2.54	3.6	5
m.p./K	1550	922	1112	1041	1000	970
b.p./K	3240	1378	1767	1654	2122	1973
E^0, M^{2+}/M/V	−1.85	−2.37	−2.87	−2.89	−2.90	

mediate in properties between Be^{2+} and Ca^{2+}, and magnesium in its +2 state forms compounds ranging from the covalent dialkyls to the ionic MgO. The oxidation state +2 is the only one of importance in the compounds of Group IIA metals; compounds of the unipositive elements are highly unstable to disproportionation.

Because their atoms are smaller, the metals are denser, and have higher *m.p.* and *b.p.* (Table 14.2) than the elements of Group IA. The E^0 values for the M^{2+}/M couples show the metals to be almost as strongly reducing as the alkali metals, but Be and Mg do not corrode easily in moist air because the oxide film adheres strongly to the surface and protects the metal beneath.

The Group IIA elements vary in structure; solid Be and Mg have h.c.p. lattices and Ba has a b.c.c. structure, whereas Ca and Sr have c.c.p. structures at ordinary temperature which change to h.c.p. at about 500 K and to b.c.c. near their respective *m.p.* The lower members of the group particularly are much harder than the alkali metals.

The metals combine directly, under appropriate temperature conditions, with oxygen, sulphur, nitrogen and the halogens. Ca, Sr and Ba also react with hydrogen on heating at normal pressure, but Mg does so only at 20 MPa, and BeH_2 cannot be made directly. Ca, Sr and Ba release H_2 from cold water but Mg does so only when amalgamated to prevent the formation of a protective oxide film. Mg burns in steam but Be does not react even at red heat. The metals generally dissolve in dilute acids but Be is rendered passive by HNO_3. Unlike the others Be is soluble in alkali-metal hydroxides, resembling Al in this respect. Ca, Sr and Ba dissolve sparingly in liquid NH_3 to give coloured solutions which, however, release hydrogen more readily than those of the alkali metals.

Beryllium (6×10^{-4} % of lithosphere) has its only commercial source in beryl, $Be_3Al_2Si_6O_{18}$. This, when fused and quenched in water, becomes soluble in concentrated H_2SO_4, giving a solution containing the sulphates of beryllium, aluminium and the alkali metals. The addition of $(NH_4)_2SO_4$ allows the aluminium to be removed as the sparingly soluble ammonium alum. The $BeSO_4$ is then crystallised from the solution and converted to BeO (Fig. 14.1).

Alternatively, compacts of beryl powder and Na_2SiF_6 are heated at 1000 K. From the sinter, sodium fluoroberyllate, Na_2BeF_4, is extracted with water, most of the Al and Si remaining in insoluble compounds. Crude

Beryl
fused, quenched and
dissolved in H_2SO_4
$BeSO_4 + Al_2(SO_4)_3$
$(NH_4)_2SO_4$
$NH_4Al(SO_4)_2 \cdot 12H_2O$
(sparingly soluble)
BeO
heat
$Be(OH)_2$
(insoluble)
NH_4OH
$BeSO_4$

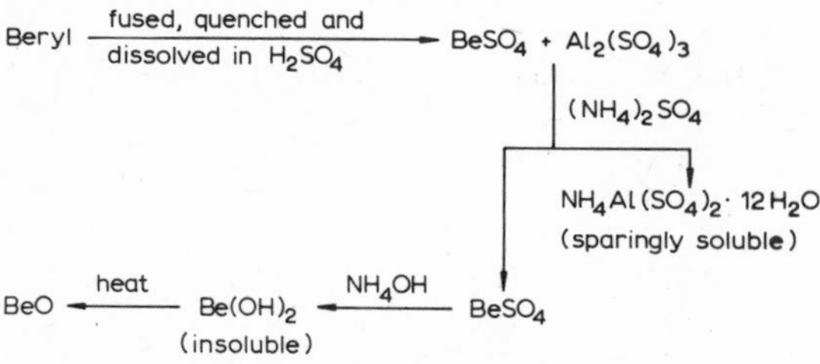

Fig. 14.1. BeO from beryl by sulphate route.

$Be(OH)_2$ is precipitated from the solution by raising the pH and purified by converting it to Na_2BeO_2 with NaOH, filtering and reprecipitating the $Be(OH)_2$ by hydrolysis.

Other beryllium compounds and the metal are obtainable from the hydroxide (Fig. 14.2).

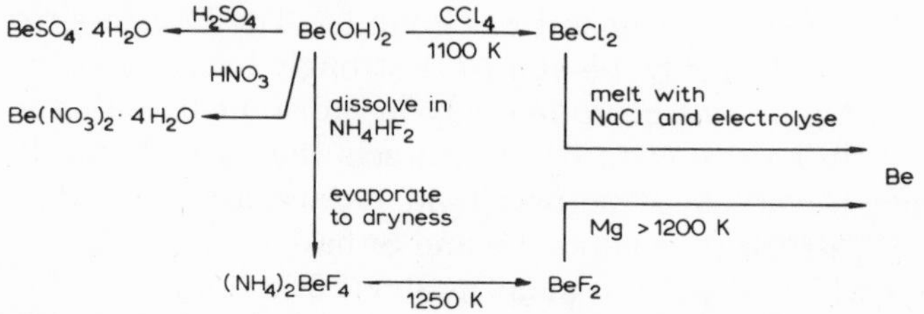

$BeSO_4 \cdot 4H_2O$
H_2SO_4
$Be(OH)_2$
CCl_4
1100 K
$BeCl_2$
HNO_3
$Be(NO_3)_2 \cdot 4H_2O$
dissolve in
NH_4HF_2
melt with
NaCl and electrolyse
Be
evaporate
to dryness
Mg >1200 K
$(NH_4)_2BeF_4$
1250 K
BeF_2

Fig. 14.2. Formation of beryllium and some of its compounds from $Be(OH)_2$.

The metal is purified by melting in a vacuum and is cast under argon. It has a high tensile strength, but is brittle even in the purest form yet obtained. Its mechanical properties are greatly improved by sintering it as a compressed powder (1420 K). Its transparency to X-rays makes Be a useful window material, and its high *m.p.* and low neutron cross-section allow its use in the nuclear energy industry, especially as it remains protected by an oxide film when heated in air up to at least 870 K. Increasingly, it is being used in alloys; beryllium copper, containing 2—2.25% Be and 0.25—0.5% Ni, is particularly hard and has a high elasticity. BeO has ceramic properties of possible use in atomic reactors.

Magnesium (1.93% of earth's crust) occurs in magnesite, $MgCO_3$, dolomite, $MgCa(CO_3)_2$, kieserite, $MgSO_4 \cdot H_2O$, carnallite, $KMgCl_3 \cdot 6\,H_2O$, and many silicates. Dolomite is not isomorphous with $CaCO_3$ or $MgCO_3$; its composition is always very close to $MgCa(CO_3)_2$ and it appears to be a distinct compound. The chlorophyll of green plants is a compound containing magnesium. Magnesium chloride is obtained from carnallite, from which it separates as a solid phase when the mineral is fused.

Magnesium is now made by the reduction of calcined dolomite with ferrosilicon in retorts of nickel—chromium steel, at 1430 K and 100 Pa. The reaction, effectively

$$2\,MgO \cdot CaO + Si \rightarrow 2\,Mg + Ca_2SiO_4$$

is of considerable thermodynamic interest. The chemical potentials of the products are kept low

(a) by working at extremely low pressure so that the partial pressure of magnesium vapour is small;

(b) by using calcined dolomite instead of MgO as starting material, which gives calcium silicate as product instead of silica.

A favourable free energy of reaction is thus achieved at the comparatively low temperature of 1430 K.

The magnesium vapour is condensed in steel-tube condensers. A much purer metal (99.98% Mg) is produced in this way than by electrolysis.

The metal is used in light alloys, particularly with aluminium, but also with Zn, Mn, Sn, Zr and Ce. Mg is a good reducing agent; turnings are heated with UF_4 in graphite-lined, closed, steel reactors to produce billets of uranium metal.

Calcium (3.4% of lithosphere) occurs as carbonate in the minerals aragonite and calcite and the rocks limestone, chalk and marble, and as sulphate in anhydrite, $CaSO_4$, and gypsum, $CaSO_4 \cdot 2\,H_2O$. All are plentiful and widely distributed. The metal is made either by electrolysing the fused chloride, $CaCl_2$, a by-product of the Solvay process, or by reducing calcined limestone with aluminium at 1470 K and 100 Pa in a retort of nickel—chromium steel. The reaction is effectively

$$6\,CaO + 2\,Al \rightarrow 3\,Ca + Ca_3Al_2O_6$$

The calcium which distils over is about 99% pure, the principal impurity being magnesium.

The high free energies of formation of CaO and CaF_2 make calcium a powerful reducing agent. It has been used in the preparation of a number of metals, examples being V, Zr, Th, U, Y and the lanthanides. It is also used to remove nitrogen from argon.

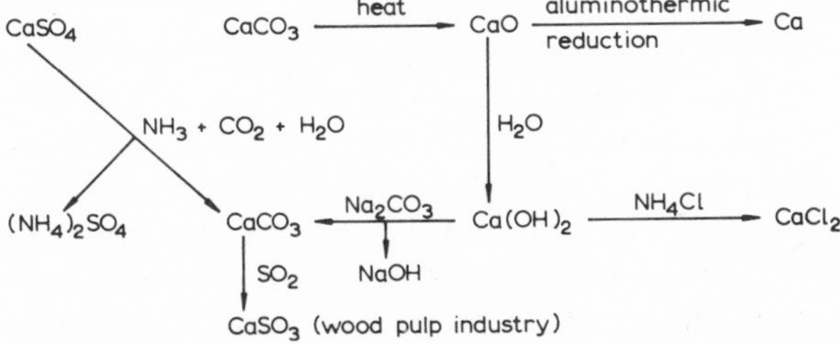

Fig. 14.3. Calcium **compounds** from anhydrite and limestone.

Strontium (0.02% of lithosphere) occurs as strontianite, $SrCO_3$, and celestite, $SrSO_4$. The metal is without economic importance except in pyrotechnics. The radioactive strontium-90 is long-lived and, being easily assimilated and incorporated with calcium in bone, is a dangerous product of uranium fission.

Barium (0.04% of lithosphere) occurs widely in veins as barytes, $BaSO_4$, which is sometimes converted to witherite, $BaCO_3$, by the atmosphere. The

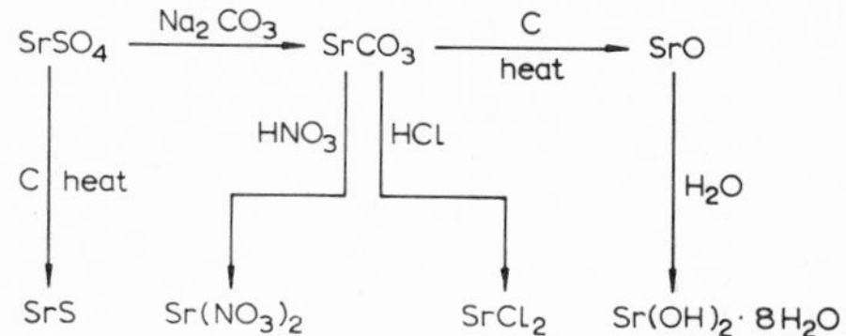

Fig. 14.4. The chemistry of some strontium compounds.

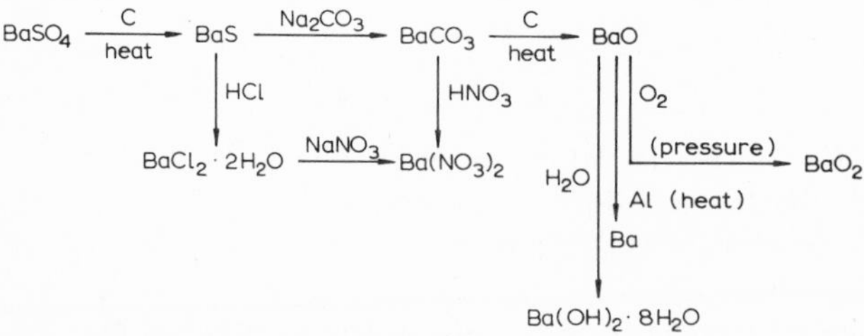

Fig. 14.5. Manufacture of barium and some of its compounds from barytes.

reactions used in the production of barium compounds from sulphate and carbonate are shown in Fig. 14.5. $BaCO_3$ is used to remove sulphate ions from solutions fed to alkali—chlorine cells. Precipitated $BaSO_4$ is used as a filler in paper and as a white pigment.

Radium is obtained from pitchblende, U_3O_8, in which it is formed by the disintegration of ^{238}U, the equilibrium ratio being 3.4×10^{-7}. The sulphate is co-precipitated with $BaSO_4$ when $BaCl_2$ is added to a sulphuric acid extract of the ore. After boiling with NaOH to remove lead, the sulphates are converted to carbonates by sodium carbonate fusion and dissolved in HCl to the chlorides. Fractional crystallisation of these removes much of the barium; the final separation of the rest is effected by the same means after conversion of the chlorides to bromides.

14.2. Halides

Unlike the Group IA halides, some of these are only feebly ionic, particularly those of beryllium. BeF_2 appears to be one of very few metal fluorides which are not completely ionised in solution. The solid often occurs in a glassy modification of random structure rather than as crystals. It is hygroscopic and very soluble in water, the very great solvation energy of Be^{2+} outweighing the effect of a high lattice energy. The complex tetrafluoroberyllate ion, BeF_4^{2-}, is stable in solution as well as in crystals; its compounds resemble sulphates in structure and solubility. Several compounds are known of the type $M^I_2M^{II}(BeF_4)_2 \cdot 6\,H_2O$, analogous with schönite, $K_2Mg(SO_4)_2 \cdot 6\,H_2O$.

The fluorides of magnesium and the alkaline earth metals are sparingly soluble; MgF_2 and KF form K_2MgF_4, but the MgF_4^{2-} ion does not exist in solution. Magnesium forms many double salts but few true complexes with halide ions.

Beryllium chloride exists as a polymeric solid containing chains of the form:

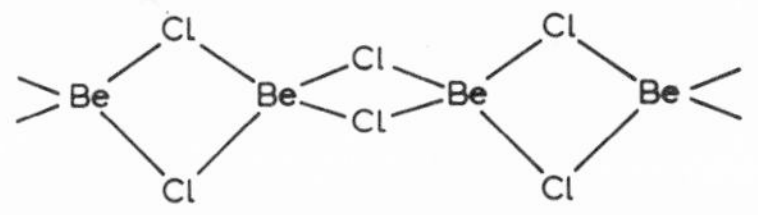

Co-ordination about the beryllium is irregular tetrahedral, the adjacent $Be{<}^{Cl}_{Cl}{>}Be$ planes being at 90° to each other but the angle ClBeCl being 98° instead of 109.5° of a regular tetrahedron. In the vapour (*b.p.* 761 K) beryllium chloride is without dipole moment, its molecule being linear, corresponding to Be in the sp valence state. In the fused state (at 678 K), $BeCl_2$ is one of the few halides with an electrical conductance intermediate between that of a characteristically ionic halide (NaCl) and a covalent one (CCl_4). The compound is soluble in organic solvents and, like $AlCl_3$, catalyses the Friedel—Crafts reaction; it thus behaves as a Lewis acid. Aldehydes, ketones and ethers co-ordinate readily to anhydrous $BeCl_2$, $BeBr_2$ and BeI_2:

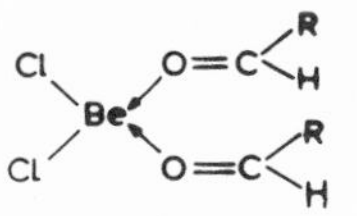

The complexes of BeX_2 (X = Cl, Br or I) with bipyridyl and *o*-phenanthroline:

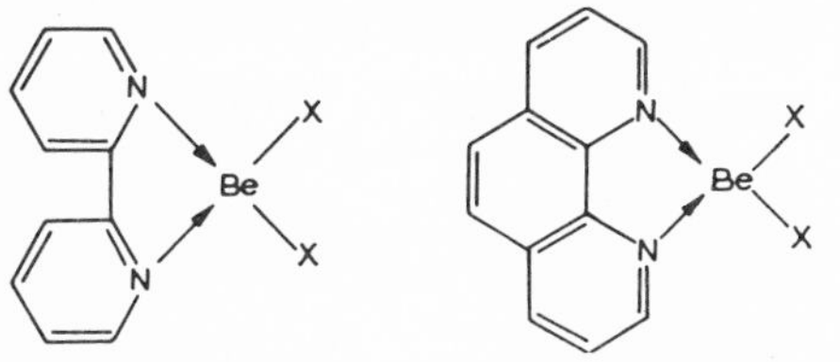

owe their colour to charge transfer (6.6.3). These complexes involve the use of sp^3 hybrid orbitals giving a tetrahedral disposition of the four bonds about the Be atom.

Beryllium chloride forms many double salts such as Na_2BeCl_4, but the $BeCl_4^{2-}$ ion does not exist in solution as does BeF_4^{2-}. The chloride has a hydrate $BeCl_2 \cdot 4\,H_2O$, an unusually stable tetra-ammine $Be(NH_3)_4Cl_2$ and a chelate ethylenediamine complex, $Be(en)_2Cl_2$. The formation of 4-co-ordinate complexes by Be^{2+} is very common; nearly all its inorganic salts have tetrahydrates containing the ion $[Be(H_2O)_4]^{2+}$. The water of hydration is tightly held; $Be(H_2O)_4Cl_2$, for instance, loses very little H_2O even when kept for several months in a vessel containing P_2O_5. The ionic mobility of the aquated beryllium ion in water is low; the solutions have high viscosities and abnormal osmotic pressures. The large ions of the group, such as Ba^{2+}, do not form co-ordinate links with water molecules, the solvation forces being of ion—dipole type (4.1.8), and the hydration numbers in aqueous solution are smaller.

Magnesium chloride forms complexes with ethers, aldehydes and ketones similar to those of $BeCl_2$. It has several hydrates.

$MgBr_2$ and $MgCl_2$ are soluble in many organic solvents and the halides (other than the fluorides) of even strontium and barium are moderately soluble in alcohol, resembling lithium in this respect.

Of the halides of Ca, Sr and Ba, the fluorides and also $SrCl_2$ have the fluorite lattice, but $CaCl_2$ and $CaBr_2$ have slightly deformed versions of the rutile structure, CaI_2 has the layer lattice of cadmium iodide, and $SrBr_2$, $BaBr_2$ and BaI_2 are more complicated; in general there is an increase in covalent character with the ratio $r_{X^-}/r_{M^{2+}}$.

The halides of calcium, strontium and barium react with their hydrides at 1170 K in hydrogen to give mica-like solids which have the PbClF structure.

$CaCl_2 + CaH_2 \rightarrow 2\ CaHCl$

14.3. Oxides

All the normal oxides have the sodium chloride structure (6 : 6 co-ordination), except BeO which has a wurtzite lattice (4 : 4 co-ordination). The compounds are strongly exothermic with ΔH_f from −550 to −700 kJ mol^{-1}. BeO is unreactive to water; even the large enthalpy of hydration of Be^{2+} is evidently insufficient to overcome the stabilising effect of the high lattice energy. MgO reacts with water only when prepared at a low temperature, but all the alkaline earth oxides slake readily:

$CaO + H_2O \rightarrow Ca(OH)_2$

The hydroxides increase in solubility with molecular weight. The monoxides, SrO and BaO are converted to peroxides when heated with oxygen under pressure:

$2\ BaO + O_2 \rightleftharpoons 2\ BaO_2$

The stability of the O_2^{2-} ion is evidently due to the low polarising power of these cations compared with those of earlier elements in the group. Peroxide hydrates of the form $MO_2 \cdot 8\ H_2O$ are formed by Ca, Sr and Ba when H_2O_2 is added to cold saturated solutions of the hydroxides; dehydration is possible without decomposition. All three anhydrous peroxides contain O_2^{2-} ions in the tetragonal calcium carbide lattice.

14.4. Compounds with Nitrogen

The metals combine with nitrogen on heating to give solid nitrides, M_3N_2. Reaction occurs at 1170 K in the case of beryllium but at much lower temperatures for the other metals. Be_3N_2 is hydrolysed slowly by water but quite quickly by aqueous acids and alkalis. The other nitrides react rapidly, however:

$$Mg_3N_2 + 6\ H_2O \rightarrow 3\ Mg(OH)_2 + 2\ NH_3$$

Magnesium and the alkaline earth metals dissolve in liquid ammonia. The solutions of calcium have been shown to have molar extinction coefficients (6.4) twice those of solutions of any Group I metal in ammonia, probably because the calcium solutions contain twice as great a concentration of solvated electrons:

$$Na \xrightarrow{NH_3} Na(NH_3)_x^{+} + e(NH_3)_y$$

$$Ca \xrightarrow{NH_3} Ca(NH_3)_n^{2+} + 2e(NH_3)_m$$

Amides can be obtained by evaporating solutions of the metals in ammonia, but $Mg(NH_2)_2$ is best made by passing NH_3 into a solution of $MgEt_2$ in ether. Magnesium amide decomposes at red heat to the nitride:

$$3Mg(NH_2)_2 \rightarrow Mg_3N_2 + 4NH_3$$

but calcium amide resembles lithium amide in giving an imide as the first product:

$$Ca(NH_2)_2 \rightarrow CaNH + NH_3$$

14.5. Sulphides

Beryllium burns in sulphur vapour forming BeS. This cannot be made in the wet way, though, like BeO, it is unreactive to water. Magnesium sulphide, similarly made from the metal and sulphur, hydrolyses rapidly:

$$2MgS + 2H_2O \rightarrow Mg(OH)_2 + Mg(SH)_2$$

The hydrosulphide is soluble and the reversible reaction

$$Mg(SH)_2 + 2H_2O \underset{\text{low temperature}}{\overset{\text{high temperature}}{\rightleftharpoons}} Mg(OH)_2 + 2H_2S$$

provides a means of purifying hydrogen sulphide. The sulphides of Ca, Sr and Ba are made by reducing the sulphates with carbon. They hydrolyse in the same way as MgS.

The structures of the sulphides, selenides and tellurides of the Group IIA metals illustrate the effect of the cation/anion size ratio on co-ordination number (Table 14.3).

Polysulphides. Studies on barium sulphide—sulphur mixtures indicate the existence of BaS_2 and BaS_3, both unstable at their respective *m.p.* When sulphur is added to boiling aqueous $Ba(SH)_2$ it dissolves, and orange-red crystals of $BaS_4 \cdot H_2O$ together with a little yellow $BaS_3 \cdot 3\ H_2O$ separate on cooling.

TABLE 14.3

STRUCTURES AND CO-ORDINATION OF SULPHIDES, SELENIDES AND TELLURIDES OF GROUP IIA ELEMENTS

	Be	Mg	Ca	Sr	Ba
Sulphide	B	R	R	R	R
Selenide	B & W	R	R	R	R
Telluride	W	B	R	R	R
Structures:	B = zinc blende		W = wurtzite		R = rocksalt
Co-ordination:	4:4		4:4		6:6

No individual polysulphide of the other members of the group has been made.

14.6. Carbides

CaO, SrO and BaO react with carbon in an electric furnace to give ionic acetylides MC_2. Magnesium metal reacts with carbon at 770 K to give MgC_2 which is also an acetylide, but at higher temperatures, with an excess of carbon, a compound Mg_2C_3 is formed which releases propyne on hydrolysis and is therefore thought to contain C_3^{4-} ions:

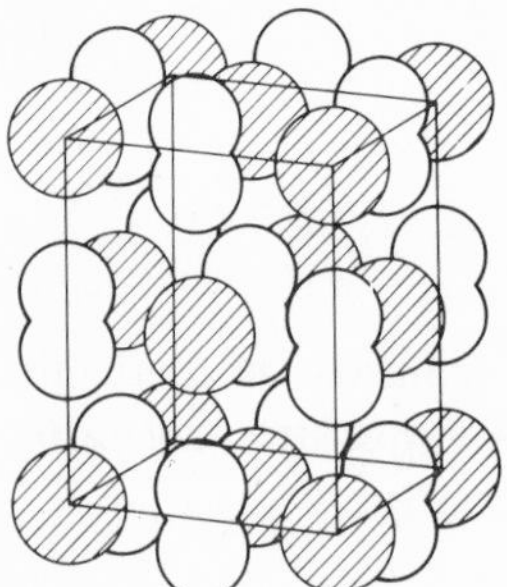

Fig. 14.6. Arrangement of C_2^{2-} ions and Ca^{2+} ions in CaC_2.

$$C_3^{4-} + 4\ H_2O \rightarrow CH_3C{\equiv}CH + 4\ OH^-$$

14.7. Oxoacid Salts

The carbonates exhibit the increase in stability with molecular weight expected to accompany increase in cationic size. Addition of a soluble carbonate to a solution of a beryllium salt gives hydroxide carbonates of indeterminate composition. Magnesium salts, similarly treated, yield the carbonates $Mg(OH)_2(MgCO_3)_4 \cdot x\ H_2O$, where $x = 4$ or 5 depending on the temperature of precipitation. Passage of CO_2 into the suspension gives a solution, probably not the hydrogen carbonate; this deposits $MgCO_3 \cdot 3\ H_2O$ on warming to 320 K. The same rhombic hydrate is obtained from solutions of $MgSO_4$ and sodium hydrogen carbonate. Sodium carbonate precipitates the anhydrous carbonates of Ca, Sr and Ba from solution. Their thermal stability is indicated by their dissociation pressures at 1070 K: $CaCO_3$ 22.2 kPa, $SrCO_3$ 93 Pa, $BaCO_3$ 4 Pa. A figure for $MgCO_3$ is not available but its thermal stability is certainly less than that of $CaCO_3$. The trend is similar to that in the Group IA carbonates, and for similar reasons (13.5).

Carbonates of bipositive cations crystallise in hexagonal form when their radii lie between 78 and 100 pm, and in rhombic form when between 100 and 143 pm. The radius of Ca^{2+} being 99 pm, both a hexagonal (calcite) and a rhombic (aragonite) form occur.

Beryllium chloride dissolved in an ethyl acetate—dinitrogen tetroxide mixture gives straw-coloured crystals of $Be(NO_3)_2 \cdot 2\ N_2O_4$. These decompose on heating in a vacuum, first to the involatile anhydrous nitrate, $Be(NO_3)_2$, then to the volatile oxide nitrate $Be_4O(NO_3)_6$. The structure of this closely resembles that of the well known oxide acetate, $Be_4O(CH_3CO_2)_6$, in which an oxygen atom is at the centre of a tetrahedron of beryllium atoms, and the acetate groups form six chelate rings along the edges of the tetrahedron, of which only one is shown in the partial formula:

In $Be_4O(NO_3)_6$, the nitrate groups form the bridges between the beryllium atoms (Fig. 14.7):

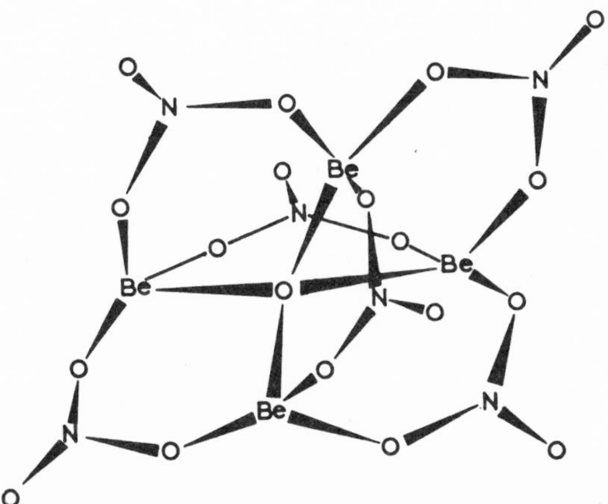

Fig. 14.7. Beryllium oxide nitrate (basic beryllium nitrate).

Like the oxide acetate and similar compounds, the oxide nitrate is volatile; however, it differs from them in being insoluble in non-polar liquids such as CCl_4 and C_6H_6.

The hydration and solubilities of the sulphates are of interest. In the $BeSO_4$—H_2SO_4—H_2O system, $BeSO_4 \cdot 4\ H_2O$ is the stable crystalline form below 362.1 K and $BeSO_4 \cdot 2\ H_2O$ above; there is no evidence of a monohydrate. $MgSO_4$ has many hydrates. $BeSO_4$ and $MgSO_4$ are freely soluble in water. $CaSO_4$ has a low but appreciable solubility. Its dihydrate, gypsum, is readily dehydrated:

$$CaSO_4 \cdot 2\ H_2O \xrightarrow{390\text{—}400\ K} CaSO_4 \cdot \tfrac{1}{2}\ H_2O \xrightarrow{410\text{—}470\ K} CaSO_4$$

The easy cleavage of gypsum crystals arises from its layer lattice structure; the Ca^{2+} and $SO_4{}^{2-}$ together make up the individual layers which are linked by hydrogen bonds between the water molecules and oxygen atoms of the sulphate ions. Every water molecule has as its nearest neighbours one Ca^{2+} and two oxygen atoms. By contrast strontium and barium sulphates are anhydrous and very sparingly soluble: $CaSO_4$ 2×10^{-3}, $SrSO_4$ 1×10^{-4}, $BaSO_4$ 2.4×10^{-6} at 288 K (all expressed as weight of solute/weight of solvent).

14.8. Organometallic Compounds

Beryllium

Beryllium alkyls are liquids or low-melting solids made by treating $BeCl_2$ with lithium alkyls or by warming the metal with a dialkylmercury:

$$Me_2Hg + Be \rightarrow Me_2Be + Hg$$

The aryls are best made in ether solution, from which LiCl is precipitated:

$$2\ PhLi + BeCl_2 \rightarrow Ph_2Be + 2\ LiCl$$

Both alkyls and aryls are spontaneously inflammable in air and are violently hydrolysed by water.

Dimethylberyllium is a solid polymer but its vapour contains the dimer

Me—Be(μ-Me)₂Be—Me

The methyl bridges can be explained in terms of three-centre bonds (11.5.2). Di-isopropylberyllium is a dimer in benzene. On heating it is converted to the polymer $[(CH_3)_2CHBeH]_n$. But pyrolysis of di-tertiarybutylberyllium at 370 K gives BeH_2 and $(CH_3)_2C{=}CH_2$. With this and the higher alkyls there seems to be no tendency to polymerisation.

Beryllium alkyls react with nitrogen ligands such as dipyridyl and ortho-phenanthroline to give highly coloured compounds:

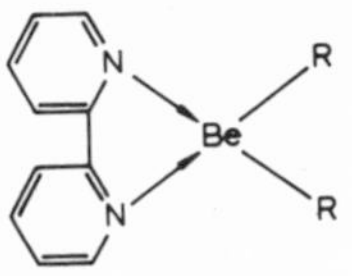

(R = alkyl)

They probably owe their colour to electron transfer from the Be—C bond to the unoccupied orbital of lowest-energy on the nitrogen ligand.

Magnesium

The best-known organomagnesium compounds are the Grignard reagents usually formulated RMgX (R = alkyl or aryl, X = Cl, Br or I). They are made by treating magnesium with a carefully dried solution of the alkyl or aryl halide in ether. Unsolvated RMgX cannot be isolated from the ether solutions. However, X-ray diffraction studies have been made on some crystalline etherates. In $C_2H_5MgBr \cdot 2\,Et_2O$ the magnesium atom is coordinated tetrahedrally to a carbon atom, a bromine atom and two oxygen atoms:

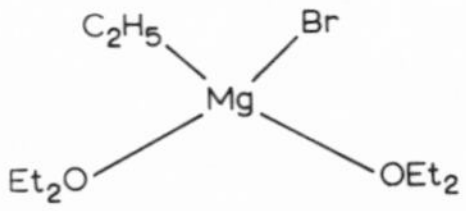

Solutions of Grignard reagents are known to contain various molecular species in equilibrium, represented in general terms, ignoring solvation, as

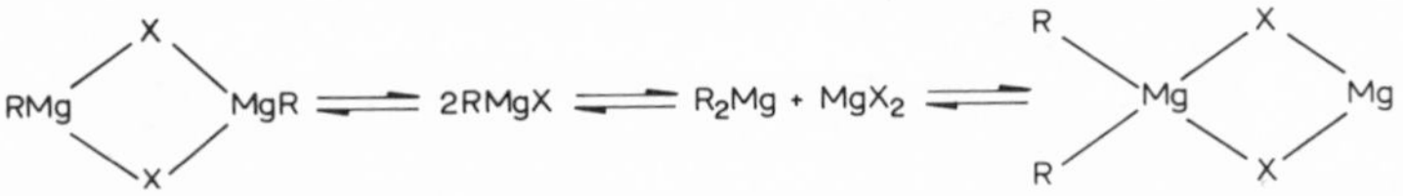

$$RMg(\mu\text{-}X)_2MgR \rightleftharpoons 2RMgX \rightleftharpoons R_2Mg + MgX_2 \rightleftharpoons R_2Mg(\mu\text{-}X)_2Mg$$

The proportions of the various species depend upon the nature of R and X, the solvent, the concentration and the temperature.

Dialkyl and diaryl magnesium compounds resemble the corresponding beryllium compounds and are made by similar methods. $(C_2H_5)_2Mg$ is a chain polymer similar in structure to $(C_2H_5)_2Be$.

Magnesium reacts with cyclopentadiene to give the colourless, ionic $(C_5H_5)_2Mg$ which has the same sandwich structure as ferrocene.

Calcium, strontium and barium

Organometallic compounds of these elements are of little importance, but the cyclopentadienyls are interesting. These can be made by the action of cyclopentadiene on the hydrides at ~570 K, or on the metals Ca and Sr in dimethylformamide.

14.9. Complexes

Beryllium forms neutral chelate complexes with oxygen ligands. An example is bis(acetylacetonyl)beryllium:

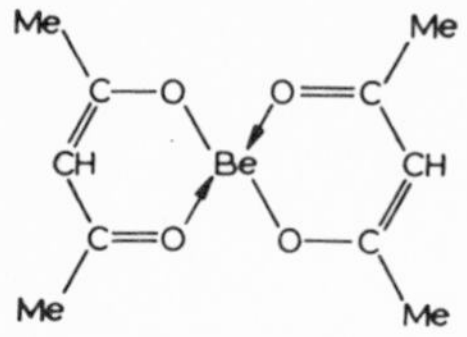

made by dissolving $Be(OH)_2$ in dilute acetic acid and treating the solution with acetylacetone. Linear chelate polymers such as

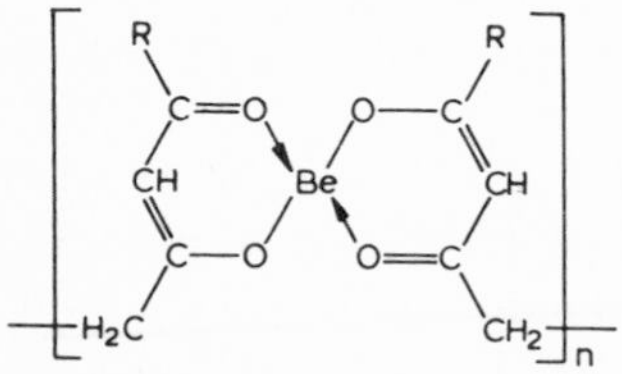

can be made.

Beryllium also forms anionic complexes with chelating oxygen ligands. An alkaline solution of catechol dissolves $Be(OH)_2$ to give a salt of

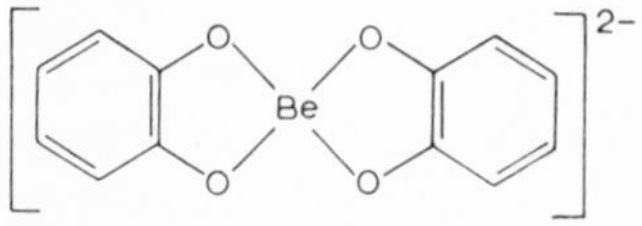

Magnesium compounds hydrate readily, and anhydrous magnesium halides form addition compounds with aldehydes, ketones and ethers, which are structurally similar to those of Be. Otherwise its complexes are few and unstable.

Calcium, strontium and barium form some poorly characterised β-diketone complexes and rather unstable ammines; the tendency to covalence, even as indicated by hydration of the ions, decreases with cation size. The determination of magnesium and calcium with sodium ethylenediamine tetraacetate, EDTA, probably involves the formation of a chelate complex whose stability is enhanced by the presence of 5-membered rings.

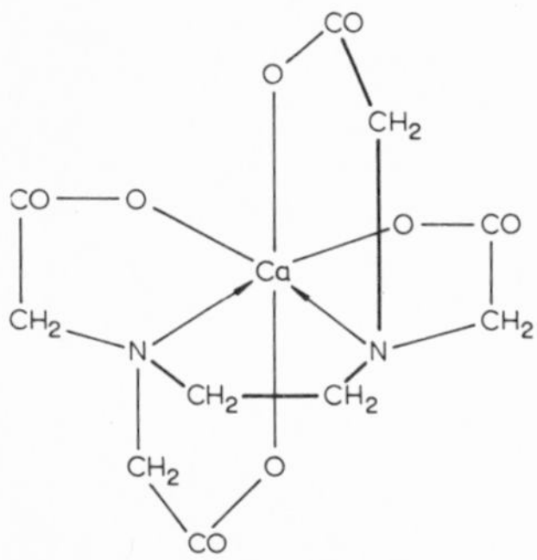

Fig. 14.8. Calcium EDTA complex (binegative ion).

14.10. Similarities between Beryllium and Aluminium

Though considerably larger than Be^{2+}, the greater charge of the Al^{3+} ion renders its polarising power of the same order. The similarities between the two elements led to an early belief that beryllium was a member of Group III. Its low atomic heat supported this idea, and not until 1871 did Mendeleev correctly place the element in Group II — an early triumph for the periodic classification. The diagonal similarity between beryllium and aluminium (cf. 13.9) remains:

(i) The standard electrode potentials of the metals are of the same order (Be^{2+}/Be −1.85 V; Al^{3+}/Al −1.66 V).

(ii) Both metals are rendered passive by nitric acid.

(iii) Both metals dissolve in caustic alkalis with hydrogen evolution.

(iv) The halides are similar in their solubilities in organic solvents and their behaviour as Lewis acids.

(v) Beryllium carbide, Be_2C, and aluminium carbide, Al_4C_3, both yield methane on hydrolysis:

$$Be_2C + 4\,H_2O \rightarrow 2\,Be(OH)_2 + CH_4$$

$$Al_4C_3 + 12\,H_2O \rightarrow 4\,Al(OH)_3 + 3\,CH_4$$

Further Reading

D.A. Everest, The chemistry of beryllium, Elsevier, Amsterdam, 1964.
R.W. Mooney and M.A. Aia, Alkaline earth phosphates, Chem. Rev., 61 (1961) 433.
N.A. Bell, Beryllium halides and pseudohalides, Adv. Inorg. and Radiochem., 14 (1972) 255.
R.J.P. Williams, The biochemistry of sodium, potassium, magnesium and calcium, Quart. Rev., 24 (1970) 331.
E.C. Ashby, Grignard reagents, Quart. Rev., 21 (1967) 259.
G.W.R. Canham and A.B.P. Lever, Bioinorganic chemistry, J. Chem. Ed., 49 (1972) 657.

Boron and Aluminium — Group III

15

15.1. Introduction

Boron is a non-metal and its chemistry is entirely that of its covalent compounds. Although it has the electron configuration $2s^2 2p^1$, the total energy release in the formation of three bonds in a compound BX_3 is always sufficient to provide for promotion and hybridisation leading to a valence state sp^2. All monomeric boron compounds of the type BX_3 have triangular molecules with X—B—X angles of 120°, as predicted by the simple electron-repulsion theory (5.2). The B—X bonds are usually rather shorter than would be expected for simple electron-pair bonds.

Because BX_3 compounds have only three pairs of valence electrons their boron atoms tend to be electron acceptors. The tendency to complete an octet is shown by the existence of tetrahedrally co-ordinated boron in compounds such as H_3BCO and ions such as BF_4^-.

The aluminium atom is much larger than the boron atom; the Al^{3+} ion can be formed, and a compound like AlF_3 is predominantly ionic, but covalent compounds of the metal are much more common than ionic ones. Unlike the boron compounds, the tribromide, tri-iodide and the lower trialkyls of aluminium are dimers, Al_2X_6, in which the aluminium atom is tetrahedrally co-ordinated (Fig. 15.1). Nevertheless these compounds resemble those of boron in being strong electron acceptors. Unlike the boron atom, for which

TABLE 15.1

ATOMIC PROPERTIES OF BORON AND ALUMINIUM

	B	Al
Z	5	13
Electron configuration	$[He]2s^2 2p^1$	$[Ne]3s^2 3p^1$
$I(1)$/kJ mol^{-1}	798	573
$I(2)$/kJ mol^{-1}	2420	1810
$I(3)$/kJ mol^{-1}	3650	2740
Covalent radius/pm	82	118
Metallic radius/pm	98	143
$r_{M^{3+}}$/pm	20	50

TABLE 15.2

PHYSICAL PROPERTIES OF BORON AND ALUMINIUM

	B	Al
ρ/g cm^{-3}	2.4	2.7
M.p./K	2570	930
B.p./K	2820	2770

the maximum co-ordination number is four, the aluminium atom is large enough to be six-co-ordinate in many of its compounds.

15.2. The Elements: Preparation and Properties

The densities of boron and aluminium are normal for the places they occupy in the Periodic Table. Boron's extremely high *m.p.* indicates very strong binding forces. The structures of several crystalline forms of pure boron have been clearly established; of these the simplest is the rhombohedral form, which contains units of nearly regular icosahedra in a slightly deformed close packing. Crystals of the purest material are very hard, 9—10 on Mohs' scale. The specific conductance increases about 100 times between 290 K and 870 K.

Aluminium has a low *m.p.* compared with neighbouring elements; its face-centred cubic lattice is characteristic of a true metal; it is soft, and its conductance is high.

Boron (5×10^{-3}% of the lithosphere) probably owes its scarcity to the readiness with which it suffers transmutation. It occurs as borates in hot springs and lakes in volcanic regions: the minerals are borax, $Na_2B_4O_7 \cdot 10\,H_2O$, kernite, $Na_2B_4O_7 \cdot 4\,H_2O$ and colemanite, $Ca_2B_6O_{11} \cdot 5\,H_2O$. Natural boron consists of two isotopes $^{10}_{5}B$ (19.6%) and $^{11}_{5}B$ (80.4%) whose nuclear spins of 3 and $\frac{3}{2}$ respectively are made use of in structure elucidation.

An amorphous form of the element can be made by reducing the oxide with magnesium:

$$Na_2B_4O_7 \xrightarrow[\text{to aqueous solution}]{\text{HCl}} H_3BO_3 \xrightarrow{\text{heat}} B_2O_3 \xrightarrow[\text{Mg}]{\text{heat with}} B$$

The brown product always contains some boron suboxide in solid solution. It is used in the manufacture of impact-resistant steels and, because of its high neutron cross-section, for alloys for atomic reactor control rods. Aluminothermic reduction of B_2O_3 yields crystalline material once thought to be pure boron but now known to contain the aluminium borides, AlB_{12} and AlB_2. Black crystalline boron has been made by reducing BBr_3 vapour with hydrogen on a tantalum filament at 1600 K. Probably because of the larger particle size and more nearly perfect lattice, this solid is much less reactive than amorphous boron. Electrolysis of KBF_4 also yields boron of high purity.

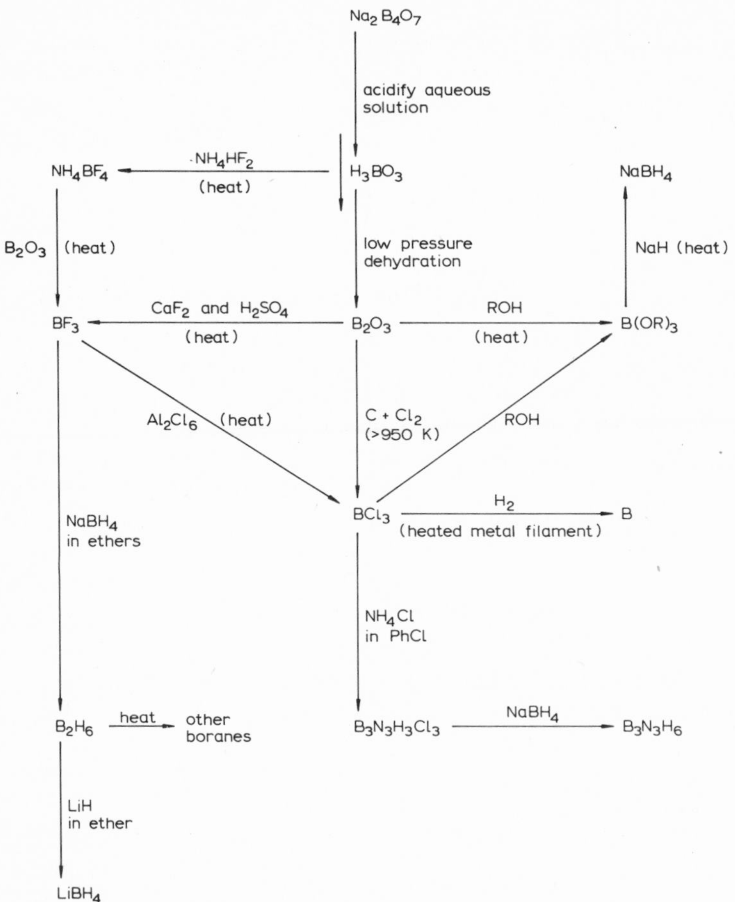

Fig. 15.1. Preparation of some boron compounds from borax.

Aluminium (7.45% of the lithosphere) is widely distributed in igneous rocks, but the only workable ore is bauxite. A solution of sodium aluminate results when bauxite is digested under pressure with caustic soda; Fe_2O_3 and other solid impurities may be removed by allowing them to settle. The clear solution is stirred, at 300—320 K, with a little crystalline $Al_2O_3 \cdot 3\,H_2O$, when much of the aluminium is thrown down as the trihydrate of alumina. This, on heating to 1500 K, becomes α-Al_2O_3; it is dissolved in a fused mixture of cryolite, Na_3AlF_6, with a little fluorspar and electrolysed for the technical production of the metal.

The metal appears unreactive because of the rapid formation, in air, of a tenacious oxide layer. Though its standard electrode potential is -1.66 V, it does not dissolve in water and, even with dilute HCl, reacts slowly until the oxide layer has been removed, after which dissolution is rapid. Amalgamation dislodges the oxide layer, causes the metal to oxidise rapidly in air, and also makes it a good reducing agent even in neutral solution. The metal reacts in a similar way to boron with oxygen, sulphur, nitrogen and the halogens. The considerable enthalpy of formation of alumina enables aluminium to be used for the reduction of metal oxides, such as MnO_2 and Cr_2O_3, by the *thermit process:*

$$2\,Al + 1\tfrac{1}{2}\,O_2 \rightarrow Al_2O_3 \qquad \Delta H = -1690\ kJ\ mol^{-1}$$
$$2\,Cr + 1\tfrac{1}{2}\,O_2 \rightarrow Cr_2O_3 \qquad \Delta H = -1140\ kJ\ mol^{-1}$$
$$\text{thus } 2\,Al + Cr_2O_3 \rightarrow Al_2O_3 + 2\,Cr \qquad \Delta H = -550\ kJ\ mol^{-1}$$

Aluminium dissolves readily in caustic alkali solutions to give the aluminate ion (Fig. 15.4).

$$2\,Al + 2\,OH^- + 10\,H_2O \rightarrow 2[Al(OH)_4(H_2O)_2]^- + 3\,H_2$$

Nitric acid renders the metal passive.

Aluminium furnishes light, strong alloys: Al—Si alloys (~12% Si) can be cast; Duralumin (4% Cu, 0.5% Mn and 0.5% Mg) and Y-alloy (4% Cu, 2% Ni and 1% Mg) can be both wrought and cast.

15.3. Halides

Boron and aluminium give trihalides. BF_3 is most conveniently made by heating B_2O_3 with concentrated H_2SO_4 and a fluoride. It reacts with aluminium chloride or bromide to produce involatile AlF_3 and volatile BCl_3 or BBr_3. The volatility of the boron trihalides decreases with molecular weight (Table 15.3).

TABLE 15.3

MELTING POINTS AND BOILING POINTS OF BORON AND ALUMINIUM TRIHALIDES

	B		Al	
	M.p./K	*B.p.*/K	*M.p.*/K	*B.p.*/K
Fluoride	146	172		1564 *
Chloride	166	285		453 *
Bromide	227	364	370	528
Iodide	316	483	453	654

* Sublimation temperatures.

The boron compounds are covalent and monomeric in the vapour phase, as is BCl_3 in benzene solution. By contrast, AlF_3 is an ionic crystalline solid which contains 6-co-ordinate aluminium and sublimes at high temperature as discrete AlF_3 molecules. The chloride also crystallises in an essentially ionic lattice but like the bromide exists as dimers both in the vapour phase and in

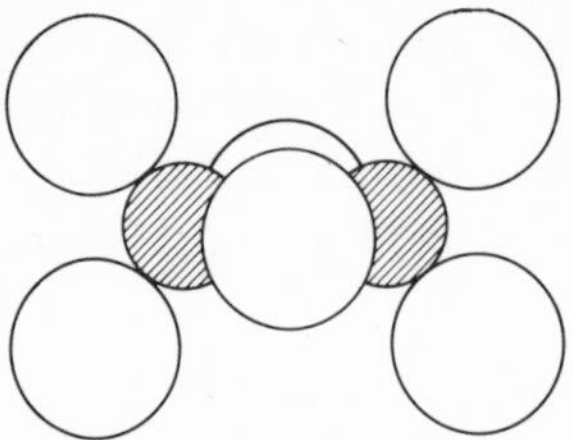

Fig. 15.2. Molecule of Al_2Cl_6.

non-polar solvents, in which the halogen atoms are tetrahedrally arranged about each aluminium atom (Fig. 15.2); this results in a 'bridge' structure of D_{2d} symmetry.

In co-ordinating solvents such as ether, however, Al_2Cl_6 gives place to a tetrahedral, mononuclear complex:

$$R_2O + AlCl_3 \longrightarrow R_2O \cdot AlCl_3$$

Electron diffraction shows that the boron trihalides are planar molecules of D_{3h} symmetry with the angle X—B—X equal to 120°; the bonds are somewhat shorter than the normal single bonds, there being fewer bond-pair repulsions.

Hydrolysis of the chloride, bromide and iodide of boron is rapid and complete. The reaction presumably begins with the donation of an oxygen lone pair to the boron:

$$H_2O + BCl_3 \longrightarrow H_2O \cdot BCl_3 \longrightarrow HO \cdot BCl_2 + HCl$$

$$\downarrow 2H_2O$$

$$B(OH)_3 + 2\,HCl$$

The trifluoride, however, forms 1 : 1 and 1 : 2 adducts with water:

$$BF_3 \xrightarrow{H_2O} H^+\left[\begin{matrix} F & & F \\ & B & \\ F & & OH \end{matrix}\right]^- \xrightarrow{H_2O} H_3O^+\left[\begin{matrix} F & & F \\ & B & \\ F & & OH \end{matrix}\right]^-$$

the latter being stable enough to be distilled without decomposition.

The relative acceptor strengths of BF_3, BCl_3 and BBr_3 have been found by measuring the enthalpies of formation and dipole moments of such compounds as their 1 : 1 adducts with pyridine. The results, together with those from infrared studies, indicate that the electron-acceptor power increases in the order

$$BF_3 << BCl_3 < BBr_3$$

This is opposite to the order suggested by the relative electronegativities of the halogens. The inductive effect in these molecules is evidently outweighed by another effect, which is a tendency for electrons from the smaller halogen atoms to be partly back-donated to the boron, to give some double-bond character to the B—X bonds and to reduce the electron deficiency on the boron.

The fact that BF_3 forms a wider range of complexes than the other trihalides is probably due to a greater difficulty of heterolysis with the B—F bond than with the B—Cl and B—Br bonds. Thus molecules of the alcohols, aldehydes and ketones, which form addition compounds with BF_3, break the boron—halogen bonds in the other halides:

$$2\,ROH + BF_3 \rightarrow ROH_2^+[BF_3OR]^-$$
$$3\,ROH + BCl_3 \rightarrow B(OR)_3 + 3\,HCl$$

The tendency of donor groups containing O, N, S and P to co-ordinate with BF_3 makes the compound an extremely useful catalyst in organic chemistry.

Boric acid dissolves in 50% HF to give tetrafluoroboric acid:

$$H_3BO_3 + 4\,HF \rightarrow HBF_4 + 3\,H_2O$$

The compound has not been isolated in the pure state. The BF_4^- ion in crystalline fluoroborates is shown, by X-ray analysis, tc be tetrahedral, KBF_4 being isomorphous with $KClO_4$. Tetrafluoroborates also resemble perchlorates in solubility. Other tetrahalogenoborates have been made: $C_5H_5NHBCl_4$ and $C_5H_5NHBBr_4$, and the corresponding iodo-complexes.

Anhydrous $AlCl_3$ resembles BF_3 in its acidity and catalytic power. Anhydrous AlF_3 is insoluble in water, though soluble stable hydrates may be prepared; the other halides dissolve with considerable hydrolysis. Their solutions in organic solvents have low conductance.

Unstable oxide halides of both boron and aluminium are known. When B_2O_3 and BF_3 are heated together a cyclic compound is formed:

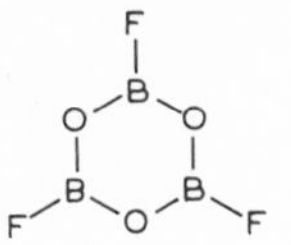

Aluminium oxyhalides result from the thermal degradation of corresponding AlX_3—ether complexes or by heating an excess of trihalide with an oxide

$$Et_2O \cdot AlCl_3 \rightarrow AlOCl + 2\,EtCl$$
$$3\,AlBr_3 + As_2O_3 \rightarrow 3\,AlOBr + 2\,AsBr_3$$

Boron subhalides B_2X_4 have been reported. B_2Cl_4 was first prepared by passing BCl_3 vapour at 150—300 Pa through a glow discharge between mercury electrodes:

$$2\,BCl_3 + Hg \rightarrow B_2Cl_4 + HgCl_2$$

A more convenient preparative route to diboron compounds is based on a Wurtz type of reaction:

$$2(Me_2N)_2BCl + 2\,Na \rightarrow (Me_2N)_2B\text{—}B(NMe_2)_2 + 2\,NaCl$$

$$\downarrow 8\,HCl$$

$$B_2Cl_4 + 4[Me_2NH_2]^+Cl^-$$

Structurally, in the vapour and liquid phase the two BCl_2 groups are in planes perpendicular to one another (D_{2d}):

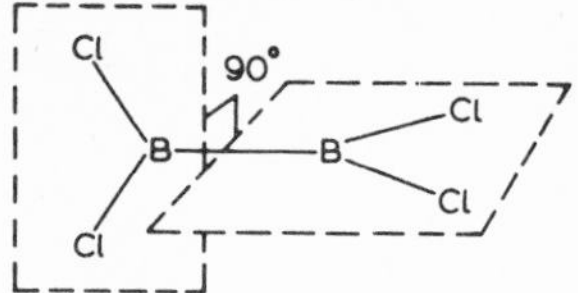

In the solid, however, the arrangement is coplanar (D_{2h}). The compound is reduced by $LiBH_4$ (11.5.3) to diborane. It decomposes slowly at room temperature to B and BCl_3, and reacts with water to give boric acid and hydrogen. The molecule is converted to an ion by some donor molecules: Me_3N reacts with it to give $[(Me_3N)_2B_2Cl_4]^{2-}$. Gaseous B_2F_4 has been obtained by fluorinating B_2Cl_4 with SbF_3. The thermally unstable liquid B_2Br_4 is made by treating B_2Cl_4 with BBr_3 and the yellow solid B_2I_4 by passing a radiofrequency discharge through BI_3 vapour.

The yellow solid B_4Cl_4 and the red solid B_8Cl_8, formed by the spontaneous decomposition of B_2Cl_4, have interesting structures. The molecule of B_4Cl_4, consists of a tetrahedron of boron atoms surrounded by four chlorine atoms so arranged as to maintain T_d symmetry. In B_8Cl_8, the molecule is a distorted dodecahedron (D_{2d}) of boron atoms with a chlorine atom bound to every one of them.

15.4. Oxides

A glassy form of boric oxide, B_2O_3, results from dehydrating H_3BO_3 at red heat:

$$H_3BO_3 \xrightarrow[-H_2O]{370\text{ K}} HBO_2 \xrightarrow[-H_2O]{\text{red heat}} B_2O_3$$

A crystalline form, *m.p.* 720 K, appears after dehydrating HBO_2 by keeping it under reduced pressure and slowly raising the temperature to 670 K over a period of some weeks. It contains two-dimensional sheets (Fig. 15.3). Boric oxide, though more acidic than Al_2O_3, is amphoteric; it combines with metal oxides to give metaborates as in the borax-bead test:

$$CuO + B_2O_3 \rightarrow Cu(BO_2)_2$$

and with phosphorus pentoxide to give a phosphate:

$$B_2O_3 + P_2O_5 \rightarrow 2\ BPO_4$$

Boron phosphate is slightly soluble when freshly prepared, but becomes insoluble on heating, and is stable enough to be sublimed (1720 K). In the crystal lattice both the boron and phosphorus atoms are surrounded, tetrahedrally, by oxygen atoms. There is also an arsenate, $BAsO_4$.

A white oxide with the empirical formula BO which contains both B—B

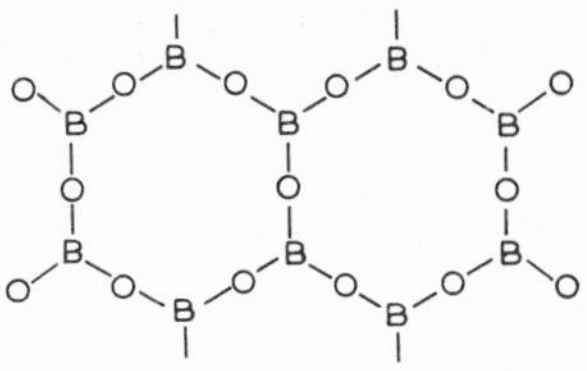

Fig. 15.3. Two-dimensional sheet in B_2O_3.

and B—O—B bonds is obtained by heating $B_2(OH)_4$

$$(Me_2N)_2BB(NMe_2)_2 \xrightarrow{H_2O} (HO)_2BB(OH)_2 \xrightarrow[<100\ Pa]{520\ K} (BO)_n$$

$$(Me_2N)_2BB(NMe_2)_2 \xrightarrow{ROH} (RO)_2BB(OR)_2 \xrightarrow{H_2O} (HO)_2BB(OH)_2$$

The solid turns brown on prolonged heating. It exists mainly as B_2O_2 in the vapour, but is highly polymeric in the solid. With water it gives white crystals of hypoboric acid, $H_4B_2O_4$, as well as boric acid and hydrogen.

Alumina, Al_2O_3, just as boric oxide, may be prepared by dehydrating one of the hydrous oxides. These exist in four well defined forms: the monohydrate AlO(OH), as boehmite (γ-monohydrate) and diaspore (α-monohydrate), and the trihydrate $Al(OH)_3$, as gibbsite (γ-trihydrate) and bayerite (α-trihydrate). Of these, all but bayerite occur naturally in bauxite. On dehydration diaspore passes directly to corundum (α-alumina); the others yield a series of anhydrous aluminas, e.g. γ-alumina, which pass into α-alumina only at higher temperatures (about 1120 K). α-Al_2O_3 has a well-defined, close-packed O^{2-} structure with Al^{3+} distributed symmetrically in octahedral spaces, whereas γ-Al_2O_3 has a distorted, badly organised, microcrystalline structure of the spinel type. In consequence, α-Al_2O_3 is dense, hard, and resistant to chemical attack and can be brought into solution only after fusion with a flux such as $KHSO_4$, whereas γ-Al_2O_3 is less dense and soft, and has a high surface area, so that it is relatively soluble in aqueous alkalis and acids as well as an excellent and selective absorbent used in dehydration, decolorisation and chromatography.

Ruby and blue or white sapphire are α-alumina with traces of specific impurities. Synthetic forms are made by fusing finely powdered alumina with a trace of the colouring oxide (Cr_2O_3 for ruby) in an oxyhydrogen flame.

Aluminium hydroxide acts primarily as a base, but it also ionises weakly as an acid ($pK_a = 12.2$). The osmotic properties of sodium aluminate solutions are identical with those of NaOH; thus both must have the same number of ions. Furthermore, conductance and pH measurements indicate that hydrolysis to NaOH and colloidal alumina is slight. Sodium aluminate appears to ionise as a 1 : 1 electrolyte:

$$NaAl(OH)_4(H_2O)_2 \rightleftharpoons Na^+ + [Al(OH)_4(H_2O)_2]^-$$

The high viscosity can be explained by the linking of hydrated aluminate

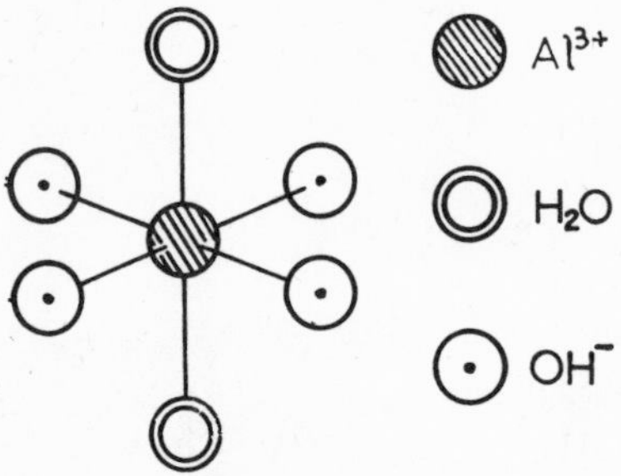

Fig. 15.4. Aluminate ion present in sodium aluminate.

ions (Fig. 15.4) with each other and with water molecules by hydrogen bonding.

15.5. Boric Acid, Borates and Aluminium Alkoxides

Phase studies of the system B_2O_3—H_2O show that HBO_2 and H_3BO_3 are the only stable boric acids, though salts of greater complexity exist such as $Na_2B_4O_7$, $Ca_2B_6O_{11}$ and $NaCaB_5O_9$. The flaky crystals of orthoboric acid have a layer structure formed by triangular BO_3 groups joined by hydrogen bonds (Fig. 15.5). The compound is volatile in steam. It is a very weak acid,

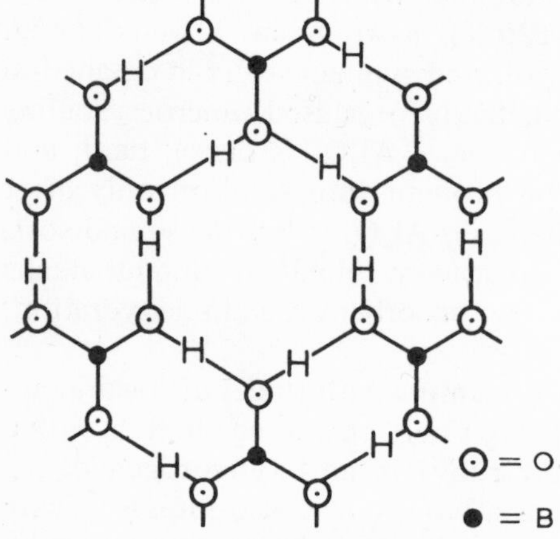

Fig. 15.5. Hydrogen-bonded structure of boric acid.

which behaves in a solution as an electron-acceptor (i.e. a Lewis acid) rather than as a proton-donor:

$$H_3BO_3 + 2\ H_2O \rightleftharpoons B(OH)_4^- + H_3O^+ \qquad pK_a = 9.2$$

Crystal analysis of borates shows they bear some similarity to the silicates, the unit being the triangular BO_3 in place of the tetrahedral SiO_4. The rather uncommon orthoborates, e.g. $Mg_3(BO_3)_2$, possess discrete BO_3^{3-} ions. The pyroborates such as $Co_2B_2O_5$ contain ions of the form:

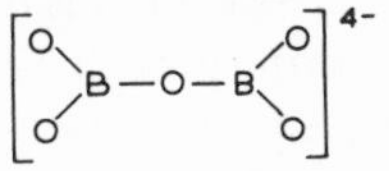

Of the borates much the most common are the metaborates; these have chain and ring forms. Sodium metaborate, correctly formulated $Na_3B_3O_6$, contains the six-membered ring ions:

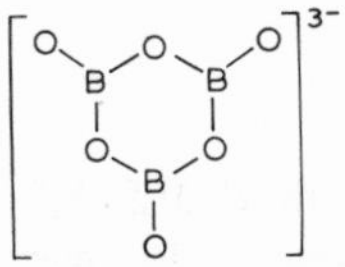

Calcium metaborate, CaB_2O_4, typical of the bipositive metal borates, has negatively charged chains held together by Ca^{2+} ions between them:

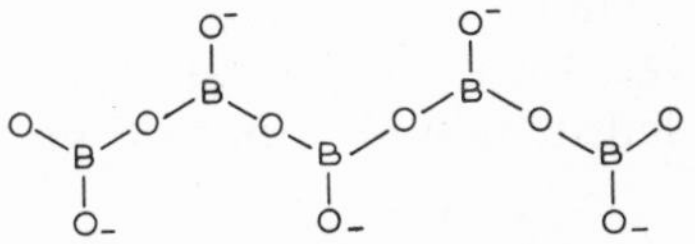

A wide variety of polyborate anions resulting from combinations of both trigonal BO_3 and tetrahedral BO_4^- groups have been identified in the crystal structures of hydrated borates. Thus borax, $Na_2B_4O_7 \cdot 10\ H_2O$, has an anion made up of two BO_4^- and two BO_3 units while colemanite $Ca_2B_6O_{11} \cdot 5\ H_2O$ has four BO_4^- and two BO_3 groups.

Esters of orthoboric acid can be made by treating boric acid with H_2SO_4 and the alcohol. The esters are monomeric liquids with normal Trouton constants. Polyols (1,2- or 1,3-*cis* diols) like glycerol are capable of reacting with boric acid in 1 : 1 stoichiometry and enhancing the acidity of monobasic boric acid to such an extent that it can be titrated with standard aqueous NaOH.

Borate esters however are now more commonly referred to as alkoxyboron compounds. From this standpoint it is more usual to think of their preparation and isolation under anhydrous conditions from a halide

$$BX_3 + 3\ ROH \rightarrow B(OR)_3 + 3\ HX$$

Aluminium alkoxides can be formed analogously, and the metal is even soluble in certain alcohols, activated with a trace of iodine:

$$2\ Al + 6\ Pr^iOH \rightarrow 2\ Al(OPr^i)_3 + 3\ H_2$$

$Al(OPr^i)_3$ is a white solid, melting in the range 291—311 K, soluble in organic solvents, and applied in organic syntheses of aldehydes and ketones.

15.6. Oxo-compounds of Aluminium

The Al^{3+} ion (radius 50 pm) is sufficiently similar in size to the Si^{4+} ion (radius 41 pm) to replace it in the SiO_4 tetrahedra of silicate structures. The deficiency of positive charge created by the change is made up by incorporating other positive ions. Such explanations account for the wide distribution of Al in the earth's crust and also the vast complexity of aluminium silicate chemistry.

15.7. Alums

Potash alum, $KAl(SO_4)_2 \cdot 12\ H_2O$, contains K^+, Al^{3+} and tetrahedral $SO_4{}^{2-}$ ions. Six of the water molecules are octahedrally co-ordinated to the Al^{3+} and six are used to link these $[Al(H_2O)_6{}^{3+}]$ ions to neighbouring sulphate ions. It is thus a lattice compound rather than a complex. A unipositve ion which is smaller than K^+ does not form a very stable alum, and the still smaller Li^+ ion does not form one at all. Moreover, for alum formation the radius of the M^{3+} ion must be small; the large lanthanide M^{3+} ions do not form alums.

15.8. Nitrogen Compounds

White, insoluble, refractory boron nitride, BN, is simply made by fusing together B_2O_3 and NH_4Cl, by strongly heating compounds such as borazine

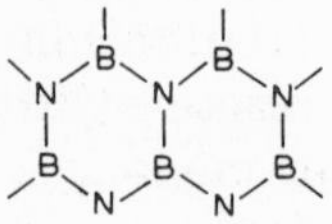

Fig. 15.6. Graphite-like structure of boron nitride.

and $BF_3 \cdot NH_3$, and by burning boron in ammonia. It has a graphite-like structure (Fig. 15.6) in which the bonding within the layers is by sp^2 hybrids of both B and N, the remaining electrons being in delocalised π orbitals above and below the plane. The B—N distance in the plane is 145 pm but the sheets are more than 330 pm apart. The structure differs from that of graphite in having the hexagons directly under one another (B under N). Boron nitride is very stable and unreactive. It is, however, decomposed by steam at red heat:

$$BN + 3\ H_2O \rightarrow NH_3 + H_3BO_3$$

and by fluorine and HF at lower temperatures:

$$2\ BN + 3\ F_2 \rightarrow 2\ BF_3 + N_2$$
$$BN + 4\ HF \rightarrow NH_4BF_4$$

Under pressures of about 7 GPa, boron nitride at 3300 K changes into the adamantine form (7.2.14), borazon, claimed to be harder than diamond. The conversion is similar to that of graphite into diamond.

When heated aluminium combines directly with nitrogen to give AlN with a wurtzite structure. It is much more reactive than BN, being hydrolysed by cold water:

$$AlN + 3\ H_2O \rightarrow NH_3 + Al(OH)_3$$

A wide variety of derivatives containing B—N bonds can be prepared from boron halide precursors:

$$BCl_3 + 6\ R_2NH \rightarrow B(NR_2)_3 + 3\ R_2NH_2^+Cl^-$$

A similar range of aluminium—nitrogen derivatives is accessible by appropriate routes. The compound $(PhAlNPh)_4$ has a structure based on an Al_4N_4 cube.

$$4\ Ph_3Al + 4\ PhNH_2 \xrightarrow{-PhH} (Ph_2AlNHPh)_2 \xrightarrow{-PhH} (PhAlNPh)_4$$

15.9. Organometallic Compounds

Boron trialkyls can be made from the halides (often $Et_2O \cdot BF_3$ for ease of handling) with an appropriate Grignard reagent:

$$Et_2O \cdot BF_3 + 3\ RMgX \rightarrow R_3B$$

Hydroboration is the name given to the addition of B—H bonds to carbon multiple bonds:

$$B_2H_6 + 6\ CH_2{=}CH_2 \rightarrow 2\ Et_3B$$

For this reaction the B_2H_6 can be prepared in situ by reduction of a boron compound, ($NaBH_4/H_2SO_4$, for example) and a suitable solvent chosen to facilitate product separation.

Trimethylboron is a gas and Et_3B and Pr_3^iB are liquids. Both the latter compounds are monomeric in the vapour phase, and their molecules are planar at boron. The lower alkyls of boron inflame spontaneously in air or chlorine gas and form ammines with NH_3, but do not react with water. By contrast, the aryls are stable in air. The trialkyls of boron are oxidised, for example by aqueous HBr, to dialkyl boric acids, R_2BOH, also called boronous acids. Their esters can also be made by the action of trialkyls on aldehydes:

$$R \cdot CHO + BEt_3 \rightarrow RCH_2OBEt_2 + C_2H_4$$

Corresponding aromatic derivatives are known.

The monoalkyl boric acids, or boronic acids, $RB(OH)_2$ can be made by hydrolysis of their esters; these esters are obtained by treating methyl or ethyl borates with a Grignard reagent:

$$(MeO)_3B + MeMgI \rightarrow (MeO)_2BMe + MeOMgI$$

$$\downarrow \text{hydrolysis}$$

$$(HO)_2BMe$$

Anhydrides of these acids can be made by treating them with P_2O_5. The methyl compound $(CH_3 \cdot BO)_3$ has a cyclic structure:

Compounds containing BR_4^- ions can be made. The most important is $Na[B(C_6H_5)_4]$, sodium tetraphenylborate, which is soluble in water and causes the precipitation of large cations like Rb^+ and Cs^+ for gravimetric determination.

Magnesium—aluminium alloy reacts with an alkyl halide in ether to give $R_3Al \cdot OEt_2$. Trialkyls are also made on a large scale by direct synthesis from α-olefins, aluminium and hydrogen. The process is based on the fact that Et_3Al for example will react directly with Al and H_2 to form the diethylaluminium hydride:

$$2\ Et_3Al + Al + \tfrac{3}{2}H_2 \rightarrow 3\ Et_2AlH$$

The hydride then adds olefin:

$$3\ Et_2Al{-}H + \rangle C{=}C\langle \rightarrow 3\ Et_3Al$$

Thus although 2 R_3Al is used in the cycle, 3 R_3Al is produced.

Trimethylaluminium, *m.p.* 288 K, is a dimer (Fig. 15.7). The structure was determined by X-ray and is supported by spectroscopic evidence. The ^1H

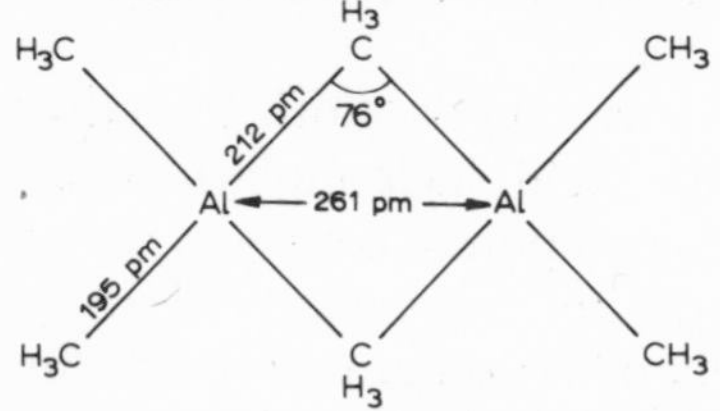

Fig. 15.7. Trimethylaluminium dimer.

n.m.r. spectrum of $Al_2(CH_3)_6$ shows separate resonances for the bridging groups and the terminal groups at 198 K but towards room temperature the resonances coalesce to a single sharp peak. Evidently the groups can exchange places by some process akin to the ring-whizzing described in Section 18.2.5. All the lower trialkyls are spontaneously inflammable in air and react violently with water:

$$Et_3Al + 3\ H_2O \rightarrow Al(OH)_3 + 3\ EtH$$

Triphenylaluminium, $(C_6H_5)_3Al$, is best made from aluminium and diphenylmercury, $(C_6H_5)_2Hg$.

Some of the trialkyls form adducts with alkali-metal halides:

$Na[AlEt_3F]$ $K[AlEt_3Cl]$ $Na[Al_2Et_6F]$

Salts of the $Al_2Et_6F^-$ ion are much stronger electrolytes than those of the $AlEt_3F^-$ ion because the unipositive ion cannot approach close enough to the halogen in the complex ion

```
┌    Et        Et     ┐ -
│    |         |      │
│ Et—Al — F — Al—Et   │
│    |         |      │
└    Et        Et     ┘
```

to form a dipole.

Both the trialkyls and triaryls of aluminium act as Lewis acids and form a variety of compounds with electron-donating reactants. The trialkyls, for example, form 1 : 1 adducts with amines which, on pyrolysis, give alkanes and organometallic Al—N polymers. Triphenylaluminium reacts with aliphatic amines in toluene to give polymers such as $[Me_2N \cdot Al(C_6H_5)_2]_n$ and $[MeN \cdot Al(C_6H_5)]_n$; but with non-orthosubstituted aromatic amines like *p*-toluidine it gives crystalline tetramers $(C_6H_5Al \cdot NAr)_4$.

A variety of compounds of the type R_2AlX are known, where R = alkyl and X = CN, halogen, NMe_2, or SMe.

15.10. Borides

Metal borides can be made by direct combination of the elements in a vacuum at high temperature (2300 K), by reduction of a metal oxide with B_4C and carbon,

$$2\ TiO_2 + B_4C + 3\ C \rightarrow 2\ TiB_2 + 4\ CO$$

or by electrolysis of a fused borate. In the last method, the liberated metal reduces borate to boron and combination then occurs. The borides are hard, have high *m.p.* and are good conductors; they resemble the interstitial carbides and nitrides of metals. They are usually fairly stable and resistant to attack but basic oxidising agents such as Na_2O_2 decompose them on heating. The structures, like those of intermetallic compounds, are determined more by the requirements of the metal and boron lattices than by valency relationships.

Metal borides have arrangements based on:

(i) isolated boron atoms (M_2B, M_3B and M_4B),
(ii) zig-zag chains (MB),
(iii) double chains (M_3B_4),
(iv) hexagonal layers (MB_2),
(v) three-dimensional frameworks (MB_6 and MB_{12}).

The iron boride FeB (Fig. 15.8) is an example of (ii); the Fe atoms are at

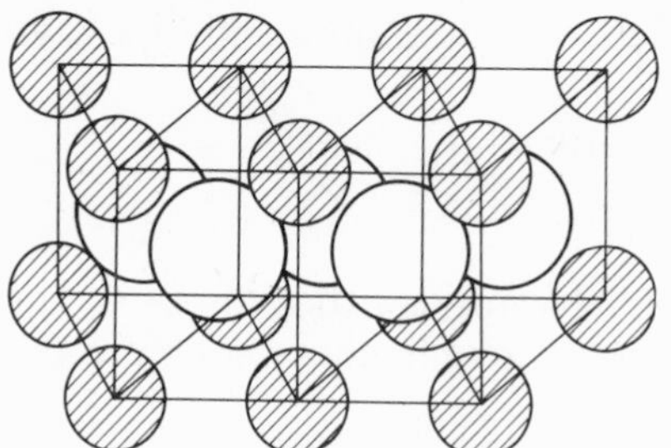

Fig. 15.8. Structure of FeB.

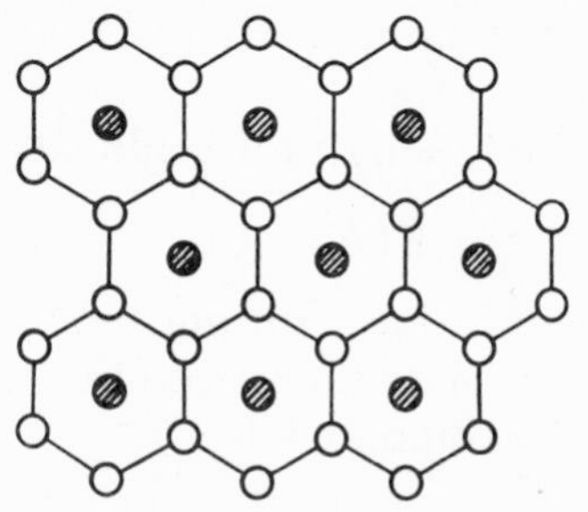

Fig. 15.9. Structure of AlB_2, with separate layers c aluminium atoms and boron atoms.

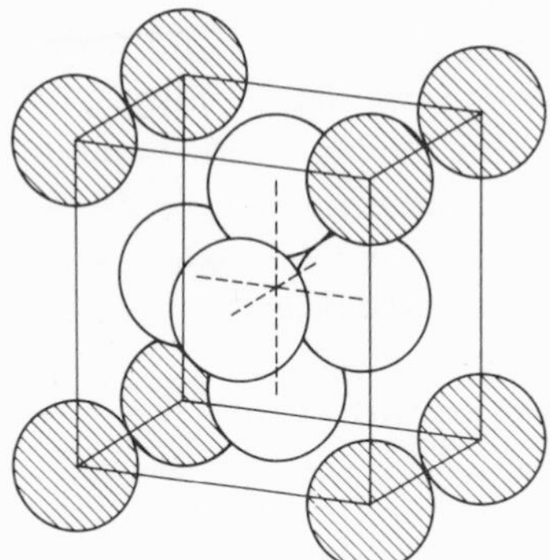

Fig. 15.10. Cubic structure of CaB_6.

the corners of trigonal prisms and the B atoms at their centres, the latter being covalently bound in zig-zag chains. Aluminium boride, AlB_2, is an example of (iv); the borons are joined in hexagons to form infinite layers with the aluminiums in layers between them (Fig. 15.9). The borides of alkaline earth metals, CaB_6, SrB_6, BaB_6, and several isomorphous borides of the lanthanides have cubic CsCl-type structures with an octahedron of boron atoms occupying the centre of the unit cell (Fig. 15.10).

15.11. Complexes

Boron

Boron's strong tendency to 4-covalence is shown in its complexes. The BF_4^- ion, like the isoelectronic BeF_4^{2-} ion and CF_4 molecule, is tetrahedral. The borohydride ion is also tetrahedral, the boron atom again being in the sp^3 valence state. The trihalides co-ordinate readily with ethers:

$Et_2O \cdot BF_3$	*b.p.* 401 K
$Et_2O \cdot BCl_3$	*m.p.* 329 K

The trifluoride also forms co-ordination compounds with esters, aldehydes and ketones; the compound shown is a stable solid:

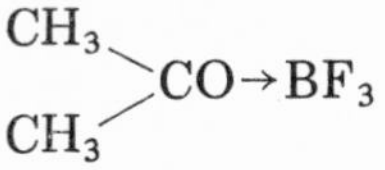

$$(CH_3)_2CO \rightarrow BF_3$$

Chelate oxo-anions containing 4-covalent boron are formed by the borate ion with some cis-diols. With catechol:

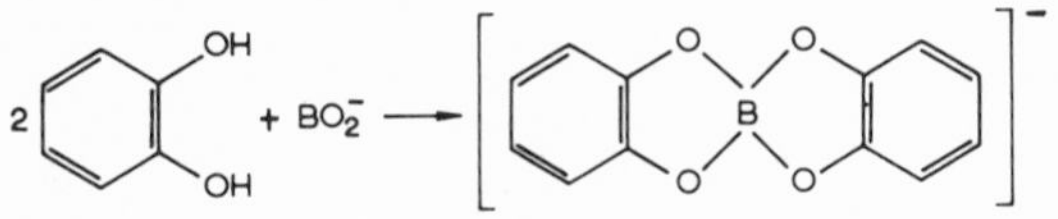

$$2\ C_6H_4(OH)_2 + BO_2^- \longrightarrow [(C_6H_4O_2)_2B]^-$$

With mannitol:

$$2\ C_6H_8(OH)_6 + BO_2^- \rightarrow [C_6H_8(OH)_4]_2BO_4^-$$

Boron has the environment

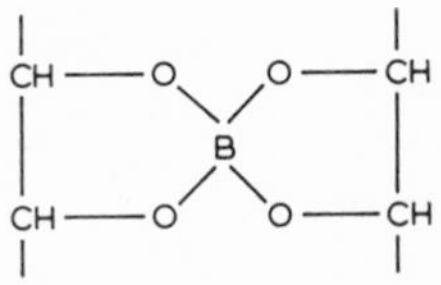

CH—O, O—CH / B / CH—O, O—CH

in this ion, which does not form a bond with a proton since the boron atom is already exerting its maximum covalence of four, but does bind H_3O^+ electrostatically. For this reason, boric acid can be titrated against NaOH to a definite end-point in the presence of mannitol or glycerol.

BF_3 and BCl_3 differ from one another in their reactions with β-diketones in benzene solution:

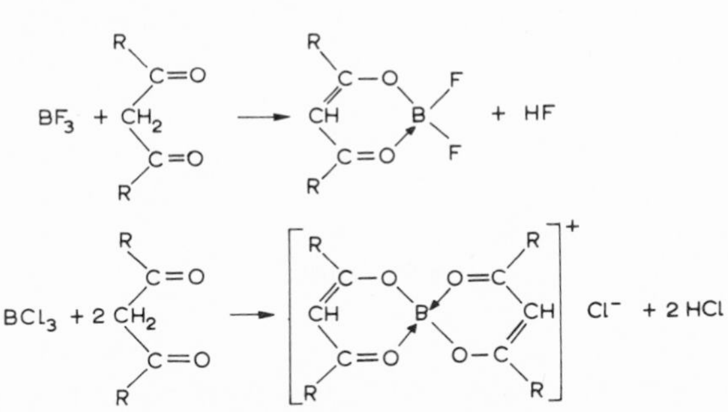

$$BF_3 + CH_2(COR)_2 \longrightarrow (RCO\cdot CH\cdot COR)BF_2 + HF$$

$$BCl_3 + 2\ CH_2(COR)_2 \longrightarrow [(RCO\cdot CH\cdot COR)_2B]^+Cl^- + 2\ HCl$$

The second complex is a positive boronium ion. Another example of a boronium compound is $[(C_6H_5)_2B(\text{dipyridyl})]^+ClO_4^-$.

Aluminium

Aluminium forms fluoro-, chloro- and bromo-complexes, containing tetrahedral AlX_4^- ions and, in the case of fluorine, octahedral AlF_6^{3-} ions. In the cubic cell of cryolite, Na_3AlF_6, the corners and centre are occupied by dis-

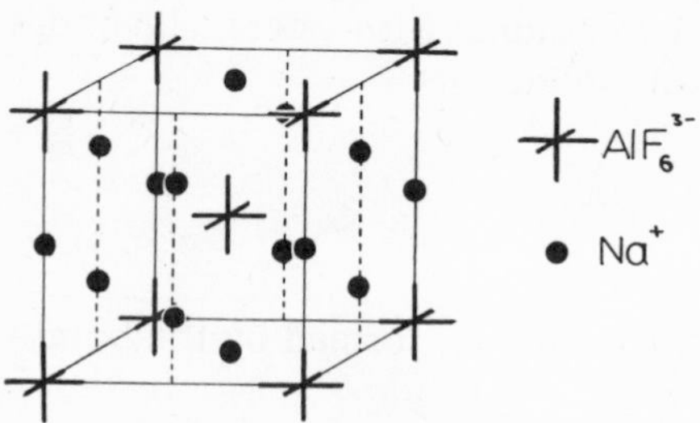

Fig. 15.11. Structure of Na_3AlF_6.

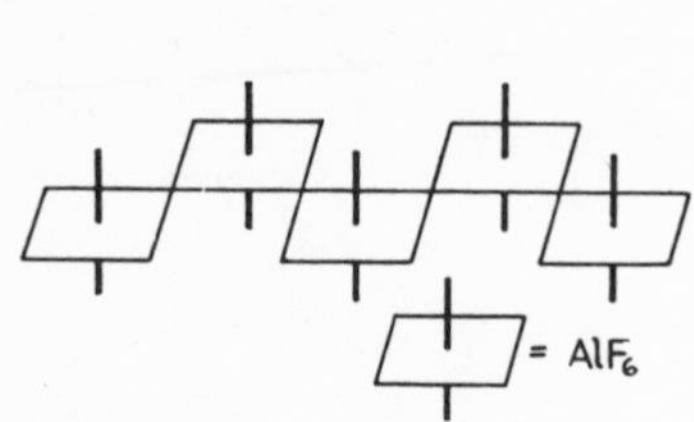

Fig. 15.12. Infinite chain ions in Tl_2AlF_5.

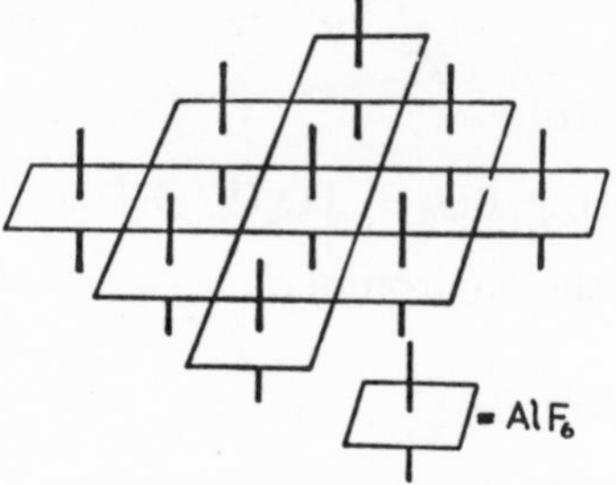

Fig. 15.13. Infinite planar ions joined at corners in $NaAlF_4$.

torted AlF_6^{3-} octahedra, and the Na^+ ions are arranged as in Fig. 15.11. In Tl_2AlF_5, infinite chain ions $(AlF_5)_n^{2n-}$ are formed by the sharing of corners of AlF_6 octahedra (Fig. 15.12), whereas in $NaAlF_4$, the planar $(AlF_4)_n^{n-}$ ions are formed by sharing corners with four other octahedra (Fig. 15.13).

The trihalides of aluminium form 1:1 adducts with ethers, aldehydes, ketones and alcohols, such as:

$$\begin{matrix} R \\ & \rangle O \rightarrow AlX_3 \\ R \end{matrix}$$

and with ammonia they give solid, covalent monoammines:

$$H_3N \rightarrow AlX_3$$

Octahedral 6-co-ordinate chelate oxo-complexes are common. They may be described in terms of sp^3d^2 hybridisation, the 3d orbitals being energetically only slightly higher than the 3p. Such orbitals are not available in boron and octahedral complexes do not occur. This is also true of the β-diketone complexes and the trioxalato-aluminates such as

$$K_3\left[Al\begin{pmatrix} COO \\ | \\ COO \end{pmatrix}_3\right]$$

Further Reading

N.N. Greenwood, The chemistry of boron, Pergamon, Oxford, 1975.
A.J. Banister and K. Wade, The chemistry of aluminium, gallium, indium and thallium, Pergamon, Oxford, 1975.
R. Thompson, Uses of boron compounds, Chem. in Britain, (1970) 140.
K. Wade, Electron deficient compounds, Nelson, London, 1971.
E.L. Muetterties, The chemistry of boron and its compounds, Wiley, New York, 1967.
Borax to boranes, Advances in Chemistry Series, No. 32, American Chemical Society, Easton, Pa., 1961.
N.N. Greenwood and R.L. Martin, Boron trifluoride co-ordination compounds, Quart. Rev., VIII (1954) 1.
W. Gerrard and M.F. Lappert, Reactions of boron trichloride with organic compounds, Chem. Rev., 58 (1958) 1081.
H.G. Heal, Recent advances in boron chemistry, Royal Institute of Chemistry monograph, London, (1960) No. 1.
A.K. Holliday and A.G. Massey, Boron subhalides and related compounds with boron—boron bonds, Chem. Rev., 62 (1962) 303.
T.G. Pearson, The chemical background of the aluminium industry, Royal Institute of Chemistry Monograph, No. 3, London, 1955.
N.N. Greenwood, R.V. Parish and P. Thornton, Metal borides, Quart. Rev., XX (1966) 441.
R. Thompson, Borides: Their chemistry and applications, Royal Institute of Chemistry Monograph, No. 5, London, 1965.
B.R. Currell and M.J. Frazer, Inorganic polymers, RIC Reviews, 2 (1969) 13.

Gallium, Indium and Thallium — Group IIIB

16

16.1. Introduction

The metals gallium, indium and thallium all have the $ns^2 np^1$ configuration and ^2P ground state met with in boron and aluminium. The atomic and ionic radii of the elements begin with values close to those of aluminium and increase somewhat with increasing atomic number (Table 16.1).

The chemistry of gallium is very similar to that of aluminium in that there is little tendency to form unipositive compounds. A tendency to retain the $5s^2$ shell appears in indium with the compound InCl. This contains unipositive indium, although as the redox potential data indicate, it disproportionates in water to give the trichloride. With thallium the $6s^2$ shell is very stable and the Tl^+ ion (radius 140 pm) appears in many well-characterised salts (see below).

The M^I state increases in stability in going down the sub-group, but the M^{III} state decreases in stability. The chlorides MCl_3 are essentially covalent. Whereas $GaCl_3$ and $InCl_3$ are stable, $TlCl_3$ begins to lose chlorine above 340 K.

16.2. Preparation and Properties of the Elements

The metals are remarkable for the difference between their melting and boiling points (Table 16.2). The extreme example, gallium, is liquid at ordinary pressure over a range of two thousand degrees and has been employed for high-temperature (1250 K) thermometry, in a quartz envelope. It has a strong tendency to superfusion and will remain liquid at room temperature for a considerable period.

Gallium (10^{-4}% of the earth's crust) is present in zinc blende, ZnS, which may contain up to 0.5%, and also in certain coal ash. The metal, which is deposited by electrolysis from alkaline solutions of its salts is silvery-white, hard and brittle. High-purity gallium has been made by the hydrogen reduction of zone-refined gallium trichloride in a quartz vessel. The orthorhombic crystal has a complex lattice. The co-ordination is strictly one-fold, a given gallium atom having one atom situated 243 pm from it and six others at dis-

TABLE 16.1

ATOMIC PROPERTIES OF GALLIUM, INDIUM AND THALLIUM

	Ga	In	Tl
Z	31	49	81
Electron configuration	[Ar]$3d^{10}4s^24p^1$	[Kr]$4d^{10}5s^25p^1$	[Xe]$5d^{10}6s^26p^1$
$I(1)$/kJ mol^{-1}	578	559	588
$I(2)$/kJ mol^{-1}	1960	1810	1950
$I(3)$/kJ mol^{-1}	2940	2690	2860
Covalent radius/pm	126	144	148
Metallic radius/pm	141	166	171
$r_{M^{3+}}$/pm	62	81	95

tances varying from 270 to 279 pm. This co-ordination is denoted by $1 + \bar{6}$; the notation applies only when the six atoms are separated from the reference atom by a distance not greater than 1.2 times that of the nearest one. The metal is stable in dry air and does not decompose water. It dissolves in caustic alkalis and in mineral acids, other than nitric which renders it passive. Its chemistry (Fig. 16.1) is, in fact, very similar to that of aluminium.

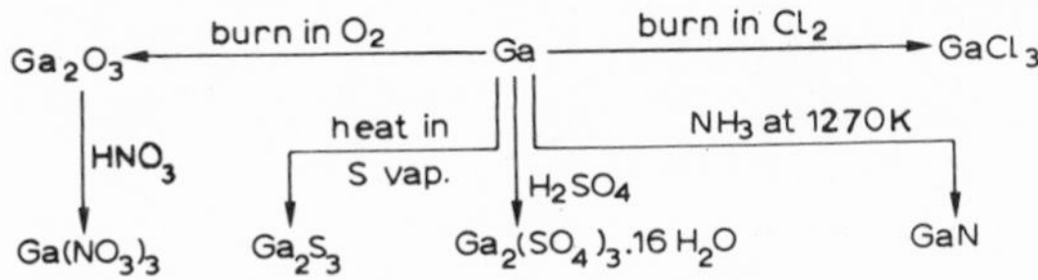

Fig. 16.1. Outline chemistry of gallium.

Indium (10^{-5} % of the earth's crust) is also present in zinc blende, but rarely above 0.1%. It can be precipitated from solution by zinc and purified electrolytically. Its tetragonal unit cell is a very slightly distorted version of the f.c.c. of a true metal (c/a = 1.08). The metal is soft. It differs from aluminium and gallium in its insolubility in boiling caustic alkalis; otherwise its reactions are similar.

TABLE 16.2

PHYSICAL PROPERTIES AND ELECTRODE POTENTIALS

	Ga	In	Tl
ρ/g cm^{-3}	5.93	7.29	11.85
M.p./K	303	429	577
B.p./K	2340	2370	1660
E^0, M^{3+}/M/V	−0.52	−0.34	+0.72
E^0, M^+/M/V		−0.25	−0.34

Thallium (10^{-5}% of earth's crust) is recovered principally from the flue dust of pyrites burners. The soft, grey metal, which has a hexagonal close-packed structure, is rather more reactive than gallium and indium because of the ease with which it forms a unipositive ion. It oxidises in moist air, decomposes steam at red heat and dissolves readily to form thallium(I) compounds in dilute mineral acids other than HCl, because of the insolubility of TlCl.

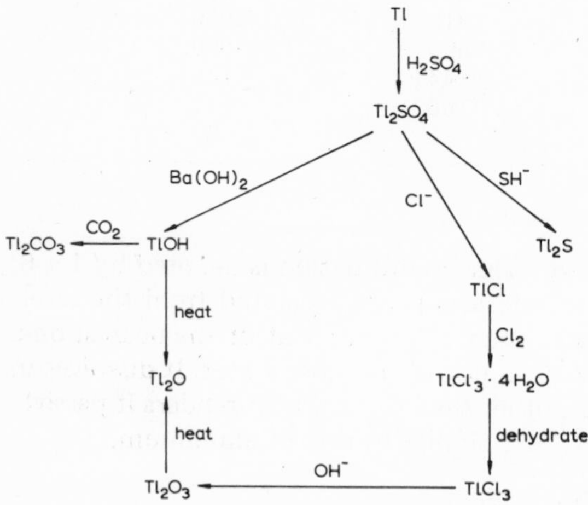

Fig. 16.2. Outline chemistry of thallium.

16.3. Halides

The trifluorides are ionic solids of high *m.p.* and low solubility similar to AlF_3. The other trihalides are soluble, though strongly hydrolysed, and largely covalent when anhydrous. The trichloride of gallium is a white solid, *m.p.* 351 K. The vapour at 770 K consists largely of the dimeric molecules Ga_2Cl_6; the liquid, on the evidence of its Raman spectrum, appears to be dimeric also. The dichloride, $GaCl_2$, can be made by heating $GaCl_3$ with the metal at 450 K. At a higher temperature $GaCl_2$ disproportionates:

$$3\ GaCl_2 \rightleftharpoons 2\ GaCl_3 + Ga$$

However, the corresponding iodide disproportionates into the triiodide and monoiodide:

$$2\ GaI_2 \rightleftharpoons GaI_3 + GaI$$

Although the empirical formula $GaCl_2$ appears to indicate bipositive gallium, the material is not paramagnetic, as it would be with a $GaCl_2$ unit containing an odd electron. Its structure is ionic, containing the tetrahedral $GaCl_4^-$ anion (which is well known) and the Ga^+ cation. The crystals are isomorphous with those of $GaAlCl_4$.

The monochlorides of both gallium and indium have been made by heating the metals in argon containing 1% Cl_2.

Indium trichloride, $InCl_3$, has a much higher *m.p.* (841 K) than the other trichlorides of the group. There is no evidence of dimerisation and the fused material is a fairly good conductor. The tribromide, too, is far more ionic in character than the tribromides of Ga and Tl.

Thallium(III) chloride hydrate, $TlCl_3 \cdot 4\,H_2O$, can be made by passing Cl_2 into an aqueous suspension of TlCl and evaporating at 330 K. The anhydrous chloride melts at 330—340 K; it is unstable, and further heating converts it to TlCl. It resembles BCl_3 in giving stable addition compounds with NH_3 and Et_2O and in not forming a dimer. Material of the composition TlI_3, made by treating TlI with iodine, is not thallium(III) iodide but a polyiodide containing the Tl^+ ion, and is isomorphous with RbI_3 and CsI_3. On heating it decomposes in two stages:

$$TlI_3 \rightarrow Tl_3I_4 \rightarrow TlI$$

Thallium(I) chloride, bromide and iodide are made by precipitation with the appropriate halide ion from a thallium(I) sulphate solution. TlCl resembles AgCl in solubility, structure and sensitivity to light but is insoluble in ammonia; the Tl^+ ion is evidently too large to form stable ammonia complexes. TlF is yellow and resembles AgF in colour, structure and solubility.

16.3.1. Halogen complexes

Halogen complexes are formed by the three metals. The fluorogallates have the octahedral GaF_6^{3-} ion but the solids are hydrates and differ in structure from cryolite. Hexafluorogallic acid, H_3GaF_6, has been made by the action of HF on $Ga[GaCl_4]$. Complexes of the general formula $[GaL_4]^+[GaX_4]^-$ (X = Cl or Br), with such ligands (L) as ethers and thioethers, are formed by allowing these volatile compounds to diffuse into $Ga[GaX_4]$ dissolved in benzene. Chelating ligands also form complexes of gallium(I) tetrachlorogallate(III) such as $[Ga(acac)_2]^+[GaCl_4]^-$ and $[Ga(dipy)_2]^+[GaCl_4]^-$.

Indium gives hydrated 6 co-ordinate complexes, $M_3[InCl_6] \cdot H_2O$.

Thallium has four types of chlorocomplexes:

$M^I[TlCl_4] \cdot x\,H_2O$ $\quad M^I_2[TlCl_5] \cdot x\,H_2O$ $\quad M^I_3[TlCl_6] \cdot x\,H_2O$ $\quad M^I_3[Tl_2Cl_9] \cdot x\,H_2O$

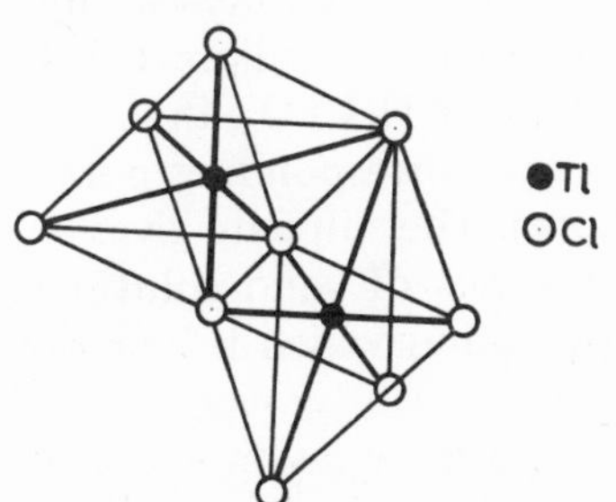

Fig. 16.3. Structure of the $Tl_2Cl_9^{3-}$ ion.

The $[Tl_2Cl_9]^{3-}$ ion is composed of two octahedra fused together by sharing three corners and a common face (Fig. 16.3). Bromocomplexes are generally similar to the chlorocomplexes. Iodocomplexes are known for thallium but not for the others.

The luminescence of solutions of thallium(I) chloride with alkali metal chlorides is due to the absorption of light by the $[TlCl_2]^-$ ion.

16.4. Oxides

The oxides, M_2O_3, can all be made by heating the metals in oxygen. The heats of formation, though high, are much lower than that of Al_2O_3, and the oxides are easily reduced.

TABLE 16.3

HEATS OF FORMATION OF Al_2O_3, Ga_2O_3 AND In_2O_3

	Al_2O_3	Ga_2O_3	In_2O_3
ΔH_f/kJ mol^{-1}	−1680	−1080	−930

The gallium(III) oxide system is very similar to that of aluminium oxide. There are high-temperature α-forms and low-temperature γ-forms of Ga_2O_3, GaO(OH) and $Ga(OH)_3$. Yellow indium(III) oxide has only one form, as has its hydrate, $In(OH)_3$. The dark-brown thallium(III) oxide can be precipitated from solutions of Tl^{III} salts by OH^- ions. It decomposes at about 370 K to give the black oxide Tl_2O. Thallium(I) hydroxide, made by adding baryta water to Tl_2SO_4 solution and evaporating, is a yellow crystalline solid which dissolves to give an alkaline solution. TlOH is almost as strong a base as KOH.

16.5. Thallium(I) Salts

The single charge and large size of the Tl^+ ion endows its compounds with many properties reminiscent of those of the alkali metals. For instance, alkaline TlOH absorbs CO_2 to form a solution of Tl_2CO_3, which hydrolyses similarly to K_2CO_3. The hydroxide is evidently a strong base. Thallium(I) sulphate, Tl_2SO_4, the orthophosphates, Tl_3PO_4, Tl_2HPO_4 and TlH_2PO_4, the chlorate and the perchlorate are all isomorphous with the corresponding potassium salts (K^+ radius = 133 pm, Tl^+ radius = 140 pm). The sulphate gives an alum $TlAl(SO_4)_2 \cdot 12\,H_2O$. It forms a continuous series of solid solutions with $(NH_4)_2SO_4$ and K_2SO_4 and a double salt with $CuSO_4$ which is isomorphous with $(NH_4)_2SO_4 \cdot CuSO_4 \cdot 6\,H_2O$.

In its halides and sulphide, unipositive thallium bears some resemblance to silver (Ag^+ radius = 126 pm); Tl_2S is precipitated by H_2S only from slightly alkaline solutions. The variable charge number of thallium, according as its

$6s^2$ shell is disturbed or not, is largely responsible for the diversity of its properties.

16.6. Other Compounds

The sulphides M_2S_3 are all made by direct combination of the elements. But GaS is also known and has an unusual layer lattice containing Ga_2^{4+} ions. The nitride GaN, unreactive to water and acids, is made by heating gallium in ammonia at 1250 K or Ga_2O_3 in ammonia at 750 K. The corresponding indium compound is best made by heating $(NH_4)_3InF_6$. Both nitrides have the wurtzite lattice.

Gallium resembles aluminium in not forming a carbonate, but a basic carbonate of indium can be precipitated from solution. Gallium(III) sulphate, $Ga_2(SO_4)_3 \cdot 16\ H_2O$, and indium(III) sulphate, $In_2(SO_4)_3$, both form alums, but thallium(III) sulphate, $Tl_2(SO_4)_3 \cdot 7\ H_2O$, obtained by dissolving Tl_2O_3 in dilute sulphuric acid, does not. The Tl^{3+} ion (radius 95 pm) is evidently too large for the lattice, resembling those of the lanthanides in this respect.

Hydrated salts are common and they usually contain octahedral $M(H_2O)_6^{3+}$ ions. The hydrated ions in aqueous solution are moderately strong acids. For the protolysis reactions

$$M(H_2O)_6^{3+} + H_2O \rightleftharpoons M(H_2O)_5OH^{2+} + H_3O^+$$

the pK_a values are 2.6 (Ga), 3.7 (In) and 1.2 (Tl), respectively. Consequently the salts of the terpositive ions are extensively hydrolysed, and their salts of weak acids, sulphides, carbonates, acetates and cyanides for example, cannot exist in the presence of water.

Bridge bonds of the kind formed by boron (the first member of Group III) are also formed by both aluminium and gallium, though less frequently. An example is

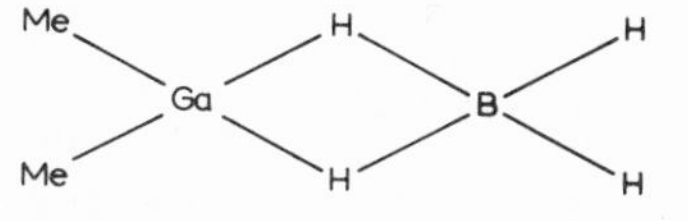

Me H H
Ga B
Me H H

16.7. Organometallic Compounds

Trialkylgalliums, R_3Ga, have been made from gallium trihalides and aluminium alkyls in the presence of KCl, or by a prolonged heating of gallium metal with mercury dialkyls. They are spontaneously inflammable in air and are hydrolysed by water:

$$R_3Ga + 2\ H_2O \rightarrow RGa(OH)_2 + 2\ RH$$

But they form very stable complexes with ethers and amines. Triethylgallium (*b.p.* 416 K) is monomeric in the vapour but dimeric in benzene.

Trimethylindium (*m.p.* 361 K) is a polymeric solid made from indium and Me_2Hg at 373 K. The structure of the solid is an unusual one; it contains unsymmetrical bridges. In the vapour and in benzene, however, the compound is monomeric. It is hydrolysed by water to $MeIn(OH)_2$, but dilute mineral acids, HX, give InX_3. The other known trialkyls of indium are monomeric, as is also $(C_6H_5)_3In$, made from diphenylmercury and indium.

The trialkyls and triaryls of thallium are thermally unstable. The solid Me_3Tl is made on treating a thallium(I) halide with methyl-lithium and methyl iodide:

$$2\,MeLi + MeI + TlX \rightarrow LiX + LiI + Me_3Tl$$

It is spontaneously inflammable in air, but hydrolyses in water only as far as Me_2TlOH. It dissolves in ether and benzene in which it behaves as a monomer.

When thallium trihalides are treated with Grignard reagents, peculiarly stable salts are formed:

$$2\,RMgX + TlX_3 \rightarrow [R_2Tl]^+X^- + 2\,MgX_2 \qquad (R = \text{alkyl or aryl})$$

These salts are not hydrolysed by water. The Me_2Tl^+ ion in aqueous solution has been shown by Raman spectroscopy to have a linear arrangement C—Tl—C, as in the isoelectronic Me_2Hg molecule.

16.8. Complexes

In their terpositive states the Group IIIB metals figure in a number of complexes. The trihalides of gallium and indium form many ammines, $MX_3(NH_3)_n$ in which n can be as large as seven. The corresponding thallium(III) ammines hydrolyse rapidly in water, but the ethylenediamine-complexes $Tl(en)_nX_3$ (n = 1, 2 or 3), are hydrolysed only slowly. The indium trihalides form 1:2 complexes with some sulphur-containing ligands such as Me_2S, MeSH and tetrahydrothiophen.

The most stable oxo-complexes are chelate compounds. Gallium and indium form tris-β-diketone complexes, soluble in alcohol and benzene, and structurally similar to those of aluminium (Fig. 16.4). Gallium and thallium

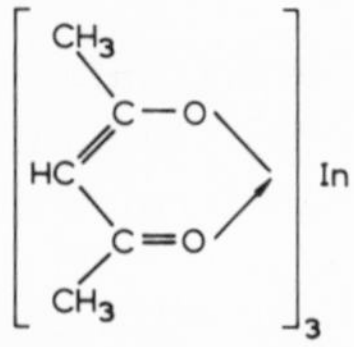

Fig. 16.4. Tris-β-diketone complex of indium.

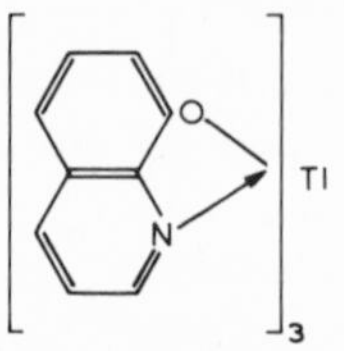

Fig. 16.5. Tris-complex of thallium with oxine

form trioxalato compounds like the oxalato-aluminates: $M^I_3[Ga(C_2O_4)_3] \cdot H_2O$. All the metals form tris-complexes with oxine (Fig. 16.5).

The three elements all form alkoxides similar to those of aluminium. These can be considered to be complexes because they are all polymeric even when dissolved in inert solvents.

Further Reading

A.J. Banister and K. Wade, The chemistry of aluminium, gallium, indium and thallium, Pergamon, Oxford, 1975.

I.A. Sheka, I.S. Chaus and T.T. Mityureva, The chemistry of gallium, Elsevier, Amsterdam, 1966.

A.G. Lee, The chemistry of thallium, Elsevier, Amsterdam, 1971.

Carbon and Silicon—Group IV

17

17.1. Introduction

The electronic structure of the carbon atom gives the element a remarkable diversity of chemical properties. The four valence electrons occupy the configuration $2s^22p^2$ in the ^3P ground state but promotion and hybridisation to $2s2p^3$ occurs very freely and accounts for the characteristic valency of four which is never exceeded.

Silicon differs from carbon in that 3d orbitals are accessible, giving a greater variety and an increased number of valency states. Si can thus exhibit a covalency greater than 4, as in the SiF_6^{2-} ion, by the use of sp^3d^2 hybrid orbitals. Although the ionisation energies (Table 17.1) indicate that silicon is capable of forming positive ions more easily than carbon, both elements form mainly covalent compounds. Carbon also exhibits the remarkable property of catenation, the ability of an element to bond to other atoms of the same element. The reason for this property is that the strength of the C—C single bond (Table 17.3) is exceptionally high for a homopolar single bond and together with the very large C—H bond strength accounts for the occurrence of the numerous compounds described as organic. In this context however it is worth noting that ΔH_f for n-alkanes are —ve but ΔG_f values become +ve after n-hexane. Clearly neither carbon nor hydrogen in the alkanes has suitable orbitals vacant, nor can their co-ordination numbers be increased,

TABLE 17.1

ATOMIC PROPERTIES OF THE ELEMENTS

	C	Si
Atomic number	6	14
Electron configuration	$[He]2s^2 2p^2$	$[Ne]3s^2 3p^2$
$I(1)/kJ\ mol^{-1}$	1090	786
$I(2)/kJ\ mol^{-1}$	2360	1575
$I(3)/kJ\ mol^{-1}$	4620	3220
$I(4)/kJ\ mol^{-1}$	6220	4350
Covalent radius/pm	77	117

TABLE 17.2

PHYSICAL PROPERTIES OF THE ELEMENTS

		C		Si
	Diamond		Graphite	
ρ/g m^{-3}	3.51		2.25	2.33
M.p./K		3820		1690
B.p./K		5100		2970

hence the energies of activation for hydrolysis and similar reactions are high. This is not so true of 3-co-ordinate carbon.

The single bond Si—Si occurs quite commonly but it is very much weaker, particularly when compared with the bonds which silicon forms with oxygen or halogens. Consequently silicon to silicon chains are short and are easily broken, so that compounds containing them are difficult to characterise.

17.2. The Elements

Elementary carbon occurs in two crystalline forms. In diamond the C atoms are arranged tetrahedrally and equidistant (Fig. 17.1), C—C = 154 pm. They are bound covalently by electron pairs which occupy localised molecular orbitals formed by an overlapping of the sp^3 hybrids. This structure confers great hardness on the crystal but permits of four well-defined cleavages.

In a graphite crystal the layers of carbon atoms are in a regular hexagonal network with a C—C bond length of 142 pm, indicating a bond order of 1.33, the different layers being 340 pm apart (Fig. 17.2). Every carbon atom is bound to three others in the layer by covalent bonds which can be described by localised molecular orbitals, built up by overlapping sp^2 hybrids, holding two electrons in each orbital. The electrons in the unhybridised p orbitals form a mobile system of metallic type. In single crystals the material

TABLE 17.3

MEAN BOND ENERGIES E(M—X)/kJ mol^{-1} FOR CARBON AND SILICON

C—H	416	Si—H	293
C—C	348	Si—C	291
C=C	620	Si—Si	200
C≡C	810		
C—N	293		
C=N	617		
C≡N	880		
C—O	344	Si—O	370
C=O	710		
C—Cl	331	Si—Cl	360

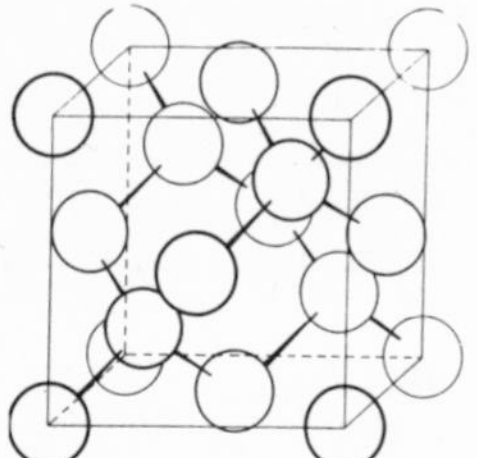

Fig. 17.1. Diamond structure.

Fig. 17.2. Graphite structure.

is soft and the cleavage well developed, one layer of atoms sliding easily over another. Carbon has particularly high *m.p.* and *b.p.*

Carbon (0.08% of the earth's crust) occurs in the elementary state as diamond, graphite and other, less ordered, forms of the element; in solid, liquid and gaseous hydrocarbons, and in mineral carbonates such as limestone, magnesite and dolomite. The atmosphere contains 0.03% of CO_2 by volume, from which source a great variety of compounds of biological generation are derived.

Clear diamonds, occasionally in coloured form, are used as gems because the high index of refraction (~2.42) enhances internal reflexion and brilliance in the cut stone. Diamonds are of great industrial importance: larger, opaque, natural diamonds are mounted in tools for cutting metal and rock, and diamond powder is a widely used abrasive for arming grinding and cutting wheels. Some of the smaller sized material is manufactured by heating graphite to temperatures of ~3000 K under pressures exceeding 10^4 MPa, in the presence of a little metal (e.g. Mn, Fe or Co) a film of which is presumed to dissolve carbon as graphite and allow it to crystallise as diamond.

Graphite, of which the natural supply is limited, is manufactured in various ways depending upon the purpose for which it is intended. Finely divided material, pure and soft for lubrication, is produced in the Acheson process. In this, powdered coke is heated for about a day to temperatures reaching ~2800 K, by letting it serve as the resistance in an electric furnace. The conversion to graphite is probably catalysed by the presence of a little silicon derived from the coke or the furnace walls.

Massive graphite required for electrodes and other refractory purposes is made by mixing powdered coke (see below for varieties) with pitch, moulding, pressing, or extruding the plastic material to shape, heating to ~1500 K in order to drive off volatiles, and finally graphitising at ~2800 K by making the pieces resistance elements in an electric furnace. The purity depends on the raw materials used and the precautions taken during the processing; it must be high when the products are intended for electrodes or for building into atomic reactors. The graphite blocks for the latter purpose are formed by an extrusion process which orients the crystallites and renders the finished material anisotropic.

Graphite is singular in being the only known material the mechanical properties of which improve with rising temperature; accordingly it is used for the dies and plungers in high-temperature (>1750 K) hot pressing of metal powders such as those of beryllium and of the refractory 'hard metals' such as TiC.

Other artificial carbons are:

(i) Charcoal, made by carburising wood, cellulose or sugar. The ash in the product depends on the starting material and can be very low.

(ii) Coke, made by carburising coal, varies widely in composition and mechanical properties; that made by carburising residues from the distillation of pitch and natural oil is a more uniform product.

(iii) Gas carbon, found in the upper part of retorts used for gas manufacture.

(iv) Animal charcoal, made by charring treated bones, consists of finely divided carbon supported on calcium phosphate. The carbon has a high surface area and is used to decolorise solutions by adsorbing the colouring matter.

(v) Carbon black, made by burning natural gas in a deficiency of air and collecting the soot on cooled metal plates, is low in ash, but contains tars and liquid and gaseous hydrocarbons. It also has a large surface area and is added as a catalyst in rubber vulcanisation and also as a filler in rubber manufacture.

(vi) Carbon fibres, made by carefully controlled pyrolysis of selected polymers such as polyacrylate esters and polyacrylonitrile. This material has a particularly great strength/weight ratio.

The first five are usually described as amorphous but most of them show some crystallinity, the more so the higher the temperature of preparation.

Diamond does not ignite in oxygen below ~1050 K and is only slowly attacked by sulphur vapour at 1250 K. Graphite reacts a little more readily, igniting in oxygen at 960 K. Both burn with a bright, flameless glow to carbon dioxide. The structure of diamond renders it chemically unreactive but that of graphite allows penetration between the layer planes of carbon atoms. Thus, though not attacked by dilute acids, it is converted to graphite oxide by a mixture of concentrated H_2SO_4 and HNO_3 to which a little $KClO_3$ has been added. Other lamellar compounds (17.10) are formed when alkali metals penetrate between the planes in graphite.

Silicon (25.7% of the earth's crust), though only about half as plentiful as oxygen, is the second most abundant element. It occurs extensively in many forms of SiO_2, silicates and aluminosilicates. The element, hard, grey and crystalline, is made commercially by heating silica with carbon or CaC_2 in an electric furnace. Crystalline silicon, made by the reduction of a silicon tetrahalide with hydrogen in a hot tube, has the diamond structure with an interatomic distance of 234 pm. It can be purified by zone refining (19.3) until the impurity content is less than 10^{-7}%, when it may be employed in semiconductor devices such as transistors.

Silicon dissolves, generally with the formation of a silicide, in most metals, exceptions being Bi, Pb and Tl. It is used as a deoxidiser and an alloying constituent in steel making, and in massive proportions in the manufacture of acid-resistant iron. For these purposes the silicon is usually added as ferrosilicon prepared by the electrochemical reduction of SiO_2 and Fe_2O_3 with carbon.

Silicon is chemically more reactive than carbon. It burns in oxygen at 670 K, the reaction being strongly exothermic:

$$Si + O_2 \rightarrow SiO_2 \qquad \Delta H = -800 \text{ kJ mol}^{-1}$$
$$C + O_2 \rightarrow CO_2 \qquad \Delta H = -393 \text{ kJ mol}^{-1}$$

It combines directly with all the halogens at temperatures ranging from 570 K upwards, with sulphur vapour at 870 K, with nitrogen at 1570 K and with carbon at 2250 K. Though acid resistant (except to HF) it is attacked by hot alkalis:

$$Si + 2\,NaOH + H_2O \rightarrow Na_2SiO_3 + 2\,H_2$$

and by steam at red heat:

$$Si + 2\,H_2O \rightarrow SiO_2 + 2\,H_2$$

17.3. Halides

The halides of carbon are very numerous because of the tendency of carbon atoms to form chains. A limited number of halides containing Si chains have been made.

Of the simple tetrahalides, CF_4 and SiF_4 are gases; CCl_4, $SiCl_4$ and $SiBr_4$ liquids; CBr_4, CI_4 and SiI_4 solids.

Carbon tetrachloride is made on a large scale by passing Cl_2 into CS_2 in the presence of a little iodine:

$$CS_2 + 3\,Cl_2 \rightarrow \underset{(b.p.\ 411\ K)}{S_2Cl_2} + \underset{(b.p.\ 350\ K)}{CCl_4}$$

Silicon tetrafluoride is conveniently made by treating a mixture of fluorite and silica with concentrated H_2SO_4:

$$2\,CaF_2 + 2\,H_2SO_4 + SiO_2 \rightarrow 2\,CaSO_4 + SiF_4 + 2\,H_2O$$

The water is retained by the sulphuric acid, and the SiF_4 is freed from HF by passing it over dry NaF, whereby solid $NaHF_2$ is formed. The other silicon tetrahalides are generally made by direct combination with the halogen. Mixed tetrahalides such as SiF_3Cl and $SiCl_2Br_2$ have also been obtained.

The fluorides of carbon are interesting. Fluorine enters the graphite lattice at 470 K to form the interstitial compound $(CF)_n$. At higher temperatures the elements give a mixture of CF_4, C_2F_4, C_2F_6 and C_3F_8. Fluorocarbons are conveniently made by passing hydrocarbons over cobalt(III) fluoride at about 450 K or chlorocompounds over SbF_3 (Swarts' reaction). Under these conditions when CCl_4 replaces the hydrocarbon a mixture of CCl_3F, CCl_2F_2, CCl_3F and CF_4 is obtained. The mixed fluorochloro-carbons, known as freons, are useful refrigerants, being volatile, non-toxic and non-corrosive. Higher boiling fluorocarbons form important lubricants which are less reactive and less sensitive to heat than corresponding hydrocarbons.

The fluorocarbons are inert and their derivatives often have very different properties from those of the hydrocarbons because of the high electronegativity of fluorine. For instance, $(CF_3)_3N$ is not basic; the attraction of electrons by the fluorine atoms prevents the nitrogen acting as a donor.

Carbon halides resist hydrolysis because only s and p orbitals are available for bond formation. This restricts the maximum covalency to four and precludes the donation of electrons by the oxygen atom in a water molecule to a carbon atom. But silicon halides do hydrolyse, since the unoccupied silicon 3d orbitals lie not far above the 3s and 3p:

$$SiX_4 + 2\,H_2O \rightarrow SiO_2 + 4\,HX$$

However, with silicon tetrafluoride, the HF formed reacts with some of the tetrafluoride to form fluorosilicic acid:

$$2\,SiF_4 + 4\,H_2O \rightarrow SiO_2 + 2\,H_3O^+ + SiF_6^{2-} + 2\,HF$$

The octahedral SiF_6^{2-} ion is the only halogeno-complex of silicon; the bonding involves sp^3d^2 hybrids and accordingly carbon does not form such a compound. Fluorosilicic acid, H_2SiF_6, known only in solution, is a strong acid. Its heavy-metal salts are soluble, and those of Na, K, Ba and the lanthanides sparingly soluble.

Silicon forms a limited number of halides containing Si chains; there is one of these with every halogen element in compounds of the general formula Si_2X_6, additional chlorides up to Si_6Cl_{14}, and bromides up to Si_3Br_8. The chloride Si_5Cl_{12} made by the action of trimethylamine on Si_2Cl_6 is probably $Si(SiCl_3)_4$, with a neopentane structure.

Differing from carbon, silicon resembles Ge, Sn and Pb in forming dihalides. The relation between pressure and temperature in the $Si/SiCl_4$ system at high temperatures shows an equilibrium to exist:

$$Si + SiCl_4 \rightleftharpoons 2\,SiCl_2$$

But the dichloride is present in appreciable quantities only above 1400 K.

Carbon forms oxohalides: COF_2 and $COCl_2$ are colourless gases made by union of carbon monoxide and the halogen. The molecules are planar, a form which suggests sp^2 hybridisation. In $COCl_2$ the small angle of 112° between the C—Cl bonds is due to a strong repulsion exerted on each by the spin-paired electrons of the C—O bond. Carbonyl bromide, $COBr_2$, is a colourless liquid best made by dropping concentrated H_2SO_4 on to CBr_4:

$$CBr_4 + H_2SO_4 \rightarrow 2\ HBr + SO_3 + COBr_2$$

All three compounds are easily hydrolysed:

$$COX_2 + H_2O \rightarrow 2\ HX + CO_2$$

Silicon oxohalides of structure,

$$X\left[-\underset{X}{\overset{X}{\underset{|}{\overset{|}{Si}}}}-O-\underset{X}{\overset{X}{\underset{|}{\overset{|}{Si}}}}-\right]_n X$$

(n = 1, 2 or 3) can be obtained either by treating SiO_2 with a mixture of oxygen and chlorine or bromine, or by partially hydrolysing the silicon tetrahalide with moist ether. The fluoride Si_2OF_6 is made by fluorinating the corresponding chloride:

$$Si_2OCl_6 \xrightarrow[SbCl_5]{SbF_3} Si_2OF_6$$

17.4. Oxides of Carbon

17.4.1. Carbon suboxide

Carbon suboxide, C_3O_2, is a gas formed when malonic acid, or one of its esters, is heated with P_2O_5. When dry, it is fairly stable at room temperature but polymerises readily on warming; the liquid (*b.p.* 279 K) so produced further polymerises to a dark red, water-soluble solid. C_3O_2 behaves as the anhydride of malonic acid:

$$C_3O_2 + 2\ H_2O \longrightarrow CH_2(COOH)_2$$

$$C_3O_2 + 2NH_3 \longrightarrow CH_2(CONH_2)_2$$

A mixture with oxygen explodes when sparked. The molecule is linear, but the structure

$$O{=}C{=}C{=}C{=}O$$

is an over-simplification; the bonds are all somewhat shorter than normal double bonds, the carbon—carbon distance being 128 pm (cf. 133 pm) and the carbon—oxygen, 116 pm (cf. 122 pm). There are two π-bond systems (cf. N_2 triple bond) but these are non-localised, and contribute only *fractional* π bonds.

17.4.2. Carbon monoxide

Carbon monoxide, CO, can be made by the dehydration of formic acid, but is too insoluble to be considered as the acid anhydride. The gas is produced industrially in large quantities by the 'producer-gas' reaction between carbon and CO_2 at high temperatures:

$$CO_2 + C \rightleftharpoons 2\ CO$$

Carbon monoxide is also one of the products of the 'water-gas' reaction:

$$H_2O + C \rightarrow CO + H_2$$

It reacts with chlorine and bromine in sunlight, and with molten sulphur and heated selenium in the dark:

$$CO + Br_2 \rightarrow COBr_2 \qquad CO + S \rightarrow COS \qquad CO + Se \rightarrow COSe$$

Several transition metals give carbonyls (17.12) and some salts and complexes of these metals also combine with carbon monoxide:

$$2\ CO + 2\ PtCl_2 \rightarrow [PtCl_2CO]_2$$

The reaction of hydrogen with CO provides the basis of an industrial process for making methanol; it is carried out at high pressure over a mixed copper—zinc catalyst. Though CO is insoluble in, and unreactive with, water at ordinary pressures, formic acid is produced at high pressures. Under similar conditions CO and aqueous NaOH combine to give sodium formate.

The CO molecule is isoelectronic with N_2 and its molecular orbitals are formally the same:

$$C(1s^2\ 2s^2\ 2p^2) + O(1s^2\ 2s^2\ 2p^4) \rightarrow CO(KK(z\sigma)^2(y\sigma)^2(x\sigma)^2(w\pi)^4)$$

However the CO^+ ion has a shorter carbon—oxygen bond than the neutral molecule, indicating that the highest occupied orbital is non-bonding, whereas the N_2^+ ion has a longer nitrogen—nitrogen bond than the neutral molecule. It is likely that in carbon monoxide the oxygen 2s electrons form a lone pair ($z\sigma$) and that one carbon sp hybrid holds a second lone pair ($y\sigma$). The bonding orbital $x\sigma$ would then be formed by overlap of a carbon sp hybrid and an oxygen p orbital. Lateral overlap of the remaining (singly occupied) 2p orbitals results in two π-type molecular orbitals, leaning somewhat towards the oxygen. Ionisation is probably from the carbon lone pair and not from a π-bond orbital. The dipole moment of the π electrons is strongly offset by the carbon lone pair; this lone pair is also responsible for the coordinating power displayed in the carbonyls.

17.4.3. Carbon dioxide

Carbon dioxide is the most stable oxide of carbon at room temperature. Industrially it is recovered from flue gases and lime kilns, the SO_2 and H_2S present being removed from the gases by scrubbing with aqueous $KMnO_4$ and Na_2CO_3. Carbon dioxide dissolves in water giving, at S.T.P., a 0.04 molar solution in which the carbonic acid is only slightly ionised:

$$H_2O + CO_2 \rightleftharpoons H_2CO_3 \overset{H_2O}{\rightleftharpoons} H_3O^+ + HCO_3^-$$

The equilibrium moves to the left as the temperature is raised and the gas can be completely expelled from the water. The pH of a saturated solution of the gas at 10^2 kPa is 3.7. Commercially the gas is solidified as *dry ice*, added to oxygen as a respiratory stimulant, and to flavoured water for carbonated drinks.

The molecule is linear with carbon—oxygen distances of 115 pm, which is considerably shorter than that calculated for a double bond (122 pm). Two π-bond systems are expected, in perpendicular planes, but each would extend over all centres (cf. carbon suboxide) contributing slightly more than half a π bond in each link.

17.4.4. Thermodynamics of the oxidation of carbon

The standard molar entropies of the carbon oxides and their elements in $J\ K^{-1}\ mol^{-1}$ are:

	C	O_2	CO	CO_2
S^0	6	204	197	213

Thus the standard entropy changes for the following reactions at 298 K are:

(a) $2\ CO(g) + O_2(g) \rightarrow 2\ CO_2(g)$

$$\Delta S^0 = (2 \times 213) - [(2 \times 197) + 204] = -172\ J\ K^{-1}\ mol^{-1}$$

(b) $C(s) + O_2(g) \rightarrow CO_2(g)$

$$\Delta S^0 = 213 - (204 + 6) = +3\ J\ K^{-1}\ mol^{-1}$$

(c) $2\ C(s) + O_2(g) \rightarrow 2\ CO(g)$

$$\Delta S^0 = (2 \times 197) - [204 + (2 \times 6)] = +178\ J\ K^{-1}\ mol^{-1}$$

Although these values relate to 298 K, they are a guide to the thermodynamics of reactions in the carbon—oxygen system at higher temperatures. For any reaction

$$-\Delta S = \left(\frac{\partial \Delta G}{\partial T}\right)_p$$

is the gradient of the graph of free energy versus temperature. The slope is very nearly linear so long as there is no change in the physical states of the reactants or products (states shown in brackets in the equations) such as the melting of a solid or the condensation of a gas.

Fig. 17.3 shows the free energy—temperature plots of the three oxidation reactions. This composite graph, in which the number of moles of the common reactant (here oxygen) is the same for every reaction, is known as an Ellingham diagram.

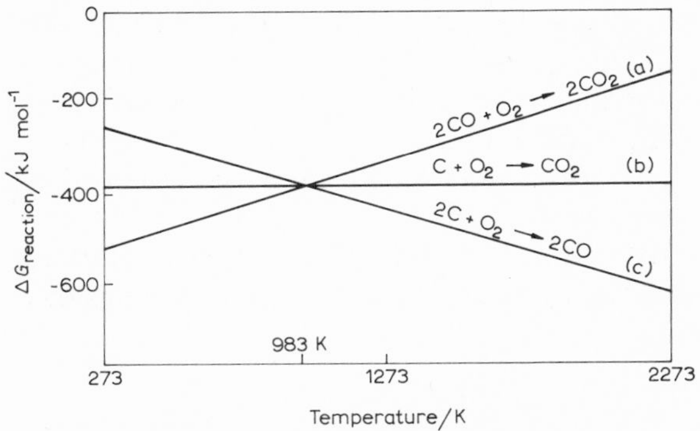

Fig. 17.3. Ellingham diagram showing the variation of the free energy of reaction with temperature for the three reactions which are formulated.

For reaction (a) ΔS is negative and ΔG increases with temperature; for (b) ΔS is almost zero and ΔG hardly changes; and for (c) ΔS is positive and ΔG decreases.

The free-energy change for the reaction

$$2\,CO \rightarrow CO_2 + C$$

is given by ΔG(c) $-$ ΔG(b). For this reaction ΔG is positive above 983 K, but negative below that temperature. Thus higher temperatures favour the

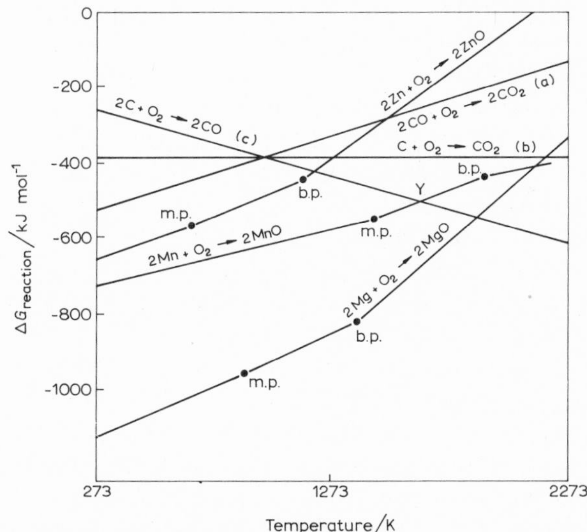

Fig. 17.4. Free energy—temperature plots illustrating the reduction of metallic oxides.

formation of CO and lower temperatures that of CO_2.

By adding to Fig. 17.3 lines representing the free-energy changes for the oxidation of metals, a diagram is produced (Fig. 17.4) from which the effect of temperature on the reduction of the respective oxides to metal by carbon or by carbon monoxide may be deduced.

The first thing to note on the diagram is that all the lines relating to the oxidation of metals are either straight (when they refer to the metal in one phase — solid for manganese) or made up of straight sections in which direction changes at phase-change points (the melting and sublimation points for zinc and magnesium). Secondly, it is observed that all the slopes are positive, that is ΔG increases with temperature.

Normally metal oxides have their atoms in a highly-ordered crystal lattice and therefore the oxides have low entropies. But the entropy of gaseous oxygen is high; thus in the oxidation:

$$\underset{\text{solid}}{2\,M} + \underset{\text{gas}}{O_2} \rightarrow \underset{\text{solid}}{2\,MO}$$

$\Delta S_{\text{reaction}}$ is negative. When, at a higher temperature, oxidation of liquid metal by oxygen occurs, $\Delta S_{\text{reaction}}$ becomes more negative in this reaction above the melting point. Again, it becomes even more negative above the boiling point because disorder in the vapour is greater than in the liquid.

Let us consider particularly the point Y (Fig. 17.4) at which ΔG has the same values for

$$2\,Mn + O_2 \rightarrow 2\,MnO \quad (d)$$

and

$$2\,C + O_2 \rightarrow 2\,CO \quad (c)$$

Moreover, for reaction (e) $2\,MnO + 2\,C \rightarrow 2\,CO + 2\,Mn$, $\Delta G = 0$. Above this temperature $\Delta G(d) > \Delta G(c)$, and $\Delta G(e)$ is therefore negative. At equilibrium the active mass of Mn exceeds that of MnO and reduction of the metal oxide is favoured. But line (d) (Fig. 17.4) reaches line (b) at a much higher temperature. Thus the existence of the stable monoxide of carbon, with its positive entropy of formation, allows many transition-metal oxides, of which MnO is typical, to be reduced by carbon at moderate temperatures. Clearly, if CO_2 were the only oxide of carbon which could be formed, the reduction of these metal oxides by carbon would need higher temperatures. Most metal ores are either oxides or compounds easily converted into oxides, and carbon, in the form of coke, is a cheap, readily-available, reducing agent, hence these thermodynamic facts are of considerable importance.

It can be seen from Fig. 17.4 that magnesium oxide becomes reducible by coke at about 1900 K. The free energies of formation of oxides like CaO and Al_2O_3 are much lower, however, and reduction with coke would require heating the reactants to very high temperatures which would be uneconomic.

Carbon, although thermodynamically easily able to reduce a particular metal oxide, may be unsuitable for that purpose for other reasons, such as the formation of metallic carbides.

17.5. Oxides of Silicon

In contrast to the discrete molecules of carbon dioxide, silicon dioxide forms condensed, three-dimensional systems of indefinite extension which are high-melting solids. Silica has three crystalline forms, quartz, tridymite and cristobalite, all have a low-temperature (α) and a high-temperature (β) modification.

$$\begin{array}{ccccc} \alpha\text{-quartz} & & \alpha\text{-tridymite} & & \alpha\text{-cristobalite} \\ \updownarrow 846\ \text{K} & & \updownarrow 413\ \text{K} & & \updownarrow 513\ \text{K} \\ \beta\text{-quartz} & \underset{1140\ \text{K}}{\rightleftharpoons} & \beta\text{-tridymite} & \underset{1740\ \text{K}}{\rightleftharpoons} & \beta\text{-cristobalite} \end{array}$$

In cristobalite the Si atoms are arranged as are the C atoms in diamond, except that they have O atoms midway between them. In quartz and tridymite the regular structure is replaced by a screw-like arrangement of the atoms. Transitions between the forms take place slowly, and all three are found in nature. However, the transitions between their respective α and β modifications are rapid.

Silica melts at 1983 K. Even when slowly cooled the molten material sets to a vitreous, non-crystalline solid. Its plastic range allows large masses to be forged and hollow-ware to be blown; its low coefficient of expansion renders it immune to thermal shock; its transparency especially to u.v. makes it suitable for lenses and prisms. Specially pure synthetic silica is now made for laboratory ware used in preparing the pure materials required by the electrical industry (e.g. transistors), and for optical parts calling for minimum absorption.

Silica gel is a hard, granular, translucent material containing about 4% H_2O. It results from the removal of salts and water from the continuous gel formed by acidifying the solution of an alkali-metal silicate, and has a large surface area. Silica gel is employed as a drying agent, as a catalyst in the hydrolysis of aryl halides to phenols, and as a support for other catalysts, such as for example the V_2O_5 used in the oxidation of naphthalene to phthalic acid or of SO_2 to SO_3. A silica—alumina gel with 10 to 13% of Al_2O_3 is made by the cogelation of silicate and aluminate solutions with a mineral acid; it is a catalyst for the cracking of petroleum.

The monosilicic acid, $Si(OH)_4$, is probably the only silicic acid species to be found in dilute solutions. However, the isothermal dehydration of silicic acid, prepared by the action of damp air on SiS_2, leads to an acid $H_8Si_4O_{12}$ for which has been suggested the structure:

```
        HO    OH
          \  /
           Si
          /  \
HO      O      O      OH
  \   /          \   /
   Si              Si
  /   \          /   \
HO      O      O      OH
          \  /
           Si
          /  \
        HO    OH
```

Further dehydration by dioxan or SO_2Cl_2 provides less definite evidence for other silicic acids.

17.6. Other Silicon—Oxygen Compounds

When trichlorosilane is hydrolysed with steam at 720 K, a compound, $[(HSiO)_2O]_n$, with strong reducing properties is formed. Its structure resembles that of mica, having large sheets of connected hexagons with Si atoms at the corners and O atoms midway along the sides (Fig. 17.5). When heated to

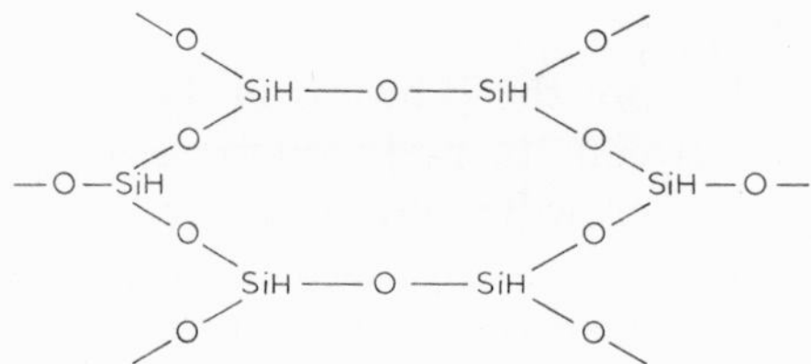

Fig. 17.5. Structure of $[(HSiO)_2O]_n$.

780 K, the compound is converted to Si_2O_3 with a loss of the hydrogen which presumably had provided hydrogen bonds between the sheets; bonds which are now replaced by Si—Si interplanal links.

Siloxene, a similar compound in the form of a flaky, white solid, is made by the action of HCl gas and ethyl alcohol on $CaSi_2$. It is spontaneously inflammable in air and a strong reducing agent. The H atoms can be replaced partly or wholly by halogens:

$$(Si_2H_2O)_n \xrightarrow{Br_2} (Si_2Br_2O)_n \begin{cases} \xrightarrow{H_2O} [Si_2(OH)_2O]_n \\ \xrightarrow{NH_3} [Si_2(NH_2)_2O]_n \end{cases}$$

Siloxene

17.7. Silicones

Hydrolysis of the alkyl and aryl substituted silicon halides produces silicones. The halogen derivatives themselves are made by passing an alkyl or aryl halide over a copper—silicon alloy at about 570 K. Evidence has been found for the mechanism:

$$2\,Cu + CH_3Cl \rightarrow CuCl + CuCH_3$$
$$Si + CuCl \rightarrow Cu + SiCl \text{ (active intermediate)}$$
$$SiCl + CuCH_3 \rightarrow CH_3SiCl + Cu$$

and so on, through $(CH_3)_2SiCl$ and $(CH_3)SiCl_2$ etc., until the four valencies of silicon are saturated. Hydrolysis of $(CH_3)_3SiCl$ gives a disiloxane:

$$(CH_3)_3SiCl + H_2O \rightarrow (CH_3)_3SiOH \text{ (trimethylsilanol)} + HCl$$

$$\downarrow \text{condensation}$$

$$(CH_3)_3Si{-}O{-}Si(CH_3)_3$$

Hydrolysis of $(CH_3)_2SiCl_2$ gives a chain compound:

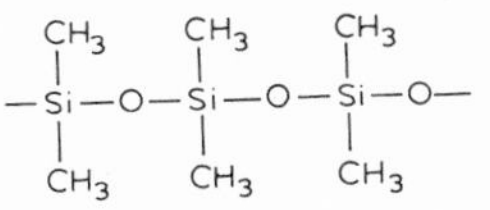

```
  CH3      CH3      CH3
  |        |        |
—Si—O—Si—O—Si—O—
  |        |        |
  CH3      CH3      CH3
```

Hydrolysis of CH_3SiCl_3 gives a cross-linked chain system:

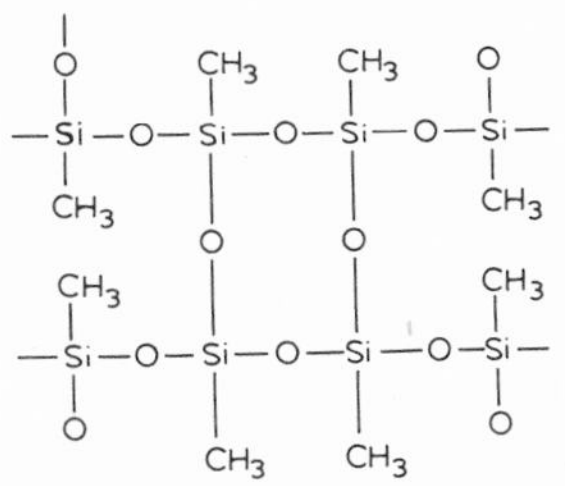

```
  |
  O        CH3      CH3      O
  |        |        |        |
—Si—O—Si—O—Si—O—Si—
  |        |        |        |
  CH3      |        |        CH3
           O        O
  CH3      |        |        CH3
  |        |        |        |
—Si—O—Si—O—Si—O—Si—
  |        |        |        |
  O        CH3      CH3      O
```

The extent of the cross-linking and the nature of the alkyl or aryl substituent determines the nature of the polymers. They range from oily liquids to rubbery solids. All are water-repellant, thermally and electrically insulating and chemically inert. These properties render them widely useful in industry; they appear as lubricants, antifoams, low-temperature hydraulic fluids, and in cosmetics.

Though the copper silicide method is preferred for the production of methyl-substituted compounds, a Grignard-type synthesis is commonly used for other alkyl and aryl substituted substances:

$$Mg + RCl \xrightarrow[\text{ether}]{\text{dry}} RMgCl \xrightarrow{SiCl_4} R_4Si \xrightarrow{SiX_4} R_n\,SiX_{4-n}$$

17.8. Sulphides

17.8.1. Sulphides of carbon

Carbon disulphide, CS_2, is a volatile liquid (*b.p.* 319 K), highly refractive, insoluble in water but soluble in ethyl alcohol and ether. The molecule is linear, with C—S distances of 155 pm, indicating that the bonds are effectively double bonds. The single-bond distance for the same pair of atoms is 179 pm.

It is made by the action of sulphur vapour on electrically heated coke. Its main uses are as a solvent and in the manufacture of CCl_4, viscose rayon and thiocarbanilide. Alkali celluloses combine with CS_2 to give cellulose xanthate:

$$\text{Cellulose} + NaOH + CS_2 \rightarrow S{=}C\begin{cases} O{-}(\text{Cel}) \\ SNa \end{cases} \qquad (\text{Cel} = \text{cellulose residue})$$

This when extruded through a jet into an acid bath gives rayon. Thiocarbanilide, $S{=}C(NH \cdot C_6H_5)_2$, results when CS_2 is passed into boiling aniline. It is used in the manufacture of dyes and pharmaceuticals, and in the vulcanisation of rubber.

A brown polymer, $(CS)_n$, is formed when CS_2 is exposed to light and a subsulphide, C_3S_2, when an arc is struck between carbon poles beneath the liquid. If the arc is between a carbon cathode and a selenium or tellurium anode then the liquids CSSe or CSTe are formed.

17.8.2. Sulphides of silicon

Silicon disulphide forms fibrous crystalline macromolecules with a structure intermediate between that of CO_2, with its individual molecules, and the three-dimensional SiO_2. In it the sulphur atoms are arranged tetrahedrally round the Si atoms:

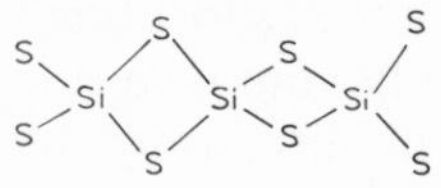

```
S      S            S
 \   /   \   S    /
  Si      Si   Si
 /   \   /   S    \
S      S            S
```

Though solid SiO cannot be obtained from SiO_2, the monosulphide, SiS, can be made by heating SiS_2 with silicon:

$$SiS_2 + Si \rightarrow 2\,SiS$$

17.9. Compounds with Nitrogen

Cyanogen, C_2N_2, is readily evolved when mercury(II) cyanide is heated with mercury(II) chloride:

$$Hg(CN)_2 + HgCl_2 \rightarrow Hg_2Cl_2 + C_2N_2$$

The colourless, very poisonous, gas (*b.p.* 252 K) reacts with alkaline solutions to give a mixture of cyanide and cyanate:

$$C_2N_2 + 2\,OH^- \rightarrow CN^- + CNO^- + H_2O$$

The molecule is linear and without dipole moment, the intermolecular distances being C—N, 116 pm and C—C, 137 pm. Since the lengths are 115 pm for the C≡N bond and 154 pm for the C—C single bond, it suggests that the π bonds of the C≡N groups are sufficiently delocalised to reduce the electron density between carbon and nitrogen and increase it between carbon and carbon.

Silicon nitride, Si_3N_4, is a refractory material made by direct combination of the elements above 1600 K. Another method is to allow the hydrides or halides of silicon to react with ammonia and to heat the amino- and imino-silanes produced. These give the polymer $[Si(NH)_2]_n$ which yields Si_3N_4.

Organosilicon compounds containing Si—N bonds are numerous. Examples are formulated as follows:

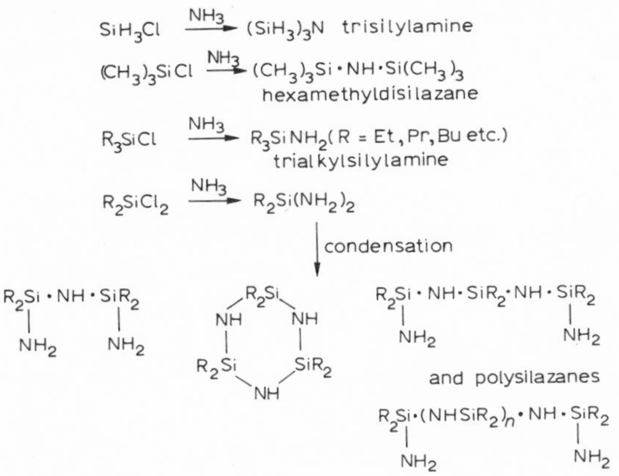

$$SiH_3Cl \xrightarrow{NH_3} (SiH_3)_3N \text{ trisilylamine}$$

$$(CH_3)_3SiCl \xrightarrow{NH_3} (CH_3)_3Si \cdot NH \cdot Si(CH_3)_3 \text{ hexamethyldisilazane}$$

$$R_3SiCl \xrightarrow{NH_3} R_3SiNH_2 (R = Et, Pr, Bu \text{ etc.}) \text{ trialkylsilylamine}$$

$$R_2SiCl_2 \xrightarrow{NH_3} R_2Si(NH_2)_2$$

↓ condensation

$R_2Si(NH_2) \cdot NH \cdot SiR_2(NH_2)$; cyclic $(R_2SiNH)_3$; $R_2Si(NH_2) \cdot NH \cdot SiR_2 \cdot NH \cdot SiR_2(NH_2)$

and polysilazanes

$R_2Si(NH_2) \cdot (NHSiR_2)_n \cdot NH \cdot SiR_2(NH_2)$

When R is a small group (H, Me, Et) the principal products are polysilazanes, but when R is butyl or aryl, the 6-membered ring compounds predominate.

Trihalogenosilanes react with ammonia to give, almost entirely, polysilazanes. Alkylsilanes require much more nucleophilic reagents than do the halogen compounds to convert them to silicon—nitrogen compounds. A solution of lithium in liquid ammonia converts Et_3SiH into a mixture of Et_3SiNH_2, triethylsilylamine, and $(Et_3Si)_2NH$, hexaethyldisilazane.

17.10. Graphite Compounds

Graphite absorbs liquid potassium and at the same time swells in a direction perpendicular to the cleavage. When the excess of potassium is evaporated, there remains a copper-coloured material with the composition KC_8 which is converted, on further heating, to KC_{24}. X-ray examination shows KC_8 to have potassium atoms inserted between every layer of carbon atoms in the graphite structure and KC_{24} to have potassium atoms between alternate layers only. The diamagnetism of the original graphite is absent from KC_8 which is, in fact, a better electrical conductor than solid potassium. Equilibrium pressure measurements have also disclosed the entity KC_{24}. Lamellar compounds MC_8 and MC_{24} (where M represents Rb or Cs) have also been made, their stability increasing with the size of the metal atom. The bonding in these compounds is essentially ionic.

Graphite also swells when heated for a period in a solution of $KClO_3$ in HNO_3, with the formation of graphite oxide, the normal interplanar distance of 340 pm in graphite being increased. Graphite oxide absorbs water. The interplanar distance is ~600 pm when the material is dried over P_2O_5, ~900 pm when kept in air, and ~1.1 nm when soaked in water. A rapid method of preparation is to heat graphite, with an anhydrous mixture of H_2SO_4, $NaNO_3$ and $KMnO_4$ at below 320 K for less than 45 min. Thus prepared, the 'dry' oxide is said to have the composition $C_7O_4H_2$, whatever the graphite used. The C : O ratio has been variously reported, but always as less than 2 : 1. Preparations have ranged from yellow to dark brown. The oxide is unstable, decomposing at ~470 K to CO_2, CO and carbon.

The constitution and structure of graphite oxide has been much investigated and finality has not been reached. But it is probable that the planar aromatic rings of the graphite (Fig. 17.2) are puckered and partly broken to allow both C—OH and C=O bonds to be formed. Although the oxide shows some acidic character ($C{-}OH \rightarrow C{-}O^- + H^+$) the bonding of the interplanar species, in contrast to that in the alkali-metal compounds, is predominantly covalent.

'Graphitic salts' can also be made. For instance sulphuric acid in the presence of strong oxidising agents forms the compound $C_{24}HSO_4 \cdot 2\,H_2SO_4$. Phosphoric, selenic and perchloric acids behave somewhat similarly. Lamellar compounds are also formed by CrO_2Cl_2 and CrO_2F_2; the product of reaction between graphite and fluorine at 470 K, $(CF)_n$, is of the same type, with an inter-layer distance of 817 pm.

17.11. Carbides

The carbides can be divided into four groups: (1) salt-like, (2) interstitial, (3) iron-type, and (4) covalent.

(1) Salt-like carbides

Members of this class can usually be made by heating the metal, its oxide or hydride with carbon, carbon monoxide or a hydrocarbon. The salt-like carbides are easily hydrolysed by water and are classified according to the aliphatic hydrocarbon they give.

(a) The acetylides, such as CaC_2, CrC_2, BaC_2 and MgC_2, give acetylene when treated with water. They are made by heating the oxide and carbon and have tetragonal crystals containing M^{2+} and $C_2{}^{2-}$ ions arranged as are the Na^+ and Cl^- ions in rock salt but with the c axis parallel to the C—C bonds lengthened (c/a ~1.2). In contrast, the compounds Cu_2C_2, Ag_2C_2 and Au_2C_2 are precipitated from aqueous solution, the first by passing acetylene into ammoniacal copper(I) chloride, and the last by passing acetylene into gold(I) thiosulphate. They are formally acetylides, but are not hydrolysed by water and are probably more covalent than CaC_2.

(b) The methanides, such as Al_4C_3 and Be_2C, yield methane on hydrolysis.

$$Al_4C_3 + 12\,H_2O \rightarrow 4\,Al(OH)_3 + 3\,CH_4$$

Both are made by combination of the elements at about 1800 K, and are much harder materials than the acetylides.

(c) Magnesium carbide, Mg_2C_3, which is formed when MgC_2 is heated, and is believed to contain C_3^{2-} ions, yields propyne on hydrolysis:

$$Mg_2C_3 + 4\ H_2O \rightarrow 2\ Mg(OH)_2 + CH_3{-}C{\equiv}CH$$

Thorium and the lanthanides also form carbides, MC_2, when their oxides are heated with carbon in an electric furnace. They have been reported as giving a mixture of acetylene, olefins and hydrogen on hydrolysis. But later work suggests that pure ThC_2 yields only acetylene, making it an acetylide.

(2) Interstitial carbides

These are made by the direct union of metal and carbon or the reduction of the oxide with carbon at about 2300 K. They are very high-melting materials (particularly TaC, *m.p.* 4170 K), good electrical conductors, very hard but brittle, and inert chemically except under oxidising conditions. Cemented carbides, based on WC, are used as hard facings for tools and dies; the metallic binder, usually cobalt, permits fabrication by sintering at 1470 K and gives increased strength. Additions of TiC and TaC serve to vary the properties. Self-bonded carbides are made by hot-pressing the powder without the addition of binder metal at 2300 K.

(3) Carbides of the iron type

These carbides are formed by metals which have a metallic radius below 130 pm. Iron (126 pm), chromium (127 pm) and manganese (127 pm) form carbides, M_3C, with properties intermediate between the salt-like and the interstitial. Structurally they can be regarded as being distorted metal lattices enclosing carbon chains. Hydrolysis is facile with water or dilute acids and gives rise to a variety of hydrocarbons.

(4) Covalent carbides

The binary compounds of carbon with elements of higher electronegativity, e.g. hydrogen, sulphur, chlorine, are gases or volatile liquids, e.g. CH_4, CS_2, CCl_4; they are formally classified as hydrides, sulphides, halides etc., not as carbides. The carbides of silicon and boron are quite different, being thermally stable, hard, chemically-inert solids. Silicon carbide, SiC, exists in three forms related structurally to one another as are diamond, zinc blende and wurtzite. It is made by reducing SiO_2 with carbon in an electric furnace. Boron carbide, B_4C, also made by reducing the oxide with carbon, has a complicated structure in which icosahedra of 12 boron atoms alternate with C_3 chains.

Silicon carbide, widely employed as an abrasive (carborundum), is finding increasing use as a refractory. It has a better thermal conductivity at high temperatures than any other ceramic and is very resistant to abrasion and corrosion especially when bonded with silicon nitride. Hot-pressed, self-bonded SiC may be suitable as a container for the fuel elements in high-temperature, gas-cooled atomic reactors and also for the structural parts of the reactors. Boron carbide, which is even harder than silicon carbide, becomes now readily available commercially because of its value as a radiation shield, and is being increasingly used as an abrasive.

17.12. Metal Carbonyls

Many transition metals of the later groups form volatile, diamagnetic carbonyls in which the charge number of the metal atom is zero. In addition to the mononuclear and binuclear carbonyls listed in Table 17.4 there are the polynuclear carbonyls

$M_3(CO)_{12}$ where M = Fe, Ru and Os,
$M_4(CO)_{12}$ where M = Co, Rh and Ir and
$M_6(CO)_{16}$ where M = Co and Rh.

TABLE 17.4

MONONUCLEAR AND BINUCLEAR METAL CARBONYLS

Group			
V	*$V(CO)_6$ Dark green crystals *d.* 200 K		
VI	$Cr(CO)_6$ Colourless crystals which decompose ~400 K	$Mo(CO)_6$	$W(CO)_6$
VII	$Mn_2(CO)_{10}$ Golden crystals *m.p.* 427 K	$Tc_2(CO)_{10}$ Colourless crystals	$Re_2(CO)_{10}$ Colourless crystals *m.p.* 450 K
VIIIA	$Fe(CO)_5$ Yellow liquid *b.p.* 376 K	$Ru(CO)_5$ Colourless liquid *f.p.* 251 K	$Os(CO)_5$ Colourless liquid *f.p.* 258 K
	$Fe_2(CO)_9$ Bronze plates *d.* 373 K		
VIIIB	$Co_2(CO)_8$ Orange crystals *m.p.* 324 K		
VIIIC	$Ni(CO)_4$ Colourless liquid *b.p.* 316 K		

* This compound is unlike the others in being paramagnetic.

17.12.1. Preparation of carbonyls

Tetracarbonylnickel, first reported as early as 1890, can be made from the freshly reduced metal and carbon monoxide at ordinary temperature and pressure:

$$Ni + 4\ CO \rightarrow Ni(CO)_4$$

A good yield is obtained by passing CO over nickel pellets, freezing out the $Ni(CO)_4$, and recycling the stripped gas with additional carbon monoxide.

Finely divided iron reacts less easily, carbon monoxide must be passed over the metal at 470 K and 10 MPa:

$$Fe + 5\ CO \rightarrow Fe(CO)_5$$

$Co_2(CO)_8$ is made by the action of an H_2/CO mixture on cobalt carbonate at 420 K and 25—30 MPa.

Many metal carbonyls can be made from metal halides suspended in organic solvents, treated with CO at 570 K and 20—30 MPa, in the presence of a reducing agent. Among the reducing agents which have been used are triethylaluminium in ether, zinc or magnesium powder in pyridine, and sodium in diglyme (diethyleneglycol diethyl ether). Thus VCl_3, CO and Na in diglyme give $Na(diglyme)_2V(CO)_6$ which yields $V(CO)_6$ on treatment with phosphoric acid.

Hexacarbonylchromium, $Cr(CO)_6$, can be made from $CrCl_3$, CO, Al_2Cl_6 and powdered Al in benzene.

Reduction of metal oxides with CO under pressure is another preparative route, e.g.:

$$Re_2O_7 + 17\ CO \rightarrow Re_2(CO)_{10} + 7\ CO_2$$
$$3\ OsO_4 + 24\ CO \rightarrow Os_3(CO)_{12} + 12\ CO_2$$

Pentacarbonyliron, $Fe(CO)_5$, is useful for converting metal salts to carbonyls by metathetical reactions. By this means WCl_6 can be converted to $W(CO)_6$, and $MoCl_5$ to $Mo(CO)_6$.

The binuclear and trinuclear carbonyls are often made by thermal or photochemical treatment of the mononuclear ones. Thus $Fe(CO)_5$ in solution in petrol is converted to $Fe_2(CO)_9$ by ultraviolet light, and $Co_2(CO)_8$ is converted to $Co_4(CO)_{12}$ by warming to 320 K.

Mixed metal carbonyls such as $(CO)_4CoMn(CO)_5$ have been made by treating a carbonyl halide of one metal with a sodium carbonylate of another, e.g.:

$$NaCo(CO)_4 + Mn(CO)_5Br \rightarrow NaBr + (CO)_4CoMn(CO)_5$$

17.12.2. Properties of the simple carbonyls

The simple carbonyls of the $M(CO)_n$ type are colourless liquids or low-melting solids, except $V(CO)_6$ which is a black solid and $Fe(CO)_5$ which is a yellow liquid. Carbonyls with more than one metal atom per molecule are usually coloured solids, but $Re_2(CO)_{10}$ forms colourless, monoclinic crystals. Physical properties are listed in Table 17.4 and representative reactions of $Fe(CO)_5$ are shown in Fig. 17.6.

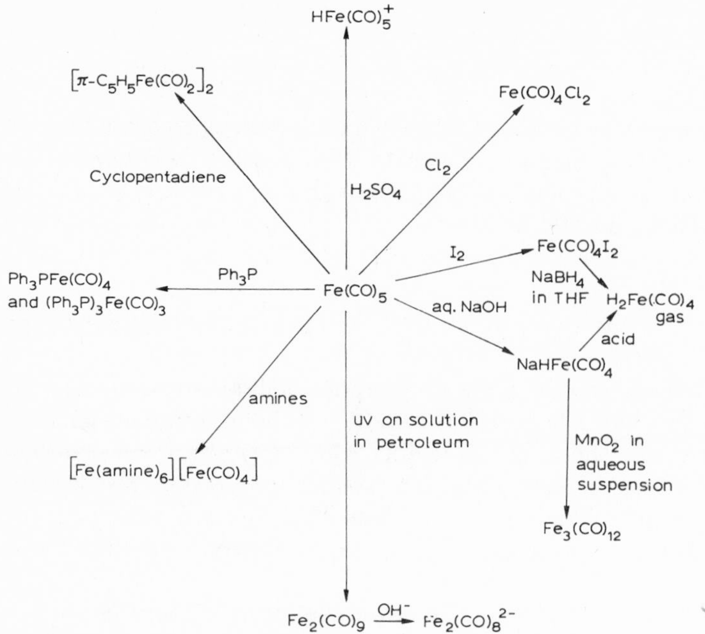

Fig. 17.6. Reactions of $Fe(CO)_5$.

The carbonyls are readily combustible; $V(CO)_6$ and the cobalt carbonyls are sensitive to air. Some, such as $Fe(CO)_5$ and $Ni(CO)_4$ are also very toxic, and must be treated with caution.

17.12.3. Structure of the simple carbonyls

In simple carbonyls of the $M(CO)_n$ type, the M—C—O bonds are collinear or nearly so. The carbon—oxygen distances are but slightly greater than the 113 pm observed in carbon monoxide, and the carbon—metal distances are somewhat less than that for a single bond. The donor power of the lone pair on the carbon atom in carbon monoxide is slight, for CO forms only very weak complexes with a few Lewis acids; one of these complexes is borine carbonyl, H_3BCO. In contrast, the electrons accepted from the carbon of CO by transition-metal ions create a ligand field which lowers the energies of just those d orbitals whose symmetry properties enable them to form bonding molecular orbitals with the vacant antibonding π orbitals of the CO molecule. The resulting 'back-donation' effect which produces some double-bond character in the M—C bond is analogous to that described for the isoelectronic CN^- ion in 6.6.1. The explanation accounts for the strength of the M—C bond and also for the weakness of the C—O bond in carbonyls relative to that bond in carbon monoxide.

Apart from vanadium ($Z = 23$) the metals which form simple mononuclear carbonyls have even atomic numbers, and the pairs of electrons donated by the CO ligands, added to the valence electrons of the metal always total eighteen. Thus Fe ($3d^6 4s^2$) has eight valence electrons, and five CO groups in $Fe(CO)_5$ donate a further ten. Similarly Ni ($3d^8 4s^2$) has ten valence electrons and these are augmented to 18 by the 8 electrons of the four CO groups in $Ni(CO)_4$. This *18-electron rule* also applies, in a modified form, to polynuclear carbonyls, as explained in a later paragraph. The carbonyls must therefore be considered to be spin-paired complexes and, with the exception of hexacarbonyl vanadium, they are all diamagnetic.

Although other metals of odd atomic number do not form mononuclear carbonyls, they can form compounds containing NO groups — nitrosylcarbonyls. Thus cobalt forms $Co(CO)_3NO$ which can be considered to obey the 18-electron rule if it is accepted that the NO group contributes three electrons to valence orbitals. Similarly $HCo(CO)_4$, a carbonyl hydride, obeys the rule, a hydrogen atom donating one electron to bonding orbitals.

The mononuclear carbonyls have molecules of high symmetry. $V(CO)_6$, $Cr(CO)_6$, $Mo(CO)_6$ and $W(CO)_6$ belong to the group O_h, $Fe(CO)_5$, and $Ru(CO)_5$ and $Os(CO)_5$ to D_{3h} and $Ni(CO)_4$ to T_d.

17.12.4. Polynuclear carbonyls

17.12.4.1. Bridging CO groups

In addition to linear M—C—O groups, some binuclear and polynuclear carbonyls of transition metals also contain bridging CO groups bonded to two metal atoms. Such a bridging group contributes one electron to orbitals on each of the metal atoms to form a σ bond with each of them. But in addition to the σ bonds some π bonding also occurs, as is evident from the low values of the C—O stretching frequencies. These are usually about 1850 cm^{-1}, which is only a little larger than the 1828 cm^{-1} of the $>C=O$ group in $COCl_2$ but much lower than the 2146 cm^{-1} of carbon monoxide itself. For example $Fe_2(CO)_9$ has a strong absorption band at 1830 cm^{-1} due to the bridging CO groups and another at about 2000 cm^{-1} due to the terminal CO groups. There is also a weak metal—metal bond, with an Fe—Fe distance of 249 pm.

Examples are also known of CO groups attached to three metal atoms. The $Rh_6(CO)_{16}$ molecule consists of an octahedron of Rh atoms each attached to the two terminal CO groups. The other four CO groups are situated above the centres of alternate faces of the octahedron and can each be considered to bridge three Rh atoms.

17.12.4.2. Metal—metal bonds

Metal—metal bonds occur in carbonyls either in conjunction with bridging CO groups, as in enneacarbonyldi-iron, $Fe_2(CO)_9$, or as the sole bond between two metal atoms in $Mn_2(CO)_{10}$ which has D_{4d} symmetry (Fig. 17.7). The Mn—Mn distance is 293 pm; this is longer than the Fe—Fe distance in $Fe_2(CO)_9$ where there is also carbonyl bridging between the metal atoms (Fig. 17.8). In dodecacarbonyltriosmium the Os atoms are arranged at the corners of an equilateral triangle and each is connected to four terminal carbonyl groups (Fig. 17.9).

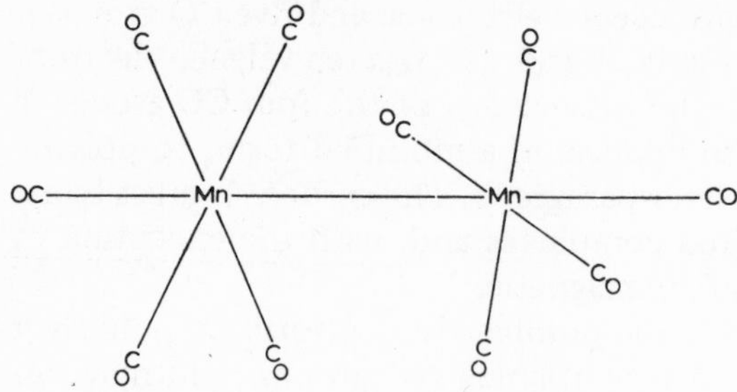

Fig. 17.7. Structure of $Mn_2(CO)_{10}$ molecule.

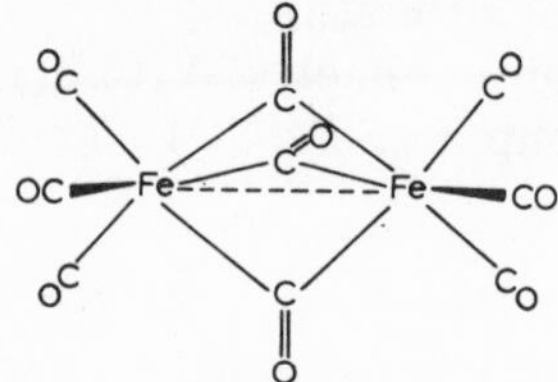

Fig. 17.8. Structure of enneacarbonyldi-iron, $Fe_2(CO)_9$.

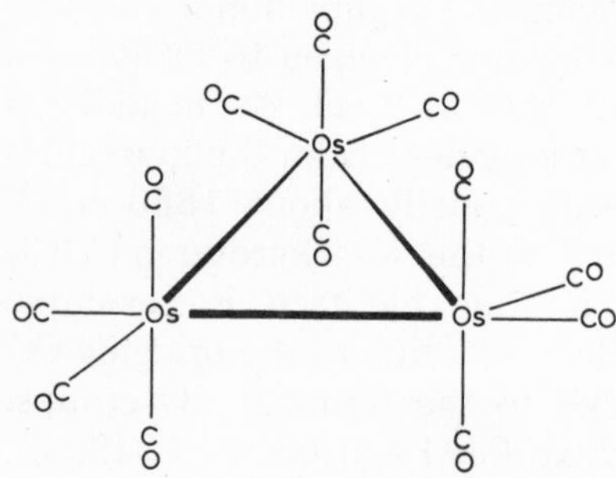

Fig. 17.9. Structure of $Os_3(CO)_{12}$ molecule.

There are two major factors which favour the formation of bonds between transition-metal atoms. Firstly the metal should be in a low oxidation state so that the effective nuclear charge acting on the outermost electrons is low, the orbitals are expanded and the interaction between neighbouring atomic cores is slight. Secondly there should be a small number of electrons in the valence shell since the more there are the more they need to occupy anti-bonding orbitals. However, with ligands like CO which are capable of with-drawing electrons from antibonding orbitals, metal—metal bonds become possible even between atoms such as Fe and Co. Polynuclear metal car-bonyls form one of the largest classes of compounds containing metal—metal bonds.

17.12.4.3. The 18-electron rule applied to polynuclear carbonyls

By considering every CO bridge bond and every metal—metal bond to represent the donation of one electron to the atom to which it is attached, the rule can be extended to polynuclear carbonyls. Thus either Fe atom in $Fe_2(CO)_9$ has

8 electrons in its valence shell ($3d^6$ $4s^2$)
6 'terminal CO' electrons
3 'bridging CO' electrons
1 metal—metal bond electron

18

It is of interest that $Co_2(CO)_8$, which has a bridged structure in the solid, exists in solution also in an unbridged form, and that in both the 18-electron rule is obeyed.

17.13. Derivatives of the Metal Carbonyls

17.13.1. Salts of carbonylate anions

Iron pentacarbonyl dissolves in aqueous alkali to give a yellow solution of sodium hydrogen iron carbonylate, $NaHFe(CO)_4$. Other carbonylates can be made from metal carbonyls by treating them with aqueous or alcoholic alkalis, with amines, or with Lewis bases such as sulphoxides. The anions of these salts are usually oxidised by air. The sodium and potassium salts are water-soluble. Addition to their aqueous solutions of Hg^{2+} and Cd^{2+} ions precipitates the monomeric $Hg[Co(CO)_4]_2$ and $Cd[Co(CO)_4]_2$. These compounds are soluble in organic solvents, are covalent in character, and have the structure:

```
CO      CO        CO      CO
  \    /   \     /   \   /
   Co        Hg        Co
  /    \   /     \   /   \
CO      CO        CO      CO
```

It is of interest to note that in an isoelectronic, isostructural series made up of carbonylate and carbonyl ions, the C—O stretching frequencies rise by about 100 cm^{-1} for every unit increase of positive charge:

Species	$Fe(CO)_4^{2-}$	$Co(CO)_4^-$	$Ni(CO)_4$
C—O frequency/cm^{-1}	1786	1886	2057

This shows that back-donation from metal to carbon in these ions is strengthened as the species becomes more negative.

Polynuclear carbonylates have also been made from polynuclear carbonyls by the action of bases. The structures of these compounds are not necessarily related to those of the parent carbonyls. The $Fe_2(CO)_8^{2-}$ ion, for example, which is derived from an enneacarbonyl with bridging CO groups, has only

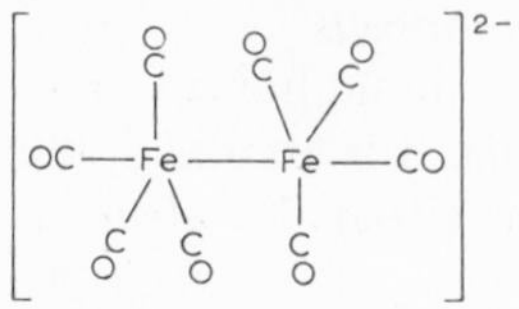

Fig. 17.10. Structure of $Fe_2(CO)_8^{2-}$ ion.

terminal CO groups and there is a metal—metal bond. The six CO groups out of the axis of the molecule, which are staggered in relation to one another (Fig. 17.10), are in planes perpendicular to the Fe—Fe bond.

17.13.2. Carbonyl hydrides

Acidification of aqueous or alcoholic $NaCo(CO)_4$ and $NaHFe(CO)_4$ in the absence of air gives, respectively, the thermally unstable yellow liquids $HCo(CO)_4$ and $H_2Fe(CO)_4$. The colourless liquid $HMn(CO)_5$ can be made from $Mn_2(CO)_{10}$ and hydrogen under 20 MPa. Electron diffraction shows the $Fe(CO)_4$ and $Co(CO)_4$ skeletons in these compounds to be slightly distorted from tetrahedral, but it has been established by means of n.m.r. spectra that the hydrogen atoms are not attached to the oxygen atoms and are only about 110—120 pm from the iron nucleus. The nature of the bonding is uncertain. In $HRe(CO)_5$ there is absorption in the infrared at 1832 cm^{-1}, a figure consistent with stretching frequencies found in known metal hydrides.

$HCo(CO)_4$ is probably the principal catalytic agent in the oxo reaction.

17.13.3. Carbonyl halides

Iron pentacarbonyl forms unstable addition compounds with the halogens:

$$Fe(CO)_5 + X_2 \rightarrow Fe(CO)_5X_2$$

These lose carbon monoxide to give more stable substances of the formula $Fe(CO)_4X_2$. The iodide, $Fe(CO)_4I_2$, can even be sublimed in a vacuum without decomposition. Reduction of the iodide by hydrogen produces $Fe(CO)_2I_2$; this can also be made by the reaction between $Fe(CO)_5$ and I_2 in boiling benzene. Both $Ni(CO)_4$ and $Co_2(CO)_8$ are decomposed by halogens and do not form carbonyl halides. There are carbonyl halides of ruthenium, rhodium, rhenium, osmium and iridium, and also of elements which do not form a simple carbonyl; examples are $[Pd(CO)Cl_2]_n$, $[Pt(CO)X_2]_n$ and Cu(CO)X (X = Cl, Br or I).

All the polynuclear carbonyl halides of which the structures are known are bridged not through carbonyl groups but through halogen atoms:

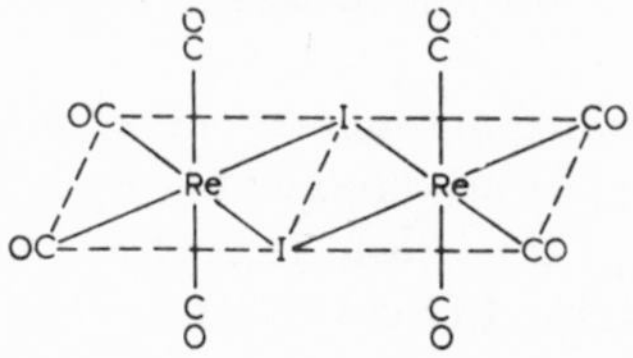

17.13.4. Nitrosyl carbonyls

When CO is passed into a suspension of $Co(CN)_2$ in aqueous potassium hydroxide at 273 K, a solution of $KCo(CO)_4$ results. Passing NO into this solution produces $Co(CO)_3NO$, a volatile liquid with physical properties very similar to $Ni(CO)_4$. Dry nitric oxide reacts with $Fe_3(CO)_{12}$ to give a mixture of $Fe(CO)_2(NO)_2$ and $Fe(CO)_5$.

The compounds $Mn(CO)(NO)_3$, $Fe(CO)_2(NO)_2$, $Co(CO)_3NO$ and $Ni(CO)_4$ are isoelectronic, since the NO molecule contains one more electron than CO. If the extra electron is imagined to be located in an MO which is close to the metal d-orbitals in energy, the structures approximate to the formal oxidation states $Mn^{-III}(CO)(NO^+)_3$, $Fe^{-II}(CO)_2(NO^+)_2$ and $Co^{-I}(CO)_3(NO^+)$. This is, however, a highly artificial representation. There is undoubtedly strong back-donation to the π^* orbitals of the NO^+; in evidence of this, the N—O stretching frequencies are in the range 1600—1900 cm^{-1} compared with about 2250 cm^{-1} for the NO^+ in nitrosonium salts. Nevertheless, the four isoelectronic compounds above can all be considered to obey the 18-electron rule in a formal sense.

17.13.5. Other derivatives

Electron-donor molecules can replace the CO groups in carbonyls. In this way neutral compounds and ions in great variety are obtainable: examples are $Fe_2(CO)_4(en)_3$, $Fe(CO)_3(PPh_3)_2$, $Mo(CO)_3(AsCl_3)_3$, $[HFe(CO)_3(PPh_3)_2]^+$, $[Co(CO)_3PPh_3]^-$. Complexes with π-bonding ligands such as cyclopentadiene are also common.

17.14. Infrared Spectra of Metal Carbonyls and their Derivatives

Although the crystal structures of many metal carbonyls and their derivatives have been determined, infrared spectroscopy is the routine method of obtaining structure information. The presence of bands in the range 1900—2050 cm^{-1} is indicative of terminal carbonyl stretching frequencies and the number and intensities of these bands can give information concerning the stereochemical arrangement of groups round the metal. Clearly this is a situation whereby the total symmetry of the molecule will determine how the vibrations of the carbonyl groups will interact.

To calculate how many carbonyl stretching vibrations would be expected for a molecule $LM(CO)_5$ (Fig. 17.11), determine its total symmetry and carry out the operations of its point group on the molecule, noting for each operation how many carbonyl groups remain unchanged in position:

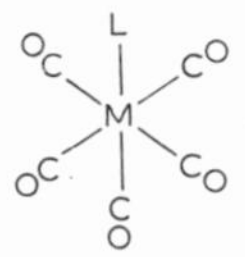

Fig. 17.11. L = ligand, M = transition metal.

C_{4v}	E	C_4	C_2	σ_v	σ_d
Γ_{CO} unchanged	5	1	1	3	1

It follows that this result is some combination of the irreducible representations of the point group C_{4v} (1.6) and can be solved by substituting the relevant values in the equation:

$$n_\Gamma = \frac{1}{g}\sum g_R \gamma_\Gamma^R \gamma_{DP}^R$$

$$n_{A_1} = \tfrac{1}{8}(1 \cdot 1 \cdot 5 + 2 \cdot 1 \cdot 1 + 1 \cdot 1 \cdot 1 + 2 \cdot 1 \cdot 3 + 2 \cdot 1 \cdot 1) = 2$$
$$n_{B_1} = \tfrac{1}{8}(1 \cdot 1 \cdot 5 + 2 \cdot -1 \cdot 1 + 1 \cdot 1 \cdot 1 + 2 \cdot 1 \cdot 3 + 2 \cdot -1 \cdot 1) = 1$$
$$n_E = \tfrac{1}{8}(1 \cdot 2 \cdot 5 + 2 \cdot 0 \cdot 1 + 1 \cdot -2 \cdot 1 + 2 \cdot 0 \cdot 3 + 2 \cdot 0 \cdot 1) = 1$$

Thus the carbonyl stretching vibrations for $LM(CO)_5$ in C_{4v} symmetry transpose as $2A_1 + B_1 + E$. To find out what combination of stretching modes each vibration represents, it is necessary to label the carbonyl groups in the molecule (Fig. 17.12). Carry out all the operations of point group C_{4v} on carbonyl group A:

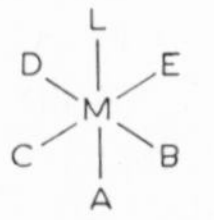

L
D E
M
C B
A

Fig. 17.12.

C_{4v}	E	$2C_4$	C_2	$2\sigma_v$	$2\sigma_d$
Γ_A	A	A A	A	A A	A A

Multiply this result by the characters of each of the irreducible representations A_1, B_1, and E individually, and follow by normalisation:

$$\Gamma_{A_1} = A + A + A + A + A + A + A + A = \frac{1}{\sqrt{8}} A$$

$$\Gamma_{B_1} = A - A - A + A + A + A - A - A = 0$$

$$\Gamma_E = 2A + 0 \quad - 2A + 0 + 0 \qquad = 0$$

Thus carbonyl at position A can only be involved in the A_1 mode. Repeat this procedure by carrying out all the operations of C_{4v} on carbonyl group B:

C_{4v}	E	$2C_4$	C_2	$2\sigma_v$	$2\sigma_d$
Γ_B	B	C E	D	B D	C E

Multiply this result as before:

$$\Gamma_{A_1} = B + C + E + D + B + D + C + E = \frac{1}{\sqrt{4}}(B + C + D + E)$$

$$\Gamma_{B_1} = B - C - E + D + B + D - C - E = \frac{1}{\sqrt{4}}(B - C + D - E)$$

$$\Gamma_E = 2B + 0 - 2D + 0 + 0 = \frac{1}{\sqrt{2}}(B - D)$$

Now carry out this procedure on carbonyl C:

C_{4v}	E	$2C_4$	C_2	$2\sigma_v$	$2\sigma_d$
Γ_C	C	B D	E	C E	B D

$$\Gamma_{A_1} = C + B + D + E + C + E + B + D = \frac{1}{\sqrt{4}}(B + C + D + E)$$

$$\Gamma_{B_1} = C - B - D + E + C + E - B - D = \frac{1}{\sqrt{4}}(C - B - D + E)$$

$$\Gamma_E = 2C + 0 - 2E + 0 + 0 = \frac{1}{\sqrt{2}}(C - E)$$

This procedure can be further repeated for carbonyls D and E and gives similar results. Thus substituting ⟵⟶ for positive signs and ⟶⟵ for negative signs (in any of the above equations) to signify in or out of phase stretching, A, B, and E modes can be illustrated as in Fig. 17.13.

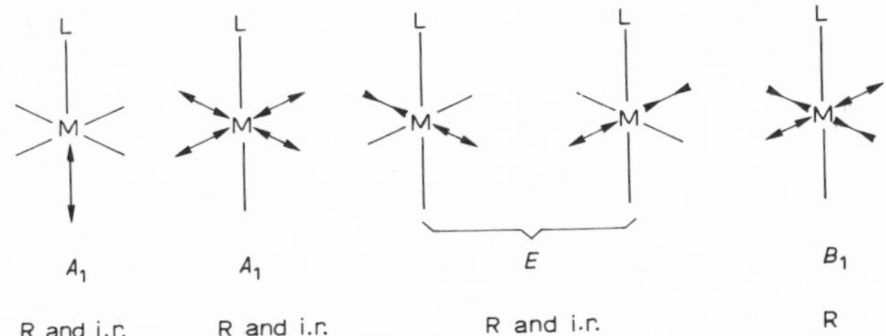

Fig. 17.13. Vibrational modes of $M(CO)_5L$ molecule. R = Raman-active mode; i.r. = infrared-active mode.

Examination of these vibrations reveals that in B_1 there is no change in dipole, hence it will be infrared inactive. Raman activity is indicated in the character table for those labelled α, and infrared activity by T, the translational modes. Thus the general pattern of the infrared spectrum would show an A_1 mode at high frequency of weak to medium intensity; the E mode very strong due to a big change in dipole; and the A_1 mode due to stretching of CO *trans* to L also weak to medium intensity. Raman spectra are consistent with such predictions showing all four modes ($2A_1 + B_1 + E$) active, the B_1 mode appearing as a strong bond.

Further Reading

A.K. Holliday, G. Hughes, S.M. Walker, M.L.H. Green and P. Powell, The chemistry of carbon: Organometallic chemistry, Pergamon, Oxford, 1975.

E.G. Rochow, The chemistry of silicon, Pergamon, Oxford, 1975.

H. Bürger, Anomalies in the structural chemistry of silicon, Angew. Chem. Internat. Ed., 12 (1973) 474.

H. Nowotny, Crystal chemistry of complex carbides and related compounds, Angew. Chem. Internat. Ed., 12 (1973) 906.

P.L. Timms, Silicon subhalides, Accounts Chem. Res., 6 (1973) 118.

S.R. Miller and L.E. Orgel, The origins of life on the earth, Chapman and Hall, London, 1974.

D.R. Williams, Life's essential elements, Ed. Chem., 10 (1973) 56.

18

Organometallic Compounds

18.1. Definition of 'Organometallic'

The term organometallic is usually restricted to compounds in which a metal atom is bonded directly to a carbon atom of a hydrocarbon radical or molecule. In this context, however, the term metal is conventionally extended to include elements less electronegative than carbon, such as boron, arsenic and silicon. Thus trimethylboron, $(CH_3)_3B$, is classified as an organometallic compound. However, trimethyl orthoborate, $B(OCH_3)_3$, is not so classified because it does not contain a boron—carbon bond. Thus although a compound contains many hydrocarbon radicals it is not classified as organometallic when all the carbon atoms are linked to the metal through atoms such as oxygen, nitrogen and sulphur. But when even one of the carbon atoms is directly linked to metal, as in $CH_3 \cdot B(OCH_3)_2$, the compound is termed organometallic.

Moreover such compounds as carbides, cyanide complexes and carbonyls which of course contain metal—carbon bonds, are not generally considered as organometallic compounds. Many organometallic compounds of the transition metals can however be most easily prepared from carbonyls.

It is sometimes convenient to divide organometallic compounds into simple and mixed. A simple organometallic compound is one, such as $(C_2H_5)_4Pb$ or $(CH_3)_3SnH$, which has only hydrocarbon radicals or hydrogen atoms attached to the metal. Simple organometallic compounds can be subdivided into the symmetrical, like $(C_2H_5)_2Hg$, and the unsymmetrical, like $C_2H_5HgC_4H_9$. A mixed organometallic compound, such as C_6H_5MgBr, $(C_4H_9)_2SnCl_2$ or $C_6H_5SbO(OH)_2$, has groups other than hydrocarbon radicals and hydrogen atoms attached directly to the metal.

18.2. Types of Metal—Carbon Bonds

In dealing with their structure and reactivity, organometallic compounds are best considered in relation to the character of the metal—carbon bond. It is possible to distinguish five types of compound: (i) ionic compounds, (ii) multicentre-bonded compounds, (iii) volatile σ-bonded compounds,

(iv) ylides and (v) π-bonded compounds, though the classification is not rigid.

18.2.1. Ionic organometallic compounds

Most of the compounds of Na, K, Rb and Cs belong to this class. They are usually colourless, salt-like solids which do not dissolve in non-polar solvents. In general the most ionic are those in which the cation is large (e.g. Cs^+) and the carbanion is small (e.g. CH_3^-) but extra thermal stability is conferred on compounds in which the carbanion contains either an acetylenic triple bond, as in $R{-}C{\equiv}C^-$, in which the sp-hybridised carbon atom is strongly electronegative, or a conjugated system, as in $CH_2{=}CH{-}CH_2^-$, in which a non-localised π-bond can stabilise the anion. These compounds containing large, resonance-stabilised carbanions are often highly coloured because energy in the visible range is absorbed when an electron is promoted from a π-orbital to a π*-orbital within the ion. For example the compound $Na^+CPh_3^-$, made by treating Ph_3CCl with sodium in ether, is red, and its solution in that solvent is electrically conducting.

18.2.2. Organometallic compounds containing multicentre bonds

The compounds, somewhat loosely called electron-deficient, are exemplified by Me_4Li_4 (Fig. 13.2), $(Me_2Be)_n$, Me_6Al_2 and the carboranes (11.5.4.1). They can be considered as intermediate in their bonding characteristics between the ionic compounds of the sodium group and the σ-bonded compounds of elements like Si, Sn and Pb. The elements with the greatest tendency to form multicentre-bonded organometallic compounds are lithium, beryllium, magnesium, boron and aluminium. With the exception of Li these are elements which form electron-deficient hydrides (11.5.1).

18.2.3. Compounds containing metal—carbon σ-bonds

Typical of these are the alkyls and aryls of the elements Zn—Hg in Group IIB, Ga—Tl in Group IIIB, Si—Pb in Group IVA, P—Bi in Group VB, Se and Te in Group VIB. Mean bond energies for the M—C bonds tend to decrease

TABLE 18.1

D(Me—M) FOR SOME METHYL DERIVATIVES

	D/kJ mol^{-1}		D/kJ mol^{-1}
Me_4C	348	Me_3N	313
Me_4Si	293	Me_3P	276
Me_4Ge	246	Me_3As	230
Me_4Sn	217	Me_3Sb	217
Me_4Pb	155	Me_3Bi	142

down a group (Table 18.1) and the thermal stabilities therefore decrease also. Thus $\Delta H_f(Me_4Si) = -238$ kJ mol^{-1}, $\Delta H_f(Me_4Sn) = -19$ kJ mol^{-1} and $\Delta H_f(Me_4Pb) = +135$ kJ mol^{-1}.

Sigma bonding between carbon and transition metals is less common than between carbon and main-group elements. Where it occurs the metal is likely to be in a low oxidation state and typically near the end of a transition series. Experiments on compounds of the type

$(R_3P)_2M^{II}(R)X$

show the kinetic stabilities to increase in the order M = Ni < Pd < Pt. Generalising, σ-organometallic compounds of main-group elements are most stable for elements early in their respective groups and in high oxidation states, whereas those of transition metals are most stable for the heavier elements of their groups and in low oxidation states.

In the transition-metal compounds the substitution of halogen, particularly fluorine, for hydrogen in the hydrocarbon increases the thermal and kinetic stabilities. In the main-group compounds however the presence of a fluorine atom tends to encourage elimination reactions:

$$F\cdot CH_2\cdot CH_2\cdot B\langle \rightarrow F\cdot B\langle + CH_2{=}CH_2$$

18.2.4. Ylides

These are compounds which contain metal—carbon double bonds. They are formed both by main-group elements and transition metals. An example of the former is the Wittig reagent $Ph_3P{=}CH_2$. Typical of the latter are the compounds

$(Ph)(MeO)C{=}M(CO)_5$

in which M = Cr, Mo, W, Mn or Fe. In these the C—O bond is rather short, suggesting that the compounds are stabilised by the withdrawal of electron density from the double bond.

18.2.5. π-Bonded compounds

These have alkene, alkyne or some other carbon-containing compound or group which has a system of electrons in π-orbitals. Overlap of these π orbitals with vacant orbitals of the metal atom give rise to arrangements in which the metal atom is bound to several carbon atoms instead of to one. One of the best examples is bis(cyclopentadienyl)iron $(\pi\text{-}C_5H_5)_2Fe$. Its synthesis in 1951 marked the beginning of extensive discoveries in the organometallic chemistry of the transition metals. It can be made by the action of cyclopentadienylmagnesium bromide on $FeCl_3$ in ether or by passing cyclopentadiene vapour over finely divided iron:

2 CH=CH–CH₂ ring + Fe ⟶ (CH ring)₂ Fe + H_2

and it has a sandwich structure in which the iron atom lies between two C_5H_5 rings.

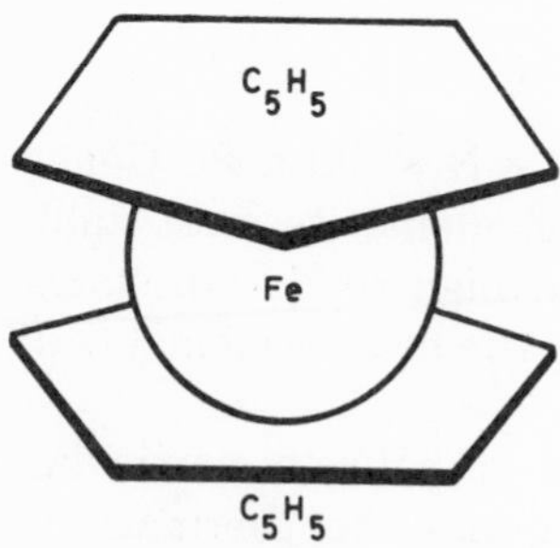

Fig. 18.1. 'Sandwich' structure of the ferrocene molecule.

The delocalized π-electrons of the cyclopentadienyl rings overlap with the unfilled d-orbitals of the metal — a typical example of a π-bonded organometallic compound of a transition metal. In this case each C_5H_5 ring supplies five electrons for bonding. Examples of other π-bonding molecules and radicals, with the number of π electrons they provide, are given in Table 18.2. The π-bonding in a typical alkene complex is described in Section 18.4.1. Some of these ligands can form either a sigma bond or π bonds with a metal atom. A compound of particular interest is π-cyclopentadienyldicarbonyl-ironcyclopentadienyl:

b, c, c, b, a — CO, Fe, CO

TABLE 18.2

SOME EXAMPLES OF π-BONDING LIGANDS

Ligand	Number of π electrons
Alkene, alkyne, PhCNS, CF_3CN	2
π-Allyl	3
Cyclobutadiene, butadiene	4
π-Cyclopentadienyl	5
Arene, pyrrole	6
Cycloheptatrienyl	7
Cyclooctatetraenyl	8

Its proton magnetic resonance spectrum at 300 K has only two lines, of equal intensity, but at 200 K the spectrum is more complicated; the absorption at high value of τ, due to the protons of the π-bonded C_5H_5, remains unchanged, but the protons on the three different sites a, b and c now give rise to three distinct peaks in the n.m.r. Studies of the variation of the spectrum with temperature suggest that the π-bonded C_5H_5 can change its point of attachment to the Fe through repeated 1,2 shifts:

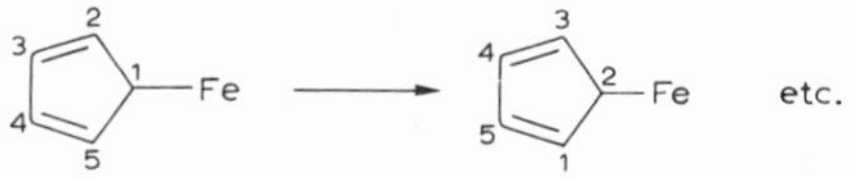

(ring 1–5, Fe on C1) → (ring 1–5, Fe on C2) etc.

It is this *site exchange*, or "ring-whizzing", which makes the protons of the σ-bonded ring appear to be equivalent at 300 K.

This compound is one example of a *fluxional molecule.* Such molecules are characterised by three properties. (i) They have two or more chemically equivalent nuclear configurations. (ii) One of these can pass to another under normally accessible experimental conditions. (iii) Interconversion proceeds rapidly enough to have detectable consequences in experimental measurements such as nuclear magnetic resonance absorption.

Many allyl complexes of both main-group elements and transition metals have *instantaneous* structures which have either localised σ-bonding:

$CH_2{=}CH{-}CH_2{-}M{-}$

or delocalised bonding:

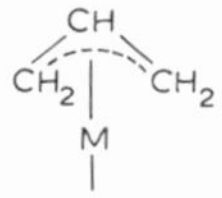

$CH_2{\cdots}CH{\cdots}CH_2$ (CH bonded to M)

The fluxional character of these molecules is such that each of the four terminal hydrogen atoms passes rapidly between the two environments available to them.

Among the other types of fluxional molecules are those in which one or more polyene is π-bonded, or both σ- and π-bonded, to several metal atoms.

18.3. Organometallic Compounds of Non-transition Elements

The differences between the σ-bonded compounds of the non-transitional elements and the generally π-bonded organometallic derivatives of transition elements are considerable. For this reason we shall consider the two types separately.

18.3.1. Methods of preparation

(I) There are three preparative methods of wide application. These can be formulated (for metals in the +2 state) as

$$(a)\ 2\,RX + 2\,M \rightarrow R_2M + MX_2$$
$$(b)\ R_2M' + M'' \rightarrow R_2M'' + M'$$
$$(c)\ R_2M' + M''X_2 \rightarrow R_2M'' + M'X_2$$

(i) Examples of (a), which is usually carried out in inert solvents such as petroleum ether, are

$$CH_3Cl + 2\,Li \rightarrow CH_3Li + LiCl$$
$$C_2H_5Cl + 2\,Na \rightarrow C_2H_5Na + NaCl$$

A slight variation of the same reaction gives rise to compounds such as the Grignard reagents, organozinc and organogermanium halides:

$$C_2H_5I + Mg \xrightarrow{\text{ether}} C_2H_5MgI$$
$$2\,C_2H_5I + 2\,Zn \rightarrow (C_2H_5)_2Zn + ZnI_2$$
$$2\,C_2H_5Cl + Ge \xrightarrow[\text{over copper}]{600\text{ K}} (C_2H_5)_2GeCl_2$$

The metal is sometimes more reactive when in the form of an alloy or an amalgam.

$$4\,C_2H_5Cl + Pb/Na\text{ alloy} \rightarrow (C_2H_5)_4Pb + 4\,NaCl$$

$$2\,C_2H_5Cl + Na/Hg\text{ amalgam} \xrightarrow[\text{acetate}]{\text{in ethyl}} (C_2H_5)_2Hg + 2\,NaCl$$

(ii) Examples of the metal-exchange reactions (b) are

$$R_2Hg + 2\,Na \rightarrow 2\,RNa + Hg$$
$$R_2Zn + Ca \rightarrow R_2Ca + Zn$$

(iii) Examples of alkyl—halogen exchange reactions (c) are

$$4\,RLi + GeCl_4 \rightarrow R_4Ge + 4\,LiCl$$
$$R_3Al + BF_3 \rightarrow R_3B + AlF_3$$
$$3\,R_2Hg + 2\,SbCl_3 \rightarrow 2\,R_3Sb + 3\,HgCl_2$$

The Grignard reagents form an important group of reagents suitable for similar exchange processes:

$$2\,RMgI + BeCl_2 \rightarrow R_2Be + MgI_2 + MgCl_2$$
$$C_6H_5MgI + CuI \rightarrow C_6H_5Cu + MgI_2$$

(II) An interesting type of preparative reaction, of less general application, is called *metallation*; it consists of a hydrogen—metal exchange:

$$2\,C_5H_6 + 2\,Na \rightarrow 2\,C_5H_5Na + H_2$$

$$C_4H_4S + HgCl_2 \longrightarrow C_4H_3S{-}HgCl + HCl$$

(III) The use of α-olefins as synthetic reagents forms the basis of a newer method of making covalent organometallic compounds, particularly of aluminium, boron and silicon.

$$AlH_3 + 3\,CH_2{=}CH \cdot R \xrightarrow{400\text{ K}} (R \cdot CH_2 \cdot CH_2)_3Al$$

$$B_2H_6 + 6\,CH_2{=}CH \cdot R \rightarrow 2(R \cdot CH_2 \cdot CH_2)_3B$$

$$(C_6H_5)_3SiH + CH_2{=}CH \cdot R \xrightarrow[\text{catalyst}]{\text{benzoyl peroxide}} (C_6H_5)_3Si \cdot CH_2 \cdot CH_2 \cdot R$$

A similar sort of reaction, termed *hydroboration*, is effected by adding an olefin to a mixture of $NaBH_4$ and $AlCl_3$ in diglyme; this gives good yields of trialkylboranes. *Methylenation* by diazomethane is yet another way of inserting carbon atoms next to the metal:

$$GeCl_4 + CH_2N_2 \xrightarrow[210\text{ K}]{\text{copper catalyst}} Cl \cdot CH_2 \cdot GeCl_3 + N_2$$

(IV) Electrolytic methods of preparation are gaining in importance. Industrially the most widely-used of all organometallic compounds, Et_4Pb, can be made by the electrolysis of $NaAlFEt_3$ between an aluminium cathode and a lead anode.

18.3.2. Relative reactivity of metal—carbon bonds

The reactivity of a compound refers to the rate at which it reacts with another substance. In general the metal is more important than the organic moiety in determining the reactivity of organometallic compounds; the lower the ionisation energy of the metal the more reactive the compound. The reactivity in which we are interested is not necessarily related to thermal stability or a tendency to inflame in air. Thus trimethylboron is very inflammable, that is very readily oxidised, but not otherwise very reactive; conversely, an alkyl of gold is thermally extremely unstable but not particularly reactive.

Orders of reactivity

The reactivity of organometallic compounds is determined by their rates of addition to unsaturated bonds such as $\rangle C{=}C\langle$, $\rangle C{=}O$ and $-C{\equiv}N$. For metals of the first three A sub-groups, this reactivity increases down the group:

$$R_2Be < R_2Mg < R_2Ca < R_2Sr < R_2Ba$$

But for the B sub-groups reactivity decreases in the same direction:

$$R_2Zn > R_2Cd > R_2Hg$$

Across the first three groups of a period reactivity decreases:

$$RNa > R_2Mg > R_3Al$$

The least reactive compound of an A sub-group is more reactive than the most reactive compound of the corresponding B sub-group:

$$R_2Be > R_2Zn$$

In Groups V, VI and VIII the reactivities decrease down the A sub-groups and increase down the B sub-groups. Across a period there is an increase in reactivity:

$R_3As < R_2Se < RBr$

This regularity is not found in Group IV, where the order of reactivity is rather variable.

It is difficult to generalise on the effect of the organic group on reactivity for a given metal, but a radical or anion which can be stabilised by π-electron delocalisation is particularly easily formed in a bond-cleavage process:

$CH_2{=}CH{-}CH_2{-}M \rightarrow [CH_2{=}CH{-}CH_2]^- + M^+$

For compounds of the same metal, an unsymmetrical compound RMR′ is usually more reactive than a symmetrical compound R_2M.

Solvents are often of importance in determining reactivity, because solvated species are frequently formed and it is their reactivities which are displayed.

18.3.3. Typical reactions of organometallic compounds of the non-transition elements

(I) With inorganic reagents

(i) Many ionic and σ-bonded organometallic compounds are oxidised spontaneously by oxygen or air.

(ii) Cleavage by hydrogenation often occurs even at room temperature and in the absence of a catalyst:

$C_6H_5K + H_2 \rightarrow C_6H_6 + KH$

(iii) Many are rapidly and violently hydrolysed by water:

$(C_2H_5)_3Al + 3\ H_2O \rightarrow Al(OH)_3 + 3\ C_2H_6$

But some, such as tetra-alkylsilanes, hydrolyse extremely slowly.

(iv) Acids, such as hydrogen halides in benzene, react to give hydrocarbons and metal salts:

$(C_2H_5)_2Mg + 2\ HX \rightarrow 2\ C_2H_6 + MgX_2$

But sometimes the reactions are slow:

$$(C_6H_5)_4Sn + HX \rightarrow (C_6H_5)_3SnX + C_6H_6$$

$$\downarrow$$

$(C_6H_5)_2SnX_2$, and so on, stepwise.

(v) Halogens such as iodine usually cleave the M—C bonds:

$RM + I_2 \rightarrow RI + MI$

(vi) Active metals usually extract halogens:

$$2\ R_3SnCl + 2\ Na \xrightarrow{\text{inert solvent}} R_3Sn \cdot SnR_3 + 2\ NaCl$$

or they may replace less active metals:

$R_2Hg + 3\ Na \rightarrow 2\ RNa + NaHg$ (amalgam)

(vii) Metal halides react in the various ways shown:

$$4\ RMgX + SnCl_4 \rightarrow R_4Sn + 2\ MgCl_2 + 2\ MgX_2$$
$$3\ R_4Sn + SnCl_4 \rightarrow 4\ R_3SnCl$$
$$2\ RLi + AuBr_3 \rightarrow R_2AuBr + 2\ LiBr$$

(II) With organic reagents

(i) Most σ-bonded organometallic compounds form addition compounds with unsaturated organic compounds:

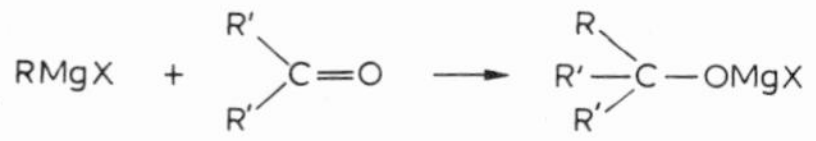

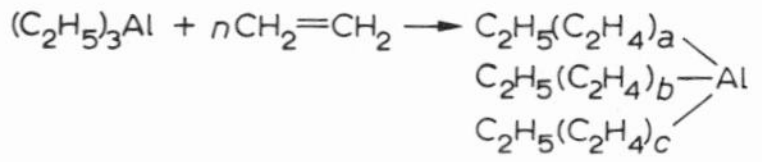

$$RMgX + R'_2C{=}O \longrightarrow RR'_2C{-}OMgX$$

$$(C_2H_5)_3Al + nCH_2{=}CH_2 \longrightarrow [C_2H_5(C_2H_4)_a][C_2H_5(C_2H_4)_b][C_2H_5(C_2H_4)_c]Al$$

$[a + b + c = n]$

The last reaction is used to produce polyethylenes which can have molecular weights up to three million. Transition-metal halides such as $TiCl_4$ catalyse the process.

(ii) Alkyls of metals usually react with aryl halides:

$ArX + AlkM \rightarrow ArM + AlkX$

the halogen being more cationic in the aryl halide than in the alkyl halide.

(iii) Many organometallic compounds behave as Lewis acids and form adducts with amines. The Lewis acid strengths, based on the strengths of the metal—nitrogen bonds, increase along the series:

$RNa < R_2Mg < R_3Al$

But they first rise and then fall again down a group:

$R_3B < R_3Al > R_3Ga > R_3In > R_3Tl$

(iv) Organic compounds containing a labile hydrogen atom which is bound directly to a carbon atom are metallated by some organometallic compounds:

$$\text{2-methylpyridine } (C_5H_4N{-}CH_3) + C_6H_5Li \longrightarrow C_5H_4N{-}CH_2^-\ Li^+ + C_6H_6$$

$$\text{thiophene } + n{-}C_4H_9Li \longrightarrow \text{2-thienyl}{-}Li + C_4H_{10}$$

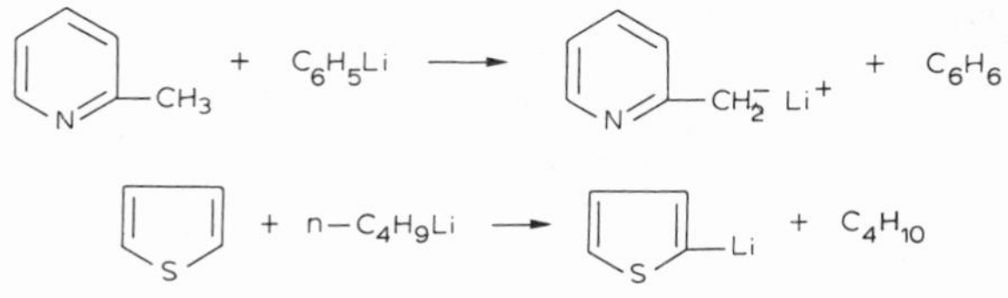

18.4. Organometallic Compounds of the Transition Elements

18.4.1. The nature of π-bonding in olefin complexes

Two compounds made by Zeise in 1827 from ethylene and potassium chloroplatinate have been shown to have the structures:

$$K^+\left[\mathrm{Cl_3Pt(CH_2{=}CH_2)}\right]^- \qquad \mathrm{(CH_2{=}CH_2)ClPt(\mu\text{-}Cl)_2PtCl(CH_2{=}CH_2)}$$

There is a planar, approximately square, bond arrangement round the platinum atoms, but it is the bonding from the metal atom to the perpendicular carbon—carbon bond which is of greatest interest. The highest-energy electrons of ethylene lie in the π-bond; molecular orbitals of only slightly higher energy are the unfilled, antibonding π* (Fig. 18.2a). Bonding with the metal can be described as due to combination of a π orbital with a hybrid AO on the metal atom, the system being stabilised by back-donation from a metal d orbital of correct symmetry (Fig. 18.2b).

18.4.2. Methods of preparing π-bonded organometallic compounds

The preparations are nearly always carried out in the absence of air and in inert solvents such as hydrocarbons and ethers. Starting materials are metal halides, carbonyls or complexes like acetylacetonates which are soluble in the organic solvents.

(i) A common general method is to treat a metal salt, usually a halide, with a reducing agent in the presence of the ligand. The reducing agent is necessary because the final product contains the metal in a low oxidation

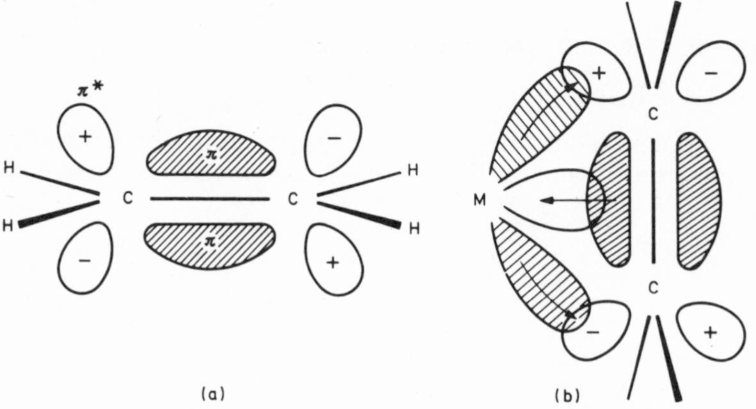

Fig. 18.2. (a) Ethylene molecule, with filled π orbitals and unfilled π* orbitals. (b) Bonding of ethylene to a transition-metal ion by overlap of filled π orbital with metal d orbital and by back-donation from the metal to the unfilled π* orbital.

state. An example is

$$\underset{\text{metal salt}}{3\,CrCl_3} + \underbrace{Al + Al_2Cl_6}_{\text{reducing agent}} + \underset{\text{ligand}}{6\,C_6H_6} \rightarrow 3(\pi\text{-}C_6H_6)_2Cr^+AlCl_4^-$$

In this preparation the dibenzenechromium $(\pi\text{-}C_6H_6)_2Cr$ is obtained by converting the chloroaluminate to the perchlorate and then reducing the $(\pi\text{-}C_6H_6)_2Cr^+$ with dithionate.

Sigma-bonded alkyls of main-group metals are often used as reducing agents. Thus $NiCl_2$, reduced with a trialkylaluminium in the presence of butadiene, gives the compound cyclododeca-1,5,9-trienenickel

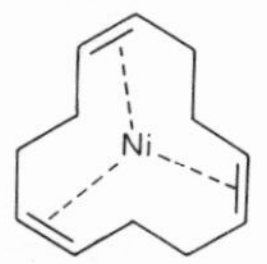

Ni

This reaction is also an example of *oligomerisation*; three butadiene molecules condense to give the organic moiety of the organometallic compound.

(ii) Ligand replacement reactions are useful routes. Thus Zeise's salt is made by the replacement by ethylene of one of the chloride ligands in $PtCl_4^{2-}$:

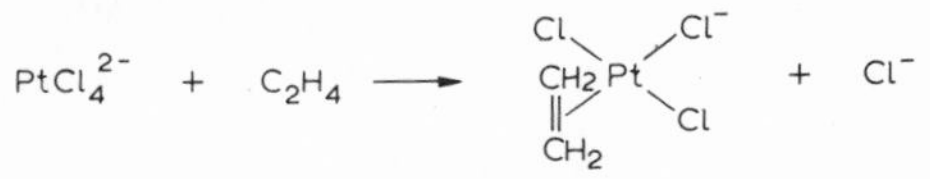

$$PtCl_4^{2-} + C_2H_4 \longrightarrow [(CH_2{=}CH_2)PtCl_3]^- + Cl^-$$

Similarly

$$2\,RhCl_3 + 4\,C_2H_4 \longrightarrow [(CH_2{=}CH_2)_2Rh(\mu\text{-}Cl)_2Rh(CH_2{=}CH_2)_2] + 2\,Cl_2$$

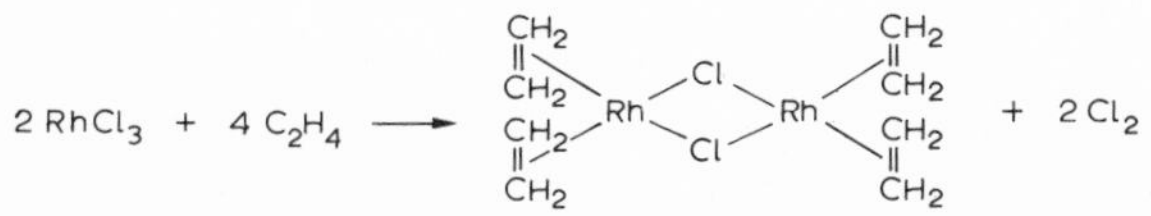

Metal carbonyls are particularly good starting materials for ligand replacement reactions. Thus cyclooctadiene reacts with $[Rh(CO)_2Cl]_2$ at room temperature:

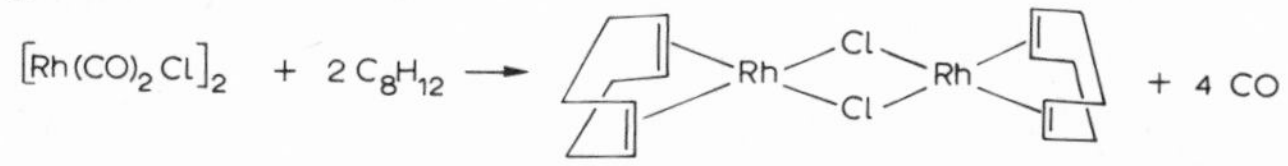

$$[Rh(CO)_2Cl]_2 + 2\,C_8H_{12} \longrightarrow [(C_8H_{12})Rh(\mu\text{-}Cl)_2Rh(C_8H_{12})] + 4\,CO$$

Sometimes a higher temperature is necessary:

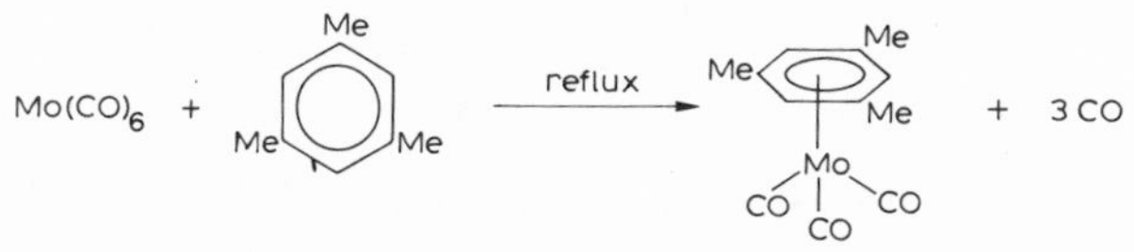

$$Mo(CO)_6 + C_6H_3Me_3 \xrightarrow{\text{reflux}} (\pi\text{-}C_6H_3Me_3)Mo(CO)_3 + 3\,CO$$

It may be necessary to prepare the compound under pressure if the ligand is highly volatile:

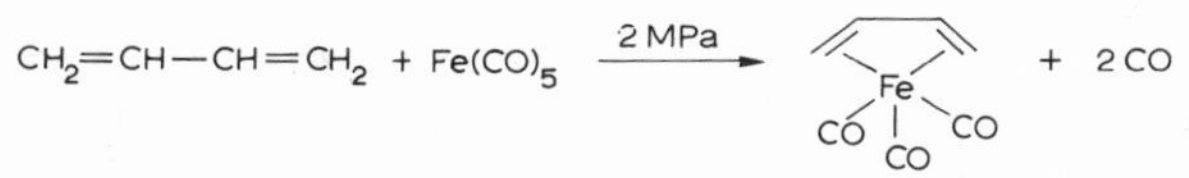

$$CH_2{=}CH{-}CH{=}CH_2 + Fe(CO)_5 \xrightarrow{2\ MPa} (C_4H_6)Fe(CO)_3 + 2\,CO$$

or it may be sufficient to irradiate the reactants with ultraviolet radiation:

$$C_2H_4 + \pi\text{-}C_5H_5Mn(CO)_3 \xrightarrow[\text{(petroleum)}]{h\nu} \pi\text{-}C_5H_5MnC_2H_4(CO)_2 + CO$$

Replacement of one hydrocarbon by another is often possible. Thus $C_6H_6Cr(CO)_3$ reacts with cycloheptatriene:

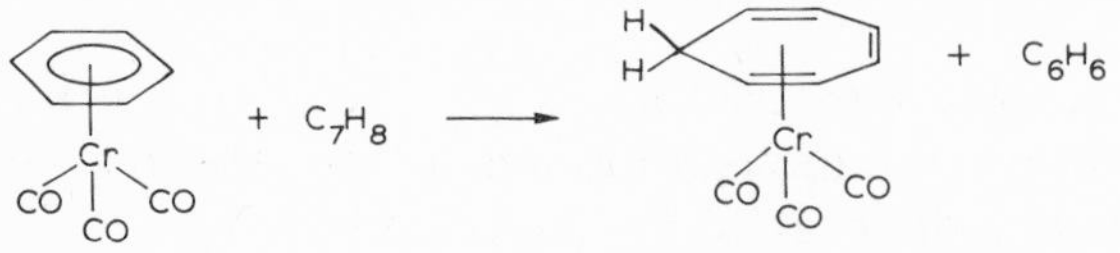

Ligand-transfer reactions between π-complexes can also be used. Thus $[\pi\text{-}C_4Ph_4PdBr_2]_2$ reacts with $(\pi\text{-}C_5H_5)_2Co$ to give the compound

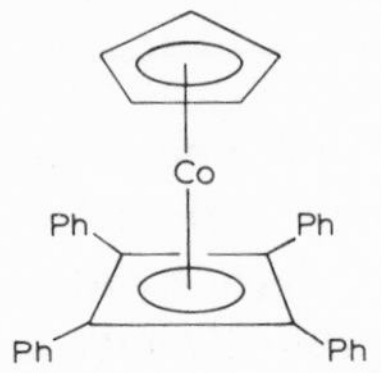

(iii) Carbonyl anions (17.13.1) are also useful starting materials because they react with halogen-substituted hydrocarbons:

$$Mn(CO)_5^- + ClCH_2{-}CH{=}CH_2 \rightarrow (CO)_5Mn - CH_2{-}CH{=}CH_2 + Cl^-$$

This σ-complex first formed is converted to the π-allyl compound

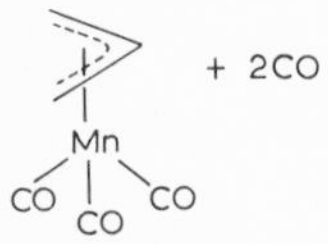

by ultraviolet irradiation.

(iv) Grignard reagents can often be used in the preparation of π-bonded organometallic compounds. Thus π-allyls are usually made by the action of allylmagnesium chloride on a metal halide in ether:

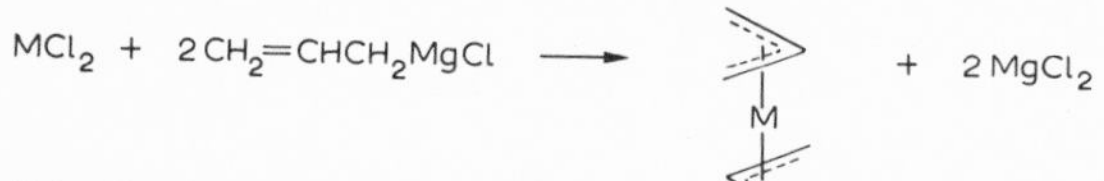

(v) Cyclopentadienyl compounds are often obtainable by the treatment of a metal halide with the ionic sodium cyclopentadienide in a solvent such as tetrahydrofuran:

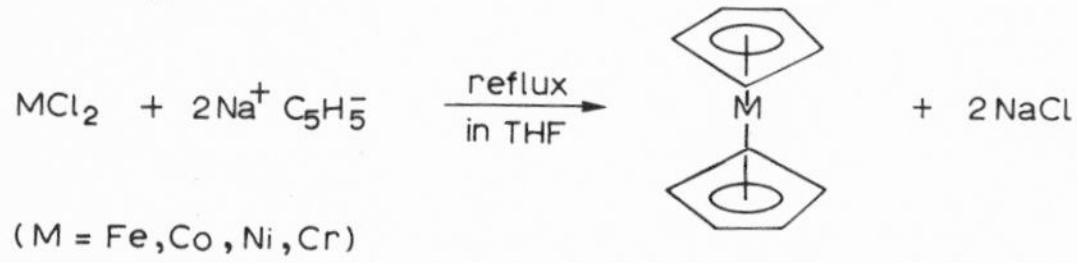

18.4.3. Reactions of π-bonded organometallic compounds

The reactivity of a ligand molecule may be modified by co-ordination to a metal. For example, a polyene ring may be made much more susceptible to attack by a nucleophile. Thus the compound π-cyclo-octatetraenetricarbonyliron is converted by $POCl_3$ in dimethylformamide to an aldehyde:

$$C_8H_8Fe(CO)_3 \xrightarrow[\text{in DMF}]{POCl_3} C_8H_7(CHO)Fe(CO)_3$$

In this case the aldehyde cyclo-C_8H_7CHO itself can be obtained from this reaction product by oxidation with Ce^{4+}. There are now many examples known of reactions which become possible only when the organic ligand is co-ordinated to a metal.

Reactions between two organic ligands can be effected when both ligands are brought into proximity by being co-ordinated to the same metal atom. Furthermore, two ligands *cis* to one another on a metal site can undergo *insertion* reactions which are features of many industrially important examples of homogeneous catalysis. In the Wacker process, acetaldehyde is made by the air-oxidation of ethylene in aqueous $PdCl_2$ in the presence of $CuCl_2$. The probable course of the reaction is outlined below:

$PdCl_4^{2-}$ + $CH_2{=}CH_2$ → $Cl_2Pd(Cl^-)(CH_2{=}CH_2)$

$Cl_2Pd(Cl^-)(CH_2{=}CH_2)$ —hydrolysis→ $Cl_2Pd(OH^-)(CH_2{=}CH_2)$

$Cl_2Pd(OH^-)(CH_2{=}CH_2)$ ⇌ (*cis* insertion) $Cl_2Pd(\square)(CH_2{\cdot}CH_2OH)$ (vacant site)

$Cl_2Pd(\square)(CH_2{\cdot}CH_2OH)$ —proton shift→ $Cl_2Pd(\square)H$ + $CH_3{\cdot}CHO$

$Cl_2Pd(\square)H$ + H_2O + C_2H_4 + $CuCl_2$ → $Cl_2Pd(OH^-)(CH_2{=}CH_2)$ + HCl + CuCl

CATALYTIC CYCLE

The CuCl is oxidised by air back to $CuCl_2$:

$$4\ CuCl + 4\ HCl + O_2 \rightarrow 4\ CuCl_2 + 2\ H_2O$$

The overall reaction is therefore

$$2\,C_2H_4 + O_2 \rightarrow 2\,CH_3\text{—}CHO$$

but the formation of the carbon—oxygen bond probably arises in the *cis* insertion process.

The hydroformylation of olefins, the oxo process, almost certainly proceeds by insertion reactions too. The olefins are converted to aldehydes and hence to alcohols by treatment with carbon monoxide and hydrogen under pressure in the presence of cobalt. The active catalyst is probably $HCo(CO)_4$ and the probable course of reaction is outlined below:

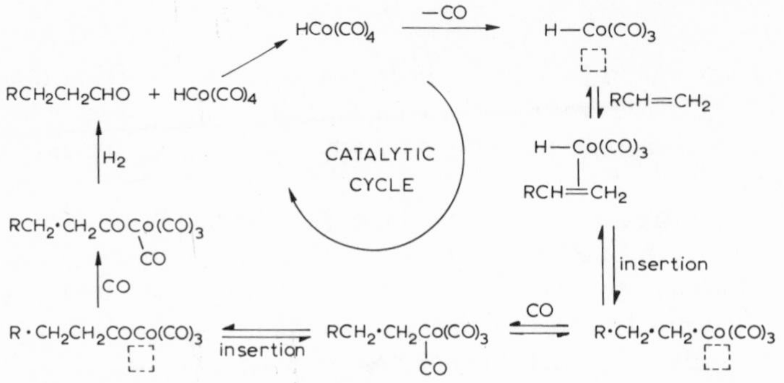

HCo(CO)4 —(−CO)→ H—Co(CO)3 [ ]
RCH2CH2CHO + HCo(CO)4
H2
CATALYTIC CYCLE
RCH=CH2
H—Co(CO)3 | RCH=CH2
RCH2·CH2COCo(CO)3 | CO
CO
insertion
R·CH2CH2COCo(CO)3 [ ] ⇌ (insertion) RCH2·CH2Co(CO)3 | CO ⇌ (CO) R·CH2·CH2·Co(CO)3 [ ]

This reaction is the route to long-chain alcohols for the synthesis of detergents.

Ziegler—Natta polymerisation of ethylene to polyethylene with $TiCl_4$ and Et_3Al as catalyst may also involve *cis* insertion:

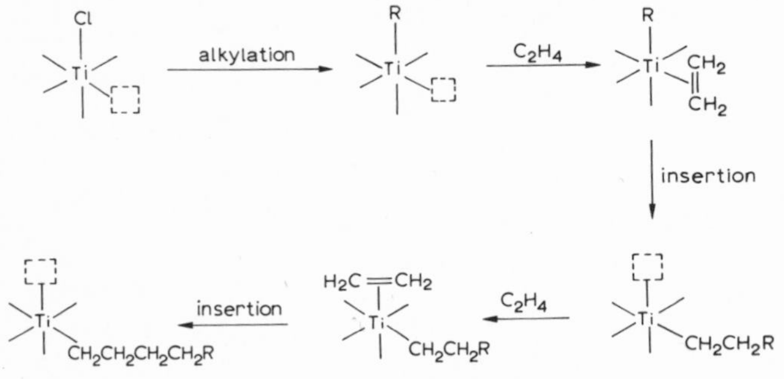

Cl–Ti [ ] —(alkylation)→ R–Ti [ ] —(C_2H_4)→ R–Ti(CH2=CH2)
insertion
[ ]–Ti–CH2CH2CH2CH2R ←(insertion)— H2C=CH2–Ti–CH2CH2R ←(C_2H_4)— [ ]–Ti–CH2CH2R

and so on.

Further Reading

G.E. Coates, M.L.H. Green and K. Wade, Organometallic compounds, 3rd edition, Vol. 1, The main group elements, 1967; Vol. 2, Transition elements, 1968, Methuen, London.
F.G.A. Stone and R. West (Eds.), Advances in organometallic chemistry, 1964—date.
E.I. Becker and M. Tsutsui, Organometallic reactions, Vol. 1 and 2, Wiley, New York, 1970.

B.L. Shaw and N.I. Tucker, Organo-transition metal compounds and related aspects of homogeneous catalysis, Pergamon, Oxford, 1975.

D.S. Matteson, Organometallic reaction mechanisms of the non-transition elements, Academic Press, New York, 1974.

R.F. Heck, Organotransition metal chemistry, a mechanistic approach, Academic Press, New York, 1974.

Germanium, Tin and Lead — Group IVB

19

19.1. Introduction

The atoms of the Group IVB elements resemble those of carbon and silicon in having the ns^2np^2 electron configuration and ^3P ground state, and in forming the tetrahedral bonds associated with sp^3 hybridisation. But downwards there is an increasing tendency to form instead an 'inert-pair' ion; and, in its most stable salts, lead preserves an ns^2 'core', appearing as Pb^{2+}. Since the atoms have rather low electronegativities, the bonds in many of their compounds are fairly strongly ionic. The usual practice is to regard the metals formally as ions, with charge number +4 or (for Pb) +2, and to assign ionic radii (as distinct from the atomic or 'metallic' radii) on this basis.

The Pb^{2+} ion is recognisable in PbS, which has the typical rock salt structure, and in the solid PbF_2 (*m.p.* 1091 K). But Sn^{2+} is not an apt description of tin in its bipositive compounds, these being predominantly covalent. The ionisation energies and the electronegativities of the elements are shown in Table 19.1.

The first ionisation energies of Sn and Pb may be contrasted with those of Cd and Hg atoms which have comparable radii (~148 pm) but also have the closed shell structure ns^2. Their similar electrode potentials indicate similar free energies (and hence, roughly, heats) of hydration for the M^{2+} ions. The electrode potentials for M^{4+}/M^{2+} are, however, very different; consequently, lead(IV) compounds in acid solution are much stronger oxidising agents than tin(IV) compounds. Tin(II) is a convenient laboratory reducing agent. The free energies, relative to the element, of the common oxidation states in solution at pH = 0 are shown graphically in Fig. 19.1.

Germanium(II) is unstable, tending to disproportionation:

$$2\ Ge^{2+} + 2\ H_2O \rightarrow Ge + GeO_2 + 4\ H^+$$

but Sn^{II} and Pb^{II} are both quite stable in this respect.

19.2. Properties of the Elements

The elements themselves show a marked transition from the rather nonmetallic germanium to the distinctly metallic lead. Germanium has a dia-

TABLE 19.1

ATOMIC PROPERTIES OF THE ELEMENTS

	Ge	Sn	Pb
Atomic number	32	50	82
Electron configuration	$[Ar]3d^{10}4s^2 4p^2$	$[Kr]4d^{10}5s^2 5p^2$	$[Xe]5d^{10}6s^2 6p^2$
$I(1)$/kJ mol^{-1}	760	707	715
$I(2)$/kJ mol^{-1}	1540	1415	1450
$I(3)$/kJ mol^{-1}	3310	2950	3090
$I(4)$/kJ mol^{-1}	4420	3830	4080
Metallic radius/pm	137	162	175
Covalent radius/pm	122	141	154
$r_{M^{2+}}$/pm	73	93	120
$r_{M^{4+}}$/pm	53	71	84
Electronegativity	2.02	1.72	1.55

mond-type lattice and a *m.p.* intermediate between the very high values of carbon and silicon and the low ones of tin and lead. Tin exists in three solid forms:

$$\text{grey } (\alpha) \text{ tin} \underset{}{\overset{286.5\,K}{\rightleftharpoons}} \text{white } (\beta) \text{ tin} \underset{}{\overset{434\,K}{\rightleftharpoons}} \text{brittle tin}$$

(diamond-type lattice) (body-centred tetragonal) (rhombic)

Lead, however, has only the characteristically metallic c.c.p. form. Both tin and lead resemble gallium and indium in their long temperature ranges of liquidity, respectively 2035 and 1430 K.

Germanium has remarkable electrical properties. In the purest form it has

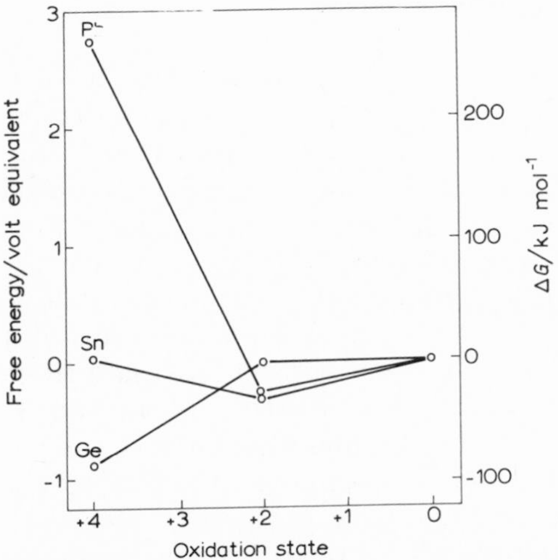

Fig. 19.1. Free energy diagram for Group IVB metals.

TABLE 19.2

PHYSICAL PROPERTIES OF THE ELEMENTS

	Ge	Sn	Pb
ρ/g cm^{-3}	5.36	5.77 (α) 7.29 (β)	11.34
M.p./K	1210	505	600
B.p./K	3100	2540	2030

a specific resistance, at room temperature, of about 50 Ω cm^{-1}. But when the temperature is raised increasing numbers of electrons are excited and pass from a filled energy band, over a narrow energy gap, into an empty 'conduction band' and the resistance falls. The metal is an 'intrinsic semiconductor'. Access to the conduction bands can also be attained by the introduction of 'impurities'. The presence of one part per million of a Group III or Group V metal in pure germanium can reduce the specific resistance by a factor of 50 or more. Antimony, for instance, with its five valence electrons, when incorporated in the 4-co-ordinate lattice contributes the extra electron to the germanium conduction band and provides *negative* current carriers or 'N-type' semiconduction. A Group III impurity such as Al creates 'holes' in a lattice of 4-co-ordination, the localised electron deficiency causes the migration of electrons from the conduction band. As these 'holes' behave as *positive* current carriers, the effect is termed 'P-type' semiconduction. Combinations of such n- and p-types of germanium function as electronic devices like rectifiers and transistors.

19.3. Preparation of the Elements

Germanium is recovered principally from flue dusts and coal ash. The only ore used commercially is germanite, (Cu, Ge, Fe, Zn, Ga) (S, As), containing about 6% Ge. When these substances are strongly heated with HCl the chloride $GeCl_4$, distils off. It is hydrolysed to GeO_2, which is dried and reduced by hydrogen, carbon or a mixture of C and KCN under a molten salt flux. The element is silvery-white, hard, brittle and without allotropes.

Very pure germanium is made by zone refining, usually achieved by slowly traversing a rod of the solid element with a molten zone by means of a moving radio-frequency heater. Impurities are thus concentrated in a short section at one end of the bar, which is discarded. The success of the process depends on the difference between the solid and liquid solubilities of the impurity elements present in the germanium at the point of solidification.

Tin (4×10^{-3}% of the earth's crust) is remarkable for its three allotropes and its ten stable isotopes, ranging in mass from 112 to 124. The only important ore is cassiterite, SnO_2. When the ore is roasted, the impurities sulphur and arsenic are oxidised to SO_2 and As_2O_3 respectively; the latter sublimes into cooled chambers and is a source of arsenic. The residual SnO_2 is reduced with carbon in a blast or reverberatory furnace, and the metal is tapped off. Its principal use is in the manufacture of tinplate, but the alloys with copper (bronzes), with lead (solders) and with lead and antimony (type metals) are important. The metal, ordinarily in its tetragonal form, is silvery, with a slight blue tinge, soft and malleable. Grey or α-tin, stable below 286.5 K, has a diamond-like structure and is brittle. The change from white to grey, which is accompanied by a fall in density, is slow above 220 K unless some grey tin is present to catalyse the change. Above 434 K, γ-tin, a more brittle metallic modification, is stable.

Lead (2×10^{-4}% of the earth's crust) occurs principally as galena, PbS, and cerussite, $PbCO_3$. The metal is obtained by ore-hearth smelting in which furnace conditions are controlled so that (a) in which about 2/3 of the sulphide reacts is followed by (c). Reactions (b) and (d) play a minor part.

(a) $2\,PbS + 3\,O_2 \rightarrow 2\,PbO + 2\,SO_2$
(b) $[PbS + 2\,O_2 \rightarrow PbSO_4]$
(c) $PbS + 2\,PbO \rightarrow 3\,Pb + SO_2$
(d) $[PbS + 2\,PbSO_4 \rightarrow Pb + 2\,PbO + 3\,SO_2]$

Alternatively, (a) may be followed by blast furnace smelting:

$PbO + CO \rightarrow CO_2 + Pb$

A later method of extraction is by the electrolysis of PbS dissolved in molten $PbCl_2$; Pb and S are liberated at cathode and anode respectively.

Lead is without allotropes and the pure metal is very soft; the hardness is increased by the presence of Sb, Cu and the other elements. The resistance of lead to atmospheric corrosion and to attack by acids leads to its employment in chemical plant and for pipes and cable sheathing. The presence of 0.050—0.065% of tellurium improves the properties desirable for these purposes, such as the grain-size, hardness, tensile strength and corrosion resistance. Alloyed with a little Sb, it is used in large quantities for the electrode grids in lead—acid storage batteries.

Silver occurs in lead ores and appears in the metal from the smelter; its removal and recovery from crude lead is usually commercially worthwhile. Parkes' process, used for this purpose, is based on the low solubility of zinc in molten lead and the very high distribution coefficient for silver between the zinc and lead layers. After stirring zinc into the molten lead, the Zn—Ag alloy, which freezes and floats to the top when the molten metal cools, is skimmed off and processed for the recovery of silver.

19.4. Reactions of the Metals

The principal reactions of germanium and tin are shown in Fig. 19.2 and involve the elements in their quadrivalent states. There are compounds of

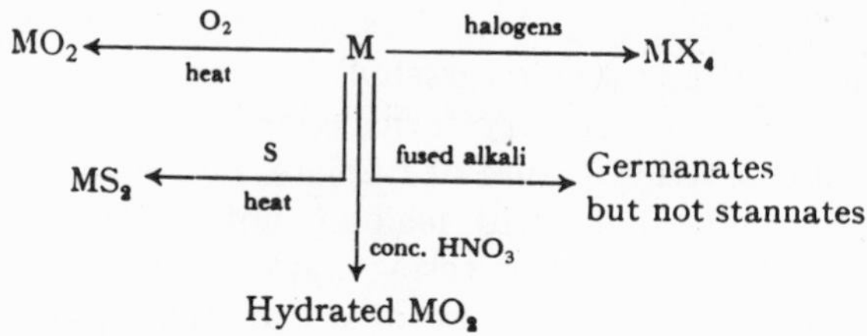

Fig. 19.2. Outline chemistry of Ge and Sn (M).

bivalent germanium such as $GeCl_2$; they suffer disproportionation when heated and are readily oxidised to the quadrivalent state.

However, lead behaves differently from germanium and tin because its 6s electrons have little tendency to participate in sp^3 hybridisation and the quadrivalent state is thermally unstable except with ligands of very high electronegativity. Hence the element usually features in its +2 oxidation state (Fig. 19.3). The three elements are rather unreactive toward acids.

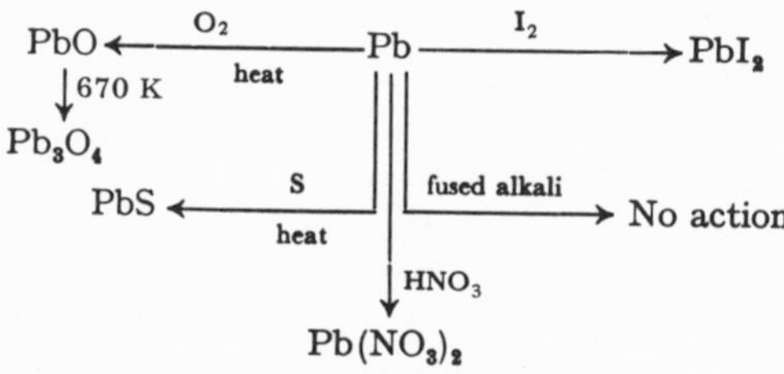

Fig. 19.3. Outline chemistry of Pb.

Germanium is not attacked by HCl or HF. Both Ge and Sn are oxidised by concentrated HNO_3 to the hydrated dioxides:

$$3\,M + 4\,HNO_3 \rightarrow 3\,MO_2 + 4\,NO + 2\,H_2O$$

In contrast, lead is dissolved, forming the nitrate:

$$3\,Pb + 8\,HNO_3 \rightarrow 3\,Pb(NO_3)_2 + 4\,H_2O + 2\,NO$$

The high hydrogen overvoltage at tin is responsible for the metal's lack of reactivity towards cold dilute acids. However, on heating, concentrated HCl and H_2SO_4 both react:

$$Sn + 2\,HCl \rightarrow SnCl_2 + H_2$$
$$Sn + 2\,H_2SO_4 \rightarrow SnSO_4 + SO_2 + 2\,H_2O$$

The action of these acids on lead is limited by the low solubilities of $PbCl_2$ and $PbSO_4$ which form coatings on the metal; but on heating both acids penetrate the coatings and dissolve the metal. Some H_2 is obtained, as well as SO_2, from concentrated H_2SO_4.

19.5. Halides

19.5.1. Tetrahalides

The physical properties of the tetrahalides, except those of SnF_4 and PbF_4, correspond to those of the covalent halides of carbon and silicon. GeF_4 is a gas similar to SiF_4, the tetrachlorides of the three elements and the bromides of germanium and tin are liquids, and GeI_4 and SnI_4 solids of low *m.p.* The compounds SnF_4, which sublimes at 978 K, and PbF_4, which melts at 870 K, are both more ionic in character than the other tetrahalides.

The tetrahalides of Ge and Sn are usually made by the action of the halogen itself on the heated element; the fluorides are obtained by the reactions:

$$GeO_2 + 4\,HF \rightarrow GeF_4 + 2\,H_2O$$
$$SnCl_4 + 4\,HF \rightarrow SnF_4 + 4\,HCl$$

The tetrahalides PbF_4 and $PbCl_4$ result from oxidising the dihalide with the corresponding halogen, the latter at a moderate temperature because of the instability of $PbCl_4$. Neither bromine nor iodine is sufficiently strongly electron-accepting to withdraw electrons from the $6s^2$ shell of Pb^{2+}; hence the absence of these tetrahalides. And PbF_4 and $PbCl_4$ dissociate on warming:

$$PbX_4 \rightarrow PbX_2 + X_2$$

Lead tetrafluoride is consequently a good fluorinating agent.

The known tetrahalides with their physical characteristics are set out in Table 19.3.

TABLE 19.3
TETRAHALIDES OF GERMANIUM, TIN AND LEAD

Halides	Ge	Sn	Pb
Tetrafluorides	Colourless gas *subl. p.* 236 K	Colourless solid *subl. p.* 978 K	Pale-yellow crystals *m.p.* 870 K
Tetrachlorides	Colourless liquid *b.p.* 356 K	Colourless liquid *b.p.* 387 K	Yellow, oily liquid thermally unstable
Tetrabromides	Pale-grey crystals *m.p.* 299 K	White crystals *m.p.* 304 K	
Tetraiodides	Red crystals *m.p.* 419 K	Orange crystals *m.p.* 417 K	

The Raman spectra of mixtures of $SnCl_4$, $SnBr_4$ and SnI_4 indicate that mixed halides with all the possible halogen combinations are present.

The tetrahalides are hydrolysed irreversibly by water, complexes often being produced. Thus $SnCl_4$, hydrolysed in the presence of cineol, produces a salt of the ion $[SnCl_3(OH)_3]^{2-}$, indicating that hydrolysis is initiated by the co-ordination of water molecules to the metal atom:

$$SnCl_4 + 2\ H_2O \rightarrow SnCl_4(H_2O)_2 \rightleftharpoons H_2[SnCl_4(OH)_2] \rightleftharpoons H[SnCl_3(OH)_2] + HCl$$
$$H[SnCl_3(OH)_2] + H_2O \rightleftharpoons H_2[SnCl_3(OH)_3]$$

A crystalline hydrate, $SnCl_4 \cdot 5\ H_2O$, is obtained when a limited amount of water is added to tin(IV) chloride. The ionic character of this solid, as compared with $SnCl_4$, is probably due to the presence of the complex ion $[Sn(H_2O)_4]^{4+}$.

GeF_4 hydrolyses similarly to SiF_4 which gives fluorosilicic acid, H_2SiF_6:

$$3\ GeF_4 + 2\ H_2O \rightarrow 2\ H_2GeF_6 + GeO_2$$

Octahedral complex ions, MX_6^{2-}, are generally very stable; GeF_6^{2-}, like SiF_6^{2-}, is particularly so. Ammonium hexachloroplumbate(IV) is precipitated when NH_4Cl is added to the solution produced by passing Cl_2 into a suspension of $PbCl_2$ in HCl:

$$PbCl_2 + Cl_2 \rightarrow PbCl_4$$
$$PbCl_4 + 2\ NH_4Cl \rightarrow (NH_4)_2PbCl_6$$

19.5.2. Dihalides

The dihalides are rather more ionic in character. Those of germanium are colourless-to-yellow solids made by passing the tetrahalide vapour over the heated metal:

$$GeX_4 + Ge \rightarrow 2\ GeX_2$$

At higher temperatures this reaction is reversed and there is disproportionation. Germanium di-iodide has the layer-lattice structure of CdI_2. The germanium dihalides are all strong reducing agents. Trichlorogermane, $GeHCl_3$, is formed when $GeCl_2$ reacts with HCl gas at 313 K; it is a colourless, fuming liquid.

The dihalides of tin and lead are monomeric in the vapour; the molecules are angular. In $PbBr_2$ the Br—Pb—Br angle is 86°. These anhydrous compounds cannot strictly be considered ionic, but tetratin(II) hydroxide dichloride, $Sn_4(OH)_6Cl_2$, the only definite compound crystallising from aqueous solutions, is the salt $[Sn_3(OH)_4]^{2+}[Sn(OH)_2Cl_2]^{2-}$. The lead dihalides are all sparingly soluble in cold water and can be precipitated:

$$Pb(NO_3)_2 + 2\ HF \rightarrow PbF_2 + 2\ HNO_3$$

The difluoride, PbF_2, crystallises with the rutile structure. Yellow PbI_2 dissolves in hot water to give a colourless solution containing hydrated Pb^{2+} ions.

Complex halides such as $KSnCl_3$, K_2SnCl_4, K_2PbCl_4 and $(NH_4)_2PbCl_4$ can be made from tin(II) and lead(II) chlorides. The formulae of the alkali metal fluoroplumbates(II) depend on the ionic radius of the Group I metal. Potassium fluoride gives K_4PbF_6 with PbF_2, but rubidium and caesium fluorides give the perovskite-type compounds $MPbF_3$. Potassium and rubidium also form non-stoichiometric compounds $M_nPb_{1-n}F_{2-n}$ where $n = 0.2$ to 0.3.

These have an anti-αAgI structure with additional F^- ions fitting into the lattice.

19.6. Oxides

The dioxides, GeO_2 and SnO_2, are the ultimate oxidation products of the metals and are stable at high temperatures. Lead dioxide, PbO_2, is much less thermally stable and is made by the oxidation of Pb^{2+} in alkaline solution. All three dioxides have structures of the rutile type but GeO_2 has also a cristobalite structure stable above 1306 K.

Though freshly precipitated germanium dioxide was formerly referred to as germanic acid there is no X-ray evidence for a definite hydrate. However, GeO_2 reacts with basic oxides to form germanates usually isomorphous with the silicates. Sodium metagermanate, Na_2GeO_3, is soluble in water. Tetraethoxygermanium, $Ge(OEt)_4$, *b.p.* 458 K, made by refluxing $GeCl_4$ with NaOEt in alcohol, is converted into a gel of hydrated GeO_2 on the addition of water to the alcoholic solution. Removal of the alcohol under reduced pressure leaves a hard material with adsorbent characteristics similar to silica gel.

The alkali-metal stannates and plumbates form trihydrates containing octahedral anions, e.g. $Sn(OH)_6{}^{2-}$. These are isomorphous with one another and with the platinates. The dioxides dissolve in acids only in the presence of F^- and Cl^- ions.

X-ray examination of the products of the thermal decomposition of PbO_2 discloses two intermediate, non-stoichiometric phases between PbO_2 and Pb_3O_4. The α-phase is close to Pb_7O_{11} and the β-phase to Pb_2O_3. Red lead itself, Pb_3O_4, formed by heating either litharge, PbO, or lead dioxide, PbO_2, in air, consists of chains of PbO_6 octahedra sharing opposite edges linked by Pb atoms each co-ordinated to three oxygens.

The monoxides, though amphoteric, are only weakly so on the acidic side, particularly SnO and PbO. Black GeO can be made by heating in an inert atmosphere the yellow precipitate of hydrated GeO obtained when a solution of $GeCl_4$ is warmed with hypophosphorous acid. Tin(II) oxide results when the oxalate SnC_2O_4 is heated in the absence of oxygen:

$$SnC_2O_4 \rightarrow SnO + CO + CO_2$$

Crystalline black SnO and yellow PbO are both tetragonal with layer lattices in which the metal atom is bonded to four oxygen atoms arranged in a square on one side of it. The adjacent layers are held together by metal—metal bonds.

19.7. Sulphides

The sulphides, GeS_2 and SnS_2, unlike SiS_2, are not hydrolysed by water. GeS_2 is made by direct combination of the elements, but the reaction between Sn and S goes as far as SnS_2 only in the presence of NH_4Cl. GeS_2 has a quartz-like structure, SnS_2 a layer lattice of the cadmium iodide type. The compounds are precipitated in a somewhat impure condition by adding H_2S to acidified germanate and stannate solutions. The precipitates dissolve in alkali sulphide solutions to give thiogermanates and thiostannates: $Na_2SnS_3 \cdot 8\ H_2O$ and $Na_4SnS_4 \cdot 18\ H_2O$ have been isolated but a thiogermanate has not been obtained from solution.

Black GeS is the most thermally stable of the Ge^{II} compounds. It is best made by reducing a suspension of GeS_2 in 6 *M* HCl with hypophosphorous acid:

$$GeS_2 + 2\ HCl + H_3PO_2 + H_2O \rightarrow GeCl_2 + H_3PO_3 + 2\ H_2S$$

neutralising the solution produced and treating it with H_2S:

$$GeCl_2 + S^{2-} \rightarrow GeS + 2\ Cl^-$$

SnS is obtained as a grey solid by heating the metal with sulphur at 1150 K, or as a brown precipitate when H_2S is passed into a tin(II) salt solution. It does not dissolve in alkali sulphides; but in polysulphides, such as yellow ammonium sulphide, it gives solutions of thiostannates from which the higher sulphide is precipitated on acidification.

The structure of PbS is of particular interest as the compound crystallises with the typically ionic NaCl lattice in marked contrast to the layer lattice of PbO. It may well be the least ionic compound to do so.

19.8. Nitrogen Compounds

Germanium di-iodide reacts with liquid ammonia to give a yellow product with the empirical formula GeNH which is converted into a dark-brown nitride, Ge_4N_3, when heated to 600 K in a vacuum

$$GeI_2 + 3\ NH_3 \rightarrow GeNH + 2\ NH_4I$$

Tin(IV) iodide reacts with potassium amide to give potassium ammonostannate and not an amide or imide,

$$SnI_4 + 6\ KNH_2 \rightarrow K_2Sn(NH_2)_6 + 4\ KI$$

and metallic tin reacts with KNH_2 in liquid NH_3 to give $KSn(NH_2)_3$. This is easily oxidised, in the presence of an excess of KNH_2, to the ammonostannate shown above. Lead iodide reacts with potassium amide in liquid ammonia to give an imide:

$$PbI_2 + 2\ KNH_2 \rightarrow PbNH + 2\ KI + NH_3$$

19.9. Oxoacid salts

Tin(IV) nitrate is formed by the action of N_2O_5 or $ClNO_3$ on $SnCl_4$. The white volatile $Sn(NO_3)_4$ contains 8-co-ordinate tin, with each nitrate group behaving as a bidentate ligand.

Tin(II) sulphate, an excellent source of Sn(II), is best prepared by the displacement of Cu from a copper sulphate solution by metallic tin. Its solubility in water decreases with temperature, and basic salts such as $Sn_3(OH)_2OSO_4$ are obtained in ammoniacal solution. Hygroscopic $Sn(SO_4)_2 \cdot 2\,H_2O$ is obtained from hydrous tin(IV) oxide in hot dilute sulphuric acid and is best stored in sealed ampoules. Tin(II) and tin(IV) carboxylates can be prepared from the corresponding oxide or halide with the requisite carboxylic acid. Tin(II) acetate hydrolyses slowly over several hours but tin(IV) carboxylates are in general sensitive to hydrolysis.

The other Group IVB oxoacid salts of importance are those of lead. Most of the lead(II) salts are sparingly soluble. The solubilities of $PbSO_4$, $PbCrO_4$ and PbC_2O_4 resemble those of the corresponding barium salts; the Pb^{2+} ion (120 pm) and the Ba^{2+} ion (135 pm) are similar in size. Lead acetate, $Pb(CH_3COO)_2 \cdot 3\,H_2O$, is easily soluble in water but ionises very slightly.

Lead tetra-acetate, the only stable lead(IV) oxoacid salt, is deposited in white needles on cooling solutions of red lead, Pb_3O_4, in hot acetic acid (PbO_2 is insoluble). It is used as an oxidising agent:

$$\begin{array}{l} R\cdot CH{-}OH \\ \quad | \\ R'\cdot CH{-}OH \end{array} \xrightarrow{Pb(CH_3COO)_4} \begin{array}{l} R\cdot CH{=}O \\ \\ R'\cdot CH{=}O \end{array}$$

and is also employed as a methylating agent:

This is probably a free-radical reaction in which the CH_3 radical, derived from the pyrolysis of the tetra-acetate, takes part:

$$Pb(CH_3COO)_4 \rightarrow CH_3^{\bullet} + CH_3COO^{\bullet} + CO_2 + Pb(CH_3COO)_2$$

19.10. Organometallic Compounds

The alkyl and aryl compounds of germanium and tin are numerous. They are generally rather stable, thermally and hydrolytically, as might be expected for organic derivatives of the congeners of carbon. Lead alkyls are less thermally stable, and fewer are known, but they include the most commercially important of all organometallic compounds, Me_4Pb and Et_4Pb.

19.10.1. Organogermanium compounds

Germanium alkyls were first made from germanium tetrahalides and zinc alkyls, but the Grignard method is the more versatile:

$GeBr_4 + 2\,Et_2Zn \rightarrow Et_4Ge + 2\,ZnBr_2$
$GeCl_4 + 4\,RMgCl \rightarrow R_4Ge + 4\,MgCl_2$

Tetra-ethylgermanium is monomeric in the vapour and in benzene solution; it is not oxidised by hot nitric acid. The simplest tetra-arylgermanium, $(C_6H_5)_4Ge$, can be made from $GeCl_4$, C_6H_5Br and sodium. The solid, *m.p.* 506 K, volatilises without decomposition and is not attacked by boiling aqueous alkalis. Like the alkyls, the aryls are soluble in organic solvents and tend to be somewhat more reactive than the corresponding silicon compounds.

Organogermanium compounds containing Ge—Ge bonds are common; they can be made by a reaction analogous to the Würtz reaction:

$2\,Et_3GeBr + 2\,Na \rightarrow 2\,NaBr + Et_3Ge{-}GeEt_3$
(*b.p.* 488 K, monomeric in benzene)
$2(C_6H_5)_3GeBr + 2\,Na \rightarrow 2\,NaBr + (C_6H_5)_3Ge{-}Ge(C_6H_5)_3$
(*m.p.* 613 K, monomeric in benzene)

Polygermane derivatives are also known, and it is of interest that a mixed silicon—germanium alkyl has been made, also by Würtz coupling:

$Me_3GeBr + Me_3SiBr + 2\,Na \rightarrow 2\,NaBr + Me_3Ge{-}SiMe_3$

Germanium forms many organo-compounds analogous to carbon and silicon compounds by reactions similar to those of organic chemistry.

19.10.2. Organotin compounds

Tin tetra-alkyls and their halogen derivations are formed in reactions such as:

$SnCl_4 + 4\,RMgX \rightarrow R_4Sn + 4\,MgXCl$
$n\,R_4Sn + (4-n)SnX_4 \rightarrow 4\,R_nSnX_{4-n}$

The tetra-alkyls are colourless liquids, insoluble in water but soluble in organic solvents. They neither polymerise nor co-ordinate. Derivatives of distannane and the polystannanes up to about Sn_6 are known but become progressively more difficult to characterise.

$2\,Et_3SnBr + 2\,Na \rightarrow Et_3SnSnEt_3 + 2\,NaBr$

$$n\,Ph_2SnH_2 + n\,Et_2Sn(NEt_2)_2 \longrightarrow \left(\begin{array}{cc} Ph & Et \\ | & | \\ -Sn- & -Sn- \\ | & | \\ Ph & Et \end{array} \right)_n + 2n\,Et_2NH$$

Organotin halides are the primary starting materials in organotin derivative chemistry and undergo many facile substitution reactions

$4\,Me_3SnCl + LiAlH_4 \rightarrow 4\,Me_3SnH + LiAlCl_4$
$2\,Bu_3SnCl + 2\,NaOH \rightarrow (Bu_3Sn)_2O + H_2O + 2\,NaCl$
$Ph_3SnCl + LiNEt_2 \rightarrow Ph_3SnNEt_2 + LiCl$
$Et_3SnCl + NaOEt \rightarrow Et_3SnOEt + NaCl$

The major uses of organotin compounds are in the application of organotin carboxylates as stabilizers in PVC and as catalysts in polyurethane and silicone reactions.

19.10.3. Organolead compounds

The alkyls of lead are much less thermally stable than those of germanium and tin. A few organolead compounds containing Pb—Pb bonds have been obtained, but they are unstable. Although plumbane itself cannot be made, unstable dialkyl plumbanes are produced by the reduction of the corresponding halide by KBH_4 in ether:

$$R_2PbX_2 \xrightarrow{KBH_4} R_2PbH_2$$

Tetramethyl- and tetraethyllead react together:

$$Me_4Pb + Et_4Pb \rightarrow 2\ Me_2PbEt_2$$

Further Reading

E.G. Rochow and E.W. Abel, The chemistry of germanium, tin and lead, Pergamon, Oxford, 1975.

F. Glockling, The chemistry of germanium, Academic Press, New York, 1969.

R.W. Weiss (Ed.), Organometallic compounds, Vol. II, Compounds of germanium, tin and lead, Springer, New York, 1967.

A.K. Sawyer (Ed.), Organotin compounds, Vol. 1, 1971; Vol. 2, 1971; Vol. 3, 1973; Marcel Dekker, New York.

20

Nitrogen and Phosphorus—Group V

20.1. Introduction

These elements with five valence electrons form bonds which are almost exclusively covalent in character. For this reason it is not generally profitable to invoke an ionic description, however formal, and the definition of an ionic radius is of little value except perhaps for certain phosphorus complexes where an empirical P^{5+} may be defined.

The chemistry of nitrogen and phosphorus is dominated by the tendency of the atoms to complete their octets. As nitrogen only accepts electrons from the most electropositive elements, bonding occurs predominantly by the formation of single or multiple bonds which may lead to ionic species by subsequent gain or loss of an electron. However the existence of stable compounds containing an odd number of electrons (NO, NO_2) should not be ignored. The chief differences between nitrogen and phosphorus arise from the availability of d orbitals in the shell with principal quantum number 3 which give phosphorus the capacity for valency expansion to 5- and 6-co-ordination associated with trigonal bipyramidal and octahedral hybridisation (5.2). This together with the tendency for nitrogen to form p_π—p_π multiple bonds means that there are few isostructural species.

In the ground state, both atoms have three singly occupied p orbitals. With this configuration, which gives a ^4F state, is associated a spherically symmetrical distribution of electrons and a high ionisation energy. Since the

TABLE 20.1

ATOMIC PROPERTIES OF NITROGEN AND PHOSPHORUS

	N	P
Atomic number	7	15
Electronic configuration	[He] $2s^2 2p^3$	[Ne] $3s^2 3p^3$
Covalent radius/pm	70	110
Ionic radius M^{5+}/pm		34
Ionisation energy/MJ mol^{-1}	3.82	2.52
Electronegativity	3.07	2.06

compounds are essentially covalent, however, ionisation energies are less important than electronegativities (Table 20.1).

20.2. Structures and Bonding

Structural data for both nitrogen and phosphorus compounds are tabulated in the relevant sections (Tables 20.2 and 20.3). In general their shapes are as would be predicted on the basis of the Sidgwick—Powell approach. However no one method of describing the bonding is entirely adequate for all situations. In addition to molecular orbital and valence bond methods, the Linnett double quartet theory will be used. This is a modification of the usual Lewis octet description (which most nitrogen compounds tend to satisfy), and emphasises the tendency of electrons with similar spin to repel each other while electrons with opposed spins do not. Hence the process of electron pairing in the formation of bonds or lone pairs is considered to involve electrons of opposed spin.

The electron configuration of the nitrogen molecule may be described in molecular orbital terms as

$$N_2[KK(\sigma^b 2s)^2(\sigma^* 2s)^2(\sigma^b 2p_z)^2(\pi^b 2p_{xy})^4]$$

The presence of eight electrons in bonding orbitals and two in antibonding orbitals gives a difference of three electron pairs and hence a bond order of 3, as determined by the usual LCAO/MO method. However, because the s and p_z atomic orbitals belong to the same symmetry species, they can be hybridised to give two new levels, σ_1, predominantly an s orbital with some p_z character, and σ_2, predominantly p_z with some s character. Thus hybridisation has the effect of increasing the bonding power of σ_2 leaving σ_1 as filled equivalent non-bonding levels on each nitrogen as shown in Fig. 20.1. The resultant MO energy level diagram indicates the relative positions of bonding

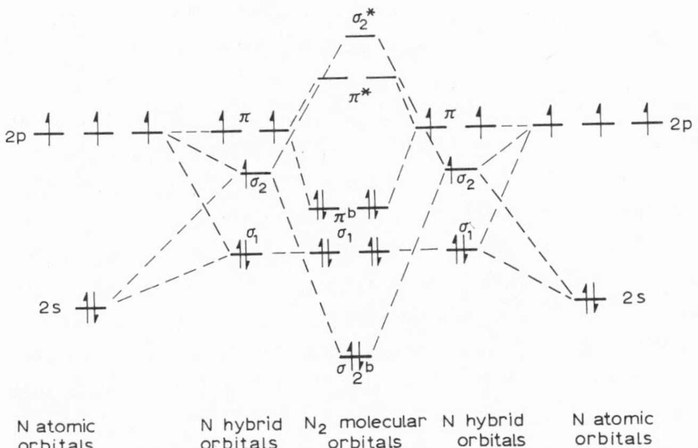

Fig. 20.1. Molecular orbital energy level diagram for molecular nitrogen.

and antibonding orbitals, the triple bond being consistent with a bond length of 109 pm and a dissociation energy of 945 kJ mol^{-1}.

The idea that four electron pairs (lone pairs or bond pairs) dispose themselves in the vicinity of a nitrogen atom is used extensively to explain molecular geometry. Examples are given in Table 20.2. Distortions from regular geometries due to interactions between electron pairs are also observed. For example, bond angles in NH_3, NMe_3 and NF_3 decrease as the electronegativity of the atom or group attached to nitrogen increases, and the electrons of the bond pairs are drawn away from the nitrogen so that they interact less with one another.

TABLE 20.2

HYBRIDISATION OF THE NITROGEN ATOM

Hybrid	Bonding pairs	Lone pairs	Molecular shape	Molecular symmetry	Example
sp^3	4σ	0	tetrahedral	T_d	NH_4^+
			tetrahedral	C_{3v}	F_3NO
	3σ	1	pyramidal	C_{3v}	NH_3
sp^2	$3\sigma\ 1\pi$	0	trigonal planar	D_{3h}	$[O{-}N(O)O]^-$
				C_{2v}	$Cl{-}N(O)O$
	$2\sigma\ 1\pi$	1	angular	C_s	$Cl{-}N{=}O$
sp	$2\sigma\ 2\pi$	0	linear	$D_{\infty h}$	$[O{=}N{=}O]^+$

In the gas phase, phosphorus, unlike nitrogen, exists as a tetrahedral P_4 molecule. The σ bonds (predominantly 3p in character), in spite of their being 'bent', are apparently preferred to the π bonds of a nitrogen-like configuration. The P—P—P bond angle of 60° as compared with the pure p-orbital angle of 90° represents a strain energy of 95.7 kJ mol^{-1} for P_4 and accounts for its high reactivity.

Phosphorus, like nitrogen, can complete an octet of four tetrahedrally arranged electron pairs either by forming three covalent bonds which have a pyramidal arrangement (PCl_3, NH_3) or by forming four covalent bonds in ions such as PH_4^+ which is akin to NH_4^+. But because 3d orbitals are also available, valency expansion to a co-ordination maximum of 6 is possible (Table 20.3). Distortions similar to those mentioned for nitrogen compounds are also operative with phosphorus. For example the 93° bond angle in PH_3 is much less than 107° in NH_3 as nitrogen forms stronger and hence shorter bonds with hydrogen, whereby the repulsive effect of the lone pair is less than in the phosphorus compound. In phosphoryl halides OPX_3, the X—P—X angle is always less than the tetrahedral angle because the P=O bond repels the P—X bonds more than single bonds repel each other (5.2).

TABLE 20.3
HYBRIDISATION OF PHOSPHORUS ATOM

Orbital or hybrid	Bonding pairs	Lone pairs	Molecular shape	Molecular symmetry	Example
p^3	3σ	1	trigonal pyramid	C_{3v}	PH_3
sp^3	4σ	0	tetrahedral	T_d	PCl_4^+
	3σ	1	trigonal pyramid	C_{3v}	PCl_3
	$4\sigma\ 1\pi$	0	tetrahedral	T_d	PO_4^{3-}
				C_{3v}	$OPCl_3$
sp^3d	5σ	0	trigonal bipyramid	D_{3h}	PF_5
sp^3d^2	6σ	0	octahedral	O_h	PCl_6^-

20.3. Thermodynamic Stability of the Oxidation States in Nitrogen and Phosphorus

The free energies, relative to the element, of the various oxidation states of nitrogen and phosphorus in solution at pH = 0 are shown graphically in Fig. 20.2.

For nitrogen there is a pronounced minimum at the zero oxidation state, which emphasises the stability of the nitrogen molecule with respect to NH_2OH and $H_2N_2O_2$. The points for all its positive oxidation states between 0 and +5 lie above the straight line connecting G(V) and G(0). These compounds are all thermodynamically unstable towards disproportionation. The only species represented in the graph which has a lower free energy than nitrogen itself is NH_4^+:

$$N_2 + 8\,H^+ + 6\,e \rightarrow 2\,NH_4^+ \qquad \Delta G^0 = -158\ \text{kJ mol}^{-1}$$

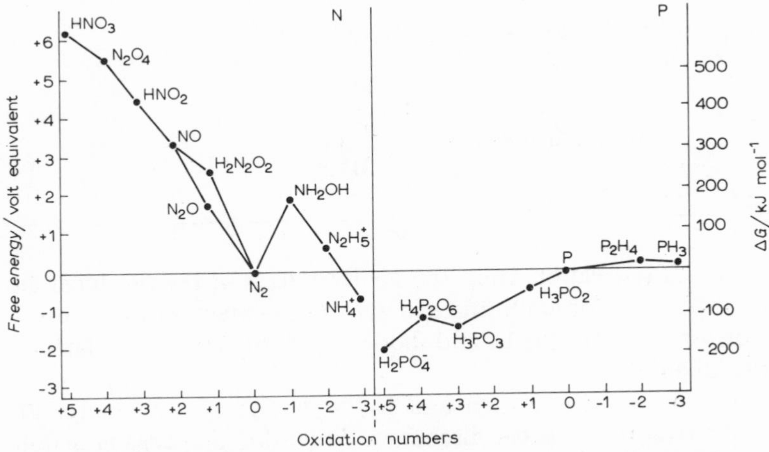

Fig. 20.2. Free energies of oxidation states of nitrogen and phosphorus, relative to the elements, at pH = 0.

For phosphorus compounds at pH = 0 the picture is entirely different. The phosphorus molecule itself is unstable with respect to disproportionation into the +1 and −3 states:

$$4\,P + 6\,H_2O \rightarrow PH_3 + 3\,H_3PO_2 \qquad \Delta G^0 = -128\ \text{kJ mol}^{-1}$$

At pH = 14 the free energy release is even greater:

$$4\,P + 3\,OH^- + 3\,H_2O \rightarrow 3\,H_2PO_2^- + PH_3 \qquad \Delta G^0 = -336\ \text{kJ mol}^{-1}$$

Hypophosphorous acid itself is unstable:

$$3\,H_3PO_2 \rightarrow 2\,H_3PO_3 + PH_3 \qquad \Delta G^0 = -127\ \text{kJ mol}^{-1}$$

The decomposition of phosphorous acid into phosphine and phosphoric acid has a small positive free-energy change:

$$4\,H_3PO_3 \rightarrow PH_3 + 3\,H_3PO_4 \qquad \Delta G^0 = +4\ \text{kJ mol}^{-1}$$

But in hot solutions, from which the PH_3 volatilises, the decomposition proceeds to completion. Thus, through the various steps, phosphorus is unstable with respect to its hydrolysis to H_3PO_4 and PH_3. The low free energies of the positive oxidation states of phosphorus indicate that even weak oxidising agents will oxidise the element all the way to H_3PO_4.

20.4. The Elements: Preparation and Properties

20.4.1. Nitrogen

Nitrogen, the principal gas of the atmosphere (78 vol.%), is separated industrially from liquid air by fractional distillation (Table 20.4). The gas is

TABLE 20.4

PHYSICAL PROPERTIES OF NITROGEN AND PHOSPHORUS

	N	P (white)
Density/g cm^{-3}	1.027 (solid at 20.6 K)	1.828
M.p./K	63.1	317.2
B.p./K	77.3	553.6

rather inert. Chemical separation from the air by means of the producer-gas reaction provides nitrogen for ammonia and cyanamide manufacture. Terrestrial nitrogen consists of nitrogen-14 and 0.365% of nitrogen-15; the latter is used for isotopic labelling.

Nitrogen, in common with oxygen and hydrogen, gives a characteristic colour in the low-pressure electrical discharge. When the discharge is switched off there is no 'afterglow' in hydrogen, it is exceedingly brief in oxygen, but persists for several seconds in nitrogen. This glowing gas is commonly known as 'active nitrogen' as a consequence of its higher chemical reactivity when compared with molecular nitrogen. It seems reasonably certain that the discharge produces nitrogen atoms in the ground state. The 'afterglow' accompanies a pre-association of atoms into excited molecules from which they return to stable ground-state N_2.

$$N_2 \xrightarrow{\text{discharge}} 2\,N(^4S) \rightarrow N_2{}^* \rightarrow N_2 + h\nu$$

The nitrogen atoms have a relatively long life as recombination involves a three-body collision.

$$N(^4S) + N(^4S) \xrightarrow{M} N_2{}^* \rightarrow N_2 + h\nu \text{ (yellow)} \qquad (1)$$

M can represent a surface or, at slightly higher pressures, atomic or molecular nitrogen, but both reactions are partly responsible for producing excited molecular states, and it is these excited molecules returning to the ground state that give rise to the light emission.

The nitrogen atom content of active nitrogen may be estimated by following the colour changes in a series of chemiluminescent association reactions with nitric oxide. With no added nitric oxide, active nitrogen in a flow system emits the characteristic yellow afterglow downstream from the discharge as in (1). Addition of NO at flow rates less than the available concentration of $N(^4S)$ atoms in the active nitrogen stream results in the very fast reaction (2):

$$N(^4S) + NO \rightarrow N_2 + O(^3P) \qquad (2)$$

Oxygen atoms so formed may react with an excess of atomic nitrogen to produce excited nitric oxide molecules which emit the blue β and γ bands of NO as in (3):

$$N(^4S) + O(^3P) \rightarrow NO^* \rightarrow NO + h\nu \text{ (blue)} \qquad (3)$$

With relatively small amounts of NO, reaction (3) occurs simultaneously with (1), and the gas stream appears purple. When the flow rate of NO exceeds that of $N(^4S)$, reaction (2), being the fastest, eliminates all nitrogen atoms, hence (1) and (3) are not possible. Instead, excited nitrogen dioxide is formed which emits greenish-yellow NO_2 bands as in (4):

$$NO + O(^3P) \rightarrow NO_2{}^* \rightarrow NO_2 + h\nu \text{ (greenish yellow)} \qquad (4)$$

If NO is added at a flow rate equivalent to that of atomic nitrogen, both NO and $N(^4S)$ atoms are destroyed, and the slow, light-producing reactions (1), (3) and (4) cannot occur. Thus in the 'NO titration' as the flow rate of NO is slowly increased, the colours can change progressively through yellow, purple, blue to darkness and to greenish yellow. The flow rate at the dark 'end point' indicates the concentration of nitrogen atoms.

The chemically inert character of nitrogen is largely due to the high bonding energy (945 kJ mol^{-1}) of the molecule. Nevertheless it should not be forgotten that the natural process of nitrogen fixation takes place at ambient temperature and pressure while the present alternative ways of reacting nitrogen are energy consuming.

Combination occurs with both oxygen and hydrogen under suitable conditions; both reactions have been extensively studied and used commercially:

$$N_2 + O_2 \rightarrow 2\,NO \qquad N_2 + 3\,H_2 \rightarrow 2\,NH_3$$

Of the metals, only lithium combines at moderate temperatures, $6Li + N_2 \rightarrow 2Li_3N$; the Group IIA metals combine at about red heat, $3Ca + N_2 \rightarrow Ca_3N_2$; and boron and aluminium at bright red heat, $B + N \rightarrow BN$. Silicon and some elements of higher groups (20.13) react at temperatures above 1500 K. The nitrides of the Group I metals other than lithium cannot be made by direct combination (13.6).

Nitrogen reacts with a wider variety of compounds. One of the earliest reactions to be put to commercial use involved calcium carbide:

$$CaC_2 + N_2 \xrightarrow{1000\ K} CaCN_2 + C$$

More recently reactions with organometallic reagents have been investigated with a view to establishing a catalytic fixation of nitrogen. Titanium alkoxides are used in one such method:

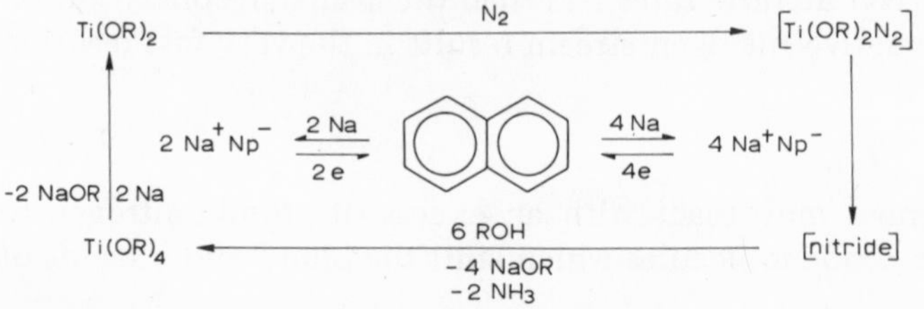

$Ti(OR)_2 \xrightarrow{N_2} [Ti(OR)_2N_2]$

$[Ti(OR)_2N_2] \rightarrow [nitride]$

$[nitride] \xrightarrow{6\ ROH,\ -4\ NaOR,\ -2\ NH_3} Ti(OR)_4$

$Ti(OR)_4 \xrightarrow{-2\ NaOR,\ 2\ Na} Ti(OR)_2$

$2\,Na^+Np^- \underset{2e}{\overset{2\ Na}{\rightleftharpoons}}$ (naphthalene) $\underset{4e}{\overset{4\ Na}{\rightleftharpoons}} 4\,Na^+Np^-$

Molecular nitrogen has also been found to act as a ligand (somewhat analogous to carbon monoxide) in forming complexes with some transition-metal compounds.

$$Co(acac)_3 + 3\,Ph_3P + N_2 \xrightarrow{Et_2AlOEt} CoH(N_2)(PPh_3)_3$$

$$FeCl_2 + 3\,PEtPh_2 + N_2 \xrightarrow{NaBH_4/EtOH} FeH_2(N_2)(PEtPh_2)_3$$

$$MoCl_4(PPhMe_2)_2 + 2\,PPhMe_2 + 2\,N_2 \xrightarrow[THF]{Na/Hg} cis\text{-}Mo(N_2)_2(PPhMe_2)_4$$

The N_2 ligand can also be introduced into the complex by means of another nitrogen compound like hydrazine or an azide

$$K_2[RuCl_5(H_2O)] + N_2H_4 \cdot H_2O \rightarrow [Ru(NH_3)_5N_2]Cl_2$$

$$trans\text{-}IrCl(CO)(PPh_3)_2 + RCON_3 \rightarrow trans\text{-}IrCl(N_2)(PPh_3)_2 + RCONCO$$

As might be expected, the bonding in $M{-}N_2$ compounds is qualitatively similar to that of terminal metal carbonyls. The differences stem from the relative energies of the molecular orbitals in free N_2 and CO which is consistent with N_2 being weaker than CO in both its σ-donor and π-acceptor character.

20.4.2. Phosphorus

Phosphorus (0.11% of the lithosphere) is found mainly in minerals based on calcium phosphate, such as collophanite, the monohydrate, $Ca_3(PO_4)_2 \cdot H_2O$, and apatite, $Ca_5F(PO_4)_3$. About 90% of the phosphate rock mined is converted into fertilisers, the rest is used for making elementary phosphorus, phosphorus compounds and such alloys as phosphor bronze. For the manufacture of fertilisers, rock phosphate is finely ground and treated with sufficient concentrated sulphuric acid to convert it to the soluble dihydrogen phosphate:

$$Ca_3(PO_4)_2 + 2\ H_2SO_4 \rightarrow Ca(H_2PO_4)_2 + 2\ CaSO_4$$

Elementary phosphorus is extracted from the mineral by heating with sand and coke, generally electrically, and condensing the vapour to give the white variety of the element:

$$Ca_3(PO_4)_2 + 3\ SiO_2 \rightarrow 3\ CaSiO_3 + P_2O_5$$
$$P_2O_5 + 5\ C \rightarrow 2\ P + 5\ CO$$

White phosphorus when pure melts to a colourless liquid which crystallises in the cubic system; the solid contains individual P_4 tetrahedra held together by Van der Waals forces. It can be changed to a hexagonal form by high pressure, the transition temperature being 196 K.

The black allotrope, similar to graphite in appearance and conductance, first made only at very high pressure, was later made from white phosphorus at ordinary pressure in the presence of mercury and a 'seed' of the black variety. Transformation was complete after 8 days during which the temperature was gradually raised from 490—640 K. The atoms in black phosphorus are covalently linked to form buckled sheets in an extended network of 3-coordinated atoms.

Red phosphorus is produced when either the white or the black allotrope of the element is heated at 700 K. It is an intermediate form in which the structural order is incompletely organised. Its physical properties are those of a polymeric material: the P_4 tetrahedra have been opened up and interlinked during its formation from white phosphorus, whereas the puckered layers of P atoms have been broken down during its formation from black phosphorus. The vapour pressure of red phosphorus is surprisingly high in view of its structure.

When phosphorus is cooled rapidly from 1300 K to 77 K a dark brown solid is obtained, stable indefinitely at that temperature and probably containing P_2 molecules. At 173 K the brown phosphorus changes irreversibly to a mixture of approximately 20% red and 80% white phosphorus; as the temperature is raised the rate of conversion increases but the ratio of the allotropes alters little.

A vitreous form of phosphorus has been prepared by heating the white form with mercury. It is stable in moist air and is harder than black phosphorus.

20.5. Halides

20.5.1. Nitrogen halides

Trihalides of nitrogen with fluorine and chlorine have been isolated, but for bromine and iodine only the ammonia complexes, $NBr_3 \cdot 6\,NH_3$ and $NI_3 \cdot NH_3$ are known.

Nitrogen trifluoride is a colourless gas (*b.p.* 144 K) of normal vapour density, made by electrolysing fused NH_4HF_2 in a copper vessel. Being an exothermic compound ($\Delta H_f = -119.4$ kJ mol^{-1}), NF_3 is much more thermally stable and chemically inert than the other nitrogen halides. It is not hydrolysed by water or alkali and is non-explosive, though it reacts violently with hydrogen on sparking:

$$2\,NF_3 + 3\,H_2 \rightarrow N_2 + 6\,HF$$

Nitrogen trichloride is a yellow, oily liquid made by the action of chlorine on NH_4Cl in concentrated aqueous solution. By contrast NCl_3 is an endothermic compound ($\Delta H_f = 230$ kJ mol^{-1}), it explodes above its boiling point or on impact, and is easily hydrolysed:

$$NCl_3 + 3\,OH^- \rightarrow NH_3 + 3\,OCl^-$$

Difluorodiazine, N_2F_2, a minor product of the electrolysis of NH_4HF_2, has well defined isomers (*cis* *m.p.* $<$ 78.15 K, *b.p.* 167.45 K; *trans* *m.p.* 101.15 K, *b.p.* 161.75 K). Pure *trans*-isomer can be prepared from N_2F_4:

$$N_2F_4 + AlCl_3 \xrightarrow{160\text{—}200\ K} trans\text{-}N_2F_2$$

The *trans* isomer is converted to the *cis* on heating; it is appreciably less reactive than the *cis* with glass and mercury.

Tetrafluorohydrazine, N_2F_4, is produced when NF_3 is passed over hot copper and other metals which abstract fluorine. The compound boils at 200 K, has a critical temperature of 309 K, and has the D_{2h} structure

```
F       F
 \     /
  N — N
 /     \
F       F
```

Its vapour contains NF_2 free radicals as well as N_2F_4 molecules (*cf.* N_2O_4, 20.6.5). With chlorine under ultraviolet irradiation it gives the monochlorofluoride:

$$Cl_2 + N_2F_4 \rightarrow 2\,NClF_2$$

20.5.2. Derivatives of nitrogen halides

Monochloramine, NH_2Cl, is formed when NH_3 and NaOCl react in aqueous solution in equimolar quantities:

$NH_3 + NaOCl \rightarrow NH_2Cl + NaOH$

Distillation of the mixture at low pressure, followed by drying and condensing the vapour, gives colourless crystals *m.p.* 207 K, unstable at higher temperatures when dry. Dichloramine, $NHCl_2$, is formed when chlorine is passed into ammonium sulphate solution buffered to pH 5 ± 0.5. NH_2F can be made pure. It is more reactive than NF_3, being hydrolysed by alkalis; NH_2Cl, on the contrary, is less reactive than NCl_3.

TABLE 20.5

STRUCTURAL DATA FOR NITROGEN—HALOGEN COMPOUNDS

			Bond length/pm		*Bond angles*/°	
			N—X	N—N	XNX	NNX
NF_3	pyramidal	C_{3v}	137		103	
N_2F_4	skew	C_2	139	148		102—105
					dihedral angle 66	
$\cdot NF_2$	angular	C_{2v}	136.5		103	
cis-N_2F_2	planar	C_{2v}	141	121		114
trans-N_2F_2	planar	C_{2h}	140	122		106
NCl_3	pyramidal	C_{3v}				
N_2H_4	gauche	C_2		145		112
			N—X	N—H	XNX	HNX
HNF_2	pyramidal	C_2	137	108	104.3	103.5
$HNCl_2$	pyramidal	C_2	175		106	102
			N—X	N—H	HNH	HNX
H_2NCl	pyramidal	C_2	176		106.8	102
H_3N	pyramidal	C_{3v}		102	109.1	

Purple $NBr_3 \cdot 6\,NH_3$ resulting from the action of bromine vapour on an excess of ammonia, and black $NI_3 \cdot NH_3$ made by treating iodine with strong aqueous ammonia, are both highly explosive solids.

20.5.3. Phosphorus halides

The halides of phosphorus are, with the exceptions indicated in Table 20.6, colourless.

The trihalides are usually prepared by direct combination under control-

TABLE 20.6

PHOSPHORUS HALIDES

	Trihalides	Pentahalides	Others
Fluorides	PF_3 *b.p./*K 171	PF_5 *b.p./*K 188.6	
Chlorides	PCl_3 *b.p./*K 349	PCl_5 *sublimes/*K 433	P_2Cl_4 *b.p./*K 453
Bromides	PBr_3 *b.p./*K 446	PBr_5 *m.p./*K 373 (yellow)	
Iodides	PI_3 *m.p./*K 334 (dark red)		P_2I_4 *m.p./*K 398 (orange)

led conditions, but PF_3 is best made by the action of AsF_3 or CaF_2 on PCl_3:

$$AsF_3 + PCl_3 \rightarrow AsCl_3 + PF_3$$

Hydrolysis gives phosphorous acid and the halogen hydracid; the ease of hydrolysis increases with the atomic weight of the halogen:

$$PX_3 + 3\ H_2O \rightarrow H_3PO_3 + 3\ HX$$

They all react with oxygen and sulphur:

$$2\ PX_3 + O_2 \rightarrow 2\ POX_3 \qquad PX_3 + S \rightarrow PSX_3$$

and with halogens except when X = I:

$$PX_3 + X_2 \rightarrow PX_5$$

By using their lone pairs, the molecules can act as ligands in complexes:

$$Ni(CO)_4 + 4\ PCl_3 \rightarrow Ni(PCl_3)_4 + 4\ CO$$

Mixed halides such as PF_2Br have been made; they decompose on heating:

$$3\ PF_2Br \rightarrow 2\ PF_3 + PBr_3$$

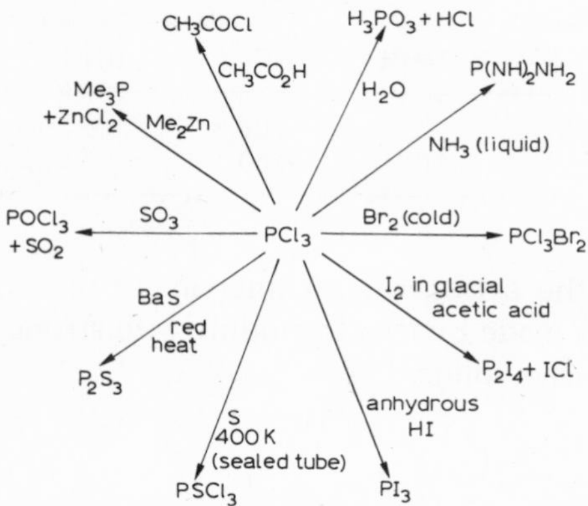

Fig. 20.3. Typical reactions of phosphorus trichloride.

The pentahalides are produced by the addition of another molecule of

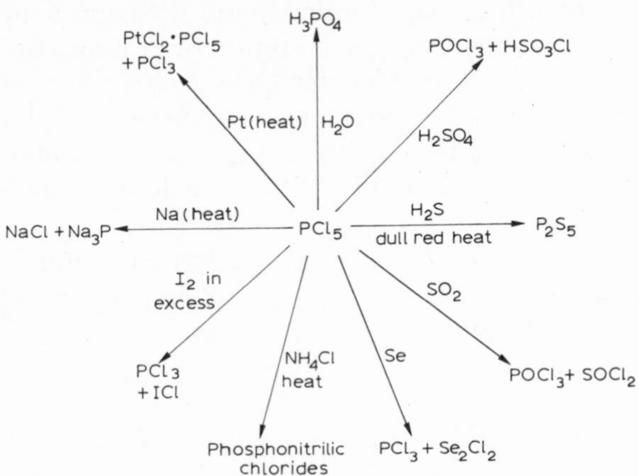

Fig. 20.4. Typical reactions of phosphorus pentachloride.

halogen to the trihalide, one 3d orbital of the phosphorus atom being now occupied. Solid PCl_5 is made by dropping liquid PCl_3 into dry chlorine. The thermal stability of the pentahalides decreases as the molecular weight rises. PCl_5 dissociates on slight heating,

$$PCl_5 \rightleftharpoons PCl_3 + Cl_2$$

and is readily reduced by some metals on warming,

$$PCl_5 + Cd \rightarrow PCl_3 + CdCl_2$$

When the vapour from phosphorus pentachloride is passed over CaF_2 at 575—675 K, the pentafluoride is formed. In the vapour phase PF_5 and PCl_5 have trigonal bipyramidal molecules with axial and equatorial bond lengths of 157.7 and 153.4 pm in PF_5 and 214 pm and 201 pm respectively in PCl_5.

Solid PCl_5 contains tetrahedral PCl_4^+ and octahedral PCl_6^- ions arranged in a CsCl lattice, but PBr_5 has PBr_4^+ and Br^-. A solution of PCl_5 in nitromethane conducts electricity, and PBr_5 gives similar conducting solutions due to the species PBr_4^+ and PBr_6^-, the anion being stabilised by solvation while the cation is relatively stable.

Hydrolysis of the pentahalides occurs in two stages:

$$PX_5 + H_2O \rightarrow POX_3 + 2\ HX$$
$$POX_3 + 3\ H_2O \rightarrow H_3PO_4 + 3\ HX$$

Of the tetrahalodiphosphines, P_2F_4 is the most recently discovered and was prepared by the coupling reaction:

$$2\ PF_2I + Hg \rightarrow P_2F_4 + HgI_2$$

Hydrolysis is thought to occur by:

$$2\ P_2F_4 + H_2O \rightarrow 2\ HPF_2 + F_2POPF_2$$

Tetrachlorodiphosphine is a colourless, oily, fuming liquid with a *m.p.* of 245 K made by passing a silent discharge through a mixture of PCl_3 and hydrogen. The heat of formation of Cl_2PPCl_2 is estimated to be $\Delta H_f = -446$ kJ mol^{-1} and the P—P bond energy is 244 kJ mol^{-1}. It decomposes at room temperature and rapidly at its *b.p.* 453 K to PCl_3, phosphorus and a non-volatile solid; it is hydrolysed to HCl, H_3PO_3 and P; and it is liable to inflame spontaneously in air.

P_2I_4, *m.p.* 398 K, is deposited as orange crystals when solutions of iodine and phosphorus in CS_2 are mixed. It decomposes to PI_3 and P on heating and is readily hydrolysed to H_3PO_3, H_3PO_2, and HI.

20.6. Oxides of Nitrogen

It is convenient to classify the oxides of nitrogen in the traditional manner in terms of the oxidation state of nitrogen (Table 20.7). Thus although the

TABLE 20.7

PHYSICAL PROPERTIES OF THE OXIDES OF NITROGEN

	N_2O	NO	N_2O_3	N_2O_4	N_2O_5
Oxidation state of N	+1	+2	+3	+4	+5
Molecular weight	44	30	76	92	108
M.p./K	182.3	109.5	172.4	261.9	305.6
B.p./K	184.7	121.4	233 to 276	294.3	(sublimes)
Colour of solid	colourless	colourless	light blue	colourless	colourless

oxides are neutral species their redox potentials (see Fig. 20.2) will be the same in acidic or basic media and they can readily be related to both anions and cations. Structural data are included in Table 20.8.

20.6.1. Nitrous oxide

Nitrous oxide, N_2O, is made by heating NH_4NO_3. Colourless and unreactive at room temperature, the gas is, through decomposition, an oxidising agent above 900 K; $2\,N_2O \rightarrow 2\,N_2 + O_2$. It is not the anhydride of hyponitrous acid and, though but slightly soluble, gives a neutral solution. With regard to the linear structure of N_2O, if both the nitrogen and oxygen atoms are considered to use sp hybrid orbitals the structure can be explained in terms of two σ bonds and two pairs of non-localized π orbitals extending over all three centres. The lowest orbitals in each pair, which differ only by rotation through 90° about the axis, give two partial π bonds in the N—N and N—O bonding (Fig. 20.5a). The higher orbitals will be weakly bonding in

TABLE 20.8

STRUCTURAL DATA FOR NITROGEN—OXYGEN COMPOUNDS

	Shape	Point group	*Bond length*/pm		*Bond angles*/°	
			N—O	N—N	NNO	ONO
N_2O	linear	$C_{\infty v}$	119	112	180	
NO	linear	$C_{\infty v}$	115			90
$[NO]_2$ (cryst)	planar	C_{2h}	112			
NO^+	linear	$C_{\infty v}$	240			
$[NO]_2^{2-}$ (*trans*)	planar	C_{2h}				
N_2O_3	planar	C_s	112	186	*NO* 105	
			118		NO_2 113	129
					118	
NO_2	angular	C_{2v}	119.7			134
NO_2^+	linear	$D_{\infty h}$	115			180
NO_2^-	angular	C_{2v}	124			115
NO_3^-	planar	D_{3h}	122			120
N_2O_4	planar	D_{2h}	118	175		134
N_2O_5	planar	D_{2h}	120			
			130			

one link, N—N, and weakly antibonding in the other, N—O (Fig. 20.5b). Both bonds in the molecule will have some triple-bond character, this being higher in the N—N link.

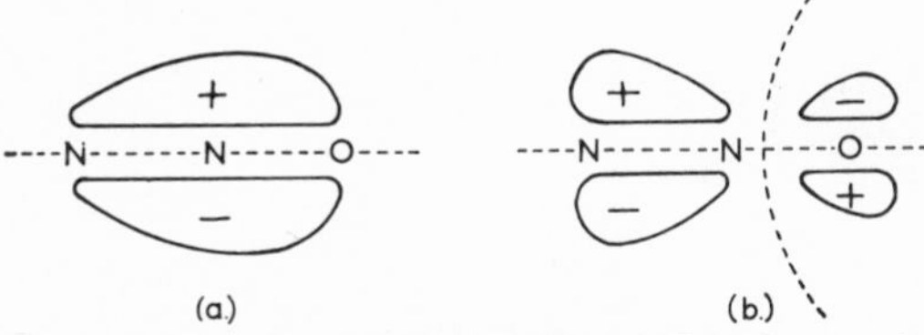

Fig. 20.5. Delocalised π orbitals (showing nodes) in N_2O: (a) bonding in both regions, N—N and N—O; (b) weakly bonding in N—N, weakly anti-bonding in N—O.

Perhaps the bonding in N_2O is most simply represented by the Linnett non-pairing structure

$$:\dot{N}\dot{=}N\dot{-}\ddot{O}:$$

The molecule is isoelectronic and isosteric with CO_2 and the physical properties of the two compounds show similarities (*crit. temp.*, N_2O, 309.6 K; CO_2, 304.2 K). Other related 16-electron triatomic systems include NO_2^+, N_3^-, NCO^-, CN_2^{2-}. The chief commercial use of nitrous oxide is as an anaesthetic. Probably its most remarkable reaction is with molten sodium or potassium amide at 500 K to yield azides:

$$2\ KNH_2 + N_2O \rightarrow KN_3 + KOH + NH_3$$

20.6.2. Nitric oxide

Nitric oxide, NO, is a colourless gas which condenses to a dark-blue liquid. The molecule is paramagnetic, having one odd electron in an antibonding π orbital. This is consistent with the fact that the cation NO^+ (that is NO with the odd electron removed) has a shorter and stronger bond. The solid contains rectangular dimers with the dimensions indicated, although the X-ray evidence does not distinguish between similar and opposite orientations of the N—O groups:

```
   238 pm
N ---------- O
|            | 110 pm
O ---------- N
```

The liquid has a high Trouton constant and a low dielectric constant; being very stable, association to N_2O_2 is endothermic at ordinary temperature. The gaseous molecule is monomeric and has a small dipole moment of 0.15 debye.

Nitric oxide is made on a large scale by oxidising ammonia with air on the surface of platinum above 750 K:

$$4\ NH_3 + 5\ O_2 \rightarrow 4\ NO + 6\ H_2O$$

The gas is oxidised to N_2O_4 by oxygen and to nitrosyl halides, NOX, by F_2, Cl_2 and Br_2. The reactions have empirical third-order kinetics but are probably bimolecular, involving the dimer even though there is no direct evidence for its presence in the vapour. For instance, the formation of nitrosyl chloride may be represented thus:

$$\text{(i)}\quad NO + NO \rightleftharpoons N_2O_2 \left(\text{with } K = \frac{[N_2O_2]}{[NO]^2}\right)$$

$$\text{(ii)}\quad N_2O_2 + Cl_2 \rightarrow 2\ ClNO$$

The rate is proportional to $[N_2O_2][Cl_2]$, i.e. to $K[NO]^2[Cl_2]$.

Paramagnetism of nitric oxide

The paramagnetic moment of NO increases up to 520 K but above that temperature it remains almost constant; the variation is not in accordance with the Curie law. The two lowest electronic energy levels of the NO molecule are separated by only 121 cm^{-1}, which is comparable with kT at normal laboratory temperature. The distribution of the molecules between two states of similar energy but different J value is responsible for the anomalous variation of magnetic moment with temperature.

20.6.3. Compounds from nitric oxide

Nitric oxide forms compounds in three ways:

(i) Electron sharing, to form highly coloured, covalent substances, typified by the volatile, easily hydrolysed nitrosyl halides (see below).

(ii) Electron gain, to form NO^-, as found in sodium nitrosyl, made by the action of nitric oxide on the metal in liquid ammonia: $Na + NO \rightarrow NaNO$.

(iii) Electron loss, to form NO^+, the nitrosonium ion (see below).

The covalent nitrosyl halides have bent molecules (see Table 20.9) with long N—X bonds suggesting much polar character. FNO is made by mixing F_2 with an excess of NO in a copper tube, and ClNO by passing N_2O_4 over moist KCl at room temperature:

$$N_2O_4 + KCl \rightarrow ClNO + KNO_3$$

BrNO is formed when NO is passed into bromine at 258 K. The iodide has not been made. Nitrosyl halides react with hydroxylic compounds to give nitrites:

$$ClNO + ROH \rightarrow RONO + HCl$$

TABLE 20.9

STRUCTURAL DATA FOR OXYGEN—NITROGEN—HALOGEN COMPOUNDS

	State	Point group	Bond length/pm		Bond angles/°	
			N—X	N—O	XNO	ONO
F_3NO	tetrahedral	C_{3v}	148	115		
FNO	angular	C_s	152	113	110	
ClNO	angular	C_s	198	114	113	
BrNO	angular	C_s	214	115	117	
FNO_2	planar	C_{2v}	135	123		125
$ClNO_2$	planar	C_{2v}	184	120		131
$FONO_2$	planar	C_s		129, 129, 139		125

FNO and IrF_6 mixed in a copper vessel at 77 K and warmed to room temperature produce ONF_3:

$$3\ FNO + 2\ IrF_6 \rightarrow 2\ NOIrF_6 + F_3NO$$

Trifluoramine oxide is a colourless gas with infrared and n.m.r. spectra consistent with the C_{3v} structure $F_3\overset{+}{N}\overset{-}{O}$, one which renders the compound moderately stable.

The ionisation potential for the process $NO \rightarrow NO^+ + e$ is 910 kJ mol^{-1} much lower than that for $N_2 \rightarrow N_2^+ + e$ (1510 kJ mol^{-1}) or $O_2 \rightarrow O_2^+ + e$ (1160 kJ mol^{-1}). Just as happens with the odd electron in an atom, the odd electron in NO is more readily lost than one from a doubly occupied orbital since such an electron moves outside a closed shell and is more effectively screened from the nucleus (cf. 3.4.4.1). Nitrosonium compounds such as $NO^+ClO_4^-$, $NO^+HSO_4^-$ and $NO^+BF_4^-$, are prepared in non-hydroxylic solvents; they are usually isomorphous with the hydroxonium and ammonium salts $H_3O^+ClO_4^-$ and $NH_4^+ClO_4^-$. The nitrosonium ion is intermediate in size between H_3O^+ and NH_4^+.

Though stable in non-hydroxylic solvents, nitrosonium compounds react with hydroxylic solvents:

$$NO^+ + OH^- \rightleftharpoons HNO_2 \rightleftharpoons NO_2^- + H^+$$

This equilibrium is displaced to the left in strongly acid media.

In metal complexes containing the NO group, nitrogen is the donor atom. Usually NO is bound as NO^+, an ion which is isoelectronic with CN^- and CO. The NO group can thus donate one extra electron to the metal; this accounts for the stability of tricarbonyl-nitrosylcobalt, $Co(NO)(CO)_3$, which has the same electronic pattern as $Ni(CO)_4$ (17.12.3). Co-ordination complexes of this type do not usually contain more than one NO group.

Unstable metal nitrosyls are formed by Fe, Ru and Ni. Black $Fe(NO)_4$, made by heating iron carbonyl with NO under pressure at 320 K, is the most stable. The ionic formula $NO^+[Fe(NO)_3]^-$ has been suggested to explain its low volatility. Ruthenium tetranitrosyl, $Ru(NO)_4$, is made as cubic, red crystals when NO is passed into $Ru_3(CO)_{12}$. A compound of empirical formula $Ni(NO)_2$ is obtained as a blue powder when NO is passed into $Ni(CO)_4$ dissolved in $CHCl_3$.

The nitrosyl carbonyls such as $Co(NO)(CO)_3$ are much more stable than the nitrosyls themselves. Nitrosyl halides, $Fe(NO)_2X$, $Co(NO)_2X$ and $Ni(NO)X$ are known; their stability falls from Fe to Ni and from I to Cl. Fluorides are unknown. The most stable nitrosyl halide, $Fe(NO)_2I$, results from passing NO over FeI_2 at 370 K.

$$2\ FeI_2 + 4\ NO \rightarrow 2\ Fe(NO)_2I + I_2$$

20.6.4. Dinitrogen trioxide

Pure N_2O_3 can only be obtained as a pale-blue solid or as an intense blue liquid just above its freezing point (172 K). At higher temperatures, dissociation (the reverse of the preparative route) becomes important as indicated by colour changes:

$$\underset{\text{blue}}{N_2O_3} \rightleftharpoons \underset{\text{colourless}}{N_2O} + \underset{\text{brown}}{NO_2}$$

X-ray data on single crystals of N_2O_3 suggest a disordered structure, while electron-diffraction data indicate, for the vapour phase, a planar molecule containing non-equivalent nitrogen atoms consistent with the nitroso-nitro structure O_2NNO. The N—N bond length of 186 pm is remarkably long when compared with a conventional single N—N bond length of 145 pm in hydrazine. Such a bond length would not suggest π-interaction, nor would σ-bonding be consistent with molecular planarity and its implied restriction of rotation. A similar situation occurs in N_2O_4 (see 20.6.5) and MO calculations have suggested bond orders of less than 1 consistent with hybrid structures (Fig. 20.6).

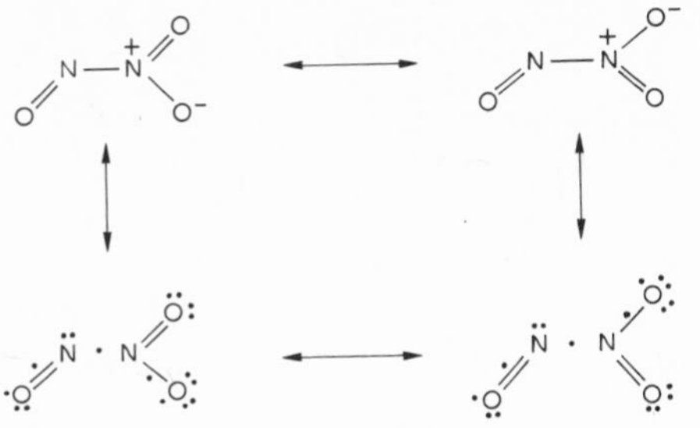

Fig. 20.6. Nitroso-nitro structures of N_2O_3.

Reactions involving water support the idea of N_2O_3 as the formal anhydride of nitrous acid:

$$2\ HNO_2 \rightarrow N_2O_3 + H_2O$$

as alkaline solutions of N_2O result in nitrite formation.

$$N_2O_3 + 2\ OH^- \rightarrow 2\ NO_2^- + H_2O$$
$(NO + NO_2)$

In concentrated acids, such as tetrafluoroboric, perchloric, sulphuric and selenic, the blue colour disappears as fully ionised nitrosonium salts are formed:

$$N_2O_3 + 3\ H_2SO_4 \rightarrow 2\ NO^+ + H_3O^+ + 3\ HSO_4^-$$

20.6.5. Dinitrogen tetroxide

Dinitrogen tetroxide is best made by heating $Pb(NO_3)_2$ and condensing the compound from the gases evolved. It is colourless and diamagnetic in the dimeric form, N_2O_4, which is found pure only in the solid state. This melts to a pale-yellow liquid, which contains about 1% of the brown, paramagnetic monomer, NO_2, with one unpaired electron. The vapour darkens progressively on heating and, at 373 K, has about 90% of the monomer. The monomer, NO_2, an odd-electron molecule, has many of the characteristics of a free radical since it (i) associates with other radicals, (ii) abstracts hydrogen from saturated hydrocarbons, (iii) adds to unsaturated hydrocarbons. Its photolysis, decomposition and oxidising action owe little, however, to its radical character.

There are two dimeric modifications of NO_2, each of which can exist in several conformations. The dimer $ONONO_2$ is stable at the temperature of liquid helium, while the O_2NNO_2 dimer is stable at higher temperatures. There are five curious features about the structure (Table 20.8):

(i) the very long N—N bond, almost 3 pm longer than that in hydrazine,
(ii) the large ONO angles,
(iii) the high barrier to internal rotation,
(iv) the planarity of the molecule,

(v) the stability of the molecule relative to the $ONONO_2$ and ONOONO conformations.

The N—N bond has been variously described as a π-only bond (Coulson, 1957) and a 'splayed' single bond (Bent, 1963) implying that the electron density is not so strongly concentrated in the line of the nuclei as in a σ bond. Valence bond or Linnett non-pairing structures explain the long N—N bond on the basis of repulsion due to formal positive charges lying on adjacent atoms or one-electron σ bonds but fail to account for the planarity.

MO calculations suggest that although the N—N bond is predominantly σ in character, there is a fairly large π contribution from orbitals which lie in the plane of the molecule. Vicinal interactions between the unshared electrons of the oxygen atoms and the N—N antibonding orbital are probably important factors in the stabilisation of the structure.

As mentioned previously NO_2 is an angular molecule and can be considered to be intermediate in structure between CO_2 and O_3. Only with Linnett valence bond structures is it possible to maintain an octet of electrons round each atom for the 17 electron molecule:

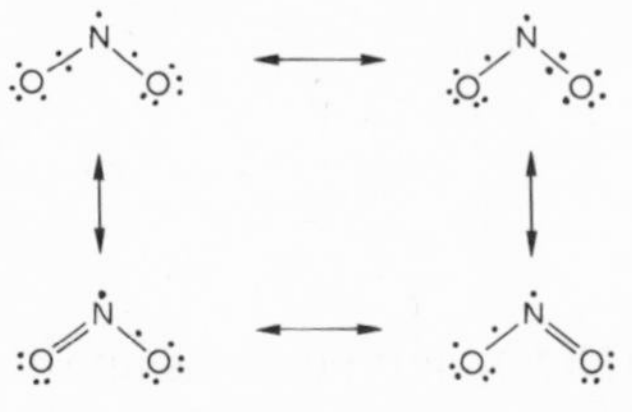

20.6.5.1. General reactions of dinitrogen tetroxide

The ionisation of N_2O_4 is encouraged by

(i) solvents of high dielectric constant such as $HClO_4$ and H_2SO_4; in concentrated sulphuric acid both nitrosonium and nitronium ions are produced:

$$N_2O_4 + 3\ H_2SO_4 \rightleftharpoons NO^+ + NO_2^+ + H_3O^+ + 3\ HSO_4^-$$

(ii) the removal of NO^+ as complexes by Lewis bases,

(iii) the removal of NO_3^-, as, for example, when $Zn(NO_3)_2$ dissolves to form a complex ion:

$$2\ N_2O_4 + Zn(NO_3)_2 \rightarrow 2\ NO^+ + Zn(NO_3)_4^{2-}$$

Very many electron donors, such as amines and ethers, combine with N_2O_4 to produce compounds of the types:

$[Base\ NO^+]NO_3^-$ and $[(Base)_2NO^+]NO_3^-$

Metals which react with liquid N_2O_4 (Na, K, Zn, Ag, Pb and Hg) all liberate nitric oxide:

$M + N_2O_4 \rightarrow MNO_3 + NO$

Salts react to produce nitrates:

$N_2O_4 + KCl \rightarrow KNO_3 + NOCl$
$2\ N_2O_4 + 2\ KI \rightarrow 2\ KNO_3 + 2\ NO + I_2$
$N_2O_4 + NaClO_3 \rightarrow NaNO_3 + NO_2 + ClO_2$

Hydroxylic and amino solvents are usually nitrosated by N_2O_4:

$N_2O_4 + 2\ ROH \rightarrow RONO + ROH_2^+ + NO_3^-$
$N_2O_4 + 2\ R_2NH \rightarrow R_2N.NO + R_2NH_2^+ + NO_3^-$

but with primary amines nitrogen is evolved in a two-stage reaction:

$RNH_2 + N_2O_4 \rightarrow R.NH_2NO^+ + NO_3^-$
$RNH_2NO^+ + RNH_2 \rightarrow RNH_3^+ + N_2 + H_2O$

Mixtures of N_2O_4 with organic solvents have proved particularly useful for the preparation of anhydrous nitrates. Copper does not dissolve in liquid N_2O_4 but it reacts with an N_2O_4/ethyl acetate mixture to give the blue compound $Cu(NO_3)_2 \cdot N_2O_4$ from which anhydrous $Cu(NO_3)_2$ is obtained by heating to 358 K. The ethyl acetate acts as an electron donor towards the N_2O_4 molecule, the energy of interaction being about 8 kJ mol^{-1}, so the ionisation of the N_2O_4 is enhanced.

$n\ EtOAc + N_2O_4 \rightleftharpoons [(EtOAc)_n NO]^+ + NO_3^-$

20.6.6. Dinitrogen pentoxide

This is made by the dehydration of nitric acid with P_2O_5. The molecular structure (Table 20.8) of the compound in non-ionising solvent is

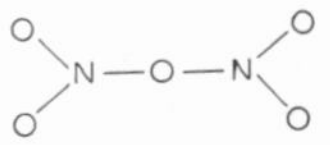

$$O_2N{-}O{-}NO_2$$

with the central NON bond angle close to 180°. But in concentrated sulphuric acid it ionises to the nitronium ion, NO_2^+:

$N_2O_5 + 3\ H_2SO_4 \rightleftharpoons 2\ NO_2^+ + 3\ HSO_4^- + H_3O^+$

The NO_2^+ ion, with a Raman absorption at 1400 cm^{-1}, is also present in sulphuric, selenic and perchloric acid solutions of HNO_3. The same Raman line is given by solid nitronium perchlorate, $NO_2^+ClO_4^-$, which can be separated from a solution made by dissolving HNO_3 and $HClO_4$ in nitromethane. Compounds such as $NO_2^+BF_4^-$ and $NO_2^+PF_6^-$ have been made by dissolving the appropriate fluoride or oxide in bromine and treating the solution with N_2O_4 followed by BrF_3.

Solid N_2O_5 consists of NO_2^+ and NO_3^- ions, having linear and planar shapes respectively (Table 20.8). Colourless crystals of N_2O_5 are stable in diffuse light below 280 K but on warming, or exposing to sunlight, decomposition occurs. Being the formal anhydride of nitric acid, N_2O_5 dissolves in water with a hissing noise to give HNO_3, while with hydrogen peroxide, pernitric acid is also formed:

$$N_2O_5 + H_2O \rightarrow 2\ HNO_3$$
$$N_2O_5 + H_2O_2 \rightarrow HNO_3 + HNO_4$$

The strong oxidising action of N_2O_5 can cause violent reactions with reducing agents.

$$N_2O_5 + I_2 \rightarrow I_2O_5 + N_2$$

20.6.7. Nitryl halides

Nitryl fluoride, FNO_2, is made when fluorine reacts with a 2 : 1 excess of NO_2, or when fluorine is passed over heated $NaNO_2$:

$$NaNO_2 + F_2 \rightarrow NaF + FNO_2$$

Nitryl chloride, $ClNO_2$, is made by treating ClNO with ozone. Both the fluoride and chloride are colourless gases. They have planar molecules (Table 20.9), consistent with their hydrolysis to HNO_3 and HX.

20.6.8. Pernitryl fluoride

This compound, $FONO_2$, also known as fluorine nitrate (Table 20.9), is formed as a colourless, explosive gas by the action of fluorine on concentrated nitric acid. It is purified by freezing with liquid air followed by fractional distillation at 230 K. $FONO_2$ explodes to give FNO and O_2 when sparked, but a first-order decomposition into the same products proceeds slowly at 350—380 K. The gas hydrolyses slowly with water yielding oxygen, fluoride and nitrate ions:

$$2\ FONO_2 + 4\ OH^- \rightarrow 2\ F^- + 2\ NO_3^- + 2\ H_2O + O_2$$

It oxidises iodides:

$$FONO_2 + 2\ I^- \rightarrow NO_3^- + F^- + I_2$$

20.7. Oxides of Phosphorus

The three oxides of phosphorus are P_4O_6, $(PO_2)_n$ and P_4O_{10} (Table 20.10). The first and last have structures based on the P_4 tetrahedron.

In the P_4O_6 molecule (Fig. 20.7) oxygen atoms bridge the edges of the tetrahedron, the P—O distance is 165 pm and the angles POP and OPO are 127.5° and 99° respectively. In the P_4O_{10} molecule (Fig. 20.8) an extra oxy-

TABLE 20.10

PHYSICAL PROPERTIES OF THE OXIDES OF PHOSPHORUS

	P_4O_6	$(PO_2)_n$	P_4O_{10}
Oxidation state of P	+3	+4	+5
Molecular weight	220	293—721	284
Melting point/K	297	sublimes	693
Boiling point/K	448		713

gen atom is attached to each P atom at a distance of 140 pm. It is interesting that a compound $P_4O_6S_4$, made by heating P_4O_6 with sulphur, has a similar structure.

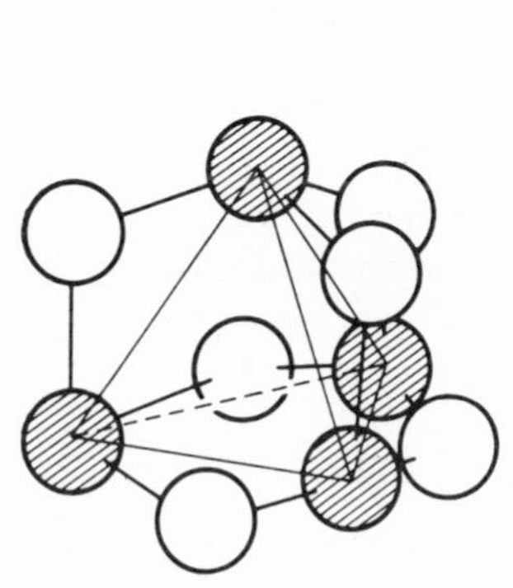

Fig. 20.7. Structure of P_4O_6 molecule.

Fig. 20.8. Structure of P_4O_{10} molecule.

Phosphorus trioxide, P_4O_6, is made by passing a mixture of oxygen and nitrogen, 3 : 1 by volume, over white phosphorus at 320 K. It is converted to phosphorous acid by cold water:

$$P_4O_6 + 6\ H_2O \rightarrow 4\ H_3PO_3$$

but hot water produces a mixture of phosphoric acid, phosphorus and phosphine.

When P_4O_6 is heated above 480 K the polymeric oxide $(PO_2)_n$ and red phosphorus are formed. Solid $(PO_2)_n$ sublimes at about 450 K and the density of the vapour indicates the molecular formula P_4O_8. The oxide is not the anhydride of hypophosphoric acid, $H_4P_2O_6$.

Phosphoric oxide, P_4O_{10}, produced when phosphorus burns in an excess of oxygen, has three crystalline forms of increasing structural complexity; they are hexagonal, orthorhombic and tetragonal. It is the anhydride of the phosphoric acids:

$$P_4O_{10} \xrightarrow{2\,H_2O} 4\,HPO_3 \xrightarrow{2\,H_2O} 2\,H_4P_2O_7 \xrightarrow{2\,H_2O} 4\,H_3PO_4$$

phosphoric oxide — metaphosphoric acid — pyrophosphoric acid — orthophosphoric acid

20.8. Oxohalides of Phosphorus

Compounds of the POX series have not been made but those of the POX_3 series are represented by:

	m.p./K	*b.p.*/K	*Distance*/pm P—O	*Distance*/pm P—X	*Bond angle*/° X—P—X
POF_3	244	244/110 kPa	156	152	107
$POCl_3$	274.3	378.2	158	202	106
$POBr_3$	328	464.8	141	206	

and by the 'mixed' halides POF_2Cl, $POFCl_2$, POF_2Br and $POFBr_2$.

Phosphoryl chloride is best made by heating PCl_5 with anhydrous oxalic acid:

$$PCl_5 + (CO_2H)_2 \rightarrow POCl_3 + 2\,HCl + CO + CO_2$$

The bromide can be obtained similarly from PBr_5. Phosphoryl fluoride and the 'mixed' halides are made by treating the chloride or bromide with SbF_3.

$POCl_3$ hydrolyses rather slowly in cold water:

$$POCl_3 + 3\,H_2O \rightarrow H_3PO_4 + 3\,HCl$$

It is used in organic chemistry as a rather milder reagent than PCl_5 for replacing —OH groups by —Cl.

The POX_3 molecules all have C_{3v} symmetry:

O
|
P
X X X

with phosphorus to oxygen very close to the double-bond distance. The P—X distances are almost the same as in the corresponding PX_3 molecules. The compounds are monomeric when liquid; the liquids have normal Trouton constants, but there is slight self-ionisation in $POCl_3$:

$$2\,POCl_3 \rightleftharpoons POCl_2^+ + POCl_4^-$$

which acts as an ionising solvent.

More complex pyrophosphoryl and polyphosphoryl halides are known and prepared by dehydration using P_4O_{10} or PCl_5.

$$4\ HOP(O)F_2 + P_4O_{10} \longrightarrow 2\ F_2P(O)OP(O)F_2 + 4\ HPO_3$$

$$2\ HOP(O)Cl_2 + PCl_5 \longrightarrow Cl_2P(O)OP(O)Cl_2 + POCl_3 + 2\ HCl$$

20.9. Oxoacids of Nitrogen

20.9.1. Hyponitrous acid

Hyponitrous acid, $H_2N_2O_2$, has been made from its salts. A solution of sodium hyponitrite is produced by reducing concentrated aqueous $NaNO_2$ with sodium amalgam. The insoluble silver salt, precipitated from a neutral solution of sodium hyponitrite, when treated with anhydrous HCl in dry ether gives a solution of hyponitrous acid:

$$Ag_2N_2O_2 + 2\ HCl \rightarrow H_2N_2O_2 + 2\ AgCl$$

The white crystals of the acid obtained by evaporating the ether solution, after removing the silver chloride, decompose spontaneously:

$$H_2N_2O_2 \rightarrow N_2O + H_2O$$

For this reason, few physical constants have been recorded but the silver salt has been shown by infrared spectroscopy to have anions with the *trans* configuration.

$$^{-}O{-}N{=}N{-}O^{-}$$

20.9.2. Nitroxylic acid

Nitroxylic acid, H_2NO_2, has not been made; but the yellow sodium salt Na_2NO_2 is produced by the electrolysis of $NaNO_2$ in liquid ammonia, or by the reduction of $NaNO_2$ with sodium in the same solvent. The salt decomposes at 373 K, giving Na_2O, $NaNO_2$, $NaNO_3$ and nitrogen.

20.9.3. Nitrous acid

Nitrous acid, HNO_2, can be made in aqueous solution by dissolving a mixture of NO and NO_2 in ice-cold water. Though the acid is known only in solution, its salts and esters are moderately stable. The Group IA and Group IIA nitrites are thermally stable, being made by heating the nitrate alone or with a reducing agent like carbon. Nitrous acid is a fairly strong acid (pK_a = 3.3 at 291 K). The solution decomposes slowly:

$$3\ HNO_2 \rightarrow HNO_3 + 2\ NO + H_2O$$

It is easily oxidised to nitric acid; but it is also easily reduced to NO, N_2O, $H_2N_2O_2$, NH_2OH, N_2 or NH_3, depending on the reducing agent used:

Fe^{2+} and Ti^{3+} $\rightarrow$ NO
Sn^{2+} $\rightarrow H_2N_2O_2$
H_2S (at pH 8—9) $\rightarrow NH_3$

The reaction

$$2\ HNO_2 + 2\ I^- + 2\ H_3O^+ \rightarrow 2\ NO + I_2 + 4\ H_2O$$

is quantitative and can be used for the determination of NO_2^- ion. The nitrite ion is angular (Table 20.11) in which the nitrogen atom has three sp^2 orbitals, one containing the lone-pair electrons:

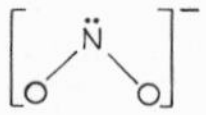

Nitrite esters can usually be made by treating the alcohols with cold acidified sodium nitrite.

TABLE 20.11

STRUCTURAL DATA FOR OXOACIDS OF NITROGEN AND RELATED SPECIES

Compound	Shape	Point group	*Bond lengths*/pm		*Bond angles*/°	
			N—O	N—OH	ONO	HON
$N_2O_2^{2-}$	*trans* planar	C_{2h}				
H_2NNO_2		C_{2v}	118		129	
HNO_2	*trans* planar	C_s	120	146	116	104
NO_2^-	angular	C_{2v}	113—123		116—132	
HNO_3	planar	C_s	122	141	130	
NO_3	planar	D_{3h}	122		120	

20.9.4. Hyponitric acid

Hyponitric acid, $H_2N_2O_3$, is very unstable, but its sodium salt can be made as the monohydrate by treating free hydroxylamine in methanol with ethyl nitrate in the presence of sodium methoxide:

$$NH_2OH + EtNO_3 \rightarrow \underset{+\ EtOH}{HONHNO_2} \xrightarrow{NaOMe} Na_2N_2O_3 \cdot H_2O$$

On acidification decomposition occurs:

$$H_2N_2O_3 \rightarrow H_2O + 2\ NO$$

20.9.5. Nitric acid

Nitric acid, HNO_3, is made industrially from synthetic ammonia. The ammonia is oxidised in a stream of air on the surface of a hot platinum gauze to nitric oxide:

$$4\ NH_3 + 5\ O_2 \rightarrow 4\ NO + 6\ H_2O$$

As the nitric oxide/air mixture is cooled, the nitric oxide is oxidised to an equilibrium mixture of nitrogen dioxide and dinitrogen tetroxide:

$$2\ NO + O_2 \rightarrow N_2O_4 \text{ and } NO_2 \text{ (equilibrium mixture)}$$

The equilibrium mixture is dissolved in water in the presence of air, and consecutive reactions occur until all the oxides of nitrogen are converted to nitric acid:

$$3\ N_2O_4 + 2\ H_2O \rightarrow 4\ HNO_3 + 2\ NO$$
$$2\ NO + O_2 \rightarrow N_2O_4$$

The pure acid exists as white crystals, *m.p.* 231.5 K. Melting these gives a faintly yellow liquid which shows weak Raman lines at 1050 cm^{-1} and 1400 cm^{-1}, indicating some dissociation to NO_2^+ and NO_3^- has occurred. The liquid, which has a density of 1.54 at 273 K, boils at 356 K. Crystalline hydrates are formed, such as $HNO_3 \cdot H_2O$, *m.p.* 235 K, sometimes called the ortho-acid, and $HNO_3 \cdot 3\ H_2O$, *m.p.* 254.6 K. The acid is extracted from 6—16 *M* aqueous solution into toluene as the hemihydrate:

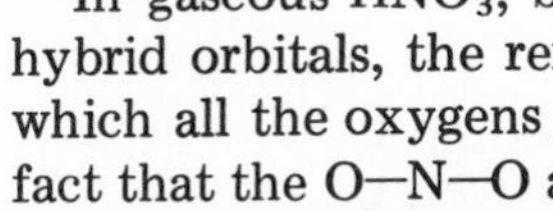

```
        O---H                 O
       /     \               /
HO—N          O---HO—N
       \     /               \
        O---N                 O
```

Nitric acid forms a constant-boiling mixture with water, of composition 68.4% HNO_3 and *b.p.* 395 K. Concentration can be effected by distillation with H_2SO_4.

In gaseous HNO_3, bonding to the nitrogen atom is evidently through sp^2 hybrid orbitals, the remaining p electrons being used in π bond formation in which all the oxygens participate. There is no satisfactory explanation of the fact that the O—N—O angle nearest the hydrogen atom is rather larger (116°) than that on the other side (114°). Any intramolecular hydrogen-bonding would be expected to reduce the angle a little.

Liquid nitric acid and its strong aqueous solutions decompose, particularly on exposure to light, to give N_2O_4, oxygen and water. A solution of N_2O_4 in anhydrous nitric acid constitutes fuming nitric acid. When HNO_3 acts as an oxidising agent the usual products are NO from the dilute acid and N_2O_4 from the concentrated acid, but other products, such as N_2O, NH_2OH and NH_3, can be obtained under appropriate conditions. The oxidising action is catalysed by the presence of N_2O_4.

20.10. Oxoacids of Phosphorus

The acids can be classified according to the formal charge number of the phosphorus (Table 20.12). In them the phosphorus atoms always have four sp^3 bonds.

TABLE 20.12

OXOACIDS OF PHOSPHORUS

Charge number of P	Formula	Name	
+1	H_3PO_2	hypophosphorous acid	
+3	HPO_2	meta-	
	$H_4P_2O_5$	pyro-	phosphorous acids
	H_3PO_3	ortho-	
+4	$H_4P_2O_6$	hypophosphoric acid	
+5	HPO_3	meta-	
	$H_4P_2O_7$	pyro-	phosphoric acids
	H_3PO_4	ortho-	

The peroxo-acids are dealt with elsewhere (24.6.3).

20.10.1. Hypophosphorous acid

Hypophosphorous acid, H_3PO_2, is made by acidifying, with dilute sulphuric acid, the barium hypophosphite solution obtained when white phosphorus is dissolved in hot baryta water:

$$Ba(H_2PO_2)_2 + H_2SO_4 \rightarrow 2\ H_3PO_2 + BaSO_4$$

It can be separated as deliquescent, colourless crystals, *m.p.* 299.6 K. Raman spectra, X-ray analysis of its salts and a uniformly monobasic character show it to have the structure represented in Fig. 20.9. In this the hydrogen atom of the OH is ionisable and the anion is $H_2PO_2^-$; this is tetrahedral.

The acid ionises fairly strongly in water ($pK_a = 2$). Both acid and salts are strong reducing agents:

$$H_3PO_3 + 2\ H_3O^+ + 2\ e \rightleftharpoons H_3PO_2 + 3\ H_2O \qquad E^0 = -0.59\ V$$
$$HPO_3^{2-} + 2\ H_2O + 2\ e \rightleftharpoons H_2PO_2^- + 3\ OH^- \qquad E^0 = -1.65\ V$$

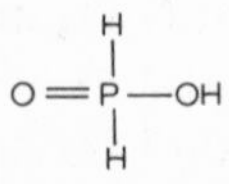

Fig. 20.9. Structure of hypophosphorous acid.

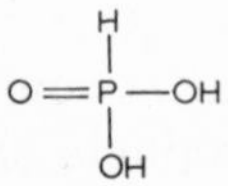

Fig. 20.10. Suggested structure of orthophosphorous acid.

20.10.2. Orthophosphorous acid

Orthophosphorous acid, H_3PO_3, can be made by hydrolysing PCl_3 with ice-cold water. The solid (*m.p.* 343 K) can be crystallised from solution. Nuclear magnetic resonance spectroscopy of the acid and of its salts and esters shows that there is one hydrogen atom directly linked to phosphorus (Fig. 20.10). This hydrogen, like the corresponding hydrogen atoms in hypophosphorous acid, is not ionised.

The first ionisation constant of H_3PO_3 is higher than that of H_3PO_4 ($pK_a = 2$). Both orthophosphorous acid and its salts are strong reducing agents:

$$H_3PO_4 + 2\,H_3O^+ + 2\,e \rightleftharpoons H_3PO_3 + 3\,H_2O \qquad E^0 = -0.20\text{ V}$$

$$PO_4^{3-} + 2\,H_2O + 2\,e \rightleftharpoons HPO_3^{2-} + 3\,OH^- \qquad E^0 = -1.05\text{ V}$$

20.10.3. Hypophosphoric acid

Hypophosphoric acid, $H_4P_2O_6$, can be prepared as a dihydrate from the solution obtained when $Na_2H_2P_2O_6$ is poured through a column of cation-exchange resin in its hydrogen form:

$$2\,H(\text{resin}) + Na_2H_2P_2O_6 \rightarrow 2\,Na(\text{resin}) + H_4P_2O_6$$

The sodium salt is made by the action of alkaline, aqueous NaOCl on red phosphorus.

The acid is tetrabasic, but the commonest salts are those in which only two of the hydrogens are replaced. Both acid and salts are remarkably stable towards oxidising and reducing agents. For the reaction

$$2\,H_3PO_4 + 2\,H_3O^+ + 2\,e \rightleftharpoons H_4P_2O_6 + 4\,H_2O \qquad E^0 = -0.8\text{ V}$$

and for

$$H_4P_2O_6 + 2\,H_3O^+ + 2\,e \rightleftharpoons 2\,H_3PO_3 + 2\,H_2O \qquad E^0 = +0.4\text{ V}$$

Support for the formula $H_4P_2O_6$ comes from cryoscopic and Raman evidence and from the diamagnetism of the salts. The ^{31}P n.m.r. spectrum shows the structure to contain one H—P bond and a P—O—P linkage. Accordingly, the structure

```
 ┌    O       O    ┐3-
      |       ‖
 H — P — O — P — O
      ‖       |
 └    O       O    ┘
```

is proposed for the ternegative anion, $HP_2O_6^{3-}$.

20.10.4. Phosphoric acids

The P_4O_{10}—H_2O system is complex. Analysis of mixtures containing more than 72.4% P_4O_{10} shows that the relative amounts of condensed acids, such as $H_4P_2O_7$, $H_5P_3O_{10}$ and $H_6P_6O_{18}$, vary considerably with composition. The mixtures are not composed simply of H_3PO_4 and P_4O_{10}.

20.10.4.1. *Orthophosphoric acid*

Orthophosphoric acid, H_3PO_4, a colourless solid, *m.p.* 315.5 K, is made by removing water from syrupy phosphoric acid under reduced pressure. The arrangement of oxygen atoms round the phosphorus atom in this tribasic acid is only approximately tetrahedral (Fig. 20.11).

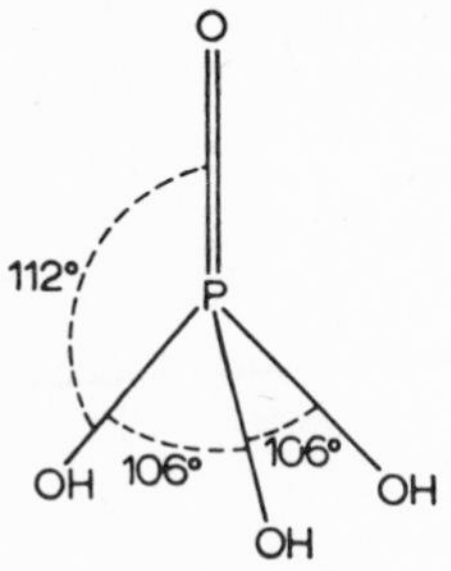

Fig. 20.11. Structure of orthophosphoric acid.

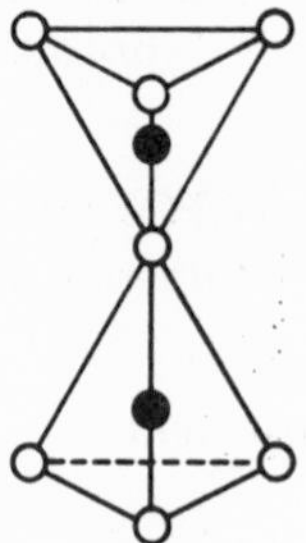

Fig. 20.12. PO_4 tetrahedra in $P_2O_7^{4-}$ ion.

The 'keto' oxygen is attached to an adjacent 'hydroxo' oxygen by a hydrogen bond.

20.10.4.2. *Phosphates*

These salts form the phosphorus minerals of which apatite, $3\ Ca_3(PO_4)_2 \cdot CaF_2$, is the most important. It is the source of phosphorus, superphosphate and phosphoric acid. The last is made by treating apatite with concentrated sulphuric acid and filtering off the insoluble calcium sulphate. Much sodium phosphate is used in water treatment. Most of the heavy metals give insoluble phosphates, and can be removed from water in that form.

There are many esters of orthophosphoric acid of which tri-*n*-butyl phosphate (TBP) is the best known because it has been extensively used in the solvent extraction of metal salts from aqueous solution. TBP is made by treating $POCl_3$ with the alcohol:

$$POCl_3 + 3\ C_4H_9OH \rightarrow (C_4H_9O)_3PO + 3\ HCl$$

It is used in a kerosene solution and extracts such nitrates as $UO_2(NO_3)_2$ and $PuO_2(NO_3)_2$ from dilute nitric acid as $MO_2(NO_3)_2 \cdot 2$ TBP. By washing this loaded organic phase with 0.1 *M* HNO_3 containing a reducing agent, Pu^{VI} is reduced to Pu^{III} and enters the aqueous phase, leaving the U^{VI} still unreduced as the TBP solvate in the kerosene. From this the uranium can be removed by washing the organic phase with water in which the salt $UO_2(NO_3)_2$, essential to the TBP solvate, is ionised and goes into solution.

20.10.4.3. *Pyrophosphoric acid*

Pyrophosphoric acid, $H_4P_2O_7$, is the colourless solid crystallised from an 80% P_4O_{10} solution which contains H_3PO_4 and $H_5P_3O_{10}$ in addition to $H_4P_2O_7$. The first dissociation constant of $H_4P_2O_7$ is higher than that of the ortho acid ($pK_a = 0.8$, cf. $pK_a = 2.1$ for H_3PO_4). All the hydrogens are replaceable, but the commonest pyrophosphates are $M^I{}_2H_2P_2O_7$ and $M^I{}_4P_2O_7$. These contain the anion $P_2O_7{}^{4-}$ (Fig. 20.12).

20.10.4.4. *Metaphosphoric acids*

Metaphosphoric acids of empirical formula HPO_3 may be obtained on further dehydration of H_3PO_4 by heating at 590 K. Their nature is in doubt; the existence of a monomer is unlikely, for the vapour is believed to be dimeric even at white heat. The metaphosphates of sodium have been studied in some detail. The sparingly soluble Maddrell's salt, made by heating NaH_2PO_4 at 590 K, is thought to be a mixture of two salts of similar, but not identical, structures. A trimetaphosphate $Na_3(PO_3)_3$ is obtained by heating $Na_2H_2P_2O_7$. The $P_3O_9{}^{3-}$ anion consists of three $PO_4{}^{3-}$ tetrahedra joined through common oxygen atoms (Fig. 20.13).

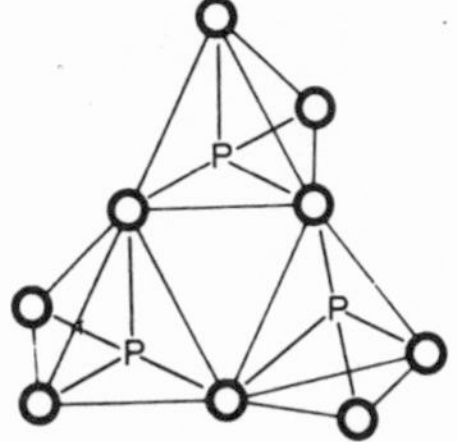

Fig. 20.13. Structure of $P_3O_9{}^{3-}$ ion.

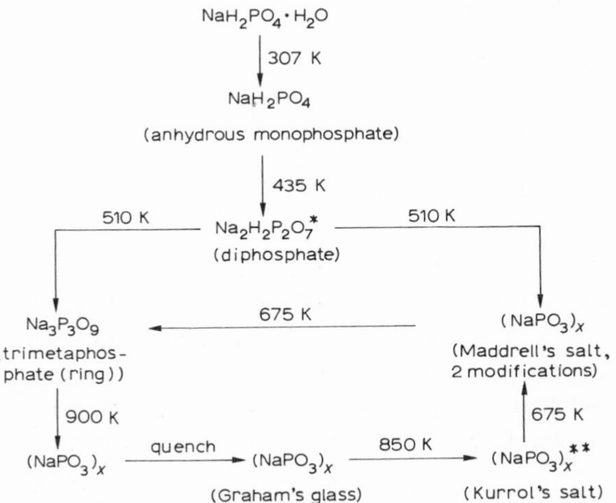

Fig. 20.14. Condensed phosphates obtained by heating $NaHPO_4 \cdot H_2O$. * Proportions depend on water vapour pressure. ** Tempering just below the melting point and inducing crystallisation by local cooling.

The compound known as sodium hexametaphosphate is made by rapidly cooling molten metaphosphate. It is best considered as a metaphosphate glass. The material is soluble and acts as a water softener, removing Ca^{2+} ion from solution by chelation to the colloidal polyanions:

—P(=O)(O)—O—P(O)(=O)—O—P(=O)(O)—O—P(O)(=O)—O— (chelating Ca, Ca)

20.10.4.5. *Polyphosphates*

The term polyphosphate is usually applied to salts of the chain-like ions $[P_nO_{3n+1}]^{-n-2}$:

MO—P(O)(OM)—O—P(O)(OM)—O—P(O)(OM)—OM
tripolyphosphate

MO—P(O)(OM)—O—P(O)(OM)—O—P(O)(OM)—O—P(O)(OM)—OM
tetrapolyphosphate

The ^{31}P n.m.r. spectra have been of value in elucidating the structures of these ions.

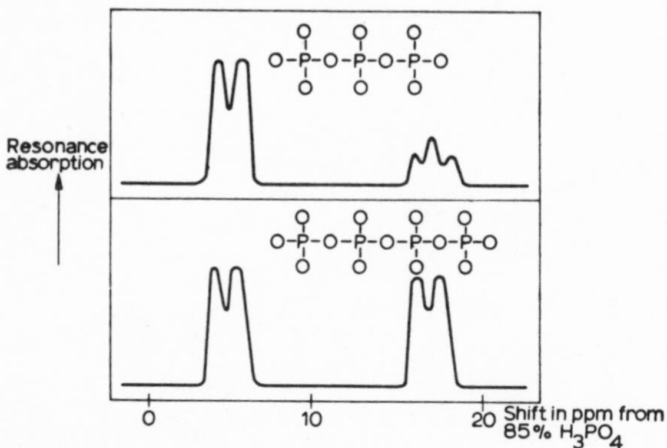

Fig. 20.15. ^{31}P n.m.r. spectra of tripolyphosphate and tetrapolyphosphate ions.

The absorption peaks on the left of Fig. 20.15 are due to ^{31}P atoms in end groups of the tri- and tetrapolyphosphate ions. They are doublets, indicating that each ^{31}P atom is attached to one other ^{31}P neighbour in an environment different from its own. The absorption peak at the top right encloses only half the area of the one below it and is a 1.2.1 triplet, indicating that one ^{31}P atom is attached to two others which are in different chemical environments from itself.

The peak on the right of the tetrapolyphosphate curve covers the same area as that on the left, indicating that there are as many middle ^{31}P groups as end ones in this ion, namely two. The doublet form indicates that each is attached to only one other ^{31}P atom which has an environment different from its own.

Polyphosphates are added to detergents for the removal of surface dirt. They are strongly adsorbed on the dirt particles which they thus render hydrophilic.

20.10.5. Fluorophosphoric acids

Phosphoryl fluoride, OPF_3, hydrolyses in alkaline solutions in the following manner:

$$O{-}PF_3 \xrightarrow[\text{(fast)}]{OH^-} \left[O{-}P(O)(F)F\right]^- \xrightarrow[\text{(slow)}]{OH^-} \left[O{-}P(O)(O)F\right]^{2-} \xrightarrow[\text{(slow)}]{OH^-} PO_4^{3-}$$

The last two stages of hydrolysis are so slow that salts of the fluorophosphoric acid, HPO_2F_2, can be separated from the solution, but the acid itself has not been made. The ammonium salt of the acid, $NH_4PO_2F_2$, can also be obtained by fusing together NH_4F and P_2O_5.

The acid H_2PO_3F is produced as an oily liquid from a mixture of P_2O_5 and concentrated aqueous HF:

$$P_2O_5 + 2\ HF + H_2O \rightarrow 2\ H_2PO_3F$$

20.11. Sulphur Compounds of Nitrogen and Phosphorus

20.11.1. Sulphur nitrides

Sulphur lies between nitrogen and phosphorus in electronegativity. Thus binary sulphur—nitrogen compounds are correctly named sulphur nitrides. On the other hand, sulphur—phosphorus compounds are named phosphorus sulphides.

Tetrasulphur tetranitride, S_4N_4, is produced when dry NH_3 is passed into a solution of S_2Cl_2 in dry ether, separating as orange crystals, *m.p.* 451 K.

$$6\ S_2Cl_2 + 16\ NH_3 \rightarrow S_4N_4 + 12\ NH_4Cl + 8\ S$$

The compound is diamagnetic and strongly endothermic. It explodes on heating, is soluble in many organic solvents, and is attacked slowly by water which does not wet it easily. Boiling alkalis cause hydrolysis:

$$S_4N_4 + 6\ OH^- + 3\ H_2O \rightarrow S_2O_3^{2-} + 2\ SO_3^{2-} + 4\ NH_3$$

Electron and X-ray diffraction show all the bonds to be equal in length and the interbond angles NSN = 104° and SNS = 113°. The probable form of the molecule is an eight-membered 'cradle' ring (Fig. 20.16). It is the parent substance of other sulphur nitrides and derivatives of sulphur nitrides. Thus the ring can be split by heating to give white disulphur dinitride, S_2N_2, also soluble in organic solvents, which is converted into the insoluble dark-blue

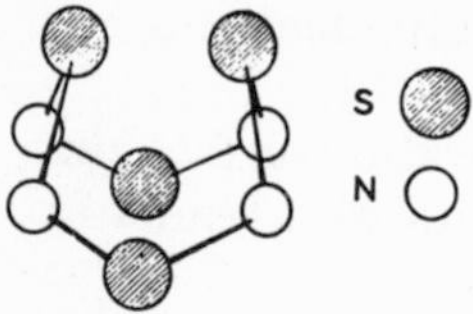

Fig. 20.16. Cradle-shaped structure of S_4N_4 molecule.

polymer, $(SN)_n$, at room temperature. Moreover, S_4N_4 can be reduced by tin(II) chloride to tetrasulphur tetraimide:

```
H     H
|     |
N—S—N
|     |
S     S
|     |
N—S—N
|     |
H     H
```

An interesting reaction of S_4N_4 is that with PCl_3 to give the compound $P_4N_2Cl_{14}$ which has a complex cation:

$[Cl_3P \cdot N \cdot PCl_2 \cdot N \cdot PCl_3]^+ PCl_6^-$

20.11.2. Phosphorus sulphides

There are four sulphides of phosphorus, P_4S_3, P_4S_5, P_4S_7 and P_4S_{10}. The last is the only sulphide with a structure similar to that of an oxide. The structural relationship of the sulphides to the P_4 molecule and to one another is clearly shown in Fig. 20.17 where the dark circles stand for phosphorus atoms.

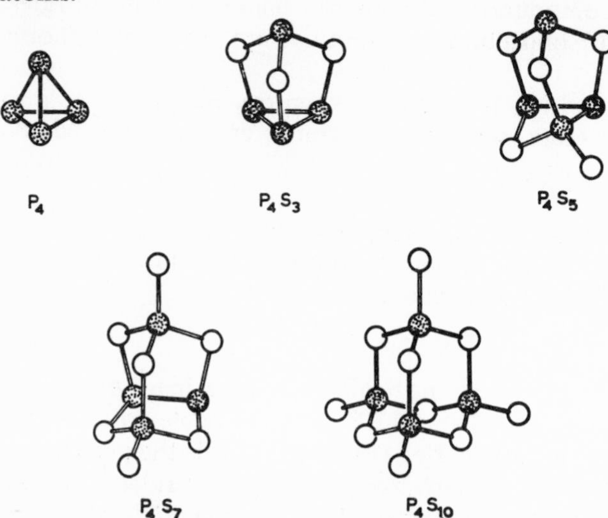

Fig. 20.17. Structures of phosphorus sulphides. (By courtesy of S. van Houten.)

The sulphide P_4S_3, made by heating red phosphorus with sulphur in an atmosphere of CO_2, is the yellow crystalline solid, *m.p.* 447 K, used in 'strike-anywhere' matches. The other sulphides have a similar physical appearance, somewhat higher melting points and much lower stabilities. P_4S_3 is attacked only very slowly by cold water and cold HCl, but decomposed by hot water to give H_2S, PH_3 and phosphorus oxo-acids. Cold HNO_3 converts it to H_3PO_4, H_2SO_4 and sulphur; aqueous KOH produces PH_3, hydrogen, and the phosphite, hypophosphite and sulphide ions.

20.12. Thiohalides

The compound thiazyl chloride, SNCl, the sulphur analogue of nitrosyl chloride, ClNO, is made from a suspension of NH_4Cl in refluxing S_2Cl_2:

$$NH_4Cl + 2\ S_2Cl_2 \rightarrow 3\ S + SNCl + 4\ HCl$$

and by other methods. Phosphorus does not form a corresponding compound, but the thiophosphoryl halides are known:

	B.p./K
PSF_3	220
$PSCl_3$	398
$PSBr_3$	448 (with decomposition)

Also there are 'mixed' halides corresponding to the phosphoryl halides such as OPF_2Cl.

$SPCl_3$ is made by heating P_2S_5 with PCl_3, and $SPBr_3$ by the action of bromine on a solution of sulphur and phosphorus in CS_2. Both can be converted to 'mixed' fluorides or to SPF_3 by the action of SbF_3, according to the conditions. The molecules are similar to those of OPX_3 (20.8) with P—S 194 pm, indicating a double bond, in $SPCl_3$ and $SPBr_3$, but down to 185 pm in SPF_3, indicating some triple-bond character in the fluorine compound.

20.13. Nitrides

The nitrides are of three types; these are the ionic, the covalent and the interstitial.

20.13.1. Ionic nitrides

These contain the N^{3-} ion, of radius 171 pm; this, incidentally, is the only simple ternegative ion known. Lithium nitride, Li_3N, made from the elements, with a heat of formation of 198 kJ mol^{-1}, is the only exothermic and thermally stable nitride of Group I. The explosive Na_3N, K_3N and Rb_3N are produced when an arc is struck between a Pt cathode and an alkali metal anode under liquid nitrogen. Beryllium, magnesium and other Group IIA elements form ionic nitrides, like lithium, when heated in nitrogen. Be_3N_2 and Ca_3N_2 have the highest heats of formation, 570 and 434 kJ mol^{-1} respectively. The ionic nitrides are hydrolysed to the hydroxides and ammonia.

20.13.2. Covalent nitrides

These include volatile compounds of nitrogen with hydrogen, the halogens and carbon respectively and also involatile, high-melting, adamantine compounds formed by nitrogen with Group III elements, as typified by aluminium nitride, AlN. Boron nitride, BN, has an adamantine form, borazon, produced at high pressure, as well as the usual form which is similar in structure to graphite. The latter is used as an anti-sticking compound in glass-makers' moulds and also for coating crucible linings, since it is not wetted by molten iron. Boron nitride is readily made by heating B_2O_3 and NH_4Cl together, the product being a white powder. Silicon nitride, Si_3N_4 ($\Delta H_f^\circ = -752$ kJ mol^{-1}), is a commercial product manufactured by converting silicon tetrachloride to the amide with ammonia and then decomposing the amide by heat:

$$SiCl_4 \xrightarrow{NH_3} Si(NH_2)_4 \xrightarrow{heat} Si_3N_4$$

20.13.3. Interstitial nitrides

The transition metals, including those of Group IIIA, form true interstitial compounds (7.2.13), many with the composition MN and the NaCl structure. Other interstitial nitrides, Mo_2N, W_2N, Fe_4N and Mn_4N, are hard, high-melting, good conductors of metallic appearance, commonly made by heating the powdered metal in nitrogen or ammonia at about 1500 K. They vary considerably in thermal stability, reactivity and ease of hydrolysis, and are less thermally stable than the corresponding metal oxides. For some elements the nitride, oxide and carbide are isomorphous, an example being TiN, TiO and TiC. Heats of formation are often high in spite of the high dissociation energy of molecular nitrogen. Some interstitial nitrides are conveniently made by ammonolysis of halides in liquid ammonia, the amides first formed being converted to nitrides on heating. Another method is to effect simultaneous reduction and nitriding by heating the oxide, for instance iron oxide, in ammonia.

20.14. Phosphides

A mixture of sodium phosphides is made by direct combination between the elements. It appears to contain Na_3P and Na_5P_2 and is used for sea flares, since with water it produces a mixture of hydrogen phosphides which is spontaneously inflammable in air. Also used for the same purpose are Mg_3P_2 and Zn_3P_2, made by warming intimate mixtures of the elements, and Ca_3P_2, made by heating CaO in phosphorus vapour. Thus prepared, calcium phosphide always contains some phosphate.

Ferrophosphorus contains the phosphides Fe_2P and FeP. These phosphides and those of the other transition metals are dark-coloured, metal-like solids which are not attacked by water. They conduct electricity.

20.15. Formal Derivatives of Ammonia

20.15.1. Hydrazoic acid

Hydrazoic acid, HN_3, is present in the aqueous distillate from the action of NaN_3 on dilute H_2SO_4. Sodium azide for the purpose is made by passing N_2O over sodium amide at 460 K:

$$N_2O + 2\,NaNH_2 \rightarrow NaN_3 + NaOH + NH_3$$

Pure HN_3, *b.p.* 308.8 K, is an extremely explosive material and aqueous solutions should not be concentrated. Hydrazoic acid is a colourless, mobile liquid, easily exploded by shock. The vapour is monomeric.

The acid dissociates only slightly in aqueous solution ($pK_a = 4.7$). Metals dissolve with the evolution of NH_3 and N_2:

$$Zn + 3\,HN_3 \rightarrow Zn(N_3)_2 + NH_3 + N_2$$

Crystalline salts are formed with ammonia, NH_4N_3 (empirically N_4H_4), and with hydrazine, $N_2H_5N_3$ (empirically N_5H_5). Other azides are the explosive chloroazide ClN_3 (gas), bromoazide BrN_3 (orange liquid), and iodoazide IN_3 (yellow solid). The first and second can be made from NaN_3 by treating it with sodium hypochlorite and bromine respectively.

Fully ionised azides like those of the Group I metals and barium are non-explosive and yield nitrogen and the metal on heating. But LiN_3 is an exception; it is converted to the nitride. The covalent azides, such as those of organic radicals, together with the heavy metal azides including those of Ag, Cu, Tl, Pb and Hg, are explosive, lead azide being widely used as a detonator.

In covalent azides such as HN_3 and those of organic radicals, the nitrogen atoms are arranged collinearly but the molecule has C_s symmetry.

In solid inorganic azides, the N_3^- ion is linear with $D_{\infty h}$ symmetry. In general it behaves like a halide ion and, being both isoelectronic and isostructural with cyanate (CNO^-) and isocyanate (NCO^-), it is regarded as a true pseudohalide (26.11). The azide ion has an approximately ellipsoid shape with similar steric properties to the chloride ion (diameter 362 pm). Its $D_{\infty h}$ symmetry and its 16 electrons make it both isoelectronic and isostructural with CO_2, NO_2^+, CNO^- the fulminate ion, NCO^- the cyanate ion, and NCN^{2-} the cyanamide ion. Satisfactory valence bond and molecular orbital structures have been described. The individual gaseous ion, which has been detected by mass spectrometry, probably has the same structure.

115pm 115pm
N N N
350 pm
510 pm

20.15.2. Hydroxylamine

Hydroxylamine, NH_2OH, is a colourless solid, *m.p.* 305.2 K, very soluble in water and the lower alcohols, less so in other organic liquids. It has a *trans* conformation with C_s symmetry, and a single covalent bond length for the N—O distance. The hydrogen sulphate, $[NH_3OH]HSO_4$, is made industrially by adding concentrated H_2SO_4 to refluxing nitromethane at 380—395 K:

$$CH_3NO_2 + H_2SO_4 \rightarrow [NH_3OH]HSO_4 + CO$$

This hydrogen salt is converted to the normal sulphate $[NH_3OH]_2SO_4$ by the action of 85% methanol, and the sulphate is converted to the chloride by dissolving it in hot, concentrated HCl and crystallising the chloride from the solution:

$$[NH_3OH]_2SO_4 + 2\,HCl \rightarrow 2[NH_3OH]Cl + H_2SO_4$$

When the chloride is added to sodium methoxide in methanol and the precipitated NaCl is filtered off, a solution is obtained from which free NH_2OH can be crystallised at 253—263 K. Before reaching its m.p., the compound decomposes into NH_3, H_2O, N_2O and NO. It is a weak monoacid base without acidic properties. Like its aquo analogue, H_2O_2, and its ammono analogue hydrazine, N_2H_4, hydroxylamine can act either as an oxidising or a reducing agent according to the circumstances.

20.16. Phosphonitrilic Compounds

Compounds of phosphorus and nitrogen based on the π-bonded unit

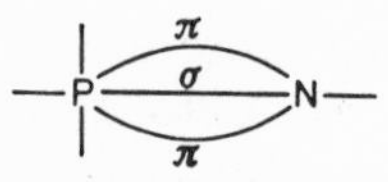

have attracted attention because they form polymers with properties which might be useful at high temperatures. The materials range from fluids possibly useful for heat transfer and lubrication, to solids with some mechanical characteristics of organic polymers. They show similarities to aromatic compounds, although the π-bonding is somewhat different.

Phosphonitrilic chlorides were first prepared by the ammonolysis of phosphorus pentachloride; dry distillation gives a mixture of $(PNCl_2)_3$ and $(PNCl_2)_4$:

$$n\,PCl_5 + n\,NH_4Cl \rightarrow (PNCl_2)_n + 4n\,HCl$$

The method was improved by dissolving the PCl_5 in tetrachloroethane; this increases the yield of higher polymers. The product is a liquid mixture of polymers from which the excess of NH_4Cl is removed by filtration. The filtrate is mainly $(PNCl_2)_3$, but other polymers, up to $(PNCl_2)_6$, are also present. Concentrating the liquid under reduced pressure gives a material, part of which is insoluble in petroleum ether and is believed to be a mixture of linear polymers, $PCl_4(PNCl_2)_nCl$, but most of which is soluble in that solvent and consists of cyclic $(PNCl_2)_n$ polymers.

The preparative reaction is very versatile; the phenyls, $(PNPh_2)_{3,4}$, can be obtained from Ph_2PCl_3 and the methyls, $(PNMe_2)_{3,4}$, from Me_2PCl_3. Bromides and mixed halides can also be made.

The trimeric compounds have a planar P_3N_3 ring in which the hybrid P—N

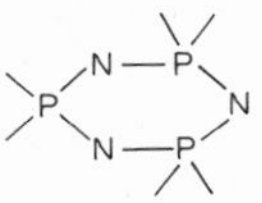

bonds are all equal in length (160 pm) and shorter than the P—N bond (178 pm) in the phosphoramidate ion, $NH_2PO_3^{2-}$. They are comparatively inert compounds, thermally stable and not readily hydrolysed; they can be steam-distilled without serious loss. The crystalline trimer (and also the tetramer) can be further condensed to rubber-like materials, of molecular weight >20,000, by heating at ~600 K. Unfortunately, however, the elastic properties are lost on exposure to damp air.

Although the chloro-compounds are much less reactive than is usual for phosphorus—chlorine compounds, they react smoothly with alcohols and alkoxides:

$$(PNCl_2)_3 + 6\ NaOC_2H_5 \rightarrow [PN(OC_2H_5)_2]_3 + 6\ NaCl$$

This product is a clear, viscous oil which changes to a gel on standing. Ammonia and amines produce compounds such as $[PN(NH_2)_2]_3$ and $[PN(NH \cdot C_6H_5)_2]_3$, and some or all of the chlorine in $(PNCl_2)_3$ can be replaced to give such compounds as $[PNCl(Me_2N)]_3$, $P_3N_3Cl_4Ph_2$, and $[PN(MeO)_2]_3$. At 470 K, $[PN(OC_2H_5)_2]_3$ isomerises thus:

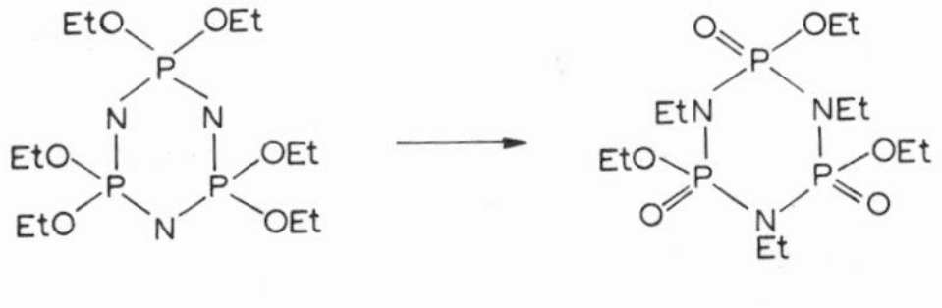

The bromo-compounds $(PNBr_2)_3$ and $(PNBr_2)_4$ are formed when NH_4Br replaces NH_4Cl in the preparative reaction; the fluoro-analogues are, however, made by the action of potassium fluorosulphite on $(PNCl_2)_3$ at 400 K:

$$(PNCl_2)_3 + 6\ KSO_2F \rightarrow (PNF_2)_3 + 6\ KCl + 6\ SO_2$$

The fluoro-trimer, *m.p.* ~300 K, *b.p.* 324 K, is insoluble in polar liquids, even in concentrated H_2SO_4. By contrast, $(PNMe_2)_3$, *m.p.* 468 K, dissolves readily in water.

The tetramer formed in the preparative reaction with NH_4Cl is also cyclic, but the ring is puckered, not planar as with the trimer $(PNCl_2)_3$:

Cl Cl
Cl P N—P N
Cl | | Cl
N P—N P Cl
Cl Cl

However, the compound $(PNF_2)_4$, made by treating $(PNCl_2)_4$ with KSO_2F, is planar; this indicates that the substituents, as well as the ring size, are important in determining whether the ring is puckered or flat. A chlorofluoropolymer, $(PNFCl)_4$, is made by heating the trimer $(PNCl_2)_3$ with PbF_2. This is the only reaction yet reported by which a phosphonitrilic ring can be enlarged. All the possible trimeric and tetrameric phosphonitrilic fluoride chlorides have now been obtained. Chlorine in the tetramer $(PNCl_2)_4$ can be substituted to give $(PNPhCl)_4$ by treatment with PhMgBr, and $[PN(NHPh)_2]_4$ by treatment with $PhNH_2$.

Bonding in the trimeric cyclic phosphonitrilic halides is interesting. Four electrons of the individual nitrogen atoms may be considered to occupy approximately sp^2 hybrid orbitals, one in every P—N bond and two in a lone pair. The fifth electron, in a p_z orbital, is available for π bond formation. The individual phosphorus atoms have four electrons arranged approximately tetrahedrally in σ bonds, leaving the fifth electron, in a d orbital, available for π bonding. Overlap results in d_π—p_π bonding; this is unlike the π bonds in benzene which arise from an overlap of p orbitals only.

Further Reading

K. Jones, The chemistry of nitrogen, Pergamon, Oxford, 1975.

A.D.F. Toy, The chemistry of phosphorus, Pergamon, Oxford, 1975.

C.B. Colburn (Ed.), Developments in inorganic nitrogen chemistry, Elsevier, Amsterdam, Vol. 1, 1966; Vol. 2, 1972.

A.N. Wright and C.A. Winkler, Active nitrogen, Academic Press, New York, 1968.

R. Brown and C.A. Winkler, The chemical behaviour of active nitrogen, Angew. Chem. Internat. Ed., 12 (1973) 181.

A.D. Allen and F. Bottomley, Nitrogen compounds of the transition metals, Accounts Chem. Res., 1 (1968) 360.

E.E. van Tamelen, Chemical modification of molecular nitrogen under mild conditions, Accounts Chem. Res., 3 (1970) 361.

R. Eisenberg and C.D. Meyer, Co-ordination chemistry of nitric oxide, Accounts Chem. Res., 8 (1975) 26.

C.A. McAuliffe (Ed.), Transition metal complexes of phosphorus, arsenic and antimony ligands, Macmillan, London, 1973.

S.W. Schneller, Nitrogen fixation, J. Chem. Ed., 49 (1972) 787.

Arsenic, Antimony and Bismuth — Group VB

21

21.1. Introduction

The elements of Group VB, like nitrogen and phosphorus, have the $ns^2\,np^3$ electron configuration and the ^4S ground state which indicates three singly-occupied orbitals. Arsenic, antimony and bismuth differ from nitrogen and phosphorus, however, in having a lower electronegativity, which decreases with increasing atomic number. They are dominantly tervalent elements and tend to form three electron-pair bonds roughly at right angles, leaving an ns^2 lone pair, but some hybridisation occurs and results in distortion towards more nearly tetrahedral bonds. All five valence electrons may, nevertheless, participate in bonding, d-hybridisation yielding bipyramidal, square pyramidal and octahedral sets of bonds. In many of these compounds the metal is described formally as having charge number +5 or +3, but the bonding is often mainly covalent; only in bismuth compounds is there any real approach to a simple cation. In solution the only ions which occur are M^{3+}, moreover they are hydrated (Table 21.1).

The ionisation energies are high, as would be expected from the electron configuration, and the standard electrode potentials M^{3+}/M are all quite close together, diminishing ionisation energies apparently being offset by the increased ionic radii and reduced heats of hydration.

The 3rd, 4th and 5th ionisation energies of bismuth are larger than those of antimony. This, coupled with the greater size of the Bi^{3+}, Bi^{4+} and Bi^{5+}

TABLE 21.1

ATOMIC PROPERTIES OF GROUP VB ELEMENTS

	As	Sb	Bi
Atomic number	33	51	83
Electron configuration	[Ar] $3d^{10}\,4s^2\,4p^3$	[Kr] $4d^{10}\,5s^2\,5p^3$	[Xe] $4f^{14}\,5d^{10}\,6s^2\,6p^3$
Covalent radius/pm	121	141	146
Ionic radius, M^{3+}/pm	69	90	120
Metallic radius/pm	139	159	170

TABLE 21.2

IONISATION ENERGIES, ELECTRODE POTENTIALS AND ELECTRONEGATIVITIES FOR GROUP VB ELEMENTS

	As	Sb	Bi
$I(1)$/kJ mol^{-1}	947	834	703
$I(2)$/kJ mol^{-1}	1798	1592	1609
$I(3)$/kJ mol^{-1}	2734	2450	2466
$I(4)$/kJ mol^{-1}	4834	4255	4370
$I(5)$/kJ mol^{-1}	6040	5400	5403
$I(6)$/kJ mol^{-1}	12300	10420	8520
E^0, M^{3+}/M/V	0.25	0.21	0.32
Electronegativity (Allred—Rochow)	2.20	1.82	1.67

ions, is the prime cause of the diminished stability of the +5 oxidation state of bismuth.

21.2. Thermodynamic Stability of the Oxidation States

The free energies, relative to the element, of the various oxidation states in solution at pH = 0 are shown graphically in Fig. 21.1.

For the three elements, the +3 state is stable with respect to disproportionation to the +5 state and the zero state. The redox potential for the

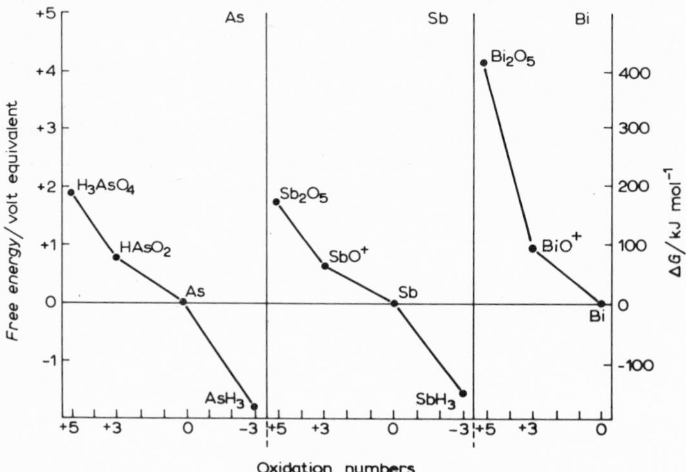

Fig. 21.1. Free energies, relative to the element, of the oxidation states of As, Sb and Bi at pH = 0.

As^V/As^{III} couple is strongly dependent on pH:

$$H_3AsO_4 + 2\,H^+ + 2\,e \rightarrow HAsO_2 + 2\,H_2O \qquad E^0 = +0.56\ V$$
$$AsO_4^{3-} + 2\,H_2O + 2\,e \rightarrow AsO_2^- + 4\,OH^- \qquad E^0 = -0.67\ V$$

Thus in strongly acidic solution arsenic acid oxidises iodide to iodine (E^0, I_2/I^- = 0.54 V) but in neutral and alkaline solutions iodine oxidises arsenite to arsenate.

Although the metals are most commonly depicted with charge numbers +5 or +3, their tendency towards non-metallic character is revealed in compounds with true metals. In such compounds as Na_3As, Mg_3Sb_2 and K_3Bi, many of which resemble the corresponding phosphides, the Group VB metal apparently accepts electrons to achieve a formal charge of −3.

21.3. The Elements

The elements well exemplify the trend towards a more metallic character with increasing atomic number; but even bismuth is not a true metal in the structural sense. The metallic allotropes of As, Sb and Bi consist of puckered sheets in which each atom is covalently bonded to three neighbouring atoms, the different sheets being held together by metallic binding. The $8-N$ rule, which applies to most non-metallic solid elements, continues to operate. It is significant too that the elements are polyatomic in the vapour. Arsenic, for example, exists principally as tetrahedral As_4 molecules at 1100 K; at higher temperatures dissociation to As_2 occurs but is incomplete even at 1870 K. Antimony behaves similarly. Bismuth vapour is an equilibrium mixture of Bi_2 and Bi, the former being in an appreciable proportion even at 2300 K.

Both arsenic and antimony have yellow cubic α-forms, soluble in CS_2. Yellow arsenic consists of tetrahedral As_4 units (cf. P_4); the α-form of antimony is probably similar in structure but is unstable above 190 K. Black β-forms, analogous to amorphous phosphorus, result from rapid cooling of the vapours. They are metastable, changing rapidly at 620—670 K into the γ-forms which are moderately good conductors of heat and electricity and exhibit metallic lustre, though brittle, easily fractured and of low ductility. Antimony and bismuth differ from normal metals in having lower electrical conductivity as solids than as liquids; for Bi the conductance of the solid is

TABLE 21.3

PHYSICAL PROPERTIES OF GROUP VB METALS

	As	Sb	Bi
Density/g cm^{-3}	5.7	6.6	9.8
Melting point/K	1090 (3.6 MPa)	903	544
Boiling point/K	889 (sublimes)	1908	1853

only 0.48 of the liquid conductance. Bismuth also has the lowest thermal conductivity of any metal at the ordinary temperature.

The physical properties given in Table 21.3 are for the metallic forms. Arsenic sublimes at the ordinary pressure; the *m.p.* given is that for the element at 3.6 MPa pressure.

21.3.1. Occurrence, extraction and uses

The principal ore of arsenic (5×10^{-4}% of earth's crust) is arsenical pyrites, FeAsS, but the element occurs commonly with nickel, copper and tin. The oxide, As_4O_6, is recovered from flue-dusts collected during the extraction of these metals. Sublimation of the crude material in the presence of galena, which prevents the formation of arsenites, purifies the oxide; this is then reduced to arsenic by carbon in a cast iron retort. The element itself has few uses; about 0.5% added to lead increases the surface tension of the molten metal and allows spherical lead-shot to be produced. The principal commercial form is the so-called white arsenic, As_4O_6. Arsenic compounds have been used mainly for their toxicity; arsenical insecticides were much used.

Antimony (5×10^{-5}%) occurs as stibnite, Sb_2S_3, which is converted to the volatile oxide by roasting in air:

$$2\,Sb_2S_3 + 9\,O_2 \rightarrow Sb_4O_6 + 6\,SO_2$$

The sublimate can be reduced in a blast furnace similar to that used for smelting lead. The metal is alloyed with lead for electrical storage-battery plates and with tin in pewter, in both cases conferring greater hardness and mechanical strength. It is also used with lead in corrosion-resistant piping, with lead and tin in type metals, and with tin, copper and lead in bearing metals. The oxide, Sb_4O_6, is employed in vitreous enamels and as a pigment.

Bismuth (10^{-5}%) occurs as Bi_2S_3, associated with the sulphide ores of lead and copper and also with SnO_2. The flue-dusts from the roasting of lead, copper and tin ores, and the anode sludge from copper refining, are worked-up for bismuth. The oxide can be reduced at ~800 K with iron and carbon in the presence of a flux. Electrolytic purification is possible from a solution of $BiCl_3$ in HCl. The metal is used particularly in fusible alloys of which Wood's metal (Bi, Pb, Sn and Cd in the wt. ratio 4 : 2 : 1 : 1), *m.p.* 344 K, and type metal are examples. Bismuth and many of its alloys expand on solidification and give sharp impressions. The salts are used in pharmaceutical preparations, showing a marked contrast in toxicity to those of arsenic.

21.3.2. Reactions of the elements

The elements combine on heating with oxygen, sulphur and the halogens (X):

$$4\,M + 3\,O_2 \rightarrow M_4O_6\ (2\,Bi_2O_3)$$
$$2\,M + 3\,S \rightarrow M_2S_3$$
$$2\,M + 3\,X_2 \rightarrow 2\,MX_3$$

Antimony gives in addition to the trihalide some pentahalide with both F_2 and Cl_2, but arsenic does this with fluorine only. The elements all dissolve in hot, concentrated H_2SO_4 with the evolution of SO_2. Antimony and bismuth yield sulphates, arsenic forms As_4O_6. Bismuth dissolves readily in nitric acid to give $Bi(NO_3)_3$; the other two elements are converted to mixtures of oxides. Hydrochloric acid has little action on any of the elements. Cold aqua regia dissolves antimony, producing a solution containing $SbCl_6^-$ ions. Arsenic dissolves in fused NaOH but the others do not:

$$2\,As + 6\,NaOH \rightarrow 2\,Na_3AsO_3 + 3\,H_2$$

21.4. Halides

AsF_3 is a colourless, fuming liquid made by heating As_4O_6 with CaF_2 and H_2SO_4 in a lead retort. The gaseous fluoride, AsF_5, is the chief product of the treatment of arsenic with fluorine. In common with the structure of all the corresponding halides of this group, the trihalides are pyramidal (C_{3v}) and the pentahalides are trigonal bipyramidal (D_{3h}).

Pentahalides of arsenic, other than AsF_5, are unknown. The trichloride, $AsCl_3$, a colourless liquid, is formed by the chlorination of arsenic. White crystalline $AsBr_3$ and red crystalline AsI_3 are conveniently obtained by treating the element with the halogen in a CS_2 solution. The hygroscopic solid $AsCl_2F_3$ is made by passing chlorine into ice-cold AsF_3. The conductance of the liquid rises with the addition of Cl_2, suggesting the formation of an ionic compound, possibly $[AsCl_4]^+[AsF_6]^-$. Hydrolysis of the arsenic trihalides becomes more difficult with increasing atomic weight of the halogen. Unlike phosphorus and antimony, arsenic does not form a well-characterised oxohalide.

The white, solid SbF_3 hydrolyses but slightly and can be made by the action of HF on Sb_2O_3. It forms complexes such as K_2SbF_5. The viscous SbF_5, made by refluxing $SbCl_5$ with anhydrous HF followed by fractional distillation, freezes to a non-ionic solid. In this it differs from PCl_5 which contains

TABLE 21.4

HALIDES OF GROUP VB ELEMENTS

	F	Cl	Br	I
As	AsF_3	$AsCl_3$	$AsBr_3$	AsI_3
	AsF_5			
Sb	SbF_3	$SbCl_3$	$SbBr_3$	SbI_3
	SbF_5	$SbCl_5$		
Bi	BiF_3	$BiCl_3$	$BiBr_3$	BiI_3

PCl_4^+ and PCl_6^- ions in the crystal, it is composed of covalent trigonal bipyramidal molecules in both solid and vapour states. The white crystalline $SbCl_3$, made by heating antimony with mercury(II) chloride and recrystallising the product from CS_2, forms complexes such as K_2SbCl_5. The trichloride hydrolyses in two stages to well-characterised oxide chlorides:

$$SbCl_3 + H_2O \rightarrow SbOCl + 2\ HCl$$
$$4\ SbCl_3 + 5\ H_2O \rightarrow Sb_4O_5Cl_2 + 10\ HCl$$

Unlike arsenic, antimony forms a pentachloride, $SbCl_5$, which is made in a similar way to PCl_5:

$$SbCl_3 + Cl_2 \rightarrow SbCl_5$$

An acid $(HSbCl_6)_2 \cdot 9\ H_2O$ can be crystallised from an HCl solution of $SbCl_3$ into which chlorine has been passed. The tribromide and tri-iodide of antimony are made from the elements. SbI_3 hydrolyses to give $Sb_4O_5I_2$.

The white powder BiF_3, like SbF_3, is made from the oxide and HF; with an excess of oxide, the oxide fluoride BiOF is formed. The trichloride, tribromide and tri-iodide of bismuth can all be made by direct combination of the elements: $BiBr_3$ is yellow and BiI_3, black. The last is hydrolysed by hot water to bronze crystals of BiOI. Bismuth pentafluoride is a white solid made by treating molten bismuth with low-pressure fluorine. It is a very strong fluorinating agent.

21.5. Oxides

The trioxides of arsenic, antimony and bismuth are of structural interest, because they show a transition from the molecular lattice characteristic of covalent compounds to an ionic lattice. Arsenic oxide contains As_4O_6 molecules, similar in structure to P_4O_6 molecules and based on the As_4 tetrahedron. The cubic form of Sb_4O_6 is similar, but above 840 K this is converted to the macromolecular valentinite form containing infinite chains (Fig. 21.2).

Bismuth oxide has a number of forms of which two are important: the low-temperature α-form with a complex structure; and a simple cubic form

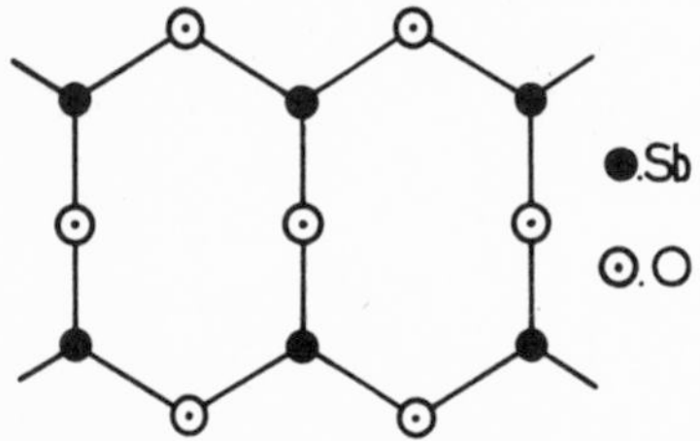

Fig. 21.2. Valentinite structure of Sb_2O_3.

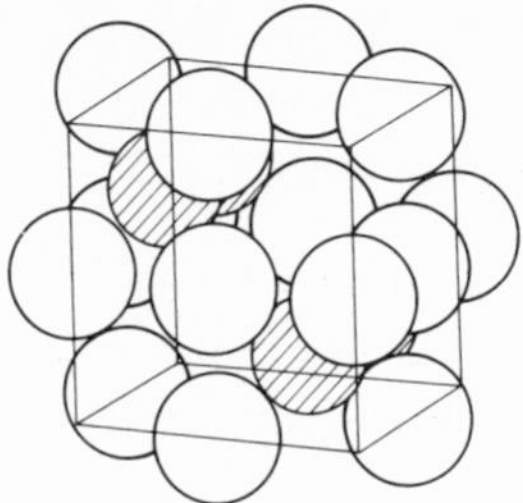

Fig. 21.3. Unit cell of Bi_2O_3.

obtained after the oxide has been fused for a long time in a porcelain crucible and probably stabilised by traces of impurities. The latter has the same ionic structure as Mn_2O_3, the bismuth ions being octahedrally co-ordinated (Fig. 21.3). The trioxides thus range in structure from the molecular, through the macromolecular, to the ionic.

Structural evidence on the pentoxides of arsenic and antimony is lacking. They are not formed from the elements but are produced when the elements are oxidised with nitric acid and the products are dehydrated. Prolonged dehydration of a hydrated oxide of Sb^V gives a compound, Sb_2O_4, which is structurally analogous to $SbTaO_4$. Oxidation of Bi_2O_3 by chlorine, bromine and persulphates takes place, but the nature of the products is unknown. A non-stoichiometric, buff powder, called sodium bismuthate, is made by fusing Bi_2O_3 with NaOH and often formulated $NaBiO_3$. It is insoluble in water and moderately concentrated nitric acid, in which it will oxidise Mn^{2+} to MnO_4^- at room temperature.

Arsenic, like phosphorus and vanadium, forms finite oxo-anions. The condensed arsenates formed when $NaH_2AsO_4 \cdot H_2O$ is heated are shown in Fig. 21.4. There is not an arsenic analogue of sodium trimetaphosphate. The con-

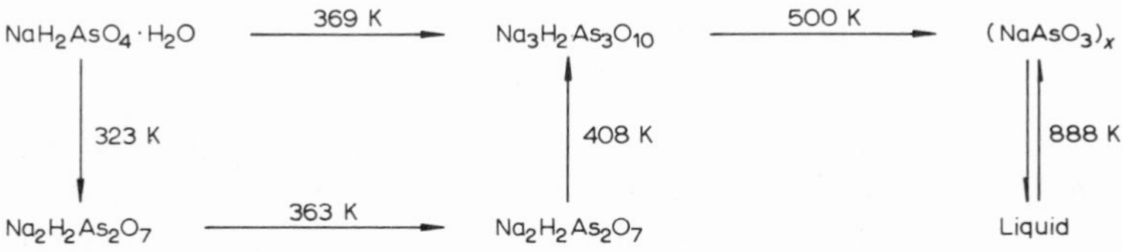

Fig. 21.4. Condensed arsenates obtained from $NaH_2AsO_4 \cdot H_2O$ by heating.

densed ions are much less stable to hydrolysis than are the ions of the condensed phosphates; they change rapidly to AsO_4^{3-} ions.

The oxygen chemistry of Sb^V is quite different, being based not on the tetrahedral but the octahedral co-ordination of Sb^V by oxygen. There are two main groups of these oxo-complexes:

(i) Salts containing $Sb(OH)_6^-$ ions. A well-known example is $Na[Sb(OH)_6]$, formerly called sodium pyroantimonate. Hexahydroantimony ammines can be made by replacing K^+ from $K[Sb(OH)_6]$ with cobalt, chromium and copper ammines:

$[Co(NH_3)_5Cl][Sb(OH)_6]_2 \cdot H_2O$ $[Co(NH_3)_6][Sb(OH)_6]_3 \cdot 3\,H_2O$

(ii) Mixed oxides based on SbO_6 octahedra, such as M^ISbO_3, $M^{II}Sb_2O_6$, $M^{III}SbO_4$ and $M_2{}^{II}Sb_2O_7$. Examples are $NaSbO_3$, structurally similar to ilmenite, $FeTiO_3$, and $FeSbO_4$, with a rutile structure. Obviously none of these compounds is an antimonate.

21.6. Sulphides

As with the oxides, there is a transition from molecular to ionic structure in passing from arsenic sulphide to bismuth sulphide. Reaction of the elements gives a vapour containing As_4S_4 molecules with a 'cradle' structure and As—S = 233 pm, As—As = 249 pm; ∠ As—S—As = 101°, ∠ S—As—S = 93° (Fig. 21.5), and As_4S_6 molecules with the P_4O_6 structure and As—S = 225 pm; ∠ As—S—As = 100°, ∠ S—As—S = 114°. The compound is precipitated by passing H_2S into the acidified solution of an arsenite; it dissolves in aqueous alkali-metal sulphides to give a solution from which thioarsenites, for instance Na_3AsS_3, may be crystallised:

$$As_2S_3 + 3\,S^{2-} \rightarrow 2\,AsS_3{}^{3-}$$

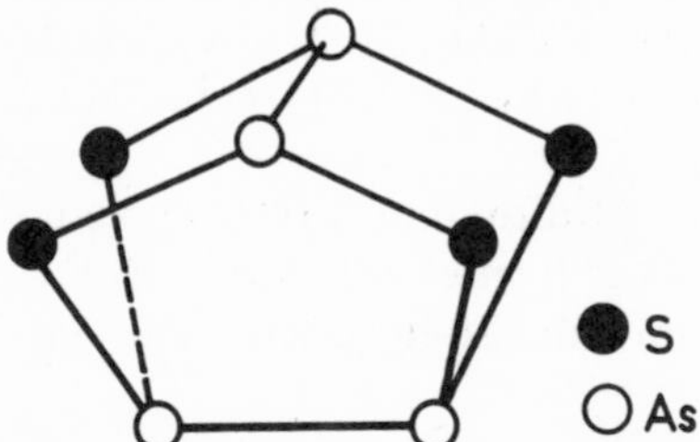

Fig. 21.5. Structure of As_4S_4 molecule.

Thioarsenates, such as $Na_3AsS_4 \cdot 8\,H_2O$ and $(NH_4)_3AsS_4$, are obtained from solutions made by the action of alkali-metal polysulphides on As_2S_3. Acidification of a thioarsenate solution gives a yellow precipitate containing a sulphide of arsenic, but there is no structural evidence that it is As_2S_5. A yellow solid of this composition is obtained by fusing As with S, extracting the cold mixture with aqueous NH_3 and then precipitating with HCl at 273 K *or* by passing H_2S *quickly* through a *strongly* acidified arsenate solution. Although these methods give material which analyses as As_2S_5, yet again there is no structural evidence for such a compound.

Antimony trisulphide occurs as the mineral stibnite. An orange form is precipitated when H_2S is passed into a solution of $SbCl_3$ in hydrochloric acid. When heated at 470 K in CO_2, this changes to a dark-grey, rhombic modification of greater density. Both forms dissolve in solutions containing sulphide ion to give thioantimonites; and, though they are soluble in polysulphide solutions, crystalline thioantimonates cannot be obtained from the solutions. The existence of an antimony pentasulphide is doubtful.

The commercial product used in rubber vulcanisation is made by boiling Sb_2S_3 with S in aqueous NaOH and precipitating the solution with HCl. Its composition is variable, it always contains free sulphur and there is no structural evidence for a compound of the composition Sb_2S_5.

Bi_2S_3 is formed as grey, rhombic crystals when Bi is fused with sulphur, and as a dark brown precipitate when H_2S is passed into the solution of a bismuth salt. It is insoluble in solutions containing sulphide ion. The compounds $KBiS_2$ and $NaBiS_2$ are made by fusing Bi_2S_3 with the appropriate Group IA sulphide; they oxidise rapidly in air.

21.7. Oxoacid Salts of Bismuth

Bismuth, alone of the elements of the sub-group, forms stable salts with oxo-anions. The deliquescent nitrate, $Bi(NO_3)_3 \cdot 5\ H_2O$, can be crystallised from a solution of the metal or Bi_2O_3 in 20% HNO_3. It hydrolyses successively to $BiO(NO_3)$ and $BiO(OH) \cdot BiO(NO_3)$, the latter being stable in boiling water. There is no evidence of a significant quantity of $Bi(OH)^{2+}$ ion in solutions of these basic nitrates in $M/2$ to M HNO_3.

Bismuth sulphate, $Bi_2(SO_4)_3$, a white solid produced by evaporating a solution of the metal in concentrated H_2SO_4, gives on hydrolysis an insoluble hydroxide sulphate, $Bi(OH)_3 \cdot Bi(OH)SO_4$, which heating converts to the yellow bismuthyl sulphate $(BiO)_2SO_4$.

The addition of an ammonium carbonate solution to a bismuth nitrate solution precipitates the oxide carbonate, $[(BiO)_2CO_3]_2 \cdot H_2O$, which readily loses water and CO_2 on heating.

21.8. Organometallic Compounds

21.8.1. Arsenic

Organoarsenic compounds are numerous: research on them was stimulated by the early discovery that some of them were trypanocides. The trialkylarsines, R_3As, are colourless liquids made by treating arsenic trihalides with dialkyls of zinc or with Grignard reagents:

$$AsX_3 + 3\ RMgX \rightarrow R_3As + 3\ MgX_2$$

They oxidise rapidly in air and react with anhydrous hydrogen halides to form salts which are composed by water. But with alkyl halides they give tetra-alkylarsonium salts which dissolve in water to give neutral solutions:

$$R_3As + RX \rightarrow R_4As^+X^-$$

The trialkylarsines act as electron donors in forming co-ordination compounds such as $(R_3As)_2PdCl_2$. The triarylarsines resemble the alkyl analogues.

Monoalkyl and dialkyl arsines are readily prepared. When sodium arsenite is treated with an alkyl halide a primary arsonic acid is formed. $RAsO(OH)_2$, which can be reduced to $RAsH_2$ by zinc and hydrochloric acid. Reduction of a secondary arsonic acid, R_2AsOOH, gives R_2AsH. These compounds liberate hydrogen from hydrogen halides:

$$R_2AsH + HCl \rightarrow R_2AsCl + H_2$$

The products can react with the parent compounds to give diarsine derivatives:

$$R_2AsCl + R_2AsH \rightarrow R_2As{-}AsR_2 + HCl$$

Large numbers of complexes of tertiary arsines with transition-metal compounds have been isolated.

21.8.2. Antimony

The trialkyls of antimony are colourless liquids, soluble in organic solvents but not in water. The lower ones are oxidised spontaneously in air. They are made by treating $SbCl_3$ with zinc alkyls or Grignard reagents. They form adducts with halogens, sulphur and selenium:

$$R_3Sb + S \rightarrow R_3SbS$$

There are also monoalkyl and dialkyl stibines and corresponding aryls.

21.8.3. Bismuth

Bismuth does not form monoalkyl or dialkyl derivatives, but rather unstable trialkyls can be made from the action of a Grignard reagent on bismuth trihalides. They are spontaneously inflammable and decompose above 425 K, but they are stable to water, in which they are insoluble.

Further Reading

J.D. Smith, The chemistry of arsenic, antimony and bismuth, Pergamon, Oxford, 1975.

G.O. Doak and L.D. Freedman, Organometallic compounds of arsenic, antimony and bismuth, Wiley, New York, 1970.

Oxygen, Sulphur, Selenium, Tellurium and Polonium — Group VIB

22

22.1. Introduction

As in the preceding group, Group VB, the elements of Group VIB show a gradual transition to more metallic character with increasing atomic number. Oxygen resembles nitrogen in being the only member of its group to be a gas at room temperature. The small atomic radius of oxygen probably plays a major part in determining its chemical behaviour, by contrast with fluorine where the chief factor is the low dissociation energy (155 kJ mol^{-1}) of the molecule. For the oxygen molecule this is 492 kJ mol^{-1}. The elements of Group VIB have six electrons in their ns^2np^4 valence shells and all show bivalence in compounds in which they have two covalent links and two lone pairs. But in the elements below oxygen d hybridisation commonly occurs, giving complexes in which the atom is usually shown with the charge numbers +4 or +6.

In general the chemistry of oxygen is not cationic but even with the high ionisation energies for oxygen, compounds with polyatomic anions are known in which oxygen is formally cationic; an example is the O_2^+ ion in $O_2^+PtF_6^-$.

The first ionisation energy of oxygen is high. The electron affinity $O + 2\,e \rightarrow O^{2-}$, $A = -752$ kJ mol^{-1}, has a surprisingly large negative value; in other words, O^- resists the introduction of the second electron needed to form the O^{2-} ion. Yet ionic oxides are common, owing to the large lattice energy of the crystals. The electronegativity ascribed to oxygen is extremely high, being exceeded only by that of fluorine. This accords with oxides having a more ionic character than the corresponding sulphides.

22.2. Properties of the Elements

22.2.1. General

The gradation of physical properties follows a similar pattern to that of the preceding group and is similarly related to the atomic structure. Gaseous oxygen consists of diatomic molecules, sulphur of 'puckered' rings of 8 atoms in the rhombic, and probably also in the monoclinic crystalline form. The

TABLE 22.1

ATOMIC PROPERTIES OF GROUP VIB ELEMENTS

	O	S	Se	Te	Po
Atomic number	8	16	34	52	84
Electronic configuration	$[He]2s^2 2p^4$	$[Ne]3s^2 3p^4$	$[Ar]3d^{10} 4s^2 4p^4$	$[Kr]4d^{10} 5s^2 5p^4$	$[Xe]4f^{14} 5d^{10} 6s^2 6p^4$
Covalent radius/pm	73	104	117	137	164
Ionic radius M^{2-}/pm	140	184	198	221	(230)
Ionic radius M^{4+}/pm				89	102
$I(1)$/kJ mol^{-1}	1410	1070	1005	930	870
$I(2)$/kJ mol^{-1}	3640				
Electron affinity/kJ mol^{-1} ($A(1) + A(2)$)	−752	−353	−435		
Electronegativity (Allred—Rochow)	3.50	2.44	2.48	2.01	1.76

TABLE 22.2

SOME PROPERTIES OF GROUP VIB ELEMENTS

	O	S	Se (grey)	Te	Po
Density of solid/g cm^{-3}	1.300	2.060 (α)	4.820	6.250	9.196 (α)
Melting point/K	54.3	387.6	493.6	722.9	527
Boiling point/K	90.1	717.7	958	1663	1235

red allotrope of selenium also contains 8-membered rings but the usual grey form which is rather metallic-looking has zig-zag chains of atoms. Tellurium (Fig. 22.1) is rather similar to, but more metallic than, grey Se. It is monotropic in contrast to S, Se and Po. Polonium is dimorphic; the low-temperature α-form has a simple cubic lattice and the high-temperature β-form a simple rhombohedral one, the transition temperature being about 309 K. The density is similar to that of bismuth. Polonium has a low *m.p.*, resembling that of Bi, but its *b.p.* is more in accord with a Group VI element.

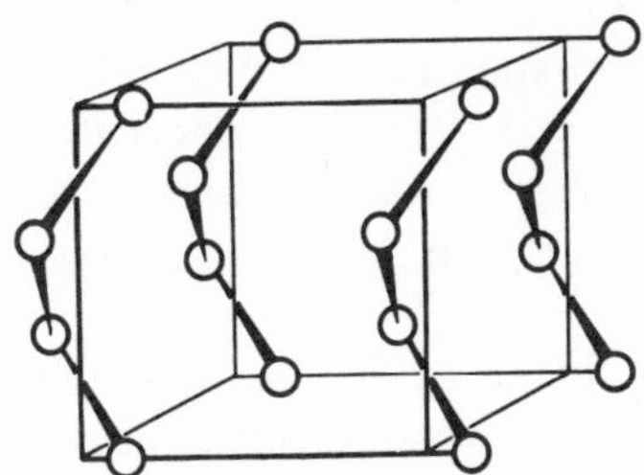

Fig. 22.1. Unit cell of tellurium, showing spiral arrangement of atoms.

Chemically, the elements of Group VIB range from the very electronegative, non-metallic oxygen to the decidedly metallic polonium. Reactivity towards metals and hydrogen decreases down the group and becomes slight after selenium.

Otherwise the middle elements, sulphur, selenium and tellurium show considerable chemical similarity. They burn in air to give SO_2, SeO_2 and TeO_2. They all react energetically with fluorine to give hexafluorides, XF_6, and with chlorine to give tetrachlorides, XCl_4. The three elements are attacked by hot nitric and sulphuric acids but not by non-oxidising acids such as HCl and HF. Sulphur and selenium dissolve in aqueous solutions of alkali sulphides and selenides to form the respective polysulphides and polyselenides. The elements react with most metals, on heating, to give binary compounds:

$$Fe + S \rightarrow FeS$$
$$Cu + S \rightarrow CuS$$

22.2.2. Oxygen

The lithosphere contains about 47% oxygen by weight and the oceans about 89%. The atmosphere has 20.95% oxygen by volume. Taking hydrosphere and lithosphere together it constitutes nearly 50% by weight, almost twice the abundance of silicon and seven times that of aluminium. The element is a large-scale industrial product, stored and transported mainly as a liquid, and obtained by the fractional distillation of liquid air.

Ordinary oxygen contains, in addition to the most abundant isotope oxygen-16, about 0.2 vol. % ^{18}O and 0.04 vol. % ^{17}O. Water can be enriched in ^{18}O by fractional distillation. This isotope has been used as a non-radioactive tracer to show, for example, that the oxygen liberated when H_2O_2 is oxidised by Ce^{4+} or MnO_4^- ions comes wholly from the hydrogen peroxide.

The oxygen molecule is paramagnetic. It contains two more electrons than N_2, and these, according to molecular orbital theory (4.1), are in the antibonding orbitals:

$$2\,O(1s^2\,2s^2\,sp^4) \rightarrow O_2[KK(z\sigma)^2(y\sigma)^2(x\sigma)^2(w\pi)^4(v\pi)^2]$$

As there are two such orbitals, they are singly occupied by these electrons which have parallel spin. The pale-blue colour of O_2 in the gas phase is consistent with the presence of unpaired electrons as in free radicals.

At room temperature and pressure, oxygen shows spectroscopic evidence of the diamagnetic species O_4; the equilibrium concentration of O_4 decreases as the temperature is raised. The bond between the two O_2 molecules is weaker than an electron-pair bond but stronger than a Van der Waals attraction.

The solubility coefficient of oxygen in water is 0.029 at 293 K. There is optical evidence for the hydrate $O_2 \cdot H_2O$ which is believed to have the structure:

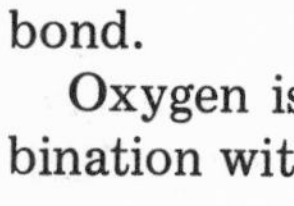

96 pm
177 pm
121 pm

The hydrogen bonds differ from those between water molecules in that they have a proton sharing three electrons instead of four, one from an oxygen atom in the biradical form of the oxygen molecule and two from the O—H bond.

Oxygen is decidedly reactive, particularly at higher temperatures its combination with other elements being often strongly exothermal:

$$2\,Ca + O_2 \rightarrow 2\,CaO \qquad \Delta H = -1270\ kJ\ mol^{-1}$$
$$4\,Al + 3\,O_2 \rightarrow 2\,Al_2O_3 \qquad \Delta H = -3360\ kJ\ mol^{-1}$$
$$Si + O_2 \rightarrow SiO_2 \qquad \Delta H = -\ 885\ kJ\ mol^{-1}$$
$$4\,P + 5\,O_2 \rightarrow P_4O_{10} \qquad \Delta H = -3020\ kJ\ mol^{-1}$$

It does not react readily with nitrogen or directly with the halogens.

Atomic oxygen is produced by passing an electric discharge through the gas at 130 Pa pressure, by radio-frequency excitation at higher pressures and by irradiating the gas with ultraviolet light of wavelength less than 190 nm. The atoms constitute a very strong oxidising agent:

$$O + 2\,H^+ + 2\,e \rightarrow H_2O \quad E^0 = +2.2\ V$$

22.2.3. Ozone

Ozone is made by passing a silent discharge through oxygen or by electrolysing a strong aqueous perchloric acid at 220 K between a lead cathode and a platinum anode, when anodic oxidation of the water occurs. Gaseous ozone is deeper blue than oxygen; it condenses at 161 K to a dark blue liquid which freezes at 80 K to a dark purple solid. Surprisingly, the liquid is not completely miscible with liquid oxygen.

Microwave and electron diffraction studies show the molecule to be angular with C_{2v} symmetry.

O
127.8 pm
116.8°
O O

It may be considered to have two σ bonds and a delocalised π orbital stretching over the three atoms. Every atom may be regarded as roughly sp^2 hybridised, and the end atoms to have two lone pairs and the central atom one.

Ozone is one of the strongest oxidising agents. In acid solution:

$$O_3 + 2\,H^+ + 2\,e \rightarrow O_2 + H_2O \qquad E^0 = +2.07\ V$$

Only fluorine, atomic oxygen, and F_2O have higher redox potentials. Ozone oxidises moist sulphur to H_2SO_4, raises silver(I) compounds to the +2 state, and converts olefinic compounds to ozonides. The reaction $2\,O_3 \rightarrow 3\,O_2$, which is catalysed by many metals and metal oxides, is exothermic and rapid above 470 K.

Ozone is an important constituent of the upper atmosphere, and current interest in its chemistry is related to the problems associated with atmospheric pollution.

22.2.4. Sulphur

Sulphur (0.052% of the lithosphere) occurs mainly as the element and in sulphides and sulphates.

The element has several allotropes. The form stable below 370 K is α-sulphur, whose rhombic crystals are built up from 8-membered, puckered rings of S atoms with S—S bond lengths of 212 pm and bond angles of 105.4° (Fig. 22.2). They are packed into 'crankshafts' stacked in crossed layers.

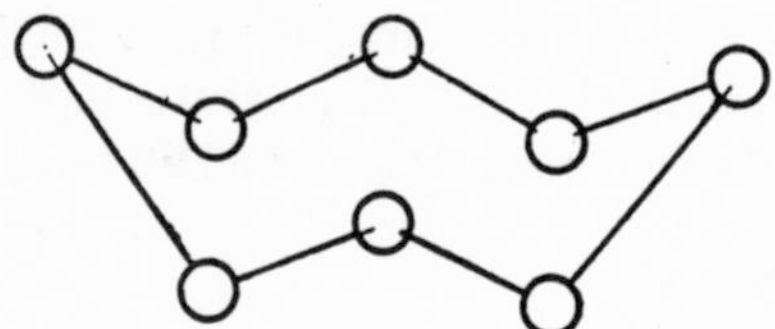

Fig. 22.2. 8-membered-ring of sulphur atoms.

Between 370 and 395 K β-sulphur is stable. Its monoclinic crystals also contain 8-membered rings. The two modifications (with a range of stability and a definite transition temperature) are the best-known example of enantiotropic allotropy:

$$S_\alpha \underset{370\text{ K}}{\rightleftharpoons} S_\beta$$

Conversion from S_α to S_β is accompanied by a small evolution of heat (0.4 kJ per g. atom) and a slight increase in volume. Although β-sulphur has a different crystal form, it is probably not very different in structure from α-sulphur.

However, there is a second monoclinic form, γ-sulphur, which separates as needle-like crystals from certain solvents and also from melts cooled so as to avoid nucleation by α-sulphur. This has also 8-membered rings stacked in 'sheared-penny rolls' which give a close-packed hexagonal arrangement in two dimensions.

Rhombohedral, or ρ-sulphur, like the α-, β- and γ-forms, has properties typical of a solid composed of small, covalent molecules. It is made by pouring an aqueous solution of $Na_2S_2O_3$ into concentrated HCl at 273 K and extracting the mixture with toluene; from the toluene solution solid S_ρ can be crystallised. Although this form is metastable with respect to S_α, the transformation is slow. X-ray analysis shows S_ρ to contain 6-membered puckered rings with the chair configuration (Fig. 22.3).

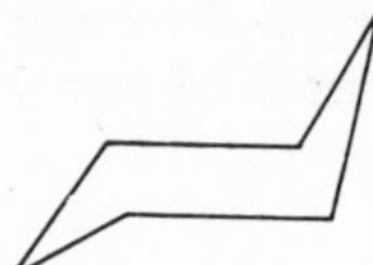

Fig. 22.3. Chair configuration of S_6 rings in ρ-sulphur.

Ordinary sulphur, S_α, melts at 388 K to a pale yellow liquid, S_λ, of low viscosity; it consists of S_8 rings, a structure which is maintained up to 433 K. Above this temperature long spiral chains of μ-sulphur are formed and increase in proportion up to the *b.p.* (717.6 K). Above 433 K there is also present a third form, π-sulphur, probably S_6 and a precursor of S_μ; it increases to a maximum at about 453 K. At this temperature the viscosity is also at a maximum and the liquid is almost unpourable, probably because of cross-linking between the chains.

When molten sulphur above 433 K is poured into water, plastic sulphur is formed. Much of it consists of long spiral chains of S_μ. On standing it slowly hardens, becoming a mixture of S_μ and solid S_α, the latter, although amorphous, containing S_8 groups.

Paramagnetic forms have been made (cf. oxygen above). When sulphur vapour at 770—970 K and 130 Pa, consisting mainly of S_2 molecules, is passed over a surface at 77 K, a purple paramagnetic solid is obtained which may contain S_2 molecules. At 193 K it reverts to S_α.

Gaseous sulphur has been shown by vapour density measurements to contain S_8, S_6, S_4 and S_2, the relative proportions depending on the temperature; the last is paramagnetic, like O_2.

22.2.5. Selenium

This element (9×10^{-6}% of the earth's crust) occurs in small quantities in sulphide ores, particularly FeS_2. It is extracted from the flue-dusts produced in the roasting of sulphide ores and from the 'lead-chamber mud' formed in sulphuric acid manufacture, as a solution in aqueous KCN. From the filtered solution it is precipitated by the addition of HCl.

$$KCN + Se \rightarrow KCNSe$$
$$KCNSe + HCl \rightarrow Se + HCN + KCl$$

Selenium is used for decolorising glass and in photoelectric cells. Its electrical conductance in the metallic form is increased as much as 200 times by light. Another application is in the iron-selenium barrier-layer cell which generates a current when illuminated. This system is extensively used in rectifiers because current flows more readily from iron to selenium than in the opposite direction when an external alternating potential is applied.

The selenium precipitated from KCNSe by acid and from H_2SeO_3 by SO_2 is red and amorphous. It dissolves in CS_2 and slow evaporation of the solution below 545 K gives red, α-monoclinic Se, whereas rapid evaporation gives red β-monoclinic Se. Se_α contains puckered Se_8 rings with Se—Se—Se bonds of 234 pm and angles of 105.5°. Hexagonal crystals of grey 'metallic' Se, the stable allotrope at room temperature, are made by keeping the other forms at 470—490 K for a time. Present evidence suggests that liquid Se has only one form. Several compounds, with the formula Se_nS_{8-n} and analogous to S_8 and Se_8, have been isolated from fused sulphur—selenium mixtures, but only one, TeS_7, from fused sulphur—tellurium mixtures.

22.2.6. Tellurium

The element (2×10^{-7}% of the lithosphere) occurs in sulphide ores, particularly those of copper, and as the tellurides of silver and gold. Its source is the anode sludge from the electrolytic refining of copper. The sludge is treated with fuming sulphuric acid and the tellurium precipitated from the diluted solution with zinc. The element is added to lead to improve resistance to heat, mechanical shock and corrosion. The stable metallic form consists of hexagonal grey crystals with low electrical conductance little affected by light. Tellurium and grey selenium form a continuous range of solid solutions in which are chains of randomly arranged Se and Te atoms. A black, amorphous form is precipitated from telluric acid by SO_2.

22.2.7. Polonium

The metal, known in trace quantities since 1898, is now made as the polonium-210 isotope in milligram quantities by the neutron irradiation of ^{209}Bi.

$$^{209}_{83}Bi(n,\gamma)^{210}_{83}Bi \xrightarrow[5\ \text{days}]{\beta^-} {}^{210}_{84}Po(\alpha, 138.4\ \text{days})$$

The isotope is virtually a pure α-emitter, but its high specific activity, 4.5 curies per mg, makes it a dangerous material, the maximum permissible body burden for ingested ^{210}Po being only 4×10^{-12} g. (The 100-year polonium-209 would be preferred for chemical work but its production is expensive.) Polonium-210 is separated from the irradiated Bi by electrochemical replacement with Ag from solution in 0.5—2 M HCl at 340—350 K: E^0, Po^{4+}/Po = +0.75 V. The elements Au, Hg, Pt and Te must first be removed by reduction with hydrazine. The polonium is sublimed from the surface of the silver in a vacuum; if the separation is delayed, sublimation becomes more diffi-

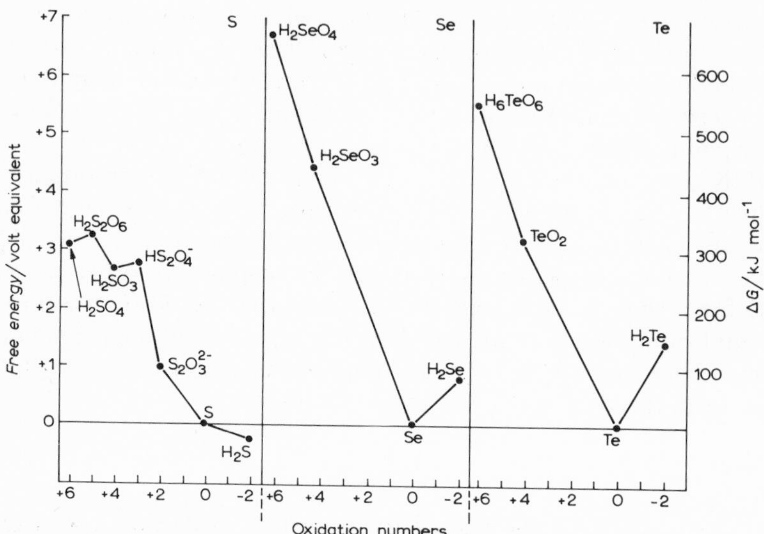

Fig. 22.4. Oxidation states of sulphur, selenium and tellurium. Free energies relative to the elements at pH = 0.

cult, possibly through compound formation. The metal is obtained from solution by precipitation as PoS followed by the decomposition of the sulphide in a vacuum at 800 K, when the metal sublimes leaving a residue of decay lead sulphide. Polonium intimately mixed with beryllium forms a useful, weak source of neutrons.

22.3. Oxidation States of Sulphur, Selenium and Tellurium

The most common oxidation states in Group VI are −2, +4 and +6, represented respectively by the compounds H_2Se, H_2SeO_3 and H_2SeO_4. The free energies of some of the oxidation states, relative to the element, in aqueous solution at pH = 0, are illustrated in Fig. 22.4.

Features to be noted are (i) the extremely unstable character of Se^{VI}, and therefore the strong oxidising power of the element in this state, relative to its lower oxidation states, (ii) the stability of compounds of the +4 state towards disproportionation for the three elements, (iii) the instability of the hydrides of selenium and tellurium, and (iv) the instability of the S^{V} and S^{III} states towards disproportionation.

22.4. Stereochemistry

22.4.1. Oxygen

The oxygen atom, with a $1s^22s^22p_x^22p_y^12p_z^1$ electronic configuration, forms two bonds in nearly all its compounds; the next orbital (3s) lies too high in energy for the promotion of an electron to it and the consequent increase in valency to be feasible. Most of the simple inorganic covalent oxides and the aliphatic ethers seem to owe their structure to approximately sp^3 hybrid orbitals, two of which are occupied by lone pairs. Examples are:

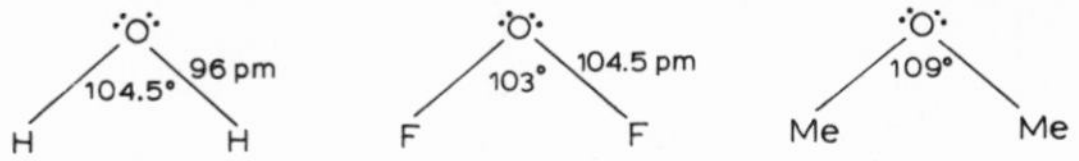

In ozone and in aromatic ethers, trigonal sp^2 hybridisation is involved, the angle at the central oxygen being approximately 120°. The molecule $(SiH_3)_2O$ is linear, however, suggesting that sp hybridisation can occur at the oxygen.

The ion O^+ is isoelectronic with the atom N. It forms three covalent bonds in H_3O^+; these are arranged similarly to the N—H bonds in ammonia and are due basically to sp^3 hybridisation, with one position occupied by a lone pair. In the ion,

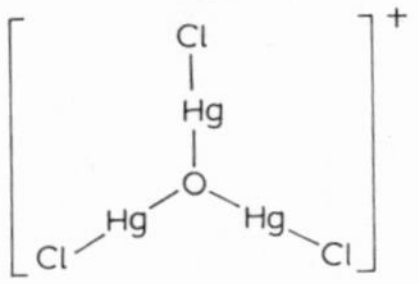

TABLE 22.3

STEREOCHEMISTRY OF S, Se, Te

σ pairs	π pairs	Lone pairs	Hybrid	Shape	Point group	Example
2	2	1	sp^2	V-shaped	C_{2v}	* SO_2
3	3	0	sp^2	Triangular	D_{3h}	* SO_3
4	2	0	sp^3	Tetrahedral	C_{2v}, T_d	SO_2Cl_2, $[SeO_4]^{2-}$
3	1	1	sp^3	Trigonal pyramidal	C_{2v}	$SOCl_2$, $SeOF_2$
2	0	2	sp^3	V-shaped	C_{2v}	H_2S, Cl_2S
4	0	1	sp^3d	Distorted tetrahedral	C_{2v}	$TeCl_4$
6	0	0	sp^3d^2	Octahedral	O_h	SF_6, SeF_6, $Te(OH)_6$
5	0	1	sp^3d^2	Square pyramidal	C_{4v}	$[MeTeI_4]^-$

* In SO_2 and SO_3 there are also π lone-pair and non-localised bonds.

however, the O—Hg—Cl groups are nearly linear and the whole structure is approximately planar. There is no unambiguous example of 4-covalence in oxygen, those usually quoted being ice and basic beryllium nitrate.

22.4.2. Sulphur, selenium and tellurium

The stereochemistry of these elements is tabulated in Table 22.3. Sulphur, when it is two-bonded, differs from oxygen in having a valency angle less than the tetrahedral angle, 109.5°, except in SO_2. This lesser angle persists when aromatic groups are attached; for instance in $(p\text{-}CH_3C_6H_4)_2S$ the C—S—C angle is 109°. The small bond angle in H_2S (92.1°) suggests pure p orbitals, but nuclear quadrupole coupling constants indicate there are both s and d contributions to the bonding orbitals.

22.4.3. Polonium

The crystal structures of several polonium compounds have been determined. Both forms of PoO_2 have ionic lattices; ZnPo has a zinc-blende, and PbPo a rock-salt structure. In $PoBr_4$, the Po is octahedrally co-ordinated with Br. Ammonium hexachloropolonate(IV), $(NH_4)_2PoCl_6$, is isomorphous with $(NH_4)_2PtCl_6$: the Po—Cl distance, 238 pm, suggests a basically covalent bond.

22.5. Halides

Although the compounds of oxygen and fluorine should be considered here as oxygen fluorides, all halogen—oxygen compounds will be discussed subsequently in 25.7.

TABLE 22.4

HALIDES OF GROUP VIB ELEMENTS

	O	S	Se	Te	Po
F	F_2O_2, F_2O	S_2F_2, SF_4, S_2F_{10}, SF_6	SeF_4 SeF_6	TeF_4 Te_2F_{10}, TeF_6	
Cl	Cl_2O, ClO_2 Cl_2O_6, Cl_2O_7	S_xCl_2, S_2Cl_2, SCl_2, SCl_4	Se_2Cl_2, $SeCl_4$	$TeCl_2$ $TeCl_4$	$PoCl_2$ $PoCl_4$
Br	Br_2O, BrO_2, BrO_3	S_2Br_2.	Se_2Br_2, $SeBr_4$	$TeBr_2$ $TeBr_4$	$PoBr_2$ $PoBr_4$
I	I_2O_4, I_4O_9 I_2O_5			TeI_4	PoI_4

22.5.1. Hexahalides

Sulphur, selenium and tellurium all form the hexafluoride by direct combination; other hexahalides are unknown. SF_6 and SeF_6 are chemically inert, colourless gases. The former is non-toxic and at once the most inert sulphur compound and possibly the most inert non-ionic fluorine compound; it resists attack by fused KOH. The somewhat more reactive SeF_6 is reduced by NH_3 above 450 K. TeF_6, also a colourless gas, is even more reactive, being hydrolysed by water:

$$TeF_6 + 6\ H_2O \rightarrow 6\ HF + H_6TeO_6$$

This difference in reactivity arises from the higher maximum covalence possible with tellurium. The hexafluoride molecules are octahedral, with six sp^3d^2 hybrid orbitals, and the low *b.p.* of the compounds are ascribable to the non-polarisable F atoms sheathing the molecules.

In addition to SF_6, some S_2F_{10} is obtained during the direct fluorination of sulphur. It is a liquid, *b.p.* 302 K, and is like SF_6 in being unreactive to alkalis. Unlike SF_6, however, it is highly toxic. It reacts with Cl_2 at 570 K to give $ClSF_5$, a colourless gas which differs from SF_6 and S_2F_{10} in being rapidly hydrolysed by alkalis. The S—S bond in S_2F_{10} is about 6% longer than the usual single-bond distance and the S—F distances are rather short.

22.5.2. Tetrahalides

Sulphur tetrafluoride, SF_4, *b.p.* 233 K, which can be made by the direct fluorination of sulphur at 200 K, is more conveniently prepared by the action of sulphur dichloride on a suspension of sodium fluoride in acetonitrile or tetramethylene sulphone at 350 K, a medium offering some chance of ionisation of the NaF is necessary.

$$3\ SCl_2 + 4\ NaF \rightarrow SF_4 + S_2Cl_2 + 4\ NaCl$$

The structure of SF_4 has been deduced from its infrared, Raman, and ^{19}F n.m.r. spectra. At 180 K the fluorine resonance spectrum has two resonance triplets of equal intensity (Fig. 22.5).

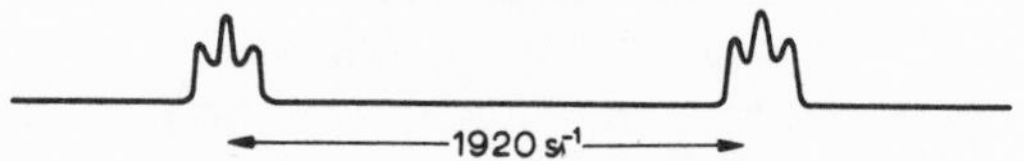

Fig. 22.5. Pair of resonance triplets in ^{19}F n.m.r. spectrum of SF_4 at 180 K.

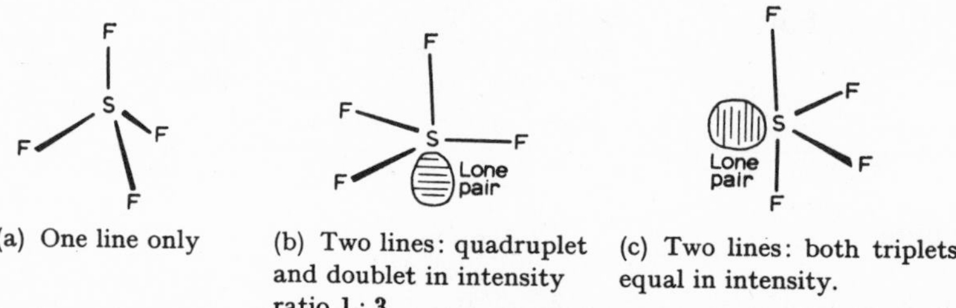

(a) One line only

(b) Two lines: quadruplet and doublet in intensity ratio 1 : 3

(c) Two lines: both triplets, equal in intensity.

Fig. 22.6. Types of n.m.r. spectra consistent with three hypothetical structures.

Accordingly, the molecule must contain two pairs of equivalent fluorine nuclei. The spectrum could not arise from a regular tetrahedral structure (T_d) (a), or a triangular pyramidal structure (C_{3v}) (b), but is consistent with the trigonal bipyramidal structure (C_{2v}) (c) with one equatorial position occupied by a lone pair (Fig. 22.6). This is just the structure predicted by the Sidgwick—Powell rule.

Sulphur tetrafluoride is a useful fluorinating agent. It converts mixtures of alkali-metal fluorides and metal oxides into complex fluorides:

$$NaF + TiO_2 \xrightarrow{SF_4} Na_2TiF_6$$

$$NaF + Sb_2O_3 \xrightarrow{SF_4} NaSbF_6$$

$$NaF + UO_2 \xrightarrow{SF_4} Na_2UF_6$$

Selenium tetrafluoride is conveniently made by passing diluted fluorine over a selenium surface at 273 K. The liquid, *b.p.* 379 K, gives with NaF, KF, RbF and CsF the complex fluorides M^ISeF_5, unlike the other halogeno complexes of selenium which are $M^I_2SeF_6$. Both SF_4 and SeF_4 form solid 1 : 1 addition compounds with BF_3, AsF_5 and SbF_5. Pure, crystalline TeF_4 is obtained by the reaction of TeO_2 with SeF_4 at 350 K followed by evaporation of the excess of reagent and the $SeOF_2$.

$$TeO_2 + 2\ SeF_4 \rightarrow TeF_4 + 2\ SeOF_2$$

Tetrachlorides of S, Se, Te and Po are formed by direct chlorination. SCl_4 is a yellow liquid, stable only at low temperatures; $SeCl_4$ is a colourless, crystalline solid subliming at 469 K, and $TeCl_4$ a white, hygroscopic solid, *m.p.* 497 K. Fused $TeCl_4$ is an even better conductor of electricity than BrF_3 probably because

$$2\ TeCl_4 \rightleftharpoons TeCl_3^+ + TeCl_5^-$$

It gives with HCl the acid H_2TeCl_6; the salt K_2TeCl_6 is isomorphous with K_2SiF_6 and K_2SnCl_6. The anion $TeCl_6^{2-}$ is a regular octahedron. The tetrachlorides of selenium and tellurium have the same structure as that of SF_4 shown in Fig. 22.6(c).

Yellow $PoCl_4$ combines with NH_4Cl to give a compound which structural evidence suggests to be $(NH_4)_2PoCl_6$. A series of compounds M_2PoX_6 where M = NH_4 or Cs and X = Cl, Br or I have been made, and X-ray diffraction indicates that they are iso-structural with one another and with their tellurium analogues.

Selenium, tellurium and polonium form tetrabromides. $SeBr_4$, a yellow solid, loses Br_2 to form Se_2Br_2 even at room temperature, and hydrolyses to a clear solution of H_2SeO_3 and HBr. Orange-red crystals of $TeBr_4$, along with green $TeBr_2$, result when tellurium reacts with bromine. Polonium forms carmine-red $PoBr_4$ at 470 K. A salmon-pink $PoBr_2Cl_2$ is also known. The diffraction pattern of $PoBr_4$ indicates a face-centred cubic cell of edge 560 pm; the intensities suggest this contains only one $PoBr_4$ unit with the Po atoms randomly distributed over the sites normally occupied by four cations. The Po atom is surrounded octahedrally by six Br atoms at a Po—Br distance of 280 pm; this lies between the covalent octahedral distance (260 pm) and the theoretical ionic distance (297 pm).

Neither sulphur nor selenium forms an iodide; Te and Po form only tetra-iodides. TeI_4 consists of iron-grey crystals, made either by direct union or by reaction in aqueous solution:

$$TeO_2 + 4\ HI \rightarrow TeI_4 + 2\ H_2O$$

PoI_4 is a black solid.

22.5.3. Dihalides

The best-characterised dihalides are SCl_2, a red liquid made by saturating S_2Cl_2 with chlorine at room temperature, $TeCl_2$, a black solid made by the action of CCl_2F_2 on tellurium at 775 K and $TeBr_2$, already mentioned. In SCl_2 the bond angle is about 103°, in $TeBr_2$ 98°, suggesting considerable p-character in the bonding orbitals of the centre atom. $PoCl_2$ has been made by reducing $PoCl_4$ with SO_2.

22.5.4. Lower halides

Disulphur difluoride can be made by heating sulphur with a mixture of HgF and CaF_2 or by passing a mixture of SO_2 and S_2Cl_2 over heated KF. There is evidence for two isomeric forms, S=SF_2 and FS—SF.

Sulphur and selenium form dimeric chlorides and bromides. S_2Cl_2, made by passing dry Cl_2 over molten sulphur, is an amber liquid whose vapour dissociates a little and becomes red. Hydrolysis by water is slow; HCl, sulphur, SO_2 and oxyacids, principally $H_2S_5O_6$, are formed. Electron diffraction on the vapour indicates a non-planar structure analogous to that of hydrogen peroxide (24.1):

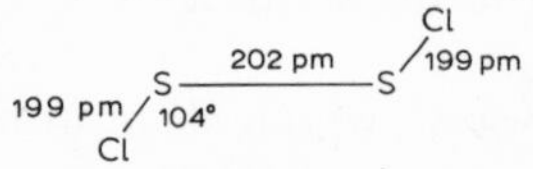

Cl
202 pm
199 pm
S——S
199 pm
104°
Cl

The S—S bond is appreciably shorter than in the S_8 ring (208 pm). Se_2Cl_2 has a similar structure. It is made by the reaction between Se and $SeCl_4$, into which it disproportionates on warming:

$$2\ Se_2Cl_2 \rightarrow 3\ Se + SeCl_4$$

The brown, oily liquid hydrolyses readily to H_2SeO_3, Se and HCl. The bromides S_2Br_2 and Se_2Br_2, both red liquids, are products of direct combination of the elements.

Lower chlorides of sulphur, S_xCl_2, have been made with chains of up to five S atoms by reducing S_2Cl_2 vapour with hydrogen at a hot surface and passing the products over a cold surface.

22.6. Oxides

22.6.1. Monoxides

Disulphur monoxide, S_2O, is made by subjecting a mixture of S and SO_2 at 420—470 K to an electric discharge, or by heating certain heavy metal oxides with sulphur vapour at a low pressure. It has the structure SSO with S—S 188 pm, S—O 146 pm, and angle SSO 118°. It has many of the properties of a free radical, can exist for several days at low pressure, and reacts immediately with Hg, Fe, and Cu. Alcoholic KOH converts it to K_2S, $K_2S_2O_4$ and K_2SO_3.

Although SO and SeO are unknown, the black oxide, TeO, is obtained when $TeSO_3$ is heated. When heated, $PoSO_3$ also produces a black oxide believed to be PoO.

Powdered sulphur reacts vigorously with liquid SO_3 forming a blue-green solid which decomposes slowly at room temperature, rapidly on warming, to give S, SO_2 and SO_3. Se and Te also dissolve in SO_3 producing solutions of green $SeSO_3$ and red $TeSO_3$.

TABLE 22.5
OXIDES OF GROUP VIB ELEMENTS

S	Se	Te	Po
S_2O			
		TeO	PoO
S_2O_3			
SO_2	SeO_2	TeO_2	PoO_2
SO_3	SeO_3	TeO_3	
SO_4			

22.6.2. Dioxides

The four elements of the sulphur family form dioxides; these are structurally very different from one another. The SO_2 molecule is angular (119.5°), owing its shape to sp^2 hybrid orbitals about the sulphur, one occupied by a lone pair, but there is a little d-orbital contribution, too. The solid SeO_2, made by burning Se in air, is colourless and crystalline, sublimes at 588 K and, under pressure, melts to an orange liquid; the colour is lost on cooling. X-ray analysis shows it to contain macromolecular chains with oxygen atoms projecting alternately on opposite sides and at 90° to the plane of the chain.

The links to the shared oxygen and to the projecting oxygen both possess some double-bond character. SeO_2 was used in organic chemistry to oxidise aldehydes and ketones containing the —CH_2—CO— system to —CO—CO—, the SeO_2 being reduced to Se. The fact that only one —CH_2— group is affected suggests that a complex is formed, but none has been isolated.

TeO_2, also made by combustion of the element, exists in two colourless forms, one having a rutile and the other a brookite type of structure, both indicating that the oxide has a dominantly ionic character. PoO_2, of which there are two forms, red tetragonal and yellow face-centred cubic, decomposes at 770 K in a vacuum, surprisingly leaving the metal.

The reactions of SO_2, SeO_2 and TeO_2 with water are of interest. Gaseous SO_2 dissolves, but the acid H_2SO_3 cannot be isolated. Liquid SO_2 has a limited miscibility with water but it is completely miscible with benzene. SeO_2 gives an acidic aqueous solution from which colourless hexagonal H_2SeO_3 can be crystallised. TeO_2 is almost insoluble in water but dissolves not only in alkalis, but also in H_2SO_4, HCl and HNO_3. The rhombic oxide hydroxide nitrate, $Te_2O_3(OH)NO_3$, crystallises from a solution in nitric acid, an indication of the amphoteric character of TeO_2.

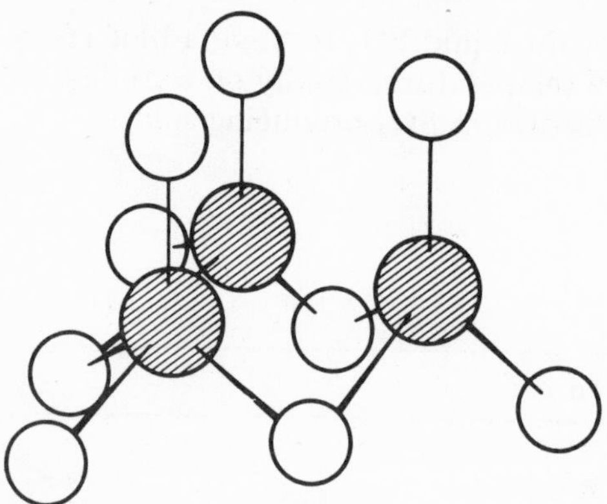

Fig. 22.7. Cyclic trimeric form of sulphur trioxide.

22.6.3. Trioxides

Sulphur trioxide, made by the catalytic oxidation of SO_2 with oxygen, is a colourless, readily volatile solid. The vapour consists of planar, D_{3h}, monomeric molecules of zero dipole moment, O—S—O angle 120° and with bonds 143 pm. The partly double-bond character is ascribable to a d-orbital contribution, stronger than that in SO_2. Solid SO_3 exists in three forms; the two having asbestos-like structures with long chains of SO_2 groups linked by oxygen atoms probably owe their existence to traces of moisture; the ice-like variety is a cyclic trimer (Fig. 22.7). Bond lengths indicate the oxygen atoms in the ring to be joined to the sulphur atoms by bonds which are essentially single. The projecting oxygen atoms are held by double bonds.

Sulphur trioxide dissolves in liquid SO_2. There is an exchange of oxygen but not of sulphur atoms between the two oxides, suggesting equilibria in the solution:

$$SO_2 + SO_3 \rightleftharpoons SO^{2+} + SO_4{}^{2-} \text{ or } SO_2 + 2\,SO_3 \rightleftharpoons SO^{2+} + S_2O_7{}^{2-}$$

Though the means of comparing the strengths of Lewis acids is not entirely satisfactory, SO_3 is clearly one of the strongest.

Sulphur trioxide, made industrially by the oxidation of SO_2 (22.7.1), is an intermediate in the manufacture of sulphuric acid and is itself an important sulphonating reagent in the detergent industry.

Selenium trioxide is formed, together with much SeO_2, when an electric discharge is passed through selenium vapour in oxygen at 520 Pa pressure. The colourless, deliquescent solid is the anhydride of selenic acid. TeO_3 is an orange solid made by heating telluric acid, H_6TeO_6, very strongly.

Sulphur is the only element in the group to form polyperoxides. The white polymer is produced when dry SO_2 and O_2 are passed through an ozoniser to give a solid with empirical formula SO_{3-4}. The compounds are derived from β-SO_3 by the substitution of peroxo bridges. Such materials are powerful oxidising agents, converting aniline to nitrobenzene, and they can be hydrolysed to mixtures of H_2SO_5 and H_2SO_4.

22.7. Oxoacids

22.7.1. Oxoacids of sulphur

Sulphurous acid	H_2SO_3	In aqueous solution and in salts
Disulphurous acid	$H_2S_2O_5$	Exemplified only in its salts
Dithionous acid	$H_2S_2O_4$	Exemplified only in its salts
Sulphuric acid	H_2SO_4	*m.p.* 283.6 K
Disulphuric acid	$H_2S_2O_7$	*m.p.* 308 K
Thiosulphuric acid	$H_2S_2O_3$	Isolated as solid stable at ~195 K
Thionic acids	$H_2S_xO_6$	(x = 2—5) In aqueous solution and in salts

Sulphurous acid cannot be isolated from aqueous solutions of SO_2. However, the solution ionises fairly strongly:

$$H_2O + H_2SO_3 \rightleftharpoons H_3O^+ + HSO_3{}^- \qquad pK_a = 1.77$$

Standard redox potentials are:

$$SO_4{}^{2-} + 4\,H^+ + 2\,e \rightleftharpoons H_2SO_3 + H_2O \qquad E^0 = +0.20\text{ V}$$
$$H_2SO_3 + 4\,H^+ + 4\,e \rightleftharpoons S + 3\,H_2O \qquad E^0 = -0.45\text{ V}$$

It has thus rather strong reducing properties, being itself reduced to sulphur only by very strong reducing agents. The dibasic acid forms two series of salts. The hydrogen sulphites, M^IHSO_3, yield disulphites of the type $M^I{}_2S_2O_5$ on heating:

$$2\,NaHSO_3 \rightarrow Na_2S_2O_5 \text{ (the so-called sodium metabisulphite)} + H_2O$$

All of these compounds are reducing agents.

The sulphite ion in salts is pyramidal with C_{3v} symmetry; the S—O distance (139 pm) is shorter than the S—O bonds in SO_2 or $SOCl_2$.

Dithionites are made by reducing bisulphites, usually with zinc, or by shaking sodium or potassium amalgam with dry SO_2 in the absence of oxygen. The salts reduce nitrocompounds to amines. An $Na_2S_2O_4$ solution is used to reduce and dissolve vat dyes such as indigo. X-ray analysis shows the $S_2O_4^{2-}$ ion in sodium dithionite to consist of two SO_2^- groups attached by a very long S—S bond (239 pm).

Sulphuric acid, commercially the most important acid of sulphur, is made by way of the catalytic oxidation of SO_2 to SO_3; this is done by air, either at a surface, such as V_2O_5 or special mixtures of oxides, in the contact process, or formerly with oxides of nitrogen as catalyst in the lead chamber process. The SO_3 produced in the contact process is dissolved in sulphuric acid which is diluted as required. In the chamber process nitrosyl sulphuric acid was probably formed from sulphurous acid and nitrogen dioxide and was then hydrolysed by water thus:

$$2\,NOHSO_4 + H_2O \rightarrow 2\,H_2SO_4 + NO_2 + NO$$
$$2\,NO + O_2 \rightarrow 2\,NO_2$$

Though a strong acid, the second ionisation of sulphuric acid is slight in any but the most dilute solutions. In the absence of water, the acid contains HSO_3^+ ions, the active sulphonating agent. X-ray diffraction data show the SO_4^{2-} ion to be almost tetrahedral, and the dimensions indicate considerable double bond character. The ion often holds 'anion water' in crystal structures, examples being $CuSO_4 \cdot 5\,H_2O$ and $FeSO_4 \cdot 7\,H_2O$. In some complex compounds the SO_4 group is bound covalently to the central atom. Examples are $[(NH_3)_5Co(SO_4)]^+$, in which the SO_4 group is attached by a single bond, and $[(NH_3)_4Pt(SO_4)]^{2+}$, in which the group is bound by two covalent bonds.

Disulphuric (pyrosulphuric) acid, $H_2S_2O_7$, is formed when SO_3 dissolves in H_2SO_4. It is an excellent sulphonating agent. Sodium disulphate (pyrosulphate), $Na_2S_2O_7$, can be made by heating sodium hydrogen sulphate strongly:

$$2\,NaHSO_4 \rightarrow H_2O + Na_2S_2O_7$$

The salt is hydrolysed in water to give HSO_4^- ions. Evidence for higher isopolysulphates exists in the compound $(NO_2)_2S_3O_{10}$.

Anhydrous thiosulphuric acid, which is very unstable, has been obtained by the reaction of H_2S with SO_3 at 195 K, either alone or in a liquid such as freon. It can also be made by the action of H_2S on chlorosulphuric acid, HSO_3Cl, at the same temperature. Thiosulphates are made by boiling alkali-metal sulphite solutions with sulphur:

$$SO_3^{2-} + S \rightarrow S_2O_3^{2-}$$

Selenium also dissolves in sulphite solutions to give selenosulphates:

$$SO_3^{2-} + Se \rightarrow SSeO_3^{2-}$$

Thiosulphates are formed by many elements and radicals; some are fairly stable and soluble in water, those of the heavier metals are less soluble and less stable. The $S_2O_3^{2-}$ ion is structurally analogous to SO_4^{2-} (Fig. 22.8).

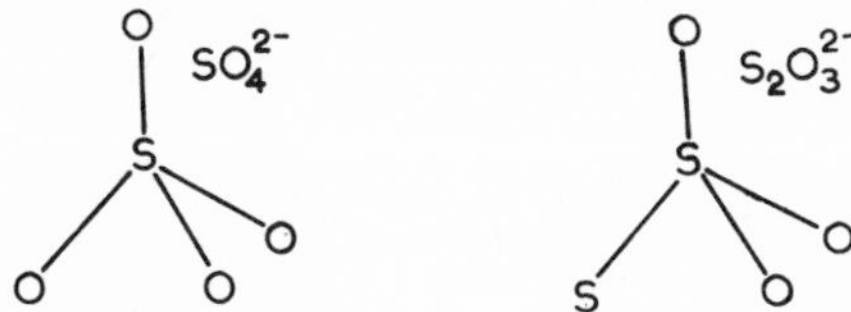

Fig. 22.8. Analogy between $S_2O_3^{2-}$ and SO_4^{2-} ions.

$Na_2S_2O_3$, made by boiling ^{35}S with inactive Na_2SO_3, yields a precipitate of active sulphur on acidification but no activity remains in the SO_3^{2-} solution:

$$^{35}S + {}^{32}SO_3^{2-} \xrightarrow{\text{boil}} {}^{35}S^{32}SO_3^{2-} \xrightarrow{\text{acid}} {}^{35}S + {}^{32}SO_3^{2-}$$

Hence there is no exchange between the two S atoms, and thus they cannot be equivalent in the molecule. When $H_2{}^{35}S$ is used in the reaction:

$$4\ SO_2 + 2\ H_2S + 6\ NaOH \rightarrow 3\ Na_2S_2O_3 + 5\ H_2O$$

similar evidence is obtained on acidifying the salt produced. All the central S atoms of the $S_2O_3^{2-}$ ions come from SO_2, but the ligand atoms partly from SO_2 and partly from H_2S.

Addition of acid to a thiosulphate solution does not cause decomposition until a pH of 4.6 is reached. The $HS_2O_3^-$ ion then formed, being much less stable than the $S_2O_3^{2-}$ ion, decomposes quickly:

$$HS_2O_3^- \rightarrow HSO_3^- + S$$

Thus thiosulphates exist only as neutral salts.

Dithionic acid appears in a solution of sulphurous acid which is being oxidised by finely divided MnO_2. From this solution, baryta precipitates all the sulphur-containing anions except dithionate, and, when the excess of Ba^{2+} has been precipitated by H_2SO_4, only $H_2S_2O_6$ is left in the solution. Although conductance measurements show the acid to be dibasic, acid salts are unknown. Dithionates are soluble in water and, unlike salts of the higher thionic acids, are not decomposed by sulphites and sulphides.

Potassium trithionate, $K_2S_3O_6$, can be made by passing SO_2 into a solution of $K_2S_2O_3$. The salt crystallises on standing, leaving in solution $K_2S_4O_6$ and $K_2S_5O_6$ formed along with it. When aqueous $K_2S_3O_6$ is acidified, decomposition occurs:

$$H_2S_3O_6 \rightarrow H_2SO_4 + SO_2 + S$$

Sodium tetrathionate, $Na_2S_4O_6$, is formed when $Na_2S_2O_3$ is oxidised with iodine. Although this, and other tetrathionates, can be isolated as solids, they are thermally unstable. Aqueous tetrathionic acid results from adding dilute H_2SO_4 to a solution of PbS_4O_6; attempts at concentration eventually cause decomposition:

$$H_2S_4O_6 \rightarrow H_2SO_4 + SO_2 + 2\ S$$

Sodium pentathionate, $Na_2S_5O_6$, is made by treating a concentrated $Na_2S_2O_3$ solution with very dilute hydrochloric acid in the presence of As_4O_6 at 263 K:

$$5\ S_2O_3^{2-} + 6\ H^+ \rightarrow 2\ S_5O_6^{2-} + 3\ H_2O$$

The liquid deposits crystals of $Na_2S_5O_6$ on standing. The acid itself is stable in fairly strong aqueous solution but eventually decomposes on concentration:

$$H_2S_5O_6 \rightarrow H_2SO_4 + SO_2 + 3\ S$$

X-ray diffraction shows the dithionate ion to consist of two triangular pyramids joined at their apices (Fig. 22.9). In the trithionate ion the third sulphur atom lies between the other two with an S—S—S angle of 103°. In the tetrathionate and pentathionate ions the additional sulphur atoms are attached to one another in zig-zag chains.

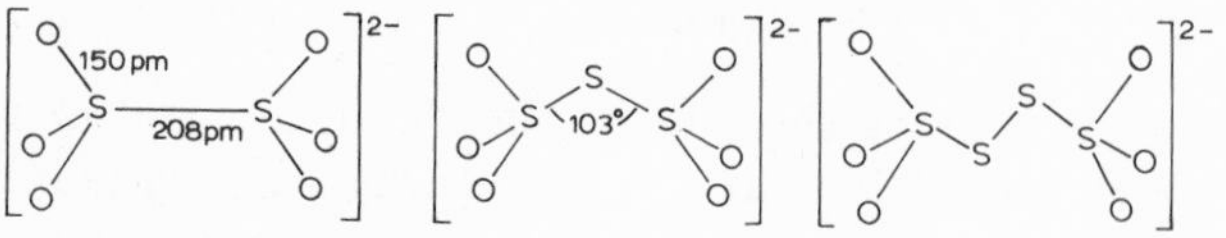

Fig. 22.9. Structures of dithionate, trithionate and tetrathionate ions.

22.7.2. Oxoacids of selenium

Selenium, in contrast to sulphur, forms only the two oxoacids:

Selenious acid, H_2SeO_3, colourless solid
Selenic acid, H_2SeO_4, colourless solid, *m.p.* 332 K

When a solution of SeO_2 is evaporated the acid, H_2SeO_3, separates as hexagonal prisms. The Raman spectra of aqueous solutions of selenious acid show its dissociation to be very slight. There are normal and acid selenites, which ionise strongly in solution, and also superacid salts such as $KH_3(SeO_3)_2$. Heteropolyacids are formed with vanadic, molybdic and uranic acids.

Selenious acid is converted to selenic acid, H_2SeO_4, when refluxed with 30% H_2O_2. A 97.4% solution can be made by vacuum desiccation; the pure acid crystallises on cooling. This resembles sulphuric acid in (i) ionising strongly; (ii) forming selenates isomorphous with the sulphates; (iii) forming a nitroso-acid $(NO)HSeO_4$. It differs from H_2SO_4 by losing oxygen when heated above 470 K and by oxidising chlorides to chlorine.

$$SeO_4^{2-} + 4\ H^+ + 2\ e \rightarrow H_2SeO_3 + H_2O \qquad E^0 = +1.15\ V$$

cf. $$SO_4^{2-} + 4\ H^+ + 2\ e \rightarrow H_2SO_3 + H_2O \qquad E^0 = +0.17\ V$$

22.7.3. Oxoacids of tellurium

Tellurous acid has not been prepared, the dioxide being insoluble in water. Tellurites may, however, be crystallised from solutions of TeO_2 in aqueous alkali-metal hydroxides.

Telluric acid, H_6TeO_6, can be made as colourless crystals by dissolving tellurium in aqua regia, adding a chlorate, evaporating in a vacuum, precipitating with HNO_3 and, finally, recrystallising the product from water. The compound is a weak dibasic acid which gives the salts $NaTeO(OH)_5$ and $Na_2TeO_2(OH)_4$. Although telluric acid was thought of as the dihydrate of H_2TeO_4, in the crystal it has a slightly distorted octahedral structure consistent with the formulation $Te(OH)_6$, further implied by the preparation of the alkoxide $Te(OMe)_6$.

22.7.4. Halogen derivatives of the oxoacids

22.7.4.1. Sulphur

The principal oxohalides of sulphur are set out below:

	M.p./K	*B.p.*/K	S—X/pm	S—O/pm
Sulphinyl (thionyl) halides				
OSF_2	163	229	158.5	141.2
$OSCl_2$	172	347	207	145
$OSBr_2$ (red)	223	413 (7 Pa)	227	(145)
Sulphonyl (sulphuryl) halides				
O_2SF_2	153	218	153	140.5
O_2SCl_2	219	204	199	143
O_2SFCl	148	280		
O_2SFBr	187	314		

Other oxohalides such as OSF_4, O_3SF_2, $FOSF_5$ and $F_2S_2O_5$ are known.

Sulphinyl chloride, $OSCl_2$, is made by the action of SO_2 on PCl_5:

$$SO_2 + PCl_5 \rightarrow OSCl_2 + POCl_3$$

The colourless liquid is obtained by fractionating the mixture. The molecule is pyramidal and owes its shape to the use of sp^3 hybrid orbitals round an S atom with one lone pair.

Sulphinyl fluoride, OSF_2, a colourless gas which does not attack glass or mercury at room temperature, results from the action of SbF_3 on $OSCl_2$. The red liquid $OSBr_2$ is made by treating $OSCl_2$ with HBr. OSF_2 hydrolyses slowly in water, in which it is sparingly soluble; the chloride hydrolyses violently:

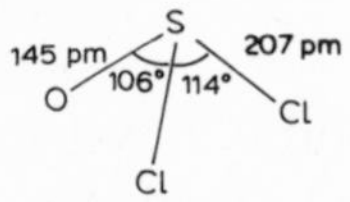

Fig. 22.10. Structure of sulphinyl chloride.

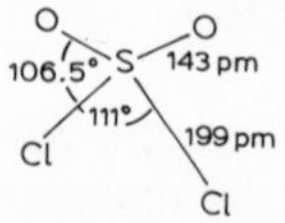

Fig. 22.11. Structure of sulphonyl chloride.

$OSCl_2 + H_2O \rightarrow SO_2 + 2\ HCl$

When SO_2 and Cl_2 are passed alternately over camphor they combine to give sulphonyl chloride, O_2SCl_2, which can be distilled from the product. The liquid is rather slowly hydrolysed by water. The molecule forms a distorted tetrahedron (Fig. 22.11). Sulphonyl fluoride, O_2SF_2, made by heating SO_2 and F_2 together, is an inert gas, unaffected by water but decomposed by hot alkalis:

$O_2SF_2 + 4\ OH^- \rightarrow SO_4^{2-} + 2\ F^- + 2\ H_2O$

Important halogen-containing acids are:

	M.p./K	*B.p.* /K
Chlorosulphonic (chlorosulphuric) $ClSO_3H$	193	428
Fluorosulphonic (fluorosulphuric) FSO_3H	184	436

When dry HCl is passed into fuming H_2SO_4, which contains SO_3, the compound $ClSO_3H$, chlorosulphuric acid, is formed. Its relation to H_2SO_4 and SO_2Cl_2 is shown; all the compounds have tetrahedral co-ordination:

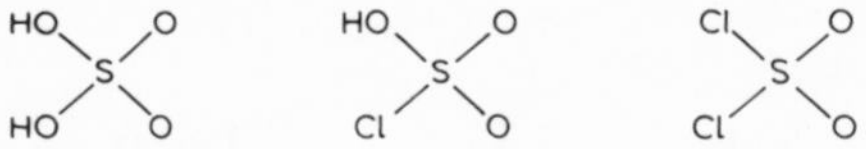

It is hydrolysed violently by water:

$(HO)SO_2Cl + H_2O \rightarrow H_2SO_4 + HCl$

Fluorosulphuric acid, $(HO)SO_2F$, made by distilling fuming H_2SO_4 with CaF_2, is a colourless liquid which in excess with water achieves equilibrium

$H_2O + FSO_3H \rightleftharpoons HF + H_2SO_4$

Fluorosulphuric acid also acts as a fluorinating agent in inorganic and organic chemistry, e.g.

$B(OH)_3 + 3\ FSO_3H \rightarrow BF_3 + 3\ H_2SO_4$

22.7.4.2. Selenium

The seleninyl (SeO^{2+}) compounds contain Se^{IV} and are thus related to H_2SeO_3.

	$M.p./K$	$B.p./K$
$OSeF_2$	288	398
$OSeCl_2$	284	450
$OSeBr_2$	315	490/95 kPa (decomp)
O_2SeF_2	173	265

$OSeCl_2$ is made by heating SeO_2 and $SeCl_4$ together in a sealed tube. It is converted to $OSeF_2$ when passed over AgF, and to $OSeBr_2$ when distilled from NaBr.

The selenonyl (SeO_2^{2+}) derivatives of H_2SeO_4 include O_2SeF_2, an easily hydrolysable gas made by heating $BaSeO_4$ with $FSeO_3H$, itself obtained by the action of HF on SeO_3.

22.7.5. Oxoacid salts of tellurium and polonium

Tellurium and polonium are the only elements of Group VIB which are sufficiently metallic to form oxoacid salts. An oxide sulphate, $Te_2O_3SO_4$, is obtained as white crystals when TeO_2 is heated with concentrated H_2SO_4. A corresponding compound of polonium can be crystallised from a solution of $PoCl_4$ in 0.02—0.025 N H_2SO_4. With stronger H_2SO_4, however, a colourless hydrate of $Po(SO_4)_2$ is formed. This loses water on standing or heating, becoming pink and finally purple. The purple anhydrous salt is stable up to 670 K.

Tellurium dissolves in 8 M HNO_3 to give orthorhombic 2 $TeO_2 \cdot 2\ HNO_3$ which degrades to the dioxide at 460 K. Decomposition of the dinitrogen tetroxide adduct $Po(NO_3)_4 \cdot N_2O_4$ gives $Po(NO_3)_4$, but this decomposes further even in vacuo to a substance probably derived from the unknown $PoO(NO_3)_2$.

Further Reading

E.A.V. Ebsworth, J.A. Connor and J.J. Turner, The chemistry of oxygen, Pergamon, Oxford, 1975.

M. Schmidt, W. Siebert and K.W. Bagnall, The chemistry of sulphur, selenium, tellurium and polonium, Pergamon, Oxford, 1975.

E. Nickless (Ed.), Inorganic sulphur chemistry, Elsevier, Amsterdam, 1968.

K.W. Bagnall, The chemistry of selenium, tellurium and polonium, Elsevier, Amsterdam, 1966.

W.C. Cooper (Ed.), Tellurium, Van Nostrand, New York, 1971.

K.C. Mills, Thermodynamic data for inorganic sulphides, selenides and tellurides, Butterworths, London, 1974; Sulphur research trends, ACS Advances in Chemistry No. 110, 1972.

A.A. Kudryavtsev, The chemistry and technology of selenium and tellurium, Collet, London, 1974.

M.J. McEwan and L.F. Phillips, Chemistry of the atmosphere, Arnold, London, 1975.

The Oxides

23

23.1. Introduction

Of the known elements the lighter noble gases alone are without oxides.

In this chapter are considered only what we term the *normal* oxides, namely those with separate oxygen atoms or ions attached directly and only to the atom or ion of another element. Besides these there are the peroxides and superoxides in which the oxygen atoms are attached in pairs. In the hydrogen peroxide molecule (24.1) they are connected to each other by a single bond, H—O—O—H; in the lattice of ionic peroxides, such as Na_2O_2, they appear as O_2^{2-} ions; and in superoxides, such as KO_2, as O_2^- ions.

Normal oxides fall broadly into three classes according to their behaviour in water: (i) *basic* which are invariably ionic oxides of metals and which, when soluble, give alkaline solutions; (ii) *acidic*, essentially covalent compounds, which are roughly the soluble oxides of the non-metals and the higher oxides of the transition metals, and which give acidic solutions; (iii) *neutral* which include water and the relatively insoluble gases CO and N_2O.

The distribution through the Periodic Table of representative basic and acidic oxides is shown in Table 23.1, the oxides adjacent to the dividing line are commonly amphoteric; they show basic or acidic properties according to the conditions.

Oxygen, owing to its high electronegativity, readily forms ionic compounds, and most metallic oxides have simple ionic structures. Some metallic oxides, however, show considerable covalent character when the metal (e.g. a transition metal) can have a high charge number and a correspondingly greater electronegativity. The charge effect is well shown by manganese; MnO is an ionic solid with a rock-salt lattice, whereas Mn_2O_7 is a covalent liquid.

Oxides of the non-metals range from the volatile monomers (e.g. CO_2, N_2O, SO_2) to involatile macro-molecules (e.g. B_2O_3, SiO_2). Intermediately there are many oxides which, through association, are less volatile than might be expected (e.g. H_2O, SO_3, P_2O_5).

The classification of oxides as acidic and basic is not rigid. Generally the molecular oxides of the non-metals are acidic, many being acid anhydrides, and so too are the higher oxides of transition metals(e.g. CrO_3, Mn_2O_7). Such

TABLE 23.1

DISTRIBUTION OF OXIDES THROUGH THE PERIODIC TABLE

Non-transition elements

Basic				Acidic		
Li_2O	BeO	B_2O_3	CO_2	N_2O_5		F_2O
Na_2O	MgO	Al_2O_3	SiO_2	P_2O_5	SO_3	Cl_2O
K_2O	CaO	Ga_2O_3	GeO_2	As_2O_3	SeO_2	Br_2O
Rb_2O	SrO	In_2O_3	SnO	Sb_2O_3	TeO_2	I_2O_5
Cs_2O	BaO	Tl_2O	PbO	Bi_2O_3	PoO_2	

Transition elements and Groups IB and IIB

	Basic		Acidic				Basic		
Sc_2O_3	TiO_2	V_2O_5	CrO_3	Mn_2O_7	Fe_2O_3	CoO	NiO	Cu_2O	ZnO
Y_2O_3	ZrO_2	Nb_2O_5	MoO_3	Tc_2O_7	RuO_4	Rh_2O_3	PdO	Ag_2O	CdO
Ln_2O_3 *	HfO_2	Ta_2O_5	WO_3	Re_2O_7	OsO_4	IrO_2	PtO	Au_2O	HgO
Ac_2O_3									

* Ln = lanthanide atom.

neutral oxides as H_2O, CO and N_2O are exceptional. The more or less covalent CuO is without acid character and the similar PbO is weakly basic. Although the ionic metal oxides are usually basic, ZnO and Al_2O_3 are amphoteric.

23.2. Methods of Preparation

Many normal oxides are formed on burning the element in air or oxygen. This is true not only of the non-metals boron, carbon, sulphur and phosphorus, but also for the volatile metals zinc, cadmium, indium and thallium, the transition metals cobalt and iron, when in a finely divided condition, and the noble metals osmium, ruthenium and rhodium. With some elements, limiting the supply of oxygen produces the lower oxide (e.g. P_4O_6 in place of P_4O_{10}).

Lower oxides have been made by reducing a higher oxide with carbon or hydrogen (e.g. MnO_2 to MnO and V_2O_5 to V_2O_3 by hydrogen).

A few elements are oxidised by steam at red heat:

$$3\,Fe + 4\,H_2O \rightarrow Fe_3O_4 + 4\,H_2$$
$$C + H_2O \rightarrow CO + H_2$$
$$Mg + H_2O \rightarrow MgO + H_2$$

In the last reaction magnesium burns brilliantly.

Metal oxides are commonly prepared, and often manufactured, by the thermal decomposition of hydroxide, carbonate or nitrate:

$$Cu(OH)_2 \rightarrow CuO + H_2O$$
$$CaCO_3 \rightarrow CaO + CO_2$$
$$2\,Pb(NO_3)_2 \rightarrow 2\,PbO + 4\,NO_2 + O_2$$

Certain precipitated hydroxides are easily converted to oxide, $Tl(OH)_3$ to Tl_2O_3 even in boiling water, but others, for instance $Cu(OH)_2$, which gives a hydrated black oxide, cannot be completely dehydrated without producing a non-stoichiometric oxide. Gold(III) hydroxide, $Au(OH)_3$, is changed only to AuO(OH) after prolonged standing over P_2O_5.

Precipitation by alkali from soluble salts usually gives a hydroxide:

$$Ca^{2+} + 2\,OH^- \rightarrow Ca(OH)_2$$

Sometimes a complex hydrated oxide is formed:

$$2\,Pb^{2+} + 4\,OH^- \rightarrow Pb_2O(OH)_2 + H_2O$$
$$H_2PtCl_6 + 6\,OH^- \rightarrow H_2Pt(OH)_6 + 6\,Cl^-$$

But in other instances the oxide itself results:

$$Hg^{2+} + 2\,OH^- \rightarrow HgO + H_2O$$
$$2\,Au^+ + 2\,OH^- \rightarrow Au_2O + H_2O$$

Other oxidising agents, especially nitric acid and the oxides of nitrogen, convert some elements to oxides. Nitric acid, for example, oxidises sulphur to SO_2 and SO_3, and germanium and tin to GeO_2 and SnO_2 respectively.

The preparation of pure, single-phase oxides of true stoichiometry is often difficult, sometimes impossible. FeO and MnO_2 are examples, the former being always metal-deficient, the latter always oxygen-deficient (36.8, 35.5).

23.3. Oxide Structures

23.3.1. Metallic oxides of predominantly ionic character

23.3.1.1. MO type

Metal oxides of MO type are generally simple structurally, having O^{2-} and M^{2+} ions arranged in 4 : 4 or 6 : 6 co-ordination, depending on the ratio of the ionic sizes. The Be^{2+} ion (31 pm) has only four O^{2-} ions (140 pm) round it in BeO; but most MO-type ionic oxides, with M^{2+} ranging from 50—100 pm, have the 6 : 6 co-ordination and rock-salt lattices. Examples are MgO, CaO, SrO, BaO, CdO, VO, MnO and CoO; however NiO has a slightly distorted version of the lattice which gives it a rhombohedral structure. The normal oxides of the Group IA metals are ionic oxides of the M_2O type. They have the antifluorite crystal lattice arising from 4 : 8 co-ordination; a structure so named because the positions of anions and cations are the reverse of those in CaF_2 (Fig. 7.19).

23.3.1.2. MO_2 type

There are two main types of ionic oxides which are empirically formulated MO_2. Where the metal ion is large (Th^{4+}, 95 pm; Ce^{4+}, 101 pm; U^{4+}, 89 pm) the crystals are built up of fluorite-type unit cells with 8 : 4 co-ordination. But where the metal ion is smaller (Sn^{4+}, 71 pm; Ti^{4+}, 68 pm) the structure

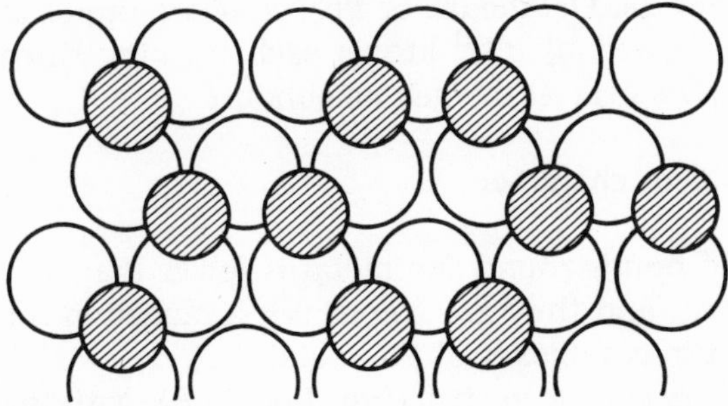

Fig. 23.1. Positions of Al^{3+} ions in α-Al_2O_3 (shaded) relative to hexagonally close-packed O^{2-} ions (unshaded).

is based on the rutile lattice with 6 : 3 co-ordination. Other examples of this structure are VO_2, RuO_2, PbO_2 and TeO_2. The rutile lattice is slightly deformed in MoO_2 and WO_2.

23.3.1.3. MO_3 type

The structurally simplest MO_3 type oxide is rhenium oxide, ReO_3 (Fig. 7.23). WO_3 is like it, but slightly deformed.

23.3.1.4. M_2O_3 type

The usual lattices of the M_2O_3 group of ionic oxides are those of corundum (α-Al_2O_3) and of the A- and C-type lanthanide sesquioxides. The positions of Al^{3+} and O^{2-} in the corundum lattice are shown in Fig. 23.1. The A-type lanthanide sesquioxides, La_2O_3 is an example, have a hexagonal lattice and unusual co-ordination. Each La^{3+} ion has, as nearest neighbours, four O^{2-} at 242 pm and three O^{2-} at 269 pm.

The C-type oxide (Fig. 23.2), Sc_2O_3 is an example, has the simpler symmetry of the cubic lattice and 6 : 4 co-ordination.

23.3.1.5. M_3O_4 type

In oxides with the empirical composition M_3O_4, the metal necessarily appears with two charge numbers. As a result, the compounds often resemble complex oxides (see below). Red lead, Pb_3O_4, which is isomorphous with

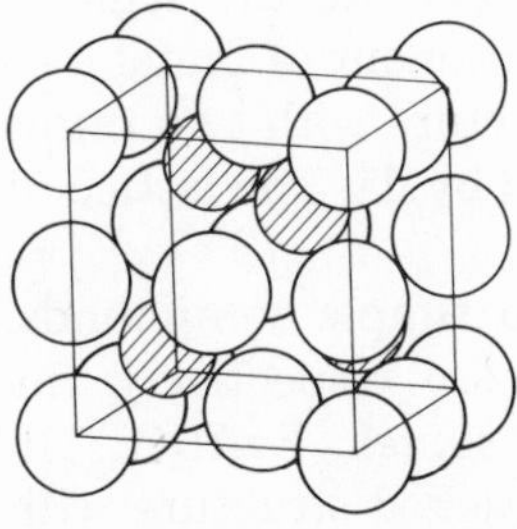

Fig. 23.2. C-type lanthanide oxide.

$ZnSb_2O_4$, has a complicated structure. Octahedra of $Pb^{IV}O_6$ share opposite edges to form chains; these are linked by Pb^{II} atoms each forming three bonds, pyramidally arranged, with its nearest oxygen neighbours.

23.3.2. Metal oxides of more covalent character

In these oxides the number of bonds round the metal is fewer than size considerations alone would suggest, and they are frequently arranged in the same manner as the bonds present in complexes of the metal.

Silver(I) and copper(I) oxides are similar in structure (Fig. 38.8), with the M—O bonds collinear and 2 : 4 co-ordination. But in PdO and PtO each metal atom has four coplanar bonds and each oxygen four tetrahedral ones in a tetragonal unit cell (Fig. 37.9).

Layer lattices, when they occur in oxides, indicate a predominantly covalent character in the metal—oxygen bonds. The oxides SnO and PbO have layer lattices. Each metal atom is joined to four oxygens arranged in a square on one side of it (Fig. 23.3).

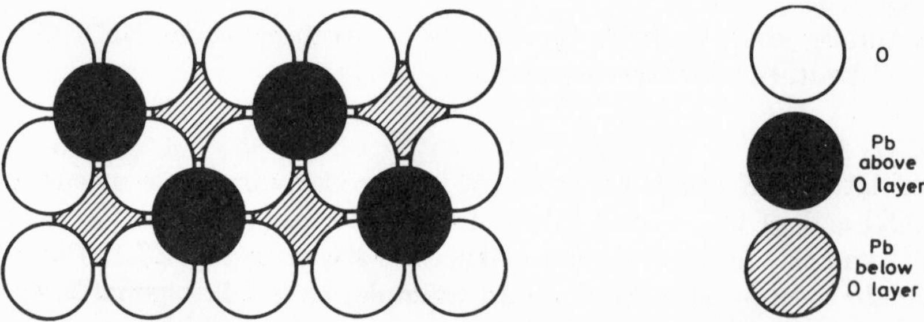

Fig. 23.3. Layer lattice of PbO (shown in plan in depth).

Molybdenum(VI) oxide, MoO_3, also has a layer structure; in it MoO_6 octahedra share two edges and two corners with one another.

23.3.3. Complex oxides

Some complex oxides have complex ions (e.g. CO_3^{2-} and NO_3^-) and others are without complex ions. Only the latter will be discussed here. They are of two main types. In the first, the lattice structures are the same as those of simple compounds, but in them random replacement of metal ions has occurred. Thus Li_2TiO_3 has a random rock-salt structure with two thirds of the metal ion positions occupied by Li and one third by Ti. The lattice of $FeSbO_4$ has a random rutile structure.

In the second type, the structures are not those of simple compounds. Three of these structures are of importance: they are respectively the *perovskite*, $CaTiO_3$; the *ilmenite*, $FeTiO_3$; and the *spinel*, $FeAl_2O_4$. In perovskite the large Ca^{2+} and O^{2-} ions form a close-packed structure with smaller Ti^{4+} ions uniformly arranged in some of the interstices (Fig. 7.27).

Other complex oxides with a perovskite structure are $SrTiO_3$, $CaZrO_3$ and $LaAlO_3$.

In ilmenite, the oxygen ions are in close-packed hexagonal arrangement with both the Fe^{2+} and Ti^{4+} ions severally occupying one third of the octahedral holes. The compounds $MnTiO_3$, $CoTiO_3$ and $NiTiO_3$ also have the ilmenite structure and there is no justification for calling them titanates. Ilmenite is isomorphous with α-Fe_2O_3 which has the corundum structure. It differs from α-Fe_2O_3 only in having Ti^{4+} ions alternating with Fe^{2+} ions instead of ions of the same metal, namely Fe^{3+}.

A slightly more complicated arrangement is that in the spinels (Fig. 7.28). In these the O^{2-} ions are arranged as in cubic close-packing, the M^{2+} ions occupy some of the tetrahedral spaces and the M^{3+} ions some of the octahedral ones. Compounds with this structure include $CoAl_2O_4$ and $MnAl_2O_4$; these again have the M^{2+} ions in positions of tetrahedral, and the M^{3+} ions in positions of octahedral co-ordination. There is, however, a second rather more complicated type of spinel structure with a partly random arrangement of the metal ions. Examples are $Fe(MgFe)O_4$ and $Fe(TiFe)O_4$, so formulated because one half of the Fe atoms are symmetrically arranged in octahedral positions and the other half randomly arranged, with the ions of the second metal, in tetrahedral positions.

23.3.4. Oxides with chain structures

Some elements of the B sub-groups form oxides with chain-like structures which can be considered as intermediate in valency character between the infinite three dimensional metallic oxides and the molecular, non-metallic oxides. Examples are SeO_2:

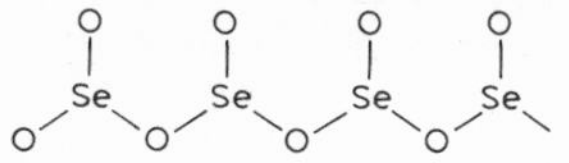

```
   O       O       O       O
   |       |       |       |
   Se      Se      Se      Se
 /    \  /    \  /    \  /    \
O      O       O       O
```

and valentinite, which is a form of Sb_2O_3:

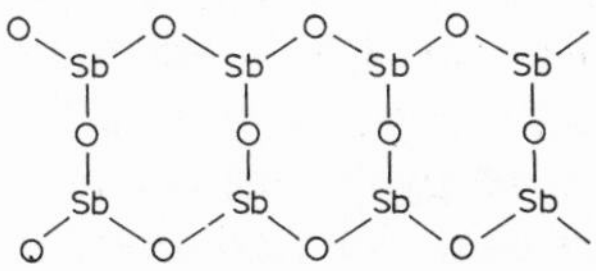

```
O       O       O       O
  \   /   \   /   \   /   \   /
   Sb      Sb      Sb      Sb
   |       |       |       |
   O       O       O       O
   |       |       |       |
   Sb      Sb      Sb      Sb
  /   \   /   \   /   \   /   \
O       O       O       O
```

In this the Sb atoms form pyramidal bonds with bond angles about 90°.

23.3.5. Solid oxides containing individual molecules

The covalent oxides of non-metals form solids in which individual molecules are present, sometimes as the simple monomer, for instance CO_2 and SO_2, and sometimes as a dimer or trimer. Dimeric forms include P_4O_6 and As_4O_6. One of the forms of sulphur trioxide is trimeric (Fig. 22.7).

23.4. Comparison of Oxides and Sulphides

Oxides are generally more ionic in character than the corresponding sulphides. The electronegativity assigned to oxygen is 3.50, that to sulphur only 2.44. Accordingly, elements of low electronegativity form markedly ionic bonds with oxygen but not necessarily with sulphur. This difference is clearly brought out when the structures of oxides and sulphides are compared. The only unequivocally ionic sulphides are those of the Group IA and Group IIA metals and of a few of the transition metals, such as MnS. The existence of any truly ionic M_2S_3 sulphide is doubtful, although the ready hydrolysis of Al_2S_3 suggests that it may be one. No ionic MS_2 sulphide is known. The S^{2-} ion is evidently too easily polarised to exist in the neighbourhood of a small quadripositive cation. Instead, most of the solid disulphides have layer lattices, a type of structure very uncommon among the oxides:

TiO_2, SnO_2 } Rutile structure TiS_2, SnS_2 } Layer lattices

Both the pyrites and marcasite forms of FeS_2 contain covalently bound S_2 units which are not ionic.

23.5. Thermochemistry of the Oxides

That many elements react spontaneously with oxygen is remarkable when it is realised that the process of converting molecular oxygen into oxide ions is strongly endothermic [$\Delta H_f^0\ O^{2-}_{(g)} = 924$ kJ mol^{-1}]. The same is also true of the process of oxidation of the element [e.g. $\Delta H_f^0\ Fe^{3+}_{(g)} = 5733$ kJ mol^{-1}]. Thus the driving force in the process is the high lattice energy of the element's oxide [e.g. $U(Fe_2O_3) = 15062$ kJ mol^{-1}].

In a series of oxides such as those of the elements of Period 3 differences between the enthalpies of formation of its members (Fig. 23.4) depend largely on the differences between the ionisation energies of the elements combining with oxygen and on the differences between the lattice energies of the oxides so formed. When a positive noble-gas type ion is replaced by one of higher charge but of the same structure, the enthalpy of formation is less negative, and consequently the stability decreases. The apparently anomalous fall in enthalpy of formation per oxygen equivalent from Na_2O to MgO is understandable when the lower crystal energy of Na_2O, caused by the mutual repulsion between unipositive metal ions, is taken into account. The effect is demonstrated more emphatically by the nitrides, where the stable AlN has a large crystal energy and the unstable Na_3N a small one. Nevertheless Li_3N is stable enough to be formed by direct combination of the elements, because the Li^+ ion is much smaller than the Na^+ ion and thus is smaller in relation to the size of the N^{3-} anion.

Enthalpies of formation of the oxides usually become more negative down

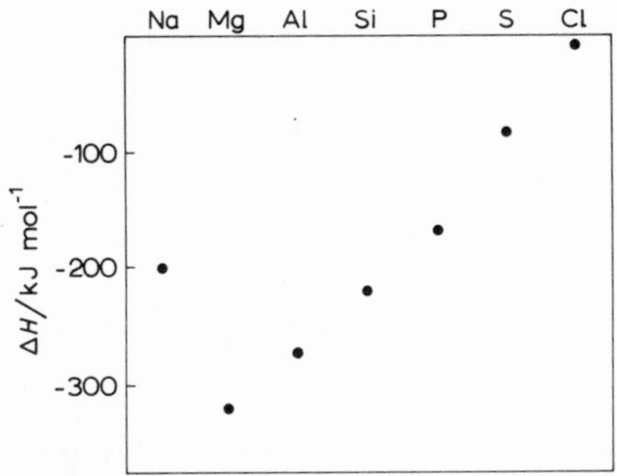

Fig. 23.4. Enthalpy of formation (ΔH) per 8 g of oxygen for Na_2O, MgO, Al_2O_3, SiO_2, P_4O_{10}, SO_3, Cl_2O_7.

a group in the Periodic Table:

	CO_2	SiO_2	TiO_2	ZrO_2	HfO_2
ΔH_f per oxygen equivalent/kJ mol^{-1}	−98.5	−208	−227	−269	−349

Ionisation energies decrease with increase in size of the M^{4+} ions in passing from Ti^{4+} to Hf^{4+}. Lattice energies also decrease as the sum of the radii r_+ and r_- increases, but the effect is overshadowed by the change in ionisation energy, particularly where the radius of the positive ion (r_+) is small compared with that of the oxide ion (r_-).

23.6. Non-stoichiometry

The solid oxides and sulphides all show, to a greater or lesser extent, homogeneous phases which vary in composition from that represented by the stoichiometric formula of the named compound. These phases often retain the same structure but, nevertheless, show a certain range of differences in their other properties.

23.6.1. Oxides of metals as semiconductors

Non-stoichiometric oxides of metals can behave as impurity semiconductors (7.2.23.1) with discrete energy levels between the highest filled band of electrons and the conduction band. In *excess* or *n-type* semiconducting oxides such as $Zn_{1+x}O$ the interstitial metal atoms act as sources of electrons whose energy levels lie only slightly below that of the conduction band (Fig. 23.5).

The levels are discrete rather than connected into a band because the

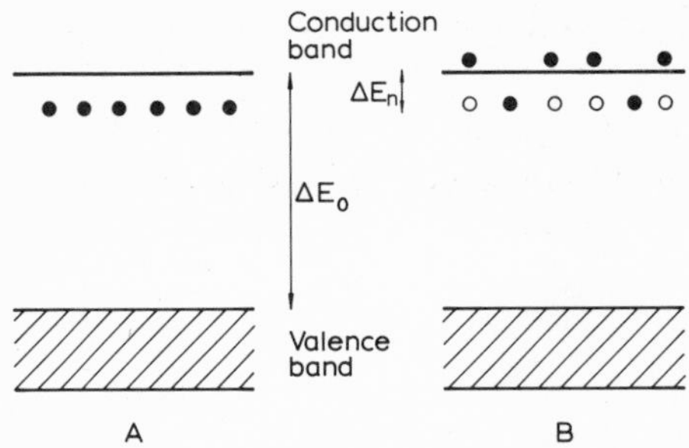

Fig. 23.5. Energy levels in n-type extrinsic semiconductor. A. Occupancy at very low temperature. B. Occupancy at higher temperature.

metal atoms are far apart. As ΔE_n is small, electrons can be promoted into the conduction band of the crystal either thermally or by irradiation with light. For non-stoichiometric zinc oxide ΔE_n is only about 5 kJ mol^{-1} whereas ΔE_0, the intrinsic energy gap (7.2.23.1), is 310 kJ mol^{-1}. In *deficit* or *p-type* semiconducting oxides such as $Cu_{2-x}O$ the metal ions of higher charge act as acceptors of electrons promoted from the highest valence band (Fig. 23.6).

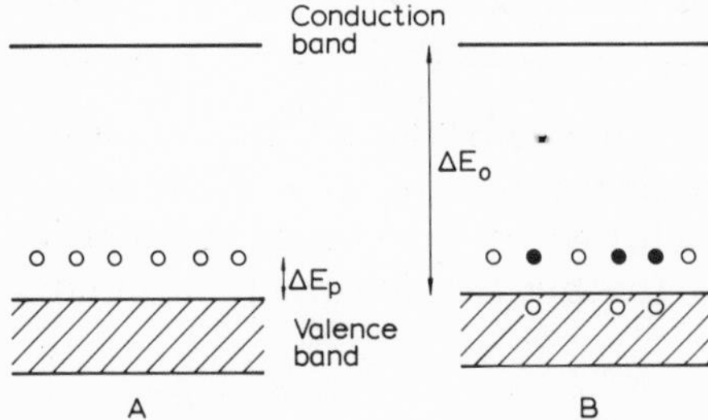

Fig. 23.6. Energy levels in p-type extrinsic semiconductor. A. Occupancy at very low temperature. B. Occupancy at higher temperature.

When the temperature is raised, electrons are promoted through a small energy gap ΔE_p and trapped, in the case of $Cu_{2-x}O$, by Cu^{2+} ions which are thus converted to Cu^+. As a consequence 'positive hole' conduction becomes possible through the valence band.

The conductivities of both types of non-stoichiometric oxides are strongly influenced by oxygen pressure. An increase in pressure of oxygen decreases the number of impurity centres in an n-type oxide (e.g. Zn atoms in $Zn_{1+x}O$) and hence reduces its conductivity. With metal-deficit semiconducting oxides an increase in oxygen pressure creates more defects (e.g. Cu^{2+} ions in $Cu_{2-x}O$) and raises the conductivity.

23.6.2. Semiconducting metal oxides as catalysts

The n-type semiconducting oxides such as ZnO, CdO, Fe_2O_3 and V_2O_5 are often used as heterogeneous catalysts for oxidation reactions. A possible sequence of reactions for the oxidation of SO_2 is as follows:

$$SO_2(g) + 2\,O^{2-}\text{ (in lattice)} = SO_4^{2-}\text{ (on surface)} + 2\,e$$
$$Zn^{2+}\text{ (in lattice)} + 2\,e = \text{Zn atom}$$
$$\text{Zn atom} + \tfrac{1}{2}\,O_2(g) = Zn^{2+} + O^{2-}\text{ (in solid)}$$
$$SO_4^{2-} = O^{2-}\text{ (in lattice)} + SO_3(g)$$

$$\text{Adding: } SO_2(g) + \tfrac{1}{2}\,O_2(g) = SO_3(g)$$

The p-type semiconducting oxides such as Cu_2O and NiO often catalyse the decomposition of gaseous oxides like N_2O. A possible sequence of reactions for the decomposition of N_2O by Cu_2O is as follows:

$$N_2O\text{ (adsorbed)} + 2\,Cu^{2+}\text{ (in lattice)} = N_2(g) + 2\,Cu^{2+} + O^{2-}\text{ (in solid)}$$
$$2\,Cu^{2+} + O^{2-} = 2\,Cu^{+} + \tfrac{1}{2}\,O_2$$

$$\text{Adding: } N_2O = N_2 + \tfrac{1}{2}\,O_2$$

However, p-type oxides can also behave as catalysts for oxidation, usually rather selectively. A possible sequence for reactions for the oxidation of CO to CO_2 on an NiO surface is as follows:

$$\tfrac{1}{2}\,O_2(g) + 2\,Ni^{2+} = 2\,Ni^{3+} + O^{2-}\text{ (in solid)}$$
$$CO(g) + O^{2-} = CO_2(g) + 2\,e$$
$$2\,e + 2\,Ni^{3+} = 2\,Ni^{2+}$$

$$\text{Adding: } CO(g) + \tfrac{1}{2}\,O_2(g) = CO_2(g)$$

Further Reading

E.A.V. Ebsworth, J.A. Connor, and J.J. Turner, The chemistry of oxygen, Pergamon, Oxford, 1975.

D.A. Johnson, Some thermodynamic aspects of inorganic chemistry, Cambridge University Press, 1968.

G.V. Samsonov (Ed.), The oxide handbook, Plenum, New York, 1972.

24

Peroxides and Peroxo-compounds

24.1. Hydrogen Peroxide

The geometry and probable electronic structure of H_2O_2 in the vapour phase are shown in Fig. 24.1. The lengths of the O—O and O—H bonds are 147.5 pm and 95 pm respectively. The HOO bond angle suggests only a slight distortion from localised single bonds between O and H involving 2p and 1s orbitals. However the dihedral angle (111.5° for the molecule in the vapour) is easily distorted (just over 90° in crystalline H_2O_2), so perhaps the σ-bonds from oxygen might better be described as between pure p and sp^3 in character. This configuration balances minimum lone pair—lone pair repulsions with contributions to bonding which can arise from slight delocalisation of the lone pairs. H_2O_2 is the simplest molecule to show restriction of rotation about a single bond. Pure liquid hydrogen peroxide is almost colourless and is syrupy in consistency. It freezes at 272.7 K to tetragonal crystals, and is much denser than water (1.643 g cm^{-3} at 269 K). Like water, it is extensively associated even close to the boiling point of 423.3 K. The entropy of vaporisation (114.5 J K^{-1} mol^{-1}) is high, indicating the destruction of a considerable 'local order' (11.8.1) when the liquid vaporises, which is characteristic of polar liquids. The dipole moment is in fact high (2.01 Debyes); the dielectric constant is large (84.2 at 273 K); and the autoprotolysis constant (9.3), $a(H_2O_2^+)a(HO_2^-) = 1.55 \times 10^{-12}$ at 298 K, indicates that the liquid is a good ionising solvent.

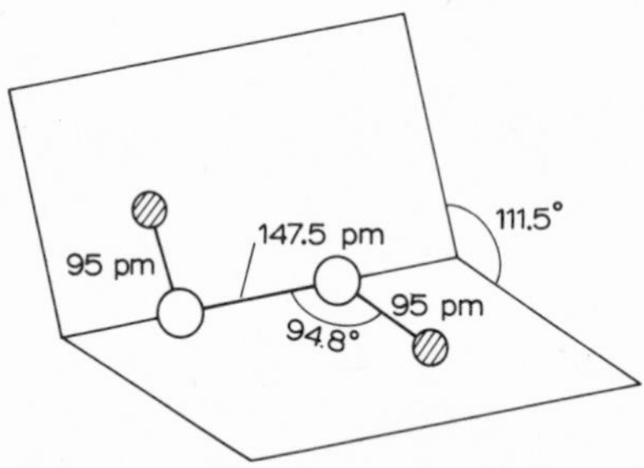

Fig. 24.1. Structure of H_2O_2 molecule.

The enthalpy and free energy of formation are much less negative than those for water:

$$H_2(g) + O_2(g) \rightarrow H_2O_2(g) \qquad \Delta H_f = -141 \text{ kJ mol}^{-1}$$
$$\Delta G_f = -123 \text{ kJ mol}^{-1}$$

Liquid hydrogen peroxide is thermodynamically unstable:

$$H_2O_2(l) \rightarrow H_2O(l) + \tfrac{1}{2} O_2(g) \qquad \Delta H = -98.5 \text{ kJ mol}^{-1}$$
$$\Delta G^0 = -123 \text{ kJ mol}^{-1}$$

but its decomposition at 298 K is not rapid in the absence of catalysts. Catalysts are silver, the platinum group metals, cobalt, iron, copper, MnO_2 and several other oxides. As usual, the rate of reaction depends on their surface area.

Hydrogen peroxide is a strong oxidising agent in both acid and alkaline solutions:

$$H_2O_2 + 2\,H^+ + 2\,e \rightleftharpoons 2\,H_2O \qquad E^0 = +1.77 \text{ V}$$
$$HO_2^- + H_2O + 2\,e \rightleftharpoons 3\,OH^- \qquad E^0 = +0.87 \text{ V}$$

But with some strong oxidising agents, such as Cl_2, or MnO_4^- in acid solution, hydrogen peroxide is itself oxidised:

$$2\,MnO_4^- + 5\,H_2O_2 + 6\,H^+ \rightarrow 2\,Mn^{2+} + 8\,H_2O + 5\,O_2$$
$$Cl_2 + H_2O_2 \rightarrow 2\,HCl + O_2$$

In acid solution:

$$2\,H^+ + O_2 + 2\,e \rightleftharpoons H_2O_2 \qquad E^0 = +0.68 \text{ V}$$

but in alkaline solution

$$\tfrac{1}{2}\,O_2 + H^+ + e \rightleftharpoons OH \rightleftharpoons \tfrac{1}{2}\,H_2O_2 \qquad E^0 = +0.22 \text{ V}$$

and oxidation can proceed more readily.

The standard redox potential of the cyanoferrate(III)—cyanoferrate(II) couple (+0.356 V) lies between these two values. As a result H_2O_2 will oxidise a cyanoferrate(II) in acid solution:

$$2\,Fe(CN)_6^{4-} + H_2O_2 + 2\,H^+ \rightarrow 2\,Fe(CN)_6^{3-} + 2\,H_2O$$

but in alkaline solution cyanoferrate(III) oxidises H_2O_2:

$$2\,Fe(CN)_6^{3-} + H_2O_2 + 2\,OH^- \rightarrow 2\,Fe(CN)_6^{4-} + 2\,H_2O + O_2$$

In the oxidation of aqueous $H_2{}^{18}O_2$ by chlorine and MnO_4^-, and in the catalytic decomposition of hydrogen peroxide by Fe^{3+} and MnO_2, all the molecular oxygen that is released comes from the H_2O_2, none from the water. Evidently the O—O bond of the peroxide is not broken.

24.2. Production of Hydrogen Peroxide

In a formal sense, H_2O_2, with its oxygen atoms in the oxidation state −1, can be prepared by oxidising O^{2-} or reducing molecular O_2. Both methods have been used.

Formerly, its production was based on the electrolysis of ammonium sulphate in an excess of sulphuric acid between platinum electrodes at high current density; this produces a solution of the ammonium salt of perdisulphuric acid by oxidation at the anode and liberates hydrogen at the cathode:

$$2\ NH_4HSO_4 \rightarrow (NH_4)_2S_2O_8 + H_2$$

The solution, on heating at 5 kPa, hydrolyses and a distillate of water and H_2O_2 is obtained:

$$(NH_4)_2S_2O_8 + 2\ H_2O \rightarrow 2\ NH_4HSO_4 + H_2O_2$$

The aqueous H_2O_2 is concentrated to 98% by low-pressure fractionation, water being more volatile than hydrogen peroxide.

Nowadays, an autoxidation process is used in the manufacture of hydrogen peroxide. Air is passed through a 10% solution of 2-ethylanthrahydroquinol in a mixture of benzene and 7—11 carbon-atom alcohols, or an alternative mixed solvent of a safer character.

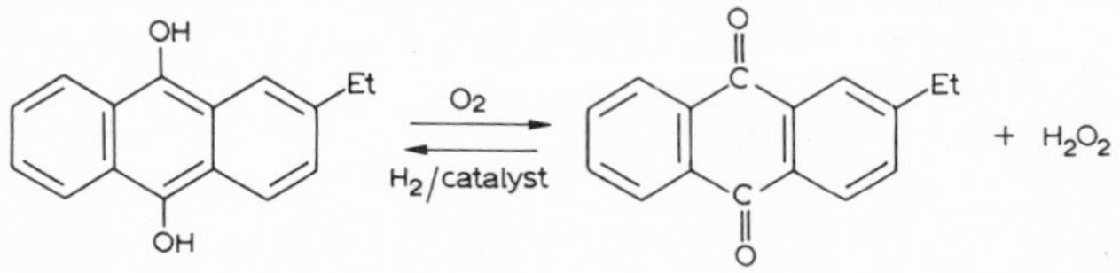

When the concentration of H_2O_2 reaches about 5.5 g/l, the organic phase is extracted with water to give an 18% aqueous solution which is concentrated as described above. The 2-ethylanthraquinone which remains in the organic phase is reduced to the original compound with hydrogen in the presence of palladium on an inert support, the catalyst being suspended in the liquid by the stream of gas and recycled. Very pure and anhydrous H_2O_2 is obtained by fractional crystallisation, but care must be taken to avoid contamination with organic materials which form explosive mixtures.

24.3. Peroxides

A peroxide contains, by definition, the anion $(O—O)^{2-}$ and produces H_2O_2 when acidified. It is thus incorrect to apply the term to dioxides such as NO_2 and PbO_2. Peroxides are formed by all Group IA and several Group IIA metals: Li_2O_2, Na_2O_2, K_2O_2, Rb_2O_2, Cs_2O_2, and MgO_2, CaO_2, SrO_2, BaO_2.

The best known of these, sodium peroxide, Na_2O_2, is made by heating sodium in oxygen. The pale yellow commercial material contains about 10% NaO_2. The compound $Li_2O_2 \cdot H_2O_2 \cdot H_2O$ is precipitated when alcoholic H_2O_2 is added to LiOH; careful drying gives the anhydrous compound. The other peroxides of this group are made by controlled oxidation; either by using the calculated amount of air on the metal at 600 K, or by passing the required amount of air through a solution of the metal in liquid ammonia at 220 K. The compounds all give H_2O_2 on acid hydrolysis:

$$Na_2O_2 + H_2SO_4 \xrightarrow{\text{cold}} Na_2SO_4 + H_2O_2$$

Belief in the existence of $(O—O)^{2-}$ ions in the Group I compounds is based on their reactions, not on their crystal structure which is unknown.

Of the Group IIA peroxides, that of barium is most stable. It is made by heating BaO in air or oxygen at 700 K. SrO_2 is made less easily; at 700 K and 10 MPa the equilibrium mixture $2SrO + O_2 \rightleftharpoons 2SrO_2$ contains about 15% of the strontium as SrO_2. Octahydrates of these two peroxides, and also $CaO_2 \cdot 8\,H_2O$, are precipitated by adding H_2O_2 to the appropriate aqueous hydroxide. All the octahydrates can be dehydrated to the anhydrous peroxides. The monohydrate, $BaO_2 \cdot H_2O$, consists of Ba^{2+} ions and helical chains of peroxide groups, held together by hydrogen bonds. Anhydrous MgO_2 is made by adding ~0.15 *M* NaOH to $MgSO_4$ dissolved in 30% H_2O_2 and drying the precipitate over phosphorus pentoxide.

Of the Group IIB peroxides, anhydrous ZnO_2 is prepared in a similar way to MgO_2, and CdO_2 is made by the action of 30% H_2O_2 on an ammoniacal solution of $CdSO_4$ and drying the precipitate produced at 390 K. Like MgO_2, the zinc and cadmium peroxides have the pyrites structure with an O—O distance, in the peroxide ion, of 150 pm.

24.4. Superoxides

The alkali metals other than lithium form coloured compounds of the formula MO_2, containing the anion O_2^-; and termed superoxides. NaO_2 is formed only at high oxygen pressures, those of the other metals more easily. In NaO_2 the arrangement of oxygen anions is not strictly ordered. The superoxide KO_2 is similar to BaO_2 in structure and stability:

Ionic radius		O—O distance	
K^+	133 pm	O_2^-	132—135 pm
Ba^{2+}	135 pm	O_2^{2-}	148—149 pm

The paramagnetism of the superoxides suggests an ionic picture featuring the O_2^- anion; the moment of 2.04 μ_B observed for KO_2 is characteristic of one unpaired electron.

Canisters containing potassium superoxide are a source of oxygen for high-altitude climbing. In the presence of a little $CuCl_2$, it reacts with expired CO_2, one gram setting free 236 cm^3 O_2 at S.T.P. The compound reacts even with ice-cold water:

$$2\,KO_2 + 2\,H_2O \rightarrow 2\,KOH + H_2O_2 + O_2$$

and with acetic acid in diethyl phthalate solution:

$$2\,CH_3COOH + 2\,KO_2 \rightarrow O_2 + H_2O_2 + 2\,CH_3COOK$$

Potassium, rubidium and caesium also form oxides of the composition M_2O_3. When white K_2O_2 is heated in oxygen at very low pressure red K_2O_3 is produced. The magnetic susceptibilities of Rb_2O_3 and Cs_2O_3 are consistent with the presence of superoxide ions, O_2^-, and peroxide ions, O_2^{2-}, in the proportion 2 : 1.

24.5. Hydroperoxides

When sodium peroxide is added to ethyl alcohol containing a little sodium ethoxide, sodium hydroperoxide is produced:

$$Na_2O_2 + EtOH \rightarrow NaOEt + NaOOH$$

In the absence of NaOEt the compound reacts with the alcohol:

$$NaOOH + EtOH \rightarrow NaOEt + H_2O_2$$

Following this further reaction occurs:

$$2\ NaOOH + H_2O_2 \rightarrow (NaOOH)_2 \cdot H_2O_2$$

A compound in the same class, potassium hydroperoxide peroxohydrate $(KOOH)_2 \cdot 3\ H_2O_2$, has also been isolated.

24.6. Peroxoacids

A peroxoacid is one which contains the peroxide linkage and either (i) is formed by the action of $H_2O_2^-$ on the normal acid, or (ii) gives H_2O_2 on treatment with dilute H_2SO_4. The empirical Riesenfeld test involves the addition of the acid or a salt to a 30% KI solution buffered to pH 7.5—8 with $NaHCO_3$. It is claimed that only true peroxoacids oxidise the KI to iodine under these conditions. However, the only unambiguous evidence for the peroxide link must come from determinations of structure.

24.6.1. The peroxosulphuric acids

The best-characterised peroxoacids are those of Group VI. There are two peroxoacids of sulphur, peroxodisulphuric acid, $H_2S_2O_8$, and peroxomonosulphuric acid, H_2SO_5. The former is made by the electrolysis of 50% H_2SO_4 at 273 K with a Pt anode and a high current density; the HSO_4^- ions are oxidised at the anode to $H_2S_2O_8$. The acid can also be made by the action of 100% H_2O_2 on chlorosulphuric acid:

$$\begin{matrix} O-H \\ | \\ O-H \end{matrix} + \begin{matrix} Cl \cdot SO_2 \cdot OH \\ \\ Cl \cdot SO_2 \cdot OH \end{matrix} \longrightarrow \begin{matrix} O-SO_2 \cdot OH \\ | \\ O-SO_2 \cdot OH \end{matrix} + 2\ HCl$$

The HCl is removed in a vacuum, leaving the colourless, crystalline $H_2S_2O_8$, *m.p.* 328 K, which is stable when dry but hydrolyses easily:

$$H_2S_2O_8 + H_2O \rightarrow H_2SO_4 + H_2SO_5$$
$$H_2SO_5 + H_2O \rightarrow H_2SO_4 + H_2O_2$$

Ammonium peroxodisulphate, $(NH_4)_2S_2O_8$, is produced at the anode when a mixture of $(NH_4)_2SO_4$ and H_2SO_4 is electrolysed. The less soluble potassium salt can be precipitated by adding $KHSO_4$ to the solution:

$$(NH_4)_2S_2O_8 + 2\ KHSO_4 \rightarrow 2(NH_4)HSO_4 + K_2S_2O_8$$

Peroxomonosulphuric acid, H_2SO_5, is made by
(i) grinding $K_2S_2O_8$ with concentrated H_2SO_4, allowing the slurry to stand, and then pouring it on to ice:

$$H_2S_2O_8 + H_2O \rightarrow H_2SO_4 + H_2SO_5$$

(ii) treating SO_3 with 100% H_2O_2:

$$H_2O_2 + SO_3 \rightarrow H_2SO_5$$

(iii) adding the calculated amount of anhydrous H_2O_2 to well-cooled chlorosulphuric acid:

$$\begin{array}{l}OH\\ |\\ OH\end{array} + Cl{\cdot}SO_2{\cdot}OH \longrightarrow \begin{array}{l}O{-}SO_2{\cdot}OH\\ |\\ OH\end{array} + HCl$$

Like $H_2S_2O_8$, it is a colourless crystalline solid, but differs in being monobasic. The salts themselves are unstable, although a benzoyl derivative of the potassium salt has been made:

$$\begin{array}{l}O{-}SO_2{\cdot}OK\\ |\\ O{-}CO{\cdot}C_6H_5\end{array}$$

Both acids give the Riesenfeld reaction and oxidise Fe^{2+} to Fe^{3+}. Unlike H_2O_2, however, neither acid has any action on $KMnO_4$, CrO_3 or TiO^{2+}. The two acids differ in their reactions with aniline and with KI at low pH:

	Potassium iodide	Aniline
$H_2S_2O_8$:	I_2 set free slowly.	Oxidised to aniline black.
H_2SO_5:	I_2 set free quickly.	Oxidised to nitrosobenzene.

24.6.2. Sulphur—fluorine compounds containing peroxo groups

Peroxocompounds have been made from complex oxofluorides of sulphur. The compound F_5SOF, made by treating OSF_2 with an excess of fluorine, is converted by ultraviolet irradiation into bis(pentafluorosulphur) peroxide, $F_5S{-}O{-}O{-}SF_5$, a dense, colourless liquid, *b.p.* 322 K, which liberates iodine rather slowly from KI.

Peroxodisulphuryl difluoride, $FO_2S{-}O{-}O{-}SO_2F$, *b.p.* 340 K, is produced in the electrolysis of fluorosulphuric acid and also in the catalytic fluorination of SO_3 vapours by F_2 when these are passed over AgF_2 coated on copper ribbon. It can behave as an oxygenating agent (a), a fluorosulphonating agent (b) or as both (c):

(a) $CO + S_2O_6F_2 \rightarrow CO_2 + S_2O_5F_2$
(b) $HgO + 2\ S_2O_6F_2 \rightarrow Hg(SO_3F)_2 + O_2 + S_2O_5F_2$
(c) $2\ SOClF + 3\ S_2O_6F_2 \rightarrow 4\ S_2O_5F_2 + Cl_2$

It liberates iodine rapidly from aqueous KI at room temperature.

24.6.3. Group V peroxoacids

Peroxonitric acid, HNO_4, is reported to be formed when anhydrous H_2O_2 is mixed with N_2O_5 at 200 K. At high concentration the compound decomposes explosively below 273 K; a 70% aqueous solution is moderately stable and weaker solutions hydrolyse rapidly. Unlike a solution of HNO_3, H_2O_2 or HNO_2, that of HNO_4 liberates bromine from KBr. It oxidises aniline to nitrobenzene.

When HCl is added to sodium nitrite in 5% aqueous H_2O_2 a fugitive, light-brown colour is produced which disappears with the evolution of oxygen. When the still brown liquid is run into NaOH, a bright yellow colour develops which may persist for upwards of 12 hours. The brown liquid is a dilute solution of peroxonitrous acid, H—O—O—N=O. Among other evidence for the presence of this compound is that the cold solution gives, with aromatic compounds, *o*-hydroxy and *m*-nitro derivatives. Their production can be explained by the following reactions:

$$HO{-}O{-}N{=}O \rightarrow HO\cdot + NO_2$$

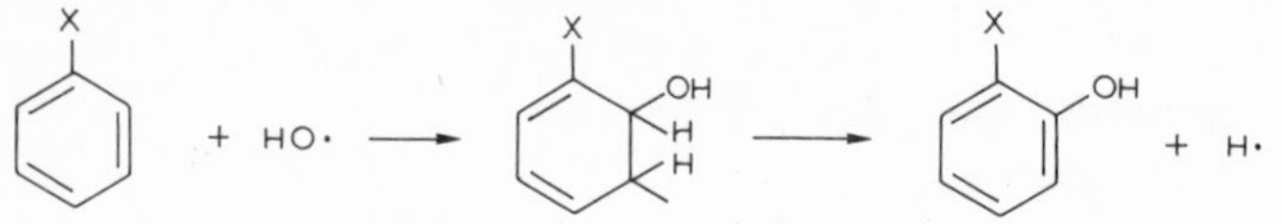

and

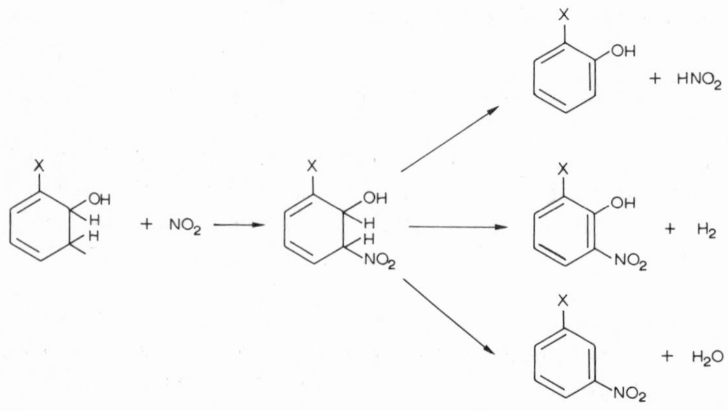

A peroxomonophosphoric acid, H_3PO_5, presumably

```
     OH
     |
O—P—O—OH
     |
     OH
```

is obtained by the addition of ~14% H_2O_2 to P_2O_5 in acetonitrile which moderates the reaction. The aqueous solution oxidises aniline to a mixture of nitrosobenzene and nitrobenzene, and manganese(II) salts to permanganates in the cold. Salts of peroxodiphosphoric acid can be made by the electrolysis of concentrated solutions of orthophosphates: $2K_2HPO_4 \rightarrow K_4P_2O_8$.

24.6.4. Group IV peroxoacids

When a concentrated solution of K_2CO_3 is electrolysed, at about 260 K, with a high current density, the pale blue peroxocarbonate, $K_2C_2O_6$, is deposited at a smooth platinum anode. The compound responds to the Riesenfeld test, reacts with dilute H_2SO_4 to give H_2O_2 and with MnO_2 to give oxygen, and oxidises PbS to $PbSO_4$. This electrolytically produced peroxocarbonate is believed to contain the ion

```
[O—C—O—O—C—O]2-
    ||      ||
    O       O
```

Several alkali-metal peroxocarbonates are claimed to have been made by passing CO_2 into suspensions of their peroxides in ice-cold water. Later Group IA peroxocarbonates, for instance $M_2CO_4 \cdot x\,H_2O$, $MHCO_4$ and $M_2C_2O_6$, were obtained by treating saturated carbonate solutions with 30% H_2O_2 and precipitating with alcohol. Some of these liberate iodine and oxygen, some oxygen only, from a neutral solution of KI.

24.7. Some Peroxo-compounds of Transition Metals

There is increasing interest in peroxocompounds of the transition metals and in organometallic peroxides, but only a few of the transition-metal peroxocompounds can be dealt with here.

When hydrogen peroxide is added to a dichromate in dilute H_2SO_4 a blue colour is produced which fades rapidly unless extracted with ether. The ether extract contains $Et_2O \cdot CrO_5$. Organic bases give more stable co-ordination compounds such as $C_5H_5N \cdot CrO_5$. The addition of alcoholic H_2O_2 and KOH to an ethereal solution of CrO_5 yields a blue salt, $KCrO_6 \cdot H_2O$. The compound has 5 peroxo groups for every 2 chromium atoms and it has been suggested that the diamagnetic anion is

```
     O2           O2
     |            |
[O—Cr—O—O—Cr—O]2-
     |            |
     O2           O2
```

A series of red peroxochromates, $M^I_3CrO_8$, can be made by adding H_2O_2 to soluble chromates under slightly alkaline conditions. Red magnesium peroxochromate, $Mg_3(CrO_8)_2 \cdot 13\ H_2O$, is also known, as are the double salts of calcium, strontium and barium peroxochromates with alkali-metal peroxochromates, $K_2Ca_5(CrO_8)_4 \cdot 19\ H_2O$ and $K_2Ca_2(CrO_8)_2 \cdot 7\ H_2O$. The CrO_8^{3-} anion has four peroxide groups per Cr atom, the centres of the O—O bonds being tetrahedrally arranged about the central Cr atom.

Hydrogen peroxide reacts with acidified molybdate solutions. The acid $H_2Mo_2O_{11}$ is present in solutions of molybdate in concentrated perchloric acid to which H_2O_2 has been added. The so-called 'permolybdic acid' is regarded as a salt of this acid with the cation $[HMo_2O_6]^+$, not itself peroxidic.

Vanadates, niobates and tantalates react with alkaline H_2O_2 to give peroxo-salts formulated as $M^I_3VO_8$, $M^I_3NbO_8$ and $M^I_3TaO_8$. In strongly acid solution the vanadyl ion, VO^{3+}, gives a red colour with H_2O_2:

$$VO^{3+} + H_2O_2 \rightarrow \underset{\text{(red)}}{VO_2^{3+}} + H_2O$$

In weaker acid:

$$VO_2^{3+} + H_2O_2 + 2\ H_2O \rightleftharpoons \underset{\text{(yellow)}}{VO_6^{3-}} + 6\ H^+$$

Titanium, zirconium and hafnium peroxo-compounds are produced by adding ammoniacal H_2O_2 to their salts. The oxide salt, $TiOCl_2$, gives a compound formulated $Ti(OH)_3OOH$. This, with KOH, produces $K_4TiO_8 \cdot 6\ H_2O$ which has been isolated as a solid; it does not, however, give the Riesenfeld reaction. In acid solution the $(TiO)^{2+}$ ion reacts with H_2O_2 to give the yellow colour ascribed to the hydrated $(TiO_2)^{2+}$ ion. The discharge of this colour by fluoride ions is almost certainly due to TiF_6^{2-} formation. A solid, $K_2(TiO_2)(SO_4)_2 \cdot 3\ H_2O$, has been isolated which gives the Riesenfeld reaction.

24.8. Peroxohydrates

A compound of the empirical formula $NaBO_3 \cdot 4\ H_2O$ is obtained by the action of Na_2O_2 on a cold solution of borax, $Na_2B_4O_7 \cdot 10\ H_2O$, and also by the electrolysis of a solution of borax and Na_2CO_3 with a platinum gauze anode. It does not free iodine from KI (see below). The substances $LiBO_4 \cdot H_2O$ and $KBO_5 \cdot H_2O$ have been made by adding ~14% H_2O_2 to metaborate solutions and precipitating with alcohol.

There are so many similarities in the physical properties of H_2O_2 and water that the existence of salts in which H_2O_2 forms peroxohydrates just as H_2O forms hydrates is not surprising. It is now known that the preparation from borax is such a compound, and that it is correctly formulated $NaBO_2 \cdot H_2O_2 \cdot 3\ H_2O$. The material is used in tooth powders and for rapid oxidation of vat dyes. A peroxohydrate of sodium carbonate,

$2\ Na_2CO_3 \cdot 3\ H_2O_2$, made by adding H_2O_2 to Na_2CO_3 solution, is used in washing powders.

Peroxohydrates can be distinguished from true peroxosalts by the ease with which ether extracts H_2O_2 from their aqueous solutions, and by their failure to give the Riesenfeld reaction. Distinctions based on such empirical tests cannot, however, be considered wholly satisfactory, and the chemistry of peroxo-compounds will not be fully understood until there is more structural information.

Further Reading

E.A.V. Ebsworth, J.A. Connor and J.J. Turner, The chemistry of oxygen, Pergamon, Oxford, 1975.

A.G. Sykes and J.A. Weil, The formation, structure and reactions of binuclear complexes of cobalt. Prog. Inorg. Chem., 13 (1969) 1.

I. Vol'nov, Peroxides, superoxides and ozonides of alkali and alkaline earth metals, Plenum, New York, 1966.

The Halogens — Group VIIB

25

25.1. Introduction

Fluorine, chlorine, bromine, iodine and astatine are the terminal members of their respective periods and form a group in which the elements show a strong family relationship and also a very regular modification of properties. (The alkali metals, separated from the halogens by the noble gases, are the first members of these periods after Group 0 and have a similarly strong family relationship among themselves). The molecules of all the halogens are diatomic at room temperature and thus differ from the elements of Groups V and VI in which the first elements, nitrogen and oxygen alone are diatomic under these conditions. The *m.p.* of chlorine, bromine and iodine are very low compared with those of the previous members of their respective periods, sulphur, selenium and tellurium.

The halogens range in physical properties from an almost colourless gas (F_2) through a dark-red, volatile liquid (Br_2) to an almost black, crystalline solid made up of I_2 molecules. The fall in volatility is due to an increase down the group of the Van der Waals force of attraction between the molecules. The elements are all chemically active and, because of that, none is found in the free state in nature. Their activity makes them powerful disinfectants but dangerous when inhaled in any but very low concentrations.

As their electron configurations show, each element is a single p electron short of a noble-gas structure, and for this reason the dominating feature of their chemistry is the ease with which their atoms acquire an electron and become uninegative ions. That these anions are large has important structural consequences. Fluorine is always formally uninegative; however, the other elements can exhibit positive charge numbers in binary compounds, but only when in combination with strongly electropositive elements. Iodine forms a number of compounds in which unipositive iodine is stabilised by co-ordination, for example:

$$I_2 + AgNO_3 + 2\ C_5H_5N \rightarrow I(NC_5H_5)_2{}^+NO_3{}^- + AgI$$

Ionisation energies in the group are generally high, but fall markedly with atomic number. Electron affinities show a maximum at chlorine. Nevertheless fluorine is the better oxidising agent in aqueous solution, and even when dry

TABLE 25.1

ATOMIC AND PHYSICAL PROPERTIES OF THE HALOGENS

	F	Cl	Br	I	At
Atomic number	9	17	35	53	85
Electronic configuration	[He] $2s^2 2p^5$	[Ne] $3s^2 3p^5$	[Ar] $3d^{10} 4s^2 4p^5$	[Kr] $4d^{10} 5s^2 5p^5$	[Xe] $4f^{14} 5d^{10} 6s^2 6p^5$
$I(1)$/kJ mol^{-1}	1685	1255	1145	1010	(935)
Electron affinity/kJ mol^{-1}	340	356	342	303	(298)
Electronegativity	4.10	2.83	2.74	2.21	
Covalent radius/pm	71	99	114	133	(~140)
Ionic radius/pm	133	181	195	216	(~230)
M.p./K	40	172.16	265.9	386.75	
B.p./K	85	239.10	332	458	
Energy of dissociation of X_2/kJ mol^{-1}	155	243	193	151	(~116)
Hydration energy of X^-/kJ mol^{-1}	485	350	320	280	(~277)
E^0, $X_2/2X^-$/V	2.87	1.36	1.07	0.53	(~0.3)

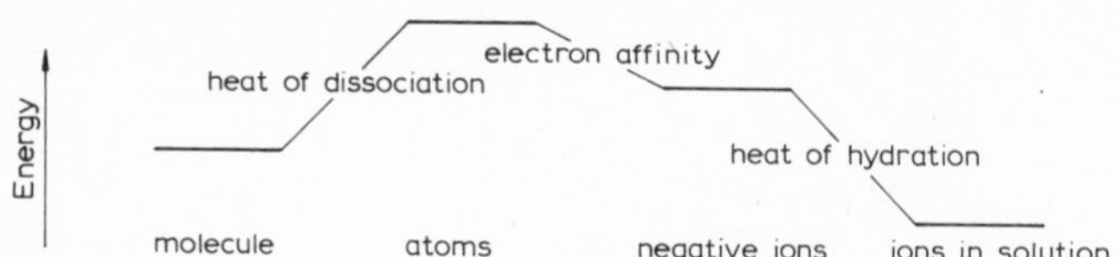

Fig. 25.1. Energetics of the change from free molecule to ion in solution.

will usually replace chlorine from its compounds.

The redox potential, E^0, $X_2/2X^-$, measures a free-energy change, usually dominated by the ΔH term, but depending, as may be seen from the energy diagram (Fig. 25.1), on (a) the energy needed to break the molecule into atoms (the enthalpy of dissociation), (b) the energy liberated when the atom is converted into a negative ion (the electron affinity), and (c) the energy set free on the hydration of the ion. For the fluorine molecule (b) is less than for the chlorine molecule but, since the energy needed to break the F—F bond is also less and the hydration energy more, the total energy drop is much greater. Molecular fluorine is, in fact, an extremely powerful oxidising agent. In spite of their lower dissociation energies, bromine and iodine are weaker oxidising agents than chlorine; this is due to their smaller electron affinities and smaller hydration energies.

The great reactivity of fluorine largely stems from the low energy of the F—F bond. The figure of 155 kJ mol^{-1} is derived from a study of its dissociation between 800 and 1100 K and from a Born-type treatment of heats of dissociation and heats of formation of alkali-metal fluorides together with their heats of sublimation and those of the metals concerned. The F—F bond is weak because of repulsion between the non-bonding electrons; the stronger X—X bond, actually in Cl_2 and Br_2, and relatively in I_2, is due to hybridisation of p and d orbitals. The weakness of the σ bond between atoms of the second period elements is also evident in the molecules $H_2N{-}NH_2$ and HO—OH.

Another factor contributing to the exothermicity of many of the reactions of elementary fluorine is the short, strong bonds formed by its atoms with those of most other elements. When crystalline fluorides are formed their lattice energies are high because the F^- ion is comparatively small.

25.2. Oxidation States of the Halogens

The only oxidation state of fluorine which is stable in aqueous solution is the —1. The principal oxidation states of chlorine, bromine and iodine are —1, +1, +3, +5, and +7. The oxoacids are all powerful oxidising agents with respect to their reduction to the aqueous hydrogen halide, and even in alkaline solutions the higher oxidation states are easily reduced to —1. All the points lie close to the straight line between the highest state and the lowest. This means that all except the —1 state are oxidising and all except

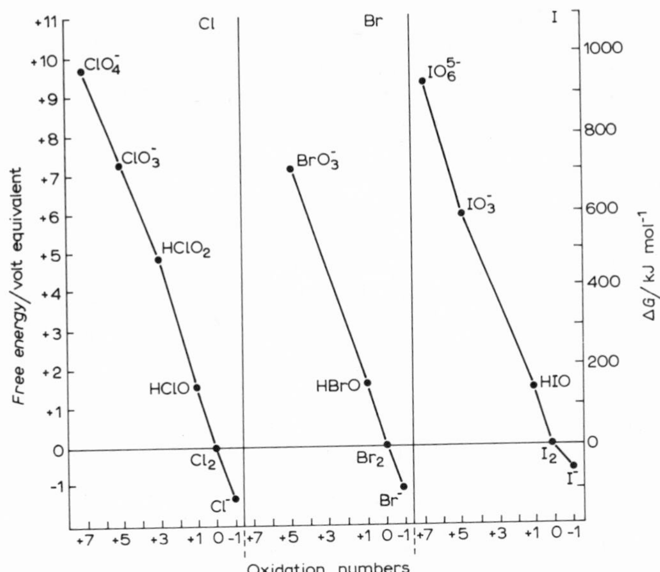

Fig. 25.2. Free energies of oxidation states relative to the elements at pH = 0.

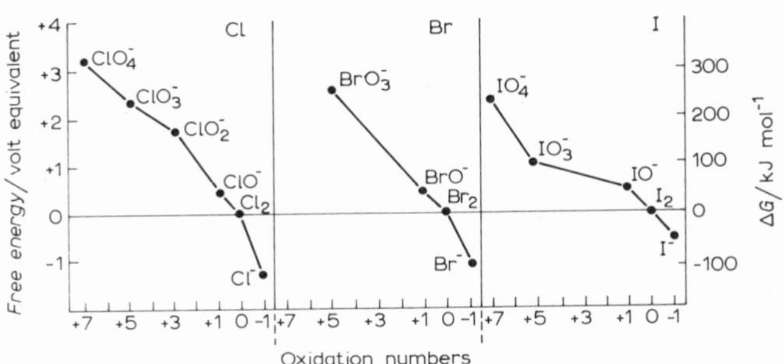

Fig. 25.3. Free energies of oxidation states relative to the elements at pH = 14.

the —1 and +7 states are liable to disproportionate.

At pH 14 the redox potentials, as indicated by the gradients of the graphs (Fig. 25.3), are lower. Features of these diagrams are the instability of Cl_2 and Br_2 (and, to a lesser extent, I_2), to disproportionation in alkaline solution, and the comparative stability of the IO_3^- ion. As in acid solution, the +3 state is an unstable one.

25.3. Stereochemistry of Halogen Complexes

Halogens with a formally positive charge number occur in complexes in which the co-ordination numbers range from 2 to 7 and the other element present is either oxygen or another halogen which is more electronegative

TABLE 25.2

STEREOCHEMISTRY OF HALOGEN COMPLEXES

Electron pairs round central atom	Hybrids	Electrons used in π bonding	Lone pairs	Charge number of central atom	Shape of molecule or ion	Point group	Examples
4	sp^3	1	2	+3	V-shaped	C_{2v}	ClO_2^-
		2	1	+5	Trigonal pyramid	C_{3v}	ClO_3^-, BrO_3^-
		3	0	+7	Tetrahedron	T_d	ClO_4^-, IO_4^-
5	sp^3d	0	3	+1	Linear	$D_{\infty h}$	ICl_2^-, I_3
		0	2	+3	T-shaped	C_{2v}	ClF_3
		1	1	+5	Distorted tetrahedron	C_{2v}	$IO_2F_2^-$
6	sp^3d^2	0	2	+3	Square	D_{4h}	ICl_4^-
		0	1	+5	Square pyramid	C_{4v}	IF_5
		1	0	+7	Octahedron	O_h	IO_6^{5-}
7	sp^3d^3	0	0	+7	Pentagonal bipyramid	D_{5h}	IF_7

than the first. The complexes may be neutral molecules, but are usually ions. In complexes with oxygen, the halogen atom is surrounded by oxygen atoms; in interhalogen compounds the larger halogen atom is surrounded by the smaller halogen atoms. The number of bond pairs around the central atom can be increased by either (a) the use of some of its electrons to form π bonds with surrounding atoms, or (b) the promotion of some of its electrons to nd levels followed by the formation of σ bonds with the surrounding atoms.

The stereochemistry of typical complexes involving a formally positive halogen is summarised in Table 25.2.

All the molecules and ions employing four electron pairs are basically tetrahedral, those employing five pairs, trigonal bipyramidal and those with six pairs, octahedral. Bond directions are determined by the strong repulsive forces exerted by the lone-pair electrons on the bond pairs.

In an unsymmetrical molecule some distortion of the simple shapes can occur. Thus ClF_3 is not exactly T-shaped. Microwave spectroscopy shows a C_{2v} structure with the bond lengths and inter-bond angles to be as indicated in Fig. 25.4. This is consonant with the idea that lone pairs occupy two positions in a trigonal bipyramid arising from sp^{3}d hybridisation.

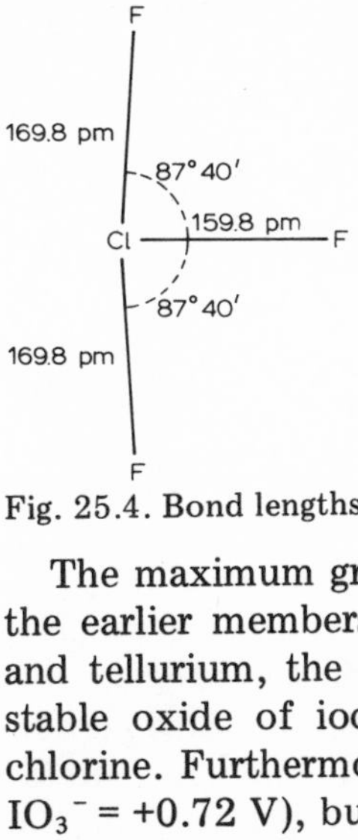

Fig. 25.4. Bond lengths and interbond angles of ClF_3.

The maximum group charge number of +7 is less stable in iodine than in the earlier members of the group; in this respect iodine resembles bismuth and tellurium, the two elements which precede it in the period. The most stable oxide of iodine is I_2O_5, whereas Cl_2O_7 is the only stable oxide of chlorine. Furthermore, the periodates are less stable than iodates (E^0, IO_4^-/IO_3^- = +0.72 V), but perchlorates are more stable than chlorates (E^0, ClO_4^-/ClO_3^- = +0.17 V).

25.4. Occurrence and Separation of the Halogens

25.4.1. Fluorine

The principal source of fluorine (0.08% of the lithosphere) is fluorspar, CaF_2, but much fluorine is recovered from industrial effluent gases arising from the aluminium, iron and other industries.

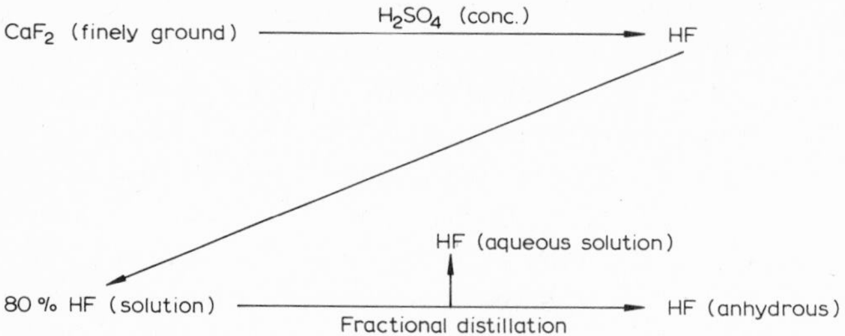

Fig. 25.5. Preparation of HF from fluorspar.

High grade fluorspar is used for the manufacture of anhydrous hydrofluoric acid. The anhydrous acid is a source of fluorine itself, of the fluorides of many metals, and also of freons such as CF_2Cl_2.

Industrial cells for preparing fluorine are rectangular steel vessels 1.5 m × 0.5 m and 80 cm deep. The electrolyte, KHF_2 together with up to 0.6 molar parts of HF kept molten at 378—398 K, is electrolysed by a current of about 0.5 A cm^{-2} between carbon anodes and sheet-steel cathodes. A monel metal diaphragm, perforated below the surface of the liquid but continuous above it, serves to keep separate the fluorine and hydrogen produced. Sealing and electrical insulation is provided by Teflon (CF_2 polymer). There are small 10 A cells convenient for laboratory work and very reliable in operation. Electrolysis depletes the cell of HF which is added from time to time in the anhydrous form. Liquid fluorine is now available in industrial quantities. Interest in the production of fluorine and its compounds has been stimulated by the use made of uranium hexafluoride for separating, by gaseous diffusion, the ^{235}U isotope from the 99.28% of ^{238}U in natural uranium, and of uranium tetrafluoride in the production of uranium metal (31.3).

25.4.2. Chlorine

The element (0.19% of lithosphere) is derived mainly from NaCl which is either crystallised from brines or mined. The gas is a product of the electrolysis of aqueous sodium chloride for caustic soda production, with carbon anodes and a mercury cathode. It is also a by-product of the manufacture of metallic sodium, and also of magnesium and calcium, by electrolysing the appropriate fused chloride. Its chief uses are as a bleach, a bactericide, and an industrial chemical.

25.4.3. Bromine

This element (0.01% of lithosphere), once largely derived from salt deposits, is now obtained chiefly from sea-water, by passing chlorine into it at pH 3.5:

$$2\,Br^- + Cl_2 \rightarrow Br_2 + 2\,Cl^-$$

The bromine is blown out with air and absorbed in Na_2CO_3 solution to give bromate and bromide. On acidification the element can be distilled off:

$$6\ H^+ + BrO_3^- + 5\ Br^- \rightarrow 3\ H_2O + 3Br_2$$

25.4.4. Iodine

Iodine (10^{-4} % of lithosphere) is obtained mainly from iodates present in Chilean nitrates. Iodate-rich solutions are reduced with sodium bisulphite to liberate the element:

$$2\ IO_3^- + 5\ HSO_3^- \rightarrow 3\ HSO_4^- + 2\ SO_4^{2-} + H_2O + I_2$$

25.4.5. Astatine

The longest lived of the 21 isotopes of astatine which have been defined are ^{210}At (8.3 h) and ^{211}At (7.21 h). Astatine-211 is made by bombarding a cooled ^{209}Bi target with high-energy α-particles, and separating it from the target by vacuum distillation at 600—900 K. Because of its short half-life and consequent high activity, its chemistry has been studied by tracer methods only. In spite of the difficulties, at least four oxidation states have been established. These closely resemble iodine, the element itself being extracted into organic solvents and At^- coprecipitating with AgI. A +5 state is carried by insoluble iodates and at least one other state between 0 and 5) appears to exist in aqueous solution, possibly AtO^- or AtO_2^-.

25.5. Reactions of the Halogens

Most metals combine directly with all the halogens and particularly readily with fluorine. Some non-metals also react:

$$H_2 + X_2 \rightarrow 2\ HX$$
$$2\ P\ (As) + 3\ X_2 \rightarrow 2\ PX_3\ (AsX_3)$$

The reactivity of the halogen decreases with atomic number. Fluorine, particularly, and chlorine to a lesser extent, often oxidise both metals and non-metals to higher states of oxidation than do bromine and iodine. Thus both fluorine and chlorine raise the charge number of phosphorus and arsenic to +5 in such compounds as PCl_5 and AsF_5. Sulphur is converted to SF_6 by fluorine, to SCl_2 by chlorine and to S_2Br_2 by bromine. The formation of SF_6 from its elements may be pictured in terms of the energy changes shown in Fig. 25.6. The energy set free in the formation of SF_6 is large; except in the F_2 molecule itself, fluorine forms stronger bonds than any of the other halogens. On the other hand the dissociation energy of fluorine is only 155 kJ mol^{-1}, and there is therefore a net binding energy available, in spite of the considerable energy of the sulphur sp^3d^2 valence state. Chlorine is more

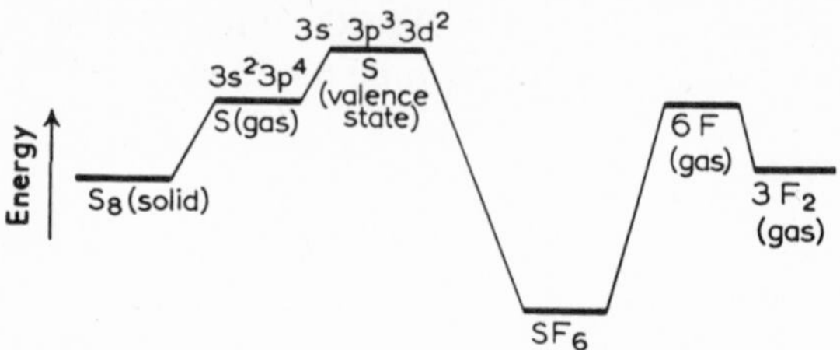

Fig. 25.6. Energetics of the formation of SF_6 from its elements.

difficult to dissociate into atoms than fluorine and forms weaker bonds with sulphur. Insufficient energy is available to offset the valence state promotion energy for SCl_6 formation, and such a compound is not known.

A halogen of lower atomic number oxidises the ion of another halogen of higher atomic number, both when the ion is in solution and in the crystal lattice.

$$Cl_2 + 2\,Br^- \rightarrow Br_2 + 2\,Cl^-$$

The respective reactions of the halogens with water also illustrate the decrease in oxidising power with atomic number. For the reaction

$$4\,H^+ + O_2 + 4\,e \rightleftharpoons 2\,H_2O \qquad E^0 = +\,0.81\ V$$

Since E^0, $F_2/2F^-$ is +2.87 V, the redox potential for

$$2\,F_2 + 2\,H_2O \rightleftharpoons 4\,H^+ + 4\,F^- + O_2 \text{ is } +\,2.06\ V$$

equivalent to a standard free energy change ΔG of —798 kJ mol^{-1}. Fluorine accordingly sets free oxygen from water.

Since E^0, $I_2/2I^-$ is +0.53 V, the redox potential for

$$2\,I_2 + 2\,H_2O \rightleftharpoons 4\,H^+ + 4\,I^- + O_2 \text{ is } -0.28\ V$$

equivalent to a standard free energy change, ΔG, of +105 kJ mol^{-1}. Here the reaction takes the opposite direction and oxygen can oxidise the iodide ion to iodine. With chlorine and bromine the oxidation of water to oxygen is thermodynamically possible but has so high an activation energy that another course is followed:

$$X_2 + 2\,H_2O \rightleftharpoons H_3O^+ + X^- + HOX$$

This reaction is naturally strongly dependent on pH; the addition of alkali favours the formation of halide and hypohalite.

25.6. Interhalogen Compounds

These are made by direct combination of the elements in a nickel tube; the product depends on the conditions, thus for instance:

$$Cl_2 + F_2 \text{ (equal volumes)} \xrightarrow{470\ K} 2\,ClF$$

$$Cl_2 + 3\,F_2 \text{ (excess } F_2) \xrightarrow{550\ K} 2\,ClF_3$$

TABLE 25.3

INTERHALOGEN COMPOUNDS

Type AX	ClF	BrF	IF	BrCl	ICl	IBr
M.p./K	117.5	240	Disproportionates	207	300.5	314
B.p./K	173	293		278	370—3	389
Internuclear distance/pm	162.8	175.6	190.8	213.8	232.1	248.5

Type AX_3	ClF_3	BrF_3	IF_3	I_2Cl_6
Shape	T-shape C_{2v}	T-shape C_{2v}		Planar D_{2h}
A—X distance/pm	169.8	181.0		238
	159.8	172.1		268
Bond angle X—A—X	87.5°	86.25°		84°, 94°
M.p./K	196.8	281.9	Yellow solid dec. 245 K	Dissociates 384

Type AX_5, AX_7	ClF_5	BrF_5	IF_5	IF_7
Shape	Square pyramidal C_{4v}	Square pyramidal C_{4v}	Square pyramidal C_{4v}	Pentagonal bipyramidal D_{5h}
A—X distance/pm	162	168.9	184.4	178.6
	172	177.4	186.9	185.8
Bond angle X—A—X		84.8°	81.9°	
M.p./K	170	212.6	282.5	277.9 subl.
B.p./K	260	314.4	377.6	

Bromine vapour diluted with nitrogen reacts with a limited supply of fluorine to give mainly BrF_3. With an excess of F_2, BrF_5 is the chief product. There are eleven interhalogen compounds which fall into the four classes shown in Table 25.3. For every class with more than one member, the boiling points increase as the difference between the electronegativities of the two halogens increases. The greatest increase is that between BrF_5 and IF_5. Interhalogen compounds containing either three or four of the elements are not known, though the polyhalide ions $ClIBr^-$ and $FICl_3^-$ exist.

Interhalogen compounds of the AX type resemble the halogens themselves in physical properties (Fig. 25.7); the divergences are naturally greatest where differences in electronegativity are marked. In general compounds containing fluorine are more volatile than those in which it is replaced by chlorine, and so on down the group.

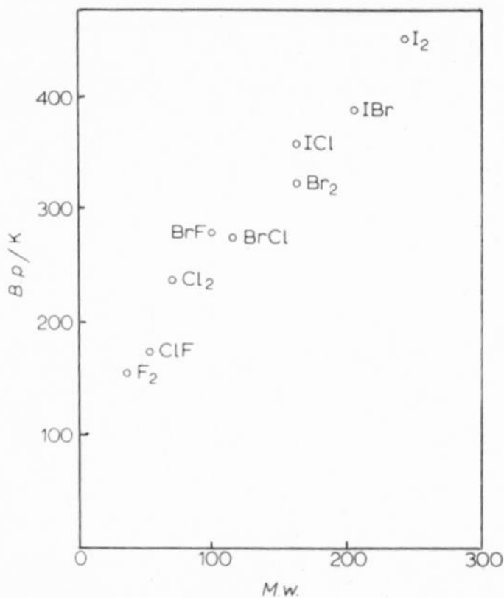

Fig. 25.7. Relation between molecular weight and boiling point for halogens and interhalogen compounds.

The thermal stabilities of the diatomic compounds fall in the order IF > BrF > ClF > ICl > IBr > BrCl, corresponding to the difference in electronegativity between the two-atoms. The more polar the bond, the more thermally stable the molecule. The bond lengths, in the gaseous molecules, relate, in that the greater the difference in electronegativity the shorter the bonds compared with the sum of the two covalent radii (Table 25.3).

The AX compounds usually convert metals to mixed halides. The compounds are more reactive than the elements because the A—X bond energy is less than the X—X bond energy (X being the more electronegative element involved). Hydrolysis of these interhalogens usually proceeds:

$$AX + 2\,H_2O \rightarrow HOA + H_3O^+ + X^- \qquad \text{(X the more electronegative)}$$

They form addition compounds with olefins:

$$\begin{array}{c}\text{—CH=CH—}\\ + \\ \text{A—X}\end{array} \rightarrow \begin{array}{c}\text{—CH—CH—}\\ |\quad\;\; | \\ \text{A}\quad\;\; \text{X}\end{array}$$

And they often do the same with alkali-metal halides:

$NaBr + IBr \rightarrow NaIBr_2$

Of the AX_3 compounds, ClF_3 is the most reactive but BrF_3 is more useful in preparative work. The liquid has a higher conductance:

$2\ BrF_3 \rightleftharpoons BrF_2^+ + BrF_4^-$

It is a valuable fluorinating agent, converting many metals, their oxides and very many of their chlorides, bromides and iodides to fluorides. Some metal fluorides dissolve to give tetrafluorobromites:

$KF + BrF_3 \rightarrow KBrF_4$

or hexafluorobromates:

$SbF_3 + BrF_3 \rightarrow SbBrF_6$

Chlorine trifluoride does not form corresponding compounds. Both ClF_3 and BrF_3 have high entropies of vaporisation, suggesting there is association in the liquid. ICl_3 is much less reactive than the other two AX_3 compounds. Thermal stabilities of the AX_3-type compounds are in the order: $BrF_3 > ClF_3 > ICl_3$.

Iodine trichloride appears to be unique in that the crystals are composed of planar symmetrical dimeric molecules separated by normal Van der Waals' distances. Within each I_2Cl_6 molecule, the terminal I—Cl bonds are similar in length to those in ICl, but the bridging I—Cl bonds are longer.

Of the AX_5 compounds, bromine pentafluoride is the most reactive, resembling ClF_3 in acting very violently, too violently to be used undiluted for the preparation of fluorides. Liquid IF_5 is a good conductor:

$2\ IF_5 \rightleftharpoons IF_4^+ + IF_6^-$

It reacts with KI at its boiling point to give KIF_6.

Iodine heptafluoride, the only example of the AX_7 type, can be made by heating IF_5 with F_2 at 520—540 K. It is comparable with ClF_3 and BrF_5 in its violent fluorinating action. In shape it is the unusual pentagonal bipyramid.

Several oxofluorides are known. ClO_2 reacts with F_2 to give chloryl fluoride, $FClO_2$. The compound is produced by the action of BrF_3 on potassium chlorate:

$6\ KClO_3 + 10\ BrF_3 \rightarrow 6\ KBrF_4 + 2\ Br_2 + 3\ O_2 + 6\ FClO_2$

It forms solid additives with BF_3 and SbF_5, regarded as chloronium salts $ClO_2^+BF_4^-$ and $ClO_2^+SbF_6^-$. Perchloryl fluoride, $FClO_3$, can be made by treating a perchlorate with fluorosulphuric acid. It is a colourless, inert, thermally stable gas in contrast with the reactive $FClO_2$. Structurally $FClO_3$ is a tetrahedral molecule centred on the chlorine atom. BrO_2F can be made by the direct fluorination of BrO_2 at a low temperature. IO_2F and IO_3F are obtained by the reactions:

$2\ I_2O_5 + 2\ F_2 \xrightarrow{HF} 4\ IO_2F + O_2$

$$2\,HIO_4 + 2\,F_2 \xrightarrow{HF} 2\,IO_3F + 2\,HF + O_2$$

25.6.1. Polyhalide ions

Halide ions, either in solution or in crystalline salts, frequently react with halogens and interhalogen compounds:

$$KI + I_2 \rightleftharpoons KI_3$$
$$CsBr + IBr \rightleftharpoons CsIBr_2$$

Many crystalline polyhalides contain solvent molecules, removal of which causes decomposition of the polyhalide:

$KI_3 \cdot H_2O$ $KI_7 \cdot H_2O$ $CsI_{10} \cdot 2\,C_6H_6$ $HICl_4 \cdot 4\,H_2O$

Polyhalides are unstable towards dissociation into monohalides and halogens or interhalogen compounds. The lighter halide atoms remain in the metal halide:

$$CsICl_2 \xrightarrow{heat} CsCl + ICl$$

The products are not CsI and Cl_2, presumably because the former course is favoured energetically by the higher lattice energy of CsCl. For trihalides formed by the same metal, the order of thermal stability, based on dissociation pressures, is:

$$I_3^- > IBr_2^- > ICl_2^- > I_2Br^- > Br_3^- > BrCl_2^- > Br_2Cl^-$$

The polyhalide anions have interesting structures. The trihalides are almost linear:

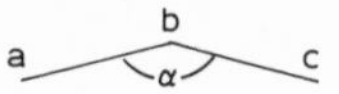

a b c α

with α in the range 171—179°. Surprisingly the I_3^- and Br_3^- ions are both slightly unsymmetrical: ab ≠ bc. The ICl_4^- ion in $KICl_4 \cdot H_2O$ and the BrF_4^- ion in $KBrF_4$ are both square, but I_5^- in NMe_4I_5 is approximately L-shaped:

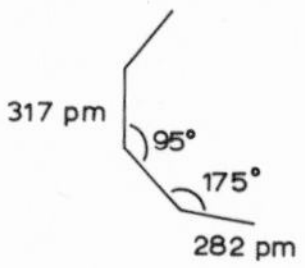

317 pm 95° 175° 282 pm

The compound Cs_2I_8 has a Z-shaped I_8^{2-} ion:

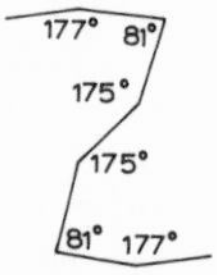

177° 81° 175° 175° 81° 177°

The L.C.A.O. method applied to polyhalide ions indicates that the bond angles should be either ~90° or ~180°, since only the p-functions of different atoms will overlap satisfactorily.

Polyhalogen cations are also known. Examples exist in the compounds $BrF_2^+SbF_6^-$, $IF_4^+SbF_6^-$ and $ICl_2^+SbCl_6^-$. The cation in the last has a Cl—I—Cl angle of 95°, again presumably mainly due to p-orbital overlap.

Acids formally corresponding to the polyhalide ions are not usually preparable, but the orange-yellow hydrate $HICl_4 \cdot 4\,H_2O$ can be crystallised from a solution of ICl_3 in aqueous HCl. The solution dissolves RbF and CsF to give $RbFICl_3$ and $CsFICl_3$.

25.7. Oxides

There are nineteen oxides of the halogens. The fluorine compounds are better considered as oxygen fluorides.

TABLE 25.4

PROPERTIES OF HALOGEN OXIDES

Compound	*M.p.*/K	*B.p.*/K	Shape	*X—O*/pm	*Bond angle*
OF_2	49.4	127.9	Angular C_{2v}	140.5	103°
O_2F_2	119	(216)			
O_4F_2	82				
Cl_2O	153	275	Angular C_{2v}	170	110°
ClO_2	214	283	Angular C_{2v}	149	118°
$ClOClO_3$	156	317			
Cl_2O_6	277	476			
Cl_2O_7	182	354			
Br_2O	255				
BrO_2	d				
BrO_3	d				
I_2O_5	d				

25.7.1. Fluorine

The colourless gas, OF_2, is made by the action of F_2 with 2% NaOH solution:

$$2\,F_2 + 2\,OH^- \rightarrow 2\,F^- + OF_2 + H_2O$$

but hydrolyses in an excess of NaOH:

$$OF_2 + 2\,OH^- \rightarrow O_2 + 2\,F^- + H_2O$$

The bonding is essentially covalent because of the similar electronegativities of oxygen and fluorine and the structure is as indicated in Table 25.4. The oxygen valence state is roughly tetrahedral (sp^3), but in OF_2 the F—O—F angle is less than the tetrahedral angle because the bond-pair—bond-pair repulsion is less than that caused by the lone pairs (5.2). OF_2 is neither explosive nor an acid anhydride; but is very poisonous. The OF radical has been observed by matrix isolation techniques involving photolysis of OF_2.

The compound O_2F_2 is an orange-red solid produced at 108 K by passing an electric discharge through an oxygen—fluorine mixture at low pressure. It decomposes into its elements above 150 K. O_4F_2 can be made similarly at lower temperatures. Likewise, other fluorine—oxygen compounds of various stoichiometry have been reported. O_3F_2 for example is now thought to be a mixture of O_2F_2 and O_4F_2.

25.7.2. Chlorine

Orange dichlorine monoxide, Cl_2O, is made by passing Cl_2 over precipitated HgO:

$$2\,Cl_2 + 2\,HgO \rightarrow HgO \cdot HgCl_2 + Cl_2O$$

The liquid can be distilled at its *b.p.*, 275 K, but at higher temperatures the gas explodes. The molecule is V-shaped and the repulsion between the bond pairs is greater than in the fluorine compound, in accordance with the closer proximity of the electrons to the oxygen. The compound is formally the anhydride of hypochlorous acid.

The photochemical decomposition of Cl_2O has been shown to proceed through a free-radical chain reaction:

$$Cl_2O + h\nu \rightarrow ClO + Cl^{\bullet}$$
$$Cl_2O + Cl^{\bullet} \rightarrow ClO + Cl_2$$
$$ClO + ClO \rightarrow Cl_2 + O_2$$

Cl_2O_3 with the assigned structure $OClClO_2$ has been identified as a photolysis product of ClO_2 at 195 K.

Chlorine dioxide, ClO_2, is best made by treating silver chlorate at 360 K with dry chlorine and condensing the ClO_2 by cooling:

$$2\,AgClO_3 + Cl_2 \rightarrow 2\,AgCl + 2\,ClO_2 + O_2$$

It is a gas with little tendency towards dimerisation, showing the paramagnetism expected for the monomeric ClO_2. The bonds in the molecule are appreciably shorter than those in Cl_2O, having much more double-bond character.

The odd-electron molecule is very reactive; detonations can occur even at 170 K and when warmed the gas explodes unless diluted. It is a powerful oxidising agent, since for

$$ClO_2 + 4\,H_3O^+ + 5\,e \rightarrow Cl + 6\,H_2O \qquad E^0 = +1.50\ V$$

The gas, ClO_2, can be looked upon as a mixed anhydride:

$$2\,ClO_2 + 2\,OH^- \rightarrow ClO_2^- + ClO_3^- + H_2O$$

Chlorine perchlorate is prepared by the reaction:

$$CsClO_4 + ClOSO_2F \xrightarrow{230\ K} CsSO_3F + ClOClO_3$$

Dichlorine hexoxide, Cl_2O_6, results from mixing ClO_2 with ozonised oxygen at 273 K. It is a dark-red liquid, less explosive than ClO_2, and reacts with alkalis to give chlorate and perchlorate:

$$Cl_2O_6 + 2\ OH^- \rightarrow ClO_3^- + ClO_4^- + H_2O$$

Though the molecular weight in carbon tetrachloride agrees with the formula Cl_2O_6, the weak paramagnetism of the aqueous solution suggests some dissociation to ClO_3. The dissociation energy $Cl_2O_6 \rightarrow 2\ ClO_3$ is only 6 kJ mol^{-1}.

Dichlorine heptoxide, Cl_2O_7, is made by dehydrating perchloric acid with P_2O_5 at a low temperature and distilling the product. The colourless, oily liquid is not so strong an oxidising agent as the other oxides of chlorine. It is the anhydride of perchloric acid. The gaseous molecule has a $O_3ClOClO_3$ structure. It is more thermodynamically stable than any of the other oxides of chlorine.

25.7.3. Bromine

Dibromine monoxide, Br_2O, is a dark-brown solid and is made in a similar way to Cl_2O. It is formally the anhydride of hypobromous acid.

Bromine dioxide, BrO_2, is a yellow solid below 230 K; above it is unstable. It can be prepared by quantitative ozonolysis of Br_2 in a fluorocarbon solvent

$$Br_2 + 4\ O_3 \xrightarrow[195\ K]{CF_3Cl} 2\ BrO_2 + 4\ O_2$$

It is less explosive than ClO_2, and hydrolyses in 5 *M* alkali to give bromide and bromate:

$$6\ BrO_2 + 6\ OH^- \rightarrow Br^- + 5\ BrO_3^- + 3\ H_2O$$

The action of a glow discharge on a mixture of bromine and oxygen between 260 and 290 K produces a white solid of crystalline appearance, stable below 200 K, which is BrO_3.

25.7.4. Iodine

The only true oxide of iodine is di-iodine pentoxide, I_2O_5, made by dehydrating iodic acid at 500 K in a stream of dry air:

$$2\ HIO_3 \rightarrow I_2O_5 + H_2O$$

The white powder decomposes to iodine and oxygen above 650 K. It is a fairly strong oxidising agent. The reaction

$$I_2O_5 + 5\ CO \rightarrow I_2 + 5\ CO_2$$

is quantitative at 340 K, and is used for determining CO in gaseous mixtures. The oxides, empirically I_2O_4 and I_4O_9, are of unknown structure.

25.8. Oxoacids

Although fluorine is more electronegative than oxygen, hypofluorous acid has recently been prepared.

The oxoacids of the halogens

HOF	HOCl	HOBr	HOI *
	$HClO_2$ *	$HBrO_2$	HIO_2 *
	$HClO_3$ *	$HBrO_3$ *	HIO_3
	$HClO_4$	$HBrO_4$	HIO_4
			$H_7I_3O_{14}$
			H_5IO_6

* Known only in aqueous solution.

In the chlorine, bromine and iodine compounds the halogen atom in the oxoanion is positive in relation to the oxygen atoms, as indicated by the δ^-:

$$H_2O + HClO_3 \rightarrow [ClO_3]^- \text{ (Cl}^{\delta+}\text{, each O}^{\delta-}) + H_3O^+$$

Increase of the charge number of the halogen atom from +1 to +7 is accompanied by (i) increased thermal stability, (ii) decreased oxidising capacity, (iii) increased acid strength.

Periodic acid is exceptional in being more strongly oxidising than iodic acid:

	Cl	I
E^0, $HOX/\frac{1}{2}X_2$/V	+1.63	+1.45
E^0, $HXO_3/\frac{1}{2}X_2$/V	+1.47	+1.19
E^0, $HXO_4/\frac{1}{2}X_2$/V	+1.34	+1.38

Bromic acid is a particularly strong oxidising agent (E^0, $HBrO_3/\frac{1}{2}Br_2$ = +1.52 V). Perbromic acid is a recent discovery. The increasing stability of the oxoanions as the charge number of the halogen rises has already been discussed.

25.8.1. Hypohalous acids

These are weak acids ($pK_a \sim 8$) which exist principally in aqueous solution; indeed HOI is more correctly considered as iodine hydroxide. Their aqueous solutions, except that of HOF, are made by shaking precipitated

HgO in water with the particular halogen:

$$2\ X_2 + 2\ HgO + H_2O \rightarrow HgO \cdot HgX_2 + 2\ HOX$$

Sodium hypochlorite, used commercially in cotton bleaching, is made by the electrolysis of brine, the electrolyte being agitated to mix the anode and cathode products. At the cathode, the reaction $2\ H^+ + 2\ e \rightarrow H_2$ increases the concentration of OH^-, while at the anode, the reaction $2\ Cl^- \rightarrow Cl_2 + 2\ e$ releases chlorine. These combine:

$$Cl_2 + 2\ OH^- \rightarrow OCl^- + Cl^- + H_2O$$

and the solution becomes progressively stronger in NaOCl without chlorine being evolved.

HOF has been obtained pure by the fluorination of ice.

Hypohalite ions are unstable to the disproportionation

$$3\ OX^- \rightarrow XO_3^- + 2\ X^-$$

Because the activation energy for the hypochlorite reaction is rather large, conversion to ClO_3^- occurs rapidly only in hot solutions. The change is fast at about 325 K for the hypobromite, OBr^-, and at room temperature for the hypoiodite, OI^-.

25.8.2. Halous acids

Of these chlorous acid, $HClO_2$, is best known, and then only in solution. It is a stronger acid than hypochlorous ($pK_a \sim 2$). Chlorites are best made by the reaction of ClO_2 with peroxides:

$$Na_2O_2 + 2\ ClO_2 \rightarrow 2\ NaClO_2 + O_2$$

Heating converts an alkali-metal chlorite to chloride and chlorate:

$$3\ NaClO_2 \rightarrow 2\ NaClO_3 + NaCl$$

Halites are thought to disproportionate in alkaline media:

$$2\ XO_2^- \rightarrow XO_3^- + XO^-$$

25.8.3. Halic acids and halates

Chloric and bromic acids are obtainable only in aqueous solution, but HIO_3 separates, as white crystals, when iodine is oxidised with fuming nitric acid. The acids are strong oxidising agents and fairly strong acids ($pK_a \sim -2$).

Alkali-metal chlorates are much more soluble than bromates and iodates, and are conveniently made by the electrolysis of hot chloride solutions. Aqueous KCl (25%) is electrolysed at 343—348 K till it is saturated with chlorate; it is then cooled and $KClO_3$ crystallises.

Chlorate crystals contain the pyramidal ClO_3^- ion. Formally, the chlorine has charge number +5 and forms σ bonds like nitrogen. But a lone pair from each O ligand overlaps a chlorine 3d orbital and can form a π bond by donation; thereby charge returns to the chlorine.

Sodium iodate, $NaIO_3$, has orthorhombic symmetry, and discrete IO_3^- ions. Iodates follow the CsCl structural arrangement and chlorates and bromates that of NaCl. However, the crystalline acid HIO_3, in its orthorhombic α-form, appears to exist as separate tetrahedral molecules.

The thermal decomposition of the alkali-metal halates is complex:

$$2\ KClO_3 \rightarrow 2\ KCl + 3\ O_2$$
$$\text{or } 4\ KClO_3 \rightarrow 3\ KClO_4 + KCl$$
$$2\ KBrO_3 \rightarrow 2\ KBr + 3\ O_2$$

Bromates of the heavier metals give mixtures of oxide, bromide and oxygen, others give oxide, bromine and oxygen. The alkali-metal iodates give periodate and iodide; other iodates react in a similar way to the bromates.

25.8.4. Perhalic acids

Perchloric acid distils as a colourless oily liquid when a perchlorate is heated with concentrated H_2SO_4 at 1—3 kPa pressure. The hot, concentrated liquid is liable to detonate in the presence of a trace of reducing agent, particularly a carbon compound. The cold, aqueous acid gives hydrogen with Zn and Fe, without any reduction of the perchlorate ion:

$$Zn + 2\ HClO_4 \rightarrow Zn(ClO_4)_2 + H_2$$

Perchloric acid is strongly ionised ($pK_a \sim -11$). The tetrahedral ClO_4^- ion is the least polarised of anions and perchlorates are much used for adjusting the ionic strength of solutions without the risk of forming complexes with the cations present. Of the many crystalline hydrates, the monohydrate crystal is composed of H_3O^+ and ClO_4^- ions and is isomorphous with NH_4ClO_4 whereas the dihydrate contains $H_5O_2^+$ and ClO_4^- ions.

Perchlorates can be made by heating chlorates under controlled conditions:

$$4\ KClO_3 \xrightarrow[\text{in silica flask}]{750\text{ K}} 3\ KClO_4 + KCl$$

and by the electrolytic oxidation of cooled chlorate solutions at high current densities. Any chlorate remaining can be decomposed by HCl which does not react with the ClO_4^- ion. The potassium perchlorate is separated by fractional crystallisation. Perchlorates are often isomorphous with the permanganates and perrhenates. Most perchlorates are very soluble, but those of potassium, rubidium, caesium and ammonium are not.

Perbromic acid and perbromates were first obtained in studies of β-decay of $^{83}SeO_4^{2-}$. Bromate ions have now been oxidised in aqueous solution electrolytically, with XeF_2, and by the action of fluorine on strongly basic solutions.

The periodate picture is more complicated because of the stability of paraperiodic acid, H_5IO_6. Thus when a stream of chlorine is passed through a boiling solution of iodine in an excess of caustic soda, a white precipitate of $Na_2H_3IO_6$ is formed:

$$18\ NaOH + I_2 + 7\ Cl_2 \rightarrow 2\ Na_2H_3IO_6 + 14\ NaCl + 6\ H_2O$$

Treatment of a suspension of this with $AgNO_3$ solution gives a black precipitate of silver paraperiodate:

$$Na_2H_3IO_6 + 5\ AgNO_3 \rightarrow Ag_5IO_6 + 2\ NaNO_3 + 3\ HNO_3$$

When chlorine is passed into a suspension of this salt, avoiding an excess of the gas, a solution is obtained from which, after filtration, paraperiodic acid, H_5IO_6, can be crystallised:

$$4\ Ag_5IO_6 + 10\ Cl_2 + 10\ H_2O \rightarrow 4\ H_5IO_6 + 20\ AgCl + 5\ O_2$$

The colourless crystals contain octahedral molecules, corresponding approximately to sp^3d^2 hybridisation with secondary π bonding. HIO_4 is obtained by heating H_5IO_6 in a vacuum at 373 K. The acid is a powerful oxidising agent. In aqueous solution HIO_4 is converted back to H_5IO_6, which is rather weakly ionised:

$$H_5IO_6 \rightleftharpoons H^+aq + H_4IO_6^- \qquad pK_a = +3.3$$

The $H_4IO_6^-$ ion is in equilibrium with IO_4^- in the solution:

$$H_4IO_6^- \rightleftharpoons IO_4^- + 2\ H_2O$$

To sum up, the periodates are of four formula types:

KIO_4	from metaperiodic acid,	HIO_4
$Na_4I_2O_9$	from dimesoperiodic acid,	$H_4I_2O_9$
$Pb_3(IO_5)_2$	from mesoperiodic acid,	H_3IO_5 (hypothetical)
Ag_5IO_6	from paraperiodic acid,	H_5IO_6

The salts, like the acids, are also powerful oxidisers, converting Mn^{2+} ions to MnO_4^-, and iodides to iodine.

25.9. Halogen and Interhalogen Cations

The existence of I_3^+ and I_5^+ in 100% H_2SO_4 has been confirmed by conductimetric and cryoscopic measurements. When I_2 and HIO_3 are dissolved in the acid in the molar ratio 7 : 1 the equilibrium:

$$HIO_3 + 7\ I_2 + 8\ H_2SO_4 \rightleftharpoons 5\ I_3^+ + 3\ H_3O^+ + 8\ HSO_4^-$$

is established. Furthermore the conductivity remains the same if more I_2 is added, evidently because I_5^+ is formed. The dark-brown solution of I_5^+ has absorption peaks at 240, 270, 345 and 450 nm in fluorosulphuric acid at 190 K whereas the red-brown solution of I_3^+ has absorption peaks at 305 and 470 nm.

The paramagnetic, blue ion I_2^+ is obtained when solutions of I_2 in HSO_3F are oxidised with $S_2O_6F_2$. The blue colour is most intense when the $I_2/S_2O_6F_2$ ratio is 2.0.

$$2\,I_2 + S_2O_6F_2 \rightarrow 2\,I_2^+ + 2\,SO_3F^-$$

There are peaks in the absorption spectrum at 410, 490 and 640 nm, but these peaks slowly fade in HSO_3F while peaks appear at 305 and 470 pm, evidently as a result of the disproportionation

$$8\,I_2^+ + 3\,SO_3F^- \rightarrow I(SO_3F)_3 + 5\,I_3^+$$

But fresh solutions of I_2^+ in HSO_3F become bright red near the freezing point of the acid. The change has been shown by spectroscopic, magnetic and conductimetric measurements to be due to the formation of I_4^{2+}.

Blue crystalline solids formulated as $I_2^+[Sb_2F_{11}]^-$ and $I_2^+[TaF_{11}]^-$ have been made by treating I_2 with SbF_5 and TaF_5 respectively.

Bromine cations are also known. Brown Br_3^+ can be formed by oxidation of Br_2 with $S_2O_6F_2$ in super-acid (9.10). Further oxidation changes the colour to cherry-red as Br_2^+ is formed. A full X-ray study on the scarlet solid $Br_2^+[Sb_3F_{16}]^-$ shows it to contain Br_2^+ ions with a bond length of 213 pm.

Evidence for Cl_3^+ and Cl_2^+ is weaker, but a yellow solid $Cl_3^+[AsF_6]^-$ has been made from Cl_2, ClF and AsF_5 at 200 K; it decomposes rapidly at room temperature.

Some triatomic interhalogen cations have been established, and structural data have been obtained for ClF_2^+, BrF_2^+, ICl_2^+ and Cl_2F^+. All are V-shaped; for example the ClF_2^+ ion in $ClF_2^+[SbF_6]^-$ has an angle F—Cl—F of 96° and a Cl—F bond length of 158 pm. The Cl_2F^+ ion is unsymmetrical. The Raman frequencies in $Cl_2F^+[A_5F_6]^-$ are 744 cm^{-1} for the Cl—F stretching and 528 and 535 for the Cl—Cl stretching.

Further Reading

R.J. Gillespie and M.J. Morton, Halogen and interhalogen cations, Q. Rev., 25 (1971) 553.

T.A. O'Donnell, The chemistry of fluorine, Pergamon, Oxford, 1975.

A.J. Downs and C.J. Adams, The chemistry of chlorine, bromine, iodine and astatine, Pergamon, Oxford, 1975.

V. Gutmann (Ed.), Halogen chemistry, Academic Press, New York, 1967.

E.W. Lawless and I.C. Smith, Inorganic high-energy oxidisers, Arnold, London, 1968.

H.J. Emeléus, The chemistry of fluorine and its compounds, Academic Press, New York, 1969.

Z.E. Jolles (Ed.), Bromine and its compounds, Academic Press, New York, 1966.

E. Appelman, Nonexistent compounds, two case histories (BrO_4^- and HOF), Accounts Chem. Res., 6 (1973) 112.

The Halides and Pseudohalides

26

THE HALIDES

26.1. General

26.1.1. Classes of halides

The halides may be divided broadly into two main classes:
(i) those formed by metals which have ions of low charge;
(ii) those formed by non-metals and by many B sub-group metals.
The first class comprises the saline halides with three dimensional, ionic lattices; they are salts with high *m.p.* and *b.p.* and are good conductors when fused. Most of the halides of elements of the first three A sub-groups belong to this class. The second class comprises the volatile, non-conducting halides which usually have molecular lattices. There are a few halides with some of the properties of each class, examples are those of beryllium and aluminium.

The classes are readily distinguished by considering the boiling points of the chlorides of the first three elements in Group I to IV (Table 26.1). Those

TABLE 26.1

BOILING POINTS OF CHLORIDES

	B.p./K		*B.p.*/K		*B.p.*/K		*B.p.*/K
Monochlorides		Dichlorides		Trichlorides		Tetrachlorides	
LiCl	1650	$BeCl_2$	760	BCl_3	285	CCl_4	349
NaCl	1710	$MgCl_2$	1670	$AlCl_3$	456 (sub)	$SiCl_4$	330
KCl	1650	$CaCl_2$	1870	$ScCl_3$	1270	$TiCl_4$	409

to the left of the full lines are saline and those to the right are covalent. Beryllium and aluminium chlorides which are shown in boxes display both ionic and covalent character, the first being the more ionic the second the more covalent. Solid aluminium chloride has a layer-lattice structure, with a greatly distorted octahedral co-ordination of chlorine atoms around the aluminium atoms, but the vapour formed on sublimation at 455 K contains Al_2Cl_6 molecules in which there is tetrahedral arrangement of chlorine atoms

TABLE 26.2

MOLAR CONDUCTANCES OF FUSED CHLORIDES AT THEIR *m.p.*

(*M.p.*/K) *Molar conductance*/S

Monochlorides			Dichlorides			Trichlorides			Tetrachlorides		
LiCl	(887)	166	$BeCl_2$	(678)	0.086	BCl_3	(166)	0	CCl_4	(250)	0
NaCl	(1073)	134	$MgCl_2$	(987)	29	$AlCl_3$	(466 *)	1.5×10^{-5}	$SiCl_4$	(203)	0
KCl	(1043)	104	$CaCl_2$	(1055)	52	$ScCl_3$	(1212)	15	$TiCl_4$	(243)	0

* Under pressure.

around the aluminium atoms (Fig. 15.2). The solid can be liquified only under pressure.

The change from ionic to covalent halides is least abrupt in the lithium period. In it the boiling point of beryllium chloride is 890 K below that of lithium chloride, whereas in the next period the sublimation point of aluminium chloride is 1217 K below the boiling point of magnesium chloride.

The molar conductances of these compounds, measured at their melting points, also bring out the change from ionic to covalent character which has just been described. Incidentally, the least readily melted halides are, broadly speaking, the best conductors when fused, and usually the transportation of current is mainly by the cations.

As $BeCl_2$ shows, a rigid division into saline and volatile halides is not possible. Many metals which form ions such as M^{2+}, M^{3+} or M^{4+}, give halides with intermediate properties. These compounds usually have either layer lattice structures ($CdCl_2$, $FeBr_2$, BiI_3) or chain structures ($PdCl_2$).

For a series of halides of the same metal, the covalence also increases with the size of the halogen atom; thus CaF_2 has an ionic crystal, and $CaCl_2$ has a very slightly deformed rutile structure, but CaI_2 has the CdI_2 layer lattice. Fluorides, as might be expected, tend to be saline, for instance AuF_3. TiF_3 and PbF_4 are saline though the corresponding chlorides are volatile. An example of a halide combining saline and covalent characteristics is iron(III) chloride which is volatile (*b.p.* 588 K), soluble in organic solvents, dimeric in the vapour and readily hydrolysed, but, nevertheless, a good conductor of electricity when fused.

Halides frequently form hydrates which differ in properties from the anhydrous materials. Thus many M^{III} fluorides (e.g. AlF_3) are quite insoluble when made by dry methods but the hydrates (e.g. $AlF_3 \cdot 3.5\ H_2O$) produced from solutions dissolve readily in water. This is because the co-ordination of water molecules round cations greatly reduces the lattice energies and facilitates solution.

26.2. Methods of Preparation

Dry methods of preparation are obligatory when the product is easily hydrolysed.

(i) Direct halogenation is a particularly versatile method, useful for all types of halide:

$$2\ Fe + 3\ Br_2 \rightarrow 2\ FeBr_3$$
$$Sn + 2\ Cl_2 \rightarrow SnCl_4$$
$$S + 3\ F_2 \rightarrow SF_6$$

(ii) Heating oxide or sulphide with carbon and chlorine, with carbon tetrachloride, or with phosgene, is usually a good method for non-saline metal halides:

$$Cr_2O_3 + 3\ C + 3\ Cl_2 \rightarrow 2\ CrCl_3 + 3\ CO$$

$$ZrO_2 + 2\,C + 2\,Cl_2 \rightarrow ZrCl_4 + 2\,CO$$
$$2\,BeO + CCl_4 \rightarrow 2\,BeCl_2 + CO_2$$

(iii) A mixture of S_2Cl_2 and chlorine often converts an oxide, or even a sulphate, to a chloride. This is a common method of making the trichlorides of the lanthanides:

$$4\,Lu_2O_3 + 9\,Cl_2 + 3\,S_2Cl_2 \rightarrow 8\,LuCl_3 + 6\,SO_2$$

(iv) Chlorine gives a mixture of chlorides with metallic uranium. The tetrachloride UCl_4 results from heating the dioxide UO_2 with CCl_4:

$$UO_2 + CCl_4 \rightarrow UCl_4 + CO_2$$

Probably UCl_6 is first formed and decomposes to give UCl_4.

(v) Fluorides are made by treating the chlorides with anhydrous HF:

$$CrCl_3 + 3\,HF \rightarrow CrF_3 + 3\,HCl$$
$$TiCl_4 + 4\,HF \rightarrow TiF_4 + 4\,HCl$$

by treating covalent halides with SbF_3 in the presence of $SbCl_5$ as catalyst:

$$SiCl_4 \rightarrow SiF_4 \qquad BCl_3 \rightarrow BF_3$$

or by heating the oxides with CaF_2 or a mixture of CaF_2 and H_2SO_4:

$$ZrO_2 + 2\,CaF_2 \rightarrow 2\,CaO + ZrF_4$$
$$GeO_2 + 2\,CaF_2 + 2\,H_2SO_4 \rightarrow GeF_4 + 2\,CaSO_4 + 2\,H_2O$$

Wet methods are feasible when the halide is not hydrolysed, either because it is ionic or because it is insoluble. The products are frequently hydrated:

(i) The metal may be dissolved in the aqueous halogen acid:

$$Zn + 2\,HCl \rightarrow ZnCl_2 + H_2$$

(ii) The oxide or hydroxide may be dissolved in the halogen acid:

$$MgO + 2\,HCl \rightarrow MgCl_2 + H_2O$$
$$Bi_2O_3 + 6\,HF \rightarrow 2\,BiF_3 + 3\,H_2O$$

(iii) Precipitation methods are frequently useful for non-hydrated metal halides:

$$M(NO_3)_n + nX^- \rightarrow \downarrow MX_n + nNO_3^-$$

The insoluble halides include the chlorides and bromides of the Ag^+, Cu^+, Au^+ and Tl^+ ions, of the Pb^{2+}, Pt^{2+} and the mercury(I) ion, Hg_2^{2+}. The iodides of these metals, together with HgI_2, PdI_2, BiI_3 and AuI_3, are also insoluble.

Certain ions are precipitated as fluoride from solutions of the corresponding chloride, for example, those of Mg^{2+}, Ca^{2+}, Sr^{2+}, Ba^{2+}, Pb^{2+}, and Cu^{2+} and Al^{3+}; but AgF, Hg_2F_2 and TlF are soluble.

26.3. The Structures of Halides

26.3.1. AB structures

The alkali-metal halides have either NaCl type (6 : 6) or CsCl type (8 : 8) structures (Figs. 7.11 and 7.12). The ratios of the ionic radii, r_+/r_- in

KF	RbF	CsF
0.98	1.09	1.24

suggest that the structures should have 8 : 8 co-ordination (higher co-ordination is not consistent with overall electrical neutrality), yet they have the 6 : 6 NaCl structure. Ammonium chloride, bromide and iodide all possess two forms, that with the CsCl structure being stable below the respective transition temperatures and that with the NaCl structure above; the transition temperatures are:

NH_4Cl	NH_4Br	NH_4I
457 K	410 K	256 K

At a low temperature NH_4Br and NH_4I have a third form which is tetragonal. Hydrogen bonding imposes a wurtzite structure on NH_4F (Fig. 7.17). The copper(I) halides have the zinc blende structure.

26.3.2. AB_2 structures

Difluorides usually have fluorite or rutile structures. Large ions such as Hg^{2+}, Sr^{2+}, Pb^{2+} and Ba^{2+} form a fluorite (8 : 4) lattice (Fig. 7.19); the smaller Ni^{2+}, Co^{2+}, Zn^{2+} and Mn^{2+} cations give rise to the rutile (6 : 3) structure (Fig. 7.20). The crystals of some chlorides and bromides of metals with large bipositive cations are ionic. $CaCl_2$ has a deformed rutile structure with four Cl^- ions 276 pm, and two Cl^- ions 270 pm, from the Ca^{2+}. However, CaI_2 has the same layer lattice as CdI_2 (Fig. 7.22).

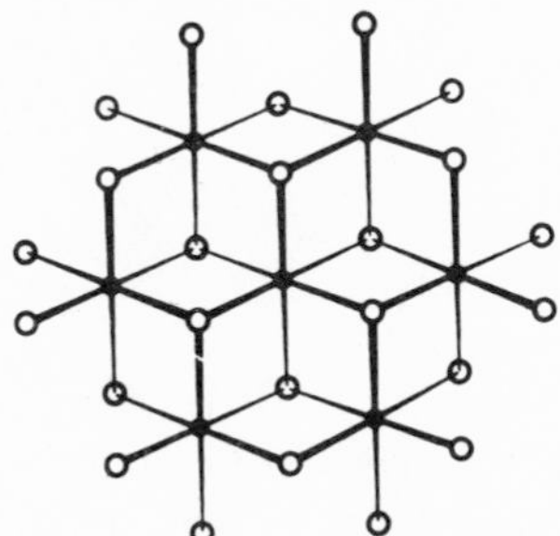

Fig. 26.1. Arrangement of octahedra in $CdCl_2$.

In $PbCl_2$ the lead ion occupies the centre of a trigonal prism with six Cl^- ions at the corners and three outside the face centres. $SrBr_2$ is similar. But most dihalides form layer lattices. $CdCl_2$ can be considered as an NaCl structure in which half the octahedral holes between the Cl atoms are unoccupied (Fig. 26.1).

CdI_2 can be considered as being made up of hexagonally close-packed iodine ions in which only half the octahedral holes are filled with cadmium ions.

Other halides of these layer types are:

$CdCl_2$ type: $FeCl_2$, $CoCl_2$, $NiCl_2$, NiI_2, $ZnCl_2$, $MnCl_2$
CdI_2 type: CaI_2, MgI_2, PbI_2, MnI_2, $MgBr_2$, $FeBr_2$

The compounds $PdCl_2$ (Fig. 26.2), $CuCl_2$ and $CuBr_2$ have chain structures:

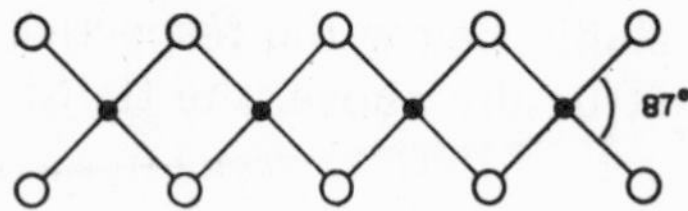

Fig. 26.2. Chain structure of $PdCl_2$.

The well-characterised PbFCl has an interesting layer lattice (Fig. 26.3). Other compounds with similar structure are PbFBr and several oxochlorides such as LaOCl and BiOCl in which there are O^{2-} ions instead of the F^- ions of similar size.

Fig. 26.3. Unit cell of PbFCl.

26.3.3. AB_3 types

The trifluorides of many lanthanides and actinides have very slightly distorted rhenium trioxide structures (Fig. 7.23). Examples are CeF_3, PrF_3, NdF_3, ScF_3, EuF_3, AcF_3, UF_3 and NpF_3.

Bismuth trifluoride has the CaF_2 structure with twelve extra fluorine atoms at the mid-points of the edges and one at the centre of the unit cell (Fig. 7.25).

The chlorides and bromides of the lanthanides and actinides usually have

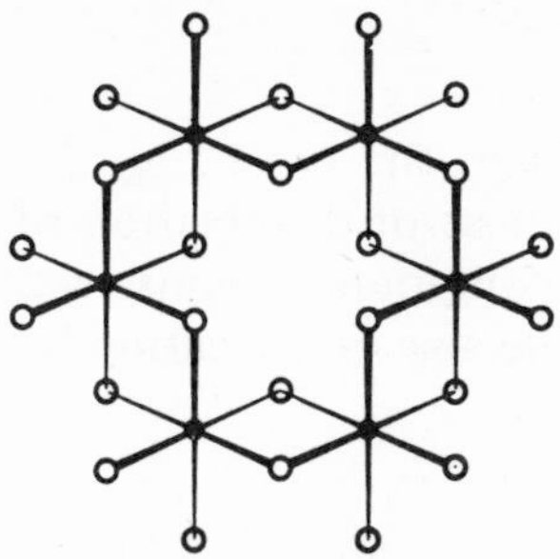

Fig. 26.4. Arrangement of octahedra in $CrCl_3$.

the same structure as $PbCl_2$ but with one third of the prisms lacking metal ions. Examples are $LaCl_3$, $CeCl_3$, $NdCl_3$, $LaBr_3$, $CeBr_3$, $AcCl_3$, $NpCl_3$.

The structure of $CrCl_3$ (Fig. 26.4) is that of $CdCl_2$ with the places of one third of the Cd atoms not filled. Crystalline $AlCl_3$ is similar in structure to $CrCl_3$.

BiI_3 is similarly related to CdI_2. It can also be considered as a structure of close-packed iodine atoms in which only one third of the octahedral holes are filled. Other examples of this structure are SbI_3, AsI_3, $FeCl_3$ and $CrBr_3$.

26.4. Halides Containing Individual Molecules

The halides of B, C, N, O, Si and S and the Group IVB, VB and VIB elements consist of finite molecules. GeI_4 and SnI_4 also contain individual tetrahedral molecules. Al_2Br_6 molecules are present in aluminium bromide. PCl_5 and PBr_5 are interesting; the former contains tetrahedral PCl_4^+ ions and octahedral PCl_6^- ions, the latter tetrahedral PBr_4^+ ions and Br^- ions.

The hexachlorides of tungsten and uranium have a deformed hexagonal close-packing of chlorine atoms with metal atoms filling only one sixth of the octahedral holes.

26.5. Halide Molecules in the Vapour State

The shapes of the free molecules in halide vapours are determined by mutual repulsion between electron pairs and may bear little relation to the disposition of atoms in the respective solid structures. The shapes of the molecules in the vapour depend largely on the electron configuration of the less electronegative element. Thus the PCl_5 molecule is trigonal bipyramidal, AsF_3 has a pyramidal molecule (C_{3v}). Aluminium chloride exists principally as the dimer, Al_2Cl_6, with D_{2d} symmetry, in the vapour just above the *b.p.* (Fig. 15.2). The trihalides of gold are planar dimers (Fig. 38.5) with D_{2h} symmetry.

26.6. Enthalpies of Formation

Of the alkali-metal halides, the fluorides have much the most negative enthalpies of formation. This is ascribed to the low heat of dissociation of the F_2 molecule and the high lattice energies of the compounds themselves. In the fluorides the enthalpies of formation rise as the size of the cation increases, in the other halides they fall:

	Li	Na	K	Rb	Cs	
Fluorides	—608	—575	—567	—558	—554	kJ mol^{-1}
Chlorides	—407	—412	—437	—442	—449	kJ mol^{-1}

This is principally because the fraction $1/(r_{\text{anion}} + r_{\text{cation}})$ which occurs in the lattice energy formula decreases more rapidly in the fluoride series on account of the small size of the anion; this change has more effect than have the decrease of ionisation energy and sublimation energy of the respective metals.

Lattice energy considerations are important in fluorination of an organic halide by alkali-metal fluorides:

$$\rangle\!\!-C-Cl + MF \rightarrow \rangle\!\!-C-F + MCl$$

Here ΔG is largely dependent on the difference between the lattice energies of MF and MCl. As this difference is proportional to

$$\frac{1}{r_{M^+} + r_{F^-}} - \frac{1}{r_{M^+} + r_{Cl^-}} = \frac{r_{Cl^-} - r_{F^-}}{(r_{M^+} + r_{F^-})(r_{M^+} + r_{Cl^-})}$$

the greater the size of the M^+ ion the smaller the energy needed to produce MCl from MF and the less the energy which has to be supplied by the C—Cl → C—F change. Fluorine-exchanging ability therefore increases with cation size for the fluorides of metals which form chlorides isomorphous with the fluorides. Incidentally, the difference between the lattice energy of AgF and AgCl is very small because of the partly covalent bonding in AgCl and this accounts for the specially high fluorinating power of AgF.

Trends in enthalpies of formation of halides through the Periodic Table are often surprisingly regular, as is shown for chlorides of elements of the third period in Fig. 26.5.

The enthalpy of formation (per halogen atom) of a halide plotted against the logarithm of the atomic number of its non-halogen atom displays a straight line relationship for compounds in which the maximum group valency is achieved. This is shown for the chlorides of Groups I to IV (Fig. 26.6).

Enthalpies of formation of chlorides, bromides and iodides decrease for the elements in going down Groups I, II, IIIA, IVA and VA. Elsewhere in the Periodic Table, enthalpies of formation usually increase down a group, but there are some irregularities, particularly in the first two periods, as appears from:

BF_3	-1104 kJ mol^{-1}	CCl_4	-138 kJ mol^{-1}
AlF_3	-1305 kJ mol^{-1}	$SiCl_4$	-643 kJ mol^{-1}

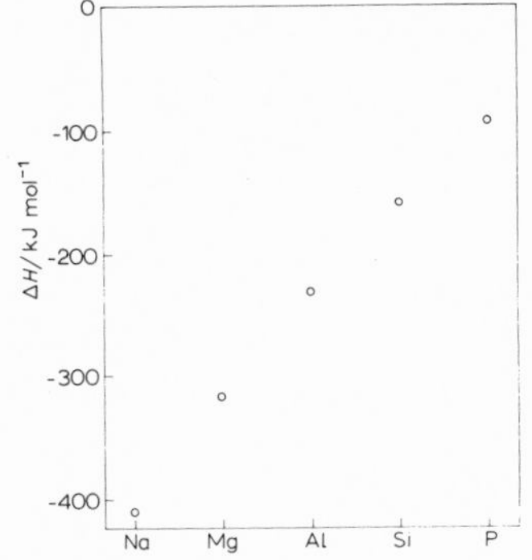

Fig. 26.5. Enthalpies of formation per 35.5 g Cl for NaCl, $MgCl_2$, $AlCl_3$, $SiCl_4$ and PCl_5.

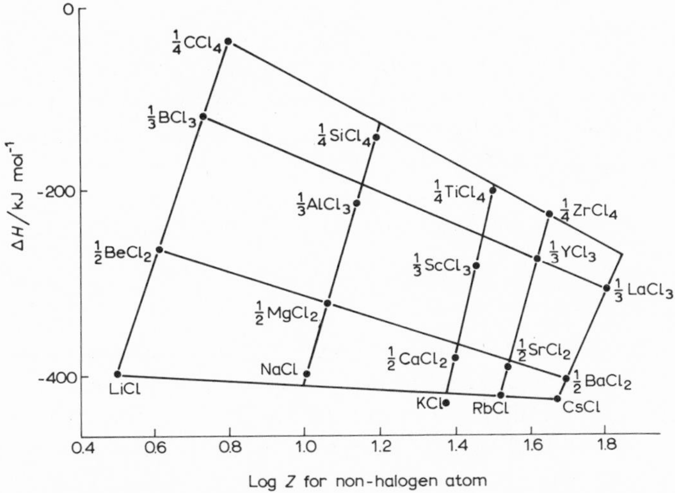

Fig. 26.6. Approximately linear relationships between enthalpies of formation and log Z for some simple chlorides.

The fluorides of Group VIB are interesting, their enthalpies of formation, ΔH_f, being SF_6 -1100 kJ mol^{-1}, SeF_6 -1030 kJ mol^{-1} and TeF_6 -1320 kJ mol^{-1}. Fluorine apparently forms particularly strong bonds with atoms having orbitals available for π-bonding, such as the 4f orbitals in Te; in this kind of situation the halogen lone pairs could make a small contribution to bonding by back-donation.

The enthalpies of formation of hypothetical halides can be calculated by the Born—Haber treatment. For the halides for elements of variable charge

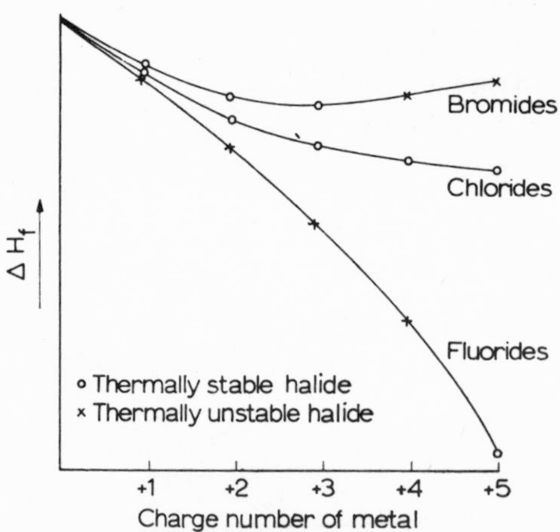

Fig. 26.7. Enthalpies of formation of halides related to thermal stability.

number, the highest curve in Fig. 26.7 is the type commonly found for bromides and iodides.

According to simple theory, the reduction to be expected in enthalpy between a bromide and a chloride, or a chloride and a fluoride, is proportional to the square of the charge number. This accounts for the form of the two lower curves. Both MBr_4 and MBr_5 are unstable and must liberate heat on conversion to MBr_3, whereas all the five chlorides are stable. When the curve becomes concave towards the charge number axis, as often happens for fluorides, the lower fluorides are metastable and heat must be liberated in the reaction

$$2\,MF_3 \rightarrow MF + MF_5.$$

It is stressed that in halides of the non-metals, the charge numbers shown by the non-metal change by steps of two because of the necessity of forming singly occupied orbitals. The implied promotion of electrons in the valence state will, of course, occur only when the formation of the bonds results in a net binding energy.

26.7. Hydrolysis of Halides

Most of the alkali-metal and barium halides dissolve in water without changing the pH, although there is undoubtedly some hydration of the ions. The alkali-metal fluorides give slightly alkaline reactions, however, because of the comparatively low ionisation of HF:

$$F^- + H_2O \rightleftharpoons HF + OH^-$$

PbF_4 hydrolyses to give PbO_2 and HF. This reaction is made possible by the high lattice energy of the PbO_2 formed, and by the large energy of hydration of the four F^- ions.

Most covalent halides hydrolyse readily:

$BCl_3 + 3\ H_2O \rightarrow H_3BO_3 + 3\ HCl$
$SiCl_4 + 2\ H_2O \rightarrow SiO_2 + 4\ HCl$
$3\ SiF_4 + 2\ H_2O \rightarrow SiO_2 + 2\ H_2SiF_6$
$TeF_6 + 6\ H_2O \rightarrow H_6TeO_6 + 6\ HF$

Halides of layer-lattice type often hydrolyse in stages and reversibly:

$FeCl_3 + H_2O \rightleftharpoons Fe(OH)Cl_2 + HCl$, etc.

giving the solution a weakly acid reaction. Some hydrolyse reversibly to an oxide halide:

$BiCl_3 + H_2O \rightleftharpoons BiOCl + 2\ HCl$

Hydrolysis is probably initiated by the donation of an electron pair to the central atom from the oxygen atom of the water molecule. Thus for example:

$$Cl_4Si \leftarrow 2\,OH_2 \rightleftharpoons SiO_2 + 4\ HCl$$

That Si—Cl bonds have some double bond character and involve the use of the 3d orbitals of Si, with donation from the chlorine lone pairs, is suggested by the very negative enthalpy of formation of $SiCl_4$. One d orbital is occupied in sp^2d hybridisation, but the empty 3d orbitals can also accept electrons from the oxygen of water. CCl_4, with its higher enthalpy of formation, is bound by simple σ-bonds and is not hydrolysed because there are no low-lying carbon orbitals of suitable symmetry to accept electrons from the water molecule.

The fluoride TeF_6 is much more readily hydrolysed than SeF_6; it also has a much more negative enthalpy of formation than SeF_6 which may be due, at least in part, to π-bonding in the Te—F bonds, made possible by the use of 4f electrons.

The hydrolysis of covalent halides in which the central atom has attained its maximum covalency but still has one or more lone pairs is initiated by the donation of electrons to a proton of the water molecule:

$$Cl_3N \rightarrow H{-}OH \rightleftharpoons NHCl_2 + HOCl$$

$$Cl_2O \rightarrow H{-}OH \rightleftharpoons 2\ HOCl$$

Such chlorides always yield HOCl instead of HCl on hydrolysis. NF_3 and F_2O do not hydrolyse because the fluorine is too electronegative to permit the donation of electrons to another molecule.

It should be noted that stability in the presence of water is not entirely dependent on the thermodynamic properties of a halide. The free energy changes for the following reactions show that they are thermodynamically feasible:

$$CF_4(g) + 2\ H_2O(g) \rightleftharpoons CO_2(g) + 4\ HF(g) \qquad \Delta G^\circ = -151\ \text{kJ mol}^{-1}$$
$$SF_6(g) + 3\ H_2O(g) \rightleftharpoons SO_3(g) + 6\ HF(g) \qquad \Delta G^\circ = -302\ \text{kJ mol}^{-1}$$

The failure of appreciable hydrolysis to occur under ordinary conditions must be due to the magnitude of the activation energies; this would be expected to be considerable since in each case the central atom is exhibiting its maximum valency, and the formation of an activated complex would require complete reorganisation of the electronic structure.

26.8. Colour of the Halides

Halides, whether solid or in solution, are usually colourless unless the metal ion itself has a characteristic colour. The principal exceptions are certain anhydrous iodides: AgI, yellow; PbI_2, bright yellow; BiI_3, dark brown; HgI_2, scarlet.

26.9. Hydrates

Sodium fluoride does not form a hydrate. The hydrate of sodium chloride, $NaCl \cdot 2\ H_2O$, separates from a saturated solution only below 263 K, whereas NaBr and NaI form both di- and penta-hydrates at the ordinary temperature. CaF_2 is anhydrous but the other calcium halides form hydrates. By contrast, potassium and silver fluorides have hydrates, $KF \cdot 2\ H_2O$, $AgF \cdot 2\ H_2O$ and $AgF \cdot 4\ H_2O$, although their other halides are anhydrous. In $KF \cdot 2\ H_2O$ each K^+ and each F^- is surrounded approximately octahedrally by two H_2O molecules and four ions of opposite sign, each H_2O having two K^+ and two F^- ions arranged tetrahedrally round it.

In very few instances is the degree of hydration of fluoride and chloride the same; where it is, as in

$FeF_2 \cdot 4\ H_2O$ $\qquad$ $FeCl_2 \cdot 4\ H_2O$
$CoF_2 \cdot 2\ H_2O$ $\qquad$ $CoCl_2 \cdot 2\ H_2O$

the structures are not yet known.

Electrostatic considerations provide a useful guide to the likelihood of hydrate formation. For a pair of ions A^+ and B^- an over-all energy decrease is most likely to happen when r_{A^+} and r_{B^-} are large. Conversely when the ionic radii are small, as in LiF and NaF, a hydrate is not usually formed. As, however, the energy decrease in hydrate formation also depends on the energy change due to the hydration of individual ions, and as the smallest ions often have the greatest hydration energies, the formation of a hydrate depends on a rather fine energy balance. Thus AgF, for instance, with two small ions, forms a hydrate.

Many metallic dihalides form hexahydrates containing $M(H_2O)_6^{2+}$ ions. Examples are $MgCl_2 \cdot 6\ H_2O$, $FeCl_2 \cdot 6\ H_2O$, $CoCl_2 \cdot 6\ H_2O$ and $NiCl_2 \cdot 6\ H_2O$. The radius-ratio rule suggests that these hydrates should have a fluorite structure, but $Mg(H_2O)_6Cl_2$ is, in fact, less symmetrical. However, $Mg(NH_3)_6Cl_2$ has a fluorite lattice.

Solubilities are closely related to the ease of hydrate formation. In the ionic halides, solubilities increase with the size of the halide ion; fluorides are often particularly insoluble. In the more covalent halides this order of solubility is usually reversed.

26.10. Complex Halides

These generally contain MX_4^{n-4} ions and MX_6^{n-6} ions, where n is the charge number of the metal. Many new fluoro-complexes containing metals with unusual charge numbers have been made by the use of BrF_3 as a non-aqueous solvent and fluorinating agent: examples are Cs_2CoF_6, K_3CuF_6 and $KIrF_6$.

Where the central metal ion is small the fluoro-complexes are less stable in solution than the chloro-complexes, perhaps because appreciable π-bonding is possible only in the latter. The complex fluorides of base metals with large positive ions are more stable than the corresponding complex chlorides. Hydration energies also play a part in determining relative stabilities in solution.

Complex fluorides may be grouped into different structural types:

(i) Those with structures like simple halides:
$KLaF_4$ and K_2UF_6 have the CaF_2 structure (Fig. 7.19) with random arrangement of metal ions in the Ca^{2+} positions. $BaUF_6$ and $BaThF_6$ have the random lanthanum fluoride lattice (LaF_3 has the ReO_3 structure; see Fig. 7.23).

(ii) Those with perovskite-type lattices (Fig. 7.27):
$KMgF_3$, $KZnF_3$ and $KNiF_3$ are examples.

(iii) Uranium and thorium complexes with a fluorine deficiency:
Na_3UF_7 has a fluorite lattice with U^{4+} ions filling one quarter of the Ca^{2+} positions and Na^+ the other three quarters; one eighth of the F^- positions are unoccupied. There are many uranium and thorium complexes with lattices of similar type.

(iv) Complexes of Group IVA and VA elements of unusual formula:
K_2NbF_7 contains NbF_7^{2-} ions which have six fluorines at the corners of a trigonal prism and a seventh in a position normal to the centre of one face; $(NH_4)_3SiF_7$ contains octahedral SiF_6^{2-} ions and discrete F^- ions.

The TaF_8^{3-} ion has the rather unusual square antiprism structure (Fig. 33.1).

(v) Fluoroaluminates:

These may contain AlF_6 octahedra linked in chains or layers, for instance Tl_2AlF_5 (Fig. 26.8).

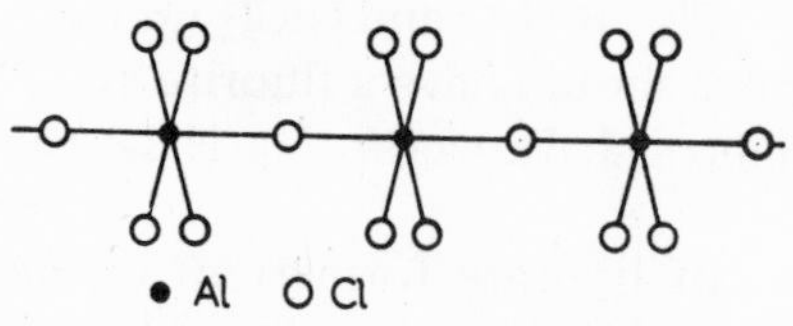

Fig. 26.8. Chain ions in Tl_2AlF_5.

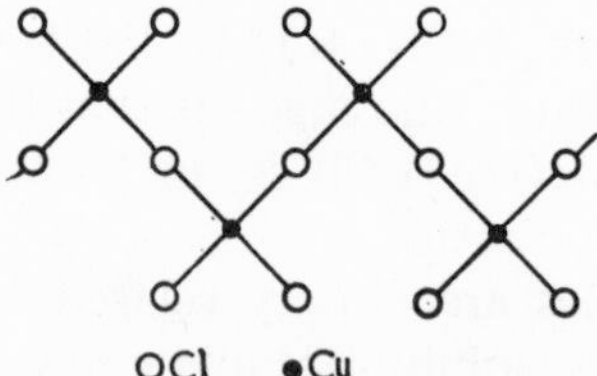

Fig. 26.9. Double-chain ions in $CsCuCl_3$.

The chloro-complexes usually contain finite complex anions. None has the perovskite structure of $KMgF_3$. The complex $CsAuCl_3$ contains equal numbers of square $AuCl_4^-$ ions and linear $AuCl_2^-$ ions in addition to Cs^+ ions. Cs_2CuCl_4 contains finite $CuCl_4^{2-}$ ions, but $CsCuCl_3$ has a double chain (Fig. 26.9).

In K_2HgCl_4 and K_2SnCl_4 there are chains based on octahedral arrangement round the noble metal (Fig. 26.10), and in NH_4CdCl_3, double chains (Fig. 26.11).

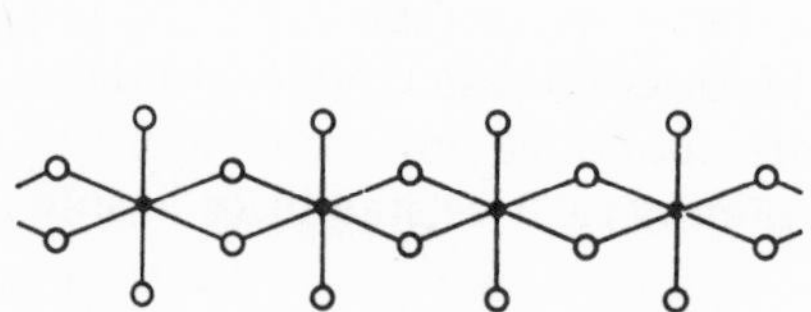
Fig. 26.10. Chain ions in K_2HgCl_4.

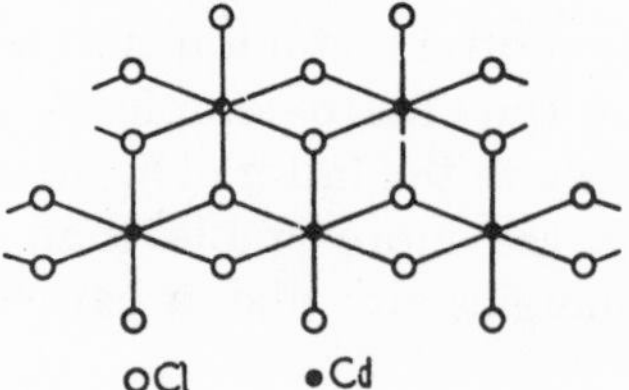

Fig. 26.11. Double chain ions in NH_4CdCl_3.

THE PSEUDOHALOGENS AND PSEUDOHALIDES

26.11. Pseudohalides

Several uninegative groups show a similarity to halide ions in both their ionic and covalent compounds. The corresponding pseudohalogen covalent dimer of four of these entities, cyanogen, $(CN)_2$, thiocyanogen, $(SCN)_2$, selenocyanogen, $(SeCN)_2$, and azidocarbondisulphide, $(SCSN_3)_2$, have been isolated (Table 26.3).

Characteristic reactions of pseudohalogens and pseudohalides are given below:

(1) The parent pseudohalogens are volatile molecular compounds involving a symmetrical combination of two radicals X—X. They react in a

TABLE 26.3
PSEUDOHALIDE IONS AND THE PSEUDOHALOGENS

Pseudohalide	Ion	Pseudohalogen
Cyanide	CN^-	cyanogen, $(CN)_2$
Cyanate	OCN^-	
Thiocyanate	SCN^-	thiocyanogen, $(SCN)_2$
Selenocyanate	$SeCN^-$	selenocyanogen, $(SeCN)_2$
Tellurocyanate	$TeCN^-$	
Azide	N_3^-	
Azidothiocarbonate	$SCSN_3^-$	azidocarbondisulphide, $(SCSN_3)_2$
Isocyanate	ONC^-	

similar manner to the halogens with alkalis:

$$(CN)_2 + 2\ OH^- \rightleftharpoons CN^- + OCN^- + H_2O$$
$$(SCSN_3)_2 + 2\ OH^- \rightleftharpoons SCSN_3^- + OSCSN_3^- + H_2O$$
$$[Cl_2 + 2\ OH^- \rightleftharpoons Cl^- + OCl^- + H_2O]$$

and in forming addition compounds with olefins and other unsaturated molecules:

$$C_2H_4 + (SCN)_2 \rightarrow NCSC_2H_4SCN$$

(2) Inter-pseudohalogen compounds and pseudohalogen—halogen compounds are known. Cyanogen chloride and bromide, CNCl and CNBr, result from the action of Cl_2 and Br_2 on HCN. Cyanogen fluoride has been identified spectroscopically among products of fluorination of cyanogen.

Chloroazide, ClN_3, bromoazide, BrN_3, and iodoazide, IN_3, have also been made. In chloroform, thiocyanogen, $(SCN)_2$, combines with chlorine to give white crystals of SCNCl. Cyanogen bromide reacts with sodium azide to give cyanogenazide

$$CNBr + NaN_3 \rightarrow CNN_3 + NaBr$$

Such compounds as CN(SCN) and CN(SeCN) have been obtained in crystalline form.

(3) The pseudohalogens react with various metals to give salts containing X^- anions; the silver, mercury(I) and lead(II) salts are like the corresponding halides in being sparingly soluble in water,

(4) With hydrogen, the uninegative groups form acids HX which are, however, very weak compared with halogen acids

$$\begin{aligned} pK_a \text{ for } HN_3 &= 4.4 \\ HCN &= 8.9 \\ HCl &= -7.4 \end{aligned}$$

The difference in acidities must arise from the differences in (a) H—X bond energy, and (b) hydration energy of the X^- ion.

(5) The pseudohalides form ions analogous to polyhalide ions, thus $NH_4(SCN)_3$ and $K(SeCN)_3$ resemble KCl_3, and there are also complexes with

metals similar to the less numerous halide complexes: e.g. $Fe(CN)_6^{3-}$ (cf. FeF_6^{3-}). The stabilities of analogous halide and pseudohalide complexes often differ widely.

(6) Molecular pseudohalogen compounds analogous to molecular halides, e.g. CH_3NCO and $Si(NCO)_4$, can be formed, though in contrast with the halogens, isomers may result depending on the position of attachment as in CH_3SCN and CH_3NCS.

(7) In common with halide ions, a pseudohalide ion is oxidised to the parent pseudohalogen by suitable oxidising agents

$$2\ Fe^{3+} + 2\ SCN^- \rightarrow 2\ Fe^{2+} + (SCN)_2$$

or simply by heating the lead(IV) salts:

$$Pb(SeCN)_4 \rightarrow Pb(SeCN)_2 + (SeCN)_2$$
$$[\text{cf. } PbCl_4 \rightarrow PbCl_2 + Cl_2]$$

26.12. Pseudohalogens

Cyanogen, $(CN)_2$, is made by heating AgCN alone or $Hg(CN)_2$ with $HgCl_2$:

$$2\ AgCN \rightarrow 2\ Ag + (CN)_2$$
$$Hg(CN)_2 + HgCl_2 \rightarrow Hg_2Cl_2 + (CN)_2$$

Thiocyanogen, $(SCN)_2$, is released by the action of bromine on AgSCN suspended in ether:

$$2\ AgSCN + Br_2 \rightarrow 2\ AgBr + (SCN)_2$$

The yellow solid polymerises irreversibly at room temperature giving an insoluble brick-red material. The dimer oxidises iodides to iodine and Cu^+ to Cu^{2+} ions.

Selenocyanogen, $(SeCN)_2$, also yellow and crystalline, is displaced by iodine from AgSeCN:

$$2\ AgSeCN + I_2 \rightarrow 2\ AgI + (SeCN)_2$$

Azidocarbondisulphide, $(SCSN_3)_2$, is formed as white crystals when $KSCSN_3$ is oxidised with hydrogen peroxide:

$$CS_2 + KN_3 \xrightarrow{310\ K} KSCSN_3 \xrightarrow{H_2O_2} (SCSN_3)_2$$

The compound decomposes at room temperature:

$$(SCSN_3)_2 \rightarrow 2\ N_2 + 2\ S + (SCN)_2$$

Further Reading

V. Gutmann (Ed.), Halogen chemistry, Vol. 3, Academic Press, New York, 1967.

J.H. Canterford and R. Colton, Halides of the second and third row transition metals, Wiley—Interscience, New York, 1968.

R. Colton and J.H. Canterford, Halides of the first-row transition metals, Wiley—Interscience, New York, 1969.

The Transition Metals

27

The elements which constitute Groups IIIA to IB in the fourth, fifth and sixth periods of the Periodic Table are classified as transition metals. Their atoms, either in their neutral state, or, in the case of Cu, Ag and Au in some common oxidation state, have partly filled d or f shells.

27.1. 'd-Block' and 'f-Block' Elements

The 'd-block' elements, or outer transition elements, are those of Groups IVA to IB together with Sc and Y. They all have partly filled d shells in their neutral atoms or in common oxidation states (Cu^{II}, Ag^{II}, Au^{III}), but their f shells are either completely empty (Periods 4 and 5) or fully occupied as in the 4f shell of the d-block elements of Period 6.

The 'f-block' elements are the inner transition elements, the lanthanides of Period 6 and the actinides of Period 7. Most of them have partly filled f shells either in their neutral atoms or their common oxidation states. Exceptions are La($5d^16s^2$) and Lu($4f^{14}5d^16s^2$) but these are so similar to the metals which lie between them that it is usual to consider all fifteen elements together, and likewise the actinides from Ac to Lr. The lanthanides and actinides have some general similarity to the Group IIIA elements; accordingly Sc and Y are described with the lanthanides in Chapter 30.

The lanthanides resemble one another far more closely than do the 'd-block' metals. In the latter the unfilled d orbitals are near the periphery of the atom and are consequently strongly influenced by atoms or groups around them. By contrast, the unfilled f orbitals of a lanthanide are overlaid with 5s and 5p electrons and are thus thoroughly shielded from external influences; consequently the chemistry of the lanthanides is affected little by the number of f electrons. The actinides are somewhat similar, but in them the 5f orbitals are less thoroughly shielded from external influences, and differences are rather greater.

27.2. The Position of the Transition Metals in the Periodic Table

It is of interest to consider why the transition metals occur where they do in the Periodic Table. Fig. 27.1 illustrates, not to scale, the relative energy

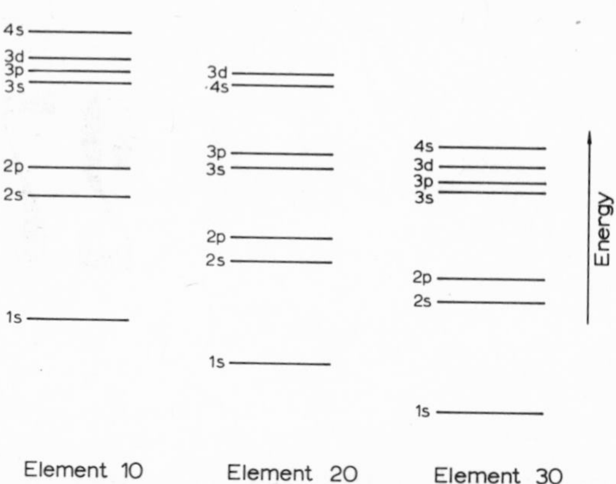

Fig. 27.1. Variation of energy levels of orbitals with nuclear charge and occupancy of shells.

levels of the lowest orbitals of elements 10, 20 and 30. As the charge on the nucleus is increased from +10 in neon to +20 in calcium, the orbital energies fall, but the 4s orbital, which penetrates the argon core ($Z_{Ar} = 18$) to some considerable extent, falls rapidly in energy relative to the 3d orbitals, which penetrate the argon core very slightly. In fact the 4s level at calcium lies lower than the 3d, and Ca has the ground-state configuration $1s^2\ 2s^2\ 2p^6\ 3s^2\ 3p^6\ 4s^2$. However the 3d lies below the 4p and is the next orbital to begin filling with electrons. As the charge on the nucleus increases from +20 (at Ca) to +30 (at Zn) and the 3d shell is occupied, its electrons tend to become more effective in shielding the 4s electrons from the nucleus, a consequence of the relative radial probability densities of s and d orbitals (3.2). The pattern is repeated in the second d-block, but here the 5s and 4d are the orbitals principally involved. In the case of the f-block elements the 4f, 5d and 6s lie extremely close together in energy at element 57, and one electron enters the 5d shell before the 4f starts to fill; La has a $5d^1\ 6s^2$ configuration. In the later elements the f shell is stabilised relative to the 5d just as the 3d is stabilised relative to the 4s when the 3d orbitals begin to fill up.

27.3. Physical Properties of the Elements

Nearly all the transition elements have the simple h.c.p., c.c.p. or b.c.c. lattices characteristic of true metals (4.1.7) and display the typical metallic properties — high tensile strength, ductility, malleability, high thermal and electrical conductivity, and metallic lustre. Their *m.p.* and *b.p.* are high. In any row of d-block metals the melting points rise to a maximum at the Group VIA metal and then, except for anomalous values in Mn and Tc, fall regularly again (Fig. 27.2). In Fig. 27.3 are represented the boiling points

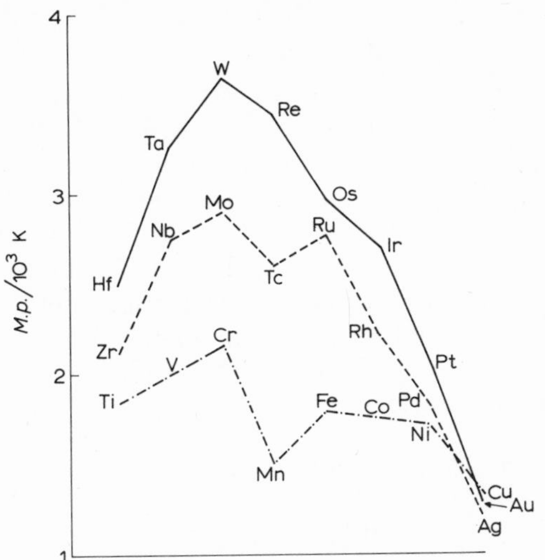

Fig. 27.2. Melting points of d-block metals.

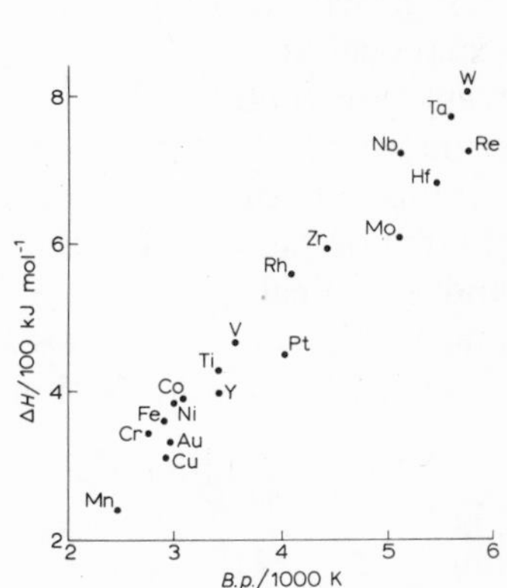

Fig. 27.3. Relation between enthalpy of vaporisation and boiling point for some transition metals.

and the enthalpies of vaporisation of some d-block metals. Clearly these quantities are related in an approximately linear manner and, as the enthalpy of vaporisation is an important factor in determining the standard electrode potential of a metal (8.2), the metals of highest boiling point will tend to be noble metals.

Transition metals form many alloys; the best known are the ferrous alloys constituting the many varieties of steel (36.3.1). Alloys of transition metals with non-transition metals are also of great industrial importance; examples are brass (copper/zinc) and bronze (copper/tin).

27.4. Chemical Properties of the Metals

Most of these metals dissolve in acids. For the lanthanides the E°, Ln^{3+}/Ln values are about −2.3 V, and the actinides are about as active. The top row of d-block elements, except copper, all dissolve in non-oxidising acids. Only the metals of Group IB and the platinum metals (37.1) show noble character.

Because unpaired d electrons are a characteristic of many transition-metal compounds, paramagnetism is common in them. The promotion of an electron from one d orbital to another usually requires a quantum of energy appropriate to absorption in the visible spectrum; thus many transition-metal compounds are coloured, but not as a rule strongly, because d—d transitions are forbidden ones in the quantum-mechanical sense (6.4).

27.5. Variable Oxidation States in Ionic Compounds

The most important chemical characteristic of the transition metals is that nearly all of them exhibit several oxidation states (8.4). Although the existence of more than one oxidation state is not unknown in the non-transition metals (e.g. Tl^{I}, Tl^{III}; Sn^{II}, Sn^{IV}), the phenomenon is much less common, and it is rare for adjacent states (e.g. +2 and +3) to occur in the compounds of any non-transition metal (an exception is Hg). To illustrate the major factors affecting oxidation state in a typical non-transition metal and a typical transition metal, let us consider first the ionic compounds of magnesium and manganese.

The first three ionisation energies are respectively:

	$I(1)$/kJ mol^{-1}	$I(2)$/kJ mol^{-1}	$I(3)$/kJ mol^{-1}
Mg	737	1450	7731
Mn	717	1509	3251

Clearly the sum $I(1) + I(2)$ is about the same in each case, but $I(3)$ is far greater for Mg than for Mn, because in Mg it refers to the removal of a p electron which is subject to a very high effective nuclear charge (3.4), whereas in Mn it refers to the removal of a d electron which is rather strongly shielded from the nucleus. The enthalpy of formation of, say, the difluoride of either metal will be given by

$$\Delta H_f(MF_2) = \Delta H_s(M) + \Delta H_d(F_2) + (I(1) + I(2))M - 2A(F) - U(MF_2) \quad (4.2.3)$$

The terms $(I(1) + I(2))M$ and $U(MF_2)$ are much larger than the others, and the enthalpy of formation is largely determined by them. Both compounds are exothermic because U is large. But for the hypothetical compound MgF_3, for which ΔH_f would be

$$\Delta H_f(MgF_3) = \Delta H_s(Mg) + \tfrac{3}{2}\Delta H_d(F_2) + (I(1) + I(2) + I(3))Mg - 3A(F) - U(MgF_3)$$

the probable value of U, though large compared with that for MgF_2, would not be big enough to compensate for the very large third ionisation energy $I(3)$, and the compound would be highly endothermic. In the case of MnF_3, however, the value of $I(3)$ is so much smaller that the compound is exothermic. The lattice energy of a salt MX_3 is greater than that for MX_2 for three reasons: the increase in the product $z_+ \times z_-$, the increase in the Madelung constant (4.2.3) and the decrease in the interionic distance because $r_{M^{3+}} < r_{M^{2+}}$. Another aspect of the lattice energy consideration is that since r_{F^-} is small in comparison with r_{Cl^-}, for example, the ratio $(r_{M^{3+}} + r_{F^-})/(r_{M^{2+}} + r_{F^-})$ is smaller than $(r_{M^{3+}} + r_{Cl^-})/(r_{M^{2+}} + r_{Cl^-})$. Thus the greatest increase in U in going from the MX_2 halide to MX_3 occurs in the fluoride, and this is the major reason why metals often reach higher oxidation states in their ionic fluorides than in their ionic chlorides; examples are MnF_3, CoF_3 and AgF_2.

If the foregoing argument about ratios of interionic distances is extended to the consideration of the variation of cationic radii through the first transition series (27.1), we should expect the ratio $(r_{M^{3+}} + r_{Cl^-})/(r_{M^{2+}} + r_{Cl^-})$ for example, to fall through the series and the trichlorides to become increasingly exothermic. In fact none of the metals after iron forms a trichloride, and the major reason why this is so is probably the increase in $I(3)$ through the latter part of the first row of the d block.

Before we proceed to consider variations of oxidation state in the more covalent transition-metal compounds, we shall consider a little further the variation of ionisation energies in the d-block.

27.6. Variations in Ionisation Energies

Fig. 27.4 illustrates that the first ionisation energies of the Period 6 elements lie higher than those of the other two periods, a consequence of the greater effective nuclear charge acting on outer electrons because of the weak shielding afforded by 4f electrons. The ionisation energies of the Period 4 and Period 5 elements are irregular; alternations between the relative values are not due to any single cause, but one factor must be the lack of consistency in the ground-state configurations. For example, the occupancy of the outer orbitals is $3d^3 4s^2$ in V and $3d^4 4s^1$ in Nb, but both their unipositive ions have a $d^4 s^0$ configuration. It will be noticed that there is a considerable increase in $I(1)$ for the first-row elements after Cr. There is a corresponding increase in $I(3)$ after Fe:

	V	Cr	Mn	Fe	Co	Ni	Cu
$I(3)$/MJ mol^{-1}	2.83	2.99	3.26	2.96	3.24	3.40	3.56

This is almost certainly the major reason for the failure to make trichlorides of metals beyond iron which is referred to in the foregoing paragraph. For both the ionisations

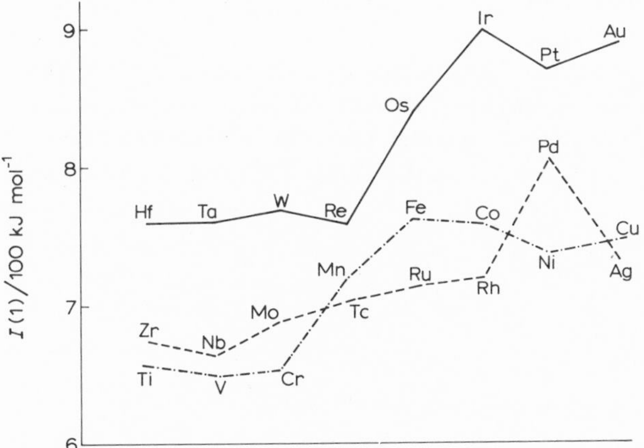

Fig. 27.4. Variation of first ionisation energies in the d-block metals.

$Cr(3d^5 4s^1) \rightarrow Cr^+(3d^5)$

and

$Fe^{2+}(3d^6) \rightarrow Fe^{3+}(3d^5)$

there is no loss of exchange stabilisation (3.4.3), almost certainly the major reason for the low $I(1)$ for Cr and the low $I(3)$ for Fe.

27.7. Transition Metals in Covalent Molecules

In the formation of covalent molecules, it is unrealistic to imagine electrons being completely removed from a metal ion, and the oxidation number represents its formal rather than its actual charge. Nevertheless the total ionisation energy required to produce an isolated metal ion is a useful indication of the energy needed to raise the metal to that particular oxidation state in a compound. In this connection it is interesting to compare consecutive ionisation energies for first-row d-block elements with those of their heavier congeners. A comparison of the first four ionisation energies of Ni with those of Pt shows that while less energy is required to produce Ni^{2+} than Pt^{2+}, more energy is required to produce Ni^{4+} than Pt^{4+}.

	$[I(1) + I(2)]$ /MJ mol^{-1}	$[I(3) + I(4)]$ /MJ mol^{-1}	*Total*/MJ mol^{-1}
Ni	2.49	8.80	11.29
Pt	2.66	6.70	9.36

Thus Ni^{II} compounds tend to be more thermodynamically stable than Pt^{II} compounds, but compounds containing Pt^{IV} are more stable than those containing Ni^{IV}; for example K_2PtCl_6 is a well known compound but K_2NiCl_6 has not been made. Trends are similar elsewhere in the d-block. In general, oxidation states of +4 and above are more stable in the second and third rows than in the first. Another way of expressing the same idea is that second and third row elements have a greater tendency to form covalent compounds, because a high formal charge on the metal means more polarisation of the ligands.

A quantitative treatment of the factors influencing oxidation states in covalent molecules presents a far greater problem than that for ionic crystals. To illustrate the difficulties, let us consider the conversion of MX_2(solid) to MX_4(vap.) in a series of steps, assuming the temperature to be constant:

		Energy requirement
$MX_2(s)$	$= M^{2+}(g) + 2X^-(g)$	$U(MX_2)$
$M^{2+}(g) + 2\,e$	$= M(g)$	$-[I(1) + I(2)]$
$2X^-(g)$	$= 2X(g) + 2\,e$	$2A(X)$
$X_2(g)$	$= 2X(g)$	$D(X_2)$
$M(g) + 4X(g)$	$= MX_4(g)$	$-4E(M{-}X)$
$MX_2(s) + X_2(g)$	$= MX_4(g)$	

The total energy required for the process is therefore

$$U(MX_2) - [I(1) + I(2)] + 2A(X) + D(X_2) - 4E(M{-}X)$$

Though conversion of a lower ionic halide into a higher covalent one is common in practice, it is clear that the thermodynamic treatment presents a complicated problem. For example, the halogen which forms the strongest M—X bonds in the covalent compound, fluorine, is also the one which forms the ionic compound MX_2 of highest lattice energy, thus two of the largest terms included in the energy requirement equation will always tend to compensate one another. There are three others factors which complicate the issue. The value of E(M—X) in a covalent compound falls as the oxidation state of M is increased because of the energy dissipated in promoting electrons to produce a higher 'valence state' (4.1.1.2). Secondly, entropy effects may become important, particularly in the conversion of a solid halide to a gaseous one. Finally, the distinction between ionic and covalent bonds is not a sharp one. For all of these reasons, a quantitative examination of stabilisation of oxidation states is extremely difficult; in particular very few data on promotion energies are available. Thus chemists have sought more general, qualitative considerations to provide explanations. One of the most useful of these is discussed in the paragraphs which follow.

27.8. Pauling's Electroneutrality Principle

This principle, which can be applied to explain the relation of the stability of a particular oxidation state in an atom to the character of atoms which surround it, was introduced by Pauling in 1948. It states that stable molecules have electronic structures such that no single atom carries a charge greater than $\pm e$.

The principle can be applied effectively to transition-metal complexes as well as to molecules. Let us consider the character of bonds between metal and oxygen in an aquated ion $M(H_2O)_6{}^{n+}$. For a d-block metal of the first row, the electronegativity of M is about 1.7 and the difference in electronegativity between the metal and the oxygen is therefore about 1.8. A reference to Fig. 27.5, which is also due to Pauling, shows that the M—O bond has about 50% covalent character. In other words, each of the six oxygens surrounding M will donate about half an electronic unit of charge to it, making $3e$ in all. Thus if M is to be approximately electrically neutral in the aquated ion, it must carry a charge of about +3 originally. The electroneutrality principle thus indicates that a first-row transition metal ion should carry a charge of about +3 for the hexa-aquocomplex to be stable towards oxidation and reduction. In fact, the most common oxidation states shown by transition metals in such ions are +2, +3 and +4. The charge is not held by the oxygen atoms, but is largely transferred to the twelve H atoms. Thus the charge possessed by the hydrated ion becomes spread over its surface, in accordance with elementary electrostatic theory. Clearly the greater the posi-

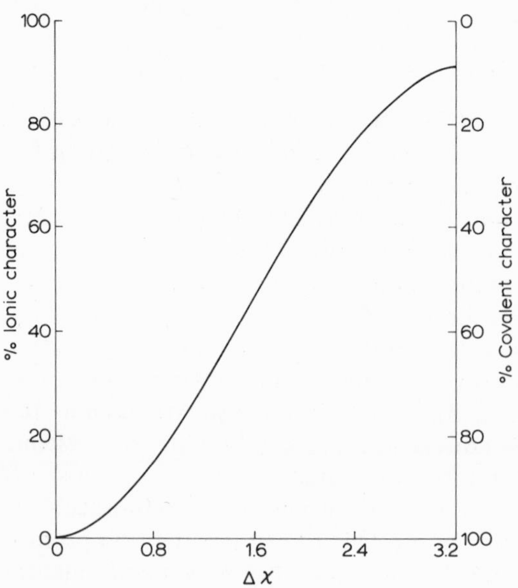

Fig. 27.5. Relation between the ionic character of a bond and the difference between the electronegativities of the atoms which it connects.

tive charge transferred to the hydrogen atoms, the greater the tendency for one to be lost as a proton, and the greater the acidity of the hydrated ion. As an example the equilibrium constant for the reaction

$$Fe(H_2O)_6^{3+} + H_2O \rightleftharpoons Fe(H_2O)_5OH^{2+} + H_3O^+$$

is about 2×10^{-3} whereas for

$$Fe(H_2O)_6^{2+} + H_2O \rightleftharpoons Fe(H_2O)_5OH^+ + H_3O^+$$

it is only about 10^{-7}.

27.8.1. Co-ordination numbers

The electroneutrality principle is of importance in deciding the co-ordination number of a metal in a particular oxidation state. Thus in the isoelectronic series of cations, Ag^+, Cd^{2+}, In^{3+}, the chloride ion forms complexes $AgCl_2^-$, $CdCl_4^{2-}$ and $InCl_6^{3-}$, as would be expected from the principle if the metal—chlorine bonds are about 50% covalent. Considering different ligands, a metal ion M^{n+} can be converted into an atom of roughly zero charge by accepting a large negative charge from each of a few easily polarisable ligands, or, alternatively, a small negative charge from each of a larger number of ligands which are less easily polarised. Thus Fe^{3+} forms a 4-co-ordinate ion, $FeCl_4^-$, with Cl^- ions but a 6-co-ordinate ion, FeF_6^{3-}, with the less polarisable F^- ions. When the ligands are large enough, the co-ordinate ion number may be affected by steric considerations, but these are usually thought to be of subsidiary importance. In general, the more electronegative the ligand the higher will be the most favoured co-ordination number.

Some readily polarisable groups such as the CN^- ion are, however, exceptions to the generalisation. Unlike the Cl^- ion, the CN^- ion forms the 6-co-ordinate Fe^{III} complex $Fe(CN)_6{}^{3-}$. Although electrons attached to the carbon atoms are directed strongly towards the iron atom, the symmetry and the energy of the t_{2g} electrons of the metal are so favourable for the formation of $d_\pi-p_\pi$ bonds with the ligand (6.6.1) that much of the negative charge is 'back-donated' to the empty π^* orbitals of the CN^- ligands.

27.8.2. Ligands stabilising oxidation states

A further consequence of the electroneutrality principle is that the highest oxidation states of the metals are stabilised by strongly electronegative, non-π-accepting ligands. In the $MnO_4{}^-$ ion, for example, the oxygen atoms exert sufficient attraction on the bonding electrons to prevent them from being attracted wholly to the metal. Charge numbers of +5 and +6 are particularly common among the oxo-complexes of transition metals ($VO_4{}^{3-}$, $CrO_4{}^{2-}$); and charge numbers of +7 ($MnO_4{}^-$, $ReO_4{}^-$) and even +8 (OsO_4) are known.

High oxidation states are also stabilised by the strongly electronegative fluorine in fluoro-complexes, but chlorine and the less electronegative bromine and iodine have decreasing power to do this. Thus although Fe^{III} is stable in chloro- and bromo-complexes, in its iodo-complexes the highest oxidation state of iron is +2. The iron atom can be considered to capture completely the two electrons of one of the Fe—I bonds; and, as a consequence, Fe^{III} is reduced to Fe^{II} and iodine is liberated.

Low oxidation states are stabilised by ligands which have vacant orbitals of suitable symmetry and energy for the formation of strong π bonds with the metal. It is possible for the carbon atom of CO to direct electrons strongly towards a metal even in its zero oxidation state (for instance Ni^0) because the negative charge added to the metal atom can be dissipated by back-donation into the π orbitals.

TABLE 27.1

OXIDATION STATES OF RUTHENIUM

Charge number	Compound or ion
0	$Ru(CO)_5$
+2	$[Ru(NH_3)_6]^{2+}$
+3	$[RuCl(NH_3)_5]^{2+}$
+4	$RuCl_6{}^{2-}$
+5	RuF_5
+6	RuF_6
+7	$RuO_4{}^-$
+8	RuO_4

A particularly good example of a metal exhibiting many different oxidation states is ruthenium; some of its compounds are listed in Table 27.1. These compounds illustrate the principle outlined above; the highest oxidation states here are stabilised by oxide and fluoride ions, the intermediate states are stabilised by chloride ions and ammonia molecules, and the lowest states by CO molecules. The O^{2-} ion stabilises even higher oxidation states than the F^- ion, probably because it uses its second electron for π bonding to the metal. A neutral molecule like ammonia cannot stabilise as high an oxidation state as can a negative ion in which the co-ordinating atom has similar electronegativity; thus the substitution of Cl^- for NH_3 in $Ru(NH_3)_6^{2+}$ enables the oxidation state of the ruthenium to be raised by one. In an ion like $Ru(NH_3)_6^{3+}$ the hydrogens tend to become so positive that the complex ion undergoes considerable dissociation as an acid.

27.8.3. d-Shell occupancy and the stability of oxidation states

Complexes in which the transition metal has a more than half-filled d shell tend to be more covalent, for a particular oxidation state and ligand, than complexes in which the transition metal has fewer d electrons, because of the increasing effective nuclear charge felt at the surface of the metal atom. The stability of such complexes is considerably affected by the number of d electrons present. Complexes in which the metal has 6, 8, or 10 d electrons are much more numerous than those in which the metal has a d^7 or d^9 configuration; a single unpaired electron appears to react particularly freely with electrons on other atoms.

d^6 complexes

Low-spin complexes of d^6 ions are especially common and range over many oxidation states:

V^{-I}	Cr^0	Mn^I	Fe^{II}	Co^{III}	Ni^{IV}
$V(CO)_6^-$	$Cr(CO)_6$	$Mn(CO)_6^+$	$Fe(CN)_6^{4-}$	$Co(NH_3)_6^{3+}$	NiF_6^{2-}

The low-spin d^6 configuration is particularly favourable for octahedral arrangement.

d^7 complexes

In agreement with the generalisation above, the d^7 configuration is rather an uncommon one. Strikingly, for ruthenium, which exhibits every other oxidation state between 0 and 8, no $Ru^I(d^7)$ compound has yet been prepared. But the configuration does occur in a few compounds of other metals with oxidation states as far apart as −1 in $Cr_2(CO)_{10}^{2-}$ and +3 in K_3NiF_6.

d^8 complexes

This again, like d^6, is a very common configuration, ranging over a wide range of oxidation states; 5-co-ordination is often observed in the low-spin complexes:

Cr^{-II}	Mn^{-I}	Fe^{0}	Co^{I}	Ni^{II}
$[Cr(CO)_5]^{2-}$	$[Mn(CO)_5]^-$	$Fe(CO)_5$	$[Co(PhNC)_5]^+$	$[Ni(CN)_5]^{3-}$

d^9 complexes

Oxidation states with a d^9 configuration are moderately common, but in most cases the spare electron is paired off to achieve, in effect, a d^8 arrangement. In $Fe_2(CO)_8{}^{2-}$, for example, there is a metal—metal bond. Paramagnetism in d^9 complexes is consequently comparatively rare.

d^{10} complexes

This is almost as common as the d^6 configuration. The most usual co-ordination number is four.

Fe^{-II}	Co^{-I}	Ni^{0}	Cu^{I}	Zn^{II}
$[Fe(CO)_4]^{2-}$	$[Co(CO)_4]^-$	$Ni(CO)_4$	$[Cu(CN)_4]^{3-}$	$[Zn(CN)_4]^{2-}$

27.8.4. Geometry stabilising oxidation states

There is another way in which d-shell occupancy can affect the stability of an oxidation state. A particular geometrical arrangement of ligands, enforced by the environment, can stabilise an otherwise unusual oxidation state. For example d^3 and low-spin d^6 configurations are favourable for octahedral co-ordination but not for tetrahedral.

We have outlined above some of the considerations which have a bearing on the stabilities of transition-metal compounds to oxidation and reduction. The subject is still somewhat empirical; very much more information is required about such things as promotion energies and ligand-field strengths before there can be a unified, quantitative theory.

27.9. Oxidation States of d-Block Metals in Aqueous Solution

The value of E°, M^{2+}/M, i.e. the *emf* of the cell in which the reaction

$$2\,H^+(aq) + M(s) \rightarrow M^{2+}aq + H_2(g)$$

occurs under reversible conditions, is determined by the factors which are explained in Section 8.3. The observed variation for those first-row transition elements for which measurements can be made is

Element	V	Cr	Mn	Fe	Co	Ni	Cu
E^0, $M^{2+}/M/V$	−1.18	−0.9	−1.18	−0.44	−0.28	−0.25	+0.34

Probably the major factor responsible for the variation is the sum $I(1) + I(2)$, but sublimation energies vary erratically from the value of 240 kJ mol^{-1} for Mn to 470 kJ mol^{-1} for V. Not surprisingly, there is no simple trend for E° values.

For the M^{3+}/M^{2+} couples, however, the problem is simplified because the sublimation energy of M is no longer a factor; the third ionisation energy of M and the hydration energies of the two ions alone determine the E° value. The measured values are:

Element	V	Cr	Mn	Fe	Co
E^0, M^{3+}/M^{2+}/V	−0.26	−0.41	+1.51	+0.77	+1.97

The variations are similar in general to the variations in $I(3)$ from V to Co, but the anomalously low value for Cr compared with V is almost certainly due to the fact that conversion of a high-spin d^4 (e.g. Cr^{2+}) ion into a d^3 ion in an octahedral field is accompanied by a large increase in ligand field stabilisation energy, whereas conversion of a d^3 ion (e.g. V^{3+}) into a d^2 results in a loss of LFSE (7.2.19).

27.10. Ionic Radii in d-Block Series

Through any of the d-block series there is a general downward trend in ionic radii because there is an increase in effective nuclear charge. The trend is illustrated here by the bipositive ions of the first-row elements in octahedral co-ordination.

Ion	Ti^{2+}	V^{2+}	Cr^{2+}	Mn^{2+}	Fe^{2+}	Co^{2+}	Ni^{2+}	Cu^{2+}
$r_{M^{2+}}$/pm	90	88	84	80	76	74	69	72

Minor variations are probably due to ligand-field effects caused by differences in the occupancy of the t_{2g} and e_g orbitals. The more uniform 'lanthanide contraction' in the first 'inner-transition' series is discussed in Section 30.1.

27.11. Hydrated Ions of Transition Metals

Six co-ordinate, hydrated, bipositive and terpositive ions are particularly common among the transition metals. The energies of hydration of these free ions are of considerable interest, and the variation of these energies from one metal to another has been explained by the use of crystal-field theory. The values for the bipositive ions of the fourth-period metals are shown in Fig. 27.6.

It might be expected that the hydration energies would increase smoothly from Ca^{2+} to Zn^{2+} with the effective nuclear charge exerted at the surface of the free ion. But although the total occupancy of d orbitals increases steadily through the series the occupancies of the sets of t_{2g} and e_g orbitals do not. In $Ti(H_2O)_6{}^{2+}$ for example, there are two t_{2g} electrons but there is no e_g electron. The field created round the free ion by the approach of the

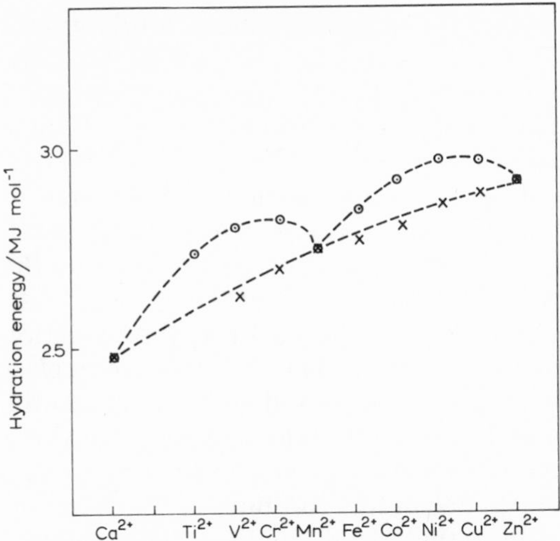

Fig. 27.6. Hydration energies of bipositive ions of Period 4 transition metals: ⊙, experimental values; ×, after subtraction of ligand-field stabilisation energies.

water molecules releases a stabilisation energy of $2 \times (2\Delta/5)$, where Δ is the crystal-field splitting. Thus the heat set free when the free Ti^{2+} ion is converted into its hexa-aquo complex is $4\Delta/5$ greater than would be expected if the two electrons were distributed in a spherically symmetrical manner. For the $V(H_2O)_6{}^{2+}$ (t_{2g}^3) ion, the extra energy released is $6\Delta/5$, but in the $Cr(H_2O)_6{}^{2+}$ ($t_{2g}^3\ e_g$) ion it is only $3\Delta/5$, because the occupancy of the higher-energy e_g orbital reduces the stabilisation energy by $3\Delta/5$ (7.2.19), and in the $Mn(H_2O)_6{}^{2+}$ ion the stabilisation is zero because the occupancy is $t_{2g}^3\ e_g^2$, with a spherically symmetrical arrangement of d electrons.

The second part of the series from Mn^{2+} to Zn^{2+} repeats the pattern as first the t_{2g} and then the e_g orbitals become doubly occupied. In $Ni(H_2O)_6{}^{2+}$, for example, ($t_{2g}^6\ e_g^2$), the stabilisation energy released when the aquo-complex is formed is $6\Delta/5$ as in $V(H_2O)_6{}^{2+}$.

The ligand-field splittings for the ions can be determined from visible absorption spectra and the stabilisation energies can be calculated. When these stabilisation energies are subtracted from the experimental hydration energies a fairly smooth curve is obtained (Fig. 27.6).

Curves of very similar appearance to those for the hydration energies of the ions are obtained by plotting the stability constants (28.6.2) of M^{II} complexes of a particular ligand for the same series of metals. The lattice energies of a series of salts such as the dichlorides also fall on a similar curve.

27.12. Metal—Metal Bonds in Transition-metal Compounds

In some compounds, in which the metals exhibit low oxidation numbers, strong interactions occur between metal atoms. These compounds can be divided into two classes:

(i) Halides and some oxides and sulphides of Period 5 and Period 6 transition metals in low oxidation states.

(ii) Metal carbonyls and related compounds (17.12).

If we consider a thermochemical cycle for the production of a halide M_yX_z, where M is a d-block element of the second or third row, we find that a major consideration is the large enthalpy of atomisation of the metal. In a compound where the ratio z/y is small, the energy released in M—X bond formation is not enough to compensate, hence stability is achieved by the formation of other strong bonds, metal—metal bonds. But this type of interaction is only possible if two conditions are fulfilled. First, the oxidation state of the metal must be low enough to ensure that the effective nuclear charge acting on the outer orbitals is insufficient to reduce their size to the point where they do not overlap effectively. As the relevant orbitals of the heaviest atoms are largest, these naturally have a greater chance to overlap than those of the first-row elements. Secondly, the number of X atoms around any metal atom must be small so that M—M bonding is not inhibited by steric considerations. This condition is, in any case, a consequence of a low formal charge in the metal.

Metal—metal bonds are formed by first-row metals only in oxidation states near zero, as in carbonyls and carbonylate ions (17.12), but some M—M interaction can occur even up to oxidation state +4 in compounds of third-row metals. However, strong interactions are uncommon in metals of groups beyond VIIA because of the contraction of the outer orbitals as the effective nuclear charge increases and also because the filling of the d orbitals makes valence states of high multiplicity unattainable. In carbonyl compounds, however, neither of these strictures apply to the same extent because orbital contraction is least when the formal charge on the metal is least and also because back-bonding to π^* orbitals of the CO molecules removes electrons from the antibonding M—M orbitals.

The effect of reducing formal charge number in enhancing M—M interactions can be illustrated by referring to the metals Re and Mo. In its +7, +6 and +5 states rhenium does not form Re—Re bonds. The Re_2Cl_{10} molecule has a long Re—Re distance of 374 pm, compared with 275 pm in the metal itself. But the $Re_2Cl_9^-$ ion containing Re^{IV} consists of two $ReCl_6$ octahedra sharing a face and strongly distorted to make the Re—Re distance only 271 pm (Fig. 27.7), and in the +3 state, strong metal—metal bonds appear in Re_3Cl_9 (Fig. 35.4) and in $Re_2Cl_8^{2-}$ (Fig. 27.8).

The pattern is similar for Mo. In the Mo_2Cl_{10} molecule the Mo—Mo distance is long, but in MoO_2 the rutile structure is strongly distorted to give a metal—metal distance of only 251 pm, and Mo^{II} forms strongly bound clusters of Mo atoms (Fig. 34.2).

Transition-metal compounds containing metal—metal bonds can be divided into three types:

(i) Cluster compounds contain arrangements of metal atoms each of which forms several single bonds to other atoms in the cluster (e.g. $Mo_6Cl_8^{4+}$ in Fig. 34.2).

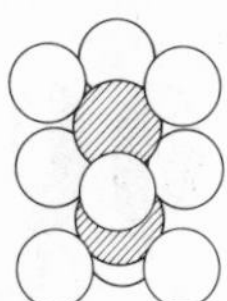

Fig. 27.7. $Re_2Cl_9^-$ ion.

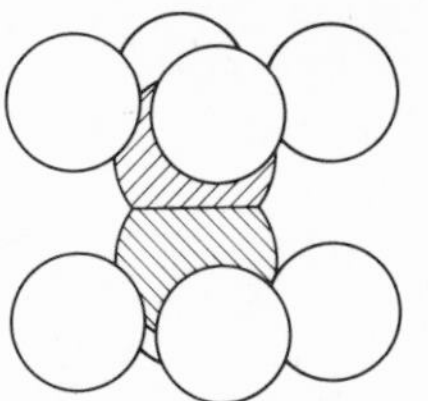

Fig. 27.8. $Re_2Cl_8^{2-}$ ion.

(ii) Some molecules and ions contain a pair of metal atoms with a bond of high order between them. In the $Re_2Cl_8^{2-}$ ion (Fig. 27.8) the rhenium—rhenium bond has an order of about 4, the internuclear distance is only 224 pm, and another indication of the strength of the bond is that the chlorines are eclipsed, evidently because at least one component of the metal—metal bond has a strong rotational dependence which overcomes the tendency for repulsions between the Cl orbitals to produce a staggered configuration.

(iii) Some molecules and ions contain small clusters, for example the triangle of Re atoms in Re_3Cl_9, in which the metals form multiple bonds to a few others. The rhenium—rhenium distance in Re_3X_9 is in the range 240—250 pm. Each rhenium atom employs five of its atomic orbitals to form bonds to halogens, thus four valence orbitals containing a total of four electrons are available to form two double bonds, one to each metal neighbour.

Further Reading

R.S. Nyholm and M.L. Tobe, The stabilisation of oxidation states of the transition metals, Adv. Inorg. and Radiochem., 5 (1963) 1.

H.D. Kaesz and R.B. Saillant, Hydride complexes of the transition metals, Chem. Rev., 72 (1972) 231.

R. Mason, Valence in transition-metal complexes, Chem. Soc. Rev., 1 (1972) 431.

M.C. Baird, Metal—metal bonds in transition-metal compounds, Prog. Inorg. Chem., 9 (1968) 1.

R.B. King, Transition metal cluster compounds, Prog. Inorg. Chem., 15 (1972) 287.

R.D. Johnston, Transition metal clusters with π-acid ligands, Adv. Inorg. and Radiochem., 13 (1970) 471.

R. Colton and J.H. Canterford, Halides of the first-row transition elements, Wiley—Interscience, New York, 1969.

J.H. Canterford and R. Colton, Halides of the second and third-row transition elements, Wiley—Interscience, New York, 1968.

L.E. Orgel, An introduction to transition-metal chemistry: Ligand field theory, 2nd edition, Methuen, London, 1966.

I.F. Roberts, Absorption spectra of the d-block elements, Ed. Chem., 8 (1971) 178.

D.L. Kepert and K. Vrieze, Compounds of transition elements involving metal—metal bonds, Pergamon, Oxford, 1975.

G.L. Eichhorn Ed., Inorganic biochemistry, Elsevier, Amsterdam, 1973.

Bioinorganic chemistry, ACS Advances in Chemistry No. 100, 1971.

M.N. Hughes, The inorganic chemistry of biological processes, Wiley, New York, 1972.

C.A. McAuliffe Ed., Techniques and topics in bioinorganic chemistry, Macmillan, London, 1975.

R.J.P. Williams, Role of transition metal ions in biological processes, RIC Reviews, 1 (1968) 13.

Complex or Co-ordination Compounds and Ions

28

28.1. Introduction

Various particular aspects of the subject of complex compounds and ions have been discussed earlier in the book; here we deal broadly with the subject as a whole.

28.1.1. Definitions

A sufficient indication of what is meant by a complex is presented in the following definition: a complex molecule or ion is one in which an atom (A) is attached to other atoms (B) or groups of atoms (C) to a number in excess of the charge or oxidation number of the atom (A). In this definition (A) is the *central* or *nuclear atom* and (B) and (C) are *ligands.* An atom whether it be the atom (B) alone, or an atom in the group (C), which is directly attached to (A), is a *co-ordinating atom.* A ligand (C) with more than one potential co-ordinating atom is *multidentate* (*uni-*, *bi-*, *terdentate*).

A *chelate ligand* is one using more than one of its co-ordinating atoms. A complex with more than one nuclear atom is *polynuclear.* A *bridging group* is a group attached to two nuclear atoms in a polynuclear complex.

28.1.2. Examples

Two or more compounds capable of independent existence often combine:

$$Fe(CN)_2 + 4\ KCN \rightarrow K_4Fe(CN)_6$$
$$2\ KCl + PtCl_4 \rightarrow K_2PtCl_6$$
$$K_2SO_4 + Al_2(SO_4)_3 + 24\ H_2O \rightarrow 2\ KAl(SO_4)_2 \cdot 12\ H_2O$$
$$AgCl + 2\ NH_3 \rightarrow Ag(NH_3)_2Cl$$
$$KF + MgF_2 \rightarrow KMgF_3$$

These products differ widely in their behaviour, particularly in water.

Potassium hexacyanoferrate(II), $K_4Fe(CN)_6$, dissolves in water to give a solution with none of the reactions of the Fe^{2+} ion; the metal is present as the $Fe(CN)_6^{4-}$ ion. A solution of K_2PtCl_6 contains $PtCl_6^{2-}$ ions and a solution of $Ag(NH_3)_2Cl$ contains $Ag(NH_3)_2^+$ ions. These are all complex ions.

The solution of an alum $KAl(SO_4)_2 \cdot 12\,H_2O$, on the other hand, gives reactions characteristic of K^+, Al^{3+} and SO_4^{2-} ions. The brown caesium rhodium alum, $CsRh(SO_4)_2 \cdot 12\,H_2O$, behaves similarly in water. However, it can be partially dehydrated to the red $CsRh(SO_4)_2 \cdot 4\,H_2O$, and this does not produce a precipitate with $BaCl_2$ immediately after it is dissolved because $[Rh(H_2O)_4(SO_4)_2]^-$ ions, not SO_4^{2-} ions, are formed. The alums and other hydrates are often distinguished from the complex salts and are called lattice compounds, but clearly a sharp distinction is not always possible.

In solid K_2PtCl_6 there are octahedral $PtCl_6^{2-}$ ions. But $KMgF_3$ is without discrete complex ions; the compound has the perovskite structure. There is thus no simple way of defining a complex in terms of solid structure.

Not all complexes contain complex ions. The well-known 'nickel dimethylglyoxime', bis(dimethylglyoximato)nickel(II), $[Ni(C_4H_7O_2N_2)_2]$ and bis(acetylacetonato)copper(II) are uncharged molecules.

Oxo-anions and halogeno-anions can be regarded as complexes. For example, the sulphate ion is formally an S^{6+} ion co-ordinated tetrahedrally to four O^{2-} ions, and the fluorosilicate ion is an Si^{4+} ion co-ordinated octahedrally to six F^- ions. These complex anions exist as such in crystals of their salts.

28.2. Methods of Preparing Complexes

Here are classified and described some of the more important methods of preparing complexes. The reactions of complexes are dealt with in Chapter 29.

28.2.1. Substitution reactions in aqueous solution

This, the commonest way of synthesis, involves the replacement of water molecules round the metal atom by other ligands. Thus when an excess of NH_3 is added to an aqueous solution of $CuSO_4$, the co-ordinated water round the copper cation is replaced by ammonia molecules to give $[Cu(NH_3)_4]^{2+}$ ions. Dark-blue $Cu(NH_3)_4SO_4$ crystallises from the solution when ethanol is added. In certain cases the addition of a second anion to an aqueous solution enables an insoluble complex to be formed. Thus tetrapyridinenickel(II) nitrite is precipitated as dark-blue crystals by treating aqueous $NiSO_4$ with pyridine and then slowly adding a solution of sodium nitrite:

$$[Ni(H_2O)_6]SO_4 \xrightarrow{\text{pyridine}} [Nipy_4]SO_4 \xrightarrow{NO_2^-} [Nipy_4](NO_2)_2$$

When a product is non-ionic it is often rather insoluble and separates as it is formed, for instance:

$$K_2[PtCl_4] + en \rightarrow \downarrow PtCl_2en + 2\,KCl$$

Here substitution of the chloroligands by ethylenediamine is incomplete. Generally, however, completely substituted complexes are easier to prepare than mixed complexes. In theory it should be possible to obtain intermediate products by limiting the concentration of the substituent, but in practice such products are usually difficult to isolate from the reaction mixture.

28.2.2. Substitution reactions in non-aqueous solvents

If the aquo-complex of the metal is too stable, either kinetically or thermodynamically, for easy substitution by the required ligand, or if the ligand is not water-soluble, a non-aqueous solvent can frequently be used. Solvents such as diethyl ether and dimethylformamide, $HCONMe_2$, have proved useful. Chromium(III) complexes with amine ligands are often made in dimethylformamide; in aqueous solution, however, the bases would precipitate hydrated chromium(III) oxide.

$$CoCl_3 + 3\,HCONMe_2 \rightarrow Co(HCONMe_2)_3Cl_3 \xrightarrow{2\,en} [CoCl_2en_2]Cl$$

Many tetrahedral complexes formerly thought to be incapable of preparation on thermodynamic grounds have been made in solvents other than water. Thus compounds of VCl_4^- have been prepared in acetonitrile.

28.2.3. Direct reaction between solid salt and liquid ligand

Metal ammines can be made by allowing a solution of a metal salt in liquid ammonia to evaporate:

$$NiCl_2 + 6\,NH_3 \rightarrow [Ni(NH_3)_6]Cl_2$$

Similarly the addition of $PtCl_2$ to liquid ethylenediamine gives $[Pten_2]Cl_2$. The direct reaction between a metal ion and a basic solvent is particularly useful when the presence of water would cause hydrolysis.

28.2.4. Thermal decomposition of complexes

What are effectively substitution reactions in the solid state may occur when a solid complex is heated. A very simple example is the conversion of pale-pink $[Co(H_2O)_6]Cl_2$, by heating, to blue $CoCl_2$ (the 'invisible ink' reaction). The Cl^- ion can be considered to replace water molecules which are driven out of the co-ordination sphere when the compound is heated. Sometimes the loss of water from a complex ion occurs even at room temperature. Aquopenta-amminecobalt(III) nitrate, $[CoH_2O(NH_3)_5](NO_3)_3$, made by dissolving nitratopenta-amminecobalt(III) nitrate (below) in warm aqueous ammonia, loses water on standing in air:

$$[CoH_2O(NH_3)_5](NO_3)_3 \rightarrow [CoNO_3(NH_3)_5](NO_3)_2 + H_2O$$

A similar transfer of anions to the co-ordination sphere can often be used to prepare complexes such as chloroammines:

$$[Pt(NH_3)_4]Cl_2 \xrightarrow{520\ K} PtCl_2(NH_3)_2 + 2\ NH_3$$

28.2.5. Substitution reactions accompanied by oxidation

Cobalt(III) complexes can be made from aqueous Co^{II} salts because the redox potential of the Co^{III}/Co^{II} couple falls markedly in the presence of most complexing agents. Cobalt(II) nitrate reacts with concentrated aqueous ammonia and NH_4NO_3 in the presence of H_2O_2 to give nitratopenta-ammine-cobalt(III) nitrate:

$$2[Co(H_2O)_6](NO_3)_2 + 8\ NH_3 + 2\ NH_4NO_3 + H_2O_2$$

$$\rightarrow 2[Co(NO_3)(NH_3)_5](NO_3)_2 + 14\ H_2O$$

Sometimes the course of reaction is altered by the presence of a heterogeneous catalyst such as activated charcoal. Thus $[CoCl(NH_3)_5]Cl_2$ is obtained when cobalt(II) chloride is treated with aqueous ammonia, hydrogen peroxide and ammonium nitrate, but when activated charcoal is also added the product is $[Co(NH_3)_6]Cl_3$. The action of the catalyst here is not fully understood; it may catalyse an otherwise very slow intermediate reaction such as:

$$[Co(NH_3)_5OH]^{2+} \xrightarrow{NH_3} [Co(NH_3)_6]^{3+} + OH^-$$

28.2.6. Reduction reactions

Complexes in which the metal is in an unusually low oxidation state must be prepared in the absence of oxygen. An alkali metal in liquid ammonia is often useful as a reducing agent in these circumstances. Thus potassium tetracyanonickelate(II) is reduced to the tetracyanonickelate(0):

$$K_2[Ni(CN)_4] + 2\ K \xrightarrow{NH_3} K_4[Ni(CN)_4]$$

28.2.7. More specialised methods

The application of the *trans*-effect principle to the preparation of *cis*- and *trans*-isomers is treated in Section 29.8. The resolution of optically active isomers can be carried out by making diastereoisomeric salts. For non-ionic racemates, however, separation is usually made by chromatographic means. Thus triglycinecobalt(III) has been resolved by pouring a solution of it through a column packed with starch. One isomer is preferentially absorbed on the starch because the N atoms are more favourably orientated for bond formation with the —OH groups of the starch molecules.

28.3. Methods of Investigating the Structure of a Complex

The principles of some of the methods used for investigating structure are outlined briefly below. The list of methods is not complete, but it includes those of most general application.

28.3.1. Stoichiometry of reaction

The formulae of the ions in an ionic complex may be determined by means of precipitation reactions. Thus from the complex $[CoCl_2en_2]Cl \cdot H_2O$ only one-third of the chlorine is precipitated by $AgNO_3$ solution, but from its isomer $[CoCl(H_2O)en_2]Cl_2$ two-thirds of the chlorine is so precipitated. Whence the distribution of the chlorine anions follows.

28.3.2. Molar conductance

Ions with unit charge have ionic conductances at infinite dilution of 65 ± 15 S with very few exceptions (H_{aq}^+ and OH_{aq}^- are the main ones); doubly charged ions have conductances of about 130 S and triply charged ones, 195 S. Thus ionic salts of the types MX, MX_2 and MX_3 (X = uninegative ion) have molar conductances of about 130, 260 and 390 S respectively. Whence a complex empirically $PtCl_4(NH_3)_5$, which has a molar conductance of 404 S is probably $[PtCl(NH_3)_5]Cl_3$.

28.3.3. Abundance of the isomeric forms

The classical experiments of Werner on the cobaltammines, in which he used methods (28.3.1) and (28.3.2) above, convinced him that these compounds of Co^{III} always had six groups in a co-ordination shell around the cobalt. He deduced the geometrical arrangement of the six groups from the number of isomers of a particular formula type he was able to make. Thus for compounds containing ions of the type $[Coa_4b_2]^{n+}$ he was often able to make two isomers but never three. A planar hexagonal arrangement would permit three isomers:

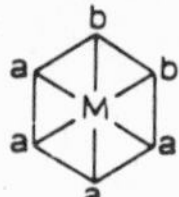

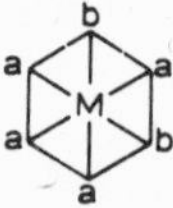

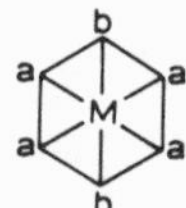

and so would a triangular prismatic one:

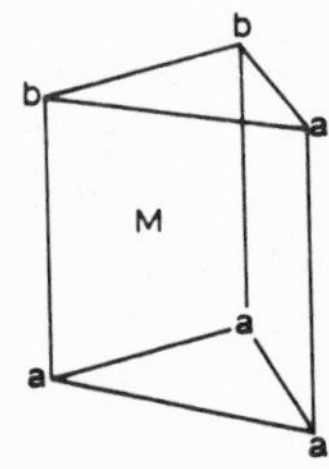

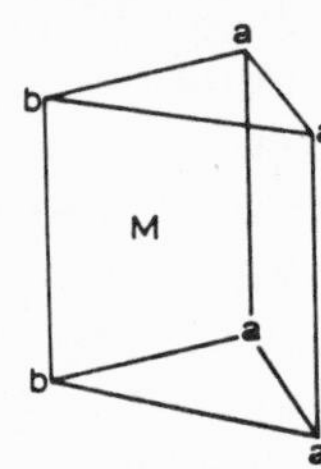

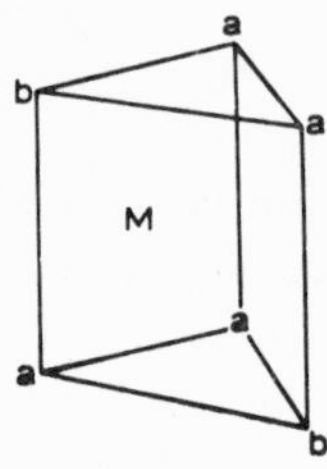

The only other symmetrical arrangement, the octahedral one, permits of only two:

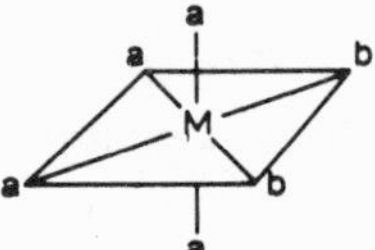

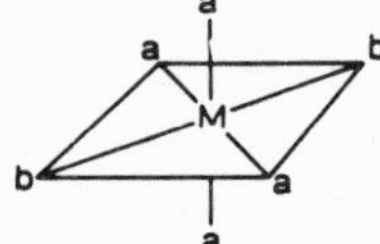

The octahedral arrangement of the co-ordinated ligands in Co^{III} complexes thus deduced was supported by the fact that a complex ion like $[CoCl_2en_2]^+$ had one *trans*-isomer and two enantiomorphic *cis*-isomers (28.4.6). No other arrangement around the cobalt could explain the number and character of these isomeric ions.

Similarly for $PtCl_2(NH_3)_2$, the fact that it exists in two forms (below) rules out the possibility of a tetrahedral arrangement of the four groups around the platinum.

28.3.4. Dipole moment

Measurement of dipole moment (5.3.7) is often of value in deciding between two forms. Of the two forms of dichlorodiammineplatinum(II) the *cis*-form has a dipole moment but the *trans*-form has not:

Cl NH$_3$ / Pt / Cl NH$_3$	Cl NH$_3$ / Pt / NH$_3$ Cl
cis	*trans*

Incidentally, the fact that the *trans*-form has zero moment indicates that the four ligands are arranged in a square planar configuration. A square pyramidal arrangement, for example, would be expected to have a moment, because Pt—Cl and Pt—NH_3 have different polarities, as is shown by the existence of a moment in the *cis*-form of the compound.

28.3.5. Optical rotatory dispersion

Many complex ions are capable of being resolved into optical enantiomers. The effect of a particular enantiomer on the plane of polarisation of light depends not only on the arrangement of its atoms, but also on the wavelength of the light. The same isomer may rotate the plane to the right at one frequency and to the left at another. A rotatory dispersion curve is shown in Fig. 28.1.

Rotatory dispersion curves have been described for a large number of cobalt(III) and chromium(III) complexes, and the absolute configurations of some of the complexes have been assigned from the relation of their curves to the curve for the (+)$Coen_3{}^{3+}$ ion, for which the absolute configuration had been determined by means of X-ray diffraction.

28.3.6. Infrared spectra

Information about metal—ligand bonds can often be obtained from infrared absorption spectra. Comparison of the spectra of the urea complexes of Pt^{II}, Pd^{II}, Cr^{III}, Fe^{III}, Cu^{II} and Zn^{II} shows that in the first two the urea ligands are co-ordinated through the nitrogen atoms and in the other four through the oxygen atoms. When co-ordination is through the oxygen the infrared

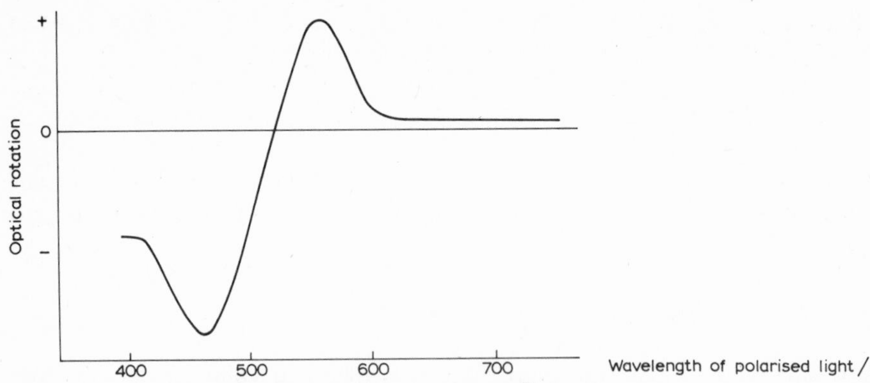

Fig. 28.1. Rotatory dispersion curve for (+)$Coen_3^{3+}$ ion.

absorption spectrum is similar to that of urea itself, although the C—O stretching frequency is a little lower in the complexes. But when co-ordination is through the nitrogen, the N—H stretching and deformation frequencies are reduced and the C—O stretching frequency increased; the spectra of the M—N complexes are quite different from the urea spectrum.

The infrared spectra of the two forms of $[CoNO_2(NH_3)_5]^{2+}$ show that the stable, brown form is a nitro complex and the unstable red form a nitrito complex, $[CO\cdot ONO\cdot (NH_3)_5]^{2+}$. The symmetric and antisymmetric stretching frequencies of the $Co—N{<}^{O}_{O}$ group are 1315 and 1430 cm^{-1}; those of the Co—O—N—O group are 1065 and 1460 cm^{-1}.

28.3.7. Visible absorption spectra

With the increasing application of ligand-field and molecular-orbital theories to the interpretation of visible spectra, the measurement of the frequency of absorption peaks and their extinction coefficients has become an extremely valuable method of gaining information about structure. A series of compounds of a metal in a particular oxidation state have spectra of similar general form.

For octahedral complexes of nickel(II) there are absorption peaks at about 10,500, 13,000, 18,000, and 28,000 cm^{-1}; the extinction coefficients are usually less than 1 m^2 mol^{-1}. Examples are:

Complex	*Absorption peak*/cm^{-1}	*E*/ m^2 mol^{-1}	Complex	*Absorption peak*/cm^{-1}	*E*/ m^2 mol^{-1}
$Ni(NH_3)_6^{2+}$	10 750	0.40	$Nien_3^{2+}$	11 200	0.73
	13 500	0.50		12 400	0.50
	17 500	0.48		18 350	0.67
	28 200	0.63		29 000	0.86

The square complexes of nickel(II), such as bis(dimethylglyoxime)nickel(II) have at most three bands, the one of lowest energy is at 15,000–18,000 cm^{-1}; the second and third bands are not often observable because of the high absorption of the ligands in the ultraviolet. The extinction coefficients of the bands are relatively high, usually in the range 10–30 m^2 mol^{-1}.

The tetrahedral Ni^{II} complex $NiCl_4^{2-}$, present in a mixture of $NiCl_2$ and LiCl which has been melted, absorbs at 15,250 and 14,200 cm^{-1}, with both extinction coefficients about 16 m^2 mol^{-1}. The spectra of octahedral, square, and tetrahedral nickel(II) are thus sufficiently different from one another to enable the three possible structures to be distinguished; minor variations between spectra of the same type which arise from differences in ligand-field strengths are not large enough to mask the major characteristics.

28.3.8. Magnetic measurements

The measurement of magnetic moments indicates the number of unpaired d electrons in a transition-metal complex, but this information is not always a reliable guide to the stereochemistry, as was once believed. The paramagnetic bis(acetylacetonato)nickel(II), $Ni(acac)_2$, was formerly assumed to be 4-co-ordinate and tetrahedral. It was argued that a square planar arrangement of four ligands, with dsp^2 hybrid orbitals, would be a diamagnetic, spin-paired complex in which the eight electrons of Ni^{II} were accommodated in the four d orbitals not occupied by ligand electrons. Thus the square structure was ruled out. The tetrahedral structure, however, would be expected to be paramagnetic, since if sp^3 hybrid orbitals were used for bonding there would be two unpaired electrons in the d orbitals. When the structure was determined by X-ray analysis $Ni(acac)_2$ was found to have a trimeric structure in which the nickel was 6-co-ordinated (Fig. 28.2).

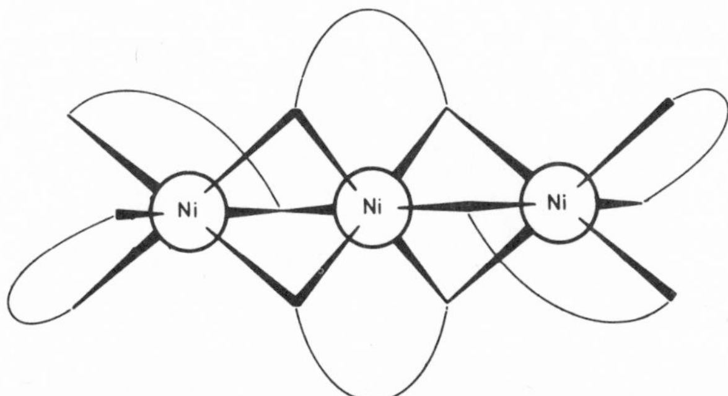

Fig. 28.2. Structure of $Ni_3(acac)_6$, the bis(acetylacetonato)nickel(II) trimer.

This example has been mentioned to draw attention to the fact that there is no unequivocal magnetic criterion of stereochemistry based on the number of unpaired electrons, as was thought in the early days of the valence-bond theory.

However, information about the stereochemical arrangement of ligands can be derived from a consideration of deviations from spin-only values for the magnetic moments. In octahedral complexes the orbital contributions to the moments are quenched in low-spin d^6 and d^7 complexes and in high-spin d^3, d^4, d^5, d^8 and d^{10} complexes, but the other configurations retain some orbital contribution to the moment. In tetrahedral complexes orbital contribution to the moment is absent for d^1, d^2, d^6 and d^7 and may be present for d^3, d^4, d^8 and d^9. For other arrangements with lower symmetry, orbital contribution is possible for fewer configurations:

Square pyramidal and square planar	d^2 and d^4
Trigonal bipyramidal	d^1, d^3, d^4, and d^5

Examples of the use of the deviation from spin-only values for diagnosing stereochemical arrangement are provided by tetrahedral Ni^{II} complexes which have higher moments than octahedral Ni^{II} complexes, as expected for a d^8 arrangement; and by $Co^{II}(d^7)$ tetrahedral complexes which have moments closer to the spin-only value than do their octahedral counterparts.

28.3.9. Nuclear magnetic resonance spectra

For certain complexes n.m.r. spectroscopy is useful in the diagnosis of structure. The compounds $RuXH[C_2H_4(PEt_2)_2]_2$, where X is Cl, Br, I, SCN, NO_2 or CN, are found to have five sharp, equally spaced bands with intensities in the approximate ratios 1 : 4 : 6 : 4 : 1 in the absorption region associated with the unique hydrogen atom. The hydrogen atom ($I = \frac{1}{2}$) is evidently coupled to four other atoms for which $I = \frac{1}{2}$, all in the same position relative to itself, confirming the structure shown in Fig. 28.3.

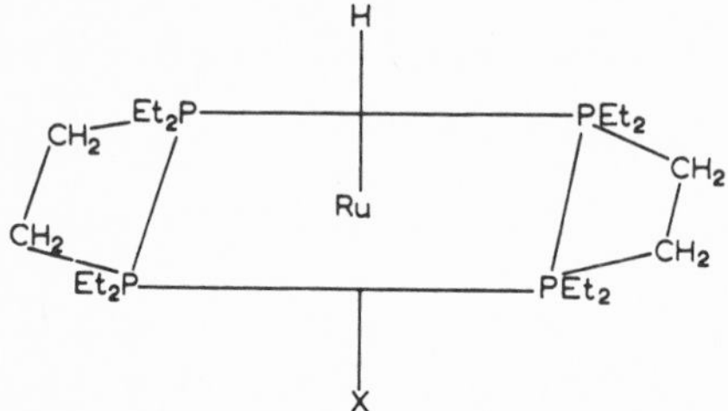

Fig. 28.3. Structure of $RuXH[C_2H_4(PEt_2)_2]_2$.

28.3.10. X-ray analysis

This is also referred to as the absolute method, because it is the only one capable of supplying unequivocal evidence of structure. Its disadvantage is that it is much more laborious than magnetic and spectroscopic determinations. However, as more X-ray determinations of the structures of complexes are made, more examples of incorrect predictions from magnetic and spectroscopic data are disclosed.

For bis(trimethylphosphine oxide)cobalt(II) nitrate, X-ray analysis showed that the cobalt is 6-co-ordinate, with a very irregular arrangement of ligand atoms; there is a bidentate nitrate ion in the co-ordination sphere and the whole is a molecule with no strict symmetry elements whatever. None of the previous indirect evidence suggested such an anomalous structure. For the given paramagnetic bis(benzyldiphenylphosphine)nickel(II) bromide, X-ray analysis reveals that the triclinic unit cell contains three molecules, two tetrahedral and one square planar, a stereochemical complication which could not have been disclosed by spectroscopic or magnetic evidence.

28.4. Isomerism in Complex Compounds

Various types of isomerism occur in complex compounds.

28.4.1. Ionisation isomerism

This kind of isomerism occurs when there is an interchange of groups between the co-ordination sphere of the nuclear atom and ions outside this sphere. Sulphatopenta-amminecobalt(III) bromide, $[Co(SO_4)(NH_3)_5]Br$, and bromopenta-amminecobalt(III) sulphate, $[CoBr(NH_3)_5]SO_4$, are such a pair of isomers. The former is red and in aqueous solution gives a precipitate of AgBr with $AgNO_3$ but no precipitate with barium chloride; the latter is violet and gives an immediate precipitate with $BaCl_2$. It should be noticed that in these compounds an SO_4 group occupies only one co-ordination position though it neutralises two ionic charges on the Co^{3+}; thus the number of co-ordinate links formed by a ligand ion is not necessarily the same as the number of charges it carries.

An even more obvious example of ionisation isomerism is afforded by $[PtCl_2(NH_3)_4]Br_2$ and $[PtBr_2(NH_3)_4]Cl_2$.

28.4.2. Hydration isomerism

There are three isomers of $CrCl_3 \cdot 6\,H_2O$. The violet-grey form is shown, by its conductance and the fact that all the chlorine is precipitated immediately with $AgNO_3$, to be the hexa-aquochromium(III) chloride, $[Cr(H_2O)_6]Cl_3$. The dark green substance obtained from hot solutions is the dichloro-tetra-aquo salt, $[CrCl_2(H_2O)_4]Cl \cdot 2\,H_2O$, from which $AgNO_3$ removes only one third of the chlorine. The third compound, the chloropenta-aquochromium(III) dichloride hydrate, $[CrCl(H_2O)_5]Cl_2 \cdot H_2O$, also green, yields two-thirds of its chlorine by precipitation with $AgNO_3$.

Other examples of hydration isomerism are found in the compounds:

$[CoCl(H_2O)en_2]Cl_2$ and $[CoCl_2en_2]Cl \cdot H_2O$
$[CrCl_2(H_2O)_2py_2]Cl$ and $[CrCl_3(H_2O)py_2]H_2O$

The exact nature of the bonds holding water molecules in the co-ordination sphere and outside it is not always known. In some hydrates those outside the co-ordination sphere are held in the interstices of the crystal lattice, in others they are attached to the simple ions.

28.4.3. Co-ordination isomerism

This isomerism occurs when both cation and anion are complex. Typical examples are (a) $[Co(NH_3)_6][Cr(CN)_6]$ and $[Cr(NH_3)_6][Co(CN)_6]$; and (b) $[Pt(NH_3)_4][PtCl_4]$ and $[PtCl(NH_3)_3][PtCl_3(NH_3)]$.

Co-ordination isomerism also occurs where a nuclear element is present in two states of different charge, as platinum is in $[Pt(NH_3)_4][PtCl_6]$ and $[PtCl_2(NH_3)_4][PtCl_4]$.

A special type of co-ordination isomerism is co-ordination position isomerism. It can occur in bridged complexes such as the ions (a) and (b) below, in which ammonia and chloro ligands are differently placed relative to the two cobalt atoms:

(a) $[(NH_3)_4Co(\mu\text{-}OH)_2Co(NH_3)_2Cl_2]^{2+}$

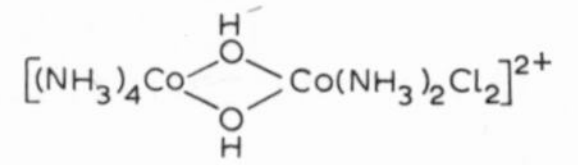

(b) $[Cl(NH_3)_3Co(\mu\text{-}OH)_2Co(NH_3)_3Cl]^{2+}$

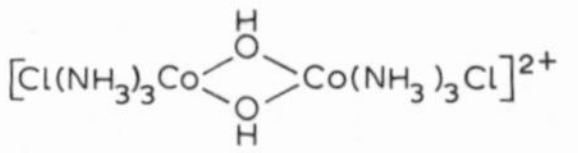

28.4.4. Linkage isomerism

The NO_2 group may be co-ordinated to a metal atom either through the nitrogen atom to act as a nitro ligand ($-NO_2$) or through one of the oxygens to act as a nitrito ligand ($-ONO$). Examples are known of isomerism arising from this difference in linkage, for instance:

$[CoNO_2(NH_3)_5]^{2+}$ (the nitropenta-amminecobalt(III) ion)
$[Co(ONO)(NH_3)_5]^{2+}$ (the nitritopenta-amminecobalt(III) ion)

Other groups are capable of co-ordinating through different member atoms; for example the CNS^- ion can co-ordinate either through nitrogen or through sulphur. Linkage isomers of this type have been prepared:

$(CO)_5Mn{-}SCN$ and $(CO)_5Mn{-}NCS$

28.4.5. Geometrical isomerism

In the planar complexes of metals showing the co-ordination number four there is a possibility of *cis*- and *trans*-isomerism:

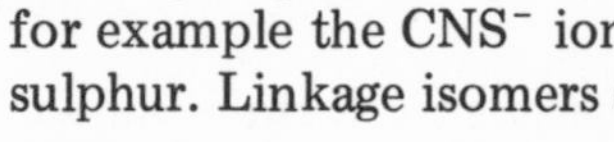

Examples are $Pt(NH_3)_2Cl_2$ and $Pd(NH_3)_2(NO_2)_2$; both have *cis*- and *trans*-isomers. The type of isomerism also arises in planar chelate compounds containing unsymmetrical bidentate ligands. But it is not necessary for the two co-ordinating atoms of the bidentate ligand to be different; it is only necessary that the chelate ring have dissimilar halves:

CH_2—NH_2 Pt NH_2—CH_2 / CO—O O—CO (*cis*) and CH_2—NH_2 Pt O—CO / CO—O NH_2—CH_2 (*trans*)

Geometrical isomerism cannot occur in the Ma_3b type of square complex, but the Mabcd type gives rise to three isomers:

a, b, M, d, c; a, c, M, d, b; a, b, M, c, d

The first complex of this type to be obtained in three isomeric forms was $[PtNO_2pyNH_3(NH_2OH)]^+$.

With 6-co-ordinate octahedral complexes, geometrical isomerism is also possible. *Cis-* and *trans-*isomerism is found in an ion such as $[CoCl_2(NH_3)_4]^+$:

cis(blue-violet) *trans*(green)

Complexes of the type Ma_3b_3 exist in only two isomeric forms:

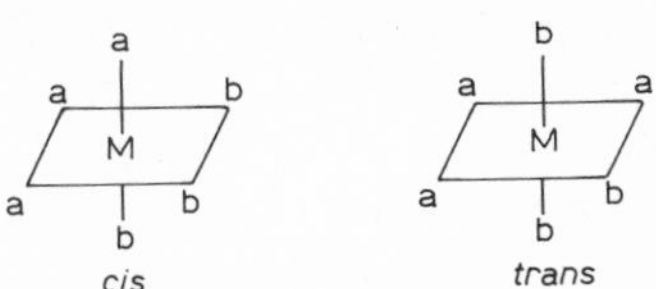

trans

An example is trichlorotripyridinerhodium(III).

Unsymmetrical bidentate ligands can give rise to *cis-* and *trans-*isomers in a similar way:

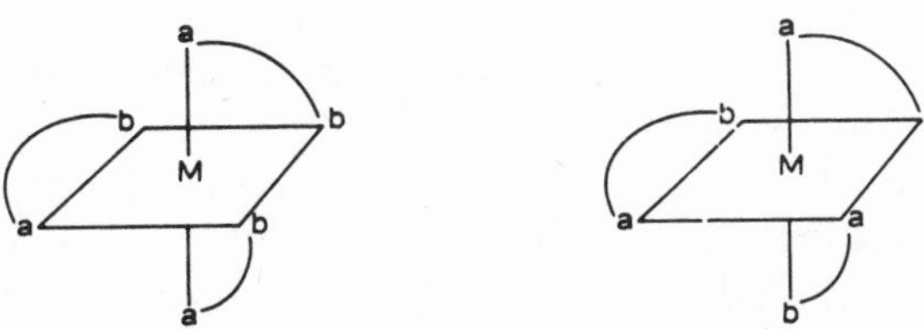

The letters a and b represent different ends of an unsymmetrical bidentate ligand, not necessarily different co-ordinating atoms. An example is triglycinecobalt(III) (see 28.7.3.1). In such an instance, both the *cis-* and *trans-*form each have a pair of optical isomers.

28.4.6. Optical isomerism

Werner showed (1908) that the element carbon is not a necessary constituent of an optical isomer when he resolved the compound

$$\left[Co\left\{\begin{matrix}OH\\OH\end{matrix}Co(NH_3)_4\right\}_3\right]Br_6$$

This kind of isomerism occurs when a compound can be represented by two asymmetrical structures, one of which is the mirror image of the other.

Resolution into optical enantiomers is usually achieved by methods similar to those of organic chemistry. Diastereoisomers of different solubility are made by forming a salt from a racemic mixture of active cations with an active anion such as, for example, d-bromocamphorsulphonate.

Six-co-ordinate complexes containing bidentate ligands afford many examples of optical isomerism. A particularly common type is $M(aa)_2b_2$ (aa = bidentate ligand) for which the *cis*- but not the *trans*-form can be resolved into optical isomers.

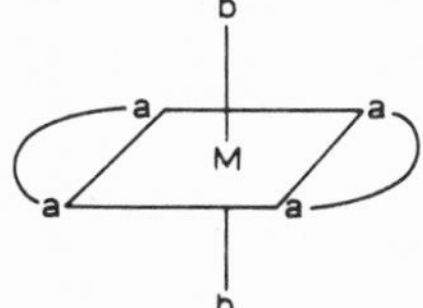

Trans-form (symmetrical), not resolvable.

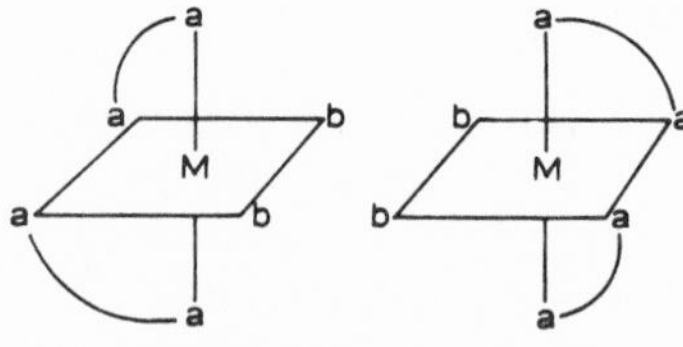

Two *cis*-forms (unsymmetrical), resolvable.

An example is the dichloro-bis(ethylenediamine)cobalt(III) cation, $[CoCl_2en_2]^+$.

Octahedral complexes with three bidentate ligands also show optical isomerism. An example is the tris(ethylenediamine)cobalt(III) cation.

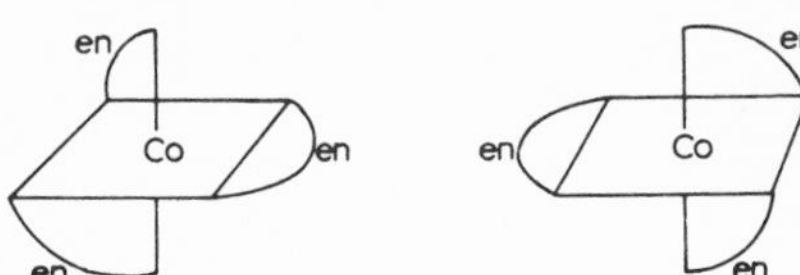

Multidentate ligands can also give rise to optical isomerism in octahedral complexes. A $(Co\ edta)^-$ complex ion has been resolved.

Optical isomerism should be possible in octahedral complexes Mabcdef which contain only unidentate ligands. Such a complex of Pt^{IV} has been made — $PtIBrClNO_2pyNH_3$. It should have fifteen geometrical isomers, each with a pair of optical enantiomers; a few of the different geometrical isomers of this compound have been made but none has been resolved.

By analogy with carbon, optically active tetrahedral complexes Mabcd are to be expected, but none has yet been prepared. However, tetrahedral complexes of Be^{II}, B^{III} and Zn^{II} with unsymmetrical bidentate ligands have been made and resolved. An example is bis(benzoylacetonato)beryllium(II).

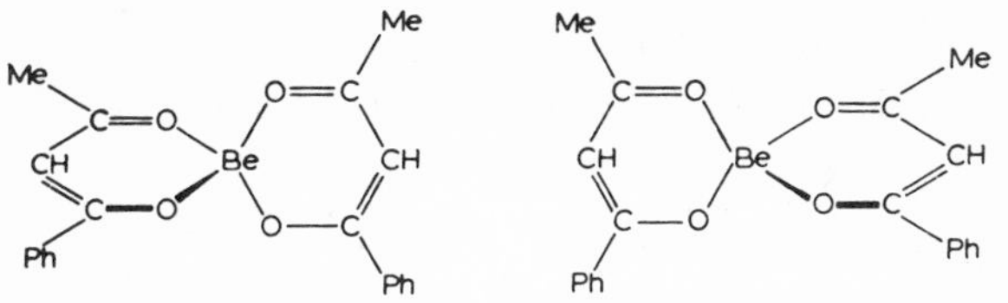

Optical isomerism in square complexes is at present limited to a few examples. One of these is the ion

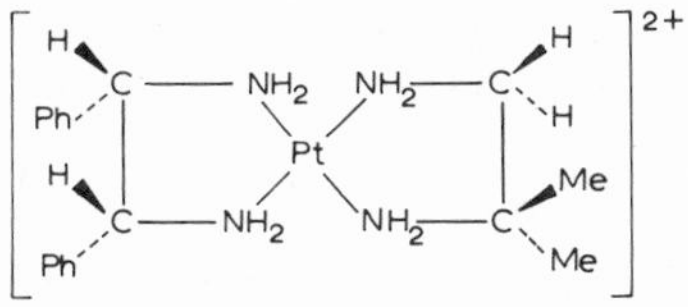

which can be seen to be without either a plane or an axis of symmetry.

28.5. Nomenclature of Complex Compounds

The formulae and names adopted for complex ions and compounds follow the main recommendations in *Nomenclature of Inorganic Chemistry*, 2nd edition, 1970, issued by the International Union of Pure and Applied Chemistry; and in *Handbook for Chemical Society Authors*, 1960, which has much other useful information. Those referring to mononuclear complexes are summarised below.

28.5.1. Formulae

(i) Place the symbol for the nuclear atom first, follow with the symbols for the ligands, and enclose the complex in square brackets.

(ii) Indicate the charge number of the nuclear atom, when required, by Stock notation: Ni^{0}, Cu^{I}, Fe^{II}.

(iii) Place the ligands in the order (a) anionic, (b) neutral and cationic. Arrange the ligands within these classes in the sequences:

(a) (1) H^-, O^{2-}, OH^-, I^-, Br^-, Cl^-, F^-;
(2) other inorganic anions containing two or more elements, that with the smaller number of atoms first, and, when the number is the same, that with the central atom of larger atomic number first;
(3) organic anions in alphabetical order.

(b) (1) H_2O, NH_3;
(2) other inorganic ligands in this order of their central atoms: B, Si, C, Sb, As, P, N, Te, Se, S, I, Br, Cl;
(3) organic ligands in alphabetical order.

28.5.2. Names

(i) Cite names of ligands first, the name of the nuclear atom last.

(ii) Names of complex cations and neutral molecules have no distinguishing termination; names of complex anions end in -ate.

(iii) Cite ligands in their order in the formula. Names of the anionic ligands end in -o; those of the neutral, except aquo for H_2O and ammine for NH_3, and of the cationic are the same as for the molecule or cation.

(iv) The anionic groups CN, O·NO, NO_2, NO_3, are named cyano, nitrito, nitro and nitrato. Anions derived from hydrocarbons are given the name of the radical, sometimes with an -o ending.

(v) The groups NO, NS, CO, CS are termed nitrosyl, thionitrosyl, carbonyl and thiocarbonyl, and are treated as neutral in computing the oxidation number of the nuclear atom.

(vi) The prefixes mono, di, tri, tetra, penta, hexa, hepta, octa, ennea, deca, hendeca and dodeca are used (without hyphen, except when two vowels are brought together as in the first example below) to indicate the numbers of the individual ligands in a complex. When, however, the name includes a numerical prefix with a different significance, then bis, tris and tetrakis are used, the group to which they refer being often placed in parentheses:

$[Ni(CO)_2(Ph_3P)_2]$ dicarbonyl-bis(triphenylphosphine)nickel

and sometimes in cases like these:

$[Fe(C_5H_5)_2]Cl$ bis(cyclopentadienyl)iron(III) chloride, and
$Ca(PCl_6)_2$ calcium bis(hexachlorophosphate)

Details about formulae and names of polynuclear complexes can be found in *Nomenclature of Inorganic Chemistry* or the *Handbook for Chemical Society Authors*. It need only be mentioned here that a bridging group is indicated in the formula by separating it from the rest of the complex by hyphens, and in the name by prefixing it with μ-:

$[(NH_3)_5Cr—OH—Cr(NH_3)_5]Cl_5$

μ-hydroxo-bis{penta-amminechromium(III)} chloride.

28.5.3. Examples of formulae and names

$[Co(NH_3)_6]^{3+}$	the hexa-amminecobalt(III) ion
$[Co(O{\cdot}NO)_3]^{3-}$	the hexanitritocobaltate(III) ion
$[Fe(CN)_6]^{4-}$	the hexacyanoferrate(II) ion
$K_3[Fe(CN)_6]$	potassium hexacyanoferrate(III)
$Co[Cl_2(NH_3)_4]Cl$	dichlorotetra-amminecobalt(III) chloride
$K[Co(CN)(CO)_2NO]$	potassium cyanodicarbonylnitrosylcobaltate(0)

$[Co(NH_3)_2(en)_2]Cl_3$	diammine-bis(ethylenediamine)cobalt(III) chloride
$[Co(NO_2)_3(NH_3)_3]$	trinitrotriamminecobalt(III)
$[Pt(NH_3)_6]Cl_4$	hexa-ammineplatinum(IV) chloride.
$[CrCl(H_2O)(en)_2]^{2+}$	the chloroaquo-bis(ethylenediamine)chromium(III) ion
$\left[(NH_3)_4Co\genfrac{}{}{0pt}{}{NH_2}{OH}Co(NH_3)_4\right]^{4+}$	the octa-ammine μ-amido-μ-hydroxo-dicobalt(III) ion
$\left[\begin{matrix} CH_3\cdot C{=}N(OH) \\ CH_3\cdot C{=}N(OH) \end{matrix} CoCl_2\right]$	dichlorodimethylglyoxime-*N*,*N′* cobalt(II)

28.6. Complexes in Aqueous Solution

28.6.1. Aquo-complexes

Dissolving a salt in water rarely gives unhydrated ions; the perchlorate ion, ClO_4^-, is a possible exception. The first step in the process of solution is probably hydration of the ions; it will be assumed in discussion that the number of water molecules in the solvated shell corresponds with the co-ordination number of the ion, although this is not always true. Relatively few salts give neutral solutions, and when $FeCl_3$ or $CuSO_4$ dissolve they must do more than yield $Fe(H_2O)_6^{3+} + 3\,Cl_{aq}^-$ or $Cu(H_2O)_4^{2+} + SO_4{}^{2-}{}_{aq}$, for their solutions are acid. A hydrated cation behaves as an acid, e.g.:

$$Fe(H_2O)_6^{3+} + H_2O \rightarrow Fe(OH)(H_2O)_5^{2+} + H_3O^+$$

The loss of protons in this way will be the easier the greater the positive field about the cation. Elements of Period 3 exemplify this: the electrostatic field about Na^+ being slight, it is only weakly hydrated; but the progressive increase in field about the following elements leads, successively, to the formation of $Mg(H_2O)_6^{2+}$, $Al(H_2O)_6^{3+}$, $Si(OH)_4$, $PO(OH)_3$, $SO_3(OH)^-$, ClO_4^- in acid solution, and of $Mg(OH)(H_2O)_5^+$, $Al(OH)_4(H_2O)_2^-$, $SiO_2(OH)_2^{2-}$, PO_4^{3-}, SO_4^{2-}, ClO_4^- in alkaline solution.

These are mononuclear complexes, but the cation $Cr_2(OH)_2^{4+}$ exists in solution, and polynuclear complexes have proved to be frequent products of hydrolysis. Indeed, mercury(II) is one of the few metals which always gives mononuclear species: $Hg(H_2O)_2^{2+}$, $Hg(OH)(H_2O)^+$, $Hg(OH)_2$. Beryllium produces mainly $Be_3(OH)_3^{3+}$ which is presumed to have a cyclic structure. Iron(III) appears as the ions $Fe(OH)_2^+$ and $Fe_2(OH)_2^{4+}$, the first being paramagnetic and the second diamagnetic.

It is instructive to follow the behaviour of potassium alum which dissolves to give the ions K_{aq}^+, $[Al(H_2O)_6]^{3+}$ and $SO_4{}^{2-}{}_{aq}$. The hexa-aquoaluminium ion is immediately soluble but tends in time, especially when the temperature is raised, to hydroxylate in stages by proton loss:

$$[Al(H_2O)_6]^{3+} \rightarrow [Al(OH)(H_2O)_5]^{2+} + H_{aq}^+$$

$$[Al(OH)(H_2O)_5]^{2+} \rightarrow [Al(OH)_2(H_2O)_4]^+ + H_{aq}^+$$

A water molecule in these aquo-complexes can be replaced by $SO_4{}^{2-}$:

$$[Al(OH)(H_2O)_5]^{2+} + SO_4{}^{2-} \rightarrow [Al(OH)(SO_4)(H_2O)_4] + H_2O$$

The passage from solution to sol probably occurs by hydroxo-bridging

$$[Al(OH)(H_2O)_5]^{2+} + [Al(OH)(SO_4)(H_2O)_4] \rightarrow \left[(H_2O)_5Al\begin{matrix}\overset{H}{O}\\ \underset{H}{O}\end{matrix}Al(SO_4)(H_2O)_4\right]^{2+}$$

This and similar kinds of bridging produce polymers of indefinite size which eventually separate as precipitates. The change from soluble complexes, through sols, to precipitates is accelerated when the pH is raised. It is characteristic of the precipitate that repeated washing with water fails to remove all the $SO_4{}^{2-}$ anions; this is because they occupy a place in the co-ordination sphere of the Al^{3+} cation and are not merely adsorbed on the surface of the particles.

28.6.2. Stability constants

Complex formation by the replacement of molecules of water from the solvated shell of a metal cation in aqueous solution, with, for example, unidentate ligands.

$$M(H_2O)_n + (n-x)L \rightleftharpoons ML_{(n-x)}(H_2O)_x + (n-x)H_2O$$

usually produces an equilibrium of two or more complexes, the reaction proceeding stepwise with an equilibrium constant for every step:

$$k_1 = \frac{a\{M(H_2O)_{n-1}\,L\}}{a\{M(H_2O)_n\}a(L)}$$

$$k_2 = \frac{a\{M(H_2O)_{n-2}L_2\}}{a\{M(H_2O)_{n-1}L\}a(L)}$$

$$k_n = \frac{a(ML_n)}{a\{M(H_2O)L_{n-1}\}a(L)}$$

The constants k_1, k_2 ... k_n are the *consecutive stability constants.*

For most complexes the values gradually fall from k_1 to k_n, for example in the Cu^{2+}—NH_3 system, log $k_1 = 4.3$, log $k_2 = 3.6$, log $k_3 = 3.0$ and log $k_4 = 2.3$. This decrease in value is to be expected on statistical grounds. Let us consider the two equilibria

$$M(H_2O)_4 + L \rightleftharpoons M(H_2O)_3L + H_2O$$

and

$$M(H_2O)_3L + L \rightleftharpoons M(H_2O)_2L_2 + H_2O$$

In the first, there are four positions at which $M(H_2O)_4$ can be substituted by L but only one site at which $M(H_2O)_3L$ can be substituted by H_2O in the reverse reaction. In the second there are three positions in which $M(H_2O)_3L$ can be substituted by L but only two positions in which $M(H_2O)_2L_2$ can be substituted by H_2O in the reverse reaction. On purely statistical considerations therefore we should expect the ratio k_1/k_2 to be $\frac{4}{1}/\frac{3}{2}$, i.e., $2\frac{2}{3}$. Thus log k_2 should be less than log k_1, by about log 2.66 = 0.45. A similar argument can be applied to show the ratio k_2/k_3 to be $\frac{3}{2}/\frac{2}{3} = 2.25$ and so on, and it can also be extended to 6-co-ordinate species.

The product of consecutive stability constants is the *cumulative stability constant*, β.

Thus, for the foregoing general example

$$k_1 \times k_2 \times \ldots k_n = \frac{a(ML_n)}{a\{M(H_2O)_n\} \times \{a(L)\}^4} = \beta_n$$

The symbols pk (for $\log_{10}k$) and pβ (for $\log_{10}\beta$) are usually employed in tabulating stability constants. For the $Cu(NH_3)_4$ complex already mentioned, pβ_4 will be 4.3 + 3.6 + 3.0 + 2.3 = 13.2.

The gradual fall in consecutive stability constants predicted by the statistical argument is not universal. In the Hg^{2+}—Cl^- system, for example, pk_1 is 6.74, pk_2 is 6.48 but pk_3 and pk_4 are both about 1.0. The sudden large decrease is probably due to the fact that $HgCl_2$ is linear but $HgCl_4^{2-}$ is tetrahedral. The energy needed to change the sp hybridisation of the Hg orbitals in the linear complex to sp^3 hybridisation in the tetrahedral complex is evidently responsible.

In a few cases there is a reversal of the usual order of k values. Thus for the Fe^{2+}—orthophenanthroline system k_3 is much larger than k_1 and k_2. The addition of the third ligand molecule converts the complex from a high-spin $t_{2g}^4e_g^2$ configuration to a low-spin t_{2g}^6 one. The concomitant gain in ligand field stabilisation makes the *tris* complex particularly thermodynamically stable.

Fig. 28.4 is a plot of the percentage of the various aquo-ammine complexes of copper(II) present in an aqueous solution containing different concentrations of free ammonia. For example, when the concentration of free ammonia is 10^{-1} mol m^{-3} the solution contains about 5% of the triammine, 45% of the diammine, 40% of the mono-ammine and 10% of aquated copper ions.

Stability constants are calculated from the concentrations of the species

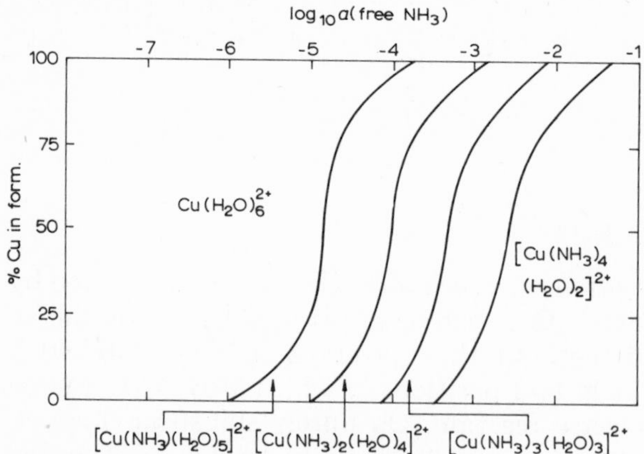

Fig. 28.4. Percentage of various copper(II) aquo-ammine complexes at different concentrations of free ammonia.

present in equilibrium mixtures containing the metal ion and the ligand in a wide range of proportions. Activity coefficients are kept constant by appropriate additions of a salt, usually sodium perchlorate, whose ions do not compete with those of the cation and ligand. Concentrations at different ionic strengths are extrapolated to zero ionic strength. It may be necessary to find the number of water molecules displaced at each step; the total of these is not necessarily the same as the co-ordination number of the cation in the solid compound. Particularly in a polar solvent such as water, the ligands may not displace all the solvent molecules.

Ignoring hydration and charge, the equilibrium of a complex with its cation and ligands is expressed by:

$$M + n\,L \rightleftharpoons ML_n$$

The total concentration of the metal, in the solvated cations and the complex, $[M]_t$, and the total concentration of the ligand, free and in the complex, $[L]_t$, can be found by analysis. The method of determining the concentration of the complex, $[ML_n]$, depends upon the system. When either the free ligand or the complex is coloured, or has a convenient absorption elsewhere in the spectrum, optical densities, i.e. log (intensity of transmitted light/intensity of incident light), at a specific wave length are measured. Sometimes the concentrations of the uncomplexed metal ions are obtained potentiometrically with a suitable electrode. Polarography and extraction methods are also used, and occasionally the required information may be deduced from pH measurements.

To determine the stoichiometry of the reaction, Job's method of continuous variation is used: $[ML_n]$ is plotted against $[M]_t/\{[M]_t + [L]_t\}$ and a maximum in the curve indicates complex formation. Its composition corresponds to the position of the maximum.

With a complex of rather low stability, formation is not complete when M and L are in the proportion indicated by the plot and, as a result, the values of β drift as the ratio of metal to complexing agent is altered. But something very near the true value can be obtained by successive approximations.

Distribution methods for determining [M] depend on knowing the distribution coefficient for the metal ion or the ligand between two immiscible solvents. Thus free ammonia in equilibrium with the ammine complex of a metal can be determined from the ammonia concentration of a chloroform layer in equilibrium with the aqueous solution.

When the ligand is protonated, the equilibrium,

$$M + n\ HL \rightleftharpoons ML_n + n\ H^+$$

may be set up, then:

$$\beta = \frac{a\{ML_n\} \times \{a(H^+)\}^n}{a(M) \times \{a(HL)\}^n}$$

Concentrations at equilibrium can be obtained from pH measurements. The method is applicable only when L is the anion of a weak acid; obviously it cannot be applied to the co-ordination of chloride and similar ions, for which the conjugate acid of the ligand is completely ionised before complex formation occurs.

28.7. Factors Affecting the Stability of Complexes

28.7.1. Electrostatic field round the cation

Of the complexes formed by an individual ligand with a metal anion in two different oxidation states, those in which the anion has the higher charge number are nearly always the more stable, for instance with hexacyanoferrate(II), $Fe(CN)_6^{4-}$, $p\beta_6$ is 24, but with hexacyanoferrate(III), $Fe(CN)_6^{3-}$, $p\beta_6$ is 31.0. This is understandable if the ligands are held by the electrostatic charge on the central ion; the smaller, more highly charged, iron(III) ion will exert a stronger attraction than the larger, lower charged, iron(II) ion.

When measured, the order of stability of complexes of these bipositive metal ions in Period 4 proved to be related, irrespective of the ligand, to the ionic radius:

Order of stability	Mn^{2+} <	Fe^{2+} <	Co^{2+} <	Ni^{2+} <	Cu^{2+} >	Zn^{2+}
Ionic radius M^{2+}/pm	91	83	82	78	69	74

28.7.2. Distribution of charge

If complex formation be imagined to involve the donation of electron pairs from ligand to cation, and if the bonds were perfectly homopolar, then one unit of charge would be transferred for every bond formed. In these circumstances, a bipositive cation would accumulate a negative charge of four in becoming 6-co-ordinate, and the result would be a very unstable condition unlikely to persist. The stability of a complex increases as the Pauling 'postulate of neutrality' condition (27.8) is approached, that is when each atom has a net charge between $-\frac{1}{2}$ and $+\frac{1}{2}$. Consequently ions such as Mg^{2+}, for which E^0, M^{n+}/M has a large negative value, attract electrons weakly and therefore form their strongest complexes with co-ordinating atoms of high electronegativity, for instance oxygen. At the other extreme, the noble metal ions with positive redox potentials accept electrons more readily, and form their strongest co-ordinate bonds with donor atoms such as sulphur and iodine which are easily polarised.

Acceptor metal ions are classified into two types:

Class (a) are acceptors which form their most stable complexes with the co-ordinating atoms N, O and F of the second period.

Class (b) are acceptors which form their most stable complexes with the co-ordinating atoms P, S, Cl, As, Br, I of later periods.

The class (b) acceptors lie in a roughly triangular area in the middle of the lower part of the Periodic Table:

		Cu		
Rh	Pd	Ag	Cd	
Ir	Pt	Au	Hg	Tl

Neighbouring elements to right and left of this area have some class (b) character; the other metals, and those non-metals which fill the position of central atoms in complexes, are class (a) acceptors.

The class (b) acceptors are metals with positive redox potentials which accept electrons readily; they also have large complements of d electrons available for dative π-bonding. Their most stable complexes are formed with ligands like PMe_3, S^{2-} and I^- which have vacant d orbitals, or like CO and CN^- which have vacant molecular orbitals of low energy.

A class (a) acceptor is a synonym for hard acid (9.15) when applied to a metal ion, and a class (b) acceptor is the same as a soft acid.

28.7.3. Chelation

The stability of complexes is greatly increased by chelation; that is the formation of rings by polydentate ligands. Thus, for example, the stability constants of 6-co-ordinate Ni^{2+} with NH_3 and $NH_2CH_2CH_2NH_2$ are, respectively, 10^9 and 10^{19}. And the sexidentate EDTA ligand forms such stable soluble complexes that, for instance, the precipitation of Ca^{2+} by oxalate from alkaline solution is entirely prevented.

Measurements of the free energy changes and the heats of reaction associated with chelation usually show that there has been a considerable increase in entropy. This occurs because several molecules of solvent in the solvated ion are replaced by a smaller number of multidentate ligands, or even by one:

$$M(H_2O)_n + L \rightleftharpoons ML + n\ H_2O$$

The reaction represented increases the number of entities in the solution $\frac{1}{2}(n + 1)$ times; consequently it increases the disorder when n is greater than one.

The enhanced stability of complexes containing chelated ligands is known as the *chelate effect.*

The size of the chelate ring is also of importance in the stabilisation of complexes; the optimum number of atoms in the ring is five, unless some of them are joined by double bonds. The following are examples:

28.7.3.1. *Bidentate ligands*

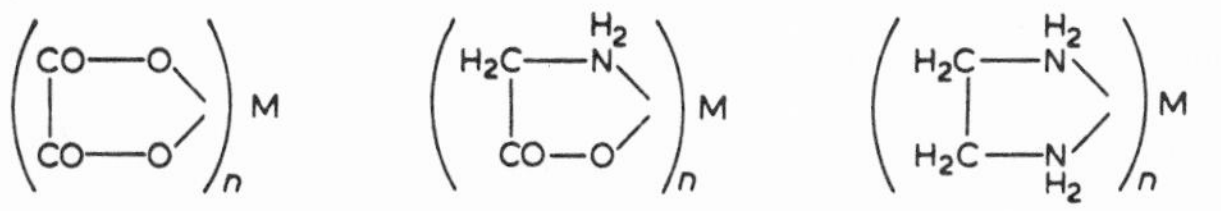

Oxalato complexes **Glycine complexes** **Ethylenediamine complexes**

Dimethylglyoxime can act as a bidentate ligand in two ways: (i) as the whole molecule with a Cu^{2+} ion to form the bipositive bis(dimethylglyoxime)-copper(II) ion (a), or (ii) as the ion derived from the dimethylglyoxime molecule (b) with a Ni^{2+} ion to form the uncharged bis(dimethylglyoximato)-nickel(II) molecule (c):

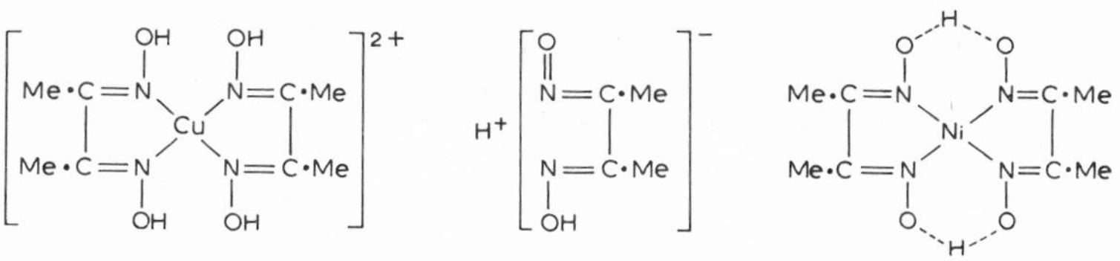

(a) (b) (c)

Another example of a complex molecule is bis(acetylacetonato)copper(II):

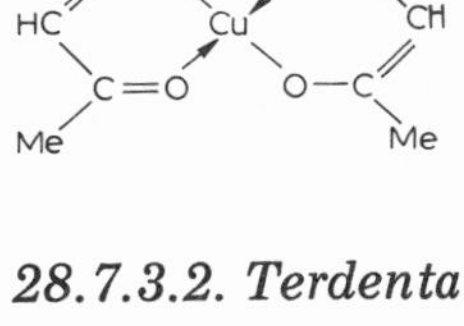

28.7.3.2. *Terdentate ligands*

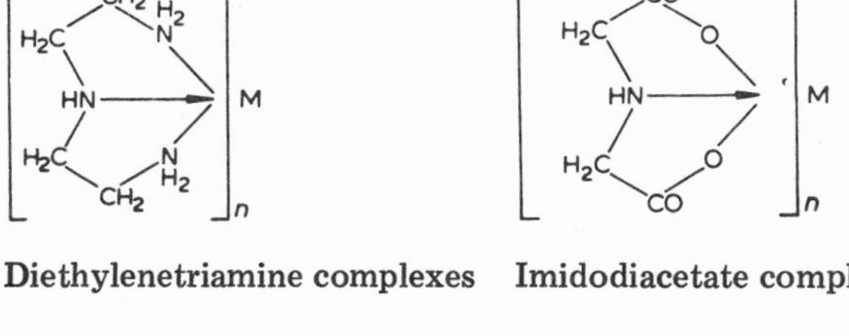

Diethylenetriamine complexes **Imidodiacetate complexes**

When some of the atoms in the ring are joined by double bonds the optimum size is six atoms. Examples:

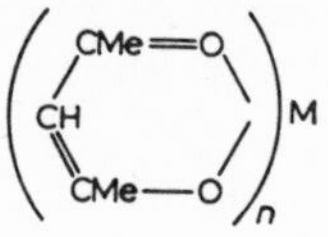

Acetylacetonato complexes 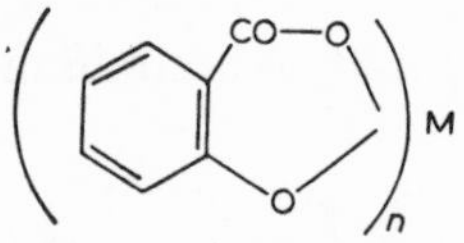 Salicylato complexes

28.7.3.3. Quadridentate ligands, e.g.

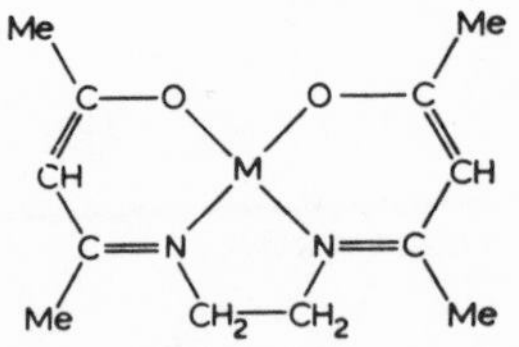

Bis(acetylacetone)ethylenedi-imine complex

In this the double-bonded rings are six-membered and the single-bonded ring five-membered.

The stability of a complex is affected by steric strain in some multidentate ligands. For example, ethylenediamine forms more stable complexes than its *N,N'*-tetramethyl derivative, $Me_2N \cdot CH_2 \cdot CH_2 \cdot NMe_2$; this is because electron donation by the nitrogen atoms to the metal brings the Me groups close together.

Further Reading

L.E. Orgel, An introduction to transition metal chemistry: Ligand field theory, 2nd edition, Methuen, 1966.

M.M. Jones, Elementary co-ordination chemistry, Prentice-Hall, Englewood Cliffs, New Jersey, 1964.

M.C. Day and J. Selbin, Theoretical inorganic chemistry, 2nd edition, Reinhold, New York, 1969.

J. Lewis and R.G. Wilkins (Eds.), Modern co-ordination chemistry, Interscience, New York, 1960.

F. Basolo and R.C. Johnston, Co-ordination chemistry, Benjamin, New York, 1964.

F.J.C. Rossotti and H. Rossotti, The determination of stability constants, McGraw-Hill, New York, 1961.

J. Bjerrum, G. Schwarzenbach and L.G. Sillen (Eds.), Stability constants of metal-ion complexes, The Chemical Society, London, Part I: Organic ligands, 1957; Part II: Inorganic ligands, 1958.

J.J. Fortman, Optical and geometrical isomerization of β-diketone complexes, Co-ord. Chem. Rev., 6 (1971) 331.

G.B. Kauffman, Alfred Werner's research on polynuclear co-ordination compounds, Co-ord. Chem. Rev., 9 (1972–3) 363.

G.B. Kauffman, Alfred Werner's research on structural isomerism, Co-ord. Chem. Rev., 11 (1973) 161.
G.B. Kauffman, Alfred Werner's research on optically active co-ordination compounds, Co-ord. Chem. Rev., 12 (1974) 105.
S.F.A. Kettle, Co-ordination compounds, Nelson, London, 1969.

Reactions of Metal Complexes

29

29.1. Classes of Substitution Reactions

Substitution reactions in inorganic chemistry include the replacement of a ligand (L) in a co-ordination complex by another ligand (Y), and the replacement of a metal ion (M) by another metal (M′). In the terminology developed for organic reactions these are called S_N (substitution-nucleophilic) and S_E (substitution-electrophilic) reactions, respectively:

$$ML_n + Y \rightarrow ML_{n-1}Y + L \qquad (S_N)$$
$$ML_n + M' \rightarrow M'L_n + M \qquad (S_E)$$

Although S_E reactions, such as that between Hg^{2+} and $[CoCl(NH_3)_5]^+$, are known, they are much less common than S_N and will not be considered here.

Turning to S_N reactions, we shall deal first with simple substitution reactions in which no change occurs in the oxidation state of the metal. In these reactions the substitution process can be considered as an acid—base reaction with the metal ion (or positive species) acting as a Lewis acid (9.14) and the ligands acting as bases.

Two different paths of nucleophilic substitution, each with a slow and a fast step, are generally recognised. The first can be formulated:

$$ML_n \rightarrow ML_{n-1} + L \qquad \text{(slow)}$$
$$ML_{n-1} + Y \rightarrow ML_{n-1}Y \qquad \text{(fast)}$$

This is the S_N1 (indicating substitution-nucleophilic-unimolecular) or dissociation mechanism. The reaction is termed unimolecular because only one species (ML_n) is involved in the formation of the activated complex (ML_{n-1}) with which further reaction occurs. As normally the activation energy for the first step will be high and that for the second step will be low, the rate of the overall reaction will depend upon $[ML_n]$, not upon $[Y]$. Thus the reaction will be first order with respect to ML_n and of zero order with respect to Y. (Of course, if there are other competing mechanisms the order will be complex.) During the formation of an activated complex in an S_N1 reaction, the co-ordination number of the metal is reduced.

It should be noted that rate equations in kinetics are expressed in terms of concentrations, unlike equilibrium equations in thermodynamics in which relative activities are used.

The second path of nucleophilic substitution, also with two steps, can be formulated:

$$ML_n + Y \rightarrow ML_nY \text{ (slow)}$$
$$ML_nY \quad \rightarrow ML_{n-1}Y + L \text{ (fast)}$$

This is the substitution-nucleophilic-bimolecular, S_N2, or displacement mechanism. The rate of reaction depends on both $[ML_n]$ and $[Y]$; in fact, it is first order with respect to ML_n, first order with respect to Y, and thus second order overall. The activated complex is made by the addition of another ligand, and during its formation the co-ordination number of the metal is increased.

Unequivocal evidence in favour of an S_N1 or an S_N2 mechanism is usually difficult to obtain because the activated complex can rarely be detected directly. Furthermore there is always the possibility of a mechanism which permits an almost simultaneous admission of Y to, and ejection of L from, the co-ordination sphere of M.

29.2. Methods of Measuring Rates of Substitution

In a slow reaction, the time available allows rates to be measured by ordinary analytical methods. For instance, in the replacement:

$$\underset{\text{purple}}{[CoCl(NH_3)_5]^{2+}} + H_2O \rightarrow \underset{\text{pink}}{[CoH_2O(NH_3)_5]^{3+}} + Cl^-$$

the progress of reaction may be followed

(i) by determining the Cl^- ion in solution analytically;

(ii) by observing the change in the optical density of the solution at an appropriate wavelength;

(iii) by observing the change in electrical conductance;

(iv) by measuring the pH; the complex ion $[CoH_2O(NH_3)_5]^{3+}$ acts as an acid, with $pK_a = 5.7$.

When complexes are formed from optically active ligands, rates of reaction can be followed polarimetrically. The compound 1,2-propylenediaminetetra-acetic acid (PDTA), also written H_4pdta:

$$\begin{matrix} HO_2C\cdot CH_2 & & CH_3 & & CH_2\cdot CO_2H \\ & \searrow & | & \nearrow & \\ & & N\cdot CH_2\cdot CH\cdot N & & \\ & \nearrow & & \searrow & \\ HO_2C\cdot CH_2 & & & & CH_2\cdot CO_2H \end{matrix}$$

is resolvable and acts as a quinquedentate or sexidentate ligand in the same way as EDTA. If a metal complex containing one active form and the free optical antipode are mixed in equimolar quantities, the optical rotation tends towards zero as the reaction proceeds and the distribution of the isomers becomes random:

$[M(d\text{-pdta})] + l\text{-}H_4pdta \rightleftharpoons [M(l\text{-pdta})] + d\text{-}H_4pdta$

Radioisotopes are useful in the determination of reaction rates, particularly in electron-exchange reactions such as:

$Fe^{II}L_6 + {}^*Fe^{III}(H_2O)_6 \rightarrow Fe^{III}L_6 + {}^*Fe^{II}(H_2O)_6$

Flow techniques, relaxation spectrometry and line broadening n.m.r. methods are available for the study of fast reactions.

29.3. Factors Affecting Rates of Substitution

Rates of substitution cover a wide range, and may depend on such variables as:

(i) the charge number of the metal;
(ii) the electronic configuration of the metal ion;
(iii) the nature and geometrical arrangement of the ligands;
(iv) the nucleophilic reagent;
(v) the solvent;
(vi) the influence of steric hindrance on the reaction.

For a metal ion of high charge number, substitution is usually slower than for an isoelectronic ion of lower charge number. The complex MnH_2pdta, made from the *d*-form of the acid, reacts with the *l*-form of the acid in aqueous solution at pH 6.3:

$Mn(d\text{-}H_2pdta) + l\text{-}H_4pdta \rightarrow Mn(l\text{-}H_2pdta) + d\text{-}H_4pdta$

and the optical activity is reduced to zero in 5 minutes at 293 K. The Fe^{III} complex (which is d^5 like Mn^{II}), formulated FeHpdta, does not undergo any measurable exchange with the optically isomeric acid in 2 days under the same conditions.

Complexes which react quickly in ligand replacement processes are termed *labile*, those which react slowly or not at all are called *inert*. Whether there is kinetic inertness or kinetic lability depends upon the activation energy of the reaction and has no connection with the thermodynamic stability of the species involved; this stability depends upon the free energy of formation. For

$Ni^{2+} + 4\,CN^- \rightarrow Ni(CN)_4{}^{2-} \qquad \Delta G = -126\ \text{kJ mol}^{-1}$

but the replacement of $^{12}CN^-$ by $^{14}CN^-$, when the latter is added to a solution of the complex, takes place too quickly to be measured by ordinary radiochemical techniques.

29.4. The Influence of Electronic Configuration on Rate of Substitution

Of the octahedral complexes of the d-block elements in the first row the most inert are those which have d^3, d^8 or low-spin d^6 configurations. Those

TABLE 29.1

CHANGES IN LFSE FOR THE PROCESS

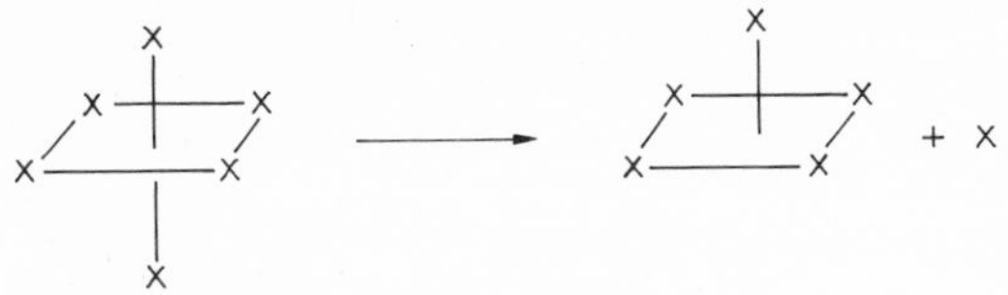

Electronic configuration of high-spin complex	(*a*) LFSE of octahedral complex	(*b*) LFSE of square pyramidal complex	Difference $(b - a)$
d^0	0	0	0
d^1, d^6	$0.400\ \Delta_0$	$0.457\ \Delta_0$	$+0.057\ \Delta_0$
d^2, d^7	$0.800\ \Delta_0$	$0.914\ \Delta_0$	$+0.114\ \Delta_0$
d^3, d^8	$1.200\ \Delta_0$	$1.000\ \Delta_0$	$-0.200\ \Delta_0$
d^4, d^9	$0.600\ \Delta_0$	$0.914\ \Delta_0$	$+0.314\ \Delta_0$
d^5, d^{10}	0	0	0

with low-spin d^4 and d^5 systems are rather more labile, and all other arrangements of d electrons are associated with kinetic lability. The pattern is similar for second- and third-row elements as far as electronic configurations are concerned, though these complexes are generally less labile than those of the Period 4 metals.

To explain these results, Basolo and Pearson (1958) calculated the changes which would occur in ligand field stabilisation energy when a particular octahedral complex was converted to the most probable activated complexes which might be formed in substitution reactions.

Consider first an octahedral d^n system which is converted to a square pyramidal activated complex by an S_N1 mechanism. The ligand field stabilisation energies for the various high-spin d^n systems in octahedral and square pyramidal environments are shown in Table 29.1 together with the change in LFSE which accompanies the formation of a square pyramidal complex in the course of reaction, making the assumption that Δ_0 does not change in the process.

For the d^3 and d^8 ions there is a loss of stabilisation energy when a square pyramidal activated complex is formed; for the other ions there is either no change or a gain. If we can assume that the change in LFSE is an important factor in determining how quickly the activated complex is formed, we should expect the d^3 and d^8 systems to be inert and all the others to be labile. Similar calculations applied to the process:

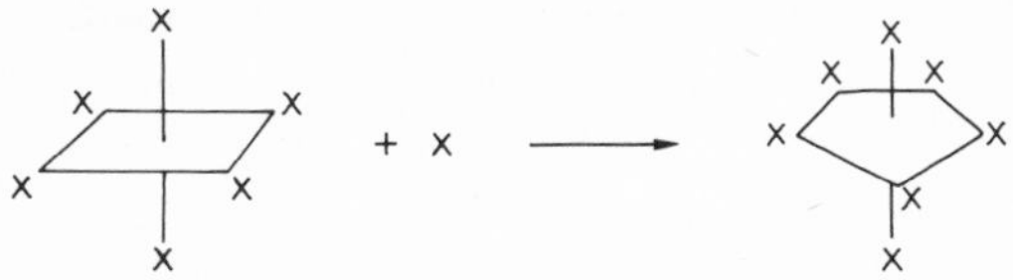

in which a pentagonal bipyramid is the activated complex, show that once again the d^3 and d^8 systems should be inert, with a loss of $0.426\Delta_0$ of LFSE when the activated complex is formed. In this case there is also a slight loss of LFSE in d^4 and d^9 systems, but all other systems should be labile.

Regarding low-spin complexes, similar calculations to the foregoing ones predict that d^6 systems should be very inert, and d^4 and d^5 systems rather inert. Thus the argument based on ligand field stabilisation can account for the known variations of lability with electron configuration.

29.5. Effect of an Inert Substituent on the Mechanism of Replacement

The rates of acid hydrolysis (the reaction between the complex and solvent water at pH < 3) of a series of complexes $[CoClXen_2]^+$ (X = NH_2, OH, Cl, NCS, NO_2) have been measured. The rate constants for the reactions:

$$[CoClXen_2]^+ + H_2O \rightarrow [CoX(H_2O)en_2]^{2+} + Cl^-$$

decrease in the order of the fall in electron-attracting power of X^-

$$NH_2^- > OH^- > Cl^- > NCS^- > NO_2^-$$

A basic group like NH_2 tends to form π bonds with the metal by the donation of p electrons. This has the effect of reducing the positive charge on the metal atom, and facilitates breaking of the Co—Cl bond. Furthermore, the π bonding stabilises the 5-co-ordinate intermediate which is formed. Both effects weaken with the decreasingly basic OH^-, Cl^- and NCS^- ligands; but, certainly in the cases where X is OH or Cl, the S_N1 mechanism is favoured.

However the NO_2^- ligand is different, it tends to withdraw electrons and thus promotes an S_N2 reaction by the formation of a 7-co-ordinate complex:

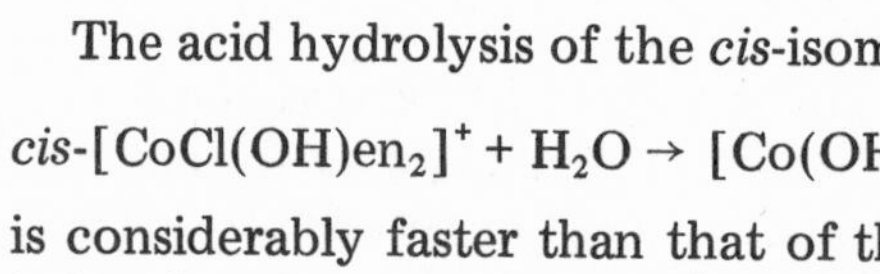

L L, O_2N—Co—Cl, L L → (electrons) L L OH_2, O_2N—Co—Cl, L L

29.6. Effect of the Stereochemical Arrangement of the Ligands

The acid hydrolysis of the *cis*-isomer:

$$cis\text{-}[CoCl(OH)en_2]^+ + H_2O \rightarrow [Co(OH)H_2O \cdot en_2]^{2+} + Cl^-$$

is considerably faster than that of the *trans*-isomer. The energy of activation is less by about 35 kJ mol^{-1}. This is believed to be due to the OH^- ligand, which is *cis* to the leaving group, Cl^-, and is in the most favourable position to donate electrons to an empty metal orbital by π-bonding. For complexes with electron-attracting ligands like NO_2, which do not form such π bonds, the hydrolysis of the *trans*-isomer is faster.

The position of an inert, electron-donating ligand also influences the configuration of the product of reaction. For the acid hydrolysis of *cis*- and *trans*-$[CoX(NCS)en_2]^+$ (X = Cl or Br) the rate constants, k, the activation energies, E, and the percentage of *cis*-$[Co(NCS)H_2Oen_2]^{2+}$ in the product are as shown:

	k/s^{-1} at 298 K	$E/kJ\ mol^{-1}$	% *cis* product
cis-Cl-compound	1.1×10^{-5}	87	100
trans-Cl-compound	5×10^{-8}	127	~60
cis-Br-compound	2.3×10^{-5}	97	100
trans-Br-compound	5×10^{-7}	126	~70

When the NCS group is *cis* to the halogen, the orbitals of the two ligands overlap sufficiently for electrons to be directed towards the space vacated by the electrons originally employed in bonding M to X, without the need for a rearrangement of the electrons around the metal. This results in the formation of a square pyramidal transition state, and retention of the original configuration.

But if a *trans*-NCS group is to provide electrons to fill up the emptied orbital, a change of configuration to trigonal bipyramid or distorted square pyramid is necessary; the activation energy is high and the original configuration is not retained.

29.7. Effect of the Nucleophile on Rates of Replacement

29.7.1. Octahedral complexes

Attempts have been made to establish an order of increasing reactivity for nucleophiles by studying the base hydrolysis of cobaltammines (hydrolysis in aqueous solution at pH > 10). The rates of reaction are commonly a million times faster than in acid hydrolysis; the reactions usually exhibit overall second-order kinetics:

Rate $\propto$ $[OH^-]$ [cobaltammine]

One explanation is that an S_N2 mechanism operates; for instance:

$$[CoCl(NH_3)_5]^{2+} + OH^- \xrightarrow{\text{slow}} [CoCl(OH)(NH_3)_5]^+ \xrightarrow{\text{fast}} Cl^- + [CoOH(NH_3)_5]^{2+}$$

But an S_N1CB (substitution, nucleophilic, unimolecular, conjugate-base) mechanism also accounts for the overall order of reaction:

(a) $\underset{\text{complex (acid)}}{[CoCl(NH_3)_5]^{2+}} + OH^- \overset{\text{fast}}{\rightleftharpoons} \underset{\text{conjugate base}}{[CoNH_2Cl(NH_3)_4]^+} + H_2O$

(b) $[CoNH_2Cl(NH_3)_4]^+ \xrightarrow{\text{slow}} [CoNH_2(NH_3)_4]^{2+} + Cl^-$

(c) $[CoNH_2(NH_3)_4]^{2+} + H_2O \xrightarrow{fast} [CoOH(NH_3)_5]^{2+}$

The speed of the rate-determining step (b) depends on [conjugate base]. If K = the equilibrium constant for the equilibrium (a), then

[conjugate base] = K[complex] [OH^-]

and the rate of the step (b) is proportional to [complex] [OH^-].

Both S_N2 and S_N1CB mechanisms would therefore give second-order kinetics, and the same product in aqueous solution; but in a non-hydroxylic solvent different products should be obtained. It has been found that for:

$CoXClen_2 + Y^- \rightarrow CoXYen_2 + Cl^-$

(X = inert ligand and Y = NO_2, N_3 or CNS) the reaction is slow in dimethylsulphoxide, but fast in the presence of small amounts of OH^-, although the products are the same in both cases. Also, the rate depends on [OH^-], not on [Y] or the nature of Y. These results support the idea of the initial formation of a 5-co-ordinate $CoXen_2$ which then reacts quickly with Y. The S_N1CB mechanism should lead to extensive rearrangement for both *cis*- and *trans*-isomers, in agreement with experimental evidence:

trans-$[CoBr(NCS)en_2]^+ \rightarrow [CoOH(NCS)en_2]^+$ 81%
cis-$[CoCl(NO_2)en_2]^+ \rightarrow [CoOH(NO_2)en_2]^+$ 34%

Thus attempts to find a scale of nucleophilic reactivities applicable to octahedral complexes have proved unsuccessful. For square complexes, however, the establishment of such a series has been possible.

29.7.2. Square complexes

The so-called square complexes are, through solvation, effectively tetragonal in most solvents (Fig. 29.1). The solvent molecules, S, are labile, and

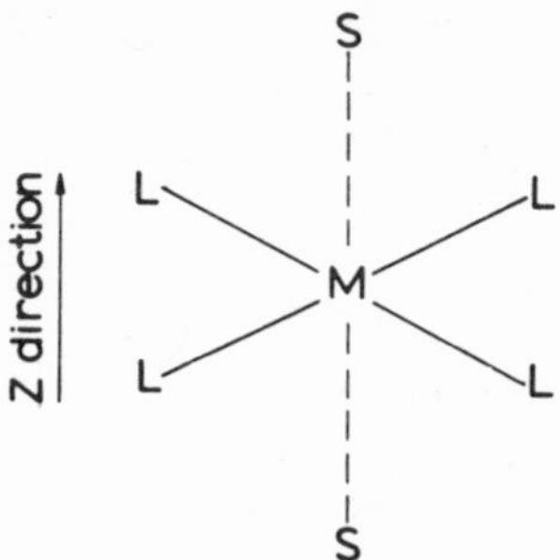

Fig. 29.1. Square complex converted to one of tetragonal symmetry by solvent molecules entering the co-ordination sphere (M—S distance long; d_{z^2} orbital contains two electrons in a typical square complex).

are replaceable by nucleophiles which can approach closer to M and cause the ejection of one of the ligands lying in the plane of the square.

There are two possible mechanisms for this. The first involves the formation of a square-pyramidal, activated complex from which the most labile ligand (L_4) is ejected:

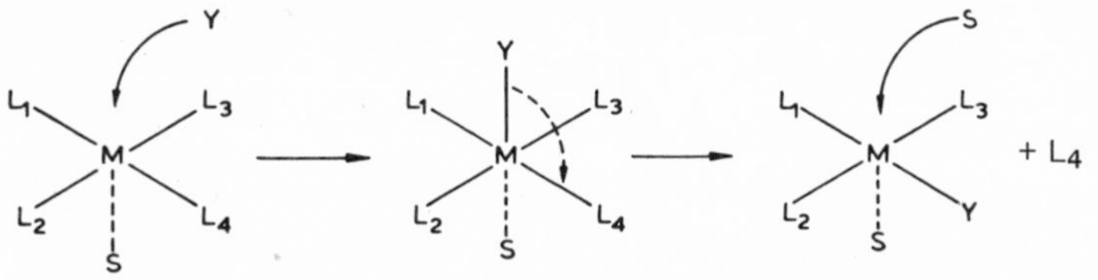

The second mechanism by either path shown below, involves the formation of a 5-co-ordinate, trigonal bipyramidal activated complex:

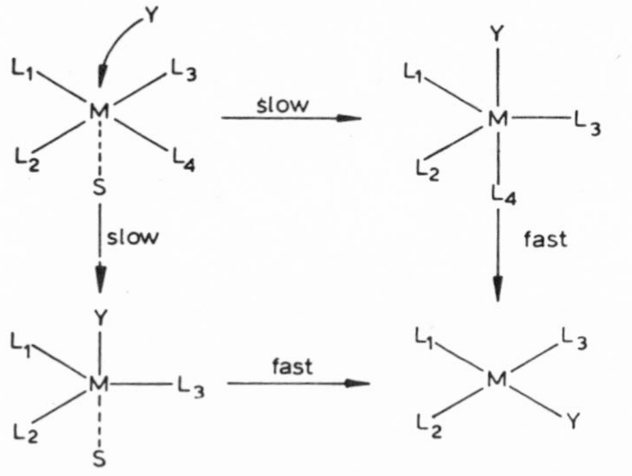

Experiments on Pt^{II} and Pd^{II} complexes of the type PtA_2LX show the rate law for the reaction:

$PtA_2LX + Y \rightarrow PtA_2LY + X$

to have the general form:

$\text{Rate} = k_1[\text{complex}] + k_2[\text{Y}][\text{complex}]$

Indeed, k_1 is interpreted as a first-order rate constant for an S_N2 reaction between complex and solvent. The values of k_2 for different nucleophiles (Y) enable these reagents to be placed in the following order of reactivity towards platinum:

$OH^- < Cl^- < Br^- \sim NH_2 <$ aniline $<$ pyridine

$< NO_2^- < N_3^- > I^- \sim SCN^- \sim$ thiourea

This order, with little variation from one Pt^{II} complex to another, is similar (except for OH^-) to that for the reactivity of the same groups towards a typical organic substrate such as an alkyl halide.

29.8. Effect of Ligands already in the Complex: The *trans* Effect

Ligand replacement reactions in square complexes such as those of Pt^{II} have some special features. For the reaction:

$$PtL_3X + Y \rightarrow PtL_2XY + L$$

the reaction product may have either of the orientations:

L X / Pt / L Y (*cis*) or L X / Pt / Y L (*trans*)

Experiment shows that the proportions of *cis-* and *trans*-isomers depend on the nature of the ligand X. Ligands vary in their power of directing an incoming nucleophilic ligand preferentially into a position *trans* to themselves.

The order of the *trans*-directing strength for common ligands is indicated:

$$H_2O < OH^- < NH_3 < Cl^- < Br^- < I^- \sim NO_2^- << CO \sim C_2H_4 \sim CN^-$$

The discovery of the *trans*-effect has enabled the synthesis of many platinum complexes to be rationalised. Thus *cis*-$PtCl_2(NO_2)(NH_3)^-$ is made by the action of ammonia on $PtCl_4^{2-}$, and of nitrite on the product:

$$[PtCl_4]^{2-} \xrightarrow{NH_3} [PtCl_3(NH_3)]^- \xrightarrow{NO_2^-} [PtCl_2(NO_2)(NH_3)]^-$$

The *trans*-isomer is obtained by reversing the order of reagent addition; this is because NO_2^- is more strongly *trans*-directing than Cl^-, but NH_3 is weaker than Cl^-:

$$[PtCl_4]^{2-} \xrightarrow{NO_2^-} [PtCl_3(NO_2)]^{2-} \xrightarrow{NH_3} [PtCl_2(NH_3)(NO_2)]^-$$

There are two approaches to the explanation of *trans*-directing character in a ligand; but these are not mutually exclusive. The first is a thermodynamic approach. A large, highly polarisable ligand, like I, distorted by the positive charge on the central atom, is imagined to polarise that atom itself to some extent and thus weaken the bond between it and the opposite ligand.

The second approach is a kinetic one and assumes an S_N2 reaction. It uses the idea that a ligand which can accept electrons donated back from the

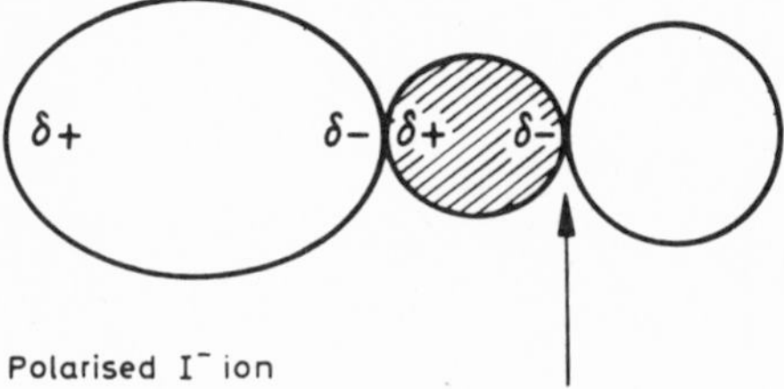

Fig. 29.2. Representation of polarisation of metal ion induced by highly polarisable ligand (signs show only polarisation effects; the central atom normally carries positive charge and the ligands are electron-donating).

metal through d_π—p_π or d_π—d_π bonds will tend to reduce the electron density both above and below the bond situated on the other side of the metal atom, and thus open up the position for nucleophilic attack. This view is in better accord with the high position of ethylene in the *trans*-directing series, since this ligand cannot have a strong electrostatic effect. However, there is at present insufficient experimental evidence to enable a comprehensive theory to be formulated.

29.9. Effect of the Solvent

The following isotopic exchange has been studied in a number of solvents:

$$trans\text{-}PtCl_2py_2 + 2\,{}^{36}Cl^- \rightleftharpoons Pt^{36}Cl_2py_2 + 2\,Cl^-$$

The results show that the solvents can be divided into two categories with respect to their influence on k_1 and k_2 in the rate equation for the forward reaction:

$$\text{Rate} = k_1[\text{complex}] + k_2[Cl^-][\text{complex}]$$

In water, ethanol, dimethylsulphoxide and methyl nitrite the rate is almost independent of $[Cl^-]$; that is $k_1 \gg k_2[Cl^-]$. But in carbon tetrachloride, benzene, tertiary butanol, ethylene dichloride and ethyl acetate the reaction is of first order with respect to free chloride ion; $k_2[Cl^-] \gg k_1$. As already explained, k_1 can be considered as a rate constant for an S_N2 reaction between complex and solvent.

The solvents of the first group are, in general, strongly co-ordinating and ionising ones, those in the second group co-ordinate weakly and do not promote ionisation. Apparently, the platinum(II) complex suffers nucleophilic attack by solvents like water and ethanol, but not by those like benzene and carbon tetrachloride. The co-ordinating solvent molecules all possess vacant orbitals capable of bonding with the filled d_{xz} and d_{yz} orbitals of Pt^{II}. Thus their electron-donating oxygen atoms are able to approach more closely to the Pt and create favourable conditions for the displacement of chloride.

29.10. Steric Effect in Replacement Reactions

The importance of solvation above and below the plane in certain reactions of Pt^{II} complexes has been verified by blocking these positions by groups attached to ligands lying in the plane (Fig. 29.3).

There is a marked reduction in reaction rates as steric shielding of the platinum by R groups is increased.

As an example, for the reactions in ethanol at 298 K:

$$trans\text{-}[PtRCl(PEt_3)_2] + py \rightarrow trans\text{-}[PtRpy(PEt_3)_2]^+ + Cl^-$$

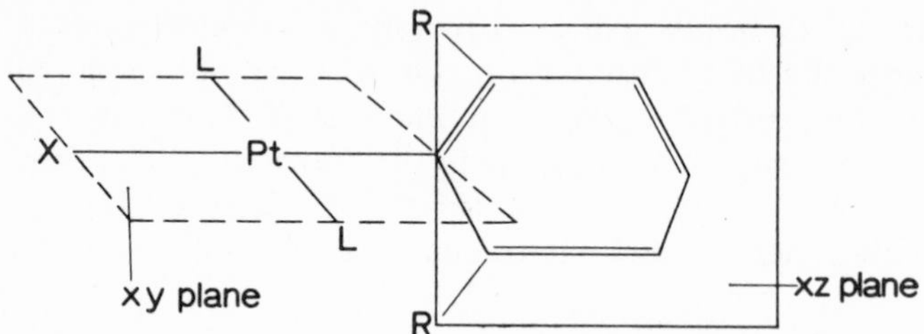

Fig. 29.3. R groups above and below Pt sterically hinder the attachment of solvent molecules.

the rate constants, k_1 are:

	k_1/s^{-1}
R = phenyl	3×10^{-5}
R = *o*-tolyl	6×10^{-6}
R = mesityl	1×10^{-6}

A similar but even more pronounced effect is observed for the analogous Pd^{II} and Ni^{II} complexes.

29.11. Reactions of Tetrahedral Complexes

Tetrahedral complexes are generally labile, and relatively little kinetic work has been done on them. The mechanism of replacement reactions in the tetrahedral complexes of Si, Ge and Sn differ from those of carbon in that 5-co-ordinate intermediates can be formed with much less expenditure of energy because the co-ordination numbers of the atoms are more easily expanded. The dominant mechanism is S_N2. Thus for optically active compounds of silicon of the type R′R″R‴SiX some substitutions occur with complete retention, some with complete inversion, of configuration; in the circumstances racemisation would result if the reaction were S_N1.

Solvent molecules seem to participate in some of the reactions of tetrahedral complexes. For instance, the bis(cyclopentadienyl)titanium(IV) halides, which have distorted tetrahedral structures, react with ionic halides in benzene or tetrahydrofuran:

$$(C_5H_5)_2TiBr_2 + 2\ R_4NCl \rightarrow (C_5H_5)_2TiCl_2 + 2\ R_4NBr$$

The rate of reaction is proportional to $[R_4NCl]$, suggesting an S_N2 process. But when LiCl is used instead of a quaternary ammonium chloride the rate is independent of [LiCl] and is slower. Presumably the rate-determining step is nucleophilic displacement by solvent molecules.

29.12. Photochemistry of Metal Complexes

Metal—ligand systems usually show two kinds of absorption band:

(i) High-intensity ultraviolet absorption attributed to electron transfer.

(ii) Low-intensity absorption in the visible range due to d—d transition.

Thus irradiation with ultraviolet rays should encourage redox processes, whereas visible light should promote substitution reactions by causing the promotion of electrons into antibonding orbitals where they cause repulsion between metal and ligand.

There is evidence that in Co^{III} complexes absorption in the charge-transfer band leads to reduction of Co^{III} to Co^{II}, but that absorption in the d—d transition band leads to substitutions. Thus for $[CoNCS(NH_3)_5]^{2+}$ in aqueous solution, the ratio of reduction to substitution (hydration to $[CoH_2O(NH_3)_5]^{3+}$) is 2.1 at 370 nm (ultraviolet) but only 0.24 at 550 nm (visible).

29.13. Redox Reactions

Let us consider a reaction between a cobalt(III) complex and a chromium(II) complex which results in the formation of a cobalt(II) complex and a chromium(III) complex. The changes in oxidation state imply the transfer of an electron:

$$Cr^{II} \rightarrow Cr^{III} + e$$
$$Co^{III} + e \rightarrow Co^{II}$$

Two kinds of mechanism are recognised for electron transfers between species in solution. In one kind, the *outer-sphere mechanism*, the electron simply hops across from one species to the other; the co-ordination spheres of the metals remain intact. In the other kind, the *inner-sphere mechanism*, the oxidant and reductant are attached to one another, at some stage in the process, by a bridging molecule, atom or ion through which the electron can pass.

29.13.1. Outer-sphere reactions

The reaction

$$^*Fe(CN)_6^{4-} + Fe(CN)_6^{3-} \rightarrow {}^*Fe(CN)_6^{3-} + Fe(CN)_6^{4-}$$

in which $^*Fe$ represents radioactively labelled iron, is a very rapid one. But both cyanoferrate(III) ions and cyanoferrate(II) ions are kinetically inert; they exchange cyanide ligands very slowly with solutions containing CN^- ions: The fact that the redox reaction is so fast but the substitution reactions of both complexes are so slow eliminates the possibility that electron exchange occurs through a bridged activated complex, because the very formation of such an activated complex is itself a substitution reaction. The other possibility is a direct electron transfer from one intact co-ordination sphere to the other. Even if one of the partners in the reaction is labile, an outer sphere mechanism is favoured if the other partner does not offer a suitable site to engage the metal ion of the labile partner. Thus the reduction of $Co(NH_3)_6^{3+}$ by $Cr(H_2O)_6^{2+}$ appears to be an outer-sphere reaction in spite of the labile character of $Cr(H_2O)_6^{2+}$; the co-ordinated NH_3 molecule does not offer unshared electrons for attachment to the Cr^{2+}. As a rule, outer-sphere reactions are of first order with respect to each reactant at constant ionic composition. Salt effects are important, however. The exchange of radioactive iron from $^*Fe(CN)_6^{4-}$ to $Fe(CN)_6^{3-}$ is thirty times faster for the Me_4N^+ salts than for the $^nPr_4N^+$ salts.

For an electron-transfer mechanism there is the critical requirement that reductant and oxidant should have similar structures. The Franck-Condon principle states that an electron moves so quickly compared with an atom that appreciable rearrangement of atoms is impossible during an electronic transition. If an electron is transferred from an $Fe(CN)_6^{4-}$ ion to an $Fe(CN)_6^{3-}$ ion the result would be an $Fe(CN)_6^{4-}$ ion in which the Fe—C bonds are too short and an $Fe(CN)_6^{3-}$ in which the Fe—C bonds are too long — both complexes being in high-energy states. Hence the process would create energy, and this description of the mechanism is therefore incorrect. Clearly the readjustment of the bond lengths must take place before the electron-jump occurs. The activation energy required to shorten the bonds in the Fe^{II} compound and lengthen those in the Fe^{III} compound is fairly small because the geometries of the two anions are similar.

Again in MnO_4^{2-}—MnO_4^- and $IrCl_6^{3-}$—$IrCl_6^{2-}$ exchanges, where again both complexes are inert, the exchange rates are far faster than could possibly be explained by dissociation and atom transfer.

When, however, two ions are very different in their geometry or in the metal—ligand distances, electron-exchange reactions between them are usually slow. Thus for $Co(NH_3)_6^{2+}$ and $Co(NH_3)_6^{3+}$, although both are octahedral, the Co—N distances and the electronic configuration round the cobalt atoms, $t_{2g}^5e_g^2$ and t_{2g}^6 respectively, are quite dissimilar. Both must change before reaction can occur: this is the reason for the slowness of the redox process.

29.13.2. Inner-sphere reactions

To prove that a reaction proceeds by an inner-sphere mechanism there must be evidence either that a particular ligand is transferred from one coordination sphere to another or that a binuclear intermediate is formed. An elegant series of experiments on the reduction of monosubstituted ammines of cobalt(III) with the hexa-aquochromium(II) ion has demonstrated ligand transfer. The ion $Cr(H_2O)_6^{2+}$, however, is labile; the half-period for H_2O exchange is about a nanosecond. But when the dipositive chromium ion is oxidised, the product containing chromium(III) (d^3) is inert. The reaction between $[(NH_3)_5CoCl]^{2+}$ and $Cr(H_2O)_6^{2+}$ is very rapid and nearly all the chromium(III) appears as $Cr(H_2O)_5Cl^{2+}$. Furthermore, if free radioactive Cl^- is present in the solution very little radioactivity is incorporated in the chromium complex. Thus the chlorine atom is transferred direct from Co to Cr in the redox process and not by way of the solvent. These results suggest a mechanism

$$[(NH_3)_5Co^{III}Cl]^{2+} + [Cr^{II}(H_2O)_6]^{2+}$$

$$\downarrow$$

$$[(NH_3)_5Co \cdots Cl \cdots Cr(H_2O)_5]^{4+} + H_2O$$

(the bridged activated complex)

$$\downarrow$$

$$[(NH_3)_5Co^{II}]^{2+} + [ClCr^{III}(H_2O)_5]^{2+}$$

The penta-amminecobalt(II) complex is quickly destroyed because the high-spin d^7 Co^{II} complex is labile and hydrolyses quickly to $Co(H_2O)_6^{2+}$ in acidic solution.

This type of mechanism seems to be general for reactions of $Cr(H_2O)_6^{2+}$ with cobalt(III) and chromium(III) complexes which provide suitable bridging groups and also for some iron(III) complexes and vanadium(III) complexes. Other reducing agents which can be demonstrated to react by inner-sphere processes in particular cases are $V(H_2O)_6^{2+}$, $Fe(H_2O)_6^{2+}$ and $Co(CN)_5^{3-}$.

Reductions of several different cobalt(III) complexes, $[Co(NH_3)_5X]^{2+}$, with Cr^{II} have been studied. Transfer of the ligand X to the chromium has been demonstrated for X = NCS^-, N_3^-, PO_4^{3-}, $CH_3CO_2^-$, Cl^-, Br^- and SO_4^{2-}. Rates of reaction differ, for example the reaction is faster for X = Br than for X = Cl. Presumably the ions which provide the best path for electron transfer produce the fastest reactions. Interestingly, organic ligands containing co-ordinated systems of double bonds (i.e. delocalised π electrons) usually conduct electrons particularly well, thus the complex

$$[(NH_3)_5Co{-}O{-}\overset{\overset{\displaystyle O}{\|}}{C}{-}CH{=}CH{-}\overset{\overset{\displaystyle O}{\|}}{C}{-}OH]^{2+}$$

is rapidly reduced by $Cr(H_2O)_6^{2+}$ whereas

$$[(NH_3)_5Co{-}O{-}\overset{\overset{\displaystyle O}{\|}}{C}{-}CH_2{-}CH_2{-}\overset{\overset{\displaystyle O}{\|}}{C}{-}OH]^{2+}$$

is reduced slowly.

The electron configurations of reactants and products are often important in determining whether a reaction proceeds by an inner-sphere mechanism or an outer-sphere one. The reactions of V_{aq}^{2+} with oxidising agents such as $Co^{III}(NH_3)_5L$ (L = H_2O, Cl, NH_3) proceed through an outer-sphere complex rather than a bridged-complex mechanism such as in the Cr_{aq}^{2+} oxidation. A t_{2g} electron on the V_{aq}^{2+} is transferred, and its orbital overlaps an orbital on the oxidant. But with Cr_{aq}^{2+} the electron to be transferred is in an e_g orbital and is badly situated for overlap with a receiver orbital; consequently transference of the electron by means of a bridge provides a path of lower energy. When L above is $C_2O_4H^-$ however, $V(H_2O)_6^{2+}$ reduces the Co^{III} complex by an inner-sphere reaction. The low-lying π orbital of the oxalato group can engage the donor orbital of V^{II}, which has π symmetry, by an inner-sphere activated complex.

Binuclear intermediate products can sometimes be identified. Ruthenium(III) (d^5) has its electron vacancy in a π orbital; thus when it forms the bond Ru—L with a bridging ligand the complex is nothing like so labile as one containing Co^{II}—L in which the transferred electron occupies an antibonding σ^* orbital. When ruthenium(III) is used as an oxidant, binuclear intermediates are often observed.

There are some reactions of aquo ions in which the most probable mechanism is the transfer of hydrogen atoms. The evidence is indirect; the reaction between $^{59}Fe(H_2O)_6{}^{2+}$ and $FeOH(H_2O)_5{}^{2+}$ is about twice as fast in H_2O as in D_2O, the usual ratio of reaction rates for processes in which a hydrogen *atom* must move — for example k_H/k_D for the reactions of alkali metals with H_2O and D_2O.

Further Reading

R.G. Wilkins, Mechanisms of ligand replacement in octahedral nickel(II) complexes, Acc. Chem. Res., 3 (1970) 408.

M.L. Tobe, Base hydrolysis of octahedral complexes, Acc. Chem. Res., 3 (1970) 377.

J. Halpern, Oxidative-addition reactions of transition-metal complexes, Acc. Chem. Res., 3 (1970) 386.

R.F. Heck, Addition reactions of transition-metal compounds, Acc. Chem. Res., 2 (1969) 10.

F. Basolo and R.G. Pearson, Mechanisms of substitution reactions of metal complexes, Adv. Inorg. and Radiochem., 3 (1961) 1.

A.G. Sykes, Further advances in the study of mechanisms of redox reactions, Adv. Inorg. and Radiochem., 10 (1967) 153.

F.R. Hartley, The *cis*- and *trans*-effects of ligands, Chem. Soc. Rev., 2 (1973) 163.

F. Basolo and R.G. Pearson, Mechanism of inorganic reactions, 2nd edition, Wiley, New York, 1967.

H. Taube, Electron transfer reactions of complex ions in solution, Academic Press, New York, 1970.

R.F. Gould (Ed.), Reactions of co-ordinated ligands, Amer. Chem. Soc., 1963.

L.E. Bennett, Metalloprotein redox reactions, Prog. Inorg. Chem., 18 (1973) 1.

The Lanthanides, Scandium and Yttrium — Group IIIA

30

THE LANTHANIDES

30.1. Electronic Structures and General Properties

The lanthanides, for which the general symbol Ln is used, have electron configurations with $6s^2$ in common and a variable occupation of the 4f level. Although lanthanum itself has not an f electron in its ground state, the element has strong similarities with those which follow it, and it will be treated in what follows as a lanthanide. Classically called the Rare Earths, these elements are also referred to as the inner-transition elements because the 4f electron build-up takes place in the fourth quantum level, below the 5s, 5p and 6s electrons which are also present in these atoms. As the electronic diversity is in this lower shell, the elements are very similar chemically.

The small differences in properties arise principally from the "lanthanide contraction". Since, for every additional proton in a nucleus the corresponding electron goes into the 4f shell which is too diffuse to screen the nucleus as effectively as a more localised inner shell, the attraction of the nucleus for the outermost electrons increases steadily with the atomic number of the lanthanide. This causes a fall in atomic size from lanthanum to lutetium. The contraction is quite regular for the terpositive ions, from 106 pm for La^{3+} to 85 pm for Lu^{3+}. The graph of metallic radii (Fig. 30.1A), however, illustrates that although there is a general downward trend the values for Eu and Yb lie much higher than the rest. These are the two lanthanides which have the greatest tendency to form bipositive ions. In the solids the atoms probably donate only two electrons to the conduction bands; the 2+ ions which are produced will be larger and less strongly bound than the 3+ ions in the other metals. Europium and ytterbium metals have notably lower densities, lower melting points (Fig. 30.1B) and lower energies of sublimation than their neighbours in the table.

The first ionisation energies of the lanthanides are about 600 kJ mol^{-1} and the second about 1.2 MJ mol^{-1}, comparable with those of calcium. The standard electrode potentials vary smoothly from -2.52 V for the La^{3+}/La couple to -2.25 V for Lu^{3+}/Lu.

The oxidation states exhibited by the lanthanides are shown in Table

TABLE 30.1

THE LANTHANIDES

Name	Symbol	Z	Configuration			*Abundance/* p.p.m.
Lanthanum	La	57		$5d^1$	$6s^2$	18.3
Cerium	Ce	58	$4f^1$	$5d^1$	$6s^2$	46.1
Praseodymium	Pr	59	$4f^3$		$6s^2$	5.5
Neodymium	Nd	60	$4f^4$		$6s^2$	23.9
Promethium	Pm	61	$4f^5$		$6s^2$	0.0
Samarium	Sm	62	$4f^6$		$6s^2$	6.5
Europium	Eu	63	$4f^7$		$6s^2$	1.1
Gadolinium	Gd	64	$4f^7$	$5d^1$	$6s^2$	6.4
Terbium	Tb	65	$4f^9$		$6s^2$	0.9
Dysprosium	Dy	66	$4f^{10}$		$6s^2$	4.5
Holmium	Ho	67	$4f^{11}$		$6s^2$	1.1
Erbium	Er	68	$4f^{12}$		$6s^2$	2.5
Thulium	Tm	69	$4f^{13}$		$6s^2$	0.2
Ytterbium	Yb	70	$4f^{14}$		$6s^2$	2.7
Lutetium	Lu	71	$4f^{14}$	$5d^1$	$6s^2$	0.8

30.2. All the metals form Ln^{3+} ions evidently because of a favourable combination of ionisation energies and lattice energies in the case of the solids, and a favourable combination of ionisation energies and hydration energies in aqueous solution. In addition, the Sm^{2+}, Eu^{2+} and Yb^{2+} ions can exist in aqueous solution, but all are strong reducing agents. The Ce^{4+} ion can exist in aqueous solution but it is a strong oxidising agent. Only isolated solid compounds in which the oxidation states are +2 or +4 are formed by Pr, Nd,

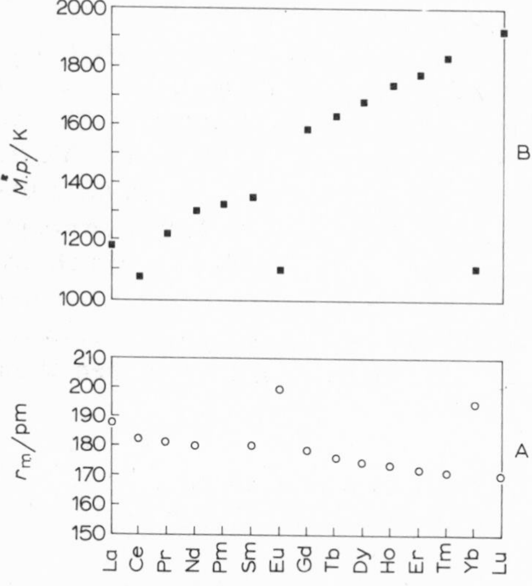

Fig. 30.1. The lanthanide metals. A. Metallic radii. B. Melting points.

TABLE 30.2

OXIDATION STATES AND CONFIGURATIONS

	+2	+3	+4
La		All form 3+ ions (configurations regular)	
Ce	$CeCl_2$ ($4f^2$)		Ce^{4+} ($4f^0$)
Pr			PrO_2 ($4f^1$)
Nd	NdI_2 ($4f^4$)		Cs_3NdF_7 ($4f^2$)
Sm	Sm^{2+} ($4f^6$)		
Eu	Eu^{2+} ($4f^7$)		
Gd			
Tb			TbF_4 ($4f^7$)
Dy			Cs_3DyF_7 ($4f^8$)
Ho			
Er			
Tm	TmI_2 ($4f^{13}$)		
Yb	Yb^{2+} ($4f^{14}$)		
Lu			

Tb, Dy and Tm (Table 30.2). These two states are most commonly associated with f^0, f^7 or f^{14} ions such as Ce^{4+}, Eu^{2+} and Yb^{2+} respectively.

30.2. Sources and Separations of the Lanthanides

Though called rare, the elements are not particularly so. Three of them, namely Ce, La and Nd, are more common than lead, and thulium is about as abundant as iodine. The lanthanides tend to be in minerals which crystallise late from a magma since their ions are too large to replace other terpositive ions; hence they separate in the pegmatites. The 'light' lanthanides are extracted chiefly from monazite, predominantly phosphates of thorium, cerium, neodymium and lanthanum, and also from orthite. Monazite is a poor source of europium, which is more commonly associated as Eu^{2+} with alkaline earth minerals, and does not carry much of the 'heavy' lanthanides which are obtained from gadolinite, $FeBe_2Y_2Si_2O_{10}$, and xenotime, YPO_4.

30.2.1. Extraction of lanthanides from monazite

Monazite is digested with concentrated H_2SO_4 to produce a paste of the sulphates containing phosphoric acid and an excess of sulphuric acid. The paste, separated by centrifugation, is dissolved in cold water, the sulphates being less soluble in hot water. The solution is neutralised with a previously prepared mixture of lanthanide oxides to precipitate thorium, zirconium and titanium. The addition of Na_2SO_4 to the clear mother liquor throws down the light lanthanides (La to Sm) as double sulphates of variable composition; the heavy lanthanides (Gd to Lu), which comprise less than 4% of those present, remain in solution. Hot NaOH is added to the double sulphate precipitate of light fraction to give a mixture of hydrated oxides. This is washed free from Na_2SO_4 and dried in air at 373 K, the cerium being thus completely oxidised to CeO_2. The composition of the solid is roughly: CeO_2 50%, Nd_2O_3 20%, La_2O_3 17%, Pr_2O_3 8%, Sm_2O_3 5%.

This mixture is extracted with dilute HNO_3 to dissolve the more basic sesquioxides, Ln_2O_3, and leave a residue of CeO_2. The crude CeO_2 is dissolved in 85% HNO_3 and added to an excess of dilute H_2SO_4, whereby the cerium is precipitated as the red, basic nitrate, $Ce(OH)(NO_3)_3 \cdot 3\,H_2O$. The dilute solution is used for leaching more of the dried hydroxide.

Separation of the light lanthanides, after removal of the Ce, has been accomplished in various ways, based mainly on solubility differences. Fractional crystallisation of the double magnesium nitrates, $2\,Ln(NO_3)_3 \cdot 3\,Mg(NO_3)_2 \cdot 24\,H_2O$, was an early method. The heavy lanthanides from the double sulphate solution (above) and from ores such as xenotime have been separated by fractional crystallisation of either the bromates or the double ammonium oxalates.

30.2.2. Separation by ion exchange

When a solution containing a mixture of lanthanide ions is brought on to an exchange resin in its hydrogen form, the order of absorption follows the atomic number of the elements. Affinity for the resin decreases with radius of the hydrated ion:

$$3\,HR + Ln^{3+} \rightleftharpoons 3\,H^+ + LnR_3$$

Orderly displacement of the cations from the resin cannot be achieved by a concentrated solution of an ammonium salt. However, the addition of an acid which complexes with the lanthanide ions improves the separation. The distribution coefficients of adjacent lanthanides between citrate buffer solutions and Amberlite IR-1 differ by a factor of about two. The eluant is most efficient at low concentration and high pH.

On a large scale, light lanthanides have been absorbed on a cation resin in a 24-unit battery of Pyrex columns 3 m by 10 cm, and displaced with 0.1% citric acid buffered to pH 6 with ammonia. The eluate, collected in 40—50 litre fractions (about a day's run), contained up to 0.4 g/l of Ln_2O_3. Over 80% of each element present was obtained in a high state of purity from a single passage through the columns. Under these conditions, the elements were displaced in the reverse order of the atomic number. EDTA has proved an even more satisfactory complexing agent than citric acid and gives purer specimens than any other agent.

30.2.3. Separation by solvent extraction

Solutions of tri-*n*-butyl phosphate in kerosene flowing counter to a nitric acid solution of lanthanides remove them in the organic phase. Differences in the extraction coefficient are approximately equal from one element to the next. Gadolinium of about 95% purity has been separated from other lanthanides on a kilogram scale in this way.

30.3. The Metals

30.3.1. Production

The metals have been produced by methods which include:
(i) electrolysis of the fused chlorides (cf. Ca),
(ii) electrolysis of CeO_2 in fused CeF_3 (cf. Al),
(iii) reduction of the anhydrous chlorides with Na (cf. Ti),
(iv) reduction of the anhydrous fluorides with Mg (cf. U).
Metallothermic reduction by method (iii) has been used for the lighter lanthanides such as La, Ce and Ga, and by method (iv) for the heavier lanthanides because the fluorides are less volatile than the chlorides and they restrict loss by evaporation at the temperatures (>1300 K) employed. It is noteworthy that Sm, Eu and Yb, which have a reasonably stable +2 state are reduced only as far as the dichlorides and their separation from other lanthanides is facilitated. These metals themselves are obtained by reducing the oxides with lanthanum.

Metals produced in the various ways described may be purified by distillation and in other ways, but for all of them there are limits beyond which purification has not been carried.

30.3.2. Physical properties

The first three lanthanides, La, Ce and Pr, are dimorphous metals with h.c.p. and c.c.p. forms. The rest have h.c.p. structures, except Eu (b.c.c.) and Yb (c.c.p.). Densities increase with atomic number from La (6.17 g cm^{-3}) to Lu (9.84 g cm^{-3}) except for Eu (5.26 g cm^{-3}) and Yb (6.98 g cm^{-3}). The low densities of these two are a consequence of the large atomic volumes (cf. metallic radii in Fig. 30.1A). In fact Eu and Yb resemble Sr more than they do their nearest neighbours.

The metals are rather soft and are malleable and ductile, but they are of low tensile strength. They are all silvery-white; the heaviest lanthanides remain bright at room temperature but the lighter ones rapidly tarnish. All except Yb are rather strongly paramagnetic; Gd is ferromagnetic up to 290 K.

Although the metals themselves have poor mechanical characteristics they are being used as alloy materials for the hardening of steel and to increase the strength and resistance to creep in magnesium at high temperatures. Their addition to the nickel–chromium alloys used for resistance wires increases their useful life.

30.3.3. Chemical character

The metals are similar to magnesium in their reducing properties, as expected from the values of E^0 for the Ln^{3+}/Ln couples. The major factor which leads to the easy formation of Ln^{3+} ions in solution is almost certainly the large enthalpies of hydration of these ions. Dry air attacks the metals slowly at room temperature but moist air attacks Eu very quickly and most of the light lanthanides quite quickly. All the metals burn in air on warming. A mixture of Ce with some of the other lighter lanthanides, called mischmetal, is pyrophoric and is used as lighter 'flints'. Eu and Yb resemble the alkali metals and the alkaline-earths in that they dissolve in ammonia to give blue solutions.

Some of the reactions of the lanthanides are illustrated in Fig. 30.2. It will be seen that in their high-temperature reactions as well as their solution properties they are of similar reactivity to magnesium.

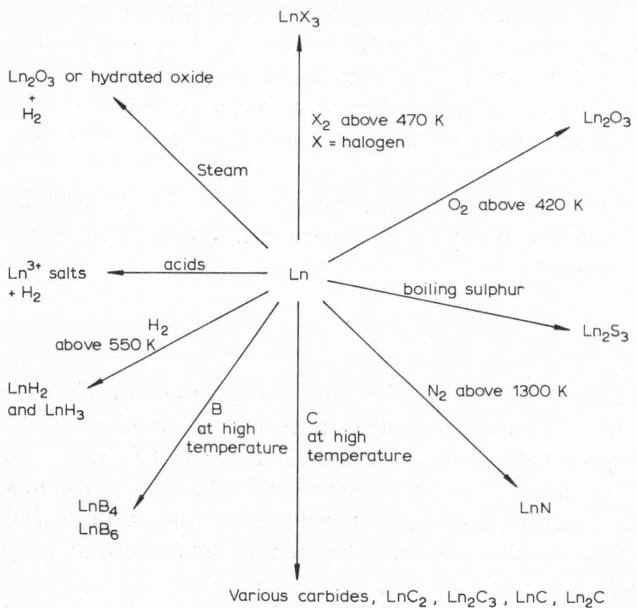

Fig. 30.2. Reactions of the lanthanides (Ln = lanthanide atom).

30.4. Absorption Spectra and Magnetic Moments

30.4.1. Absorption spectra

Neither La^{3+} nor Lu^{3+} show absorption bands in the ultra-violet, visible or near infrared, but the rest do. The bands, which are narrow compared with those of the normal transition ions, move from red at Pr^{3+} to ultraviolet at Gd^{3+} and back to red again at Tm^{3+}. The energy changes involved are probably due to excitations within the 4f shell, since complexing agents, which alter the absorption spectra of normal transition ions by modifying their outer structure, have little effect on the lanthanide ions. An ion with n electrons in the 4f level generally has a similar absorption spectrum to one with $14 - n$, but the pairs Nd^{3+} ($4f^3$), Er^{3+} ($4f^{11}$), and Pm^{3+}, Ho^{3+} are anomalous.

TABLE 30.3

COLOURS OF LANTHANIDE 3+ IONS

$4f^1$	Ce^{3+}	(colourless)	Yb^{3+}	(colourless)	$4f^{13}$
$4f^2$	Pr^{3+}	(green)	Tm^{3+}	(green)	$4f^{12}$
$4f^3$	Nd^{3+}	(blue-violet)	Er^{3+}	(pink)	$4f^{11}$
$4f^4$	Pm^{3+}	(rose)	Ho^{3+}	(yellow)	$4f^{10}$
$4f^5$	Sm^{3+}	(cream)	Dy^{3+}	(cream)	$4f^9$
$4f^6$	Eu^{3+}	(colourless; u.v. absorption)	Tb^{3+}	(colourless; u.v. absorption)	$4f^8$
→	$4f^7$ Gd^{3+} (colourless; u.v. absorption)				←

30.4.2. Paramagnetism

The lanthanide ions, other than the $4f^0$ type, La^{3+} and Ce^{4+}, and the $4f^{14}$ type, Yb^{2+} and Lu^{3+}, are all paramagnetic, with both spin and orbital moments. For the $J_0 \rightarrow J_1$ transition in Sm^{3+} and Eu^{3+}, $h\nu \sim kT$ and there is no simple expression for μ, the magnetic moment, which receives contributions from some of the lower excited states of the ion as well as from the ground state. But in all other cases μ is given to a near approximation by calculation on the basis of Russell—Saunders coupling since $h\nu$ is much greater than kT, and the Curie law is closely obeyed. The moments of isoelectronic ions (Eu^{2+}, Gd^{3+}, Tb^{4+} — all $4f^7$) are very similar.

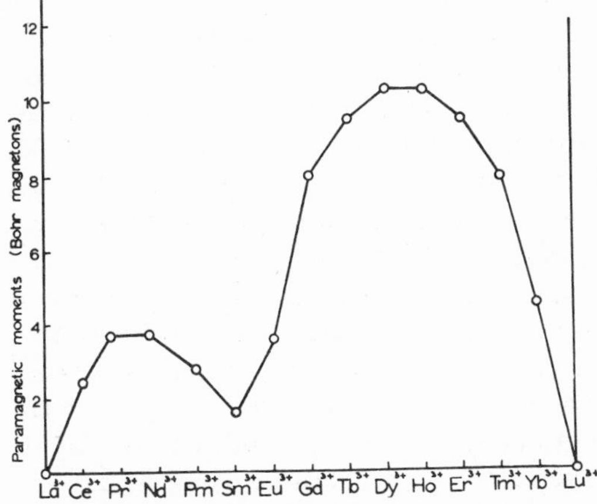

Fig. 30.3. Paramagnetic moments of terpositive lanthanide ions.

30.5. Compounds of the Lanthanides

30.5.1. The bipositive state

Sm, Eu, Tm and Yb form dihalides when the trihalides are reduced either with hydrogen or the metal:

$$2\ SmCl_3 + H_2 \rightarrow 2\ SmCl_2 + 2\ HCl$$
$$2\ TmI_3 + Tm \rightarrow 3\ TmI_2$$

In their structures the compounds usually resemble the corresponding barium compounds; for example SmF_2 and YbF_2 have the fluorite structure, as has BaF_2, and $SmCl_2$ is isostructural with $BaCl_2$.

Dihalides of Eu are made by reducing aqueous Eu^{3+} solutions with amalgamated zinc (E^0, Eu^{3+}/Eu^{2+} = —0.43 V) and Yb^{3+} ions can be reduced at a mercury cathode (E^0, Yb^{3+}/Yb^{2+} = —1.15 V). But Yb^{2+} solutions and Sm^{2+} solutions (E^0, Sm^{3+}/Sm^{2+} = —1.55 V) quickly reduce water, and the hydrates of Yb^{II} and Sm^{II} salts are oxidised by their own water of crystallisation. $EuCl_2 \cdot 2\ H_2O$ and the other Eu^{II} hydrates are stable to oxidation, however. Water-insoluble salts of Sm^{II} and Yb^{II}, the sulphates, carbonates and fluorides, resist oxidation even when moist

Sm, Eu and Yb have monoxides, obtained by reducing the Ln_2O_3 compounds with the metal in an atmosphere of argon at 1300—1600 K. SmO and YbO have the NaCl structure as have the corresponding sulphides EuS and YbS.

Aqueous Eu^{2+} solutions show intense, diffuse, absorption bands attributed to 4f—5d transitions.

30.5.2. The terpositive state

The 3+ ions of the lanthanides are found combined in crystalline compounds with a great range of anions. The anhydrous trichlorides are best made by heating the oxides in $COCl_2$ or CCl_4 vapour. They are high-melting solids, often deliquescent, very soluble in water and fairly soluble in alcohol. Their high conductivities in the molten state show them to be ionic in character. They form hydrates, usually with either 6 or 7 H_2O, e.g. $LaCl_3 \cdot 7\ H_2O$, $NdCl_3 \cdot 6\ H_2O$. The bromides and iodides are similar to the chlorides but the fluorides are much less soluble and can be made by precipitation reactions. Structures are usually rather complicated and the Ln^{3+} ions tend to have high co-ordination numbers. Thus the tysonite structure of LaF_3 which is common to many of the trifluorides and is one of the structurally simpler of the lanthanide halides consists of layers made of hexagonal nets of La^{3+} and F^- ions with F^- ions on each side of the layer, giving each La^{3+} five nearest F^- neighbours at 236 pm arranged in a trigonal bipyramid. But the layers on either side are so arranged that each La^{3+} has six more F^- ions at 270 pm arranged in a triangular prism, so the structure can be considered to be transitional between one in which the co-ordination number of the metal ion is 5 and one in which it is 11.

The lanthanide oxides, Ln_2O_3, are refractory materials which can be made by oxidising the metals or by heating the carbonates, hydroxides, oxalates, nitrates or sulphates. Unlike Al_2O_3 they dissolve in acids even after prolonged ignition. They are strongly exothermic compounds, for instance ΔH_f-(La_2O_3) = —1790 kJ mol^{-1}, ΔH_f(Sm_2O_3) = —1810 kJ mol^{-1}. Lanthanide(III) oxides are of two structural types: La_2O_3, Ce_2O_3, Pr_2O_3 and Nd_2O_3 are of the A-type (23.3.1.4), the rest are of the C-type.

The lanthanide(III) hydroxides are precipitated from solutions of Ln^{III} salts by aqueous NaOH, and most of them are insoluble in an excess of the reagent, the exceptions being $Yb(OH)_3$ and $Lu(OH)_3$ which can be converted to the compounds $Na_3Ln(OH)_6$ by heating with concentrated aqueous NaOH in an autoclave. The hydroxides are correctly formulated $Ln(OH)_3$; they are not hydrated oxides. Their solubility products are so low that they can be precipitated by ammonia even in the presence of NH_4Cl. There is a gradual reduction in basic character from $La(OH)_3$ to $Lu(OH)_3$, as expected from the decrease in size of the Ln^{3+} ions.

Normal carbonates of the lanthanides can be precipitated from solutions of Ln^{III} salts by $NaHCO_3$; in this behaviour the lanthanides resemble magnesium. Many of the carbonates form hydrates. They dissolve in alkali-carbonate solutions, and compounds such as $K_2CO_3 \cdot Ce_2(CO_3)_3 \cdot 3\,H_2O$ have been crystallised.

The nitrates are very soluble indeed but they form isomorphous series of double nitrates $3\,M(NO_3)_2 \cdot 2\,Ln(NO_3)_3 \cdot 24\,H_2O$, where M = Mg, Zn, Ni or Mn, in which solubilities are low for the lightest lanthanides and increase through the series.

The sulphates do not form alums. Small ions such as Al^{3+} (r = 50 pm) and Cr^{3+} (r = 64 pm) can be surrounded octahedrally by six H_2O molecules, but the lanthanide(III) ions (r = 85—106 pm) are too large to co-ordinate water molecules in this way. Nevertheless there are many double sulphates, $NH_4Ln(SO_4)_2 \cdot 4\,H_2O$ being the most common type.

Isomorphism is common in the hydrated salts of the terpositive lanthanides. Co-ordination numbers of the Ln^{3+} ions are frequently high, for instance in the $Ln_2(SO_4)_3 \cdot 9\,H_2O$ series the metal ions are surrounded by nine H_2O molecules arranged at the corners of a tricapped trigonal prism.

The solubilities of the lanthanide(III) compounds closely resemble those of the Group IIA metals:

Considerable solubility: $LnCl_3$, $LnBr_3$, LnI_3, $Ln(NO_3)_3$, $Ln(ClO_4)_3$, $Ln(BrO_3)_3$, $LnAc_3$.

Low solubility: Ln_2O_3, $Ln(OH)_3$, $Ln_2(Ox)_3$, $Ln_2(CO_3)_3$, $LnPO_4$, LnF_3.

Variable solubility: $Ln_2(SO_4)_3$.

Solubility in any series of salts is not simply related to ionic radii but the following generalisations can be made. The heavy lanthanides (Z = 63—71) and yttrium differ from the light lanthanides in having (a) oxalates which dissolve in an excess of $C_2O_4^{2-}$, (b) carbonates which dissolve in an excess of CO_3^{2-}, (c) basic nitrates which are more soluble in water.

Except for aquated ions, complexes of the terpositive lanthanides are rather few in number, and only those formed by strong chelating agents can be considered thermodynamically stable. In fact, the lanthanides resemble Ca and Sr rather than the 'd-block' metals in their complex chemistry. Lanthanide ions are effectively 'noble gas' ions because the f electrons are too well shielded to behave as d electrons do in the outer transition metals. Interaction between metal and ligand therefore has considerable ionic character, the metal ions behave as hard acids, and the most common co-ordinating atoms are oxygen and nitrogen. Typical chelate complexes are the neutral species $Ln(oxinate)_3 \cdot n\, H_2O$ and $Ln(\beta\text{-diketonate})_3 \cdot n\, H_2O$ and the anions $Ln(R{\cdot}CHOH{\cdot}COO)_6^{3-}$ and $Ln(edta)^-$. Co-ordination numbers often exceed 6; for example, in $Y(acac)_3 \cdot H_2O$ there are seven oxygen atoms arranged at the corners of a monocapped trigonal prism around the yttrium atom, and in $La(acac)_3 \cdot 2\, H_2O$ eight oxygens are arranged at the corners of a square antiprism.

For the EDTA complexes, pβ values range fairly smoothly from 15.5 for the La^{III} complex to 19.8 for that of Lu^{III}; for the nitrilotriacetic acid complexes the range is smaller, 10.4 to 12.5. Differences in the pβ values of the EDTA complexes form the basis for separation procedures (30.2.2).

30.5.3. The quadripositive state

The white oxide CeO_2 can be made by heating $Ce(OH)_3$ or the corresponding carbonate, oxalate or nitrate in air or oxygen. Above 918 K the solid has the fluorite structure which it retains even when the existence of anion vacancies reduces the composition as low as $CeO_{1.72}$. At room temperature the situation is more complicated; there are several two-phase regions in the same composition range. Thus at compositions between CeO_2 and $CeO_{1.8}$ there are two phases in equilibrium, an α-phase of almost stoichiometric CeO_2 and a β-phase $Ce_{32}O_{58}$ derived from the superlattice ordering of six O^{2-} vacancies from eight unit cells of CeO_2. It is interesting that re-oxidation of such a reduced phase causes a decrease in the lattice dimensions because the additional O^{2-} ions merely fill vacant anion sites whereas the change $Ce^{3+} \rightarrow Ce^{4+}$ creates smaller cations which also have greater polarising power.

Oxosalts of Pr^{III} can be converted to a black higher oxide by heating in air. There is a range of ordered phases defining a "homologous series" of oxides Pr_nO_{2n-2} with $n = 4, 7, 9, 10, 11, 12$ and ∞. Within each phase at ordinary temperature there is a limited range of composition because the ordering of defects soon ensures the emergence of the next homologue. The terbium—oxygen system is also complex; as normally prepared by the ignition of oxo-salts the brown higher oxide of terbium has a composition in the range $TbO_{1.71}$ to $TbO_{1.81}$.

Of the tetrafluorides, CeF_4 and TbF_4 can be made by the direct fluorination of the trifluorides. PrF_4 cannot be so made, but when PrF_3 is heated with NaF and F_2 at 700 K compounds such as $NaPrF_5$ and Na_2PrF_6 can be obtained which yield PrF_4 on treatment with HF. Although tetrafluorides of Nd and Dy have not been made, the compounds Cs_3NdF_7 and Cs_3DyF_7 can be prepared by the action of F_2 on CsCl and the respective lanthanide trichlorides at 600—800 K.

With the exception of Ce^{3+}, the Ln^{3+} ions resist oxidation in aqueous solution. The standard redox potential of the Ce^{4+}/Ce^{3+} couple varies greatly with the nature of the anion present.

TABLE 30.4

STANDARD REDOX POTENTIALS IN DIFFERENT MEDIA

Medium	E°, Ce^{4+}/Ce^{3+}
HCl	+1.28 V
HNO_3	+1.61 V
$HClO_4$	+1.70 V

The figures suggest that the simple Ce^{4+} ion does not exist in aqueous solution, and the variations in redox potential seem too great to be caused merely by differences in its hydration. Because of its considerable charge and moderate size (r = 101 pm) the ion has appreciable polarising power and may well form complexes with suitable anions. In accordance with this idea it is interesting that the soluble double nitrate formulated $Ce(NO_3)_4 \cdot 2\ NH_4NO_3$ has been shown to contain the ion $Ce(NO_3)_6{}^{2-}$ in which the $NO_3{}^-$ ions act as bidentate ligands and the co-ordinating oxygen atoms are arranged in an icosahedron around the cerium atom.

30.5.4. Organometallic compounds

Unlike 'd-block' transition metals the lanthanides form few π-complexes with olefins. The cyclopentadienides $Ln(C_5H_5)_3$ form the only important series of organometallic compounds. These are ionic, and have magnetic moments very similar to those of the corresponding lanthanide(III) sulphates. They are thermally stable, and sublime in vacuo at about 500 K, but they are easily oxidised by air and hydrolysed by water.

30.5.5. Chelate complexes

In spite of their high charge, lanthanide ions are too large to cause much polarisation and they behave as rather hard acids, forming their most stable complexes by co-ordination of small, strongly electronegative, donor atoms. There are a few fluorocomplexes, but much the most important complexes are those with chelating oxygen ligands. The value of anions such as the citrate ion and the ethylenediaminetetra-acetate ion in the separation of terpositive lanthanide ions has already been mentioned.

The preparation of β-diketonates by traditional methods always produces solvated species such as $[Ln(acac)_3] \cdot EtOH \cdot 3\ H_2O$ from which it is extremely difficult to remove the solvent molecules, especially the H_2O, without causing decomposition.

Both β-diketones and anions of the edta type give rise to anionic species as well as neutral complexes with lanthanide ions. Examples are $[Ln(\beta\text{-diketone})_4]^-$ ions and $Ln(edta)^-$ ions (here $edta^{4-}$; represents the ion $[(CH_2 \cdot N(CH_2CO_2)_2)_2]^{4-}$).

Other polydentate ligands which form moderately stable complexes with the lanthanides include oxine (8-hydroxyquinoline), in which both oxygen and nitrogen function as donor atoms, and nitrogen ligands such as orthophenanthroline and terpyridine. Examples of complexes formed by these ligands are $[Ln(terpyridine)_3](ClO_4)_3$ and $[Lnphen_3](SCN)_3$. Where the structures of these compounds have been studied the lanthanide atoms are found to have high co-ordination numbers.

30.6. Promethium

Discovery of the sequence of atomic numbers showed that there are 14 possible lanthanides. Attempts to find element 61, promethium, which lies between Nd and Sm, in natural occurrence have been unsuccessful because there is apparently no stable isotope.

Promethium-147 was separated by ion exchange from products of the slow-neutron fission of ^{235}U, which contain 2.6% of the isotope. It has a half life of 2.3 years and emits a soft β-radiation (0.23 MeV) but no γ-rays:

$$^{147}_{61}Pm \xrightarrow[2.3\ y]{\beta^-} {}^{147}_{62}Sm\ \text{(stable)}$$

The same nuclide results from an (n, γ) reaction on $^{146}_{60}Nd$ in the pile, followed by β^- emission from the ^{147}Nd formed:

$$\underset{\text{(stable)}}{^{146}_{60}Nd} \xrightarrow{n,\gamma} {}^{147}_{60}Nd \xrightarrow{\beta^-} {}^{147}_{61}Pm$$

Though its specific activity is fairly high, the low β-energy and the absence of γ-radiation make it fairly safe to handle, and chemical studies are possible. The chloride and nitrate have been made in mg amounts; their solutions are pink.

SCANDIUM AND YTTRIUM

These elements are respectively the first outer transition elements of Periods 4 and 5. Their outermost electrons have a similar pattern to the corresponding electrons of the lanthanides and their properties greatly resemble those of the lanthanides. For this reason these elements of Group IIIA are included in this chapter.

30.7. Scandium

Its $3d^14s^2$ structure gives element 21 properties similar to the lanthanides and to lanthanum ($5d^16s^2$) in particular. The covalent and ionic radii, 144 pm and 81 pm respectively, are however much smaller than those of the lanthanides. In consequence the Sc^{3+} ion has a greater polarising power and more readily forms complexes: for instance crystalline K_3ScF_6 can be obtained. The ionisation energies are not much larger than those of the lanthanides so far as they are known, and the metal itself is almost as reactive.

Scandium is present in some lanthanide minerals, but thortveitite, $ScSi_2O_7$, is the usual source. The metal is made by the electrolysis of a fused mixture of $ScCl_3$, KCl and LiCl on a zinc cathode followed by volatilisation of the Zn at low pressure from the Zn—Sc alloy so formed. The metal is dimorphous, with f.c.c. and h.c.p. forms. Its *m.p.* is rather high, ~1700 K.

Some important differences between scandium and the lanthanides are:

(i) Scandium oxide is a weaker base.

(ii) The chloride is more volatile.

(iii) The nitrate, $Sc(NO_3)_3 \cdot 4\ H_2O$, is more easily decomposed by heat.

(iv) The sulphate, $Sc_2(SO_4)_3 \cdot 5\ H_2O$, is very soluble in both cold and hot water.

(v) The chelate complexes with acetylacetone and oxine are more covalent in type than those of the lanthanides. Thus $Sc(acac)_3$ can be sublimed without decomposition; similar lanthanide acetylacetonates decompose at about 500 K. Scandium oxinate is extracted almost quantitatively from aqueous solution by carbon tetrachloride in a single operation, again unlike the lanthanide oxinates.

Recently Sc^0 and Y^0 compounds, $Sc(dipy)_3(THF)_3$ and $Y(dipy)_3(THF)_3$ have been obtained by reducing the trichlorides with lithium in tetrahydrofuran (THF) containing 2,2′-dipyridyl.

30.8. Yttrium

Element 39, with $4d^15s^2$ electron configuration, is also similar to the lanthanides. It occurs with the lanthanides in minerals; the best source is xenotime, YPO_4. Yttrium itself has properties approximately midway between those of Sc and La; its compounds resemble those of the heavy earths dysprosium and holmium, the ionic radius ($r_{Sc^{3+}} = 90$ pm) being similar.

Further Reading

T. Moeller, Chemistry of the lanthanides, Van Nostrand—Reinhold, Princeton, 1963.

N.E. Topp, The chemistry of the rare-earth elements, Elsevier, Amsterdam, 1965.

T. Moeller, D.F. Martin, L.C. Thompson, R. Ferrús, G.R. Feistel and W.J. Randall, The co-ordination chemistry of yttrium and the rare-earth metal ions, Chem. Rev., 65 (1965) 1.

M.D. Taylor, Preparation of anhydrous lanthanide halides, Chem. Rev., 62 (1962) 503.
K.W. Bagnall (Ed.), Lanthanides and actinides, Butterworth, London, 1972.
R.J. Callow, The rare earth industry, Pergamon, London, 1966.
L.B. Asprey and B.B. Cunningham, Unusual oxidation states of some actinide and lanthanide elements, Prog. Inorg. Chem., Vol. II, Interscience, New York, 1960.
T. Moeller, Periodicity and the lanthanides and actinides, J. Chem. Ed., 47 (1970) 417.
D.G. Karraker, Co-ordination of trivalent lanthanide ions, J. Chem. Ed., 47 (1970) 424.
O. Johnson, Role of f electrons in chemical bonding, J. Chem. Ed., 47 (1970) 431.
H. Gysling and M. Tsutsui, Organolanthanides and organoactinides, Adv. Organometallic Chem., 9 (1970) 361.
D.P. Schumacher and W.E. Wallace, Magnetic properties of some lanthanide nitrides, Inorg. Chem., 5 (1966) 1563.
G.R. Choppin, Structure and thermodynamics of lanthanide and actinide complexes in solution, Pure Appl. Chem., 27 (1971) 23.
T. Moeller, The Chemistry of the Lanthanides, Pergamon, Oxford, 1975.

The Actinides

31

31.1. Introduction

Until the properties of neptunium and plutonium became known (1941), actinium, thorium, protactinium and uranium were treated as the last members of Sub-groups IIIA, IVA, VA and VIA respectively. There were chemical reasons for this: uranium forms complexes indicating variable charge and a particularly stable +6 oxidation state which is typical of molybdenum and tungsten; and thorium nearly always occurs in a +4 oxidation state. Thus they seemed to fit into Groups VI and IV. The existence of actinium(III) salts and their isomorphism with those of lanthanum, suggested membership of Group III. Very little was known of the properties of protactinium.

However, with the production of neptunium by McMillan (1940) and of plutonium and further trans-uranic elements by Seaborg and others (1944 to present), evidence for a different classification has accumulated. The similarity of the elements actinium and its successors to lanthanum and its succes-

TABLE 31.1

THE ACTINIDES

Name	Symbol	Z	Configuration		
Actinium	Ac	89		$6d^1$	$7s^2$
Thorium	Th	90		$6d^2$	$7s^2$
Protactinium	Pa	91	$5f^2$	$6d^1$	$7s^2$
Uranium	U	92	$5f^3$	$6d^1$	$7s^2$
Neptunium	Np	93	$5f^4$	$6d^1$	$7s^2$
Plutonium	Pu	94	$5f^6$		$7s^2$
Americium	Am	95	$5f^7$		$7s^2$
Curium	Cm	96	$5f^7$	$6d^1$	$7s^2$
Berkelium	Bk	97	$5f^9$		$7s^2$
Californium	Cf	98	$5f^{10}$		$7s^2$
Einsteinium	Es	99	$5f^{11}$		$7s^2$
Fermium	Fm	100	$5f^{12}$		$7s^2$
Mendelevium	Md	101	$5f^{13}$		$7s^2$
Nobelium	No	102	$5f^{14}$		$7s^2$
Lawrencium	Lr	103	$5f^{14}$	$6d^1$	$7s^2$

sors became increasingly apparent. Actinium to lawrencium are indeed a closely related family and known as the *actinides* (cf. the lanthanides which begin thirty-two places earlier in the Periodic Table).

The actinides are characterised by the

(i) increasing stability of the 3+ ion with atomic number;
(ii) isomorphism of trichlorides, dioxides and many salts with corresponding lanthanide compounds;
(iii) decrease in ionic radii with atomic number, analogous to the lanthanide contraction;
(iv) character of the absorption spectra of the ions;
(v) magnetic moments of the ions.

All the actinides are believed to have the outer electron configuration $7s^2$ and to vary irregularly in the electron occupation of the 5f and 6d shells (Table 31.1).

31.2. Ionic Radii

TABLE 31.2

IONIC RADII OF THE ACTINIDES/pm

	Ac	Th	Pa	U	Np	Pu	Am	Cm
M^{3+}	111	—	—	103	101	100	99	98
M^{4+}	—	99	96	93	92	90	89	—

These ionic radii, derived from X-ray diffraction data, should be compared with those of the lanthanides (30.1). The size of an ion depends largely upon the quantum number of the outermost electrons and the effective nuclear charge. In the 3+ ions of these elements the outermost electrons are in a completed 6p shell; the effective nuclear charge rises with atomic number because the screening effect of extra electrons in the 5f level fails to compensate entirely for the increased nuclear charge. Thus there is a contraction in the size of the ions, similar to the lanthanide contraction, from actinium to mendelevium. The contraction is, however, greater from element to element in the actinides.

31.3. Occurrence and Separation of the Metals

Thorium, the most abundant actinide, comprises about 1.5×10^{-3}% of the lithosphere. The principal source is monazite (30.2) which contains up to 10% of thoria. To extract the metal, the ore is treated either with hot, 95% H_2SO_4 or hot, 45% NaOH solution. In one version of the acid extraction process the "monazite sulphate" is dissolved in water, separated from the insolubles and adjusted to pH 1 with ammonia. A precipitate is obtained containing thorium sulphate and thorium phosphate, together with the corresponding salts of cerium(IV) as the main impurities. The precipitate is dissolved in HNO_3, the thorium is then reprecipitated as oxalate and calcined to ThO_2. Further purification is effected by dissolving again in HNO_3 and extracting with tributyl phosphate. The $Ce(NO_3)_4$ which dissolves with the $Th(NO_3)_4$ can be removed from the organic phase by reducing to Ce^{III} and back-extracting into an aqueous layer.

When the monazite is digested with aqueous NaOH in the alkaline extraction process, thorium, uranium and the lanthanides are left as insoluble oxides which are then dissolved in HCl. The oxides of thorium and uranium are reprecipitated by adjusting the pH to 5.8. Further purification is achieved by dissolving in HNO_3 and extracting with tributyl phosphate.

The pure metal is difficult to obtain because it combines readily with hydrogen, oxygen, nitrogen and carbon. The principal method used for its manufacture is the reduction of ThO_2 with calcium in an enclosed vessel for one hour at 1200 K. The excess of Ca and the CaO are leached out with water, and the thorium powder is sintered to give billets of the metal. The powder can be purified if necessary by the iodine method which has a wide application to heavy metals. The crude powder is converted to ThI_4 by heating in iodine vapour. The volatile ThI_4 can be decomposed on a filament at 1400 K, the iodine which is liberated being recirculated to react with more of the crude metal. Thorium-232 can be converted to the fissionable nuclide uranium-233 by the sequence of processes:

$$^{232}_{90}Th + ^{1}_{0}n \rightarrow ^{233}_{90}Th \xrightarrow[22\ min]{\beta^-} ^{233}_{91}Pa \xrightarrow[27\ d]{\beta^-} ^{233}_{92}U\ (1.6 \times 10^5\ y)$$

A nuclear reactor used to initiate the production of fresh nuclear fuel in this way is known as a breeder reactor. This particular process is of interest because thorium is available in much greater quantities than uranium.

Uranium is extracted from pitchblende, which is essentially U_3O_8, but the available ores rarely contain more than 2 kg of uranium per Mg. The methods of extraction are numerous; they involve the leaching of the ore with either acid or alkali and the recovery of the metal from solution by methods including precipitation, solvent extraction and ion-exchange. The final stage is the extraction of the nitrate with tributyl phosphate. The metal is obtained by reduction of the tetrafluoride:

$$UO_2(NO_3)_2 \xrightarrow{heat} UO_2 \xrightarrow[in\ HF]{heat} UF_4 \xrightarrow[Mg\ in\ pressure\ vessel]{heat\ with} U + MgF_2\ slag$$

The fissionable isotope U-235 (0.72% of natural uranium) is separated from U-238 (99.2%) by methods including gaseous diffusion of UF_6, and is used as fuel in nuclear reactors (31.4.1).

The elements actinium and protactinium have been isolated in small amounts. The longest-lived isotope of actinium, Ac-227 (21.8 y), is made by irradiating Ra-226 in a nuclear reactor:

$$^{226}_{88}Ra(n,\gamma)^{227}_{88}Ra \xrightarrow[41\ min]{\beta^-} ^{227}_{89}Ac$$

and the actinium is separated by solvent extraction. Protactinium-233 has been obtained in 100-mg amounts by neutron-irradiation of Th-232:

$$^{232}_{90}Th(n,\gamma)^{233}_{90}Th \xrightarrow[22\ min]{\beta^-} {}^{233}_{91}Pa\ (27.4\ d)$$

The longer-lived Pa-231 (3.3×10^4 y) has been separated from the residues of the processes used for refining uranium.

The element plutonium occurs in pitchblende to the extent of about 10^{-11} of the uranium present. The quantity, though extremely small, is nevertheless too high to be accounted for by neutrons from the ^{235}U present acting on ^{238}U. The extra neutrons required may result from (α, n) reactions on the lighter elements present.

The production of plutonium and its separation from uranium are described below. The metal is obtained (cf. uranium) by the reduction of PuF_4 with calcium.

31.4. Transuranic Elements from the Nuclear Reactor

31.4.1. The nuclear reactor

An air-cooled graphite pile consists essentially of a stack of blocks of very pure graphite through which run many parallel channels arranged in a pattern calculated to make the most effective use of the available neutrons. In the channels are placed metallic uranium slugs sealed in magnox or steel cylinders and through the channels is blown cooling air. To control the neutron flux, rods of cadmium or other materials which absorb neutrons are automatically inserted into or withdrawn from the pile, and thus the power output is governed. A thick shield of concrete surrounds the pile.

Research reactors have openings through which materials for irradiation may be introduced and placed in a required neutron flux. Through others the graphite protrudes to form a 'thermal column' which makes the slower, 'thermal' neutrons available for experimental use. Both types of opening in the concrete are suitably shielded with cadmium and lead.

31.4.2. Plutonium

The element is produced almost entirely from uranium in the nuclear reactor.

The sequence of nuclear changes is:

1. A single neutron is absorbed by a ^{235}U nucleus and causes instability leading to (i) fission fragments which lose most of their kinetic energy in the uranium metal itself, (ii) γ-rays which are absorbed mainly by the structure and shielding, (iii) fast neutrons with energies of 0.5 to 1 MeV.

2. The fast neutrons lose energy by every collison with a carbon nucleus; they are thus continuously retarded and deflected back as slower neutrons into the uranium rods.

3. As the neutrons slow down their chance of capture increases. They are captured (i) by the structural materials of the pile and lost to the process; (ii) by the ^{238}U, through resonance absorption, leading to the production of ^{239}Pu; and (iii) by the ^{235}U, leading to fissions which generate about 2.5 fresh neutrons for every neutron captured. The geometry of the pile ensures that one of the fresh neutrons is captured by another ^{235}U, thus maintaining a steady rate of fission of these nuclei and a steady neutron flux.

The sequence of nuclear reactions leading to the formation of $^{239}_{94}Pu$ is:

$$^{238}_{92}U + ^{1}_{0}n \rightarrow ^{239}_{92}U + \gamma$$

$$^{239}_{92}U \xrightarrow[24\ min]{} ^{239}_{93}Np + \beta^-$$

$$^{239}_{93}Np \xrightarrow[2.3\ days]{} ^{239}_{94}Pu + \beta^-$$

The nuclide $^{239}_{94}Pu$ is an α-emitter with a half-life of 2.4×10^4 years.

Accumulation in the uranium of fission products, many of which absorb neutrons strongly, would eventually reduce the neutron flux below the intensity required to maintain the nuclear chain reaction. Accordingly the uranium is removed after a certain period and processed chemically, the plutonium and fission products such as caesium-137 and strontium-90 being extracted.

31.4.3. The large scale separation of plutonium

One method for separating Pu from reactor fuel makes use of a 30% solution in kerosene of the viscous liquid tributyl phosphate (TBP). This compound is decomposed only very slightly by the action of α-particles. The uranium slugs from the reactor, containing fission products and plutonium, are dissolved in nitric acid. The plutonium is first reduced to Pu^{4+} by the addition of nitrite to the aqueous solution, which is then agitated with the TBP in kerosene.

The fission products remain in the aqueous layer while the Pu^{4+} and $UO_2{}^{2+}$ dissolve in the organic phase, which is separated and washed with aqueous SO_2 or hydroxylamine. The Pu^{4+} is then reduced to Pu^{3+}, which enters the aqueous phase, but the $UO_2{}^{2+}$ remains unchanged in the organic layer.

In the lanthanum fluoride cycle, another method of Pu—U separation, a solution of Pu^{4+}, $UO_2{}^{2+}$ and fission products in aqueous HNO_3 is extracted first with hexone (isobutyl methyl ketone). The $UO_2{}^{2+}$ enters the organic phase. The Pu^{4+} in the aqueous phase is then reduced to Pu^{3+} which is coprecipitated on LaF_3 by the addition of HF and LaF_3. The mixed fluoride is boiled with aqueous NaOH to convert it to hydroxides which are separated and dissolved in HNO_3. Oxidation of the Pu^{3+} to $PuO_2{}^{2+}$ with a strong oxidising agent such as persulphate is followed by addition of HF which reprecipitates LaF_3 but leaves the plutonium(VI) in solution. The Pu^{VI} is then reduced to Pu^{IV} and the cycle is repeated. Decreasing amounts of the LaF_3 carrier are needed as the plutonium concentration increases.

31.4.4. Neptunium and other transuranics

Some neptunium, the long-life $^{237}_{93}Np$, is also formed in the reactor:

$$^{238}_{92}U + ^{1}_{0}n \rightarrow ^{237}_{92}U + 2^{1}_{0}n$$

$$^{237}_{92}U \xrightarrow[6.8\ days]{} ^{237}_{93}Np + \beta^-$$

This, too, is an α-emitter, with a half-life of 2.25×10^6 years and, unlike the highly active β-emitter $^{239}_{93}Np$, with a half-life of only 2.3 days, is particularly suitable for chemical work. Curium and americium also appear as the result of (n, γ) reactions involving plutonium.

$$^{239}_{94}Pu + ^{1}_{0}n \rightarrow ^{240}_{94}Pu\ (\alpha,\ 6000\ years) + \gamma$$

$$^{240}_{94}Pu + ^{1}_{0}n \rightarrow ^{241}_{94}Pu + \gamma$$

$$^{241}_{94}Pu \rightarrow ^{241}_{95}Am\ (\alpha,\ 500\ years) + \beta^-$$

$$^{241}_{95}Am + ^{1}_{0}n \rightarrow ^{242}_{95}Am + \gamma$$

$$^{242}_{95}Am \rightarrow ^{242}_{96}Cm\ (\beta,\ 162\ days) + \beta^-$$

Plutonium-241 captures further neutrons, and the product from the reactor is a mixture of several isotopes of the element. Pile-produced neptunium is, however, almost entirely neptunium-237.

The separation of the actinides up to americium can be achieved by the use of appropriate oxidation—reduction cycles coupled with solvent extraction. For example, Pu^{3+} in aqueous nitric acid can be oxidised to $PuO_2{}^{2+}$ by an oxidising couple of $E^0 > 1.02$ V and thereby extracted into hexane, but Am^{3+} is unchanged and remains in the aqueous layer.

31.5. Transamericium Actinides

Curium-242 was first made by bombarding Pu-239 with α-particles accelerated in a cyclotron to 32 MeV:

$$^{239}_{94}Pu + ^{4}_{2}He \rightarrow ^{242}_{96}Cm + ^{1}_{0}n$$

Once it became possible to isolate americium and curium in milligram quantities, high-energy α-particles could also be used for the production of berkelium (1949) and californium (1950):

$^{241}_{95}Am + ^{4}_{2}He \rightarrow ^{243}_{97}Bk + 2^{1}_{0}n$ (K-capture, 4.5 h)

$^{242}_{96}Cm + ^{4}_{2}He \rightarrow ^{245}_{98}Cf + ^{1}_{0}n$ (K-capture, 44 min)

Later, heavy ions such as C^{6+}, accelerated in cyclotrons, were used as bombarding particles:

$^{246}_{96}Cm + ^{12}_{6}C \rightarrow ^{254}_{102}No + 4^{1}_{0}n$ (α, 3 s)

$^{250}_{98}Cf + ^{11}_{5}B \rightarrow ^{257}_{103}Lr + 4^{1}_{0}n$ (α, 8 s)

Since oxidation above the +3 state becomes increasingly difficult in actinides beyond americium, separation by oxidation—reduction cycles were not possible. Instead, an ion-exchange technique was used, similar in principle to that used in the lanthanide separations (30.2.2). A cationic resin, kept at 360 K by boiling trichloroethylene, formed a column on which was poured 0.1 ml of a solution of transuranic elements in 0.05 *M* HCl. Elution with 0.4 *M* ammonium citrate, or ammonium α-hydroxybutyrate, gave peaks of activity; the order of the removal of the respective 3+ ions was the reverse of their atomic number (Fig. 31.1). In this way, einsteinium, element 99 (1952), fermium, element 100 (1954) and mendelevium, element 101 (1955) were separated by Seaborg and his collaborators. The nuclide $^{256}_{101}Md$, which undergoes spontaneous fission, has a half-life of 3.5 hours. That this process could be studied so effectively on a sample consisting of 17 atoms gives an idea of the powerful experimental methods now available.

31.6. The Metals

The actinides are base metals and have to be made by methods such as the reduction of their oxides or fluorides with calcium or barium. Thus thorium is usually obtained by reducing ThO_2 with Ca in a steel vessel, and plutonium by reducing PuF_4 with Ca.

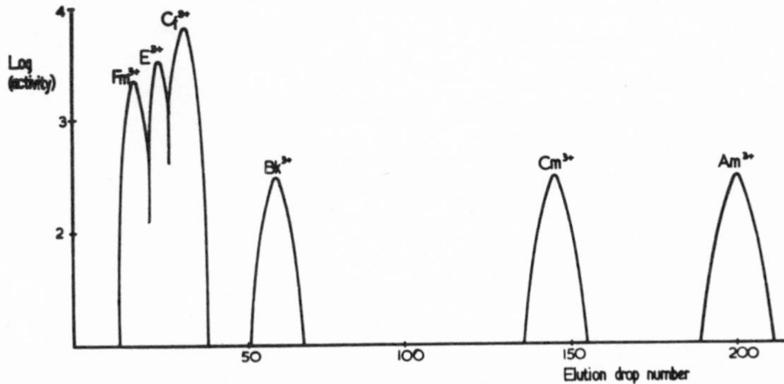

Fig. 31.1. Order of elution of trans-plutonium terpositive ions.

TABLE 31.3

PHYSICAL PROPERTIES OF SOME ACTINIDES

Metal	*M.p.*/K	Phase	Symmetry	*Density*/g cm^{-3}
Ac	1320		c.c.p.	10.0
Th	2020	α (<1700 K)	c.c.p.	11.72
		β (1700—*m.p.*)	b.c.c.	
Pa	2150		tetragonal	15.37
U	1405	α (<941 K)	orthorhombic	19.04
		β (941—1047 K)	tetragonal	18.11
		γ (1047—*m.p.*)	b.c.c.	18.06
Np	910	α (<553 K)	orthorhombic	20.45
		β (553—850 K)	tetragonal	19.36
		γ (850—*m.p.*)	cubic	18.00
Pu	913	α (<395 K)	monoclinic	19.73
		β (395—475 K)	body-centred monoclinic	17.77
		γ (475—590 K)	orthorhombic	17.19
		δ (590—726 K)	c.c.p.	15.92
		δ' (726—750 K)	tetragonal	15.99
		ϵ (750—*m.p.*)	b.c.c.	16.48
Am	1269	α (<600 K)	hexagonal	13.67

The metals are silvery white. Those near uranium have very high densities, but americium resembles the corresponding lanthanide, europium, in being less dense than its neighbours. Polymorphism is common (Table 31.3). Plutonium, for example, assumes six different crystalline forms between room temperature and its melting point, and there is not one of them for which the temperature coefficients of electric resistance and of expansion are both positive; it is indeed a most unusual metal structurally.

The metals are chemically reactive. The redox potentials for the M^{3+}/M couples, where they are known, are about −1.9 V, the metals tarnish in air, and, when finely powdered, some of them (e.g. Th and Pu) ignite spontaneously. Thorium dissolves slowly and uranium more quickly in aqueous HCl and H_2SO_4; both metals are rendered passive by HNO_3. The reactions of

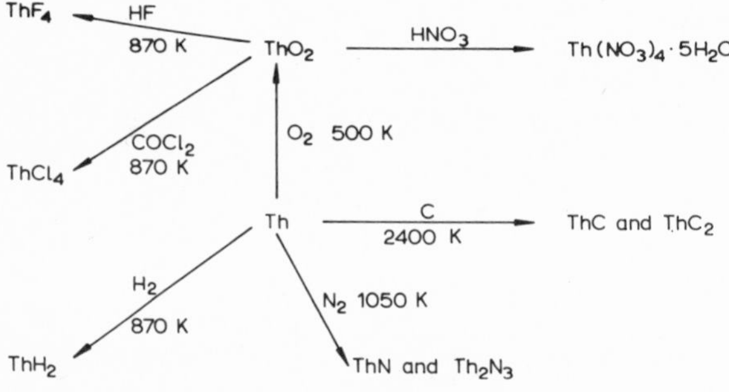

Fig. 31.2. Reactions of thorium.

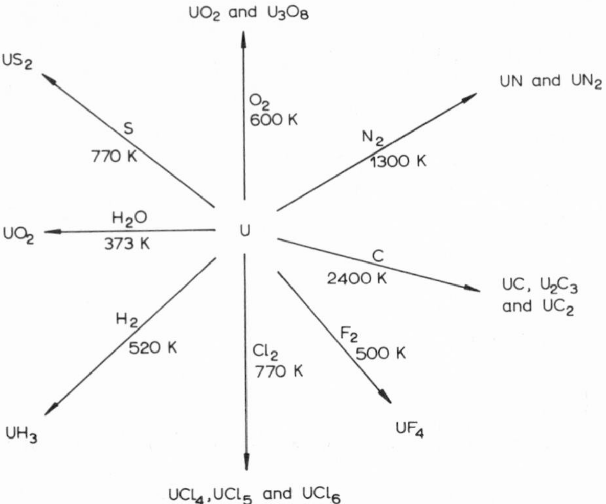

Fig. 31.3. Reactions of uranium.

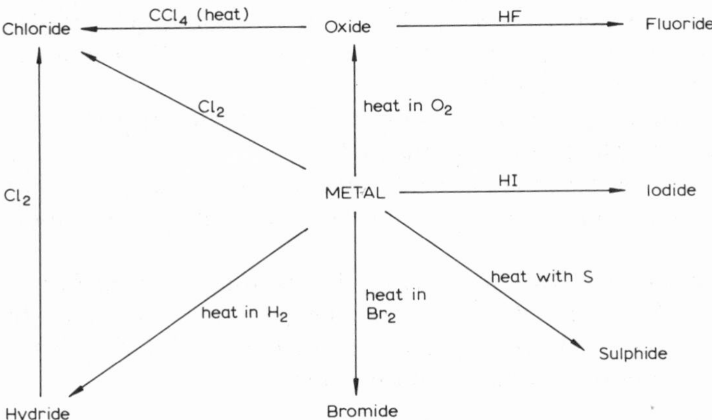

Fig. 31.4. General methods for preparing compounds of the actinides.

thorium and uranium with other elements are outlined in Figs. 31.2 and 31.3, and some general methods of preparing compounds of the other actinides are shown in Fig. 31.4.

31.7. Oxidation States

Comparison of Table 31.4 with Table 30.2, which gives the charge numbers of the lanthanides, indicates that high oxidation numbers are achieved much more easily in the actinide series. In americium and the elements

TABLE 31.4

OXIDATION STATES OF THE ACTINIDE ELEMENTS *

Ac	Th	Pa	U	Np	Pu	Am	Cm	Bk	Cf
+3	(+3)		+3	+3	+3	+3	+3	+3	+3
	+4	+4	+4	+4	+4	+4	+4	+4	
		+5	+5	+5	+5	+5			
			+6	+6	+6	+6			
				+7	+7				

* Bold print indicates principal oxidation state.

which follow it, however, the dominant oxidation state is +3 as in the lanthanide series.

The use of e.s.r. spectroscopy provides evidence that Am^{2+} ions can replace Ca^{2+} ions in the CaF_2 lattice, and the +2 state is also reported as Cf^{2+}, Es^{2+}, Fm^{2+} and Md^{2+} in aqueous solution.

The +3 state is common. Trihalides are known for actinium and for uranium and the elements which follow it. There are oxides Ac_2O_3, Pu_2O_3, Am_2O_3 and Cm_2O_3. The terpositive ions of Cm, Bk, Cf, Es and Fm cannot be oxidised or reduced in aqueous solution.

The +4 state is much more common than in the lanthanides. It is the principal oxidation state for Th, and the hydrated Pa^{4+}, U^{4+} and Pu^{4+} ions can exist in solution. Dioxides from ThO_2 to BkO_2 have the fluorite lattice. Tetrafluorides are known up to CmF_4 and tetrachlorides and bromides up to those of neptunium.

The +5 state is the most important one for Pa. It also occurs in fluoroanions such as MF_6^-, MF_7^{2-} and MF_8^{3-} for the elements from Pa to Pu. The cations UO_2^+, NpO_2^+, PuO_2^+ and AmO_2^+ can exist in aqueous solution. Their stabilities towards disproportionation are in the order

$$AmO_2^+ \sim NpO_2^+ > PuO_2^+ > UO_2^+ \qquad \text{(Fig. 31.5)}$$

TABLE 31.5

ION TYPES AND COLOURS IN AQUEOUS SOLUTION

	M^{3+}	M^{4+}	MO_2^+	MO_2^{2+}
Ac	Colourless	—	—	—
Th	—	Colourless	—	—
Pa	—	Colourless	Colourless	—
U	Red	Green	Colour unknown	Yellow
Np	Purple-blue	Yellow-green	Green	Pink
Pu	Violet	Orange-brown	Red-purple	Orange
Am	Pink	Pink	Yellow	Brown
Cm	Colourless	Unknown	—	—

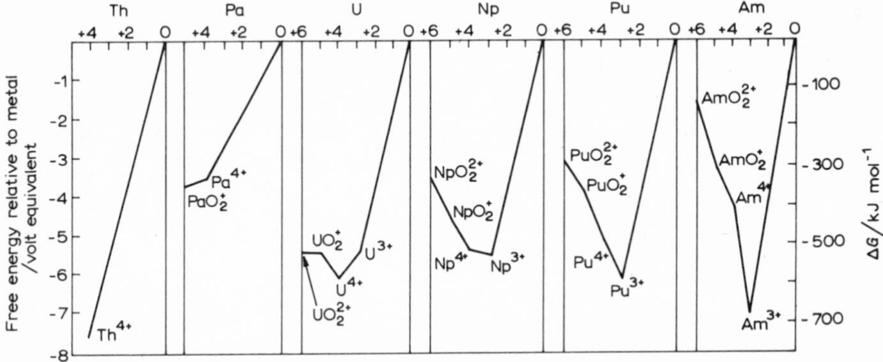

Fig. 31.5. Stabilities of oxidation states, elements 90—95. Free energies relative to the metal in aqueous solution at pH = 0.

The +6 state is represented by the hexafluorides of U, Np and Pu, and the ions MO_2^{2+} for the same three elements and americium. The UO_2^{2+} ion is very weakly oxidising but its reduction polarographically gives (a) a first stage at a half-wave potential which is little affected by pH, presumably representing a simple electron transfer:

$$UO_2^{2+} + e \rightarrow UO_2^+$$

and (b) a second wave, with an inflexion at half its height, indicating a stepwise reduction

$$UO_2^+ + 4\,H^+ + e \rightarrow U^{4+} + 2\,H_2O$$
$$U^{4+} + e \rightarrow U^{3+}$$

In support of this view the reduction of U^{VI} at a mercury cathode is reversible, whereas the reduction of U^V is irreversible and heavily dependent on pH. The behaviour of PuO_2^{2+} on polarographic reduction is similar to that of UO_2^{2+}. The couples Pu^{VI}/Pu^V, Pu^V/Pu^{IV} and Pu^{IV}/Pu^{III} have similar $E°$ values, thus the four oxidation states from +3 to +6 can exist together in measurable amounts in aqueous solution.

Recently, compounds of Np^{VII} and Pu^{VII} have been prepared. Thus when PuO_2 and Li_2O are heated together in oxygen at 700 K the green compound Li_5PuO_6 is obtained.

31.8. Halides

Actinium occurs only in the +3 state and thorium only with certainty in the +4 state in their halides. The white anhydrous ThX_4 compounds can be prepared by reactions such as:

$$ThO_2 + CCl_4 \xrightarrow{900\ K} ThCl_4 + CO_2$$

The hydrated tetrafluoride can be precipitated from Th^{4+} solutions with HF. It is insoluble in dilute acids and can be converted to the involatile anhydrous ThF_4 by heating in HF gas. The other three tetrahalides sublime without decomposition, dissolve in dilute acids, and hydrolyse in water. All four form adducts with Lewis bases. A golden diamagnetic compound, empirically ThI_2, is obtained by heating ThI_4 with Th. It has a layer lattice and is not a true Th^{II} compound.

TABLE 31.6

HALIDES * OF SOME ACTINIDES

	Ac	Th	Pa	U	Np	Pu	Am	Cm
III	AcF_3 $AcCl_3$ $AcBr_3$			UX_3	NpX_3	PuX_3	AmX_3	CmF_3
IV		ThX_4	PaF_4 $PaCl_4$	UX_4	NpF_4 $NpCl_4$ $NpBr_4$	PuF_4	AmF_4	CmF_4
V			PaX_5	UF_5 U_2Cl_{10}				
VI				UF_6 UCl_6	NpF_6	PuF_6		

* X = F, Cl, Br and I.

The pentahalides of protactinium have all been made. $PaCl_5$ is obtained by the action of Cl_2 and CCl_4 on Pa_2O_5 at 650 K. It can be reduced by hydrogen at 1000 K to the pale-green $PaCl_4$. The pentabromide and pentaiodide can be made by heating hydrated Pa_2O_5 with the corresponding aluminium halide. Brown, anhydrous PaF_4 has been made by the action of H_2 and HF at 650 K on hydrated Pa_2O_5. It can be oxidised to white PaF_5 by F_2 at 1000 K.

The halides of uranium have been extensively studied. Green, anhydrous UF_4 is made by heating UO_2 with a freon. It can be reduced to UF_3 by heating with aluminium. These involatile compounds can be converted to the white volatile UF_6 by treating them with F_2 at about 650 K. Uranium hexafluoride (*m.p.* 337 K) is the only convenient uranium compound which is sufficiently volatile to be used for the separation of U-235 from U-238 by gaseous diffusion, and it has been made on a large scale for this purpose. The molecule is octahedral and its dipole moment is zero. Pale-blue UF_5, and the black compounds U_2F_9 and U_4F_{14}, have been made by the action of heat on mixtures of UF_4 and UF_6.

The most important chloride of uranium is UCl_4 which can be made by refluxing UO_3 with certain chloro-olefins. The green solid is soluble in water and in polar organic solvents. Red UCl_3 can be made only in anhydrous conditions; one method is to treat uranium hydride with HCl.

The halides of neptunium, plutonium and americium which have been made are isostructural with the corresponding halides of uranium. Preparative methods are usually similar, too. In accord with what has been said about the stabilities of oxidation states in the actinide series, the orange NpF_6 and particularly the dark-brown PuF_6 are less thermally stable than UF_6.

The uranyl halides UO_2X_2 are of great thermal stability. They are salts of the UO_2^{2+} ion and not covalent molecules like SO_2Cl_2 and CrO_2Cl_2. There are corresponding neptunium and plutonium compounds.

31.9. The Oxides

White Ac_2O_3, made by heating the oxalate which is precipitated by addition of oxalate ions to Ac^{3+} solutions, has the A-type lanthanide oxide structure.

The only oxide of thorium, ThO_2, is a white solid with the fluorite structure, made by ignition of the hydrated oxide precipitated from Th^{4+} solution by OH^- ions.

The white oxide Pa_2O_5 is the principal compound of protactinium. It can be precipitated from Pa^V solutions as a hydrate which is then dehydrated at 750 K. The compound loses oxygen on heating in a vacuum and gives rise to non-stoichiometric phases; the limiting composition of the product appears to be PaO_2, a black solid which is also obtained by reducing Pa_2O_5 with hydrogen at high temperature.

The oxygen chemistry of uranium is complicated. A trioxide, UO_3, is obtained by decomposing $UO_2(NO_3)_2$ at 600 K. It has an orange α-form which has a layer structure linked through U—O—U bonds and there is also a red form, UO_3-II, which is usually somewhat oxygen-deficient, in which the arrangement of atoms is like that in ReO_3. UO_3 decomposes at 1000 K to give the black U_3O_8, which can be reduced with carbon monoxide to the dark-brown dioxide UO_2 which has the fluorite structure. Deviations from stoichiometry are often large, as might be expected; at high temperatures UO_2 can be oxidised to U_4O_9, without a phase change, by the incorporation of additional oxide ions in the lattice. A hydrated tetroxide has been made by precipitation from UO_2^{2+} solutions with H_2O_2. It is best considered as a U^{VI} compound containing O_2^{2-} ions and UO_2^{2+} ions.

For the elements from Np to Cm, moderate heating of the nitrate yields the dioxide, with CaF_2 structure. An oxide Np_3O_8 has been made which is isostructural with U_3O_8. Reduction of AmO_2 with hydrogen gives the pink, dimorphic Am_2O_3, and curium also has a sesquioxide, white Cm_2O_3. As in the halides the oxidation state +3 becomes increasingly important beyond plutonium.

31.10. Other Compounds

The sulphides are made by dry methods and are all compounds of lower charge number.

The disulphides of thorium and uranium are essentially covalent compounds; but materials of the formula M_2S_3 appear to be semi-metallic, except for Pu_2S_3 which is covalent. This again emphasises the increasing stability of compounds with the charge number +3 as the atomic number increases. Some of the sulphides, notably US, can be used as refractories.

TABLE 31.7

SULPHUR COMPOUNDS OF SOME ACTINIDES

Ac	Th	U	Np	Pu
	ThS	US		
Ac_2S_3	Th_2S_3	U_2S_3	Np_2S_3	Pu_2S_3
		UOS	NpOS	
	ThS_2	US_2		

Plutonium, like uranium, forms both a mononitride and a monocarbide; all have the sodium chloride structure:

$$PuCl_3 + NH_3 \xrightarrow[\text{temp.}]{\text{high}} PuN + 3\ HCl$$

The hydrolysis of uranium monocarbide, UC, yields hydrated UO_2 and a gas containing 85% CH_4, 11% H_2 and some C_2 and C_3 hydrocarbons.

Uranium also forms a dinitride, UN_2, with a fluorite structure.

Uranium usually dissolves in acids to give U^{IV} salts. The sulphate crystallises as the tetrahydrate which heating converts to the hemihydrate, but further loss of water leads to decomposition. Plutonium sulphate is similar. Double sulphates of formula $M^I_4M^{IV}(SO_4)_4$ (where $M^I = NH_4$, K or Rb and $M^{IV} = Th^{IV}$, U^{IV} or Pu^{IV}) have been made. Migration experiments indicate that these are true complexes.

Plutonium forms a nitrate of Pu^{IV} but the only nitrates of uranium are those of U^V and U^{VI}, namely $UO_2(NO_3)$ and $UO_2(NO_3)_2$.

31.11. Structural Relations in the Actinide Series

Isomorphism is common among compounds of the actinides. The dioxides ThO_2, UO_2, NpO_2, PaO_2 and AmO_2 all have the fluorite lattice. The trihalides $AcCl_3$, UCl_3, $NpCl_3$, $PaCl_3$, $AmCl_3$, $AcBr_3$, UBr_3 and α-$NpBr_3$ have a regular lattice incorporating 9-co-ordinate actinide atoms; this is the structure of $LaCl_3$ and many other lanthanide trichlorides and tribromides. The tysonite structure of LaF_3 (30.5.2) is shared by AcF_3, UF_3, NpF_3, PuF_3 and AmF_3. Isomorphism is also exhibited in the complex halides; thus Th, U and Pu form the isomorphous series KMF_5. Plutonium has a number of complex nitrates, $M^I_2Pu(NO_3)_6$, of which the ammonium salt $(NH_4)_2Pu(NO_3)_6$ is isomorphous not only with $(NH_4)_2Th(NO_3)_6$ but with $(NH_4)_2Ce(NO_3)_6$ also. In these $M(NO_3)_6^{2-}$ ions the NO_3 groups are arranged as bidentate ligands around the central metal atom, with the twelve co-ordinating oxygen atoms at the corners of a slightly distorted icosahedron.

High co-ordination numbers are common, as they are in the lanthanide series. Eight co-ordination, for example, is exemplified by cubic symmetry in $[U(NCS)_8]^{4-}$, dodecahedral in $[ThOx_4]^{4-}$, bicapped trigonal prismatic in $PuBr_3$ and square antiprismatic in $U(acac)_4$.

31.12. Absorption Spectra

Further evidence for electronic structures of the actinides is furnished by the absorption spectra of their compounds. The absorption bands are sharply defined and may be explained, as they are for the lanthanides, by characteristic atomic transitions involving 5f electrons rather than those electrons engaged more actively in bond formation. Extinction coefficients in the actinide series are higher than in the lanthanides, commonly by a factor of ten. It is noteworthy that the spectra of the 3+ ions show particular resemblances to their lanthanide analogues, for example U^{3+} to Nd^{3+}, Pu^{3+} to Sm^{3+}, and particularly Am^{3+} to Eu^{3+}.

31.13. Magnetic Properties

The paramagnetism of appropriate ions of the transuranic elements and that of the corresponding lanthanide ions discloses a remarkable parallelism (Fig. 31.6).

The moments of the lanthanide ions agree closely with theoretical prediction but those of the transuranic ions are somewhat lower than expected, possibly because the 5f electrons of the transuranic ions are less effectively screened from the crystal field which quenches the orbital contribution than

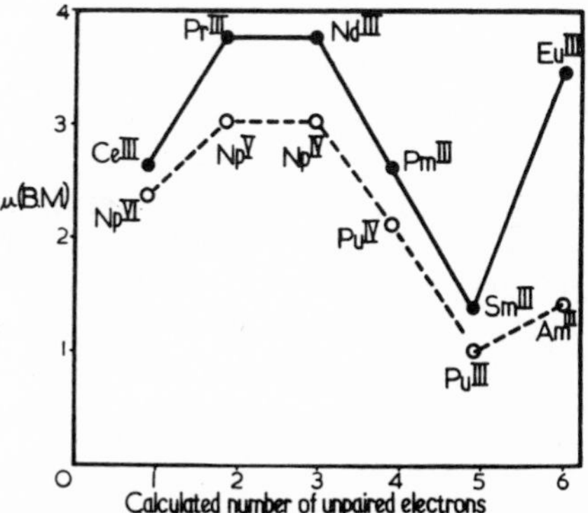

Fig. 31.6. Paramagnetism of transuranic and lanthanide ions.

are the 4f electrons of the lanthanide ions. The variation of paramagnetic moment with the number of unpaired electrons differs radically from the variation which occurs in elements of the first and second transition series where the maximum is five.

The magnetic properties lend credence to a 5f structure: those of UCl_3 indicate $5f^3$ for U^{III} (cf. Nd^{III}) and those of PuO_2^{2+}, with two unpaired electrons, indicate $5f^2$ for Pu^{VI}. Some moments of U^{IV} compounds appear to be so close to spin-only values however, as to suggest an ion with a $6d^2$ rather than a $5f^2$ structure (cf. ligand field theory). But more evidence is required before the absence of 5f electrons in U^{IV} can be safely accepted. The ions Pu^{3+} and Am^{3+} resemble Sm^{3+} and Eu^{3+}, the corresponding lanthanides; the lower excited states lie so close to the ground states that they contribute to the magnetic moment at ordinary temperatures.

31.14. Complexes

Most actinide halides form adducts with alkali-metal halides; thus thorium forms $MThCl_5$, M_2ThCl_6 and M_3ThCl_7. Many actinide tetrafluorides form adducts with NH_4F; the $NH_4F : MF_4$ ratios can be 4 : 1, 2 : 1, 7 : 6, 1 : 1 and 1 : 3. Structural evidence for complex ions is however lacking. But the compound $UCl_3 \cdot PCl_5$, made by treating UO_3 with PCl_5 at room temperature, ionises in $POCl_3$ to PCl_4^+ and UCl_6^-.

As in the lanthanide series, complexes are not numerous and nearly all contain chelating ligands. Tetrakis(acetylacetonato)thorium, $Th(acac)_4$, has α and β forms. In the α form the Th atom is surrounded by 8 oxygens arranged in an Archimedian antiprism. Uranium forms $U(acac)_4$ isomorphous with α-$Th(acac)_4$; $Pu(acac)_4$ is isomorphous with β-$Th(acac)_4$.

Uranyl compounds also react with β-diketones to give $UO_2(RCOCH{=}COR^1)_2$ complexes, often solvated. The bis(acetylacetonato) complex is dimeric in benzene. Thenoyltrifluoroacetone (TTA):

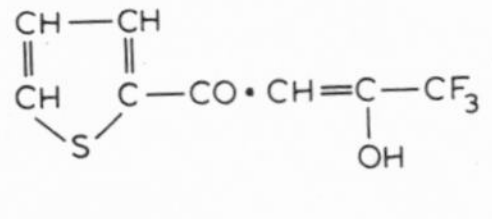

```
CH—CH
||   ||
CH   C—CO·CH=C—CF3
 \  /       |
  S         OH
```

in benzene can be used for the solvent-extraction of UO_2^{2+}, Np^{4+} and Pu^{4+}. A crystalline $UO_2(TTA)_2 \cdot 2\,H_2O$ has been isolated.

Nitrogen-ligands are also of interest. $ThCl_4$ and $ThBr_4$ form monopyridino complexes. Of the chelate complexes those from EDTA and oxine are well known. Th^{IV} and U^{IV} EDTA complexes $M(C_{10}H_{12}N_2O_8) \cdot 2\,H_2O$ can be crystallised. UO_2^{2+} forms rather unstable complexes with EDTA and nitrilotriacetic acid:

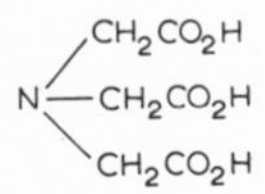

```
   CH2CO2H
  /
N—CH2CO2H
  \
   CH2CO2H
```

31.15. Organometallic Compounds

The dark-green compound (π-C_5H_5)$_3$UCl, made by the action of TlC_5H_5 on UCl_4, has a structure in which the essentially ionic U—Cl bond and C_5H_5—U bonds of sandwich type are arranged so that the C_5H_5 rings and the Cl atom form a distorted tetrahedron around the uranium atom. A bright-red (π-C_5H_5)$_4$U has also been made for which the infrared spectrum indicates the presence of symmetrical cyclopentadienyl ligands. Other cyclopentadienyl compounds include the colourless Th(C_5H_5)$_4$, for which the proton n.m.r. spectrum in $CDCl_3$ consists of a single, sharp absorption at τ 3.6, the M(C_5H_5)$_4$ compounds for M = Pa and Np, and the M(C_5H_5)$_3$ compounds for M = Am and Cm.

An interesting compound 'uranocene' (π-C_8H_8)$_2$U, made by the action of $C_8H_8^{2-}$ on UCl_4 in THF at 240 K, appears to have a ferrocene-like structure with planar eight-membered rings on either side of the U atom. There are similar compounds of thorium and neptunium.

Further Reading

J.J. Katz and I. Sheft, Halides of the actinide elements, Adv. Inorg. and Radiochem., 2 (1959) 195.

D. Brown, Some recent preparative chemistry of protactinium, Adv. Inorg. and Radiochem., 12 (1969) 1.

J.M. Cleveland, The chemistry of plutonium, Gordon and Breach, New York, 1970.

E.H.P. Cordfunke, The chemistry of uranium, Elsevier, Amsterdam, 1969.

G.T. Seaborg, Man-made transuranium elements, Prentice-Hall, Englewood Cliffs, N.J., 1963.

J. Selbin and J.D. Ortego, The chemistry of uranium(V), Chem. Rev., 69 (1969) 657.

K.W. Bagnall, The halogen chemistry of the actinides, in Halogen chemistry, Vol. 3, Ed. V. Gutmann, Academic Press, London and New York, 1967.

A.E. Comyns, The co-ordination chemistry of the actinides, Chem. Rev., 60 (1960) 115.

L.E.J. Roberts, The actinide oxides, Quart. Rev., 15 (1961) 442.

K.W. Bagnall (Ed.), Lanthanides and actinides, Butterworth, London, 1972.

C. Keller, The chemistry of protactinium, Angew. Chem. Internat. Ed., 5 (1966) 23.

C.K. Jørgensen, The loose connection between electron configuration and chemical behaviour of the heavy elements, Angew. Chem. Internat. Ed., 12 (1973) 12.

D. Brown, Halides of the lanthanides and actinides, Wiley, New York, 1968.

A.H.W. Aten, Jr. and J. Kooi (Eds.), Actinides reviews, Elsevier, Amsterdam, Vol. 1 (1967) etc.

L.B. Asprey and R.A. Penneman, The chemistry of the actinides, Chem. Eng. News, 45/32 (1967) 75.

B.B. Cunningham, Co-ordination chemistry and physical properties of the trans-plutonium actinides, Pure Appl. Chem., 27 (1971) 43.

H. Gysling and M. Tsutsui, Organolanthanides and organoactinides, Adv. Organometallic Chem., 9 (1970) 361.

S. Ahrland, K.W. Bagnall, S.H. Eberle, C. Keller, J.O. Liljenzin, J. Rydberg, D. Brown, R.M. Dell, J.A. Lee, P.G. Mardon, J.A.C. Marples, G.W.C. Milner, G. Phillips and P.E. Potter, The chemistry of the actinides, Pergamon, Oxford, 1975.

Titanium, Zirconium and Hafnium — Group IVA

32

32.1. Introduction

The elements Ti, Zr and Hf differ from the other transition metals which follow them in their strong tendency to exhibit the group's maximum charge number, +4, almost to the exclusion of lower oxidation states; especially is this true of Zr and Hf. As in the latter transition subgroups, there is a distinct break in properties between the first member and the two heavier congeners. In fact Zr and Hf are so strikingly similar in their chemistry that the elements are extremely difficult to separate. Because of the effect of the lanthanide contraction which accompanies the filling of the 4f quantum level, the metallic radii and the estimated radii of the quadripositive ions are very close despite the difference of 32 in atomic number (Table 32.1).

TABLE 32.1

ATOMIC PROPERTIES OF Ti, Zr AND Hf

	Ti	Zr	Hf
Z	22	40	72
Electron configuration	$[Ar]3d^2 4s^2$	$[Kr]4d^2 5s^2$	$[Xe]4f^{14} 5d^2 6s^2$
$I(1)$/kJ mol^{-1}	658	670	530
$I(2)$/kJ mol^{-1}	1315	1345	1425
Metallic radius/pm	147	160	158
$r_{M^{4+}}$/pm	68	80	81

32.2. The Elements

The elements are lustrous, silvery metals of high *m.p.* (Table 32.2). Titanium has a particularly low density for a transition metal and this, coupled with its strength and its resistance to corrosion, makes it of interest for the construction of supersonic aircraft. Ti and Zr have the h.c.p. structure at ordinary temperature but both have a high-temperature b.c.c. form; for Ti the transition temperature of the change is 1150 K. Hf has a structure very close

TABLE 32.2

PHYSICAL PROPERTIES OF Ti, Zr AND Hf

	Ti	Zr	Hf
ρ/g cm^{-3}	4.54	6.53	13.3
M.p./K	1950	2130	2470

to h.c.p. but six of the Hf—Hf distances are said to be about 2% longer than the other six. This metal also exists in a high-temperature b.c.c. modification.

The metals are extremely resistant to corrosion at ordinary temperatures, almost certainly because they are so easily rendered passive by the formation of an oxide layer. However, the metals react with most non-metals on heating; they combine with O_2, N_2, H_2, C, S and the halogens. Ti reacts particularly vigorously with nitrogen.

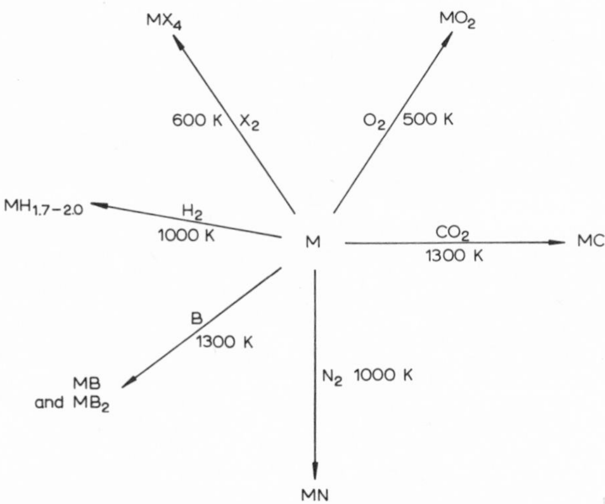

Fig. 32.1. Reactions of Ti, Zr and Hf (X = halogen).

The metals decompose steam on strong heating. Titanium dissolves in hot HCl to give $TiCl_3$, but resists attack by cold acids. Zirconium and hafnium dissolve even less readily in mineral acids, but all three metals dissolve in the presence of F^- ions, titanium giving salts of the Ti^{3+} ion, Zr and Hf giving zirconyl and hafnyl salts. The metals are remarkably resistant to corrosion by weakly acid solutions such as those of H_2S, SO_2, $FeCl_3$ (hot and cold) and even of H_2CrO_4. Hot and cold alkalis do not react with them.

32.3. Extraction of the Metals

32.3.1. Titanium

The element (0.6% of the earth's crust) is abundant but difficult to extract. It is commonly associated with siliceous rocks, but the principal workable ores are ilmenite, $FeTiO_3$, and rutile, a tetragonal form of TiO_2. Reduction of TiO_2 with carbon is unsatisfactory because the very stable carbide is produced. The ease with which the metal combines with both oxygen and nitrogen at high temperatures makes other reduction methods difficult. The commercial production of titanium is by the metallothermic reduction of titanium tetrachloride.

In this process chlorine is passed over ilmenite or rutile heated to 1200 K with carbon, and the $TiCl_4$ vapour formed is condensed.

$$2\ TiO_2 + 3\ C + 4\ Cl_2 \rightarrow 2\ TiCl_4 \rightarrow 2\ TiCl_4 + 2\ CO + CO_2$$

The liquid (*b.p.* 409 K) is purified by fractional distillation.

The tetrachloride vapour at atmospheric pressure is reduced with molten Mg at 1070 K, air being excluded by purging with argon.

$$\underset{\text{(liquid)}}{TiCl_4} + 2\ \underset{\text{(solid)}}{Mg} \rightarrow \underset{\text{(solid)}}{Ti} + 2\ \underset{\text{(solid)}}{MgCl_2} \qquad \Delta H = -483\ \text{kJ mol}^{-1};\ \Delta G = -448\ \text{kJ mol}^{-1}$$

Molten $MgCl_2$ is tapped from the reactor when 60% of the Mg has been used, in order to allow the $TiCl_4$ vapour to continue to attack the Mg until 85% of the metal has reacted. The cold solid may be chipped from the reactor mechanically and the Mg and $MgCl_2$ leached from the chips with dilute acid. Alternatively, the Mg and $MgCl_2$ may be distilled from the Ti under a high vacuum.

The fragments of titanium are melted in a water-cooled copper crucible, by striking an arc between them and a compressed titanium-sponge cathode, in an atmosphere of argon.

Very pure titanium can be made by the iodine method in which TiI_4 vapour is decomposed on a hot wire.

The metal is unusual in igniting spontaneously in oxygen, at 2.4 MPa and room temperature, when a fresh surface is exposed by fracture. The reaction is self-propagating and leads to the complete combustion of massive pieces, through a superficial melting and rapid diffusion of the oxide into the metal, leaving a new surface of metal for oxidation. Similarly ZrO_2 dissolves in molten Zr, and this metal also behaves like Ti, but Mg, Al, Nb and Ta, whose oxides are not soluble in the metal, do not show the phenomenon.

The mechanical properties of titanium are comparable with those of steel; but it is more difficult to fabricate owing to the readiness with which it takes up, and is hardened and embrittled by, oxygen and nitrogen.

32.3.2. Zirconium

The principal ores of zirconium (0.025% of the lithosphere) are the silicate zircon, $ZrSiO_4$, and the oxide baddeleyite, ZrO_2. Treatment of these with carbon and chlorine at red heat gives crude $ZrCl_4$ which, after purification, can be reduced with Mg in a modification of the titanium process. For metal of higher purity the decomposition of ZrI_4 is used.

The metal is much softer than titanium. Its principal uses at present are in bullet-proof alloy steels and, because of its low cross-section for neutron capture, in alloys for cladding the metallic fuel elements used in some atomic reactors. Hafnium must be separated as completely as possible from zirconium which is to be used in reactors, because of the high cross-section for neutron capture of hafnium.

32.3.3. Hafnium

This element, predicted from atomic number sequence, was the first element to be discovered by X-ray methods. It was found in zirconium minerals which usually contain about 0.1% but occasionally up to 7% of hafnium. Because of its very close resemblance to zirconium, separation is difficult.

On a commercial scale a mixture of $ZrCl_4$ and $HfCl_4$ is dissolved in aqueous NH_4CNS, which causes some hydrolysis and some complex formation, and then agitated with hexone, which extracts the hafnium preferentially as $HfOCl_2$. The small amount of zirconium which enters the organic phase can be stripped from it with HCl. Finally the hafnium is recovered from the hexone with H_2SO_4, and the solvent is recycled.

32.4. Oxidation States

The dipyridyl complexes $Li[Ti(dipy)_3] \cdot 3.5\ C_4H_8O$ and $Ti(dipy)_3$, which are both obtained by reducing $TiCl_4$ with lithium in solvent tetrahydrofuran in the presence of 2,2′-dipyridyl, are examples of titanium in the formal oxidation states −1 and 0 respectively. A compound of Zr^0, $Zr(dipy)_3$, has been made similarly.

Compounds of Ti^{II} and Zr^{II} are known. The dihalides $TiCl_2$, $TiBr_2$, TiI_2, $ZrCl_2$ and ZrI_2 have been made by reducing the tetrahalide vapours over the heated metals. These compounds are all oxidised by water, thus the +2 states have no aqueous chemistry. An oxide of approximate composition TiO can be made by heating TiO_2 with Ti; it has a defect NaCl lattice. Titanium forms a cyclopentadienyl, $(\pi\text{-}C_5H_5)_2Ti$, and a cyclopentadienyl carbonyl, $(\pi\text{-}C_5H_5)_2Ti(CO)_2$, both Ti^{II} compounds.

The +3 state is represented in titanium by all four halides, the oxide Ti_2O_3, alums like $RbTi(SO_4)_2 \cdot 12\ H_2O$, halogen complexes like the hexafluorotitanate(III) ion, and complexes of $TiCl_3$ with oxygen and nitrogen donors.

Aqueous solutions containing the violet $Ti(H_2O)_6{}^{3+}$ ion are useful reducing agents in volumetric analysis. They reduce Fe^{3+} to Fe^{2+}, chlorates and perchlorates to chlorides and aromatic nitro-compounds to amines; the redox potential for the $Ti^{IV}aq/Ti^{III}aq$ couple is about +0.05 V. However, the solutions are rapidly oxidised in air and must be used in an atmosphere of N_2 or CO_2. The +3 state also occurs in trihalides of Zr and Hf, several of which have been prepared, but aquated Zr^{3+} and Hf^{3+} are not known.

The oxidation state +4 is dominant in the group; all compounds of the metals in lower oxidation states are easily oxidised to it.

32.5. Halides

Tetrafluorides result from the action of anhydrous HF on the chlorides:

$$TiCl_4 + 4\ HF \rightarrow TiF_4 + 4\ HCl$$

The white solids form stable complexes:

$$ZrF_4 + 2\ KF \rightarrow K_2ZrF_6$$

The $TiF_6{}^{2-}$ ion is unstable to hydrolysis. It is converted rapidly to $TiOF_4{}^{2-}$. Surprisingly, some hexafluorozirconates, such as K_2ZrF_6, do not contain $ZrF_6{}^{2-}$ ions, but instead ZrF_8 units formed by the sharing of F^- ions.

The tetrachlorides are made by passing chlorine over the dioxides heated with carbon. $TiCl_4$ is a colourless, strongly fuming liquid but $ZrCl_4$ and $HfCl_4$ are solids. The vapours are monomeric. Water hydrolyses them:

$$TiCl_4 + 2\ H_2O \rightarrow TiO_2 + 4\ HCl$$
$$ZrCl_4 + H_2O \rightarrow ZrOCl_2 + 2\ HCl$$

This oxide chloride crystallises from the solution as the octahydrate, $ZrOCl_2 \cdot 8\ H_2O$, which has a tetragonal structure containing $Zr_4(OH)_8{}^{8+}$ ions. Phase studies of zirconium tetrachloride—alkali-metal chloride systems indicate the formation of chlorozirconates $M^I_2ZrCl_6$. Ammonium chlorotitanate, $(NH_4)_2TiCl_6$, is precipitated when NH_4Cl is added to a solution of $TiCl_4$ in concentrated HCl.

The tetrabromides and tetraiodides are made by direct combination of the elements. Some are coloured; $TiBr_4$ is yellow and TiI_4 red-brown, in accordance with the position of the ligands in the spectrochemical series (6.2.1.2). They are solids of low *m.p.*; the crystals have a cubic lattice and contain tetrahedral molecules. Some bromo-complexes have been made, for instance $(NH_4)_2TiBr_6 \cdot 2\ H_2O$, but they are much less stable than the fluoro-compounds; iodo-complexes are unknown; the stability falls rapidly with the more easily polarisable halogens.

Titanium trichloride, $TiCl_3$, is the most important trihalide. Its hexahydrate, like $CrCl_3 \cdot 6\ H_2O$ (28.4.2) exhibits hydration isomerism. Rubidium and caesium chlorides give complexes containing $TiCl_5{}^{2-}$ ions. The crystals, $M^I_2TiCl_5 \cdot H_2O$, are green but the solutions are violet. The colour change is associated with a change in the number of co-ordinated water molecules in the complex (cf. 34.4). TiF_3 is best made from titanium hydride and HF at 1000 K. It is a blue solid, stable to air, water and even concentrated H_2SO_4. Its magnetic susceptibility (1.75 B.M.) is appropriate to the Ti^{3+} ion with one d electron.

The trihalides ZrX_3 (X = Cl, Br or I), $HfBr_3$ and HfI_3 are known. They are oxidised by water to the +4 state, and therefore are without an aqueous solution chemistry.

Pure titanium dichloride is a dark-brown powder spontaneously inflammable in air. It is made by passing an electrode-less discharge through $TiCl_4$ mixed with hydrogen at low pressure. It sets free hydrogen from water. Chlorocomplexes $MTiCl_3$ and M_2TiCl_4 can be made by melting $TiCl_2$ with alkali-metal halides.

Titanium dibromide, which is made by reducing $TiBr_4$ with Ti, has been shown to have a CdI_2 structure. $ZrBr_2$ and ZrI_2 are obtained, mixed with the corresponding trihalides, when $ZrBr_4$ and ZrI_4 are reduced with hydrogen or the metal.

32.6. Oxides

Both TiO_2 and ZrO_2 are manufactured for use as white pigments, TiO_2 from ilmenite by conversion to the sulphate, followed by hydrolysis (Fig. 32.2). The usual method is, however, the vapour-phase oxidation of $TiCl_4$, attained by passing the vapour with air through a flame produced by burning a hydrocarbon in an excess of oxygen.

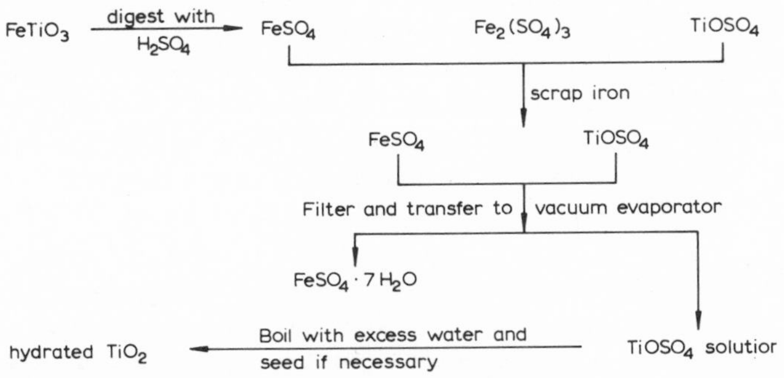

Fig. 32.2. Manufacture of TiO_2 from ilmenite.

The refractory oxide is white when cold but resembles SnO_2 in becoming yellow when hot. It has three crystalline forms, the tetragonal rutile, the slender tetragonal prisms of anatase and the flat plates of orthorhombic brookite. The pigment grades are either anatase or rutile. Rutile has 6 : 3 co-ordination and is isomorphous with cassiterite, SnO_2 (Fig. 7.20). In anatase, linear molecules of TiO_2 are present.

Zirconia (*m.p.* 2970 K) is used as a refractory and also as a pigment, mainly for white enamels. It is made from zircon (Fig. 32.3a) or baddeleyite (Fig. 32.3b).

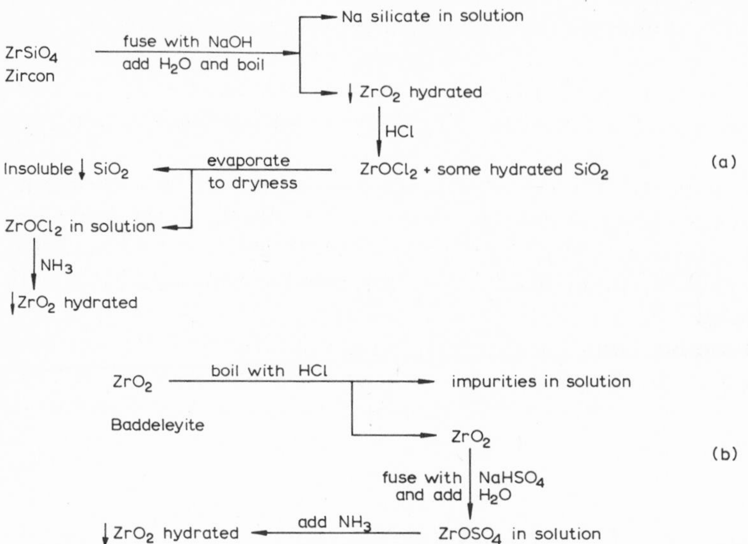

Fig. 32.3. Manufacture of zirconia, (a) from zircon, (b) from baddeleyite.

Hydrated oxides can be obtained by adding OH^- ions to solutions of the M^{IV} salts. They dissolve in acids. Hydrated TiO_2 also dissolves in concentrated aqueous alkalis to give hydrated compounds such as $M^I_2TiO_3 \cdot n\ H_2O$, of unknown structure, but the hydrated ZrO_2 is almost insoluble in alkalis. When the hydrated oxides are strongly heated the MO_2 compounds which are produced are extremely resistant to attack by acids.

White TiO_2 can be converted to the dark-violet Ti_2O_3, which has the corundum structure, by strong heating in hydrogen at atmospheric pressure. It is not attacked by non-oxidising acids but it dissolves in some oxidising acids to give Ti^{IV} compounds. A non-stoichiometric TiO, with a defect-NaCl lattice, can be made by heating TiO_2 strongly with Ti.

32.6.1. Complex oxides

Most compounds formulated as titanates are complex oxides rather than salts. One of these, perovskite, $CaTiO_3$, gives its name to the perovskite structure (Fig. 7.27) of which other examples are $SrTiO_3$, $BaTiO_3$ and $CaZrO_3$. The barium compound, however, appears in four other crystalline forms: hexagonal, tetragonal, orthorhombic and trigonal. The tetragonal form has a very high dielectric constant which varies markedly with temperature. It is used for high value capacitances. The Ba^{2+} ion is so large that the O^{2-} lattice is expanded to the extent that the Ti^{4+} is not quite large enough to fill its octahedral space. Thus the Ti^{4+} ion is easily displaced within this space by a strong electric field, with a consequent polarisation of the crystal. One form has piezoelectric properties and is employed in transducers to convert electrical into mechanical energy, in ultra-sonic applications and in gramophone pick-ups. The material is made by heating $BaCO_3$ with TiO_2. It is ground, made to shape with a binding material, and fired like a ceramic.

There are also complex oxides containing titanium which have the ilmenite structure, which is effectively a corundum lattice in which the M^{3+} ions are replaced by alternate M^{2+} and M^{4+} ions. Examples are $MgTiO_3$, $MnTiO_3$, $NiTiO_3$ and ilmenite itself, $FeTiO_3$.

Other of the complex oxides of titanium have the spinel structure (Fig. 7.28); examples are Mg_2TiO_4 and Zn_2TiO_4, made by heating the respective oxides with TiO_2, and $Mn^{II}Ti^{III}{}_2O_4$, produced when MnO_2 and TiO_2 are heated together at 1700 K. In Ba_2TiO_4, however, titanium is tetrahedrally co-ordinated to oxygen; the compound is a true titanate.

A number of complex oxides of zirconium are known; they are produced by strongly heating metal oxides, or nitrates, with ZrO_2. The calcium compound, $CaZrO_3$, is a perovskite; but several of the $(M^{II})_2ZrO_4$ compounds are known to be spinels.

32.7. Binary Compounds with Other Non-metals

Hard, refractory borides such as TiB and TiB_2 can be made by heating the metals strongly with boron. The carbides TiC and ZrC can be made similarly. These, and the mononitrides TiN and ZrN, also made by direct combination at high temperature, have the NaCl structure; the Ti atoms are in a cubic close-packed arrangement with the C or N atoms occupying octahedral spaces. Since the metals themselves have h.c.p. lattices these have not been simply expanded to accommodate the non-metal atoms.

Both Ti and Zr form nitrides M_3N_4. The titanium compound is made by the reaction between $TiBr_4$ and NH_2^- ions in liquid ammonia, and the zirconium compound by thermal decomposition of $Zr(NH_3)_4Cl_4$ obtained by the action of NH_3 on $ZrCl_4$. Both Ti_3N_4 and Zr_3N_4 lose nitrogen on heating and are converted to the mononitrides.

The three metals combine with sulphur to give disulphides, which are semi-conducting solids with metallic lustre. TiS_2 has been shown to have a structure similar to that of CdI_2.

32.8. Aqueous Solution Chemistry

Titanium(III) exists in aqueous solution as the blue-violet $Ti(H_2O)_6{}^{3+}$ ion; its absorption spectrum is discussed in Section 6.2.1.1. The hydrated ion is an acid of moderate strength; for the equilibrium

$$Ti(H_2O)_6{}^{3+} + H_2O \rightleftharpoons Ti(H_2O)_5OH^{2+} + H_3O^+$$

the pK_a value is 3.9. The hexa-aquotitanium(III) ion occurs in some alums such as $RbTi(SO_4)_2 \cdot 12\ H_2O$ and also in the violet form of $TiCl_3 \cdot 6\ H_2O$.

The Ti^{4+} ion is too strongly polarising to form simple aquo-complexes. Although salts of the Group IVA metals have formulae such as $TiOSO_4 \cdot H_2O$, $ZrO(NO_3)_2 \cdot 2\ H_2O$ and $Zr(C_2O_4)_2$ it is unlikely that MO^{2+} ions exist either in the aqueous solutions or in the crystal lattices. Solutions of titanium(IV) perchlorate contain $Ti(OH)_2(H_2O)_4{}^{2+}$ ions which are approximately octahedral, and solid $TiOSO_4 \cdot H_2O$ has been shown to contain zig-zag $(TiO)_n{}^{2n+}$ chains. The ions in zirconium(IV) crystals and their aqueous solutions are of even greater complexity. Solid $ZrOCl_2 \cdot 8\ H_2O$ contains $[Zr_4(OH)_8(H_2O)_{16}]^{8+}$ ions in which the Zr atoms are linked in a distorted square by pairs of OH bridges. These ions are also thought to be the principal Zr^{IV} species present in molar aqueous $HClO_4$.

32.9. Complexes

The halogen complexes are of interest. The compounds Na_3ZrF_7 and Na_3HfF_7 are reported to contain pentagonal bipyramidal $MF_7{}^{3-}$ ions, but the $ZrF_7{}^{3-}$ ion in $(NH_4)_3ZrF_7$ is a side-capped trigonal prism (Fig. 32.4). Eight-co-ordination of zirconium is also observed; the compound $Cu_2ZrF_8 \cdot 12\ H_2O$ contains $ZrF_8{}^{4-}$ ions which are square antiprisms, and $Cu_3Zr_2F_{14} \cdot 16\ H_2O$ contains $Zr_2F_{14}{}^{6-}$ ions in which two antiprisms share an edge.

The tetrachlorides and tetrabromides of Ti and Zr act as Lewis acids and form adducts MX_4L and MX_4L_2 with oxygen donors such as carboxylates and also with various phosphorus and arsenic donors. The MX_4L adducts seem to be predominantly dimers, the metal atoms being connected by pairs of halogen bridges. As in the MX_4L_2 monomers there is approximately octahedral co-ordination of the metal atoms.

The tetrachloride and tetrabromides of all three metals react with β-diketones in inert solvents. The octahedral chloro-acetyl-acetonates $MCl_2(acac)_2$ have *cis*-chlorines. Zr and Hf also form compounds $MCl(acac)_3$ in which the metal atoms are 7-co-ordinate.

Complexes of bidentate ligands with donor atoms other than oxygen are also known. Orthophenylenebis(dimethylarsine) forms complexes MCl_4(diars) and $MCl_4(diars)_2$ with all three tetrachlorides. The compound $TiCl_4(diars)_2$

Fig. 32.4. $ZrF_7{}^{3-}$ ion in $(NH_4)_3ZrF_7$.

has a dodecahedrally co-ordinated Ti atom; it was an early example of the rare 8-co-ordination of a first-row d-block metal. The alkoxides of titanium have been extensively studied. They can be made by reactions of the type:

$$TiCl_4 + 4\ ROH + 4\ NH_3 \rightarrow Ti(OR)_4 + 4\ NH_4Cl$$

The compounds are used in heat-resisting paints in which they eventually hydrolyse to TiO_2. These alkoxides are liquids or solids which can be easily distilled or sublimed, and dissolve in non-polar organic solvents. Solid $Ti(OEt)_4$ has been shown to be a tetramer in which the four Ti atoms are co-ordinated approximately octahedrally to the oxygens of the OEt groups, some bridging and some terminal.

Titanium and zirconium both form volatile, anhydrous tetranitrates which are complex covalent compounds, not salts. $Ti(NO_3)_4$ is obtained when the hydrated nitrate is treated with N_2O_5, and $Zr(NO_3)_4$ by gently heating the adduct formed by the action of N_2O_5 on $ZrCl_4$. The metal atom in these molecules is co-ordinated to bidentate NO_3 groups through oxygen atoms arranged at the corners of a dodecahedron; the N atoms are thereby arranged in a slightly distorted tetrahedron. Zirconium also forms tetracarboxylates, a borohydride, $Zr(BH_4)_4$, and an anhydrous oxalate which are all covalent molecular complexes.

32.10. Organometallic Compounds

Unlike Ge, Sn and Pb in Group IVB, the metals of Group IVA do not form strong metal—carbon σ-bonds. The solid $C_6H_5Ti(OPr^i)_3$ is obtained by treating $Ti(OPr^i)_4$ with phenyllithium. Although rapidly oxidised by air or water, it is thermally stable at room temperature in an inert atmosphere. CH_3TiCl_3 and $Ti(CH_2Ph)_4$ are other moderately stable compounds containing Ti—C σ-bonds.

Although the Group IVA metals do not form many sigma-bonded organometallic compounds they show general resemblance to the transition metals in their ability to form π-bonded organometallic compounds. Deep-red, diamagnetic bis(cyclopentadienyl)titanium dichloride, $(C_5H_5)_2TiCl_2$ is made by treating $TiCl_4$ with C_5H_5Na in tetrahydrofuran. The corresponding fluoride, bromide and iodide are known; so are $(\pi\text{-}C_5H_5)_2ZrBr_2$ and $(\pi\text{-}C_5H_5)_2HfCl_2$; all are stable in air.

A tetracyclopentadienyl $(C_5H_5)_4Ti$ has been made which has the structure:

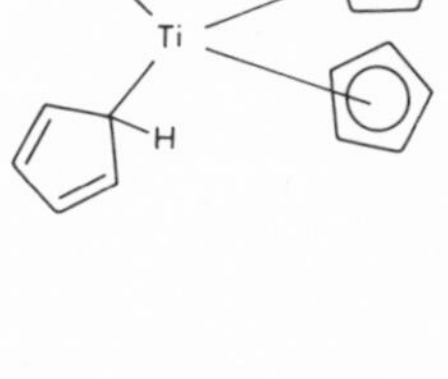

This molecule is fluxional in two senses. Firstly the σ-bonded C_5H_5 rings undergo 'ring-whizzing' (18.2.5) and secondly the σ-bonded rings and the π-bonded rings rapidly change their roles. In consequence the ^1H n.m.r. spectrum at 298 K consists of a single broad absorption.

When $(\pi\text{-}C_5H_5)_2TiCl_2$ is reduced with sodium in tetrahydrofuran a compound of empirical formula $C_{10}H_{10}Ti$ is formed. This has two isomers, one of which has been shown to be a dimer with each Ti atom attached to a carbene-like C_5H_4 group, a $\pi\text{-}C_5H_5$ group, and two bridging hydrogens. The compound is converted by carbon monoxide to $(\pi\text{-}C_5H_5)_2Ti(CO)_2$, the only well characterised carbonyl compound in this sub-group. Interestingly $C_{10}H_{10}Ti$ also absorbs molecular nitrogen; the product releases much of it as NH_3 on hydrolysis (20.4.1).

Further Reading

R.J.H. Clark, The chemistry of titanium and vanadium, Elsevier, Amsterdam, 1968.

I. Shiihara, W.T. Schwartz and H.W. Post, The organic chemistry of titanium, Chem. Rev., 61 (1961) 1.

W. Fischer, The separation of zirconium and hafnium by liquid—liquid partition, Angew. Chem. Internat. Ed., 5 (1966) 15.

T.E. MacDermott, The structural chemistry of zirconium compounds, Co-ord. Chem. Rev., 11 (1973) 1.

E.M. Larsen, Zirconium and hafnium chemistry, Adv. Inorg. and Radiochem., 13 (1970) 1.

A.D. McQuillan and M.K. McQuillan, Titanium, Butterworth, London, 1956.

W.B. Blumenthal, The chemical behaviour of zirconium, Van Nostrand, Princeton, 1958.

G.L. Miller, Zirconium, Butterworth, London, 1954.

R.J.H. Clark, D.C. Bradley and P. Thornton, The chemistry of titanium, zirconium and hafnium, Pergamon, Oxford, 1975.

P.C. Wailes, R.S.P. Coutts and H. Weigold, Organometallic chemistry of titanium, zirconium, and hafnium, Academic Press, New York, 1974.

Vanadium, Niobium and Tantalum — Group VA

33

33.1. The Metals

Vanadium, niobium and tantalum are typical transition metals. Their first ionisation energies are in the range 650—675 kJ mol^{-1} (Table 33.1), they all

TABLE 33.1

ATOMIC PROPERTIES OF VANADIUM, NIOBIUM AND TANTALUM

	V	Nb	Ta
Z	23	41	73
Electron configuration	[Ar]3d^{3}4s^2	[Kr]4d^{4}5s^1	[Xe]4f^{14}5d^{3}6s^2
$I(1)$/kJ mol^{-1}	651	654	675
Metallic radius/pm	134	146	146
$r_{M^{5+}}$/pm	56	70	73

TABLE 33.2

PHYSICAL PROPERTIES AND ELECTRODE POTENTIALS OF VANADIUM, NIOBIUM AND TANTALUM

	V	Nb	Ta
ρ/g cm^{-3}	6.11	8.57	16.6
M.p./K	2190	2740	3270
E°, M$^{IV}{}_{aq}$/M/V	−1.5		
E°, M$^{V}{}_{aq}$/M/V		−0.6	−0.7

crystallise with the b.c.c. lattice, they have high *m.p.* (Table 33.2) and they exhibit a wide variety of oxidation states in their compounds. The standard electrode potentials are extremely difficult to measure because the metals are so easily rendered passive that truly reversible electrodes cannot be prepared. Although the calculated values show the metals to be strong reducing agents, reaction does not occur with cold, dilute, non-oxidising acids. The metallic radius of Ta is very close to that of Nb, a consequence of the lanthanide contraction, and the two metals are therefore very similar in their chemistry, like the elements Zr and Hf which immediately precede them. Another consequence of the small metallic radius of Ta is that the metal has a high density, almost twice that of Nb.

Although unreactive at room temperature, the metals combine with most non-metals when heated. All three burn in oxygen to give the pentoxides. In the reactions of vanadium with halogens, the products depend upon the polarisability of the halide ion; thus V reacts with F_2 at 570 K to give VF_5, with Cl_2 at 770 K to give VCl_4 and with Br_2 at 420 K to give VBr_3. However, most of the pentahalides of Nb and Ta can be made by direct combination; even I_2 reacts with Nb at 570 K to give NbI_5. The metals also form nitrides, arsenides, carbides, silicides and borides by direct combination; these compounds are mainly of interstitial type. The metals do not dissolve readily in cold mineral acids but they can be easily taken into solution as fluorocomplexes by a hot mixture of HF and HNO_3. Surprisingly, all three react with fused alkalis; H_2 is liberated and oxoacid salts such as niobates are formed.

33.2. The Occurrence and Separation of the Elements

33.2.1. Vanadium

The element (0.02% of the lithosphere) is widely distributed — more than sixty vanadium minerals have been described — but there are few workable ores. Carnotite, $K(UO_2)VO_4 \cdot 1.5\ H_2O$, is a source of both uranium and vanadium; vanadinite, $Pb_5(VO_4)_3Cl$, which is isomorphous with apatite, can also be worked up for the element, but the metal is being extracted to an increasing extent from residues obtained in iron and titanium production.

In the usual commercial process the vanadinite or vanadium residue is roasted with NaCl which converts the vanadium to sodium vanadate. This is leached out with water, and when the solution is acidified a red polyvanadate is precipitated. This can be fused at 950 K to give a commercial grade of V_2O_5, or dissolved in aqueous Na_2CO_3 from which NH_4VO_3 is precipitated by NH_4^+ ions; the ammonium vanadate is converted to a better grade of V_2O_5 by heating to 700 K.

Vanadium metal can be made by reducing V_2O_5 with calcium in a pressure vessel in the presence of a little iodine, some $CaCl_2$ being added to flux the lime which is formed:

$$V_2O_5 + 5\ Ca \rightarrow 5\ CaO + 2\ V$$

The metal is normally refined by electrolysis of a fused $NaCl/LiCl/VCl_2$ mixture, a process particularly effective for removing interstitial oxygen, nitrogen and hydrogen. Pure metal can also be made by the iodine process – the sublimation and thermal decomposition of VI_2.

The material ferrovanadium, which is made by aluminothermic reduction of V_2O_5 in the presence of iron, is manufactured for use in steel for high-speed tools. The alloy refines the grain and carbide structure of the steel, and improves its hardness at high temperature, by combining with the carbon which is present to form V_4C_3. Pure vanadium has little tensile strength and is a soft, ductile metal.

33.2.2. Niobium and tantalum

The elements are both rare, niobium being about 10^{-4}% and tantalum 10^{-5}% of the lithosphere. A mineral which is a mixed niobate and tantalate of iron and manganese, (Fe,Mn)(Nb,Ta,O_3)$_2$, is called columbite when it contains more Nb than Ta, otherwise tantalite. Another source of the metals is pyrochlore, which is a carbonatite — a mixed carbonate and silicate mineral — containing in this case some niobate and tantalate in addition. The columbite or pyrochlore is leached with aqueous HCl at 350 K for ten hours to take the Nb and Ta into solution. The metals are usually separated by a solvent-extraction procedure; tantalum can be extracted first from weakly acidic solutions into methyl isobutyl ketone, the solution is then made strongly acidic and niobium is extracted by more of the same pure solvent. Niobium is recovered from the organic phase by agitation with de-ionised water, it is recovered by precipitation as $NbOF_3$, filtered, dried and calcined to Nb_2O_5. The metal can be obtained by reducing Nb_2O_5 with carbon or $NbCl_5$ with sodium. Tantalum is recovered from its organic phase similarly, and precipitated with KF as K_2TaF_7, which is dried and reduced to metal with sodium.

Niobium is used to inhibit intergranular corrosion in austenite steels, and as carbide in hard carbide tool compositions. An alloy with tin, of composition Nb_3Sn, is a particularly good superconducting material.

Tantalum is resistant to corrosion and for this reason is employed in both chemical research and plant construction. The element has minimal foreign body reactions in human tissue and finds a place in surgery. It is used in electrolytic rectifiers and in capacitors. These applications are possible because of the thin anodic film formed on the metal in oxoacid electrolytes.

33.3. Oxidation States

The lowest states are stabilised, as is usual, by π-acceptors such as CO. All three carbonylate anions $M(CO)_6^-$ are made by reduction of pentachlorides with sodium in diglyme under CO at 30 MPa. The vanadium compound $[Na(diglyme)_2]V(CO)_6$ can be oxidised to the dark-green, paramagnetic hexacarbonyl, which has an octahedral molecule and is an exception to the 18-electron rule (17.12.3).

The +1 state is also stabilised by ligands such as carbonyl and bipyridyl. The +2 state is unimportant, though there are well-established compounds, such as double sulphates, containing the $V(H_2O)_6^{2+}$ ion. All the trihalides of vanadium are known, there are V^{III} alums and many anionic complexes such as $V(C_2O_4)_3^{3-}$ and $V(CN)_6^{3-}$. The +4 state is an important one in vana-

TABLE 33.3

COMPOUNDS AND IONS REPRESENTATIVE OF THE OXIDATION STATES OF VANADIUM, NIOBIUM AND TANTALUM

	V	Nb	Ta
−1	$V(CO)_6^-$	$Nb(CO)_6^-$	$Ta(CO)_6^-$
0	$V(CO)_6$		
+1	$V(dipy)_3^+$	$(\pi\text{-}C_5H_5)Nb(CO)_4$	$(\pi\text{-}C_5H_5)Ta(CO)_4$
+2	$V(CN)_6^{4-}$		
+3	$V(NH_3)_6^{3+}$	$NbCl_3$	$TaBr_3$
+4	K_2VCl_6	$NbCl_4$	TaO_2
+5	$VOCl_3$	NbF_5	Na_3TaF_8

dium. It is represented by halides such as VCl_4 and VBr_4, by vanadyl compounds such as $VOCl_2$ and $VOSO_4$ and by the dark-blue VO_2. Niobium and tantalum form some tetrahalides, of which the iodides are the best known, and also complexes of these with neutral ligands such as pyridine and diarsine.

As in the preceding group, however, the most common oxidation state is the maximum one, in this case +5. Niobium and tantalum form pentahalides with all four halogens and also have oxides M_2O_5. There is a great variety of halogen complexes as well as many types of vanadates, niobates and tantalates containing the metals in the +5 state. Vanadium has a fluoride VF_5, an oxide V_2O_5 and oxohalides such as $VOCl_3$ and $VOBr_3$. The relation between the four highest oxidation states of vanadium in aqueous solution is shown in Table 33.4.

TABLE 33.4

VANADIUM COMPOUNDS IN SOLUTION

Vanadium charge number	+5		+4		+3		+2
Most common corresponding species and appropriate reducing agents	VO_3^-	$\xrightarrow{Fe^{2+}}$	VO^{2+}	$\xrightarrow[\text{or } SO_2]{Sn^{2+},\, Ti^{3+}}$	V^{3+}	$\xrightarrow[\text{or } Cr^{2+}]{Zn}$	V^{2+}
Colour in aqueous solution	colourless		blue		green		violet
Redox potential		+1.0 V		+0.3 V		−0.2 V	
Typical compounds	NH_4VO_3		$VOCl_2$ $VOSO_4$		$V_2(SO_4)_3$		VSO_4
Typical complexes			$VO(SCN)_4^{2-}$		$V(NH_3)_6^{3+}$		$V(CN)_6^{4-}$

33.4. Halides

Vanadium forms the halides and oxohalides shown in Table 33.5. There is only one pentahalide, VF_5, made by direct combination at 600 K, but there are vanadyl trihalides in which the metal is formally in an oxidation state of +5:

$$2\,VF_3 + O_2 \xrightarrow{\text{heat}} 2\,VOF_3 \text{ (white solid)}$$

$$V_2O_5 + \text{carbon} \xrightarrow[\text{in } Cl_2]{\text{heat}} VOCl_3 \text{ (liquid, } b.p. \text{ 400 K)}$$

$$V_2O_5 + \text{carbon} \xrightarrow[\text{in } Br_2]{\text{heat}} VOBr_3 \text{ (liquid, } b.p. \text{ 403 K)}$$

The brown liquid VCl_4 is made by direct combination at 800 K, and the lime-green solid VF_4 is obtained from it by the action of HF in a solvent freon.

The trihalides VBr_3 and VI_3 can be made from the elements but VCl_3 is obtained by treating the metal with HCl gas. The dark-coloured solids crystallise from water as green hexahydrates. Yellow crystalline VF_3 is obtained by prolonged heating of VCl_3 with dry HF.

TABLE 33.5

HALIDES AND OXOHALIDES OF VANADIUM

Charge number	Fluorides	Chlorides	Bromides	Iodides
+2	VF_2	VCl_2	VBr_2	VI_2
+3	VF_3	VCl_3	VBr_3	VI_3
	$VF_3 \cdot 6\,H_2O$	$VCl_3 \cdot 6\,H_2O$	$VBr_3 \cdot 6\,H_2O$	$VI_3 \cdot 6\,H_2O$
+4	VF_4	VCl_4		
	VOF_2	$VOCl_2$	$VOBr_2$	
+5	VF_5			
	VOF_3	$VOCl_3$	$VOBr_3$	
	VO_2F	VO_2Cl		

When VCl_3 is heated to 800 K in nitrogen, disproportionation occurs,

$$2\,VCl_3 \rightarrow VCl_2 + VCl_4$$

and the VCl_4 distils off, leaving VCl_2. VBr_3 and VI_3 both lose halogen on heating, giving the brown VBr_2 and the violet VI_2 respectively. This last compound can be used in the preparation of pure vanadium by the iodine method. The blue solid VF_2 can be made by the action of HF gas on VCl_2 at 900 K.

The important halides of niobium and tantalum are the pentahalides, all made by direct combination, and the oxohalides such as $NbOCl_3$, $NbOBr_3$ and $TaOBr_3$. The pentachlorides and pentabromides are monomeric in the vapour phase, but X-ray diffraction studies on the crystals show them to be made up of M_2X_{10} units containing two distorted octahedra joined along one edge:

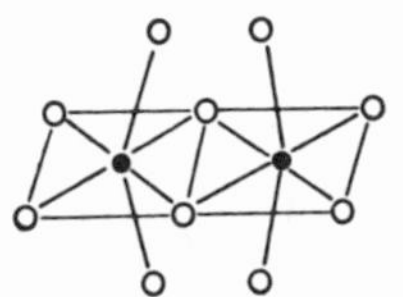

The oxohalides are usually made by heating the pentahalides in oxygen. They are less volatile than the halides themselves which can thus be separated from them by sublimation. The oxofluorides NbO_2F and TaO_2F are of structural interest, having the ReO_3 lattice (Fig. 7.23) with a random distribution of O and F in the anion positions.

With the exception of TaF_4, tetrahalides of Nb and Ta can be made by reduction of the pentahalides at high temperatures with Nb or Ta themselves, hydrogen or aluminium. They are diamagnetic, evidently because they contain metal—metal bonds.

Further reduction leads to a range of compounds containing Nb and Ta in formal oxidation states from +2 to +3. The majority of them are known to be cluster compounds, the most common cations in them being $M_6X_{12}^{n+}$ ions consisting of octahedra of metal atoms with halogens bridging all twelve edges. These cluster ions are joined in the crystals by doubly bridging halide ions. Thus the compound which is empirically Ta_2Cl_5 is formulated $[Ta_6Cl_{12}]Cl_{6/2}$, meaning that each cationic cluster is connected to three others by a total of six Cl^- ions. There is also an extensive solution chemistry of $M_6X_{12}^{n+}$ ions, where $n = 2$, 3 or 4. Salts such as $Nb_6Cl_{15} \cdot 7\,H_2O$ and $Ta_6Cl_{16} \cdot 7\,H_2O$ have been isolated from solution. The former contains the paramagnetic $Nb_6Cl_{12}^{3+}$ ion and the latter the diamagnetic $Ta_6Cl_{12}^{4+}$ ion. Anionic complexes containing metal clusters have also been obtained; for example tetraethylammonium compounds of $Nb_6Cl_{18}^{n-}$, where $n = 2$, 3 or 4, can be made from aqueous solutions containing $Nb_6Cl_{12}^{2+}$ ions and Cl^- ions.

The halides form numerous complexes. In particular the pentafluorides and the oxide fluorides combine with other metal fluorides:

$$VF_5 \xrightarrow{\text{KF in HF}} KVF_6$$

$$TaF_5 \rightarrow KTaF_6,\ K_2TaF_7 \text{ and } K_3TaF_8$$

$$NbOF_3 \rightarrow Na_3NbOF_6,\ ZnNbOF_5 \cdot 6\,H_2O$$

The TaF_8^{3-} ion in Na_3TaF_8 has the form of a slightly distorted square antiprism (Fig. 33.1). Salts containing NbF_7^{2-} have been made, but not NbF_8^{3-}, and the highest fluoroanions which appear to exist in anhydrous HF are the MF_6^- species.

Many of the pentahalides form adducts with donor molecules. Examples are $NbF_5 \cdot OEt_2$, $TaF_5 \cdot SEt_2$, $NbCl_5 \cdot (C_5H_{11}N)_6$ and $TaCl_5 \cdot (C_5H_5N)_2$.

Fig. 33.1. Square antiprism. Structure of TaF_8^{3-} in Na_3TaF_8.

33.5. Oxides

Orange-yellow V_2O_5 can be made by heating the metal in an excess of oxygen or by heating the "metavanadate" which has the empirical formula NH_4VO_3:

$$2\ NH_4VO_3 \rightarrow V_2O_5 + 2\ NH_3 + H_2O$$

The oxide dissolves in strong alkalis to form orthovanadates $M^{I}_{3}VO_4$. Some of these ($Na_3VO_4 \cdot 12\ H_2O$, $K_3VO_4 \cdot 6\ H_2O$) can be crystallised from solution at $pH > 12$. The addition of NH_4Cl to one of these solutions precipitates NH_4VO_3. When this is boiled for some time in 10% acetic acid it is converted to the golden $NH_4V_3O_8$, which is only one of a large number of polyvanadates. The decavanadates contain $V_{10}O_{28}^{6-}$ ions in which ten VO_6 octahedra share edges — these are isopolyanions similar to those found in Group VIA (34.6).

Various polyvanadate ions exist in alkaline aqueous solutions of V_2O_5. In solutions of concentration 10—100 mol m^{-3} of V_2O_5 it has been shown by spectrophotometric and potentiometric analysis that at $pH < 2$ the major species is VO_2^+, but between pH 2 and pH 6.5 the ions $H_2V_{10}O_{28}^{4-}$, $HV_{10}O_{28}^{5-}$ and $V_{10}O_{28}^{6-}$ are present in proportions depending on pH. At high pH the major species are $V_2O_7^{4-}$ and VO_3^{3-}. But at low concentration (<0.1 mol m^{-3} of V_2O_5) the ions are mononuclear and the species which occur are VO_4^{3-}, HVO_4^{2-}, $H_2VO_4^-$, H_3VO_4 and VO_2^+, in proportions depending again on pH.

The pentoxide is converted by reduction into the other oxides:

$$\underset{\text{(blue-black)}}{VO_2} \xleftarrow[\text{(heat)}]{SO_2} V_2O_5 \xrightarrow[\text{(heat)}]{H} \underset{\text{(black)}}{V_2O_3} \xrightarrow[\text{at low pressure}]{\text{heat with V}} VO$$

Of these VO has the rock-salt structure, V_2O_3 the corundum structure and VO_2 the rutile structure, but the composition range associated with a particular lattice is always large, for example compositions as low as $VO_{1.35}$ are said to retain the corundum structure, and the rock-salt structure can persist through a range of composition from about $VO_{0.85}$ to $VO_{1.15}$.

It is probable that oxides with compositions between $VO_{1.66}$ and $VO_{1.86}$ contain phases which form the homologous series V_nO_{2n-1} with n = 3, 4, 5, 6, 7 or 8, based on the rutile structure. The reason V_2O_5 itself is such a good surface catalyst, as for example in the oxidation of SO_2 to SO_3, is almost certainly the ease with which vacant oxygen sites can be produced on heating. The structure of V_2O_5 is complicated; double chains of distorted VO_5 bipyramids are linked into a layer structure by V—O—V bonds.

Hydrated oxides have been made. Rose-red $VO(OH)_2$ is made by concentrating, in an inert atmosphere, the acidic solution of a vanadate after reduction with SO_2. Black, crystalline $V_3O_5(OH)_4$ is obtained by reducing vanadic acid with zinc in the presence of concentrated NH_4Cl. A hydrated V^{III} oxide is precipitated by OH^- ions from vanadium(III) solutions but it is rapidly oxidised in air.

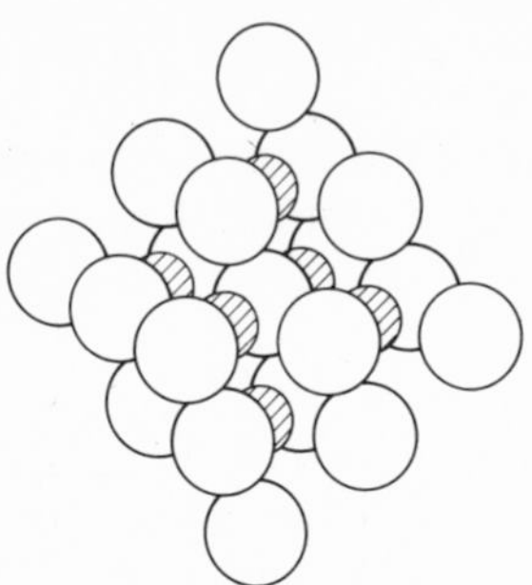

Fig. 33.2. $Ta_6O_{19}^{8-}$ ion.

The white oxides Nb_2O_5 and Ta_2O_5 are much more difficult to reduce than V_2O_5. Nb_2O_5 can be reduced with hydrogen at high temperature to NbO_2 which has a rutile-like structure, but similar treatment of Ta_2O_5 never reduces all the tantalum to the +4 state, though an electrically conducting phase containing a few interstitial metal atoms can be produced.

The pentoxides of niobium and tantalum are acidic. Fusion of Ta_2O_5 with caustic alkalis gives polytantalates of the form $M^I_8Ta_6O_{19}$. The anion $Ta_6O_{19}^{8-}$ can best be described as an octahedron of TaO_6 octahedra which share edges and have one oxygen atom common to all six (Fig. 33.2). Meta- and pyro-tantalates, e.g., $Ca(TaO_3)_2$ and $Ca_2Ta_2O_7$, are also known. A compound $NaNbO_3$, with the perovskite structure, has been obtained by heating Nb_2O_5 with Na_2CO_3, and most of the compounds called niobates and tantalates are best considered as mixed metal oxides. However, the tetrahedral NbO_4^{3-} ion does exist in $ScNbO_4$.

33.6. Binary Compounds with Other Non-metals

33.6.1. Carbides

Very hard carbides result from strongly heating the oxides with carbon. They are empirically MC and have the NaCl structure. Vanadium also forms V_4C_3 which has a defect structure, carbon atoms being missing from some of the lattice positions.

33.6.2. Nitrides

The elements combine with nitrogen at high temperatures to give the hard, very stable nitrides VN, NbN and TaN. They are not attacked by cold acids, but steam reacts with them at high temperatures to give the oxides and NH_3. The unstable higher nitrides Ta_3N_5 and VN_2 are made by heating the appropriate halides with ammonia.

33.6.3. Sulphides

The most stable sulphide of vanadium is V_2S_3, made by passing CS_2 over V_2O_5. It can be converted into V_2S_5 and VS.

$$V_2S_5 \xleftarrow[700\,K]{S} V_2S_3 \xrightarrow[1470\,K]{H_2} VS$$

Reduction of Ta_2O_5 with CS_2 at white heat gives TaS_2, the only sulphide of tantalum. NbS_2 is made by direct combination of the elements. The VS_4^{3-} ion is of interest because of its strong purple colour which is like that of MnO_4^-; presumably it also owes its colour to charge-transfer absorption, as there are no partly filled d orbitals.

33.7. Oxoacid Salts of Vanadium and Niobium

The only oxoacid salts of importance are the sulphates of vanadium. A solid sulphate has not been obtained from a solution of V_2O_5, that is vanadium with charge number +5, in H_2SO_4, but reduction of the solution gives sulphates of vanadium with charge numbers +4, +3 and +2 respectively:

$$V^{V}{}_2O_5 \text{ in } H_2SO_4 \xrightarrow{SO_2} \underset{\text{blue solution}}{V^{IV}OSO_4} \xrightarrow[\text{sulphates}]{\text{alkali}} \underset{\text{dark blue double salts}}{M^{I}{}_2SO_4 \cdot V^{IV}OSO_4 \cdot xH_2O}$$

$$\xrightarrow{Zn} V^{III}{}_2(SO_4)_3 \rightarrow \text{alums}$$

$$\xrightarrow[\text{reduction}]{\text{cathodic}} V^{II}SO_4 \cdot 7\,H_2O \text{ (isomorphous with } FeSO_4 \cdot 7\,H_2O)$$

A nitrate of niobium, $NbO(NO_3)_3$, has been made by treating $NbCl_5$ with N_2O_5 at 300 K.

$$NbCl_5 + 4\,N_2O_5 \rightarrow NbO(NO_3)_3 + 5\,NO_2Cl$$

33.8. Organometallic Compounds

As with the other transition metals, the π-bonded organometallic compounds are more numerous than the σ-bonded ones. The cyclopentadienyl compounds are of interest. Pale-green, paramagnetic $(\pi\text{-}C_5H_5)_2VCl_2$ is made by treating VCl_4 with sodium cyclopentadienide; the dark-purple, paramagnetic $(\pi\text{-}C_5H_5)_2V$ is made by treating VCl_3 with the same reagent. A 'sandwich' structure has been proposed for the latter as a result of X-ray examination. The compound is soluble in most organic solvents and is very sensitive to air.

Reduction of $(\pi\text{-}C_5H_5)_2VCl_2$ with amalgamated zinc gives $(\pi\text{-}C_5H_5)_2VCl$; this reacts with phenyllithium to give $(\pi\text{-}C_5H_5)_2VC_6H_5$ in which the phenyl group is σ-bonded to the metal. From $NbBr_5$ the bis(cyclopentadienyl)niobium(V) tribromide $(\pi\text{-}C_5H_5)_2NbBr_3$ has been made by treatment with sodium cyclopentadienide.

The compound $(\pi\text{-}C_5H_5)_2TaH_3$ is of interest because its proton n.m.r. spectrum shows it to contain non-parallel C_5H_5 rings; there is no evidence, however, that the bonding between rings and metal is fundamentally different from that in such compounds as ferrocene. The very reactive compound $(\pi\text{-}C_5H_5)_2NbH$ forms with ethylene a particularly stable adduct $(\pi\text{-}C_5H_5)_2NbH(C_2H_4)$, one of very few hydrido alkenes to have been made.

33.9. Complexes

The metals of this sub-group form many complexes. In the oxidation state +5, high co-ordination numbers often occur, as in the TaF_8^{3-} ion already mentioned and in the NbF_7^{2-} and TaF_7^{2-} ions which have the same shape as ZrF_7^{3-} (Fig. 32.4). Many of the pentahalides react with oxygen and nitrogen donors:

$$NbCl_5 \xrightarrow{NHEt_2} NbCl_3(NEt_2)_2 \cdot NHEt_2$$

$$TaBr_5 \xrightarrow[\text{in ethanol}]{\text{acetylacetone}} TaBr_2(acac)(OEt)_2$$

$NbCl_5$ and $TaCl_5$ form 1 : 1 adducts with $POCl_3$ which are monomeric in benzene. In some organic solvents, $NbCl_5$ reacts with HCN to give the complex acid $HNbCl_5CN$; the triethylammonium salt of this acid, $Et_3NH[NbCl_5CN]$, has been isolated.

Vanadium in the +4 state forms chelate complexes with salicylic acid, catechol and β-diketones. Examples are $M^I_2[VO(OC_6H_4CO_2)_2]$, $M^I_2[VO(C_6H_4O_2)_2]C_6H_6O_2 \cdot x\,H_2O$ and $VO(acac)_2$. In the last compound the vanadium atom is near the space centre of a square pyramid of oxygen atoms:

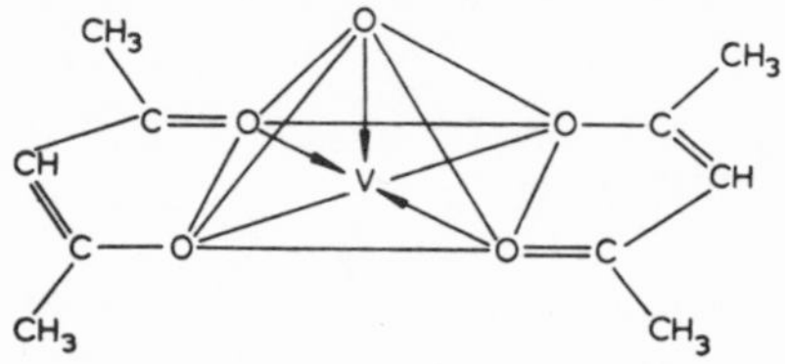

The 5-co-ordinate complexes of vanadium(IV) such as the acetylacetonate form adducts with nitrogen and phosphorus donors in which the metal becomes octahedrally co-ordinated. The introduction of a strong donor opposite to the V=O bond reduces the stretching frequency of that bond markedly, evidence that the metal is less able to accept electrons from the oxygen. Other vanadium(IV) complexes include adducts VCl_4L_2 in which L is an alkylphosphine, as well as alkoxides $V_2(OEt)_8$ and dialkylamides $V(NR_2)_4$. The tetrahalides of Nb and Ta form adducts with nitrogen, oxygen and sulphur donors. Those with unidentate donors are usually of the MX_4L_2 type and probably have *cis* octahedral configurations. Those with bidentate ligands such as o-$C_6H_4(AsMe_2)_2$ are believed to have a dodecahedral arrangement of atoms around the metal.

Vanadium(III) halides form 5-co-ordinate 1:2 adducts with some nitrogen and sulphur donors:

$VCl_3(SMe_2)_2$ $VBr_3(NMe_3)_2$

This oxidation state is also represented in chelate oxalate, β-diketone and catechol complexes:

$V(C_2O_4)_3^{3-}$ $V(C_6H_4O_2)_3^{3-}$ $V(acac)_3$

The last compound is a dark-green solid, insoluble in water, but soluble in many organic solvents. Cyano and thiocyanato complexes are also known:

$K_3V(CN)_6$ and $K_3V(CNS)_6$

Vanadium(II) is represented by $K_4V(CN)_6$. There are double sulphates of the schönite type, such as $K_2V(SO_4)_2 \cdot 6\ H_2O$, but these are lattice compounds rather than complexes.

Further Reading

M.T. Pope and B.W. Dale, Isopoly-vanadates, niobates and tantalates, Q. Rev., 22 (1968) 527.

F. Fairbrother, The chemistry of niobium and tantalum, Elsevier, Amsterdam, 1967.

J. Selbin, The chemistry of oxovanadium(IV), Chem. Rev., 65 (1965) 153.

F. Fairbrother, The halides of niobium and tantalum, in Halogen Chemistry, Vol. 3, Ed. V. Gutmann, Academic Press, London and New York, 1967.

W. Rostoker, The metallurgy of vanadium, Wiley, New York, 1958.

R.J.H. Clarke, The chemistry of titanium and vanadium, Elsevier, Amsterdam, 1968.

G.W.A. Fowles, Halide and oxyhalide complexes of elements of the titanium, vanadium and chromium subgroups, in Preparative inorganic reactions, Interscience, New York, 1964, p. 126.

R.J.H. Clark and D. Brown, The chemistry of vanadium, niobium and tantalum, Pergamon, Oxford, 1975.

34

Chromium, Molybdenum and Tungsten — Group VIA

34.1. Introduction

Although both chromium and molybdenum have the ground-state configuration d^5s^1 in contrast to d^4s^2 in tungsten, the main break in properties comes, as in the previous sub-groups of d-block elements, after Period 4. Chromium differs particularly from the other two in forming aquated bipositive and terpositive ions, which are similar to those formed by the other transition metals of Period 4.

As a consequence of the lanthanide contraction, the metallic radii of molybdenum and tungsten are the same, as are the radii of their 4+ cations (Table 34.1). The first ionisation energy of chromium is rather low for the

TABLE 34.1

ATOMIC PROPERTIES OF CHROMIUM, MOLYBDENUM AND TUNGSTEN

	Cr	Mo	W
Z	24	42	74
Electron configuration	$[Ar]3d^54s^1$	$[Kr]4d^55s^1$	$[Xe]4f^{14}5d^46s^2$
$I(1)$/kJ mol^{-1}	653	692	770
Metallic radius/pm	127	139	139
Ionic radius M^{3+}/pm	64		
$r_{M^{4+}}$/pm	55	68	68

metal's place in the Periodic Table but those of Mo and W are unremarkable.

The three metals are silvery white; they have b.c.c. structures at room temperature and are rather soft when carefully purified. The *m.p.* are high (Table 34.2), that of each metal being at the peak of its respective transition

TABLE 34.2

PHYSICAL PROPERTIES OF CHROMIUM, MOLYBDENUM AND TUNGSTEN

	Cr	Mo	W
ρ/g cm^{-3}	7.14	10.8	19.3
M.p./K	2180	2890	3680

series; thus tungsten has the highest *m.p.* of any metal and is consequently used for electric light filaments. Judged potentiometrically, chromium is a strong reducing agent ($E°$, Cr^{3+}/Cr = −0.74 V) but reactivity is inhibited by the formation of a tough oxide coating. The resistance of the metal to corrosion makes it a valuable protective plating on steel.

The metals are rather unreactive at laboratory temperature. Neither Mo nor W is easily dissolved by non-oxidising acids, though Mo is attacked readily by a mixture of concentrated HNO_3 and HF. Chromium dissolves in cold HCl but is rendered passive by HNO_3. The metals are much more reactive at high temperatures; they combine with oxygen, sulphur, nitrogen, carbon, silicon and boron; with the last four elements most of the products are interstitial compounds. With oxygen, Mo and W combine directly to form the trioxides MoO_3 and WO_3 at red heat, but Cr gives Cr_2O_3 in this way. Fluorine attacks the metals in the cold but the other halogens do so only on strong heating.

34.2. Occurrence, Separation and Properties of the Metals

34.2.1. Chromium

The element (0.02% lithosphere) is produced from chromite, $FeCr_2O_4$, which is a spinel. Direct reduction of this compound with carbon in an electric furnace yields ferrochrome, an iron—chromium alloy used in making stainless steel (12—26% Cr) and tool steels (3—6% Cr). Chromium itself can be obtained 99.8% pure by the electrolysis of aqueous chrome alum prepared from a solution of ferrochrome in sulphuric acid. The metal can be electroplated on to other surfaces to provide decoration, but steel is best plated first with either nickel or copper on which the Cr is deposited, because the decorative plating is rather porous. The purest chromium is made by the decomposition of chromium iodide, produced from the elements at 1050 K, on a filament heated to 1500 K.

The metal dissolves slowly in cold HCl and H_2SO_4, rapidly in the same acids when they are hot, but not at all in HNO_3. It reacts with chlorine at 900 K and bromine at red heat, with oxygen at 2300 K and with steam at red heat:

$$2\,Cr + 3\,X_2 \rightarrow 2\,CrX_3 \ (X = Cl \text{ or } Br)$$
$$4\,Cr + 3\,O_2 \rightarrow 2\,Cr_2O_3$$
$$2\,Cr + 3\,H_2O \rightarrow Cr_2O_3 + 3\,H_2$$

34.2.2. Molybdenum

The element (1.5×10^{-4}% of the lithosphere) is extracted from molybdenite, MoS_2, a substance which, because of its laminar structure, is also used as a lubricant, particularly for conditions of high temperature and high pressure. The concentration of MoS_2 in the ore is low, usually about 0.3%, but it can be recovered as a 90% concentrate by a flotation process. Roasting converts the MoS_2 to MoO_3, which can be either purified by sublimation or converted to ammonium molybdate by ammonia treatment. Molybdenum powder is obtained by reducing either the MoO_3 or the ammonium molybdate with hydrogen at 1000 K. The metal in its massive form is produced by sintering the powder in hydrogen at about 2500 K and 2 MPa. It is used for toughening steel and for supporting tungsten filaments in electric lamps. Annual world production is about 60 thousand tons.

Molybdenum reacts with fluorine at room temperature, with Cl_2 and Br_2 above 600 K, but not with iodine except under pressure. Thus the iodine method cannot be used, as in the case of chromium, for the production of the purest molybdenum, which must be made by the hydrogen reduction of carefully prepared MoO_3. The metal is not attacked by HCl, HF or H_2SO_4; it reacts initially with HNO_3, but quickly becomes passive. It is, however, attacked rapidly by fused oxidising agents such as Na_2O_2, $KClO_3$ and KNO_3.

34.2.3. Tungsten

The element (1.6×10^{-4}% of the lithosphere) is made principally from scheelite, $CaWO_4$, or from wolframite, a mixed crystal of $FeWO_4$ and $MnWO_4$. About 90% of the world's production of the metal (15 thousand tons per annum) is used in the manufacture of ferrous alloys, mainly tool steels (18—20% W). For this purpose ferrotungsten is made by reducing high grade concentrates of $FeWO_4$, low in manganese, with coke in an electric furnace. Tungsten for electric lamps is made from $CaWO_4$ which can be converted by HCl to tungstic acid subsequently purified by dissolving in aqueous alkali and reprecipitating. The acid is calcined to WO_3 which is reduced with hydrogen to powdered metal that can be converted to massive metal in the same way as molybdenum. Industrially, the most important compound of tungsten is its carbide, produced by direct combination at high temperature. Sintered with cobalt powder, it produces a material which is almost as hard as diamond and is applied where great resistance to wear is important, as in rock-drilling tools.

The reactivity of tungsten is very similar to that of molybdenum, but whereas Mo is converted to $MoCl_5$ by chlorination and to $MoBr_3$ by bromination, the products in the case of tungsten are the hexahalides.

34.3. Oxidation States

For Cr, as for Ti and V in the earlier groups, the highest oxidation state is that corresponding to the periodic group number, but Cr^{VI} compounds differ from those of Ti^{IV} and V^{V} in having strong oxidising power, for example $E°$, $Cr_2O_7^{2-}/Cr^{3+}$ is +1.33 V. For the higher congeners Mo and W the +6 state is much more stable to reduction, however (Fig. 34.1). This relation between

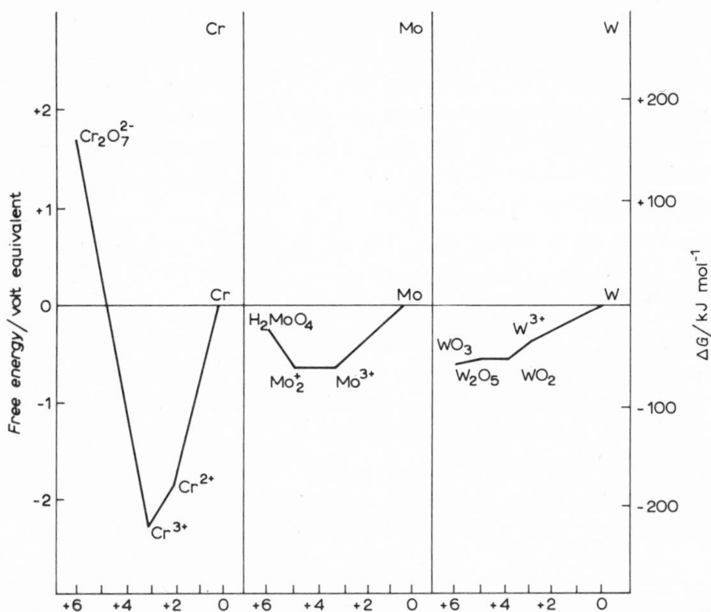

Fig. 34.1. Free energies of the oxidation states of Group VIA metals relative to the metals in aqueous solution at pH = 0.

the heavier d-block elements and those in the first row is repeated in the later transition groups also.

Cr, Mo and W all form diamagnetic hexacarbonyls in which the formal oxidation state of the metal is zero, and some carbonylate anions are known which contain the metals in negative oxidation states. For chromium the Cr^{3+} ion has an extensive aqueous chemistry, but the Cr^{2+} ion is a strong reducing agent (E°, Cr^{3+}/Cr^{2+} = −0.4 V). Mo and W do not form simple hydrated cations in aqueous solution.

Representative compounds of the various oxidation states are shown in Table 34.3. As is general in this part of the Periodic Table, the lowest oxida-

TABLE 34.3

COMPOUNDS AND IONS REPRESENTING THE OXIDATION STATES OF CHROMIUM, MOLYBDENUM AND TUNGSTEN

Oxidation state	Cr	Mo	W
−2	$Cr(CO)_5^{2-}$	$Mo(CO)_5^{2-}$	
−1	$Cr_2(CO)_{10}^{2-}$		
0	$Cr(C_6H_6)_2$	$Mo(CO)_3py_3$	$W(CO)_6$
+1	$Cr(dipy)_3^+$	$(C_6H_6)_2Mo^+$	
+2	$CrCl_2$	Mo_6Cl_{12}	$W_6Cl_8^{4+}$
+3	$Cr_2(SO_4)_3$	$Mo(CN)_6^{3-}$	$K_3W_2Cl_9$
+4	Ba_2CrO_4	MoS_2	WO_2
+5	CrF_5	$Mo(CN)_8^{3-}$	WF_6^-
+6	CrO_4^{2-}	MoO_2Cl_2	WF_8^{2-}

tion states are stabilised by π-acceptors such as CO, the intermediate states by ligands such as CN^-, Cl^-, Br^- and H_2O, and the highest states by the hard bases F^- and O^{2-}, though in tungsten the +6 state is of particular importance and is exemplified by the hexahalides WCl_6 and WBr_6 as well as WF_6, the oxide WO_3 and its corresponding acid and salts. Although the +6 state is also important in molybdenum, the +5 state is stabilised by many common ligands; the highest chloride of Mo, for example, is $MoCl_5$.

34.4. Halides

The halides of Cr, Mo and W are shown in Table 34.4, which illustrates the usual trends to be found in the halides of d-block congeners. The metals of

TABLE 34.4

HALIDES OF CHROMIUM, MOLYBDENUM AND TUNGSTEN

Charge numbers	Cr		Mo		W	
+2	CrF_2,	$CrCl_2$		Mo_6Cl_{12} *		W_6Cl_{12}
	$CrBr_2$,	CrI_2	Mo_6Br_{12},	Mo_6I_{12}	W_6Br_{12},	W_6I_{12}
+3	CrF_3,	$CrCl_3$	MoF_3,	$MoCl_3$		WCl_3
	$CrBr_3$,	CrI_3	$MoBr_3$,	MoI_3	WBr_3,	WI_3
+4	CrF_4		MoF_4,	$MoCl_4$	WF_4,	WCl_4
			$MoBr_4$		WBr_4,	WI_4
+5	CrF_5		MoF_5,	Mo_2Cl_{10}		WCl_5
					WBr_5	
+6	CrF_6		MoF_6		WF_6,	WCl_6
					WBr_6	

* The brown dichloride, $MoCl_2$, which is distinct from yellow Mo_6Cl_{12}, is also known.

Periods 5 and 6 form more of the higher halides than does the Period 4 element. Thus Mo and W, like the Group IV and Group V metals which precede them, form several halides in which all the outer d and s electrons are used in bonding. The +6 state is particularly strongly represented in tungsten, a fact of some interest because this is an unusual oxidation state for metals to exhibit in their chlorides and bromides.

The chromium(II) halides are made by the action of hydrogen halides on the metal at 1000 K (CrF_2, $CrCl_2$ and $CrBr_2$) or by direct combination of the elements (CrI_2). The d^4 configuration gives rise to Jahn—Teller distortion; in the CrF_2 crystal for example there is a tetragonal arrangement of F^- ions around the Cr^{2+} ion, with four of the fluoride ions at 200 pm and two at 243 pm. Although a true dihalide of molybdenum, $MoCl_2$, can be made by the action of chlorine on $Mo(CO)_6$, the more common halides containing Mo^{II} and W^{II} are cluster compounds of formula $(M_6X_8)^{4+}(X^-)_4$. The complex

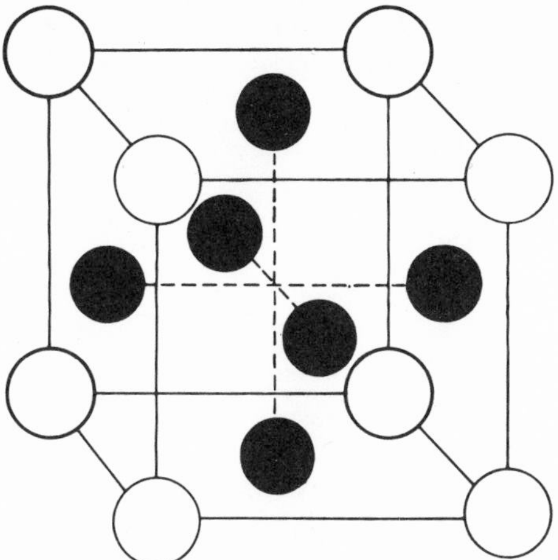

Fig. 34.2. Metal cluster ion $Mo_6Cl_8^{4+}$.

cation $Mo_6Cl_8^{4+}$ has been shown to be cubic, with Cl atoms at the eight corners and Mo atoms at the six face-centres (Fig. 34.2). The $M_6X_8^{4-}$ ion can co-ordinate six electron-pair donors, one to each metal atom along a fourfold axis of the octahedral cluster.

In Mo_6Cl_{12} each $Mo_6Cl_8^{4+}$ ion is co-ordinated thus to four bridging Cl^- ions and two non-bridging ones. This compound is prepared by the reduction of $MoCl_4$ with Mo at about 900 K, while W_6Cl_{12} and W_6Br_{12} can be made by heating the pentahalides with aluminium.

Red-violet $CrCl_3$ can be made by heating the metal with chlorine or the oxide Cr_2O_3 with CCl_4. The colour of an aqueous solution varies with temperature and chloride-ion concentration. In a cold, dilute solution the octahedral $Cr(H_2O)_6^{3+}$ ion is violet, but as the temperature is raised and the Cl^- concentration is increased, the colour changes to green because of the reaction:

$$Cr(H_2O)_6^{3+} + Cl^- \rightarrow CrCl(H_2O)_5^{2+} + H_2O$$

Three isomers of $Cr(H_2O)_6Cl_3$ can be crystallised from aqueous solutions depending on the conditions (28.4.2). The tribromide, however, has only two hydration isomers and the tri-iodide has only one form.

Of the trihalides of Mo and W, those of most interest are the brown solid MoF_3, made by the reduction of MoF_6 with Mo at about 700 K, and WCl_3, which is made by treating W_6Cl_{12} at 370 K with Cl_2. It is really a cluster compound $(W_6Cl_{12})Cl_6$ in which the $W_6Cl_{12}^{6+}$ ion consists of an octahedron of W atoms bridged along every edge by Cl atoms.

The higher fluorides of chromium are made by direct combination of the elements, CrF_4 at about 600 K, CrF_5 between 650 K and 750 K, and CrF_6 at 700 K in a bomb at 20 MPa.

The tetrahalides of Mo and W are not, in general, easy to obtain pure; the fluorides can be made by reducing the hexafluorides with benzene at about

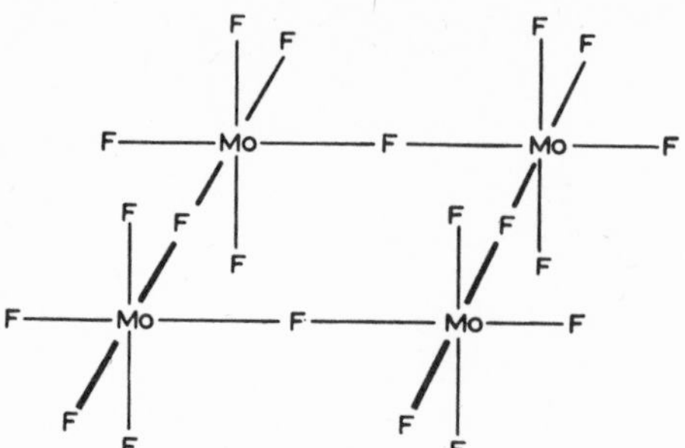

Fig. 34.3. Tetramer of molybdenum(V) fluoride.

380 K, and WCl_4 results from the reduction of WCl_6 with aluminium in a thermal gradient. It disproportionates to W_6Cl_{12} and WCl_5 at about 750 K. The most stable pentahalides are Mo_2Cl_{10} and WBr_5, both made by direct combination of the elements at high pressure.

The dark-green crystals of molybdenum(V) chloride contain Mo_2Cl_{10} molecules similar in structure to Nb_2Cl_{10} (33.4). The compound is paramagnetic, with $\mu_{eff} = 1.6\ \mu_B$ at 293 K, indicating a very slight coupling of the electron spins of the two metal atoms. In the vapour and in solvents such as benzene the compound exists as $MoCl_5$. The pentafluoride, MoF_5, a yellow solid, is made by reducing MoF_6 with $Mo(CO)_6$:

$$5\ MoF_6 + Mo(CO)_6 \rightarrow 6\ MoF_5 + 6\ CO$$

It has a tetrameric structural unit (Fig. 34.3) in which the arrangement of F atoms around the metal atoms is, however, only approximately octahedral.

The hexafluorides of Mo and W are colourless, volatile, diamagnetic compounds, both made by direct fluorination of the metal. The chloride WCl_6 and the bromide WBr_6 can also be made by direct halogenation of the metals; they are dark-blue solids. These covalent hexahalides are all readily hydrolysed by water.

34.4.1. Oxohalides

The three metals form many oxohalides of the type MO_2X_2. In the case of Cr and Mo, X can be F, Cl or Br, in the case of W, X can be Cl, Br or I. Chromyl chloride, CrO_2Cl_2, familiar as a distinguishing test for a chloride, is the yellow distillate obtained when a chloride is heated with $K_2Cr_2O_7$ and concentrated H_2SO_4:

$$Cr_2O_7^{2-} + 4\ Cl^- + 6\ H_2SO_4 \rightarrow 6\ HSO_4^- + 3\ H_2O + 2\ CrO_2Cl_2$$

The compounds are mostly yellow or orange liquids or low-melting solids; those of molybdenum and tungsten can usually be made by treating the trioxide with the hydrogen halide or the halogen. There are also many compounds of general formula MOX_4. Examples are $MoOCl_4$, which is made by the chlorination of MoO_2Cl_2 with a mixture of S_2Cl_2 and Cl_2, and $MoOF_4$

which is made from it:

$MoOCl_4 + 4\ HF \rightarrow MoOF_4 + 4\ HCl$

These oxohalides of the metals in their +6 states form many complexes (34.10).

There are also oxohalides of the metals in their +5 states, and compounds such as $CrOCl_3$ and $MoOCl_3$ have been made. The chromium compound disproportionates above 273 K into CrO_2Cl_2 and a chromium(III) compound. CrOCl, an oxochloride of Cr^{III}, is obtained when Cr_2O_3 is heated strongly with $CrCl_3$.

34.5. Oxides

All three metals form trioxides. White MoO_3 and yellow WO_3 are produced when the metals, their other oxides or their sulphides are heated in air or oxygen. The dark-red, crystalline CrO_3 is less thermally stable and cannot be made by direct combination of the elements; it is usually obtained by treating a saturated aqueous solution of a dichromate with concentrated H_2SO_4:

$Cr_2O_7^{2-} + 3\ H_2SO_4 \rightarrow 3\ HSO_4^- + H_3O^+ + 2\ CrO_3$

It differs from MoO_3 and WO_3 in being very soluble in water, but it does not appear to form any crystalline hydrates. It is widely used in organic chemistry as a strong oxidising agent, usually in the form of a solution in acetic acid. The structure consists of chains of CrO_4 tetrahedra linked at corner oxygens; not surprisingly the *m.p.* is low (470 K). MoO_3 has a layer lattice and melts at 1070 K but WO_3 has an only slightly distorted version of the ReO_3 lattice (Fig. 7.23) and melts at 1450 K.

Hydrates of these oxides, the yellow $MoO_3 \cdot 2\ H_2O$ and the colourless, isomorphous $WO_3 \cdot 2\ H_2O$, can be crystallised from cold solutions of molybdates and tungstates when they are made strongly acidic. The molybdenum compound has been shown to contain sheets of MoO_6 octahedra sharing corners, with one H_2O covalently bound to Mo and the other hydrogen-bonded in the lattice. From hot solutions of molybdates and tungstates monohydrates $MO_3 \cdot H_2O$ are precipitated rapidly by acids.

When CrO_3 is melted it loses oxygen to give a series of lower oxides, including the ferromagnetic CrO_2 used in magnetic recording tapes, which has an undistorted rutile structure, and eventually the green, thermally stable α-Cr_2O_3 which has the corundum lattice. This compound is more conveniently made by the action of heat on $(NH_4)_2Cr_2O_7$:

$(NH_4)_2Cr_2O_7 \rightarrow N_2 + 4\ H_2O + Cr_2O_3$

The oxides MoO_3 and WO_3 are reduced by heating with mild reducing agents. The products are of great structural interest and have been extensively studied. For example, the yellow WO_3 changes to deep blue as a phase

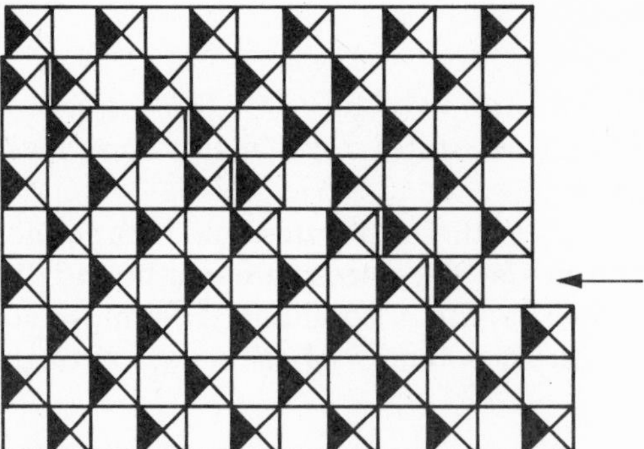

Fig. 34.4. Diagram indicating shearing in MO_3-type oxide consisting of MO_6 octahedra sharing corners.

$W_{50}O_{148}$ crystallises. Fig. 34.4 illustrates the type of crystallographic shearing which is responsible for this phenomenon. As oxygen is lost from the trioxide, oxygen vacancies are removed by slabs of corner-linked octahedra moving into positions where they share edges with the octahedra of adjacent slabs. It has been shown that shearing of this type occurs at regular intervals throughout the solid phase; the reasons for this long-range ordering of shear lines is not yet understood. Further reduction of both MoO_3 and WO_3 gives rise to phases of general formula M_nO_{3n-1} with $n = 14, 12, 11, 10, 9, 8$, and in the case of molybdenum $n = 5$ and 4 also. These last have more complicated structures in which the shearing is of a different kind, but long-range ordering again appears within a particular phase.

Further reduction with hydrogen at temperatures below 750 K, or with ammonia, produces the violet MoO_2 and brown WO_2. These compounds have much distorted rutile structures in which the short distances between metal atoms indicates bonding between them.

34.6. Chromates, Molybdates and Tungstates

Yellow solutions of the tetrahedral CrO_4^{2-} ion are turned orange below pH 6 because the dichromate ion, $Cr_2O_7^{2-}$ is produced:

$$2\,CrO_4^{2-} + 2\,H^+_{aq} \rightleftharpoons Cr_2O_7^{2-} + aq$$

The equilibrium is labile, however, and insoluble chromates like Ag_2CrO_4, $BaCrO_4$ and $PbCrO_4$ are precipitated even from mildly acidic solution. The last of these is known as chrome yellow, a useful pigment. A salt $KCrO_3Cl$ can be crystallised from a solution of $K_2Cr_2O_7$ in strong aqueous HCl. Despite the fact that $E°$, $Cr_2O_7^{2-}/Cr^{3+}$ is higher than $E°$, X_2/X^- for X = Cl, Br or I, all three compounds $KCrO_3X$ can be made similarly because the redox reactions are so slow.

Normal molybdates and tungstates containing discrete, tetrahedral MoO_4^{2-} and WO_4^{2-} ions can be crystallised from solutions within limited pH ranges.

The salt $(NH_4)_2MoO_4$ separates from a solution of MoO_3 in strong aqueous ammonia, but from a near-neutral solution the salt which is obtained is $(NH_4)_6Mo_7O_{24} \cdot 4\ H_2O$. The $Mo_7O_{24}{}^{6-}$ ion is an example of an isopolyanion. Octamolybdates containing $Mo_8O_{26}{}^{4-}$ can also be made, and the most important species obtained in tungstate solutions are $HW_6O_{21}{}^{5-}$ and $W_{12}O_{41}{}^{10-}$. These ions are probably hydrated to some extent both in solution and in crystals. It has been shown, for example, that the ion $H_2W_{12}O_{42}{}^{10-}$ is present in the salt usually formulated $(NH_4)_{10}W_{12}O_{41} \cdot 11\ H_2O$.

34.6.1. Heteropolyacids

The yellow precipitate obtained in the ammonium molybdate test for a phosphate is $(NH_4)_2H(PMo_{12}O_{40}) \cdot H_2O$. When it is washed with dilute NH_4NO_3 it becomes $(NH_4)_3(PMo_{12}O_{40})$; it is in fact a useful inorganic cation exchanger. This 12-molybdophosphate is an example of a large number of heteropolysalts formed by molybdates and tungstates when acidified in the presence of other oxo-anions. Some typical heteropolyacids are

$H_4(SiMo_{12}O_{40})$	12-molybdosilicic acid
$H_3(AsMo_{12}O_{40})$	12-molybdoarsenic acid
$H_5(BW_{12}O_{40})$	12-tungstoboric acid

They are known as 12-acids because there are twelve Mo or W atoms to one hetero-atom. More than thirty elements are known to function as hetero-atoms in these compounds; examples are Ti, Ge, Sn, Zr, Hf in their +4 states. The most common structure is shown in Fig. 34.5. It consists, in effect, of a

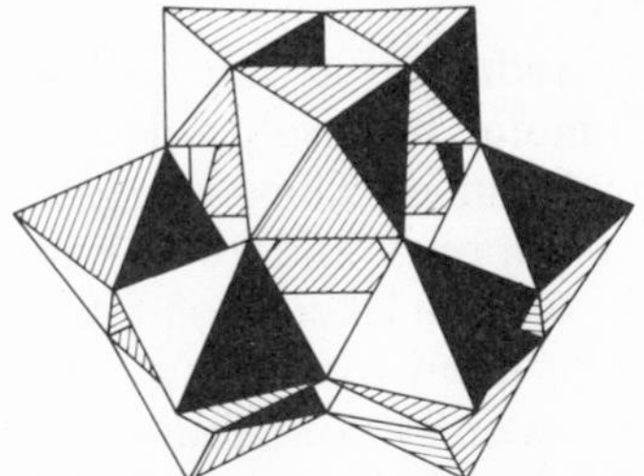

Fig. 34.5. Most common structure of a 12-acid heteropolyanion.

tetrahedron around the hetero-atom, every corner of which is shared by three WO_6 or MoO_6 octahedra each of which shares an oxygen with its neighbours. The four resulting Mo_3O_{13} groups, for example, share corners to form $(MMo_{12}O_{40})^{n-8}$ ions where n is the charge number of M. The large open spaces in ions of this type allow the inclusion of water molecules and cations. Hydrates are, in fact, common and the insoluble salts are often excellent cation-exchangers. There are a few 12-acids containing the ions $(MW_{12}O_{42})^{n-12}$ and $(MMo_{12}O_{42})^{n-12}$ where M is, for example, Ce^{IV} or Th^{IV}, in which the hetero-atoms appear to be octahedrally co-ordinated to oxygen atoms instead of tetrahedrally.

Fig. 34.6. Arrangement of MoO_6 octahedra in a 6-acid.

6-acids of molybdenum are also known; they have the general formula $H_m(MMo_6O_{24})$, where M can be I, Te, Fe, Co, Al, Rh, Cr, Cu or Mn, and its charge number $12-m$. In the 6-acids, six MoO_6 octahedra are joined by sharing edges to form a hexagonal annulus which provides 6 oxygen atoms to co-ordinate the central hetero-atom (Fig. 34.6). The structure of the isopoly-anion $Mo_7O_{24}^{6-}$ is just this with an Mo atom replacing the hetero-atom at the centre.

Molybdenum also forms salts of 9-acids, e.g. $H_6(MnMo_9O_{32})$ and salts of the more complex 10- and 11-acids with Mo: hetero-atom ratios of 10 and 11 respectively. Dimeric ions derived from the 9-acids, e.g. $(P_2Mo_{18}O_{62})^{6-}$, have also been studied.

34.7. Tungsten Bronzes

When alkali-metal tungstates are reduced with hydrogen at red heat, or when WO_3 is treated with the vapour of the alkali metal, intensely coloured, unreactive substances of bronze-like lustre are obtained. These tungsten bronzes have the general formula M_xWO_3 (M = Li, Na or K and $x < 1$) and are of considerable interest structurally. When $x > 0.3$ the lattice consists of cubic ReO_3 unit cells (Fig. 7.23) with a fraction x of them containing alkali-metal atoms at their body centres. These atoms evidently contribute their 3s electrons to delocalised energy bands rather like those in a metal, and the compounds are good electrical conductors. For values of x below 0.3 the cubic lattice is distorted towards triclinic, as in WO_3 itself, the conductivity falls, and the colour tends towards violet. It has not been found possible to make compounds with $x > 0.95$. These compounds rich in alkali metal are golden-yellow; this structure approximates closely to that of perovskite (Fig. 7.27), and the cell dimensions are greatest when x is greatest.

34.8. Binary Compounds with Other Non-metals

Of the Group VI metals, only chromium forms a hydride, which can be made by electrolytic reduction of certain Cr^{VI} solutions. A phase with composition range $CrH_{0.5-1.0}$ has a hexagonal structure; a second less thermally stable phase with a composition $CrH_{1.0}$ to $CrH_{1.7}$ has an anion-deficient modification of the fluorite structure.

Molybdenum and tungsten react at high temperatures with carbon and nitrogen to give hard, refractory carbides and nitrides. These generally have the character of interstitial compounds, but there are phases of stoichiometric composition M_2C, M_2N, MC and MN. These monocarbides and mononitrides have c.c.p. arrangements of metal atoms with C or N atoms in octahedral holes. The metallic radius of Cr is too small for the metal to incorporate interstitial C atoms in its structure, and the carbide Cr_3C_2 can be described as having carbon chains, with rather large C—C distances (ca. 165 pm), running through a greatly distorted metal lattice. It is hydrolysed by dilute acids to give a mixture of hydrocarbons, and can be considered to be intermediate between the truly ionic carbides and the interstitial ones.

There are several types of interstitial borides. MoB and WB have structures like FeB (Fig. 15.8) whereas MoB_2 and WB_2 resemble AlB_2.

The chromium—sulphur system is complex. Dark-green, paramagnetic Cr_2S_3 can be made by the action of H_2S on Cr_2O_3. When it is heated to a high temperature it can be converted eventually into CrS, but there is evidence for several intermediate phases forming a homologous series Cr_nS_{n+1}.

The most important sulphide of molybdenum is molybdenite, MoS_2, which has a structure rather like that of CdI_2, but with the layers of S atoms eclipsed instead of staggered (Fig. 34.7). Molybdenite resembles graphite in being

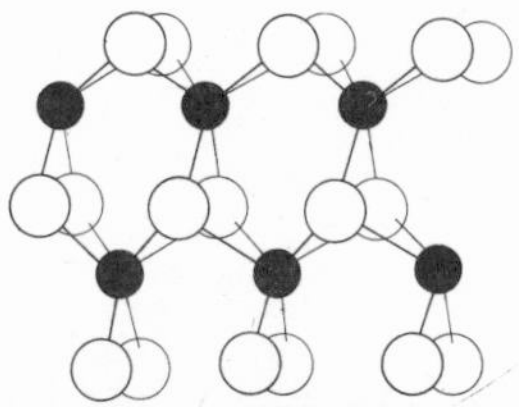

Fig. 34.7. Eclipsed arrangement of S atoms in molybdenite, MoS_2.

an excellent lubricant, and is added to engine oils for that purpose. A trisulphide, brown MoS_3, is precipitated when H_2S is passed into acidified aqueous molybdate. The sulphides WS_2 and WS_3 resemble the molybdenum compounds.

34.9. Organometallic Compounds and π-complexes

Dibenzenechromium, which has a sandwich structure with a Cr atom between parallel benzene rings, can be made by heating $CrCl_3$ with aluminium, Al_2Cl_6 and benzene to obtain $[(\pi\text{-}C_6H_6)_2Cr]^+(AlCl_4)^-$ and then reducing this with aqueous sodium dithionite. The dark-brown, diamagnetic solid can be sublimed in vacuo at 420 K but is much more sensitive to air than is ferrocene (18.2.5). Although $(C_6H_6)_2Cr$ contains the same number of bonding electrons as ferrocene, it is much less stable to electrophiles; attempts to effect aromatic substitution always cause decomposition. Bisbenzenoid complexes of Mo and W have also been made; they too are unstable towards electrophiles.

Sodium cyclopentadienide reacts with anhydrous $CrCl_2$ in tetrahydrofuran to give scarlet $(\pi\text{-}C_5H_5)_2Cr$. Compounds containing a blue cation, probably $[(\pi\text{-}C_5H_5)CrCl(H_2O)_n]^+$, are easily obtained on oxidation in the presence of HCl. The compound $(\pi\text{-}C_5H_5)_2Cr$ is, in fact, very air-sensitive. The metals of this group form many π-cyclopentadienyl carbonyl complexes. The infrared spectrum of $[(\pi\text{-}C_5H_5)Cr(CO)_3]_2$ shows no bridging carbonyl frequency, and the analogous Mo compound has been shown by X-ray analysis to have an Mo—Mo bond. The group $(\pi\text{-}C_5H_5)Mo(CO)_3$ is surprisingly inert; the dimer can be converted to $Na[(\pi\text{-}C_5H_5)Mo(CO)_3]$ by sodium amalgam in tetrahydrofuran and to $(\pi\text{-}C_5H_5)Mo(CO)_3H$ by hydrogen at 17 MPa. This hydride can be converted to the bromo-derivative $(\pi\text{-}C_5H_5)Mo(CO)_3Br$ with *N*-bromosuccinimide and into $(\pi\text{-}C_5H_5)Mo(CO)_3CF_2CF_2H$ with tetrafluoroethylene. The compound $(\pi\text{-}C_5H_5)_2MoH_2$, obtained from $MoCl_5$, C_5H_5Na and $NaBH_4$ in tetrahydrofuran, has the structure shown in Fig. 34.8. The two C_5H_5 rings are inclined at 34° to one another.

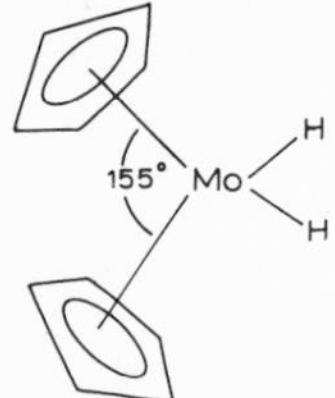

Fig. 34.8. Structure of $(\pi\text{-}C_5H_5)_2MoH_2$.

Chromium and molybdenum compounds in which the metals are π-bonded to seven-membered and eight-membered cyclic hydrocarbons have also been made. Thus $(\pi\text{-}C_5H_5)(\pi\text{-}C_7H_7)Cr$ has been made by the method which is outlined:

$$CrCl_3 \xrightarrow[C_6H_5MgBr]{C_5H_5MgBr} (\pi\text{-}C_5H_5)(\pi\text{-}C_6H_6)Cr \xrightarrow[Al_2Cl_6]{C_7H_8} Cr^+ \xrightarrow{S_2O_4^{2-}} Cr$$

The compound $C_8H_8Mo(CO)_3$ has been shown to have the structure shown in Fig. 34.9, in which the cyclo-octatetrene acts as a 6-electron donor.

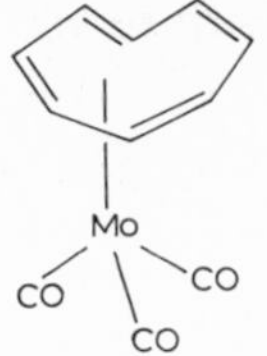

Fig. 34.9. Structure of $C_8H_8Mo(CO)_3$.

34.10. Complexes

There are not many complexes containing the metals in the +2 state. Chromium forms ammines, for instance $Cr(NH_3)_6Cl_2$, which are high-spin d^4 complexes and, as such, exhibit Jahn—Teller distortion to a tetragonal symmetry. There are some low-spin cyanocomplexes such as $K_4Cr(CN)_6$. Molybdenum(II) is represented by high-spin orthophenylenebisdimethyldiarsine complexes, $Mo(diars)_2X_2$ (X = Cl, Br, I).

Chromium(III) complexes (d^3 configuration) are very common; there are many ammine complexes of this oxidation state. The purple chloropenta-amminechromium(III) dichloride, $[Cr(NH_3)_5Cl]Cl_2$, is made by bubbling air through a solution of $CrCl_2$, NH_4Cl and NH_3 in water. It can be converted into hexa-amminechromium(III) trichloride, $[Cr(NH_3)_6]Cl_3$ (yellow), by treating its cold, concentrated solution with ammonia. A violet dichlorotetra-amminechromium(III) chloride, $[Cr(NH_3)_4Cl_2]Cl$, exists and also the triammine, $Cr(NH_3)_3Cl_3$. Werner (1910) made the latter by the reactions:

$$CrO_3 \xrightarrow[\text{dilute } H_2SO_4]{H_2O_2\text{, pyridine and}} \text{pyridinium perchromate} \xrightarrow{NH_3} CrO_4 \cdot (NH_3)_3 \xrightarrow{\text{cold concentrated HCl}} Cr(NH_3)_3Cl_3$$

Cyano- and thiocyanato-complexes of chromium are also common. Among the more interesting of these is Reinecke's salt, $NH_4[Cr(NH_3)_2(SCN)_4]H_2O$, made by adding $(NH_4)_2Cr_2O_7$ slowly to melted NH_4SCN, washing with, and recrystallising from, alcohol. The octahedral ion has the form shown in Fig. 34.10.

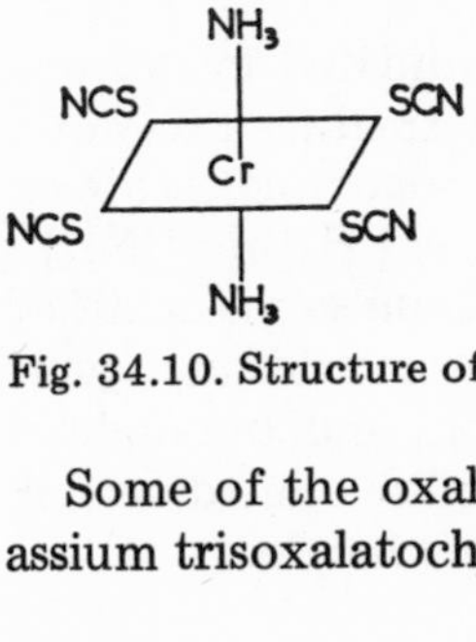

Fig. 34.10. Structure of $[Cr(NH_3)_2(SCN)_4]^-$ ion present in Reinecke's salt.

Some of the oxalato-complexes of chromium(III) are also of interest. Potassium trisoxalatochromate(III), $K_3Cr(C_2O_4)_3 \cdot 3\,H_2O$, is obtained by adding potassium oxalate to the solution obtained by reducing $K_2Cr_2O_7$ with oxalic acid:

$$K_2Cr_2O_7 + 7\,H_2C_2O_4 \rightarrow \underbrace{K_2C_2O_4 + Cr_2(C_2O_4)_3} + 6\,CO_2 + 7\,H_2O$$

$$\downarrow 2\,K_2C_2O_4$$

$$2\,K_3[Cr(C_2O_4)_3]$$

The anion of the blue crystalline compound was resolved by Werner (1912) into dextrorotatory and laevorotatory forms (Fig. 34.11A). Potassium dioxalatodiaquochromate, $K[Cr(C_2O_4)_2(H_2O)_2]$, exists in *cis*- and *trans*-forms, the former (Fig. 34.11B (i)) showing purple-green dichroism and the latter (Fig. 34.11B (ii)) being mauve.

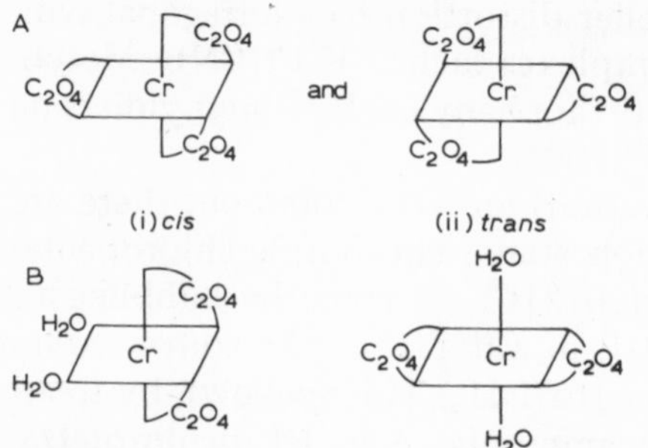

Fig. 34.11. A. *d*- and *l*-forms of $[Cr(C_2O_4)_3]^{3-}$ ion. B. *cis*- and *trans*-forms of $[Cr(C_2O_4)_2(H_2O)_2]^-$ ion.

Molybdenum(III) exists in neutral, anionic and cationic complexes. $MoCl_3$ and $MoBr_3$ form the adducts with pyridine $Mo(py)_3X_3$, but the halides undergo solvolysis in liquid ammonia, methylamine and dimethylamine to give products such as $MoBr_2NH_2$, $MoBr_2NHMe$ and $MoBr_2NMe_2$. The complex anions include many octahedral ones of the MoX_6^{3-} type in which X is a halogen or pseudohalogen. The complex Mo^{III} cations include $Mo(dipy)_3^{3+}$ and $Mo(o\text{-phen})_3^{3+}$. Tungsten(III), however, forms few complexes; these are mainly complex halides. The $W_2Cl_9^{3-}$ ion in $K_3W_2Cl_9$ has the same structure as the $Tl_2Cl_9^{3-}$ ion (Fig. 16.3).

Mo^{IV} and W^{IV} are represented by interesting 8-co-ordinated cyanocomplexes containing $M(CN)_8^{4-}$ ions. $K_4Mo(CN)_8$ is made by treating a K_3MoCl_6 solution with KCN in the presence of air. These octacyanocomplexes are remarkably thermally and hydrolytically stable. The free acid $H_4[Mo(CN)_8] \cdot 6\,H_2O$, which can be isolated as crystals when the potassium salt is acidified with HCl, can be oxidised to $H_3[Mo(CN)_8] \cdot 3\,H_2O$ by strong oxidising agents such as MnO_4^-. The Mo^{5+} ion has only one 4d electron, and the dodecahedral arrangement of CN^- ligands around it should stabilise one 4d orbital, capable of accommodating these electrons, relative to the other four:

Energy ↑

d_{xy}, d_{yz} — —

d_{z^2} —

$d_{x^2-y^2}$ —

d_{xz} —

The structure of the $Re(CN)_8^{3-}$ ion is similarly stabilised. In solid $K_4Mo(CN)_8$ the complex ion is dodecahedral (Fig. 34.12).

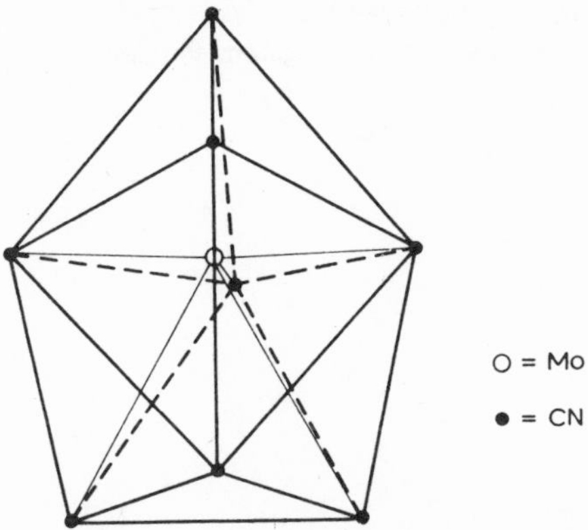

Fig. 34.12. Dodecahedral structure of $Mo(CN)_8^{4-}$ ion.

The oxidation state +5 is represented by some fluoro- and oxohalide complexes of molybdenum and tungsten, such as M^IMoF_6, M^IWF_6, $M^I_2[MoOCl_5]$ and $M^I[MoOBr_4]$. There are also the cyanocomplexes like $K_3W(CN)_8$, made by the oxidation of aqueous solutions of the corresponding +4 complexes (above).

Molybdenum(VI) and tungsten(VI) exist in anionic complexes containing oxygen and fluorine as ligands. Examples are $[MoOF_5]^-$ and $[WOF_5]^-$, $[MoO_3F_3]^{3-}$ and $[WO_3F_3]^{3-}$.

Further Reading

R.V. Parish, The inorganic chemistry of tungsten, Adv. Inorg. and Radiochem., 9 (1966) 315.

P.G. Dickens and M.S. Whittingham, The tungsten bronzes and related compounds, Q. Rev., 22 (1968) 30.

G.D. Rieck, Tungsten and its compounds, Pergamon, Elmsford, N.Y., 1967.

D.L. Kepert and K. Vrieze, Halides containing multicentred metal—metal bonds, in Halogen chemistry, Vol. 3, Ed. V. Gutmann, Academic Press, London and New York, 1967.

J.E. Fergusson, Halide chemistry of chromium, molybdenum and tungsten, in Halogen chemistry, Vol. 3, Ed. V. Gutmann, Academic Press, London and New York, 1967.

F.A. Cotton, Transition-metal compounds containing clusters of metal atoms, Quart. Rev., 20 (1966) 389.

P.C.H. Mitchell, Oxo-species of molybdenum(V) and molybdenum(VI), Quart. Rev., 20 (1966) 103.

K.F. Jahr and J. Fuchs, New methods and results in the study of polyacids, Angew. Chem. Internat. Ed., 5 (1966) 689.

K. Schwabe, The passivity of metals, Angew. Chem. Internat. Ed., 5 (1966) 185.

W.H. Hartford and M. Darrin, Chemistry of chromyl compounds, Chem. Rev., 58 (1958) 1.

E.N. Simons, Chromium, in Guide to uncommon metals, Hart, New York, 1967, p. 49.

J.D. Corbett, Metal halides in low oxidation states, in Preparative inorganic reactions, Vol. 3, Interscience, New York, 1966, p. 7.

C.D. Rieck, Tungsten and its compounds, Pergamon, Oxford, 1967.

D.L. Kepert, Isopolytungstates, Prog. Inorg. Chem., 4 (1967) 199.

C.L. Rollinson, The chemistry of chromium, molybdenum and tungsten, Pergamon, Oxford, 1975.

H.J.M. Bowen, Trace elements in biochemistry, Academic Press, London, 1966.

Manganese, Technetium and Rhenium — Group VIIA

35

35.1. Introduction

As in the preceding groups the properties of the first member stand apart from those of the higher congeners. All three elements have seven d and s electrons available for valency orbitals and the +7 state is clearly characterised in the MnO_4^-, TcO_4^- and ReO_4^- ions. Tc and Re differ from Mn in failing to form simple +2 ions. In this respect they resemble the elements of corresponding series in the earlier transition-metal groups.

TABLE 35.1

ATOMIC PROPERTIES OF Mn, Tc AND Re

	Mn	Tc	Re
Z	25	43	75
Electron configuration	$[Ar]3d^5 4s^2$	$[Kr]4d^5 5s^2$	$[Xe]4f^{14}5d^5 6s^2$
$I(1)$/kJ mol^{-1}	717	703	760
Metallic radius/pm	126	136	137

Physically, the metals are typical d-block metals; their densities (Table 35.2) are normal for the positions they occupy in the Periodic Table. The *m.p.* of Tc and Re are only slightly less than those of Mo and W which precede them, but that of manganese is low compared with the metals on either side of it, Cr and Fe. Although technetium and rhenium have h.c.p. struc-

TABLE 35.2

PHYSICAL PROPERTIES OF Mn, Tc and Re

	Mn	Tc	Re
ρ/g cm^{-3}	7.21—7.44	11.5	21.0
M.p./K	1517	2600	3450

tures, manganese exists in three forms, not one of which has the simple 8-co-ordination or 12-co-ordination typical of the transition metals. In α-Mn, for example, there are four kinds of crystallographically non-equivalent atoms and the Mn—Mn distances vary from 224 pm to 296 pm. Its brittleness and its resistance to wear may both be due to this peculiarity.

In Group VIIA the first ionisation energy of the second element, Tc, is rather less than that of the first element, Mn. In this respect the group is different from Group VIA but similar to the iron and cobalt groups. The effect on the chemistry is minimal, however. The value of E°, Mn^{2+}/Mn (Table 35.3) indicates manganese to be a better reducing agent than its neighbours, a fact which must be largely due to the lower boiling point and correspondingly low sublimation energy of the metal.

TABLE 35.3

IONISATION ENERGIES AND ELECTRODE POTENTIALS OF GROUP VIIA ELEMENTS

		Mn	Tc	Re
Ionisation energy	$I(1)$/kJ mol^{-1}	717	703	760
	$I(2)$/kJ mol^{-1}	1492	1470	
E^0, $M^{2+}/M/V$		−1.18		
E^0, $MO_2/M/V$			+0.27	+0.25
E^0, $MO_4^-/M/V$		+0.79	+0.47	+0.34

The redox potentials for technetium and rhenium (Table 35.3) are for the reactions:

$$MO_2 + 4\,H_3O^+ + 4\,e \rightleftharpoons M + 6\,H_2O$$
$$MO_4^- + 8\,H_3O^+ + 7\,e \rightleftharpoons M + 12\,H_2O$$

The free energies of some of the oxidation states relative to the metal in aqueous solution at pH 0 are given in Fig. 35.1. A striking point is the very great stability of Mn^{II} to both oxidation and reduction. The Mn^{VI} state is unstable to disproportionation in acid solution but is more stable at high pH. Thus K_2MnO_4 (which is isomorphous with K_2CrO_4) is converted by as weak an acid as H_2CO_3 into $KMnO_4$ and MnO_2:

$$3\,K_2MnO_4 + 2\,CO_2 \rightarrow 2\,KMnO_4 + MnO_2 + 2K_2CO_3$$

Technetium and rhenium salts corresponding to Mn^{2+} salts are not known. The formal charge +7 is dominant in Tc and Re; compounds in this state have far less oxidising power than MnO_4^-. Free permanganic acid exists only in aqueous solution and the oxide decomposes explosively above 273 K. The corresponding compounds of Tc and Re are, however, stable.

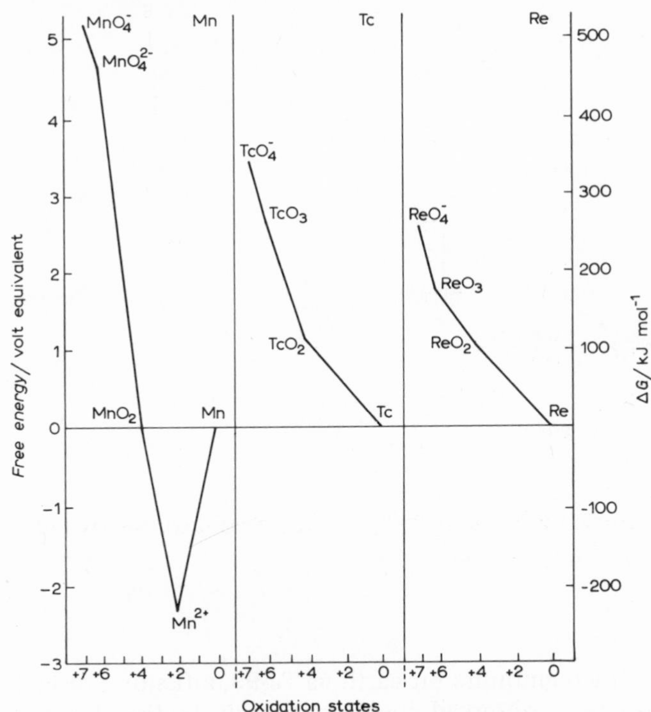

Fig. 35.1. Free energies of oxidation states relative to the metal in aqueous solution at pH = 0.

35.2. Occurrence and Separation of the Elements

35.2.1. Manganese

The element (0.085% of the lithosphere) is the most common transition metal after iron and titanium. Four of its five oxides exist in nature but the most important ore is pyrolusite, a tetragonal form of MnO_2. The principal metallurgical form of manganese is ferromanganese (ca. 80% Mn) which is made by reducing MnO_2 and Fe_2O_3 in a furnace with coke in the presence of dolomite which removes the SiO_2 as a slag. Spiegeleisen (5—20% Mn, 3—5% C) is made by a modification of the process. Purer manganese (99.9% Mn) is manufactured when required by the electrolysis of aqueous $MnSO_4$. Almost every grade of steel contains manganese. It combines with sulphur which would otherwise remain as FeS and make the steel brittle when hot; the MnS forms harmless inclusions, thereby improving the rolling and forging properties of the alloy. Furthermore, Mn acts as a deoxidiser when the metal is molten and it also improves the strength, toughness and response to heat treatment after solidification.

The metal is reactive; it combines with the halogens, oxygen, sulphur, car-

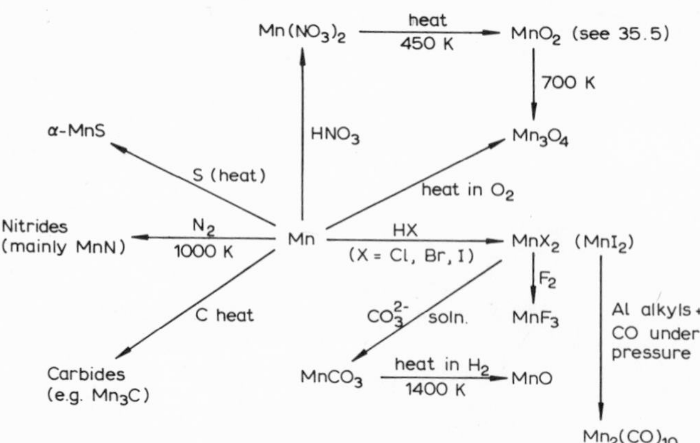

Fig. 35.2. An outline of manganese chemistry.

bon, nitrogen and many metalloids. It liberates H_2 from cold dilute HCl and H_2SO_4 and from steam at red heat.

35.2.2. Technetium

The element occurs in trace amounts on earth as Tc-99, a fission product of uranium, and has also been observed spectroscopically in the sun and some stars. In 1939 the 90-day ^{97m}Tc was made by bombarding molybdenum with high-energy deuterons from a cyclotron. The long-lived ^{99}Tc ($t_{1/2}$ = 2.12×10^5 y) forms about 6% of the fission products of uranium, and a reactor operating at 100 MW yields about 2.5 g per day. This, separated from the uranium and its fission products, is the principal source of the metal.

The hot, acidic solution of fission products is treated with tetraphenylarsonium chloride in the presence of an excess of $HClO_4$, and a precipitate of Ph_4AsTcO_4 in carrier Ph_4AsClO_4 is obtained, leaving most of the other fission products (e.g. Cs, Sr and lanthanides) in solution. The Tc is recovered by dissolving the precipitate in H_2SO_4 and electrolysing between Pt electrodes. The black TcO_2 which is deposited is converted to Tc_2O_7 by dissolving in $HClO_4$, and separated by precipitating as Tc_2S_7 with H_2S.

$$2\ HTcO_4 + 7\ H_2S \rightarrow Tc_2S_7 + 8\ H_2O$$

The metal can be made by reducing Tc_2S_7 with hydrogen at 1400 K. It is bright and silvery, tarnishes in moist air, burns in O_2 to give Tc_2O_7, in fluorine to give TcF_5 and TcF_6, and combines with S to give TcS_2 and with C to give TcC at high temperatures. It dissolves in HNO_3 and concentrated H_2SO_4, but not in HCl.

Interestingly technetium, though its isotopes are all radioactive and none has a particularly long half-life, is becoming more readily available than its heavier congener, rhenium, which, although it exists as stable nuclides ^{185}Re and ^{187}Re, is a very rare element indeed.

35.2.3. Rhenium

The element remained undiscovered until 1925 largely because it was sought in manganese ores. In fact its only appreciable occurrence is in molybdenite, but even then the mineral rarely contains more than 20 p.p.m. of Re. The element is recovered mainly from the flue dusts obtained in the roasting of MoS_2, because the Re_2O_7 obtained by the oxidation of rhenium compounds is a volatile substance. The metal is usually made by reducing NH_4ReO_4 with hydrogen at 700 K. Like Tc it dissolves in HNO_3 and H_2SO_4,

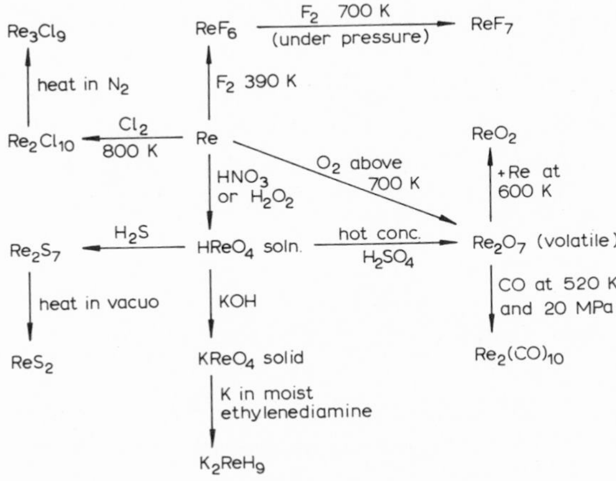

Fig. 35.3. An outline of rhenium chemistry.

but not in HCl. Unlike Tc it dissolves easily in H_2O_2, the product being $HReO_4$, but in its other reactions it resembles Tc closely. Rhenium is a very expensive metal and has found little commercial use, though it may have value as a catalyst for hydrogenation and dehydrogenation reactions.

TABLE 35.4

COMPOUNDS AND IONS REPRESENTING OXIDATION STATES OF Mn, Tc AND Re

Charge number	Mn	Tc	Re
−1	$Mn(CO)_5^-$		$Re(CO)_5^-$
0	$Mn_2(CO)_{10}$	$Tc_2(CO)_{10}$	$Re_2(CO)_{10}$
+1	$Mn(CO)_5Cl$	$K_5Tc(CN)_6$	$Re(CO)_5Cl$
+2	$Mn(H_2O)_6^{2+}$	$(\pi\text{-}C_5H_5)_2Tc$	$ReCl_2$ (diars)
+3	MnF_3	$Tc(diars)_2Cl_2^+$	Re_3Cl_9
+4	MnO_2	$TcCl_4$	K_2ReCl_6
+5	Na_3MnO_4	$TcOBr_3$	$ReOCl_4^-$
+6	MnO_4^{2-}	TcF_6	ReO_3
+7	MnO_4^-	Tc_2O_7	ReO_3Cl

35.3. Oxidation States

The representative compounds and ions shown in Table 35.4 illustrate the usual trends shown by transition metals near the middle of the d block. The highest possible oxidation state, numerically equal to the group number, is shown by all three elements. There are several indications in the table that the higher oxidation states of Tc and Re are more resistant to reduction than those of Mn, however. For example, the comparatively easily polarised Br^- ion features as a ligand in $TcOBr_3$, a compound of technetium(V), whereas the only ligands which stabilise manganese(V) are hard bases such as O^{2-}. The lowest oxidation states are all stabilised, as usual, by π-acceptor ligands such as the CO molecule.

The free energies of the reactions for the reduction of H^+aq at pH 0 are shown in Fig. 35.1. Tc^{VII} and Re^{VII} are clearly not nearly such strong oxidising agents as MnO_4^-; another feature of the diagram is that it shows the +6 state in Tc, and particularly in Re, to be much more thermodynamically stable than Mn^{VI} in aqueous solution.

35.4. Halides

The halides and oxohalides are listed in Table 35.5. Features of note are (a) the absence of higher iodides, (b) the absence of high oxidation states of

TABLE 35.5

HALIDES AND OXOHALIDES OF MANGANESE, TECHNETIUM AND RHENIUM

Charge number	Mn	Tc	Re
+2	MnF_2, $MnCl_2$, $MnBr_2$, MnI_2		
+3	MnF_3		Re_3Cl_9, Re_3Br_9, Re_3I_9
+4	MnF_4	$TcCl_4$	ReF_4, $ReCl_4$, $ReBr_4$, ReI_4
+5		TcF_5	ReF_5, $ReCl_5$, $ReBr_5$
		$TcOCl_3$, $TcOBr_3$	$ReOF_3$
+6		TcF_6	ReF_6, $ReCl_6$
		$TcOF_4$	$ReOF_4$, $ReOCl_4$, $ReOBr_4$
+7			ReF_7
			$ReOF_5$
			ReO_2F_3
	MnO_3F	TcO_3F, TcO_3Cl	ReO_3F, ReO_3Cl, ReO_3Br

Mn except in MnO_3F, (c) the existence of many oxofluorides of rhenium.

The dihalides of Mn, except MnF_2 which is insoluble and is usually made by treating $MnCO_3$ with HF, are obtained as rose-pink hydrates when the metal is dissolved in the appropriate aqueous hydrogen halide.

The red-purple MnF_3, the only trihalide of Mn, is made by the action of F_2 on $MnCl_2$ or MnI_2. With water it gives MnO_2 and MnF_2. The hygroscopic, blue MnF_4 may be made by fluorinating powdered MnF_3 at 800 K. At room temperature it slowly decomposes to MnF_3 and fluorine.

The rhenium(III) halides are of some interest structurally. Re_3Cl_9, made by heating $ReCl_5$ in nitrogen, is a dark-red solid containing the trimeric units shown in Fig. 35.4a. Half of the terminal Cl atoms are used in bridging to ad-

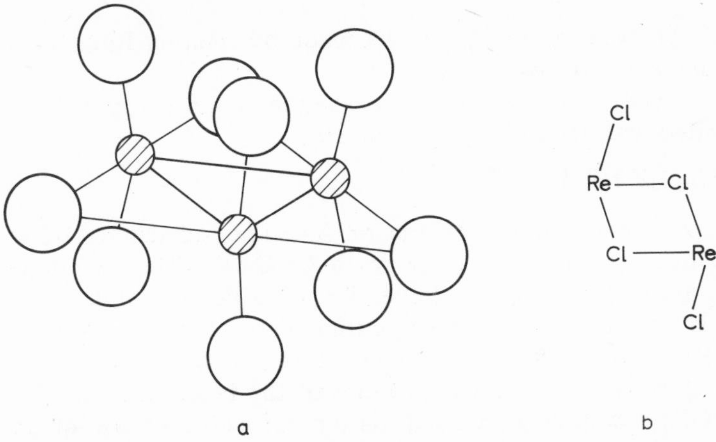

Fig. 35.4. Re_3Cl_9. (a) Trimeric unit in the solid. (b) Showing how terminal Cl atoms are used as bridges to adjacent clusters.

jacent clusters (Fig. 35.4b); Re_3Br_9 is similar, but Re_3I_9, made by the decomposition of ReI_4 in a sealed tube at 600 K, is rather different, only two rhenium atoms in each cluster being chlorine-bridged to neighbouring clusters.

$TcCl_4$ is the only thermally stable chloride of technetium yet made. The red crystals can be obtained by heating Tc_2O_7 to 700 K with CCl_4 in a sealed tube. The slightly distorted $TcCl_6$ octahedra which constitute the structure are joined into a chain by sharing edges. Of the tetrahalides of Re, the bromide and iodide can be made by treating $HReO_4$ with the respective hydrogen halide, and ReF_4 by reducing ReF_6 with Re. The amorphous, black compound known as α-$ReCl_4$ is made by treating ReO_2 with $SOCl_2$. A crystalline material known as β-$ReCl_4$, which may be the same compound, has been found to contain Re_2Cl_8 units linked into chains by single chlorine bridges.

When technetium powder reacts with fluorine the major product is the golden-yellow solid TcF_6 and the minor product the yellow solid TcF_5. There is no evidence for a heptafluoride. Rhenium, however, gives a mixture of ReF_6 and ReF_7. The former, a low-melting yellow solid, is very reactive; it hydrolyses in water:

$$3\,ReF_6 + 10\,H_2O \rightarrow 2\,HReO_4 + ReO_2 + 18\,HF$$

and reacts with silica:

$$2\,ReF_6 + SiO_2 \rightarrow 2\,ReOF_4 + SiF_4$$

The pale-yellow ReF_7 can be made in good yield by heating the foregoing ReF_6/ReF_7 mixture for some hours with fluorine under 300 kPa at 670 K. It hydrolyses to give only $HReO_4$ and HF.

The major product of the reaction between Re and Cl_2 is $ReCl_5$, a dark-brown solid which melts at 495 K. Its hydrolysis, like that of ReF_6, is accompanied by disproportionation:

$$3\ ReCl_5 + 8\ H_2O \rightarrow 2\ ReO_2 + HReO_4 + 15\ HCl$$

A compound which analyses as $ReCl_6$ can be made by treating ReF_6 with BCl_3, but its structure is not known.

The compound MnO_3F is obtained as a dark-green solid (*m.p.* 195 K) when $KMnO_4$ is treated with HF at low temperature:

$$KMnO_4 + 2\ HF \rightarrow MnO_3F + KF + H_2O$$

It decomposes explosively at room temperature. A compound claimed to be MnO_3Cl has been made by treating Mn_2O_7 with $ClSO_2OH$. The oxofluorides of Tc and Re are more thermally stable. TcO_3F, a yellow solid, is obtained when TcO_2 is heated with F_2. The colourless liquid ReO_3Cl, made by reaction between Re_2O_7 and $ReCl_5$, can be converted to the yellow solid ReO_3F by treatment with HF. A minor product of the reaction is $ReOF_5$. The compound $ReOF_4$ is well characterised; its unit of structure is a square pyramid:

The elements display a wide range of complex halides. Manganese(II) forms complexes of the types M^IMnF_3, $M^I_2MnCl_4$ and $M^I_4MnCl_4$. Moreover, $MnCl_2$ and $MnBr_2$ react with salts like the pyridinium and tetramethylammonium halides to give tetrahedral MnX_4^{2-} complexes. Manganese(III) complexes such as $M^I_2MnF_5$ and $M^I_2MnCl_5$ are known. When MnO_2 is added to a solution of HCl in CCl_4, a dark-green colour is produced and the salt $(Et_4N)_2MnCl_5$ can be extracted after addition of Et_4NCl.

Rhenium in its +2 and +3 states forms a number of complex halide ions in which there are strong metal—metal interactions. The ions $Re_2Cl_8^{2-}$ and $Re_3Cl_9^{3-}$ are discussed in Paragraph 27.12. The ion in the compound $CsReCl_4$ is a trimeric one, $Re_3Cl_{12}^{3-}$, in which there is a triangle of Re atoms each of which is seven-co-ordinate (Fig. 35.5).

The most common complex halides of Mn and Re are those in which the metal has a +4 charge; examples are $M^I_2MnCl_6$, $M^I_2ReCl_6$ and, particularly, the corresponding fluoro-complexes. This +4 state of Re is obtained from states of higher charge in the formation of these compounds:

$$2\ ReCl_5 + 4\ KCl \rightarrow 2\ K_2ReCl_6 + Cl_2$$

The fluoro-compound K_2ReF_6 is made in good yield by the action of anhydrous HF on K_2ReI_6, but the acid H_2ReF_6 cannot be isolated.

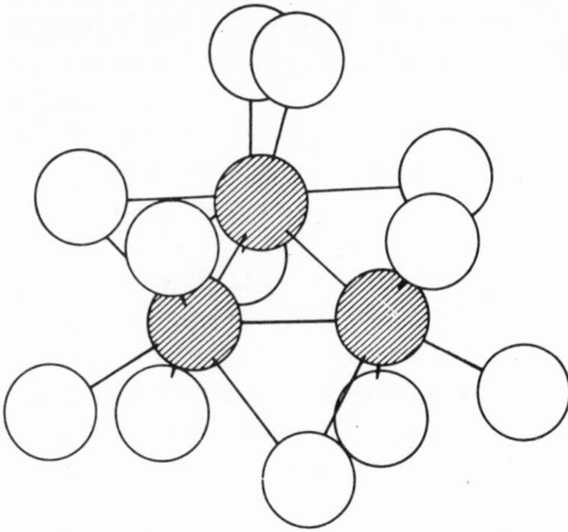

Fig. 35.5. The $Re_3Cl_{12}^{3-}$ anion in $CsReCl_4$.

The Mn^{IV} complexes are spin-free with 3 unpaired electrons. The magnetic susceptibilities of the complex chlorides of Re^{IV} and Tc^{IV} increase with temperature and the magnetic moment approaches that expected for 3 unpaired electrons. There is evidently some slight metal—metal interaction even in the +4 state (27.12).

35.5. Oxides

Manganese occurs as manganosite, MnO, with the NaCl structure, hausmannite, Mn_3O_4, with a normal spinel structure which is distorted towards the tetragonal, braunite, Mn_2O_3, with the structure of a C-type lanthanide oxide (Fig. 23.2) and pyrolusite, MnO_2, which has a distorted rutile lattice. Gross nonstoichiometry is common, for example the composition of MnO can be increased to $MnO_{1.13}$ without a new phase being formed. The element also occurs as pyrochroite, $Mn(OH)_2$, and as manganite, MnO(OH). The dull-green monoxide is usually made in the laboratory by decomposing the car-

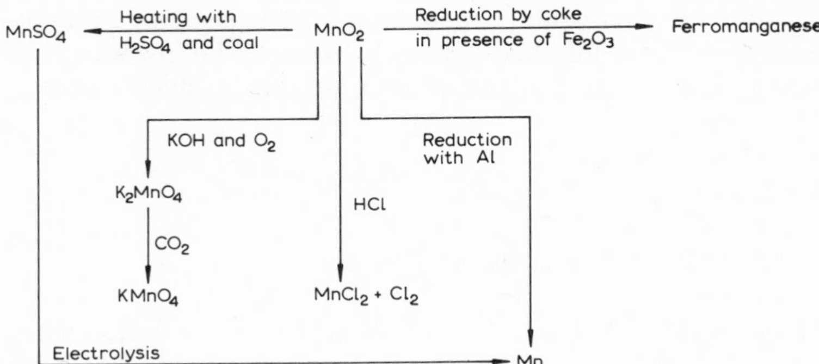

Fig. 35.6. Reactions of manganese dioxide.

bonate in a nitrogen atmosphere. The corresponding hydroxide, precipitated by OH^- from Mn^{2+} solutions, rapidly darkens in colour in air to give MnO(OH) which can be converted at 500 K into Mn_2O_3, which is believed to contain Mn^{II} and Mn^{IV} but not Mn^{III}. Black Mn_3O_4 is made by heating any of the other oxides to 1300 K. The principal ore, pyrolusite, is used (a) for making ferromanganese and manganese itself, (b) as a dipolariser in dry cells, (c) for rendering glass colourless, (d) as a drier in paint and (e) for making $MnSO_4$ which is used in treating manganese-deficient soils. Pure MnO_2 is made by decomposing $Mn(NO_3)_2$ at 450 K and then dehydrating in oxygen at 750 K, but the product is always stoichiometrically deficient in oxygen. Dimanganese heptoxide is an olive-green liquid which separates when powdered $KMnO_4$ is added to concentrated H_2SO_4 at 250 K. Very much the least stable oxide of manganese, it explodes at 285 K, but can be distilled below 270 K.

Black TcO_2 is the most thermally stable oxide of technetium, being formed when either Tc_2O_7 or NH_4TcO_4 is heated. A dihydrate is precipitated when TcO_4^- is reduced with Zn/HCl and the solution is then made alkaline. Rhenium dioxide results from heating Re_2O_7 with Re at 900 K or, as the dihydrate, from a solution of $ReCl_6^{2-}$ treated with OH^-. ReO_2 is less thermally stable than TcO_2, being converted to Re_2O_7 and Re at high temperatures. Both dioxides have the same distorted rutile structure as MnO_2.

TcO_3 has been reported as a product of heating TcO_3Br but little is known about it. However ReO_3, a red solid, is well established. It is obtained in almost quantitative yield by the thermal decomposition of $Re_2O_7(C_4H_8O_2)_3$, the addition compound of Re_2O_7 with dioxan. It disproportionates on heating:

$$3\,ReO_3 \rightarrow Re_2O_7 + ReO_2$$

The structure is a very simple one. Every Re atom is octahedrally co-ordinated and every O atom linearly co-ordinated in a cubic lattice (Fig. 7.23). Surprisingly in view of its simplicity, this is not a common lattice, though distorted versions like that in MoO_3 do occur more frequently.

The yellow Tc_2O_7 is driven off from acidic solutions of TcO_4^- on heating. Re_2O_7 is not lost at the boiling point of a dilute aqueous solution but does distil from hot solutions of ReO_4^- in strong H_2SO_4. The yellow crystals melt at 490 K; their structure is interesting — there is an infinite array of ReO_6 octahedra alternating with ReO_4 tetrahedra sharing corners. The lower-melting Tc_2O_7 contains molecules in which a pair of TcO_4 tetrahedra share a corner:

35.6. Manganates and Rhenates

A manganate(V), $Na_3MnO_4(\frac{1}{4}\ NaOH) \cdot 12\ H_2O$ separates from solution when $KMnO_4$ is reduced with alkaline Na_2SO_3. It is thermally unstable, but more stable salts such as $Ba_3(MnO_4)_2$ and K_3MnO_4 have also been prepared.

The well-known green MnO_4^{2-} ion is formed when MnO_2 is fused with an alkali-metal hydroxide and an oxidising agent such as KNO_3. The salts are unstable to disproportionation in even slightly acidic solution:

$$3\ MnO_4^{2-} + 4\ H^+ \rightarrow 2\ MnO_4^- + MnO_2 + 2\ H_2O$$

One of the most stable manganates(VI), $BaMnO_4$, can be made by adding concentrated aqueous $KMnO_4$ to boiling, saturated baryta water:

$$4\ Ba(OH)_2 + 4\ MnO_4^- \rightarrow 4\ OH^- + 2\ H_2O + O_2 + 4\ BaMnO_4$$

The room-temperature magnetic moment of $BaMnO_4$ is $1.80\mu_B$.

Rhenates(VI) of various types have been made by fusing together a perrhenate and rhenium with alkali-metal oxides. The best known are those containing the ReO_6^{6-} ion; Li_6ReO_6, Ca_3ReO_6, and the corresponding Sr and Ba salts have all been made.

35.7. Permanganates, Pertechnetates and Perrhenates

35.7.1. Permanganates

The stability of the permanganate ion MnO_4^- over a wide range of pH makes the permanganates very useful oxidising agents. Potassium permanganate, $KMnO_4$, is manufactured by the electrolytic oxidation of the alkaline manganate solution. It is used in volumetric oxidimetry, in the industrial production of such things as saccharin and benzoic acid, and for bleaching waxes. The MnO_4^- anion is much used in the volumetric determination of manganese since Mn^{2+} is quantitatively oxidised to it in dilute nitric acid by insoluble oxidising agents like sodium bismuthate.

In alkali MnO_4^- reacts as an oxidising agent thus:

$$MnO_4^- + 2\ H_2O + 3\ e \rightarrow MnO_2 + 4\ OH^- \qquad E^0 = +1.23\ V$$

and in acid oxidation occurs thus:

$$MnO_4^- + 8\ H^+ + 5\ e \rightarrow Mn^{2+} + 4\ H_2O \qquad E^0 = +1.51\ V$$

A solution of $HMnO_4$ can be made by adding H_2SO_4 to aqueous $Ba(MnO_4)_2$ below 272 K. A hydrate $HMnO_4 \cdot 2\ H_2O$ has been isolated from the solution. The acid is very soluble in water and ionises strongly.

Permanganates exhibit weak, temperature-independent magnetism which is due to a second-order Zeeman effect between the higher molecular orbital levels and the ground level.

35.7.2. Pertechnetates and perrhenates

A colourless solution of $HTcO_4$ is obtained when yellow Tc_2O_7 is dissolved in water, but as the solution is concentrated it becomes pink and then red; finally dark-red crystals of $HTcO_4$ separate from solution. This colour phenomenon is not yet explained. Colourless $KTcO_4$ has high thermal stability but NH_4TcO_4 decomposes to give TcO_2:

$$2\ NH_4TcO_4 \rightarrow N_2 + 2\ TcO_2 + 4\ H_2O$$

Solutions of $HReO_4$, which are also colourless, are made by dissolving Re_2O_7 in water or oxidising either Re itself or ReO_2 with aqueous H_2O_2. On evaporation the solution becomes greenish-yellow; careful drying over P_2O_5 eventually yields a yellow, hygroscopic solid which is a true molecular hydrate, $Re_2O_7 \cdot 2\ H_2O$. Like $HMnO_4$ and $HTcO_4$, aqueous $HReO_4$ is a strong acid. The only anion present in the solution is the tetrahedral ReO_4^- ion, but heavy-metal ions precipitate mesoperrhenates, $M_3^{II}(ReO_5)_2$. The acid is not a particularly strong oxidising agent but it oxidises HBr to Br_2.

The alkali-metal perrhenates are colourless, thermally stable, and, except for $NaReO_4$, rather insoluble. $KReO_4$, made by adding KCl to aqueous $HReO_4$, is less soluble than $KClO_4$. NH_4ReO_4 decomposes on strong heating, the products being ReO_2, N_2 and H_2O.

35.8. Sulphides

Manganese(II) sulphide has three forms. The stable, green α-MnS, alabandite, has the NaCl structure, but there are metastable red forms with the blende and wurtzite structures respectively. The compound MnS_2 contains Mn^{2+} and S_2^{2-} ions in a structure closely related to iron pyrites (36.10). MnS_2 and all three forms of MnS are semiconductors and are antiferromagnetic at low temperatures. Technetium and rhenium form the isomorphous pairs TcS_2, ReS_2 and Tc_2S_7, Re_2S_7. The black heptasulphides are precipitated when H_2S is passed under pressure into acid solutions of TcO_4^- and ReO_4^-. The sulphur which is co-precipitated is removed by leaching with CS_2. The disulphides are obtained from the heptasulphides by heating in vacuo.

35.9. Binary Compounds with Other Non-metals

Manganese combines with nitrogen at 1000 K to give a mixture of nitrides; the composition of the product depending on the pressure of the gas. Rhenium does not react with N_2 to give a nitride, but NH_4ReO_4 reacts with H_2 at 580 K to give a mixture of the metal with Re_3N.

The Mn—C system is complicated, and the reactions are difficult to study because manganese is so reactive towards oxygen and nitrogen. Orthorhombic Mn_3C hydrolyses to give a mixture of gaseous products, typically 75% H_2, 15% CH_4 and 10% C_2H_6. There is some doubt about the existence of rhenium carbides.

Both manganese and rhenium form mixtures of borides by direct combination at high temperature. They resemble other transition-metal borides in appearance and properties.

35.10. Potassium Enneahydridorhenate(VII)

When potassium perrhenate is reduced with a solution of potassium in slightly moist ethylenediamine, colourless, diamagnetic, hexagonal crystals are obtained. The positions of the Re and K atoms were found by X-ray diffraction and those of the H atoms by neutron diffraction. Proton magnetic resonance shows a single peak at high field indicating equivalence or vibrational deformation of the H atoms. The anion, ReH_9^{2-}, has the structure shown in Fig. 35.7. The Re atom at the centre of a triangular prism is surrounded by 9 H atoms, six at the corners and three on extensions of perpendiculars through the centres of the vertical sides. The Re—H distances are all 168 pm. If the hydrogens are considered to be co-ordinated as H^- ions, the Re is in a formal oxidation state of +7, and the diamagnetism is thereby explained. In the crystal the K^+ ions lie in straight lines through the H and Re atoms and form a larger tricapped triangular prism at 60° about a common vertical axis to the hydrogen one.

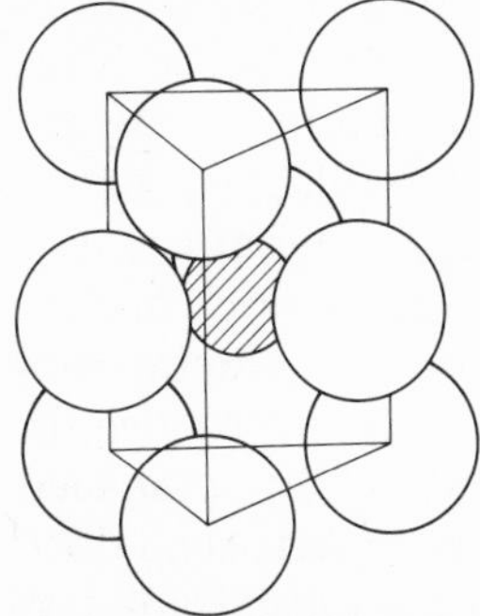

Fig. 35.7. The ReH_9^{2-} ion.

35.11. Organometallic Compounds and π-Complexes

Only one binary metal carbonyl of manganese is known, $Mn_2(CO)_{10}$, best made by the action of CO under pressure on MnI_2 in the presence of a reducing agent such as an alkylaluminium. The golden-yellow crystals are stable in air at room temperature but decompose quickly above 380 K. The molecule belongs to the symmetry group D_{4d}, as confirmed by its vibrational spectrum. The carbonyl groups have been substituted by a variety of ligands usually to give compounds of the types $Mn_2(CO)_8L_2$ and $Mn_2(CO)_9L$.

Sodium pentacarbonylmanganate(—I), $NaMn(CO)_5$, made by the reduction of $Mn_2(CO)_{10}$ with sodium amalgam in tetrahydrofuran, is a useful starting material for the preparation of many compounds containing bonds between Mn and other metals such as Ge, Pb, Mo, Fe, Co, Cu, Ag and Hg which are a particular feature of the chemistry of manganese. Typically, a halide derivative of the other metal reacts with $NaMn(CO)_5$:

$$L_mMX_n + n\ NaMn(CO)_5 \rightarrow L_mM[Mn(CO)_5]_n + n\ NaX$$

Manganese—metal bonds have also been produced by insertion reactions, e.g.:

$$Mn_2(CO)_{10} + SnCl_2 \rightarrow [(CO)_5Mn]_2SnCl_2$$

and by reactions with carbonylate cations:

$$NaCo(CO)_4 + ClMn(CO)_5 \rightarrow NaCl + (CO)_4CoMn(CO)_5$$

Manganocene, $(C_5H_5)_2Mn$, is of some interest. The amber crystals can be made by treating manganese dihalides in tetrahydrofuran with C_5H_5Na. X-ray studies show the molecule to have a sandwich structure like ferrocene and magnetic studies show the magnetic moment to be $5.86 \pm 0.05\mu_B$, consistent with the presence of an $Mn^{2+}(d^5)$ ion. Thus the indications are that the metal—ring bonds are ionic; the Mn compound is unique among d-block metal—cyclopentadienyl compounds in this respect; in all the others the metal—ring bonds are covalent. Manganocene reacts with CO under pressure to give the yellow solid $C_5H_5Mn(CO)_3$ in which the C_5H_5 undergoes attack by some electrophiles to give organic derivatives in much the same way as ferrocene does. Substitution of the CO groups can be effected under ultraviolet irradiation by amines, phosphines and arsines as well as olefins and acetylenes. In $(C_5H_5)Mn(CO)_2PhC{\equiv}CPh$ the metal is bonded by π-electrons of the acetylenic bond occupying the position vacated by the CO group.

Tc and Re form carbonyls $M_2(CO)_{10}$ with structures like that of manganese carbonyl; $Re_2(CO)_{10}$ is obtained in good yield by the action of CO on Re_2O_7 at 520 K and 20 MPa without the need for an additional reducing agent.

The reaction between C_5H_5Na and $ReCl_5$ in tetrahydrofuran yields not $(C_5H_5)_2Re$ but the hydride $(C_5H_5)_2ReH$. $TcCl_4$ reacts similarly with C_5H_5Na to give $(C_5H_5)_2TcH$. The rhenium compound has been carefully studied. It is diamagnetic and basic, its hydrogen being hydridic, not protonic. The lemon-yellow crystals dissolve in dilute HCl to give a salt:

$$(C_5H_5)_2ReH + HCl \rightarrow [(C_5H_5)_2ReH_2]^+Cl^-$$

It was in this compound that the characteristics of a bond between a transition-metal atom and a hydrogen atom were first recognised.

At 370 K and 25 MPa, $(C_5H_5)_2ReH$ reacts with CO to give pale-yellow crystals of composition $C_{12}H_{11}O_2Re$ which have been shown by i.r. and n.m.r. studies to contain molecules with the structure

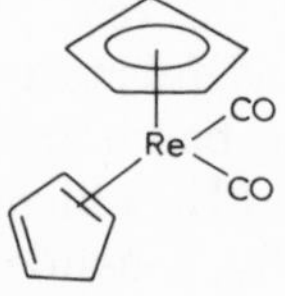

Interestingly, the compound obtained by the action of $Re_2(CO)_{10}$ on C_5H_6 is quite different; it is π-$C_5H_5Re(CO)_3$.

35.12. Complexes

The +1 state is represented by hexacyano-complexes, $K_5M(CN)_6$, for all three metals. The preparation of the olive-green $K_5Tc(CN)_6$ is typical; it is made by reducing TcO_4^- with potassium amalgam in the presence of CN^-. Exposure to air converts it to the Tc^{IV} complex $[Tc(OH)_3(CN)_4]^{3-}$.

The +2 state is common in manganese, rather uncommon in rhenium and rare in technetium. The d^5 configuration is rather unfavourable for the formation of low-spin octahedral complexes (6.3.2) but they exist in $Mn(CN)_6{}^{4-}$ and $Mn(CNR)_6{}^{2+}$. The high-spin ammines of Mn^{II} are rather unstable, but there are octahedral complexes with chelating ligands such as ethylenediamine and the oxalate ion. Mn^{II} has some tetrahedral complexes, mainly salts of $MnX_4{}^{2-}$ (X = Cl, Br, I) with large cations such as Me_4N^+.

Tc^{II} and Re^{II} exist in the compounds MCl_2(diarsine)$_2$, made by reducing M^{VII} compounds with hypophosphite in methanol in the presence of Cl^- ions and *o*-$C_6H_4(AsMe_2)_2$.

The +3 state is moderately common in this sub-group. Several ligands lower $E°$, Mn^{III}/Mn^{II} sufficiently for Mn^{II} to be oxidised in air. Examples of these are CN^- in $K_3Mn(CN)_6$ and acetylacetone in $Mn(acac)_3$, both of which can be made by atmospheric oxidation of solutions of Mn^{2+} and the ligand. Tc^{III} and Re^{III} compounds $[MX_2(diarsine)_2]ClO_4$ have been made by a method similar to that for the foregoing diarsine complexes, but with the use of weaker reducing agents.

The +4 state for manganese is not common — it exists in some hexachloro- and hexafluorocomplexes. Technetium(IV), in addition to its halide complexes, forms some cyanocomplexes. Thus $TcO_2 \cdot 2\,H_2O$ dissolves in alkali cyanides to give $Tc(OH)_3(CN)_4{}^{3-}$ which can be isolated as the dark-brown thallium salt. K_2TcI_6 and KCN react in methanol to give the dark-red $K_2Tc(CN)_6$. But similar reactions with the corresponding rhenium compound produce the Re^V complexes $Re(OH)_4(CN)_4{}^{3-}$ and $K_3Re(CN)_8$.

The +5 state in technetium is represented by $[TcCl_4(diars)_2]ClO_4$; this provided the first example of 8-co-ordinate technetium.

Higher oxidation states than +5 in this sub-group are represented mainly by the oxo-complexes and complex halides. But another particularly interesting complex is the rhenium(VI) compound tris(*cis*-1,2-diphenylethene-1,2-dithiolato)-rhenium:

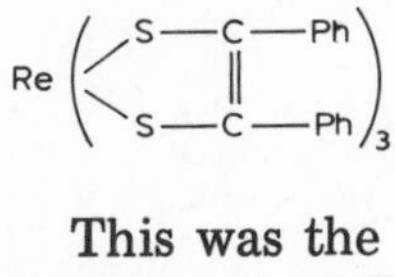

$$\mathrm{Re}\left(\begin{array}{l} \mathrm{S{-}C{-}Ph} \\ \quad\ \| \\ \mathrm{S{-}C{-}Ph} \end{array}\right)_3$$

This was the first known example of a trigonal prismatic complex. The arrangement of S and C atoms is shown in Fig. 35.8; the phenyl rings, not shown in the figure, are twisted out of the S—C—C—S planes. Though other examples of this arrangement of donor atoms have now been discovered, it remains a very rare form of 6-co-ordination compared with octahedral symmetry.

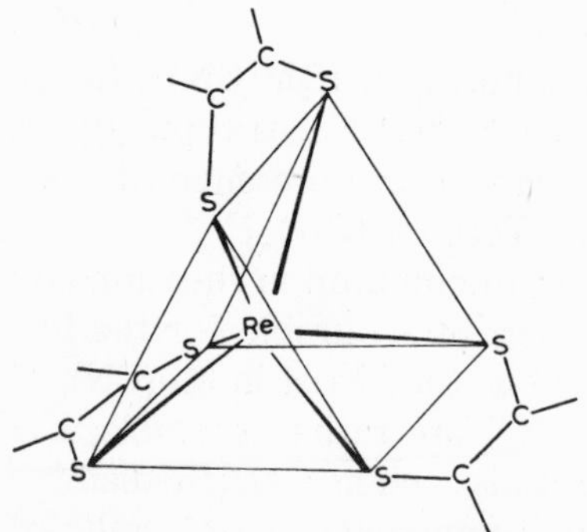

Fig. 35.8. Trigonal prismatic arrangement of bidentate sulphur ligands around rhenium atom in $Re\left(\begin{array}{l}S{-}C{-}Ph\\ \quad\ \ \| \\ S{-}C{-}Ph\end{array}\right)_3$.

Further Reading

F.A. Cotton, Strong homonuclear metal—metal bonds, Acc. Chem. Res., 2 (1969) 240.

K.V. Kotegow, O.N. Pavlov and V.P. Shvedov, Technetium, Adv. Inorg. and Radiochem., 11 (1968) 1.

R.D. Peacock, Chemistry of technetium and rhenium, Elsevier, Amsterdam, 1966.

T.A. Zordan and L.G. Hepler, Thermochemistry and oxidation potentials of manganese and its compounds, Chem. Rev., 68 (1968) 737.

K.B. Lebedev, The chemistry of rhenium, Butterworth, London, 1962.

R. Colton, The chemistry of rhenium and technetium, Interscience, London, 1965.

R.D.W. Kemmitt and R.D. Peacock, The chemistry of manganese, technetium and rhenium, Pergamon, Oxford, 1975.

Iron, Cobalt and Nickel — The First Triad of Group VIII

36

36.1. Introduction

The nine elements of Group VIII carry the three d-block series from the manganese group, VIIA, to the copper group, IB. Although there are the usual vertical similarities, as for example in Fe, Rh and Os, the horizontal similarities, as in Fe, Co and Ni, are particularly strong. The metals Fe, Co and Ni replace hydrogen from non-oxidising acids, form aquated ions with charges of either +2 or +3, and behave, in general, as typical active metals. But the heavier congeners, Ru, Rh, Pd, Os, Ir and Pt, are generally unreactive towards acids, though Pd is converted to $Pd(NO_3)_2$ and Os to OsO_4 by concentrated HNO_3, and their typical ions are anionic complexes such as $PtCl_6^{2-}$ in which the metal atom is in a high oxidation state and is covalently bound to the ligands which surround it. These metals, often known collectively as the platinum metals, are very similar in most physical and chemical properties. To simplify the correlation of properties in Group VIII, the active metals Fe, Co and Ni will be treated first and the platinum metals will be considered in the next chapter.

36.2. The Elements

The three metals resemble Cr and Mn, which precede them, in their physical properties and reactivity. All three combine with oxygen on heating; finely divided Fe and Ni are, in fact, pyrophoric. The metals also combine with the halogens, sulphur, boron, carbon, silicon and phosphorus. It is known that carbide, silicide and nitride phases are important in the metallurgy of iron. Nickel is the only transition element which combines with carbon monoxide at atmospheric pressure, tetracarbonyl nickel, $Ni(CO)_4$, being formed at 325 K. Iron and nickel dissolve easily in dilute, non-oxidising acids, but with cobalt reaction is slow even though E°, Co^{2+}/Co is negative (Table 36.2). Hot HNO_3 renders all three metals passive. Iron is oxidised quickly in moist air, the hydrated oxide which is formed being of no protection to the underlying metal because it flakes off easily. Co and Ni resist atmospheric corrosion at ordinary temperatures, however, and nickel plating is used as a

TABLE 36.1
ATOMIC PROPERTIES OF Fe, Co AND Ni

	Fe	Co	Ni
Z	26	27	28
Electron configuration	$[Ar]3d^6 4s^2$	$[Ar]3d^7 4s^2$	$[Ar]3d^8 4s^2$
$I(1)$/kJ mol^{-1}	762	758	736
$I(2)$/kJ mol^{-1}	1561	1644	1752
Metallic radius/pm	126	125	124
$r_{M^{2+}}$/pm	76	74	72
$r_{M^{3+}}$/pm	64	63	

TABLE 36.2
PHYSICAL PROPERTIES AND REDOX POTENTIALS

	Fe	Co	Ni
M.p./K	1808	1760	1728
ρ/g cm^{-3}	7.9	8.9	8.9
E^0, M^{2+}/M/V	−0.44	−0.27	−0.24
E^0, M^{3+}/M^{2+}/V	+0.77	+1.84	

protective coating on steel. The metals all have low-temperature α-forms, b.c.c. in the case of Fe and h.c.p. in Co and Ni, and also high-temperature modifications, γ-Fe, β-Co and β-Ni, all with the c.c.p. lattice. All three metals are ferromagnetic; the Curie temperature for Co is particularly high, near 1400 K.

36.3. Occurrence and Preparation of the Elements

36.3.1. Iron

Iron (5.1% of the lithosphere) is widespread, many minerals owing their tints to its presence, but the important ores are haematite, Fe_2O_3, magnetite, Fe_3O_4, and siderite, $FeCO_3$. Iron pyrites, FeS_2, is a common vein mineral, but is not worked up for the metal because its sulphur content is so high. The core of the Earth is thought to consist mainly of iron, and the metal is the major constituent of metallic meteorites.

Pure iron, which can be made by hydrogen reduction of the oxides, has little industrial use, but iron alloys are of great importance. Annual world production probably exceeds 300 million tons, which represents more than ten times the total of all other metals. The first stage in the industrial production is to reduce the calcined ores (Fe_2O_3) with coke in a blast furnace. Limestone is used to remove silica as slag. In the cooler part of the furnace

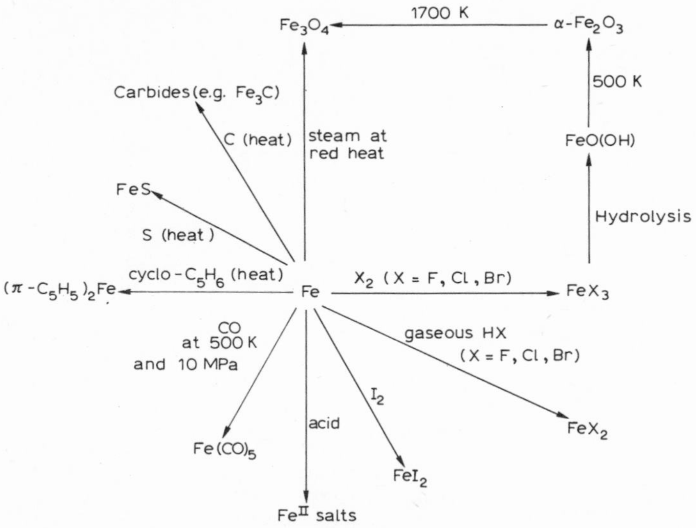

Fig. 36.1. An outline of the chemistry of iron.

the reducing agent is CO, and typical reactions are:

$$3\ Fe_2O_3 + CO \rightarrow 2\ Fe_3O_4 + CO_2$$
$$Fe_3O_4 + 4\ CO \rightarrow 3\ Fe + 4\ CO_2$$
$$Fe_2O_3 + CO \rightarrow 2\ FeO + CO_2$$

In the lower, hotter parts of the furnace (>1800 K) the reduction is mainly by carbon:

$$FeO + C \rightarrow Fe + CO$$

Most of the silica in the original ore is tapped off as a slag of calcium silicate, but some silicon, produced by the reduction of SiO_2, is incorporated in the metal. A typical pig iron, tapped from the furnace, contains about 4% C, 2.5% Si, 2.5% Mn, 1% P and 0.1% S. To convert pig iron to steel, the carbon and silicon must be largely removed by oxidation, and the phosphorus and sulphur by the use of lime to form a slag. Oxidation is now effected mainly by oxygen gas. Steel usually contains 0.2—1.5% combined carbon, which must be added to the melt, after oxidation, in the form of carbides of iron and manganese.

36.3.2. Cobalt

Cobalt (10^{-3}% of the lithosphere) is widely distributed, but its workable ores are mainly sulphides and arsenides which are associated with other metals, particularly nickel and copper. Examples are smaltite, $CoAs_2$, and cobaltite, CoAsS. The extraction of cobalt is usually ancilliary to the production of the other metal in the ore, but in a typical process the roasted ore is dissolved in H_2SO_4, iron is removed with lime, and cobalt is precipitated as

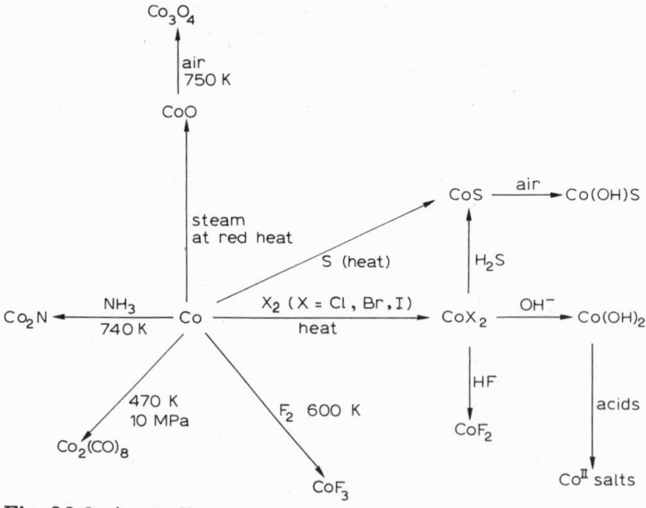

Fig. 36.2. An outline of cobalt chemistry.

CoO(OH) by addition of NaOCl. Heating converts the precipitate to Co_3O_4, which can be reduced with carbon.

36.3.3. Nickel

Nickel (0.016% of the lithosphere) occurs mainly as sulphides and arsenides. The important deposit at Sudbury, Ontario is mainly pentlandite, (Ni,Fe)S. The concentrate is roasted and smelted to form a matte, which, on heating with $NaHSO_4$ and coke, separates into two layers, the lower being mainly NiS. This is oxidised to NiO which is used for making nickel steels. The pure metal is obtained by the Mond process; first crude nickel is obtained by reducing NiO with water gas, this metal is then converted to volatile

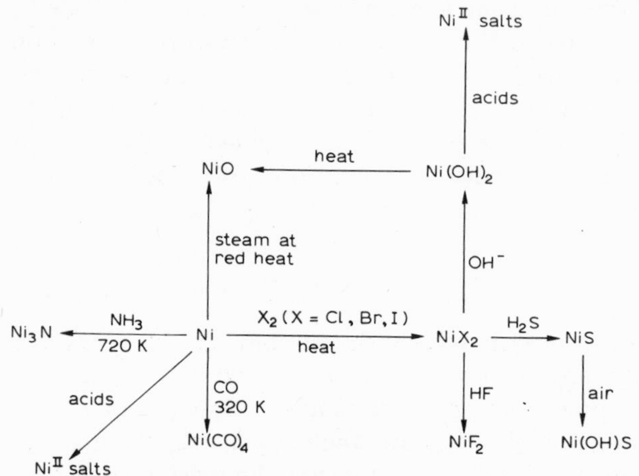

Fig. 36.3. An outline of nickel chemistry.

$Ni(CO)_4$ by the action of CO at 330 K, finally the $Ni(CO)_4$ is decomposed by passing it through nickel pellets at 450 K, the CO which is released being recycled over the impure metal.

36.4. Oxidation States

The lowest oxidation states occur as usual in π-bonded complexes such as carbonyls and carbonylate anions, and the highest in fluorocomplexes, or in the case of Fe, in an oxoanion (Table 36.3). However, the highest oxidation state falls progressively with atomic number from the value of +7 reached in manganese.

Zero oxidation state is particularly common in Ni, for example the carbonyl groups in $Ni(CO)_4$ can be replaced by other neutral, π-acceptor ligands such as isonitriles and phosphorus trihalides. The charge number +1 is uncommon, but all three elements in their dipositive states form very large numbers of compounds; Ni^{II} often forms square complexes which are diamagnetic, but high-spin tetrahedral and also octahedral complexes are common.

The terpositive state occurs much more in Fe and Co than in Ni. The Fe^{3+}aq ion is a moderately strong oxidising agent (E°, Fe^{3+}/Fe^{2+} = +0.77 V) and the Co^{3+}aq ion a powerful one (E°, Co^{3+}/Co^{2+} = +1.84 V) but in co-ordination with ligands of only moderate field strength, like ammonia and

TABLE 36.3

REPRESENTATIVE COMPOUNDS AND IONS OF OXIDATION STATES −2 TO +6 OF IRON, COBALT AND NICKEL

Charge	Fe	Co	Ni
−2	$Fe(CO)_4^{2-}$	—	—
−1	—	$Co(CO)_4^-$	$Ni_2(CO)_6^{2-}$
0	$Fe(CO)_5$	$K_4Co(CN)_4$	$Ni(PF_3)_4$, $K_4Ni(CN)_4$
+1	$(\pi\text{-}C_5H_5)_2Fe_2(CO)_4$,	$Co(dipy)_3^+$, $Co(NCC_6H_5)_5ClO_4$	$K_4Ni_2(CN)_6$
+2	$Fe(H_2O)_6^{2+}$, $FeCl_2$, $(\pi\text{-}C_5H_5)_2Fe$	$Co(H_2O)_6^{2+}$, $CoCl_4^{2-}$	$Ni(H_2O)_6^{2+}$, $Ni(en)_3^{2+}$
+3	FeF_6^{3-}, $FeO(OH)$, $K_3Fe(CN)_6$	$Co(NH_3)_6^{3+}$, $K_3Co(CN)_6$	$Ni(diars)_2Cl_2^+$, $NiBr_3(PMe_3)_3$
+4	$Fe(diars)_2Cl_2^{2+}$, Ba_2FeO_4	CoF_6^{2-}	K_2NiF_6
+5	FeO_4^{3-}	—	—
+6	K_2FeO_4	—	—

ethylenediamine, Co^{III} (d^6) forms diamagnetic complexes, many of considerable thermodynamic stability.

Charge +4 is rather uncommon, but occurs in some diarsine complexes of iron and in CoF_6^{2-} and NiF_6^{2-}. Charge +6 occurs in the ferrates(VI) which are made by the oxidation of a suspension of hydrated iron(III) oxide in aqueous alkali:

$$Fe_2O_3 + 3\ OCl^- + 4\ OH^- \rightarrow 2\ FeO_4^{2-} + 3\ Cl^- + 2\ H_2O$$

The carmine-red $BaFeO_4$ is similar to $BaCrO_4$ in structure; though fairly stable to reduction in aqueous alkali, the FeO_4^{2-} ion is a stronger oxidising agent than MnO_4^- in acidic solutions.

36.5. Halides

White FeF_2 can be made by passing HF over iron at red heat or over $FeCl_2$ at a lower temperature. It has the rutile structure. The yellow dichloride and greenish-yellow dibromide, made by passing HCl or HBr over the strongly heated metal, both have layer lattices, $FeCl_2$ the $CdCl_2$ structure, and $FeBr_2$ the CdI_2 structure. The iodide, which also has the CdI_2 lattice, can be made from iron and iodine. The halides form many hydrates. Interestingly, $FeCl_2 \cdot 6\ H_2O$ has been shown to contain *trans*-$FeCl_2(H_2O)_4$ units and not hexa-aquoiron(II) ions.

There are three trihalides. Green FeF_3, which is a rather insoluble compound, can be made by the action of fluorine on the metal. The very soluble, black $FeCl_3$ can also be made by direct combination, but $FeBr_3$ is difficult to obtain pure because it decomposes rather easily to $FeBr_2$ and bromine. The chloride and bromide have layer structures similar to that of BiI_3 (26.3.3).

Cobalt forms all four dihalides. CoF_2, best made by heating $CoCl_2$ in a stream of HF, is a pink solid with the rutile structure. The other three dihalides, which are unlike CoF_2 in being readily soluble in water, and in having layer lattices, can be made by direct combination of the elements. There are several hydrated halides. Blue $CoCl_2$ turns pink in moist air and it is used as an indicator in desiccants such as silica gel. The pink hexahydrate contains *trans*-$CoCl_2(H_2O)_4$ units. A violet dihydrate, $CoCl_2 \cdot 2\ H_2O$, crystallises from aqueous solution above 325 K, and a blue-violet monohydrate at 363 K.

Anhydrous CoF_3 can be made as a light-brown powder by treating Co with F_2 at 500 K, but it has not been possible to prepare the other trihalides.

Green, hydrated nickel dihalides can be crystallised from solutions of the metal, its oxide or carbonate in the appropriate aqueous hydrogen halides. The anhydrous chloride and bromide, both yellow solids with the $CdCl_2$ structure, can be made by direct combination of the elements. The yellow NiF_2, which has the rutile structure, is best made by treating $NiCl_2$ with HF at 750 K, and the black NiI_2 results from the mixing of ethanolic solutions of NaI and $NiCl_2$. Its hexahydrate contains $Ni(H_2O)_6^{2+}$ ions, unlike $NiCl_2 \cdot 6\ H_2O$ and $NiBr_2 \cdot 6\ H_2O$ which have *trans*-$NiX_2(H_2O)_4$ units.

36.6. Complex Halides

Complex anions of general formula MX_4^{2-} are very common. Fe^{II} and Ni^{II} halides react in ethanol with salts such as quaternary ammonium halides to give compounds $(R_4N)_2MX_4$, for example. The CoX_4^{2-} complexes are even easier to prepare: the chloro-, bromo- and iodo-complexes will crystallise from aqueous mixtures of CoX_2 with MX (M = alkali metal). The MX_4^{2-} ions are all tetrahedral or very nearly so.

High oxidation states are common in the fluorocomplexes. Yellow Cs_2CoF_6, for example, is obtained by treating Cs_2CoCl_4 with fluorine at 550 K. Red K_2NiF_6 is made in similar way; a mixture of 2 KCl + $NiCl_2$ is fluorinated. The diamagnetic compound liberates oxygen from water.

36.7. Cyanides and Cyanocomplexes

The hexacyanoferrate(II) ion, which is non-poisonous, occurs with many cations; the potassium salt $K_4Fe(CN)_6 \cdot 3\,H_2O$ can be crystallised from the solution after treating an Fe^{II} salt with an excess of KCN. Many oxidising agents convert it to hexacyanoferrate(III), e.g.:

$$Ce^{4+} + Fe(CN)_6^{4-} \rightarrow Ce^{3+} + Fe(CN)_6^{3-}$$

Hexacyanoferrate(II) reacts with an excess of Fe^{3+} to give the pigment Prussian blue, whereas $Fe(CN)_6^{3-}$ reacts with Fe^{2+} to give a precipitate known as Turnbull's blue. The two compounds have been found to give the same X-ray powder diffraction pattern and the same Mössbauer spectrum. If 1 : 1 molar proportions of Fe^{3+} and $Fe(CN)_6^{4-}$ react together, soluble Prussian blue is obtained. This has the formula $KFe[Fe(CN)_6] \cdot H_2O$. The electronic and Mössbauer spectra indicate a formulation $KFe^{III}[Fe^{II}(CN)_6]$ in which high-spin Fe^{3+} ions are octahedrally surrounded by N atoms and low-spin Fe^{2+} by C atoms (Fig. 36.4).

Unlike iron(II) cyanide, $Co(CN)_2$ can be made from the dipositive ion and CN^- in aqueous solution. When the compound is dissolved in aqueous KCN and ethanol is added, a violet salt $K_6[Co_2(CN)_{10}] \cdot 4\,H_2O$ is precipitated. Its

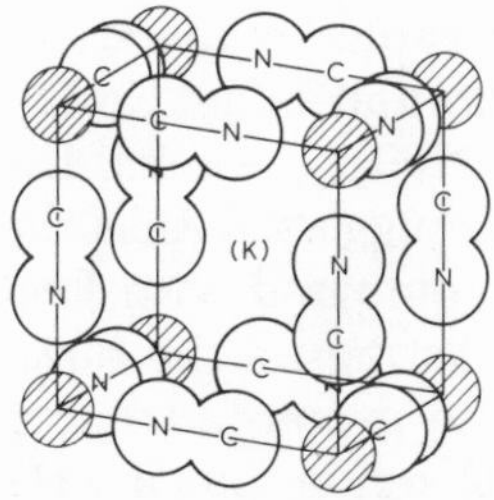

Fig. 36.4. Unit cell of soluble Prussian blue. K^+ ion in alternate cells.

anion exhibits three strong bands in the CN stretching region, and is thought to be structurally similar to $Mn_2(CO)_{10}$.

$Ni(CN)_2$ can be precipitated as a grey hydrate from Ni^{2+} solutions treated with CN^-. It dissolves in an excess of aqueous KCN, and potassium tetracyanonickelate(II), in the form of an orange monohydrate, can be crystallised from the solution. The diamagnetic compound contains a square $Ni(CN)_4{}^{2-}$ ion. It can be reduced with potassium in liquid ammonia to potassium tetracyanonickelate(0), $K_4Ni(CN)_4$, a yellow solid which liberates hydrogen from water.

36.8. Oxides

There are three oxides of iron, with the ideal compositions FeO, Fe_3O_4, Fe_2O_3, but they all show gross nonstoichiometry. The black powder made by heating iron(II) oxalate in the absence of air at ordinary pressure is always iron-deficient, its iron-rich limit being $Fe_{0.95}O$. However, it is claimed that stoichiometric FeO can be made at 5 GPa.

The brown compound FeO(OH), which is made by hydrolysis of $FeCl_3$ at high temperature, is converted at 470 K into red-brown α-Fe_2O_3; the same compound occurs in nature as haematite. This has the corundum structure in which the oxide ions are arranged as in hexagonal close-packing but there is another form, γ-Fe_2O_3, in which the oxide ions are arranged as in cubic close-packing. The black compound Fe_3O_4, which occurs as magnetite, can be made in the laboratory by heating Fe_2O_3 above 1700 K.

FeO, Fe_3O_4 and γ-Fe_2O_3 are closely related structurally. In FeO most of the octahedral spaces between the oxide ions are filled with Fe^{2+} ions in a defect NaCl lattice. In Fe_3O_4 there are both Fe^{2+} and Fe^{3+} ions in an inverse spinel structure and in γ-Fe_2O_3 there are Fe^{3+} ions but not Fe^{2+}; however the cubic arrangement of O^{2-} ions is preserved throughout. As expected, the cubic unit containing 32 O^{2-} ions becomes smaller as the iron ions are reduced in both number and size. The edge of the cube diminishes from 860 pm to 828 pm as the atomic ratio Fe : O falls from 0.486 in $Fe_{0.95}O$ to 0.400 in Fe_2O_3.

Cobalt forms only two oxides, CoO and Co_3O_4. The olive-green CoO, conveniently made by heating $CoCO_3$, has the NaCl structure and a composition very close to the ideal one. It is antiferromagnetic at room temperature. When heated to 700 K in oxygen it is converted to the normal spinel Co_3O_4. The green nickel(II) oxide, made by heating $NiCO_3$ or $Ni(NO_3)_2$, also has the NaCl lattice and almost ideal composition. It has been suggested that the very narrow variation of composition in CoO and NiO compared with FeO and VO is due to the greater effective nuclear charge acting on the outer electrons of the dipositive ions, making them less easy to promote to nonlocalised orbitals. There is no evidence for a nickel(III) oxide, but a well-defined compound, β-NiO(OH), is obtained as a black precipitate by oxidising aqueous Ni^{2+} with potassium hypobromite solution. A similar compound of cobalt is similarly made.

The oxides of iron, cobalt and nickel form compounds with oxides of other metals; the ferrites, which are dealt with in the following paragraph, are of particular technological interest.

36.9. Ferrites

These compounds, which are made by heating the appropriate metal carbonate with Fe_2O_3, have the general formula $M^{II}Fe_2^{III}O_4$. The normal ferrites, such as $ZnFe_2O_4$ and $CdFe_2O_4$, with normal spinel structure, are diamagnetic; the inverse ferrites, with inverse spinel structure, exhibit *ferrimagnetism*. In ferrimagnetic spinels all the ions occupying tetrahedral sites have parallel electronic spins and all the unpaired ions occupying octahedral sites have parallel spins, but the second set are antiparallel to the first set. Thus at low temperatures the inverse ferrite $Fe^{III}(Fe^{III}Ni^{II})O_4$ exhibits the magnetic susceptibility arising from two unpaired spins (i.e. 7—5) per formula unit:

Fe^{III} (tetrahedral) — ↑↑↑↑↑ — 5 unpaired electrons

Fe^{III} (octahedral) and Ni^{II} (octahedral) — ↓↓↓↓↓ ↓↓↑↓↑↓↑↓ — 7 unpaired (opposite in spin to those on tetrahedral sites)

The ferrimagnetism of an inverse ferrite can be enhanced by lattice substitution. In a mixed ferrite of composition $Zn_{\frac{1}{2}}Ni_{\frac{1}{2}}Fe_2O_4$, the distribution of ions among tetrahedral sites will be:

$$(Zn^{II}{}_{1/2}Fe^{III}{}_{1/2})_{tetr}(Fe^{III}{}_{3/2}Ni^{II}{}_{1/2})_{oct}O_4$$

Thus the number of unpaired spins per formula unit might be expected to increase to six (i.e. $7\frac{1}{2} + 1 - 2\frac{1}{2}$). In practice this enhancement of the magnetic susceptibility proceeds to only a limited extent because the exchange forces are weakened as the diamagnetic Zn^{2+} ions are added (Fig. 36.5).

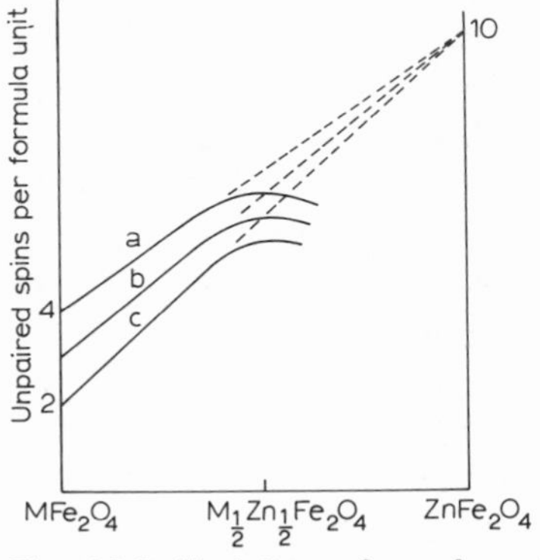

Fig. 36.5. Variation of number of unpaired spins per formula unit for mixed ferrites. In (a) M = Fe, in (b) M = Co and in (c) M = Ni.

The employment of ferrites in electromagnetic devices is associated with (a) their very square hysteresis loops, (b) their high-speed response, and (c) their high flux output. Thus the direction of an induced magnetic field in a ferrite can be sharply reversed by an electric impulse of the right size, but remains unaffected by a smaller impulse. This property is used particularly in the storage and retrieval of information, using binary notation, in the memory banks of computers.

36.10. Sulphides

The iron-sulphur system has been extensively studied but is not yet fully characterised. FeS_2 occurs in two forms, pyrites and marcasite, both brassy in appearance. Pyrites has a cubic lattice, with Fe^{2+} and $S_2{}^{2-}$ ions arranged like the ions in rock salt. The S—S bonds are parallel to trigonal axes. Both forms of FeS_2 are diamagnetic, indicating that the Fe^{2+} ions are spin-paired (t_{2g}^6). Grey FeS, which has the nickel arsenide structure, is made by direct combination. It also occurs as pyrrhotite, which is usually iron-deficient; a phase Fe_7S_8, which has an NiAs structure with one-eighth of the metal positions unoccupied, is a ferromagnetic material.

In the cobalt—sulphur system the compounds which are identified are CoS_2, with the pyrites structure, Co_3S_4, with a spinel structure, and the metal-deficient $Co_{1-x}S$. CoS_2 is ferromagnetic below 120 K; its paramagnetic moment at room temperature indicates that the Co^{2+} ion has the spin-paired $t_{2g}^6e_g^1$ configuration, an indication that the $S_2{}^{2-}$ ions exert a strong crystal field. In the nickel—sulphur system, which is very similar to the Co—S system, there are NiS_2, Ni_3S_4 and $Ni_{1-x}S$ phases with pyrites, spinel and nickel arsenide structures respectively.

36.11. Organometallic Compounds

Iron compounds have been of special importance in the development of organometallic chemistry. The use of $Fe(CO)_5$ in organic synthesis from acetylene was developed in 1949, and in 1951 two groups of workers discovered ferrocene and began the now extensive study of organometallic compounds containing π-bonded aromatic rings. Ferrocene, $(C_5H_5)_2Fe$, was made (a) by treating cyclopentadienyl magnesium bromide with $FeCl_3$ and (b) by passing cyclopentadiene with nitrogen at 570 K over finely divided iron:

$$2\ C_5H_6 + Fe \rightarrow (C_5H_5)_2Fe + H_2$$

It is now made most conveniently by treating $FeCl_2$ with sodium pentadienide in tetrahydrofuran or diglyme:

$$FeCl_2 + 2\ C_5H_5Na \rightarrow (C_5H_5)_2Fe + 2\ NaCl$$

The orange crystals are insoluble in water but soluble in most organic solvents. X-ray diffraction studies show that in the crystal the iron atom is sandwiched between two C_5H_5 rings which are staggered relative to one another. All Fe—C distances are the same (Fig. 18.1). In the vapour the two rings are eclipsed, as shown by electron-diffraction studies.

To construct a molecular orbital scheme to explain the bonding we must first consider the C_5H_5 rings. In these the five 2pπ atomic orbitals combine to form five π molecular orbitals which fall into three groups. The MO of lowest energy is one of symmetry designation A_1, there are two degenerate E_1 orbitals with a nodal plane perpendicular to the ring, and those of highest energy are the E_2 orbitals with two such nodal planes. These localised molecular orbital are then combined to give a set of ten molecular orbitals encompassing both rings; these molecular orbitals have symmetry classifications A_{1g}, A_{2u}, E_{1g}, E_{1u}, E_{2g}, and E_{2u}. In D_{5d} symmetry, the atomic orbitals of the iron atom, the 3d, 4s and 4p, have the following representations: $A_{1g}(4s, 3d_{z^2})$, $A_{2u}(4p_z)$, $E_{1g}(3d_{xz}, 3d_{yz})$, $E_{2g}(3d_{xy}, 3d_{x^2-y^2})$ and $E_{1u}(4p_x, 4p_y)$. As in all MO constructions, only those orbitals with the same symmetry properties can give rise to net overlap. Calculation shows that twelve bonding electrons can be accommodated in strongly bonding orbitals, A_{1g}, A_{2u}, E_{1u} and E_{1g} in ascending order of energy, two more in a non-bonding A_{1g} state related to the d_{z^2} orbital and four in weakly bonding E_{2g} molecular orbitals derived from metal d_{xy} and $d_{x^2-y^2}$ orbitals. Thus, as would be expected purely from symmetry considerations, the $3d_{xy}$ and $3d_{yz}$ are the metal atomic orbitals which contribute most to metal—ring bonding.

The chemical reactions of ferrocene are largely those which arise from the aromatic properties of the C_5H_5 rings, and electrophilic attack on them has been the subject of extensive study. There is evidence that electrophiles interact first with the iron:

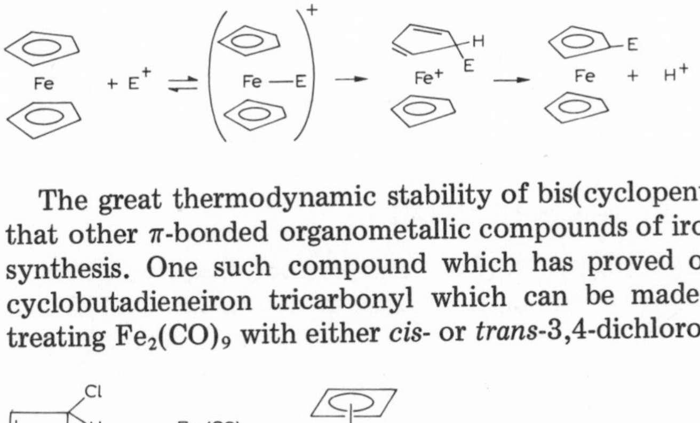

The great thermodynamic stability of bis(cyclopentadienyl)iron suggested that other π-bonded organometallic compounds of iron should be capable of synthesis. One such compound which has proved of particular interest is cyclobutadieneiron tricarbonyl which can be made as yellow crystals by treating $Fe_2(CO)_9$ with either *cis*- or *trans*-3,4-dichlorocyclobutene:

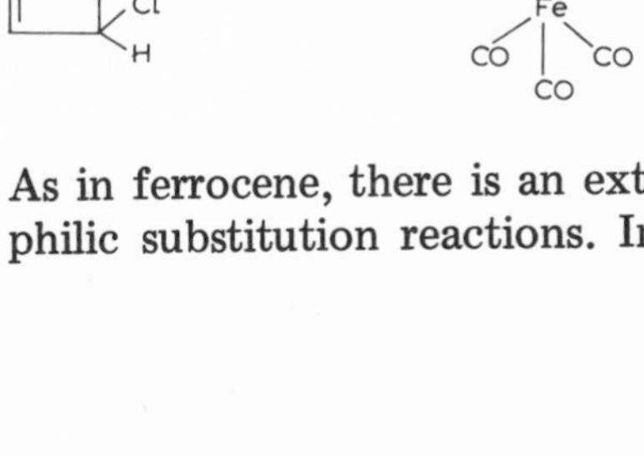

As in ferrocene, there is an extensive chemistry associated with the electrophilic substitution reactions. In Friedel—Crafts acylation, for example, the compound is much more reactive than benzene and almost as reactive as ferrocene itself.

Compounds derived from iron carbonyls and olefins are numerous. In 1930 a compound $C_4H_6Fe(CO)_3$ was made by the action of butadiene on $Fe(CO)_5$, but not until 1960 was it proved to be a π-complex. More recently it has been found that protonation of this compound gives rise to a π-allyl complex:

CH—CH, CH$_2$, CH$_2$, Fe, CO, CO, CO $\xrightarrow{H^+}$ CH, CH$_2$, CH—CH$_3$, Fe$^+$, CO, CO, CO

which, in the form of its iodide, is the starting material for yet another extensive area of organometallic chemistry.

The σ-bonded organometallic compounds of iron are far fewer than the π-complexes. The alkyls and aryls themselves are not stable enough to exist, but many compounds are known containing π-bonding ligands such as CO, Ph_3P, or π-C_5H_5 in addition to an Fe—C σ-bond. An example is $(\pi\text{-}C_5H_5)$-$Fe(CO)_2CF_2CF_2H$, made by the action of perfluoroethylene on cyclopentadienyliron dicarbonyl hydride.

The organometallic chemistry of cobalt is rather similar to that of iron. Purple, air-sensitive crystals of cobaltocene can be made by treating $Co(CNS)_2$ with C_5H_5Na in liquid NH_3. Unlike ferrocene, it is paramagnetic ($\mu = 1.76\ \mu_B$) and is readily oxidised to diamagnetic $(C_5H_5)_2Co^+$ salts. Olefin, allyl and acetylene complexes of cobalt have also been made. Among the σ-bonded organometallic compounds of cobalt the alkyl cobalt carbonyls are important in synthetic organic chemistry. The most important reaction is the reversible carbonylation:

$$RCo(CO)_4 + CO \rightleftharpoons RCOCo(CO)_4$$

The acyls so formed react with alcohols at about 320 K to give esters; thus alkyl halides can be converted to esters by reactions formulated:

$$RX + CO + Co(CO)_4^- \rightarrow RCOCo(CO)_4 + X^-$$
$$RCOCo(CO)_4 + R'OH \rightarrow RCOOR' + HCo(CO)_4$$

Although the catalytic effect of $Ni(CO)_4$ and its phosphine derivatives on the polymerisation of olefin has been studied in depth, few olefin complexes of nickel have been isolated. Perhaps the most important group of π-bonded organonickel compounds is that of π-allyls. Bis(π-allyl)nickel itself is made as yellow, pyrophoric crystals by the action of allyl magnesium bromide on $NiBr_2$ in ether at 263 K. Its sandwich structure has been confirmed by X-ray analysis:

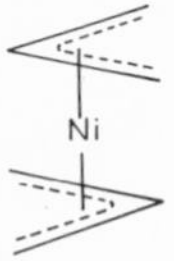

It is an active catalyst for the cyclotrimerisation of butadiene to cyclododeca-1,5,9-triene; π-cyclobutadiene and π-cyclopentadiene compounds of nickel are also known. Emerald green nickelocene, $(\pi\text{-}C_5H_5)_2Ni$, is best made by addition of a solution of $NiCl_2$ in dimethyl sulphoxide to C_5H_5K in diglyme.

Of the σ-bonded organonickel compounds, those made by the action of aryl magnesium halides on square $(R_3P)_2NiX_2$ complexes (X = halogen) are well established. These are yellow, diamagnetic, compounds of general formula $(R_3P)_2NiX(Ar)$ and $(R_3P)_2NiAr_2$ which have *trans*-square configurations.

36.12. Complexes

Although some of the complexes of these metals have already been discussed their chemistry is so extensive that other aspects are worth attention.

36.12.1. Bipositive states

36.12.1.1. Iron(II)

Most of the complexes of iron(II) (d^6) are octahedral; but diamagnetic compounds are far less common than among the isoelectronic Co^{III} compounds, not unnaturally in view of the smaller charge on the iron. The six-co-ordinate, diamagnetic complexes include the hexacyanoferrates(II) and the tris(*ortho*-phenanthroline)iron(II) ions. The octahedral $[Fe(phen)_3]^{2+}$ ion (Fig. 36.6) is blood-red; it is oxidised to pale blue $[Fe(phen)_3]^{3+}$ without any structural change. $E°$ for the system = 1.14 V, making the compound, also known as ferroin, a most useful redox indicator for the oxidation of Fe^{2+} ion ($E°$, Fe^{3+}/Fe^{2+} = 0.77 V) by cerium(IV) ion ($E°$, Ce^{4+}/Ce^{3+} = 1.45 V).

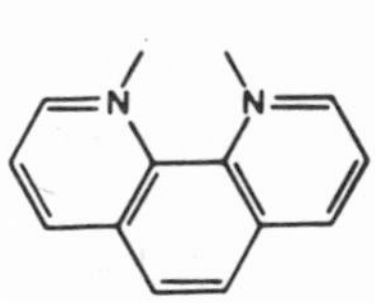

Fig. 36.6. The (phen) group in octahedral $[Fe(phen)_3]^{2+}$ ion.

The pentacyanonitrosylferrates, (the 'nitroprussides') are of interest. Cyano complexes in general do not easily form mixed complexes by replacement of CN^- groups and, in the hexacyanocomplexes of iron, only one CN^- can be replaced by NH_3, H_2O, CO, NO_2 (nitro) or NO (nitroso). Acidification of a $K_4Fe(CN)_6$—KNO_2 mixture gives first the pentacyanonitroferrate(II) ion, $[Fe(CN)_5NO_2]^{4-}$:

$$Fe(CN)_6^{4-} + NO_2^- \rightleftharpoons [Fe(CN)_5NO_2]^{4-} + CN^-$$

and then the pentacyanonitrosylferrate(II) ion, the 'nitroprusside' ion:

$$[Fe(CN)_5NO_2]^{4-} + 2\,H_3O^+ \rightleftharpoons [Fe(CN)_5NO]^{2-} + 3\,H_2O$$

The red sodium salt, $Na_2[Fe(CN)_5NO] \cdot 2\,H_2O$ is diamagnetic; surprisingly since the NO group is an odd-electron group which must confer paramagnetism should it be co-ordinated in the normal way. If the group is considered as co-ordinating in the form NO^+, however, not only is the diamagnetism comprehensible but the charge on the iron becomes clear: it is, in fact, +2.

The SH^- ion in alkaline solution converts the nitroprusside ion to purple $[Fe(CN)_5NOS]^{4-}$:

$$[Fe(CN)_5NO]^{2-} + OH^- + SH^- \rightarrow [Fe(CN)_5NOS]^{4-} + H_2O$$

High-spin octahedral iron(II) complexes are common. The pale-green hexa-aquoiron(II) ion, $Fe(H_2O)_6{}^{2+}$, present in dilute aqueous solution, owes its colour to a single absorption band with a peak at about 10,000 cm^{-1} (120 kJ mol^{-1}) in the i.r. which spreads into the visible spectrum. A few tetrahedral iron(II) complexes are known; examples are the $FeX_4{}^{2-}$ (X = halogen) anions already discussed, and neutral complexes such as $(Ph_3PO)_2FeX_2$. There are also some low-spin, square pyramidal complexes like the $[FeClO_4(OAsMe_3)_4]^+$ ion and a few square complexes, e.g. $Fe(C_6Cl_5)_2(PPhEt_2)_2$.

36.12.1.2. Cobalt(II)

The Co^{2+} ion is the only d^7 ion of common occurrence. The complexes formed from it are of several types; octahedral and tetrahedral are most common but square and 5-co-ordinate ones also occur. For a d^7 ion the ligand-field stabilisation should favour the octahedral configuration only slightly relative to the tetrahedral one, and tetrahedral symmetry proves to be more common in Co^{II} complexes than in those of any other transition metal. Tetrahedral complexes of cobalt(II) with halide ions have already been mentioned, and there are others with monodentate pseudohalide ligands such as SCN^-. Bidentate β-diketonate ions with large alkyl groups also give rise to tetrahedral co-ordination, but with small alkyl groups, as in acetylacetonate, an octahedral arrangement is favoured. With several singly-charged bidentate ions such as dimethylglyoximate and dithioacetylacetonate a square configuration is observed.

With strong-field ligands the tendency for Co^{II} complexes to oxidise to Co^{III} is very strong. Though low-spin octahedral cobalt(II) complexes such as $Co(diars)_3{}^{2+}$ have magnetic moments very close to spin-only values, the high-spin ones have large moments of 4.7 to 5.2 μ_B owing to large orbital contributions (3.4.4.4). The square complexes are all low-spin ones with magnetic moments in the range 2.2 to 2.7 μ_B.

36.12.1.3. Nickel(II)

Nickel in its dipositive state forms complexes of five structural types, octahedral, trigonal bipyramidal, square pyramidal, tetrahedral and square. The aquo ligands in the octahedral $Ni(H_2O)_6{}^{2+}$ ion can be replaced by ammines to give ions such as $Ni(NH_3)_6{}^{2+}$ and $Nien_3{}^{2+}$. Their magnetic moments are in the range 2.9 to 3.4 μ_B, indicating that there are two unpaired electrons and that the orbital contribution to paramagnetism is variable but sometimes large. The absorption spectrum contains four bands in the visible and near ultraviolet, with ϵ_{max} less than 1.0 $m^2\ mol^{-1}$.

Hydrated nickel salts are apple-green, but the octahedral complexes with nitrogen ligands are usually blue or purple because the peaks shift towards the u.v. in the stronger ligand field:

$Ni(H_2O)_6^{2+}$		$Ni(NH_3)_6^{2+}$	
Wave number/ cm^{-1}	$\epsilon/m^2\ mol^{-1}$	*Wave number/* cm^{-1}	$\epsilon/m^2\ mol^{-1}$
8 500	0.20	10 750	0.40
13 500	0.18	15 150	0.50
15 400	0.15	17 500	0.48
25 300	0.52	28 200	0.63

It should be noted that where there are several absorption peaks a general shift to shorter wavelength does not necessarily mean a colour shift towards the red as in the simple d^1 case (6.2.1.1). In the foregoing examples the ammine complex is deep blue because the principal absorption peak is in the ultraviolet whereas the aquo complex is green because there is strong absorption in the violet.

An important class of complexes incorporating 5-co-ordinate nickel(II) is exemplified by the ion $NiN[CH_2CH_2N(CH_3)_2]_3Br^+$:

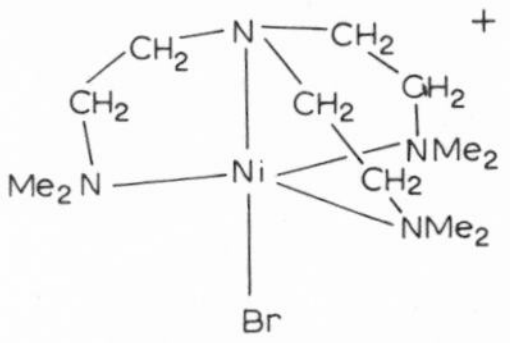

Among other quadridentate 'tripod' ligands which give rise to this symmetry are $N(CH_2CH_2PPh_2)_3$ and $P(o\text{-}C_6H_4SMe)_3$. However the ion $Ni(CN)_5^{3-}$, which contains only unidentate ligands, is also known to exist in a trigonal bipyramidal form.

Several of the paramagnetic nickel(II) complexes, once thought to be tetrahedral on stoichiometric grounds, have been found to be associated and octahedral (28.3.8). Nevertheless, increasing numbers of authentic tetrahedral complexes are being discovered: examples are $NiI_2(PPh_3)_2$, $NiI_2(py)_2$ and the chelate isopropyl- and sec-butyl-salicylaldimine complexes. They are usually blue compounds and their absorption in the red part of the spectrum ($\sim$15000 cm^{-1}) is fairly strong ($\epsilon \sim 16\ m^2\ mol^{-1}$).

Typical square planar nickel(II) complexes are uncharged complexes:

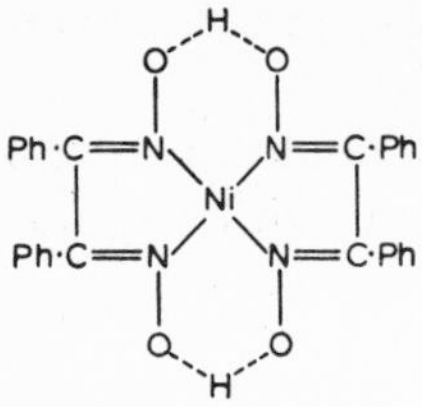

Bis(diphenylglyoxime)nickel(II)

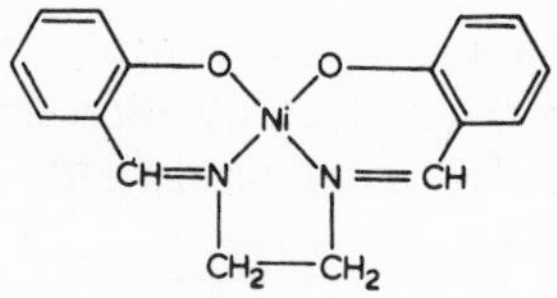

Salicylaldehydo-ethylenediaminenickel(II)

The compounds are usually pink: there are at most three bands in the visible absorption spectrum, the strongest being at about 22,000 cm^{-1} ($\epsilon \sim 15$ to 35 $m^2\ mol^{-1}$).

Many square planar Ni^{II} complexes which are diamagnetic in the solid state become paramagnetic in solution. Bis(*N*-methylsalicylaldimine)nickel(II) shows weak absorption bands which indicate that some of the nickel atoms are in a triplet ground state even in the diamagnetic solid. When the solid is dissolved in chloroform these bands become stronger, and the compound becomes paramagnetic ($\mu = 2.3\ \mu_B$). The solvent weakens the ligand field and uncoupling of spin occurs. In pyridine the same compound has the type of absorption spectrum characteristic of a paramagnetic octahedral Ni^{II} complex and the magnetic moment reaches 3.1 μ_B. Molecules of the base can enter the fifth and sixth co-ordination positions, thereby raising the energy of the d_{z^2} orbital and causing the ligands in the xy plane to be repelled:

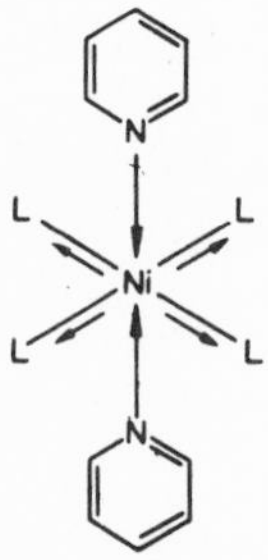

Another way in which planar nickel(II) complexes are converted into tetragonal or octahedral ones is by association either in solution or in the crystal.

There is evidence for equilibria between diamagnetic planar and paramagnetic tetrahedral isomers in solution and melts. The phenomenon is shown by a number of bis(*N*-sec-alkylsalicylaldimine)nickel(II) complexes:

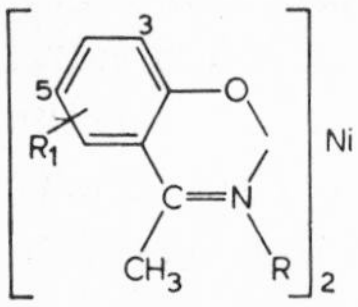

R = sec. alkyl

R_1 = alkyl at 3 *or* 5 position

Of the solids, some are diamagnetic and some paramagnetic ($\mu \sim 3.3\ \mu_B$).

When dissolved in toluene the diamagnetic ones become paramagnetic, the paramagnetic ones less so.

The existence of equilibria between conformational isomers:

tetrahedral (paramagnetic) $\rightleftharpoons$ planar (diamagnetic)

is evident from the absorption spectra.

36.12.2. Terpositive states

36.12.2.1. Iron(III)

Most iron(III) complexes are octahedral, but tetrahedral anions such as $FeCl_4^-$ are known and there is the pentagonal bipyramidal $[Fe(edta)H_2O]^-$. Of the octahedral complexes, those with ligands such as oxalate, phosphate and β-diketones, which co-ordinate through oxygen atoms, are much more common than those formed from nitrogen donors. Addition of ammonia to an iron(III) salt, even in the presence of ammonium chloride, merely causes the precipitation of hydrated iron(III) oxide. But dipyridyl and *ortho*-phenanthroline, which exert ligand fields strong enough to cause spin-pairing, form stable complexes. Among the halogen and pseudohalogen complexes are FeF_6^{3-}, $Fe(CN)_6^{3-}$, $Fe(CNS)_6^{3-}$; anions containing five of one type of ligand and one of another are rather common; $[Fe(CN)_5H_2O]^{2-}$; and $[FeF_5(H_2O)]^{2-}$; are examples.

Iron(III) (d^5) compounds might be expected to have very similar absorption spectra to those of the isoelectronic Mn^{II}, but the weak d—d transitions are in fact masked by the spread into the visible spectrum of strong charge-transfer peaks belonging to the near ultraviolet.

Binuclear, oxygen-bridged iron(III) compounds are of interest. They are of two kinds, the dihydroxo-bridged complexes exemplified by the $[Fe(H_2O)_4(OH)_2Fe(H_2O)_4]^{4+}$ ion which exists in aqueous solutions of Fe^{3+} in the pH range 2—3:

$$(H_2O)_4Fe(\mu\text{-}OH)_2Fe(OH_2)_4$$

and the monoxo-bridged complexes exemplified by the $[Fe(Hedta)]_2O^{2-}$ ion in which the two iron atoms are bound to a single bridging oxygen:

$$[\text{Hedta Fe}\!-\!O\!-\!\text{Fe edtaH}]^{2-}\quad (\angle \text{Fe–O–Fe} = 165°)$$

36.12.2.2. Cobalt(III)

There is a wide range of cobalt(III) complexes with nitrogen donors, and most of the pioneer investigation of transition-metal complexes was done on cobaltammines. Although cobalt(III) salts are oxidised only with extreme difficulty (E°, Co^{3+}/Co^{2+} = +1.84 V), H_2O_2 or even air will oxidise Co^{II} to Co^{III} in the presence of cyanide or ammonia (E°, $Co(CN)_6^{3-}/Co(CN)_6^{4-}$ = −0.83 V). Hexa-amminecobalt(III) chloride, $Co(NH_3)_6Cl_3$, is readily made by oxidising a solution of $CoCl_2$, NH_3 and NH_4Cl containing active charcoal as a catalyst:

$$2\,CoCl_2 + 2\,NH_4Cl + 10\,NH_3 + H_2O_2 \rightarrow 2\,Co(NH_3)_6Cl_3 + 2\,H_2O$$

This compound contains the $Co(NH_3)_6^{3+}$ ion. Treatment of $Co(NO_3)_2$, NH_4NO_3 and NH_3 with H_2O_2 in the absence of charcoal gives a nitratopenta-amminecobalt nitrate however, $[CoNO_3(NH_3)_5](NO_3)_2$. Other series can be made by suitable methods.

The compound $K_6Co_2(CN)_{10}$ is easily oxidised in weak acid to the very stable $K_3Co(CN)_6$. The so-called cobaltinitrites are also Co^{III} complexes. Addition of aqueous KNO_2 to a Co^{2+} solution containing acetic acid gives a yellow precipitate of potassium hexanitrocobaltate(III), $K_3Co(NO_2)_6$.

$$Co^{2+} + 7\,NO_2^- + 2\,H^+ \rightarrow Co(NO_2)_6^{3-} + NO + H_2O$$

When the violet salt, chloroaquotetra-amminecobalt(III) sulphate, $[Co(NH_3)_4(H_2O)Cl]SO_4$, is treated with cold alkali the asymmetric ion

$$\left(Co\left[\begin{array}{c} H \\ O \\ \langle\;\;\rangle Co(NH_3)_4 \\ O \\ H \end{array}\right]_3\right)^{6+}$$

is produced. The chloride of this ion was the first purely inorganic (non-carbon) compound to be resolved (Werner, 1914).

Another type of polynuclear cobalt complex is produced by aerial oxidation of ammoniacal $Co(NO_3)_2$ solution. This is the μ-peroxo-bis{penta-amminecobalt(III)} ion, $[(NH_3)_5Co{-}O{-}O{-}Co(NH_3)_5]^{4+}$. Further oxidation

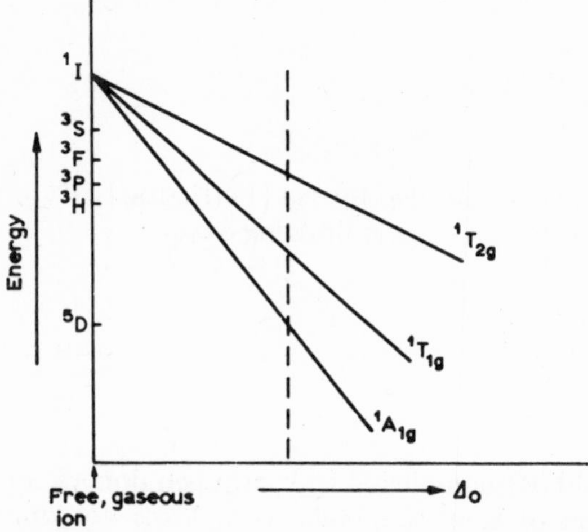

Fig. 36.7. Energy level diagram for d^6 ion in an octahedral field (singlet states only).

gives the quinquepositive ion $[(NH_3)_5CoO_2Co(NH_3)_5]^{5+}$. It was shown, by electron-spin resonance, to contain cobalt atoms of identical oxidation number, not Co^{III} and Co^{IV} as previously thought. X-ray studies indicated the axis of O—O to be perpendicular to the Co—Co axis.

All cobalt(III) complexes except CoF_6^{3-} have singlet ground states and are diamagnetic. The energy level diagram for the d^6 ion in the octahedral field, with only the singlet states shown, has the form illustrated in Fig. 36.7.

There are usually two absorption bands in the visible spectrum. For symmetrical complexes, CoX_6, each of these bands, when plotted on a wavenumber scale, is almost perfectly symmetrical about its peak. But for less symmetrical complexes such as CoX_4Y_2 there is considerable splitting of the $^1T_{1g}$ state. Furthermore, the splitting is greater for the *trans* form than for the *cis* form of such a complex. For the *trans* the absorption peak at the lower energy ($^1A_{1g} \rightarrow {}^1T_{1g}$) transition is split into two peaks, for the *cis* the peak differs from that of the symmetrical CoX_6 in having a 'shoulder'. Moreover because a *cis* octahedral isomer is without a centre of symmetry the intensity of absorption is stronger. Typical absorption curves for *cis* and *trans* isomers of the CoX_4Y_2 type are illustrated in Fig. 36.8.

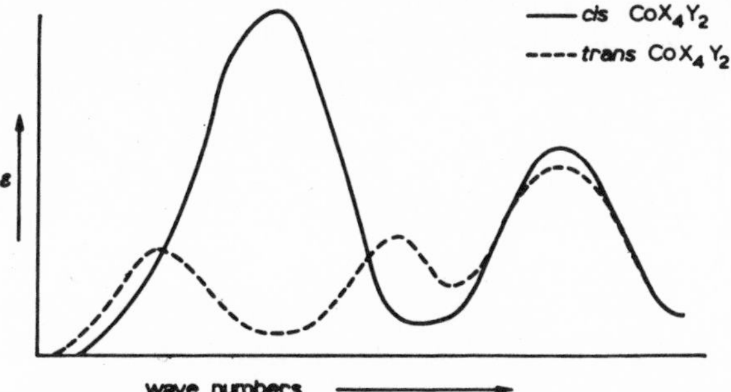

Fig. 36.8. Typical absorption curves for *cis* and *trans* CoX_4Y_2 isomers.

36.12.2.3. Nickel(III)

This is an unusual oxidation state in nickel, but a few complexes with the element in this state are known. The trimethylphosphine complexes of nickel(II) halides, $NiX_2(PMe_3)_2$, are oxidised to nickel(III) complexes even by atmospheric oxygen. When the appropriate nitrosyl halide is used as the oxidising agent the compounds formed are $NiX_3(PMe_3)_2$; they are monomeric in solution.

36.12.3. Higher oxidation states

Treatment of alcoholic $FeCl_3$ with $o\text{-}C_6H_4(AsMe_2)_2$(diarsine) gives a brilliant red precipitate of $[Fe^{III}Cl_2(\text{diarsine})_2]FeCl_4$ which, when warmed in nitrobenzene with concentrated HNO_3, gives a compound $[Fe^{IV}Cl_2(\text{diarsine})_2](FeCl_4)_2$. The magnetic moment of 2.98 μ_B is consistent with the presence of two unpaired electrons in a tetragonally distorted set of t_{2g} orbitals. Other compounds of Fe^{IV}, Co^{IV} and Ni^{IV} are principally fluoroanions which have already been discussed, but some diamagnetic Ni^{IV} heteropoly-anions such as $[NiMo_9O_{32}]^{6-}$ and $[NiNb_{12}O_{38}]^{12-}$ contain nickel in this formal oxidation state.

The charge number +6 does not occur in Co and Ni, and in Fe only as the ferrates such as $BaFeO_4$ which have already been discussed.

Further Reading

M. Rosenblum, The chemistry of the iron group metallocenes, Wiley, New York, 1965.

T.W. Thomas and A.E. Underhill, Metal—metal interactions in transition-metal complexes containing infinite chains of metal atoms, Chem. Soc. Rev., 1 (1972) 99.

J.M. Pratt and R.G. Thorpe, *Cis* and *trans* effects in cobalt(III) complexes, Adv. Inorg. and Radiochem., 12 (1969) 375.

R.S. Young (Ed.), Cobalt. Its chemistry, metallurgy and uses, Reinhold, New York, 1960.

R.W. Thomas, Revolution in iron and steel making, Ed. Chem., 2 (1965) 167.

U.R. Evans, The mechanism of rusting, Q. Rev., 21 (1967) 29.

E.K. Barefield, D.H. Busch and S.M. Nelson, Iron, cobalt and nickel complexes having anomalous magnetic moments, Q. Rev., 22 (1968) 457.

D. Nicholls, The chemistry of iron, cobalt and nickel, Pergamon, Oxford, 1975.

P. Hambright, The co-ordination chemistry of metalloporphyrins, Co-ord. Chem. Rev., 6 (1971) 247.

D.G. Brown, The chemistry of vitamin B_{12} and related inorganic model systems, Prog. Inorg. Chem., 18 (1973) 177.

The Platinum Metals

37

37.1. Introduction

The second triad of Group VIII, ruthenium, rhodium and palladium, and the third triad, osmium, iridium and platinum, are sufficiently alike in physical and chemical character to be known collectively as the platinum metals. Vertical similarities are much stronger between members of these triads than between them and iron, cobalt and nickel, but some correlations can be observed. Thus Fe, Ru and Os form monomeric carbonyls, $M(CO)_5$, the ions MO_4^{2-} and the very stable complex ions $M(CN)_6^{4-}$, $M(phen)_3^{2+}$ and $M(bipy)_3^{2+}$. Co, Rh and Ir all form the dimeric carbonyls $M_2(CO)_8$ and the ions $M(CN)_6^{3-}$ and $M(C_2O_4)_3^{3-}$, and Ni, Pd and Pt have the very stable $M(CN)_4^{2-}$ ions in common as well as neutral complexes with oximes, the best known being the dimethylgloxime complexes.

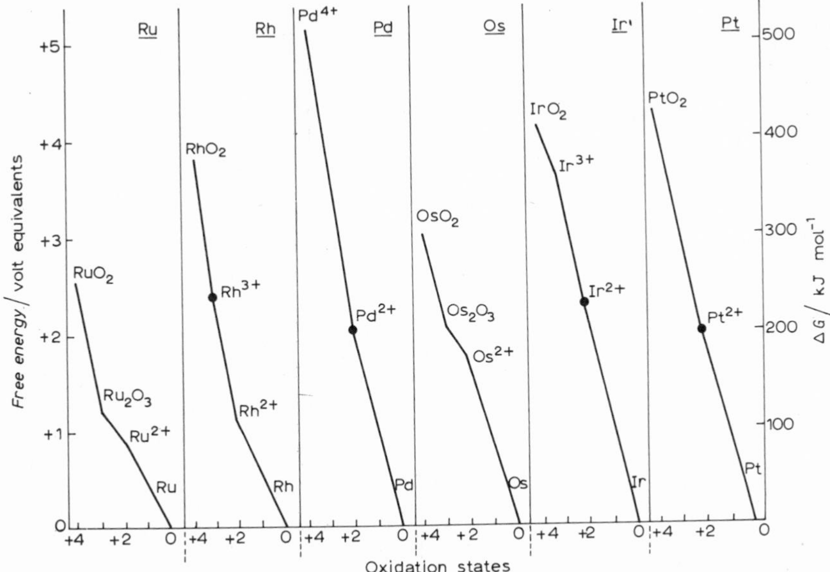

Fig. 37.1. Free energies of oxidation states of the platinum family, relative to the metal, in aqueous solution at pH = 0.

TABLE 37.1

ATOMIC PROPERTIES OF Ru, Rh, Pd, Os, Ir and Pt

	Ru	Rh	Pd	Os	Ir	Pt
Z	44	45	46	76	77	78
Electron configuration	$[Kr]4d^7 5s^1$	$[Kr]4d^8 5s^1$	$[Kr]4d^{10}$	$[Xe]4f^{14} 5d^6 6s^2$	$[Xe]4f^{14} 5d^7 6s^2$	$[Xe]4f^{14} 5d^9 6s^1$
$I(1)/kJ\ mol^{-1}$	720	720	804	840	840	870
Metallic radius/pm	134	134	137	135	136	138
$r_{M^{2+}}$/pm		86	80			80
$r_{M^{3+}}$/pm	69	69				
$r_{M^{4+}}$/pm	65			67	66	

Metallic radii in the platinum metals are almost uniform (Table 37.1); not surprisingly the elements show great physical similarities. The first ionisation energies are not much greater than those of the d-block elements which precede them but the electrode potentials are strongly positive, a major factor being undoubtedly the very high sublimation energies.

37.2. Stability of Oxidation States in Aqueous Solution

The free energies of oxidation states up to +4, relative to the metal, are illustrated in Fig. 37.1. Notable features of the diagram are the relative stability of the +3 state in the elements Ru and Os, the tendency of this state to disproportionation in Rh and Ir (cf. their congeners), and the absence of the +3 state in Pd and Pt (cf. Ni).

Higher oxidation states of Ru and Os exist in solution, for instance in RuO_4^- and $OsO_4(OH)_2^{2-}$, but they are unstable at low pH and their redox potentials are in doubt.

37.3. The Elements: Preparation and Properties

The six metals together comprise about $2 \times 10^{-6}\%$ of the lithosphere and are often found native. Osmiridium is a natural alloy of osmium and iridium.

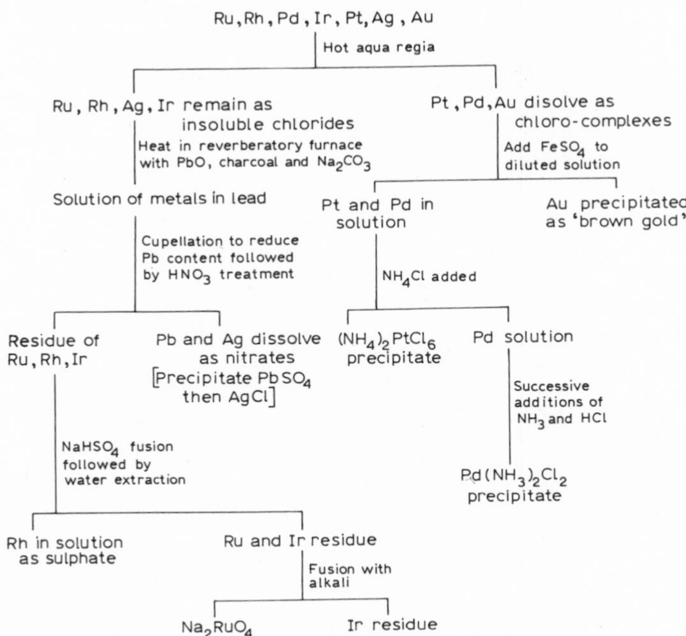

Fig. 37.2. Separation of the platinum metals.

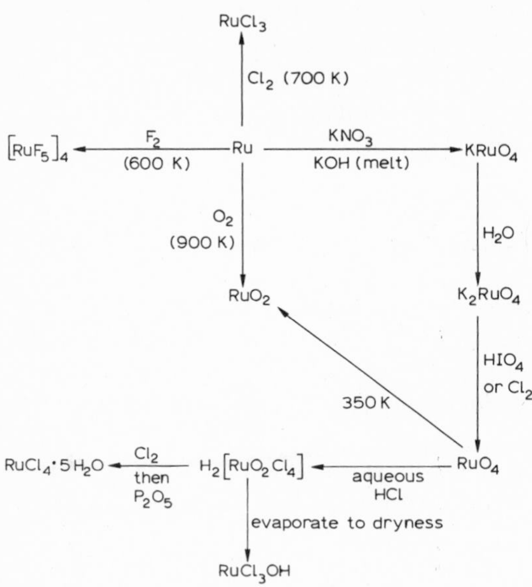

Fig. 37.3. Ruthenium, some reactions and preparations.

Important ores are sperrylite, $PtAs_2$, cooperite, PtS, and braggite, (Pt, Pd, Ni)S. The Sudbury (Canada) nickel and copper sulphide deposits contain about 2 p.p.m. of the platinum metals, other than osmium. The metals, along with copper and gold, are segregated in the undissolved sludges from the electro-refining of Ni and Cu and in the involatile residues after the removal of $Ni(CO)_4$ in the carbonyl process. The Sudbury separation (Fig. 37.2)

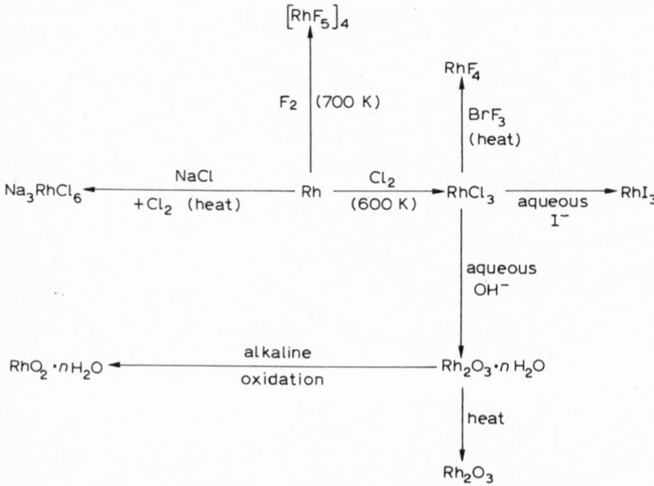

Fig. 37.4. Rhodium, some reactions and preparations.

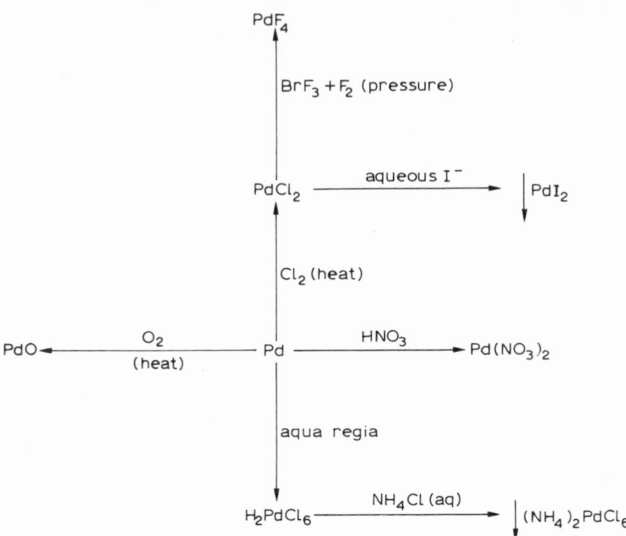

Fig. 37.5. Palladium, some reactions and preparations.

begins with an extraction by hot aqua regia in which only platinum, palladium and gold dissolve; the various metals are recovered as indicated in Fig. 37.2.

Total world production of platinum metals exceeds 100 tons per annum.

The densities of the metals fall into two groups, those in the third triad being the highest of all the elements. Ru and Os have h.c.p. structure, the

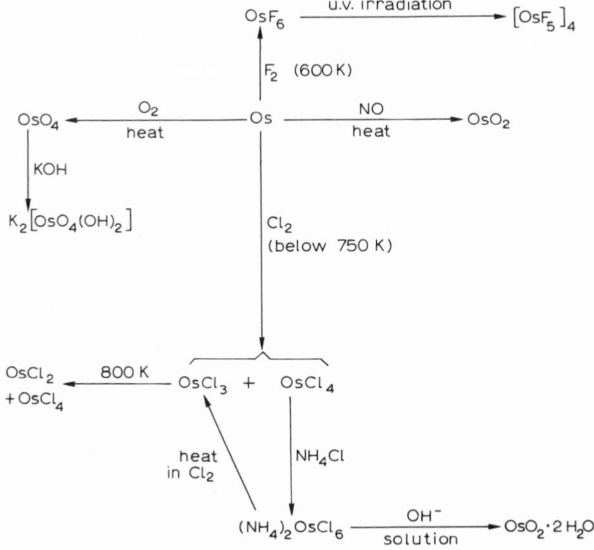

Fig. 37.6. Osmium, some reactions and preparations.

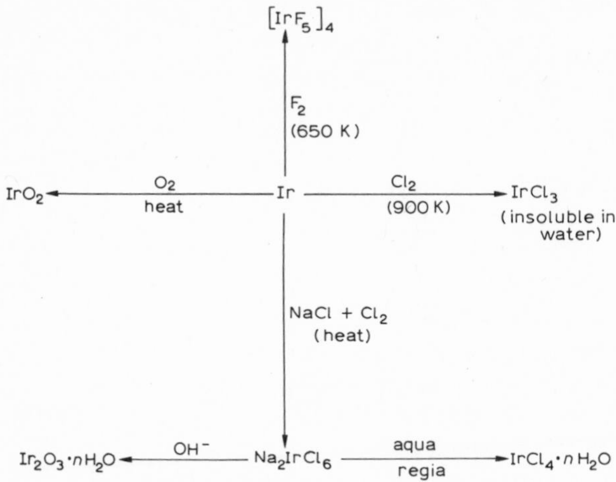

Fig. 37.7. Iridium, some reactions and preparations.

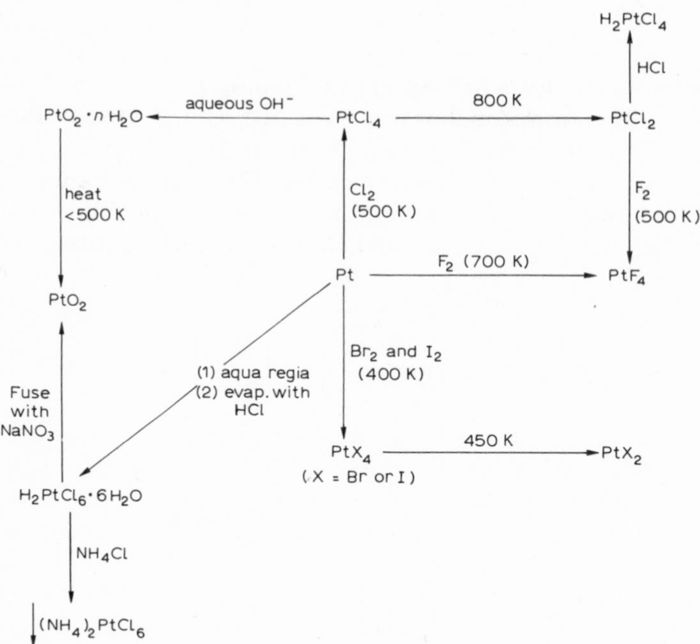

Fig. 37.8. Platinum, some reactions and preparations.

rest are cubic close-packed. Melting points, hardness and mechanical strength decrease progressively in each period. These differences have been attributed to the decreasing availability of electrons for bonding among the atoms in the solid state, since electrons which become paired in atomic d orbitals fail to participate in metallic bonding.

TABLE 37.2

PHYSICAL PROPERTIES OF THE PLATINUM METALS

	Ru	Rh	Pd	Os	Ir	Pt
M.p./K	2720	2240	1830	2970	2730	2040
ρ/g cm^{-3}	12.45	12.41	12.02	22.61	22.65	21.45

Several of the metals are useful in catalysis. Platinum is used as a surface catalyst in (a) the hydrogenation of olefins and acetylenes, (b) the oxidation of NH_3 to HNO_3, (c) the reforming of hydrocarbons and (d) the oxidation of hydrocarbons in fuel cells. Rhodium supported in a finely divided state on alumina is particularly useful for catalysing the hydrogenation of benzene. Some of the intermetallic compounds of iridium, notably Ti_3Ir, $ZrIr_2$ and Nb_3Ir, are superconductors. Many of the elements are resistant to corrosion: platinum is employed for electrodes and crucibles and extensively for spinnerets to extrude rayon, and rhodium provides a surfacing material for searchlight mirrors. Palladium is used for purifying hydrogen which diffuses rapidly and selectively through the warm metal.

The metals are relatively unreactive. Palladium alone among them dissolves in HNO_3 and Pt, Os and Pd only dissolve in aqua regia. Platinum does not react with oxygen. Some of the reactions of the metals are illustrated in the foregoing diagrams. The oxides, sulphides, selenides, tellurides, phosphides, a few of the sulphates and some of the halides are the only uncomplexed compounds.

37.4. Oxidation States

Only a few examples of negative oxidation states are known; rhodium forms the carbonylate anion $Rh(CO)_4^-$, and there is some evidence for the $Ru(CO)_4^{2-}$ ion. In the case of iridium a tetrahedral ion $[Ir(CO)_3PPh_3]^-$ has been made.

The highest oxidation state of +8 is exhibited by ruthenium and osmium in their oxides, RuO_4 and OsO_4, but in rhodium and iridium the maximum charge number is only +6. Indeed the general charge number pattern of the preceding transition elements changes at this point in the Periodic Table; thereafter the elements tend to show less diversity of oxidation state. This is evinced by the closer similarity between Ni, Pd and Pt than between the respective members of the two preceding vertical groupings.

The charge numbers +1 and +5 are uncommon. All the elements except palladium form hexafluorides, but in palladium the highest known oxidation state is +4.

TABLE 37.3

COMPOUNDS AND IONS REPRESENTATIVE OF THE OXIDATION STATES OF Ru, Rh, Pd, Os, Ir AND Pt

Oxidation number	Ru	Rh	Pd	Os	Ir	Pt
−1		$Rh(CO)_4^-$			$[Ir(CO)_3PPh_3]^-$	
0	$Ru(CO)_5$	$Rh_2(CO)_8$	$Pd(NCR)_2$	$Os(CO)_5$	$Ir_4(CO)_{12}$	$Pt(PR_3)_4$
+1		$[RhCl(CO)_2]_2$		$Os(NH_3)_6Br$	$IrClCO(PEt_3)_2$	·
+2	$Ru(NH_3)_6^{2+}$	$RhCl(dipy)_2^+$	PdO	$Os(CN)_6^{4-}$	$Ir(NH_3)_4Cl_2$	PtS
+3	$RuCl(NH_3)_5^{2+}$	$RhCl_6^{3-}$		$OsCl_6^{3-}$	$IrCl_6^{3-}$	
+4	$RuCl_6^{2-}$	RhF_6^{2-}	$PdCl_6^{2-}$	$OsCl_6^{2-}$	IrO_2	$PtCl_6^{2-}$
+5	RuF_5			OsF_6^-	IrF_6^-	PtF_6^-
+6	RuF_6	RhF_6		OsF_6	IrF_6	PtF_6
+7	RuO_4^-			$OsOF_5$		
+8	RuO_4			OsO_4		

37.5. Halides

The hexafluorides are known for all the platinum metals except palladium. They are volatile, reactive substances which are normally kept in nickel or monel apparatus because they attack glass, some of them even at room temperatures. The compounds are normally made by direct combination of the elements followed by rapid condensation on a cold finger. Dark-red crystals of PtF_6 so made have extraordinary oxidising powers. The vapour was discovered to oxidise O_2 to O_2^+, a result which stimulated the work which led to the preparation of the first compound of cationic xenon, $XePtF_6$. The hexafluorides have octahedral molecules.

The five metals which form hexafluorides also form tetrameric compounds $(MF_5)_4$ in which the octahedrally co-ordinated metal atoms are linked by M—F—M bridges similar to those in the corresponding compounds of Nb, Ta and Mo. They are normally obtained by direct fluorination at temperatures rather higher than those used for the preparation of hexafluorides.

All six metals form tetrahalides. Brick-red PdF_4 is made by fluorination of a compound formerly called PdF_3 but now known to be $Pd^{II}(Pd^{IV}F_6)$ which is made by treating $PdBr_2$ with BrF_3 and heating to 450 K the adduct $Pd_2F_6 \cdot 2\,BrF_3$ so formed. RhF_4 and PtF_4 can also be made from BrF_3 adducts:

$$RhCl_3 \xrightarrow{BrF_3} RhF_4 \cdot 2\,BrF_3 \xrightarrow{heat} RhF_4$$

$$Pt \xrightarrow{BrF_3} PtF_4 \cdot 2\,BrF_3 \xrightarrow{heat} PtF_4$$

Red-brown $PtCl_4$ is obtained by heating to 570 K the compound $H_2PtCl_6 \cdot 6\,H_2O$ obtained by dissolving Pt in aqua regia. The other two tetrahalides of platinum are dark-brown compounds made by direct combination of the elements.

Of the trifluorides, RuF_3 is best made by reducing RuF_5 with iodine at 520 K:

$$5\,RuF_5 + I_2 \rightarrow 5\,RuF_3 + 2\,IF_5$$

and IrF_3 by reducing IrF_6 with Ir. Most of the other trihalides shown in Table 37.4 are made by direct combination of the elements or by precipitation from solution:

$$RhCl_3 + 3\,I^- \rightarrow RhI_3 + 3\,Cl^-$$

The dark-red $PdCl_2$ and $PdBr_2$ are made from the elements at red heat, but PdF_2 is best made by refluxing $Pd^{II}[Pd^{IV}Cl_6]$ with SeF_4:

$$Pd_2Cl_6 + SeF_4 \rightarrow 2\,PdF_2 + SeCl_6$$

Dark-red $PtCl_2$ can be obtained by thermal decompositions of $PtCl_4$, but the dibromide and di-iodide are so obtained only with difficulty.

TABLE 37.4

HALIDES OF THE PLATINUM METALS

Oxidation state	Ru	Rh	Pd	Os	Ir	Pt
+2	$RuBr_2$		PdF_2, $PdCl_2$ $PdBr_2$, PdI_2	OsI_2		$PtCl_2$ $PtBr_2$, PtI_2
+3	RuF_3, $RuCl_3$ $RuBr_3$, RuI_3	RhF_3, $RhCl_3$ $RhBr_3$, RhI_3		$OsCl_3$ $OsBr_3$ OsI_3	IrF_3, $IrCl_3$ $IrBr_3$, IrI_3	
+4	RuF_4	RhF_4	PdF_4	OsF_4, $OsCl_4$ $OsBr_4$		PtF_4, $PtCl_4$ $PtBr_4$, PtI_4
+5	$(RuF_5)_4$	$(RhF_5)_4$		$(OsF_5)_4$	$(IrF_5)_4$	$(PtF_5)_4$
+6	RuF_6	RhF_6		OsF_6	IrF_6	PtF_6

37.6. Halogen Complexes

The principal fluoro- and chloro-anions are shown in Table 37.5.

TABLE 37.5
PRINCIPAL FLUORO- AND CHLORO-ANIONS OF THE PLATINUM METALS

Charge number	Ru	Rh	Pd	Os	Ir	Pt
+2			$PdCl_4^{2-}$			$PtCl_4^{2-}$
+3	RuF_6^{3-}	RhF_6^{3-}			IrF_6^{3-}	
	$RuCl_6^{3-}$	$RhCl_6^{3-}$		$OsCl_6^{3-}$	$IrCl_6^{3-}$	
+4	RuF_6^{2-}	RhF_6^{2-}	PdF_6^{2-}	OsF_6^{2-}	IrF_6^{2-}	PtF_6^{2-}
	$RuCl_6^{2-}$	$RhCl_6^{2-}$	$PdCl_6^{2-}$	$OsCl_6^{2-}$	$IrCl_6^{2-}$	$PtCl_6^{2-}$
+5	RuF_6^-	RhF_6^-		OsF_6^-	IrF_6^-	PtF_6^-

The chloro-complexes can sometimes be made by heating the metal with an alkali-metal chloride in a stream of chlorine:

$2\ Rh + 6\ NaCl + 3\ Cl_2 \rightarrow 2\ Na_3RhCl_6$

Another common method of preparation is to add NH_4Cl or KCl to the complex chloro-acid, ammonium and potassium salts being sparingly soluble as a rule:

$H_2PtCl_6 + 2\ NH_4Cl \rightarrow (NH_4)_2PtCl_6 + 2\ HCl$
$H_2OsCl_6 + 2\ NH_4Cl \rightarrow (NH_4)_2OsCl_6 + 2\ HCl$

The ammonium salt is frequently used in purification processes as it gives the metal on heating. The addition of KCl to a Na_3RhCl_6 solution precipitates the 5-co-ordinate K_2RhCl_5.

Ruthenium(III) exists in $RuCl_4^-$, $RuCl_5^{2-}$ and $RuCl_7^{4-}$, ions of unknown structure, as well as in $RuCl_6^{3-}$.

Platinum resists attack by concentrated aqueous HCl except in the presence of KCl or RbCl which, by forming insoluble chloroplatinates, disturb the equilibrium which is normally in favour of the metal. Similarly, gaseous HCl reacts with Pt in the presence of KCl, K_2PtCl_4 is first formed and subsequently disproportionates to Pt and K_2PtCl_6, on cooling.

Of all the chloro-complexes, only the chloroplatinates(II) and (IV) are not hydrolysed in aqueous solution at pH 7. The rest give precipitates of hydrated oxides under these conditions.

Of the fluorides, K_3RuF_6 is made by fusing $RuCl_3$ with KHF_2, and K_3RhF_6 by fusing $K_3Rh(NO_2)_6$ with KHF_2. Fluorocomplexes of +5 states can be made by fluorination of mixtures of alkali-metal halides with platinum-group halides:

$RuCl_3 + MCl + 3\ F_2 \rightarrow MRuF_6 + 2\ Cl_2$
$2\ OsCl_4 + 2\ MCl + 6\ F_2 \rightarrow 2\ MOsF_6 + 5\ Cl_2$

Bromine trifluoride is also used as the fluorinating agent:

$$IrBr_3 + MCl \xrightarrow{BrF_3} MIrF_6$$

These +5 complexes are unstable in water or dilute aqueous alkali. Thus when $KIrF_6$ is treated with dilute KOH it is converted to K_2IrF_6, an Ir^{IV} complex, which remains unchanged by the water. The hexafluororuthenates and hexafluoro-osmates(IV) are made similarly.

There are bromocomplexes corresponding to some of the chlorocomplexes listed above; among the most important are the salts of $IrBr_6^{2-}$, $OsBr_6^{2-}$ and $PtBr_4^{2-}$. Few iodocomplexes are known.

37.7. Oxides

The principal oxides of the platinum metals are shown in Table 37.6. A number of other solids of uncertain composition appear to exist and the oxygen chemistry of the group needs further investigation. Black PdO can be made by direct combination of the elements or by fusing $PdCl_2$ with Na_2CO_3, leaching out the soluble sodium salts and then dehydrating. The oxide, which is always somewhat oxygen-deficient, has a tetragonal lattice (Fig. 37.9) in which the metal atoms have four coplanar bonds as in Pd^{II} complexes, and the oxygens are tetrahedrally co-ordinated.

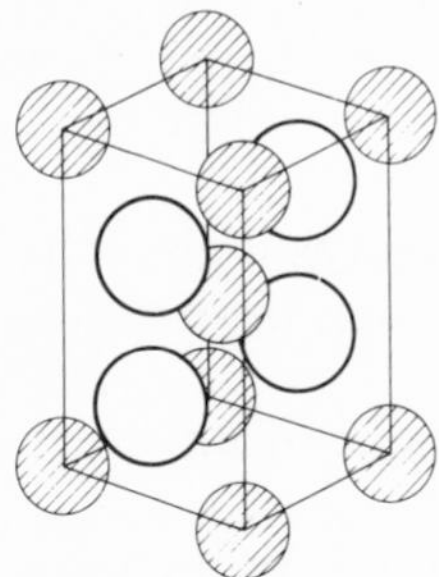

Fig. 37.9. Unit cell of PdO.

A yellow, hydrated oxide $Rh_2O_3 \cdot n\,H_2O$ is precipitated when a slight excess of alkali is added to an $RhCl_3$ solution. The anhydrous oxide, which has the corundum structure, is best made from it by converting it with nitric acid to $Rh(NO_3)_3 \cdot 2\,H_2O$ which is then decomposed by heating. If the solid so obtained is heated in oxygen under pressure it is converted to black RhO_2. A grey, hydrated iridium(III) oxide can be obtained by heating K_3IrCl_6 with Na_2CO_3 and removing the soluble products with water, but it is at least partly oxidised in air to a hydrated IrO_2. Black, anhydrous IrO_2 and black RuO_2 can both be made by heating the metals in oxygen, but OsO_2 is best made by the action of nitric oxide on the metal at 920 K. The foregoing dioxides all have the rutile structure. Platinum does not react directly with oxygen, being in this respect the most noble of the six metals. A brown oxide of approxi-

TABLE 37.6

ANHYDROUS OXIDES OF THE PLATINUM METALS

Charge number	Ru	Rh	Pd	Os	Ir	Pt
+2			PdO			
+3		Rh_2O_3				
+4	RuO_2	RhO_2		OsO_2	IrO_2	PtO_2
+8	RuO_4			OsO_4		

mate composition PtO_2 but unknown structure is formed by fusing H_2PtCl_6 and $NaNO_3$ at 750 K, washing free of sodium salts and then drying the product.

The tetroxides of ruthenium and osmium are solids of low *m.p.* (RuO_4, 298 K and OsO_4, 314 K). Orange RuO_4 volatilises when a stream of chlorine is passed through an acidified solution of a ruthenate. It sublimes in a vacuum, but decomposes explosively into RuO_2 and oxygen at about 450 K. Colourless OsO_4 is much more thermally stable; it can be made by heating the finely divided metal in oxygen. Both compounds contain tetrahedral molecules; they are extremely soluble in CCl_4, which can be used to extract them from aqueous solutions. They are both strong oxidising agents and both can be considered to be acidic oxides. They differ in their behaviour towards alkalis, however. OsO_4 dissolves to give the ion $[OsO_4(OH)_2]^{2-}$ but RuO_4 liberates oxygen from the water and gives eventually a ruthenate(VI) ion:

$$2\ RuO_4 + 4\ OH^- \rightarrow 2\ RuO_4^{2-} + 2\ H_2O + O_2$$

37.8. Sulphides

Both palladium and platinum form monosulphides. PdS can be obtained as a brown precipitate by passing H_2S into a solution of $PdCl_4^{2-}$, and the grey PtS is made by heating together $PtCl_2$, Na_2CO_3 and sulphur. In both, the

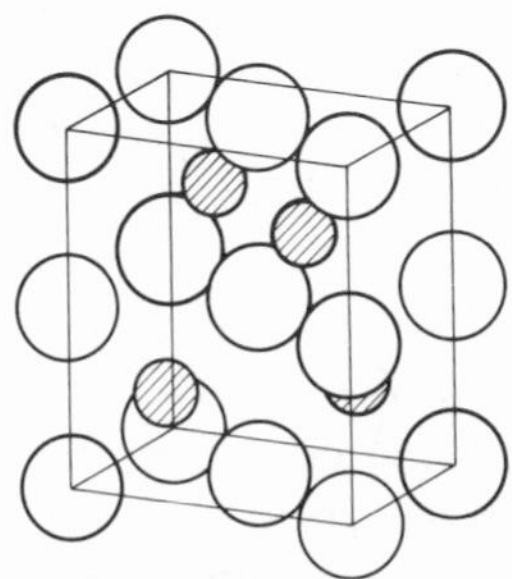

Fig. 37.10. Unit cell of PtS.

metal atom is surrounded by sulphur atoms in approximately square-planar co-ordination (Fig. 37.10). All the metals except palladium form disulphides which are obtained by direct combination of the elements. RuS_2, RhS_2 and OsS_2 have the cubic, pyrites structure containing S_2^{2-} ions and are therefore compounds of the metals in their +2 states, but PtS_2 has the CdI_2 structure. Iridium forms a trisulphide IrS_3 when $IrCl_3$ is heated with an excess of sulphur in a sealed tube, and rhodium forms a compound Rh_2S_5 in a similar reaction. The platinum metals also react with selenium and tellurium, the principal products being diselenides and ditellurides. In general they resemble the sulphides structurally, thus the compounds of Ru, Rh and Os have the pyrites structure whereas $PtSe_2$ and $PtTe_2$ have the CdI_2 lattice.

37.9. Oxoacid Salts

Various yellow hydrates of $Rh_2(SO_4)_3$ are known. The red sulphate $Rh_2(SO_4)_3 \cdot 6\,H_2O$ obtained by evaporating the yellow solutions at 370 K gives no precipitate with $BaCl_2$ and is evidently complex. Yellow $Ir_2(SO_4)_3$aq, made by treating Ir_2O_3 with II_2SO_4 in the absence of air, is also of unknown structure. Both Rh^{III} and Ir^{III} form alums, however, with the sulphates of K, Rb, Cs, NH_4 and Tl^{I}.

A hydrated palladium(II) nitrate, $Pd(NO_3)_2 \cdot 2\,H_2O$, obtained from a solution of the metal in nitric acid, has an infrared spectrum which indicates the existence of unidentate nitrato groups. A brown, volatile, anhydrous compound $Pd(NO_3)_2$, made by treating the dihydrate with liquid N_2O_5, is shown by the same method to contain bridging nitrato groups.

Clearly the platinum-group metals form very few compounds which can be considered as true salts.

37.10. Organometallic Compounds and π-Complexes

Platinum, unlike the other Group VIII metals, forms some thermally stable alkyl compounds in which there is little or no π-bonding. Orange crystals of Me_3PtI are obtained from the reaction between $PtCl_4$ and MeMgI by treating the product with water and afterwards extracting with benzene. The solid is insoluble in water and is not attacked by concentrated acids and alkalis, but it reacts with potassium to give Pt_2Me_6:

$$2\,Me_3PtI + 2\,K \rightarrow Me_3Pt{-}PtMe_3 + 2\,KI$$

Tetramethylplatinum has an interesting structure. The molecule is a cubic tetramer in which every platinum is octahedrally co-ordinated to methyl groups; three of these are terminal groups and three bridging groups to other platinum atoms (Fig. 37.11). Octahedral arrangements in Pt^{IV} compounds are frequently preserved by such unusual types of co-ordination.

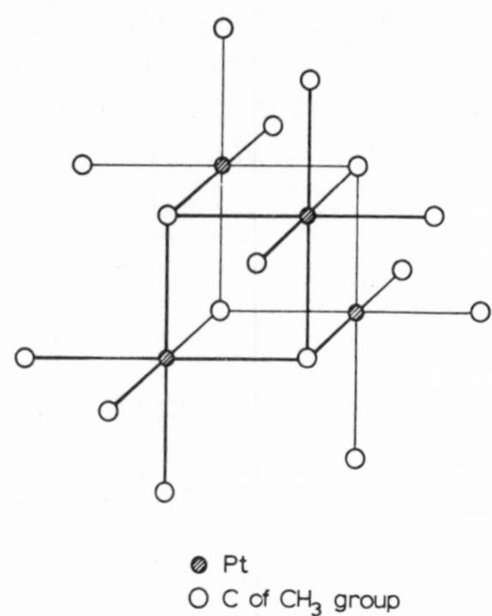

Fig. 37.11. Tetramethylplatinum: cubic structure of the tetrameric molecules.

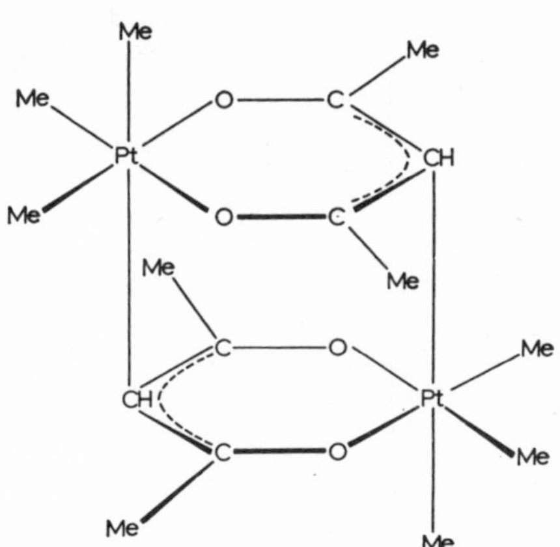

Fig. 37.12. Dimeric molecule of trimethylplatinum acetylacetonate.

Trimethylplatinum acetylacetonate, made by the action of Me_3PtI on acetylacetone, has a dimeric molecule (Fig. 37.12). This dimer reacts with dipyridyl to give the monomeric $Me_3Pt(dipy)(O_2C_5H_7)$ in which the octahedral arrangement around the platinum is preserved through co-ordination, not, however, to the two oxygens of the β-diketone, but to the methylenic carbon atom (Fig. 37.13).

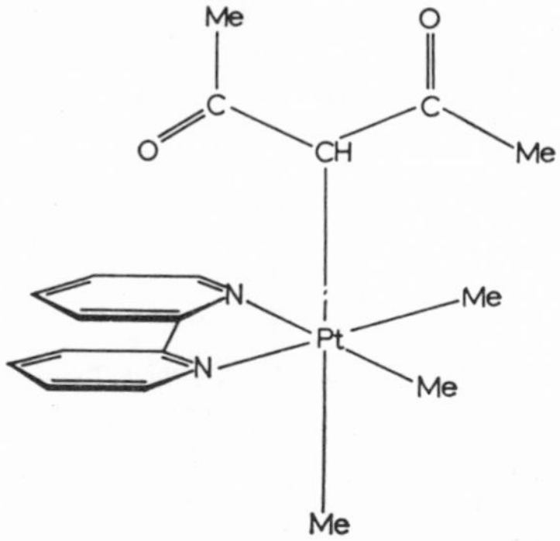

Fig. 37.13. Co-ordination of platinum to methylenic carbon atom in $Me_3Pt(dipy)(O_2C_5H_7)$.

Of the many olefin complexes formed by the platinum metals some have already been discussed. Complex acetylides are also known. Those of Pd^{II} and Pd^0 have been prepared in liquid ammonia by the reactions:

$$K_2Pd(CN)_4 + 2\,KC{:}CR \rightarrow K_2Pd(CN)_2(C{:}CR)_2 + 2\,KCN$$
$$K_2Pd(CN)_2(C{:}CR)_2 + 2\,K \rightarrow K_2Pd(C{:}CR)_2 + 2\,KCN$$

Similar platinum complexes are similarly made. These are all true acetylido-complexes containing metal—carbon σ bonds similar to those in cyano-complexes. In another interesting group of acetylene complexes, $(R_3P)_2Pt(R_1C⋮CR_2)$, the structures are best explained by considering the metal to be σ-bonded to both the carbons of the acetylene group:

```
R3P       C—R'
   \     / ||
    Pt     ||
   /     \ ||
R3P       C—R''
```

The cyclopentadienyls of ruthenium and osmium resemble ferrocene in being stable in air and capable of aromatic substitution without decomposition. Yellow $(\pi\text{-}C_5H_5)_2Rh$ and colourless $(\pi\text{-}C_5H_5)_2Os$ are made by treating $RuCl_3$ and $OsCl_4$ respectively with C_5H_5Na.

The ions $(C_5H_5)_2Rh^+$ and $(C_5H_5)_2Ir^+$, made by treating the acetylacetonates with C_5H_5MgBr, can be reduced to the biscyclopentadienyls by $NaBH_4$.

37.11. Complexes

37.11.1. Zero oxidation states

This state is exhibited in the many carbonyls (17.12) and also in complexes containing phosphine and arsine ligands. Examples are $Ru(CO)_3(PPh_3)_2$ and $Os(CO)_3(PPh_3)_2$. The Pd^0 compound $K_4Pd(CN)_4$ is obtained by reducing $K_2Pd(CN)_4$ with potassium in liquid ammonia. The brown diamagnetic $Pd(RNC)_2$ compounds (R = Ph, p-$CH_3 \cdot$ Ph and p-CH_3OPh) react with phosphines to give compounds such as $Pd(PPh_3)_3$ and $Pd(PPh_3)_4$. Similar complexes of Pt^0 are obtained by reducing $PtX_2(PR_3)_2$ with hydrazine in the presence of an excess of the phosphine. The compound $Pt(PF_3)_4$ is obtained by treating $PtCl_2$ with PF_3 at 370 K under pressure.

37.11.2. Unipositive states

The polymer $[RuCOBr]_n$ is obtained as one of the products of the reaction between $RuBr_3$ and CO under pressure. Bright-yellow $[Os(NH_3)_6]Br$, with a magnetic moment of 1.5 μ_B, can be made by reducing $Os(NH_3)_6Br_3$ with potassium in liquid ammonia.

Rhodium(I) complexes are quite numerous; most of them contain π-bonding ligands. The compound

```
OC        Cl        CO
   \    /    \    /
    Rh        Rh
   /    \    /    \
OC        Cl        CO
```

reacts with many donor ligands to give monomeric $Rh(CO)_2ClL$ compounds. The compound $Rh(PPh_3)_3Cl$ can be made by treating $RhCl_3 \cdot 3\,H_2O$ with PPh_3 in ethanol. It dissociates to give $Rh(PPh_3)_2Cl$, which catalyses the homogeneous hydrogenation of olefins and acetylenes; the compound takes up H_2 to give *cis*-$H_2RhCl(PPh_3)_2$ to which the olefin attaches itself in the sixth co-ordination position and is thereby activated for hydrogenation to occur.

Iridium(I) is also a common species, but again nearly all the complexes contain π-bonding ligands. Like their Rh^I analogues, many of the square d^8 complexes undergo oxidative addition reactions to give octahedral d^6 complexes, thus $Ir(CO)Cl(PPh_3)_2$ reacts with H_2, HCl and Cl_2 to give octahedral Ir^{III} complexes. The yellow compound of iridium made by reducing $[Ir(NH_3)_6]Br_3$ with potassium in liquid ammonia was first thought to be $Ir(NH_3)_5$ but it is diamagnetic and is more probably the Ir^I compound $HIr(NH_3)_5$.

37.11.3. Bipositive states

37.11.3.1. Ru^{II} and Os^{II}

The d^6 complexes of these are diamagnetic, octahedral, and usually inert. They are obtained by the direct action of the ligands on compounds in which the metals are in higher oxidation states:

$$RuCl_3 \xrightarrow[\text{evaporation on steam bath}]{\text{KCN solution}} K_4Ru(CN)_6$$

$$(NH_4)_2OsBr_6 \xrightarrow[\text{540 K}]{\text{dipyridyl}} Os(dipy)_3Br_2$$

$$RuCl_3 \xrightarrow[\text{in ethanol}]{C_6H_4(AsMe_2)_2} Ru(diars)_2Cl_2$$

In these reactions the KCN, dipyridyl and diarsine all act as reducing agents.

37.11.3.2. Rh^{II} and Ir^{II}

There are few genuine examples of these states. Many of the compounds which have been reported have been shown later to be hydrido complexes of the +3 states.

37.11.3.3. Pd^{II} and Pt^{II}

These complexes are formally of d^8 configuration; they are diamagnetic, and usually square planar. Pd^{II} and Pt^{II} are typical soft acids. Their most stable complexes are formed with donors such as phosphorus, arsenic, sulphur and chlorine. Oxygen-donor ligands produce only a few unstable complexes, although there are many complexes with nitrogen ligands.

There are a few octahedral complexes of Pt^{II} such as $[PtCl_5NO]^{2-}$ and $[PtCl(NO)(NH_3)_4]^{2+}$.

37.11.4. Terpositive states

37.11.4.1. Ru^{III} and Os^{III}

These d^7 complexes are all low-spin, with one unpaired electron. Ruthenium(III) complexes are much more common than those of osmium(III).

Rutenium halides react with ammonia to give hexa-ammines, which are changed to acidopenta-ammines by boiling with acid:

$$RuCl_3 + 6\ NH_3 \rightarrow [Ru(NH_3)_6]Cl_3 \xrightarrow{\text{HCl (boil)}} [RuCl(NH_3)_5]Cl_2$$

Complexes of osmium(III) with nitrogen ligands are not common, but $[Os(NH_3)_6]^{3+}$, $[OsBr(NH_3)_5]^{2+}$ and $[Os(dipy)_3]^{3+}$ occur.

Only a few Ru^{III} and Os^{III} complexes with oxygen donors are known — mainly β-diketone and oxalato chelates.

In keeping with the soft acid character of these states they form stable complexes with phosphines and arsines. Examples are $RuX_3(PPh_3)_3$ and $[Os(diars)_2X_2]^+$.

37.11.4.2. Rh^{III} and Ir^{III}

There are many stable complexes of these d^6 states.

The cationic complexes resemble those of Co^{III} (also d^6), and the structures, where known, are octahedral. They are all diamagnetic; even a weak-field ligand like F^- causes spin-pairing in the $RuF_6{}^{3-}$ ion.

There is a well-defined $Rh(H_2O)_6{}^{3+}$ ion in rhodium alums and in the perchlorate $Rh(H_2O)_6(ClO_4)_3$; this is notable because aquo ions of this type are uncommon among second and third row d-block elements. Salts of *cis*-$[Rhen_2Cl_2]^+$ have been resolved into optical isomers. The $[Rh(dipy)_3]X_3$ salts (X = Cl, Br, I, SCN, ClO_4) are yellow because of charge-transfer bands in the blue region of the spectrum.

Rhodium(III) forms many more anionic complexes than does cobalt(III); they are usually more labile than the cationic and neutral complexes.

37.11.4.3. Palladium and platinum

The +3 state probably does not exist in the true sense in either of these metals, for instance, PdF_3 is actually $Pd^{II}[Pd^{IV}F_6]$ and $PtBr_3en$ contains equal numbers of $PtBr_2en$ and $PtBr_4en$ groupings.

37.11.5. Quadripositive states

Ru^{IV}, Rh^{IV}, Pd^{IV}, Os^{IV} and Ir^{IV} occur principally in complex halides but there is a greater range of Pt^{IV} complexes, some of which can be made by *trans* addition:

$$\text{(A,B,C,D)Pt} + Cl_2 \longrightarrow \text{(A,B,C,D)PtCl}_2\ (\text{Cl above and below})$$

In the product the equatorial positions of the octahedral Pt^{IV} complex have the same arrangement as in the square Pt^{II} complex. The Pt^{IV} complexes are invariably octahedral; they provide a particularly extensive range of ammines, from $Pt(NH_3)_6^{4+}$ through $PtX(NH_3)_5^{3+}$, all the way to PtX_6^{2-} (X = Cl, Br, SCN, NO_2). Ethylenediamine, hydrazine and hydroxylamine are other nitrogen donors which appear in these complexes.

37.12. Hydridocomplexes

When ligands which exert a strong field are attached to a transition metal it acquires some of the σ-bonding character of a B sub-group metalloid. Thus it can form metal—metal bonds, as in $Mn_2(CO)_{10}$, and strong σ-bonds to hydrogen, as in carbonyl hydrides, e.g. $MnH(CO)_5$, cyclopentadienyl hydrides, e.g. $ReH(C_5H_5)_2$, and carbonylcyclopentadienyl hydrides, e.g. $MoH(C_5H_5)(CO)_3$.

The platinum metals form a particularly interesting series of molecular hydrides in which stabilisation is due to tertiary phosphines and amines. Examples are:

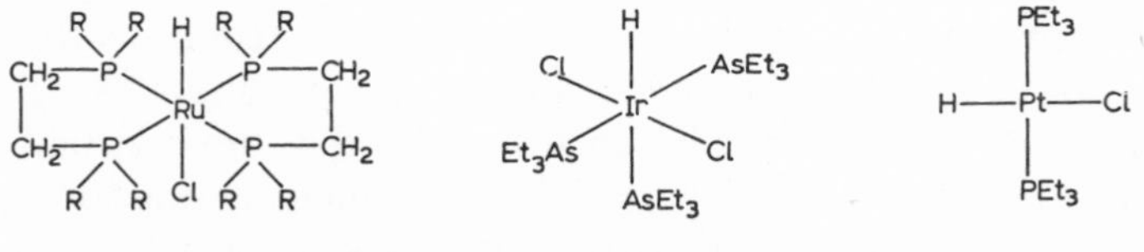

These hydrides are usually made by reduction of the corresponding halogen complexes:

$PtCl_2(PEt_3)_2 \rightarrow PtHCl(PEt_3)_2$

The reducing agents which have been used include $LiAlH_4$ in tetrahydrofuran, H_3PO_2 in alcohol and NH_2NH_2 in water.

In these compounds hydrogen acts as an anionic ligand, the metal retains the same charge number, co-ordination number and stereochemistry as it has in the halogen compound from which the hydride is made. Thus *trans*-$PtHBr(PEt_3)_2$ has the slightly distorted square planar structure (Fig. 37.14).

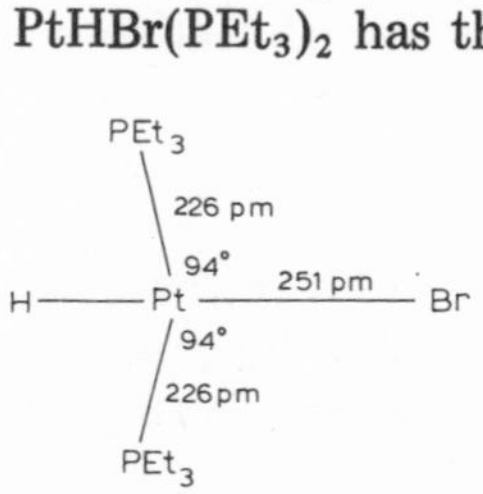

Fig. 37.14. Bond lengths and bond angles in *trans*-$PtHBr(PEt_3)_2$.

The chemical shifts in the proton magnetic resonance spectrum are very large (20—30 p.p.m.) — well removed from those due to organic substituents on the phosphorus atoms. In a compound such as $PtHBr(PEt_3)_2$ the hydrogen resonance is split into a triplet by the two equivalent ^{31}P nuclei and there is further large splitting by ^{195}Pt.

The infrared absorption spectra show a strong, sharp band due to metal—hydrogen bond stretching, but the position of the band varies a great deal from one compound to another (1726—2242 cm^{-1}). In platinum complexes of the *trans*-$PtHX(PEt_3)_2$ type the M—H stretching frequency is reduced by X ligands of increasing *trans* effect.

X	NO_3	Cl	Br	I	NO_2	SCN	CN
M—H stretching frequency/cm^{-1}	2241	2183	2178	2156	2150	2112	2041

The hydrogen itself exerts a strong *trans* effect. In the process:

trans-$PtXCl(PEt_3)_2 + py \rightarrow PtXpy(PEt_3)_2^+ + Cl^-$

the rate of reaction is 10^6 times faster for X = H than for X = Cl. The hydrogen produces a strong ligand field at the metal; the d—d transition bands in the spectra of these compounds are shifted right into the ultraviolet range, and usually masked by charge-transfer bonds, but the available evidence is that H^- lies near CN^- in the spectrochemical series.

The dipole moments of these compounds are of interest. For *trans*-$PtHCl(PEt_3)_2$ the moment is 4.2 D. As the usual moment associated with a transition-metal to chlorine bond is about 2 D the M—H moment appears to be about 2.5 D, with H positive, in conflict with the idea that the hydrogen is co-ordinated as hydride ion. However, some of the moment is due to the distortion from square symmetry (Fig. 37.14) which places positive P atoms towards the hydrogen; furthermore the presence of the hydrogen has the effect of lengthening the metal—chlorine bond.

Thermal stabilities of analogous hydridocomplexes in any group of transition elements usually rise with the atomic mass of the element. Thus in the nickel group $PtHCl(PEt_3)_2$ is stable enough to be distilled at 400 K (1.5 Pa pressure); the palladium compound $PdHCl(PEt_3)_2$ is rather unstable even in the solid state; but the corresponding nickel compound has not been isolated, although its n.m.r. spectrum has disclosed its presence in the solution made by reducing $NiCl_2(PEt_3)_2$ with $LiAlH_4$. The order of thermal stability of these transition-metal hydridocomplexes is thus the reverse of that found for hydrides of a B-sub-group:

Ni complex < Pd complex < Pt complex *cf.* $NH_3 > PH_3 > AsH_3 > SbH_3$

The arsine- and phosphine-stabilised hydridocomplexes of the platinum metals are usually octahedral or square, but the trigonal bipyramidal $RhH(CO)(PPh_3)_3$ has been made by reducing the square planar $RhCl(CO)(PPh_3)_2$ with hydrazine in alcohol.

The π-bonding ligands PH_3 and AsR_3 which stabilise the metal—hydrogen σ bonds are also effective in stabilising metal—carbon σ bonds and there are some organometallic compounds similar to these hydrides. *Trans*-$PtHCl(PEt_3)_2$ reacts reversibly with ethylene to give *trans*-$PtClEt(PEt_3)_2$.

Further Reading

J.R. Miller, Recent advances in the stereochemistry of nickel, palladium and platinum, Adv. Inorg. and Radiochem., 4 (1962) 133.

M.L.H. Green and D.J. Jones, Hydride complexes of the transition metals, Adv. Inorg. and Radiochem., 7 (1965) 115.

W.P. Griffith, The chemistry of the rare platinum metals, Wiley-Interscience, New York, 1967.

R.N. Goldberg and L.G. Hepler, Thermochemistry and electrode potentials of the platinum-group metals and their compounds, Chem. Rev., 68'(1968) 229.

W.P. Griffith, Osmium and its compounds, Quart. Rev., 19 (1965) 254.

F.R. Hartley, Metal—olefin and —acetylene bonding in complexes. Angew. Chem. Internat. Ed., 11 (1972) 596.

K. Krogmann, Planar complexes containing metal—metal bonds, Angew. Chem. Internat. Ed., 8 (1969) 35.

J. Chatt, Hydrido- and related organo-complexes of transition metals, Proc. Chem. Soc. London, (1962) 318.

R.N. Goldberg and L.G. Helper, Thermochemistry and oxidation potentials of the platinum group metals and their compounds, Chem. Rev., 68 (1968) 229.

S.E. Livingstone, The chemistry of ruthenium, rhodium, palladium, osmium, iridium and platinum, Pergamon, Oxford, 1975.

Copper, Silver and Gold — Group IB

38.1. The elements

Traditionally the coinage metals because of their resistance to corrosion, Cu, Ag and Au form the sub-group which marks the ends of the respective 'd-block' series. The atoms have the configurations $(n-1)d^{10}ns^1$ but the metals are nevertheless considered as transitional because ions with incomplete d shells can be formed by all three. The first ionisation energy (Table 38.1) is much greater than that of the preceding alkali metal with its $(n-1)p^6ns^1$ configuration, because the ten d electrons are rather ineffective in shielding the outer s electron from the coulombic field of the nucleus. However, d electrons can either be lost in the formation of 2+ ions (Cu^{2+} and Ag^{2+}) since the second ionisation energies are not high, or can take part in covalent bonding to give formal oxidation states as high as +3 (particularly Au^{III}). The compounds of the metals in these higher oxidation states are coloured and paramagnetic like those of typical transition metals.

Metallic binding in the elements is strong because electrons of the d shells are involved as well as the outer s electrons; consequently the *m.p.* and enthalpies of sublimation are high (Fig. 27.3); a further consequence is that conversion of the metals to aquated cations is energetically unfavourable, and the metals do not corrode. Conversely, they are easily deposited electrolytically from aqueous solutions of their cations.

Again because of the high effective nuclear charges acting on the outer

TABLE 38.1

ATOMIC PROPERTIES OF THE ELEMENTS

	Cu	Ag	Au
Z	29	47	79
Electron configuration	$[Ar]3d^{10}4s^1$	$[Kr]4d^{10}5s^1$	$[Xe]4f^{14}5d^{10}6s^1$
$I(1)$/kJ mol^{-1}	745	731	889
$I(2)$/kJ mol^{-1}	1958	2072	1980
Ionic radii M^+/pm	128	143	144
Metallic radii/pm	96	126	137

TABLE 38.2

PHYSICAL PROPERTIES OF THE ELEMENTS

	Cu	Ag	Au
$\rho/\mathrm{g\ cm^{-3}}$	8.93	10.5	19.3
M.p./K	1357	1234	1338
B.p./K	2868	2485	3000

electrons, the internuclear distances in the solids are low, and the metals have high densities, particularly Au, in which, following the lanthanide contraction, the metallic radius is almost identical with that of Ag. The ions are also small, for example the radius of Cu^+ is only 96 pm compared with 133 pm for K^+. Thus the compounds tend to have high lattices energies, and many of them are insoluble in water as a consequence.

38.2. Extraction and Uses

The metals all occur native, gold almost exclusively so. Copper makes up about 7×10^{-3}% of the lithosphere mainly as copper pyrites $CuFeS_2$, cuprite, Cu_2O, and malachite, $Cu_2(OH)_2CO_3$; silver about 2×10^{-5}% mainly as argentite, Ag_2S, horn silver, AgCl, and pyrargyrite, Ag_3SbS_3; gold about 5×10^{-7}% mainly as metal. Their extraction from ores is chemically relatively easy.

38.2.1. Copper

Oxide and carbonates ores are readily reduced by heating with coke and a flux:

$$CuCO_3 \xrightarrow{\text{heat}} CuO \xrightarrow{C} Cu$$

The main source is copper pyrites which is relieved of volatile arsenic and antimony by roasting, after which it is slagged and reduced.

(i) Partial oxidation: $CuFeS_2 \xrightarrow{O_2} Cu_2S + FeS + SO_2$

(ii) Slagging off iron: $Cu_2S + FeS + SiO_2 \rightarrow \underset{\text{(less dense)}}{FeSiO_3\text{ slag}} + \underset{\text{(more dense)}}{Cu_2S\text{ liquid sulphide}}$

(iii) Oxidation—reduction: $Cu_2S \xrightarrow{O_2} Cu_2O + Cu_2S \xrightarrow{\text{heat alone}} Cu + SO_2$

Copper is also extracted from its ores by leaching with a solution of H_2SO_4 and $Fe_2(SO_4)_3$, which converts the CuS or $CuCO_3$ to $CuSO_4$ solution. Copper is refined electrolytically, the silver and gold present separating as an anode sludge.

38.2.2. Silver and gold

Whether present as metal or compound, these elements can be extracted from the finely divided ore by means of aqueous sodium cyanide. The cyanide ion reduces the oxidation potential of the noble metal so that atmospheric oxygen brings it into solution as a soluble complex:

$$4\ Ag + 8\ NaCN + H_2O + O_2 \rightarrow 4\ NaAg(CN)_2 + 4\ NaOH$$
$$Ag_2S + 4\ NaCN \rightarrow 2\ NaAg(CN)_2 + Na_2S$$

The sodium sulphide is largely oxidised to Na_2SO_4 by air and the reverse reaction is thereby impeded. The silver and gold are precipitated from their solution as cyano-ions, after this has been filtered, by the addition of zinc:

$$2\ Ag(CN)_2^- + Zn \rightarrow 2\ Ag + Zn(CN)_4^{2-}$$
$$2\ Au(CN)_4^- + 3\ Zn + 4\ CN^- \rightarrow 2\ Au + 3\ Zn(CN)_4^{2-}$$

Silver and gold are recovered during the purification of lead and nickel.

38.2.3. Uses

World production of copper is about 6 million tons per annum. It is therefore a metal of great commercial importance, being used particularly as an electrical conductor, in casting alloys and in coinage. The biochemical behavior of copper is of interest. It is a constituent of many oxidases — enzymes which catalyse redox reactions — in which it functions through the shifting of the Cu^{II}/Cu^{I} equilibrium during ligand replacements. Copper deficiency in plants gives rise to die-back, inability to form seeds, chlorosis, and the reduction of photosynthetic activity. Copper deficiency in soils is made good by spraying with aqueous $CuSO_4$.

World production of silver is about 10 thousand tons per annum. Its alloys with copper, particularly Sterling silver (92.5% Ag, 7.5% Cu), are used in tableware. The metal is also used in silvering mirrors and, increasingly, in storage batteries.

The annual world production of gold is about 15 hundred tons — three-quarters of this in South Africa.

About half of the world's gold is used for adjusting trade balances. Its use in jewellery is well known, and it is becoming of increasing importance in electronics.

38.3. Physical Properties

The metals all have the c.c.p. structure and are rather soft when pure, presumably because of the regularity of the packing in the lattice. They are particularly malleable and ductile and are excellent conductors of electricity and heat. The boiling points are high, in excess of 2400 K. The elements form alloys with one another and with many other metals; of particular importance are brass (copper—zinc) and bronze (copper—tin), which are much harder than pure copper.

38.4. Reactions

Gold is rather unreactive, though it is attacked by BrF_3 and aqua regia in the cold and also by warm Cl_2 and Br_2. Copper and silver are more reactive; the former is converted to CuO by oxygen at red heat. The halogens react with the heated metal to give dihalides, and sulphur combines with it to give a product of approximate composition Cu_2S. Non-oxidising acids have no

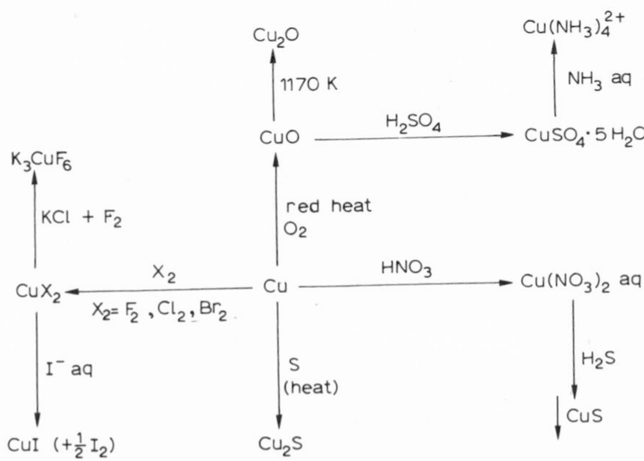

Fig. 38.1. Reactions of copper.

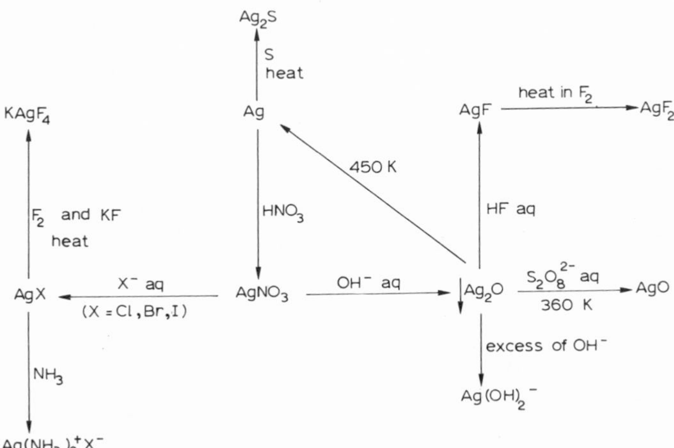

Fig. 38.2. Reactions of silver.

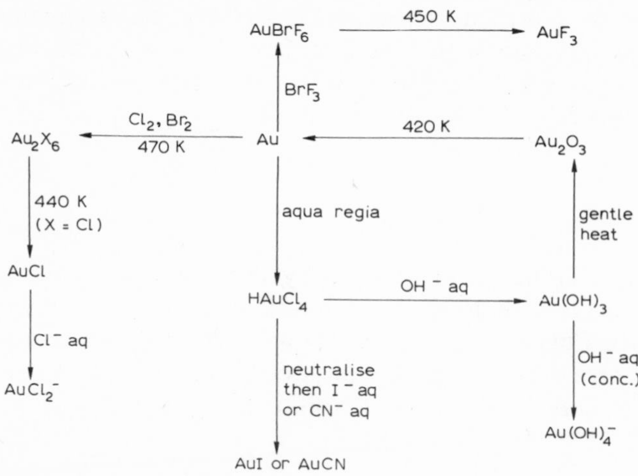

Fig. 38.3. Reactions of gold.

action, but HNO_3 and H_2SO_4 dissolve copper, being themselves reduced. Silver is less reactive than copper except towards S and H_2S which both convert it to black Ag_2S. Alkaline cyanide solutions containing oxidising agents such as H_2O_2 are good solvents for Ag and Au, which react to form the $Ag(CN)_2^-$ and $Au(CN)_4^-$ ions respectively.

38.5. Oxidation States

Charge numbers in the sub-group are limited to +1, +2 and +3. There is no evidence for the zero state or negative states. In the simple compounds of the unipositive metals the bonding is rather covalent in character; lattice energies are therefore enhanced and solubilities are low. In the presence of complexing agents such as CN^- and NH_3, however, it is possible to obtain water-soluble complex ions like $Cu(CN)_4^{3-}$ and $Ag(NH_3)_2^+$. These colourless complexes contain the unipositive metal with a non-bonding d^{10} shell, and

TABLE 38.3

REPRESENTATIVE COMPOUNDS AND IONS OF THE OXIDATION STATES OF Cu, Ag AND Au

	Cu	Ag	Au
+1	Cu_2O, CuI, $Cu(CN)_4^{3-}$	AgCl, $Ag(CN)_2^-$ $Ag(NH_3)_2^+$	$Au(CN)_2^-$
+2	CuO, $CuCl_2$, CuF_4^{2-}	$Ag(py)_4^{2+}$	
+3	CuF_6^{3-}	AgF_4^-	$AuBr_4^-$, $Au(CN)_4^-$

the arrangement of ligands resembles that in the corresponding non-transition metal compound; thus $Ag(NH_3)_2^+$ is linear.

The charge +2 is common, particularly in copper. The d^9 arrangement in Cu^{II} gives rise to square arrangements of ligands around the metal atom or to distorted octahedral arrangements in which two ligands are attached, by somewhat longer bonds, above and below the plane of the other four. Compounds of Ag^{II} are mostly strong oxidising agents; it has not yet been possible to make a compound of Au^{II}.

All three metals form M^{III} complexes. Pale-green K_3CuF_6 has a paramagnetic moment of 2.8 μ_B, consistent with the presence of a d^8 configuration containing two unpaired electrons. Gold(III) complexes are common; they are oxidising agents in which the Au is usually at the centre of a square formed by four ligands.

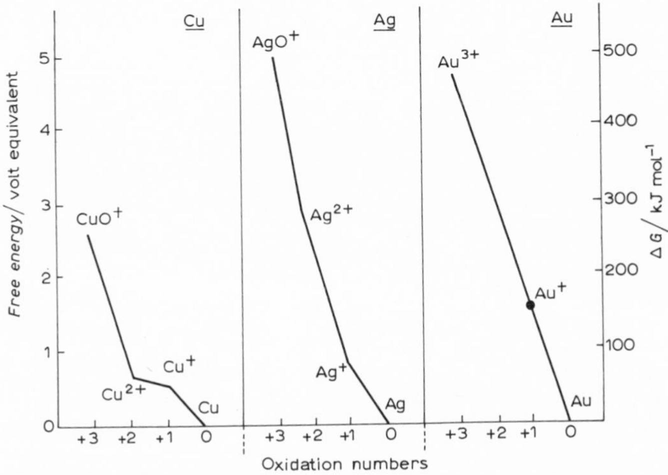

Fig. 38.4. Free energies of the oxidation states of the group IB elements relative to the metals at pH = 0.

The free energies of the oxidation states of Cu, Ag and Au relative to the metals in aqueous solutions at pH 0 are represented graphically in Fig. 38.4. Although $I(2)$ for Cu is rather high, the enthalpy of hydration of Cu^{2+} is so much more negative than that of Cu^+ that the dipositive ion is much the more stable in aqueous solution. Thus soluble Cu^+ salts disproportionate in water. Silver(I) salts do not do so however.

38.6. Halides

White CuCl or CuBr can be precipitated from aqueous $CuCl_2$ or $CuBr_2$ by a variety of reducing agents including aqueous SO_2, $SnCl_2$ and $N_2H_5HSO_4$, but CuI is obtained merely by adding I^- to a solution of Cu^{2+} salt;

$$2\ Cu^{2+} + 2\ I_2 \rightarrow 2\ CuI + I_2$$

a reaction which is used for the volumetric determination of Cu^{2+}. All three copper(I) halides have the zinc blende structure, and they have considerable covalent character, the experimental values for the lattice energies being markedly larger than the values calculated on the basis of ionic models. The halides are moderately soluble in solutions containing the corresponding halide ion because complex halides are formed.

The silver(I) halides, white AgCl, cream AgBr and pale yellow AgI are made by precipitation reactions, but the more soluble AgF is best prepared by dissolving Ag_2O in HF and evaporating until yellow AgF crystallises. AgCl, AgBr and anhydrous AgF have the NaCl structure but AgI has both wurtzite and blende structures, the former being used to cause nucleation of ice crystals in supercooled clouds to induce rainfall, presumably because its crystal structure is similar to that of ice. Above 419 K, AgI is converted to a b.c.c. form which has an unusual type of defect structure in which the Ag^+ ions form what may be called an interstitial fluid, being apparently able to move through the rigid network of I^- ions. At 419 K this α-form has a specific conductance of 1.3 S cm^{-1} compared with 3.5×10^{-4} S cm^{-1} for the adamantine form at 416 K. The silver(I) halides are all sensitive to light; the bromide particularly is important in photography.

A subfluoride of silver, Ag_2F, can be made by treating a solution of AgF in aqueous HF with colloidal silver. Structurally it is composed of successive layers of Ag, Ag and F atoms with a distance between Ag layers (299 pm) similar to the Ag—Ag distances in the metal, and a distance between Ag and F as in the monofluoride (245 pm). The bronze solid is a good conductor and can be considered as intermediate between a metal and a salt.

Gold(I) chloride can be made by heating $AuCl_3$ to 430 K. It decomposes at a slightly higher temperature.

Anhydrous CuF_2, made from the elements, is an ionic compound with a distorted rutile structure. But the more covalent anhydrous $CuCl_2$ and $CuBr_2$ have polymeric chain structures formed by planar CuX_4 groups sharing opposite edges, the chains being packed so that each Cu has two more halogens at a rather greater distance. In $CuCl_2$ the Cu—Cl distances are 230 pm within the chain and 295 pm between chains. In $CuCl_2 \cdot 2\ H_2O$ however there are finite groups with the structure:

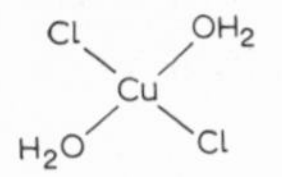

Cl OH2
Cu
H2O Cl

Silver(II) fluoride, made by the action of F_2 on Ag or AgCl; although it is thermally stable, the vapour pressure of the fused compound being only 10 kPa at 970 K, is a powerful fluorinating agent, particularly useful in the preparation of fluorocarbons from hydrocarbons.

AuF_3, made by treating Au with BrF_3 and then warming the $AuBrF_6$ so

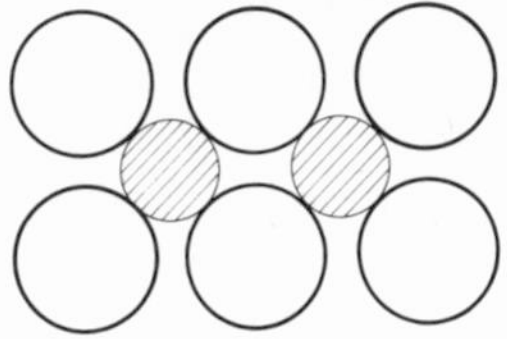

Fig. 38.5. Planar molecule of Au_2Cl_6.

formed, has a polymeric structure in which each Au is surrounded by a square of F atoms and each square is linked to the next by *cis* bridges to give an infinite hexagonal helix. It is a vigorous fluorinating agent.

Gold(III) chloride contains planar Au_2Cl_6 molecules in both the solid and the vapour (Fig. 38.5). It is best made by direct chlorination at 470 K.

38.7. Complex Halides

The halogen complexes of copper(I), with general formulae $M^I{}_2CuX_3$ and $M^ICu_2X_3$, contain 4-co-ordinate Cu^I, the former containing single chains, and the latter double chains of CuX_4 tetrahedra sharing corners. In Cs_2AgI_3 there are chains of AgI_4 sharing corners, but in $Me_4NAg_2I_3$ similar tetrahedra are joined by sharing edges.

Among the halide complexes of the dipositive metals, $K_2CuCl_4 \cdot 2\,H_2O$ (Fig. 38.6) has a tetragonal unit cell with features deriving from both $CuCl_2 \cdot 2\,H_2O$ and $CuCl_2$. In anhydrous Cs_2CuCl_4 the $CuCl_4{}^{2-}$ ion is not planar but a flattened tetrahedron; in aqueous solution the absorption spectrum is different from that of the solid, possibly because the anion reverts to a planar configuration on hydration.

The Cu^{II} complex $CsCuCl_3$ consists of Cs^+ ions and infinite chain ions $(CuCl_3{}^{2-})_n$ (Fig. 38.7).

The compound formulated $CsAuCl_3$, however, contains both $[AuCl_2]^-$ and $[AuCl_4]^-$ ions; it contains both Au^I and Au^{III}. There is an isostructural compound $Cs_2[AgCl_2][AuCl_4]$.

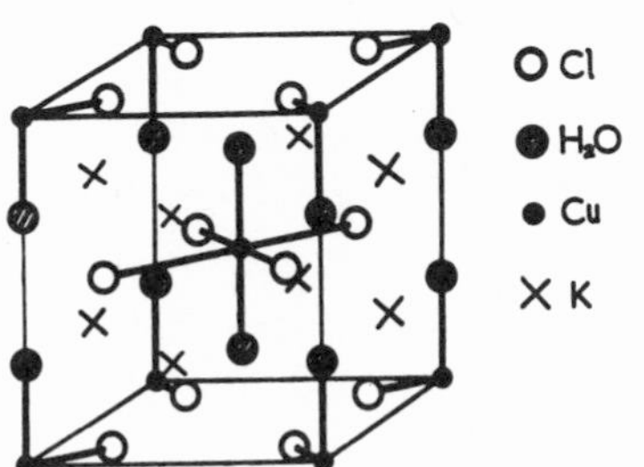

Fig. 38.6. Structure of $K_2CuCl_4 \cdot 2\,H_2O$.

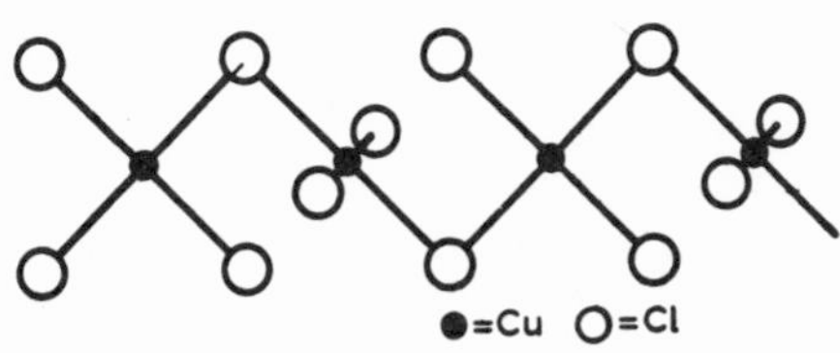

Fig. 38.7. Spiral arrangement of atoms in $(CuCl_3{}^{2-})_n$ ions.

Of halogen complexes of the terpositive metals, K_3CuF_6 is known, but corresponding chloro- and bromo-complexes have not been made. As so often happens in transition-metal chemistry the F^- ion is the only halide ion capable of stabilising the highest oxidation state. The paramagnetic moment of K_3CuF_6 indicates the presence of two unpaired electrons, suggesting that copper(III) has the $t_{2g}^6e_g^2$ configuration.

Gold(III) halogen complexes are particularly stable. The $AuBr_4^-$ ion in $KAuBr_4 \cdot 2\,H_2O$ is square, as is the AuF_4^- ion present in $KAuF_4$.

38.8. Oxides

Copper(I) oxide occurs as the red mineral cuprite. It is best made in the laboratory by reducing Fehling's solution — an alkaline solution of a Cu^{II} salt containing just sufficient tartrate to keep the copper in solution — with an aldehyde. A hydrated Ag_2O is obtained as a dark-brown precipitate when aqueous OH^- is added to aqueous Ag^+. It is difficult to remove all the water without causing decomposition but anhydrous Ag_2O can be obtained by heating Ag powder in oxygen. The oxides Cu_2O and Ag_2O are isomorphous and of unusual structure. The metal atoms have two collinear bonds and the oxygen four tetrahedral bonds in a cubic structure similar to that of β-cristobalite. The low co-ordination 4:2 indicates covalent character. The structure represented in Fig. 38.8 is incomplete because an identical framework, in which the oxygens marked B take up positions A, interpenetrates it — a structure unique in crystal chemistry.

Ag_2O is decomposed completely into silver and oxygen at 570 K under normal pressure. It gives an alkaline reaction with water; the solubility, though slight, is in conformity with the low lattice energy. Cu_2O has much greater thermal stability, a much more negative enthalpy of formation (ΔH_f is —170 kJ mol^{-1} compared with —30 kJ mol^{-1} for Ag_2O), and a lower solubility than Ag_2O.

There is some doubt about the existence of a true gold(I) oxide, though the corresponding sulphide has been characterised.

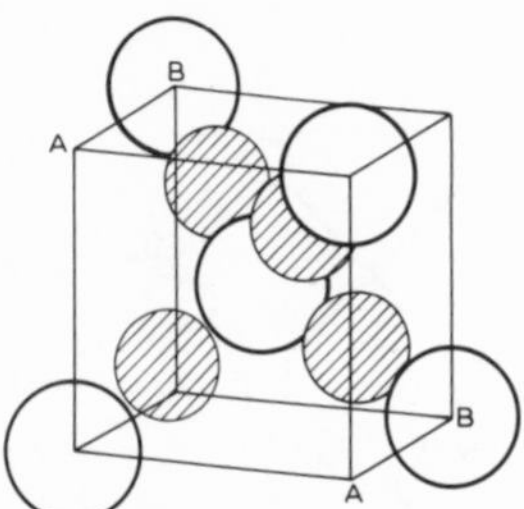

Fig. 38.8. Cu_2O structure. An identical framework in which positions A move back to occupy positions B interpenetrates the framework which is illustrated.

Copper(II) oxide can be made by heating powdered copper in air or oxygen or by decomposition of the pale-blue $Cu(OH)_2$ made by adding OH^-aq to a solution of Cu^{2+} in just sufficient aqueous NH_3 to hold the cations in solution. The structure of CuO resembles that of PdO; it involves 4:4 co-ordination, with coplanar bonds around the copper atoms and a tetrahedral arrangement of copper atoms around the oxygens. The Cu—O distance of 195 pm indicates a large degree of covalency.

A compound with the composition AgO can be obtained as a black precipitate by treating aqueous $AgNO_3$ with alkaline $S_2O_8{}^{2-}$. It is not paramagnetic as a true Ag^{II} compound should be. Neutron diffraction studies have shown that it contains Ag^{I} co-ordinated to two oxygens and Ag^{III} co-ordinated to four oxygens. The compound is a semiconductor and a very strong oxidising agent. It decomposes at 370 K to give Ag_2O and oxygen, and dissolves in aqueous HNO_3, with some loss of oxygen, to give a paramagnetic solution which presumably contains Ag^{2+}.

Addition of OH^- to aqueous $AuCl_4{}^-$ gives a light-brown precipitate of composition $Au(OH)_3$ which is converted to brown Au_2O_3 by heating to constant weight at 420 K. The anhydrous oxide is soluble in concentrated mineral acids and also in hot aqueous alkalis; yellow needles of $KAuO_2 \cdot 3\,H_2O$ can be crystallised from a solution of Au_2O_3 in aqueous KOH.

38.9. Sulphides

Black Cu_2S occurs as chalcocite. It can be made by passing dry H_2S over Cu at 700 K. Silver(I) sulphide, which is also black and also occurs naturally, can be made by numerous methods but most conveniently by precipitation with H_2S from aqueous Ag^+ solutions. Black gold(I) sulphide is best made by passing H_2S into an acidified solution of $Au(CN)_2{}^-$.

The compound CuS, which occurs as covellite, is not a true Cu^{II} compound. One-third of the Cu atoms are co-ordinated trigonally to S atoms and two-thirds are co-ordinated tetrahedrally, furthermore two-thirds of the sulphur is present as S_2 groups. When covellite is heated with sulphur a dark-purple compound is obtained with the approximate composition CuS_2 and the pyrites structure. Gold(III) sulphide, made by passing H_2S into a solution of Au_2Cl_6 in ether, is a black, insoluble solid.

38.10. Other Binary Compounds

A nitride Cu_3N can be made by heating CuF_2 with NH_3. It has a reversed ReO_3 structure with Cu—N distances of 190 pm. It is evidently a true Cu^{I} compound; it reacts with acids to give a Cu^{2+} salt, an $NH_4{}^+$ salt and a precipitate of copper:

$$2\,Cu_3N + 8\,H^+ \rightarrow 2\,NH_4{}^+ + 3\,Cu^{2+} + 3\,Cu$$

There is also an explosive azide CuN_3 made by reducing aqueous Cu^{2+} with HSO_3^- in the presence of N_3^- ions. An explosive silver(I) azide is precipitated by hydrazine from aqueous $AgNO_3$. There is some doubt about the existence of a true nitride or azide of gold; the grey, explosive compound made by the action of NH_3 on aqueous $HAuCl_4$ contains $Au_2O_3 \cdot 3\,NH_3$ and $HNAuCl \cdot NH_3$.

The explosive solid Cu_2C_2 is obtained as a dark-red precipitate when C_2H_2 is passed into a solution of a Cu^+ salt in aqueous ammonia. Yellow Ag_2C_2 is made similarly from acetylene and ammoniacal $AgNO_3$. The dry solid detonates violently at 400 K.

38.11. Oxoacid salts

Copper(I) sulphate is the only ionic Cu^I compound. It is made by heating Cu_2O with dimethyl sulphate:

$$Cu_2O + Me_2SO_4 \rightarrow Cu_2SO_4 + Me_2O$$

It is decomposed immediately in water:

$$Cu_2SO_4 \rightarrow Cu + CuSO_4$$

This is to be expected from the redox potentials of the Cu^+/Cu and Cu^{2+}/Cu couples. The Cu^+ ion is, however, stabilised by complexing and appears as colourless crystals of $[Cu(NH_3)_2]_2SO_4$; these are produced when ethyl alcohol is added to a solution made by dissolving Cu_2O in an aqueous solution of ammonium sulphate and ammonia.

Copper(II) sulphate pentahydrate, $Cu(H_2O)_4SO_4 \cdot H_2O$, has four water molecules and an oxygen from each of two SO_4^{2-} anions octahedrally arranged about the Cu^{2+} cation. The fifth water molecule is hydrogen-bonded to oxygen atoms. The positions of all the water molecules and hydrogen bonds have been determined by neutron refraction. The H—O—H angles of all the water molecules are near the tetrahedral angle, the O—H····O angles between 154—176°, and the O—O—O angles between 105—130°; so that some of the hydrogen bonds are bent, one by as much as 26°. The corresponding ammine hydrate, $Cu(NH_3)_4SO_4 \cdot H_2O$, is known, but not a penta-ammine.

Copper(II) nitrate has the hydrates $Cu(NO_3)_2 \cdot 6\,H_2O$ and $Cu(NO_3)_2 \cdot 9\,H_2O$. Attempts to dehydrate them produce basic salts. However, copper reacts vigorously with N_2O_4 in ethyl acetate, and a solid $Cu(NO_3)_2 \cdot N_2O_4$ can be crystallised from the solution. Gentle heating distils off the N_2O_4 leaving the green, volatile $Cu(NO_3)_2$. This compound is monomeric in the vapour, the Cu atom being co-ordinated to NO_3 groups acting as bidentate ligands.

The most important soluble salt of silver, $AgNO_3$, can be prepared by dissolving Ag in HNO_3. It is stable up to 620 K but at 700 K it decomposes into the metal, nitrogen, nitrogen oxides and oxygen. Unlike the halides, the pure nitrate is not photosensitive.

Copper(II) acetate monohydrate is interesting for its magnetic properties.

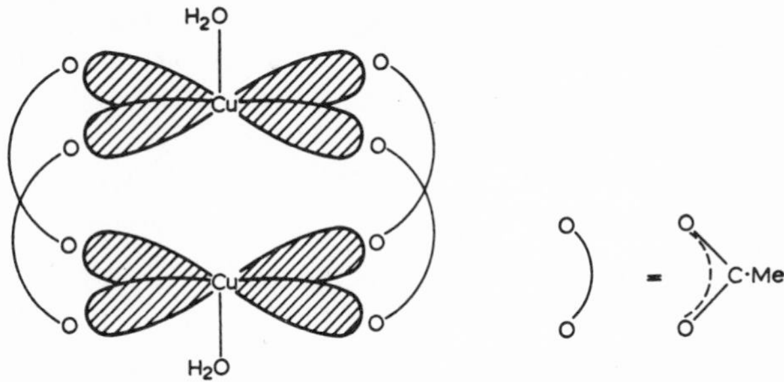

Fig. 38.9. Dimer of copper(II) acetate monohydrate, illustrating δ-bonding.

The moment decreases as the temperature falls, suggesting a temperature-dependent equilibrium between the diamagnetic d^{10} and the paramagnetic d^9 configurations, with a Cu—Cu bond energy of only about 4 kJ mol^{-1}. The compound has a dimeric structure (Fig. 38.9). The copper atoms have square pyramidal arrangements of oxygens around them. Interaction between the two copper atoms is pictured as being due to the face-to-face overlap of the $d_{x^2-y^2}$ orbitals, shown in the diagram. This type of bonding, called δ-bonding, though it is extremely weak compared with σ- and π-bonding, can lead to diamagnetism through the coupling of spins.

Pale-yellow Ag_2CO_3 can be precipitated from aqueous $AgNO_3$ with alkali carbonates. In contrast only basic carbonates can be precipitated from Cu^{2+} solutions, an indication that Ag_2O is more basic than CuO.

Silver perchlorate is not only deliquescent and very soluble but forms a monohydrate. It also dissolves in benzene and toluene, recrystallising from the latter as $AgClO_4 \cdot C_7H_8$. Solubility in organic solvents is thus shown to be an unreliable guide to covalent character, for silver perchlorate is considerably ionised both in water and nitromethane.

Oxoacid salts of gold are uncommon; one of the more stable, yellow $Au_2(SeO_4)_3$, crystallises from a solution of gold in hot selenic acid.

38.12. Organometallic Compounds

Gold resembles platinum in forming numerous σ-bonded alkyl derivatives. The halides $AuCl_3$ and $AuBr_3$ react with Grignard reagents to give colourless solids, R_2AuX:

$$AuBr_3 + 2\ RMgBr \rightarrow R_2AuBr + 2\ MgBr_2$$

The compounds are insoluble in water but soluble in organic solvents. The molecules are dimeric and halogen-bridged:

R Br R
Au Au
R Br R

They react with halogens to give red monoalkyl derivatives:

$R_2AuBr + Br_2 \rightarrow RAuBr_2 + RBr$

The cyanides R_2AuCN are tetrameric:

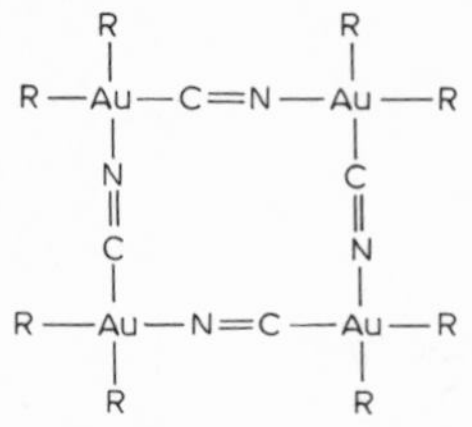

```
     R               R
     |               |
R —Au—C≡N—Au—R
     |               |
     N               C
     ‖               ‖
     C               N
     |               |
R —Au—N≡C—Au—R
     |               |
     R               R
```

Aryl compounds cannot be made from Grignard reagents, but yellow solids of the general formula $ArAuCl_2$ are prepared by dissolving $AuCl_3$ in the aromatic hydrocarbon:

$AuCl_3 + C_6H_6 \rightarrow C_6H_5AuCl_2 + HCl$

The known alkyl and aryl compounds of silver are of low thermal stability. The usual method of preparation is the treatment of $AgNO_3$ with a tetra-alkyl or aryl of lead in ethanol:

$Ag^+ + R_4Pb \rightarrow R_3Pb^+ + RAg$

38.13. Complexes

38.13.1. The +1 state

The three metals of the sub-group form ammines containing linear $M(NH_3)_2^+$ ions. There are similar alkylamine and pyridine complexes. Phosphine, and substituted phosphines, and arsines react with monohalides of the metals to form rather unstable complexes:

H_3PCuBr Et_3PCuI Et_3AsCuI

These are tetrameric in benzene. A full structural determination on $(Et_3AsCuI)_4$ has shown it to have a structure based on interpenetrating tetrahedra of copper and arsenic atoms (Fig. 38.10).

The silver compounds are similar, but Ph_3PAuCl has been shown to be monomeric.

Complexes of the unipositive metals with oxygen donors are few and unimportant but there are many with sulphur donors. Copper(I) forms thiourea complexes with 1, 2, 3 and 4 thiourea molecules per copper atom; silver(I) complexes are similar. Gold(I) halides form two series of compounds with ethylenethiourea, the salts:

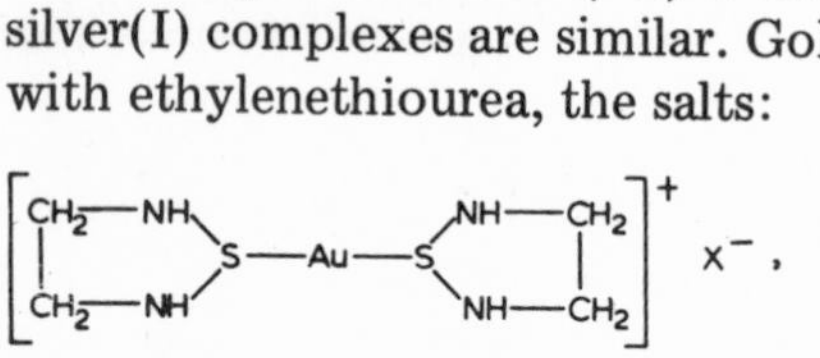

```
⎡ CH2—NH              NH—CH2 ⎤+
⎢ |        \          /      |   ⎥
⎢ |         S—Au—S          |   ⎥  X⁻ ,
⎢ |        /          \      |   ⎥
⎣ CH2—NH              NH—CH2 ⎦
```

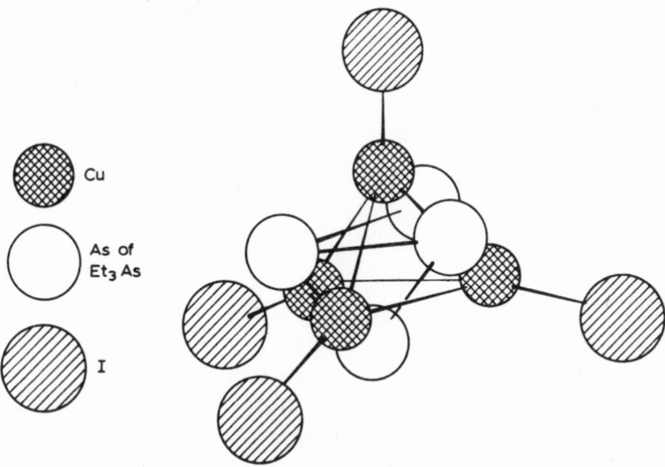

Fig. 38.10. Structure of $[Et_3AsCuI]_4$.

and the covalent compounds:

```
CH2—NH
|        \
|         S—Au—X ·
|        /
CH2—NH
```

The unipositive metals form cyanocomplexes of empirical formula $M(CN)_2^-$. The silver and gold complexes are monomeric but $KCu(CN)_2$ has been shown to contain spiral, polymeric ions:

```
  |                                |
  Cu                               Cu
 /   \                           /    \
      C                        N
        \                    /
         N                 C
           \             /
             Cu
             |
             C
             |
             N
```

Copper(I) ammines react with alkynes to give yellow or red polymers in which the copper atom of one RC≡CCu unit is π-bonded to the —C≡C— bond of another. Both the ammines and the chlorocuprates(I) absorb carbon monoxide.

38.13.2. The +2 state

Copper(II) forms many complexes with nitrogen ligands. The four-co-ordinate complexes are usually much more stable than the six-co-ordinate ones, although many of these exist:

$Cu(NH_3)_4^{2+}$	$Cu(en)_2^{2+}$ and $Cu(dipy)_2^{2+}$	(very stable)
$Cu(NH_3)_6^{2+}$	$Cu(en)_3^{2+}$ and $Cu(dipy)_3^{2+}$	(rather unstable)

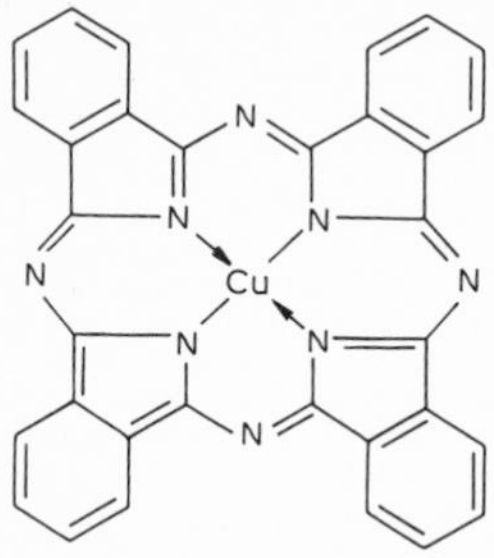

Fig. 38.11. Copper(II) phthalocyanine.

Glycine reacts with Cu^{II} salts to give square planar diglycinecopper, $(NH_2CH_2CO_2)_2Cu$, co-ordinated through both oxygen and nitrogen. There are also numerous complexes with chelating oxygen donors, such as β-diketones, β-ketoesters, catechol and salicylic acid.

Among the copper compounds of technological interest the copper(II) phthalocyanine derivatives are of particular note. The parent compound (Fig. 38.11) is made by heating a Cu^{2+} salt with phthalic anhydride and urea. The derivatives have given colour technologists a series of blue and green pigments of outstanding brightness and colour fastness. Considering the size of the molecule, copper phthalocyanine is remarkably thermally stable; decomposition is not observed below 800 K. The best-known Ag^{II} complexes are those formed by oxidising Ag^+ with persulphate in the presence of nitrogen ligands such as pyridine and bipyridyl: $Ag(py)_4S_2O_8$ is isomorphous with its Cu^{II} analogue which is known to have square planar symmetry. Gold(II) complexes are not known.

38.13.2.1. Effect of solvent on the spectra of copper(II) complexes

Bipositive copper (d^9) can be considered to form square planar complexes in which the odd electron occupies the $d_{x^2-y^2}$ orbital. When additional ligands are available they occupy the fifth and sixth co-ordination positions, but as the d_{z^2} orbital is doubly occupied, these ligands will not approach so closely as those in the plane; the complex thus becomes tetragonal. With the compound in solution the apical positions can be occupied by solvent molecules; the more basic the solvent the stronger the ligand field so created:

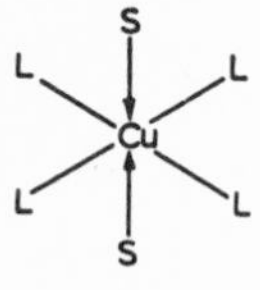

The square planar energy diagram for a d^9 system is given on the left of the diagram shown in Fig. 38.12. As solvent molecules approach the Cu atom the d_{z^2}, d_{xz} and d_{yz} orbitals increase in energy; at the same time the ligands in the xy plane move away from the copper and the $d_{x^2-y^2}$ and d_{xy} energies fall

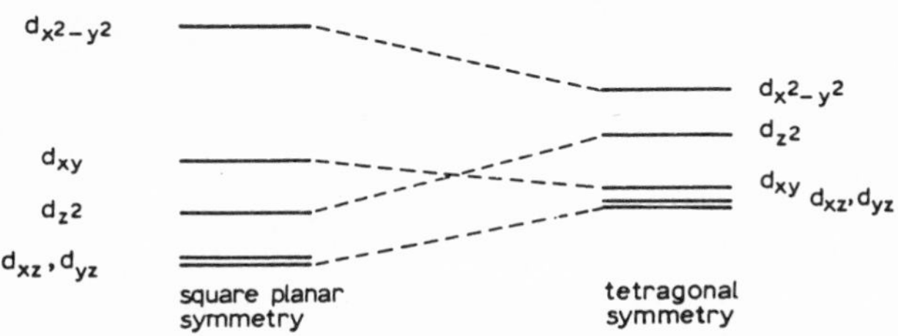

Fig. 38.12. Changes in energy levels as square planar symmetry is converted to tetragonal.

(right of diagram). The extent of this effect depends on the particular solvent.

The absorption spectrum of copper(II) acetylacetonate in a variety of solvents illustrates this (Table 38.4). As the solvent becomes more basic, the

TABLE 38.4

EFFECT OF SOLVENT ON THE SPECTRUM OF $Cu(acac)_2$

Wave number/ cm^{-1}	Transition	Dioxan	n-Pentanol	Pyridine	Piperidine
ν_1	$\left.\begin{matrix}d_{xz}\\d_{yz}\end{matrix}\right\} \rightarrow d_{x^2-y^2}$	17,500	17,100	15,900	15,100
ν_2	$d_{xy} \rightarrow d_{x^2-y^2}$	15,100	15,200	14,800	14,800
ν_3	$d_{z^2} \rightarrow d_{x^2-y^2}$	13,500	13,000	12,100	11,300

$d_{xz} \rightarrow d_{x^2-y^2}$ transition involves a smaller energy change, as does also the $d_{z^2} \rightarrow d_{x^2-y^2}$ transition, but the $d_{xy} \rightarrow d_{x^2-y^2}$ transition is almost unchanged. However, the identification of absorption bands in terms of orbital transitions is rarely as simple as in this d^9 case.

38.13.3. The +3 state

Copper(III) is uncommon, but occurs in K_3CuF_6 and in the blue-grey, diamagnetic $KCuO_2$ made by heating CuO with KO_2 in oxygen. Silver(III) occurs in the diamagnetic $KAgF_4$, made by heating $AgNO_3$ and KCl together with fluorine, and in periodate complexes containing $Ag(IO_6)_2^{7-}$ made by boiling Ag_2O with aqueous IO_4^- and OH^-. There are also complexes containing $Ag(TeO_6)_2^{9-}$ obtained by oxidising Ag^+ with $S_2O_8^{2-}$ in the presence of OH^- and TeO_2.

Gold(III) complexes are more common than Cu^{III} and Ag^{III} complexes. Four-, five- and six-co-ordination occurs, as in $[Au(diars)_2]^{3+}$, $[Au(diars)_2I]^{2+}$ and $[Au(diars)_2I_2]^+$ respectively. Chelating nitrogen, phosphorus, arsenic and sulphur donors are of particular importance.

Further Reading

C.C. Addison and N. Logan, Anhydrous metal nitrates, Adv. Inorg. and Radiochem., 6 (1964) 72.

J.A. McMillan, Higher oxidation states of silver, Chem. Rev., 62 (1962) 65.

B.R. James and R.J.P. Williams, Oxidation—reduction potentials of cupric compounds, J. Chem. Ed., 38 (1961) 2007.

B. Armer and H. Schmidbaur, Organogold chemistry, Angew. Chem. Internat. Ed., 9 (1970) 101.

J.V. Huxley, Recent developments in copper production, Ed. Chem., 10 (1973) 94.

A.G. Massey, N.R. Thompson, B.F.G. Johnson and R. Davis, The chemistry of copper, silver and gold, Pergamon, Oxford, 1975.

Zinc, Cadmium and Mercury — Group IIB

39

39.1. The Elements

The atoms of Zn, Cd and Hg have the electronic structure $(n-1)d^{10}ns^2$ and differ from the transition metals which immediately precede them in that they do not form ions with incomplete d shells. Nevertheless they form some complexes, particularly with nitrogen, sulphur, oxygen and halogen donors, which resemble those of the later d-block metals. The possibility of d_π bonding is very low, however, and the elements do not form complexes with CO, NO and olefins.

TABLE 39.1

ATOMIC PROPERTIES OF Zn, Cd AND Hg

	Zn	Cd	Hg
Z	30	48	80
Electron configuration	$[Ar]3d^{10}4s^2$	$[Kr]4d^{10}5s^2$	$[Xe]4f^{14}5d^{10}6s^2$
$I(1)$/kJ mol^{-1}	906	867	1008
$I(2)$/kJ mol^{-1}	1729	1624	1800
Metallic radius/pm	138	154	157
Covalent radius/pm	131	148	149
$r_{M^{2+}}$/pm	74	97	110

The metallic radii (Table 39.1) are rather larger than those of the metals which precede them, evidently because only the s electrons are involved in

TABLE 39.2

PHYSICAL PROPERTIES AND ELECTRODE POTENTIALS OF Zn, Cd AND Hg

	Zn	Cd	Hg
ρ/g cm^{-3}	7.1	8.6	13.6
M.p./K	692	594	234
B.p./K	1180	1041	630
E°, M^{2+}/M/V	−0.763	−0.402	+0.854
E°, M_2^{2+}/M/V			+0.789

metallic bonding; single-bond covalent radii are also rather large. The M^{2+} ions vary a great deal in size, $r_{Zn^{2+}}$ being much less than $r_{Hg^{2+}}$. The first two ionisation energies of Hg are unusually high, and, as its large bipositive ion has a small hydration energy, the standard electrode potential, $E°$, Hg^{2+}/Hg (Table 39.2) is much more positive than the values for Zn and Cd. Thus mercury is a noble metal, whereas its two congeners liberate hydrogen from dilute acids.

39.2. Physical Properties

The solids do not have the close-packed structures characteristic of true metals; Zn and Cd have distorted versions of the h.c.p. structure in which the distances between layers of atoms are about 15% larger than the ideal, and solid mercury has an even more distorted modification of the hexagonal pattern. Consequently the metals are less dense than the corresponding metals of Group IB, their *m.p.* are distinctly lower and their boiling points are strikingly low. Furthermore the solid metals are of low tensile strength compared with the d-block metals.

39.3. Chemical Properties

Because their enthalpies of vaporisation are small (Fig. 27.3) Zn and Cd tend to form aquated ions more readily than do Cu and Ag. The metals dissolve easily in non-oxidising acids and combine directly with oxygen, sulphur, phosphorus and the halogens on heating. Zinc, unlike its heavier congeners, dissolves in strong aqueous alkalis; solid zincates such as $Na_2Zn(OH)_4$ can be crystallised from the solutions. Mercury resembles Cu and Ag in its reactions with acids, being dissolved only by those with oxidising properties; it combines rather slowly with oxygen at 620 K but the HgO which is formed decomposes at 670 K. The metal reacts readily with sulphur and the halogens, however. All three metals form alloys, many of which are of technical importance. Solutions of other metals in mercury are called amalgams; truly stoichiometric compounds such as $NaHg_2$ are formed in some cases.

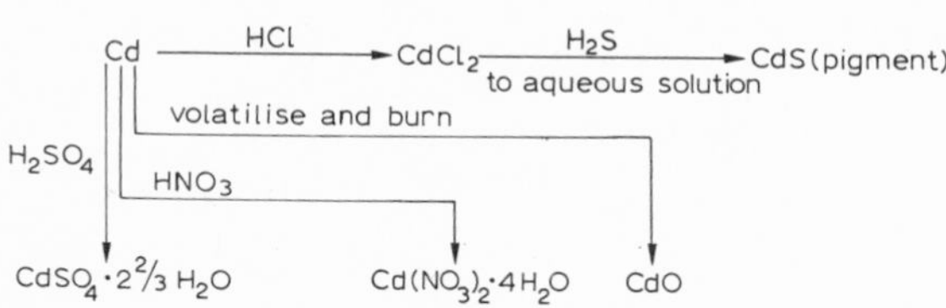

Fig. 39.1. Outline chemistry of Cd.

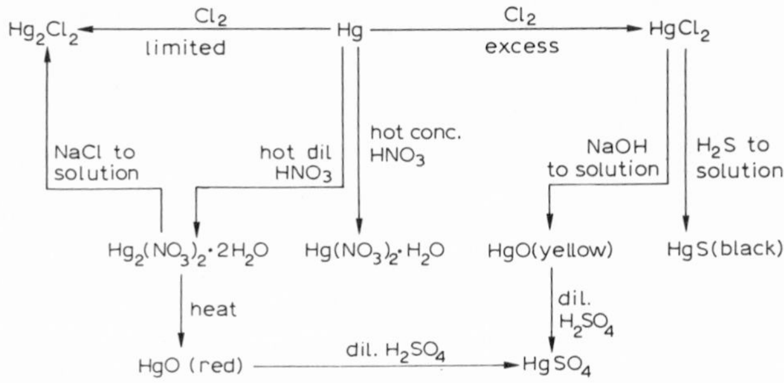

Fig. 39.2. Outline chemistry of Hg.

39.4. Oxidation States

The +2 state is the only one of importance for Zn and Cd, though solid $Cd_2(AlCl_4)_2$, which has been isolated from solutions of Cd and $CdCl_2$ in molten $NaAlCl_4$, has been shown to contain the ion Cd_2^{2+}; the Raman spectrum indicates a Cd—Cd bond with $\nu = 183 \pm 3\ cm^{-1}$ and $k = 111\ N\ m^{-1}$. Mercury(I) compounds containing the Hg_2^{2+} ion, such as $Hg_2(NO_3)_2$ and $Hg_2(ClO_4)_2$, can be made by reducing aqueous solutions of the corresponding Hg^{2+} compounds with mercury. For the reaction

$$Hg^{2+}(aq) + Hg(l) \rightleftharpoons Hg_2^{2+}(aq)$$

the equilibrium constant at 298 K is about 160, independent of the concentration of Hg^{2+}aq. The sparingly soluble Hg^{I} halides, halates and sulphate can be made from the aqueous nitrate or perchlorate by precipitation reactions. There is evidence from X-ray analysis, Raman spectroscopy and magnetic susceptibility measurements that the mercury(I) ion is always $Hg—Hg^{2+}$; the length of the bond varies from 243 pm in solid Hg_2F_2 to 269 pm in Hg_2I_2, and the force constant for the Hg—Hg bond is about $168\ N\ m^{-1}$. The fact that the equilibrium constant of the foregoing reaction between Hg^{2+}aq and Hg is independent of concentration is itself an indication that one Hg_2^{2+} ion, not two Hg^+ ions, is formed from each Hg^{2+} ion which reacts.

It is not possible to make an oxide, sulphide or cyanide of unipositive mercury. Attempts to precipitate these compounds always result in the formation of the Hg^{II} compound and the metal:

$$Hg_2^{2+} + 2\ OH^- \rightarrow HgO + Hg + H_2O$$

Mercury(II) exists in its ionic compounds as the simple, bipositive ion Hg^{2+}.

39.5. Occurrence, Extraction and Uses

39.5.1. Zinc

The element (0.02% of the lithosphere) occurs principally as the sulphide, which has two forms, the commoner, cubic zinc blende or sphalerite and the less common, hexagonal wurtzite. Iron and cadmium are nearly always present as substitutional impurities, and the ore also serves to concentrate a number of rarer elements such as indium, germanium and gallium. The carbonate smithsonite is another commercially important source of zinc. The ores are concentrated by selective sedimentation or froth flotation and roasted to give ZnO which is reduced with anthracite or coke at about 1400 K; the zinc vapour which escapes is condensed to give solid metal. The metal can also be extracted electrolytically using aqueous $ZnSO_4$ obtained from ZnO as the electrolyte, aluminium as the cathode and lead as the anode. The purest zinc is best made by zone refining however.

World production of zinc is about 5 million tons per annum, of which about one-third is used for galvanising steel. Alloys such as brass are also important commercially. A newly-developed alloy containing 80% Zn and 20% Al is of great potential. It becomes super-plastic at 540 K and can be pressed into complex shapes, but its hardness and strength at room temperature are comparable with those of steel — it seems a particularly suitable material for car bodies.

39.5.2. Cadmium

The element is rare (2×10^{-5}% of the earth's crust) and is derived exclusively from zinc ores; the mineral greenockite, CdS, is widely dispersed and of no commercial value. Its lower *b.p.* enables cadmium, which may amount to 0.5% of crude zinc, to be separated by distillation. It can also be precipitated from Zn^{2+}/Cd^{2+} solutions by addition of zinc powder. It is used in electroplating, in some bearing metals and fusible alloys, and in the control rods of nuclear reactors, because ^{113}Cd has a particularly large cross-section for the capture of thermal neutrons.

39.5.3. Mercury

The element (5×10^{-5}% of the lithosphere) has only one important ore, red, rhombohedral cinnabar, HgS. The ore is crushed and concentrated by selective washing and flotation. Roasting in air at 900 K releases the vapour

$$HgS + O_2 \rightarrow Hg + SO_2$$

which is condensed to the liquid metal. World production is about 8 thousand tons per annum. The metal is used in mercury vapour lamps and in many kinds of scientific equipment such as thermometers, gauges, diffusion pumps and electric relays.

39.6. Halides and Halogen Complexes

The fluorides of zinc and cadmium, made by treating the carbonates with HF and dehydrating in a vacuum the hydrates so formed, are high-melting, rather insoluble, ionic compounds. ZnF_2 has the rutile structure, but CdF_2 has the fluorite lattice; 8 : 4 co-ordination is made possible by the greater size of the Cd^{2+} ion. The other dihalides of zinc and cadmium have much lower *m.p.*, are freely soluble in water, and have layer lattices, those of the

TABLE 39.3

PHYSICAL PROPERTIES OF Cd^{II} HALIDES

	CdF_2	$CdCl_2$	$CdBr_2$	CdI_2
M.p./K	1383	841	840	660
ΔH_{fusion}/kJ mol^{-1}	22.7	22.3	20.9	33.0
Solubility w/w at 298 K	0.045	1.40	1.15	0.86
Crystal lattice	fluorite	←	layer	→

zinc halides being based upon 4-co-ordination of the metal atoms and those of the cadmium halides upon 6-co-ordination. The chlorides are made in their anhydrous forms by passing dry HCl over the heated metal, the bromides and iodides can be made by direct combination of the elements. Anhydrous $ZnCl_2$ crystallises in three different modifications, α, β and γ in which the Zn—Cl distances are respectively 234, 231 and 227 pm, but in all three the zinc atoms are tetrahedrally co-ordinated to chlorine atoms.

Complex halides of the types M^IZnX_3 and $M^I_2ZnX_4$ are known (X = F, Cl and Br), and also $M^I_3ZnCl_5$. The compounds M^IZnF_3 do not contain ZnF_3^- ions but have either true perovskite structures (Fig. 7.27) or slightly distorted modifications. $M^I_2ZnF_4$, however, and all the chlorides and bromides, do contain complex ions. Iodo-complexes of zinc have low stability contants, and no solid species has yet been isolated.

Hg_2F_2 can be made by treating Hg_2CO_3 with HF. Unlike the other mercury(I) halides which are made from aqueous $Hg_2(NO_3)_2$ by precipitation reactions, it is appreciably soluble in water. The ionic compound HgF_2, made by the action of F_2 on $HgCl_2$, has the fluorite structure and is strongly hydrolysed by water. Hg^{II} chloride however is predominantly covalent, the solid containing linear Cl—Hg—Cl molecules. Like $HgCl_2$, the bromide and iodide are low-melting solids, but they have layer lattices and are much less soluble. The very poisonous, corrosive, colourless $HgCl_2$ is made by the action of chlorine on the metal. The colourless $HgBr_2$ and the red HgI_2 are much less soluble in water and can be precipitated from aqueous solution.

Cd^{II} and Hg^{II} form complex anions with halides; their soft-acid character is shown by the fact that the thermodynamic stabilities increase from the chloro- to iodo-complexes (Table 39.4).

Cadmium halides form autocomplexes in concentrated aqueous solution. The anomalous transport number of cadmium in the electrolysis of aqueous CdI_2 was noticed by Hittorf and correctly attributed to the equilibrium

$$2\,CdI_2 \rightleftharpoons Cd^{2+} + CdI_4^{2-}$$

TABLE 39.4

THERMODYNAMIC STABILITIES OF HALOGEN COMPLEXES OF MERCURY(II)

System	pK_1	pK_2	pK_3	pK_4
Hg^{2+}/F^-	1.03	—	—	—
Hg^{2+}/Cl^-	6.74	6.48	0.85	1.00
Hg^{2+}/Br^-	9.05	8.28	2.41	1.26
Hg^{2+}/I^-	12.87	10.95	3.67	2.37

The Raman spectra of CdI_4^{2-} and $CdBr_4^{2-}$ are similar to that of the tetrahedral molecule $SnBr_4$.

The great thermodynamic stability of the HgI_4^{2-} ion is illustrated by the fact that aqueous solutions can be made quite strongly alkaline without causing precipitation (Nessler's reagent). Furthermore, aqueous iodides dissolve HgO:

$$HgO + 3\,I^- + H_2O \rightarrow HgI_3^- + 2\,OH^-$$

$$HgO + 4\,I^- + H_2O \rightarrow HgI_4^- + 3\,OH^-$$

The chlorocomplexes of Hg^{II} are of structural interest. The anions usually contain distorted $HgCl_6$ octahedra linked in ribbons and layers.

39.7. Oxides and Hydroxides

39.7.1. Zinc

Zinc oxide occurs as zincite, which has the wurtzite structure, but it is made industrially by burning the metal in air. It is used on a large scale as a filler in rubber and as a pigment (Chinese white). ZnO is an intrinsic semiconductor with an energy gap of 316 kJ mol^{-1} but when heated in zinc vapour the solid becomes a metal-excess n-type semiconductor with an energy gap of only 5 kJ mol^{-1}. These doped crystals, which are shown by X-ray analysis to contain extra Zn atoms, are yellow, green or brown depending on the proportion of zinc.

Addition of aqueous OH^- to a Zn^{2+} solution produces various crystalline forms of $Zn(OH)_2$, the proportions depending on concentrations, pH and temperature. Of the six modifications, the ϵ-form is the only one stable in contact with water below 312 K. It is completely converted to ZnO by heating at 1150 K, it dissolves easily in dilute mineral acids and in aqueous alkalis; analytical results support the existence of species such as $Zn(OH)_3^-$ and $Zn(OH)_4^{2-}$ in the alkaline solutions. Ammonia solutions also dissolve $Zn(OH)_2$:

$$Zn(OH)_2 + 4\,NH_3 \rightarrow Zn(NH_3)_4^{2+} + 2\,OH^-$$

39.7.2. Cadmium

Cadmium oxide is obtained as a brown powder by the pyrolysis of $CdCO_3$ or $Cd(NO_3)_2$, but strong ignition in oxygen converts it to deep-red, cubic crystals. It has the NaCl structure, an example of the tendency of Cd^{II} to co-ordinate octahedrally, as distinct from Zn^{II} which is usually tetrahedrally co-ordinated.

White $Cd(OH)_2$ is precipitated by the addition of aqueous OH^- to a solution of Cd^{2+} ions. The solid, which has a layer structure resembling that of CdI_2, is soluble in dilute acids and aqueous NH_4Cl but only slightly soluble in OH^- solutions. It is, in fact, a much stronger base than $Zn(OH)_2$. The solid decomposes to give CdO at 470 K.

39.7.3. Mercury

The normal, orthorhombic form of HgO is obtained as a yellow precipitate when OH^- is added to $Hg(NO_3)_2$ in the cold, or as a red precipate when hot aqueous Na_2CO_3 is added to mercuric nitrate solution. The red form, which gives the same X-ray pattern as the yellow but contains larger crystals, is also the product of the pyrolysis of mercury(II) nitrate. A hexagonal modification of HgO is obtained by precipitation with OH^- from aqueous K_2HgI_4 containing a slight excess of I^-. It is usually obtained as small, red crystals but it too has a more finely-divided yellow modification. Both the orthorhombic and the hexagonal forms of HgO are based on Hg—O—Hg—O—Hg chains, which are planar in the former and spiral in the latter. There is no evidence for a mercury(II) hydroxide, nor is there a mercury(I) oxide.

39.8. Sulphides

39.8.1. Zinc

Zinc sulphide occurs principally as blende; the hexagonal wurtzite is comparatively rare. The white sulphide precipitated when ammonium sulphide is added to a Zn^{2+} solution is amorphous and is readily soluble in dilute mineral acids, but it gradually changes to a less soluble form. This, heated with aqueous H_2S under pressure, gives blende, but heated with H_2S gas gives wurtzite.

39.8.2. Cadmium

Cadmium sulphide occurs as the mineral greenockite. When made by heating cadmium oxide and sulphur, it has the wurtzite form; but the precipitate obtained by adding H_2S to a weakly acidic Cd^{2+} solution has the blende structure.

39.8.3. Mercury

The stable modification of HgS, the mineral cinnabar, is isostructural with the hexagonal form of HgO. In the Hg—S—Hg chains the bond distance, 236 pm, is equal to the sum of the single-bond covalent radii; every Hg has two other pairs of S atoms, at 310 and 330 pm respectively, which give it a distorted octahedral environment. The black HgS precipitated from solution is a metastable form with a blende structure. A mineral metacinnabar which also has the blende structure, is much less common than cinnabar itself.

39.9. Mercury—Nitrogen Compounds

Mercury forms direct covalent as well as co-ordinate links with nitrogen. Mercury(II) chloride reacts with gaseous ammonia to give 'fusible white precipitate':

$$HgCl_2 + 2\ NH_3 \rightleftharpoons Hg(NH_3)_2Cl_2$$

The $[Hg(NH_3)_2]^{2+}$ ions are arranged at the face centres of cubes of Cl^- ions with their axes randomly arranged along the a, b and c axes of the crystal. The crystal contains finite $[Hg(NH_3)_2]^{2+}$ ions (Fig. 39.3).

Treatment of $HgCl_2$ with aqueous ammonia gives 'infusible white precipitate', $HgNH_2Cl$, containing indefinitely long chains of $[—Hg—NH_2—]^+$ with Cl^- ions between them in an orthorhombic structure (Fig. 39.4). The infusibility and low solubility are ascribed to chain structures.

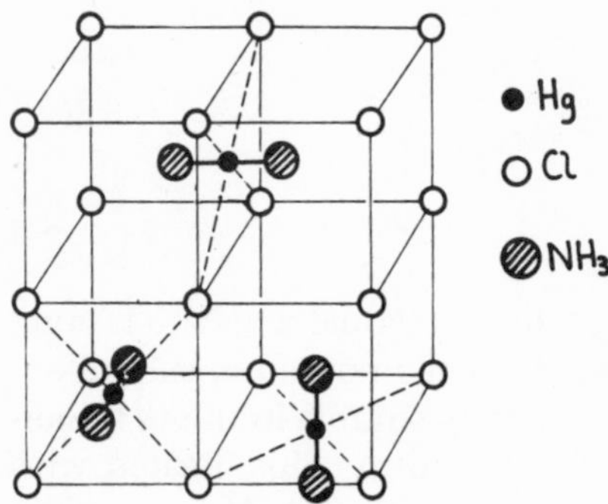

Fig. 39.3. Random arrangement of $Hg(NH_3)_2{}^{2+}$ ions in 'fusible white precipitate', $Hg(NH_3)_2Cl_2$.

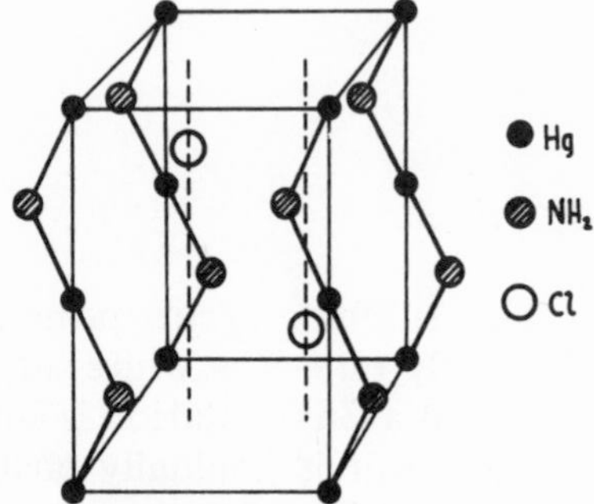

Fig. 39.4. Orthorhombic cell of 'infusible white precipitate', $HgNH_3Cl$, showing zig-zag chains $(—Hg—NH_2—)_n$.

When HgO is warmed with aqueous ammonia the compound $Hg_2NOH \cdot 2\ H_2O$, Millon's base, is produced. This has a Hg_2N^+ network of cristobalite type with OH^- ions and water molecules held in the network by ionic, hydrogen bond and dispersion forces (Fig. 39.5).

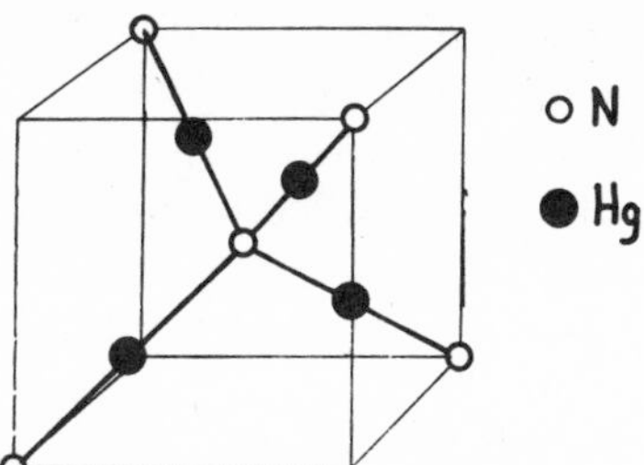

Fig. 39.5. Arrangement of Hg_2N^+ network in Millon's base, $Hg_2NOH \cdot 2\ H_2O$.

The Hg—N distances in all these compounds are similar (206 pm). The iodide corresponding to Millon's base, $Hg_2NI \cdot H_2O$, is the precipitate produced from Nessler's reagent by ammonia. The nitrogen atom uses sp^3-hybrid orbitals in these compounds; the collinear bonds of the Hg probably involve sp hybridisation.

When ammonia reacts with $Hg(ClO_4)_2$, a compound of much greater ionic character than $HgCl_2$ is formed, a complex tetra-ammine in which there are co-ordinate links between nitrogen and mercury:

$$Hg(ClO_4)_2 + 4\ NH_3 \rightleftharpoons [Hg(NH_3)_4]^{2+}(ClO_4)_2^-$$

A mercury(I) compound reacts with aqueous ammonia to give free mercury together with the ammine formed by the corresponding mercury(II) compound under the same conditions. Thus Hg_2Cl_2 gives a black precipitate which is a mixture of finely divided mercury and infusible white precipitate.

39.10. Oxo-acid Salts

The oxo-acid salts of Zn^{2+} and Cd^{2+} are frequently similar to those of Mg^{2+} in solubility, degree of hydration and crystal form; for example $ZnSO_4 \cdot 7\ H_2O$ is isomorphous with $MgSO_4 \cdot 7\ H_2O$. They are much less thermally stable than magnesium salts, however; $ZnCO_3$ is converted to ZnO and CO_2 at 570 K, and $Zn(NO_3)_2$ to ZnO, N_2O_4 and O_2 at 410 K; even the sulphate decomposes above 1000 K. Oxo-salts of Cd^{2+} and of both Hg_2^{2+} and Hg^{2+} are also easy to decompose by heating. One of the most thermally stable of the oxo-acid salts in the subgroup is the pyrophosphate $Zn_2P_2O_7$, which is used in the gravimetric determination of zinc.

Cadmium sulphate forms a hydrate $CdSO_4 \cdot 2\frac{2}{3}\ H_2O$ which has an interesting structure. There are two sets of Cd^{2+} with slightly different environments; all are surrounded, approximately octahedrally, by two H_2O molecules and four sulphate oxygens. There are four kinds of crystallographically nonequivalent H_2O molecules; three-quarters of them are attached to Cd^{2+} ions and one-quarter have as neighbours two other H_2O molecules and two sulphate oxygen atoms.

Zinc forms a basic acetate $Zn_4O(OCOCH_3)_6$ isomorphous with that of beryllium when the normal acetate is distilled at low pressure. Unlike the Be salt it is hydrolysed rapidly by water, presumably because Zn^{II} can co-ordinate easily to two H_2O molecules, thereby raising its co-ordination number to six, whereas the co-ordination number for Be^{II} surrounded by oxygen atoms does not exceed four.

39.11. Organometallic Compounds

Zinc dialkyls can be made by treating the metal or, preferably, a zinc-copper couple, with the appropriate alkyl iodide in an autoclave at 420 K:

$$2\ Zn + 2\ RI \rightarrow ZnR_2 + ZnI_2 \qquad (R = Me, Et, n\text{-}Pr, n\text{-}Bu, n\text{-}pentyl)$$

Other preparative methods are:

$$ZnCl_2 + 2\ LiR \xrightarrow[\text{solution}]{\text{ether}} ZnR_2 + 2\ LiCl \qquad (R = Ph,\ C_{10}H_7,\ C_6F_5)$$

$$ZnCl_2 + 2\ RMgX \xrightarrow[\text{solution}]{\text{ether}} ZnR_2 + MgX_2 + MgCl_2 \qquad (R = Et, vinyl, t\text{-}Bu)$$

$$Zn + HgR_2 \rightarrow ZnR_2 + Hg \qquad (R = Me, i\text{-}Bu, p\text{-}C_6H_4F)$$

Dimethyl zinc and diethyl zinc are volatile liquids which can be distilled unchanged in the absence of air but are oxidised immediately if air is present. The compounds are monomeric in the vapour and in benzene solution. The X-ray analysis of $ZnMe_2$ indicates a linear C—Zn—C arrangement. Diaryl compounds of zinc are white, crystalline solids; Ph_2Zn melts at 330 K.

An increasingly important use of organozinc compounds is as polymerisation catalysts. Thus diethyl zinc in the presence of $TiCl_4$ promotes the polymerisation of vinyl monomers, propene, butadiene and isoprene.

Cadmium alkyls are best made by treating $CdBr_2$ with Grignard reagents in ether, the aryls by treating $CdBr_2$ with lithium aryls in the same solvent. Cadmium dialkyls are uniformly less volatile and less reactive than the corresponding zinc compounds. Thus they do not usually inflame in air and they react quietly with water and alcohols. With acid chlorides they yield ketones — a useful method of synthesis:

$$2\ RCOCl + R'_2Cd \rightarrow 2\ RCOR' + CdCl_2$$

Organomercury compounds, which are stable to air and water, are nevertheless sufficiently reactive towards other metals to be of considerable value in the synthesis of their organometallic compounds. The mercury—carbon bond is a rather weak one (~65 kJ mol^{-1}), and the lack of reactivity towards oxygen is probably due mainly to the weakness of the Hg—O bond. Organomercury compounds can, in fact, be made by treating aromatic compounds, as well as alkenes and alkynes, with mercury—oxygen compounds such as basic salts. This process of *mercuration* goes particularly smoothly with substituted benzene derivatives:

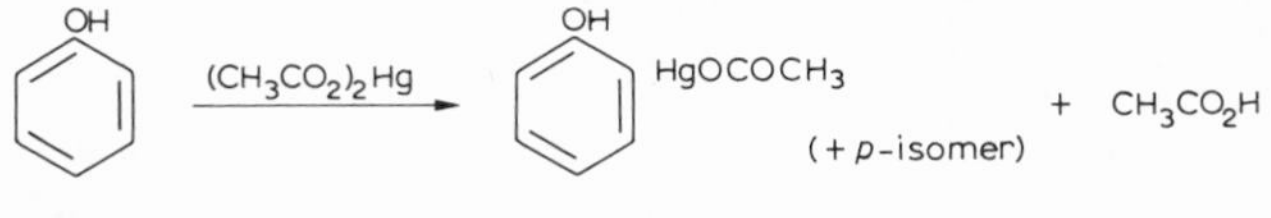

$$\text{C}_6\text{H}_5\text{OH} \xrightarrow{(CH_3CO_2)_2Hg} o\text{-HOC}_6\text{H}_4\text{HgOCOCH}_3 \; (+ \, p\text{-isomer}) + CH_3CO_2H$$

Alkyls of mercury, as well as alkylmercury halides, can be made by treating mercury(II) bromide with Grignard reagents in ether:

$$CH_3MgBr + HgBr_2 \rightarrow CH_3HgBr + MgBr_2$$
$$CH_3HgBr + CH_3MgBr \rightarrow CH_3HgCH_3 + MgBr_2$$

The alkylmercury halides are well-characterised compounds containing a linear arrangement C—Hg—X. Dimethyl mercury, *b.p.* 365 K, and diethyl mercury, *b.p.* 432 K are monomers in the vapour.

Zinc and mercury both form cyclopentadienyl compounds but they are quite different in character. Biscyclopentadienylzinc is made in low yield by the action of C_5H_5Na on $ZnCl_2$ in ether. It is air-sensitive, rapidly hydrolysed by water, and only slightly soluble in organic solvents. The similarity of its infrared spectrum to that of bis(cyclopentadienyl)magnesium suggests that it too has a 'sandwich' structure in which the bonding is largely ionic. Bis(cyclopentadienyl)mercury, made from C_5H_5Na and $HgCl_2$ in tetrahydrofuran, is stable to water, soluble in most organic solvents, and capable of forming Diels—Alder adducts. Sigma bonding between the mercury atom and the cyclopentadiene groups is indicated.

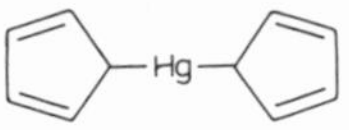

$$C_5H_5\text{—Hg—}C_5H_5$$

39.12. Complexes

The Zn^{2+} ion is a harder acid than Cd^{2+} and Hg^{2+}, and resembles Mg^{2+} in forming strong complexes with oxygen donors such as β-diketones; these complexes usually have tetrahedrally co-ordinated zinc atoms. Other arrangements do occur, however. The acetylacetonate monohydrate $Zn(acac)_2 \cdot H_2O$ contains 5-co-ordinate zinc; the oxygen atoms are arranged around the zinc at the corners of a distorted triangular bipyramid. Six-co-ordinate zinc complexes are uncommon; one interesting example is hydrazinium sulphatozincate $(N_2H_5)_2Zn(SO_4)_2$. In the solid, Zn is co-ordinated to one N atom of each cation and doubly co-ordinated to each SO_4 group acting as a bidentate ligand.

Cadmium(II) and particularly mercury(II) are softer acids than Zn^{2+} and tend to form stronger bonds with oxygen and nitrogen. However the dihalides of all three metals form addition compounds with thioethers:

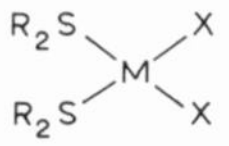

$$(R_2S)_2MX_2$$

(R = alkyl, M = Zn, Cd or Hg and X = halogen)

The dihalides, except the fluorides, of all three metals form complexes with *o*-phenylenebisdimethylarsine. They are non-electrolytes in which the metal atom is tetrahedrally co-ordinated to two arsenic and two halogen atoms:

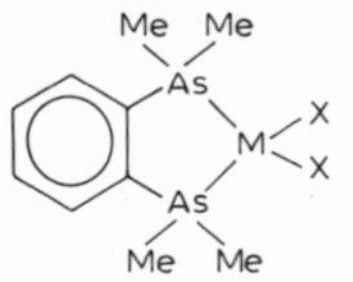

Mercury dihalides also form bridged structures with substituted phosphines and arsines, e.g.:

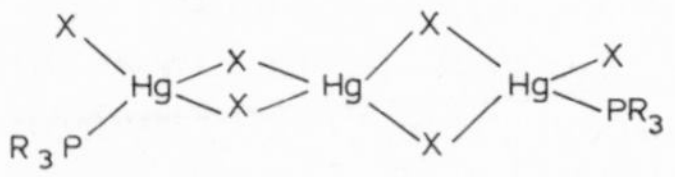

Salts of all three metals form ammines, principally tetrahedral complex cations $M(NH_3)_4^{2+}$, but in addition mercury(II) salts form linear diammines, and zinc and cadmium octahedral hexa-ammines, $[M(NH_3)_6]^{2+}$. Ethylenediamine, however, produces octahedral $[Men_3]^{2+}$ ions with all three metals.

The Hg_2^{2+} ion forms few complexes presumably because the Hg^{2+} ion forms stronger bonds with most donors, so that Hg_2^{2+} disproportionates and the products of reaction between mercury(I) and a ligand are the mercury(II) complex and the metal. Aniline, however, does give the fairly stable $[Hg_2(NH_2Ph)_2]^{2+}$ ion.

Further Reading

H.L. Roberts, Some general aspects of mercury chemistry, Adv. Inorg. and Radiochem., 11 (1968) 309.

D. Breitinger and K. Brodersen, Development of and problems in the chemistry of mercury—nitrogen compounds, Angew. Chem. Internat. Ed., 9 (1970) 357.

W.N. Lipscomb, Structures of mercury—nitrogen compounds, Anal. Chem., 25 (1953) 737.

D.H. Klein, Some general and analytical aspects of environmental mercury contamination, J. Chem. Ed., 49 (1972) 7.

B.J. Aylett, The chemistry of zinc, cadmium, and mercury, Pergamon, Oxford, 1975.

Problems

The arrangement of problems follows the order of the chapters but, particularly in the later problems, the student will need to have recourse to topics raised earlier.

1. Place the following molecules and ions in their point groups: (a) SO_2ClF; (b) pyridine, C_5H_6N; (c) *trans*-N_2F_2; (d) ethylene, C_2H_4; (e) CO_3^{2-}; (f) benzene; (g) $Mn_2(CO)_{10}$; (h) NO_2^+; (i) NO_2^-; (j) B_2H_6; (k) $P_3N_3Cl_6$; (l) $Ni(CO)_4$; (m) $SiHCl_3$; (n) COS.

2. Assign the following molecules to point groups:

(a) 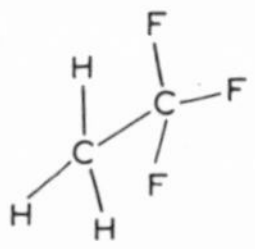 (b) 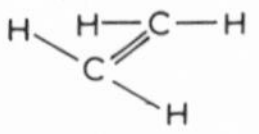 (c) 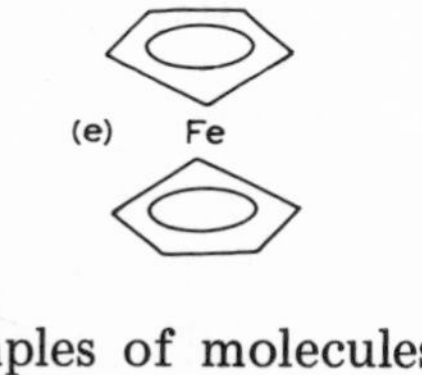 (d) 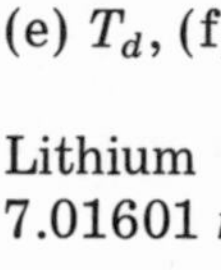 Cr (e) 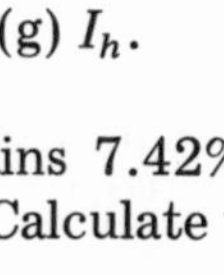 Fe (f) 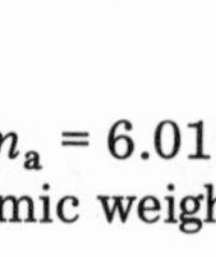 Ru

3. Give examples of molecules or ions, with the following point groups, containing either boron or aluminium: (a) C_{3h}, (b) D_{2h}, (c) D_{3h}, (d) D_{3d}, (e) T_d, (f) O_h, (g) I_h.

4. Lithium contains 7.42% ${}^6_3Li$ ($m_a = 6.01513\ m_u$) and 92.58% ${}^7_3Li$ ($m_a = 7.01601\ m_u$). Calculate the atomic weight.

5. There are three natural isotopes of magnesium: ${}^{24}_{12}Mg$ (m_a = 23.98504 m_u, abundance = 78.70%), ${}^{25}_{12}Mg$ (m_a = 24.98584 m_u, abundance = 10.13%) and ${}^{26}_{12}Mg$ (m_a = 25.98259 m_u, abundance = 11.17%). Calculate A_r(Mg).

6. Calculate the total binding energy of the ${}^{4}_{2}He$ nucleus (m_a = 4.00260 m_u). Use the data on the proton, neutron and electron given in Table 2.1.

7. The *B.E.* per nucleon for a nuclide is given with fair accuracy for many nuclides by the equation:

   B.E. per nucleon/pJ = $2.26 - 2.08\ A^{-1/3} - 0.096\ Z^2\ A^{-4/3}$

   Use the equation to calculate the *B.E.* per nucleon for ${}^{64}_{30}Zn$ (A = 64, Z = 30).

8. Sulphur-35 (m_a = 34.96903 m_u) emits a β^--particle but no γ-ray. The product is chlorine-35 (m_a = 34.96885 m_u). Calculate the maximum energy of the β^--particle emitted.

9. ${}^{32}P$ is pure β^--emitter. Calculate the maximum β energy from: $m_a({}^{32}P)$ = 31.97391 m_u, $m_a({}^{32}S)$ = 31.97207 m_u.

Interpret the following data in Examples 10—13

10. At 40 MHz, the ${}^{19}F$ n.m.r. spectrum of selenium(IV) fluoride above its melting point of 264 K consists of a single band, but a solution at 73 K displays two bands which merge when the temperature is raised above 83 K.

11. The 40 MHz ${}^{19}F$ n.m.r. spectrum of SbF_5 at 263 K consists of 3 broad peaks with intensities in the ratio 2 : 2 : 1. The fine structure due to F—F spin coupling is not completely resolved.

12. The ${}^{31}P$ n.m.r. spectrum of P_4S_3 shows two peaks of intensity ratio 3 : 1. The more intense peak is a doublet and the less intense a quadruplet.

13. At 200 K the 60 MHz ${}^{1}H$ n.m.r. spectrum of ${}^{205}TlMe_3$ in solution comprises two sharp equally intense bands. At 100 MHz and the same temperature a similar spectrum results but at 300 K a single broad band is recorded in a position midway between the bands observed at lower temperature.

14. Using the information in 2.7.1. predict the following ${}^{1}H$ n.m.r. spectra: (a) PH_3, (b) ${}^{14}NH_4$, (c) ${}^{15}NH_3$, (d) ${}^{10}BH_3$, (e) ${}^{11}BH_4^-$.

15. Predict the following ^{19}F n.m.r. spectra: (a) $^{32}SF_6$, (b) $^{10}BF_3$, (c) BeF_4^-, (d) HPF_2, (e) PF_3.

16. What do the following quadrupole coupling constants for the ^{127}I atom indicate about I—X bonds in the compounds?

	(eqQ/h)/MHz
I_2	−2156
ICl	−3131
CH_3I	−1739
CHI_3	−2064
CI_4	−2102
DI	−1823
LiI	−198.5
KI	−60

17. What do the following sets of isomer shift data indicate about the Mössbauer atom?

	δ/mm s^{-1}
KIO_4	−2.34
KI	−0.51
KIO_3	+1.56

	δ_{SnO_2}/mm s^{-1}
$SnCl_2$	4.15
α-Sn	2.10
$SnCl_4$	0.80
$(C_5H_5)_2Sn$	3.73
$(C_6H_5)_2Sn$	1.42

18. Calculate the energy of radiation which has a frequency 200.0 MHz.

19. Calculate the wavelength of light which has a frequency of 5.00×10^{14} Hz.

20. Calculate the wavelength of a free electron moving at a constant velocity of 3.00×10^6 m s^{-1}.

21. Calculate the kinetic energy of a free electron moving at 2.00×10^5 m s^{-1}.

22. Calculate the energy gained by an electron in being accelerated through an electric potential of 100.0 V.

23. Calculate the velocity of an electron, originally at rest, after it has been accelerated through an electric potential of 100.0 V.

24. Calculate the force of attraction between a proton and an electron separated by a distance of 100.00 pm in a vacuum.

25. Calculate the potential energy of a proton and an electron separated by a distance of 200.0 pm in a vacuum.

26 Calculate the wavelength of the second line in the visible hydrogen spectrum.

27. Use Slaters' rules to calculate the effective nuclear charge acting on the 3s electron in the atom of Na in its ground state.

28. Calculate the effective nuclear charge acting on a 3p electron in the atom of Cl in its ground state.

29. The electron configuration of Pd in its ground state is $[Kr]4d^{10}$, not $[Kr]4d^8 5s^2$. Suggest an explanation.

30. Use the law of second differences to calculate the fourth ionisation energy for sulphur from the following data: $I(1)$ Al = 577 kJ mol^{-1}, $I(2)$ Si = 1577 kJ mol^{-1}, $I(3)$ P = 2911 kJ mol^{-1}.

31. Estimate the electron affinity of magnesium from the data given in Question 30.

32. Calculate the magnetic susceptibility, χ_m, of a substance X from the following data. The force acting on a tube of X with an internal cross-section of 1.24 cm^2, in a magnetic flux of density 1.05 tesla, is 2.35×10^{-3} newtons.

33. Calculate μ_{eff} for the Fe^{2+} ion in iron(II) sulphate given that for $FeSO_4 \cdot 7\,H_2O$, $\chi_m = 8.71 \times 10^{-4}$ at 298 K, $M = 0.278$ kg mol^{-1} and the density $\rho = 1.90 \times 10^3$ kg m^{-3}.

34. Use the spin-only formula to calculate the magnetic moment of the Mn^{2+} ion in its ground state.

35. Calculate the value of the Landé splitting factor g for the free Co^{2+} ion in its ground state ($^4F_{9/2}$).

36. Calculate value of μ_{eff} for the Co^{2+} ion in an octahedral ligand field of strength 125 kJ mol^{-1} using the value of the spin—orbit coupling constant given in Table 3.5.

37. What is the bond order of the bond in each of the following diatomic molecules and ions? (a) NO^+, (b) ICl, (c) H_2^+, (d) OCl^-.

38. Calculate E(C—N) from the following data:

		ΔH/kJ mol^{-1}
C(s)	= C(g)	+715
N_2(g)	= 2 N(g)	+950
2 C(s) + N_2(g) + 5 H_2(g)	= 2 CH_3NH_2(g)	−56

Use the values for E(H—H), E(C—H) and E(N—H) given in Table 4.4.

39. Calculate E(N—F), given that the enthalpy of dissociation of N_2 gas is 950 kJ mol^{-1} and the enthalpy of formation of NF_3 gas is −113 kJ mol^{-1}. Use the value of E(F—F) given in Table 4.4.

40. Calculate E(C—H) from the following data:

D(H—C) = 340 kJ mol^{-1}
D(H—CH) = 452 kJ mol^{-1}
D(H—CH_2) = 436 kJ mol^{-1}
D(H—CH_3) = 436 kJ mol^{-1}

41. Use Pauling's method and the data in Table 4.4 to calculate the electronegativity of iodine.

42. Use the value of the single-bond covalent radius of iodine in Table 4.3 to calculate the electronegativity of the element on the Allred—Rochow scale.

43. Use Table 4.3, Table 4.5, the figure for χ_I calculated in Question 41 and the Schomaker—Stevenson equation to calculate the probable bond length in the I—Cl molecule.

44. Calculate the lattice energy of KBr using the Born—Landé equation: $A_a = 1.748$, $r_0 = 328$ pm, $n = 9.5$.

45. Use the Born—Landé equation to calculate the lattice energy of MgO, which has the rocksalt lattice. $n = 7$ and $r_0 = 210$ pm.

46. Calculate the lattice energy of RbF from the following enthalpy data:

		ΔH^0/kJ mol^{-1}
Rb(s)	= Rb(g)	+78
Rb(g)	= Rb^+(g) + e	+402
F_2(g)	= 2 F(g)	+160
F(g) + e	= F^-(g)	−350
F_2(g) + 2 Rb(s)	= 2 RbF(s)	−1104

47. Calculate the electron affinity of the hydrogen atom using the following enthalpy data:

		ΔH^0/kJ mol^{-1}
$H_2(g)$	$= 2\ H(g)$	+436
$H_2(g) + 2\ K(s)$	$= 2\ KH(s)$	−118
$K(s)$	$= K(g)$	+83
$K(g)$	$= K^+(g) + e$	+417
$H^-(g) + K^+(g)$	$= KH(s)$	−742

48. Use Pauling's method to calculate the radius of the Br^- ion given that the internuclear distance in the crystal of RbBr is 342 pm.

49. Calculate the radius of the Cs^+ ion for eight-co-ordination if $r_6 = 169$ pm and $n = 12$.

50. Calculate the radius of Li^+ for tetrahedral co-ordination; $r_6 = 60$ pm and $n = 5$.

51. Calculate the lattice energy for K_2SeO_4 using the Kapustinskii equation with $r_{SeO_4^{2-}} = 243$ pm and $r_{K^+} = 133$ pm.

52. Calculate the thermochemical radius of the BF_4^- ion from Kapustinskii's equation given that U is 665 kJ mol^{-1} for $NaBF_4$ and $r_{Na^+} = 95$ pm.

53. For Rb_2SO_4 the lattice energy is 1729 kJ mol^{-1} and ΔH(hydration) = +24 kJ mol^{-1}. Use this information and the enthalpy of hydration of Rb^+ given in Table 4.10 to calculate the enthalpy of hydration of SO_4^{2-}.

54. Use the Sidgwick—Powell method to predict the shapes of the following molecules and ions: (a) TeF_4, (b) ICl_2^-, (c) BH_4^-, (d) IO_4^-, (e) SeO_4^{2-}, (f) ClO_2, (g) I_3^-, (h) I_3^+, (i) SeO_2Cl_2, (j) $AsOCl_3$, (k) $SbCl_4^-$, (l) AsF_2^+, (m) NO_3^-, (n) SiF_6^{2-}.

55. In the rotational—vibrational spectrum of $^1H^{127}I$ the spacing of the rotational lines is 13.1 cm^{-1}. Calculate (a) the moment of inertia of the HI molecule, (b) the bond length.

56. Calculate the moment of inertia of the $^{14}N^{16}O$ molecule assuming the masses of the atoms to be exactly 14 m_u and 16 m_u. The N—O bond has a length of 115 pm. Use this information to calculate the spacing of the rotational lines in its infrared spectrum.

57. Calculate the force constant for the carbon—oxygen bond in $^{12}C^{16}O$ given that the band centre for the $\nu = 0$ to $\nu = 1$ transition is at 2143 cm^{-1}.

58. For ICl vapour the graph of P_M against $1/T$ is a straight line of gradient 1.78×10^{-3} m^3 mol^{-1} K. Calculate the dipole moment of the ICl molecule.

59. For He gas at 1.01×10^5 Pa and 413 K, $\epsilon_r = 1.000\ 006\ 84$. Calculate the polarisability of the He atom.

60. According to Smith and Hannay the percentage ionic character of a bond is given by the equation:

$$\%\ \text{ionic character} = 16(\Delta\chi) + 3.5(\Delta\chi)^2$$

The bond length in CO is 113 pm and the difference in electronegativity between O and C, $\Delta\chi$, is 1.0. (a) Calculate the expected dipole moment of the CO molecule. (b) The actual value is only 3.6×10^{-31} C m; suggest an explanation.

61. Predict the ligand field splitting of d orbitals in (a) a linear ML_2 complex, (b) a cubic ML_8 complex.

62. Calculate Δ_0 for an octahedral complex of a d^1 ion which has an absorption peak in the visible spectrum at 585 nm.

63. How many pairs of d electrons of parallel spin are there in (a) a high-spin d^7 ion, (b) a low-spin d^7 ion?

64. Which of the following types of ions would be expected to form undistorted tetrahedral $MX_4^{\pm n}$ complexes: d^2, d^3, d^5, d^7, d^8?

65. Most Cu^{II} complexes have a tetragonal arrangement of ligands around the copper. Suggest an explanation.

66. Tetrahedral complexes of Ni^{II} are uncommon. Suggest a possible reason.

67. Calculate the extinction coefficient for a complex at a particular wavelength if the intensity of a beam of that wavelength is reduced by 66.7% in passing through 2.00 cm of an aqueous solution containing 1.50 mol m^{-3} of the complex.

68. Anhydrous $CoCl_2$ gives a pale-pink solution in water but a deep-blue one in dimethylformamide. Suggest an explanation.

69. Consider the $d_{x^2-y^2}$ orbital. Indicate whether it is symmetric or antisymmetric to each of the symmetry operations: i, σ_{xy}, C_4^z, σ_{yz}, C_2^x.

70. Caesium bromide was examined in a Bragg spectrometer using X-rays of wavelength 57.6 pm. The maximum for first-order reflection ($n = 1$) was obtained at a reflection angle of $7°43'$, for second-order reflection ($n = 2$) at $15°35'$. Calculate the distances between the planes from which the X-rays were reflected.

71. CaO was examined with X-rays of wavelength 57.6 pm. First-order

maxima were observed for 100, 110 and 111 planes (see 7.1.1) at reflection angles of 6°54′, 9°52′ and 6°00′ respectively. What are the values of d_{100}, d_{110} and d_{111}? What further evidence does this information supply?

72. Copper has a cubic close-packed structure. The unit cell is a cube of edge 360.8 pm containing four Cu atoms. Calculate the density of the metal assuming that the packing in the crystal is ideal.

73. In solid CsCl a cube of side 413 pm contains one pair of ions. M(CsCl) = 0.169 kg mol^{-1} and the density of the solid is 3.97×10^3 kg m^{-3}. Calculate the Avogradro constant N_A assuming the structure of CsCl to be ideal.

74. Calculate the lattice energy of AlF_3(s) from the following thermochemical data:

		ΔH^0/kJ mol^{-1}
Al(s)	= Al(g)	+326
Al(g)	= Al^{3+}(g) + 3 e	+5138
2 Al(s) + 3 F_2(g)	= 2 AlF_3(s)	−2620
F_2(g)	= 2 F(g)	+160
F(g) + e	= F^-	−350

75. Calculate ΔH^0 for the reaction

CH_3Cl(g) + NaF(s) = CH_3F(g) + NaCl(s)

given that

		ΔH^0/kJ mol^{-1}
Cl(g) + e	= Cl^-(g)	−348
F(g) + e	= F^-(g)	−350
CH_3Cl(g)	= CH_3(g) + Cl(g)	+338
CH_3F(g)	= CH_3(g) + F(g)	+472
U(NaF)	= 894 kJ mol^{-1}	
U(NaCl)	= 768 kJ mol^{-1}	

76. If the energy required to produce one mole of Schottky pairs in an ionic crystal is 193 kJ, what proportion of such defects would occur in such a crystal (a) at 300 K, (b) at 1000 K?

77. The electrical conductance of germanium is increased by a factor of 10^5 when a few parts per million of arsenic are added. Explain.

78. The following are internuclear distances in halides which have the CsCl structure:

Cs—Cl	336 pm	Tl—Cl	325 pm
Cs—I	395 pm	Tl—I	364 pm

The radius ratio in TlI has the limiting value; in it the iodide ions, considered as hard spheres, just touch one another. Calculate the radii of I^-, Cl^-, Tl^+ and Cs^+ for 8-co-ordination.

79. Explain why WO_3 can incorporate small amounts of some alkali metals into its structure.

80. Suggest a reason why KCl is more subject to Schottky defects than to Frenkel defects.

81. For a metal M the calculated lattice energies of its ionic compounds with a halogen X are $U(MX)$ = 1340 kJ mol^{-1}, $U(MX_2)$ = 3840 kJ mol^{-1} and $U(MX_3)$ = 7020 kJ mol^{-1}. For the metal $I(1)$ = 360 kJ mol^{-1}, $I(2)$ = 1236 kJ mol^{-1} and $I(3)$ = 2990 kJ mol^{-1}, and its enthalpy of sublimation is 334 kJ mol^{-1}. Will either MX or MX_2 have a tendency to disproportionate?

82. Comment on the fact that Li_2TiO_3 and $LiFeO_2$ form continuous ranges of solid solutions with one another and also with MgO.

83. Use the following data to calculate the standard enthalpies of formation of (a) the proton, H^+(g), (b) the gaseous Cl^- ion.

		ΔH^0/kJ mol^{-1}
H_2(g)	= 2 H(g) +	+436
Cl_2(g)	= 2 Cl(g)	+242
H(g)	= H^+(g) + e	+1315
Cl(g) + e	= Cl^-(g)	−364

(c) Use your results and the following information to find ΔH° for the reaction $NH_3(g) + H^+(g) = NH_4^+(g)$

		ΔH°/kJ mol^{-1}
$\frac{1}{2}N_2(g) + \frac{3}{2}H_2(g)$	$= NH_3(g)$	−46
$\frac{1}{2}N_2(g) + 2\,H_2(g) + \frac{1}{2}Cl_2(g)$	$= NH_4Cl(s)$	−314
$NH_4^+(g) + Cl^-(g)$	$= NH_4Cl(s)$	−682

84. Use Table 8.1 to calculated the standard *emf*'s of the cells

(a) Zn|Zn^{2+}| |Ag^+|Ag
(b) Ni|Ni^{2+}| |Cu^{2+}|Cu
(c) Cu|Cu^{2+}| |Ag^+|Ag

85. Write down the oxidation state of sulphur in each of the following molecules and ions: SO_3, H_2S, S_2^{2-}, HSO_4^-, $S_2O_3^{2-}$, $S_4O_6^{2-}$.

86. Write down the oxidation state of manganese in each of the following compounds: MnF_3, MnO_3F, $K_4Mn(CN)_6$, MnO_2, K_2MnO_4, Mn_2O_7, $Mn_2(CO)_{10}$, $Mn(CO)_5I$.

87. Write ionic equations for the reaction between IO_3^- and I^- (a) in dilute acetic acid, (b) in concentrated aqueous HCl.

88. Calculate the equilibrium constant of the reaction:

$$Fe^{2+} + Ag^+ \rightleftharpoons Fe^{3+} + Ag \quad \text{at } 298\ K.$$

(Obtain the necessary data from Table 8.1).

89. For Fe edta$^-$ pβ = 25.1 and for Fe edta^{2-} pβ = 14.3. Calculate $E°$, Fe edta$^-$/Fe edta^{2-} at 298 K.

90. For $Co(CN)_6^{3-}$ pβ = 64 and for $Co(CN)_6^{4-}$ pβ = 19. Calculate $E°$, $Co(CN)_6^{3-}/Co(CN)_6^{4-}$ at 298 K.

91. Calculate the value for $E°$, $HBrO/Br_2$ at pH 5.

92. A solution of HCl was added from a burette to water containing exactly 1.00×10^{-3} mol KIO_3, an excess of both KI and $Na_2S_2O_3$ and a few drops of methyl orange. No iodine colour developed as the HCl was added and the methyl orange did not change colour until 30 cm^3 of acid had been added. Explain the titration. Calculate the strength of the HCl solution.

93. The IO_4^- ion is reduced by I^- in two ways, depending on pH.

(i) at pH ~ 8

$$IO_4^- + H_2O + 2\ I^- = 2\ OH^- + IO_3^- + I_2$$

(ii) at pH ~ 1

$$IO_4^- + 8\ H^+ + 7\ I^- = 4\ H_2O + 4\ I_2$$

The IO_3^- ion, however, is not reduced by I^- at pH 8, but is reduced in acid solution:

$$IO_3^- + 5\ I^- + 6\ H^+ = 3\ H_2O + 3\ I_2$$

A solution containing both IO_4^- and IO_3^- ions, of volume 25.0 cm^3, buffered at pH 8 with $NaHCO_3$, liberated iodine from an excess of KI which needed 7.60 cm^3 of a solution of arsenite containing 50.0 mol m^{-3} of AsO_2^-:

$AsO_2^- + I_2 + 2\ H_2O = AsO_4^{3-} + 2\ I^- + 4\ H^+$

A further 25.0 cm^3 were acidified, and liberated sufficient iodine from an excess of KI to require 50.0 cm^3 of a solution of thiosulphate containing 100.0 mol m^{-3} of $S_2O_3^{2-}$. Calculate the amounts of IO_4^- and IO_3^- in 1000 cm^3 of the original solution.

94. Addition of oxalic acid to an iron(III) solution gives a yellow solid thought to have the composition $FeC_2O_4 \cdot x\ H_2O$. A sample, 264.4 mg, was dissolved in dilute aqueous H_2SO_4 and titrated with MnO_4^- at 298 K until all the Fe^{II} was oxidised. The solution was then heated to 340 K, and the titration was continued until the oxalate ion was also completely oxidised. A total of 44.10 cm^3 of a permanganate solution containing 20.0 mol m^{-3} MnO_4^- was required. The Fe^{3+} ions were then reduced with zinc amalgam to Fe^{2+} and the solution then required 14.70 cm^3 of the same permanganate solution to re-oxidise the Fe^{II}. Show that the formula of the yellow precipitate corresponds to $FeC_2O_4 \cdot x\ H_2O$ and find the value of x.

95. Aqueous $CuSO_4$ give precipitates of Cu^I compounds when I^- or CN^- is added, but Cu_2SO_4 is immediately converted to $CuSO_4$ and Cu by water. Explain.

96. For $Co^{3+}aq/Co^{2+}aq$, $E° = +1.82$ at 298 K. For $Co(en)_3^{3+}$ $p\beta = 49$ and for $Co(en)_3^{2+}$ $p\beta = 14$. Calculate $E°$ for the $Co(en)_3^{3+}/Co(en)_3^{2+}$ couple at 298 K.

97. Use Table 8.1 to answer the following questions: Which of the following are capable of being oxidised by $Cr_2O_7^{2-}$ at pH 0 in aqueous solution: (a) Br^-, (b) Hg_2^{2+}, (c) Cl^-, (d) Cu, (e) HNO_2? Which of the following are capable of being reduced by HNO_2 under the same conditions of acidity: (f) H_2O_2, (g) Br_2, (h) MnO_4^-? Write balanced ionic equations for the reactions you choose.

98. Explain why FeI_3 cannot be made in aqueous solution.

99. Explain why a neutral solution of iodine in aqueous KI will oxidise As^{III}, whereas in strongly acidic solution As^V will oxidise I^- to I_2.

100. For the half-reaction $2H^+ + 2e = H_2$ at pH 7, $E° = -0.414$ V. $E°$, $Cr^{3+}/Cr^{2+} = -0.40$ V. What effect would you expect the Cr^{2+} ion to have on air-free water (a) at pH 7, (b) at pH 0?

101. Pure iron wire (521 mg) was dissolved in an excess of aqueous HCl in an inert atmosphere. KNO_3 (253 mg) was added to the hot solution. When reaction was complete the Fe^{2+} ion remaining in solution required

18.00 cm^3 of a solution of dichromate containing 16.7 mol m^{-3} $Cr_2O_7^{2-}$ for titration. Deduce the stoichiometry of the reaction between Fe^{2+} and NO_3^- ions.

102. For the metal vanadium a simplified Latimer scheme is

$$V^{V} \xleftrightarrow{+1.00\ V} V^{VI} \xleftrightarrow{+0.31\ V} V^{III} \xleftrightarrow{-0.2\ V} V^{II} \xleftrightarrow{-1.5\ V} V^{0}$$

Given that $E°$, Zn^{2+}/Zn = −0.76 V; $E°$, Sn^{4+}/Sn^{2+} = +0.15 V; $E°$, Fe^{3+}/Fe^{2+} = +0.76 V, select suitable reducing agents to reduce (a) V^{V} to V^{II}, (b) V^{V} to V^{III}, (c) V^{V} to V^{IV}.

103. $E°$, Fe^{3+}/Fe^{2+} = +0.76 V and $E°$, Fe^{2+}/Fe = −0.44 V, Calculate $E°$, Fe^{3+}/Fe.

104. Write down the formulae of the conjugate bases of the following acids: H_3PO_4, HSO_4^-, SH^-, $Cr(H_2O)_6^{3+}$, $H_2AsO_4^-$.

105. Write down the formulae of the conjugate acids of the following bases: $C_2H_5NH_2$, $H_4IO_6^-$, $B(OH)_4^-$, $VO(OH)^+$.

106. Crystals of perchloric acid monohydrate are isomorphous with those of anhydrous $KClO_4$ and NH_4ClO_4. Suggest an explanation.

107. For the $Fe(H_2O)_6^{3+}$ ion pK_a = 2.7. Calculate the pH of an aqueous solution containing 1.00 mol m^{-3} $FeCl_3$. Ignore secondary ionisation.

108. What is the pH of an aqueous solution of sodium fluoride containing 10.0 mol m^{-3} of NaF? (pK_a for HF = 3.22.)

109. A solution of KCN containing 1.00 mol m^{-3} has a pH of 10.0. What is pK_a for HCN?

110. For NH_4^+, pK_a = 9.0. What is the pH of a solution made by dissolving 10.7 g NH_4Cl in 1 kg of aqueous ammonia containing 0.100 mol kg^{-1} of NH_3?

111. Calculate the pH of a solution containing NaH_2PO_4 and Na_2HPO_4 in the mole ratio 2 : 1 (pK_a for $H_2PO_4^-$ is 7.2).

112. For H_2O, ϵ_r = 78.5 and for MeOH, ϵ_r = 32.6, both at 298 K. Calculate the approximate value for the dissociation constant of $C_2H_5NH_2$ as a base in methanol given that pK_a for $C_2H_5NH_3^+$ in aqueous solution is 10.70. Assume the charges on the ions to be separated by 300 pm.

113. Write balanced equations for the following acid—base equilibria in 100% H_2SO_4:

HNO_3(1 mol) + H_2SO_4(2 mol) $\nu = 4$
H_3BO_3(1 mol) + H_2SO_4(6 mol) $\nu = 6$
I_2(7 mol) + HIO_3(1 mol) + H_2SO_4(8 mol) $\nu = 16$

(ν is the number of ions on the right of the equation; it can be determined cryoscopically.)

114. Suggest a reason why the order of acid strength in the oxo-acids of chlorine is $HClO_4 > HClO_3 > HClO_2 > HOCl$.

115. If Na_2SO_3 labelled with ^{35}S is dissolved in liquid SO_2 the ^{35}S exchanges with the solvent, but if labelled $SOCl_2$ is dissolved in SO_2 there is no exchange with the solvent. Comment.

116. Comment on the fact that HI is a stronger acid in aqueous solution than HF even though the anhydrous HI molecule is much less polar than the HF molecule.

117. Discuss the position of hydrogen in the periodic table with regard to the alkali metals and the halogens.

118. Comment on the fact that the following compounds are not known: BH, ArH, HCl_3, TiH_4, PH_5, SH_6, IH_7.

119. Show how the physical and chemical properties of hydrides of the elements may be related to fundamental physical characteristics of the atoms of the elements.

120. Compare the properties of CH_4, NH_3, H_2O, and HF with those of SiH_4, PH_3, H_2S and HCl.

121. Suggest a reason why the O—H stretching frequency in the infrared spectrum of ice is lower than that in the spectrum of water vapour.

122. When one mole of B_2H_6 is heated with two moles of NH_3 a white solid with ionic character is formed which decomposes on further heating to give a volatile, covalent liquid. Explain the reactions.

123. Estimate the energy of the $\nu = 0$ to $\nu = 1$ transition in (a) $D^{35}Cl$, (b) $D^{79}Br$, assuming the force constants to be the same as for HCl and HBr respectively (Table 5.2).

124. Enthalpies of vaporisation at their boiling points for the simples hydrides of the elements of Group VB and Group VIB are:

	ΔH(vaporisation)/kJ mol^{-1}		ΔH(vaporisation)/kJ mol^{-1}
NH_3	23.3	H_2O	40.6

PH_3	14.6	H_2S	18.7
AsH	16.7	H_2Se	19.3
SbH_3	21.0	H_2Te	23.1

Plot molar enthalpies of vaporisation against molar masses for all but the first hydride in each group and hence estimate the enthalpies of vaporisation that NH_3 and H_2O would have if hydrogen bonding did not occur. In which of these two liquids are the hydrogen bonds evidently the stronger?

125. In the emission spectrum of He^+ in the ultraviolet the lines of longest wavelength are at 24.30 nm, 25.63 nm and 30.37 nm. Calculate the second ionisation energy of He.

126. The Group 0 elements in their solid state contain close-packed atoms, the internuclear distances being 357 pm in He, 320 nm in Ne, 382 pm in Ar and 394 pm in Kr. Suggest a reason for the anomalously large distance between nuclei in solid helium.

127. From the ionisation energies given below calculate the approximate value of the electron affinity of Xe using the law of second differences (3.4.4.1). $I(1)(\text{Cs})$ = 380 kJ mol^{-1}, $I(2)(\text{Ba})$ = 966 kJ mol^{-1}, $I(3)(\text{La})$ = 1850 kJ mol^{-1}.
Making the assumption that the lattice energy of a possible compound CsXe is similar to that of CsI, 592 kJ mol^{-1}, and given that the sublimation energy of caesium is 78 kJ mol^{-1}, calculate the approximate enthalpy of formation of an ionic caesium xenide. Does this quantity suggest that the compound might eventually be prepared?

128. Discuss the structures and gradation of properties for the series of compounds: XeF_6, XeF_4, XeF_2, $XeOF_2$, XeO_2, XeO_3.

129. Calculate the enthalpy of formation of $XeCl_2$ from the following data:

		ΔH/kJ mol^{-1}
Xe	$= Xe^+ + e$	+1170
Cl_2	= 2 Cl	+243
Cl + e	$= Cl^-$	−390
Xe^+ + Cl	$= (XeCl)^+$	−167
$(XeCl)^+ + Cl^-$	$= XeCl_2$	−217

130. A reaction between xenon and fluorine gave a solid containing XeF_2 and XeF_4. Two samples of equal weight were analysed as in (a) and (b).
(a) Treatment with water gave 60.2 cm^3 of gas at 290 K and 1.00×10^5 Pa, of which 24.1 cm^3 was shown to be oxygen and the rest xenon. The aqueous solution contained XeO_3aq which oxidised 30.0 cm^3 of an $FeSO_4$ solution containing 100.0 mol m^{-3} Fe^{2+}.

(b) Treatment with aqueous KI gave I_2 which required 35.0 cm^3 of an $Na_2S_2O_3$ solution containing 200.0 mol m^{-3} $S_2O_3^{2-}$ for titration. What was the composition of the solid?

131. Criticise the statement that the CsCl lattice is an example of body-centred cubic packing.

132. Why is an aqueous solution of LiCl very slightly acidic?

133. Predict for francium (a) its *m.p.*, (b) the type of lattice in its chloride, (c) which oxide it will form on heating in air, (d) how it will react with liquid NH_3 containing some $Fe(NO_3)_3$, (e) which of its salts are sparingly soluble.

134. Calculate the enthalpy change of the reaction:

$$KCl(s) + Na(g) = NaCl(s) + K(g)$$

from U(KCl) = 694 kJ mol^{-1}, U(NaCl) = 768 kJ mol^{-1}, I(1)(K) = 426 kJ mol^{-1}, I(1)(Na) = 501 kJ mol^{-1}. If ΔS for the reaction is zero, what is the approximate composition of the gas in equilibrium with excess of the solids at 1000 K?

135. Use the data given below to calculate the enthalpy change of the reaction:

$$2\ NaI(s) + Cl_2(g) = 2\ NaCl(s) + I_2(s)$$

		ΔH/kJ mol^{-1}
$Cl_2(g)$	$= 2\ Cl(g)$	+224
$I_2(s)$	$= 2\ I(g)$	+213
$Na^+(g) + I^-(g)$	$= NaI(s)$	−435
$Na^+(g) + Cl^-(g)$	$= NaCl(s)$	−766
$Cl(g) + e$	$= Cl^-(g)$	−363
$I(g) + e$	$= I^-(g)$	−314

136. Suggest reasons why ΔH(hydration) increases in the series Be^{2+}, Mg^{2+}, Ca^{2+}, Sr^{2+}, Ba^{2+} (i.e. the energy *release* decreases).

137. The isotopic mass of $^{40}_{20}Ca$ is 39.9751 m_u. Calculate the nuclear binding energy per nucleon for this nuclide.

138. A white solid **A**, which is insoluble in water, decomposes violently when it is warmed to give a solid **B** and a gas **C**. The solid **B** is insoluble in water and HCl, but dissolves in HNO_3 to give a solution of **D** and a colourless gas **E** which turns brown in air. The addition of HCl to a solution of **D** produces a white precipitate **F** which dissolves easily in hot water.
The gas **C** is neutral to acid—base indicators; when passed over mag-

nesium it converts the metal to a white solid G which reacts with water to produce a white solid H and a colourless gas J which has an alkaline reaction. A solution of F in hot water reacts with H_2S to give a black precipitate K and a filtrate L. The solid K dissolves in 60% HNO_3 to produce the gas E, a pale-yellow solid M and a solution of D. Identify the lettered substances.

139. Calculate the standard redox potential of the Mg^{2+}/Mg couple at 298 K from the following data on Gibbs free energy changes:

		ΔG°/kJ mol^{-1}
$Mg(s) + \frac{1}{2} O_2(g)$	$= MgO(s)$	−573
$MgO(s) + H_2O(l)$	$= Mg(OH)_2(s)$	−31
$H_2(g) + \frac{1}{2} O_2(g)$	$= H_2O(l)$	−241
$H_2O(l)$	$= H^+ + OH^-$	+80

and from $K_s(Mg(OH)_2) = 5.5 \times 10^{-12}$

140. The ionisation of Mg to Mg^{2+} requires three times as much energy as Mg to Mg^+. The formation of O^{2-} is endothermic while O^- is exothermic. Explain why magnesium oxide is formulated as $Mg^{2+}O^{2-}$ rather than Mg^+O^- and suggest experimental evidence to prove that Mg^+O^- is incorrect.

141. Account for the fact that calcium forms $CaCl_2(s)$ but not CaCl(s) given the following data:

		ΔH^0/kJ mol^{-1}
Ca(s)	$\rightarrow$ Ca(g)	+177
Ca(g)	$\rightarrow Ca^+(g) + e$	+589
$Ca^+(g)$	$\rightarrow Ca^{2+}(g) + e$	+1146
$Cl_2(g)$	$\rightarrow$ 2 Cl(g)	+242
Cl(g) + e	$\rightarrow Cl^-(g)$	−354

For $CaCl_2$ $A_a = 5.00$, $r_0 = 281$ pm and $n = 9$; for CaCl $A_a = 1.748$, $r_0 = 314$ pm and $n = 9$.

142. The solubility product of $BaSO_4 = 2.0 \times 10^{-10}$ at 293 K. Calculate the solubility of the salt at that temperature.

143. Calculate the entropy change of the reaction

$Sr(H_2O)_6^{2+} + edta^{4-} = Sr\ edta^{2-} + 6\ H_2O$

given that pβ for Sr $edta^{2-}$ is 8.88 at 273 K and 8.53 at 298 K.

144. Thermodynamic properties of CO_2 and of some compounds of Group IIA metals are given for 298 K:

	H^0/kJ mol^{-1}	S^0/J K^{-1} mol^{-1}
$CO_2(g)$	−393	214

MgO	−598	28
$MgCO_3$	−1112	66
BaO	−558	70
$BaCO_3$	−1217	112

Calculate ΔG° values for the reactions (a) $MgCO_3 = MgO + CO_2$ and (b) $BaCO_3 = BaO + CO_2$.

145. Sanderson (J. Chem. Ed., (1952) 539) devised a method for assigning electronegativities based on a quantity which he called the stability ratio, S. He calculated S from the average electronic density of an atom compared with that of a hypothetical isoelectronic inert atom. For the Group IIA metals be obtained values of S, Be, 1.88; Mg, 1.56; Ca, 1.22; Sr, 1.10; Ba, 0.98. He obtained values of χ, the electronegativities of the metals, from the equation:

$$\chi = (0.21S + 0.77)^2$$

These values are close to those of Pauling. Calculate the Sanderson electronegativity values for the Group IIA metals.

146. Estimate the decomposition temperature of $BaSO_4$ from those of the other sulphates of the Group IIA metals: $BeSO_4$, 853 K; $MgSO_4$, 1168 K; $CaSO_4$, 1422 K; $SrSO_4$, 1647 K. Use the method of Ostroff and Sanderson (J. Inorg. Nuclear Chem., 9 (1959) 45). Plot decomposition temperatures against $(r/S)^{1/2}$ where S is the stability ratio of the atom (Question 145 above) and r pm is the radius of the M^{2+} ion (Table 14.1). Draw the best straight line and find the intercept for $BaSO_4$ at the appropriate value of $(r/S)^{1/2}$.

147. The ion below has two isomers. What is the type of isomerism exhibited?

148. The ^{11}B n.m.r. spectrum of B_4H_{10} at 15.1 MHz has a low-field triplet and a high-field doublet. In B_4D_{10} the two absorptions are single peaks. Comment on these results.

149. Compare the boiling point sequence for the halogen compounds of boron and aluminium.

150. The sum of the theoretical covalent radii for boron and fluorine is about 150 pm whereas the experimentally determined B—F bond length in BF_3 is 130 pm. Comment.

Suggest reasons for the observations in Examples 151—156

151. The B—F bond length in amine—BF_3 complexes is longer than in BF_3 itself.

152. BBr_3 is a better electron acceptor than BF_3.

153. BMe_3 fails to react with 2,6-dimethylpyridine although the latter is a stronger base than pyridine.

154. BH_3 is a better electron acceptor than BMe_3.

155. The halide complexes formed by Al(III) are AlF_6^{3-}, $AlCl_4^-$, $AlBr_4^-$ and AlI_4^-.

156. AlF_3 is almost insoluble in anhydrous HF but it dissolves readily when NaF is added.

157. $(CH_3)_3CNH_2$ reacts with BCl_3 to give a white solid of composition C, 40.8; H, 7.7; N, 11.9; Cl, 30.3; B, 9.3%. The solid dissolves in non-polar solvents, and cryoscopic determinations indicate a molar mass of 0.47 kg mol^{-1}. It has zero dipole moment. The ^{11}B n.m.r. spectrum shows a single resonance and the chemical shift is that characteristic of 3-coordinate boron. Hydrolysis yields equimolar amounts of $(CH_3)_3CNH_2$, H_3BO_3 and HCl. Suggest the most probable molecular structure.

158. For the compound

```
  ,NH—BCl,
ClB       NH
  `NH—BCl'
```

in the gas phase ΔH_f° is —1012 kJ mol^{-1}. The standard enthalpies of formation of gaseous monoatomic B, Cl, N and H are respectively 590 kJ mol^{-1}, 122 kJ mol^{-1}, 472 kJ mol^{-1} and 218 kJ mol^{-1}. Calculate E(B—N) for this compound, given that E(B—Cl) = 456 mol^{-1} and E(N—H) = 390 kJ mol^{-1}.

159. Although X-ray analysis shows trimethylaluminium to be a dimer with two bridging CH_3 groups between the Al atoms, the room-temperature ^1H n.m.r. spectrum has a single peak. Suggest a possible explanation.

160. Comment on the fact that $InCl_2$ is diamagnetic.

161. Given that E°, Tl^+/Tl = 0.34 V and E°, Ti^{3+}/Tl = +0.72 V, calculate (a) E°, Tl^{3+}/Tl^+ and (b) the equilibrium constant for $3\ Tl^+(aq) \rightleftharpoons 2\ Tl + Tl^{3+}(aq)$ at 298 K.

162. Thallium(I) iodide is isomorphous with KI but insoluble in water. Suggest a reason.

163. The isotopic masses of ^{14}C and ^{14}N are respectively 14.003242 m_u and 14.003074 m_u. Calculate the maximum β-energy of carbon-14, which is a pure β-emitter.

164. The β-activity from 1.00 g carbon made from the timbers of a Saxon church registered 8.53 counts per minute in a counting device. A 1.00-g specimen of carbon made from a freshly cut sapling gave 9.80 counts per minute in the same equipment. Estimate the age of the ancient timber ($t_{1/2}$ for carbon-14 = 5600 y).

Suggest reasons for the observations in Examples 165—169

165. $SiCl_4$ is more easily hydrolysed than CCl_4.

166. Trisilylamine is planar and is a very weak base.

167. Graphite is chemically more reactive than diamond.

168. CO_2 is gaseous whereas SiO_2 is crystalline under normal conditions.

169. The reaction between water and silica to give H_2SiO_3 is endothermic yet silica gel is an excellent drying agent.

170. How far can the differences between carbon and silicon chemistry be associated with (a) the change in the relative affinity for hydrogen and oxygen, and (b) the increase in maximum covalence from 4 to 6?

171. When $(CH_3)_2SiCl_2$ is treated with an alkali metal in tetrahydrofuran the major product **A** is a crystalline solid with a molar mass of 0.290 kg mol^{-1} and the analysis: C, 41.4; H, 10.3; Si, 48.3%. The proton n.m.r. spectrum in benzene consists of a single band. The compound **A** reacts with bromine to give **B**, which contains 35.5% Br. When **B** is treated with $(CH_3)_3SiCl$ and sodium in toluene one of the products is a liquid **C** which has a proton n.m.r. spectrum consisting of four bands with areas in the ratio 3 : 2 : 2 : 1. Write structures for **A**, **B**, and **C**.

172. Suggest an explanation for the fact that $Mn(CO)_4NO$ is diamagnetic.

173. Discuss the significance of the CO stretching frequencies for the following: (a) CO 2145 cm^{-1}, (b) $Mo(CO)_6$ 2000 cm^{-1}, (c) $Mo(CO)_3(NH_3)_3$ 1855 cm^{-1}, and (d) $Mo(CO)_3(PPh_3)_3$ 1949 cm^{-1}.

174. In the series $V(CO)_6^-$. $Cr(CO)_6$, $Mn(CO)_6^+$ the C—O stretching vibrations occur at 1860, 2000 and 2090 cm^{-1} respectively. Explain.

175. The complex $Mo(py)_2(CO)_4$ has two forms, one with a single band in

the infrared at about 2000 cm^{-1}, the other with four bands. Explain.

176. How many infrared-active modes of vibration are expected for (a) CF_4 (tetrahedral), (b) XeF_4 (square)?

177. Using group theory predict the carbonyl vibrational stretching modes for the isomers of $ML_3(CO)_3$.

178. Although organometallic compounds of the type R_nM are usually thermodynamically unstable, explain why many are isolable and some show considerable resistance to thermal decomposition.

179. Describe the 18-electron rule, explain its basis and illustrate its use in organometallic chemistry.

180. Outline the factors which contribute to the change in reactivity of a molecule or ion as a result of its co-ordination to a metal ion.

181. Explain why the GeOGe angle in $(H_3Ge)_2O$ is 108°, whereas in $(H_3Si)_2O$ the SiOSi angle is 150°.

182. Account for the trends in stability of the oxidation states of Si, Ge, Sn and Pb.

183. Estimate (a) the mean bond energy E(Ge—Cl) and (b) the intrinsic bond energy for the Ge—Cl bond in $GeCl_4$, i.e. E(Ge—Cl) + $\frac{1}{4}$ (promotion energy $s^2p^2 \rightarrow sp^3$). Using this estimate and the other data below, comment on the relative thermal stabilities of the tetrachlorides of Group IV.

M	*a*	State	*b*	*c*	*d*	*e*	*f*
Si	—627	g	+452	+30	393	+398	494
Ge	—544	l	+372	+34	—	+502	—
Sn	—489	g	+301	+39	322	+472	439
Pb	—331	l	+196	+38	243	+607	402

(*a*) = $\Delta H_f^0(MCl_4)$, (*b*) = ΔH_f^0(M,g), (*c*) = $\Delta H_{vapn.}(MCl_4)$, (*d*) = E(M—Cl), (*e*) = promotion energy $s^2p^2 \rightarrow sp^3$, (*f*) = intrinsic bond energy of M—Cl in MCl_4, all in kJ mol^{-1}. ΔH for $Cl_2 \rightarrow 2$ Cl is + 242 kJ mol^{-1}.

184. When an aqueous solution containing $Sn(ClO_4)_2$ and $Pb(ClO_4)_2$ is shaken with an excess of a powdered lead—tin alloy at 298 K, it is found that the ratio of concentrations $[Sn^{2+}]/[Pb^{2+}]$ at equilibrium is 0.46. Calculate E^0, Sn^{2+}/Sn given that E^0, Pb^{2+}/Pb = —0.126 V.

185. A compound X, made by treating diphenylgermane with $Co_2(CO)_8$ in toluene, has the composition C, 42.1; H, 1.85; Co, 21.8; Ge, 13.5; O,

20.7% and a molar mass of 0.541 kg mol^{-1}. It is diamagnetic and has an infrared spectrum with five absorptions in the range 2000—2100 cm^{-1} and another at 1840 cm^{-1}. The compound reacts under pressure with carbon monoxide to give **Y**, which is also diamagnetic, has the composition C, 42.2; H, 1.7; Ge, 12.6; Co, 20.5, O, 23.0%, and shows seven infrared bands in the range 1995 cm^{-1} to 2100 cm^{-1} but none near 1850 cm^{-1}. Suggest structures for **X** and **Y**.

Suggest reasons for the observations in Examples 186—189

186. NO_2^+ is linear whereas NO_2 and NO_2^- are angular.

187. The first ionisation energy of the N_2 molecule is about 70% greater than that of the NO molecule.

188. The N—O infrared stretching vibration in NO occurs at 1876 cm^{-1} but in $NOBF_4$ it occurs at 2200 cm^{-1}.

189. Boron nitride and boron phosphide both crystallise with a graphite structure.

190. Use the following results to determine the stoichiometry of the reaction:

$x Ti^{3+} + y NH_3OH^+ + z H^+ \rightarrow$ products

When 5.00 cm^3 of a solution of $(NH_3OH)_2SO_4$ containing 58.2 mol m^{-3} were added to 25.0 cm^3 of a solution of $TiCl_3$ containing 105.0 mol m^{-3}, the unused Ti^{3+} needed 12.80 cm^3 of a solution of $Fe_2(SO_4)_3$ containing 57.2 mol m^{-3} to oxidise it. What are the products?

Suggest reasons for the observations in Examples 191—194

191. The commonest oxoanions of nitrogen and phosphorus are NO_3^- and PO_4^{3-}.

192. NF_3 is the only stable trihalide of a Group V element which is not readily hydrolysed.

193. PCl_3 is hydrolysed to HCl and H_3PO_3 by water, but NCl_3 gives HOCl and NH_4^+.

194. The bond angles in NH_3, PH_3 and NF_3 are 107°, 94° and 103° respectively.

195. Outline how nuclear magnetic resonance spectroscopy can be used to elucidate structures of oxo-anions of phosphorus.

196. On the basis of simple group theory describe the bonding in a molecule of nitrogen dioxide monomer.

Comment on the statements in Example 197—205

197. NF_3 is remarkably inert, it neither explodes like NCl_3 nor like NH_3 adds to Lewis acids.

198. Some nitrogen compounds contain an odd number of electrons yet their structures can be rationalised by the octet rule.

199. The chemical and physical properties of elemental phosphorus depend markedly on the allotropic form of the element.

200. The ring in $(PNF_2)_3$ is planar, whereas that in $[PN(NMe_2)_2]_4$ has approximate D_{2d} symmetry and that in $(NPCl_2)_4$ has approximately S_4 symmetry; the P—N (ring) bond lengths in these three compounds are 151, 158 and 158 pm respectively.

201. The 'resonance energy' per P—N (ring) is slightly greater in a cyclotetraphosphatetrazene than that in the corresponding cyclotriphosphatriazene.

202. The ionisation potential of the oxygen atom is less than that of the nitrogen atom but O is more electronegative.

203. Although aqueous H_2S is a stronger acid than water, liquid H_2S is much less ionised than water.

204. SF_6 exists but SH_6 has not been prepared.

205. TeF_6 is hydrolysed by cold water to a mixture of hydrofluoric and telluric acids but SF_6 is not hydrolysed even by boiling aqueous alkali.

206. Discuss the following observation. Radioactive $Na_2S_2O_3$ can be prepared by boiling ^{35}S-labelled powdered sulphur with a solution of Na_2SO_3. When a solution of the radioactive thiosulphate is treated with dilute HCl, radioactive sulphur is precipitated and the solution which remains is inactive.

207. How far does the fact that at room temperature oxygen is a gas while sulphur is a solid reflect the essential difference between 2nd and 3rd period elements?

208. For the SO_3 molecule, determine which sulphur orbitals are involved in σ- and which in π-bonding.

209. When S_4N_4 is dissolved in dry dimethoxyethane, a red diamagnetic solution is produced. If the solution is shaken with potassium the colour changes and an e.s.r. spectrum can be obtained consisting of nine lines of relative intensity 1 : 4 : 10 : 16 : 19 : 16 : 10 : 4 : 1 consistent with hyperfine interaction between an unpaired electron and four equivalent N atoms ($I = 1$). On further shaking with potassium the e.s.r. signal almost disappears, then reappears in the same form and finally disappears again. Suggest the series of changes responsible for these observations.

210. An aqueous solution of SO_2 (10.0 cm^3) was diluted to 500.0 cm^3, and 10.0 cm^2 of the diluted solution was added to 25.0 cm^3 of a solution of iodine. The excess of iodine required for titration 9.30 cm^3 of a thiosulphate solution containing 100 mol m^{-3} $S_2O_3^{2-}$. In a separate experiment, 25.0 cm^3 of the iodine solution required 22.5 cm^3 of the same thiosulphate solution for titration.
10.0 cm^3 of S_2Cl_2 (density 1.68 g cm^{-3}) was dissolved in 100 cm^3 of petroleum ether, and a coloured solution was obtained which was added slowly to, and shaken with, 50.0 cm^3 of the original aqueous solution of SO_2. Reaction was complete when about 68 cm^3 of the coloured solution had been added. Neutralisation of the aqueous layer with aqueous KOH gave crystals of an anhydrous potassium salt. The only other product, KCl, remained in solution.
A sample of the anhydrous potassium salt weighing 50.0 mg was dissolved in 50% aqueous HCl. It required 23.2 cm^3 of a potassium iodate solution containing 25.0 mol m^{-3} IO_3^- for complete reaction. Deduce equations for the reaction between (a) SO_2 and S_2Cl_2 in aqueous solution and (b) the anhydrous potassium salt and KIO_3 in 50% aqueous HCl.

211. Comment on the variation on M—O distances (pm) in some oxides which possess the rocksalt structure: CaO 240, TiO 212, VO 205, MnO 222, FeO 217, CoO 212, NiO 208, ZnO (calculated for co-ordination number 6) 214.

212. Discuss mechanisms for semiconducting oxides acting as catalysts for surface oxidation of CO.

213. Classify the following oxides into insulators, n- and p-type semiconductors. ZnO (wurtzite), Al_2O_3 (corundum), NiO (rock salt), Co_3O_4 (spinel), MgO (rock salt), and WO_3 (rhenium oxide structure).

Suggest reasons for the observations in Examples 214–216

214. Many peroxides are coloured.

215. The stability of peroxides and superoxides increases on passing from Li to Cs.

216. Hydrogen peroxide can be regarded as both an oxidising and a reducing agent.

217. Use simple symmetry arguments to derive a qualitative model of the electronic configuration of ozone. Account for the fact that the O—O internuclear bonding distance in the ozonide ion is greater than that in ozone.

Comment on the statements in Examples 218—224

218. The process $O + 2\,e \rightarrow O^{2-}$ is endothermic yet a large number of compounds containing this ion exist.

219. The acidic and basic character of the oxides of a metal varies with the oxidation state of the metal.

220. The dissociation energies of F_2, and I_2 are similar.

221. The interhalogen compounds so far discovered are diamagnetic and contain an even number of atoms.

222. I_2 forms purple solutions in cyclohexane or CCl_4 but brown solutions in ether or pyridine.

223. In the conductimetric titration of $SnBr_2F_{10}$ with $KBrF_4$ in liquid BrF_3, the conductivity passes through a minimum.

224. Elements tend to exhibit higher oxidation states in their fluorides than in their iodides.

225. Bromine reacts with ozone, in the solvent perfluoropentane at 220 K, to give a solid **A** which is hydrolysed by aqueous NaOH to Br^- and BrO_3^- ions. This solution, treated with an excess of KI and HCl, liberates iodine which requires for titration 5 mole $Na_2S_2O_3$ for every mole of **A** taken. **A** reacts with F_2 in perfluoropentane to give colourless crystals of **B** which contains 61.1% Br. When **B** is warmed it releases 1 mole O_2 and $\frac{1}{3}$ mole Br_2 per mole **B** and leaves **C**, a yellow liquid with a large dipole moment. **C** reacts with KF to give crystals of **D**, which contains 20.0% K, 41.0% Br and 39.1% F. Suggest structures for **A**, **B**, **C** and **D**.

226. Suggest a structure for SOF_6 which shows two fluorine resonance lines in the ^{19}F n.m.r. spectrum of relative intensities 5:1 with the structures of a doublet and a sextuplet respectively.

227. A compound of empirical formula C_3F_9NS shows a ^{19}F n.m.r. spectrum comprising a septuplet and a quartet. The relative intensities of the lines are 1, 6, 15, 20, 15, 6, 1 and 16, 48, 48, 16. Suggest a structure for the compound.

228. A metal **M** reacts with Cl_2 on heating to give a liquid compound **A** which can be converted to a solid **B** when treated with an excess of the metal. A solution of **B** in aqueous HCl reduces (a) Hg^{II} salts to Hg^{I} and then to Hg, (b) nitrobenzene to aniline, (c) Fe^{3+} to Fe^{2+}. The compound **A** reacts with C_6H_5MgBr to give a product **C** which contains only carbon, hydrogen and **M**. If **A** and **C** are heated together, a new compound **D** is formed which can be hydrolysed by water to give HCl and a polymer. If H_2S is passed through a solution of **B** in HCl a yellow precipitate **E** is obtained which dissolves in ammonium sulphide solution as a compound **F**. Identify the metal **M** and the compounds **A** to **F**.

229. Describe the spectrum you would expect to see for the ^{19}F n.m.r. resonance of BrF_5 which has a tetragonal pyramidal structure. (The apex fluorine is less shielded. Ignore the spin of the Br nuclei.)

230. Suggest some reason for the compound $(CF_3)_3N$, unlike $(CH_3)_3N$, being without basic character.

231. Suggest an explanation for the fact that the bond in ClF is about 5% shorter and that in BrF about 7% shorter than the sum of the respective single-bond covalent radii.

232. Electrolysis of a solution of NaCl produced a solution containing NaCl, NaClO, and $NaClO_3$.
To 25.0 cm^3 of this solution was added acetic acid and an excess of KI. The iodine which was liberated needed 18.75 cm^3 of a solution of sodium thiosulphate containing 1.00 kmol m^{-3} for titration. In this reaction only the ClO^- is reduced, not the ClO_3^-.
To a further 25.0 cm^3 an excess of Fe^{2+} ion was added, the solution was heated to boiling point and then cooled. It was found that 5.40×10^{-2} mol Fe^{2+} was oxidised in the process. In this both ClO^- and ClO_3^- were reduced.
A further 25.0 cm^3 was completely reduced with SO_2 and the chloride was precipitated as AgCl (2.8668 g). Calculate the concentrations of Cl^-, ClO^- and ClO_3^- in the solution.

233. The hydrolysis of $(SCN)_2$ in alkaline solution gives sulphate, cyanide and thiocyanate ions. Does this reaction suggest that $(SCN)_2$ is properly classed as a pseudohalogen?

234. From the following enthalpies of formation: $\Delta H_f^0(FeF_2) = -830$ kJ

mol^{-1}, $\Delta H_f^0(FeF_3)$ = −1058 kJ mol^{-1}, $\Delta H_f^0(MgF_2)$ = −1104 kJ mol^{-1}, $\Delta H_f^0(MgF)$ = −350 kJ mol^{-1} (the last being calculated on the assumption that MgF has the same lattice energy as NaF), calculate the standard enthalpy of the reactions

(a) 2 FeF_3 = 2 FeF_2 + F_2
(b) 2 FeF_3 + Fe = 3 FeF_2
(c) 2 MgF = MgF_2 + Mg

Use your result to illustrate one essential difference between the chemistry of a transition metal and that of a non-transition metal.

235. Deduce the d-orbital splitting diagram for a square pyramidal molecule ML_5 with C_{4v} symmetry.

236. Comment on the fact that elements of the 1st transition series possess many properties different from those of heavier transition elements.

237. Comment on the values of pβ, for the following complexes: Cu en^{2+}, 10.72; Cu $dmen^{2+}$, 9.72; Cu $tmen^{2+}$, 7.20 (en = $NH_2CH_2CH_2NH_2$, dmen = $CH_3NHCH_2CH_2NHCH_3$ and tmen = $(CH_3)_2NCH_2CH_2N(CH_3)_2$).

238. The logarithms of the stability constants for the reaction

$M^{2+} + edta^{4-} \rightleftharpoons Medta^{2-}$

(H_4edta = EDTA) are given below for several metals. Comment on the trend.

M	pK_1
Mn	13.6
Fe	14.3
Co	16.2
Ni	18.6
Cu	18.8
Zn	16.3

239. A dark-green salt formulated $CrCl_3 \cdot 6\ H_2O$, freshly dissolved in water, gives a white precipitate when $AgNO_3$ is added to it, but when the precipitate is filtered off a further precipitation of AgCl occurs when the solution is warmed. Explain.

240. An aqueous suspension of calcium oxalate is dissolved when the sodium salt of EDTA is added to it. Explain.

241. Use the spin-only formula to predict the paramagnetic moments of (a) potassium hexafluoroferrate(III), (b) sodium hexacyanomanganate(III), (c) hexa-amminechromium(III) sulphate.

242. A specimen (83.5 mg) of a diamagnetic complex, of composition $CoCl_3en_2H_2O$ and molar mass 0.303 kg mol^{-1}, was dissolved in water and poured through a cation-exchange column in its hydrogen form, releasing acid which required 11.0 cm^3 of aqueous NaOH containing 50.0 mol m^{-3} OH^- for neutralisation. Write down the ionic formula of the complex and indicate the possible structures the cation can have.

243. The intensity of a beam of monochromatic light is reduced by 80% on passing through 3.00 cm of an aqueous solution containing 2.0 kmol m^{-3} of an inorganic compound. What is the extinction coefficient?

244. The *emf* of the cell

$$\mathrm{Cu}\left|\begin{array}{ll}\mathrm{Cu(NO_3)_2} & (0.010\ m)\\ \mathrm{NH_2CH_2CH_2NH_2} & (0.170\ m)\\ \mathrm{KNO_3} & (1.00\ m)\end{array}\right|\left|\begin{array}{ll}\mathrm{Cu(NO_3)_2} & (0.010\ m)\\ \mathrm{KNO_3} & (1.00\ m)\\ & \end{array}\right|\mathrm{Cu}$$

is +0.535 V at 298 K. Calculate $p\beta_2$ for the complex $[Cuen_2]^{2+}$ assuming it to be the only ethylenediamine complex present in the left-hand compartment of the cell. Assume relative activities are proportional to molalities in the solutions of equal ionic strength.

245. The substitution reactions of the ion $[Pd(NH_2CH_2CH_2NHCH_2CH_2NH_2)Cl]^+$ are rapid, whereas those of the ion $[Pd(Et_2NCH_2CH_2NHCH_2CH_2NEt_2)Cl]^+$ are very slow. Explain.

246. Suggest an explanation of the fact that electron transfer from Fe^{2+} to Fe^{3+} is reduced by a factor of two when D_2O replaces H_2O as solvent.

247. Comment on the difference in the activation energies for the electron-transfer reactions between the following pairs:

	E_A/kJ mol^{-1}
(a) $[Fe(CN)_6]^{4-}$ and $[Fe(CN)_6]^{3-}$	19.6
(b) $[Co(NH_3)_6]^{2+}$ and $[Co(NH_3)_6]^{3+}$	56.5

Above pH 4 the first-order constant for the aquation of the ion $CoNO_3(NH_3)_4H_2O^{2+}$ is of the form $k = k_1 + k_2/[H^+]$. For the aquation of *trans*-$Co(NO_2)_2bipy^{2+}$ in the pH range 2—11, however $k = k_1$[complex] (independent of $[H^+]$). Comment on these results.

248. For the reaction

$$Co(NH_3)_6^{3+} + 6\ H_3O^+ \rightleftharpoons Co(H_2O)_6^{3+} + 6\ NH_4^+$$

ΔG° is about —200 kJ mol^{-1}. Nevertheless the hexamminecobalt(III) ion will keep almost indefinitely in acidic aqueous solution. Explain this fact.

249. Which of the following complexes do you expect to be labile and which inert? (a) $V(H_2O)_6^{2+}$, (b) $V(H_2O)_6^{3+}$, (c) FeF_6^{3-}, (d) $Co(CN)_6^{3-}$, (e) CoF_6^{3-}, (f) $Nien_3^{2+}$.

250. Why is $Co(NH_3)_6^{3+}$ reduced much less rapidly than $Co(NH_3)_5Cl^{2+}$ by hexa-aquochromium(II) ions?

251. Studies with oxygen-18 shows that in the reaction

$$SO_2 + ClO_3^- + H_2O = HSO_4^- + ClO_2^- + H^+$$

every HSO_4^- ion which is formed contains one oxygen atom derived from a chlorate ion. Explain this observation.

252. Of the complexes of lanthanide(III) ions only those of La^{3+}, Gd^{3+} and Lu^{3+} have magnetic moments which are in accordance with the "spin-only" formula. Explain.

253. Whereas octahedral Mn^{2+} (d^5) complexes have magnetic moments of either about 5.9 μ_B or 1.7 μ_B at room temperature, octahedral Sm^{3+} (f^5) complexes have moments of about 1.6 μ_B. Explain.

254. A stable isotope with $Z = 81$, $A = 205$, on irradiation with high-energy protons, gives two identical fission fragments only. Write a nuclear equation for the process and suggest a likely mode of radioactive decay for the product.

255. TiF_4 forms a six-co-ordinate complex TiF_4L_2 with the monodentate ligand L, in which the ^{19}F n.m.r. spectrum has two equally intense 1 : 2 : 1 triplets. Substitution of one L by another ligand Y gives a complex which shows a doublet of doublets and two doublets of 1 : 2 : 1 triplets each with one-half the area of the doublet of doublets. Deduce whether the isomers are *cis* or *trans*.

256. The compound $VO[(CH_3CO)_2CH]_2$ exhibits an e.s.r. spectrum in frozen $CHCl_3$ at 77 K. Predict its essential features. $I(\text{Vanadium}) = \frac{7}{2}$.

257. A blue oxalatovanadate can be made by the addition of aqueous ammonium vanadate to a hot solution of ammonium oxalate and oxalic acid. Use the following experimental results to find the values of x and y in the formula $(NH_4)_xVO(C_2O_4)_y \cdot 2\,H_2O$.
A sample of the complex, weighing 237.4 mg, when dissolved in an excess of hot, dilute H_2SO_4, required for titration 38.95 cm^3 of a solution of potassium permanganate containing 19.4 mol m^{-3} $KMnO_4$.
When a few crystals of sodium sulphite were added to the solution and it was warmed, it turned blue. The solution was boiled to expel the ex-

cess of SO_2, and then cooled. It then required 7.80 cm^3 of the same $KMnO_4$ solution for titration.
At the pH which was used the value of $E°$, H_2SO_4/H_2SO_3 lies between $E°$, vanadium(V)/vanadium(IV) and $E°$, vanadium(IV)/vanadium(III).

258. Balance these two equations by replacing the letters *a* to *k* with stoichiometric numbers:

(i) $aVO_2^+ + bSO_2 = cVO^{2+} + dSO_4^{2-}$
(ii) $eVO^{2+} + fMnO_4^- + gH_2O = hVO_2^+ + jMn^{2+} + kH^+$

A solution of ammonium vanadate (25 cm^3) was acidified with H_2SO_4, heated to boiling and reduced by passing SO_2 through it. After the excess of SO_2 had been boiled off, the solution required the addition of 23.2 cm^3 of a potassium permanganate solution containing 2.960 kg m^{-3} $KMnO_4$ to produce a permanent pink colour. Calculate the concentration of vanadium in the ammonium vanadate solution.
In another experiment, 10 cm^3 of the same ammonium vanadate was reduced by acidifying and passing through a column of amalgamated zinc. The violet solution which was washed through the column required 27.8 cm^3 of the same $KMnO_4$ solution to produce a permanent pink colour. Calculate the oxidation state of vanadium in the violet solution.

259. Comment on the following observations:
Both $NbBr_5$ and $TaBr_5$ can be reduced with Al at 750 K to give dark-green, diamagnetic products of empirical composition M_3Br_7. A similar reduction of $TaCl_5$ gives dark-brown Ta_2Cl_5 which is paramagnetic, with $\mu_{eff} = 1.46\ \mu_B$ at 293 K. $NbCl_5$ and $NbBr_5$ can be reduced with Nb metal in the presence of KX (X = Cl or Br) to give dark-green, diamagnetic $K_2Nb_3X_9$.
All the reduction products dissolve in dilute aqueous HCl to give green solutions from which compounds $M_3X_7 \cdot 4\,H_2O$ can be crystallised. These solutions, on treatment with $AgNO_3$, give only one mole AgX for every three moles of M.

260. When molybdenum(II) chloride is dissolved in aqueous HCl a complex crystalline acid A ($Mo_6Cl_{14}H_{18}O_8$) is obtained. Conductivity measurements on a solution of the crystals in nitrobenzene show A to be a 1 : 2 electrolyte. Mixing A with $^{36}Cl^-$ in solution results in exchange of six Cl atoms per molecule of A, the reaction being of first order with respect to A and zero order with respect to Cl^-. When A is heated in vacuo a compound of empirical formula $MoCl_2$ is produced. If A is treated with dilute aqueous ammonia and then dehydrated at 470 K in vacuo, a compound B, $Mo_6Cl_8H_4O_4$, is produced. Replacement of chlorine in B proceeds slowly in dilute aqueous alkali, the reaction being of first order with respect to B and first order with respect to OH^-. Suggest structures for A, and B and interpret the kinetic results.

261. A compound A was made by fusing together KCl and $CrCl_3$. The chromium in a 200-mg sample of A was oxidised to $Cr_2O_7^{2-}$ which liberated 1.1075×10^{-3} mol I_2 from an excess of KI solution. The chlorine in A was determined by precipitation. An excess of $AgNO_3$ in aqueous HNO_3 precipitated 478 mg AgCl from a solution containing 200 mg of A. Assuming the only elements in A to be K, Cr and Cl, calculate its empirical formula and suggest a possible structure for the chlorochromate ion in it.

262. Suggest probable structures for the compounds B and C

$$Mo(CO)_6 \xrightarrow{\text{cycloheptatriene}} A + MoC_{10}H_8O_3\ (B)$$

$$\downarrow Ph_3CBF_4 \text{ in inert solvent}$$

$$Ph_3C + MoC_{10}H_7O_3BF_4\ (C)$$

A is a gas of molar mass 0.028 kg mol^{-1}, B has three infrared absorption bands in the range 1880—2000 cm^{-1} and four peaks of equal intensity in the proton n.m.r. spectrum. It is a non-electrolyte. C is an electrolyte. It has three absorptions in the range 1950—2030 cm^{-1} and only one peak in the proton resonance spectrum.

263. The principal isotope of Cr (84%) has zero nuclear spin. The e.s.r. spectrum of $K_3[Cr(CN)_5NO]$ in dilute aqueous solution has as its main feature a triplet of lines of equal intensity. For N-14, $I = 1$. Comment on the oxidation state of chromium in the complex.

264. Comment on the trends in the carbonyl stretching frequencies in the following compounds.

	Absorption peaks/cm^{-1}
$py_3Mo(CO)_3$	1888 and 1746
$(Ph_3P)_3Mo(CO)_3$	1949 and 1835
$(Ph_3As)_3Mo(CO)_3$	1957 and 1847
$(Cl_3As)_3Mo(CO)_3$	2031 and 1992

265. Nitric oxide reacts with $Mn(CO)_5I$, and one of the products, A, is a green solid containing 31.8% Mn and 24.3% N. It is a non-electrolyte in benzene and shows strong infrared absorptions at 2088, 1823 and 1734 cm^{-1}. A reacts with triphenylphosphine to give a further green solid B containing 13.5% Mn, 10.3% N and 7.6% P. It is diamagnetic, a non-electrolyte in benzene, has a molar mass of 0.407 kg mol^{-1} and infrared absorptions at 1781 and 1688 cm^{-1}. Suggest structures for A and B.

266. Write a balanced ionic equation for the action of carbon dioxide upon an alkaline solution of K_2MnO_4.

267. A solution of $MnSO_4 \cdot 4\ H_2O$ containing 584 mg of the solid was found to react with 29.8 cm^3 of a $KMnO_4$ solution A in the cold and in the presence of an excess of H_2SO_4 to yield a red solution. The colour of this solution was discharged by 34.5 cm^3 of a solution B containing 37.24 kg m^{-3} of $FeSO_4 \cdot (NH_4)_2SO_4 \cdot 6\ H_2O$. In a separate titration 25 cm^3 of B were found to be equivalent to 21.6 cm^3 of solution A. Use these data to determine the oxidation state of manganese in the red solution and write an ionic equation for the reaction between Mn^{2+} and MnO_4^-.

268. What inferences can be drawn from the fact that the compound formulated $MnRe(CO)_{10}$ is diamagnetic and has no infrared carbonyl absorption below 2000 cm^{-1}?

269. Treatment of $NaMn(CO)_5$ with $CH_2{=}CHCH_2Br$ in the cold gives a product A of molar mass 0.236 kg mol^{-1} with a proton n.m.r. spectrum of four lines with relative intensities 1 : 1 : 1 : 2. On warming, A is converted to B which has a molar mass of 0.208 kg mol^{-1} and a proton resonance spectrum of three lines with relative intensities 2 : 1 : 2. Suggest structures for A and B.

270. Both $Re_2(CO)_{10}$ and $Re_2Cl_8^{2-}$ have structures in which there is an Re—Re bond. The carbonyl has a D_{4d} molecule but the chloroanion has a D_{4h} structure. Suggest a possible explanation.

271. $ReBr_3$ (0.426 g) was dissolved in acetone and treated with Ph_3PHBr (0.120 g) to give an almost quantitative yield of a dark-red compound which, after recrystallisation, was found to have the composition C, 13.3; H, 1.0; Br, 49.5; Re, 34.3; P, 1.9%. In nitrobenzene solution the compound behaves as a 1 : 1 electrolyte. Its apparent molar mass in acetone is 0.810 kg mol^{-1}. Comment on the compound and its structure.

272. $(\pi\text{-}C_5H_5)_2Mn$ reacted with CO under pressure at 390 K to give a volatile yellow solid A containing C 47.1%, H 2.5%, Mn 26.9% with strong infrared absorptions near 2000 cm^{-1}. A sample of A (0.100 g) reacted with pyridine to give 11.0 cm^3 of CO measured at 273 K and 101 kPa and a non-volatile product B with a molecular weight of 255. The compound A also reacted with butadiene to give C which was shown to have a molecular weight of 202 and a single carbonyl stretching absorption in the infrared spectrum. A also reacted with concentrated H_2SO_4 in acetic anhydride to give a solid D, a strong acid of composition $C_8H_5MnO_6S$, of which a sample weighing 0.171 g required 6.03 cm^3 of aqueous NaOH containing 100 mol m^{-3} OH^- for neutralisation. Write down structural formulae for A, B, C and D.

273. Discuss the bonding in $(Ph_3P)_2NiCl_2$, given the following information. It is blue, with two principal absorption bands at 580 nm ($\epsilon = 14.2$ m^2 mol^{-1}) and 927 nm ($\epsilon = 14.9$ m^2 mol^{-1}). It is a non-conductor in acetone and has a dipole moment of about 2×10^{-29} C m. The *ortho* and *para* hydrogens have n.m.r. chemical shifts considerably upfield relative to their positions in Ph_3P, but the *meta* hydrogen resonance is downfield. The paramagnetic moment is 3.41 μ_B at 300 K.

274. A sample containing only iron and Fe_2O_3 was dissolved in HCl and reduced with a slight excess of aqueous $SnCl_2$. After treatment with $HgCl_2$ to remove the excess of $SnCl_2$, the solution was diluted and a solution containing H^+, Mn^{2+}, SO_4^{2-}, and PO_4^{3-} ions was added. If 0.225 g of the sample, so treated, required 37.5 cm^3 of a potassium permanganate solution containing 3.132 kg m^{-3} $KMnO_4$, calculate the percentages of Fe and Fe_2O_3 in the sample.

275. Given that the mean octahedral ligand field splitting energies produced by oxygen ligands are 111, 171 and 208 kJ mol^{-1} for Co^{II}, Fe^{III} and Cr^{III} respectively, predict which type of spinel (normal or inverse) is likely to be formed by $CoCr_2O_4$ and $CoFe_2O_4$. Assume that the tetrahedral ligand field splitting is $\frac{4}{9}$ of the corresponding octahedral value.

276. The compound 'diarsine'

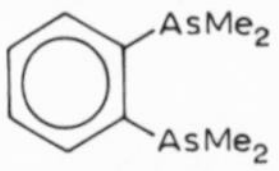

reacts with $NiCl_2$ to give a red complex Ni(diarsine)$_2$Cl$_2$, which can be oxidised by chlorine to Ni(diarsine)$_2$Cl$_3$ which is shown by magnetic measurements to contain one unpaired electron per ion. Further oxidation with fuming HNO_3 and $HClO_4$ gives a deep-green bipositive complex ion. Discuss the stereochemistry of the three compounds. What is the probable magnetic moment of the bipositive ion?

277. Room-temperature paramagnetic moments in the range 3.5 to 4.2 μ_B are typical of tetrahedral complexes of Ni^{II}. Explain.

278. Symmetrical μ-dihyroxotetrakis(ethylenediamine)dicobalt(III) tetrachloride reacts with aqueous HCl to give a product Co(en)$_2$Cl$_3$ which is resolvable into optical isomers. When it is kept for some time in acidic solution the compound changes colour and gives an isomer which is not resolvable. Explain.

279. Explain why the Mössbauer spectrum of $K_4Fe(CN)_6$ has a single peak whereas those of (a) $Na_2[Fe(CN)_5NO]$ and (b) $FeSO_4 \cdot 7\,H_2O$ have double peaks.

280. Addition of F^- ions to a weakly acidic solution of $FeCl_3$ (a) removes the colour, (b) reduces the oxidising power, and (c) interferes with the thiocyanate test for Fe^{3+}. Suggest explanations.

281. The redox potential for the Fe^{III}/Fe^{II} couple is +0.76 V in aqueous solution in the absence of complexing agent but only —0.12 V in the presence of an excess of EDTA. Explain.

282. $Fe(CO)_5$ has zero dipole moment. What does this suggest about the shape of the molecule?

283. Anhydrous iron(III)chloride reacts with anhydrous ethylenediamine to give a white compound containing 18.2% Fe, 23.1% Cl and 27.4% N. It has a paramagnetic moment of 4.0 μ_B and is resolvable into optical isomers. Suggest a possible structure for the compound.

284. Suggest one possible reason why Fe_3O_4 is an inverse spinel whereas Co_3O_4 is a normal spinel.

285. A metal M dissolves in dilute HCl to produce a cation which has a magnetic moment of 5.0 μ_B. In the complete absence of oxygen the addition of OH^- to the solution produces a white precipitate A which on exposure to air turns green and finally to a brown solid B. Upon ignition, B gives a brown solid C which on mild reduction yields a ferrimagnetic, black solid D. B is soluble in dilute HCl; the solution, E, oxidises KI to I_2, but not if an excess of F^- is added before the iodide. When Cl_2 is passed into a suspension of B in concentrated aqueous sodium hydroxide, a red solution F is obtained from which $BaCl_2$ precipitates a red-brown solid G which is a very strong oxidising agent. Identify the metal M and the compounds A to G.

286. There are two compounds of formula $CoBr(SO_4)(NH_3)_5$. One, a red compound, dissolves in water to give a solution from which $AgNO_3$ solution precipitates AgBr but $BaCl_2$ gives no precipitate. The other, a violet compound, gives a precipitate with $BaCl_2$ but not with $AgNO_3$. Explain what structures would explain the behaviour of the two compounds. What names would be assigned to the two forms in the Stock notation?

287. A compound $NiI_2(HPPh_2)_2$, obtained from the reaction between NiI_2 and diphenylphosphine, is diamagnetic in the solid state, is monomeric and a non-electrolyte in nitrobenzene, has a very low dipole moment and a ^{31}P n.m.r. spectrum which indicates that the two P atoms are in identical environments. Draw a probable structure.

288. Reaction between cobalt, carbon monoxide and hydrogen yields a

pale-yellow solid **A** (M = 0.172 kg mol^{-1}) which has a very strong infrared absorption at 2059 cm^{-1} and a single, very intense band in its proton resonance spectrum. The compound **A** loses hydrogen just above its *m.p.* and gives **B** (M = 0.342 kg mol^{-1}) which reacts with lithium in diethyl ether, liberating a gas **C** (M = 28 g mol^{-1}) and a solution from which a deep-red crystalline solid **D** can be obtained. **D** contains Co, 38.1; Li, 1.5; C, 25.9%, has three distinct CO infrared bands at 2080, 1850 and 1600 cm^{-1} and M = 0.464 kg mol^{-1}. Identify **A, B, C,** and **D.**

289. In a close-packed arrangement of N metal atoms, where N is a very large number (a) how many octahedral spaces and (b) how many tetrahedral spaces are there? Assuming the metallic radius of Co in α-Co to be 126 pm what is (c) the radius of the largest particle which can be accommodated in the tetrahedral spaces without distorting the lattice?

290. The exposure to sunlight of a solution of $Fe(CO)_5$ in acetic acid and acetic anhydride produced a bronze-coloured solid A which exhibited absorption bands in the infrared spectrum near 2000 cm^{-1} and 1830 cm^{-1} and weak peaks in its mass spectrum at mass values of 364, 336, 308, 280 and 252 as well as strong peaks at 196, 168, 140, 112, 84 and 56. The compound **A** reacted with $Me_2P{\cdot}PMe_2$ in benzene under pressure at 430 K to give an orange-coloured solid B with an empirical formula $C_6H_6FePO_4$ which shows absorptions in the infrared spectrum attributable to terminal CO groups. The mass spectrum of **B** had strong peaks attributable to fragments $Fe_2P_2Me_4(CO)_n^+$ [n = 0—8] and weak peaks corresponding to fragments $FeP_2Me_4(CO)_m^+$ [m =0—4]. In the Mössbauer spectrum of A and B there are twin peaks, separated by 0.43 mm s^{-1} in A and 2.58 mm s^{-1} in B. Suggest structures for A and B. Assume the iron to be ^{56}Fe only.

291. Pentacarbonyl iron, refluxed at 470 K with a hydrocarbon of molecular formula $C_{10}H_{12}$, yielded a compound A of molar mass 0.354 kg mol^{-1} which contained Fe, 31.57; C, 47.49; O, 18.09 and H, 2.85%. When **A** was heated it was converted to a diamagnetic orange solid **B** containing Fe, 30.03; C, 64.52 and H, 5.42%. On treatment with nitric acid **B** is converted to a blue, paramagnetic compound. Write structural formulae for A and B.

292. Osmium tetroxide reacts with an excess of di-n-butylphenylphosphine in ethanol and concentrated HCl to produce a red compound A which contains 52.4% C and 11.1% Cl, has an experimentally determined molar mass of 0.955 kg mol^{-1} and a magnetic moment at room temperature of 2.2 μ_B. When A is refluxed with CCl_4 for several days a yellow compound B is produced which contains 43.3% C and 18.3% Cl and has a molar mass of 0.783 kg mol^{-1}. Identify the compounds A and B.

293. Ruthenium forms a complex $RuClH(Et_2PCH_2CH_2PEt_2)_2$ which, when dissolved in $CHCl_3$, gives a proton n.m.r. absorption at $\tau = 31.8$ which is a symmetrical quintet with relative intensity distribution 1 : 4 : 6 : 4 : 1. Suggest a structure for the complex.

294. The complex $Pt(NH_3)_2(NO_3)_2$ has an α-form and a β-form. The α-form reacts with oxalic acid to give $Pt(NH_3)_2C_2O_4$ but the β-form gives $Pt(NH_3)_2(C_2O_4H)_2$ with the same reagent. Explain why this is so. What physical method could be used to distinguish between the α- and β-forms?

295. The complex

Cl
Ph Et_2P — Ir — Cl
Ph Et_2P — Ir — CO
Cl

which has a dipole moment of 9.65 D and a carbonyl absorption at 2100 cm^{-1} in the infrared spectrum, reacts with ethanol to give a rearranged monomeric product of formula $HIr(PEt_2Ph)_2Cl_2CO$ with a dipole moment of only 1.2 D, a carbonyl absorption near 2100 cm^{-1} and an additional i.r. absorption at 2008 cm^{-1}. Its n.m.r. spectrum includes a 1:2:1 triplet centred on $\tau = 19.0$ which is absent in the starting material. Suggest a structure for the product.

296. $IrCl_3$ reacts with $(C_6H_5)_3P$ in dimethylformamide and methanol to give a complex **A** with the analysis C, 56.9; H, 3.8; Cl, 4.6; Ir, 24.7; P, 7.9% and an observed molar mass of 0.780 kg mol^{-1}. **A** shows a single, intense, infrared absorption at 1950 cm^{-1}. The ^{31}P n.m.r. spectrum shows one peak, and the only proton n.m.r. absorption occurs at $\tau = 2.8$. The complex reacts with sodium amalgam in tetrahydrofuran under 2×10^5 Pa of CO to give an anionic complex **B** which in turn reacts with $(CH_3)_2SnCl_2$ to give a neutral complex **C**. **C** shows a single, intense, infrared peak at 2000 cm^{-1}, two absorption regions in the proton n.m.r. ($\tau = 2.58$, complex, intensity 5; $\tau = 8.96$, triplet with small coupling, intensity 1) and a single peak in the ^{31}P n.m.r. spectrum. **C** contains C, 43.1; H, 2.9; Ir, 31.3; Sn, 9.7; P, 5.1% and has a molar mass of 1.23 kg mol^{-1}. Suggest possible structures for **A**, **B** and **C**.

297. Ruthenium forms a series of complex hydrides with the general formula $RuXH(PEt_2CH_2CH_2PEt_2)_2$ in which X = I, Br, Cl, SCN, NO_2, CN. They are diamagnetic and give non-conducting solutions in nitrobenzene. The infrared spectrum of the compound in which X = Cl shows a strong absorption at 1938 cm^{-1}, for X = Br, 1945 cm^{-1} for X = SCN,

1919 cm^{-1} and for X = CN, 1803 cm^{-1}. The dipole moments are in the range 3.8—5.0 D (for comparison that of *trans*-$PtHCl(PEt_3)_2$ is 4.2 D). The proton n.m.r. spectra all show a 1 : 4 : 6 : 4 : 1 quintet at $\tau \sim 27$. Suggest a general structure and comment on the infrared data.

298. Calculate the solubility of $CuSO_4$ at 288 K from the following experimental results. A sample of saturated solution which weighed 5.092 g was made up to exactly 100 cm^3 with water. A 10.0 cm^3 portion of the solution, treated with an excess of KI, liberated sufficient iodine to oxidise 14.0 cm^3 of a solution of sodium thiosulphate containing 9.126 kg m^{-3} of $Na_2S_2O_3 \cdot 5\,H_2O$.

299. Suggest a reason why pβ_2 for $Ag(NH_3)_2^+$ is 7.2 whereas pβ for $Ag(en)^+$ is only 6.0.

300. $E°$, Au^{3+}/Au^+ = 1.2 V and $E°$, VO_2^+/VO^{2+} = 1.2 V. Suggest experimental conditions in which VO_2^+ can be used to oxidise Au^+ to Au^{3+}.

301. Account for the difference between the magnetic properties of $CuSO_4 \cdot 5\,H_2O$ and those of $Cu(OOCCH_3)_2 \cdot H_2O$. The former has $\mu_{eff} = 1.95\ \mu_B$ independent of temperature but the latter has $\mu_{eff} = 1.39\ \mu_B$ at room temperature and $\mu_{eff} \sim 0$ at 130 K.

302. The solubility of AgCl (weight for weight of water) is 2.41×10^{-7} at 291 K. Calculate the solubility product at that temperature.

303. Explain why the electrical conductance of zinc oxide is increased when the solid is heated in a vacuum.

304. A solution of $Cd(NO_3)_2$ containing 100 mol m^{-3} was added to an equal volume of a solution containing 1.00 kmol m^{-3} KCN. Find by calculation whether any CdS will be precipitated from the solution by the addition of sufficient soluble sulphide to make the concentration of S^{2-} 1.00 mol m^{-3}. The solubility product of CdS is 7×10^{-25} and the overall stability constant of $Cd(CN)_4^{2-}$ is 1.3×10^{17}.

305. For CdS, K_s, the solubility product, is 7×10^{-25}. Will CdS be precipitated from a solution of a cadmium salt containing 1.00 mol m^{-3} Cd^{2+} saturated with H_2S gas (equilibrium concentration 100.0 mol m^{-3} H_2S) at pH = 0? The first and second dissociation constants of H_2S are 10^{-7} and 10^{-14} respectively. What will happen at pH 2?

Answers to Problems

1. (a) C_s, (b) C_{2v}, (c) C_{2h}, (d) D_{2h}, (e) D_{3h}, (f) D_{6h}, (g) D_{4d}, (h) $D_{\infty h}$, (i) C_{2v}, (j) D_{2h}, (k) D_{3h}, (l) T_d, (m) C_{3v}, (n) $C_{\infty v}$.

2. (a) C_3, (b) D_2, (c) C_{4h}, (d) D_{6h}, (e) D_{5d}, (f) D_{5h}.

3. (a) $B(OH)_3$, (b) Al_2Cl_6, (c) BCl_3, (d) $H_3B \cdot NH_3$, (e) BH_4^-, (f) $AlCl_6^{3-}$, (g) $B_{12}H_{12}^{2-}$.

4. $A_r = 6.941$.

5. $A_r = 24.31$.

6. 4.54 pJ per atom.

7. 1.40 pJ per nucleon.

8. 27 fJ.

9. 275 fJ (1.71 MeV).

10. Exchange between fluorine positions. At 73 K in solution two fluorine environments.

11. SbF_5 is a viscous liquid which must be associated. Suggests association through *cis* fluorine bridges:

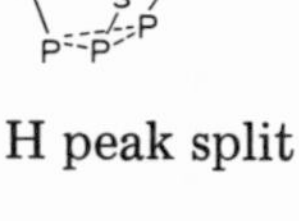

12. Two phosphorus environments in ratio 3:1,

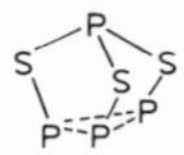

13. H peak split at 300 K by ^{205}Tl.

14. (a) Two equally intense peaks, (b) three equally intense peaks, (c) two equally intense peaks, (d) seven equally intense peaks, (e) four equally intense peaks.

15. (a) One peak, (b) seven equally intense peaks, (c) four equally intense peaks, (d) two pairs of equally intense peaks, (e) two equally intense peaks.

16. Higher value correlates with increasing polarity of I—X bond: I—Cl > I—I, I—CI_3 > I—CHI_2 > I—D > I—CH_3, Li—I > K—I.

17. Isomer shift correlates with the s electron density at the Mössbauer atom: I^{VII} (s^0p^0) < I^{-I} (s^2p^6) < I^V (s^2p^0), Sn^{II} (s^2p^0) > Sn^0 (sp^3) > Sn^{IV} (s^0p^0), Cp_2Sn is Sn^{II}, Ph_2Sn is polymeric (Sn—Sn bonds). ∴ Sn^{IV}.

18. 1.325×10^{-25} J.

19. 599.6 nm.

20. 242.5 pm.

21. 1.822×10^{-20} J.

22. 16.02 aJ.

23. 5.93×10^6 m s^{-1}.

24. 23.07 nN.

25. −1.153 aJ.

26. 468.3 nm.

27. 2.2 *e*.

28. 8.1 *e*.

29. Large exchange stabilisation.

30. 4.579 MJ mol^{-1}.

31. 89 kJ mol^{-1}.

32. 4.32×10^{-5}.

33. 4.56×10^{-23} A m^2 = 4.91 μ_B.

34. 5.92 μ_B.

35. $\frac{4}{3}$.

36. 4.14 μ_B.

37. (a) 3, (b) 1, (c) $\frac{1}{2}$, (d) 1.

38. 280 kJ mol^{-1}.

39. 274 kJ mol^{-1}.

40. 416 kJ mol^{-1}.

41. 2.74.

42. 2.21.

43. 224 pm.

44. 663 kJ mol^{-1}.

45. 3.964 MJ mol^{-1}.

46. 762 kJ mol^{-1}.

47. 35 kJ mol^{-1}.

48. 193 pm.

49. 174 pm.

50. 54 pm.

51. 1.74 MJ mol^{-1}.

52. 228 pm.

53. —1.075 MJ mol^{-1}.

54. (a) Distorted tetrahedron (trigonal bipyramid with one equatorial atom missing), (b) linear, (c) tetrahedral, (d) tetrahedral, (e) tetrahedral, (f) V-shaped, (g) linear, (h) V-shaped, (i) tetrahedral, (j) tetrahedral, (k) as (a), (l) V-shaped, (m) triangular, (n) octahedral.

55. (a) 4.27×10^{-47} kg m^2, (b) 160.4 pm.

56. 3.4 cm^{-1}.

57. 1.87×10^3 N m^{-1}.

58. 1.80×10^{-30} C m = 0.54 D.

59. 3.42×10^{-41} F m^2.

60. (a) 3.53×10^{-30} C m. (b) The C atom has a lone pair in a hybrid orbital directed away from the O atom.

61.

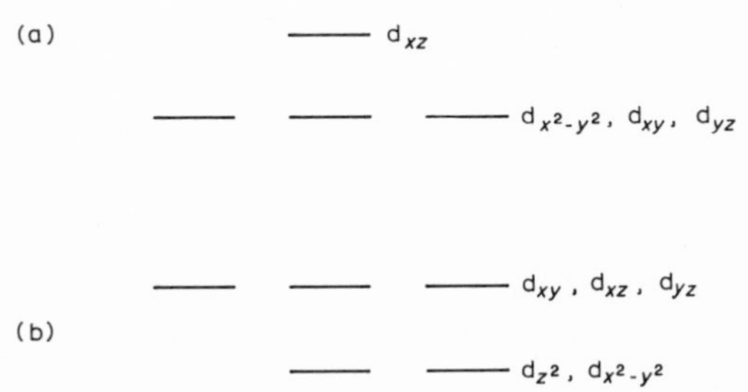

(a)

d_{z^2}

d_{xz}

$d_{x^2-y^2}$, d_{xy}, d_{yz}

(b)

d_{xy}, d_{xz}, d_{yz}

d_{z^2}, $d_{x^2-y^2}$

62. 204 kJ mol^{-1}.

63. (a) 11, (b) 9.

64. d^2, d^5 and d^7.

65. See 6.3.4.

66. See 6.3.6.

67. 15.9 $m^2\ mol^{-1}$.

68. The aquo complex is octahedral, the dmf complex is tetrahedral.

69. Antisymmetric to C_4^z, symmetric to the others.

70. 214 pm.

71. 240 pm, 168 pm and 275 pm. The ratio is $1{:}1/\sqrt{2}{:}2/\sqrt{3}$, the relative spacings in the NaCl lattice.

72. 8.99 $g\ cm^{-3}$.

73. $6.04 \times 10^{23}\ mol^{-1}$.

74. 5.964 MJ mol^{-1}.

75. −6 kJ mol^{-1}.

76. (a) 1.6×10^{-17}, (b) 9.2×10^{-6}.

77. See 7.2.23.1.

78. $r_{I^-} = 209$ pm, $r_{Tl^+} = 155$ pm, $r_{Cl^-} = 170$ pm, $r_{Cs^+} = 186$ pm.

79. See 7.2.18.

80. Interstitial spaces are too small.

81. MX will disproportionate to MX_2 and M but MX_2 is thermally stable.

82. The cations Li^+, Mg^{2+}, Fe^{3+} and Ti^{4+} are similar in size and will replace one another as long as total positive charge balances total negative charge.

83. (a) +1533 kJ mol^{-1}, (b) −243 kJ mol^{-1}, (c) −876 kJ mol^{-1}.

84. (a) 1.562 V, (b) 0.59 V, (c) 0.459 V.

85. +6, −2, −1, +6, +2, $+2\frac{1}{2}$.

86. +3, +7, +2, +4, +6, +7, 0, +1.

87. (a) $IO_3^- + 5\ I^- + 6\ H^+ = 3\ I_2 + 3\ H_2O$.
(b) $IO_3^- + 2\ I^- + 6\ H^+ + 3\ Cl^- = 3\ ICl + 3\ H_2O$.

88. $K = 0.32$.

89. +0.13 V.

90. −0.83 V.

91. 1.30 V.

92. The reactions are:

$$5\,I^- + IO_3^- + 6\,H^+ = 3\,H_2O + 3\,I_2$$
$$3\,I_2 + 6\,S_2O_3^{2-} = 6\,I^- + 3\,S_4O_6^{2-}$$

The $[H^+]$ remains low until all the IO_3^- is removed. [HCl] = 200 mol m^3.

93. 1.52×10^{-2} mol IO_4^- and 1.31×10^{-2} mol IO_3^-.

94. $x = 2$.

95. See Table 8.1.

96. —0.25 V.

97. a, c, d and e:

$$Cr_2O_7^{2-} + 14\,H^+ + 6\,Br^- = 2\,Cr^{3+} + 7\,H_2O + 3\,Br_2$$
$$Cr_2O_7^{2-} + 14\,H^+ + 3\,Hg_2^{2+} = 2\,Cr^{3+} + 7\,H_2O + 6\,Hg^{2+}$$
$$Cr_2O_7^{2-} + 14\,H^+ + 3\,Cu = 2\,Cr^{3+} + 7\,H_2O + 3\,Cu^{2+}$$
$$Cr_2O_7^{2-} + 5\,H^+ + 3\,HNO_2 = 2\,Cr^{3+} + 3\,NO_3^- + 4\,H_2O$$

f, g, h:

$$HNO_2 + H_2O_2 = H_2O + 2\,H^+ + NO_3^-$$
$$HNO_2 + Br_2 + H_2O = 2\,Br^- + 2\,H^+ + NO_3^-$$
$$5\,HNO_2 + 2\,MnO_4^- + H^+ = 5\,NO_3^- + 2\,Mn^{2+} + 3\,H_2O$$

98. Fe^{3+} oxidises I^- to I_2.

99. E^0, $As^V/As^{III} > E^0$, I_2/I^- at pH 0 but $< E^0$, I_2/I^- at high pH.

100. (a) No reaction, (b) hydrogen is liberated.

101. $3\,Fe^{2+} + NO_3^- + 4\,H^+ = 3\,Fe^{3+} + 2\,H_2O + NO$.

102. (a) Zn, (b) Sn^{2+}, (c) Fe^{2+}

103. —0.04 V.

104. $H_2PO_4^-$, SO_4^{2-}, S^{2-}, $Cr(H_2O)_5OH^{2+}$, $HAsO_4^{2-}$.

105. $C_2H_5NH_3^+$, H_5IO_6, H_3BO_3, VO^{2+}.

106. H_3O^+ is similar in size to NH_4^+ and K^+, and produces a similar ionic lattice.

107. pH = 2.85.

108. pH = 7.6.

109. $pK_a = 9.0$.

110. pH = 8.7.

111. pH = 6.9.

112. pK_a is about 4.8.

113. $HNO_3 + 2\,H_2SO_4 = NO_2^+ + 2\,HSO_4^- + H_3O^+$
$H_3BO_3 + 6\,H_2SO_4 = B(HSO_4)^- + 2\,HSO_4^- + 3\,H_3O^+$
$7\,I_2 + HIO_3 + 8\,H_2SO_4 = 5\,I_3^+ + 8\,HSO_4^- + 3\,H_3O^+$

114. See 9.13.

115. SO_2 ionises to give SO_3^{2-} ions but evidently not SO^{2+} ions.

116. The H—I bond is weak compared with the H—F bond.

117. See 10.1.2, 10.3 and 11.

118. See 11.1.

119. See 10.4 and 11.

120. See 11.6, 11.7, 11.8, 11.9.

121. Stronger hydrogen-bonding.

122. Salt-like $[(NH_3)_2BH_2]^+BH_4^-$ is converted to borazine.

123. (a) 50.4 kJ mol^{-1}, (b) 46.2 kJ mol^{-1}.

124. $NH_3 \sim 14$ kJ mol^{-1}, $H_2O \sim 18.5$ kJ mol^{-1}. H-bonding is stronger in H_2O.

125. 5.32 MJ mol^{-1}.

126. High zero-point energy.

127. $A(Xe) = 92$ kJ mol^{-1}, $H_f(CsXe) = -226$ kJ mol^{-1}.

128. See 12.6.

129. $\Delta H_f(XeCl_2) = +82$ kJ mol^{-1}.

130. 25 mole % XeF_2 and 75 mole % XeF_4.

131. Term is applied only to solids containing one kind of atom.

132. Li^+ polarises H_2O molecules slightly.

133. (a) ~280 K; (b) CsCl type; (c) FrO_2; (d) H_2 and $FrNH_2$; (e) $FrBPh_4$, $FrClO_4$, and Fr_2PtCl_6.

134. $\Delta H = +1$ kJ mol^{-1}. Equilibrium composition = 47% K, 53% Na.

135. $\Delta H = -749$ kJ mol^{-1}.

136. Increasing ionic size reduces polarising power.

137. 797 GJ mol^{-1} = 8.27 MeV.

138. **A** = $Pb(NO_3)_2$, B = Pb, **C** = N_2, **D** = $Pb(NO_3)_2$, **E** = NO, **F** = $PbCl_2$, **G** = Mg_3N_2, **H** = $Mg(OH)_2$, **J** = NH_3, **K** = PbS, **L** = HCl and **M** = S.

139. E^0, $Mg^{2+}/Mg = -2.38$ V.

140. The lattice energy calculation from thermodynamic data agrees with the Born—Meyer calculation based on the formulation $Mg^{2+}O^{2-}$.

141. $\Delta H^0(\text{CaCl}) = -154$ kJ mol^{-1}, $\Delta H^0(\text{CaCl}_2) = -2946$ kJ mol^{-1}. For $2\,\text{CaCl} = \text{CaCl}_2 + \text{Ca}$, $\Delta H^0 = -2638$ kJ mol^{-1}. ∴ CaCl disproportionates.

142. 3.3×10^{-6} (as mass/mass of water).

143. 90 J K^{-1} mol^{-1}.

144. (a) +69 kJ mol^{-1}, (b) +216 kJ mol^{-1}.

145. Be, 1.36; Mg, 1.21; Ca, 1.05; Sr, 1.00; Ba, 0.96.

146. 1850 K.

147. Optical isomerism.

148. In B_4H_{10} there is coupling of ^{11}B with (a) protons of BH_2 groups and (b) protons of BH groups but ^{11}B—^{10}B coupling is not important.

149. Volatility of BX_3 decreases with molecular weight. BX_3 are covalent and monomeric whereas AlF_3 is ionic and the chloride and bromide are dimers.

150. Planar (D_{3h}) molecule, few bond-pair repulsions, and tendency for back-donation of electrons from F to B.

151. Four-co-ordinate B where N of amine donates electrons.

152. B in BBr_3 more electron deficient than in BF_3.

153. Steric hindrance between 2,6-dimethyl groups and BMe_3.

154. B in BH_3 more electron deficient than in BMe_3.

155. F gives maximum co-ordination number and steric factors.

156. AlF_3 is ionic, HF not a good source of F^-, NaF gives Na_3AlF_6.

157.

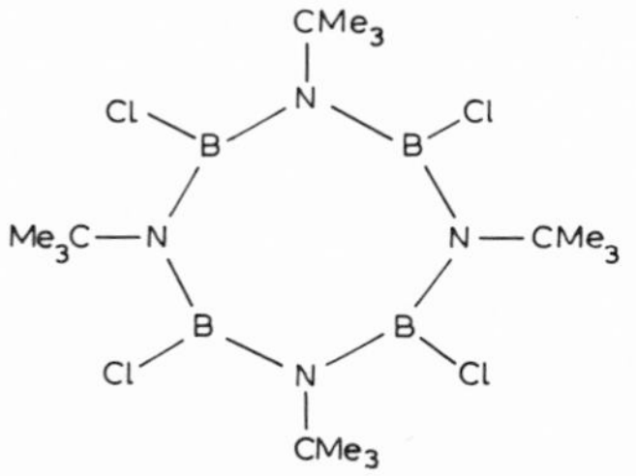

CMe3
N
Cl B B Cl
Me3C—N N—CMe3
B B
Cl N Cl
CMe3

158. E(B—N) = 447 kJ mol^{-1}.

159. The molecule is fluxional (see 18.2.5).

160. It contains In^{III} and In^{I} but not In^{II}.

161. (a) +1.25 V, (b) $K = 10^{-36}$.

162. See 4.2.8.2.

163. 0.156 MeV ≡ 15 GJ mol^{-1}.

164. 1100 years.

165. Availability of d orbitals on Si to expand co-ordination number.

166. d_π—p_π Si—N bonding removing pyramidal shape and no lone pair of electrons.

167. Graphite layer sp^2 structure, diamond sp^3 tetrahedral.

168. p_π—p_π C=O bonding in free molecule of CO_2, tetrahedral. SiO_4 units in crosslinked SiO_2 structure.

169. The drying process is not a chemical change but a physical absorption.

170. See Chapter 19.

171. A =

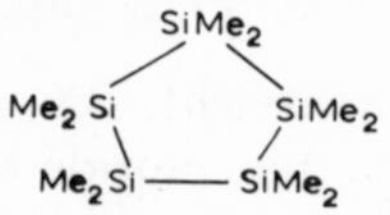

SiMe$_2$
Me$_2$Si SiMe$_2$
Me$_2$Si —— SiMe$_2$

B = $Br(SiMe_2)_5Br$, C = $Me(SiMe_2)_7Me$.

172. NO acts as a 3-electron donor.

173. CO groups in $Mo(CO)_6$ are all terminal. NH_3 as electron donor supplies electrons to Mo for back-donation to π^* orbitals of CO. PH_3 is a weaker electron donor than NH_3.

174. Back-donation to π^* orbitals of CO is weaker the more positive the charge number of the metal.

175. The first is the *trans* form, the second the *cis* form.

176. (a) 2, (b) 3.

177.

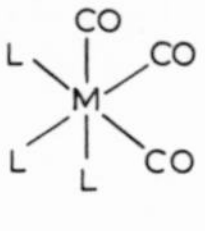

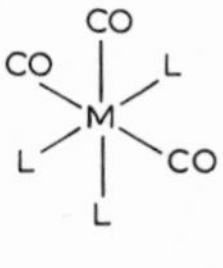

M with CO, CO, CO, L, L, L (fac)

C_{3v}	E	$2C_3$	$3\sigma_v$
Γ_{CO}	3	0	1

$A_1 + E$
R and i.r. active

M with CO, CO, CO, L, L, L (mer)

C_{2v}	E	C_2	σ_{xz}	σ_{yz}
Γ_{CO}	3	1	3	1

$2A_1 + B_1$
R and i.r. active

178. Kinetic stability versus thermodynamic stability.

179. See 18.2.5.

180. See 18.3.2.

181. d_π—p_π interaction in Si—O—Si but not in Ge—O—Ge.

182. The oxidation state (II) becomes more important down the group. See 19.1.

183. E(Ge—Ge) = 342 kJ mol^{-1}. Intrinsic energy = 467 kJ mol^{-1}.

184. —0.136 V.

185.

X = Ph_2Ge bridging two Co atoms [Co–Co bonded, each Co bearing $(CO)_3$, with a bridging CO]

Y = $Ph_2Ge[Co(CO)_4]_2$

186. NO_2^+ has no residual electrons on N, whereas NO_2 has one (ONO = 132°) and NO_2^- has two (ONO = 115°).

187. Bond order is 3 in N_2 but in NO it is only 2.5, with the extra electron in π^* and easier to remove.

188. 1876 cm^{-1} corresponds to NO with bond order 2.5. 2200 cm^{-1} corresponds to NO^+ with bond order 3.

189. BN is isoelectronic with C_2. Unit structure has B alternating in a ring with N or P, with angles of 120° implying sp^2 hybridisation at B. The distance between the sheets is greater than the interatomic distances in the rings.

190. $x = 2, y = 1$. $2\ Ti^{3+} + NH_3OH^+ + 2\ H^+ = 2\ Ti^{4+} + NH_4^+ + H_2O$.

191. The co-ordination number of 3 for oxygen and the planar structure reflect the stabilising influence of p_π—p_π bonding in NO_3^-. Tetrahedral PO_4^{3-} occurs because d_π—p_π bonding is possible for an element of Period 3.

192. Electronegative F in pyramidal NF_3 dominates, hence NF_3 has no donor properties.

193. Electronegativities suggest polarisation P^+—Cl^- and N^-—Cl^+.

194. Angle in NH_3 close to tetrahedral angle allowing for lone-pair—bond-pair repulsion. Effect of lone pair greater in NF_3 because electrons of N—F bonds lie near F atoms and repel one another less. Angle in PH_3 close to 90° suggesting large p orbital contribution.

195. See 20.10.4.5.

196.

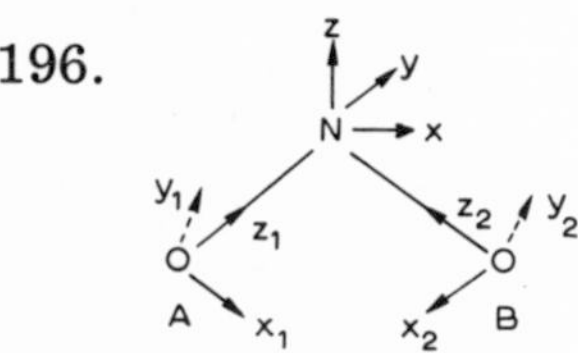

Consider how the orbitals of N (2s, $2p_x$, $2p_y$, $2p_z$) and linear combinations of the two O atomic nuclei transform according to symmetry operations of the point group (C_{2v}) for NO_2:

C_{2v}	E	C_2	σ_{xz}	σ_{yz}	
A_1	1	1	1	1	
A_2	1	1	−1	−1	
B_1	1	−1	1	−1	
B_2	1	−1	−1	1	
N_{2s}	1	1	1	1	A_1
N_{2p_x}	1	−1	1	−1	B_1
N_{2p_y}	1	−1	−1	1	B_2
N_{2p_z}	1	1	1	1	A_1
$O(2s_A + 2s_B)$	1	1	1	1	A_1
$O(2s_A - 2s_B)$	1	−1	1	−1	B_1
$O(2p_{x_A} + 2p_{x_B})$	1	−1	1	−1	B_1
$O(2p_{x_A} - 2p_{x_B})$	1	1	1	1	A_1
$O(2p_{y_A} + 2p_{y_B})$	1	−1	−1	1	B_2
$O(2p_{y_A} - 2p_{y_B})$	1	1	−1	−1	A_2
$O(2p_{z_A} + 2p_{z_B})$	1	1	1	1	A_1
$O(2p_{z_A} - 2p_{z_B})$	1	−1	1	−1	B_1

Clearly the four N atomic orbitals can combine with four combinations of O atomic nuclei with similar symmetry. These eight bonding molecular orbitals together with A_2 non-bonding orbital will contain 17 electrons leaving three empty antibonding molecular orbitals. The various combinations can be illustrated from the above transformations as in 4.1.1.7.

197. Electronegative fluorines in NF_3 balance the electron distribution round N to such an extent that the dipole moment is only 0.2 D. For NH_3 it is 1.5 D which would imply that the bond polarity of N—H is reversed. NCl_3, unlike NF_3, is endothermic.

198. Using Linnett non pairing structures; e.g. NO_2 has 17 electrons.

199. See 20.4.2.

200.

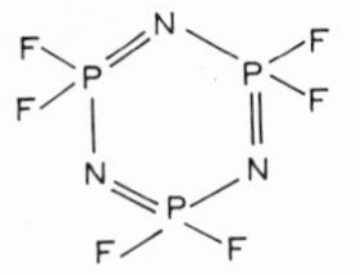

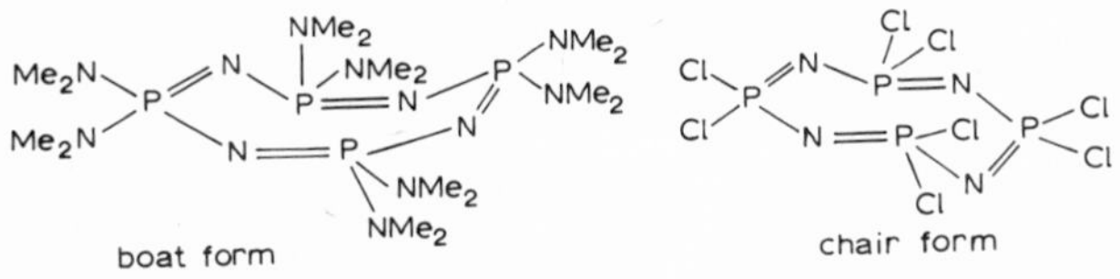

boat form   chair form

These structures are illustrated with localised double bonds to clarify the valency situation. However each P—N bond in each particular ring is the same length and their values reflect the relative extent of d_π—p_π bonding.

201. The ring in cyclotriphosphatriazene is planar, that in cyclotetraphosphatetrazene is not.

202. See basic definitions of these terms in Chapter 4.

203. In order to dissociate an acid it is not only necessary to place a negative charge on the element but also to break the bond to hydrogen.

204. Because of high S—F bond strength, co-ordination saturation and lack of polarity of the molecule.

205. Te is capable of 8 co-ordination and hence a reagent could attack in this way; S has a co-ordination maximum of 6.

206. There is no exchange between the two S atoms in the $S_2O_3^{2-}$ ion.

207. See Chapter 22, but also note the structural units are O_2 and S_8.

208. Similar to Question 196. SO_3 is a planar molecule with point group D_{3h}.

209. $S_4N_4 \rightarrow S_4N_4^- \rightarrow S_4N_4^{2-} \rightarrow S_4N_4^{3-} \rightarrow S_4N_4^{4-}$.

210. (a) $2\ SO_2 + S_2Cl_2 + 2\ H_2O = S_4O_6^{2-} + 2\ Cl^- + 4\ H^+$
(b) $S_4O_6^{2-} + 7\ IO_3^- + 7\ HCl + 3\ H^+ = 7\ ICl + 8\ HSO_4^- + H_2O$

211. Gradual increase of effective nuclear charge modified by occupancy of e_g orbitals lying opposite to the oxygens.

212. See Chapter 23.

213. n-type, ZnO and WO_3; p-type NiO and Co_3O_4; insulators, MgO and Al_2O_3.

214. Many peroxides contain some superoxide ions (O_2^-) which are paramagnetic (i.e. have an unpaired electron).

215. Stability correlates with the electropositive character of the metal. Largest metal ions polarise oxygen anions least.

216. $H_2O_2 + 2\,H^+ + 2\,e \rightarrow 2\,H_2O$
$HO_2^- + H_2O + 2\,e \rightarrow 3\,OH^-$
$H_2O_2 \rightarrow O_2 + 2\,H^+ + 2\,e$

The last occurs only with very strong oxidising agents such as MnO_4^-.

217. Similar to Question 196. O_3 has C_{2v} symmetry. Linnett structures would be:

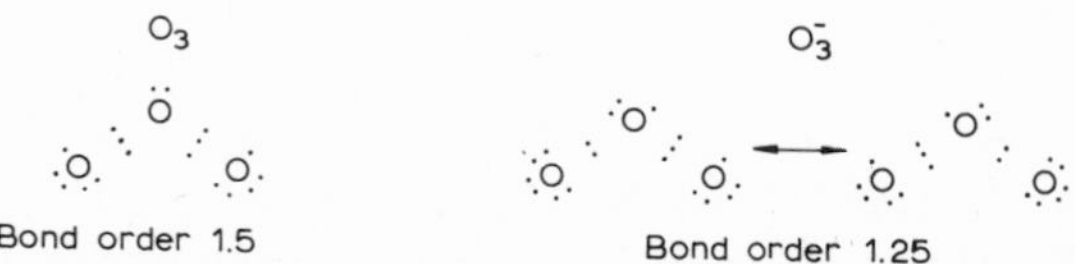

218. This energy is offset by large lattice energy terms.

219. See Chapter 23.

220. The low dissociation energy of F_2 arises from the repulsion between unpaired electrons on the two atoms which are themselves small.

221. Presumably the presence of unpaired electrons would result in further reaction.

222. Brown colour in polar solvents due to 'charge transfer' complexes of the type I_2 .... Solvent. Non-polar solvents show no such interaction.

223. $2\,BrF_3 \rightleftharpoons [BrF_2]^+[BrF_4]^-$
$[BrF_2]_2^+[SnF_6]^{2-} + 2\,KBrF_4 \rightarrow K_2SnF_6 + 4\,BrF_3$

224. See 25.5.

225. **A** = BrO_2, **B** = BrO_2F, **C** = BrF_3, **D** = $KBrF_4$.

226.
F
F F
S
F OF
F

227. $(CF_3)_2N{-}S{-}CF_3$.

228. **M** = Sn, **A** = $SnCl_4$, **B** = $SnCl_2$, **C** = $SnPh_4$, **D** = $SnPh_2Cl_2$, **E** = SnS, **F** = $(NH_4)_4SnS_2$.

229. A quintuplet: relative intensities 1:4:6:4:1 and a doublet, 32:32.

230. N is poor electron donor because F atoms draw electrons from it.

231. Increasing ionic character.

232. $[ClO^-] = 375$ mol m^{-3}, $[ClO_3^-] = 235$ mol m^{-3}, $[Cl^-] = 190$ mol m^{-3}.

233. $2(SCN)_2 + 8\,OH^- \rightarrow 3\,SCN^- + 3\,CN^- + SO_4^{2-} + 4\,H_2O$

has analogies with

$X_2 + 2\,OH^- \rightarrow X^- + XO^- + H_2O$

$$3\,XO^- \rightarrow 2\,X^- + XO_3^-$$
$$4\,XO_3^- \rightarrow X^- + 3\,XO_4^-$$

234. (a) +456 kJ mol^{-1}, (b) —374 kJ mol^{-1}, (c) —404 kJ mol^{-1}. Both FeF_2 and FeF_3 are thermally stable; MgF_2 is stable but MgF is unstable to disproportionation.

235. Consider the effect of removing one ligand from ML_6 with O_h symmetry. The degeneracy in the d-orbital splitting diagram would be reduced.

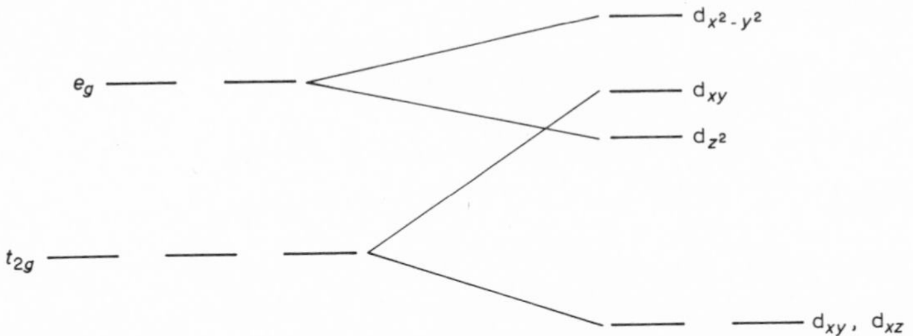

236. Points to consider include radii, oxidation states, magnetic and spectroscopic properties, structural differences. See relevant chapters.

237. The methyl groups attached to the nitrogens sterically hinder co-ordination.

238. Stabilities vary with reciprocals of ionic radii (Irving—Williams series).

239. Cl is replaced from co-ordination sphere by H_2O.

240. CaOx $\rightleftharpoons$ Ca^{2+} + Ox^{2-} moves to right as Ca^{2+} is removed as soluble Ca edta^{2-} ion.

241. (a) 5.90 μ_B, (b) 2.83 μ_B, (c) 3.88 μ_B.

242. $[CoClH_2Oen_2]Cl_2$

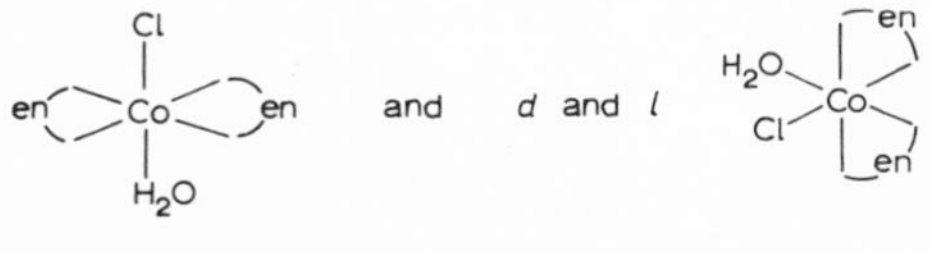

and *d* and *l*

243. 0.0116 m^2 mol^{-1}.

244. $p\beta_2 = 18.5$.

245. Polar positions relative to the plane of co-ordination are overlapped by ethyl groups which inhibit nucleophilic attack.

246. See 29.13.2.

247. (a) Both octahedral and both low-spin. (b) Both octahedral but oxidation converts high-spin complex to low-spin.

248. See 29.7.1.

249. (a) inert (d^3), (b) labile (d^2), (c) labile (d^5), (d) inert (low-spin d^6), (e) labile (high-spin d^6), (f) inert (d^8).

250. Cl acts as bridging atom.

251. An inner-sphere activated complex is formed at some stage with a structure $(O_2Cl—O—SO_2)^-$.

252. They are respectively f^0, f^7 and f^{14} ions for which $L = 0$.

253. The Mn^{2+} values correspond with d^5 (high-spin) and d^5 (low-spin). In Sm^{3+} there is strong L—S coupling.

254. $^{205}_{81}Tl + ^{1}_{1}H \rightarrow 2\ ^{103}_{41}Nb$. Will decay by β^- emission.

255. Both are *cis*-isomers.

256. Eight lines about equally spaced and of equal intensity.

257. $x = 2$ and $y = 2$.

258. b, d, f, g, and $j = 1$; a, c, and $k = 2$; l and $h = 5$. $[V] = 86.9$ mol m^{-3}. Oxidation state = +2.

259. All contain metal cluster ions (See 34.4).

260. **A** = $H_2MoCl_{14} \cdot 8\ H_2O$, **B** = $Mo_6Cl_8(OH)_4$. Cl in complex ion is replaced by OH^-.

261. **A** is $K_3Cr_2Cl_9$. Anion $Cr_2Cl_9{}^{3-}$ consists of two $CrCl_6$ octahedra sharing a face.

262. **A** = CO **B** = CH_2—$Mo(CO)_3$

C = $[Mo(CO)_3]^+$ BF_4^-

263. Cr^I is present in a low-spin d^5 state.

264. Back-donation to the π^* orbitals of the CO ligands is reduced by increasingly strong electronegative groups attached to the metal.

265. **A** = CO, Mn, NO, NO, NO **B** = $Ph_3P—Mn(NO)_3$

266. $3\ MnO_4{}^{2-} + 2\ CO_2 = 2\ MnO_4{}^- + MnO_2 + 2\ CO_3{}^{2-}$

267. Mn^{III}.
$MnO_4{}^- + 4\ Mn^{2+} + 8\ H^+ = 5\ Mn^{3+} + 4\ H_2O$.

268. There is an Mn—Re bond and 5 terminal CO groups attached to each metal atom.

269. (**A**) $CH_2{=}CH—CH_2Mn(CO)_5$, (**B**) (π-allyl)$Mn(CO)_4$.

270. The Re—Re bond in the carbonyl has a bond order of about one and the CO groups are in equilibrium in positions as far from one another as possible. In the anion the Re—Re bond has a bond order of about 4 (1σ, 2π and 1δ bond). The δ bonding imposes the eclipsed structure.

271. The compound is $Ph_3PH^+(Re_3Br_{10})^-$.

272.

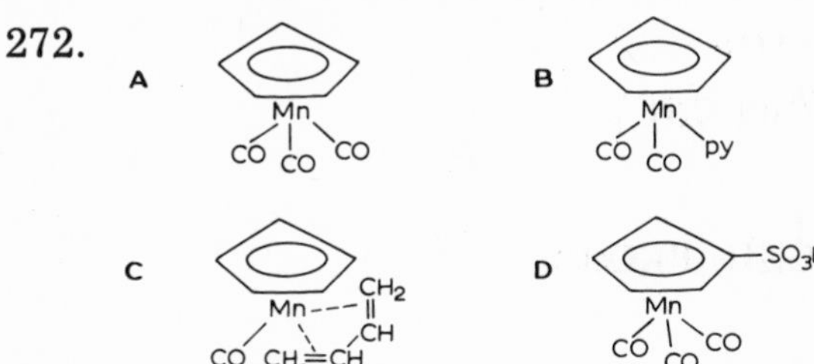

273. The compound has two P atoms and two Cl atoms tetrahedrally arranged around the Ni.

274. 74% Fe and 26% Fe_2O_3.

275. $CoCr_2O_4$ is a normal spinel, $CoFe_2O_4$ is inverse.

276. The planar $Ni(diarsine)_2^{2+}$ ion is oxidised to the octahedral $[NiCl_2(diarsine)_2]^+$ and then to the octahedral $[NiCl_2(diarsine)_2]^{2+}$. The last is diamagnetic.

277. The orbital contribution is variable.

278. The *d* and *l* forms contain *cis*-$[CoCl_2en_2]^+$ ions and the optically inactive form contains *trans*-$[CoCl_2en_2]^+$.

279. (a) Low-spin d^6 but with one hetero-ligand, (b) high spin d^6. There is thus a field gradient around both Fe nuclei.

280. The thermodynamically stable FeF_6^{3-} ion is formed.

281. See 8.6.

282. The molecule has a regular trigonal bipyramidal structure.

283. $[Fe(en)(en)(en)]^{2+}$ $2\,Cl^-$

284. Ligand field stabilisation of d^6 Fe^{2+} ion in octahedral field.

285. **M** = Fe, **A** = $Fe(OH)_2$, **B** = $Fe(OH)_3$, **C** = Fe_2O_3, **D** = Fe_3O_4, **E** = $FeCl_3$, **F** = Na_2FeO_4, **G** = $BaFeO_4$.

286. Red $[CoSO_4(NH_3)_5]Br$, sulphatopenta-amminecobalt(III) bromide. Violet $[CoBr(NH_3)_5]SO_4$, bromopenta-amminecobalt(III) sulphate.

287.

PHPh$_2$
I—Ni—I
PHPh$_2$

288. $\mathbf{A} = HCo(CO)_4$, $\mathbf{B} = Co_2(CO)_8$, $\mathbf{C} = CO$, $\mathbf{D} = LiCo_3(CO)_{10}$.

289. (a) N, (b) $2N$, (c) 27 pm.

290.

$\mathbf{A}$ = $(CO)_3Fe$—CO (three bridging CO)—$Fe(CO)_3$

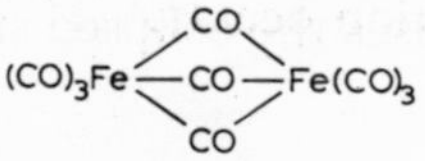

$\mathbf{B} = (CO)_4FePMe_2PMe_2Fe(CO)_4$.

291. $\mathbf{A} = (\pi C_5H_5)Fe(CO)_2Fe(CO)_2(\pi\text{-}C_5H_5)$,
$\mathbf{B} = (\pi\text{-}C_5H_5)_2Fe$.

292. $\mathbf{A} = (C_{14}H_{23}P)_3OsCl_3$, $\mathbf{B} = (C_{14}H_{23}P)_2OsCl_4$.

293.

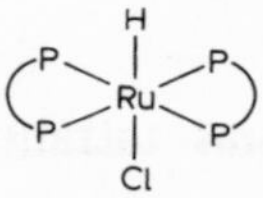

H
P P
Ru
P P
Cl

294. α is *cis* $\xrightarrow{H_2C_2O_4}$ H_3N, H_3N Pt O—CO, O—CO

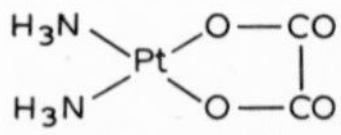

β is *trans* $\xrightarrow{H_2C_2O_4}$ H_3N Pt $O\cdot CO\cdot CO_2H$, $HCO_2\cdot CO\text{-}O$ NH_3

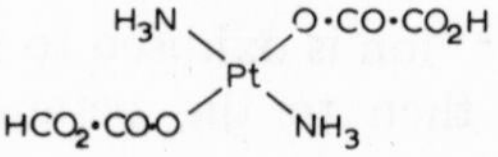

Dipole moment measurements.

295.

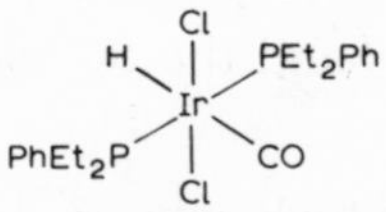

Cl
H PEt_2Ph
Ir
$PhEt_2P$ CO
Cl

296. $\mathbf{A} = (Ph_3P)_2Ir(CO)Cl$, $\mathbf{B} = [Ph_3P\ Ir(CO)_3]^-$,
$\mathbf{C} = Ph_3PIr(CO)_3Sn(Me)_2Ir(CO)_3PPh_3$.

297.

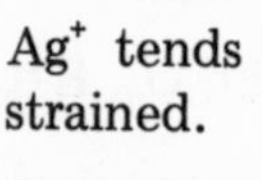

H
P P
Ru
P P
X

Ligands with the strongest *trans*-directing power have greatest weakening effect on the Ru—H bond diametrically opposite.

298. 0.193 (as $w(CuSO_4)/w(H_2O)$).

299. Ag^+ tends to form linear complexes. Hence the 5-membered ring is strained.

300. Strongly acidic conditions.

301. See 38.10.

302. 2.8×10^{-12}.

303. See 23.6.1.

304. CdS is precipitated.

305. At pH 0, no precipitate. At pH 2, precipitation occurs.

Appendix I

Character Tables for Some Chemically Important Symmetry Groups

C_s	E	σ_h		
A'	1	1	T_x, T_y, R_z	x^2, y^2 z^2, xy
A''	1	−1	T_z, R_x, R_y	xz, yz

C_i	E	i		
A_g	1	1	R_x, R_y, R_z	x^2, y^2, z^2 xy, xz, yz
A_u	1	−1	T_x, T_y, T_z	

The C_n groups

C_2	E	C_2		
A	1	1	T_z, R_z	x^2, y^2, z^2, xy
B	1	−1	T_x, T_y, R_x, R_y	xz, yz

C_3	E	C_3	C_3^2		$\epsilon = \exp(2\pi i/3)$
A	1	1	1	T_z, R_z	$x^2 + y^2, z^2$
E	$\left\{\begin{matrix}1\\1\end{matrix}\right.$	$\begin{matrix}\epsilon\\\epsilon^*\end{matrix}$	$\left.\begin{matrix}\epsilon^*\\\epsilon\end{matrix}\right\}$	$(T_x, T_y), (R_x, R_y)$	$(x^2 - y^2, xy), (xz, yz)$

C_4	E	C_4	C_2	C_4^3		
A	1	1	1	1	T_z, R_z	$x^2 + y^2, z^2$
B	1	−1	1	−1		$x^2 - y^2, xy$
E	$\left\{\begin{matrix}1\\1\end{matrix}\right.$	$\begin{matrix}i\\-i\end{matrix}$	$\begin{matrix}-1\\-1\end{matrix}$	$\left.\begin{matrix}-i\\i\end{matrix}\right\}$	$(T_x, T_y), (R_x, R_y)$	xz, yz

C_5	E	C_5	C_5^2	C_5^3	C_5^4		$\epsilon = \exp(2\pi i/5)$
A	1	1	1	1	1	T_z, R_z	x^2+y^2, z^2
E_1	$\{ 1$	ϵ	ϵ^2	ϵ^{2*}	$\epsilon^* \}$	$(T_x, T_y), (R_x, R_y)$	(xz, yz)
	1	ϵ^*	ϵ^{2*}	ϵ^2	ϵ		
E_2	$\{ 1$	ϵ^2	ϵ^*	ϵ	$\epsilon^{2*} \}$		(x^2-y^2, xy)
	1	ϵ^{2*}	ϵ	ϵ^*	ϵ^2		

C_6	E	C_6	C_3	C_2	C_3^2	C_6^5		$\epsilon = \exp(2\pi i/6)$
A	1	1	1	1	1	1	T_z, R_z	x^2+y^2, z^2
B	1	-1	1	-1	1	-1		
E_1	$\{ 1$	ϵ	$-\epsilon^*$	-1	$-\epsilon$	$\epsilon^* \}$	$(T_x, T_y), (R_x, R_y)$	(xz, yz)
	1	ϵ^*	$-\epsilon$	-1	$-\epsilon^*$	ϵ		
E_2	$\{ 1$	$-\epsilon^*$	$-\epsilon$	1	$-\epsilon^*$	$-\epsilon \}$		(x^2-y^2, xy)
	1	$-\epsilon$	$-\epsilon^*$	1	$-\epsilon$	$-\epsilon^*$		

The C_{nv} groups

C_{2v}	E	C_2	$\sigma_v(xz)$	$\sigma_v(yz)$		
A_1	1	1	1	1	T_z	x^2, y^2, z^2
A_2	1	1	-1	-1	R_z	xy
B_1	1	-1	1	-1	T_x, R_y	xz
B_2	1	-1	-1	1	T_y, R_x	yz

C_{3v}	E	$2C_3$	$3\sigma_v$		
A_1	1	1	1	T_z	x^2+y^2, z^2
A_2	1	1	-1	R_z	
E	2	-1	0	$(T_x, T_y), (R_x, R_y)$	$(x^2-y^2, xy), (xz, yz)$

C_{4v}	E	$2C_4$	C_2	$2\sigma_v$	$2\sigma_d$		
A_1	1	1	1	1	1	T_z	x^2+y^2, z^2
A_2	1	1	1	-1	-1	R_z	
B_1	1	-1	1	1	-1		x^2-y^2
B_2	1	-1	1	-1	1		xy
E	2	0	-2	0	0	$(T_x, T_y), (R_x, R_y)$	(xz, yz)

C_{5v}	E	$2C_5$	$2C_5^2$	$5\sigma_v$		
A_1	1	1	1	1	T_z	x^2+y^2, z^2
A_2	1	1	1	-1	R_z	
E_1	2	2 cos 72°	2 cos 144°	0	$(T_x, T_y), (R_x, R_y)$	(xz, yz)
E_2	2	2 cos 144°	2 cos 72°	0		(x^2-y^2, xy)

C_{6v}	E	$2C_6$	$2C_3$	C_2	$3\sigma_v$	$3\sigma_d$		
A_1	1	1	1	1	1	1	T_z	$x^2 + y^2, z^2$
A_2	1	1	1	1	−1	−1	R_z	
B_1	1	−1	1	−1	1	−1		
B_2	1	−1	1	−1	−1	1		
E_1	2	1	−1	−2	0	0	$(T_x, T_y), (R_x, R_y)$	(xz, yz)
E_2	2	−1	−1	2	0	0		$(x^2 - y^2, xy)$

The C_{nh} groups

C_{2h}	E	C_2	i	σ_h		
A_g	1	1	1	1	R_z	x^2, y^2, z^2, xy
B_g	1	−1	1	−1	R_x, R_y	xz, yz
A_u	1	1	−1	−1	T_z,	
B_u	1	−1	−1	1	T_x, T_y	

C_{3h}	E	C_3	C_3^2	σ_h	S_3	S_3^2		$\epsilon = \exp(2\pi i/3)$
A'	1	1	1	1	1	1	R_z	$x^2 + y^2, z^2$
E'	$\{1$	ϵ	ϵ^*	1	ϵ	$\epsilon^*\}$	(T_x, T_y)	$(x^2 - y^2, xy)$
	1	ϵ^*	ϵ	1	ϵ^*	ϵ		
A''	1	1	1	−1	−1	−1	T_z	
E''	$\{1$	ϵ	ϵ^*	−1	$-\epsilon$	$-\epsilon^*\}$	(R_x, R_y)	(xz, yz)
	1	ϵ^*	ϵ	−1	$-\epsilon^*$	$-\epsilon$		

C_{4h}	E	C_4	C_2	C_4^3	i	S_4^3	σ_h	S_4		
A_g	1	1	1	1	1	1	1	1	R_z	$x^2 + y^2, z^2$
B_g	1	−1	1	−1	1	−1	1	−1		$x^2 - y^2, xy$
E_g	$\{1$	i	−1	−i	1	i	−1	−i$\}$	(R_x, R_y)	(xz, yz)
	1	−i	−1	i	1	−i	−1	i		
A_u	1	1	1	1	−1	−1	−1	−1	T_z	
B_u	1	−1	1	−1	−1	1	−1	1		
E_u	$\{1$	i	−1	−i	−1	−i	1	i$\}$	(T_x, T_y)	
	1	−i	−1	i	−1	i	1	−i		

C_{5h}	E	C_5	C_5^2	C_5^3	C_5^4	σ_h	S_5	S_5^7	S_5^3	S_5^9		$\epsilon = \exp(2\pi i/5)$
A'	1	1	1	1	1	1	1	1	1	1	R_z	$x^2 + y^2, z^2$
E_1'	$\{1$	ϵ	ϵ^2	ϵ^{2*}	ϵ^*	1	ϵ	ϵ^2	ϵ^{2*}	$\epsilon^*\}$	(T_x, T_y)	
	1	ϵ^*	ϵ^{2*}	ϵ^2	ϵ	1	ϵ^*	ϵ^{2*}	ϵ^2	ϵ		
E_2'	$\{1$	ϵ^2	ϵ^*	ϵ	ϵ^{2*}	1	ϵ^2	ϵ^*	ϵ	$\epsilon^{2*}\}$		$(x^2 - y^2, xy)$
	1	ϵ^{2*}	ϵ	ϵ^*	ϵ^2	1	ϵ^{2*}	ϵ	ϵ^*	ϵ^2		
A''	1	1	1	1	1	−1	−1	−1	−1	−1	T_z	
E_1''	$\{1$	ϵ	ϵ^2	ϵ^{2*}	ϵ^*	−1	$-\epsilon$	$-\epsilon^2$	$-\epsilon^{2*}$	$-\epsilon^*\}$	(R_x, R_y)	(xz, yz)
	1	ϵ^*	ϵ^{2*}	ϵ^2	ϵ	−1	$-\epsilon^*$	$-\epsilon^{2*}$	$-\epsilon^2$	$-\epsilon$		
E_2''	$\{1$	ϵ^2	ϵ^*	ϵ	ϵ^{2*}	−1	$-\epsilon^2$	$-\epsilon^*$	$-\epsilon$	$-\epsilon^{2*}\}$		
	1	ϵ^{2*}	ϵ	ϵ^*	ϵ^2	−1	$-\epsilon^{2*}$	$-\epsilon$	$-\epsilon^*$	$-\epsilon^2$		

C_{6h}	E	C_6	C_3	C_2	C_3^2	C_6^5	i	S_3^5	S_6^5	σ_h	S_6	S_3		$\epsilon = \exp(2\pi i/6)$
A_g	1	1	1	1	1	1	1	1	1	1	1	1	R_z	x^2+y^2, z^2
B_g	1	−1	1	−1	1	−1	1	−1	1	−1	1	−1		
E_{1g}	1	ϵ	$-\epsilon^*$	−1	$-\epsilon$	ϵ^*	1	ϵ	$-\epsilon^*$	−1	$-\epsilon$	ϵ^*	(R_x, R_y)	(xz, yz)
	1	ϵ^*	$-\epsilon$	−1	$-\epsilon^*$	ϵ	1	ϵ^*	$-\epsilon$	−1	$-\epsilon^*$	ϵ		
E_{2g}	1	$-\epsilon^*$	$-\epsilon$	1	$-\epsilon^*$	$-\epsilon$	1	$-\epsilon^*$	$-\epsilon$	1	$-\epsilon^*$	$-\epsilon$		(x^2-y^2, xy)
	1	$-\epsilon$	$-\epsilon^*$	1	$-\epsilon$	$-\epsilon^*$	1	$-\epsilon$	$-\epsilon^*$	1	$-\epsilon$	$-\epsilon^*$		
A_u	1	1	1	1	1	1	−1	−1	−1	−1	−1	−1	T_z	
B_u	1	−1	1	−1	1	−1	−1	1	−1	1	−1	1		
E_{1u}	1	ϵ	$-\epsilon^*$	−1	$-\epsilon$	ϵ^*	−1	$-\epsilon$	ϵ^*	1	ϵ	$-\epsilon^*$	(T_x, T_y)	
	1	ϵ^*	$-\epsilon$	−1	$-\epsilon^*$	ϵ	−1	$-\epsilon^*$	ϵ	1	ϵ^*	$-\epsilon$		
E_{2u}	1	$-\epsilon^*$	$-\epsilon$	1	$-\epsilon^*$	$-\epsilon$	−1	ϵ^*	ϵ	−1	ϵ^*	ϵ		
	1	$-\epsilon$	$-\epsilon^*$	1	$-\epsilon$	$-\epsilon^*$	−1	ϵ	ϵ^*	−1	ϵ	ϵ^*		

The D_n groups

D_2	E	$C_2(z)$	$C_2(y)$	$C_2(x)$		
A	1	1	1	1		x^2, y^2, z^2
B_1	1	1	−1	−1	T_z, R_z	xy
B_2	1	−1	1	−1	T_y, R_y	xz
B_3	1	−1	−1	1	T_x, R_x	yz

D_3	E	$2C_3$	$3C_2$		
A_1	1	1	1		x^2+y^2, z^2
A_2	1	1	−1	T_z, R_z	
E	2	−1	0	$(T_x, T_y), (R_x, R_y)$	$(x^2-y^2, xy), (xz, yz)$

D_4	E	$2C_4$	$C_2(=C_4^2)$	$2C_2'$	$2C_2''$		
A_1	1	1	1	1	1		x^2+y^2, z^2
A_2	1	1	1	−1	−1	T_z, R_z	
B_1	1	−1	1	1	−1		x^2-y^2
B_2	1	−1	1	−1	1		xy
E	2	0	−2	0	0	$(T_x, T_y), (R_x, R_y)$	(xz, yz)

D_5	E	$2C_5$	$2C_5^2$	$5C_2$		
A_1	1	1	1	1		x^2+y^2, z^2
A_2	1	1	1	−1	T_z, R_z	
E_1	2	2 cos 72°	2 cos 144°	0	$(T_x, T_y), (R_x, R_y)$	(xz, yz)
E_2	2	2 cos 144°	2 cos 72°	0		(x^2-y^2, xy)

D_6	E	$2C_6$	$2C_3$	C_2	$3C_2'$	$3C_2''$		
A_1	1	1	1	1	1	1		x^2+y^2, z^2
A_2	1	1	1	1	−1	−1	T_z, R_z	
B_1	1	−1	1	−1	1	−1		
B_2	1	−1	1	−1	−1	1		
E_1	2	1	−1	−2	0	0	$(T_x, T_y), (R_x, R_y)$	(xz, yz)
E_2	2	−1	−1	2	0	0		(x^2-y^2, xy)

The D_{nh} group

D_{2h}	E	$C_2(z)$	$C_2(y)$	$C_2(x)$	i	$\sigma(xy)$	$\sigma(xz)$	$\sigma(yz)$		
A_g	1	1	1	1	1	1	1	1		x^2, y^2, z^2
B_{1g}	1	1	−1	−1	1	1	−1	−1	R_z	xy
B_{2g}	1	−1	1	−1	1	−1	1	−1	R_y	xz
B_{3g}	1	−1	−1	1	1	−1	−1	1	R_x	yz
A_u	1	1	1	1	−1	−1	−1	−1		
B_{1u}	1	1	−1	−1	−1	−1	1	1	T_z	
B_{2u}	1	−1	1	−1	−1	1	−1	1	T_y	
B_{3u}	1	−1	−1	1	−1	1	1	−1	T_x	

D_{3h}	E	$2C_3$	$3C_2$	σ_h	$2S_3$	$3\sigma_v$		
A_1'	1	1	1	1	1	1		$x^2 + y^2, z^2$
A_2'	1	1	−1	1	1	−1	R_z	
E'	2	−1	0	2	−1	0	(T_x, T_y)	$(x^2 - y^2, xy)$
A_1''	1	1	1	−1	−1	−1		
A_2''	1	1	−1	−1	−1	1	T_z	
E''	2	−1	0	−2	1	0	(R_x, R_y)	(xz, yz)

D_{4h}	E	$2C_4$	C_2	$2C_2'$	$2C_2''$	i	$2S_4$	σ_h	$2\sigma_v$	$2\sigma_d$		
A_{1g}	1	1	1	1	1	1	1	1	1	1		$x^2 + y^2, z^2$
A_{2g}	1	1	1	−1	−1	1	1	1	−1	−1	R_z	
B_{1g}	1	−1	1	1	−1	1	−1	1	1	−1		$x^2 - y^2$
B_{2g}	1	−1	1	−1	1	1	−1	1	−1	1		xy
E_g	2	0	−2	0	0	2	0	−2	0	0	(R_x, R_y)	(xz, yz)
A_{1u}	1	1	1	1	1	−1	−1	−1	−1	−1		
A_{2u}	1	1	1	−1	−1	−1	−1	−1	1	1	T_z	
B_{1u}	1	−1	1	1	−1	−1	1	−1	−1	1		
B_{2u}	1	−1	1	−1	1	−1	1	−1	1	−1		
E_u	2	0	−2	0	0	−2	0	2	0	0	(T_x, T_y)	

D_{5h}	E	$2C_5$	$2C_5^2$	$5C_2$	σ_h	$2S_5$	$2S_5^3$	$5\sigma_v$		
A_1'	1	1	1	1	1	1	1	1		$x^2 + y^2, z^2$
A_2'	1	1	1	−1	1	1	1	−1	R_z	
E_1'	2	2 cos 72°	2 cos 144°	0	2	2 cos 72°	2 cos 144°	0	(T_x, T_y)	
E_2'	2	2 cos 144°	2 cos 72°	0	2	2 cos 144°	2 cos 72°	0		$(x^2 - y^2, xy)$
A_1''	1	1	1	1	−1	−1	−1	−1		
A_2''	1	1	1	−1	−1	−1	−1	1	T_z	
E_1''	2	2 cos 72°	2 cos 144°	0	−2	−2 cos 144°	−2 cos 144°	0	(R_x, R_y)	(xz, yz)
E_2''	2	2 cos 144°	2 cos 72°	0	−2	−2 cos 144°	−2 cos 72°	0		

D_{6h}	E	$2C_6$	$2C_3$	C_2	$3C'_2$	$3C''_2$	i	$2S_3$	$2S_6$	σ_h	$3\sigma_d$	$3\sigma_v$		
A_{1g}	1	1	1	1	1	1	1	1	1	1	1	1		x^2+y^2, z^2
A_{2g}	1	1	1	1	−1	−1	1	1	1	1	−1	−1	R_z	
B_{1g}	1	−1	1	−1	1	−1	1	−1	1	−1	1	−1		
B_{2g}	1	−1	1	−1	−1	1	1	−1	1	−1	−1	1		
E_{1g}	2	1	−1	−2	0	0	2	1	−1	−2	0	0	(R_x, R_y)	(xz, yz)
E_{2g}	2	−1	−1	2	0	0	2	−1	−1	2	0	0		(x^2-y^2, xy)
A_{1u}	1	1	1	1	1	1	−1	−1	−1	−1	−1	−1		
A_{2u}	1	1	1	1	−1	−1	−1	−1	−1	−1	1	1	T_z	
B_{1u}	1	−1	1	−1	1	−1	−1	1	−1	1	−1	1		
B_{2u}	1	−1	1	−1	−1	1	−1	1	−1	1	1	−1		
E_{1u}	2	1	−1	−2	0	0	−2	−1	1	2	0	0	(T_x, T_y)	
E_{2u}	2	−1	−1	2	0	0	−2	1	1	−2	0	0		

The D_{nd} groups

D_{2d}	E	$2S_4$	C_2	$2C'_2$	$2\sigma_d$		
A_1	1	1	1	1	1		x^2+y^2, z^2
A_2	1	1	1	−1	−1	R_z	
B_1	1	−1	1	1	−1		x^2-y^2
B_2	1	−1	1	−1	1	T_z	xy
E	2	0	−2	0	0	$(T_x, T_y), (R_x, R_y)$	(xz, yz)

D_{3d}	E	$2C_3$	$3C_2$	i	$2S_6$	$3\sigma_d$		
A_{1g}	1	1	1	1	1	1		x^2+y^2, z^2
A_{2g}	1	1	−1	1	1	−1	R_z	
E_g	2	−1	0	2	−1	0	(R_x, R_y)	$(x^2-y^2, xy)(xz, yz)$
A_{1u}	1	1	1	−1	−1	−1		
A_{2u}	1	1	−1	−1	−1	1	T_z	
E_u	2	−1	0	−2	1	0	(T_x, T_y)	

D_{4d}	E	$2S_8$	$2C_4$	$2S_8^3$	C_2	$4C'_2$	$4\sigma_d$		
A_1	1	1	1	1	1	1	1		x^2+y^2, z^2
A_2	1	1	1	1	1	−1	−1	R_z	
B_1	1	−1	1	−1	1	1	−1		
B_2	1	−1	1	−1	1	−1	1	T_z	
E_1	2	$\sqrt{2}$	0	$-\sqrt{2}$	−2	0	0	(T_x, T_y)	
E_2	2	0	−2	0	2	0	0		(x^2-y^2, xy)
E_3	2	$-\sqrt{2}$	0	$\sqrt{2}$	−2	0	0	(R_x, R_y)	(xz, yz)

D_{5d}	E	$2C_5$	$2C_5^2$	$5C_2$	i	$2S_{10}^3$	$2S_{10}$	$5\sigma_d$		
A_{1g}	1	1	1	1	1	1	1	1		x^2+y^2, z^2
A_{2g}	1	1	1	−1	1	1	1	−1	R_z	
E_{1g}	2	2 cos 72°	2 cos 144°	0	2	2 cos 72°	2 cos 144°	0	(R_x, R_y)	(xz, yz)
E_{2g}	2	2 cos 144°	2 cos 72°	0	2	2 cos 144°	2 cos 72°	0		(x^2-y^2, xy)
A_{1u}	1	1	1	1	−1	−1	−1	−1		
A_{2u}	1	1	1	−1	−1	−1	−1	1	T_z	
E_{1u}	2	2 cos 72°	2 cos 144°	0	−2	−2 cos 72°	−2 cos 144°	0	(T_x, T_y)	
E_{2u}	2	2 cos 144°	2 cos 72°	0	−2	−2 cos 144°	−2 cos 72°	0		

D_{6d}	E	$2S_{12}$	$2C_6$	$2S_4$	$2C_3$	$2S_{12}^5$	C_2	$6C_2'$	$6\sigma_d$		
A_1	1	1	1	1	1	1	1	1	1		x^2+y^2, z^2
A_2	1	1	1	1	1	1	1	−1	−1	R_z	
B_1	1	−1	1	−1	1	−1	1	1	−1		
B_2	1	−1	1	−1	1	−1	1	−1	1	T_z	
E_1	2	$\sqrt{3}$	1	0	−1	$-\sqrt{3}$	−2	0	0	(T_x, T_y)	
E_2	2	1	−1	−2	−1	1	2	0	0		(x^2-y^2, xy)
E_3	2	0	−2	0	2	0	−2	0	0		
E_4	2	−1	−1	2	−1	−1	2	0	0		
E_5	2	$-\sqrt{3}$	1	0	−1	$\sqrt{3}$	−2	0	0	(R_x, R_y)	(xz, yz)

The S_n groups

S_4	E	S_4	C_2	S_4^3		
A	1	1	1	1	R_z	x^2+y^2, z^2
B	1	−1	1	−1	T_z	x^2-y^2, xy
E	{1	i	−1	−i}	$(T_x, T_y), (R_x, R_y)$	(xz, yz)
	{1	−i	−1	i}		

S_6	E	C_3	C_3^2	i	S_6^5	S_6		$\epsilon = \exp(2\pi i/3)$
A_g	1	1	1	1	1	1	R_z	x^2+y^2, z^2
E_g	{1	ϵ	ϵ^*	1	ϵ	ϵ^*}	(R_x, R_y)	$(x^2-y^2, xy), (xz, yz)$
	{1	ϵ^*	ϵ	1	ϵ^*	ϵ}		
A_u	1	1	1	−1	−1	−1	T_z	
E_u	{1	ϵ	ϵ^*	−1	$-\epsilon$	$-\epsilon^*$}	(T_x, T_y)	
	{1	ϵ^*	ϵ	−1	$-\epsilon^*$	$-\epsilon$}		

The groups $C_{\infty v}$ and $D_{\infty h}$ for linear molecules

$C_{\infty v}$	E	$2C_\infty^\Phi$	...	$\infty\sigma_v$		
$A_1 \equiv \Sigma^+$	1	1	...	1	T_z	x^2+y^2, z^2
$A_2 \equiv \Sigma^-$	1	1	...	−1	R_z	
$E_1 \equiv \Pi$	2	$2\cos\Phi$	...	0	$(T_x, T_y), (R_x, R_y)$	(xz, yz)
$E_2 \equiv \Delta$	2	$2\cos 2\Phi$	...	0		(x^2-y^2, xy)
$E_3 \equiv \Phi$	2	$2\cos 3\Phi$	...	0		
...	...	...	...	...		

$D_{\infty h}$	E	$2C_\infty^\Phi$	...	$\infty\sigma_v$	i	$2S_\infty^\Phi$	...	∞C_2		
$A_{1g} \equiv \Sigma_g^+$	1	1	...	1	1	1	...	1		x^2+y^2, z^2
$A_{2g} \equiv \Sigma_g^-$	1	1	...	−1	1	1	...	1	R_z	
$E_{1g} \equiv \Pi_g$	2	$2\cos\Phi$	...	0	2	$--2\cos\Phi$	...	0	(R_x, R_y)	(xz, yz)
$E_{2g} \equiv \Delta_g$	2	$2\cos 2\Phi$	...	0	2	$2\cos 2\Phi$	...	0		(x^2-y^2, xy)
...	...	...	...	...	...	...	...	...		
$A_{1u} \equiv \Sigma_u^+$	1	1	...	1	−1	−1	...	−1	T_z	
$A_{2u} \equiv \Sigma_u^-$	1	1	...	−1	−1	−1	...	1		
$E_{1u} \equiv \Pi_u$	2	$2\cos\Phi$	...	0	−2	$2\cos\Phi$	...	0	(T_x, T_y)	
$E_{2u} \equiv \Delta_u$	2	$2\cos 2\Phi$	...	0	−2	$-2\cos 2\Phi$	...	0		
...	...	...	...	...	...	...	...	...		

The cubic groups

T_d	E	$8C_3$	$3C_2$	$6S_4$	$6\sigma_d$		
A_1	1	1	1	1	1		$x^2+y^2+z^2$
A_2	1	1	1	−1	−1		
E	2	−1	2	0	0		$(2z^2-x^2-y^2, x^2-y^2)$
T_1	3	0	−1	1	−1	(R_x, R_y, R_z)	
T_2	3	0	−1	−1	1	(T_x, T_y, T_z)	(xy, xz, yz)

O_h	E	$8C_3$	$6C_2$	$6C_4$	$3C_2(=C_4^2)$	i	$6S_4$	$8S_6$	$3\sigma_h$	$6\sigma_d$		
A_{1g}	1	1	1	1	1	1	1	1	1	1		$x^2+y^2+z^2$
A_{2g}	1	1	−1	−1	1	1	−1	1	1	−1		
E_g	2	−1	0	0	2	2	0	−1	2	0		$(2z^2-x^2-y^2, x^2-y^2)$
T_{1g}	3	0	−1	1	−1	3	1	0	−1	−1	(R_x, R_y, R_z)	
T_{2g}	3	0	1	−1	−1	3	−1	0	−1	1		(xy, xz, yz)
A_{1u}	1	1	1	1	1	−1	−1	−1	−1	−1		
A_{2u}	1	1	−1	−1	1	−1	1	−1	−1	1		
E_u	2	−1	0	0	2	−2	0	1	−2	0		
T_{1u}	3	0	−1	1	−1	−3	−1	0	1	1	(T_x, T_y, T_z)	
T_{2u}	3	0	1	−1	−1	−3	1	0	1	−1		

Icosahedral

I_h	E	$12C_5$	$12C_5^2$	$20C_3$	$15C_2$	i	$12S_{10}$	$12S_{10}^3$	$20S_6$	15σ		
A_g	1	1	1	1	1	1	1	1	1	1		$x^2+y^2+z^2$
T_{1g}	3	$\frac{1}{2}(1+\sqrt{5})$	$\frac{1}{2}(1-\sqrt{5})$	0	−1	3	$\frac{1}{2}(1-\sqrt{5})$	$\frac{1}{2}(1+\sqrt{5})$	0	−1	(R_x, R_y, R_z)	
T_{2g}	3	$\frac{1}{2}(1-\sqrt{5})$	$\frac{1}{2}(1+\sqrt{5})$	0	−1	3	$\frac{1}{2}(1+\sqrt{5})$	$\frac{1}{2}(1-\sqrt{5})$	0	−1		
G_g	4	−1	−1	1	0	4	−1	−1	1	0		
H_g	5	0	0	−1	1	5	0	0	−1	1		$(2z^2-x^2-y^2, x^2-y^2, xy, xz, yz)$
A_u	1	1	1	1	1	−1	−1	−1	−1	−1		
T_{1u}	3	$\frac{1}{2}(1+\sqrt{5})$	$\frac{1}{2}(1-\sqrt{5})$	0	−1	−3	$-\frac{1}{2}(1-\sqrt{5})$	$-\frac{1}{2}(1+\sqrt{5})$	0	1	(T_x, T_y, T_z)	
T_{2u}	3	$\frac{1}{2}(1-\sqrt{5})$	$\frac{1}{2}(1+\sqrt{5})$	0	−1	−3	$-\frac{1}{2}(1+\sqrt{5})$	$-\frac{1}{2}(1-\sqrt{5})$	0	1		
G_u	4	−1	−1	1	0	−4	1	1	−1	0		
H_u	5	0	0	−1	1	−5	0	0	1	−1		

Appendix II

Some physical constants

Physical constant	Symbol	Value	
		Measure	Unit
Speed of light in a vacuum	c	$2.997\,945\,0 \times 10^{8}$	$m\ s^{-1}$
Permeability of a vacuum	μ_0	$4\pi \times 10^{7}$	$kg\ m\ s^{-2}\ A^{-2}$ (or $H\ m^{-1}$)
Permittivity of a vacuum	$\epsilon_0 = \mu_0^{-1}\ c^{-2}$	$8.854\,187\,8 \times 10^{-12}$	$kg^{-1}\ m^{-3}\ s^{4}\ A^{2}$ (or $F\ m^{-1}$)
Unified atomic mass constant	$m_u = m(^{12}C)/12$	$1.660\,565 \times 10^{-27}$	kg
Rest mass of proton	m_p	$1.672\,648 \times 10^{-27}$	kg
Rest mass of neutron	m_n	$1.674\,954 \times 10^{-27}$	kg
Rest mass of electron	m_e	$9.109\,534 \times 10^{-31}$	kg
Charge of proton	e	$1.602\,189\,2 \times 10^{-19}$	C
Boltzmann constant	k	$1.380\,662 \times 10^{-23}$	$J\ K^{-1}$
Planck constant	h	$6.626\,176 \times 10^{-34}$	J s
Rydberg constant	$R_\infty = m_e e^4/8\epsilon_0^2 h^3 c$	$1.097\,373\,18 \times 10^{7}$	m^{-1}
Bohr magneton	$\mu_B = eh/4\pi m_e$	$9.274\,078 \times 10^{-24}$	$A\ m^{2}$ (or $J\ T^{-1}$)
Avogadro constant	N_A	$6.022\,045 \times 10^{23}$	mol^{-1}
Gas constant	R	$8.314\,41$	$J\ K^{-1}\ mol^{-1}$
Faraday constant	F	$9.648\,456 \times 10^{4}$	$C\ mol^{-1}$

Appendix III

The Elements

Z	Name	Symbol	A_r
1	hydrogen	H	1.008
2	helium	He	4.002 60
3	lithium	Li	6.94
4	berylium	Be	9.012 18
5	boron	B	10.81
6	carbon	C	12.011
7	nitrogen	N	14.006 7
8	oxygen	O	15.999
9	fluorine	F	18.998 4
10	neon	Ne	20.17
11	sodium	Na	22.989 8
12	magnesium	Mg	24.305
13	aluminium	Al	26.981 5
14	silicon	Si	28.08
15	phosphorus	P	30.973 8
16	sulphur	S	32.06
17	chlorine	Cl	35.453
18	argon	Ar	39.94
19	potassium	K	39.10
20	calcium	Ca	40.08
21	scandium	Sc	44.955 9
22	titanium	Ti	47.9
23	vanadium	V	50.941
24	chromium	Cr	51.996
25	manganese	Mn	54.938 0
26	iron	Fe	55.84
27	cobalt	Co	58.933 2
28	nickel	Ni	58.7
29	copper	Cu	63.54
30	zinc	Zn	65.3
31	gallium	Ga	69.72
32	germanium	Ge	72.5
33	arsenic	As	74.921 6
34	selenium	Se	78.9
35	bromine	Br	79.904
36	krypton	Kr	83.80

The elements (continued)

Z	Name	Symbol	A_r
37	rubidium	Rb	85.467
38	strontium	Sr	87.62
39	yttrium	Y	88.905 9
40	zirconium	Zr	91.22
41	niobium	Nb	92.906 4
42	molybdenum	Mo	95.9
43	technetium	Tc	98.906 2
44	ruthenium	Ru	101.0
45	rhodium	Rh	102.905 5
46	palladium	Pd	106.4
47	silver	Ag	107.868
48	cadmium	Cd	112.40
49	indium	In	114.82
50	tin	Sn	118.6
51	antimony	Sb	121.7
52	tellurium	Te	127.6
53	iodine	I	126.904 5
54	xenon	Xe	131.30
55	caesium	Cs	132.905 5
56	barium	Ba	137.3
57	lanthanum	La	138.905
58	cerium	Ce	140.12
59	praseodymium	Pr	140.907 7
60	neodymium	Nd	144.2
61	promethium	Pm	—
62	samarium	Sm	150.4
63	europium	Eu	151.96
64	gadolinium	Gd	157.2
65	terbium	Tb	158.925 4
66	dysprosium	Dy	162.5
67	holmium	Ho	164.930 3
68	erbium	Er	167.2
69	thulium	Tm	168.934 2
70	ytterbium	Yb	173.0
71	lutetium	Lu	174.97
72	hafnium	Hf	178.4
73	tantalum	Ta	180.947
74	tungsten	W	183.8
75	rhenium	Re	186.2
76	osmium	Os	190.2
77	iridium	Ir	192.2
78	platinum	Pt	195.0
79	gold	Au	196.966 5
80	mercury	Hg	200.5
81	thallium	Tl	204.3
82	lead	Pb	207.2
83	bismuth	Bi	208.980 6
84	polonium	Po	—
85	astatine	At	—
86	radon	Rn	—
87	francium	Fr	—

The Elements (continued)

Z	Name	Symbol	A_r
88	radium	Ra	226.025 4
89	actinium	Ac	—
90	thorium	Th	232.038 1
91	protactinium	Pa	231.035 9
92	uranium	U	238.029
93	neptunium	Np	237.048 2
94	plutonium	Pu	—
95	americium	Am	—
96	curium	Cm	—
97	berkelium	Bk	—
98	californium	Cf	—
99	einsteinium	Es	—
100	fermium	Fm	—
101	mendelevium	Md	—
102	nobelium	No	—
103	lawrencium	Lr	—
104	rutherfordium	Rf	—

Appendix IV

Four-place logarithms of numbers

N	0	1	2	3	4	5	6	7	8	9	Mean differences								
											1	2	3	4	5	6	7	8	9
10	0000	0043	0086	0128	0170	0212	0253	0294	0334	0374	4	8	12	17	21	25	29	33	37
11	0414	0453	0492	0531	0569	0607	0645	0682	0719	0755	4	8	11	15	19	23	26	30	34
12	0792	0828	0864	0899	0934	0969	1004	1038	1072	1106	3	7	10	14	17	21	24	28	31
13	1139	1173	1206	1239	1271	1303	1335	1367	1399	1430	3	6	10	13	16	19	23	26	29
14	1461	1492	1523	1553	1584	1614	1644	1673	1703	1732	3	6	9	12	15	18	21	24	27
15	1761	1790	1818	1847	1875	1903	1931	1959	1987	2014	3	6	8	11	14	17	20	22	25
16	2041	2068	2095	2122	2148	2175	2201	2227	2253	2279	3	5	8	11	13	16	18	21	24
17	2304	2330	2355	2380	2405	2430	2455	2480	2504	2529	2	5	7	10	12	15	17	20	22
18	2553	2577	2601	2625	2648	2672	2695	2718	2742	2765	2	5	7	9	12	14	16	19	21
19	2788	2810	2833	2856	2878	2900	2923	2945	2967	2989	2	4	7	9	11	13	16	18	20
20	3010	3032	3054	3075	3096	3118	3139	3160	3181	3201	2	4	6	8	11	13	15	17	19
21	3222	3243	3263	3284	3304	3324	3345	3365	3385	3404	2	4	6	8	10	12	14	16	18
22	3424	3444	3464	3483	3502	3522	3541	3560	3579	3598	2	4	6	8	10	12	14	15	17
23	3617	3636	3655	3674	3692	3711	3729	3747	3766	3784	2	4	6	7	9	11	13	15	17
24	3802	3820	3838	3856	3874	3892	3909	3927	3945	3962	2	4	5	7	9	11	12	14	16
25	3979	3997	4014	4031	4048	4065	4082	4099	4116	4133	2	3	5	7	9	10	12	14	15
26	4150	4166	4183	4200	4216	4232	4249	4265	4281	4298	2	3	5	7	8	10	11	13	15
27	4314	4330	4346	4362	4378	4393	4409	4425	4440	4456	2	3	5	6	8	9	11	13	14
28	4472	4487	4502	4518	4533	4548	4564	4579	4594	4609	2	3	5	6	8	9	11	12	14
29	4624	4639	4654	4669	4683	4698	4713	4728	4742	4757	1	3	4	6	7	9	10	12	13
30	4771	4786	4800	4814	4829	4843	4857	4871	4886	4900	1	3	4	6	7	9	10	11	13
31	4914	4928	4942	4955	4969	4983	4997	5011	5024	5038	1	3	4	6	7	8	10	11	12
32	5051	5065	5079	5092	5105	5119	5132	5145	5159	5172	1	3	4	5	7	8	9	11	12
33	5185	5198	5211	5224	5237	5250	5263	5276	5289	5302	1	3	4	5	6	8	9	10	12
34	5315	5328	5340	5353	5366	5378	5391	5403	5416	5428	1	3	4	5	6	8	9	10	11
35	5441	5453	5465	5478	5490	5502	5514	5527	5539	5551	1	2	4	5	6	7	9	10	11
36	5563	5575	5587	5599	5611	5623	5635	5647	5658	5670	1	2	4	5	6	7	8	10	11
37	5682	5694	5705	5717	5729	5740	5752	5763	5775	5786	1	2	3	5	6	7	8	9	10
38	5798	5809	5821	5832	5843	5855	5866	5877	5888	5899	1	2	3	5	6	7	8	9	10
39	5911	5922	5933	5944	5955	5966	5977	5988	5999	6010	1	2	3	4	5	7	8	9	10
40	6021	6031	6042	6053	6064	6075	6085	6096	6107	6117	1	2	3	4	5	6	8	9	10
41	6128	6138	6149	6160	6170	6180	6191	6201	6212	6222	1	2	3	4	5	6	7	8	9
42	6232	6243	6253	6263	6274	6284	6294	6304	6314	6325	1	2	3	4	5	6	7	8	9
43	6335	6345	6355	6365	6375	6385	6395	6405	6415	6425	1	2	3	4	5	6	7	8	9
44	6435	6444	6454	6464	6474	6484	6493	6503	6513	6522	1	2	3	4	5	6	7	8	9
45	6532	6542	6551	6561	6571	6580	6590	6599	6609	6618	1	2	3	4	5	6	7	8	9
46	6628	6637	6646	6656	6665	6675	6684	6693	6702	6712	1	2	3	4	5	6	7	7	8
47	6721	6730	6739	6749	6758	6767	6776	6785	6794	6803	1	2	3	4	5	5	6	7	8
48	6812	6821	6830	6839	6848	6857	6866	6875	6884	6893	1	2	3	4	4	5	6	7	8
49	6902	6911	6920	6928	6937	6946	6955	6964	6972	6981	1	2	3	4	4	5	6	7	8
50	6990	6998	7007	7016	7024	7033	7042	7050	7059	7067	1	2	3	3	4	5	6	7	8
51	7076	7084	7093	7101	7110	7118	7126	7135	7143	7152	1	2	3	3	4	5	6	7	8
52	7160	7168	7177	7185	7193	7202	7210	7218	7226	7235	1	2	2	3	4	5	6	7	7
53	7243	7251	7259	7267	7275	7284	7292	7300	7308	7316	1	2	2	3	4	5	6	6	7
54	7324	7332	7340	7348	7356	7364	7372	7380	7388	7396	1	2	2	3	4	5	6	6	7

N	0	1	2	3	4	5	6	7	8	9	Mean differences 1	2	3	4	5	6	7	8	9
55	7404	7412	7419	7427	7435	7443	7451	7459	7466	7474	1	2	2	3	4	5	5	6	7
56	7482	7490	7497	7505	7513	7520	7528	7536	7543	7551	1	2	2	3	4	5	5	6	7
57	7559	7566	7574	7582	7589	7597	7604	7612	7619	7627	1	2	2	3	4	5	5	6	7
58	7634	7642	7649	7657	7664	7672	7679	7686	7694	7701	1	1	2	3	4	4	5	6	7
59	7709	7716	7723	7731	7738	7745	7752	7760	7767	7774	1	1	2	3	4	4	5	6	7
60	7782	7789	7796	7803	7810	7818	7825	7832	7839	7846	1	1	2	3	4	4	5	6	6
61	7853	7860	7868	7875	7882	7889	7896	7903	7910	7917	1	1	2	3	4	4	5	6	6
62	7924	7931	7938	7945	7952	7959	7966	7973	7980	7987	1	1	2	3	3	4	5	6	6
63	7993	8000	8007	8014	8021	8028	8035	8041	8048	8055	1	1	2	3	3	4	5	5	6
64	8062	8069	8075	8082	8089	8096	8102	8109	8116	8122	1	1	2	3	3	4	5	5	6
65	8129	8136	8142	8149	8156	8162	8169	8176	8182	8189	1	1	2	3	3	4	5	5	6
66	8195	8202	8209	8215	8222	8228	8235	8241	8248	8254	1	1	2	3	3	4	5	5	6
67	8261	8267	8274	8280	8287	8293	8299	8306	8312	8319	1	1	2	3	3	4	5	5	6
68	8325	8331	8338	8344	8351	8357	8363	8370	8376	8382	1	1	2	3	3	4	4	5	6
69	8388	8395	8401	8407	8414	8420	8426	8432	8439	8445	1	1	2	2	3	4	4	5	6
70	8451	8457	8463	8470	8476	8482	8488	8494	8500	8506	1	1	2	2	3	4	4	5	6
71	8513	8519	8525	8531	8537	8543	8549	8555	8561	8567	1	1	2	2	3	4	4	5	5
72	8573	8579	8585	8591	8597	8603	8609	8615	8621	8627	1	1	2	2	3	4	4	5	5
73	8633	8639	8645	8651	8657	8663	8669	8675	8681	8686	1	1	2	2	3	4	4	5	5
74	8692	8698	8704	8710	8716	8722	8727	8733	8739	8745	1	1	2	2	3	4	4	5	5
75	8751	8756	8762	8768	8774	8779	8785	8791	8797	8802	1	1	2	2	3	3	4	5	5
76	8808	8814	8820	8825	8831	8837	8842	8848	8854	8859	1	1	2	2	3	3	4	5	5
77	8865	8871	8876	8882	8887	8893	8899	8904	8910	8915	1	1	2	2	3	3	4	4	5
78	8921	8927	8932	8938	8943	8949	8954	8960	8965	8971	1	1	2	2	3	3	4	4	5
79	8976	8982	8987	8993	8998	9004	9009	9015	9020	9025	1	1	2	2	3	3	4	4	5
80	9031	9036	9042	9047	9053	9058	9063	9069	9074	9079	1	1	2	2	3	3	4	4	5
81	9085	9090	9096	9101	9106	9112	9117	9122	9128	9133	1	1	2	2	3	3	4	4	5
82	9138	9143	9149	9154	9159	9165	9170	9175	9180	9186	1	1	2	2	3	3	4	4	5
83	9191	9196	9201	9206	9212	9217	9222	9227	9232	9238	1	1	2	2	3	3	4	4	5
84	9243	9248	9253	9258	9263	9269	9274	9279	9284	9289	1	1	2	2	3	3	4	4	5
85	9294	9299	9304	9309	9315	9320	9325	9330	9335	9340	1	1	2	2	3	3	4	4	5
86	9345	9350	9355	9360	9365	9370	9375	9380	9385	9390	1	1	2	2	3	3	4	4	5
87	9395	9400	9405	9410	9415	9420	9425	9430	9435	9440	0	1	1	2	2	3	3	4	4
88	9445	9450	9455	9460	9465	9469	9474	9479	9484	9489	0	1	1	2	2	3	3	4	4
89	9494	9499	9504	9509	9513	9518	9523	9528	9533	9538	0	1	1	2	2	3	3	4	4
90	9542	9547	9552	9557	9562	9566	9571	9576	9581	9586	0	1	1	2	2	3	3	4	4
91	9590	9595	9600	9605	9609	9614	9619	9624	9628	9633	0	1	1	2	2	3	3	4	4
92	9638	9643	9647	9652	9657	9661	9666	9671	9675	9680	0	1	1	2	2	3	3	4	4
93	9685	9689	9694	9699	9703	9708	9713	9717	9722	9727	0	1	1	2	2	3	3	4	4
94	9731	9736	9741	9745	9750	9754	9759	9763	9768	9773	0	1	1	2	2	3	3	4	4
95	9777	9782	9786	9791	9795	9800	9805	9809	9814	9818	0	1	1	2	2	3	3	4	4
96	9823	9827	9832	9836	9841	9845	9850	9854	9859	9863	0	1	1	2	2	3	3	4	4
97	9868	9872	9877	9881	9886	9890	9894	9899	9903	9908	0	1	1	2	2	3	3	4	4
98	9912	9917	9921	9926	9930	9934	9939	9943	9948	9952	0	1	1	2	2	3	3	4	4
99	9956	9961	9965	9969	9974	9978	9983	9987	9991	9996	0	1	1	2	2	3	3	3	4

Index